Engineering Materials

This series provides topical information on innovative, structural and functional materials and composites with applications in optical, electrical, mechanical, civil, aeronautical, medical, bio- and nano-engineering. The individual volumes are complete, comprehensive monographs covering the structure, properties, manufacturing process and applications of these materials. This multidisciplinary series is devoted to professionals, students and all those interested in the latest developments in the Materials Science field, that look for a carefully selected collection of high quality review articles on their respective field of expertise.

Indexed at Compendex (2021)

More information about this series at http://www.springer.com/series/4288

Felipe A. La Porta · Carlton A. Taft
Editors

Functional Properties of Advanced Engineering Materials and Biomolecules

Springer

Editors
Felipe A. La Porta
Department de Química
Federal University of Technology—Paraná
Londrina, Paraná, Brazil

Carlton A. Taft
Centro Brasileiro de Pesquisas Físicas
Rio de Janeiro, Brazil

ISSN 1612-1317 ISSN 1868-1212 (electronic)
Engineering Materials
ISBN 978-3-030-62228-2 ISBN 978-3-030-62226-8 (eBook)
https://doi.org/10.1007/978-3-030-62226-8

This Springer imprint is published by the registered company Springer Nature Switzerland AG
The registered company address is: Gewerbestrasse 11, 6330 Cham, Switzerland

Contents

ZnO/bentonite Hybrids Obtained by a Simple Method of Synthesis and Applied as Catalyst for Biodiesel Production

Ana Flávia Félix Farias, Sandro Marden Torres, Elson Longo, Maguy Jaber, Maria Gardennia Fonseca, Liliana Fátima Bezerra Lira Pontes, and Ieda Maria Garcia dos Santos

Abstract ZnO/bentonite hybrids were obtained by the microwave-assisted hydrothermal method, in order to change the bentonite surface and improve its activity for biodiesel production. No previous treatment of the clay was used and in situ crystallization of ZnO was done by a fast single step method. A factorial design was used to optimize synthesis conditions (e.g., reaction time, solution pH, and ZnO amount). The hybrids were fully characterized to determine its structural, chemical, and morphological properties. The results demonstrated the positive contribution of the microwave-assisted hydrothermal method to the formation of ZnO/bentonite hybrids, in addition to showing that the pH was the most important parameter influencing the formation of ZnO. A pH value of 8 favored the incorporation of approximately 19.6% of ZnO onto the smectite particles even after 5 min of microwave irradiation. The ZnO/bentonite hybrid with the biggest amount of ZnO was applied as a catalyst in the transesterification reaction of soybean oil. As such, our results

A. F. F. Farias · M. G. Fonseca · I. M. G. Santos (✉)
Núcleo de Pesquisa e Extensão—Laboratório de Combustíveis e Materiais (NPE/LACOM), Universidade Federal da Paraíba, João Pessoa, PB, Brazil
e-mail: ieda@quimica.ufpb.br

A. F. F. Farias
Laboratório de Síntese de Materiais Cerâmicos (LabSMaC), Universidade Federal de Campina Grande, Campina Grande, PB, Brazil

S. M. Torres
Depto de Engenharia de Materiais (DEMAT), Laboratório de Tecnologia de Novos Materiais (TECNOMAT), Universidade Federal da Paraíba, João Pessoa, PB, Brazil

E. Longo
Laboratório Interdisciplinar de Eletroquímica e Cerâmica (LIEC)/Center for Development of Functional Materials (CDFM), Universidade Federal de São Carlos, São Carlos, SP, Brazil

M. Jaber
Sorbonne Universités, Laboratoire d'Archéologie Moléculaire et Structurale, Paris, France

L. F. B. L. Pontes
Depto de Química, Universidade Federal da Paraíba, João Pessoa, PB, Brazil

F. A. La Porta and C. A. Taft (eds.), *Functional Properties of Advanced Engineering Materials and Biomolecules*, Engineering Materials,
https://doi.org/10.1007/978-3-030-62226-8_1

demonstrated an ethyl ester conversion rate higher than 73% with better performance than that of natural bentonite (44%), which represents approximately 66% of the improvement of the catalytic activity.

1 Introduction

As is well-known, the chemical modification of clay minerals is an exciting field for the synthesis of new advanced functional materials, and it includes the immobilization of organic and inorganic substances in the interlayer space on the external surface to form new solids that combine the properties of both precursors [1–9].

The hybridization of clay minerals has the potential to improve technological applications in different fields, such as adsorption [2], antimicrobial materials [10, 11], drug delivery [12], catalysis [13, 14] and others [5, 15]. Due to their important functional properties and the application potential of these hybrid structures, new synthesis methods have been proposed, and new hybrid species have been created based on nano-oxide/clay minerals [5, 6, 11].

Oxide-clay mineral nanocomposites show several favorable characteristics. As such, oxide nanoparticles can be located on the surface of the clay matrix therefore improving its properties and reuse and decreasing the solubility of the nanoparticles due to a higher tendency of hybrids to flocculate. In addition, the physical and chemical properties of the clay mineral substrate are maintained and new properties are enhanced [11, 16].

Recently, hybrid materials composed of ZnO immobilized on silicate mesoporous materials and clay minerals have widely been studied [6, 11, 17–19]. ZnO is a semiconductor with high chemical and thermal stabilities in addition to being amphoteric, low cost, abundant, and environmentally friendly [6, 20–22]. Due to these properties, ZnO is widely used as a photocatalyst [20, 23], catalyst [21, 24], antimicrobial material [10, 25], etc.

ZnO/clay mineral hybrids are usually prepared by a conventional hydro and/or solvothermal method that involves several steps, namely, the purification of the clay mineral, pre-expansion and/or sodium ion exchange and the cation exchange with the ZnO precursor, and requires more than 24 h of treatment, and in some cases, post-calcination process is necessary [6, 8, 19, 20, 26–30]. These hybrids have also been obtained by the sol–gel method [31, 32] and by reflux of ZnO and acid-treated bentonite in aqueous media under microwave heating [33, 34].

Hur et al. [26] synthesized ZnO-montmorillonite heterostructures using a hydrothermal reaction between a colloidal dispersion of exfoliated sodium montmorillonite nanolayers and a zinc acetate solution. The heterostructures were described as exfoliated montmorillonite hybrids with a house-of-cards-like stacking of the crystallites containing ZnO nanoparticles (i.e., as guest species located inside the house-of-card mesopores or in the montmorillonite interlayer space). Suchithra et al. [19] applied a microwave-assisted hydrothermal method to obtain a hybrid nanostructure composed of ZnO intercalated into the interlayer space of aluminosilicate and deposited on its surface, through sodium ion exchange and swelling the bentonite by sonication for 30 min. Polyethylene glycol was used as a capping molecule in order to promote Zn intercalation.

Despite several studies that report obtaining hybrids based on ZnO and clay minerals, the development of a simple, low-cost, rapid method for synthesizing of these inorganic-inorganic ZnO/bentonite hybrids is needed. In a previous work [14], our research group evaluated the synthesis of oxides/bentonite hybrids by microwave-assisted hydrothermal synthesis, with interesting results for ZnO/bentonite hybrids. In the present work, optimization of the synthesis conditions was done with an emphasis on the influence of pH on the final hybrid structural and morphological properties. Pristine bentonite was used without previous purification, acid treatments or pre-expansion treatments, and using a Zn^{2+} solution without capping molecules. ZnO was semiconductor chosen due to its promising results for the synthesis of biodiesel, with high rates of conversion depending on the synthesis method and the reaction conditions during transesterification [14, 35, 36].

2 Materials and Methods

2.1 *Synthesis of the Hybrid Materials*

The ZnO/bentonite hybrids were synthesized by a microwave-assisted hydrothermal synthesis from the following analytical grade reagents: dehydrated zinc acetate-$C_4H_6O_4Zn.2H_2O$ purchased from Synth (P.A > 98%), ammonium hydroxide-NH_4OH from Vetec (P.A 99%) and bentonite (BENTONISA S.A., Brazil). The cation exchange capacity (CEC) of the pristine clay samples according to the supplier was 108 $cmol.kg^{-1}$.

A 2^3 factorial design was then performed to optimize the synthesis conditions of the hybrids. The studied factors and levels were as follows: solution pH (pH), 8.0 (−1) and 11.0 (+1); relatively proportion of Zn^{2+} to bentonite mass in the aqueous solution (C), 10% (−1) and 20% (+1); and reaction time (T), 5 min (−1) and 15 min (+1). The software Statistica 8.0 was then used to construct the planning matrix and evaluate the results obtained by X-ray fluorescence (XRF), once the desired response correlates to the percentage of ZnO in the hybrid.

The quantities of the reagents were adapted to obtain the approximately 2 g of ZnO/bentonite hybrid in each synthesis. First, solutions of the ZnO precursor salt were prepared at 0.02 and 0.04 mol L^{-1}. Bentonite was added into these solutions under stirring for 20 min, followed by an adjustment to the desired pH with added of NH_4OH. The dispersion was transferred to a Teflon vessel and hydrothermalized in a microwave reactor (RMW-1, Inove Produtos and Tecnologia/Ltda), according to the procedure previously described [14]. As a reference, pure ZnO was obtained under the same conditions without the addition of bentonite. The material was deagglomerated with a porcelain mortar followed by sieving with a 200-mesh sieve (0.074 mm).

After hydrothermalization, the supernatant solutions of the reaction media used for the syntheses were collected to quantify of Na, Ca, and Zn. The quantitative analyses were carried out by inductively coupled plasma-optical emission spectroscopy (ICP OES, Radial) using an Arco spectrometer from Spectro.

The pristine and obtained materials were characterized by an analysis of the point of zero charge (pH_{PZC}), X-ray diffraction (XRD), XRF, infrared spectroscopy (FTIR), specific surface area from nitrogen adsorption/desorption isotherms, field emission-scanning electron microscopy (FE-SEM) and transmission electron microscopy (TEM).

The pH_{PZC} was determined for bentonite according to a procedure described in the literature [37, 38]. XRD patterns of random samples were obtained using a Shimadzu XRD-6000 diffractometer at 30 kV and 30 mA with nickel-filtered Cu Kα radiation ($\lambda = 0.15418$ nm), in the range from 3 to 80°, at a step of 0.03° and a rate of 0.02°s^{-1}. XRF was performed in a Shimadzu EDX-800 X-ray fluorescence spectrometer in an attempt to determine the actual percentage of ZnO in the samples obtained after the formation of the hybrids. The samples were prepared by homogenization of bentonite with ZnO (analytical grade, Synth; used as received), in a mortar, followed by pressing of pellets. The ZnO amounts varied from 2.5 to 50 mass% in order to obtain a calibration curve from nine samples. After XRF analysis, these results were graphed and given as the intensity of Zn^{2+} peak (cps) as a function of the ZnO percentage (mass%).

The infrared spectra were obtained using the transmittance mode on a Shimadzu FTIR Prestige-21 between 4000 and 400 cm^{-1} by preparing KBr pellets at 1% (m/m), with a 4 cm^{-1} resolution and 20 iterations.

The textural properties, in all cases, were determined from the N_2 adsorption/desorption isotherms obtained using a Bel Japan Belsorp Mini II apparatus. The specific surface area was determined using the BET method, and the mesopore diameter and volume were determined by the BJH method. To remove adsorbed water from the surface, the samples were pre-heated at 200 °C for 120 min.

FE-SEM analyses were obtained using a FEG-VP Zeiss Supra 35 microscope. The samples were dispersed in isopropanol, sonicated for 1 min and then deposited on a Si substrate.

TEM analyses were obtained from a Philips CM-200 microscope operating at 200 kV. Samples in the form of bulk powders were dispersed in ethanol and then deposited on 400-mesh copper grids covered with an ultrathin carbon membrane of 2–3 nm thickness.

2.2 *Biodiesel Synthesis*

The transesterification reaction was performed in the 4842 Parr reactor using soybean oil from Campestre (Brazil), ethyl alcohol (Química Moderna, Brazil) and the catalyst, as displayed in Eq. (1).

$$
\begin{array}{ccccc}
H_2C{-}OCOR_1 & & & ROCOR_1 & H_2C{-}OH \\
| & & & + & | \\
HC{-}OCOR_2 & +\ 3ROH & \underset{\Delta}{\overset{Catalyst}{\rightleftharpoons}} & ROCOR_2 & +\ HC{-}OH \\
| & & & + & | \\
H_2C{-}OCOR_3 & & & ROCOR_3 & H_2C{-}OH \\
\text{Triacyglyceride} & \text{Alcohol} & & \text{Mixture of alkylesters} & \\
 & & & & \text{Glycerol}
\end{array}
\qquad (1)
$$

For the catalysis tests, 30 g of soybean oil with 3% m/m of the catalyst in relation to the oil amount and a molar ratio for ethanol:oil of 6:1 were used. Catalytic reaction was performed as previously described, with a reaction time of 240 min [14].

The kinematic viscosity was measured according to the ASTM D 4455 standard using a V18 Julabo bath at 40 °C and a Cannon Fenske glass capillary viscometer. ^{1}H nuclear magnetic resonance (^{1}H NMR) spectroscopy was used to evaluate the ester characteristic peak of the samples and quantify their conversion rate into biodiesel using peak integration according to the methodology described in the literature [14, 39, 40]. The ^{1}H NMR spectra were obtained at a frequency of 200 MHz from a Varian Gemini 300BB spectrometer.

The percent conversion was calculated from Eq. (2) and was adapted to the synthesis of ethylic esters, as described in the literature [41].

$$\%C_{EE} = 100\left(\frac{I_{TAG+EE} - I_{TAG}}{I_{\alpha CH_2}}\right) \qquad (2)$$

where: $\%\ C_{EE}$ = Percentage of conversion to ethyl esters; I_{TAG+EE} = integrated area of the overlapping peaks, between 4.1 and 4.2 ppm, assigned to methylenic hydrogens of glyceryl, and hydrogens ethoxy group-OCH_2 superimposed; I_{TAG} = integrated peak area between 4.25 and 4.35 ppm, attributed to methylenic hydrogens of glyceryl indicating the presence of mono-, di- and triglycerides in the sample; $I_{\alpha\text{-}CH2}$ = integrated area of the methylene hydrogen of carbonyl approximately 2.2 and 2.4 ppm.

3 Results and Discussion

3.1 Previous Evaluation of the Microwave-Assisted Hydrothermal Reaction

The pH_{PZC} measurements are shown in Fig. 1. Our results unveil two different pH_{PZC} values for bentonite: 2.0 and 7.3. Indeed, a charged surface leads to the repulsion or attraction of ions from the solution. Therefore, the adsorption of positive species from the solution is, of course, favored on a negatively charged surface, which occurs

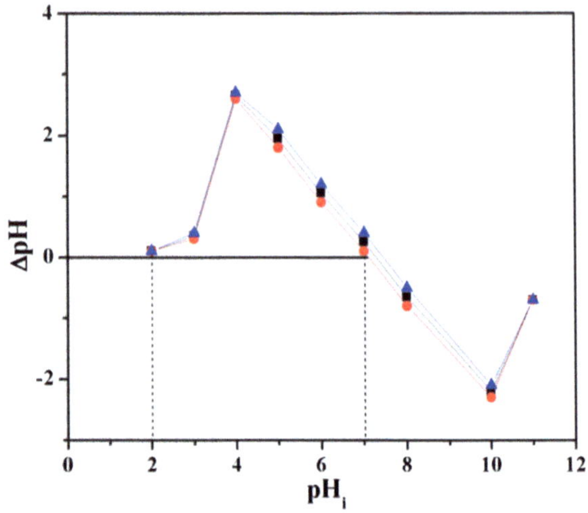

Fig. 1 Variation of the pH of zero charge depending on the initial pH of the bentonite solution

at pH values above the pH_{PZC}, e.g., 7.3 for bentonite. Therefore, the deprotonated surface of the bentonite at pH values 8.0 and 11.0 favors the interaction with the cationic species.

The ZnO synthesis by the microwave-assisted hydrothermal method was performed under alkaline conditions, as previously reported [42, 43]. In an alkaline solution obtained by the addition of NH_4OH, different zinc species were present. As such, the nature of the ionic species in solution was evaluated by the Visual Minteq 3.0 program [44], described in Table 1. Their presence and nature were dependent on the zinc precursor concentration, alkalinizing agent, pH, and temperature. As such, the influence of acetate was not considered due to a limitation of the program. For both pH values (8 and 11), the predominant zinc species were positively charged,

Table 1 Zinc species formed in solutions with concentrations of 0.02 and 0.04 mol L^{-1} at 100 °C and pH values 8.0 and 11 adjusted with NH_4Cl. The counter ion (acetate) was not considered

Ionic species (mol %)	Synthesis conditions			
	pH 8.0 [Zn] = 0.02 mol L^{-1}	pH 8.0 [Zn] = 0.04 mol L^{-1}	pH 11.0 [Zn] = 0.02 mol L^{-1}	pH 11.0 [Zn] = 0.04 mol L^{-1}
$Zn(NH_3)_4{}^{2+}$	78.55	88.36	63.62	87.87
$Zn(NH_3)_3{}^{2+}$	19.80	11.19	9.09	7.60
$Zn(NH_3)_2{}^{2+}$	1.53	0.44	0.40	0.20
$ZnNH_3{}^{2+}$	0.10	0.01	0.01	–
$ZnOH^+$	0.02	–	1.80	0.30
$Zn(OH)_2$ (aq)	–	–	16.86	2.64
$Zn(OH)_3{}^-$	–	–	7.87	1.33
$Zn(OH)_4{}^{2-}$	–	–	0.35	0.06

which is favorable for adsorption on the negatively charged bentonite. Indeed, at pH 8.0, all of the zinc complexes are cationic for both zinc concentrations used in the present work, while a small number of negative complexes (less than 10 mol%) are formed at pH 11.0, especially at lower Zn^{2+} concentrations. For instance, 8.2 mol% of negatively charged species (hydroxy-zinc complexes) may be present in solution at 0.02 mol L^{-1} zinc salt while only 1.39 mol% are present at 0.04 mol L^{-1}.

The mixtures prepared at pH = 8 or 11, as well as the material precipitated at pH 8, were analyzed by infrared spectroscopy, as displayed in Fig. 2. Two bands at 1555 and 1405 cm^{-1} in the spectrum of the precipitate were assigned to the asymmetric and symmetric modes of COO^-, respectively, which indicates that Zn^{2+} complexation by acetate is taking place [45]. These bands were observed at 1558 and 1418 cm^{-1} for both solutions.

At pH 11, one band assigned to Me-OH bonds is observed at 1102 cm^{-1} [46, 47], while a broadband assigned to NH_3 is observed at 1260 cm^{-1} [48, 49]. Other bands assigned to NH_3 are not observed due to the superposition with H_2O bands, which

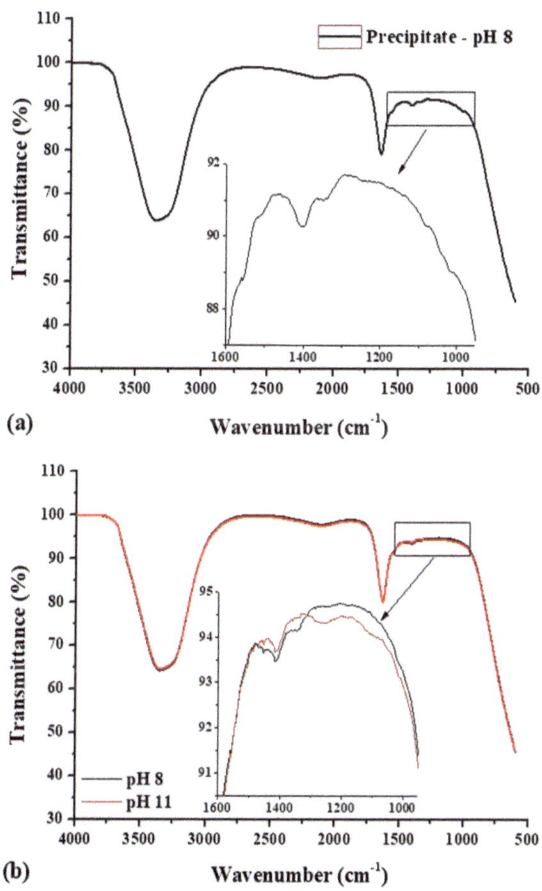

Fig. 2 Infrared spectra of the precursors before hydrothermallyzation. **a** Gel precipitated from solution at pH 8; **b** Solutions at different pH values

presents broad intense bands around 3300 cm^{-1} and 1640 cm^{-1}. These data (Table 1 and Fig. 2) agree with the literature, indicating that ammonia, used as the source of hydroxide anions in solutions at pH > 10, favors the formation of $Zn(OH)_4{}^{2-}$ species by forming stable tetrahedral complexes responsible for the growth of zinc oxide [42, 43]. Moreover, the interaction between Zn^{2+} and $RCOO^-$ has also been investigated as well as the formation of $Zn(NH_3)^{4+}$ at pH > 11 [45, 50]. For instance, Zn leaching from a metallurgical slag was recently done by the addition of $NH_4CH_3CO_2$ to a NH_3–H2O system, which leads to the formation of Zn acetate ammonium complexes [45].

3.2 *Characterization of the Hybrid Materials*

The chemical composition of the pristine bentonite was carried out by XRF, resulting in mass percentages values of SiO_2 (77%), Al_2O_3 (13%), Fe_2O_3 (6%), MgO (2%), CaO (1%), and small amounts of TiO_2, Na_2O and K_2O (< 0.5%). These data agree with the literature [51, 52].

The amount of ZnO in the hybrids was quantified by XRF and evaluated by a 2^3 factorial design (Table 2). A higher percentage of incorporated ZnO was observed for experiment 2, using the highest cation concentration (20 mass%) with a pH value of 8.0 and a 5-min reaction time. According to the analysis of the experimental design response (mass% of ZnO), the lowest pH value favored ZnO formation in the hybrid because the difference between the amount of Zn^{2+} initially present in the reaction solution and the amount of ZnO quantified in the hybrid by XRF was smaller than 10 mass%. For a pH of 11, this difference varied from 35 to 68%, as displayed in Table 2.

The effects of the variables studied are shown in the Pareto graph (Fig. 3a) obtained from the statistical analysis of the factorial design results. According to this graph,

Table 2 The 2^3 factorial design matrix used to synthesize ZnO/bentonite hybrids and the resulting quantity of ZnO present in the hybrid

Run	Level of variables [actual (coded)]			ZnO amount (mass %)
	pH	Cation concentration C (%)	Reaction time T (min)	
1	8.0 (−1)	10 (−1)	5 (−1)	10.4 ± 0.1
2	8.0 (−1)	20 (+1)	5 (−1)	19.2 ± 0.1
3	8.0 (−1)	10 (−1)	15 (+1)	10.7 ± 0.1
4	8.0 (−1)	20 (+1)	15 (+1)	18.7 ± 0.1
5	11.0 (+1)	10 (−1)	5 (−1)	6.5 ± 0.1
6	11.0 (+1)	20 (+1)	5 (−1)	6.9 ± 0.1
7	11.0 (+1)	10 (−1)	15 (+1)	4.7 ± 0.1
8	11.0 (+1)	20 (+1)	15 (+1)	6.4 ± 0.1

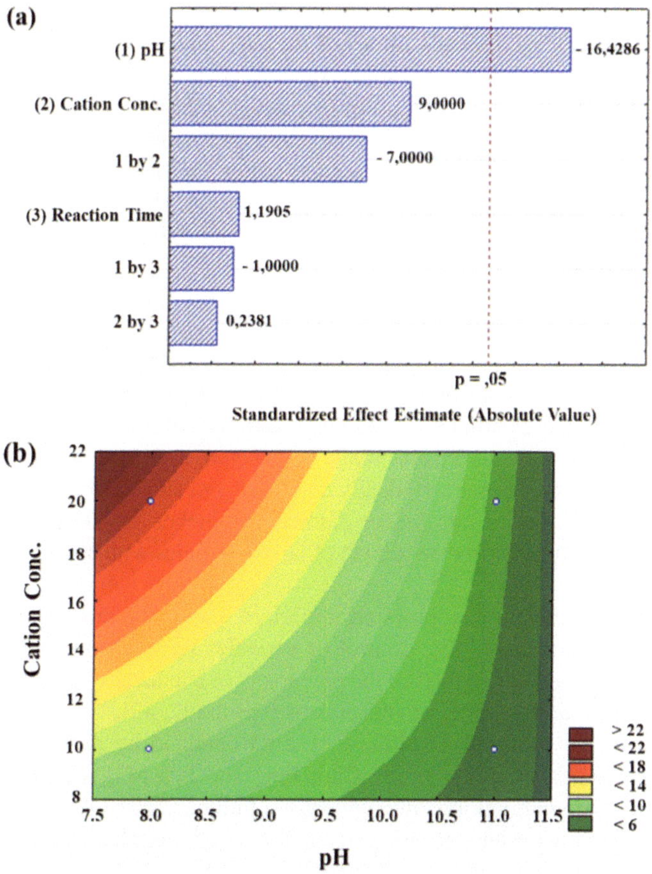

Fig. 3 **a** A Pareto graph for 2^3 factorial design obtained for the analyzed variables (the vertical line defines the 95% confidence interval); **b** The response surface of the factorial design obtained for Zn^{2+} concentration and pH

it is possible to determine the variables with a greater influence on the resulting ZnO amount incorporated in the hybrids. According to the Pareto graph and to the variance analysis (ANOVA), it was possible to observe that the pH demonstrated a significant negative effect for the present study, indicating that solutions with lower pH values resulted in higher ZnO amounts.

Considering the information displayed in Fig. 1 and Table 1, this behavior may be related to the higher number of ionic complexes with positive charges ($Zn(NH_3)_4{}^{2+}$ and $Zn(NH_3)_3{}^{2+}$) at pH 8.0. This favors the adsorption on the negatively charged clay mineral surface leading to a higher amount of ZnO—more than double the amount for some synthesis conditions. Negative and neutral complexes were formed at pH 11.0.

Analysis of the Pareto graph also showed that the ZnO concentration and the reaction time do not significantly influence the ZnO amount incorporated onto the bentonite, which may be attributed to the random effect of the crystal dissolution–recrystallization process during microwave-assisted hydrothermal synthesis.

The surface response as a function of the initial Zn^{2+} concentration and pH for the evaluation of the factorial design is presented in Fig. 3b. According to the adjusted model, the surface response confirms that a higher ZnO amount can be obtained when pH 8.0 is used, which may also be attributed to a higher dissolution–recrystallization process at higher pH.

Dissolution–recrystallization process is known as Ostwald ripening, corresponding to the nucleation and growth of crystalline particles, and is related to the microwave irradiation time during hydrothermal synthesis [53]. Different morphologies may be obtained by controlling the Ostwald ripening using solubility differences among small and large particles during the microwave-assisted hydrothermal reaction process. Larger crystals are essentially unchanged, whereas lower density particles in colloidal aggregates are gradually dissolved and subjected to mass transport recrystallization processes into higher diameter or higher density particles in the same aggregate, thus favoring the growth of larger particles with higher crystallinity [53, 54].

Microwave irradiation also plays an important role in the morphology of ZnO nano- and microstructures [42]. Control of variables such as time and irradiation intensity, in addition to the effects of the precursors of chemical species and reaction temperature, can directly influence the crystal growth [42, 43].

A study of the synthesis conditions of ZnO obtained by microwave-assisted hydrothermal synthesis was done by Cho, Jung and Lee [43], who describe that the addition of ammonium ions into aqueous solution, allows OH^- to easily react with Zn^{2+} at high pH values, forming stable $Zn(OH)_4{}^{2-}$ tetrahedral complexes that act as seeds for the growth of the ZnO structure. $Zn(NH_3)_4{}^{2+}$ is another ZnO forming complex, favoring the growth kinetics contributing to the formation of defects at the beginning of the crystal growth. The lower stability of these complexes as well as the higher number of nuclei leads to smaller seeds and favoring the Oswald Ripening effect. In the present work, the smaller amount of ZnO formed from solutions at higher pH indicates that the increase of the pH and the reaction time during synthesis of nanohybrids favor the redissolution processes of ZnO crystals.

The previously discussed dissolution process caused by the high pH values, associated with the adsorption process of Zn^{2+} complexes onto the bentonite surface, may be responsible for the lower amount of ZnO at pH 11.

Analysis of the supernatant solutions collected after the synthesis of the T05C20pH08 and T05C20pH11 hybrids was done to identify the presence of the exchangeable cations (Na^+ and Ca^{2+}) and remaining Zn^{2+} in the solution. These data were compared to the amount of Zn^{2+} in the solution before hydrothermalization and to the initial composition of bentonite, to quantify the amount of Zn^{2+} exchanged with the interlamellar cations (Na^+ and Ca^{2+}) and the amount of Zn^{2+} deposited on the bentonite as ZnO.

According to the initial composition of the pristine bentonite, the total number of exchangeable cations (0.51 mmols) was very similar to the total amount of these cations in the solution after hydrothermalization (0.46 mmols), indicating that an almost complete exchange of these cations occurred, leading to the addition of 0.32 mmols of Zn^{2+} into the interlamellar region for solutions at pH 8.0 and 11.0. The difference between the amount of ZnO in the hybrids obtained by XRF (19.2 mass% at pH 8.0 and 6.9 mass% at pH 11.0) and the amount of Zn^{2+} in the solution indicated the amount of ZnO supported on the surface of the bentonite, which was 4.4 mmol and 1.38 mmol for solutions with pH 8.0 and 11.0, respectively. These amounts correspond to 93% and 81% of the total amount of Zn incorporated to the bentonite, respectively.

The quantification of Zn in the reaction solution after hydrothermalization also allowed the calculation of the total percentage of Zn incorporated to the bentonite: 19.6 mass% at pH 8.0 and 10.1 mass% at pH 11.0. Comparing these values to the percentage of ZnO obtained by XRF (Table 2), two different results are observed. For solutions at pH 8.0, a small difference between the two values is observed (0.4%), indicating that the Zn incorporated to the bentonite during hydrothermalization was not eliminated during the washing step. For solutions at pH 11.0, the difference was much greater (3.2%), indicating that ZnO or zinc complexes formed during hydrothermalization may be eliminated during the washing step, which is likely due to its weak physical adsorption on the bentonite surface. The formation of anionic compounds was reported in Table 1 and previously determined to be responsible for the smaller amount of ZnO formed during the synthesis when $pH = 11$ is used.

An evaluation of the XRD patterns (Fig. 4) indicates that bentonite is composed of clay minerals from the smectite group, likely a mixture of montmorillonite and nontronite, as also suggested by the presence of 6 mass% of Fe_2O_3 in the XRF analyses, and the impurity of quartz.

Well-defined diffraction reflections characteristic of hexagonal wurtzite ZnO were observed for the hybrid samples (T05C10pH11, T05C10pH08, T05C20pH08, T15C20pH08 and T05C20pH11). The reflections assigned to the smectite were still observed after synthesis, confirming that the smectite structure was maintained. The basal distance for the bentonite was similar to that described in the literature for montmorillonite [20].

For all experimental conditions, the basal distances decreased to values between 1.19 and 1.27 nm, as displayed in Table 3, which may be attributed to a dehydration process of the smectite. A decrease in the intensities of the (001) reflections was also observed and may be related to the deposition of ZnO on the bentonite surface, which did not crystallize inside the interlayer region. This behavior may result from the high concentration of the zinc precursor in the solution, leading to the crystallization of larger particles on the surface and decrease the amount of Zn^{2+} in the interlayer region of the smectite, as previously discussed.

A different behavior was observed by Fatimah et al. [20], who intercalated ZnO into montmorillonite and observed an increase of the basal distance (from 1.49 to 1.62 nm), which was similar to that observed by Khaorapapong et al. [6] (from 1.24 to 1.54 nm) for ZnO/sodic bentonite hybrids.

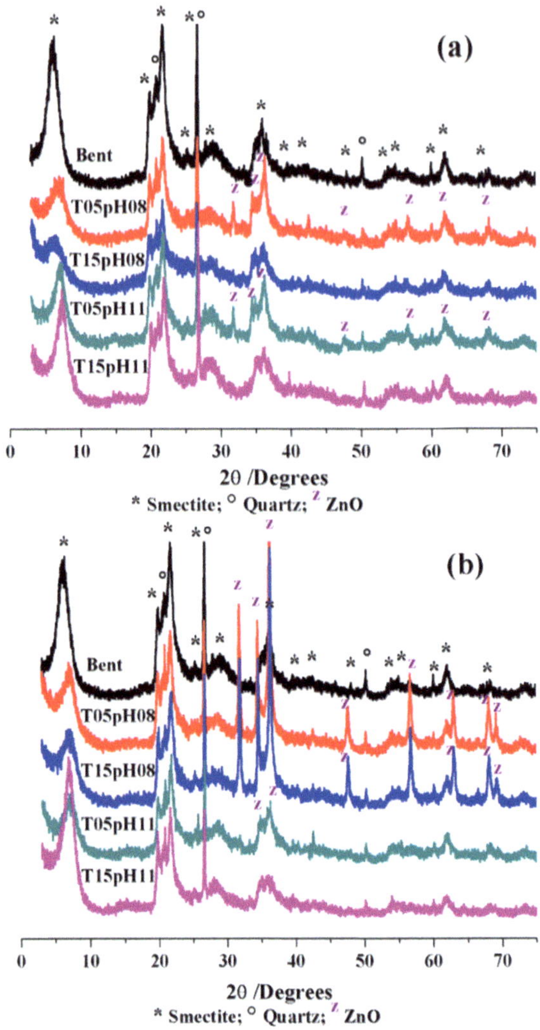

Fig. 4 XRD patterns of bentonite (Bent) and ZnO/bentonite hybrids obtained under different synthesis conditions with ZnO/bentonite mass/mass proportions of **a** 10% and **b** 20%

Table 3 Basal spacings (001 reflection) of the pristine bentonite and hybrid materials

Ample	Bent	T05 C10 pH08	T15 C10 pH08	T05 C10 pH11	T15 C10 pH11	T05 C20 pH08	T15 C20 pH08	T05 C20 pH11	T15 C20 pH11
2θ (°)	6.3	7.2	7.3	7.0	7.1	7.4	7.1	6.6	7.2
d (nm)	1.39	1.23	1.21	1.27	1.24	1.19	1.25	1.34	1.22

Infrared spectra of the bentonite precursor and the ZnO/bentonite hybrids are shown in Fig. 5 and Table 4, where are describes assignment of the bands observed in clay minerals is well described in the literature [55–58]. The main changes in the smectite octahedral layer were observed for the absorption of bands attributed to the stretching of the group (OH), which showed displacement and even disappearance for the hybrids M05B10Z11 and M15B20Z11. Also were observed for the Fe-OH out-of-plane band, which disappeared for some samples. For the tetrahedral layer, the Si–O-Si deformation bands were shifted for all of the materials. These results may be related to the interactions among the ZnO nanoparticles and the bentonite surface or at the edges of the smectite structure. Dislocations were also observed for the bands assigned to quartz, indicating that ZnO may also be formed on its surface. Bands assigned to acetate and ammonia were also observed for some samples, in agreement to the evaluation of the precursors, displayed in Fig. 2.

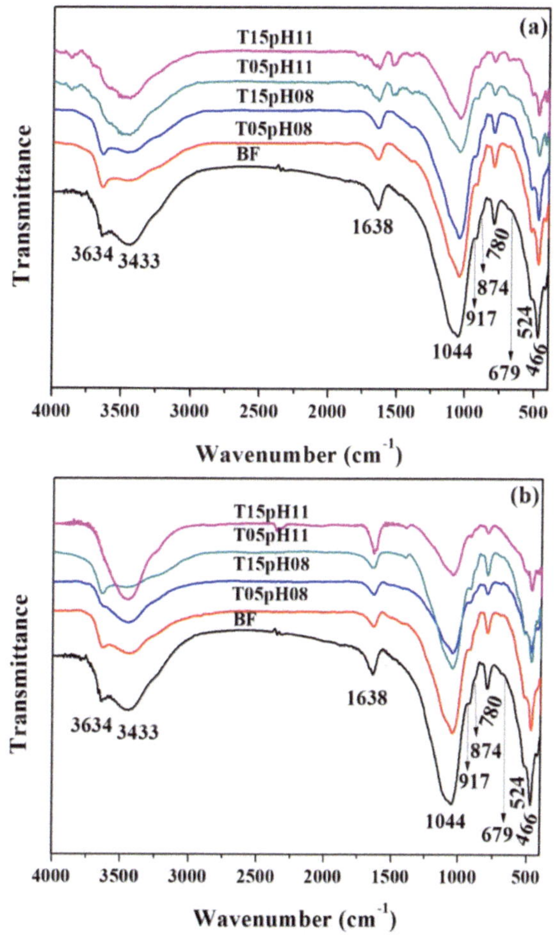

Fig. 5 IR spectra of the bentonite precursor and ZnO/bentonite hybrids. **a** 10% of ZnO and **b** 20% of ZnO

Table 4 Infrared spectra assignments for the pristine bentonite and the ZnO/bentonite hybrids

Assignments (cm^{-1})	Bent	C10				C20			
		T05 pH08	T15 pH08	T05 pH11	T15 pH11	T05 pH08	T15 pH08	T05 pH11	T15 pH11
OH octahedral stretching	3634	3634	**3641**	–	**3625**	**3641**	**3626**	**3641**	–
C = O vibration (CH_3COO^- group)	–	–	–	1529	1529	–	–	–	–
C-O vibration (CH_3COO^- group) or N–H from ammonium	–	1400	1400	1400	1400	–	–	1400	1400
Si–O stretching	1044	**1051**	**1051**	1043	**1051**	**1051**	**1051**	**1051**	1043
Al–Al–OH deformation	917	917	917	917	917	917	917	917	917
Al–Mg–OH deformation	874	873	872	872	873	873	871	876	873
Quartz	780	791	791	783	780	780	783	791	793
Fe–Fe–OH deformation with Fe-OH out of plane vibration	693	693	–	688	689	695	696	–	–
Si–O–Al deformation	524	522	522	524	522	524	522	522	522
Si–O–Si deformation	466	**460**	**470**	**470**	**477**	**477**	**470**	**470**	**477**

*Numbers in bold indicate bands with dislocation in relation to the bentonite precursor

The N_2 adsorption/desorption isotherm of the precursor bentonite is presented in Fig. 6, and the results are displayed in Table 5. No meaningful change of the curve profiles occurred after the synthesis of the ZnO/bentonite hybrids, and all of the samples presented a type-II isotherm, with H3 hysteresis, which is characteristic of mesoporous materials.

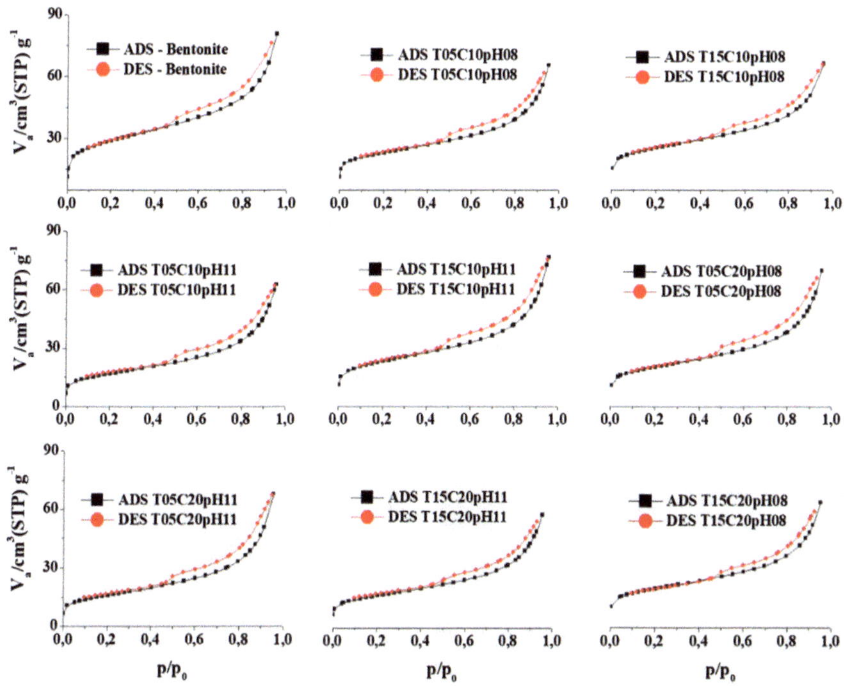

Fig. 6 N_2 adsorption/desorption isotherms for the pristine bentonite and the ZnO/bentonite hybrids

Table 5 Textural properties obtained from the BJH and BET methods applied to the isothermal N_2 adsorption/desorption curves

Sample	S (m^2g^{-1})	V_p (cm^3g^{-1})	d_p (nm)
Bent	99	0.1016	3.75
T05C10pH08	66	0.0859	3.75
T15C10pH08	72	0.0841	3.75
T05C10pH11	56	0.0854	3.75
T15C10pH11	77	0.1017	3.75
T05C20pH08	71	0.0945	3.75
T15C20pH08	67	0.0862	3.75
T05C20pH11	54	0.0961	3.75
T15C20pH11	55	0.0778	4.25

*S specific surface area; V_p pore volume and d_p pore diameter

A median pore diameter of 3.75 was determined for all samples except T15C20pH11, and it confirms the presence of mesopores [50, 59, 60]. The data showed a decrease of the specific surface area and pore volume for all of the samples except T15C10pHZ11. These results indicate that ZnO may be deposited on the surface or inside the pores.

The FE-SEM image of the pristine bentonite (Fig. 7a) indicates a heterogeneous morphology; some regions have a rough morphology, whereas others have plates with lamellar stacking. This behavior was expected due to the use of non-purified bentonite.

The SEM images (Fig. 7b and c) of ZnO obtained under the same conditions as the hybrid materials indicate different morphologies for the synthesis under different pH values. At shorter times and lower pH values, rod-like morphologies were obtained by the aggregation of small particles. Over longer times and higher pH values, flower- or star-like morphologies were formed. These morphologies indicate the existence of the Ostwald ripening effect during the crystallization of ZnO, which agrees with the studies performed by Cho [43], Ram et al. [61]. The synthesis of ZnO by microwave-assisted hydrothermal described by Ram et al. [61] was performed at different pH values (8, 10, 12 and 13.5). According to the authors, the ripening process was more pronounced when fewer nuclei are present in the reaction medium. This favors a

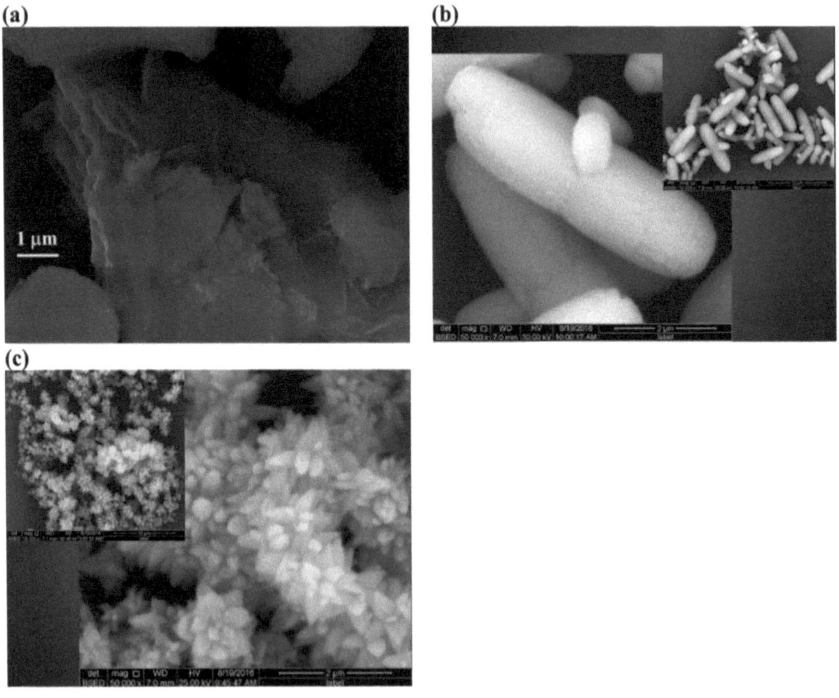

Fig. 7 FE-SEM images of the pristine bentonite **a** and SEM images of the ZnO obtained under the following conditions: **b** 05 min, pH 08 and **c** 15 min, pH 11

greater size distribution of the morphologies, showing rods of smaller lengths or the growth of tiny flakes. The highest pH favors the production of $[Zn(OH)_4]^{2-}$ nuclei, enhancing the formation of well-defined hexagonal rods.

For the hybrids, significant changes were noticed in the morphology of ZnO for all synthesis conditions used, as observed in the FE-SEM and TEM images shown in Figs. 8 and 9, respectively.

The presence of bentonite in the reaction medium has a considerable interference effect on the morphology of the ZnO nanoparticles, which is likely due to its crystallization on the smectite surface. Moreover, the solution pH and reaction time also change the distribution of these oxides on the bentonite surface. This behavior is observed in Fig. 8 and agrees with the textural results displayed in Table 4, showing a decrease in the specific surface area after the hybrid is synthesized.

The hybrids synthesized with the longest reaction time, highest Zn^{2+} concentration and highest pH (T15C20pH11) favored the formation of ZnO with a dandelion-type morphology on the bentonite surface (Fig. 8d). This may indicate that a dissolution–recrystallization process occurs, which favors the formation of the needle-like morphology. This behavior may be related to less stable complexes formed at pH 11, which favor dissolution processes, especially over longer reaction times. The same

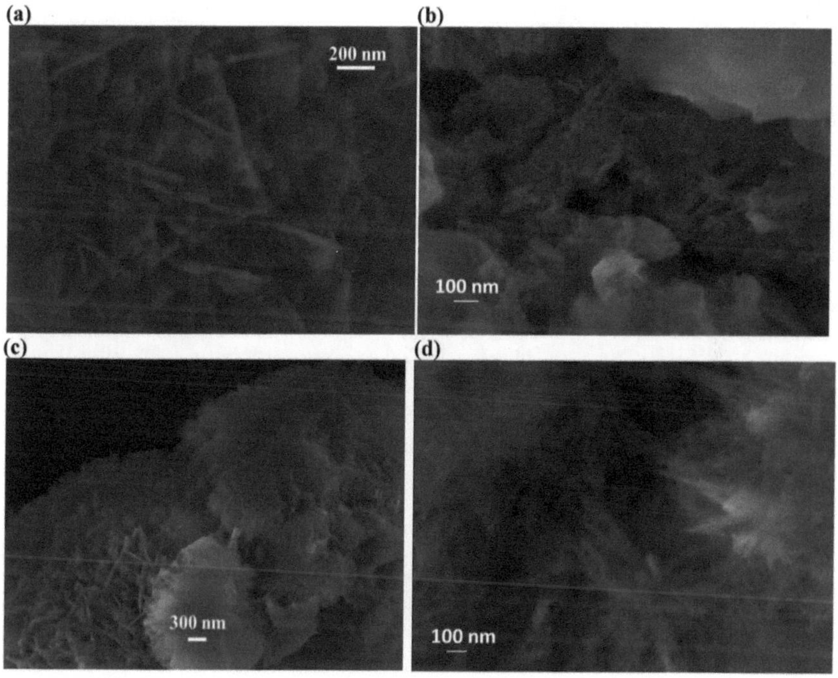

Fig. 8 FE-SEM images of the ZnO/bentonite hybrids. **a** T05C20pH08; **b** T15C10pH11; **c** T05C20pH11; **d** T15C20pH11

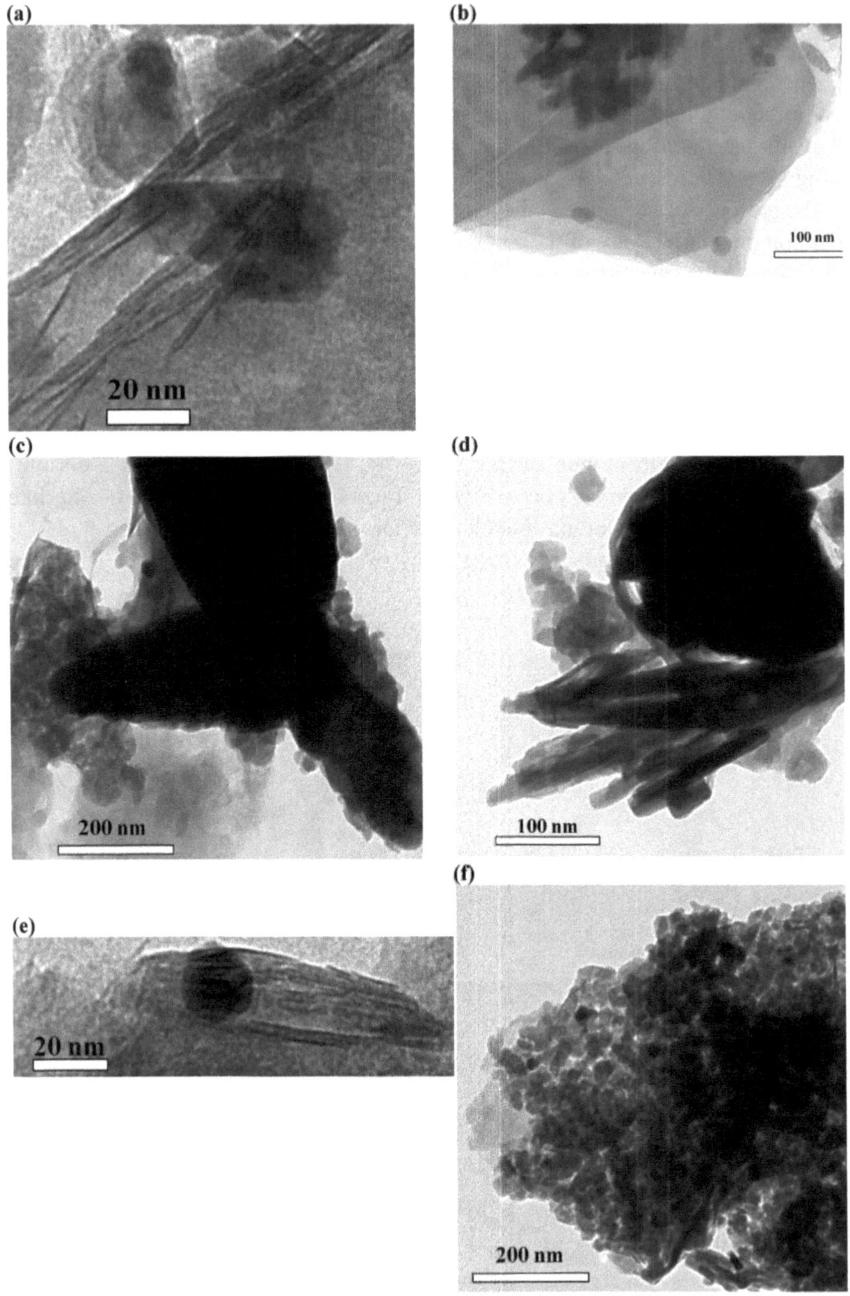

Fig. 9 TEM images of the pristine bentonite **a** and **b** of the ZnO/bentonite hybrids. **c** and **d** T05C20pH08; **e** and **f** T05C20pH11

morphology was obtained by Barreto [42] and Cho et al. [43] used a microwave-assisted hydrothermal method and NH_4OH as an alkalinizing agent to synthesize ZnO.

For some hybrids (T05C20pH08, T05C20pH11, T15C20pH11, T15C10pH11), the formation of plates, needles and nanoparticles was also observed, as shown in Fig. 8. These different ZnO morphologies formed on the bentonite surface.

TEM analysis was performed for the precursor bentonite and for the hybrid samples (Fig. 9) obtained from solutions with higher Zn2 + concentrations, shorter reaction times and lower pH values.

Figure 9a and b showed a heterogeneous morphology with lamellar regions characteristic of smectite and particles due to the presence to quartz, which agree with the observations in the literature and with FE-SEM images (Fig. 7a) [62, 63]. According to Fig. 9c–f, ZnO particles were deposited on the quartz and smectite surfaces with an almost spherical morphology or with needle-like morphology, which blocks the bentonite pores, as indicated by the decrease of specific surface area/pore volume (Table 4).

The morphology evaluated by SEM, FE-SEM and TEM images is in agreement with the literature [43, 61, 63, 64] and with the other results obtained in this work. For instance, XRD patterns suggest a decrease in the basal plane distance of the hybrids indicating that ZnO crystallization did not occur in the interlayer spaces. The BET results also suggested that crystallization occurred on the surface, with a decrease in the specific surface area of the hybrids, compared to natural bentonite. Moreover, the displacement of the bands in the IR spectra suggests the interaction among the ZnO nanoparticles and bentonite. The formation of the hybrid is schematically displayed in Fig. 10.

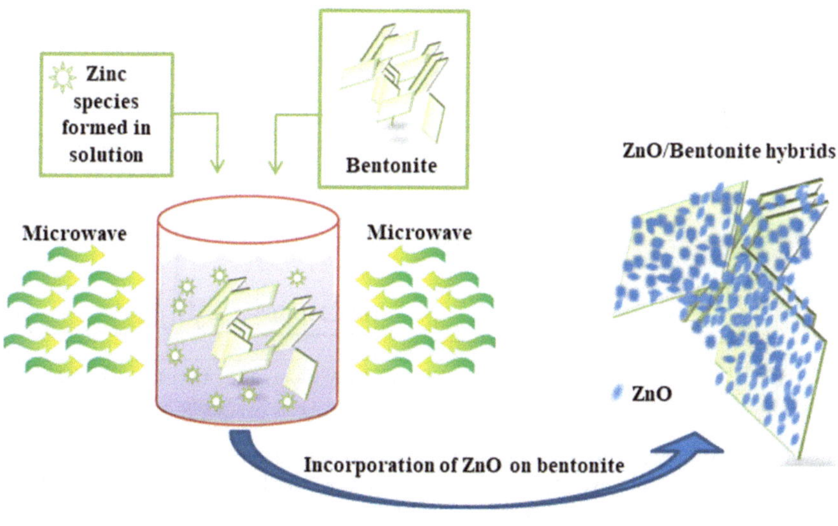

Fig. 10 Schematic diagram showing the formation of the ZnO/bentonite hybrid

3.3 Catalytic Test

The catalyst (T05C20pH08) with the highest amount of ZnO was chosen for the synthesis of biodiesel by transesterification of soybean oil at 200 °C for 240 min in a Parr reactor. The performance of three catalysts, the ZnO/bentonite hybrid, the pure ZnO and the bentonite, was evaluated by determining the kinematic viscosity of the product of the catalytic tests. A decrease of 70.5% of the viscosity was obtained compared to the standard soybean oil indicating that a change occurred in the structure of the triacylglyceride when the hybrid catalyst was tested. On the other hand, viscosity analysis of the catalytic tests using ZnO (active phase) and bentonite (used as support for the active phase) indicated a viscosity reduction of 74.5% and 32.1%, respectively. A comparison of the results should consider that only 19.2% of the active phase (ZnO) is present in the hybrid, indicating that its formation improved the catalytic effect of ZnO. This likely is due to the interaction among the active phase and the support.

Confirmation of the oil conversion into fatty acid ethyl esters was obtained by ^{1}H NMR spectroscopy (Fig. 11), which indicated the catalytic effectiveness with a 73.4% conversion to biodiesel, in spite of the use of the ethylic route, which leads to smaller conversions but brings lower toxicity. When pristine bentonite was used as catalyst, 43.9% conversion was attained, which confirms the improvement of the catalytic activity of this clay and increases its added value. These results showed the viability of the use of hybrids for biodiesel production. The profile of the ^{1}H NMR spectra of the oil before the catalytic test was characterized by the peaks assigned to triacylglycerides (two doublets between 4.0 and 4.4 ppm), but a different profile was observed after the catalytic test in the presence of the T05C20pH08 hybrid, showing a quartet between 4.0 and 4.2, which is assigned to ethyl esters [14, 39–41].

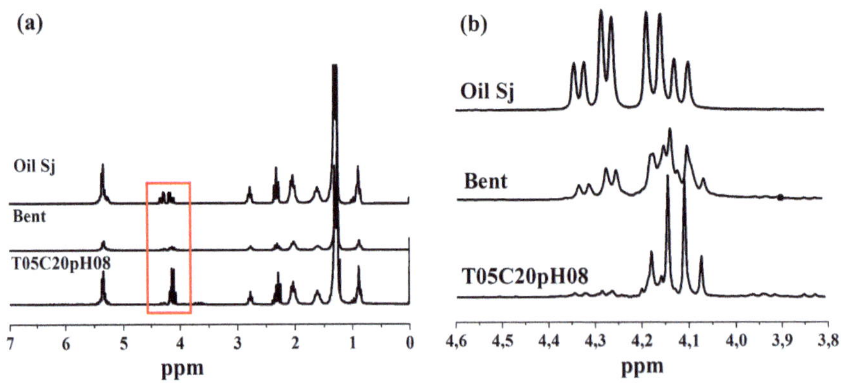

Fig. 11 ^{1}H NMR spectra of the oil and of the products of the catalytic tests performed at 200 °C using as the ZnO/bentonite hybrid catalysts (T05C20pH08) and the Bentonita (Bent), before (**a**) and after (**b**) of spectral amplification between 3.8 and 4.6 ppm

Previously, our research group evaluated the impregnation of different oxides on bentonite and obtained the highest conversion using ZnO/bentonite as catalyst for the ethylic transesterification of soybean oil [14]. Comparison with literature data shows different results for ZnO impregnated on silica, alumina, or zeolite. For instance, 1 mass% of ZnO was impregnated on fused SiO_2 for the esterification of the free fatty acids present in the Jathropa Curcas oil, with 95% of conversion [65]. 15% of ZnO was also supported on Hβ-zeolite for the esterification of acetic acid with butanol, but a decrease of activity was observed comparing to pure zeolite and to pure ZnO [66]. The influence of the support on the methylic transesterification of jathropa curcas using ZnO as catalyst was evaluated by Singh et al. [67]. The authors supported 25 mass% of ZnO on α-alumina, γ-alumina and on the ZSM-5 zeolite. All of the systems had more than 90% of conversion after reaction at 200 °C for 60 min. Besides the higher alcohol amount, which favors the transesterification reaction, the use of methanol also leads to higher conversions than ethanol. In spite of this, Navas et al. [68] obtained 41% of conversion using 0.27 mol % of ZnO supported on γ-Al_2O_3 as catalyst for the methylic transesterification of soybean oil.

4 Conclusions

Microwave-assisted hydrothermal synthesis using an as-received bentonite sample without prior treatment favors the formation of the ZnO/bentonite hybrids for all synthesis conditions applied in this work. The solution pH was the determining factor in obtaining a hybrid with a greater proportion of ZnO. The two pH values used contribute positively to the bentonite/ZnO interaction, maintaining the bentonite structure but changing its surface and morphological characteristics, as observed from the micrographs. At pH 8.0, a greater amount of ZnO crystallized on the hybrids, while a needle-like morphology may be obtained at pH 11.0, when higher concentrations and longer times were used. The synthesis conditions that lead to a greater amount of ZnO in the hybrids were pH 8.0, 20 mass% of Zn^{2+} and 5 min of reaction time. The incorporation of ZnO onto the bentonite provided a good catalytic effect on the transesterification reaction of soybean oil. The rates of conversion of soybean oil into ethyl esters reaching values higher than 73%.

Acknowledgements This work was supported by CAPES/COFECUB Project (grant number 835/2015), CT-INFRA/FINEP/MCTIC, CNPq/MCTIC (grant number 406857/2013-0—Edital 40/2013), grant 0012/2019, Paraíba State Research Foundation (FAPESQ) and CAPES/MEC (Coordenação de Aperfeiçoamento de Pessoal de Nível Superior—Finance Code: 001) as well as by contributions from the Archeology Molecular and Structural Laboratory in France.

References

1. Fournier, F., De Viguerie, L., Balme, S., Janot, J.-M., Walter, P., Jaber, M.: Physico-chemical characterization of lake pigments based on montmorillonite and carminic acid. Appl. Clay Sci. **130**, 12–17 (2016)
2. Santos, S.S.G., França, D.B., Castellano, L.R.C., Trigueiro, P., Silva Filho, E.C., Santos, I.M.G., Fonseca, M.G.: Novel modified bentonites applied to the removal of an anionic azo-dye from aqueous solution. Colloids Surf. A **585**, 124152 (2020)
3. Jaber, M., Miehe-Brendle, J., Le Dred, R.: Mercaptopropyl Al–Mg phyllosilicate: Synthesis and characterization by XRD, IR, and NMR. Chem. Lett. **9**, 954–955 (2002)
4. Jaber, M., Miehe-Brendle, J.: Synthesis, characterization and applications of 2:1 phyllosilicates and organophyllosilicates: contribution of fluoride to study the octahedral sheet Microporous. Mesoporous. Mater. **107**, 121–127 (2007)
5. Kamada, K., Kang, J.-H., Paek, S.-M., Choy, J.-H.: CeO_2—layered aluminosilicate nanohybrids for UV screening. J. Phys. Chem. Solids **73**, 1478–1482 (2012)
6. Khaorapapong, N., Khumchoo, N., Ogawa, M.: Preparation of zinc oxide–montmorillonite hybrids. Mater. Lett. **65**, 657–660 (2011)
7. Alves, L., Ferraz, E., Gamelas, J.A.F.: Composites of nanofibrillated cellulose with clay minerals: A review. Adv. Colloid Interface **272**, 101994 (2020)
8. Meshram, S., Limaye, R., Ghodke, S., Nigam, S., Sonawane, S., Chikate, R.: Continuous flow photocatalytic reactor using ZnO–bentonite nanocomposite for degradation of phenol. Chem. Eng. J. **172**, 1008–1015 (2011)
9. Varadwaj, G.B.B., Rana, S., Parida, K.: Cs salt of Co substituted lacunary phosphotungstate supported K10 montmorillonite showing binary catalytic activity. Chem. Eng. J. **215–216**, 849–858 (2013)
10. de Silva, R.T., Pasbakhsh, P., Mae, L.S., Kit, A.Y.: ZnO deposited/encapsulated halloysite–poly (lactic acid) (PLA) nanocomposites for high performance packaging films with improved mechanical and antimicrobial properties. Appl. Clay Sci. **111**, 10–20 (2015)
11. Barabaszová, K.C., Holešová, S., Šulcová, K., Hundáková, M., Thomasová, B.: Effects of ultrasound on zinc oxide/vermiculite/chlorhexidine nanocomposite preparation and their antibacterial activity. Nanomaterials **9**, 1309 (2019)
12. Cavalcanti, G.R.S., Fonseca, M.G., Silva Filho, E.C., Jaber, M.: Thiabendazole/bentonites hybrids ascontrolledrelease systems. Colloids Surf. B **176**, 249–255 (2019)
13. Cornu, D., Lin, L., Daou, M., Jaber, M., Krafft, J.-M., Herledan, V., Laugel, G., Millot, Y., Lauron-Pernot, H.: Influence of acid-base properties of Mg-based catalysts in transesterification. Role of Magnesium Silicate Hydrate formation. Catal. Sci. Tech. **7**, 1701–1712 (2017)
14. Farias, A.F.F., Moura, K.F., Souza, J.K.D., Lima, R.O., Nascimento, J.D.S.S., Cutrim, A.A., Longo, E., Araujo, A.S., Carvalho-filho, J.R., Souza, A.G., Santos, I.M.G.: Biodiesel obtained by ethylic transesterification using CuO, ZnO and CeO_2 supported on bentonite. Fuel **160**, 357–365 (2015)
15. Tae, J.-W., Jang, B.-S., Kim, J.-R., Kim, I., Park, D.-W.: Catalytic degradation of polystyrene using acid-treated halloysite clays. Solid State Ionics **172**, 129–133 (2004)
16. Wang, C., Shi, H., Zhang, P., Li, Y.: Synthesis and characterization of kaolinite/TiO_2 nano-photocatalysts. Appl. Clay Sci. **53**, 646–649 (2011)
17. Fernández, L., Garro, N., Haskouri, J., El P-Cabero, M., A-Rodríguez, J., Latorre, J., Guillem, C., Beltrán, A., Beltrán, D., Amorós, P.: Mesosynthesis of ZnO–SiO_2 porous nanocomposites with low-defect ZnO nanometric domains Nanotechnology **19**, 1–10 (2008)
18. Khumchoo, N., Khaorapapong, N., Ogawa, M.: Formation of zinc oxide particles in cetyltrimethylammonium-smectites. Appl. Clay Sci. **105–106**, 236–242 (2015)
19. Suchithra, P.S., Shadiya, C.P., Mohamed, A.P., Velusamy, P., Ananthakumar, S.: One-pot microwave mediated growth of heterostructured ZnO@AlSi as a potential dual-function eco-catalyst for treating hazardous pollutants in water resources. Appl. Catal. B Environ. **130–131**, 44–53 (2013)

20. Fatimah, I., Wang, S., Wulandari, D.: ZnO/montmorillonite for photocatalytic and photochemical degradation of methylene blue. Appl. Clay Sci. **53**, 553–560 (2011)
21. Kim, M., Lee, H.-S., Yoo, S.J., Youn, Y.-S., Shin, Y.H., Lee, Y.-W.: Simultaneous synthesis of biodiesel and zinc oxide nanoparticles using supercritical methanol. Fuel **109**, 279–284 (2013)
22. Shohany, B.G., Zak, K.: Doped ZnO nanostructures with selected elements—structural, morphology and optical properties: a review. Ceram. Int. **46**, 5507–5520 (2020)
23. Ivan, R., Popescu, C., Pino, A.P., Logofatu, C., György, E.: Carbon-based nanomaterials and ZnO ternary compound layers grown by laser technique for environmental and energy storage applications. Appl. Surf. Sci. **509**, 145359 (2020)
24. Laskar, I.B., Rokhum, L., Gupta, R., Chatterjee, S.: Zinc oxide supported silver nanoparticles as a heterogeneouscatalyst for production of biodiesel from palm oil. Environ. Prog. Sustainable Energy 1–11 (2019)
25. Piedade, A.P., Pinho, A.C., Branco, R., Morais, P.V.: Evaluation of antimicrobial activity of ZnO based nanocomposites for the coating of non-critical equipment in medical-care facilities. Appl. Surf. Sci. **513**, 145818 (2020)
26. Hur, S.G., Kim, T.W., Hwang, S.-J., Hwang, S.-H., Yang, J.H., Choy, J.-H.: Heterostructured nanohybrid of zinc oxide-montmorillonite clay. J. Phys. Chem. B **110**, 1599–1604 (2006)
27. Abdo, J., Zaier, R., Hassan, E., AL-Sharji, H., Al-Shabibi, A.: ZnO–clay nanocomposites for enhance drilling at HTHP conditions Surf. Interface Anal. **46**, 970–974 (2013)
28. Akkari, M., Aranda, P., Amara, A.B.H., R-Hitzky, E.,: Clay-nanoarchitectures as photocatalysts by in situ assembly of ZnO nanoparticles and clay minerals J. Nanosci. Nanotechnol. **18**, 223–233 (2018)
29. Khumchoo, N., Khaorapapong, N., Ontam, A., Intachai, S., Ogawa, M.: Efficient photodegradation of organics in acidic solution by ZnO–smectite hybrids. Eur. J. Inorg. Chem. 3157–3162 (2016)
30. Ahmad, S., Su, X., Yang, C., Wang, X., Liu, X., Wang, J.: Space-confined growth of layered basic zinc acetate nanosheets and their orderly fragmented ZnO nanoparticles on clay platelets. J. Hazard. Mater. **371**, 213–223 (2019)
31. Sasikala, S.P., Nibila, T.A., Babitha, K.B., Mohamed, A.A.P., Solaiappan.: A Competitive photo-degradation performance of ZnO modified bentonite clay in water containing both organic and inorganic contaminants Sustainable Environ. Res. **29**(1), 1–12 (2019)
32. Xu, H., Zhang, D., Xu, A., Wu, F., Cao, R.: Quantum sized zinc oxide immobilized on bentonite clay and degradation of C.I. acid Red 35 in aqueous under ultraviolet light. Int. J. Photoenergy. **7**, 750–869 (2015). https://dx.doi.org/10.1155/2015/750869
33. Motshekga, S.C., Ray, S.S., Onyango, M.S., Momba, M.N.B.: Microwave-assisted synthesis, characterization and antibacterial activity of Ag/ZnO nanoparticles supported bentonite clay. J. Hazard. Mater. **262**, 439–446 (2013)
34. Motshekga, S.C., Ray, S.S., Onyango, M.S., Momba, M.N.B.: Preparation and antibacterial activity of chitosan-based nanocomposites containing bentonite-supported silver and zinc oxide nanoparticles for water disinfection. Appl. Clay Sci. **114**, 330–339 (2015)
35. Antunes, W.M., Veloso, C.O., Henriques, C.A.: Transesterification of soybean oil with methanol catalyzed by basic solids. Catal. Today **133–135**, 548–554 (2008)
36. Xie, W., Yang, Z., Chun, H.: Ba-ZnO catalysts for soybean oil transesterification. Catal. Lett. **117**, 159–165 (2007)
37. Cristiano, E., Hu, Y.-J., Siegfried, M., Kaplan, D., Nitsche, H.: A comparison of point of zero charge measurement methodology. Clays Clay Miner. **59**(2), 107–115 (2011)
38. Mahmood, T., Saddique, M.T., Naeem, A., Westerhoff, P., Mustafa, S., Alum, A.: Comparison of different methods for the point of zero charge determination of NiO. Ind. Eng. Chem. Res. **50**, 10017–10023 (2011)
39. López, J.M., Cota, T.N.J.G., Monterrosas, E.E.G., Martínez, R.N., González, V.M.C., Flores, J.L.A., Ortega, Y.R.: Kinetic study by 1H nuclear magnetic resonance spectroscopy for biodiesel production from castor oil. Chem. Eng. J. **178**, 391–397 (2011)
40. Tariq, M., Ali, S., Ahmad, F., Ahmad, M., Zafar, M., Khalid, N., Khan, M.A.: Identification, FT-IR, NMR (^{1}H and ^{13}C) and GC/MS studies of fatty acid methyl esters in biodiesel from rocket seed oil. Fuel Process. Technol. **92**, 336–341 (2011)

41. Ghesti, G.F., Macedo, J.L., Resck, I.S., Dias, J.A., Dias, S.C.L.: FT-Raman spectroscopy quantification of biodiesel in a progressive soybean oil transesterification reaction and its correlation with ^{1}H NMR spectroscopy methods. Energy Fuels **21**, 5 (2007)
42. Barreto, G.P., Morales, G., Quintanilla, M.L.L.: Microwave assisted synthesis of ZnO nanoparticles: effect of precursor reagents, temperature, irradiation time, and additives on nano-ZnO morphology development. J. Mater. 1–11 (2013)
43. Cho, S., Jung, S.-H., Lee, K.-H.: Morphology-controlled growth of ZnO nanostructures using microwave irradiation: from basic to complex structures. J. Phys. Chem. C **112**, 12769–12776 (2008)
44. Gustaffson, J.P.: Visual Minteq, version 3.0 https://vminteq.lwr.kth.se/. Accessed 11, 2015 (2012)
45. Ma, A., Zheng, X., Li, S., Wang, Y., Zhu, S.: Zinc recovery from metallurgical slag and dust by coordination leaching in NH_3–CH_3COONH_4–H_2O system. R. Soc. Open Sci. **5**, 1–16 (2018)
46. Nakamoto, K.: Infrared and Raman Spectra of Inorganic and Coordination Compounds, 4th edn. Wiley, New York (1980)
47. Srivastava, O.K., Secco, E.A.: Studies on metal hydroxy compounds. II. Infrared spectra of zinc derivatives ε-$Zn(OH)_2$, β-ZnOHCl, ZnOHF, $Zn_5(0H)_8Cl_2$, and $Zn_5(0H)_8Cl_2.H_2O$. Can. J. Chem. **45**, 585–588 (1967)
48. Müller, A., Böschen, I., Baran, E.J., Aymonino, P.J.: Über Tetramminmetallchalkogenometallate. Monatsh. Chem. **104**, 836–847 (1973)
49. Williams, P.A., Ferrer, M.E.G., Baeza, N., Piro, O.E., Baran, E.J.: New saccharinato/ammonia complexes of Nickel(II) and Zinc(II). A structural and spectroscopic study. Z. Anorg. Allg. Chem. **632**, 619–623 (2006)
50. Shepherd, D., Best, S.M.: Production of zinc substituted hydroxyapatite using various precipitation routes. Biomed. Mater. **8**, 1–11 (2013)
51. H-Ashtiani, M., Characterization of nano-porous bentonite (Montmorillonite) particles using FTIR and BET-BJH analyses Part. Part. Syst. Char. **28**, 71–76 (2011)
52. Noyan, H., Önal, M., Sarikaya, Y.: The effect of sulphuric acid activation on the crystallinity, surface area, porosity, surface acidity, and bleaching power of a bentonita. Food Chem. **105**, 156–163 (2007)
53. Ismail, I.H., Ibrahim, K., Khairuddean, M., Yen, T.S., Wan, Y.C.: Time growing comparison for ZnO nanorod by using microwave irradiation method. Sol-Gel Sci. Technol. **72**, 369–374 (2014)
54. Zou, X., Fan, H., Tian, Y., Zhang, M., Yan, X.: Microwave-assisted hydrothermal synthesis of Cu/Cu_2O hollow spheres with enhanced photocatalytic and gas sensing activities at room temperature. Dalton Trans. **44**, 7811–7821 (2015)
55. Madejova, J.: FTIR techniques in clay mineral studies. Vib. Spectrosc **31**, 1–10 (2003)
56. Petit, S., Madejova, J.: Fourier Transform Infrared Spectroscopy In: Handbook of Clay Science, 2nd edn., vol. 5A. Elsevier Ltd., London, UK, Chapter 2.7
57. Zhang, Y., Liu, Q., Wu, Z., Zhang, Y.: Thermal behavior analysis of two bentonite samples selected from China. J. Therm. Anal. Calorim. **121**, 1287–1295 (2015)
58. Madejova, J., Bujdák, J., Janek, M., Komadel, P.: Comparative FT-IR study of structural modifications during acid treatment of dioctahedral smectites and hectorite. Spectrochim. Acta, Part a **54**, 1397–1406 (1998)
59. Noyan, H., Onal, M., Sarikaya, Y.: The effect of heating on the surface area, porosity and surface acidity of a bentonite. Clays Clay Miner. **54**, 375–381 (2006)
60. Sing, K.S.W.: Reporting physisorption data for gas/solid systems with special reference to the determination of surface area and porosity. Pure Appl. Chem. **54**, 2201–2218 (1982)
61. Ram, S.D.G., Kulandainathan, M.A., Ravi, G.: On the study of pH effects in the microwave enhanced rapid synthesis of nano-ZnO. Appl. Phys. A **99**, 197–203 (2010)
62. Khaorapapong, N., Khumchoo, N., Ogawa, M.: Preparation of copper oxide in smectites. Appl. Clay Sci. **104**, 238–244 (2015)
63. Kumar, J.P., Ramacharyulu, P.V.R.K., Prasad, G.K., Singh, B.: Montmorillonites supported with metal oxide nanoparticles for decontamination of sulfur mustard. Appl. Clay Sci. **116–117**, 263–272 (2015)

64. Chang, J., Ahmad, M.Z., Wlodarski, W., Waclawik, E.R.: Self-assembled 3D ZnO porous structures with exposed reactive 0001 facets and their enhanced gas sensitivity. Sensors **13**, 8445–8460 (2013)
65. Corro, G., Bañuelos, F., Vidal, E., Cebada, S.: Measurements of surface acidity of solid catalysts for free fatty acids esterification in Jatropha curcas crude oil for biodiesel production. Fuel **115**, 625–628 (2014)
66. Nagvenkar, A., Naik, S., Fernandes, J.: Zinc oxide as a solid acid catalyst for esterification reaction. Catal. Commun. **65**, 20–23 (2015)
67. Singh, D., Bhoi, R., Ganesh, A., Mahajani, S.: Synthesis of biodiesel from vegetable oil using supported metal oxide catalysts. Energy Fuels **28**, 2743–2753 (2014)
68. Navas, M.B., Lick, I.D., Bolla, P.A., Casella, M.L., Ruggera, J.F.: Transesterification of soybean and castor oil with methanol and butanol using heterogeneous basic catalysts to obtain biodiesel. Chem. Eng. Sci. **187**, 444–454 (2018)

Review: Simulation Models for Materials and Biomolecules

Carlton Anthony Taft, Jose Gabriel Solano Canchaya, Jose Divino dos Santos, and Junio Cesar Francisco Silva

Abstract We make an overview of chemical/physical computational simulation models for materials and biomolecular systems emphasizing basic philosophies, theoretical foundations and underlying limitations from Schrodinger´s equation to actual state of the art modeling as well as future trends and perspectives. We start with the ab initio models, including HF, CI, CC, MPT, MCSCF, CASSCF, MRC, offering good accuracy, whereas the need to address larger complex systems and computational limitations led to semiempirical models (ZDO, CNDO, INDO, ZINDO, NDDO, MNDO, AM1, PM3, PM6, PM7, DFTB) with inherent simplification/parametrization of integrals. Limitations of the ab initio approach and lack of accuracy for semiempirical models, when not specifically parameterized, led to density functional methods with excellent cost/performance/advantages and wide applications in both materials and biomolecular systems which resulted, however, in decades-long search for the best density functionals organized by Jacob's ladder (LSDA, GGA, mGGA, GH, LH, RSH, DH, MCKS, PT2/RPA). Methods such as TDFT, MC, MD, MM and AIMD are summarized. Localized/all-electron/PW basis functions, OPW, APW, LMTO, LAPW, pseudopotential (PAW, NCPP, USPP) approach within the framework of DFT, AI, SE, AIMD, introduced methodologies for electronic structure/properties calculations for the solid material state which, with support of efficient, versatile computer codes, made feasible simulations of a wide range of properties in materials/biomolecular systems. We introduce recent cognitive overload addressed by sharing/feedback/free access of data/software/technical experience and discuss as well methodologies, research areas, databases of the so-called second computer revolution/fourth scientific paradigm (material learning

C. A. Taft (✉)
Centro Brasileiro de Pesquisas Físicas, Rua Dr. Xavier Sigaud 150, Urca, 22290-180, Rio de Janeiro, Brazil
e-mail: catff@terra.com.br

J. G. S. Canchaya
Grupo de Métodos Computacionais Aplicados a Nanomateriais-GMCAN, Faculty of Physical Sciences, National University of San Marcos, Avenida Venezuela Cidra 11, Lima, Peru

J. D. dos Santos · J. C. F. Silva
Universidade Estadual de Goias (UEG), 75132-903, Anápolis, GO, Brazil

F. A. La Porta and C. A. Taft (eds.), *Functional Properties of Advanced Engineering Materials and Biomolecules*, Engineering Materials,
https://doi.org/10.1007/978-3-030-62226-8_2

models applied to material science). We also address Docking, Pharmacophore, Homology modeling for Biomolecular Systems as well as Coarse-Grained methodology including Force Matching, Inverse Boltzmann, Inverse Monte Carlo, Bayesian Dissipative Particle Dynamics with applications in Soft Matter, Macromolecules, Polymers, Interfacial Systems (polymer/material), Biomolecules, Water, Proteins, Carbohydrates.

1 Introduction

The Schrödinger's equation (SE) famous in wave mechanics, in turn, describes the exchange energy of attraction or repulsion among like particles, leading into consideration the electron spin (for the time-dependent solution), and the Pauli exclusion principle, basically sparked a new quantum theory. Such approaches introduced a new era, whereas the necessary physics laws for giving a more rigorous mathematical description of most chemistry and a substantial part of physics became known but with very complicated unsolvable equations. We can use a 3 N wavefunction that obeys SE to determine properties for a system of N particles; however, the $\sim 10^{23}$ number of atoms in macroscopic biomolecules/materials systems makes it nontrivial to solve the SE leading to alternatives/approximations. More specifically, computational simulations based on the approximations of the quantum–mechanical description of the underlying interactions among electrons and atomic nuclei have been long developed over several decades, leading to significant progress in the field of solid-state physics, chemistry, biology, materials science and so on. In all cases, such apporaches promote a more in-depth understanding and the possibility to contribute significantly to the emergence of new materials/biomolecular designed for future technologies [1–409].

The three main approaches used to solve SE are the presently still expensive ab initio methods, which yields a convergent path to exact solutions; semi-empirical methods with limited accuracy but more realistic modeling of larger systems and density functional theory which has good price/performance ratio. The SE Hartree–Fock self-consistent (mean-field) ab initio method to electron configurations in atoms/molecules and crystals is a well-known "zero-approximation" approach [1–197]. Post HF methods include configuration interaction, coupled cluster, many-body perturbation theory, complete active space self-consistent field, and multi-reference configuration interaction methods. The semi-empirical models, despite fundamental and parameterization sources of errors, can be parameterized from experiments to perform better for particular cases addressing larger systems as well. Density functional theory (DFT) introduces the ground state density to solve SE with sufficient accurate results for medium-size molecules, but not a systematic path for improvement. However, efficient algorithms and increasingly sophisticated codes, in general, have been devised in recent years toward solving the Kohn–Sham (KS) equations, that is, thereby making the application of DFT-based methods as a popular tool for simulation of diverse complex systems. Time-dependent DFT (TDDFT) introduces time-dependent single-particle equations with an additional gain of a wide/growing range of new applications [1–197].

As is well-known, the structure and conformational dynamics of diverse molecular systems are, in principle, governed by noncovalent-like interactions and are thus crucial to reactive properties. Molecular mechanics, used in some semiempirical simulation models, is a classical potential function for describing noncovalent interactions and hence might, in theory, be expressed as a sum set of its functional forms. Molecular dynamics calculate the forces on atoms of the system and uses Newton's laws of motion to predict time spatial dynamics and have become an important simulation tool across numerous multidisciplinary fields. The intersection of QM and Monte Carlo (MC) methods yields quantum MC (QMC) methods, which are usually capable, for example, achieving a level of very high accuracy. As such, the combined usage of MM and QM yields the QM/MM method. In ab initio molecular dynamic methods, in particular, both electronic structure and dynamics of the atoms are solved in practice concurrently through the use of Newton's equations set, that is, allowing calculations of the dynamic time evolution of structures without the need for predefined potential energy surfaces [198–243].

Material simulations use a combination including all-electron, local, plane waves, augmented, projected, muffin-tin basis functions, pseudopotential approach within the framework of SE, AI, DFT, AIMD via software codes such as VASP, CRYSTAL, GAUSSIAN, WIEN2K, CASTEP, and others. We introduce recent cognitive overload addressed by sharing/feedback/free access of data/software/technical experience. Methodology, databases, research fields, future predictions of applications in materials science are addressed [244–272].

Molecular docking among the biomolecular models is extremely useful for the predicting binding modes of molecules using a protein target. Pharmacophore modeling is basically the ensemble of steric and electronic features required, more specifically, towards ensuring optimal supramolecular interactions of a desired biological target in order to trigger/block macromolecule response [1, 273–301]. Comparative, knowledge-based, homology modeling has the objective of predicting the three-dimensional (3D) structures of a given protein sequence. This consolidated strategy has usually based on known 3D systems with one or more homologous proteins. Pharmacophore modeling represents a conceptual approach for describing structure binding affinity relationships using sets of approximate rules-of-the-thumb and can be described as molecular frameworks having the important features of a drug's biological activity [273–301].

In coarse-graining, one cluster groups of atoms into sites (beads/quasi-particles), which can interact through more computationally efficient interactions reducing the total number of degrees of freedom, whereas a combination of CG interactions allows significant jumps in accessible temporal/spatial scales. Two broad approaches are typically used to design CG models, i.e. bottom-up models based on atomic simulations and the top-down approach, which uses macroscopic experimental data to derive the parameters. CG has opened, on time scales inaccessible to all-atom models, new incredible ways to simulate large-scale material and biomolecular processes. As such, the CG models addressed in this chapter include the Martini Model, Force

Matching, iterative Boltzman inversion, inverse Monte Carlo, Bayesian Interference with Dissipative Molecular Dynamics. Applications include soft matter, macromolecules, polymers, biomolecular systems, interfaces, water, carbohydrates, and proteins [302–372].

Efficient computational algorithms and visualization tools have/are been implemented in sophisticated codes, boosting applications and opening new doors to innovative research on materials/biomolecular systems. Machine learning (ML) methods in materials science have fueled the actual second computer revolution/fourth scientific paradigm [373–409].

However, there is no simple/clear hierarchy' regarding with temporal/spatial scales involved for selecting best simulation models for each case and the best route is to probably to first understand the basics of each model (objective of this chapter) and decide first what questions/phenomena are of primary interest. This chapter summarizes the state of the art of typical multidisciplinary simulation models for materials/biomolecular systems emphasizing their basic philosophies, theoretical foundations, and underlying limitations and may thus help workers in the field understand/make good choices for the usage of simulation models.

2 Ab Initio Methods

The SE determines the properties for a system of N particles with three approaches used for the solution of the equations, i.e. ab initio, semi-empirical and density functional methods. As such, it is known that the *ab initio* or first principles approach has widely characterized as purely theoretical from the fundamental point of view. In non-relativistic quantum mechanics (QM), well-defined approximations are widely used to solve SE that can be systematically improved, that is, converting into exact solutions from using only fundamental physical constants of nature for such calculational simulations. On the other hand, as is well-known, the semi-empirical methods involve approximations, generalizations used to simplify calculations and produce results with a sufficient empirical basis. We note, however, that even the ab initio methods incorporate some approximation, such as the Born–Oppenheimer model, yielding some degree of semi-classical features. In addition, the basis sets, which must be complete, are "put in by hand"" and restricted by computational capacity, i.e., from a conceptual viewpoint, no calculation is fully ab initio [1, 9, 12–23].

From classical mechanics, it was already known the difficulty of finding exact solutions for interactions in many-body systems and, for the elucidation of the main features of molecular systems, quantum physics ended up confronted with the same n-body problem which made it necessary even for ab initio methods to introduce some type of approximation. Note that the SE used in this case is independent of time, which is also considered as an idealization. Relativistic mass effects are also ignored and most computational simulation programs usually assume the Born–Oppenheimer approximation, i.e. separating the molecule into nuclear movement and electronic structure which is an acceptable simplification since the nuclei are much heavier than the electrons, that in turn move much faster than nuclear mass, that is, as if the nucleus was in a fixed position and the nuclear and electronic coordinates could be separated [1, 12–23, 70–75].

In order to understand molecular structure and reactivity using QM, concepts of bonding and valence are central, whereas alternatives to elucidate chemical bonding are usually based on valence bond (VB) and molecular orbital (MO) theories, which are non-self-consistent field (SCF) and SCF models respectively. As such, it is well-known that the LCAO (linear combination of atomic orbitals) is the essence of the MO approach, which, from a simple electronic structure viewpoint, helps understand the relationship among atoms and molecules. More chemically intuitive, VB concepts are based on an understanding of bond/lone pairs. Thus, the progress in the computater field lead to the study of medium-sized molecules from the orthogonal MO formalism, with computationally tractable equations, whereas electron correlation is neglected in simple MO theory and overemphasized in VB theory. However, the important advantage of MO theory is that it can be improved via, for example, configuration interaction (CI), i.e., excitation yielding single/double/multiply occupied configurations, which could be described by Slater determinants.

In the case of HF method, each electron of the system interacts in particular with the average field of the other electrons and since two electrons of the same spin cannot occupy a single orbital, hence the exchange–correlation is included in formalism of HF theory. Still, the instantaneous electrostatic interactions are not considered and correlated motions of electrons are, of course, neglected. Note that the difference between exact and HF energies yields the correlation energy (CE) which is necessary for a nice description of electronic structure derived properties of interest chemical system and is widely addressed in calculation methods popularly known as post-HF methods such as SCF multiconfiguration (MCSCF); configuration interaction (CI); multireference CI (MRCI); coupled cluster (CC), many-body perturbation theory (MBPT) and complete active space self-consistent field (CASSCF) methods [1, 9, 12–23].

As such, a single Slater determinant could be reasonable in closed shell structures but otherwise Post-HF methods are necessary where HF can still determine the best determinantal trial wave function energetically to be improved via configuration of excited states in which are generated by promoting electrons to virtual (unoccupied) orbitals [1, 9, 12–23]. The MOs can be used to construct the determinants in the MCSCF method with the coefficients optimized by the variational principle. The MOs (from RHF) can be partitioned into inactive and active spaces in CASSCF in order to generate the configurations. An unperturbed Hamiltonian and a perturbation term are used in the MBPT to obtain corrections of various orders as the perturbation is, of course, increased from zero-to-finite values and written as a Taylor expansion. Constructed for interacting particles, the coupled-cluster (CC) theory uses n-particle excitation operators yielding coupled nonlinear equations, whereas amplitudes are solved through iteration techniques yielding CC energies. It is worthwhile mentioning that density functional theory (DFT), discussed in subsequent sections, includes electron correlation at a relatively low computational cost when compared to post-HF methods [1, 9, 12–23].

It is also noteworthy that the accurate description of noncovalent interactions (e.g. hydrogen bonding, stacking, sigma-hole interactions, halogen bonds, cation-π) has a pivotal role in applying of these computational methods in various fields, ranging from biochemistry to material science. In contrast, these methods accuracy has been evaluated via comparing interaction energies calculated for fixed geometries. Hence, from this perspective, the nonbonding, noncovalent interactions we refer to are weak with minimum more significant changes in the electronic structures properties for geometries of the interest. It should be noted that in more considerable interatomic distances (at about of several angstroms), these interactions operate, i.e., where overlapping is not necessary and attraction arises from electrical properties of participating molecules involving permanent multipoles, between an induced multipole and a permanent multipole as well as between induced multipoles and time-variable multipoles [12, 19, 24–27]. The stabilizing energy of the noncovalent complexes can, in principle, be attributed towards induction, coulombic, charge transfer, induction, and are attractive terms. The role of the exchange repulsion term is keeping the subsystems apart. Charged molecules, for induction terms, polarize neighboring species. Interactions between fluctuating multipoles yield London (dispersion) interactions. In the case of charge transfer interactions, in particular, the electrons flow from basically donor-to-acceptor. Repulsion and dispersion contributions are typically described by van der Waals (vdW) forces (inverse of the sixth and twelfth power terms), although there could be other long-range contributions. These interactions can involve, host, guest, surrounding (crystal lattice, solvation, and the gas phase). Ion pairing interactions can be attributed to electrostatic interactions. Nonetheless, there is a substantial challenge in the quantitative evaluation of these nonbonding interactions, whereas semiempirical methods with more approximations for large systems can be considered [1, 9, 12, 24–27].

3 Semiempirical Methods

Semi-empirical simulations use empirical parameters to simplify the resolution of their calculations, whereas some of the equation's parameters are fixed by values established experimentally compared to ab initio methods, which are based solely on the use of fundamental physics/chemistry constants [1, 9, 12, 28–88].

The history of semiempirical methods is long, with many important contributors. Quantum-chemical calculation of molecules using semi-empirical models has substantially progressed since the Huckel method in 1931 for organic molecules with conjugated π systems, which was extended by Hoffmam in 1963 for non-planar molecules [29]. The early Huckel-type methods have based on noniterative schemes, including as strategy only one-electron integrals, whereas the π-electron Hůckel method used the connectivity matrix of geometry in order to generate its MOs, and all electrons of valence were included in the extended Hůckel theory [28–88]. Early stages of quantum chemistry used the zero-differential-overlap (ZDO) approximation to address evaluation of three and four center integrals at different levels, i.e.

complete neglect of differential overlap (CNDO), intermediate neglect of differential overlap (INDO) and neglect of diatomic differential overlap (NDDO) [28–88].

Pople and co-workers were early users of CNDO and also introduced the INDO models [29–31], among which ZINDO was introduced by Zerner [32] to reproduce electronic spectra. However, one of the most sophisticated of these HF based formalisms is the neglect of diatomic differential overlap (NDDO) introduced by Pople et al. [33] in which all-electron repulsion two-center integrals involving charge distributions on two different atoms are included and could be considered as a forerunner for many current semiempirical methods. We note that many semiempirical methods emerged at a time when the processing power of computers was much slower than it is today, limiting applications to studies of systems of a few dozen atoms, whereas each new method with an additional approximation was extremely welcome.

Within this framework, the modified neglect of differential overlap (MNDO) method [34], introduced by Dewar's group, was one of the most extensively used NDDO methods, which was later extended to d orbitals by Thiel et al. [35]. Offering a good compromise between computer time and accuracy (Austin model 1) AM1 was introduced in 1985 by Dewar et al. [36]. Stewart [37] later developed PM3, whereas both AM1 and PM3 were highly popularized by the MOPAC software [38].

With a considerable increase in the processing capacity of computers, during the following two decades there was a great decrease of interest in semiempirical methods. The Recife Model 1 (RM1) [39, 40] maintained the AM1 formalism but focused on the parameterization procedure to reproduce numerous experimental properties. Note that extension to a spd basis in the PM3 method, in turn, led to the development of both PM6 [41] and PM7 [42], which covers the whole periodic table, including rara-earth elements, with specific applications in molecular properties and solid-state.

A number of hybrid approaches were introduced, based on DFT, such as the semiempirical TB (tight binding) versions of DFT methods (DFTB) and the self-consistent charge (SCC) DFT method. Conceptually, however, these semiempirical methods can, in theory, be considered as a simplified version of ab initio MO approaches, whereas extensive integral approximations and parameterizations are used [43–45]. During the long history of semi-empirical methods, the MNDO approach was mostly used in the 1980s and 1990s for quantum-chemical computations whereas AM1, PM3, MNDO methods are still used along with more recent versions such as PM6, PM7 although DFT calculations have become more commonplace [1].

Electronic structure methods typically approximate the molecular orbitals $\{\psi_i\}$ using a strategy based on the linear combination of basis, i.e. whereas all-electrons in the conventional HF theory are typically expanded in the atomic orbitals (AO) basis with SE methods treating only electrons of valence shell assumed in the field of the nuclei and the inner-shell electrons (unpolarizable) using Slater-type orbitals. Besides that, $C_{\mu i}$, the molecular orbital coefficients, can then be obtained from $FC = SC\epsilon$ (Roothaan – Hall equation), where C, F, S, ϵ are the Fock, MO-coefficients, overlap matrices and diagonal matrix containing the orbital energies, respectively.

The Fock matrix (F) is usually separated into two-electron (G) and one-electron (h) parts [58].

Yet, only two-electron integrals which in general are an integral of the type $(\mu\mu|\nu\nu)$ are, in turn, calculated in the zero-differential overlap (ZDO) approximation. Note that this approximation is also used in the SE methods of CNDO family; with however, the retained integrals parametrized using the Mataga – Nishimoto approximation [29]. Similar to CNDO is INDO (intermediate neglect of differential overlap), whereas integrals centered on the same atom are, therefore, parametrized. INDO can be further modified by including two-electron integrals $(\mu\nu|\lambda\sigma)$ centered on different atoms, leading to the NDDO approximations where simulation of the interaction is not only between two nuclei but also the electronic core [1, 29–31, 58]. MNDO is an improvement of NDDO by introducing approximate integrals from multipole interactions to replace two-center as well as improved core – core interaction terms. As such, the AM1 method introduces a complete reparameterization of MNDO and adds up to four Gaussian functions in order to the core – core repulsion term yielding improvements in hydrogen bonding predictions. PM3 is a reparameterization of the AM1 method, however, using a different computational strategy. Also, it should be emphasized that among these aspects the PM3 method only considerer two Gaussian functions in order to correct the core – core repulsion. Reparameterization of PM3 led to RM1, whereas PM6 uses a much broader set of reference data in order to fit the parameters with diverse improvements in the description of the core – core terms as well as added to certain elements d-orbitals in their atomic basis, from a scheme similar to the MNDO/d method [35]. PM7 further includes hydrogen-bond corrections, a dispersion correction, and several semiempirical integrals are modified to enforce classical longer distance convergence [58].

Using pairwise distance directed gaussians (PDDG) for a descripition empirical of core repulsions led to the further develompent of both PDDG/PM3 and PDDG/MNDO models [47]. The usage of a larger set of reference data served to introduce the general-purpose known as RM1 model. Inclusions of orthogonalization corrections in terms of the one-electron in which led to the OM1, OM2, OM3, and MSINDO models, which can, in principle, account to the important effects of Pauli exchange repulsions [48, 49]. In the OMx-D methods [48] dispersion corrections are added to the OMx methods. The MAIS (method adapted for intermolecular studies) and the PIF (parametrized interaction functions) [51–53] add longe-range functions to improve more specifically the description of nonbonded interactions. As such, the SCP-NDDO approach improves the polarization of PM3 method and in this case includes a post-SCF dispersion contribution [54].

Derived within the framework of DFT, the density functional tight-binding (DFTB) method is based on an expansion of the energy known as Taylor series from a reference density [55, 56]. In this perspective, the DFTB calculations are quite similar to the NDDO/MNDO methods, and hence are faster than standard DFT-GGA approaches at about 2–3 orders of magnitude. This is largely due to the exclusive use of a minimal AO basis set to the treatment of the electrons of valence. As such, more specifically, the high computational efficiency results are strongly dependent on integral approximations used, as well as modern computational implementations, which lead to determination beforehand of the AO Hamiltonian matrix elements, that is, so that they do not have computed during such simulations [55, 56, 59]. Thus, in this perspective, it is evident that the Taylor expansion, in principle, does not pose a crucial approximation, i.e., being, therefore, merely a formulation used to cast the DFT total energy in a form suitable for such approximations. DFTB2 (referred to as SCC-DFTB) was parameterized for many first-row transition metals, including Sc, Ti, Fe, Co, and Ni [59]. DFTB3 was also parametrized for Mg^{2+} and Zn^{2+} [60, 63, 65, 73, 74].

Summarizing, the SE approach involves minimal basis sets and approximations to electron integrals, which can lead to computational efficiency increase and consequently perform 1000 times longer sampling or larger systems. We can further improve these enhancements by integrating based on SE approaches using modern computational architectures, computational algorithms, faster diagonalizations, and extended algorithms. Nonetheless, there are sources of error for SE since there are limitations of the parent HF method (electron correlation). DFTB, uses the generalized gradient approximation (GGA) [58], which lacks electron correlation and does not describe dispersion properly. The usage of a minimal basis set also introduces some errors in calculations of van der Waals interactions, electronic polarizability, hydrogen bonding, and the integral approximations also lead to nonbonded interaction errors. However, it is well-known that empirical corrections/formal extensions at the SE approach have significantly improved the accuracy for the intermolecular interactions and even condensed-phase related problems based on the use of NDDO/MNDO/DFTB models as strategy. Future modifications should, however, attempt to further improve hydrogen-bonding, dispersion errors, Pauli repulsion, and scaling with respect to system size.

4 Density Functional Theory

The electronic energy, $E_e[\rho(r)]$ can, in principle, be conveniently written as a functional of the electron density within the Born–Oppenheimer approximation [89, 90], $E_e[\rho(r)] = T[\rho(r)] + V_{en}[\rho(r)] + J[\rho(r)] + Q[\rho(r)]$ where $Q[\rho(r)]$ is the quantum electron–electron interaction energy, $J[\rho(r)]$ is defined as the electron–electron repulsion energy, $V_{en}[\rho(r)]$ is the nuclear–electron attraction energy and $T[\rho(r)]$ is defined as the electron kinetic energy. The kinetic energy contribution is, of course, the largest unknown term and must be approximated accurately. The difference between the kinetic energy and its approximation together with $Q[\rho(r)]$ yields the exchange–correlation energy ($E_{xc}[\rho(r)]$) often represented as the sum of exchange $E_x[\rho(r)]$ and correlation $E_c[\rho(r)]$ functionals addressed in this section [1, 89–142].

DFT aims to develop for $T[\rho(r)]$ and $Q[\rho(r)]$ accurate approximate functionals [1]. The infinite uniform electron gas exact approximation to the kinetic energy is the early, well known Thomas–Fermi model [92], only applicable, however, to systems with nearly uniform densities, whereas it is difficult to design accurate kinetic energy functionals. Kohn and Sham avoided this problem by showing that a single Slater determinant of orbitals could accurately approximate the kinetic energy and describe a fictitious system of non-interacting electrons having the same density as the exact electronic wave function [90]. The properties of many-body systems are usually a function of their density in the ground state. They hence can thus be determined yielding a new independent electron problem to specifically substitute the many-body problem [1].

Besides that, Kohn–Sham (KS) DFT defines equations for a set of orbitals to be solved whose density is defined to be exactly that of the real system with basis a better cost/performance than wave-function theory, although the exact formula is not known, requiring hence the use of different approximations [1]. The sum, Exc[ρ(r)], of exchange Ex[ρ(r)] and correlation functional Ec[ρ(r)], are the only unknown terms in KS-DFT or DFT which will be interchangeably used in order to refer to modern DFT [90, 91]. During the last 3 decades, numerous [89–189] non-empirical/semi-empirical density functionals have been developed with numerous important contributions, including those of John Perdew´s in non-empirical density functional development as well as the contributions from Axel Becke´s for the development of semi-empirical density functionals which have influenced most of the popular density functionals currently in usage [1, 89–189].

We start by dividing the DFT functionals into non-empirical and semi-empirical models. The non-empirical models (TPSS, PBE, and others) [93, 94] involves the design of functional forms satisfying exact constraints like the expansion of the second-order gradient in the slowly varying limit, whereas free parameters can be determined from fitting to appropriate norms [91, 93, 94]. The semi-empirical routes, i.e., B3LYP, B97 [95, 96] and others involve a selection of flexible functional forms, i.e., power series with undetermined coefficients enhancing variables based on ρ,∇ρ, and so on. In contrast, note that the coefficients are in general fitted to accurate reference values, such as coupled-cluster with singles, doubles, and with perturbative inclusion of triplets model at the complete basis set limit. Combining the two approaches [97, 98] has also been explored (ωB97, MS1).

We note, however, that such density functionals are not systematically improvable, i.e. there is no guarantee that more exact constraints/additional ingredients/ flexible functional forms will yield improvements for all types of interactions, whereas in contrast, wave function theory allows systematically improvable models. Thus, it has been reported 'Jacob' Ladder in DFT [99] which starts with the Hartree model (zero exchange–correlation energy, classical electrostatics for the electron–electron interactions) introducing additional ingredients to move up the ladder to accuracy (~0.1 kcal/mol for nonbonded interactions and 1 kcal/mol for chemical accuracy). In a metaphorical sense, it should be noted that the Jacob's Ladder by Perdew involves five rungs that correspond to increasingly sophisticated models used for the treatment of exchange–correlation functionals, whereas each rung contains new physical content leading to improved accuracy at each higher level [99]. In this perspective, it is well-known that some functionals are based on non-empirical strategies, i.e. adopting known conditions allows fix all variables, however, are usually insufficient in power/number to fix the form of all but the simplest functionals completely [91]. In this line, semi-empirical functionals`are in large part used for more complex functional designs with increased potential for overfitting, whereas some argue that such approaches used in many modern functionals sacrifice physical rigor for flexibility of empirical fitting [1, 89–189].

The first rung of Jacob's Ladder depends only on the electron density, i.e. the simplest exchange–correlation functionals or local spin-density approximation (LSDA), which is highly inaccurate for most real systems with inhomogeneous density distributions. We note that there is an exact analytic LSDA form for exchange functional whereas the LSDA correlation functional has no exact analytic form and the three most popular parameterizations (VWN5, PZ81, and PW92 [100–102]) were fitted to accurate Quantum Monte Carlo data [103].

The second rung of the Jacob's Ladder uses the density gradient, $\nabla\rho$, to introduce the generalized gradient approximation (GGA) functionals to account for inhomogeneities in the density, improving the LSDA model significantly. In the 1960s Frank Herman and colleagues made the earliest attempts to use the density gradient to improve LSDA using X$\alpha\beta$ exchange functionals [104, 105]. Since the potential of the X$\alpha\beta$ exchange functional diverged asymptotically alternatives were introduced by Becke leading to B86, PBE and B97 density functionals [95, 96, 106]. Combining exchange and correlation, some successful GGA density functionals include PBE, B86, BLYP, PW91, PBE, revPBE, RPBE, PBEsol [93, 107–113]. Parameterization of GGA exchange–correlation by Handy and co-workers led to semi-empirical GGA functionals, such as HCTH/93, HCTH/120, HCTH/147 and HCTH/407 [114–116]. Along this line the Truhlar group and colleagues [117, 118] also introduced semi-empirical GGA functionals. Other GGA exchange–correlation functionals introduced were BOP, BPBE, mPW91, OLYP, PBEOP, rPW86PBE and SOGGA [119–123].

Laplacian of the density, as the kinetic energy density, can also be used for further improving the accuracy of density functionals, whereas the introduction of these functionals defines the third rung of Jacob's Ladder that are known as meta-GGA (meta-generalized gradient approximations). The more popular kinetic energy density is used to add flexibility to the functional form (constraint/least-squares fitting) and detect electron delocalization in molecules [124]. Note that non-empirical meta-GGA functionals include PKZB, TPSS, TPSS, MS0, MS1, MS2, MVS, SCAN, TM, BLOC [125–130]. Semi-empirical meta-GGA functionals include VSXC, τ-HCTH, TPSS), M06-L, M11-L, MN12-L, MN15-L, mBEEF [131–138]. As is well-known, the meta-GGAs tend to outperform GGAs, mainly due to more sensitive to the integration grid, lead to nice results for barrier heights and even medium-range dispersion [91]. However, this greater sensitivity also can affect accuracy for weakly interacting systems. Despite systematic improvements offered by Jacob's Ladder there are major limitations to exchange–correlation functionals described, i.e. self-interaction error (SIE) [101], strong correlation and long-range dynamic correlation (dispersion).

The exact exchange term in KS-DFT is replaced by the exchange–correlation functional, typically not one-electron SIE (self-interaction energy) free. One solution is replacing the local exchange functional with use of the exact HF exchange functional and using a local correlation functional that yields, for any one-electron system, exactly zero correlation energy, such as LYP. Early attempts to make this combination was not successful, and this approach was not continued until Becke, in the early 1990s, introduced the idea of mixing exchange–correlation functional with a global fraction of exact exchange yielding a successful solution defining Rung 4 of Jacob's Ladder. Through the adiabatic connection formula, in general, this global hybrid (GH) functionals can theoretically be justified, as well [91, 139–142].

B3PW91, was developed in 1993 by Becke [95], from fitting three linear parameters to 56 atomization energies, being known as the first global hybrid functional. Yet, the most popular density functional, B3LYP has an exact exchange mixing parameter of 20% [95]. PBE0, with 25% exact exchange, is one of the most popular non-empirical global hybrid GGA [119]. This formalism can also be extended to meta-GGAs. Similarly, among the preferred global hybrid meta-GGA density functionals for which non-empirical models include TPSSh, revTPSSh, MS2h, MVSh and SCAN0 [91, 143–146]. Truhlar has contributed numerous global hybrid meta-GGA density functionals, including MPW1B95, MPWB1K, PW6B95, PWB6K, M05, M05-2X, M06, M06-2X, M06-HF, M08-HX, M08-SO, MN15 [91, 147–154] with a fraction of exact exchange functionals varying between 27 and 100% (M06-HF). Although global hybrid functionals introduce a significant improvement for kinetics and bonded interactions, they do not address self-interaction completely.

In the range-separation [155, 156] approach the exact exchange contribution is divided into a long-range component ($EHF_{x,lr}$) and a short-range component ($EHF_{x,sr}$) each complemented by error functions whereas the short-range component is optionally scaled to give a non-zero fraction and the long-range component set to one in order to ensure that the functional is SIE free in the long-range. In should be noted that the corresponding local exchange functional has similarly partitioned with scaled short- and long-range components. Range-separated hybrid (RSH) semi-empirical GGA functionals include ωB97, ωB97X [97] less-empirical long-range corrected RSH GGA functionals LC-ωPBE08, LRC-ωPBE, and LRC-ωPBEh, screened-exchange functionals HSE-HJS, N12-SX as well as RSH meta-GGA functionals (M11, MN12-SX) and CAM-B3LYP RSH GGA functional [158–164]. It is also possible to entirely remove the HF contribution to form screened-exchange functionals (HSE-HJS) [91]. Based on the local hybrid functionals represent another class on Rung 4 of Jacob's Ladder, including e.g. B05 and PSTS functionals which explicitly depend on the exchange energy density and modulate between local and exact exchange [165–168]. Using a multi-configurational wave function to replace the single Kohn–Sham reference should, in principle help define Rung 5 of Jacobs Ladder [169]. These double hybrids functionals, adopting as strategy the use wave function expressions for incorporated the electron correlation (based on the second-order perturbation theory, called PT2, or the random phase approximation); in addition to wave function exchange in which tends a yielding higher fractions of exact exchange, i;e lower SIE [91, 169].

In addition to SIE, it is well-known that local and hybrid exchange–correlation functionals are also unable to account properly for long-range dynamic correlation and along the past decade has seen significant works with a focus on improving the description of non-covalent interactions within KS-DFT yielding the DFT-D dispersion approach by Grimme and co-workers [170–172] as well as nonlocal correlation (NLC) functionals vdW-DF2 by Lundqvist and co-workers and VV10 by Vydrov and Van Voorhis [173]. This lead to a damped, atom–atom empirical potential (DFT-D) trained as a correction for any functional (e.g., that tend to underbind non-covalent interactions) was introduced by Grimme yielding DFT-D1, DFT-D2, DFT-D3 as well as the first training of a semi-empirical functional with the inclusion of a dispersion correction simultaneously, i.e. B97-D (local GGA functional with D2 tail) [170–173] Other similar functionals include ωB97X-D, ωM05-D, CHG, ωB97X-D3, ωM06-D3 [174–176]. NLC functionals have also been attracting interest since they are less empirical than the DFT-D approaches using only two fitted parameters yielding VV10 and LC-VV10, B97M-rV, SCAN + rVV10 and rVV10 [177–179].

We note that all density functionals fail to some extent for strongly correlated multi-reference systems, which is expected since the HF and KS-DFT are single-determinant methods, whereas multiple determinants would be required to describe multi-reference systems accurately. From a practical viewpoint, the strong correlation within KS-DFT is the least solved problem requiring more local functionals and hybrids with a small fraction of exact exchange typically perform acceptably well [91]. We also note that it is possible to include multi-reference systems in the training set of a semi-empirical functional with however, worsening performance for single-reference systems.

It has been noted that there is no universally accurate present-day density functional. Notwithstanding, these functionals are computationally efficient whereas numerical errors are generally acceptable with main errors attributed to correlation and nonlocal exchange yielding delocalization errors associated with SIE reflected in non-covalent interactions (NCI), barrier heights (BH), isomerisation energies, radical-neutral complexes, ionisation energies in large molecules, mixed-valence systems. In this view, the global hybrids or range-separated hybrids with exact exchange fractions are nice functionals for SIE-sensitive problems [91]. Cohen et al. indicated that the DFT method with approximate functionals has a very nice performance for a broad range of physical and chemical properties [184]. Vibrational frequencies are also well described by different density functionals, whereas electric polarizabilities are susceptible to SIE [91]. DFT chemical shifts have been widely applied. In has been noted that the standard Hohenberg–Kohn theorem does not hold in the relation to magnetic fields requiring different functionals for each field strength or current density dependence [180–182].

Future routes of interest include the combinatorial design approach for the development of strategies known as PT2 and/or RPA-based Rung 5 density functionals as well as new descriptors substantially created an enormous impact on the development of functionals, more specifically, with lower SIE/long-standing self-interaction corrections [91]. Future trends include the usage of a multi-configurational wave function to replace the single Kohn–Sham reference, which will however face definitional ambiguities and complexities that confronts generalization of single-determinant-based wave functions [91, 183–189]. Moreover, it is known that the development of current–density-dependent functionals is also of interest. In general, the KS-DFT can be used in the TDDFT framework allows calculating electronic excited states and responses due to time-dependent perturbations as summarized in the following section.

5 Time Dependent Density Functional Theory

Extensive studies carried out with ground state stationary properties of molecules, excluding excited electronic states, the KS-DFT replaces the N-electron wave function with an effective mathematically simpler single-particle charge density, which can be solved much faster [9, 190–197]. A single-particle problem substitutes the interacting many-electron system. As is well-known, the TDDFT method has been proving that applies the same concept to time-dependent problems by replacing the complicated many-body time-dependent SE, but replacing time-dependent single-particle equations with orbitals yielding the same time-dependent density [9, 190–197]. This is possible because the known Runge-Gross theorem [190] indicating that, in theory, a given an initial wavefunction/particle statistics/interaction, at most one time-dependent external potential $V_S(r, t)$ can yield a given time-dependent density n(r, t). In addition, the TDDFT equations can therefore be used to describes a set of N non-interacting electrons evolving in V_S (r, t) and, consequently, producing in this case of the same n(r, t) for an interacting system. Applications and development of TDDFT have indicated considerable growth of predictions for the non-perturbative regime (systems in laser fields); linear and higher-order regime (optical response/electronic transitions as well as back to the ground-state approximations from the treatment of excitations) [9, 190–197].

Theoretical analysis tools, such as TDELF (time-dependent electron localization function), it should yield observation time-resolved of the processes related to formation/modulation/breaking of chemical bonds that providing a visual understanding of the dynamics of excited electrons, with a range of applications from above-threshold ionization of metal clusters to coherent control of quantum wells, multiharmonic generation in benzene and linear response to a spatially uniform electric field [190–197]. We note, however, that there are sources of error in TDDFT calculations/predictions, which are likely due to local (or gradient-corrected) approximations and properties that require non-locality in time.

6 Quantum Monte Carlo

The intersection among the QM and MC methods is, of course, referred to as quantum Monte Carlo (QMC). In contrast, the true/exact Schrodinger wave function is approximated by a varied trial wave function to get more accurate functions. QMC methods can solve the stationary SE using stochastic processes as well as sampling of wave functions of electron positions (in the space) yielding high accuracy [1, 9, 198–214].

Solving the many-body problem SE is difficult with exponential growth of the Hilbert number of particles whereas the premise of MC methods indicating that properties of interest can, in principle, be computed stochastically converging, irrespective of the dimension of the Hilbert space, as the inverse of the squared root of independent samples [202]. Moreover, MC based task is to generate independent samples as strongly interacting systems/arbitrary topologies, i.e. in the case of classical simulations based on MC of quantum systems [9, 198–214].

This QMC approaches include variational MC (VMC), fixed node diffusion Monte Carlo (FN DMC), Green's function Monte Carlo (GFMC), Fermion Monte Carlo (FMC), Self-healing diffusion Monte Carlo (SH DMC), auxiliary field quantum Monte Carlo (AFQMC), reputation quantum Monte Carlo (RQMC), full CI Quantum Monte Caro (FCI-QMC) [9, 198–214].

7 Molecular Dynamics

In molecular dynamics (MD), given the positions of all of the atoms in a system, the force exerted on each atom by all others is calculated and Newton's laws of motion used to predict spatial position of individual atoms as a function of time, whereas these one-step processes are repeated, recalculating the forces on each atom which are used to update all atomic positions and velocities. MD simulations for atomic systems is thus a numerical/step-by-step solution for the classical motion equations, whereas the forces on the atoms are typically derived from potential energies (functions of the complete set of 3 N atomic coordinates) yielding a trajectory (series of molecular configurations) [9, 215–230].

The potential energy mathematical models of molecular systems typically consist of a sum of interaction types, i.e. intramolecular/local contributions (bond stretching, angle bending, dihedral/improper torsions), repulsive, as well as Van der Waals interactions (12–6 Lennard–Jones (LJ) potential) which have improved substantially over the past decade. We have noted that need nice models to represent the interatomic forces that exist among the atoms of the system whereas the spatial and/or time-scales required are prohibitively expensive and empirical methods based on force-field are usually used to describing the dependence of the system energy based on particle coordinates [9, 215–230].

Usually, a simple harmonic function often represents bond stretching controlling length of covalent bonds, angle bending by harmonic/trigonometric potential, torsional energies usually by cosine functions; improper torsions measured by appropriate angles that evaluate deviations from planarity; Van der Waals interactions among two atoms as a result of both repulsive (overlapping clouds between atoms) and attractive interactions between induced dipoles (12–6 LJ potential); electrostatic interactions with partial atomic charges assigned to each nucleus and Coulomb's law to compute contribution to the total energy. Crossed terms can also be used to describe the coupling between stretching bending torsion. Hydrogen bonding is often described by choice of van der Waals parameters and partial charges. Polarization effects can be included explicitly using methods such as fluctuating charges, shell models, and induced point dipoles [9, 215–230].

In order to perform the simulations, we usually need to know the positions and velocities of initial each particle, which for some systems (crystals, proteins, etc.) are reported in the form of a crystallographic information framework file. For the intramolecular potentials analytical expressions for the forces can be used or determined by numerical differentiation. We can then simulate, in this case, our system based on the isolated N-particles, that is, surrounded by vacuum or imposing some boundary conditions. In order to solve the equations numerically [219] we need to discretize the trajectory over small time steps using an integrator with minimal requirements to compute the forces, i.e., nice stability if large time steps Δt are in general used to obtain nice accuracy and conservation of energy/momentum, beyond the time reversibility. Verlet [227] proposed the sum of Taylor expansions for $+\Delta t$ and $-\Delta t$, whereas some terms cancel leading to simple expressions, i.e. Verlet integrator, velocity-Verlet as well as leap-frog algorithms. One form of controlling the temperature during the simulation is based on the usage of a Berendsen thermostat, consisting of a weak coupling of the system to a heat bath [228]. Note that the modifying the motion equation , an additional term that acts as a frictional force and a coupling constant that is responsible by determines the coupling strength [9, 215–230].

There are, however, intrinsic limitations related to using empirical simulations based on the force-field, i.e., such approaches are, in principle, well-known by the inability to reveal details on the electronic structure; handle reactions (breaking/formation of bonds, electronic excitations, charge transfers) and its accuracy, in general, depends on the parameterization quality; limited transferability. These simulations are, therefore, diverse orders of magnitude faster than traditional ab initio calculations, that is, such approaches avoiding solving the electronic structure of the interest system in order to obtain the interatomic forces making it possible to easily handle systems of large complexity.

8 Ab Initio Molecular Dynamics

Ab initio molecular dynamics (AIMD), [9, 231–243] became of a simulation type of considerable interest, especially in the last decades. Despite the success of MD simulations in the field of medicional chemistry, however, because the predefined/fixed potentials can in principle be a drawback for complex systems [9, 231–243]. Car and Parrinello [231] proposed that electronic wave function parameters could a priori be treated with the use of dynamical variables and application of the steepest descent method to the classical Newtonian equations of motions, in this case, are used to solve electronic structure problems whereas the classical motions of the atom nuclei could be coupled with fictitious electron dynamics. In this context, it is known that the QM adiabatic can in principle be used to timescale separation of fast electronic and slow nuclear motion in which transform the electronic problem into a classical mechanical adiabatic energy scale separation whereas both the electronic structure problem and the dynamics of the atoms are solved concurrently by a set of Newton's equations allowing calculations of the fully dynamic time evolution of structures without the need of predefined a potential energy surfaces [9, 231–243].

This first-principles "on-the-fly" the electronic potential energy surface calculations associated with the nuclear dynamics method has opened new perspectives and research opportunities [238]. An AIMD simulation assumes that the Born–Oppenheimer approximation is valid. Hence, the system is composed of nuclei and electrons and, in this case, the nuclear dynamics is treated classically. Since they cannot solve the ground-state electronic problem exactly, in particular, approximate electronic structure methods are, in turn, needed in order to complement these findings [9, 231–243]. Whereas in modern AIMD, the electronic structure method for complex system is most commonly based on KS-DFT formalism, leading to constructing an interaction potential in order to selecting an approximation for solving the SE [9, 231–243].

9 Material Simulations

The orbitals of the methods discussed previously are represented by a linear combination of functions such as Slater-type (STOs) with an exponential decay of the radial part as well as Gaussian-type basis functions (GTOs), whereas only the orbital expansion coefficients are usually optimized, that is, yielding linear combinations of primitives that can in principle be used to define angular momentum (which is called contracted basis sets). Both STO/GTO sets are usually centered at nuclei in which leads to a well-known linear combination of AO (LCAO), i.e., so-called local methods that can be used in material simulations [9, 12, 244–272]. In addition, we note that the usage of plane waves accommodates the periodicity of the lattice/potentials imposed by Bloch's Theorem, yielding a orthonormal functions used for expansion/labeling orbitals in reciprocal space. As such, it is known that the plane waves do not depend on the positions of the nuclei and in general are delocalized (in space), which is widely interesting for solid-state material simulations.

The core levels (completely localized) could be determined by early workers (1940), whereas one method of solution for the band levels was a variational calculation using a complete set of· plane waves (due to lattice periodicity only those wave vectors which reduce to the k value of the desired Bloch state need be included). Orthogonalizing the plane waves to the core levels and performing a variational calculation using a limited number of the OPW's (orthogonalized plane waves) led to lowest roots of the secular equation converging to lowest band levels with the increasing number of OPW's, i.e. the orthogonalized plane wave (OPW) method [9, 12, 249–252].

We note that for plane waves enormous cutoff parameters could be necessary to improve the quality of basis set/core potentials. Defining plane waves in curved space yields orthonormal functions forming a complete basis set for reciprocal point sampling introducing a generalization of the plane wave concept (GPW). The cut-off energy becomes lower in vacuum regions and highest in regions close to nuclei, that is, requiring a lower number of GPW and allowing even in this case all-electron electronic structure calculations for situations where plane wave fails, as well. Also, it is known that the plane waves although they are considered simple, lack the spatial bias of GTOs, stimulating longstanding efforts to combine/exploit the mutual strengths of plane waves and localized gaussian basis functions, taking into consideration that periodicity makes possible calculations of unit cell calculations of bulk matter [9, 12, 249–252].

The so-called augmented wave methods have basis functions which are combinations of atom-like wave functions in the atomic regions and envelope type functions between the atomic regions whereas space is divided into atomic regions as well interstitial regions between the spheres leading to partial solutions in each region which are matched at the interface between interstitial and atomic regions. The muffin-tin orbital (MT) approach uses a shape approximation of the potential well in a crystalline lattice which was typically used in QM simulations of the band structure in the case of solids whereas the potential is assumed to be spherically symmetric in the muffin-tin region (atom-centered spheres, defining the atomic regions) and constant in the interstitial region. Early electronic structure methods used the MT approach, including the augmented plane wave (APW) method, the linear muffin-tin orbital method (LMTO) and the KKR (Korringa, Kohn, Rostoker) and other Green's function methods [9, 12, 249–252].

In the most straightforward MT orbital approach (MTO) non-overlapping spheres are usually centered on their respective atomic positions. In contrast, the screened potential experienced satisfactory, in this case, by an electron can, in principle, be approximated to be spherically symmetric on the nucleus and was shown as constant in the case of the remaining interstitial region. Therefore, in this perspective, continuity of the potential in the atom-centered spheres and interstitial region can, in principle, be enforced. As is well-known, the wave functions can in principle be expanded in terms of spherical harmonics and the eigenfunctions of a radial SE in the atom-centered regions. As such, the single-electron wave functions usually is expanded in plane waves, e.g., in the case of the interstitial region due to constant potential. The usage of basis functions other than plane waves was termed the augmented plane-wave method (APW) with many variations. All LMTO based methods, e.g., the wavefunctions (in the interstitial region) are also known as Hankel functions. In this regard, usually, each basis function is based on a numerical solution inside a muffin-tin-like sphere, which in turn is combined with the value and slope to a Hankel function tail on the sphere boundary [9, 12, 249–252],

In linear augmented plane wave (LAPW) model, the basis functions are constructed from the connecting sinusoidal plane wave functions in the interstitial region with LCAO-like functions in the spheres with the advantage that in this case electrons of core and semicore are explicitly included yielding similar computational effort for all materials. As such, the full-potential LMTO (defined here as FP-LMTO) method is an implementation of DFT within the LDA functional with no shape approximation to the crystal potential and the crystal can thus be divided into regions inside muffin-tin spheres, where SE is solved numerically, and using Hankel functions in an interstitial region. The KKR method used the variational method and MT approximations. Some of the methods that descended from the APW method, in general, have been considered as one of the most accurate for electronic structure calculations of materials, mainly due to introducing the schemes related to linearise the secular equation, including the LAPW method as well as the so-called APW + lo method with localized orbital extensions of the basis sets [9, 12, 249–252].

The pseudopotential (PP) theory started as an extension to the OPW method, i.e., separate the total wave function into oscillatory/smooth components (pseudo wave functions). The PP is a well-elaborated approximation in general used to simplified description of a systems with the objective of replacing the effects of the core electrons motion of an atom and its nucleus so the SE contains modified terms for the potential of core electrons and the true potential of the ions is replaced by a weaker potential for the valence electrons. The pseudopotential replaces the atomic all-electron potential, whereas core states because they are less available in a system of interest they can then be eliminated, and the valence states described by pseudo-wavefunctions with much fewer nodes using plane-wave in simulations are more practical. As such, it is known that the valence electrons for being more available are chemically active and hence in this case can in principle be dealt with explicitly and the core electrons are frozen, i.e., considered together with nuclei as rigid non-polarizable ion cores. However, it is possible to self-consistently update PP with its embedded chemical environment (relaxation of frozen core approximation) [9, 12, 249–252]. Effective core potentials (ECPs) typically only freeze core electrons for local basis functions. First-principles pseudopotentials can be obtained from an atomic reference state, in this case, the same energies/amplitude/density are needed for pseudo-and all-electron valence eigenstates outside a chosen core cut-off radius. So-called Softer PP's have a larger cut-off radius, more rapidly convergent but less transferable, i.e. less accurate reproduction of realistic features in different environments. Some objectives in these models were to reduce size/number of basis set size and inclusion of relativistic as well as other effects. This is a one-electron approximation where small-core approximation assumes no significant overlap among core and valence wave-functions and nonlinear core corrections deal with situations where the overlap is, of course, non-negligible [9, 12, 249–252].

Two common forms of PP are norm-conserving and ultrasoft for basis set have a lower cut-off (for the frequency) in describing the electron wavefunctions allowing good convergence with reasonable computing resources. The norm-conserving pseudopotentials (NCPP) are constructed such that inside the cut-off radius each pseudo-wavefunction norm is identical to that of the corresponding all-electron wavefunction, whereas the pseudo and all-electron wavefunctions are in general identical outside the cut-off radius. In the case of the Ultrasoft PP (USPP) model, the norm-conserving constraint is relaxed in order to reduce the basis-set size significantly, i.e., due to the introducing a generalized eigenvalue problem. In contrast, the electron density in this case can be usually subdivided in a soft part (throughout the unit cell) as well as a hard part which localized in core regions [9, 12, 249–252].

The linear combination of plane waves and core states led to the orthogonalized plane waves (OPW) method which possible to obtain the same eigenvalues from the secular equation using the PP method where the core electrons can be considered inert/removed and represented by ECPs to which the valence electrons are subjected. PP methods, including the projected augmented wave (PAW), norm-conserving (NCPP), ultrasoft (USPP) methods. The PAW approach maps a smooth auxiliary wave function in a true all-electron wave function, having components that are smooth and continuous everywhere with rapid oscillations only in the core region of each atom, whereas each part can be treated separately and KS DFT all-electron wavefunctions can be retrieved. PP calculations reached accuracy in which in general are comparable towards all-electron methods, became this strategy an important tool for the description of many-electron systems in the 1970s. Noteworthy, PPs can be used to describe the electron–ion interaction, and in plane-wave calculations, the Hellmann–Feynman theorem facilitates calculations of forces acting on the atoms and stresses on the unit cell is straightforward facilitating ab initio MD materials simulations [9, 12, 249–252].

The Car and Parrinello method (CP) introduced [231] a dynamically simulated annealing strategy to solve the motion equations of the coupled many-electron system using as strategy fast Fourier Transforms in order to switch among real- and momentum-space representations of the wave function minimizing simultaneously the total energy of a complex system yielding a new powerful tool for materials research. However, with all the success of modern DFT has paricularly been characterized by using the more traditional approach due to the strong requirements of the CP method.

Moreover, DFT calculations under periodic conditions for solids can be determined through choice of a basis set to expand the KS eigenfunctions (plane waves and localized basis functions); full-potential/pseudopotential to describe interactions between the ionic core and the valence electrons (also used to speed up local orbital methods); method to obtain eigenstates of the Kohn–Sham Hamiltonian; choice of exchange–correlation functional within the hierarchy of DFT as well as post-DFT corrections and efficient algorithms have been refiend for solving the KS equations which are implemented in sophisticated codes for material simulations.

A number of codes are available to materials science workers including VASP(plane wave basis sets and Pseudo/PAW potentials at atpcms.mpi.univie.ac.at; CRYSTAL(Local basis sets and all-electron potentials at www.crystal.unit.it); SIESTA(local and numerical basis sets, pseudopotentials at wwuam.es/departamentos/ciencias; Gaussian(local basis sets and all-electron potentials at www.gaussian.com); WIEN2K(LAPW basis set with all-electron potentials at www.wien2k.at); CASTEP(plane wave basis with Pseudo potential at www.tcm.phy.cam.ac.uk/castep/); CPMD(plane wave basis set with Pseudo potentials at www.cpmd.org/); Dacapo(plane wave basis sets and pseudo potentials at dcwww.camp.dtu.dk/campos/Dacapo/); FHImd(Plane wave basis functions and Pseudo potentials at www.fhi-berlin.mpg.de/th/fhimd/); PWscf (Plane waves and Pseudopotentials at www.pwscf.org/; Quickstep(Gaussian + plane wave basis functions and Pseudo potentials at cp2k.berlios.de/quickstep.html); FPLO(local basis sets and all-electron potentials at www.ifw-dresden.de/agtheo/FPLO/); ABINIT(plane wave basis and Pseudo potentials at www.abinit.org); ADF(Local basis/all-electron basis sets at www.scm.com); FLAIR; LAPW (basis sets and all electron potentials at www.uwm.edu/~weinert/flair.html); QMD-FLAPW(LAPW basis sets and all-electron potentials at flapw.com) [244–272].

The advent of computational methods such as DFT, MC, MD, AIMD simulations allowed exploration of phase and composition space very efficiently, whereas a combination of experiments and theoretical approaches has allowed substantially cut of time and cost for desiging new materials [1]. Based on the constant increase, in all cases, in computing power coupled with the development of more efficient codes that allowed to boost the emergence of computational high-throughput studies towards accelerating the development of designing new materials with tunable properties to screen for better candidates cognitive overload addressed by sharing/feedback/free access of data/software/technical experience.

One of the routes practiced nowadays for predictive capabilities of computational materials science derived from overlapping advances in computational methods and techniques, as well as best practices, with each area equally important and feeding into the other, allow workers to advance understanding combining efficiently simple models with powerful softwares. This allows the processing of complex systems with million-particle models. The cognitive overload is addressed by sharing/feedback/free access of data/software/technical experience [266]

Central to this route is based on the use of simplified models in which results in a detailed nanoscale understanding of materials. This lead to reproducibility of research, accessibility of materials simulation tools, and demand for computationally literate researchers [266]. Some free access tools are Diffractometer, Foyer, freud, RDF, mBuild, MDAnalysis, MDTraj, MorphCT, packmol, Planckton PLUMED, pyLAT, Rhaco, SSAGES, Signac-Flow, Signac, VMD, VOTCA [266].

In yet, another direction, the large-scale calculations/simulations together with experimental high throughput studies are producing a large amount of data, that is, enabling ML methods for materials science and heralding a second computational revolution since the number of possible materials is estimated to be as high as a ~ 10^{100} justifying this substantial growth of interest/investment. The Materials Genome Initiative (MGI) involves data-driven ML approaches which are gradually penetrating materials design, and have already been applied to development of excellent studies in materials science [372, 373]

Advances in computational and experimental methods have considerably increased the complexity/quantity of data generated which has to be stored/interpreted for identification/correlation of patterns via ML algorithms requiring appropriate methodologies to extract insights similar to what has been done during the last decade with biomolecular systems, i.e. usage of methodologies including screening, data mining, and ML techniques.

10 Machine Learning Methods in Material Simulations

First, we had experimental science, subsequently theoretical science and then during latter half of last century we developed simulations generating a huge amount of data leading us to big data-driven science, distinguished as the new/fourth paradigm for scientific exploration and fueling the second computational revolution in materials science [245, 255, 372–409] after the first revolution being fueled by MD, MC, DFT. The recent growth in computing power, improved algorithms and large datasets have led to a great interest in ML algorithms used for dimensionality reduction, regression, classification and clustering of high-dimensional input data. These methods have a long history in chemistry and biology (establishing a correlations for a large set of data) but have only recently emerged in materials science. Additionally, recent advances in experimental and computational approaches have increased significantly the complexity/quantity of the generated data, which needs storage/interpretation in materials science and engineering. In this sense, has been observed a high requirement on resources/equipment, made on a very limited number of materials, conducted over long time periods, traditionally, played key roles in finding/characterizing new materials and consequently great discoveries were made through serendipity/human intuition. DFT, MD, MC combined with experiments explored phase/composition space efficiently allowing substantial cuts in time and cost of materials were the fuel for the first so-called first material science computational revolution.

The development of more efficient codes, constant computing power/increase is allowing computational/experimental high-throughput screening of large material groups which producing large-scale information (big data) making possible the usage of ML methods in materials science for research among a possible large number (10^{100}) of materials, heralding the so-called second computational revolution. This process is also stimulated by projects such as the MGI [372, 373] which thrives on supporting data-intensive systematic research approach between experiment and theory and ML applications can already be found in materials science addressing the calculation of many material properties, leading to the prediction of new stable materials from first-principle calculations [255]. Such approaches have considered well more efficient than humans towards designing new material with new functional properties for many technological applications. We note, however, that not understanding this new approach and its laws is one of the major criticisms regarding ML algorithms which are often treated as machine built models, i.e., black boxes that are too alien/complex for human understanding.

As such, the main objective of ML is closely related to optimize the performance of tasks via past experience and/or examples and can be categorically divided as supervised, unsupervised, and reinforcement learning. Supervised ML aims at finding the unknown functions that connect known inputs to unknown outputs based on pattern extrapolation in labeled training data whereas unsupervised learning searches for patterns in unlabeled data. The algorithm in semi-supervised learning is provided with both labeled and unlabeled data. Supervised learning is the most used ML approach in materials science [245].

The workflow for supervised learned goes as follows. First, selection of subset of population having known values of target property which are known or can be created. This is followed by a selection of ML algorithms to fit the target quantity finding, generating, cleaning the data to consistency and accuracy. Subsequently, the next step is setting up model inputs for algorithm to be used by translating the information into appropriate inputs. This is followed by training the model via the optimization of its performance (measured/cost/ function), adjusting hyperparameters controlling the training process/structure/properties of the model. Separate validation datasets can be used for testing and training to optimize the hyperparameters. Notably, all applications of ML have to consider aspects of overfitting/underfitting when the model is unable to express the data complexity or features not adequately describing data and can lead to high training error. As such, the model also needs to be evaluated on test sets, to estimate generalization/extrapolation using methods such as simple holdout, Monte Carlo cross-validation, k-fold cross-validation, leave-one-out cross-validation. The success of ML methods for materials Science depends on the quality/amount of available databases such as the MGI [245, 255, 372–409].

Other most-used theoretical/experimental databases include COD at https://crystallography.net, ICSD at https://www2.fiz-karlsruhe.de/icsd, AFLOWlib at https://aflowlib.org/, materials Project at https://materialsproject.org/,OQDM at https://oqmd.org/, CMR at Ahttps://cmr.fysik.dtu.dk/,MaterialsWeb at https://materialsweb.org/twodmaterials, JARVIS-DFT 2D materials at https://ctcms.nist.gov/~knc6/JVASP.html, NOMAD at https://repository.nomad-coe.eu/, Materials Cloud at https://materialscloud.org/discover, Citrination at https://citrination.com, C2DB at https://cmr.fysik.dtu.dk/c2db/c2db.html. The success of databases for materials informatics depends on a findability, accessibility, interoperability, and reusability treatment of data possessing findability, accessibility, interoperability, and repurposability [255, 372–409].

As such, the representation of the data in a suitable form usually is considered as the main ingredient of a ML algorithm [255]. Thus, the features for a ML algorithm in real material science applications have to be able to capture all physical and chemical information related to materials behavior in different conditions and environments, i.e. feature extraction or engineering. Ideally, descriptors should be uncorrelated. Particularly, the fitting of potential energy surfaces is the most likely feature studied in materials science. Note that the descriptors quality is related to the differentiability to the completeness of the representation, due to the basic symmetries in terms of invariance (hence such as rotation, reflection, translation, permutation of atoms of the same species) and movement of the atoms.

Simple representations showing shortcomings as features include Z-matrices, Weyl matrices, transformations of pairwise distances including their reciprocal/exponential transformations, worked only for a number of fixed atoms. And yet, are not unique about permutation of the atoms. Some automatic feature selection algorithms include sure independence screening and sparsifying, operator (SISSO), least absolute shrinkage and selection operator (LASSO), principal component analysis (PCA), decision trees. The algorithms for material science include linear- and kernel-based regression and classification methods, variable selection and extraction algorithm linear methods, nonlinear decision tree-based methods such as random forests (RFs), highly randomized trees and neural networks, simple fully connected feed-forward networks, convolutional networks, variational autoencoders (VAEs) and generative adversarial networks (GANs) [245, 255, 372–409].

Many properties of materials can be predicted with application of ML methods, include Curie temperature, Vibrational free energy and entropy, Band gap, Dielectric breakdown strength, Lattice parameter, Debye temperature and heat capacity, Glass transition temperature, Thermal expansion coefficient, Thermal boundary resistance, Thermal conductivity, Local magnetic moments, Melting temperature, Magnetocaloric effects, Grain boundaries, Grain boundary energy, Grain boundary mobility, Interface energy Seebeck coeficient, Thermoelectric figure of merit, Bulk and shear moduli, Electrical resistivity, Density of states, Fermi energy/ Poisson ratio, Dopant solution energy, Metal–insulator classification, Topological invariants, Superconducting critical temperature Li-ion conductivity and battery state-of-charge [245, 255, 372–409].

Regarding perspectives and future trends, we note that similar to the industrial revolution, the ML will train machines in order to identify patterns/relations among material properties and features at nanoscale level. As such, the use of ML algorithms has mostly been applied in materials design in order to elucidate quantitative structure–property relationships involving numerous properties of interest to the development of modern technology. Additional needed further properties should also be predictable with the same accuracy. In future research, particularly, each methodology will be data availability dependent. Future perspectives include realistic calculation of properties for $\geq 10^5$ materials using universal deep neural networks models, predicting of fundamental properties (e.g., bandgap energies, Fermi level, bulk moduli, shear moduli, formation energies, and Poisson ratios) for a huge wide variety of functional materials as well as DFT calculations errors, in general, can be comparable or smaller than obtained in diverse experiments, that is, reducing the need for training only a single structural prototype/property which will in turn drastically reduce required investigative resources. With the availability of ab initio data fast interpolative prediction of materials electronic properties based on the use of ML techniques is now possible. ML algorithms should also excel in 2D materials, symmetry protected topological materials, and disentangling high-dimensional correlations. Superconductivity and other complex nanoscale phenomena have also begun to be addressed, and there are paths open for possibilities of novel superconductor discoveries. Magnetic phenomena is, of course, important in a variety of technologies. However, the small number of research in this field using ML techniques indicates the field is just starting to emerge. Quantum phase transitions in crystalline systems and topological materials classification are also fields of interest to ML methods [245, 255, 372–409].

The usage of neural network models will significantly increase with the popularity of such strategies to a wide materials science audience. Also, it is possible to use surrogate-based optimization (active learning), transfer learning, or shift the problem of predicting a property to predicting errors in crude models concerning training data. It is expected high-throughput searches in the coming years of all unresearched common crystal structure prototypes in order to speed up material discovery. Also, it is known that the obtain of quality experimental data can in principle be well rare/expensive to be generate, the prototypes options may be quickly exhausted. As such, the direct prediction of a crystal structure can be a challenging problem, mainly because the most available data is due to a small number of prototypes. Bayesian optimization for global structure prediction should be further explored. Neural networks built Force fields indicate efficiency parallel to classical force fields and accuracy comparable to DFT in solid-state and should completely replace classical force fields in the long term. ML methods allow us to investigate emerging/novel phenomena from complex systems mainly due to superior numerical scaling [245, 255, 372–409].

Linear kernel models and decision trees, using simple algorithms, were among the first ML applications in materials science [255]. Future predictions suggest two routes; the first is the continuation of present research via more sophisticated ML methods lacking benchmarking datasets and standards essential to measure the progress in features/algorithms. In this direction, the prediction of formation energies/band gaps should be a critical step. The usability of ML is the second direction. The electronic structure Community uses both routes addressed. Some crucial points can also be summarized as follows. In ML methods interpretability is a multifaceted concept with different priorities that depends on research goals and dataset. However, it is well-known that simulatability is usually non-existent in the field of materials science and hence should not be of concern, as well. As such, the progress in this direction of materials informatics should use post hoc knowledge techniques to improve trust in high-performing black-box models [245, 255, 272–409].

11 Molecular Docking

Notably, along the last decade, was observed a significant development of computer-aided drug design (CADD) methods. Such computational strategies have sufficient accuracy for diverse practical applications related to new drug discovery. In this regard, structure-based methods are beneficial for predicting binding modes for various molecules of interest and estimating their relative affinity yielding high-throughput docking of 10^6 small molecules classified by scoring based on force fields, which can identify binders from using a protein target. Many approved drugs have benefited from the usage of CADD, supports hit identification/optimization by predicting how small molecules bind to the protein target and estimating relative binding affinities [273–290].

Virtual screening evaluates the library of many molecules for the possibility of binding to the protein indicating the ones with the highest affinity, i.e. reducing the number of false negatives selected for in vitro assays. Numerous software has been developed with the specific purpose of performing an automatic molecular docking of small molecules into protein target structures. However, only a few have gained wide recognition/usage by a large community, including Glide, [273], GOLD [274]

Dock [275], and AutoDock [276]. The different types of interaction in docking involves include electrostatic forces (dipole–dipole, charge-dipole, charge-charge), electrodynamics forces (Van der Waals interaction), steric forces (caused by entropy), solvent-related forces (hydrogen bond and hydrophobic interactions).

As such, it is known that the molecular docking involves typically usage of search algorithms and scoring functions. The search algorithms create the desired number of configurations that determine the binding modes and include Point complementary, Monte Carlo, Fragment-based, Genetic algorithms, Systematic searches and Distance geometry. The objective of scoring functions is to rank positioning of ligands, whereas ideally the score should correspond directly to the binding affinity for indicating the best binders. Hence, the scoring functions are typically empirical. The methods applied for docking are typically Lock and Key\Rigid Docking whereas receptor and ligands are fixed during docking; induced fit\flexible docking where both ligand and receptor can be conformationally flexible and the most optimum pose is selected [278–280].

In the molecular docking process the intermolecular interaction between two molecules is studied from the using *in silico* models, i.e. involves the protein receptor (macromolecule) and the ligand molecule (inhibitor) with 3 main steps. The first step consists of preparation of protein, whereas a three dimensional structure is retrieved from the Protein data bank (PDB), pre-processed (removal of water molecules from cavity, stabilization of charges, filling missing residues, generation of side chains, etc.) [278–280]. Subsequently, the active site of the protein is identified, whereas water molecules and heteroatoms are removed. The following step comprises the preparation of ligand, including those from databases such as ZINC, Pub Chem or even sketched. The Lipinsky's rule of 5 can be used to discard non-drug like compounds, i.e., less than five hydrogen bond donors and ten hydrogen bond acceptors, molecular mass less than 500 Da, partition coefficient (LogP) not over 5 and molar refractivity between 40 and 130 [283, 286]. Finally, the ligand and protein are docked, interactions analyzed, and scoring functions used to ascertain best-docked ligand complexes.

The interactions of molecular docking may in principle lead to inhibition or activation of the protein. They can be employed in diverse areas, including optimization (drug discovery), hit identification (virtual Screening), prediction of KA (biological activity), bioremediation, de-orphaning of protein, binding site prediction, searching for lead structures for protein targets, protein–protein/nucleic acid interactions, protein engineering, mechanisms of enzymatic reactions, studies of structure–function. Current protein–ligand docking tools include Glide (Monte Carlo Sampling, Glide Score), GOLD (Genetic Algorithm, Chempl), Auto Dock (Genetic Algorithm, Lamarckian genetic algorithm, Simulated Annealing) [278–280].

Algorithms from academia and industry are being incorporated in the molecular docking software improving functionality, speed, graphic display, techniques, search engines, visualization of molecules, and also easy access to structural databases making molecular docking an essential component for the medicinal chemist. We note, however, that although the success of this approach is target-and software suite-dependent, it does not necessarily correlate with respect to the binding affinity but depends on the quality ligand–protein interactions. In contrast, molecular docking is, in turn, biased towards molecules binding to a protein target, with respect to the available crystal structure, as well as detectable affinity [273–290].

12 Pharmacophore Modeling

Pharmacophores have been considered important models/paradigms of noncovalent chemistry. First, we understood the three-dimensional (3D) nature of molecules in which are based on stereochemistry rules to determining conformations [291–296]. Subsequently, a better understanding of the key-and-lock paradigm (exhibiting complementary geometric shapes) has been explained by ligand binding to macromolecules. Excluding those that govern our universe, it was shown that the chemical nature of the noncovalent binding forces are very complex. Particularly, it is known that steric "repulsion forces" are in general useful models to provide a rationalize behavior of the electron clouds which can be well-described by interacting ideal 'atomic spheres' and leads to the introduction of a van der Waals corrective term. It is found that modern medicinal chemistry occurs, for example, at this "atom sphere" level of approximation, that is, where the charge of the cations interact with anions, the donors with acceptors, as well as hydrophobes among themselves and thus, the principle of the functional group is an essential paradigm [291–296].

Effectively, pharmacophore modeling (PM) represents a conceptual approach for describing elusive structure binding affinity relationships using sets of approximate rules-of-the-thumb and can be described as molecular frameworks having the important features of a drug's biological activity that can be represented by points in 3D space composed of functional groups such as hydrogen bond acceptors and donors (HBA and HBD), anions, cations, aromatics and hydrophobic areas (Hyp) [291–296] whereas PM does not require the 3D structure of the target protein. These models are basically the ensemble of steric and electronic features that are able to ensure optimal interactions with a macromolecule target (receptor, enzyme, ion channel) in order to trigger/block biological response. Interesting to note is that the success of pharmacophore models help us to gain a fundamental understanding in-depth of protein–ligand interactions which, if fulfill the requirements of the pharmacophore, are likely to show biological activity [291–296].

As such, the extraction of common chemical features interacting with the target proteins is important for pharmacophore generation. For pharmacophore-based virtual screening of biologically active conformations, in addition to pharmacophore features, it is necessary to generate a 3D structural database of ligands. Main advantage of this approach is the screening of millions of compounds for hit identification whereas disadvantages include the fact that may not be well represented important interactions in some specific models with the loss of meaningful information such as binding free energy contributions.

In principle, two strategies can be used in defining a pharmacophore. In the structure-based or direct approach structural 3D information is incorporated in the protein whereas the indirect/ligand based approach aligns active molecules and determines their similarities. Simulation-based affinity predictions (free energy perturbation simulations, flexible docking) are very time-consuming for very large-scale practical usage; however, are usually based on severe approximations of the physical reality [291–296].

It must be noted that the first step discussed in pharmacophore modeling can be used to the classification of the atoms in terms of their nature/chemical environment, have been taken into predefined categories associated with specific interactions such as hydrophobes, polar positives and polar negatives without explicit characterization of the strength of interactions. Nevertheless, the next step is due to the conformational sampling of known binders and non-binders, to serve in the ML process for pharmacophore extraction as well as for selection of candidate compounds from the electronic database to be confronted with pharmacophore hypothesis refined during the analysis of the virtual screening. Meanwhile, conformational flexibility can be a problem of large complexity, especially for compounds exceeding several tens of rotatable bonds. Although drug-like compounds can in principle be less complicated, however be extremely numerous, whereas reducing satisfactory the conformational sampling time to a few seconds or minutes for each compound, using in this case to simplified molecular models, does not guarantee the sampling of biologically relevant conformers. Construction of the pharmacophore hypothesis is the critical step in pharmacophore modeling, which delimits a set of space regions that harbors specified functional groups when low-energy conformers are aligned with respect to it. These space regions supposedly represent a complete map of interaction spots where favorable contacts to the protein site occur (hot spots) [291–296].

13 Homology Modeling

Considered to be the molecules that do the work in living things, proteins play key roles in many physiological processes (cell cycle, transport, metabolism, signal transduction, immune response). The consecutive order of about 20 amino acid (as building blocks), in general, determines the primary sequence of a protein polypeptide chain, which can in principle be easily deduced from its coding nucleotide sequence. Additionally, proteins fold into three-dimensional structures yielding spatial amino acid arrangements with unique functional properties making the knowledge of the refined structure of the protein of great value for recognition of its molecular function (most drugs target proteins). It is thus highly desirable to be able to deduce three-dimensional properties of proteins. Despite considerable progress in experimental structure determinations, there are still technological limitations in the number of protein structures elucidated and deposited in the PDB [297–301].

The molecular process by which the proteins fold is complex. Although some progress has been made with computational methods, they are expensive, with a prediction accuracy often not sufficient for applications in pharmacology and molecular biology. In contrast, protein models obtained from inexpensive homology modeling (bridging gap among sequences, structures and function) are sufficiently accurate for applications in numerous research fields.

Since the 3D structures are in general more conserved than sequences within families, when there is homology at the sequence level we then can assume that two proteins e.g. are structurally similar. Comparative, knowledge-based, homology modeling has the objective of predicting the 3D structure of a given protein sequence (based on known 3D structures) of one or more homologous proteins. To this end, it is known that the homology modeling given the broad range of applications, in particular, follow a four-step process, i.e., selection of a template structure, the alignment of the target sequence to a template, model building, and its complete assessment [1, 222, 282].

First, one selects the best-known template from the PDB. As such, the introduction of database searching strategies (using as a query the target sequence) has enabled a great improvement in the case of the sequence sensitivity and alignment accuracy. The quality of the template structure is very important for the quality of the model and should be selected from crystals with low free R-factors and high resolution. The presence of ligands, substrates, cofactors, domain conformation, oligomerization states play biological roles and should be considered. In general, a good criterion to select adequate template structures is good sequence similarity between targets and templates. It additionally is essential to give all due attention to the alignment phase. A model system can be mainly built for the protein-based on template structure and alignment. At the early prediction stage, a vital factor to consider is the percentage of sequence identity among two substrings. An evaluation of the predicted model can in principle be used to evaluate the quality of the predicted coordinates [297–301].

14 Coarse Grain Simulation Models

Coarse Grain (CG) simulations are a necessary/important emerging field with rapid progress during the last decade since, although the usage of parallel computing allows simulation of molecular dynamics of increasingly larger systems, there is still not sufficient gain towards longer simulation times when sequential nature of trajectories are involved. As selected examples, we note that polymer melts and modeling of biological systems are still challenging due to the broad spatial and temporal scales involved. In order, for example, to connect biology to the molecular world it is necessary to understand how molecular-scale dynamics propagate upward in scale to define the function of specific biological structures whereas, for example, traditional all-atom models are still inadequate in order to simulate the large spatio/temportal scales involved in cellular processes [302–371].

Using lower resolution to replace atomistic details, CG has opened (on time scales inaccessible to all-atom models) the route to simulate large-scale biomolecular processes. Given the broad range of applications of these strategies, we have here selected (as part of state-of-the-art) some examples, such as protein folding, membrane protein gating/self-assembly, DNA hybridization, and modeling of carbohydrate fibers [302–371]. The development and application of CG models for studied diverse biomolecular processes of interest has been an area of active research whereas in order to maintain the predictive quality of such simulations, there is a need to develop realistic CG models. The methodology for the development of CG models should be guided by experimental results based on numerous large-scale coarse grain simulations, in order to identifying significant interactions and reconnecting to the molecular scale (all-atom MD simulations) whereas the CG method can be extended refined and repeated iteratively. Among the coarse grain models included in this review are Force Matching [323], Inverse Montecarlo [323], Iterative Boltzmann inversion (IBI) [314], Trajectory Matching [304], Bayesian dissipative dynamic models [310]. Applications include soft matter, macromolecular complexes, polymers, interfacial properties, biomolecular systems, water, proteins and carbohydrates [302–371].

It is known that the atomistic degrees of freedom can be neglected in the CG models, however, the detailed physics description of a system may in principle be sensitive to small-scale phenomena. As such, there may also be challenges to develop accurate/transferable force fields (extracted from atomistic simulations/reproducing key experimental data). In most forms of coarse-graining, one clusters groups of atoms into new CG sites, i.e. beads (the type of quasi-particles), which can interact through more computationally efficient interactions. Hence, the combination of the CG interactions with a reduction in degrees of freedom allows significant jump in accessible temporal/spatial scales [302–371].

Statistical mechanics, provides a closer connection between the world of molecular interactions with the macroscopic world of thermodynamics. Coarse grain methods can be explained by the formula dr $\exp[-\beta V(r)]$, V is defined as the potential energy of the system of atoms and $\beta = 1/k_B T$ (T is the thermodynamic temperature and k_B is the Boltzmann constant). In contrast, the atomic scale and its potential energy function (QM, Born-Oppenheimer approximation) are connected with the macroscopic thermodynamics world (free energy/derivatives) yielding entropy, internal energy and heat capacity [325]. This formula yields the basis for distribution functions/properties/averages implicitly highlighting the importance of coarse-graining method since the evaluation of these integrals are a challenge for many-dimensional systems such as those for materials and biomolecules. Using CG modeling, the computational challenge can be significantly reduced, simplifying/speeding up complex simulations without loosing contact with the reality of molecular properties.

CG modeling offers computational reduction via combination of less degrees of freedom and an effective potential that is simpler/smoother than the all-atom potential, facing however, the challenges of not knowing a priori optimal CG site choice (mapping) or the CG effective potential. One route of CG simulation is actually to map groups of atoms clustered into CG sites or beads with effective interactions. For the beads, the interaction potentials are derived/parameterized in energy-based CG, in order to reproduce satisfactorily the energies for all-atom of the system. Using force-matching methods, basically, the sum of atomistic forces are in general mapped onto corresponding beads and in structure-based CG methods interactions are, for example, reproduced from atomistic simulations [351]. However, it should be noted that exist some challenges related to designing CG models. For instance, this includes choice of pseudoatom sites (i.e., groups of multiple atoms), the effective energy function (such as interactions between pseudoatoms), and also useful dynamical equations. One of the key problems is to simplify the model so as to preserve in this case the chemical-physical nature behavior of the molecule, building CG force fields that are realistic and transferability to different chemical groups and conditions [302–316].

The CG method is typically COM-based when the targets are atomistic simulations, whereas mapping determines which atomistic-level atoms contribute to each bead and hence are usually used in order to determine (at the CG level) the position of the bead. In parallel, experimental measures are used to optimize target and the mapping in this case only serves as an interpretation guide for atomistic CG simulations. The development of CG models in this case requires defining of pseudoatoms sites, derivation of the energy functions, and defining dynamical equations to the time-based study. Different mappings of coarse-graining can range from 4:1 (four heavy atoms/associated hydrogens correspond to one CG particle) to 20:1 (twenty united atoms represented by one CG particle). Force fields models include the Martini force fields (12–6 shifted LJ potential), 9–6 LJ potential, (Morse potential), (SAFT-γ) all of which are typically used as standard MC and MD methods [302–329].

First, the bottom-up models are based on atomic simulations (including iterative Boltzmann inversion (IBI), force matching (FM)). Second, the top-down approach uses macroscopic experimental data (compressibility, diffusion, surface tension, and chemical potential) to derive the parameters. The forward CG procedures, in particular, no use free parameters in order to tune/recovers the MD system faithfully. However, in the case of reverse CG procedure, it has been indicated that with properly defined parameters, the model can yield accurate effective potentials whereas an initial guess of the CG force fields are left undetermined posed (only a few parameters). In order to correct the free parameters, in this case, it was an (iterative) inverse optimization is then carried out to obtain the target (mesoscopic) properties. Most CG methods available belong to this category, i.e., IMC, IBI, FM, stochastic parametric optimization (SPO) using Bayesian interference, and minimization of relative entropy [302–317].

The typical Martini CG Model uses a 4:1 mapping in which four heavy atoms/associated hydrogens correspond to one CG particle. The model consists mainly of polar, apolar, and charged interaction sites, which are further categorized based on hydrogen bonding and polarity. The 12–6 LJ potential is used for the non-bonded interaction, and a standard potential energy function term is used for the bonded term [305, 329. 345, 356, 358]. The IBI method within the bottom-up approach is an iterative method for potential inversion, whereas potential functions can be deduced from atomistic configurations. This structure-based method, has some important limitations regarding, for example, transferability to heterogeneous systems, the dynamics of chains, and also nonunicity of the resulting potential. IBI is primarily used to get non-bonded interactions for the reproduction of reference systems from experiments or MD simulations [323, 326].

The Force Matching (FM) is an alternative strategy to the IBI method and is usually based on the analysis of instantaneous forces on the beads. This method uses least square minimization with the specific aim of defining average conservative interactions. In order to optimize the part of the non-conservative interactions (random and dissipative), trajectories with constrained dynamics can be used [304, 323].

Although MD has become a standard tool in order to investigating molecular systems, e.g. all-atoms models are still prohibitive computationally for systems with large spatial–temporal scales and is unrealistic even at the mesoscale applications of soft matter, physical and biological systems. It is known that dissipative particle dynamics (DPD) conserves the momentum of the system, in addition to provides the correct hydrodynamic behavior for the case of fluids at the mesoscale. Similar to MD, the dynamics of DPD are computed in this case by integrating Newton's equation of motion, with however, a softer potential allowing larger time steps and has found widespread application in soft matter (polymer solutions, colloidal suspensions, blood rheology, blood coagulation, and others). Describing the significant interactions between the DPD particles via functional forms is the essence of a good CG model. One strategy is in general a direct forward-path (e.g. microscopic dynamics evaluations) where the CG force fields among the DPD particles are directly constructed from available MD trajectories. A new pragmatic Bayesian method has recently been developed, which yields parameters for the CG-force fields from the reproduction of reference trajectory with the highest probability. The objective was to adapt it for polymers by considering the intramolecular bonding and bending interactions within the polymer chains. The Bayesian CG model assumes that there is a DPD CG model whose behavior conserves degrees of freedom like the higher resolution models [310, 313, 362].

15 CG of Soft Matter

Since the properties of soft condensed matter are usually determined by structure and dynamics in a wide range of time/length scales, simulation of these systems face challenges of scale bridging to link chemical structures and properties. CG models have been successful in recent years for simulation of physical phenomena on scales requiring some chemical specificity. Some concerns include protein interactions, lipid bilayers, polymer material, protein interactions, surface properties/interactions, and others. CG simulation of single interaction sites representing groups of atoms facilitates studying soft matter since the total number of particles are reduced, interaction potentials are simplified and potential energy surfaces are smoothed, leading to an acceleration of the molecular dynamics. CG of soft matter models can also use adaptive multiscale and hybrid simulations, i.e., joint atomistic and coarse-graining models [325, 333, 339, 340].

16 CG of Macromolecular Complexes

Progress in CG modeling has facilitated computational studies of biological macromolecules with their large separation of length/time scales. Numerous biological processes use macromolecules, as building blocks, for large-scale complexes (ribosomes, viruses, cytoskeletal filaments) which often contain many copies of the same macromolecule (collective aggregation via non-covalent interactions) leading to ordered and functional dynamic suprastructures whereas configurational transitions act as regulatory signals emphasizing the importance of a hierarchical CG approach [307]. An underlying assumption in a structure-based CG model is that conservation of close contacts between residues (observed in native structures) are essential for functional dynamics of macromolecules, whereas mapping is typically prescribed such that each CG site represents a different residue [307]. Among the methods used to describe the energetics of the effective CG interactions is the network model where CG molecules are described by masses connected by springs, and the predicted ground state yields experimental structures. Another model is the so-called (off-lattice) Go models [322, 323], which describes native contacts via attractive, nonbonded interactions with the other non-native interactions being purely repulsive. However, both models have limited usability when large conformational transitions or changes in the chemical environment are of interest [305, 307, 326, 334].

The self-organizing polymer (SOP) model (a variant of Go models) use softer bonded and repulsive interactions improving agreement with force-induced folding and unfolding behavior and empirical success for proteins and RNA [307]. Some CG models have opted for hybridized network-based and native-contact-based models suited for studies of large-scale macromolecular complexes [311]. Algorithms inspired by graph theory have been proposed as a means to differentiate network weights based on chemical fragments [309]. Network bonds and native-contact models have been replaced by local density kernels for flexibility analysis [307]. A good example of knowledge-based models is protein structure prediction and homology modeling [297–301]. Leverage of Bayesian inference techniques is another promising direction for knowledge-based models [362]. Recent CG models show mechanistic insights into polymeric/aggregated assemblies for exploration of collective motions with computational efficiency, dynamical processes of macromolecular complexes, and gain for hierarchical insights [307, 310].

17 CG of Polymers

Physicists often describe polymers as long, flexible, one-dimensional molecules, and for a long time many properties of polymers follow chain length scaling laws allowing them to be described by simple simulation models (bead-springs) with generic features (bonds, bending potentials, van der Waals attraction, etc.). For the physical-chemist, the chemical nature of the monomers is important in order to better explain/predict properties of specific polymers. Particularly, it is known that the MD simulations can accomplish a good enough degree of realism but are far from reaching space/time scales relevant to the slow dynamics of these large molecules, indicating the need for CG simulations [304, 314, 317, 319, 333, 339].

Typical vibrations for covalent bonds indicate length scales of angstroms with sub-picosecond time scale whereas lengths of monomer are nanometers with tens of picoseconds relevant dynamics; polymer chains are ~10–100 nm with corresponding single-chain interactions of 10–1000 ns, and polymeric networks are on the order of 1–100 μm indicating relaxation times of micro to milliseconds. In case of the bulk polymeric materials show length scales of millimeters/centimeters with aging and relaxation of polymeric materials in seconds/hours/years confirm the need to introduce coarse-grain modelling if one wants to characterize correctly the hierarchy of scales [9].

Early CG models have been developed in order to reproduce universal polymer properties (structures/relaxation times), however, in this case not reproducing polymer chains properties with specific chemistry [303, 314, 363–368]. Nonetheless, based on the more realistic CG models have been developed in this case from top-down and bottom-up parameterization to reproduce these properties. One of the most used CG methods for polymers (polystyrene) is IBI, where the time needed for atomistic MD simulations is prohibitive [307, 326]. MD of star polymer melts provided multiscale dynamics for coarse-graining with molecules represented as chains of beads connected by springs, whereas each molecule has ten arms interconnected with a center bead, and each arm has two identical monomers [314]. LJ potential was then used to describe excluded volume interactions between monomers, which are truncated for only repulsion (Weeks-Chandler-Andersen (WCA) potential) [326]. MD of star polymer melts was done to obtain atomistic fields for CG, whereas each molecule of a star polymer, including internal degrees of freedom, is grouped into a single CG particle whose interaction is evaluated from microscopic dynamics [314]. The IBI method has been successfully applied to a variety of polymer melts formed by dendrimers, vinyl polymer chains, polystyrene, and polyethylene chains [303, 363–368].

It has been recently noted that with CG we could obtain faster equilibration of large polymer lengths with time scales enabling studies macromolecular phenomena (such chain relaxation, structural rearrangements, disorder/order transitions, particle dispersion to aggregation transitions as well as small system size for a broad range of polymer design parameters and system conditions in order to identify useful parameters, conditions for the desired morphology prior to experiment/expensive atomistic simulations. With recent computing power advances, polymer simulations can guide in vitro polymer design efforts [368]. Recently, it was investigated the scale gaps of the atomistic model of polyethylene melts, using the bead–spring Kremer–Grest model (simple beads with nonbonded interactions described by repulsive LJ potential) as well as DPD yielding polymer properties (threshold values indicating the onset of static/dynamic universality of polymers). Hence, the work stated that could be quantitatively evaluated for CG models the "scale gap" of the static and dynamic properties/similarity of the polymer network/computational efficiency relative to that of the atomistic model [363].

18 CG for Interfacial Properties

In the case of liquid/vapor interface of alkanes, the Martini Model force fields perform relatively well, since such molecules are formed by identical groups interacting predominantly via repulsion-dispersion interactions. Coarse graining strategies have widely been applied with success to calculate surface tension of liquid–vapor interfaces for CH_4, SF, CO_2 [313, 363, 364, 371].

The polymer/metal interface has enormous relevance for industrial applications (durability, composite manufacturing, packaging for microelectronics, coatings, packaging for microelectronics components), whereas understanding the behavior of polymer chains near the solid surface, e.g. is in general critical for optimizing the interfacial strength. CG force field developed by IBI for a confined polymer solution was used to investigate sensitivity to changes in the degree of polymer arrangement at near the surface with constant concentration. Recently the Martini force field was used to calculate/reproduce to within ± 10 mN m^{-1} interfacial tensions for liquid/liquid interfaces. A Bayesian method was particularly tested on the pentane molecule for the bulk/interfacial properties [316], and was later extended in order to the prediction of properties of polymer melts [304, 310, 313, 325, 371].

Other types of CG potentials can in principle be used in association with methods as DPD, allowing for more extended time steps. However, most CG simulations of liquid–vapor interfaces are non-dissipative or use density-dependent many-body interactions [310, 328]. Canchaya et al. extended the CG model designed for DPD with an optimization based on trajectory matching in order to improve transferability across thermodynamic conditions [303]. For instance, a CG-DPD model of n-pentane from constant-NPT atomistic simulations of bulk liquid phases was developed and applied to the surface tension of the liquid–vapor interface over a wide temperature range [303]. According to the database used for the development of the potentials, in particular, it is possible to build a good CG model that, in turn, performs very well in the surface tension reproduction on the orthobaric curve [310].

Conchaya et al. [303] reported simulations of the interaction between a metal oxide (Cu_2O) surface and a cis-1,4 polybutadiene polymer chain at the mesoscopic scale using a CG model via dissipative particle dynamics. A bottom-up approach was used to link underlying molecular properties with structure (in a wide range of length/time scales). A recently developed Bayesian scheme was used for parameterization of the model in order to investigate local structures of polymers (confined slab geometry) via radius of gyration, orientation, and density profiles. A method is a force-matching approach using basic input atomistic simulation trajectories, whereas configurations are recorded with frequencies that match the proposed time scale of the coarse-grain simulation yielding a collection of suitable groups of atoms (coarse grain units). With big 50 fs time-separation, atomistic trajectories are converted to the corresponding

CG beads. The atomistic trajectories, with relative large time-separation (50 fs) are compared to DPD simulations in order to obtain bond and pair interactions that yield optimal reproducibility of the dynamic evolution [303, 328].

19 CG Biomolecular CG

In biological systems/processes, the length/time involved prevents the use of prohibitive fully atomistic simulations, and CG models can be used to overcome the atomistic limitations by allowing larger time steps. The availability of experimental structures in recent years has helped all-atom MD simulations to address protein structural and functional studies as well [306, 308, 311, 323, 324, 326, 327, 335, 361–364].

20 Water CG Models

Water is an important biomolecular system to study since it is a solvent present in all biological systems and shows very challenging complex collective behavior stimulating the development of implicit, explicit, and polarizable models. Yet, it is known that the water has unique properties, included a density maximum at near 277 K, increased diffusivity with compression. Particularly, this anomalous behavior of water, in general, complicates its dynamics and chemical/physics properties compared to other solvents. From a computational viewpoint, water offers great challenges since it comprises more than 80% of the particles/computational effort in MD simulations, indicating the advantages of CG solvent models. The collective effect of water generates highly cooperative hydrophobic forces, stabilize base-pair stacking in nucleic acids, drives protein folding, determines dielectrics, and is coupled with long-range electrostatic interactions [315, 332, 341, 342]. One strategy to model water implicitly utilizes the hydrophobic/charge screening between nonsolvent molecules (lipid, protein folding, DNA, carbohydrate models), among others. Another more sophisticated strategy models the van der Waals interactions and hydrophobic effects by FF depending on solute size and can be via solvent accessible surface area contributions [315, 323, 332].

In some applications, the solvent such as water, can be implicitly modelled whereas the CG model of the solute is optimized to interact correctly with modelling the solvent molecules explicitly. The implicit models can be efficient but can also miss critical physical details with a description strongly coupled to long-range electrostatic interactions. Earlier, one water CG model can limit the amount of time saved. Another straightforward strategy is representing each water molecule by a single site (1/3 of atomistic sites required), whereas spherically symmetric solvent sites reproduce water pair correlations and replace long-ranged Coulombic interactions with simple short-ranged pair potentials [315, 323, 332]. Good examples are the soft-sticky dipole model (explicit dipole and orientation-dependent hydrogen-bonding interactions) and the monatomic water model (three-body potentials to enforce the tetrahedral water ordering) [323, 332]. These models still provide a limited description of the tetrahedral structure/electrostatic water properties. Some CG models associate single sites with 3–5 water and even replace explicit long-range electrostatic interactions by short-ranged potentials, presenting however, conceptual challenges. Some low-resolution water models use explicit polarizability in order to model solvent electrostatics. The big multipole water model employs three bonded sites to describe the local dipole/quadrupole of 4-water clusters. Notably, CG models preserving anisotropy of water interactions often provides a more realistic description of hydrophobic effects [315, 323, 332].

21 Proteins and Carbohydrates

Although the availability of experimental structures in recent years has helped all-atom MD simulations address protein structural and functional studies, it is still a challenge to study protein folding processes. Methods to combat this issue include enhanced sampling methods for calculation of the free-energy landscapes, whereas the binding strength is quantified through the difference in energy between the free and the bound state of a system, i.e. umbrella sampling, parallel tempering, and multi canonical MD simulation. Also, Replica Exchange Molecular Dynamics (REMD) is used frequently for studying the folding thermodynamics of disordered proteins via concurrent simulations for numerous replicas/conditions in order to elucidate minimum/transition states. Markov state models (transition probability matrices enclosing the metastable states in system configurational space) have also emerged as a statistically sound method for tracking the dynamics of a system in thermodynamic equilibrium [302, 305, 353].

A brief survey of the recent progress of the CG models and their applications to biological complexes (with emphasis on carbohydrate and protein models) is given as follows. The Martini model uses up to five beads per amino acid residue mechanism with applications to allostery, peptide binding to GPCRs, and parameters developed for ATP molecules [305, 356–359]. The UNRES model uses two beads per residue with steered molecular dynamics, a deep feed-forward neural network for phosphorylated residues [361]. The CABS model uses four beads per residue and addresses flexible docking of peptides. The PRIMO model uses three to eight beads per residue in order to provide high resolution and transferability [322, 342]. OPEP utilizes up to six beads per residue and Replica Exchange MD and OPEP [336] in order to address protein–protein docking [361]. PRIME uses four beads per residue to investigate protein peptides [342]. The Bereau and Deserno models use four beads per residue model to analyze peptide-membranes [344]. The Kim and Hummer single bead model addresses highly flexible protein complexes and phase behavior of intrinsically disordered proteins [354, 355]. For carbohydrates, the M3B three beads per monosaccharide model was a pioneer CG method [346]. The

Bellesia model with five beads per monosaccharide investigated structural transitions in cellulose [337]. The MARTINI three beads per monosaccharide model was used to investigate the physicochemical properties of Chitosan [345]. The Srinivas single bead per monosaccharide model is a solvent-free CG model used to study cellulose fibrils [338].

Notable, It becomes clear the challenge of simulating a large number of interacting atoms in particle-based systems above ~10^7 atoms (cells, membranes, proteins) as well as the wide/large diversity in temporal scales. The still great limitations of computational power suggest as future trends/perspectives further development of new protocols for coarse grain and ML in the actual second computer revolution/fourth scientific paradigm age, which could allow larger jumps in accessible temporal/spatial scales for materials and biomolecular systems of interest. However, despite the great promise and advances of CG simulations, from foundation principles to actual force-field parametrization and algorithms/methods, there are some limitations in the model, such as the force-field parametrization description of nonbonded interactions in terms of pairwise functions, which puts a bound on accuracy and transferability. The usage of ML/kernels/neural networks for CG force fields development still has to address the challenges of more complex systems/computational efficiency. Developments/application of ML techniques in chemistry and materials science is speeding up expensive computational calculations. In contrast, CG models progress has been more limited by the absence of more improved computational scaling. Ultimately, ML models will be strongly dependent on data availability.

References

1. La Porta, F.A., Taft, C.A. (eds.): Emerging Research in Science and Engineering Based on Advanced Experimental and Computational Strategies, Engineering Materials (2009). https://doi.org/10.1007/978-3-030-31403-3_16
2. Kawano, D.F., da Silva, C.H.T.P., Taft, C.A.: Advances in medicine and biology, Chapter 4. In: Berhardt L.V. (ed.) Pharmacophore Perception by Pattern Recognition: Model Construction, Validation and Application to Virtual Screening in a Practical Exercise, Advances in Medicine and Biology, Chapter, vol 159 (2020). ISBN: 978-1-53617-184-6
3. Mattos, E.C., Silva, L.P., Valero, C., de Castro, P. A., dos Reis, T.F., Ribeiro, L.F.C., Marten, Silva-Rocha, M.R.R., Westmann, C., da Silva, C.H.T.P., Taft, C.A., Al-Furaiji, N., Bromley, M., Mortensen, U.H., Benz,J.P., Brown, N.A., Goldman, G.H.: The *Aspergillus fumigatus* phosphoproteome reveals roles of high-osmolarity glycerol mitogen-activated protein kinases in promoting cell wall damage and *Caspofungin* tolerance. Molecul. Biol. Physiol. **11**(1), 1–23 (2020). https://doi.org/10.1128/mBio.02962-19
4. Barcellos, M.P., Santos, C.B.R., Federico, L.B., Almeida, P.F., da Silva, C.H.T.P., Taft, C.A.: Pharmacophore and structure-based drug design, molecular dynamics studies to design novel potential pad4 inhibitors. J. Biomolecul. Struct. Dyn. **37**, 966–998 (2019)
5. Amorin, L.H., Suzuki, V.Y., de Paula, N.H., Duarte, J.L., da Silva, M.A., Taft, C.A., La Porta, F.A.: Electronic, structural, optical and photocatalytic properties of graphitic arbon nitride. New J. Chem. **43**, 13647–13653 (2019)
6. Ramos, R.S., Costa, J.S., Silva, R.C., da Costa, G.V., Rodrigues, A.B.L., Rabelo, E.M., Souto, R.N.P., Taft, C.A., da Silva, C.H.T.P., Rosa, J.M.C., dos Santos, C.B.R., Macedo, W.J.: C,

Identification of potential inhibitors from pyriproxyfen with insecticidal activity by virtual screening. Pharmaceuticals **12**, 20 (2019)
7. Cruz, J.V., Serafim, R.B., da Silva, G.M., Giulliati, S., Campos, J.M., Neto, M.F.A., Leite, F.H.A., Taft, C.A., da Silva, C.H.T.P., Santos, C.B.R.: Computational design of new protein kinase 2 inhibitors for the treatment of inflammatory diseases using QSAR, pharmacophore-structure-based virtual screening and molecular dynamics. J. Mol. Model. **24**, 225 (2018)
8. Silva, J.C.F., dos Santos, J.D., Taft, C.A., Martins, J.B.L., Longo, E.: Stability of rolled-up GaAS nanotubes. J. Mol. Mod. **23**, 204 (2017)
9. Taft, C.A., Canchaya, J.G.S.: Review on Simulation models for materials and biomolecular study and design. In: Longo E., La Porta F.A. (eds.) Recent advances in complex functional materials (2017). https://doi.org/10.1007/978-3-319-53898-3_15
10. da Silva, C.H.T.P., Taft, C.A.: 3d descriptor calculations and conformational search to investigate potential bioactive conformations, with applications in 3d-qsar and virtual screening in drug design. J. Biomolecul. Struct. Dyn. **35**, 2966–2974 (2016)
11. Taft, C.A, (ed.): Thematic issue, ceramics, nanotubes, advanced materials: theoretical and experimental structure-property relationships V. Curr. Phys. Chem. **6**, 1–80 (2016)
12. Canchaya, J.G.S., Furtado, N.C., Taft, C.A.: An overview of fuel cells and simulation models. Curr. Phys. Chem. **5**, 223–252 (2015)
13. Taft, C.A. (ed): Thematic issue, ceramics, nanotubes, advanced materials: theoretical and experimental structure-property relationships II, Curr. Phys. Chem. **3**, 377–508 (2014)
14. Taft, C.A., da Silva, C.H.T.P. (eds.) New Developments in Medicinal Chemistry, vol. 2. Bentham Science, UAE (2014)
15. Taft, C.A. (ed.): Thematic issue, ceramics, nanotubes, advanced materials: theoretical and experimental structure-property relationships I. Curr. Phys. Chem. **3**, 377–508 (2013)
16. Taft, C.A., da Silva, C.H.T.P. (eds.) New Developments in Medicinal Chemistry, vol. 1. Bentham Science, UAE (2010)
17. Hirata, S.: Third- and fourth-order perturbation corrections to excitation energies from configuration interaction singles. J. Chem. Phys. **122**, 094105 (2005)
18. Johnson, E.R., Becke, A.D.: A post-Hartree-Fock model of intermolecular interactions: inclusion of higher-order corrections. J. Chem. Phys. **124**, 174104 (2006)
19. Mochizuki, Y., Tanakac, K.: Modification for spin-adapted version of configuration interaction singles with perturbative doublés. Chem. Phys. Lett. **443**(2007), 389–397 (2007)
20. Garcia, J., Szalewicz, K.: Ab initio extended hartree-fock plus dispersion method applied to dimers with hundreds of atoms. J. Phys. Chem. A **124**, 1196–1203 (2020)
21. Qin,K.S., Ichibha, T., Hongo, K., Maezono, R., Qin, K.S.: Inconsistencies in ab initio evaluations of non-additive contributions of DNA stacking energies. Chem. Phys. **529**, 110554 (2020)
22. Levine, D.S., Hait, D., Tubman, N.M., Lehtola, S.K., Whaley, B., Head-Gordon, M.: CASSCF with extremely large active spaces using the adaptive sampling configuration interaction method. J. Chem. Theory Comput. **16**, 2340–2354 (2020)
23. Park, W., Al-Saadon, R., MacLeod, M.K., Shiozaki, T., Vlaisavljevich, B.: Multirerence electron c/orrelation methods: Journeys along potential energy surfaces. Chem. Rev. **120**, 5878–5909 (2020)
24. Lomas, J.S.: Steric clashes, doughnuts and exploding cigars: some comments on non-covalent interactions. Chem. Europe **4**, 4238–4244 (2019)
25. Alkorta, B., Elguero, J., Frontera, N.: Not only hydrogen bonds: other noncovalent interactions. Crystals **10**, 180 (2020)
26. Puzzarini, C., Spada, L., Alessandrini, S., Barone, V.: The challenge of non-covalent interactions: theory meets experiment for reconciling accuracy and interpretation. J. Phys. Condens. Matt. **32**, 343002 (2020)
27. Dubecky, D., Mitas, L., Jurecka, P.: Noncovalent interactions by Quantum Monte Carlo. Chem. Rev. **116**, 5188–5215 (2016)
28. Hoffmann, R.: An extended Huckel theory I Hydrocarbons. J. Chem. Phys. **39**, 1397 (1963)

29. Pople, J.A., Segal, G.A.: Approximate self-consistent molecular orbital theory II. Calculations with complete neglect of differential overlap. J. Chem. Phys. **43**, S136–S151 (1965)
30. Pople, J.A., Santry, D.P., Segal, G.A.: Approximate self-consistent molecular orbital theory. I. Invariant Procedures. J. Chem. Phys. **43**, S129–S135 (1965)
31. Pople, J.A., Beveridge, D.J.: Approximate Molecular Orbital Theory. McGraw-Hill, New York (1970)
32. Ridley, J., Zerner, M.: An intermediate neglect of differential overlap technique for spectrostroscopy: Pyrrole and the azines. Theor. Chim. Acta **32**, 111 (1973)
33. Hehre, W.J., Stewart, R.F., Pople J.A.: Self-consistent molecular orbital methods. I. Use of gaussian expansions of slater type atomic orbitals. Chem. Phys. **51**, 2657–2664 (1969)
34. Dewar, M.J.S., Thiel, W.: Ground states of molecules. The MNDO method. Approximations and parameters. J. Am. Chem. Soc. **99**, 4899–4907 (1977)
35. Thiel, W., Voityuk, A.A.: Extension of MNDO to d-orbitals parameters and results for the second-row elements and for the zinc group. J. Phys. Chem. **100**, 616–626 (1996)
36. Dewar, M.J.S., Zoebisch, E.G., Healy, E.F., Stewart, J.J.P. Development and use of quantum mechanical molecular models. 76. AM1: A new general purpose quantum mechanical molecular model. J. Am. Chem. Soc. **107**, 3902–3909 (1985)
37. Stewart, J.: Optimization of parameters for semiempirical methods V: modification of NDDO approximations and application to 70 elements. J. Mol. Model. **13**, 1173–1213 (2007)
38. Stewart, J.J.P.: MOPAC2012; Stewart computational chemistry. Colorado Springs, CO (2012) https://OpenMOPAC.net;
39. Rocha, G.B., Freire, R.O., Simas, A.M., Stewart, J.J.P.: RM1: a reparameterization of AM1 for H, C, N, O, P, S, F, Cl, Br, and I. J. Comput. Chem. **27**, 1101 (2006)
40. Lima, N.B.D., Rocha, G.B., Freire, R.O., Simas, A.M.: RM1 Semiempirical model: chemistry, pharmaceutical research, molecular biology and materials science. J. Braz. Chem. Soc. **30**, 683–716 (2019)
41. Stewart, J.J.P.: Optimization of parameters for semiempirical methods II applications. J. Comput. Chem. **10**, 221–264 (1989)
42. Stewart, J.: Optimization of parameters for semiempirical methods VI: more modifications to the nddo approximations and re-optimization of parameters. J. Mol. Model. **19**, 1–32 (2013)
43. Elstner, M., Frauenheim, T., Kaxiras, E., Seifert, G., Suhai, S.A.: Self-consistent charge density-functional based tight-binding scheme for large biomolecules. Phys. Status Solidi B **217**, 357–376 (2000)
44. Elstner, M., Seifert, G.: Density functional tight binding. Philos. Trans. R. Soc. A **372**, 20120483 (2014)
45. Gaus, M., Cui, Q., Elstner, M.: Density functional tight binding: application to organic and biological molecules. WIREs Comput. Mol. Sci. **2014**(4), 49–61 (2014)
46. Korth, M.: Third-generation hydrogen-bonding corrections for semiempirical QM Methods and force fields. J. Chem. Theory Comput. **6**, 3808–3816 (2010)
47. Repasky, M.P., Chandrasekhar, J., Jorgensen, W.L.: PDDG/PM3 and PDDG/MNDO: improved semiempirical methods. J. Comput. Chem. **23**, 1601–1622 (2002)
48. Tuttle T., Thiel, W.: OMx-D: semiempirical methods with orthogonalization and dispersion corrections. implementation and biochemical application. Phys. Chem. Chem. Phys. **10**, 2159–2166 (2008)
49. Ahlswede, B., Jug, K.: Consistent modifications of SINDO1: I approximations and parameters. J. Comput. Chem. **20**, 563–571 (1999)
50. Ferro, N., Bredow, T.: Assessment of quantum-chemical methods for electronic properties and geometry of signalling biomolecules. J. Comput. Chem. **31**, 1063 (2010)
51. Bernal-Uruchurtu, M., Ruiz-López, M.: Basic ideas for the correction of semiempirical methods describing H-bonded systems. Chem. Phys. Lett. **330**, 118–124 (2000)
52. Bernal-Uruchurtu, M.I., Martins-Costa, M.T.C., Millot, C., Ruiz-López, M.F.: Improving description of hydrogen bonds at the semiempirical level: water-water interactions as test case. J. Comput. Chem. **21**, 572–581 (2000)

53. Chang, D.T., Schenter, G.K., Garrett, B.C.: Self-consistent polarization neglect of diatomic differential overlap: application to water clusters. J. Chem. Phys. **128**, 164111 (2008)
54. Murdachaew, G., Mundy, C.J., Schenter, G.K., Laino, T., Hutter, J.: Semiempirical self-consistent polarization description of bulk water, the liquid-vapor interface, and cubic ice. J. Phys. Chem. A **115**, 6046–6053 (2011)
55. Elstner, M., Fraunheim, T., Kaxiras, E., Seifert, G., Suhai, S.: Self-consistent charge density-functional based tight-binding scheme for large biomolecules. Phys. Status Solidi B **217**, 357–376 (2000)
56. Gaus, M., Cui, Q., Elstner, M.: DFTB3: extension of the self-consistent-charge densityfunctional tight-binding method (SCC-DFTB). J Chem Theory Comput. **7**, 931–948 (2012)
57. Dewar, M.J.S., AMPAC 10; Semichem, USA, 2012. Available at https://www.semichem.com/ampac/default.php
58. Christensen, A.S., Kubar, T., Cui, Q., Elstner, M.: Semiempirical quantum mechanical methods for noncovalent interactions for chemical and biochemical applications. Chem. Rev. **116**, 5301–5337 (2016)
59. Elstner, M., Porezag, D., Jungnickel, G., Elsner, J., Haugk, M., Frauenheim, T., Suhai, S., Seifert, G.: Self-consistent-charge density-functional tight-binding method for simulations of complex materials properties. Phys. Rev. B: Condens. Matter Mater. Phys. **58**, 7260–7268 (1998)
60. Lu, X., Gaus, M., Elstner, M., Cui, Q.: Parameterization of DFTB3/3OB for magnesium and zinc for chemical and biological applications. J. Phys. Chem. B **119**, 1062–1082 (2015)
61. Polzella, M.S.: Re-evaluating semi-empirical computer simulations in quantum chemistry. Foundat. Chem. **21**(1), 83–95 (2019). https://doi.org/10.1007/s10698-018-09329
62. Wu, X., Koslowski, A., Thiel, W.: Semiempirical Quantum Chemistry. In: Walker, R.C., Götz, A.W. (eds.) Electronic Structure Calculations on Graphics Processing Units: from Quantum Chemistry to Condensed Matter Physics, pp. 239–253. Wiley, Chichester (2016)
63. Domínguez, A., Niehaus, T.A., Frauenheim, T.: Accurate hydrogen bond energies within the density functional tight binding method. J. Phys. Chem. A **119**, 3535–3544 (2015)
64. Weber, W., Thiel, W.: Orthogonalization corrections for semiempirical methods. Theor. Chem. Acc. **103**, 495–506 (2000)
65. Cui, Q., Elstner, M.: Density functional tight binding: values of semi-empirical methods in an ab initio era. Phys. Chem. Chem. Phys. **16**, 14368–14377 (2014)
66. Maia, J.D.C., Carvalho, G.A.U., Mangueira, G.P., Santana Jr, S.R., Cabral, L.A.F., Rocha, G.B.: GPU Linear algebra libraries and GPGPU programming for accelerating MOPAC semiempirical quantum chemistry calculations. J. Chem. Theory Comput. **8**, 3072–3081 (2012)
67. Wu, X., Koslowski, A., Thiel, W.: Semiempirical quantum chemical calculations accelerated on a hybrid multicore CPU-GPU computing platform. J. Chem. Theory Comput. **8**, 2272–2281 (2012)
68. Gao, J.L., Truhlar, D.G., Wang, Y.J., Mazack, M.J.M., Loffler, P., Provorse, M.R., Rehak, P.: Explicit polarization: a quantum mechanical framework for developing next generation force fields Acc. Chem. Res. **47**, 2837–2845 (2014)
69. Giese, T.J., Huang, M., Chen, H.Y., York, D.M.: Recent advances toward a general purpose linear-scaling quantum force field. Acc. Chem. Res. **47**, 2812–2820 (2014)
70. Niklasson, A.M.N., Cawkwell, M.J.: Generalized extended lagrangian born-oppenheimer molecular dynamics. J. Chem. Phys. **141**, 164123 (2014)
71. Thiel, W.: Perspectives on semiempirical molecular orbital theory. Adv. Chem. Phys. **93**, 703–757 (1996)
72. Sattelmeyer, K.W., Tirado-Rives, J., Jorgensen, W.: Comparison of SCC-DFTB and NDDO-based semiempirical molecular orbital methods for organic molecules. J. Phys. Chem. A **110**, 13551–13559 (2006)
73. Otte, N., Scholten, M., Thiel, W.: Looking at self-consistent-charge density functional tight binding from a semiempirical perspective. J. Phys. Chem. A **111**, 5751–5755 (2007)

74. Kruger, T., Elstner, M., Schiffels, P., Frauenheim, T.: Validation of the density functional based tight-binding approximation method for the calculation of reaction energies and other data. J. Chem. Phys. **122**, 114110 (2005)
75. Parr, R.G.: A method for estimating electronic repulsion integrals over LCAO MOs in complex unsaturated molecules. J. Chem. Phys. **20**, 1499–1499 (1952)
76. Li, J., Williams, B., Cramer, C.J., Truhlar, D.G.: A Class IV charge model for molecular excited states. J. Chem. Phys. **110**, 724–733 (1999)
77. Voityuk, A.A.: INDO/X: a new semiempirical method for excited states of organic and biological molecules. J. Chem. Theory Comput. **10**, 4950–4958 (2014)
78. Voityuk, A.A.: Intermediate neglect of differential overlap for spectroscopy. WIREs Comput. Mol. Sci. **3**, 515–527 (2013)
79. Sustmann, R., Williams, J.E., Dewar, M.J.S., Allen, L.C., von Ragué Schleyer, P.: Molecular orbital calculations on carbonium Ions. II. Methyl, Ethyl, and Vinyl cations. The series $C_3H_7^+$. J. Am. Chem. Soc. **91**, 5350−5357 (1969)
80. Stewart, J.J.P., Csaszar, P., Pulay, P.: Fast semiempirical calculations. J. Comput. Chem. **3**, 227–228 (1982)
81. Neymeyr, K., Seelig, F.F.: Neglect of diatomic differential overlap. In nonempirical quantum chemical orbital theories, I. On the justification of the neglect of diatomic differential overlap approximation. Int. J. Quantum Chem. **53**, 515−518 (1995)
82. Kolb, M., Thiel, W.: Beyond the MNDO model: methodical considerations and numerical results. J. Comput. Chem. **14**, 775–789 (1993)
83. Ohno, K.: Some remarks on the Pariser-Parr-Pople method. Theor. Chim. Acta **2**, 219–227 (1964)
84. Clark, T.: Quo Vadis Semiempirical MO-Theory? J. Mol. Struct. THEOCHEM, **530**, 1−10 (2000)
85. Voityuk, A.A., Rosch, N.: AM1/d parameters for molybdenum. J. Phys. Chem. A **104**, 4089–4094 (2000)
86. Jurecka, P., Cerny, J., Hobza, P., Salahub, D.R.: Density functional theory augmented with an empirical dispersion term. Interaction energies and geometries of 80 noncovalent complexes compared with Ab Initio quantum mechanics calculations. J. Comput. Chem. **28**, 555−569 (2007)
87. Podeszwa, R., Patkowski, K., Szalewicz, K.: Improved Interaction energy benchmarks for dimers of biological relevance. Phys. Chem. Chem. Phys. **12**, 5974–5979 (2010)
88. Gaus, M., Goez, A., Elstner, M.: Parametrization and benchmark of DFTB3 for organic molecules. J. Chem. Theory Comput. **9**, 338–354 (2013)
89. Hohenberg, P., Kohn, W.: Inhomogenous electron gas. Phys. Rev. **136**, B864 (1964)
90. Kohn, W., Sham, J.: Phys.,Self-consistent equations including exchange and corelation effects, L.J. Rev. **140**, A1133 (1965)
91. Mardirossian, N., Head-Gordon, M.: Thirty years of density functional theory in computational chemistry: an overview and extensive assessment of 200 density functionals. Mol. Phys. **115**, 2315 (2017)
92. Thomas, L.H.: The calculation of atomic fields. Proc. Cambridge Philos. Soc. **23**, 542–548 (1927)
93. Perdew, J.P., Burke, K., Ernzerhof, M.: Generalized gradient approximation made simple. Phys. Rev. Lett. **7**, 3865 (1966)
94. Tao, J., Perdew, J.P., Staroverov, V.N., Scuseria, G.E.: Climbing the density functional ladder: Nonemirical meta-generalized gradient approximation designed for molecules and solids. Phys. Rev. Lett. **91**, 146401 (2003)
95. Becke, A.D.: Density-functional thermochemistry, III.The role of exact exchange, J. Chem. Phys. **98**, 5648 (1993)
96. Becke, A.D.: Density-functional thermochemistry, V. Systematic optimization of exchange-correlation functionals, J. Chem. Phys. **107**, 8554 (1997)
97. Chai, J.D., Head-Gordon, M.J.: Systematic optimization of long-range corrected hybrid density functionals Chem. Phys. **128**, 084106 (2008)

98. Sun, J., Haunschild, R., Xiao, B., Bulik, I.W., Scuseria, G.E., Perdew, J.P.: Semilocal and hybrid meta-generalized gradient approximations based on the understanding of the kinetic-energy-density dependence. J. Chem. Phys. **138**, 044113 (2013).
99. Perdew, J.P., Ruzsinszky, A., Tao, J., Staroverov, V.N., Scuseria, G.E., Csonka, G.I.: Prescription for the design and selection of density functional approximations: More constraint satisfaction with fewer fits. J. Chem. Phys. **123**, 062201 (2005)
100. Vosko, S.H., Wilk, L., Nusair, M.: Accurate spin-dependent electron liquid correlation energies for local spin density calculations: a critical analysis. Can. J. Phys. **58**, 1200 (1980)
101. Perdew, J.P., Zunger, A.: Self-interaction correction to density-functional approximations for many-electron systems Phys. Rev. B **23**, 5048 (1981)
102. Perdew, J.P., Wang, Y.: Accurate and simple analytic representation of the electron-gas correlation energy. Phys. Rev. B **45**, 13244 (1992)
103. Ceperley, D.M., Alder, B.J.: Ground state of the electron gas by a stochastic method. Phys Rev. Lett. **45**, 566 (1980)
104. Herman, F., Van Dyke, J.P., Ortenburger, I.B.: Improved statistical exchange approximation for inhomogeneous many-electron systems. Phys. Rev. Lett. **22**, 807 (1969)
105. Herman, F., Ortenburger, I.B., Van Dyke, J.P.: A method for improving the physical realism of first-principles band structure calculations. Int. J. Quant. Chem. **4**, 827 (1969)
106. Becke, A.D.: Density functional calculations of molecular bond energies. J. Chem. Phys. **84**, 4524 (1986)
107. Becke, A.D.: Density-functional exchange-energy approximation with correct asymptotic behavior. Phys. Rev. A **38**, 3098 (1988)
108. Perdew, J.P., Chevary, J.A., Vosko, S.H., Jackson, K.A., Pederson, M.R., Singh, D.J., Fiolhais, C.: Atoms, molecules, solids, and surfaces: applications of the generalized gradient approximation for exchange and correlation. Phys. Rev. B **46**, 6671 (1992)
109. Zhang, Y., Yang, W.: Comment on generalized gradient approximation made simple. Phys. Rev. Lett. **80**, 890 (1998)
110. Hammer, B., Hansen, L.B., Nørskov, J.K.: Improved adsorption energetics within density-functional theory using revised Perdew-Burke-Ernzerhof functionals. Phys. Rev. B **59**, 13–74 (1999)
111. Perdew, J.P., Ruzsinszky, A., Csonka, G.I., Vydrov, O.A., Scuseria, G.E., Constantin, L.A., Zhou, X., Burke, K.: Restoring the density-gradient expansion for exchange in solids and surfaces. Phys. Rev. Lett. **100**, 136406 (2008)
112. Perdew, J.P.: Density-functional approximation for the correlation energy of the inhomogeneous electron gas. Phys. Rev. B **33**, 8822 (1986)
113. Lee, C., Yang, W., Parr, R.G.: Development of the Colle-Salvetti correlation-energy formula into a functional of the electron density. Phys. Rev. B **37**, 785 (1988)
114. Hamprecht, F.A., Cohen, A.J., Tozer, D.J., Handy, N.C.: Development and assessment of new exchange-correlation functionals. J. Chem. Phys. **109**, 6264 (1998)
115. Boese, A.D., Doltsinis, N.L., Handy, N.C., Sprik, M.: New generalized gradient approximation functionals. J. Chem. Phys. **112**, 1670 (2000)
116. Boese, A.D., Handy, N.C.: A new parametrization of exchange–correlation generalized gradient approximation functionals, J. Chem. Phys. **114**, 54–97 (2001). Cheguei 20 08 2020
117. Peverati, R., Truhlar, D.G.: Exchange-correlation functional with good accuracy for both structural and energetic properties while depending only on the density and its gradient. J. Chem. Theory Comput. **8**, 2310 (2012)
118. Yu, H.S., Zhang, W., Verma, P., He, X., Truhlar, D.G.: Nonseparable exchange–correlation functional for molecules, including homogeneous catalysis involving transition metals. Phys. Chem. Chem. Phys. **17**, 12146 (2015)
119. Tsuneda, T., Suzumura, T., Hirao, K.: A reexamination of exchange energy functionals. J. Chem. Phys. **110**, 10664 (1999)
120. Adamo, C., Barone, V.: Exchange functionals with improved long-range behavior and adiabatic connection methods without adjustable parameters: The mPW and mPW1PW models. J. Chem. Phys. **108**, 664 (1998)

121. Handy, N.C., Cohen, A.J.: Left-right correlation energy. Mol. Phys. **99**, 403 (2001)
122. Murray, E.D., Lee, K., Langreth, D.C.: Investigation of exchange energy density functional accuracy for interacting molecules. J. Chem. Theory Comput. **5**, 2754 (2009)
123. Zhao, Y., Truhlar, D.G.: Construction of a generalized gradient approximation by restoring the density-gradient expansion and enforcing a tight Lieb-Oxford bound. J. Chem. Phys. **128**, 184109 (2008)
124. Schmider, H., Becke, A.: Chemical content of the kinetic energy density. J. Mol. Struct. THEOCHEM. **51**, 527 (2000)
125. Perdew, J.P., Kurth, S., Zupan, A., Blaha, P.: Accurate density functional with correct formal properties: a step beyond the generalized gradient approximation. Phys. Rev. Lett. **82**, 2544 (1999)
126. Sun, J., Xiao, B., Ruzsinszky, A.: Nonempirical Rung 3.5 density functionals from the Lieb-Oxford bound. J. Chem. Phys. **137**, 051–101 (2012)
127. Becke, A.D.: Simulation of delocalized exchange by local density functionals. J. Chem. Phys. **112**, 4020 (2000)
128. Perdew, J.P., Ruzsinszky, A., Csonka, G.I., Constantin, L.A., Sun, J.: Workhorse semilocal density functional for condensed matter physics and quantum chemistry. Phys. Rev. Lett. **103**, 026403 (2009)
129. Sun, J., Perdew, J.P., Ruzsinszky, A.: Semilocal density functional obeying a strongly tightened bound for exchange. Proc. Natl. Acad. Sci. U.S.A. **112**, 685 (2015)
130. Sun, J., Ruzsinszky, A., Perdew, J.P.: Strongly constrained and appropriately normed semilocal density functional. Phys. Rev. Lett. **115**, 036402 (2015)
131. Zhao, Y., Truhlar, D.G.: A new local density functional for main-group thermochemistry, transition metal bonding, thermochemical kinetics, and noncovalent interactions. J. Chem. Phys. **125**, 194101 (2006)
132. Peverati, R., Truhlar, D.G.: M11-L: a local density functional that provides improved accuracy for electronic structure calculations in chemistry and physics. J. Phys. Chem. Lett. **3**, 117 (2012)
133. Peverati, R., Truhlar, D.G.: An improved and broadly accurate local approximation to the exchange–correlation density functional: the MN12-L functional for electronic structure calculations in chemistry and physics. Phys. Chem. Chem. Phys. **14**, 13171 (2012)
134. Yu, H.S., He, X., Truhlar, D.G.: MN15-L: a new local exchange-correlation functional for kohn-sham density functional theory with broad accuracy for atoms, molecules, and solids. J. Chem. Theory Comput. **12**, 1280 (2016)
135. Voorhis, T.V., Scuseria, G.E.: A novel form for the exchange-correlation energy functional. J. Chem. Phys. **109**, 400 (1998)
136. Boese, A.D., Handy, N.C.: New exchange-correlation density functionals: the role of the kinetic-energy density. J. Chem. Phys. **116**, 9559 (2002)
137. Goerigk, L., Grimme, S.: A general database for main group thermochemistry, kinetics, and noncovalent interactions—assessment of common and reparameterized (meta-)GGA density functionals. J. Chem. Theory Comput. **6**, 107 (2010)
138. Wellendorff, J., Lundgaard, K.T., Jacobsen, K.W., Bligaard, T.: mBEEF: an accurate semi-local Bayesian error estimation density functional. J. Chem. Phys. **140**, 144107 (2014)
139. Harris, J., Jones, R.O.: The surface energy of a bounded electron gas. J. Phys. F: Metal Phys. **4**, 1170 (1974)
140. Gunnarsson, O., Lundqvist, B.I.: Exchange and correlation in atoms, molecules, and solids by the spin-density-functional formalism. Phys. Rev. B **13**, 4274 (1976)
141. Langreth, D.C., Perdew, J.P.: Exchange-correlation energy of a metallic surface: Wave-vector analysis. Phys. Rev. B **15**, 2884 (1977)
142. Harris, J.: Adiabatic-connection approach to Kohn-Sham theory. Phys. Rev. A **29**, 1648 (1984)
143. Adamo, C., Barone, V.: Toward reliable density functional methods without adjustable parameters: the PBE0 model. J. Chem. Phys. **110**, 6158 (1999)
144. Staroverov, V.N., Scuseria, G.E., Tao, J., Perdew, J.P.: Comparative assessment of a new nonempirical density functional: molecules and hydrogen-bonded complexes. J. Chem. Phys. **119**, 12129 (2003)

145. Csonka, G.I., Perdew, J.P., Ruzsinszky, A.: Global hybrid functionals: a look at the engine under the hood. J. Chem. Theory Comput. **6**, 3688 (2010)
146. Hui, K., Chai, J.D.: SCAN-based hybrid and double-hybrid density functionals from models without fitted parameters. J. Chem. Phys. **144**, 044114 (2016)
147. Zhao, Y., Truhlar, D.G.: Hybrid meta density functional theory methods for thermochemistry, thermochemical kinetics, and noncovalent interactions: the MPW1B95 and MPWB1K models and comparative assessments for hydrogen bonding and van der Waals interactions. J. Phys. Chem. A **108**, 6908 (2004)
148. Zhao, Y., Truhlar, D.G.: Design of density functionals that are broadly accurate for thermochemistry, thermochemical kinetics, and nonbonded interactions. J. Phys. Chem. A **109**, 5656 (2005)
149. Zhao, Y., Schultz, N.E., Truhlar, D.G.: Exchange-correlation functional with broad accuracy for metallic and nonmetallic compounds, kinetics, and noncovalent interactions, J. Chem. Phys. **123**, 103–161 (2005).
150. Zhao, Y., Schultz, N.E., Truhlar, D.G.: Design of Density functionals by combining the method of constraint satisfaction with parametrization for thermochemistry, thermochemical kinetics, and noncovalent interactions. J. Chem. Theory Comput. **2**, 364 (2006)
151. Zhao, Y., Truhlar, D.: The M06 suite of density functionals for main group thermochemistry, thermochemical kinetics, noncovalent interactions, excited states, and transition elements: two new functionals and systematic testing of four M06-class functionals and 12 other functionals. Theor. Chem. Acc. (Theoretica Chimica Acta) **120**, 215 (2008)
152. Zhao, Y., Truhlar, D.G.: Density functional for spectroscopy: no long-range self-interaction error, good performance for rydberg and charge-transfer states, and better performance on average than B3LYP for ground states. J. Phys. Chem. A **110**, 13126 (2006)
153. Zhao, Y., Truhlar, D.G.: Exploring the limit of accuracy of the global hybrid meta density functional for main-group thermochemistry, kinetics, and noncovalent interactions. J. Chem. Theory Comput. **4**, 1849 (2008)
154. Yu, H.S., He, X., Li, S.L., Truhlar, D.G.: MN15: A Kohn-Sham global-hybrid exchange–correlation density functional with broad accuracy for multi-reference and single-reference systems and noncovalent interactions. Chem. Sci. **7**, 5032 (2016)
155. Gill, P.M.W., Adamson, R.D., Pople, J.A.: Coulomb-attenuated exchange energy density functionals. Mol. Phys. **88**, 1005 (1996)
156. Leininger, T., Stoll, H., Werner, H.J., Savin, A.: Combining long-range configuration interaction with short-range density functionals. Chem. Phys. Lett. **275**, 151 (1997)
157. Gagliardi, Truhlar, D.G., Manni, G.L., Carlson, R.K., Hoyer, C.E., Bao, J.L.: Multiconfiguration pair-density functional theory: a new way to treat strongly correlated systems, Acc. Chem. Res. **50**, 66 (2017).
158. Weintraub, E., Henderson, T.M., Scuseria, G.E.: Long-range-corrected hybrids based on a new model exchange hole. J. Chem. Theory Comput. **5**, 754 (2009)
159. Rohrdanz, M.A., Martins, K.M., Herbert, J.M.: A long-range-corrected density functional that performs well for both ground-state properties and time-dependent density functional theory excitation energies, including charge-transfer excited states. J. Chem. Phys. **130**, 054112 (2009)
160. Yanai, T., Tew, D.P., Handy, N.C.: A new hybrid exchange–correlation functional using theCoulomb-attenuating method (CAM-B3LYP). Chem. Phys. Lett. **393**, 51 (2004)
161. Henderson, T.M., Janesko, B.G., Scuseria, G.E.: Generalized gradient approximation model exchange holes for range-separated hybrids. J. Chem. Phys. **128**, 194105 (2008)
162. Peverati, R., Truhlar, D.G.: Screened-exchange density functionals with broad accuracy for chemistry and solid-state physics. Phys. Chem. Chem. Phys. **14**, 16187 (2012)
163. Peverati, R., Truhlar, D.G.: Improving the accuracy of hybrid meta-GGA density functionals by range separation. J. Phys. Chem. Lett. **2**, 2810 (2011)
164. Krukau, A.V., Vydrov, O.A., Izmaylov, A.F., Scuseria, G.E.: Influence of the exchange screening on the performance of screened hybrid functionals. J. Chem. Phys. **125**, 224106 (2006)

165. Becke, A.D.: Real-space post-Hartree–Fock correlation models, Chem. Phys. **122**, 064–101 (2005)
166. Jaramillo, J., Scuseria, G.E., Ernzerhof, M.: Local hybrid functionals. J. Chem. Phys. **118**, 1068 (2003)
167. Perdew, J.P., Staroverov, V.N., Tao, J., Scuseria, G.E.: Density functional with full exact exchange, balanced nonlocality of correlation, and constraint satisfaction. Phys. Rev. A **78**, 052513 (2008)
168. Boese, A.D.: Density functional theory and hydrogen bonds: are we there yet? Chem. Phys. Chem. **16**, 978 (2015)
169. Goerigk L., Grimme, S.: Double-hybrid density functionals, Wiley Interdiscip. Rev.: Comput. Mol. Sci. **4**, 576 (2014)
170. Grimme, S.: Accurate description of van der Waals complexes by density functional theory including empirical corrections. J. Comput. Chem. **25**, 1463 (2004)
171. Grimme, S.: Semiempirical GGA-type density functional constructed with a long-range dispersion correction. J. Comput. Chem. **27**, 1787 (2006)
172. Grimme, S., Antony, J., Ehrlich, S., Krieg, H.: A consistent and accurate ab initio parametrization of density functional dispersion correction (DFT-D) for the 94 elements H-Pu. J. Chem. Phys. **132**, 154104 (2010)
173. Vydrov, O.A., Voorhis, T.V.: Nonlocal van der Waals density functional: the simpler the better. J. Chem. Phys. **133**, 244103 (2010)
174. Chai, J.D., Head-Gordon, M.: Long-range corrected hybrid density functionals with damped atom-atom dispersion corrections. Phys. Chem. Chem. Phys. **10**, 6615 (2008)
175. Lin, Y.S., Tsai, G.D., Li, J.D.: Long-range corrected hybrid meta-generalized-gradient approximations with dispersion corrections, Chai. J. Chem. Phys. **136**, 154109 (2012)
176. Lin, Y.S., Li, G.D., Mao, S.P., Chai, J.D.: Long-range corrected hybrid density functionals with improved dispersion corrections. J. Chem. Theory Comput. **9**, 263 (2013)
177. Mardirossian, N., Ruiz Pestana, L., Womack, J.C., Skylaris, C.K., Head-Gordon, T., Head-Gordon, M.: Use of the rVV10 nonlocal correlation functional in the B97M-V density functional: defining B97M-rV and related functionals. Phys. Chem. Lett. **8**, 35 (2017)
178. Sabatini, R., Gorni, T., de Gironcoli, S.: Nonlocal van der Waals density functional made simple and eficiente. Phys. Rev. B **87**, 041108 (2013)
179. Peng, H., Yang, Z.H., Perdew, J.P., Sun, J.: Versatile van der waals density functional based on a meta-generalized gradient approximation. Phys. Rev. X **6**, 041005 (2016)
180. Grayce, C.J., Harris, R.A.: Magnetic-field density-functional theory. Phys. Rev. A **50**, 3089 (1994)
181. Vignale, G., Rasolt, M.: Density-functional theory in strong magnetic fields. Phys. Rev. Lett. **59**, 2360 (1987)
182. Vignale, G., Rasolt, M.: Current- and spin-density-functional theory for inhomogeneous electronic systems in strong magnetic fields. Phys. Rev. B **37**, 10685 (1988)
183. Verma, P., Truhlar, D.G.: status and challenges of density functional theory status and challenges of density functional theory. Trends Chem. **2**, 306–316 (2020)
184. Cohen, A.J., Sanchez, P.M., Yang, W.: Challenges for density functional theory. Chem. Rev. **112**, 289 (2011)
185. He, Q., Yu, B., Li, Z., Zhao, Y.: Density functional theory for battery materials. Energy Environ. Mater. **3**, 1 (2020)
186. Rehn, D.A., Wills, J.M., Battelle, T.E., Mattsson, A.E.: Dirac's equation and its implications for density functional theory based calculations of materials containing heavy elements. Phys. Rev. B **101**, 085114 (2020)
187. Dick, S., Fernandez-Serra, M.: Machine learning accurate exchange and correlation functionals of the electronic density. Nat. Commun. **11**, 3509 (2020)
188. Kharissova, O.V., Kharisov, B.I., González, L.T.: Recent trends on density functional theory–assisted calculations of structures and properties of metal–organic frameworks and metal–organic frameworks-derived nanocarbons. J. Mater. Res. **35**, 1424–1438 (2020)

189. Mazurek, A.H., Szeleszczuk, L., Pisklak, D.M.: Periodic DFT calculations-review of applications in the pharmaceutical sciences. Pharmaceutics **12**, 415 (2020)
190. Runge, E, Gross, E.K.U.: Density-functional theory for time-dependent systems. Phys. Rev. Lett. **52**, 997 (1984)
191. Werschnik, J., Gross, E.K.U., Burke, K.: Time-dependent density functional theory: past, present, and future. J. Chem. Phys. **123**, 062206 (2005)
192. Adamo, C., Jacquemin, D.: The calculations of excited-state properties with time-dependent density functional theory. Chem. Soc. Rev **42**, 845 (2013)
193. Casida, M.E., Huix-Rotllant, M.: Progress in time-dependent density-functional theory. Annu. Rev. Phys. Chem. **63**, 287–323 (2012)
194. Maitra, N.T., van Leeuwen, R., Burke, K.: Comment on Critique of the foundations of time-dependent density-functional theory. Phys. Rev. A **78**, 056501 (2008)
195. Suzuki, Y., Nagai, R., Haruyama, J.: Machine learning exchange-correlation potential in time-dependent density-functional theory. Phys. Rev. A **101**, 050501 (2020)
196. Suzuki, Y., Watanabe, K.: Excitons in two-dimensional atomic layer materials from time-dependent density functional theory: mono-layer and bi-layer hexagonal boron nitride and transition-metal dichalcogenides. Phys. Chem. Chem. Phys. **22**, 2908–2916 (2020)
197. Shao, Y., Mei, Y., Sundholm, D., Kaila, V.R.I.: Benchmarking the performance of time-dependent density functional theory methods on biochromophores. J. Chem. Theory Comput. **16**(1), 587–600 (2020)
198. QMC Austin, B.M., Zubarev, D.Y., Lester, Jr., W.A.: Quantum Monte Carlo and related approaches, Chem. Rev. **112**, 263–288 (2012)
199. Yang, Y., Gorelov, V., Pierleoni, C., Ceperley, D.M., Holzmann, M.: Electronic band gaps from quantum Monte Carlo methods. Phys. Rev. B **101**, 085115 (2020)
200. Li, B.H., Li, Z.X., Yao, H.: Fermion-induced quantum critical point in Dirac semimetals: a sign-problem-free quantum Monte Carlo study. Phys. Rev. B **101**, 085105 (2020)
201. Raczkowski, M., Peters, R., Phùng, T.T., Takemori, N., Assaad, F.F., Honecker, A., Vahedi, J.: Hubbard model on the honeycomb lattice: from static and dynamical mean-field theories to lattice quantum Monte Carlo simulations. Phys. Rev. B **101**, 125103 (2020)
202. Pollet, L.: Recent developments in Quantum Monte-Carlo simulations with applications for cold gases. Rept. Prog. Phys. **75**, 094501 (2012)
203. Mella, M., Anderson, J.B.: Intermolecular forces and fixed-node diffusion Monte Carlo: a brute force test of accuracies for He2 and He-LiH. J. Chem. Phys. **119**, 8225–8228 (2003)
204. Diedrich, C., Luchow, A., Grimme, S.: Weak intermolecular interactions calculated with diffusion Monte Carlo. J. Chem. Phys. **123**, 184106 (2005)
205. Ambrosetti, A., Alfè, D., DiStasio, R.A., Tkatchenko, Jr., A.: Hard numbers for large molecules: toward exact energetics for supramolecular systems. J. Phys. Chem. Lett. **5**, 849–855 (2014)
206. Benali, A., Shulenburger, L., Romero, N.A., Kim, J., von Lilienfeld, O.A.: Application of diffusion monte carlo to materials dominated by van der Waals interactions. J. Chem. Theory Comput. **10**, 3417–3422 (2014)
207. Ceperley, D., Chester, G.V., Kalos, M.H.: Monte Carlo simulation of a many-fermion study. Phys. Rev. B **16**, 3081–3099 (1977)
208. Hammond, B.L., Reynolds, P.J., Lester, W.A.: Valence quantum Monte Carlo with ab initio effective core potentials. J. Chem. Phys. **87**, 1130–1136 (1987)
209. Kosztin, I., Faber, B., Schulten, K.: Introduction to the diffusion Monte Carlo methods. Am. J. Phys. **64**, 633–644 (1996)
210. Luchow, A., Anderson, J.B.: Monte Carlo methods in electronic structures for large systems. Annu. Rev. Phys. Chem. **51**, 501 (2000)
211. Foulkes, W.M.C., Mitas, L., Needs, R.J., Rajagopal, G.: Quantum Monte Carlo simulations of solids. Rev. Mod. Phys. **73**, 33–83 (2001)
212. Lester Jr., W.A., Mitas, L., Hammond, B.: Quantum Monte Carlo for atoms, molecules and solids. Chem. Phys. Lett. **478**, 1–10 (2009)

213. Bajdich, M., Mitas, L.: Electronic structure Quantum Monte Carlo. Acta Phys. Slovaca **59**, 81–168 (2009)
214. Needs, R.J., Towler, M.D., Drummond, N.D., López Ríos, P.: Continuum variational and diffusion quantum Monte Carlo calculations. J. Phys. Condens. Matter, **22**, 023–201 (2010)
215. Hollingsworth, S.A., Dror, Ron, O.: Molecular dynamics simulation for all. Neuron, **99**, 1129 (2018)
216. Dror, R.O., Dirks, R.M., Grossman, J.P.,Xu, J., Shaw D.E.: Biomolecular simulation: a computational microscope for molecular biology. Annu. Rev. Biophys. **41**, 429–452 (2012)
217. Hospital, A., Goñi, J.R., Orozco, M., Gelpí, J.L.: Molecular dynamics simulations: advances and applications. Adv. Appl. Bioinforma. Chem. **8**, 37–47 (2015)
218. Doshi, U., Hamelberg, D.: Towards fast, rigorous and efficient conformational sampling of biomolecules: advances in accelerated molecular dynamics. Biochem. Biophys. Acta. **1850**, 878–888 (2014)
219. González, M.A.: Force fields and molecular dynamics simulations. Collection SFN **12**, 169–200 (2011)
220. Borhani, D.W., Shaw, D.E.: The future of molecular dynamics simulations in drug Discovery. J. Comput. Aided Mol. Des. **26**, 15–26 (2012)
221. Dong, Y., Li, Q., Martini, A.: Molecular dynamics simulation of atomic friction: a review and guide. J. Vac. Sci. Technol. **31**, 030801 (2013). https://doi.org/10.1116/1.4794357
222. Daddam, J.R., Sreenivasulu, B., Peddannade, K., Umamahesh, K.: Designing, docking and molecular dynamics simulation studies of novel cloperastine analogues as anti-allergic agents: homology modeling and active site prediction for the human histamine H1 receptor, RSC Adv. **46** (2020)
223. Wang, Q., Tang, Q., Tian, S.: Molecular dynamics simulation of sI methane hydrate under compression and tension. Open Chem. **18** (2020)
224. Bekker, G.J., Fukuda, I., Higo, J., Kamiya, N.: Mutual population-shift driven antibody-peptide binding elucidated by molecular dynamics simulations. Sci. Rep. **10**, 1406 (2020)
225. Bissaro, M., Sturlese, M., Moro, S.: Exploring the RNA-recognition mechanism using supervised molecular dynamics (SuMD) simulations: toward a rational design for ribonucleic-targeting molecules? Front. Chem. (2020). https://doi.org/10.3389/fchem.2020.00107
226. Genga, H., Chenb, F., Jiangac, J.Y.F.: Applications of molecular dynamics simulation in structure prediction of peptides and proteins. Computat. Struct. Biotechnol. J. **17**, 1162–1170 (2019)
227. Verlet, L.: Computer experiments on classical _uids. i. thermodynamical properties of Lennard-Jones molecules. Phys. Rev. **159**, 98–103 (1967)
228. Berendsen, H.J.C., Postma, J.P.M., DiNola, A., Haak, J.R.: Molecular dynamics with coupling to an external bath. J. Chem. Phys. **81**, 3684–3690 (1984)
229. Hata, H., Nishiyama, M., Kitao, A.: Molecular dynamics simulation of proteins under high pressure: structure, function and thermodynamics. Biochimica et Biophysica Acta (BBA) General Subjects **1864**(2), 129–395 (2020)
230. Baker, C.M.: Polarizable force fields for molecular dynamics simulations of biomolecules. WIREs Comput. Mol. Sci. **5**, 241–254 (2015). https://doi.org/10.1002/wcms.1215
231. Car, R., Parrinello, M.: Unified approach for molecular dynamics ad density functional theory. M. Phys. Rev. Lett. **55**, 2471–2474 (1985)
232. Iftimie, R., Minary, P., Tuckerman, M.E.: Ab initio molecular dynamics: concepts, recente developments, and future trends. PNAS **102**, 6654–6659 (2005)
233. Mark, D., Hutter, J.: Ab initio molecular Dynamics Basic Theory and Advanced Methods. Cambridge University Press, Cambridge (2009)
234. Kirchner, B., di Dio, P.J., Hutter, J.: Real-world predictions from ab initio molecular dynamics simulations. Top. Curr. Chem. **307**, 109 (2012)
235. Tuckerman, M.E.: Ab initio molecular dynamics: basic concepts, current trends and novel applications. J. Phys. Condens. Matter **14**, 1297 (2002)
236. Kuhne, T.D.: Ab-Initio Molecular Dynamics. WIREs Comput. Mol. Sci. **4**, 391 (2014)

237. Liu, J., Zhu, T., Wang, X., He, X., Zhang, J.Z.H.: Quantum fragment based ab initio molecular dynamics for proteins. J. Chem. Theor. sAnd Compt. **11**, 5897 (2015)
238. Tse, J.S.: Ab initio molecular dynamics with density functional theory. Annu. Rev. Phys. Chem. **53**, 249–290 (2002)
239. Heenen, H.H., Gauthier, J.A., Kristoffersen, H.H., Ludwig, T., Chan, K.: Solvation at metal/water interfaces: An ab initio molecular dynamics benchmark of common computational approaches. J. Chem. Phys. **152**, 144703 (2020). https://doi.org/10.1063/1.5144912
240. Galvez-Aranda, D.E., Seminario, J.M.: Solid electrolyte interphase formation between the Li0.29La0.57TiO3 solid-state electrolyte and a Li-metal anode: an ab initio molecular dynamics study. RSC Adv. **10**, 9000–9015 (2020)
241. Ang S.J., Wang, W., Schwalbe-Koda, D., Axelrod, S., Gomez-Bombarelli, R.: Active learning accelerates ab initio molecular dynamics on pericyclic reactive energy surfaces. ChemRev (2020). https://doi.org/10.26434/chemrxiv.11910948.v2
242. Ko, H.Y., Jia, J., Santra, B., Wu, X., Car, R., DiStasio Jr., R.A.: Enabling large-scale condensed-phase hybrid density functional theory based ab initio molecular dynamics, 1. Theory, algorithm, and performance. J. Chem. Theory Comput. **16**(6), 3757–3785 (2020)
243. Krylow, S., Hernandez, F.V., Bauerhenne, B., Garcia, M.E.: Ultrafast structural relaxation dynamics of laser-excited graphene: Ab initio molecular dynamics simulations including electron-phonon interactions. Phys. Rev. B **101**, 205428 (2020)
244. Hafner, J.: Ab-initio simulations of materials using VASP: density-functional theory and beyond. J. Comput. Chem. **29**, 2044–2078 (2008)
245. Schleder, G.R., Padilha, A.C.M., Acosta, C.M., Costa, M., Fazzio, A.: From DFT to machine learning: recent approaches to materials science–a review. J. Phys. Mater. **2**, 032001 (2019)
246. Saad, Y., Chelikowsky, J.R., Shontz, S.M.: Numerical methods for electronic structure calculations of materials. Soc. Indust. Appl. Mathe. **52**, 3–54 (2010)
247. Steinhauser, M.O., Hiermaier, S.: A review of computational methods in materials science: examples from shock-wave and polymer physics. Int. J. Mol. Sci. **10**, 5135–5216 (2009). https://doi.org/10.3390/ijms10125135
248. Holzwarth, N.A.W., Matthews, G.E., Dunning, R.B., Tackett, A.R., Zeng, Y.: Comparison of the projector augmented-wave, pseudopotential, and linearized augmented-plane-wave formalisms for density-functional calculations of solids. Phys. Rev. B **55**, 2005 (1997)
249. Cao, W., Wang, Z., Fu, L., Chi, J., Gao, X., Wang, W.: LW, The analysis of a plane wave pseudopotential density functional theory code on a GPU machine. Comput. Phys. Commun. **184**, 9–18 (2013)
250. Rohr, B., Stein, H.S., Guevarra, D., Wang, Y., Haber, J.A., Aykol, M., Suram, S.K., Gregoir, J.M.: Benchmarking the acceleration of materials discovery by sequential learning. Chem. Sci. **11**, 2696–2706 (2020)
251. Faber, F., Lindma, A., von Lilienfeld, A.O., Rickard Armiento, R.: Crystal structure representations for machine learning models of formation energies. Int. J. Quantum Chem. **115**, 1094–1101 (2015)
252. Hafner, J., Wolverton, C., Ceder, G., Hafner, J., Wolverton, C., Gerbrand, C.: Toward computational materials design: the impact of density functional theory on materials research. MRS Bulletin **31**, 659–668 (2006)
253. Zaporozhets, T.V., Taranovskyy, A., Jáger, G., Gusak, A.M., Erdélyi, Z., Tomán, János J.: The effect of introducing stochasticity to kinetic mean-field calculations: comparison with lattice kinetic Monte Carlo in case of regular solid Solutions. Computat. Mater. Sci. Computat. Mater. Sci. **171**, 109–251 (2020)
254. Azevedo, D.H.M., Fabris, G.S.L., Sambrano, J.R., Cordeiro, J.M.M.: Surface and electronic properties of rutile TiO2 thin films coated with PbO2. Comput. Mater. Sci. **171**, 109222 (2020)
255. Schmidt, J., Marques, M.R.G., Botti, S., Marques, M.A.L.: Recent advances and applications of machine learning in solid state materials Science. Computat. Mater. **5**(83) (2019). https://doi.org/10.1038/s41524-019-0221-0

256. He, L., Liu, F., Li, J., Rignanesed, G.M., Zhou, A.: First-principles investigation of monatomic gold wires under tension. Comput. Mater. Sci. **171**, 109226 (2020)
257. Silva, C.E., Pontes, R.B.: Structural, electronic and transport properties of a single 1,4-benzenediamine molecule attached to metal contacts of Au Ag and Cu. Computat. Mater. Sci. **171**, 109212 (2020)
258. Kawano, S., Iikubo, S., Ohtanib, H.: Role of substituted atoms in stacking fault formation in long-period stacking ordered system. Comput. Mater. Sci. **171**, 109210 (2020)
259. Fang, C., Fan, Z.: Prenucleation at the liquid-Al/α-Al_2O_3 and the liquid-Al/MgO interfaces. Comput. Mater. Sci. **171**, 109258 (2020)
260. Cui, W., Yu, S., Zhao, J.: Two-dimensional π-conjugated metal bis(dithiolene) nanosheet: A promising anchoring material for lithium-sulfur batteries. Comput. Mater. Sci. **171**, 109228 (2020)
261. Rafi, A., Lasir, M., Hammond, Karl, D.: Pseudopotential for plane-wave density functional theory studies of metallic Uranium. Computat. Mater. Sci. **171**, 109–221 (2020)
262. Jiang, C., van Rooyen, I.J., Meher, S.: Ab initio study and thermodynamic modeling of the Pd-Si-C system. Comput. Mater. Sci. **171**, 109238 (2020)
263. Geng, X., Wang, H., Xue, W., Xiang, S., Huang, H., Meng, L., Ma, G.: Modeling of CCT diagrams for tool steels using diferente machine learning techniques. Comput. Mater. Sci. **171**, 109235 (2020)
264. Morrissey, L.S., Handrigan, S.M., Nakhla, S.: Discrepancies in the mechanical properties of gold nanowires: the importance of potential type and equilibration method. Comput. Mater. Sci. **171**, 109234 (2020)
265. Khosravi, M., Murthy, V., Mackinnon, I.D.R.: Evaluation of DFT methods to calculate structure and partial atomic charges for zeolite N. Comput. Mater. Sci. **171**, 109225 (2020)
266. Jankowskia, E., Ellysona, N., Fothergilla, J.W., Henrya, M.M., Leibowitza, M.H., Millera, E.D., Albertsa, M., Chesserb, S., Guevarac, J.D., Jonesa, C.D., Klopfensteina, M., Nonemana, K.K., Singletond, R., Uriarte-Mendozae, R.A., Thomas, S., Estridgeg, C.E., Jones, M.L.: Perspective on coarse-graining, cognitive load, and materials simulation. Comput. Mater. Sci. **171**, 109129 (2020)
267. Goncharenko, A.V., Nazarov, V.U., Pinchuk, A.O.: Reciprocal space approach to effective constitutive parameters of periodic composites. Comput. Mater. Sci. **171**, 109257 (2020)
268. Ferreira, M.D., Santos, J.D., Taft, C.A., Longo, E., Martins, J.B.L.: Single walled MgF2 nanotubes. Comput. Mater. Sci. **46**, 233–238 (2009)
269. Santos, J.D., Ferreira, M.D., Martins, J.B.L., Taft, C.A., Longo, E.: Computational studies of [(SnO_2)n]m nanotubes. Curr. Phys. Chem. **3**, 451–476 (2013)
270. Costa, R.J., Martins, J.B.L., Longo, E., Taft, C.A., Santos, J.D.: Methodology to obtain and study geometries of single and double wall silicon carbide nanotubes. Curr. Phys. Chem. **60**, 60–80 (2016)
271. Sodré, J.M., Longo, E., Taft, C.A., Martins, J.B.L., dos Santos, J.D.: Electronic structure of GaN nanotubes. C. R. Chim. **20**, 190–196 (2017)
272. Silva, J.C.F., dos Santos, J.D., Taft, C.A., Martins, J.B.L., Longo, E.: Stability of rolled-up GaAs nanotubes. J. Mol. Model. **23**, 204–210 (2017)
273. Friesner, R.A., Banks, J.L., Murphy, R.B., Halgren, T.A., Klicic, J.J., Mainz, D.T., et al.: Glide: a new approach for rapid, accurate docking and scoring, 1. Method and Assessment of Docking acuracy. J. Med. Chem. **47**, 1739–1749 (2004)
274. Jones, G., Willett, P., Glen, R.C., Leach, A.R., Taylor, R.: Development and validation of a genetic algorithm for flexible docking. J. Mol. Biol. **267**, 727–748 (1997)
275. Allen, W.J., Balius, T.E., Mukherjee, S., Brozell, S.R., Moustakas, D.T., Lang, P.T., Case, D.A., Kuntz, I.D., Rizzo, R.C.: DOCK 6: impact of new features and current docking performance. J. Comput. Chem. **36**, 1132–1156 (2015)
276. Morris, G.M., Huey, R., Lindstrom, W., Sanner, M.F., Belew, R.K., Goodsell, D.S., Olson, A.J.: AutoDock4 and AutoDockTools4: automated docking with selective receptor flexibility. J. Comput. Chem. **30**, 2785–2791 (2009)

277. Kawano, D.F., da Silva, C.H.T.P., Taft, C.A.: Design of inhibitors of the human fibroblast activation protein as a strategy to hinder metastasis and angiogenesis. La Porta, F.A., Taft, C.A. (eds.) Emerging Research in Science and engineering Based on Advanced Experimental and Computational Strategies,Engineering Materials (2020). https://doi.org/10.1007/978-3-030-31403-3_11
278. Sledz, P., Caflisch, A.: Protein structure-based drug design: from docking to molecular dynamics. Curr. Opin. Struct. Biol. **48**, 93–102 (2018)
279. Lin, X., Li, X., Lin, X.: A review on applications of computational methods in drug screening and design. Molecules **25**, 1375 (2020)
280. Chaudhary, K.K., Mishra, N.A.: Review on molecular docking: novel tool for drug discovery. JSM Chem. **4**, 1029 (2016)
281. da Silva, C.H.T.P., Taft, C.A.: Computer-aided molecular design of novel glucosidase inhibitors for AIDS treatment. J. Biomol. Struct. Dyn. **22**, 59 (2004)
282. da Silva, C.H.T.P., Carvalho, I., Taft, C.A.: Homology modeling and molecular interaction field studies of α-glucosidases as a guide to structure-based design of novel proposed anti-HIV inhibitors. J. Comput.-Aided Molecul. Des. 1983 (2005)
283. da Silva, C.H.T.P., Taft, C.A., Carvalho, I.: Molecular dynamics, docking, density functional and admet studies of HIV-1 reverse transcaptose inhibitors. J. Theoret. Computat. Chem. **5**, 579 (2006)
284. da Silva, V.B., Andrioli, W.J., Carvalho, I., Taft, C.A., da Silva, C.H.T.P.: Computer-aided molecular design of novelHMG-CoA reductase inhibitors for the treatment of hypercholesterolemia. J. Theor. Comput. Chem. **6**, 811 (2007)
285. da Silva, V.B., Kawano, D., Gomes, A.S., Carvalho, I., Taft, C.A., da Silva, C.H.T.P.: Molecular dynamics, density functional, ADMET predictions, virtual screening, and molecular interaction field studies for identification and evaluation of novel potential CDK2 inhibitors in cancer therapy. J. Phys. Chem. A **112**, 8902–8910 (2008)
286. da Silva, C.H.T.P., da Silva, V.B., Resende, J., Rodrigues, P.F., Bononi, F.C., Benevenuto, C.G., Taft, C.A.: Computer-aided drug design and ADMET predictions for identification and evaluation of novel potential farnesyltransferase inhibitors in cancer therapy. J. Molecul. Graphics Modeling **28** (2009)
287. Semighini, E.P., Resende, J., Andrade, P., Morais, P.A.B., Carvalho, I., Taft, C.A., da Silva, C.H.T.P.: Using computer-aided drug design and medicinal chemistry strategies in the fight against diabetes. J. Biomol. Struct. Dyn. **28**, 787–796 (2011)
288. Ferreira, F.P., Couto, W.F., Fontana, F., Taft, C.A., da Silva, C.H.T.P.: Molecular dynamics, density functional theory, pharmacophore modeling, molecular interaction fields and ADME/Tox investigation of novel bioactive compunds interacting with CDK2 surfaces. Curr. Phys. Chem. **4**, 94105 (2014)
289. Hage-Melim, L., Sampaio, S., Taft, C.A., da Silva, C.H.T.P.: Phospholipase A2 inhibitors isolated from medicinal plants: alternative treatment against snakebites. MiniRev. Med. Chem. **13**, 1348 (2013)
290. Taft, C.A., da Silva, C.H.T.P.: Comments on the paper 'Levinthal' question, revisited, and answered. J. Biomol. Struct. Dyn. **31**, 1001 (2013). https://doi.org/10.1080/07391102.2012.748540
291. de Oliveira, M.G., de Souza, W.R.N., Rodrigues, R.P., Kawano, D.F., Borges, L.L., da Silva, V.B.: Pharmacophore mapping of natural products for pancreatic lipase 3 Inhibition. La Porta, F.A., Taft, C.A. (eds.) Emerging Research in Science and Engineering Based on Advanced Experimental and Computational Strategies,Engineering Materials (2020). https://doi.org/10.1007/978-3-030-31403-3_12
292. Schaller, D., Sribar, D., Noonan, T., Deng, L., Nguyen, T.N., Pach, S., Machalz, D., Bermudez, M., Wolber, G.: Next generation 3D pharmacophore modeling. WIREs Comput. Mol. Sci. **10**, 1468 (2020)
293. Horvath, D.: Pharmacophore-based virtual screening. Jeurgen B. (ed.) Chemoinformatics and Computational Chemical Biology, Methods in Molecular Biology, vol. 672. Springer Science Business Media, LLC (2011). https://doi.org/10.1007/978-1-60761-839-3_11

294. Kim, K., Kim, M.D., Seong, B.: Pharmacophore-based virtual screening: a review of recent applications. Expert Opin. Drug Discov. **5**(3), 205–222 (2010)
295. Schuster, D., Wolber, G.: Identification of bioactive natural products by pharmacophore-based virtual screening. Curr. Pharm. Des. **16**, 1666–1681 (2010)
296. Seidel, T., Ibus, G., Fabian Bendix,F., Wolber, G.: Strategies for 3D pharmacophorebased virtual screening, Drug Discovery Today **7** (2010)
297. Barcelos, M.P., Federico, L.B., Taft, C.A., da Silva, C.H.T.P.: Prediction of the three-dimensional structure of phosphate-6-mannose PMI present in the cell membrane of anthomonas citri subsp. Citri of interest for the citrus canker control. La Porta, F.A., Taft, C.A. (eds.) Emerging Research in Science and Engineering Based on Advanced Experimental and Computational Strategies, Engineering Materials (2020). https://doi.org/10.1007/978-3-030-31403-3_10
298. Kaczanowski, S., Zielenkiewicz, P.: Why similar protein sequences encode similar three-dimensional structures? Theor. Chem. Acc. **125**, 643–650 (2010)
299. Gront, D., Blaszczyk, M., Wojciechowski, P., Kolinski, A.: BioShell threader: protein homology detection based on sequence profiles and secondary structure profiles. Nucleic. Acids Res. **40**, 257–262 (2012)
300. Peng, J., Xu, J.: Low-homology protein threading. Bioinformatics **26**, i294-300 (2010)
301. Peng, J., Xu, J.: A multiple-template approach to protein threading. Proteins Struct. Funct. Bioinf. **79**, 1930–1939 (2011)
302. López, C.A., Rzepiela, A.J., de Vries, A.H., Dijkhuizen, L., Hünenberger, P.H., Marrink, S.J.: Martini, coarse-grained force field: extension to carbohydrates. J. Chem. Theory Comput. **5**, 3195–3210 (2009)
303. Solano Canchaya, J.G.S., Dequidt, A.,Garruchet, S., Latour, B., Martzel, N.,Devémy, J.,Goujon, F.,Blaak, R., Schnell, B.,Munch, E., Seeboth, N., Malfreyt,P.: Development of a coarse-grain model for the description of the metal oxide-polymer interface from a bottom-up approach. J. Chem. Phys. **151**, 064–703 (2019)
304. Kempfer, J.D., Dequidt, A., Couty, M., MaMalfrey, P.: Development of coarse-grained models for polymers by trajectory matching. ACS Omega **4**, 5955–5967 (2019)
305. Singh, N., Li, W.: recent advances in coarse-grained models for biomolecules and their applications. Int. J. Mol. Sci. **20**, 3774 (2019)
306. Bhattacharya, S., Lin, X.: Recent advances in computational protocols addressing intrinsically disordered proteins. Biomolecules **9**, 146 (2019)
307. Pak, J., Voth, G.A.: Advances in coarse-grained modeling of macromolecular complexes. Curr. Opin. Struct. Biol. **52**, 119–126 (2018)
308. Ekimoto, T., Ikeguchi, M.: Multiscale molecular dynamics simulations of rotary motor proteins. Biophys. Rev. **10**, 605–615 (2018)
309. Bramer, D., Wei, G.W.: Multiscale weighted colored graphs for protein flexibility and rigidity analysis. J. Chem. Phys. **148**, 054103 (2018)
310. Dequidt, A., Canchaya, J.G.S.: Bayesian parametrization of coarse-grain dissipative dynamics models. J. Chem. Phys. **143**, 084122 (2017)
311. Koehl, P., Poitevin, F., Navaza, R., Delarue, M.H.: The renormalization group and its applications to generating coarse-grained models of large biological molecular systems. J. Chem. Theory Comput. **13**, 1424–1438 (2017)
312. Opron, K., Xia, K., Burton, Z., Wei, G.W.: Flexibility-rigidity index for protein-nucleic acid flexibility and fluctuation analysis. J. Comput. Chem. **37**, 1283–1295 (2016)
313. Canchaya, J.C.S., Dequidt, A., Goujon, F., Patrice Malfrey, P.: Development of DPD coarse-grained models: From bulk to interfacial properties. J. Chem. Phys. **145**, 54107 (2016)
314. Li, Z., Bian, X., Yang, X., Karniadakis, G. E.: A comparative study of coarse-graining methods for polymeric fluids: Mori-Zwanzig vs. iterative Boltzmann inversion vs. stochastic parametric optimization, J. Chem. Phys. **145**, 044–102 (2016)
315. Zavadlav, J., Melo, M.N., Marrink, S.J., Praprotnik, M.: Adaptive resolution simulation of polarizable supramolecular coarse-grained water models. J. Chem. Phys. **142**, 244118 (2015)

316. Kempfer, K., Devémy, J., Dequidt, A., Couty, M., Malfreyt, P.: Multi-scale modeling of the polymer–filler interaction. Soft Matter **16**, 1538–1547 (2020)
317. Maurel, G., Goujon, F., Schnell, B., Malfreyt, P.: Multiscale modeling of the polymer-silica surface interaction: from atomistic to mesoscopic simulations. J. Phys. Chem. C **119**, 4817–4826 (2015)
318. Ingolfsson, H.I., Lopez, C.A., Uusitalo, J.J., de Jong, D.H., Gopal, S.M., Periole, X., Siewert, J.: Marrink the power of coarse graining in biomolecular simulations in biomolecular simulations. WIREs Comput. Mol. Sci. **4**, 225–248 (2014)
319. Carbone, P., Avendaño, C.: Coarse-grained methods for polymeric materials: enthalpy and entropy-driven models. WIREs Comput. Mol. Sci. **4**, 62–70 (2014)
320. Negami, T., Shimizu, K., Terada, T.: Coarse-grained molecular dynamics simulations of protein-ligand binding. J. Comput. Chem. **35**, 1835–1845 (2014)
321. Thota, N., Jiang, J.: Self-assembly of amphiphilic peptide (AF)6H5K15 derivatives: rsoles of hydrophilic and hydrophobic residues. J. Phys. Chem. B **118**, 2683–2692 (2014)
322. Kar, P., Gopal, S.M., Cheng, Y.M., Panahi, A., Feig, M.: Transferring the PRIMO coarse-grained force field to the membrane environment: simulations of membrane proteins and helix-helix association. J. Chem. Theory Comput. **10**, 3459–3472 (2014)
323. Noid, W.G., Perspective: Coarse-grained models for biomolecular systems. J. Chem. Phys. **139**, 090–901 (2013)
324. Saunders, M.G., Voth, G.A.: Coarse-graining methods for computational biology. Annu. Rev. Biophys. **42**, 73–93 (2013)
325. Brini, E., Algaer, E.A., Ganguly, P., Li, C., Rodriıguez-Ropero, F., Nico van der Vegt, Nico F.A.: Systematic coarse-graining methods for soft matter simulations—a review, Soft Matter **9**, 2108 (2013)
326. Bayramoglu, B., Faller, R.: Modeling of polystyrene under confinement: exploring the limits of iterative boltzmann inversion. Macromolecules **46**, 7057–7976 (2013)
327. Potoyan, D.A., Savelyev, A., Papoian, G.A.: Recent successes in coarse-grained modelling of DNA. WIREs Comput. Mol. Sci. **3**, 69–63 (2013)
328. Ghoufi, A., Malfreyt, P.: Recent advances in many body dissipative particles dynamics simulations of liquid-vapor interfaces. Eur. Phys. J. E **36**, 10 (2013)
329. Marrink, S.J., Tieleman, D.P.: Perspective on the Martini model. Chem. Soc. Rev. **42**, 6801–6822 (2013)
330. Coskuner, O., Wise-Scira, O.: Structures and free energy landscapes of the A53T mutant-type/-synuclein protein and impact of A53T mutation on the structures of the wild-type-synuclein protein with dynamics. ACS Chem. Neurosci. **4**, 1101–1113 (2013)
331. Kar, P., Gopal, S.M., Cheng, Y.M., Predeus, A., Feig, M.: PRIMO: a transferable coarse-grained force field for proteins. J. Chem. Theory Comput. **9**, 3769–3788 (2013)
332. Hadley, K.R., McCabe, C., Hadley, K.R., McCabe, C.: Coarse-grained molecular models of water: a review Mol. Simul. **38**, 671–681 (2012)
333. D'Adamo, G., Pelissetto, A., Pierleoni, C.: Coarse-graining strategies in polymer solutions. Soft Matter **8**, 5151 (2012)
334. Karimi-Varzaneh, H.A., Muller-Plathe, F.: Coarse-grained modeling for macromolecular chemistry. Top Curr. Chem. **307**, 295–321 (2012)
335. Riniker, S., Allison, J.R., van Gunsteren, W.F.: On developing coarse-grained models in biomolecular systems: a review. Phys. Chem. Chem. Phys. **14**, 12423 (2012)
336. Chebaro, Y., Pasquali, S., Derreumaux, P.: The coarsegrained OPEP force field for non-amyloid and amyloid proteins. J. Phys. Chem. B **116**, 8741–8752 (2012)
337. Bellesia, G., Chundawat, S.P.S., Langan, P., Redondo, A., Dale, B.E., Gnanakaran, S.: Coarse-grained model for the interconversion between native and liquid ammonia-treated crystalline cellulose. J. Phys. Chem. B **116**, 8031–8037 (2012)
338. Srinivas, G., Cheng, X., Smith, J.C.: A solvent-free coarse grain model for crystalline and amorphous cellulose fibrils. J. Chem. Theory Comput. **7**, 2539–2548 (2011)
339. Goujon, F., Malfreyt, P., Tildesley, D.J.: Interactions between polymer brushes and a polymer solution: mesoscale modelling of the structural and frictional properties. Soft Matter **6**, 3472–3481 (2010)

340. Roel-Touris, J., Bonvin, A.M.J.J.: Coarse-grained (hybrid) integrative modeling of biomolecularinteractions, Coarse-grained (hybrid) integrative modeling of biomolecular Interactions. Computat. Struct. Biotechnol. J. **18**, 1182–1190 (2020)
341. Wu, Z., Cui, Q., Yethiraj, A.: A new coarse-grained model for water: the importance of electrostatic interactions. J. Phys. Chem. B **114**, 10524–10529 (2010)
342. Gopal, S.M., Mukherjee, S., Cheng, Y.M., Feig, M.: PRIMO/PRIMONA: a coarse-grained model for proteins and nucleic acids that preserves near-atomistic accuracy. Proteins **78**, 1266–1281 (2010)
343. Mehandzhiyski, A.Y., Rolland, N., Garg, M., Wohlert, J., Linares, M., Zozoulenko, I.: A novel supra coarse-grained model for cellulose. Cellulose **27**, 4221–4234 (2020)
344. Bereau, T., Deserno, M.: Generic coarse-grained model for protein folding and aggregation. J. Chem. Phys. **130**, 235106 (2009)
345. Monticelli, L., Kandasamy, S.K., Periole, X., Larson, R.G., Tieleman, D.P., Marrink, S.J.: The MARTINI coarse-grained force field: extension to proteins. J. Chem. Theory Comput. **4**, 819–834 (2008)
346. Molinero, V., Goddard, V.A., III.: M3B: a coarse grain force field for molecular simulations of maltooligosaccharides and their water mixtures. J. Phys. Chem. B **108**, 1414–1427 (2004)
347. Izvekov, S., Parrinello, M., Burnham, C.J., Voth, G.A.: Effective force fields for condensed phase systems from ab initio molecular dynamics simulation: a new method for force-matching. J. Chem. Phys. **120**, 10896 (2004)
348. Reith, D., Pütz, M., Müller-Plathe, F.: Deriving effective mesoscale potentials from atomistic simulations. J. Comput. Chem. **24**, 1624–1636 (2003)
349. Clementi, C., Nymeyer, H., Onuchic, J.N.: Topological and energetic factors: What determines the structural details of the transition state ensemble and' en-route' intermediates for protein folding? An investigation for small globular proteins. J. Mol. Biol. **298**, 937–953 (2000)
350. Müller-Plathe, F.: Coarse-graining in polymer simulation: from the atomistic to the mesoscopic scale and back. ChemPhysChem **3**, 754–769 (2002)
351. Ercolessi, F., Adams, J.B.: Interatomic potentials from first principles calculations: the force-matching method. Europhys. Lett. **26**, 583–588 (1994)
352. Go, N.: Theoretical studies of protein folding. Annu. Rev. Biophys. Bioeng. **12**, 183–210 (1983)
353. Xiong, X., Chen, Z., Cossins, B.P., Xu, Z., Shao, Q., Ding, K., Zhu, W., Shi, J.: Force fields and scoring functions for carbohydrate simulation. Carbohydr. Res. **12**, 73–81 (2015)
354. Kim, Y.C., Hummer, G.: Coarse-grained models for simulations of multiprotein complexes: application to ubiquitin binding. J. Mol. Biol. **375**, 1416–1433 (2008)
355. Dignon, G.L., Zheng, W., Kim, Y.C., Best, R.B., Mittal, J.: Sequence determinants of protein phase behavior from a coarse-grained model. PLoS Comput. Biol. **14**, e1005941 (2018)
356. Lu, Y., Salsbury, F.R.: Recapturing the correlated motions of protein using coarse-grained models. Protein Pept. Lett. **22**, 654–659 (2015)
357. Delort, B., Renault, P., Charlier, L., Raussin, F., Martinez, J., Floquet, N.: Coarse-grained prediction of peptide binding to G-protein coupled receptors. J. Chem. Inf. Model. **57**, 562–571 (2017)
358. Hirano, R., Yabuchi, T., Sakurai, M., Furuta, T.: Development of an ATP force field for coarse grained simulation of ATPases and its application to the maltose transporter. J. Comput. Chem. **40** (2019)
359. Sieradzan, A.K., Jakubowski, R.: Introduction of steered molecular dynamics into UNRES coarse-grained simulations package. J. Comput. Chem. **38**, 553–562 (2017)
360. Sieradzan, A.K., Bogunia, M., Mech, P., Ganzynkowicz, R., Giełdon, A., Liwo, A., Makowski, M.: Introduction of phosphorylated residues into the UNRES coarse-grained model: towards modeling of signaling processes. J. Phys. Chem. B **123**, 5721–5729 (2019)
361. Kynast, P., Derreumaux, P., Strodel, B.: Evaluation of the coarse-grained OPEP force field for protein-protein docking. BMC Biophys. **9**, 4 (2016)
362. Ge, Y., Voelz, V.A.: Model selection using BICePs: a bayesian approach for force field validation and parameterization. J Phys Chem B **122**, 5610–5622 (2018)

363. Miwatani, R., Takahashi K.Z., Arai, N.: Performance of coarse graining in estimating polymer properties: comparison with the atomitic model. **12**, 382 (2020)
364. Lahmar, F., Tzoumanekas, C., Theodorou, D.N., Rousseau, B.: Onset of entanglements revisited. topological analysis macromolecules. 7474–7484 (2009)
365. Milano, G., Muller-Plathe, V.: Mapping atomistic simulations to mesoscopic models: a systematic coarse-graining procedure for vinyl polymer chains. J. Phys. Chem. B **109**, 18608 (2005)
366. Carbone, P., Negri, F., Muller-Plathe, F.: A Coarse-Grained Model for Polyphenylene Dendrimers: Switching and Backfolding of Planar Three-Fold Core Dendrimers. Macromolecules **40**, 7044 (2007)
367. Qian, H.-J., Chee, C.L., Florian, M.-P.: Effective control of the transport coefficients of a coarse-grained liquid and polymer models using the dissipative particle dynamics and Lowe-Andersen equations of motion. Phys. Chem. Chem. Phys. **11,** 1962 (2009)
368. Gartner III, T.E., Arthi, J.: Modeling and simulations of polymers: a roadmap macromolecules. **52,** 755–786 (2019)
369. Avendano, C., Lafitte, T., Galindo, A., Adjiman, C.S., Jackson, G., Muller, E.A.: SAFT-γ force field for the simulation of molecular fluids, 1. A single-site coarse grained model of carbon dioxide. J. Phys. Chem. B **115**, 11154 (2011)
370. Marrink, J., Risselada, H.J., Yefimov, S., Tieleman, D.P., de Vries, A.H.: The MARTINI force field: coarse grained model for biomolecular simulations. J. Phys. Chem. B **111**, 7812 (2007)
371. Ndao, M., Devemy, J., Ghoufi, A., Malfreyt, P.: Coarse-graining the liquid-liquid interfaces with the MARTINI force field: how is the interfacial tension reproduced? J. Chem. Theory Comput. **11**, 3818 (2015)
372. Green, M.L., et al.: Fulfilling the promise of the materials genome initiative with high-throughput experimental methodologies. Appl. Phys. Rev. **4**, 011105 (2017)
373. Jain, A., et al.: Commentary: the materials project: a materials genome approach to accelerating materials innovation. APL Mater. **1**, 011002 (2013)
374. Allen, F.H., Sievers, R. (eds.): Crystallographic databases. International Union of Crystallography, Chester (1987)
375. Saal, J.E., Kirklin, S., Aykol, M., Meredig, B., Wolverton, C.: Materials design and discovery with high-throughput density functional theory: the open quantum materials database (OQMD). JOM **65**, 1501–1509 (2013)
376. Kirklin, S., Saal, J.E., Meredig, B., Thompson, A., Doak, J.W, Muratahan, A., Stephan R., Chris, W.: The open quantum materials database (OQMD): assessing the accuracy of DFT formation energies. NPJ Comput. Mater. **1**, 15010 (2015)
377. Groom, C.R., Bruno, I.J., Lightfoot, M.P., Ward, S.C.: The Cambridge structural database. Acta Crystallogr. Sect. B Struct. Sci. Cryst. Eng. Mater. **72**, 171–179 (2016)
378. Walsh, A.: The quest for new functionality. Nat. Chem. **7**, 274–275 (2015)
379. Haastrup, S. et al.: The Computational 2D materials database: high-throughput modeling and discovery of atomically thin crystals. 2D Mater. **5**, 042–002 (2018)
380. Baldi, P., Brunak, S.: Bioinformatics: The Machine Learning Approach. The MIT Press, Cambridge, MA (2001)
381. Rouet-Leduc, B., Barros, K., Lookman, T., Humphreys, C.J.: Optimisation of GaN LEDs and the reduction of efficiency droop using active machine learning. Sci. Rep. **6**, 24862 (2016)
382. Legrain, F., Carrete, J., van Roekeghem, A., Curtarolo, S., Mingo, N.: How chemical composition alone can predict vibrational free energies and entropies of solids. Chem. Mater. **29**, 6220–6227 (2017)
383. Pilania, G., Liu, X.Y.: Machine learning properties of binary wurtzite superlattices. J. Mater. Sci. **53**, 6652–6664 (2018)
384. Ward, L., Wolverton, C.: Atomistic calculations and materials informatics: a review. Curr. Opin. Solid State Mater. Sci. **21**, 167–176 (2017)
385. Yuan, F., Mueller, T.: Identifying models of dielectric breakdown strength from high-throughput data via genetic programming. Sci. Rep. **7**, 17594 (2017)

386. Gaultois, M.W., et al.: Perspective: Web-based machine learning models for realtime screening of thermoelectric materials properties. APL Mater. **4**, 053213 (2016)
387. Kauwe, S.K., Graser, J., Vazquez, A., Sparks, T.D.: Machine learning prediction of heat capacity for solid inorganics. Integr. Mater. Manuf. Innov. **7**, 43–51 (2018)
388. Liu, Y., Zhao, T., Yang, G., Ju, W., Shi, S.: The onset temperature (Tg) of As Se1glasses transition prediction: a comparison of topological and regression analysis methods. Comput. Mater. Sci. **140**, 315–321 (2017)
389. Isayev, O., et al.: Universal fragment descriptors for predicting properties of inorganic crystals. Nat. Commun. **8**, 15679 (2017)
390. Furmanchuk, A., Agrawal, A., Choudhary, A.: Predictive analytics for crystalline materials: bulk modulus. RSC Adv. **6**, 95246–95251 (2016)
391. van Roekeghem, A., Carrete, J., Oses, C., Curtarolo, S., Mingo, N.: Highthroughput computation of thermal conductivity of high-temperature solid phases: the case of oxide and fluoride perovskites. Phys. Rev. X **6**, 041061 (2016)
392. Pham, T.-L., et al.: Learning structure-property relationship in crystalline materials: a study of lanthanide–transition metal alloys. J. Chem. Phys. **148**, 204106 (2018)
393. Pham, T.L., et al.: Machine learning reveals orbital interaction in materials. Sci. Technol. Adv. Mat. **18**, 756–765 (2017)
394. Pilania, G., Gubernatis, J.E., Lookman, T.: Structure classification and melting temperature prediction in octet AB solids via machine learning. Phys. Rev. B **91**, 214302 (2015)
395. Kiyohara, S., Oda, H., Tsuda, K., Mizoguchi, T.: Acceleration of stable interface structure searching using a kriging approach. Jpn. J. Appl. Phys. **55**, 045502 (2016)
396. Carrete, J., Mingo, N., Wang, S., Curtarolo, S.: Nanograined half-Heusler semiconductors as advanced thermoelectrics: an ab initio high-throughput statistical study. Adv. Funct. Mater. **24**, 7427–7432 (2014)
397. Evans, J.D., Coudert, F.X.: Predicting the mechanical properties of zeolite frameworks by machine learning. Chem. Mater. **29**, 7833–7839 (2017)
398. Yeo, B.C., Kim, D., Kim, C., Han, S.S. Pattern learning electronic density of states, Preprint at arXiv:1808.03383 (2018)
399. Broderick, S.R., Aourag, H., Rajan, K.: Classification of oxide compounds through data-mining density of states spectra. J. Am. Ceram. Soc. **94**, 2974–2980 (2011)
400. Tehrani, A.M., et al.: Machine learning directed search for ultraincompressible, superhard materials. J. Am. Chem. Soc. **140**, 9844–9853 (2018)
401. Meredig, B., Wolverton, C.: Dissolving the periodic table in cubic zirconia: data mining to discover chemical trends. Chem. Mater. **26**, 1985–1991 (2014)
402. Shi, S., et al.: Multi-scale computation methods: their applications in lithium-ion battery research and development. Chin. Phys. B **25**, 018212 (2016)
403. Carrasquilla, J., Melko, R.G.: Machine learning phases of matter. Nat. Phys. **13**, 431–434 (2017)
404. Ling, J., Hutchinson, M., Antono, E., Paradiso, S., Meredig, B.: High-dimensional materials and process optimization using data-driven experimental design with well-calibrated uncertainty estimates. Integr. Mater. Manuf. Innov. **6**, 207–217 (2017)
405. Waag, W., Fleischer, C., Sauer, D.U.: Critical review of the methods for monitoring of lithium-ion batteries in electric and hybrid vehicles. J. Power Sources **258**, 321–339 (2014)
406. Stanev, V., et al.: Machine learning modeling of superconducting critical temperature. Comput. Mater. **4**, 29 (2018)
407. Zhang, Y., Ling, C.: A strategy to apply machine learning to small datasets in materials science. npj Comput. Mater. **4**, 25 (2018).
408. Warren, J.A.: The Materials Genome Initiative and artificial intelligence, J A. MRS Bull. **43**, 452 (2018)
409. Zhuo, Y., Tehrani, A.M., Brgoch, J.: Predicting the band gaps of inorganic solids by machine learning. J. Phys. Chem. Lett. **9**, 1668–1673 (2018)

Perovskite-Like Quantum Dots Designed for Advanced Optoelectronic Applications

Felipe M. Pinto, Swayandipta Dey, Thiago M. Duarte, Carlton A. Taft, and Felipe A. La Porta

Abstract Many studies have focused on understanding synthesis parameters, defects, electronic and structural properties of novel functional materials, which has proven to be extremely important for the development of new technologies. This may, in principle, provide clues to elucidate some of their design rules at the nanoscale further. However, currently, it is still virtually impossible to predict the final shape of these colloidal nanocrystals, as well as their unique properties, which in turn are strongly dependent on crystal morphologies as prepared. Hence, diverse strategies have been widely adopted to obtain novel target materials with well-defined morphologies. Among emerging materials, more recently, perovskite-like materials have gained immense attention due to their outstanding properties. In this chapter, we will present the recent progress in synthesis, theory, and characterization of perovskite-like colloidal nanocrystals for the development of advanced optoelectronic applications.

F. M. Pinto
Department of Engineering, Federal University of Lavras, Lavras, Minas Gerais 37200-900, Brazil

S. Dey
Department of Physics of Complex Systems, Weizmann Institute of Science, 7610001 Rehovot, Israel

T. M. Duarte
Department of Chemistry, Federal University of Paraíba, João Pessoa, Paraíba 58051–085, Brazil

C. A. Taft
Centro Brasileiro de Pesquisas Físicas, CBPF, Rio de Janeiro, RJ 22290-180, Brazil

F. A. La Porta (✉)
Laboratory of Nanotechnology and Computational Chemistry, Federal Technological University of Paraná, Avenida dos Pioneiros 3131, Londrina, Paraná 86036-370, Brazil
e-mail: felipelaporta@utfpr.edu.br

F. A. La Porta and C. A. Taft (eds.), *Functional Properties of Advanced Engineering Materials and Biomolecules*, Engineering Materials,
https://doi.org/10.1007/978-3-030-62226-8_3

1 Introduction to Perovskites-Like Materials

Perovskite compounds are currently one of the most well-known class of functional materials. This is primarily due to the simplicity in obtaining different morphologies, sizes, and compositions. Additionally, these compounds possess attractive physical and chemical properties, including a direct bandgap with long carrier diffusion lengths and absorption region ideal for high performance in the solar-energy-conversion process. Such features make these compounds extremely versatile for numerous technological applications in the photovoltaics and optoelectronics fields [1–12]. From a historical point of view, the term perovskite is the name given to the mineral composed of calcium titanate ($CaTiO_3$), i.e., in honor of the Lev Alexeievitch Perovsky, a Russian mineralogist. For this reason, functional materials with the formula AMX_3, and similar structure, also receive such a name [2]. More recently, a highly revolutionized field of photovoltaic devices among this class of functional materials has gained considerable attention. This has occurred since its first application by Miyasaka et al. [3] in 2009, where the academia and industry realized the huge potential of such novel materials with functional properties. This is likely due to its high capacity to absorb and emit light at different wavelengths, brightness, and malleability, broadening the range of applications of these functional materials [1–12].

Particularly, perovskite-related materials can be purely inorganic, formed by a variety of elements from the periodic table, or they can be hybrid, where one of the cations is a molecule or organic ion [1–5]. In the case of the A site being formed by a monovalent cation, they need to be in an ideal range of radius size, as those that are substantially larger or smaller will not allow the formation of the three-dimensional perovskite phase (all-inorganic or hybrid) [4]; the most common cations are methylammonium (MA), formamidium (FA), and cesium (Cs) [5]. In the case of the M site, also known as forming the network, we usually have lead (Pb) as the primary representative of these compounds used in photovoltaic applications (or other Sn, Ge, Bi, Sb, Cu, and Ti, namely "Lead-free Perovskites," have emerged in this field). To end, the X site is formed by a halogen anion (Cl, Br, and I, or a mix of them) [6].

Despite the extraordinary results recently achieved for these functional materials in optoelectronic and photovoltaic applications [7–10], in particular, it is well-known that the halide perovskite designed with different sizes and morphologies, are highly air- as well as humidity-sensitive [6, 11, 12]. A range of studies have also reported that the decomposition and volatilization of organic components in hybrid perovskites (organic–inorganic) are in large part, one of the biggest challenges in improving their long-term stability [6, 13]. In this context, the halide perovskite all-inorganic has a high-stability at ordinary room temperature. In principle, this limitation may be partially overcome by using colloidal systems, which have emerged as significant alternatives to enhance the stability of perovskite-based devices [14–17]. However, identifying the optimal physicochemical properties of these colloidal nanocrystals is not easy. Regardless, recent major advances have been made from the synthesis of materials, theory/simulation, and applications. This study reviews some of the principal challenges and opportunities in the materials chemistry and engineering field of research.

2 Crystalline Structure and Electronic Structure Properties of Perovskite-Like Materials

Usually, the perovskite-like materials may exist in different stoichiometric, such as AMX_3, A_4MX_6, and AM_2X_5. Among these stoichiometries, in particular, the most studied is the AMX_3, and depending on experimental conditions, such materials can be obtained in different crystalline structures [18–21]. A detailed analysis of the high-temperature phase transitions reveal that $APbX_3$-related compounds can adopt four phases, such as α-cubic, β-tetragonal, and two γ- and δ-orthorhombic, as shown in Fig. 1a, b, depending on the tilting and rotation of the $[MO_6]$ polyhedral clusters in the lattice [22–24]. Particularly the phase diagram of the Cs-Pb-Br system is shown

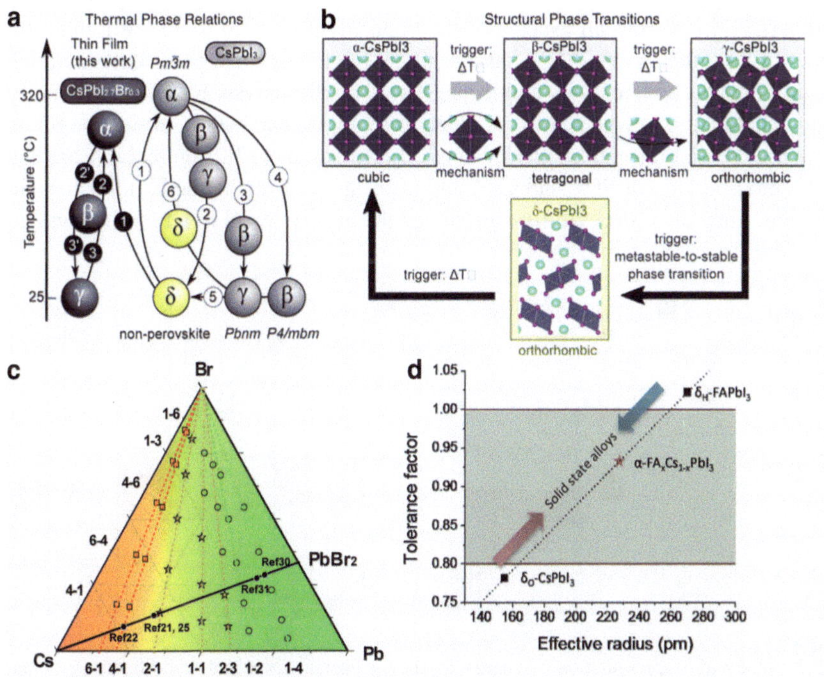

Fig. 1 **a** Representative thermal phase relation and **b** crystal structure of perovskite-like materials. Reproduced with permission from Ref. [18]. **c** Ternary phase diagram of Cs-Pb-Br system. Reproduced with permission from Ref. [21]. **d** The empirical TF index is used for predicting the stability of perovskite materials with cubic structure. Reproduced with permission from Ref. [26]

in Fig. 1c. In general, the formation of the perovskite-like structure and its stability depend on an empirical factor called Goldschmidt tolerance factor (TF) [25], defined in the following equation:

$$TF = \frac{\mathrm{Ra} + \mathrm{Rx}}{\sqrt{2}(Ra + Rm)} \tag{1}$$

where "Ra", "Rm", and "Rx" are specifically the values of ionic radius of the cation, metal ion, and halide ions, respectively, however, in this case, if the TF value is out of this range (0.8–1.0), the perovskite-like cubic structure is not formed [25]. Most known halide perovskites (with an orthorhombic, rhombohedral, or tetragonal structure) exhibit TF values in the range of 0.75–1.00 [5]. In parallel, Li et. al. [26] utilizes hybrid density functional theory (DFT) calculations to investigate the stability of diverse crystals with perovskite structure in order to provide insights on the effect of radius size of cation A, as shown in Fig. 1d. Because of their outstanding chemical and physical properties, a considerable effort has widely been devoted to the production of such perovskite-like materials with controlled size, structure, composition, and morphologies [1–17].

Goesten and Hoffman [27], carried out simulations based on density functional theory (DFT) covering all aspects of the bond in an interval of 15 eV around the gap to explore the chemical bond and bandgap of all-inorganic perovskite materials through detailed calculations and a qualitative bond analysis based on symmetry. They concluded that when cation A changes to an organic cation, it must be robust enough to persist through moderate network distortions [27]. Figure 2a, b predicts how a modification of Pb^{2+} and Cs^{2+} and a variation of the halide X influences the bandgap. They also successfully reproduced the experimental trend for the bandgap

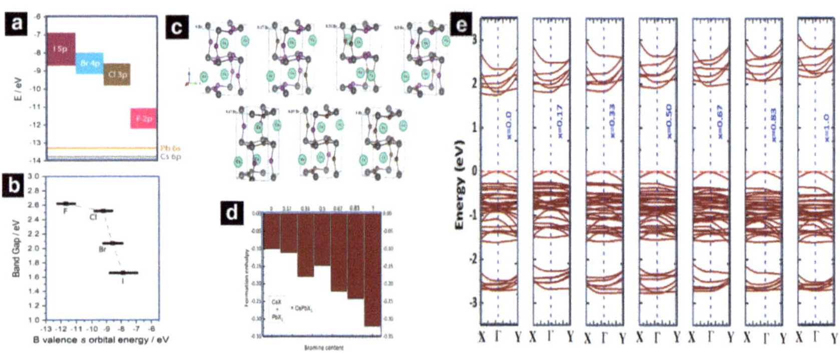

Fig. 2 Schematic representation of energy levels for Pb and Cs p and 6 s orbitals (left), and the correlation between halide p energy and band gap (right). Reproduced with permission from Ref. [27]. **a** Representation of CsPbI3-xBrx Perovskites with different concentrations. Relationship between the stability of the simulated sytems and the Br amount in the system. **b** Computed band structures for pure systems and solid-solutions of $CsPbI_{3-x}Brx$. (Reproduced with permission from Ref. [28]

of these materials with satisfactory precision (see Fig. 2b). In another study, Fadla et al. [28] used the DFT calculations to simulate the structural, electronic, and transport properties for $CsPbI_{3-x}Br_x$ solid-solution (see Fig. 2c–e). The $CsPbI_{3-x}Br_x$ solid-solution is known to improve the thermal stability and efficiency in optoelectronic and photovoltaic devices [21–23]. Figure 2 shows that the substitution of iodine for bromine leads to a reduction in the network volume of the optimized systems, with direct implications in its enthalpy of formation. These theoretical calculations were in accordance with experimental data reported in the literature [29, 30]. It was observed that all compounds have enthalpies of negative formation, and thermodynamic stability increases with the Br content [28]. All the $CsPbI_{3-x}Br_x$ solid-solution studied exhibit a direct bandgap semiconductor, where the minimum conduction band (CB) and the maximum valence band (VB) were located at the point Γ, as shown in Fig. 3e. According to these results, the bandgap energy calculated for $CsPbI_3$ was 1.75 eV (x = 0.0) end observed that the bandgap energy tends to increase and reached a value of about 2.16 eV when all 12 iodine atoms were replaced by bromine atoms (x = 1.0). Thus, they showed that the simulated systems had their electronic and structural properties altered, mainly due to the higher electronegativity of bromine compared to iodine, which evolved as the substitution occurred.

Also, as is well-known, the α-phase of the $CsPbI_3$ perovskite has an optimal value of bandgap for applications in photovoltaic devices; however, this structure is metastable in room temperature [6, 18]. In this context, theoretical calculations performed by Kye and collaborate [31] suggest that vacancy defects could stabilize the α-phase of the $CsPbI_3$ perovskite. For the models proposed in Fig. 4a, b, the chemical ionic potentials were thermodynamically analyzed using the DFT formalism. This study also found that the insertion of cationic vacancy defects induces the weakening of interactions between Cs and the $[PbI_6]$ polyhedron in the $CsPbI_3$ lattice; this decreases the energy difference between the α- and δ-phases. It was possible to calculate the bond lengths [Pb–I] in α- and δ- phases equal to 3.12 Å and 3.08–3.33 Å, respectively [31]. In this case, the energies of formation and charge transition levels of intrinsic point defects $\left(\text{such as } V_{Cs}^{-}, V_{Pb}^{-2} \text{ and } V_I^{+} \text{ with different charge states}\right)$ of α- and δ-phases of $CsPbI_3$ (see Fig. 3). It can be represented by neutral constituent species or by charged species, according to the following equations [31]:

$$\mu_{Cs} + \mu_{Pb} + 3\mu_I = \mu_{CsPbI_3} = \mu_{Cs}^{+} + \mu_{Pb}^{2+} + \mu_I^{-} \tag{2}$$

$$\mu_{Pb} + 2\mu_I = \mu_{PbI_2} = \mu_{Pb}^{2+} + 2\mu_I^{-} \tag{3}$$

$$\mu_{Cs} + \mu_I = \mu_{CsI} = \mu_{Cs}^{+} + \mu_I^{-} \tag{4}$$

Figure 3c showed the vacancy formation energies in the α- and δ- phases based on neutral species, in which the environmental condition is described by the potential chemicals of neutral species that satisfy the left side of Eqs. 2–4. In both I-poor (Pb- and Cs- rich) and I-rich (Pb- and Cs-poor) conditions studied, in particular, the δ-phase exhibits higher vacancy-forming energies (V_I, V_{Pb}, and V_{Cs}) more than

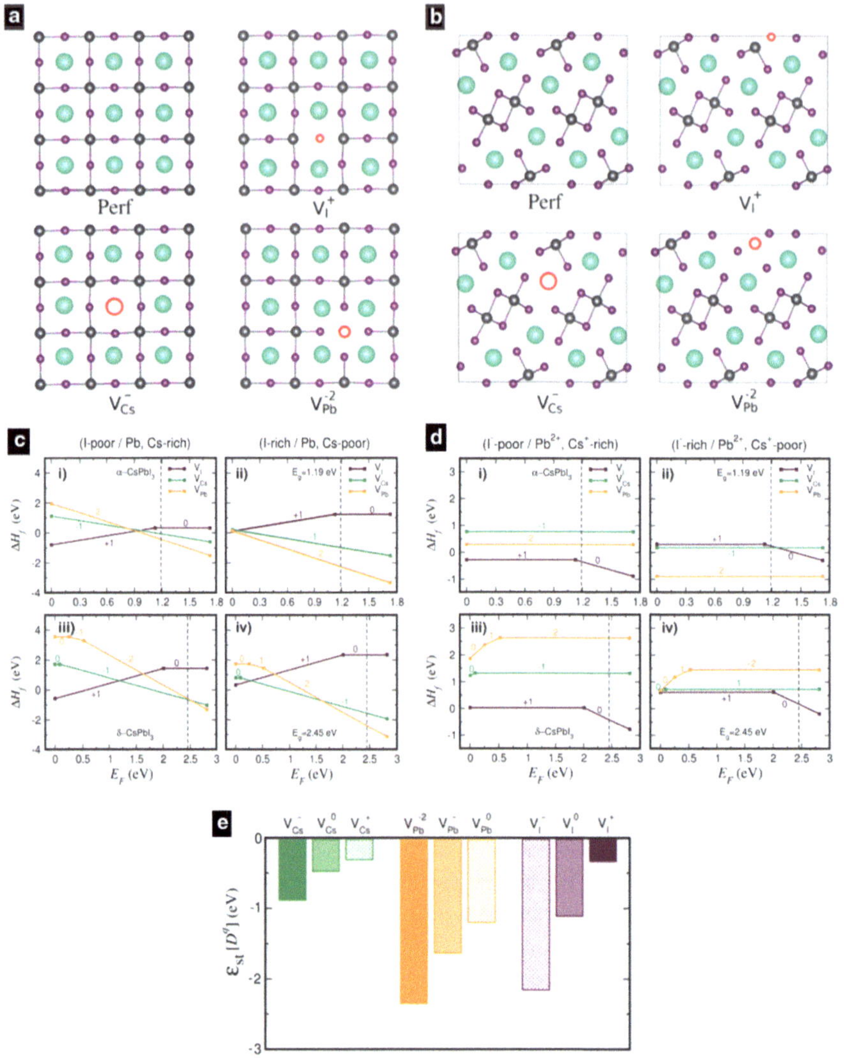

Fig. 3 Scheme of $CsPbI_3$ polymorphs with and without vacancy: **a** α-phase and **b** δ-phase. Vacancy defect formation energies as a function of Fermi level energy (E_F). **c** Representing the constituent neutral species and **d** by charged species, under the following conditions: low -I (rich-Cs and -Pb) (left panel) and rich I (poor-Cs and -Pb) (right panel) phases α-$CsPbI_3$ (i, ii) and δ-$CsPbI_3$ (iii, iv). **e** Relative stabilization energy according to type of defect (vacancy with different charge states) of the α phase over the δ phase of the $CsPbI_3$ perovskite. Solid bars represent the main load states of vacancies, while dotted bars show oxidized (for V_{Cs} and V_{Pb}) or reduced (for V_I) states. Reproduced with permission from Ref. [31]

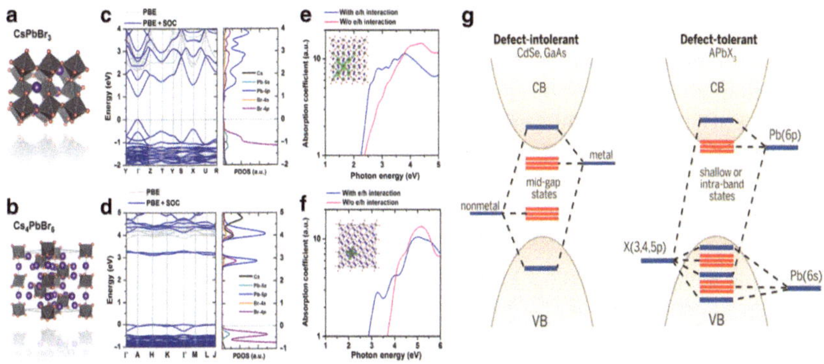

Fig. 4 **a, b** Representative crystal structure and (c,d) electronic bands and DOS calculated of $CsPbBr_3$ and Cs_4PbBr_6. **e, f** Computed absorption spectra with or without accounting for electron–hole (e/h) pair interactions concomitantly with (see inset) the exciton wave functions (green area) corresponding to the excitonic peaks. Reproduced with permission from Ref. [34]). **g** Defect tolerance in perovskite nanocrystals. Reproduced with permission from Ref. [26]

α-phase, indicating the greater stability of the δ-phase [31]. Based on the chemical potentials determined through ions, which depend on physical parameters such as concentration and temperature of the solution, as shown in Fig. 3d, the α-phase was benefited in conditions of low availability of ions Cs^+ and/or Pb^{2+} [31]. It was also possible to observe the difference between the neutral species and ionic species, as shown in Fig. 3c, d, indicating that defects of the vacancy type with their main charge states have energies of the formation independent of Fermi energy (E_F) level. As a result, the stabilization of α-$CsPbI_3$ was supported by the formation of cationic vacancies [31]. In particular, the energy barrier for internal stabilization between the α- and δ-phases of the $CsPbI_3$ perovskite, decreased with the creation of vacancies (see Fig. 3e). It has been demonstrated that the established nanometric effect originates not only from the energy gain of the surface compensating for the difference of stability, but also from the weakening of the bond between Cs and the octahedra $[PbI_6]$ clusters through the formation of vacancy (e.g., cation extraction) [31].

In addition, it is well-known that the Cs-Pb-Br system has outstanding optoelectrical properties [32, 33]. For instance, Cha et al. [32] have reported a full study on photocurrent and photoluminescence responses for both $CsPbBr_3$ and Cs_4PbBr_6 single-crystals. As a result, they observed that $CsPbBr_3$ crystals showed a highly sensitive photoresponse as well as weak PL signal; however, the Cs_4PbBr_6 crystals exhibited a strong green PL emission (e.g., with more than one order of magnitude higher PL intensity than $CsPbBr_3$), but this structure has an ultralow photoresponse under illumination [32]. Particularly the crystalline structure of these materials is presented in Fig. 4a, b. Regarding these crystalline structures, for the orthorhombic $CsPbBr_3$ structure (belong to space group Pnam), it is possible observed that all-corner-sharing the octahedra $[PbBr_6]^{4-}$ clusters formed an extended 3D network, as shown in Fig. 4a. Whereas rhombohedral Cs_4PbBr_6 structure (belong to space group R-3c), is typically formed by isolated octahedra $[PbBr_6]^{4-}$ clusters separated by Cs^+ cations [34]. Their electronic band and DOS calculated both $CsPbBr_3$ and Cs_4PbBr_6 are shown in Fig. 4c, d. According to these results, both materials have a direct bandgap at the Γ point of 2.13 eV for the $CsPbBr_3$ and of about 3.88 eV for the Cs_4PbBr_6. These results are, therefore, consistent with the experimental data [35–37]. In analogy to $CsPbBr_3$, the DOS analysis (see Fig. 4c, d) reveal that the CB of Cs_4PbBr_6 is, of course, mainly composed of Pb-6sp and Br-4p states, whereas the VB consists of both Pb-6 s and Br-4p states. On the other hand, the electronic analysis obtained by this DFT calculations, it also suggests that these Pb 6sp and Br 4p orbital states are disconnected, likely due to the octahedra $[PbBr_6]^{4-}$ clusters isolation in the Cs_4PbBr_6 lattice [34]. Consequently, these theoretical results indicated a wide-bandgap (at about 3.88 eV) for the Cs_4PbBr_6. Note that these electronic differences have a profound effect on the absorption spectra and exciton wave function features of $CsPbBr_3$ and Cs_4PbBr_6, as shown in Fig. 4e, f. In both cases, the main interband electronic transitions related to $Pb^{2+}(6s)Br^-(4p) \rightarrow Pb^{2+}(6p)$ at the Γ point contribute to the higher-energy absorption continuum [34]. In the case of $CsPbBr_3$, e.g., such results reveal an excitonic absorption is weak, which is in turn consistent with its small exciton binding energy (in range of 19 to 62 meV) [38, 39] Yet, mainly due to its electronic characteristics, the Cs_4PbBr_6 has strong quantum confinement and, consequently, that is the reason for their sharp excitonic peak at about 3.3 eV, which is in good agreement with experimental absorption measurements [34, 40].

For the case of traditional semiconductors (e.g., Si, CdSe, GaAs, and so on), defects lead to a significant decrease in the efficiency of solar cells, mainly because that they act as electronic traps or electronic dopants [41, 42]. As such, materials design for solar cell applications possess an in-principle defect intolerance; however, these perovskite materials have an unusual electronic band behavior, as shown in Fig. 4g. Such defect tolerance features may be attributed to the lack of bonding-antibonding interaction between the CB and VB, and may also potentially explain the superior optoelectronic properties of these materials [19, 43–45]. Dangling bonds at the surface of perovskite-like quantum dots (QDs) also have similar effects, leading to localized, non-bonding states [42].

Zhu et al. [46] developed a systematic theoretical study on the structural properties, lattice parameters, bond lengths, the energy of the formation, electronic band structures, and DOS in which evaluated the effect of replacing Pb by Sn in perovskites of the $CsPb_{1\text{-}x}Sn_xIBr_2$ system (x = 0, 0.25, 0.5, 0.75) in different crystalline structures, as shown in Fig. 5i. In this study, the energy of the formation was monitored for structural changes that occurred in the cubic ($Pm\bar{3}m$) and orthorhombic ($Pnma$) structures of the $CsPb_{1\text{-}x}Sn_xIBr$ solid-solution system as a function of increased Sn amount and are shown in Fig. 5a. As such, the cubic $CsPbBr_3$ structure has a lattice parameter of about 6.162 Å, bond lengths of Pb-Br are 3.081 Å, and the angles of Br-Pb-Br was 90° or 180°. On the other hand, the orthorhombic phase has lattice parameters a = 8.518 Å, b = 11.882 Å, and c = 8.240 Å, which are in agreement with the experimental values [32, 34, 46]. Finally, when evaluating the structural properties and total energy of all mixed phases $CsPb_xSn_{1-x}IBr_2$, it was observed that the lattice and volume parameters decrease with the introduction of Sn and an increase in

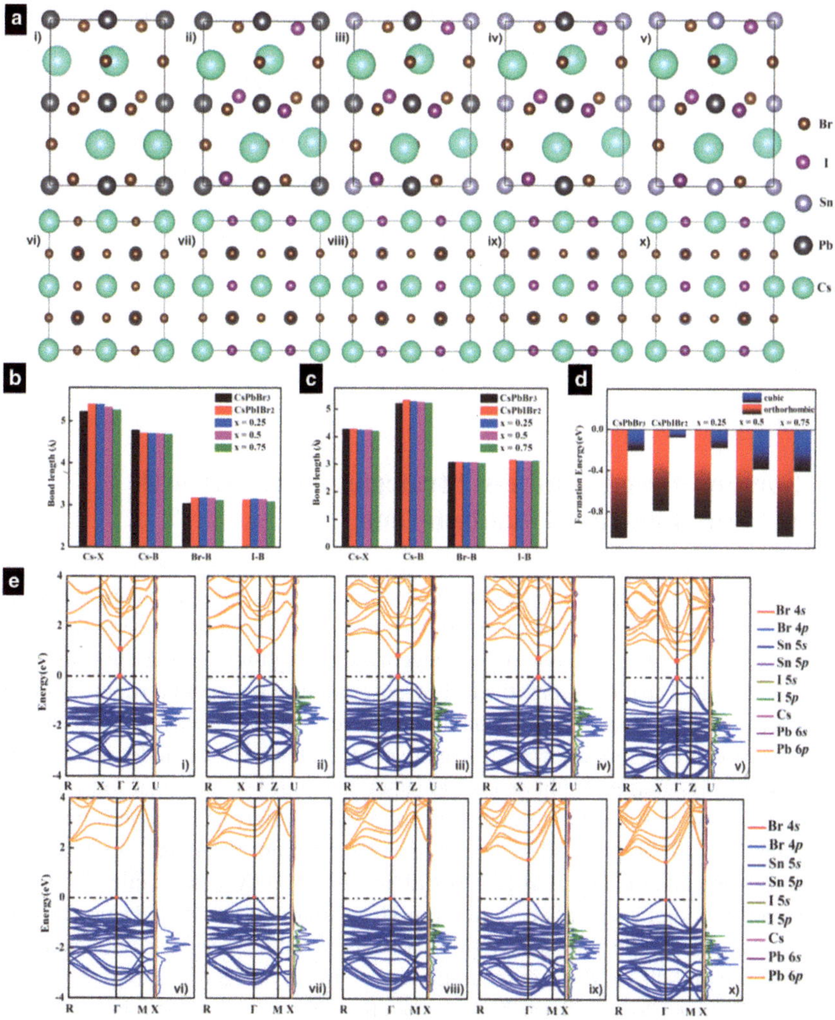

Fig. 5 Schematic representation of cubic phase **a** $CsPb_{1-x}Sn_xIBr_2$ (x = 0, 0.25, 0.5, 0.75)—(i–v) and orthorhombic (vi–x). Length bonds of the $CsPb_{1-x}Sn_xIBr$ cubic phase system (**b**) and orthorhombic (**c**). **d** Formation energy of the system $CsPb_{1-x}Sn_xIBr_2$ pure and doped. (e) Band structure and partial state density (PDOS) of the system $CsPb_{1-x}Sn_xIBr_2$. Reproduced with permission from Ref. [46]

the overlap in the of the M-X bond. This phenomenon can, in principle, be attributed to the different electronegativity values of Pb and Sn and the strong electron coupling effect of Cs that influences in the resultant structure [46]. In this case, the negative energy of the formation (ΔH) values indicated that the decomposition reactions are endothermic and gradually decreases with the increase in the proportion of Sn (x = 0.25) [46]. DFT calculation of the electronic band structure bands shows a reduction in the distance between the VB and CB for the series of lead iodides doped with Sn in the cubic phases (Fig. 5**e**-(i–v)) and orthorhombic (Fig. 5e-(vi–x)) and both systems exhibited a p-type conductivity nature. This phenomenon was observed by comparing the $CsPbIBr_2$ system doped with I and the pure $CsPbBr_3$ system. As such, the doping process allowed to verify how the *p*-orbitals influences the VB influences the system to exhibit ever smaller band gaps intervals. The halogen that has been replaced (I) has less electronegativity than the substituent (Br), which directly implies extended M-X bond lengths and covalent hybridization. In the mixed series of the two phases, the mixed series γ-$CsPb_{1-x}Sn_xIBr_2$ has smaller band intervals due to an orbital overlap much more strongly controlled by M-X bond length. Hence, for this, the analysis of the DOS calculated (see Fig. 5e) showed the increasing contribution of the 4p–4 s orbitals with the increase of the Sn amount in the lattice [46].

3 Solution-Based Synthesis Strategies for the Growth of Perovskite-Like Quantum Dots

This section will focus on the development of perovskite-like QDs using solution-based synthesis strategies. The fabrication strategies of halide-based perovskite-like QDs may be classified into two major categories: (i) high-temperature hot-injection (HI) method and (ii) a comparatively low-cost and easy-to-operate room temperature (RT)-based synthesis. The HI method requires a specific reaction environment with reaction precursors injected at a relatively high temperature under inert conditions. Both preparation methods enable superior control of the size, morphology, and composition offering high precision for a wide variety of designed functional materials [47–49]. Typical HI synthesis involves the preparation and pre-heating of a reaction precursor under inert gas conditions, followed by a quick injection of the precursor into another reaction solution mixture after an ionic metathesis reaction. This technique usually produces highly monodispersed nanocrystals with excellent crystallinity [47–49]. A distinct separation between the nucleation and growth stage reaction kinetics, along with the high temperature of reaction, enables better control over the size, shape and phase purity of the QDs. Experimental parameters, e.g., such as the operating temperature, growth time, and the speed of the precursor injection, are essential to control the size and photoluminescence emission peak position of the QDs obtained [47–49].

We begin our discussion with a strategy developed by the Kovalenko et al. [50], involving the preparation of inorganic $CsPbX_3$ colloidal nanocrystals using the HI method. In this HI synthesis, the lead and halogen precursor (PbX_2) was dissolved in three different flasks containing octadecene, oleic acid (OA), and oleylamine (OAm). Then, cesium oleate (Cs-oleate) was injected into the flask at high temperature (140–200 °C), and a few seconds later, the reaction system was cooled to room temperature using an ice bath [50]. Figure 6a, b, show the acquisition of the cubic structure for these colloidal nanocrystals, with controlled and well-defined morphologies. The

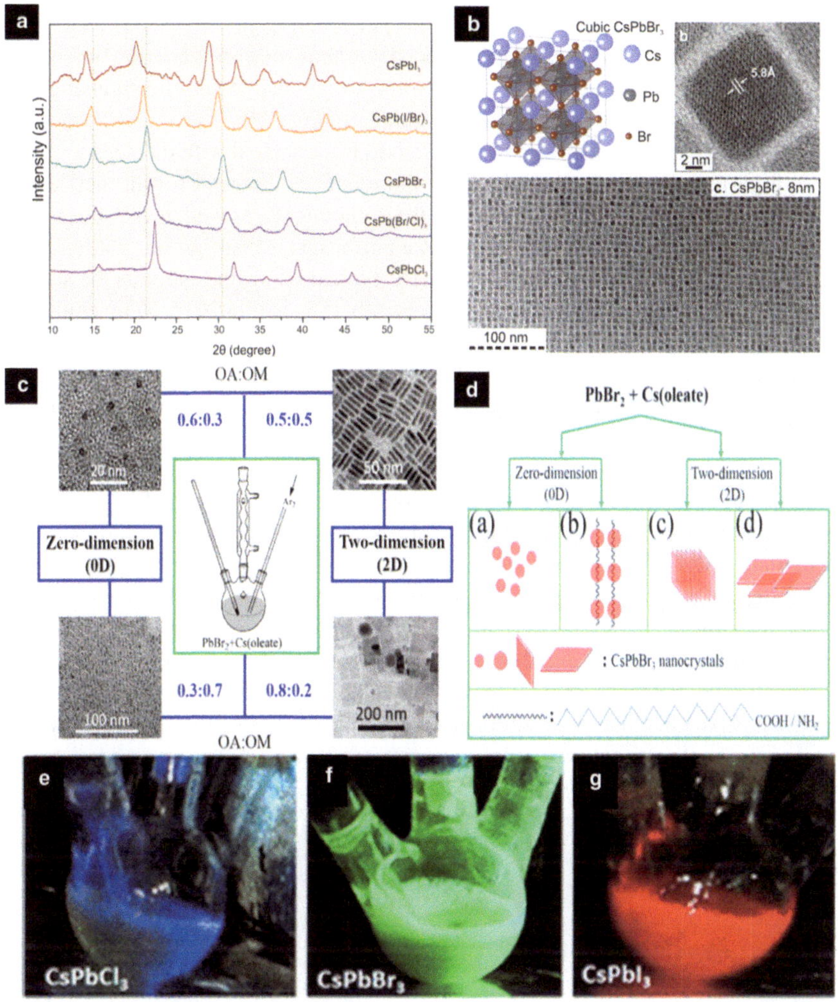

Fig. 6 **a** X-ray diffraction patterns for typical $CsPbX_3$. **b** Schematic structure cubic perovskite and typical transmission electron microscopy (TEM) images of $CsPbBr_3$ NCs. Reproduced with permission from Ref. [50]. **c** Schematic diagram of the $CsPbBr_3$ colloidal synthesis with various morphologies using different OAm/OA molar ratio as strategy. **d** Schematic synthesis method to control morphology. Reproduced with permission from Ref. [53]. **e**–**f** Photographs of the colloidal $CsPbX_3$ perovskite nanocrystals prepared by HI synthesis with illuminated reaction flask. Reproduced with permission from Ref. [54]

colloidal nanocrystals were tuned by varying the OAm and OA concentrations during HI synthesis, displaying two critical roles: (i) in the solubilization of precursors and (ii) as surfactants and stabilizers for the newly formed $CsPbX_3$ colloidal nanocrystals [51]. Additionally, the size and morphology of the perovskite-like colloidal nanocrystals are determined by controlling the polarity of the solvents in the medium (length of the ligand chain), the ratio between the proportion of OAm and OA, as well as the reaction temperature [20, 33, 52]. In Fig. 6c, d, we show how the variation of reaction conditions (i.e., different OAm/OA molar ratio), may lead to significant changes in the size and morphology of the resultant nanoparticles.

The primary disadvantages of the HI method is associated with the need to maintain a high temperature in the middle of the reaction, the need for an inert atmosphere, the frequent use of expensive and toxic chemicals, as well as the subsequent requirements for rapid cooling to maintain a uniform size distribution profile for the acquired colloidal nanocrystals. Additionally, typical HI synthesis limits the scaled-up production of solution-processed nanocrystals [55, 56]. In 2016, Zeng et al. [57] introduced a less complicated and room temperature based ligand-assisted reprecipitation (LARP) method. This method based on the use of non-precursors of Cs^+, Pb^{2+}, and halogen (in the desired stoichiometry) which are pre-dissolved in a polar solvent dimethylformamide (DMF) or dimethyl sulfoxide (DMSO), followed by rapid injection in a non-polar solvent (e.g., toluene and hexane), as shown in Fig. 7. As the solubility of these precursors, e.g., in non-polar solvents, is much less than polar solvents, $CsPbX_3$ nanoparticles quickly formed during this process [57]. Unlike the HI method, the LARP involves the co-occurrence of nucleation and crystal growth stages, initiated by the supersaturation process. Despite its simplicity and convenience for large scale productions, this preparation strategy has been known to have certain drawbacks. These drawbacks are usually associated with difficulty in purifying colloidal products, as well as its low synthetic yield and limited control on perovskite-like QDs size. The subsequent parts of Fig. 7b–f, show that the power of

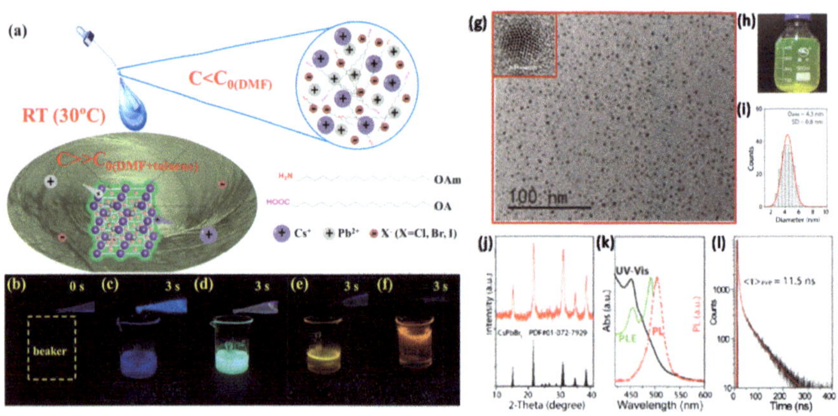

Fig. 7 **a** Schematic of Room Temperature formation of $CsPbX_3$ (X = Cl, Br, I)). Snapshots of four typical samples after the addition of precursor ion solutions for 3 s, blue (**c** Cl:Br = 1), green (**d** pure Br), yellow (**e** I:Br = 1), and red (**f** I:Br = 1.5), respectively. Reproduced with permission from Ref. [57]. **g** TEM images, **h** digital picture, **i** size distribution, **j** XRD pattern, **k** absoption and emission spectra, and **l** time-ressolved PL decay for the colloidal $CsPbBr_3$ perovskite nanocrystals prepared by LARP method Reproduced with permission from Ref. [58]

synthesis control is noticeable from the ratio of the precursor halogens and the range of light emission from the synthesized materials when exposed to light [57].

Solution-processed synthetic techniques have also been used to control the morphology of perovskite-like QDs, producing one dimensional (1D) nanowires and two dimensional (2D) nanoplatelets. Yet, three dimensional (3D) perovskite QDs, more commonly studied, can easily be tailored to lower-dimensional structures (e.g., nanosheets or nanowire-like) by carefully adjusting the amount of long-chain cations in the precursor solution during synthesis [46]. Studies have also demonstrated the direct growth of nanosheets or nanowires on a substrate surface, and it is plausible to produce nanoplatelets with lateral dimensions between 1 and 10 μm. Additionally, with careful control of the solvent composition, it is possible to grow atomically thin single-layer nanoplatelets. Zhang et al. [59] studied the morphological evolution by recording images at different growth times, finding that the resultant QDs mainly constituted a nanocuboid-like structure after 10 min of precursor injection. Specifically, large quantities of 1D nanowires (with a diameter <12 nm and lateral dimension extending up to 5 μm) were formed after 1 h of reaction. Stoumpos et al. [60, 61] also demonstrated that careful control of the injection temperature, could generate quasi-2D perovskite nanoplatelets at lower temperatures (i.e., 90–130 °C). These nanoplates with varying thicknesses are then separated by size-selective precipitation. A blue-shift in fluorescence was observed for plates with few layers (1–5 layers), indicative of the quantum confinement effect in these 2D nanocrystals. Akkerman et al. [62] prepared 3–5 monolayers thick cesium lead halide perovskite nanoplates using solution-based synthesis at room temperature. There have also been recently studies on ligand-induced engineering of morphologies in perovskite like QDs. Deng et al. [58] demonstrated the shape control in these perovskite-like QDs by altering various ligands, e.g., such as OAm, dodecylamine, hexanoic acid, and OA.

There consensus among the scientific community regarding the issues related to the presence of heavy metals in soil and water [63–67]. The toxicity of heavy metals present in perovskites (principally Pb) is directly linked to their leaching metals, which increases their mobility. Ecotoxicity tests, e.g., are used to systematically assess the toxic effects of contaminants in aquatic and terrestrial ecosystems, in which the bio-availability of the toxic substance to biota is determined [68]. Mint plant (*Mentha spicata*) stands out among the various existing ecotoxicity tests [69]. Seed germination and growth appear to be essential tools to determine the toxicity of materials, as any component present in the medium were to impact the development, this toxicity would manifest in the inhibition of germination and root growth. Abate et al. [69, 70] studied the effect of Pb substitution in perovskites for some time, to facilitate its use at a large scale, and minimize the impacts of possible Pb contamination. Figure 8 shows the adverse effects of the Pb presence on the growth of mint plants.

In addition to Pb-based perovskite, relatively less toxic Sn-based perovskite has been gaining attention in recent years due to their wide spectral response and near-infrared emission. They have been applied in solar cells, near-infrared light-emitting diodes (LEDs), and lasers [73–76]. However, there has been limited preparation of the analogs of cesium tin halide nanocrystals due to the poor solubility of the cesium

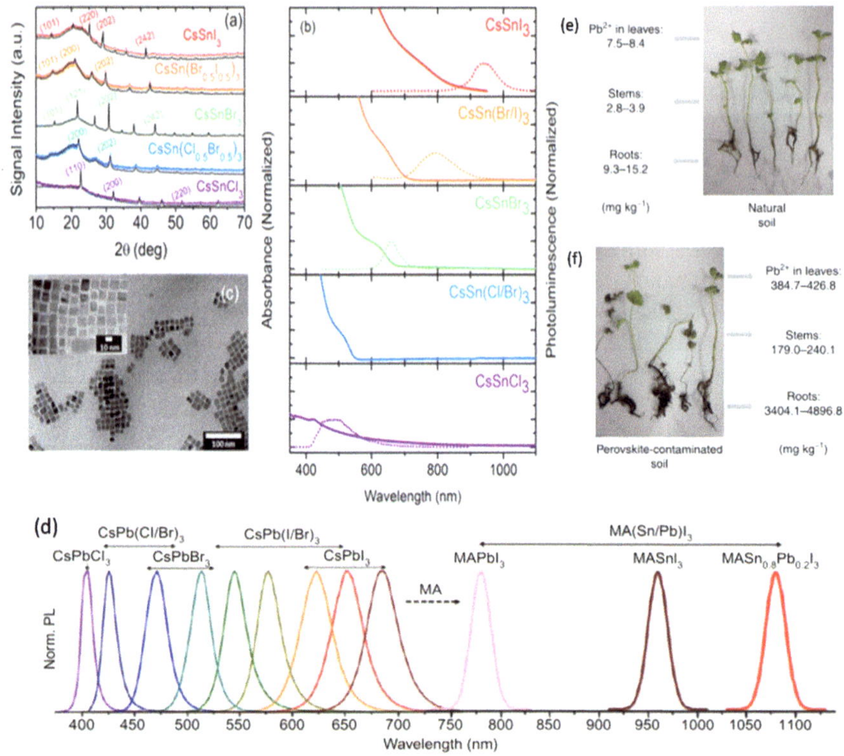

Fig. 8 **a** XRD patterns, **b** Absorbance and PL spectra of $CsSnX_3$ (X = Cl, $Cl_{0.5}Br_{0.5}$, Br, $Br_{0.5}I_{0.5}$, I). **c**TEM image of $CsSnX_3$ nanocrystals. Reproduced with permission from Ref. [71]. **d** Representative photoluminescence spectra of Blue ($CsPbX_3$) to redshifted ($MA(Sn/Pb)I_3$). Reproduced with permission from Ref. [72]). **e** The picture of mint plants grown on control soil and c 250 mg kg^{-1} Pb^{2+} perovskite-contaminated soil (**f**). The range of lead content measured in the leaves, stem, and root is reported on the side of each picture. Reproduced with permission from Ref. [69]

halides precursor. Bohm et al. [71] first synthesized these series of nanocrystals by using solvent coordinated SnX_2 precursors. Compared to Pb-based perovskite counterparts, Sn-perovskite exhibits a longer emission wavelength, as shown in Fig. 8**a–c**, enabling its usefulness for near-infrared LED devices. Although Sn-based perovskites are newer non-toxic alternatives to more popularly studied Pb-based analogs, stability is still an important issue (as Sn^{2+} is prone to oxidation to Sn^{4+} resulting in photoluminescence quenching and reduced photoluminescence quantum yields of the QDs); this needs to be addressed [77]. For perovskite colloidal nanocrystals, the position of the first absorption peak, and the fluorescence emission is modulated by controlling the size, morphology, and composition, and the imperfections of these parameters [78]. As such, these perovskite colloidal nanocrystals are designed to exhibit tunable optical properties, as shown in Fig. 8d. Perovskites may be widely used as LEDs due to their ability to absorb and to emit radiation at

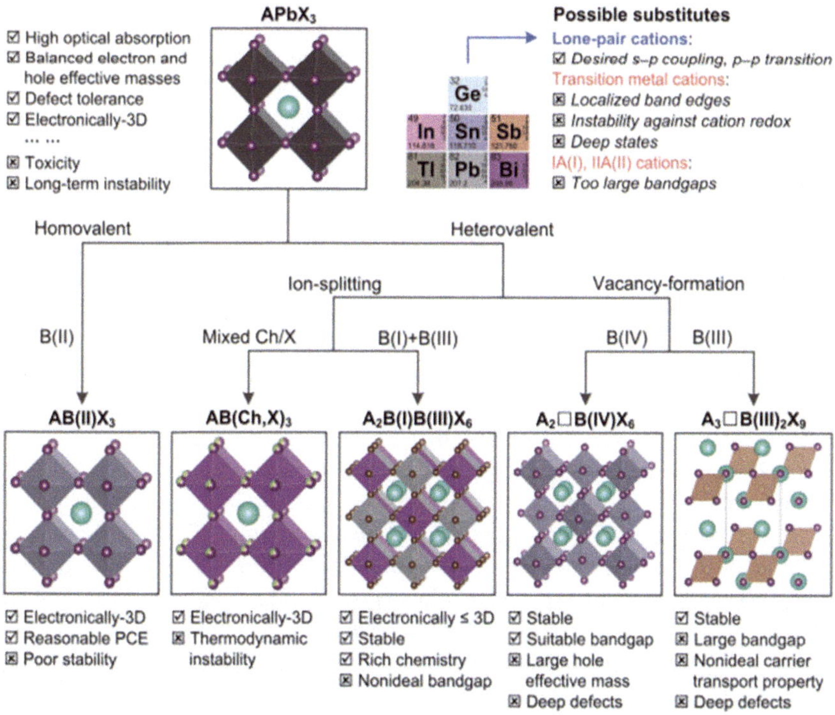

Fig. 9 Representative illustration of emerging Pb-free perovskite materials with the base on recently literature. Adapted from [79] 2019 © Wiley–VCH

different wavelengths (i.e., due to different confinement regimes for electrons and holes) [61–68]. Figure 9 shows a full scheme of the emerging Pb-free perovskite materials designed mainly for high-performance applications [79]. They can also be generated from solution-based synthesis, using a more straightforward method, and are easily replicated, offering a satisfactory cost–benefit ratio compared to other technological devices [42].

4 Advanced Optoelectronic Applications for Perovskite-Based Materials

Solution-processed colloidal materials possess advantageous properties for optoelectronics applications [47–49]. Therefore, they are promising candidates that complement "conventional" technologies when tunable color, large area, mechanical flexibility, or low-cost processing is required. The last 20 years, however, has seen the development of a wide range of electronics devices based on nanocrystals, such as lasers, LEDs, photodetectors, transistors, solar cells, and so on [41, 42]. As such, perovskite colloidal nanocrystals are gaining popularity in a wide range of technological applications, because of their versatility, simple preparation (which reduces production costs of devices significantly), and potential applications in solar cells [24–27]. The intrinsic defect tolerance nature of perovskite-like QDs and with nearly 100% photoluminescence quantum yield (PLQY) and tunable color emission make them potential candidates in solid-state lightings and displays [21–27]. Specifically, 3D metal halide perovskite, has better application prospects due to its very narrow photoluminescence bandwidth spectra and wide wavelength range. Recently, designed perovskite colloidal nanocrystals have been integrated into various low-cost solution-processed optoelectronic devices as result of their outstanding photoluminescence properties [1, 80]. These high-quality functional materials are suitable for use in LEDs, lasers, photovoltaic cells, amplifiers, photodetectors, and also biological sensors [63–87].

Zeng et al. [81] reported on the first $CsPbBr_3$ perovskite QD-based LEDs, where poly(ethylenedioxythiophene):polystyrene sulfonate (PEDOT:PSS) and poly (9-vinylcarbazole) (PVK) were used as the hole injection layers, and thin films of 2,2′,2″-(1,3,5-benzenetriyl)tris(1-phenyl-1-H-benzimidazole) (TPBi) were vacuum deposited on top of the perovskite nanocrystal films, serving as the electron injection layer. This work demonstrates the great potential of $CsPbX_3$ QDs in LED applications. As such, the LEDs based on $CsPbBr_3$ QDs had a maximum current efficiency of 0.43 ca A^{-1} and a maximum external quantum efficiency (EQE) of approximately 0.12% at a luminance of 100 cd m^{-2}. The electroluminescence of these devices could be easily tuned by changing the anion (halide) compositions in the $CsPbX_3$ QDs, and all devices with different colors exhibited an electroluminescence with a full width at half maximum (FWHM) of ~20 nm. Although the performance of this device was much lower than expected from the high solution PLQY of these QDs (50–90%) [81].

In 2016, Bakr et al. [82] passivated the surface of $CsPbX_3$ QDs using a ligand-exchange strategy, which further improved the stability of QDs and the EQE of QLEDs (3.0% for green, and 1.9% for blue). Li et al. [83] obtained nearly complete perovskite thin films using the crosslinking method, and achieved an electroluminescence yield of 5.7%. Based on the outcomes of this work, Fig. 10 shows the architecture used for LED fabrication (with imagery of the devices in operation), an energy-level diagram of the materials in the device, as well as the corresponding electroluminescence spectra.

To overcome the limited solubility of the Cs precursor, Mhaisalkar et al. [85] implemented crown-ethers to enable room temperature synthesis of $CsPbBr_3$ QDs. This ensured complete dissolution of the CsBr precursor, and rendered $CsPbBr_3$ QD inks practical for device fabrication. The resultant LEDs displayed a bright green emission, with a current efficiency and EQE of 9.22 cd A^{-1} and 2.64%, respectively. Their small size (typically from few to tens of nanometers), and high surface to volume ratio greatly impacts the electronic properties of QDs by their surface, which is typically covered with organic ligands [86]. The capping ligands are left on the surface during synthesis, and are responsible for stabilizing colloidal nanocrystals

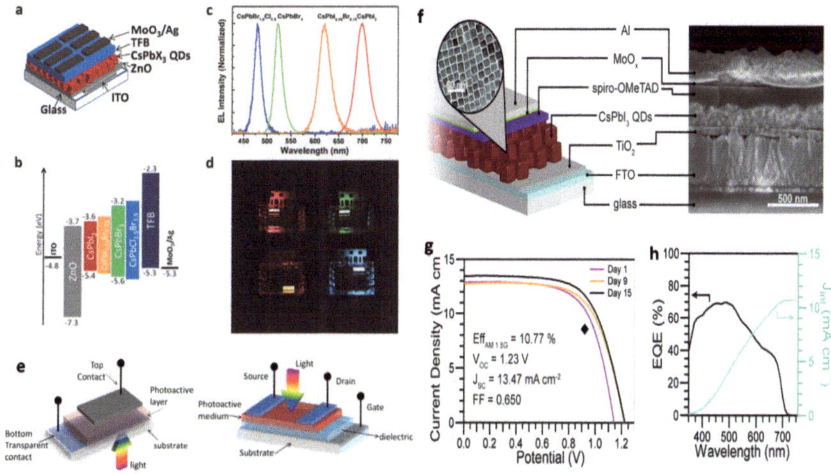

Fig. 10 Summarization of the main optoelectronic applications for colloidal perovskite-based nanocrystals. **a** Representative scheme of perovskite LEDs architecture. **b** Energy level-diagram and **c** electroluminescence spectra of perovskite LEDs. **d** Photographs of perovskite LEDs in operation. Adapted from [83]. **e** Schematic diagram of the device configurations of (left) vertical structure and (right) lateral structure photodetectors. For photodetectors with a vertical (sandwich-like) topology, the incident light window is located at the bottom, whereas, for photodetectors with the lateral topology, the incident light window is located at the top. Adapted from [84] 2013 © Wiley–VCH. **f** Schematic architecture with SEM cross-section for the colloidal $CsPbI_3$ perovskite nanocrystals solar cells. **g** Current density–voltage curves and (h) external quantum efficiency (black, left ordinate) and integrated current density (blue, right ordinate) of the device. Adapted from [87]

in the solvent. Furthermore, organic surface ligands play a dominant role in determining the interface between two or more neighboring nanocrystals in the solid-state; this strongly inhibits electrical charge injection and mobility. It has also been observed that the presence of excessive ligands may cause poor charge injection efficiency of perovskite QDs film, and insufficient ligand films exhibit low PLQY and stability. Thus, a fine equilibrium, needs to be formed between the charge injection and surface passivation. Colloidal QD-based devices, are often required to engineer their surfaces through surface functionalization via ligand exchange to improve the interfacial charge carrier transfer and tune their electronic energy levels. Over the last few years, many different ligand removals and replacement reactions (commonly referred to as ligand-exchange) have been developed to improve the electrical charge mobility of core/shell and core only nanocrystals. In a recent study by Dey et al. [87], this type of perovskite QDs and CdSe-based nano-heterojunctions was studied as a proof of concept, achieved using in-situ ligand exchange and careful manipulation of the QD surfaces. Li et al. [88] demonstrated another efficient ligand exchange method whereby a series of phenylalkylammonium bromides with different branch lengths were used for ligand-exchange. Based on the ligand-exchange method, the conductivity of the $CsPbBr_3$ QD layer had significantly improved due to the shortening of the ligand and the insertion of the π-conjugation benzene ring. As a result, high brightness (up to 12 650 cd/m^2) and low turn-on voltage (down to 2.66 V) was achieved in the $CsPbBr_3$ QD LEDs (QLEDs). This enabled, dramatic improvements in device performance with a current efficiency of 13.43 cd/A, power efficiency of 12.05 lm/W, and EQE of about 4.33%. In addition to organic–inorganic halide perovskites and all-inorganic perovskites-based QLEDs, in particular, Zhang et al. [89] reported hybrid perovskite QLEDs-based on organic–inorganic mixed cation ion FA/$CsPbBr_3$ QDs. This $FA_{1-x}Cs_xPbBr_3$ is highly dependent on the substitution content of the Cs cation. Xu et al. [90] demonstrated the $MA_{1-x}Cs_xPbBr_3$ perovskite QD-based LEDs for the first time. The optimized composition of $MA_{0.7}Cs_{0.3}PbBr_3$ QLEDs exhibited a brightness of 24 500 cd m^{-2}, current efficiency of 4.1 cd A^{-1}, and an EQE of approximately 1.3%.

The toxicity of Pb-based devices has a negative impact on the environment, marking total or partial Pb^{2+} replacement preferable. At present, the most suitable substitute is the less toxic Sn(II) ion. Sn-based perovskite semiconductors have been widely developed and applied in solar cell devices. However, only a few studies have investigated Sn-based perovskite for LEDs applications. Lai et al. [91] demonstrated QLEDs-based on $MASn(Br_{1-x}I_x)_3$ thin film exhibiting near-infrared electroluminescence spectrum with an emission wavelength of 945 nm and a maximum EQE of 0.72%. Zhang et al. [92] also reported efficient LEDs based on $CsPb_{1-x}Sn_xBr_3$ QDs with Sn(II) ion substitution and adopted a device configuration ITO/PEDOT:PSS/poly-TPD/$CsPb_{1-x}Sn_xBr_3$/TPBi/LiF/Al normal-type structure. More recently, other Pb-free halide perovskite QDs with improved stability and PLQYs, such as Cs_2AgBiX_6 [93, 94], $Cs_3Sb_2X_9$ [95, 96], and $(MA)_3Bi_2X_9$ [97] have been demonstrated, although QDLEDs based on these materials has been rarely reported.

Yet, the low electroluminescence efficiency continues to be a critical disadvantage when attempting to fabricate efficient perovskite QD-based LEDs. In addition to the general challenge of improving LED performance, the poor stability (moisture and temperature sensitivity) originated from the perovskite itself, the ion migration (and subsequent phase segregation observed in mixed halides system), and the metal electrode diffusion is, of course, the main concerns for future commercialization [90, 91, 94, 95, 97–100]. Recently, several strategies have been proposed to address these stability issues. The first is to employ all-inorganic perovskite materials to provide a potential long-term solution, while alternative approaches are based on the introduction of large organic ammonium cations to form a Ruddlesden-Popper layered perovskite, effectively hindering ion migration and improving the stability of perovskite-like QDs [101].

Within a brief timespan, diverse studies have focused on the rational design of perovskite colloidal nanocrystals, which have paved a new pathway as potential candidates to be used as solid-state QD-LEDs. Due to their functional properties, in particular, these perovskite-like QDs systems exhibit a high rate of radiative recombination, which can be precisely tuned by size, shape and composition control, as well as by the imperfections of these parameters [102]. Further research is required to improve the efficiency of device based on perovskite-like QDs. Additionally, the lifetime of perovskite QLEDs remains low for commercial applications.

Significant success was achieved in encapsulation technology, which requires further improvement, especially for flexible and stretchable perovskite-like QLEDs. Despite these hurdles, the advantages of perovskite-like QDs including facile wavelength tuning, high color purity, cost-effectiveness, and technical compatibility, render them appealing candidates for high definition LEDs [103].

Figure 10e illustrates the typical structure of a photodetector device. Therefore, in this perspective, it is well-known that for an all-optical communication greatly need the development of the ultrafast and reliable photodetectors significantly. Today's optical communication systems are usually based on silicon as well as indium gallium arsenide photodetectors, which are expensive materials [84, 104]. Based on these high costs, for this, the emergence of perovskite-based systems is, of course, considered a promising alternative for the development of a new generation of photodetectors, which may easily be integrated into the new technology of devices (which hence are mechanically flexible, lightweight, and large) [84, 104].

All-inorganic perovskite QDs are relatively more stable than those containing organic cations [6, 87]. Also, as we highlighted earlier, it is well-known that α-$CsPbI_3$ has an optimal value of bandgap as the photoactive material for solar cell applications. Particularly the Luther et al. [87] have reported a full study on the fabrication α-$CsPbI_3$ colloidal nanocrystals (of about 9 nm) photovoltaic cells with an efficiency of approximately 10.77%. Thus, the device architecture used by these authors is shown in Fig. 10f. In this study, the current density–voltage (JV) measurements were realized at room temperature and with relative humidity ~15 to 25%. These results (see Fig. 10g) is obtained for a 0.10 cm^2 cell and showed an open-circuit voltage (V_{OC}) of about 1.23 V. In additional, the EQE of the devices was also measured in this study, as shown in Fig. 10h. Finally, these results reveal a bandgap of about 1.75 eV for α-$CsPbI_3$ prepared by the HI method at 170 °C [87].

5 Summary and Outlook

In this chapter we have discussed on the crystalline structure, phase diagram, electronic properties, solution-based synthesis as well as potential integration strategies of perovskite colloidal nanocrystals into various optoelectronic devices. There are still challenges regarding the use of Pb-free technologies and increasing their stability in environmental conditions, but these are the same challenges that should guide the race for the development of better perovskite-based materials. We have demonstrated that significant improvements in the material foresight by design, including the fundamental understanding of their physical and chemical properties, are critically important and will undoubtedly provide a new technological level in the future.

References

1. Stranks, S.D., Snaith, H.J.: Metal-halide perovskites for photovoltaic and light-emitting devices. Nat. Publ. Gr. **10**, 391–402 (2015). https://doi.org/10.1038/nnano.2015.90
2. Moure, C., Peña, O.: Recent Advances in perovskites: processing and properties. Prog. Solid State Chem. **43**, 123–148 (2015). https://doi.org/10.1016/j.progsolidstchem.2015.09.001
3. Kojima, A., Teshima, K., Shirai, Y., Miyasaka, T.: Organometal halide perovskites as visible-light sensitizers for photovoltaic cells. J. Am. Chem. Soc. **131**, 6050–6051 (2009). https://doi.org/10.1021/ja809598r
4. Zhang, J., Hodes, G., Jin, Z., (Frank) Liu, S.: All-Inorganic $CsPbX_3$ perovskite solar cells: progress and prospects. Angew. Chemie - Int. Ed. **58**, 15596–15618 (2019). https://doi.org/10.1002/anie.201901081
5. Li, J., Wang, Q., Abate, A.: Perovskite solar cells. In: Thomas, S., Kalarikkal, N., Sakho, E.H.M., Wu, J., Oluwafemi, S.O. (eds.) Nanomater for Solar Cell Applications, 1st ed., pp. 417–446. Elsevier, Amsterdam (2019). https://doi.org/10.5757/vacmag.1.4.10
6. Manser, J.S., Christians, J.A., Kamat, P.V.: Intriguing optoelectronic properties of metal halide perovskites. Chem. Rev. **116**, 12956–13008 (2016). https://doi.org/10.1021/acs.chemrev.6b00136
7. Hyuck Heo, J., Hyuk Im, S., Hong Noh, J., Mandal, T.N., Lim, J. Ah Chang, Y. Hui Lee, H. Kim, A. Sarkar, A., Nazeeruddin, M.K., Grätzel, M., Il Seok, S.: Efficient inorganic-organic hybrid heterojunction solar cells containing perovskite compound and polymeric hole conductors. Nat. Photonics **7**, 486–491 (2013). https://doi.org/10.1038/NPHOTON.2013.80.
8. Burschka, J., Pellet, N., Moon, S.-J., Humphry-Baker, R., Gao, P., Nazeeruddin, M.K., Grätzel, M.: Sequential deposition as a route to high-performance perovskite-sensitized solar cells. Nature **499**, 316–319 (2013). https://doi.org/10.1038/nature12340
9. Liu, M., Johnston, M.B., Snaith, H.J.: Efficient planar heterojunction perovskite solar cells by vapour deposition. Nature **501**, 395–398 (2013). https://doi.org/10.1038/nature12509
10. Saliba, M., Matsui, T., Seo, J.-Y., Domanski, K., Correa-Baena, J.-P., Khaja, M., Zakeeruddin, S.M., Tress, W., Abate, A., Hagfeldt, A., Grä, M.: Cesium-containing triple cation perovskite solar cells: improved stability, reproducibility and high efficiency †, Cite This Energy Environ. Sci. **9** (2016). https://doi.org/10.1039/c5ee03874j
11. Deretzis, I., Smecca, E., Mannino, G., La Magna, A., Miyasaka, T., Alberti, A.: Stability and degradation in hybrid perovskites: is the glass half-empty or half-full? J. Phys. Chem. Lett. **9**, 3000–3007 (2018). https://doi.org/10.1021/acs.jpclett.8b00120
12. Niu, G., Guo, X., Wang, L.: Review of recent progress in chemical stability of perovskite solar cells. J. Mater. Chem. A. **3**, 8970–8980 (2015). https://doi.org/10.1039/c4ta04994b
13. Saidaminov, M.I., Haque, M.A., Almutlaq, J., Sarmah, S., Miao, X.H., Begum, R., Zhumekenov, A.A., Dursun, I., Cho, N., Murali, B., Mohammed, O.F., Wu, T., Bakr, O.M.: inorganic lead halide perovskite single crystals: phase-selective low-tem growth, carrier transport properties, and self-powered photodetection. Adv. Opt. Mater. **5** (2017). https://doi.org/10.1002/adom.201600704
14. Kulbak, M., Gupta, S., Kedem, N., Levine, I., Bendikov, T., Hodes, G., Cahen, D.: Cesium enhances long-term stability of lead bromide perovskite-based solar cells. J. Phys. Chem. Lett. **7**, 167–172 (2016). https://doi.org/10.1021/acs.jpclett.5b02597
15. Wang, Y., Li, X., Song, J., Xiao, L., Zeng, H., Sun, H.: All-inorganic colloidal perovskite quantum dots: a new class of lasing materials with favorable characteristics. Adv. Mater. **27**, 7101–7108 (2015). https://doi.org/10.1002/adma.201503573
16. Ling, X., Zhou, S., Yuan, J., Shi, J., Qian, Y., Larson, B.W., Zhao, Q., Qin, C., Li, F., Shi, G., Stewart, C., Hu, J., Zhang, X., Luther, J.M., Duhm, S., Ma, W.: 14.1% $CsPbI_3$ perovskite quantum dot solar cells via cesium cation passivation. Adv. Energy Mater. 1900721, 1–9 (2019). https://doi.org/10.1002/aenm.201900721

17. Swarnkar, A., Marshall, A.R., Sanehira, E.M., Chernomordik, B.D., Moore, D.T., Christians, J.A., Chakrabarti, T., Luther, J.M.: Quantum dot-induced phase stabilization of α-CsPbI3 perovskite for high-efficiency photovoltaics, Science (80-.). **354**, 92–95 (2016). https://doi.org/10.1126/science.aag2700
18. Steele, J.A., Jin, H., Dovgaliuk, I., Berger, R.F., Braeckevelt, T., Yuan, H., Martin, C., Solano, E., Lejaeghere, K., Rogge, S.M.J.: Thermal unequilibrium of strained black $CsPbI_3$ thin films, Science (80-.). **365**, 679–684 (2019). https://doi.org/10.1126/science.aax3878
19. Meggiolaro, D., Motti, S.G., Mosconi, E., Barker, A.J., Ball, J., Andrea Riccardo Perini, C., Deschler, F., Petrozza, A., De Angelis, F.: Iodine chemistry determines the defect tolerance of lead-halide perovskites, Energy Environ. Sci. **11**, 702–713 (2018). https://doi.org/10.1039/c8ee00124c
20. Shamsi, J., Urban, A.S., Imran, M., De Trizio, L., Manna, L.: Metal halide perovskite nanocrystals: synthesis, post-synthesis modifications, and their optical properties. Chem. Rev. **119**, 3296–3348 (2019). https://doi.org/10.1021/acs.chemrev.8b00644
21. Jing, Q., Xu, Y., Su, Y., Xing, X., Lu, Z.: A systematic study of the synthesis of cesium lead halide nanocrystals: does $Cs4PbBr_6$ or $CsPbBr_3$ form? Nanoscale **11**, 1784–1789 (2019). https://doi.org/10.1039/C8NR08116F
22. Brivio, F., Walker, A.B., Walsh, A.: Structural and electronic properties of hybrid perovskites for high-efficiency thin-film photovoltaics from first-principles. Apl Mater. **1**, 42111 (2013). https://doi.org/10.1063/1.4824147
23. Frost, J.M., Walsh, A.: Molecular motion and dynamic crystal structures of hybrid halide perovskites. Org. Halide Perovskite Photovoltaics 1–17 (2016). https://doi.org/10.1007/978-3-319-35114-8_1
24. Nakata, M.M., Mazzo, T.M., Casali, G.P., La Porta, F.A., Longo, E.: A large red-shift in the photoluminescence emission of Mg1—$xSrxTiO_3$. Chem. Phys. Lett. **622**, 9–14 (2015). https://doi.org/10.1016/j.cplett.2015.01.011
25. Goldschmidt, V.M.: Die Gesetze der Krystallochemie. Naturwissenschaften **14**, 477–485 (1926). https://doi.org/10.1007/BF01507527
26. Li, Z., Yang, M., Park, J.S., Wei, S.H., Berry, J.J., Zhu, K.: Stabilizing perovskite structures by tuning tolerance factor: formation of formamidinium and cesium lead iodide solid-state alloys. Chem. Mater. **28**, 284–292 (2016). https://doi.org/10.1021/acs.chemmater.5b04107
27. Goesten, M.G., Hoffmann, R.: Mirrors of bonding in metal halide perovskites. J. Am. Chem. Soc. **140**, 12996–13010 (2018). https://doi.org/10.1021/jacs.8b08038
28. Fadla, M.A., Bentria, B., Benghia, A., Dahame, T., Goumri-Said, S.: Insights on the opto-electronic structure of the inorganic mixed halide perovskites γ-CsPb(I1-xBrx)3 with low symmetry black phase. J. Alloys Compd. **832**, 154847 (2020). https://doi.org/10.1016/j.jallcom.2020.154847
29. Villars, K.C.P.: Rt-$CsPbBr_3$ ($CsPbBr_3$ Rt) crystal structure: datasheet from "PAULING FILE Multinaries Edition. SpringerMaterials. Springer, Berlin. https://materials-springer-com.ez26.periodicos.capes.gov.br/Isp/Crystallographic/Docs/Sd_1623827
30. Sutton, R.J., Filip, M.R., Haghighirad, A.A., Sakai, N., Wenger, B., Giustino, F., Snaith, H.J.: Cubic or orthorhombic? Revealing the crystal structure of metastable black-phase $CsPbI_3$ by theory and experiment. ACS Energy Lett. **3**, 1787–1794 (2018). https://doi.org/10.1021/acsenergylett.8b00672
31. Kye, Y.H., Yu, C.J., Jong, U.G., Ri, K.C., Kim, J.S., Choe, S.H., Hong, S.N., Li, S., Wilson, J.N., Walsh, A.: Vacancy-driven stabilization of the cubic perovskite polymorph of CsPbI 3. J. Phys. Chem. C. **123**, 9735–9744 (2019). https://doi.org/10.1021/acs.jpcc.9b01552
32. Cha, J.-H., Han, J.H., Yin, W., Park, C., Park, Y., Ahn, T.K., Cho, J.H., Jung, D.-Y.: Photoresponse of $CsPbBr_3$ and Cs_4PbBr_6 perovskite single crystals. J. Phys. Chem. Lett. **8**, 565–570 (2017). https://doi.org/10.1021/acs.jpclett.6b02763
33. Akkerman, Q.A., Rainò, G., Kovalenko, M.V., Manna, L.: Genesis, challenges and opportunities for colloidal lead halide perovskite nanocrystals. Nat. Mater. **17**, 394–405 (2018). https://doi.org/10.1038/s41563-018-0018-4

34. Yin, J., Maity, P., De Bastiani, M., Dursun, I., Bakr, O.M., Brédas, J.-L., Mohammed, O.F.: Molecular behavior of zero-dimensional perovskites. Sci. Adv. **3**, e1701793 (2017). https://doi.org/10.1126/sciadv.1701793
35. Stoumpos, C.C., Malliakas, C.D., Peters, J.A., Liu, Z., Sebastian, M., Im, J., Chasapis, T.C., Wibowo, A.C., Chung, D.Y., Freeman, A.J.: Crystal growth of the perovskite semiconductor $CsPbBr_3$: a new material for high-energy radiation detection. Cryst. Growth Des. **13**, 2722–2727 (2013). https://doi.org/10.1021/cg400645t
36. Nikl, M., Mihokova, E., Nitsch, K., Somma, F., Giampaolo, C., Pazzi, G.P., Fabeni, P., Zazubovich, S.: Photoluminescence of Cs_4PbBr_6 crystals and thin films. Chem. Phys. Lett. **306**, 280–284 (1999). https://doi.org/10.1016/S0009-2614(99)00477-7
37. Akkerman, Q.A., Park, S., Radicchi, E., Nunzi, F., Mosconi, E., De Angelis, F., Brescia, R., Rastogi, P., Prato, M., Manna, L.: Nearly monodisperse insulator Cs4PbX6 (X= Cl, Br, I) nanocrystals, their mixed halide compositions, and their transformation into $CsPbX_3$ nanocrystals. Nano Lett. **17**, 1924–1930 (2017). https://doi.org/10.1021/acs.nanolett.6b05262
38. Li, J., Yuan, X., Jing, P., Li, J., Wei, M., Hua, J., Zhao, J., Tian, L.: Temperature-dependent photoluminescence of inorganic perovskite nanocrystal films. RSC Adv. **6**, 78311–78316 (2016). https://doi.org/10.1039/C6RA17008K
39. Yettapu, G.R., Talukdar, D., Sarkar, S., Swarnkar, A., Nag, A., Ghosh, P., Mandal, P.: Terahertz conductivity within colloidal CsPbBr 3 perovskite nanocrystals: remarkably high carrier mobilities and large diffusion lengths. Nano Lett. **16**, 4838–4848 (2016). https://doi.org/10.1021/acs.nanolett.6b01168
40. Kondo, S., Amaya, K., Saito, T.: Localized optical absorption in Cs4PbBr 6. J. Phys. Condens. Matter. **14**, 2093 (2002)
41. Queisser, H.J. Haller, E.E.: Defects in semiconductors: some fatal, some vital, Science (80-.). **281**, 945–950 (1998). https://doi.org/10.1126/science.281.5379.945.
42. Kovalenko, M.V., Protesescu, L., Bodnarchuk, M.I.: Properties and potential optoelectronic applications of lead halide perovskite nanocrystals. Science (80-.). **358**, 745–750 (2017). https://doi.org/10.1126/science.aam7093
43. Kang, J., Wang, L.W.: High defect tolerance in lead halide perovskite CsPbBr 3. J. Phys. Chem. Lett. **8**, 489–493 (2017). https://doi.org/10.1021/acs.jpclett.6b02800
44. Brandt, R.E., Poindexter, J.R., Gorai, P., Kurchin, R.C., Hoye, R.L.Z., Nienhaus, L., Wilson, M.W.B., Polizzotti, J.A., Sereika, R., Žaltauskas, R., Lee, L.C., Macmanus-Driscoll, J.L., Bawendi, M., Stevanović, V., Buonassisi, T.: Searching for "defect-Tolerant" photovoltaic materials: combined theoretical and experimental screening. Chem. Mater. **29**, 4667–4674 (2017). https://doi.org/10.1021/acs.chemmater.6b05496
45. Steirer, K.X., Schulz, P., Teeter, G., Stevanovic, V., Yang, M., Zhu, K., Berry, J.J.: Defect tolerance in methylammonium lead triiodide perovskite. ACS Energy Lett. **1**, 360–366 (2016). https://doi.org/10.1021/acsenergylett.6b00196
46. Zhu, S., Ye, J., Zhao, Y., Qiu, Y.: Structural, electronic, stability, and optical properties of $CsPb_{1-x}$ Sn_x IBr_2 perovskites: a first-principles investigation. J. Phys. Chem. C. **123**, 20476–20487 (2019). https://doi.org/10.1021/acs.jpcc.9b04841
47. Cozzoli, P.D., Pellegrino, T., Manna, L.: Synthesis, properties and perspectives of hybrid nanocrystal structures. Chem. Soc. Rev. **35**, 1195–1208 (2006). https://doi.org/10.1039/b517790c
48. Murray, C.B., Kagan, C.R., Bawendi, M.G.: Synthesis and characterization of monodisperse nanocrystals and close-packed nanocrystal assemblies. Annu. Rev. Mater. Sci. **30**, 545–610 (2000). https://doi.org/10.1146/annurev.matsci.30.1.545
49. Rogach, A.L.: Semiconductor Nanocrystal Quantum Dots. Springer, Berlin (2008)
50. Protesescu, L., Yakunin, S., Bodnarchuk, M.I., Krieg, F., Caputo, R., Hendon, C.H., Yang, R.X., Walsh, A., Kovalenko, M.V.: Nanocrystals of cesium lead halide perovskites ($CsPbX_3$, X = Cl, Br, and I): novel optoelectronic materials showing bright emission with wide color gamut. Nano Lett. **15**, 3692–3696 (2015). https://doi.org/10.1021/nl5048779
51. Yang, D., Cao, M., Zhong, Q., Li, P., Zhang, X., Zhang, Q.: All-inorganic cesium lead halide perovskite nanocrystals: synthesis, surface engineering and applications. J. Mater. Chem. C. **7**, 757–789 (2019). https://doi.org/10.1039/c8tc04381g

52. Quarta, D., Imran, M., Capodilupo, A.L., Petralanda, U., Van Beek, B., De Angelis, F., Manna, L., Infante, I., De Trizio, L., Giansante, C.: Stable Ligand Coordination at the surface of colloidal $CsPbBr_3$ nanocrystals. J. Phys. Chem. Lett. **10**, 3715–3726 (2019). https://doi.org/10.1021/acs.jpclett.9b01634
53. Liang, Z., Zhao, S., Xu, Z., Qiao, B., Song, P., Gao, D., Xu, X.: Shape-controlled synthesis of all-inorganic $CsPbBr_3$ perovskite nanocrystals with bright blue emission. ACS Appl. Mater. Interfaces. **8**, 28824–28830 (2016). https://doi.org/10.1021/acsami.6b08528
54. Guria, A.K., Dutta, S.K., Das Adhikari, S., Pradhan, N.: Doping Mn^{2+} in lead halide perovskite nanocrystals: successes and challenges. ACS Energy Lett. **2,** 1014–1021 (2017). https://doi.org/10.1021/acsenergylett.7b00177
55. Li, H., Wang, C., Peng, Z., Fu, X.: A review on the synthesis methods of CdSeS-based nanostructures. J. Nanomater **2015**, 1–16 (2015)
56. Donegá, C.D.M., Liljeroth, P., Vanmaekelbergh, D.: Physicochemical evaluation of the hot-injection method, a synthesis route for monodisperse nanocrystals. Small **1**, 1152–1162 (2005). https://doi.org/10.1002/smll.200500239
57. Li, X., Wu, Y., Zhang, S., Cai, B., Gu, Y., Song, J., Zeng, H.: $CsPbX_3$ quantum dots for lighting and displays: roomerature synthesis, photoluminescence superiorities, underlying origins and white light-emitting diodes. Adv. Funct. Mater. **26**, 2435–2445 (2016). https://doi.org/10.1002/adfm.201600109
58. Sun, S., Yuan, D., Xu, Y., Wang, A., Deng, Z.: Ligand-mediated synthesis of shape-controlled cesium lead halide perovskite nanocrystals via reprecipitation process at room temperature. ACS Nano **10**, 3648–3657 (2016). https://doi.org/10.1021/acsnano.5b08193
59. Zhang, D., Eaton, S.W., Yu, Y., Dou, L., Yang, P.: Solution-phase synthesis of cesium lead halide perovskite nanowires. J. Am. Chem. Soc. **137**, 9230–9233 (2015). https://doi.org/10.1021/jacs.5b05404
60. Lee, B., Stoumpos, C.C., Zhou, N., Hao, F., Malliakas, C., Yeh, C.-Y., Marks, T.J., Kanatzidis, M.G., Chang, R.P.H.: Air-stable molecular semiconducting iodosalts for solar cell applications: Cs_2SnI_6 as a hole conductor. J. Am. Chem. Soc. **136**, 15379–15385 (2014). https://doi.org/10.1021/ja508464w
61. Hao, F., Stoumpos, C.C., Chang, R.P.H., Kanatzidis, M.G.: Anomalous band gap behavior in mixed Sn and Pb perovskites enables broadening of absorption spectrum in solar cells. J. Am. Chem. Soc. **136**, 8094–8099 (2014)
62. Akkerman, Q.A., Motti, S.G., Srimath Kandada, A.R., Mosconi, E., D'Innocenzo, V., Bertoni, G., Marras, S., Kamino, B.A., Miranda, L., De Angelis, F.: Solution synthesis approach to colloidal cesium lead halide perovskite nanoplatelets with monolayer-level thickness control. J. Am. Chem. Soc. **138**, 1010–1016 (2016). https://doi.org/10.1021/jacs.5b12124
63. Kumar, M., Furumai, H., Kurisu, F., Kasuga, I.: Chemosphere potential mobility of heavy metals through coupled application of sequential extraction and isotopic exchange: comparison of leaching tests applied to soil and soakaway sediment. Chemosphere **90**, 796–804 (2013). https://doi.org/10.1016/j.chemosphere.2012.09.082
64. Song, F., Gu, L., Zhu, Yuan, N.H.: Leaching behavior of heavy metals from sewage sludge solidified by cement-based binders. Chemosphere (2013). https://doi.org/10.1016/j.chemosphere.2013.01.022
65. Tsiridis, V., Petala, M., Samaras, P., Kungolos, A., Sakellaropoulos, G.P.: Environmental hazard assessment of coal fly ashes using leaching and ecotoxicity tests. Ecotoxicol. Environ. Saf. **84**, 212–20 (2012). https://doi.org/10.1016/j.ecoenv.2012.07.011
66. Zhou, X., Guo, J., Lin, K., Huang, K., Deng, J.: Leaching characteristics of heavy metals and brominated flame retardants from waste printed circuit boards. J. Hazard. Mater. **246–247**, 96–102 (2013). https://doi.org/10.1016/j.jhazmat.2012.11.065
67. Kogbara, R.B., Al-Tabbaa, A., Yi, Y., Stegemann, J.A.: pH-dependent leaching behaviour and other performance properties of cement-treated mixed contaminated soil. J. Environ. Sci. **24**, 1630–1638 (2012). https://doi.org/10.1016/S1001-0742(11)60991-1
68. An, Y.-J., Yang, C.-Y.: *Fridericia peregrinabunda* (Enchytraeidae) as a new test species for soil toxicity assessment. Chemosphere **77**, 325–329 (2009). https://doi.org/10.1016/j.chemosphere.2009.07.013

69. Li, J., Cao, H., Jiao, W., Wang, Q., Wei, M., Cantone, I., Lü, J., Abate, A.: Biological impact of lead from halide perovskites reveals the risk of introducing a safe threshold. Nat. Commun. **11**, 1–5 (2020). https://doi.org/10.1038/s41467-019-13910-y
70. Abate, A.: Perovskite solar cells go lead free. Joule **1**, 659–664 (2017). https://doi.org/10.1016/j.joule.2017.09.007
71. Jellicoe, T.C., Richter, J.M., Glass, H.F.J., Tabachnyk, M., Brady, R., Dutton, S.E., Rao, A., Friend, R.H., Credgington, D., Greenham, N.C.: Synthesis and optical properties of lead-free cesium tin halide perovskite nanocrystals. J. Am. Chem. Soc. **138**, 2941–2944 (2016). https://doi.org/10.1021/jacs.5b13470
72. Saliba, M., Correa-Baena, J.P., Grätzel, M., Hagfeldt, A., Abate, A.: Perovskite solar cells: from the atomic level to film quality and device performance. Angew. Chemie Int. Ed. **57**, 2554–2569 (2018). https://doi.org/10.1002/anie.201703226
73. Wang, H.-C., Bao, Z., Tsai, H.-Y., Tang, A.-C., Liu, R.-S.: Perovskite quantum dots and their application in light-emitting diodes. Small **14**, 1702433 (2018). https://doi.org/10.1002/smll.201702433
74. Wei, Y., Cheng, Z., Lin, J.: An overview on enhancing the stability of lead halide perovskite quantum dots and their applications in phosphor-converted LEDs. Chem. Soc. Rev. **48**, 310–350 (2019). https://doi.org/10.1039/c8cs00740c
75. Li, D., Xu, W., Zhou, D., Ma, X., Chen, X., Pan, G., Zhu, J., Ji, Y., Ding, N., Song, H.: Cesium tin halide perovskite quantum dots as an organic photoluminescence probe for lead ion. J. Lumin. **216**, 116711 (2019). https://doi.org/10.1016/j.jlumin.2019.116711
76. Xing, G., Kumar, M.H., Chong, W.K., Liu, X., Cai, Y., Ding, H., Asta, M., Grätzel, M., Mhaisalkar, S., Mathews, N., Sum, T.C.: Solution-processed tin-based perovskite for near-infrared lasing. Adv. Mater. **28**, 8191–8196 (2016). https://doi.org/10.1002/adma.201601418
77. Jena, A.K., Kulkarni, A., Miyasaka, T.: Halide perovskite photovoltaics: background, status, and future prospects. Chem. Rev. **119**, 3036–3103 (2019). https://doi.org/10.1021/acs.chemrev.8b00539
78. Talapin, D.V., Lee, J., Kovalenko, M.V., Shevchenko, E.V.: Prospects of colloidal nanocrystals for electronic and optoelectronic applications. Chem. Rev. **110**, 389–458 (2010). https://doi.org/10.1021/cr900137k
79. Xiao, Z., Song, Z., Yan, Y.: From lead halide perovskites to lead-free metal halide perovskites and perovskite derivatives. Adv. Mater. **31**, 1803792 (2019). https://doi.org/10.1002/adma.201803792
80. Pan, J., Quan, L.N., Zhao, Y., Peng, W., Murali, B., Sarmah, S.P., Yuan, M., Sinatra, L., Alyami, N.M., Liu, J., Yassitepe, E., Yang, Z., Voznyy, O., Comin, R., Hedhili, M.N., Mohammed, O.F., Lu, Z.H., Kim, D.H., Sargent, E.H.: Highly efficient perovskite-quantum-dot light-emitting diodes by surface engineering. Adv. Mater. 1–8. (2016). https://doi.org/10.1002/adma.201600784
81. Song, J., Li, J., Li, X., Xu, L., Dong, Y., Zeng, H.: Quantum dot light-emitting diodes based on inorganic perovskite cesium lead halides ($CsPbX_3$). Adv. Mater. **27**, 7162–7167 (2015). https://doi.org/10.1002/adma.201502567
82. Pan, J., Quan, L.N., Zhao, Y., Peng, W., Murali, B., Sarmah, S.P., Yuan, M., Sinatra, L., Alyami, N.M., Liu, J.: Highly efficient perovskite-quantum-dot light-emitting diodes by surface engineering. Adv. Mater. **28**, 8718–8725 (2016). https://doi.org/10.1002/adma.201600784
83. Li, G., Rivarola, F.W.R., Davis, N.J.L.K., Bai, S., Jellicoe, T.C., de la Peña, F., Hou, S., Ducati, C., Gao, F., Friend, R.H.: Highly efficient perovskite nanocrystal light-emitting diodes enabled by a universal crosslinking method. Adv. Mater. **28**, 3528–3534 (2016). https://doi.org/10.1002/adma.201600064
84. Baeg, K.J., Binda, M., Natali, D., Caironi, M., Noh, Y.Y.: Organic light detectors: photodiodes and phototransistors. Adv. Mater. **25**, 4267–4295 (2013). https://doi.org/10.1002/adma.201204979
85. Veldhuis, S.A., Ng, Y.F., Ahmad, R., Bruno, A., Jamaludin, N.F., Damodaran, B., Mathews, N., Mhaisalkar, S.G.: Crown ethers enable room-temperature synthesis of $CsPbBr_3$ quantum dots for light-emitting diodes. ACS Energy Lett. **3**, 526–531 (2018)

86. Lourenço, S.A., Silva, A.C.A., Zelaya, V.M., Cava, C.E., Rocha, G.D.A., da Silva, M.A.T., Duarte, J.L., Franchello, F., La Porta, F.A., Dantas, N.O.: Surface engineering in alloyed CdSe/$CdSe_x$ CdS_{1-x}/CdS Core-Shell colloidal quantum dots for enhanced optoelectronic applications. Emerg. Res. Sci. Eng. Based Adv. Exp. Comput. Strateg. 189–205 (2020). https://doi.org/10.1007/978-3-030-31403-3_7
87. Dey, S., Cohen, H., Pinkas, I., Lin, H., Kazes, M., Oron, D.: Band alignment and charge transfer in $CsPbBr_3$–CdSe nanoplatelet hybrids coupled by molecular linkers. J. Chem. Phys. **151**, 174704 (2019). https://doi.org/10.1063/1.5124552
88. Li, G., Huang, J., Li, Y., Tang, J., Jiang, Y.: Highly bright and low turn-on voltage $CsPbBr_3$ quantum dot LEDs via conjugation molecular ligand exchange. Nano Res. **12**, 109–114 (2019). https://doi.org/10.1007/s12274-018-2187-5
89. Zhang, X., Liu, H., Wang, W., Zhang, J., Xu, B., Karen, K.L., Zheng, Y., Liu, S., Chen, S., Wang, K.: Hybrid perovskite light-emitting diodes based on perovskite nanocrystals with organic-inorganic mixed cations. Adv. Mater. **29**, 1606405 (2017). https://doi.org/10.1002/adma.201606405
90. Xu, B., Wang, W., Zhang, X., Cao, W., Wu, D., Liu, S., Dai, H., Chen, S., Wang, K., Sun, X.: Bright and efficient light-emitting diodes based on MA/Cs double cation perovskite nanocrystals. J. Mater. Chem. C. **5**, 6123–6128 (2017). https://doi.org/10.1039/C7TC01300K
91. Lai, M.L., Tay, T.Y.S., Sadhanala, A., Dutton, S.E., Li, G., Friend, R.H., Tan, Z.-K.: Tunable near-infrared luminescence in tin halide perovskite devices. J. Phys. Chem. Lett. **7**, 2653–2658 (2016). https://doi.org/10.1021/acs.jpclett.6b01047
92. Zhang, X., Wang, W., Xu, B., Liu, H., Shi, H., Dai, H., Zhang, X., Chen, S., Wang, K., Sun, X.W.: Less-lead control toward highly efficient formamidinium-based perovskite light-emitting diodes. ACS Appl. Mater. Interfaces. **10**, 24242–24248 (2018). https://doi.org/10.1021/acsami.8b03590
93. Chen, N., Cai, T., Li, W., Hills-Kimball, K., Yang, H., Que, M., Nagaoka, Y., Liu, Z., Yang, D., Dong, A.: Yb-and Mn-doped lead-free double perovskite Cs_2AgBiX_0 (X= Cl–, Br–) Nanocrystals. ACS Appl. Mater. Interfaces **11**, 16855–16863 (2019)
94. Yang, B., Chen, J., Hong, F., Mao, X., Zheng, K., Yang, S., Li, Y., Pullerits, T., Deng, W., Han, K.: Lead-free, air-stable all-inorganic cesium bismuth halide perovskite nanocrystals. Angew. Chemie Int. Ed. **56**, 12471–12475 (2017). https://doi.org/10.1002/anie.201704739
95. Singh, A., Chiu, N.-C., Boopathi, K.M., Lu, Y.-J., Mohapatra, A., Li, G., Chen, Y.-F., Guo, T.-F., Chu, C.-W.: Lead-free antimony-based light-emitting diodes through the vapor–anion-exchange method. ACS Appl. Mater. Interfaces **11**, 35088–35094 (2019)
96. Ma, Z., Shi, Z., Yang, D., Zhang, F., Li, S., Wang, L., Wu, D., Zhang, Y., Na, G., Zhang, L.: Electrically-driven violet light-emitting devices based on highly stable lead-free perovskite $Cs_3Sb_2Br_9$ quantum dots. ACS Energy Lett. **5**, 385–394 (2019). https://doi.org/10.1021/acsenergylett.9b02096
97. Leng, M., Yang, Y., Zeng, K., Chen, Z., Tan, Z., Li, S., Li, J., Xu, B., Li, D., Hautzinger, M.P.: All-inorganic bismuth-based perovskite quantum dots with bright blue photoluminescence and excellent stability. Adv. Funct. Mater. **28**, 1704446 (2018). https://doi.org/10.1002/adfm.201704446
98. Wang, Q., Abate, A.: Strategies toward stable perovskite solar cells. Adv. Mater. Interfaces **1800264**, 1–6 (2018). https://doi.org/10.1002/admi.201800264
99. Ming, W., Yang, D., Li, T., Zhang, L., Du, M.-H.: Formation and diffusion of metal impurities in perovskite solar cell material CH_3 NH_3 PbI_3: implications on solar cell degradation and choice of electrode. Adv. Sci. **5**, 1700662 (2018). https://doi.org/10.1002/advs.201700662
100. Gualdrón-Reyes, A.F., Yoon, S.J., Barea, E.M., Agouram, S., Muñoz-Sanjosé, V., Meléndez, A.M., Niño-Gómez, M.E., Mora-Seró, I.: Controlling the phase segregation in mixed halide perovskites through nanocrystal size. ACS Energy Lett. **4**, 54–62 (2018)
101. Gao, X., Zhang, X., Yin, W., Wang, H., Hu, Y., Zhang, Q., Shi, Z., Colvin, V.L., Yu, W.W., Zhang, Y.: Ruddlesden-popper perovskites: synthesis and optical properties for optoelectronic applications. Adv. Sci. **6**, 1900941 (2019). https://doi.org/10.1002/advs.201900941

102. Longo, E., La Porta, F.A.: Recent Advances in Complex Functional Materials. Springer International Publishing, Berlin (2017). https://doi.org/10.1007/978-3-319-53898-3
103. Li, Y.F., Feng, J., Sun, H.B.: Perovskite quantum dots for light-emitting devices. Nanoscale **11**, 19119–19139 (2019). https://doi.org/10.1039/c9nr06191f
104. Bao, C., Yang, J., Bai, S., Xu, W., Yan, Z., Xu, Q., Liu, J.: High performance and stable all-inorganic metal halide perovskite-based photodetectors for optical communication applications. Adv. Mater. **1803422**, 1–8 (2018). https://doi.org/10.1002/adma.201803422

Fundamentals and Advances of the Oxidant Peroxo Method (OPM) for the Synthesis of Transition Metal Oxides

Alexandre H. Pinto, André E. Nogueira, Mayra D. Gonçalves, and Emerson R. Camargo

Abstract The ability to produce phase pure and compositionally controlled nanomaterials at temperatures lower than the ones required by solid state reaction methods is one of the most important features in a solution-chemistry synthetic method. The sol–gel based methods usually use many of organic compounds throughout the synthetic process, which can be detrimental to certain applications, as high quantities of residual carbon can be found along the final product. The Oxidant Peroxo Method, usually known by the acronym OPM, is a solution-chemistry method based on the production of peroxo complexes with hydrogen peroxide and different transition metal ions at alkaline pH. The production of these peroxo complexes leads to an amorphous material that upon calcination produces phase pure transition metal oxides with controlled composition. One special feature of the OPM method is the total absence of the organic compounds during the synthesis, which avoids the presence of undesired pyrolyzed organic molecules mixed with the metal oxide product. Additionally, the absence of organic compounds produces an oxidizing atmosphere during the synthesis, yielding very reactive powders, facilitating the production highly dense ceramic pellets for electronic applications. The production of powders with surface containing peroxo groups, also, has been beneficial for increasing the photocatalytic activity of titanium-based compounds and for use as a precursor in the solid-state reactions, which considerably decreases the processing temperature. Since its inception

A. H. Pinto
Department of Chemistry and Biochemistry, Manhattan College, 4513 Manhattan College Parkway, Riverdale, NY 10471, USA
e-mail: alex.pinto@manhattan.edu

A. E. Nogueira
Department of Chemistry, Institute of Exact and Biological Sciences (ICEB), Federal University of Ouro Preto-UFOP, Ouro Preto, MG 35400-000, Brazil
e-mail: andre.esteves@ufop.edu.br

M. D. Gonçalves · E. R. Camargo (✉)
Interdisciplinary Laboratory of Electrochemistry and Ceramics (LIEC), Department of Chemistry, Federal University of São Carlos, Rod. Washington Luis km 235, São Carlos, SP 13565-905, Brazil
e-mail: camargo@ufscar.br

F. A. La Porta and C. A. Taft (eds.), *Functional Properties of Advanced Engineering Materials and Biomolecules*, Engineering Materials,
https://doi.org/10.1007/978-3-030-62226-8_4

and first publication, back in 2001, the OPM method has been successfully applied by different research groups worldwide to produce binary oxides, i.e. TiO_2, tertiary oxides, $PbTiO_3$, $BaZrO_3$, and doped tertiary oxides $Pb_{1-x}LaTiO_3$. The variety of different metal oxides produced confirms the versatility of OPM method on yielding not only different compositions, but also different crystalline structures, like anatase, perovskite, sillenite, and spinel. Furthermore, the OPM method has yield metal oxides for many different applications, such as dielectric, optical, and photocatalytic. For instance, undoped $Bi_{12}TiO_{20}$ and Nb-doped $Bi_{12}(Ti_{1-x}Nb_x)O_{20}$ were used as efficient photocatalysts for degradation of rhodamine B under ultraviolet and visible lights, presenting better activity than TiO_2. In this chapter, the chemistry underlying the OPM method and the oxides most commonly prepared by this technique will be described, focusing how the method contributed to the advance of the synthetic, structural, and application aspects related to each one of these compounds. The future goals and applications of the method will be critically discussed. The authors hope this chapter can provide enough information to motivate a continuous dissemination of the OPM method, in view of its confirmed successful features and potential.

Keywords Chemical synthesis · Transition metal oxides · Nanomaterials · OPM method · Oxidant peroxo method

1 The Oxidant Peroxo Method (OPM)—Introduction

Some traditionally used solution-based techniques for the production of metal oxides, for instance, the sol–gel and polymerizable complex methods, use a high quantity of organic compounds. This fact can be detrimental to the final property of the material since it may be difficult to eliminate the residual carbon after the heat treatment step entirely.

Intending to have synthetic route free from organic compounds to produce oxide nanoparticles, Camargo and Kakihana developed in 2001 a synthetic route based on the in situ production of Ti-peroxo complexes by reacting metallic Ti with H_2O_2 in alkaline pH, initially used to produce lead titanate ($PbTiO_3$) [1]. Equation (1) can describe this reaction:

$$Ti_{(s)} + 3H_2O_{2(aq)} \rightarrow [Ti(OH)_3O_2]^-_{(aq)} + H_2O_{(l)} + H^+_{(aq)} \tag{1}$$

As shown in Eq. (1), the metallic Ti is oxidized by the H_2O_2, producing the peroxytitanic acid $[Ti(OH)_3O_2]^-$. The peroxytitanic acid could be used either as a precursor to produce binary oxides, such as TiO_2 or for the production of multinary oxides, like $PbTiO_3$. The production of TiO_2 nanoparticles from $[Ti(OH)_3O_2]^-$ will be discussed in depth in Sect. 2. Then, taking as an example the $PbTiO_3$ production, by adding Pb^{+2} ions, at alkaline pH, they are converted to $Pb(OH)_4{}^{-2}$. The $Pb(OH)_4{}^{-2}$ can react with the $[Ti(OH)_3O_2]^-$, according to Eq. (2):

$$Pb(OH)^{-2}_{4(aq)} + [Ti(OH)_3O_2]^{-}_{(aq)} \rightarrow PbO_{2(s)} + TiO_{2(s)} + OH^{-}_{(aq)} + H_2O_{(l)} \quad (2)$$

In this case, the peroxy group (O_2^{-2}) acts as an oxidizing agent by increasing the Pb oxidation state from +2 to +4. The product of reaction represented by Eq. (2) can be described as a mixture of amorphous PbO_2 and TiO_2. Finally, this amorphous product was calcined at 700 °C, giving rise to phase pure tetragonal $PbTiO_3$, as confirmed by XRD and Raman spectroscopy.

Although the OPM was not the first peroxide-based synthetic route, the simplicity of this method is one of its outstanding features, since the $[Ti(OH)_3O_2]^-$ is produced by merely dissolving Ti in a mixture of H_2O_2 and $NH_3{\cdot}H_2O$ in an ice bath. Then, this complex reacts with Pb^{+2} and coprecipitates the amorphous precursor. Finally, the amorphous precursor is crystallized to the desired composition by some thermal treatment. None of these steps require expensive instrumentation, a special atmosphere, or sophisticated pieces of glassware. Essentially, the $[Ti(OH)_3O_2]^-$ synthesis and amorphous precursor preparation can be carried out by using a few beakers and pipets.

Additionally, when $[Ti(OH)_3O_2]^-$ undergoes hydrolysis, an amorphous precipitate is formed, which can be heat treated using different crystallization methods. Among these methods, we can mention the traditional hydrothermal and solvothermal methods or assisted by microwave irradiation. The possibility of using different crystallization methods opens space for the study of the control of some final properties of materials such as the size and morphology of the particles.

For being an organic-free method, the OPM method eliminates the need for additional thermal treatments to remove unreacted or byproduct organic compounds. Moreover, it eliminates the possibility to have residual carbon as one of the byproducts of the process, representing an advantage when compared to sol–gel and polymerizable complex methods.

After the first paper describing the $PbTiO_3$ synthesis, the OPM method began to be successfully applied for the synthesis of many different metal oxides, either binary or multinary. Throughout this chapter, the reader will have the chance to get acquainted with several instances from the current scientific literature where these oxides were prepared. Also, to learn more about how these compounds were prepared, characterized, and applied in different situations, for instance, degradation of water pollutants or preparation of materials with the potential to be used in electronic components.

Many of the papers cited in this chapter followed strictly the procedure described in the Camargo and Kakihana first paper. Contrastingly, many of the papers cited in this chapter implemented crucial modifications to the first paper of the method. Some of them present only a distant and slight resemblance to the original method. Regardless of how similar to the original paper the procedure used was, we considered that the papers are worth to be cited.

Throughout this chapter, many examples where the OPM method was used to prepare functional materials such as photocatalysts in different wavelengths, pigments, and materials for electronic applications will be presented.

2 Binary Oxides

2.1 Titanium Dioxide (TiO_2) and Related Compounds

Titanium dioxide (TiO_2) is an n-type semiconductor having three crystalline phases: the rutile (tetragonal) thermodynamically stable phase and two metastable phases, known as anatase (tetragonal) and brookite (orthorhombic) [2, 3]. The three TiO_2 polymorphs are shown in Fig. 1.

The band gap of each polymorph varies slightly. For instance, the rutile band gap energy is equal to 3.0 eV, whereas, for anatase and brookite, it is respectively equal to 3.2 and 3.3 eV [4]. Although anatase has an indirect band gap, it usually presents better photocatalytic activity than the other two polymorphs. This is attributed to longer lifetimes of the charge carriers in anatase, and lighter effective mass of electrons and holes in anatase, resulting in an easier migration of the photogenerated charge carriers [4].

TiO_2 is a frequently studied material due to its nontoxicity, water insolubility, hydrophilicity, availability, and relatively low price [5]. Given the technological importance of TiO_2, the OPM method has been extensively used to prepare TiO_2 nanomaterials. The Ti-peroxo complex water solubility allied to the absence of organic compounds and chlorine on its structure has made the Ti-peroxo complex a versatile Ti source. Additionally, the Ti-peroxo complex has been very versatile regarding different types of thermal treatments it can be subject to in order to produce different polymorphs and morphologies of TiO_2.

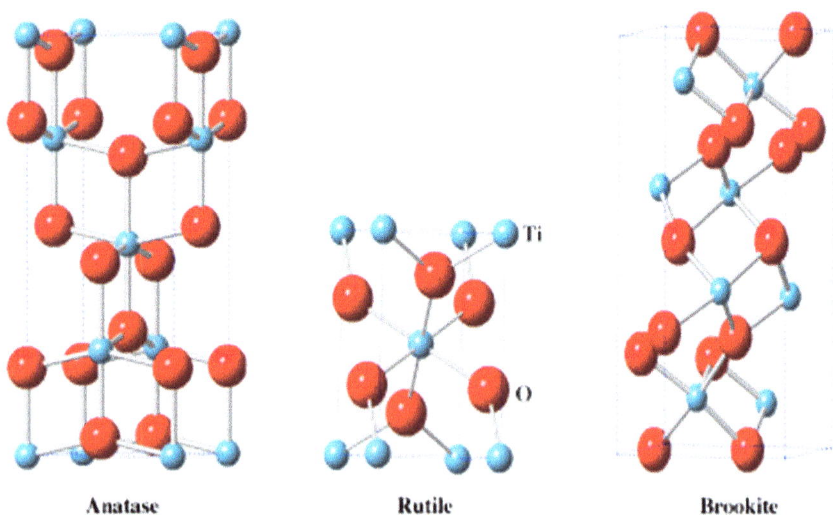

Fig. 1 Crystal structure of TiO_2 polymorphs: anatase (left), rutile (middle), and brookite (right). Adapted from Ref. [3], with permission from Copyright © 2015, Elsevier

The Ti-peroxo complex has been used as a Ti source also to prepare heterostructures of TiO_2 with other transition metal oxides and with carbon-based 2-D materials, such as graphene oxide (GO) and reduced graphene oxide (rGO). Additionally, the Ti-peroxo complex has been used as a means to insert Ti ions into intercalation compounds, such as zeolites and layered double hydroxides (LDH).

In the following sections, examples of how the Ti-peroxo complex has been used to produce these different categories of nanomaterials based on TiO_2 will be presented.

2.2 *TiO_2 Polymorphic and Morphological Controls*

One of the first papers describing TiO_2 preparation through the OPM method was authored by Ribeiro et al. [6]. There, the author studied the effect of pH adjustment of aqueous solutions containing the Ti-peroxo complex precursor on the TiO_2 preparation with controlled crystalline phases and morphologies.

They prepared the Ti-peroxo complex, which they described as being an amorphous solid, from metallic Ti, H_2O_2, and $NH_3{\cdot}H_2O$, then they use hydrothermal heating at 200 °C for 2 h, with the pH adjusted in a broad range. They figured out that at pH 0 phase pure rutile nanorods were formed, whereas, at pH values from 2 to 6, phase pure anatase nanoparticles were formed. At pH 8, these nanoparticles start to coalesce as a result of oriented attachment, leading to the formation of anatase nanorods and nanoneedles, respectively, at pH values equal to 10 and 12. Finally, at pH 14, hydrogen titanates with varied morphology are formed.

This work revealed the potential of the OPM method to tune the TiO_2 crystalline phase and morphology by varying a parameter as simple as the pH.

Taking advantage of these results, in a later paper, the same group prepared TiO_2 from the Ti-peroxo complex precursor [7]. At pH values ranging from 6 to 14, single-phase TiO_2 anatase was obtained for all samples. Except for the one at 14, which corresponded to hydrogen titanate ($H_2Ti_3O_7$). Then, they applied these powders as photocatalysts for the degradation of rhodamine B under UV light (254 nm emission maximum) illumination. The sample synthesized at pH 8 was the one that presented the best photocatalytic performance, both by weight and surface area-normalized experiments. This performance was attributed to the more isotropic morphology of the sample prepared at pH 8 compared to the anisotropic samples prepared on pH values from 10 to 14. As the anisotropic morphology leads to the formation of more defects, which can act as electron–hole recombination centers, the photocatalytic activity is decreased. Although the confirmed ability to work as photocatalysts for rhodamine B degradation, all the samples on this study were less efficient than commercial TiO_2 with estimated composition of 92% anatase and 8% rutile.

Besides powder production, the Ti-peroxo complex has proved to be successful in preparing TiO_2 thin films. Gao and co-workers [8] prepared the Ti-peroxo complex by dissolving metatitanic acid (H_2TiO_3) in a mixture of concentrated H_2O_2 and $NH_3{\cdot}H_2O$. Then, they dipped a substrate into the yellow-greenish transparent Ti-peroxo complex solution and were able to obtain an amorphous film according to

XRD deposited onto the substrate. By using XPS, TGA, and FT-IR, the authors estimated that the amorphous precipitate had the $TiO_{1.4}(O_2)_{0.5}(OH)_{0.2}\cdot 1.34H_2O$ formula when the film deposition is carried out at pH around 2.4. This film was annealed in temperatures from 400 to 700 °C for 1 h in air. This heat treatment revealed the (101) peak related to the anatase phase, from temperatures equal or higher than 600 °C.

The authors also studied powders formed from the Ti-peroxo complex, by calcining them between 300 and 900 °C. From 300 to 800 °C, phase pure anatase was formed, whereas, at higher temperatures, mixtures of anatase and rutile were obtained.

Aiming lower energy consumption and time-saving procedures, the use of microwaves has proved to be an efficient method to destabilize the Ti-peroxo complex precursor to produce TiO_2. For instance, Garcia et al. [9] prepared the Ti-peroxo complex precursor from a titanium isopropoxide solution in isopropyl alcohol by adding H_2O_2. Then they used the microwave hydrothermal method as a means to crystallize the Ti-peroxo complex amorphous precursor. Three different temperatures were studied 100, 150, and 200 °C. For each temperature, three time intervals were studied 30, 60, and 120 min. For all combinations of temperature and time, the authors obtained phase pure anatase, according to the XRD, and submicrometric agglomerates according to SEM, and specific surface area around 150 m^2/g. Then, the authors used these TiO_2 samples to test the photocatalytic degradation of methyl orange dye, under UV-A radiation. The samples prepared at 200 °C for 30 min were able to degrade the methyl orange almost completely, in an interval of 120 min, presenting photocatalytic activity comparable to the reference material TiO_2–P25.

By pursuing a deeper understanding of the microwaves on the conversion of the Ti-peroxo complex into TiO_2, Mendonça and co-workers [10] used benzyl alcohol as an additional solvent to the water in the microwave crystallization process.

The authors were aiming to gain any important information about the solvent role in the microwave crystallization process at 170 °C, and time intervals ranging from 2 to 60 min. Interestingly, the precursor obtained by the Ti-peroxo complex precursor obtained was not completely amorphous; instead, it presented broad peaks identified with a titanate anion lamellar structure, according to XRD results.

Besides the titanate anion lamellar structure, the Ti-peroxo complex was probably comprised also by ammonium cations and peroxo groups at stoichiometric ratios. This structure persisted when the precursor was heated for 4 min, and then, from 8 min and on, it was possible to observe sharps peaks related to TiO_2 anatase.

The release of these ammonium cations during hydrothermal treatment and decomposition of peroxo groups are the main steps responsible for the anatase phase formation, as shown in the scheme presented in Fig. 2. The authors confirmed the titanate anion lamellar structure of the precursor through Raman and X-ray absorption near-edge structure (XANES) spectroscopies, by comparing the precursor spectra with the potassium titanate spectra in both techniques. The anatase is the only TiO_2 polymorph observed for all heating time intervals from 8 to 60 min. The photocatalytic activity for rhodamine B degradation under UV-C radiation (254 nm emission maximum) using TiO_2 as photocatalyst increased as the TiO_2 microwave heating

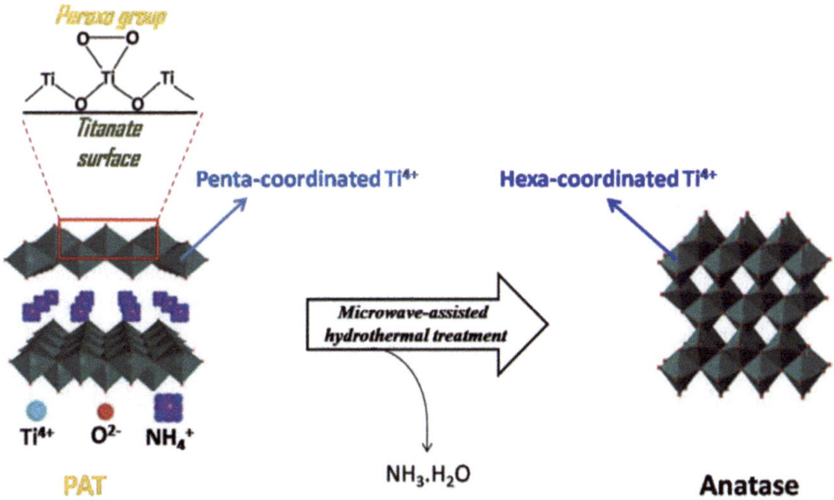

Fig. 2 Schematic representation of the NH_4^+ cations release and O_2^- degradation from the titanate lamellar structure during the microwave TiO_2 crystallization process. Adapted from Ref. [10], with permission from Copyright © 2019, Elsevier

time increased, however, no comparison with commercial TiO_2 samples were made on this paper.

2.3 Non-metal Doped TiO_2

The TiO_2 band gap around 3.2 eV for anatase and brookite, and 3.0 eV for rutile. These band gap energy values lie in the UV range of the electromagnetic spectrum, which limits the TiO_2 application in photochemical processes driven by visible light [5]. Another drawback presented by TiO_2 nanomaterials is the high recombination rate between electron and hole [5, 11]. One possible strategy to overcome these two limitations of TiO_2 is the non-metal doping.

The non-metal doping of TiO_2 narrows the band gap energy between the valence band and the conduction band. This band gap narrowing happens because the dopant ion introduces newly occupied orbitals between the valence band and conduction band [12].

Bakar et al. prepared N-doped TiO_2 using the OPM method [13]. The nitrogen source was trimethylamine, which was added to the water-soluble Ti-peroxo complex. The N added to the Ti-peroxo complex ranged from 1 to 5% by weight. After the precipitation of the solid precursors, they underwent hydrothermal treatment at 200 °C for 2 h.

The N-doped TiO_2 presented the anatase as the only crystalline phase, for all percentages of doping, revealing that the doping did not change the TiO_2 crystalline phase according to XRD and Raman spectroscopy. Doping also did not alter the morphology, since both undoped and N-doped TiO_2 presented a nanorod morphology, according to SEM and TEM micrographs. Figure 3 shows the TEM micrographs for undoped and N-doped-TiO_2, revealing that both of them have the same rice-like morphology.

Scherrer equation applied to XRD data revealed that an increasing N percentage led to a decrease in the crystallite size. In contrast, zeta potential indicated an increase in the isoelectric point with increasing N percentage. Regarding the optical properties, the band gap energy decreased systematically with the increasing N percentage. For instance, the band gap was 3.22 eV for samples doped with 1% of N and 2.85 eV for samples doped with 5% of N. This indicated that the N doping was an effective means to bring the band gap energy closer to the visible range of the electromagnetic spectrum. Besides the systematic variation in the band gap, the successful N doping was confirmed by EDS coupled to SEM and XPS.

The undoped and N-doped TiO_2 samples were used as photocatalysts in the photocatalytic degradation of methyl orange either under UV ($\lambda_{max} = 254$ nm) or visible

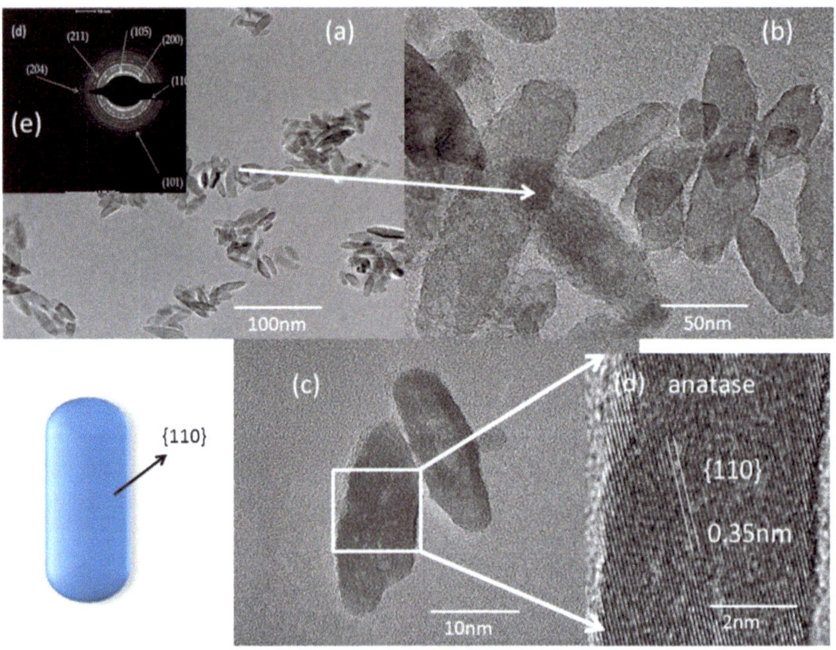

Fig. 3 TEM micrographs of the undoped TiO_2 in different magnifications (**a** and **b**), and from the N-doped TiO_2 (**c**), lattice fringes to the N-doped TiO_2, revealing the orientation according to {110} planes of the anatase phase, and selected area electron diffraction patterns (SAED) of the N-doped TiO_2 (**d**). Adapted from Ref. [13], with permission from Copyright © 2016, Elsevier

light. ($\lambda_{max} = 440$ nm). In general, higher rate constants were obtained for the process carried out under UV light illumination. Furthermore, both for UV and visible light, the sample doped with 3% of N presented the best performance. Similar results were obtained in subsequent work. There, the authors analyzed the photocatalytic degradation of rhodamine B and atrazine [14].

Following the studies regarding non-metal doping, the same group prepared TiO_2 doped with cationic and anionic sulfur [15, 16]. This research revealed exceptional flexibility of the OPM method, since by tuning the sulfur source, it was possible to make cationic doping, where S^{+6} replaced Ti^{+4} ions; or anionic doping, where S^{-2} replaced O^{-2} ions. SEM–EDS and XPS confirmed the success of the doping strategy.

To accomplish TiO_2 doping in different lattice sites, two sources of S were added straight to the Ti-peroxo complex solution, along with the Ti, H_2O_2, and $NH_3 \cdot H_2O$. For the cationic S-doped TiO_2, the source was thiourea (CH_4N_2S), whereas, for the anionic S-doped TiO_2, the source was carbon disulfide (CS_2). In both cases, the sulfur doping was done in the 1%, 2%, and 3% in a molar ratio in relation to the Ti amount. For both types of doped samples, the S doping did not cause substantial change neither in the crystalline phase nor in the morphology in comparison to undoped TiO_2. For all samples, it was observed the anatase phase and nanorods morphology.

However, interesting trends were observed about to crystallite size according to the Scherrer equation. For the anionic S-doped samples, the crystallite size increased with the increasing S content. Whereas for cationic S-doped samples, the crystallite size decreased with the increasing S content. Regarding the band gap, both types of samples were able to decrease the band gap compared to the undoped TiO_2, which was equal to 3.09 eV. The band gap of the samples containing 3% of S were 2.82 and 2.85 eV, respectively, for the anionic and the cationic doped-S samples.

Both types of samples were used for the photocatalytic degradation of methyl orange and phenol with the photocatalysts excited by visible light ($\lambda_{max} = 440$ nm). Both for methyl orange and phenol, both types of S doped samples presented photocatalytic efficiency around 80%, which was much higher than the one presented by the undoped TiO_2, around 5%. This result agrees with the fact that S-doping brings the bandgap as close as to the visible range of the electromagnetic spectrum.

2.4 TiO_2 Heterostructures

2.4.1 TiO_2—Metallic Oxides Heterostructures

A common strategy to prepare heterojunctions is the *building block* strategy. This strategy is usually based on the use of two or more pre-formed materials subject to some chemical or thermal process. This process can create an interface between the individual materials, producing a heterojunction. The heterojunction preparation is aimed to create a material presenting a positive synergistic effect between the individual components [17].

The TiO_2 has been used as a component along with other metallic oxides, using the *building block* strategy. A frequently prepared heterojunction is the TiO_2–SnO_2. Mendonça et al. prepared a TiO_2–SnO_2 heterojunction, where TiO_2 had anatase phase and SnO_2 rutile phase [18]. The TiO_2 anatase was prepared by the hydrothermal decomposition of the Ti-peroxo complex, as described by Ribeiro et al. [6]. Whereas, the SnO_2 rutile was prepared by the hydrolysis at room temperature of $SnCl_2{\cdot}2H_2O$ dissolved in ethanol.

The TiO_2–SnO_2 heterostructures were prepared with SnO_2 content equal to 10, 20, 35, and 50%, by weight. The heterojunction interface was formed by heating the pre-formed TiO_2 and SnO_2 in a microwave solvothermal procedure.

The formation of the heterojunction was confirmed by microstructural characterization and by chemical reactions tracking the rate of hydroxyl radical formation. However, the TEM-EDS elemental analyses revealed that most of the samples had a SnO_2 content lower than their nominal values.

In a subsequent paper, the authors used the same pre-formed TiO_2 and SnO_2 nanoparticles to prepare the TiO_2–SnO_2 heterostructures [19]. However, in this work, the authors studied only heterostructure composition containing 20% of SnO_2 by weight. Furthermore, the heterojunction interface was formed by regular heating hydrothermal reactions, instead of microwaves. The hydrothermal heating times ranged from 8 to 64 h. The TiO_2–SnO_2 heterostructures are shown in Fig. 4.

To probe the heterojunction formation, the authors used the heterostructures as photocatalysts for degradation of rhodamine B under UV light ($\lambda_{max} = 254$ nm) illumination.

In general, all the heterostructures presented better photocatalytic activity than pre-formed TiO_2 and the pre-formed TiO_2 subjected to hydrothermal heating for 64 h. The better activity presented by the heterostructures indicated there was a positive synergistic effect and that the heterojunction was successfully formed.

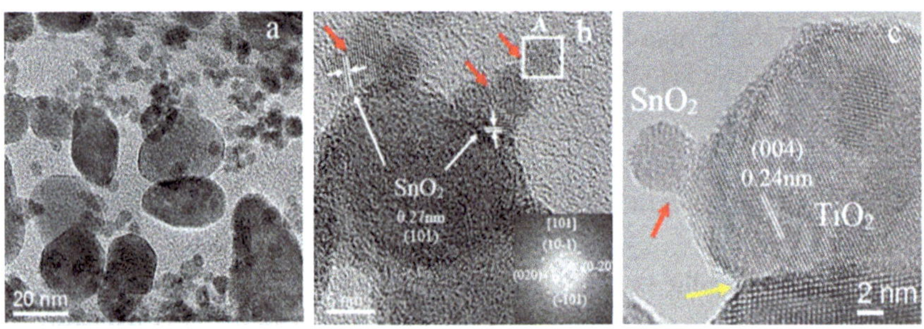

Fig. 4 **a** TEM micrographs of the TiO_2–SnO_2 heterostructures at low magnification **b** HR-TEM image showing the possible points of heterojunctions between the two oxides **c** HR-TEM showing the heterojunction between two TiO_2 particles (yellow arrow), and between a TiO_2 and a SnO_2 particles (red arrow). Besides their lattice fringes, the two oxides can be differentiated by their size difference, a SnO_2 particle is ca 5 nm, a TiO_2 particle is ca 20 nm. Adapted from Ref. [19], with permission from Copyright © 2017, Elsevier

Recently, Avansi Jr et al. prepared TiO_2–V_2O_5 heterostructures for ozone sensors [20]. The TiO_2 was prepared according to the procedures previously described by Ribeiro and Mendonça [6, 7]. Whereas the V_2O_5 was prepared by the dissolution of micrometric V_2O_5 powder in water, followed by H_2O_2 addition, the result solution underwent hydrothermal heating at 200 °C for 24 h, yielding V_2O_5 nanorods. Finally, the pre-formed V_2O_5 nanorods and Ti-peroxo complex powder were added to water, in molar ratios Ti:V equal to 5, 10, or 25%. Then the resulting mixture was subjected to hydrothermal treatment at 200 °C for 6 h. The TiO_2–V_2O_5 heterostructures produced were made up by V_2O_5 rods around 4 μm long, with TiO_2 anatase nanoparticles, with a diameter of around 20 nm, attached to the surfaces of the V_2O_5 rods. Electrical measurements revealed that the sensors made up of TiO_2–V_2O_5 heterostructures had a higher response to O_3 gas than the ones made from the pre-formed V_2O_5 by itself. The higher response to O_3 gas indicated that the heterostructure formation led to a more efficient charge carrier separation, facilitating the O_3 gas chemisorption process.

2.4.2 TiO_2—Carbon-Based Materials and TiO_2—Clays Heterostructures

Graphene is considered as being a two-dimensional material, constituted by a single layer of carbon atoms bond to each other by σ bonds, describing a honeycomb lattice formed by hexagons [21]. Graphite oxide has the same layered structure as the graphite; the difference resides in the addition of surface functional groups like hydroxyl and epoxide. In contrast, carbonyl and carboxyl groups are mostly located on the edges. The presence of these functional groups renders hydrophilicity to graphite oxide, unlike pristine graphite and graphene [22–24].

A method commonly used to prepare graphite oxide is the chemical exfoliation. It is based on the oxidation of graphite in an acidic media forming graphite oxide. The graphite oxide is then exfoliated in some solvent, for instance, water or ethanol, generating graphene oxide (GO). Finally, GO undergoes a reduction step producing reduced graphene oxide (rGO) [25].

Heterostructures made up by TiO_2 and rGO, are interesting for energy storage applications, such as batteries and capacitors [26]. In this sense, TiO_2 sub-micron sized spheres decorated with trace amounts of reduced graphene oxide (rGO) were prepared by using the OPM method [27].

The synthetic process started with titanium nitride (TiN), H_2O_2, $NH_3{\cdot}H_2O$, and graphene oxide (GO), to produce the water-soluble $[Ti(OH)_3O_2]^{-2}$ Ti-peroxo complex. Then, ethanol was added in different volume ratios in relation to the Ti-peroxo complex solution. The Ti-peroxo complex solution:ethanol volume ration were equal to 2:1, 1:1, 1:2 and 1:4. Each one of these resulting solutions were hydrolyzed at 80 °C, producing a grey amorphous powder. This powder was annealed at 450 °C for 3 h in an argon atmosphere. At the end of this process, sub-micron sized anatase TiO_2 spheres were obtained. The diameter of the spheres ranged from 1.5 μm to 100 nm, which could be easily tuned by varying the volume ratio between the Ti-peroxo complex solution and ethanol. As the amount of ethanol increased, the sphere diameter decreased. It could be explained that different volumes of ethanol led to the formation of different concentrations of the metastable $Ti(OH)_4$. The $Ti(OH)_4$ condensed in amorphous TiO_2 during the hydrolysis at 80 °C. Consequently, the $Ti(OH)_4$ concentration ended up controlling the rates of nucleation and growth, allowing to control the final diameter of the TiO_2 sub-micron sized spheres.

Then, these TiO_2 spheres were used as anodes in lithium-ion batteries (LIBs) and sodium-ion batteries (SIBs). Both for the LIBs and SIBs, the best electrochemical performance was obtained by the spheres prepared with Ti-peroxo complex:ethanol volume ration equal to 1:4. This performance could be attributed to the smaller size, more homogeneous size distribution, and a larger number of micropores.

Additionally, for comparison, TiO_2 spheres without rGO were prepared and presented worse performance than the ones containing rGO. It could be explained by the enhanced electrical conductivity of the spheres containing rGO.

A promising class of novel 2-dimensional materials is the MXenes, which are early transition metal nitrides or carbonitrides, for instance: Ti_3C_2, Nb_2C, Ti_3CN, V_2C, and Ta_4C_3 [28–31]. The MXenes are produced by selective etching of their parent 3-dimensional compounds called MAX-phases. They are compounds having the general formula $M_{(n+1)}AX_n$, where M is an early transition metal, for instance, Ti, V, Nb, or Mo. A is mainly a group IIIA or IVA (i.e., groups 13 or 14) element, for instance, Al or Si; and X is C and/or N, and $n = 1, 2$, or 3.

As examples of MAX-phases, it is possible to cite Ti_2AlC and Ti_3AlC_2. During the etching process, the A element is replaced by surface functional groups like O, OH, or F [30]. These structural features confer to MXenes a unique combination of metallic conductivity and hydrophilicity. Making them promising electrode materials for supercapacitors, and Li, Na, and K-ion batteries [31].

Aiming to prepare Na-ion capacitors, Wang et al. prepared graphene supported TiO_2 highly porous nanocomposite [32]. The synthetic process started with the MAX-phase Ti_3AlC_2, which was etched in HF to yield the MXene compound Ti_3C_2. Then, H_2O_2 was added to the Ti_3C_2 and aged for 18 h at room temperature, producing a yellow gel. The gel was dispersed in water, and GO was added. This mixture was sonicated, freeze-dried to remove water, and annealed at 500 °C for 2 h, in an argon atmosphere, producing a mesostructured TiO_2-reduced graphene oxide composite (M–TiO_2–rGO). The M–TiO_2–rGO was successfully applied as an anode in Na-ion batteries and Na-ion capacitors. The authors saw the possibility to convert at room temperature the MXene Ti_3C_2 in a water-soluble Ti-peroxo complex as one of the special features of the synthetic procedure. Furthermore, based on this feature, the authors see the potential to generalize this procedure to other MAX-phase compounds.

Layered double hydroxides (abbreviated as LDHs) are usually lamellar mixed hydroxides containing positively charged main layers and able to undergo anion exchange chemistry [33]. LDHs are used in many applications, such as adsorption of oxyanions, energy storage, photocatalysts, and water purification [34].

Taking advantage of the water solubility and organic and chlorine-free features of the Ti-peroxo complex, Jing He and co-workers used the Ti-peroxo complex as a source to insert Ti-ions inside the layered double hydroxide (LDH) [35]. Different LDHs based on M(II)-Al-NO_3, where M(II) = Mg^{+2}, Zn^{+2}, or Ni^{+2}, were prepared by co-precipitation of the metallic cation nitrate and NaOH. Then, the Ti-peroxo complex was used as an intercalating agent by adding it in an aqueous solution of the host compound M(II)-Al-NO_3. The resulting solution was stirred at room temperature for 24 h, under N_2 atmosphere. Shifts in some XRD peaks indicated that the $[Ti(OH)_3O_2]^-$ have been successfully intercalated in all three M(II)-Al-NO_3 host compounds, producing well-ordered Ti-peroxide pillared LDHs. In fact, a Ti/Al mole ratio is enough to promote the exchange of the $NO_3{}^-$ anions by the $[Ti(OH)_3O_2]^-$.

The Ti-peroxide pillared LDHs were used as catalysts for the oxidation of phenyl methyl sulfide using aqueous H_2O_2 as oxidant. All the three Ti peroxide pillared LDHs presented higher conversion than their respective parent LDH M(II)-Al-NO_3. This result revealed that the thioether conversion could be mostly attributed to the catalytic property of the Ti peroxide centers.

3 Metallic Titanates

3.1 *Lead-Based Compounds*

Since the initial work by Camargo and Kakihana, preparing lead titanate ($PbTiO_3$) in 2001 [1], the application of the OPM method for the preparation of Pb-based ceramic materials has grown. A natural pathway to increase the scope of the synthetic method was to prepare Pb-based ternary oxides, by replacing the Ti^{+4} cation by other metallic cations from the group IV B of the periodic table.

The first attempt described in the literature was about the preparation of lead zirconate ($PbZrO_3$) [36]. There, Camargo et al. used very similar chemicals used for the $PbTiO_3$ synthesis. The only difference, regarding the chemicals, was the use of $ZrO(NO_3)_2$ as Zr^{+4} source. The reaction setup is a little different as well. For the $PbTiO_3$ synthesis, a Ti-peroxo complex was prepared at pH 11 adjusted by $NH_3{\cdot}H_2O$, followed by the addition of a $Pb(NO_3)_2$ aqueous solution. Whereas for the $PbZrO_3$ synthesis, an aqueous solution containing Ti^{+4} and Zr^{+4} in 1:1 mol ratio was prepared, then, H_2O_2 and $NH_3{\cdot}H_2O$ were added into the Ti^{+4}–Zr^{+4} solution, leading to the formation of an amorphous precipitate.

The formation of the amorphous precursor in this particular reaction can be explained by the following chemical equations Eq. (3)–(5):

$$Pb(OH)^{-2}_{4(aq)} + H_2O_{2(aq)} \rightarrow PbO_{2(s)} + 2\,H_2O_{(l)} + 2\,OH^-_{(aq)} \tag{3}$$

$$ZrO^{-2}_{(aq)} + H_2O_{2(aq)} \rightarrow ZrO(O_2)H^+_{(aq)} + H^+_{(aq)} \tag{4}$$

$$Pb(OH)^{-2}_{4(aq)} + ZrO(O_2)H^+_{(aq)} \rightarrow PbO_{2(s)} + ZrO_{2(s)} + 2H_2O_{(l)} + OH^-_{(aq)} \tag{5}$$

As the H_2O_2 can not form stable complexes with Pb^{+2}, an amorphous precipitate containing Pb and O is formed, as shown in Eq. (3). Then, the stable aqueous complex $ZrO(O_2)H^+$ is formed by the addition of H_2O_2 to $ZrO(NO_3)_2$ (Eq. 4). Finally, at high pH, the $ZrO(NO_3)_2$ works as an oxidizing agent, increasing the Pb oxidation state from +2 to +4, and forming an amorphous precipitating containing Pb^{+4} and Zr^{+4} in stoichiometric condition (Eq. 5).

The amorphous precursor obtained was calcined in temperatures ranging from 300 to 1000 °C, in time intervals from 1 to 8 h. The calcination of the amorphous precursor for 2 h indicated that at a temperature as low as 300 °C, the powder is mostly amorphous with small peaks in the positions expected for PbO_2. At 550 °C, a mixture of orthorhombic $PbZrO_3$ and tetragonal ZrO_2 were obtained. Lastly, phase-pure orthorhombic $PbZrO_3$ was obtained at 700 °C.

Upon successful preparation of $PbTiO_3$ and $PbZrO_3$, the next logical step would be to prepare the solid-solution of these two compounds, which has the general formula $Pb(Ti_xZr_{1-x})O_3$ and it is known by the acronym PZT. PZT is a technologically functional material applied in sensors and actuators due to its high piezoelectric coefficient [37]. An intriguing structural property of PZT is the coexistence of the rhombohedral and tetragonal phases when x is around 0.5 [37]. This phenomenon is called the morphotropic phase boundary (MPB).

Camargo et al. made the first attempt to prepare PZT using OPM method in 2001. They prepared PZT with the $Pb(Ti_{0.40}Zr_{0.60})O_3$ composition [38]. They calcined the amorphous powder in temperatures from 700 to 1000 °C, for 2 h. When calcined at 700 °C, according to XRD, PZT powder was estimated to be made up of 53% of the rhombohedral and 47% of the tetragonal phases. Whereas at 1000 °C, it was made up of 91% of the rhombohedral and 9% of the tetragonal phases.

Three years later, Camargo and coworkers prepared PZT powders with the $Pb(Ti_{0.50}Zr_{0.50})O_3$ composition, by calcination of the amorphous powder at 700 °C for 2 h [39]. Then, the authors pressed the powders making ceramic pellets and sintered them at 1000 and 1100 °C, for 2 or 4 h. The dielectric constant of the ceramic pellets was measured at different frequencies (1, 10, and 100 kHz) versus temperature. The highest dielectric constant was equal to 17,500, and measured at 1 kHz. Whereas, the phase transition temperature from the ferroelectric to paraelectric phase was around 394 °C.

In order to study the effects of microwaves in the sintering process, Gonçalves et al. prepared $Pb(Ti_xZr_{1-x})O_3$ with the following compositions $x = 0.20$, 0.48, and 0.80 [40]. The ceramic pellets prepared from the PZT crystalline powders and sintered at 1000 °C for 2 h. Either by using an electric tubular oven or a microwave oven. The ceramic pellets with $x = 0.48$ presented densification higher than 99%, by either of the sintering methods. Figure 5 shows the SEM images of fractured $Pb(Ti_{0.48}Zr_{0.52})O_3$ ceramic pellets, displaying large and dense grains.

Despite similar densification, the dielectric constant for pellets sintered by microwave oven presented a higher dielectric constant than the ones sintered in conventional electric oven. The dielectric constant was around 17,500 for the microwave sintering, whereas it was around 12,000 for the electric oven sintering.

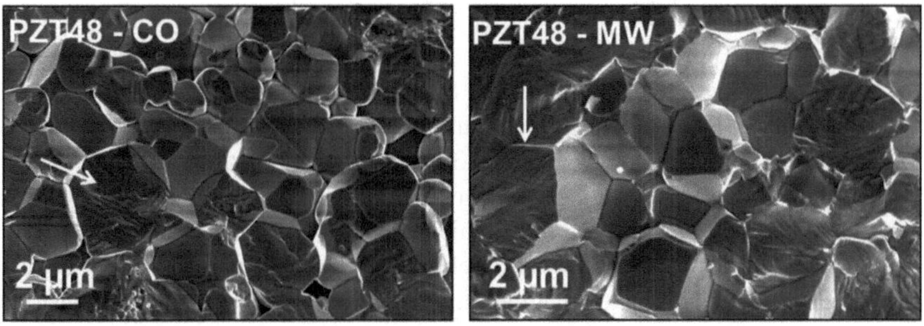

Fig. 5 **a** SEM micrographs of fractured $Pb(Ti_{0.48}Zr_{0.52})O_3$ bodies. Either sintered at an electric tubular oven (left) or a microwave oven (right). Adapted from Ref. [40], with permission from Copyright © 2016, Elsevier

Showing the versatility of the OPM method, Yamada made $Pb(Zr_{0.52}Ti_{0.48})O_3$ solid solutions with lanthanide ions La^{+3}, Nd^{+3}, and Pr^{+3} [41], symbolized by the formulas $(RE_xPb_{1-x})(Zr_{0.52}Ti_{0.48}O_3)$, with RE = La^{+3}, Pr^{+3}, or Nd^{+3} and x = 0.05, 0.10, 0.15, 0.20, 0.30, 0.38, and 0.45.The strategy to prepare Pb-based ternary oxides with + 4 ions continued by the $PbHfO_3$ preparation [42]. The source of hafnium was the $HfO(NO_3)_2$. And the amorphous precursor formation followed a similar mechanism to the one presented for the $PbZrO_3$, with the $HfO(O_2)H^+$ oxidizing Pb from +2 to +4, as shown in Eq. (6):

$$Pb(OH)^{-2}_{4(aq)} + HfO(O_2)H^+_{(aq)} \rightarrow PbO_{2(s)} + HfO_{2(s)} + 2\,H_2O_{(l)} + OH^-_{(aq)} \qquad (6)$$

The phase-pure orthorhombic $PbHfO_3$ was obtained after calcination of the amorphous precursor at 900 °C, for 1 h.

Besides the ability to prepare phase pure Pb-based ternary oxides containing +4 ions, the OPM proved to be effective in preparing $PbTiO_3$ doped by lanthanide ions. Initially, the La-substituted $PbTiO_3$ $(Pb_{(1-x)}La_x)TiO_3$ was prepared by Camargo et al., with x = 0.05, 0.10, 0.15, and 0.20 [43]. The La^{+3} was introduced in the method by dissolving La_2O_3 in a diluted nitric acid solution. The Eq. (7) explains the formation of the amorphous precipitate:

$$Pb(OH)^{-2}_{4(aq)} + La(H_2O)^{+3}_{6(aq)} + \left[Ti(OH)_3O_2\right]^-_{(aq)} \rightarrow [PbO_2 \cdot TiO_2La_2O_3] \cdot H_2O_{(s)} \qquad (7)$$

When the amorphous precursors were calcined at 800 °C, for 1 h. The powders with x = 0.05 and 0.10, presented a mixture between the tetragonal and cubic phases of $(Pb_{(1-x)}La_x)TiO_3$. Whereas, for x = 0.15 and 0.20, the samples obtained were practically only cubic phases of $(Pb_{(1-x)}La_x)TiO_3$. These results showed that doping could lower to room temperature the $PbTiO_3$ phase transition temperature from tetragonal to cubic, which is around 490 °C [44]. The La-substitution was quantified by ICP analyses and revealed experimental compositions very close to the nominal values.

These results confirmed the ability of the OPM method to precisely substitute Pb^{+2} by La^{+3} in the $PbTiO_3$ matrix.

Taking advantage of the ability of the OPM to produce $(Pb_{(1-x)}La_x)TiO_3$ powders with precise La^{+3} insertion, Pinto et al. prepared dense ceramic pellets of $(Pb_{(1-x)}La_x)TiO_3$ with x = 0.25 and 0.30 [45]. The authors measured the dielectric constant of these materials versus temperature to estimate the phase transition temperature from tetragonal to cubic. This transition is accompanied by a transition in the dielectric properties from ferroelectric (tetragonal) to paraelectric (cubic). These measurements indicated that the sample with x = 0.25 presented a transition temperature equal around 105 °C, whereas the sample with x = 0.30 presented a transition temperature equal around −8 °C. Both temperatures are below the 490 °C observed for the $PbTiO_3$ tetragonal to cubic phase transition temperature. These results confirm the hypothesis that La substitution decreases the phase transition of $PbTiO_3$ from tetragonal to cubic and that the extent of substitution correlates to how much the phase transition temperature will be decreased.

Continuing the lanthanide ions substitutions, Pinto et al. prepared praseodymium-modified lead titanate $(Pb_{(1-x)}Pr_x)TiO_3$ powders, with x = 0.20 [46]. Dense ceramic pellets sintered at 1150 °C were estimated to have a phase transition temperature from tetragonal to cubic around 208 °C, according to the dielectric constant measurements versus temperature. Although the Pr^{+3} substitution induced a decrease in the phase transition temperature, it was not so pronounced as the one induced by the La^{+3}, for a similar substitution amount. This observation opens up space for future studies aiming to investigate the properties of lanthanide ions and their ability to reduce the phase transition temperature in comparison to the one for the pure $PbTiO_3$. This type of study is necessary because of the fragility of $PbTiO_3$ ceramics. This fragility is attributed to the structural anisotropy characteristic of the tetragonal phase [47, 48]. So, to be able to stabilize the cubic phase at temperatures close to room temperature could be an option to improve the mechanical properties of $PbTiO_3$-based ceramics.

3.2 *Bismuth Titanates*

Research is currently focused on exploring materials with various types of nanostructures due to their multiple purposes in multiple fields such as energy, electronics, and the environment. Thus, bismuth titanate with its different polymorphs($Bi_4Ti_3O_{12}$, $Bi_2Ti_4O_{11}$, $Bi_{12}TiO_{20}$, Bi_8TiO_{14}, and $Bi_2Ti_2O_7$) [49] is extensively explored due to its set of exclusive properties.

These properties allow their use in the most varied applications, including high-temperature piezoelectric transducers [50], photocatalytic systems [51], hydrogen generation [52], capacitors [53], as sensors [54] among others. The various applications of bismuth titanates are possible due to their piezoelectric, electro-optical properties, high Curie temperature, high dielectric constant, low dielectric loss, high chemical stability, and photocatalytic activity [55–57].

Among the different polymorphs, the perovskite ($Bi_4Ti_3O_{12}$) and selenite ($Bi_{12}TiO_{20}$) phases were more widely studied. $Bi_4Ti_3O_{12}$ is a ferroelectric material characterized by the structure in the form of layers of the perovskite-type $(Bi_2Ti_3O_{10})^{+2}$ interspersed with layers of $(Bi_2O_2)^{-2}$ belonging to the family of materials discovered by Aurivillius in 1949, as shown in Fig. 6. $Bi_4Ti_3O_{12}$ is a material with potential applications in the electronics industry, such as capacitors, sensors, memory storage devices, optical displays, and other electro-optical devices [51].

$Bi_{12}TiO_{20}$ is a crystalline material with a body-centered cubic structure, belonging to the selenite family. The structure of $Bi_{12}TiO_{20}$ is formed by Bi–O polyhedra, in which the Bi ions are bonded to five oxygen ions forming an octahedral arrangement together with a pair of electrons from Bi^{3+}. The polyhedral network is, in turn, connected to TiO_4 tetrahedrons, as shown in Fig. 7. In the $Bi_{12}TiO_{20}$ structure, the most likely intrinsic defects are the Ti^{4+} sites occupied by a Bi^{3+} ion associated with a hole (h^+). This defect is considered the main responsible for a *p*-type conduction, in which this pristine and/or doped semiconductor showed good photocatalytic activity in the degradation of more complex organic pollutants, as reported by Nogueira et al., and Wang et al. [59, 60].

The different polymorphs of bismuth titanate can be obtained by changing the stoichiometry between bismuth and titanium (Bi:Ti). And several routes have already been used in their syntheses, such as the sol–gel method [60], hydrothermal

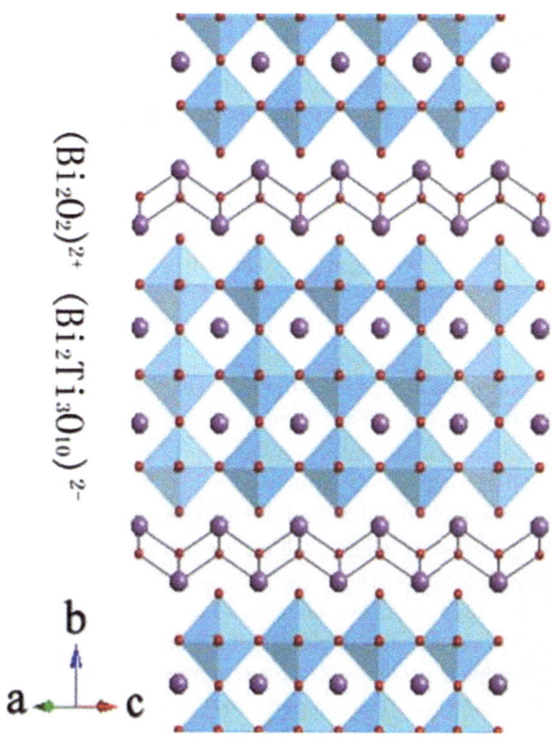

Fig. 6 Crystalline structure model of $Bi_4Ti_3O_{12}$ made up of perovskite-like layers $(Bi_2Ti_3O_{12})^{-2}$ intercalated by $(Bi_2O_2)^{+2}$ sheets. Adapted from Ref. [58], with permission from Copyright © 2017, Elsevier

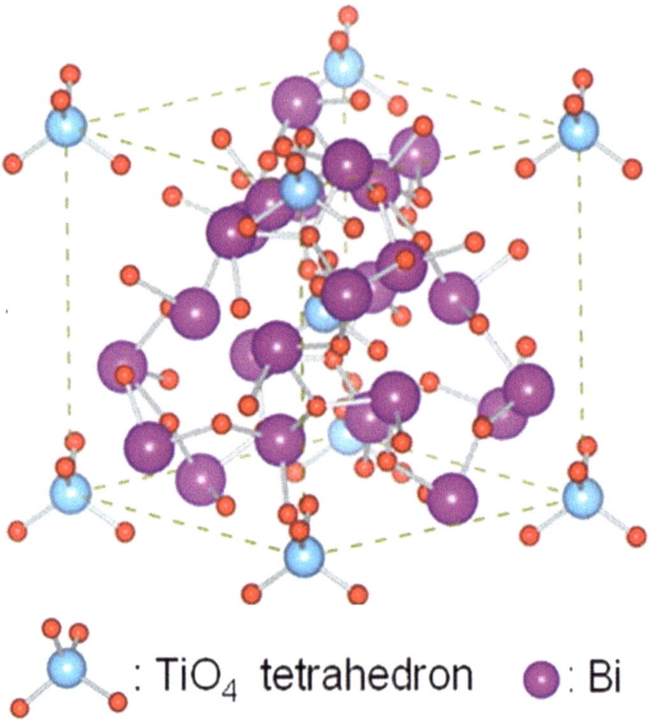

Fig. 7 Crystalline structure model of $Bi_{12}TiO_{20}$ made up of Bi–O polyhedral and TiO_4 tetrahedrons. Adapted from Ref. [61], with permission from Copyright © 2013, American Chemical Society

method, solid-state reaction [62], molten salt method [55] and even routes assisted by hydrogen peroxide [63]. Through these routes, crystalline and nanometric powders with the desired phase are obtained. However, the precursors used provide some drawbacks such as the use of halides, alkoxides, and carbonates, which are groups that are difficult to eliminate and that directly interfere in the final properties of the synthesized materials.

Nogueira et al. demonstrated that the oxidant peroxo method (OPM) is efficient for the synthesis of different structures of bismuth titanate, with potential for application in photocatalytic processes for degradation of organic pollutants [51]. In which, the main innovation in this route was the possibility of obtaining semiconductors of great technological and commercial interest, free of any type of contamination by carbon and halides, which are extremely deleterious for catalytic and electro-electronic applications. Another important feature of this synthesis method, which represented enormous energy and environmental gain is the low temperature and a shorter processing time used for the crystallization of nanometric oxides.

In this method, the Ti-peroxocomplex synthesized in situ acts in a similar way to H_2O_2 molecules. The Ti-peroxo complex is obtained from metallic titanium together with a solution of hydrogen peroxide and ammonia, forming the peroxotitanate ion $[Ti(OH)_3O_2]^-$, which has a characteristic yellow color of peroxocomplexes.

The mechanism of formation of the peroxotitanate ion was previously described in Eq. (1) (see Sect. 1).Then, when the bismuth precursor solution (bismuth nitrate) comes in contact with the Ti-peroxo complex solution, the pH rises and $Bi(OH)_5$ formation occurs.The Bi^{+5} is reduced to Bi^{+3} by an exothermic process in the presence of H_2O_2, forming amorphous bismuth oxide (Bi_2O_3) according to Eq. (8).

Simultaneously, the hydrolysis of the Ti-peroxo complex occurs, forming $Ti(OH)_4$, due to the formation of H_2O by the oxidation of hydrogen peroxide according to Eq. (9), producing a precipitate, which after a thermal treatment, resulting in crystalline bismuth titanate [59].

$$Bi(OH)_{5(s)} + H_2O_{2(aq)} \rightarrow Bi_2O_{3(s)} + H_2O_{(l)} + OH^-_{(aq)} \quad (8)$$

$$[Ti(OH)_3O_2]^-_{(aq)} + H_2O_{(l)} \rightarrow Ti(OH)_{4(s)} + O_{2(g)} + OH^-_{(aq)} \quad (9)$$

Nogueira et al. showed that the precipitate from the synthesis of bismuth titanate via OPM has irregular clusters of nanoparticles together with well-defined spherical particles. The elemental mapping through EDS microanalysis showed a high concentration of bismuth in the spherical particles, and titanium in particles with no defined morphology is related to domains of bismuth (Bi_2O_3) and $Ti(OH)_4$ oxides respectively (Fig. 8) [64, 65]. After heat treatment of the precipitate, a diffusion process occurs between the particles forming a partially sintered material. Pinto et al. reported

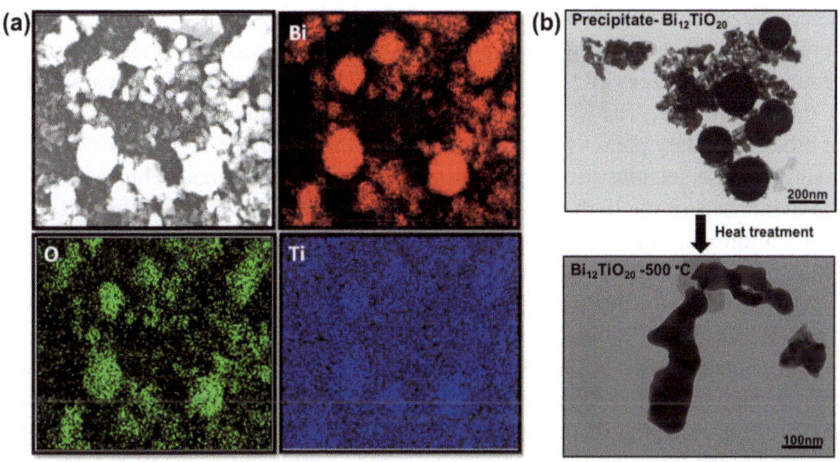

Fig. 8 **a** SEM micrograph and the corresponding distribution map of the elements bismuth (red), oxygen (green), and titanium (blue) of the precipitate of $Bi_{12}TiO_{20}$. Adapted from Ref. [64], with permission from Copyright © 2014, Springer Nature. **b** TEM image of the precipitate before (top) and after (bottom) heat treatment at 500 °C for 1 h. Adapted from Ref. [51], with permission from Copyright © 2014, Elsevier

that one of the main characteristics of the powders obtained by the OPM route is its high reactivity, which can be verified by the sintering and presence of necks between the particles [46].

Since the high reactivity of the powders from the OPM route was related to the presence of oxygenated groups on the surface of $Ti(OH)_4$, Francatto et al. [66] studied the reactivity of $Ti(OH)_4$ obtained by the OPM route in a solid-state reaction using commercial Bi_2O_3 to obtain bismuth titanate.

To obtain the $Ti(OH)_4$ nanoparticles Francatto et al. used the same principle of the OPM route, which is by the hydrolysis of the Ti-peroxo complex through the formation of water by the oxidation of H_2O_2 (Eq. 9). However, instead of the H_2O_2 oxidation process taking place through a redox reaction, the heating of the Ti-peroxo complex solution was carried out. Leading to the decomposition of H_2O_2 and the formation of O_2 and H_2O, causing the hydrolysis of the peroxo complex and the consequent formation of the precipitate [66, 67].

The $Ti(OH)_4$ precipitate formed showed a yellow color, which is related to the presence of peroxo groups on the material surface, formed by the excess of hydrogen peroxide in the reaction medium. The formation of these groups on the surface of some transition metals of groups V and VI is well known, especially in reactions in a homogeneous medium [68, 69].

Thus, to confirm this hypothesis, a solid-state reaction was carried out between commercial Bi_2O_3 with three different types of titanium precursor: commercial titanium dioxide, $Ti(OH)_4$ obtained by OPM, and a third one obtained by reducing the surface of $Ti(OH)_4$ to eliminate superficial peroxo groups [66]. Each one of the titanium precursors was mixed stoichiometrically with Bi_2O_3 to obtain the $Bi_{12}TiO_{20}$ phase. Then, the resulting powder mixture was placed in a ball mill for 24 h for homogenization and taken to a heat treatment at 600 °C for 1 h.Francato et al. demonstrated that the materials synthesized with the precursors without the peroxo groups on the surface (TiO_2-COM and TiO_2-RED) showed diffraction peaks related to the selenite ($Bi_{12}TiO_{20}$) structure of bismuth titanate together with the presence of other secondary phases [66]. However, when the precursor TiO_2-OPM was used, a single-phase selenite structure of bismuth titanate was formed. Thus confirming the hypothesis of the high reactivity of the peroxo group present on the surface of titanium dioxide. Figure 9 shows the XRD pattern of the bismuth titanate phases obtained from $Ti(OH)_4$ with peroxide groups in the surface (BT-OPM), from the $Ti(OH)_4$ without the peroxide groups in the surface (BT-RED), and the commercial TiO_2 (BT-COM).

The same behavior of the high reactivity of $Ti(OH)_4$ obtained by the OPM route was demonstrated by Ribeiro et al. in the synthesis process of lithium titanate ($Li_4Ti_5O_{12}$), in which the crystalline material was obtained in a shorter time than reported in the literature [70].

In the heat treatment process, the peroxo groups present on the surface of the titanium precursor are removed through an exothermic process, which locally releases a large amount of energy that accelerates the reaction, which facilitates the formation of the final product at lower calcination times and temperatures (Fig. 10).

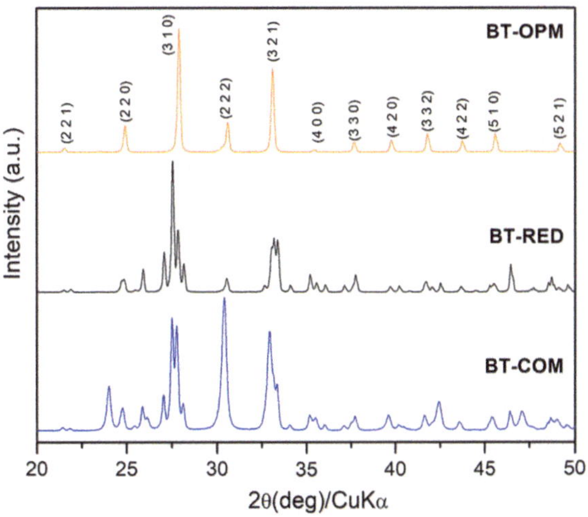

Fig. 9 XRD patterns of three different mixtures of Bi_2O_3 and titanium precursors (TiO_2-OPM, TiO_2-Red, and TiO_2-Com) calcined at 600° C for 1 h. The Miller indices shown in the BT-OPM pattern are related to the selenite phase $Bi_{12}TiO_{20}$. Adapted from Ref. [66], with permission from Copyright © 2016, Elsevier

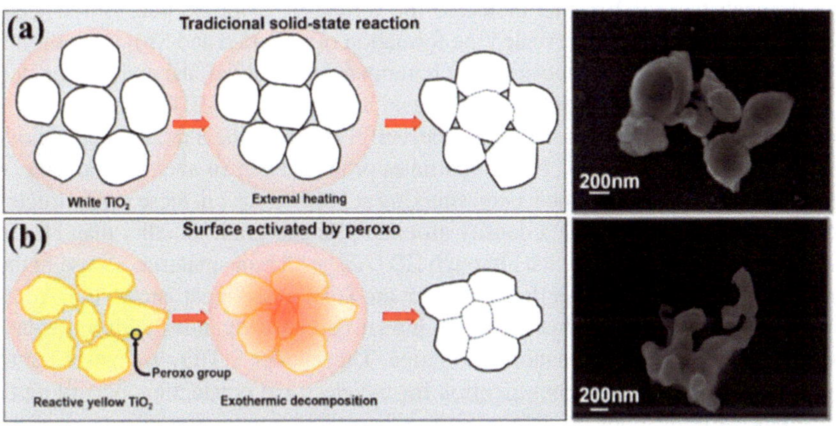

Fig. 10 A general schematics for the particle formation using **a** the solid-state reaction and **b** reactions involving peroxo-modified surfaces. In this case, the local decomposition is exothermic and hence releases enough energy to accelerate the reaction at lower temperatures. Adapted from Ref. [66], with permission from Copyright © 2016, Elsevier

In summary, obtaining bismuth titanate through the OPM route makes it possible to obtain these materials in milder conditions. It also facilitates the control of the size, morphology, and sinterability of the particles of the materials obtained. Besides, it is essential to emphasize that this approach can be extended to other elements present in the composition of various materials of technological importance, such as niobium, tungsten, vanadium, and zirconium, which are elements capable of forming "peroxo groups" on the surface of their oxides.

3.3 Other Metallic Titanates

In the area of electronic applications, strontium titanate ($SrTiO_3$) is an interesting material, since it combines the properties of ionic and electronic conductors [71]. One challenge faced in the preparation of $SrTiO_3$ ceramic materials is to make highly-dense ceramics from nanocrystalline powders. Traditionally, the sol–gel methods are used to prepare $SrTiO_3$ powders [71]. However, the high carbon content leads to the production of porous particles, due to the exothermic elimination of organic materials during the heating process. The presence of porosity renders it impossible to obtain a highly dense ceramic body [71].

To reach the goal to prepare highly dense $SrTiO_3$ ceramic bodies from nanopowders, a reasonable step is to use a synthetic method requiring little to none of the organic compounds. And this was the strategy used by Balaya and co-workers by preparing $SrTiO_3$ powders using the OPM method [71]. The authors initially prepared the Ti-peroxo complex starting from metallic Ti, H_2O_2, and $NH_3 \cdot H_2O$. Then, they added $Sr(NO_3)_2$ as a source of strontium.

Although the XRD results indicated the formation of phase pure $SrTiO_3$, the elemental analysis by ICP revealed the formation of a Ti–rich and Sr-poor material. So, to generate a more compositionally homogeneous material, the authors decided to add 1 mol of EDTA for each mole of Ti and Sr, to prepare Ti-EDTA and Sr-EDTA complexes. This strategy proved to be successful in producing a more compositionally homogeneous $SrTiO_3$. The nanoparticles produced had an average diameter of 30 nm according to TEM. Then, upon sintering at 1000 °C for 1 h, these nanoparticles formed ceramic bodies with a density around 93%, and grains smaller than 80 nm [71]. These results revealed that although EDTA is an organic material, its use in the amount specified in the paper does not harm the $SrTiO_3$ densification process. Additionally, the OPM method succeeded in preparing dense ceramics with individual grains with the size in the nanometric range. The smaller $SrTiO_3$ grain size led to the elimination of the bulk contribution for the electrical conduction, according to impedance spectroscopy, in comparison to the $SrTiO_3$ containing microcrystalline grains (around 2500 nm).

Still, about the $SrTiO_3$, Wang and Chen used the OPM method to prepare the SrO_2 [72]. Then, they used this SrO_2 to react with commercial TiO_2 to produce $SrTiO_3$ through conventional solid-state reaction, by heating the mixture of the oxides at 700 °C for 10 h. The reaction produced a small amount of the $SrCO_3$ as a byproduct, which could be eliminated from the $SrTiO_3$ by washing the powder with a 1 mol/L HNO_3 aqueous solution. After this washing process, the $SrTiO_3$ phase purity was confirmed by XRD and XPS, whereas SEM data revealed the particles had cube-like morphology with length in of the exposed facet ranging between 125 and 320 nm.

Multilayer ceramic capacitors are known as one of the most common types of components in electronics, a material commonly used to prepare these capacitors is barium titanate ($BaTiO_3$), due to its high dielectric constant and low dielectric loss [73]. Buscaglia et al. used the OPM method to prepared amorphous TiO_2 particles, which were coated by a $BaCO_3$ shell [73]. The synthesis started with the addition of $TiCl_4$ in water to produce $TiOCl_2$. Then, H_2O_2 and $NH_3{\cdot}H_2O$ were added to the $TiOCl_2$ solution to produce the Ti-peroxo complex. Although the OPM was initially designed not to use chlorides, the use of $TiCl_4$ in this paper did not cause any harm to the formation of the Ti-peroxo complex.

The $BaCO_3$ was added to the stable Ti-peroxo complex solution, with the Ba:Ti mole ration equal 1:1. This suspension was heated at 95 °C for 5 h, producing a white precipitate. The authors confirmed by XRD, TEM, and Raman spectroscopy that this precipitate was a core amorphous TiO_2 covered by a shell of $BaCO_3$ ($BaCO_3@TiO_2$). The $BaCO_3@TiO_2$ was converted to phase pure $BaTiO_3$ by heating it at 700 °C for 1 h. The TEM micrographs of the $BaCO_3@TiO_2$, of the $BaTiO_3$ and a scheme of the $BaTiO_3$ formation process, are shown in Fig. 11.

This core–shell approach presented some advantages in relation to the conventional solid-state mixture and heating of $BaCO_3$ and TiO_2 to produce $BaTiO_3$. The first one is the fact that the interface created in the nanometer scale between $BaCO_3$ and TiO_2 led to a more compositionally controlled $BaTiO_3$. Second, the $BaTiO_3$ produced via the $BaCO_3@TiO_2$ core–shell had a higher surface area than the $BaTiO_3$

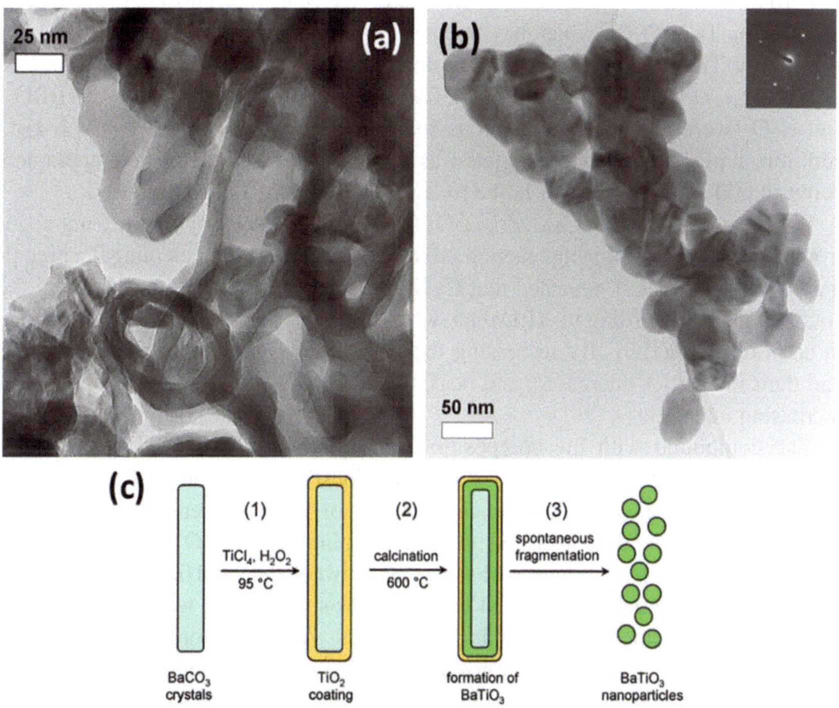

Fig. 11 TEM micrographs of **a** $BaCO_3@TiO_2$ nanostructures calcined at 550 °C for 1 h, **b** $BaTiO_3$ nanocrystals obtained from $BaCO_3@TiO_2$ calcination at 700 °C for 1 h, with a single particle electron diffraction pattern, and **c** scheme of synthetic process starting from $BaCO_3$, leading to $BaTiO_3$ nanoparticles. Adapted from Ref. [73], with permission from Copyright © 2007, American Chemical Society

produced by the conventional solid-state reaction. Finally, lower temperatures and shorter times were required to produce phase pure $BaTiO_3$ through the core–shell approach.

Zinc titanates have industrial applications like pigments, catalysts, sensors, and adsorbents for desulfurization [74]. However, to obtain compositionally pure zinc titanate is not straightforward, since the $ZnO–TiO_2$ phase diagram predicts the existence of the following compounds: $ZnTiO_3$ (cubic, hexagonal), Zn_2TiO_4 (cubic, tetragonal) and $Zn_2Ti_3O_8$ (cubic). To study the mechanism of formation of different zinc titanate phases in solution-based methods, Nikolenko et al. prepared reaction mixtures containing $TiOCl_2$, $Zn(NO_3)_2$, and H_2O_2 in distilled water, then, they added 0.1 mol/L KOH to this mixture to cause precipitation [74]. It was figured out that a pH of about 8.5 is suitable for precipitating zinc and titanium hydroxides with H_2O_2. This precipitate crystallizes as cubic $ZnTiO_3$ in temperatures around 580 to 690 °C. However, when the temperature increases to the 800 to 900 °C range, the $ZnTiO_3$ decomposes to Zn_2TiO_4 and TiO_2 [74].

Photoluminescent materials are another application field where metallic titanates are actively used. In this sense, Wanjun and Donghua took advantage of the Ti-peroxo complexes preparation to synthesize the red phosphor compound $[(Ca_{1-x}Zn_x)TiO_3{:}Pr,B]$, which has an emission peak at 614 nm [75].

Initially, $TiO(OH)_2$ was prepared in situ by the hydrolysis of tetra-*n*-butyl titanate ($Ti(OC_4H_9)_4$) (TTIP). Then, $Pr(NO_3)_3$, $Ca(NO_3)_2{\cdot}4H_2O$, $Zn(NO_3)_2{\cdot}6H_2O$, H_3BO_3, and H_2O_2 were added to the aqueous solution containing TTIP. Upon $NH_3{\cdot}H_2O$ addition, at pH 9, a yellow precipitate was formed. After calcination at temperatures between 600 and 900 °C for 1 h, the $[(Ca_{1-x}Zn_x)TiO_3{:}Pr,B]$ was produced.

For the composition $[(Ca_{1-x}Zn_x)TiO_3{:}0.001Pr,0.1B]$, by varying the Zn amount from $x = 0$ to $x = 0.3$ during reaction mixture preparation, and calcining the precipitate at 800 °C for 1 h, revealed that $CaTiO_3$ is the major phase throughout all the studied range. According to XRD data, when zinc is absent ($x = 0$), TiO_2 is formed in addition to $CaTiO_3$. By increasing the zinc content, the peaks related to TiO_2 had their intensities decreased, and peaks related to $ZnTiO_3$ started to appear with increasing intensity.

The compound with the composition $[(Ca_{0.85}Zn_{0.15})TiO_3{:}0.001Pr,0.1B]$ ($x = 0.15$) was the one present the highest photoluminescent emission peak at 614 nm.

Aiming to enhance the n-type conductivity of $LaCoO_3$ perovskites semiconductors, Robert et al. performed the Ti^{+4} doping of $LaCoO_3$ [76], producing $La(Co_{1-x}Ti_x)O_{3\pm\delta}$ compounds, with x varying between 0.01 and 0.5. The production of the Ti-peroxo complex followed by its further stabilization with citrate groups, was essential for the successful doping process. After the formation of the Ti-peroxo complex $[Ti(OH)_3O_2]^-$, citric acid ($C_6H_8O_7$) was added in the mole ratio $Ti{:}C_6H_8O_7 = 1{:}5$, converting $[Ti(OH)_3O_2]^-$ to $[Ti(C_6H_5O_7)O_2]^{-2}$. After that, the pH was further adjusted to 6–7 by adding more citric acid. Then, another solution containing $La(NO_3)_3{\cdot}6H_2O$, $Co(NO_3)_2{\cdot}6H_2O$, and citric acid, with pH 1 was added to the solution containing the $[Ti(C_6H_5O_7)O_2]^{-2}$ complex. The final mole ratio between citric acid and metallic ions was 2:1.

This step produced a gel that was isolated and heated at 300 °C, forming a brown product, which was further calcined at 1100 °C for 12 h, producing the $La(Co_{1-x}Ti_x)O_{3\pm\delta}$. The actual composition, determined by X-ray fluorescence (XRF) or EDS was very close to the nominal composition. And only for the sample with x = 0.1, TiO_2 was present as an impurity. From x between 0.1 and 0.3, the particle size increased with increasing Ti content. Whereas for x between 0.4 and 0.5 there was a decrease in the particle size. Electrical resistivity versus temperature measurements on pressed and sintered pellets revealed a semiconductor behavior for all compositions studied.

The advance in the lithium-ion batteries (LIBs) field requires the development of efficient anode materials, a potential candidate is the spinel oxide $Li_4Ti_5O_{12}$. Wang et al. prepared $Li_4Ti_5O_{12}$ starting with TiN as titanium source, with H_2O_2 and $NH_3{\cdot}H_2O$ to form the Ti-peroxo complex [77]. Then, LiOH and varying amounts of PVP were added to the peroxo complex solution. The solution was dried, and the resulting powder was calcined at 750 °C for 7 h. To test the influence of PVP, the authors also prepared samples replacing PVP by glucose, or without the addition of any capping agent. These experiments revealed that PVP was essential to keep the monodisperse microspheres morphology after the calcination process.

Later, Ribeiro and co-workers studied the influence of the peroxo groups on the surface of TiO_2 nanoparticles produced by the OPM in the production of $Li_4Ti_5O_{12}$ via, by reacting TiO_2 with LiOH via solid-state reaction in a ball-milling, followed by heating in temperatures ranging from 650 and 850 °C for time periods between 0.5 and 3 h [70].

This study was accomplished by comparing the TiO_2 produced via OPM method with commercial TiO_2 nanoparticles, and TiO_2 nanoparticles without surface peroxo groups. For these samples, the surface peroxo groups were removed by a heating treatment at 250 °C for 30 min, under H_2 atmosphere. From heat treatment temperature below 800 °C, none of the TiO_2 sources was able to produce phase pure $Li_4Ti_5O_{12}$. All of them presented some TiO_2 impurity after the end of the heating process. However, for temperatures higher than 800 °C, only the TiO_2 produced via OPM method was able to yield phase pure spinel $Li_4Ti_5O_{12}$. The success of this TiO_2 source is attributed to the higher reactivity provided by the presence of the surface peroxo groups.

4 Metallic Zirconates

4.1 Barium Zirconate ($BaZrO_3$) and Related Compounds

Finding new, environmentally friendly energy sources has become one of the main concerns of the scientific community in the last decades. Solid oxide fuel cells (SOFCs) have attracted increased attention worldwide due to their high conversion efficiency of chemical energy into electrical energy, leading to a low environmental impact [78–83]. Those cells are also a sustainable alternative to energy production, compared with other hydrogen-based technologies.

The most developed and already commercially available SOFC is based on oxygen-ion conducting solid electrolytes, and the operational temperature is within the 800–1000 °C range. The commonly used solid electrolyte is yttrium-stabilized zirconia (ZrO_2:8 mol% Y_2O_3), and it is considered state of the art regarding solid electrolytes for SOFCs operated at high temperature. This electrolyte has predominantly ionic conduction at the cell temperature, which can reach 0.1 S/cm at 1000 °C. Yet, there are several problems directly associated with the high operation temperature, which include components thermal expansion incompatibility; the cell poisoning and degradation due to CO and CO_2 formation leading to carbon deposition on the cell (coking). Those problems decrease cell performance and lifetime [79, 80, 82]. Therefore, many efforts have been made to reduce the SOFC working temperature to an intermediate range (400–700 °C, SOFC—IT) [78, 83].

Oxides with perovskite structure based on barium cerate and barium zirconate, and a combination of both (BCx, BZx and BCZx, x = trivalent cation—Y^{3+}, Yb^{3+}, Ga^{3+}, respectively) are the most studied compositions for application as solid electrolytes in SOFC- IT. Nevertheless, among the proposed compounds, Yttrium-doped barium zirconate (BZY) have demonstrated to be the most promising due to its high chemical stability under CO_2-rich atmospheres, mechanical resistance, besides it's relatively high proton conductivity at intermediate temperatures (400–600 °C) [84–89]. Indeed, the bulk conductivity at 500 °C of barium zirconate containing 20 mol% of Y^{3+} in the solid solution (BZY20) is comparable to the one achieved from gadolinium-doped ceria ($Ce_{0.9}Gd_{0.1}O_{2-\delta}$), which is among the bests known oxygen-ion conductor solid electrolytes [86, 89].

The great challenges regarding BZx solid solutions are to achieve dense ceramic bodies, with low intergranular electrical resistivity but using less aggressive sintering conditions (>24 h, >1700 °C, respectively), leading to less barium loss and higher electrical conductivities [88, 90–92].

Many alternatives synthetic and processing techniques have been used to prepare better quality samples. Among the synthetic routes, the oxidant peroxo method (OPM) is one of the most promising since it is free from common contaminants such as graphite carbon or halides [36, 38]. Besides, the OPM allows the formation of stoichiometric powders constituted by reactive nanoparticles, which improves the microstructure for the preparation of dense pellets used as protonic solid electrolytes for SOFCs [39, 45, 46].

4.2 *The Oxidant Peroxo Method for the Synthesis of $BaZrO_3$ Solid-Solutions*

Powder samples of pure and Y-doped barium zirconate have been successfully produced by a different approach to the oxidant peroxo method. Despite the changes, it will simply be referred to as OPM [36, 38]. Initially, the formation of precursor powder by the OPM occurs due to Pb^{+2} exothermic oxidation reaction when in contact with an inorganic-peroxo complex (Zr, Ti, W, among others) in alkaline solution. The formation of crystalline perovskite structure occurs by heat treatment of the amorphous precursor powder, which occurs simultaneously with the reduction of Pb^{+4} with the evolution of oxygen [1]. In the case of BZY, the amorphous precursor formation occurs due to the precipitation of the hydrated cations oxides in alkaline a H_2O_2 solution, since no oxidation state changes during either the synthesis or during further processing [93].

When dissolving $ZrO(NO_3)_2{\cdot}nH_2O$ in water, zirconila ions (ZrO^{+2}) are produced and it can form peroxo complexes.[29] Thus, the addition of $ZrO(NO_3)_2.nH_2O$ into a H_2O_2, described in Eq. (4) (Sect. 3.1), leads to the formation of a stable solution of zirconyl-peroxo complex ($ZrO(O_2)H^+$).

However, ZrO^{+2} ions form a white precipitate of insoluble [$ZrO.nH_2O$] when the pH is alkaline, making it impossible to form a stable and clear solution containing $NH_3{\cdot}H_2O$ and $ZrO(O_2)H^+$. Therefore, for BZ and BZY precursor synthesis, the zirconyl-peroxo complex has to be separately produced and then added to the Ba^{+2} and Y^{+3} cation nitrates solution. Afterward, the resulting cation solution is added to the mixture H_2O_2/ $NH_3{\cdot}H_2O$ for the precipitate of the amorphous precursor. Another change was made regarding the way of which the cation solution was added to the oxidant mixture. Aiming to control Ba^{+2} stoichiometry, the cation solution addition into the $H_2O_2/NH_3{\cdot}H_2O$ mixture was quick (rather than drop by drop in the original method) to minimize the spontaneous reaction between the barium nitrate solution with carbonic acid (H_2CO_3), favored when CO_2 dissolves in water. This reaction allows the formation of $BaCO_3$ during the precursor synthesis, which leads to a deviation of Ba stoichiometry [93].

$$CO_{2(aq)} + H_2O_{(aq)} \leftrightarrow H_2CO_{3(aq)} + 20.27\,\mathrm{kJ/mol} \tag{10}$$

$$H_2CO_{3(aq)} \leftrightarrow HCO^-_{3(aq)} + H^+_{(aq)} + 671.23\,\mathrm{kJ/mol} \tag{11}$$

$$H_2CO_{3(aq)} + Ba(NO_3)_{2(aq)} + 40.57\,\mathrm{kJ/mol} \rightarrow 2HNO_{3(aq)} + BaCO_{3(s)} \tag{12}$$

To avoid further contamination by atmospheric carbon dioxide, the synthesis can be performed in a glove box under nitrogen gas [94]. However, in this case, the water used in the synthesis and for washing the resultant precipitate has to be deionized and previously degassed, by bubbling it with nitrogen gas. Additionally, the drying step is also crucial and has been done at room temperature in a vacuum chamber before exposing the precursor powder to the atmospheric air [94].

Barium zirconate ($BaZrO_3$) and yttrium-doped barium zirconate solid solutions ($BaZr_{1-x}Y_xO_{3-\delta}$), (x = 0.1 to 0.5, where x in mol%), were successfully synthesized by the oxidant-peroxo method by this approach. This improvement in sample quality also allowed the experimental determination of enthalpy of formation insights into its defect chemistry for the above mentioned BZY solid-solutions using high-temperature drop solution calorimetry [94].

The study focused on barium zirconate composition containing 10 mol% of yttrium synthesized by this approach of the OPM demonstrated that the method is suitable for the synthesis of BZY solid-solutions consisting of nanoparticles. These nanoparticles, after appropriate processing, allowed less aggressive sintering conditions for obtaining dense bodies (>90%), with optimized proton conductivity when compared with the same material synthesized under atmospheric air [95]. These improvements were associated with the formation of a precursor powder consisting of nanoparticles in the range 20–50 nm, which has surfaces nearly free of contaminant species and where less agglomerated. Therefore, denser pellets with lower grain boundary density were obtained after sintering at 1200 °C/24 h, reaching $\sigma_T = 1.6 \times 10^{-3}$ at 536 °C [93, 94].

4.3 BZ and BZY Defect Chemistry

Along with the synthetic routes, processing parameters, and morphology optimization, it is very important to understand the BZY defect chemistry since its features are closely related to the incorporation and the mobility of the charge carriers. Thus, affecting the behavior of proton conductivity as a function of temperature, water fugacity, and electrolyte composition [96]. BZY has a perovskite structure, ABO_3, which is ideally cubic where the A-site is usually occupied by a divalent cation (A = Ba^{+2}, Pb^{+2}, Ca^{+2}). And the B-site by a tetravalent cation (B = Zr^{+4}, Ce^{+4}, Ti^{+4}), with coordination numbers 12 and 6, respectively [97].

Therefore, BZY solid-solution stoichiometry is crucial for the existence of the maximum of oxygen vacancies allowed in the structure and the capacity for its hydration for enhanced proton conductivity [98].

There are still many discrepancies related to the properties of the defect structure and hydration of BZY. These discrepancies are partly due to significant variation in the stoichiometry of the material in the published works.

A method of characterization for chemical stability and structural defects of the oxides is to obtain enthalpies of formation as a function of varying the dopant concentration. This characterization can be accomplished by using the high-temperature oxide melt solution calorimetry (*Drop solution calorimetry*) [99–101]. Previous works using this technique on ceria and thoria based fluorite structures correlate the enthalpy of formation with the conductivity behavior of these compounds [102–105].

Attempts to perform drop solution calorimetry on BZY solid-solutions produced by the solid-state method were unsuccessful, once the amount of the $BaCO_3$ contaminant could not be quantified. The high quality of BZYx ($x = 0 - 0.5$ mol% de Y^{3+}) solid-solutions synthesized by the OPM under nitrogen atmosphere, enabled the study using the Drop Solution Calorimetry to determine, for the first time, the empirical enthalpies of formation from the oxides at 25 °C [94]. The thermodynamic data also was correlated with BZY defect chemistry and its implications on the behavior of proton conductivity as a function of dopant concentration. It was found that up to 20 mol% of yttrium is still in a dilute regime in the barium zirconate matrix. The dopant content in the dilute regime, in the barium zirconate matrix, coincides with the maximum proton conductivity, showing the possibility of a defect-defect interaction as the main cause of the decrease in proton mobility above 20 mol% of yttrium [94].

The BZY solid-solutions produced by the OPM were also investigated using Water Adsorption Calorimetry. This technique consists in the combination of a surface area analyzer coupled to a Calvet microcalorimeter [106–108]. The technique allows the direct quantification of the water uptake adsorption in nanostructured materials and the simultaneous measurement of heats of reaction as a function of vapor pressure [107]. The experiments on BZYx, with x = 10 to 30 mol% of yttrium solid-solutions, were performed to systematically investigate the energetics of protonic defects environments and how their stability changes with variations in temperature and dopant concentration [108]. A three-step hydration process was observed, which elucidated a balance between surface bulk defects distribution. The results confirmed the dissociative incorporation of water in the BZY structure.

5 Other Multinary Metallic Oxides

The ability of certain metallic cations to form peroxo complexes is one of the features that signalizes for the possibility to make the OPM method a technique as general as possible. Vanadium ions (V^{+5}) are recognized to be able to form different types of peroxo complexes, depending on the relative concentration between V^{+5} and H_2O_2, and also on the pH. For instance, the mono and bisperoxo vanadium complexes are present when there is an acidic pH, the V^{+5} concentration is in the millimolar range, and there is a small excess of H_2O_2 [109] (Fig. 12).

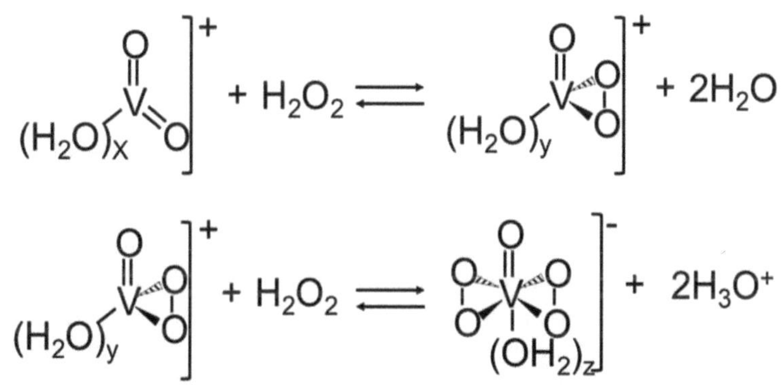

Fig. 12 Reactions of formation of the vanadium mono peroxo complex (top), and vanadium bisperoxo complex (bottom). Adapted from Ref. [109], with permission from Copyright © 2000, Elsevier

Currently, vanadium-based oxides are calling attention due to their optical properties. For instance, $BiVO_4$ is an oxide with a band gap around 2.4–2.8 eV, which lies in the visible range of the electromagnetic spectrum. This feature makes $BiVO_4$ a promising visible-light photocatalyst, for many applications. Such as photocatalytic degradation of pollutants and solar water splitting [110].

By taking advantage of the V^{+5} ions ability to form peroxo complexes, Lopes et al. prepared $BiVO_4$ using the OPM method [111]. They started the synthesis with $Bi(NO_3)_3{\cdot}5H_2O$ and NH_4VO_3 in 40 mL of deionized water, in the Bi^{+3}: V^{+5} mol ratio equal 1:1. Then, H_2O_2 was added in the H_2O_2:metallic ions mole ratio of 10:1. This reaction generated a yellow vanadium peroxo complex, which was subjected to hydrothermal treatment in temperatures ranging from 80 to 160 °C for 12 h.

All the samples crystallized in the monoclinic phase of $BiVO_4$, according to XRD, without secondary phases. SEM results indicated that the particle size decreases with increasing temperature of hydrothermal crystallization. The band gap estimation through Tauc plots revealed that all samples had a band gap of around 2.35–2.45 eV. However, no clear correlation between band gap and hydrothermal treatment time was found.

The $BiVO_4$ efficiency as photocatalyst was tested in the visible-light photocatalytic degradation of methylene blue. These results revealed that, regardless of the temperature of the hydrothermal treatment, the $BiVO_4$ samples degrade around 40 to 50% of the initial methylene blue, in 3 h. In comparison, for the same period, commercial Bi_2O_3 and the yellow vanadium peroxo complex without hydrothermal treatment were able to degrade about 8 and 20%, respectively.

In a subsequent paper, the same group managed to prepared heterostructures of the monoclinic and tetragonal phases of $BiVO_4$ (m-$BiVO_4$/t-$BiVO_4$) [112]. To be able to prepare the tetragonal phase of $BiVO_4$, the authors used V_2O_5, instead of NH_4VO_3 as V^{+5} source. Also, the H_2O_2: metallic ions mole ratio was 5:1, instead of 10:1. Under these conditions, they produced a mixture containing about 75% of the monoclinic phase, 19% of the tetragonal phase, and about 6% of the orthorhombic phase. The formation of this mixture of phases was attributed to the lower solubility of V_2O_5 in water. The presence of an interface between the monoclinic and tetragonal phases was confirmed by HR-TEM, by showing the lattice fringes related to each one of the phases in neighboring nanoparticles.

The m-$BiVO_4$/t-$BiVO_4$ heterostructures presented higher photocatalytic activity for methylene blue degradation than the phase pure monoclinic $BiVO_4$ samples. The heterostructures degrade about 60% of the initial methylene blue, against 20% of the single-phase compounds. This improved activity of the heterostructures was attributed to an efficient separation between the photogenerated electron/hole pairs.

Continuing the application of OPM prepared $BiVO_4$, Lopes et al. prepared Bi_2O_3/$BiVO_4$ heterostructures by growing $BiVO_4$ on the surface of preformed Bi_2O_3 nanoparticles [113]. The authors compared the effect of the Bi_2O_3 particle size by using commercial nanometric and micrometric Bi_2O_3 particles. The V^{+5} source was NH_4VO_3, aimed to prepare monoclinic $BiVO_4$. And the initial mole ratio between Bi^{+3} and V^{+5} varied between 1:1 and 1:2. The XRD patterns revealed a higher percentage of $BiVO_4$ in the heterostructure when the nanometric Bi_2O_3 was used.

Interestingly, the $Bi_2O_3/BiVO_4$ from nanometric Bi_2O_3 presented band gap closer to the Bi_2O_3 band gap, whereas the micrometric $Bi_2O_3/BiVO_4$ presented band gap closer to $BiVO_4$ band gap.

In general, all the heterostructures produced from the nanometric Bi_2O_3 presented better photocatalytic activity for methylene blue degradation. The best activity was obtained for the sample containing 37% of $BiVO_4$ since samples with higher $BiVO_4$ content had phase segregation according to SEM and EDS results.

Niobium oxides are technologically relevant materials due to their potential applications as transparent conductive oxides, as components in dye-sensitized solar cells, photochromic devices, and solid electrolytic capacitors [114]. In particular, $NaNbO_3$ has been applied as photocatalyst for pollutant molecules degradation [115], photocatalyst for H_2 evolution, and CO_2 reduction [116], and highly-dense piezoelectric ceramics [117].

Usually, the solid-state reaction methods tend to produce poorly homogeneous compositionally material, because of sodium volatilization in the high temperatures required to promote the reaction. In this sense, Dey et al. developed a peroxide-based route to prepare the precursor $Na[Nb(O_2)_3]0.2H_2O$, which can further be converted to $NaNbO_3$ upon heating in temperatures as low as 400 °C [118].

The $Na[Nb(O_2)_3]0.2H_2O$ is synthesized by preparing a 1.40×10^{-1} mol/L $Nb_2O_5{\cdot}3H_2O$ solution in a mixture of 40 mL H_2O_2 and 5 mL $NH_3{\cdot}H_2O$. This reaction produces the complex $[Nb(O_2)_4]^{-3}$, as described by Eq. (13):

$$Nb_2O_5 \cdot 3H_2O + 8H_2O_2 + 6NH_3 \rightarrow 2(NH_4)_3\left[Nb(O_2)_4\right] + 8H_2O \tag{13}$$

Then, Na_2CO_3 is added to the solution containing $[Nb(O_2)_4]^{-3}$, at a Na^{+1}: Nb^{+3} mol ratio of 1:1, leading to the production of the precipitate $Na[Nb(O_2)_3]0.2H_2O$, according to the reaction in Eq. (14):

$$\begin{aligned} &2\left\{(NH_4)_3\left[Nb(O_2)_4\right]\right\} + Na_2CO_3 \\ &+ H_2O \rightarrow 2\left\{Na\left[Nb(O_2)_3\right] \cdot 2H_2O\right\} \\ &+ 6NH_3 \uparrow + CO_2 \uparrow + O_2 \uparrow \end{aligned} \tag{14}$$

The authors pointed out that a pH equal 11 is essential to the successful obtaining of $Na[Nb(O_2)_3]0.2H_2O$ since highly alkaline pH can form $Na_8[Nb_6O_{19}]$. Upon $Na[Nb(O_2)_3]0.2H_2O$ calcination at 400 °C for 2 h, the $Na[Nb(O_2)_3]0.2H_2O$ was converted to phase-pure $NaNbO_3$.

Solid-solutions of potassium and sodium niobates, such as $K_xNa_{1-x}NbO_3$ are recognized for being highly piezoelectric lead-free ceramic materials [119]. However, the preparation of $K_xNa_{1-x}NbO_3$ materials via solid-state reaction has some drawbacks. For instance, the high hygroscopicity of potassium carbonate, which can make the grinding or milling steps in the solid-state reaction less efficient. Additionally, it is hard to form dense ceramic pellets from these materials because of the potassium volatilization during sintering [120].

To prepare $K_xNa_{1-x}NbO_3$ materials via solution-based chemical methods, Stavber and co-workers developed an OPM route assisted by citrate or glycine to prepare $K_{0.5}Na_{0.5}NbO_3$ powders [121]. In general, tetra and pentavalent metallic ions, such as Ti^{+4}, Zr^{+4}, Ta^{+5}, and Nb^{+5} are rarely water-soluble. So, to render water-solubility to the Nb^{+5} source, the authors prepared peroxo complexes of Nb^{+5} and bioinspired small molecules, such as citric acid and glycine.

Either for the glycine Nb^{+5} peroxo complex or the citrate Nb^{+5} peroxo complex, the Nb^{+5} source was the Niobium(V) ammonium oxo-tris(oxalate) monohydrate $[(NH_4)_3NbO(C_2O_4)_3 \cdot H_2O]$.

To prepare the citrate Nb^{+5} peroxo complex, citric acid was dissolved in H_2O_2, then, the $[(NH_4)_3NbO(C_2O_4)_3 \cdot H_2O]$ added and heated at 60 °C, after 2 h of heating, a clear yellow solution was obtained. Then $NH_3 \cdot H_2O$ was added to increase the pH to 8.50 to activate the citric acid. After an hour under stirring the citrate Nb^{+5} peroxo complex was formed. Then sodium and potassium acetate in the 1:1 mol ratio were added to the solution containing the complex, and stirred at 60 °C for 4 h, forming a yellow powder. This powder was calcined in temperatures between 450 and 700 °C for 2 h. Phase pure $K_{0.5}Na_{0.5}NbO_3$ powders were formed by calcining from 500 °C, and then, increasing calcination temperature increased the $K_{0.5}Na_{0.5}NbO_3$ particle size. The $K_{0.5}Na_{0.5}NbO_3$ powders calcined at 700 °C had a cuboidal shape, according to SEM images.

For the glycine Nb^{+5} peroxo complex preparation, a similar procedure was performed, simply by replacing citric acid with glycine, and the sources of K^+ and Na^+ were the nitrates, instead of the acetates. Also, the $NH_3 \cdot H_2O$ addition was made to increase the pH until 5.50, in order to activate the glycine. The calcination at 550 °C to form $K_{0.5}Na_{0.5}NbO_3$; however, a minimal amount of an unidentified secondary phase was observed. The $K_{0.5}Na_{0.5}NbO_3$ powders calcined at 700 °C had a mixture of cuboid and rod-like morphologies, according to SEM images.

Continuing the application of OPM method to niobates preparation, Kato and co-workers prepared niobates or tantalates from lanthanide ions (Ln). For example, $LnNbO_4$ and $LnTaO_4$ [122]. First, they prepared the ammonium peroxo complex of Nb or Ta $(NH_4)_3[M(O_2)_4]$ (M: Nb and Ta), starting from $Nb_2O_5 \cdot nH_2O$ or $Ta_2O_5 \cdot nH_2O$, then, they added H_2O_2 and $NH_3 \cdot H_2O$. Immediately after the formation of the $(NH_4)_3[M(O_2)_4]$, the lanthanide nitrates were added, and the mixture was then stirred for 1 h, obtaining amorphous precursors. Phase pure $LaTaO_4$ was obtained by calcining their respective amorphous precursor at 1000 °C. This method also allowed obtaining phase pure $ScNbO_4$, $YNbO_4$, $LaNbO_4$, $HoNbO_4$, and $YbNbO_4$. It shows how generalizable this method is.

6 Future Directions for OPM Method

Although this chapter, up to this point, has identified many different classes of materials and possible applications where the OPM method was successfully used in nanomaterials preparation. There are still opportunities to explore the OPM method further, as the nanotechnology and materials preparation and characterization are ever-evolving fields. So, in this section, we will see examples of situations where the OPM method has been sparsely to never used, and how we believe that the method can contribute to the development of certain fields.

6.1 Synthesis of Tungstates and Molybdates

According to Sergienko, elements from Columns IVB (Ti, Zr, and Hf), VB (V, Nb, and Ta), and VIB (Mo and W) can form peroxo complexes. Most of them, even with different compositions [123, 124]. In this sense, the reader can notice that throughout this chapter, we have not cited any examples where the OPM method was used to prepare tungstates or molybdates. The lack of published papers regarding these materials means that there are many opportunities where the OPM method can be explored. Maybe not all of them will work, but for sure, they worth being tried by groups performing research related to these materials.

Before we dive into the possible applications of each class of these materials, it is essential to be aware of their thermodynamically stable crystalline structures. For tungstates (AWO_4) and molybdates ($AMoO_4$), when the divalent cation A^{+2} is relatively large (i.e., ionic radius > 0.99 angstroms), these compounds crystallize in the scheelite structure. For instance, scheelite is the thermodynamically stable structure for A = Ca^{+2}, Ba^{+2}, Sr^{+2}, Pb^{+2} [125, 126].

When the ionic radius of the A^{+2} is relatively small (i.e., ionic radius < 0.77 angstroms), the AWO_4 and $AMoO_4$ crystallize in the monoclinic wolframite structure [127]. For instance, wolframite is the thermodynamically stable structure for A = Fe^{+2}, Mn^{+2}, Co^{+2}, Ni^{+2}, Mg^{+2}, and Zn^{+2} [125].

In scheelites, the W^{+6} or Mo^{+6} ions adopt tetrahedral coordinations with the oxygen atoms, by forming $[WO_4]^{-2}$ or $[MoO_4]^{-2}$ units, as shown in Fig. 13 [128].

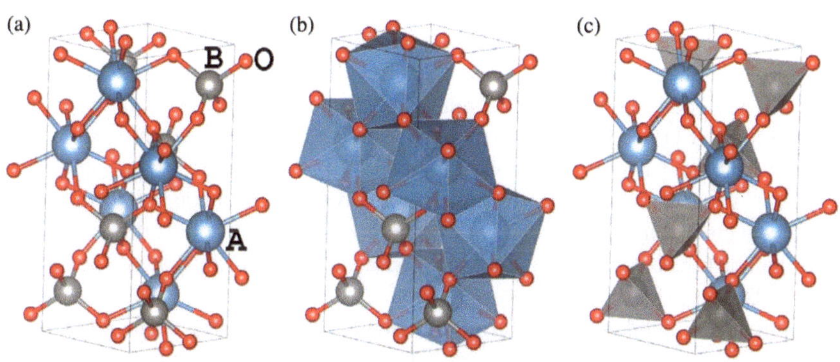

Fig. 13 The scheelite-type structure for ABO_4 (A = Ca, Pb, and B = W, Mo) compound. The $[AO_8]$ and the $[BO_4]$ approximate polyhedra in the crystalline structure are shown in the panels **b** and **c**, respectively. Adapted from Ref. [128], with permission from Copyright © 2015, Elsevier

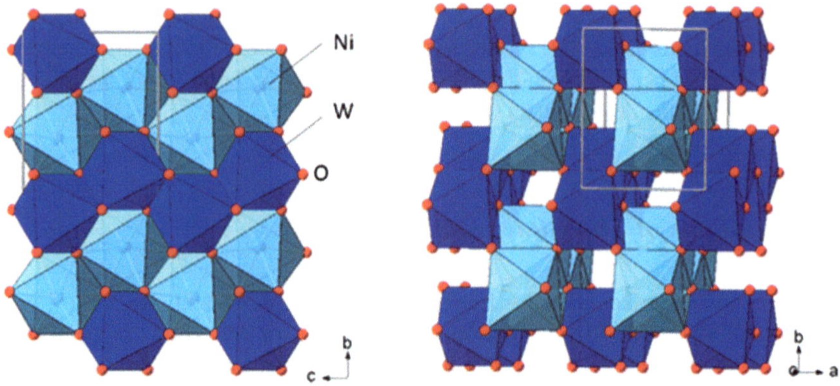

Fig. 14 Two views of $NiWO_4$, which adopts the wolframite structure. The edge-sharing chains are most evident when looking down the a-axis, as shown on the left, while corner-sharing connectivity of neighboring chains can be more clearly seen looking down the c-axis, as shown on the right. Adapted from Ref. [129], with permission from Copyright © 2014, American Chemical Society

In wolframites, the A^{+2} and W^{+6} cations are coordinated by six oxygen ligands in a distorted octahedral geometry. As it can be observed for the $NiWO_4$, in Fig. 14 [129].

In general, the wolframites present pronounced d-d electronic transitions in the visible range of the electromagnetic spectrum. Hence, the energy of this transition varies according to the type of A^{+2} cation. Consequently, it leads to compounds presenting absorption spectra in different wavelength values within the visible range, which, ultimately, opens up space for the development of pigments in a variety of colors. Figure 15 illustrates how different the colors of AWO_4 (A = Mg^{+2}, Mn^{+2}, Co^{+2}, Ni^{+2}, Cu^{+2}, and Zn^{+2}) are [129].

Another important application of the tungstates AWO_4 is as phosphors materials. For instance, Ryu et al. prepared AWO_4 (A = Ca^{+2}, Sr^{+2}, Ba^{+2}, and Pb^{+2}) nanocrystalline compounds, exhibiting blue emission when excited by a 240 nm radiation [130]. Additionally, the tungstates with the formula $L_2(WO_4)_3$, where L is a trivalent lanthanide cation, is another prevalent category of phosphor materials. The undoped $L_2(WO_4)_3$ also emit in the blue region of the visible spectrum when excited by UV radiation. However, these materials are very versatile regarding doping with other trivalent lanthanide cations besides the ones already present in the $L_2(WO_4)_3$ structure. Consequently, different doping cations and quantities may lead to different emission wavelengths. For a comprehensive review of the different types of lanthanide tungstates, the readers are encouraged to read the paper by Kaczmarek and Van Deun [131].

About the molybdates with formula $AMoO_4$, a compelling application is as cathode materials in lithium rechargeable batteries. As such, the fundamental idea around this application is the fact that molybdenum can form a Mo^{+6}/Mo^{+4} redox pair, in which the change in two oxidation state units is attractive for the development of the batteries with high energy density [132]. Additionally, the $AMoO_4$ can work

Fig. 15 Color of the pigments AWO_4 (A = Mg^{+2}, Mn^{+2}, Co^{+2}, Ni^{+2}, Cu^{+2}, and Zn^{+2}. Adapted from Ref. [129], with permission from Copyright © 2014, American Chemical Society

as host lithium-ion insertion, and the combination between A^{+2} and Li^{+1} ions can take to the production of other multinary molybdates [132].

A good starting point to adapt to develop a procedure for preparing tungstates and molybdates using the OPM method resides in the selection of an appropriate source of tungsten and molybdenum for the tungsten or molybdenum-peroxo complex. In this sense, tungstic acid (H_2WO_4) or molybdic acid (H_2MoO_4), which are the monohydrated forms of WO_3 and MoO_3, respectively, can be viable reagents since H_2WO_4 and H_2MoO_4 are soluble in $NH_3{\cdot}H_2O$, although insoluble in water [133, 134]. From the proper precursor selection, the researchers may need to optimize other stoichiometric and heat treatment parameters to obtain the pure desired crystalline phase.

6.2 Morphological Control on Multinary Oxides

Although a reasonable morphological control has been obtained for binary oxides, for instance, TiO_2, as shown in Sect. 2.2. This level of control still lacks in the case of the multinary oxides. As the high reactivity of the oxides prepared by the OPM method tends to form nearly-spherical particles, most of the time, as shown in Fig. 16, for different materials prepared by OPM.

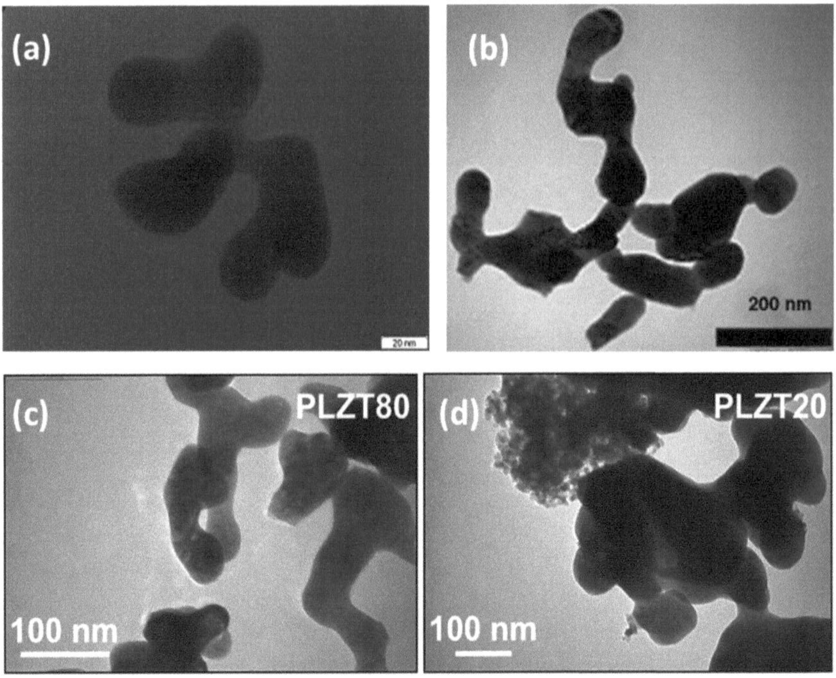

Fig. 16 TEM micrographs of **a** $(Pb_{0.75}La_{0.25})TiO_3$ calcined at 900 °C for 1 h, **b** $Pb(Zr_{0.50}Ti_{0.50})O_3$ calcined at 700 °C for 1 h, **c** $(Pb_{0.95}La_{0.05})(Zr_{0.20}Ti_{0.80})O_3$ calcined at 700 °C for 2 h, **d** $(Pb_{0.95}La_{0.05})(Zr_{0.80}Ti_{0.20})O_3$ calcined at 700 °C for 2 h. Adapted from Refs. [45, 135, 136], with permission from Copyright © 2010, 2009, 2017, Elsevier

Intending to obtain morphological control on the nanoparticle of multinary oxides, a path to be explored is the use of surfactant and capping agents during some step of the synthetic procedure. To be general about the surfactant and capping agents, from this point, we will not make a distinction between them. So, we will simply call them morphology-control agents.

For a generic multinary oxide with formula ABO_n, usually, the most recommended moment where the morphology-control agents could be added is after the formation of the peroxo complex of the cation B. For instance, in the case of the $PbTiO_3$, the Ti-peroxo complex ($[Ti(OH)_3O_2]^-$) is present in solution, then a source of Pb^{+2} ions, for instance, $Pb(NO_3)_2$, is added to form the $PbTiO_3$ amorphous precursor. So, the morphology-control agent could be added together with the $Pb(NO_3)_2$ solution.

The nature of this morphology-control agent is something that will need some optimization from the researcher. However, we can say different categories of morphology-control agents could be tested, for example, chelating agents, such as EDTA or ethylenediamine. In this case, the purpose would be that the chelating agent could coordinate Pb^{+2} and/or Ti^{+4}, then it would make the nucleation and growth rates slower during the precipitation process. Consequently, smaller particles or particles with anisotropic morphology may be obtained.

Anionic or cationic surfactants also could be used. However, to select between anionic and cationic surfactants, it is necessary a more detailed study regarding the point of zero charge (PZC) or the isoelectric point (IEP) of the amorphous precursors produced. For instance, Mendonça et al. studied the PZC of TiO_2 nanoparticles produced via hydrothermal methods at different pH values [7]. They found that the IEP ranged between 3.5 and 5.0. Considering these values as a starting point, if the amorphous precipitate had a similar surface charge than the one from TiO_2 nanoparticles, we could infer that at neutral pH, these nanoparticles would be negatively charged. So, in this case, a cationic surfactant would be more appropriate. Nevertheless, we reinforce that still there space for a detailed study about the PZC and IEP of the amorphous precursor, and this study should be a guide for the proper selection of a surfactant.

Another strategy to control the morphology of multinary oxides is the use of hard-templating methods. Hard templates are rigid materials that may control the morphology of multinary oxides by serving like nano or macro-scaled molds. In other words, the presence of the hard template would make the desired multinary oxide to acquire the morphological features determined by the shape and size of the template.

A commonly used hard template is the mesoporous carbon, which usually has a well-defined pore-size. In these procedures, the pores of mesoporous carbon are replicated in the desired material, which can lead to the production of mesoporous multinary oxides prepared from OPM method [137].

The obtaining of mesoporous multinary oxides would create opportunities for the application of these materials as catalysts for organic reaction and in nanofiltration techniques, for example. One disadvantage of using hard templates is that, in general, an additional step is necessary to remove the template without damaging the multinary oxide.

6.3 *Thin-Film Production*

The preparation and characterization of thin-films of multinary oxides from OPM method is still an underexplored area. Dispersions of the amorphous precursors could be prepared and deposited onto substrates by using conventional solution-based methods for thin-film production, for instance, spin-coating, dip-coating, and drop-casting.

An array of new opportunities and research fields can arise from this idea. Since, many steps for the selection of substrate, solvent for dispersion deposition, heat treatment steps, and thin-film characterization are necessary to obtain a thin-film. The successful thin-film obtaining would open opportunities for the application of these multinary oxides as components of electronic devices, for instance, capacitors or piezoelectric sensors.

7 Concluding Remarks

From this chapter, we can conclude that since its introduction in 2001, the OPM method has been able to produce different types of metal oxides, mainly by taking advantage of the formation of various transition metal peroxo complexes.

The original OPM route has appropriately been modified by many different ways to accommodate specific needs in the production of the materials. And these materials have been applied in varied fields, such as photocatalysis, sensors, and electronic devices.

Although a wide range of materials and applications have been explored, there are opportunities to explore many other materials and applications. Some ideas we presented in this chapter, and we will be glad to see them as the outcome of future papers and patents by any research group willing to take these ideas to fruition. We are sure that the opportunities are unlimited, in such a way that would be impossible to cover all of them in a book chapter.

In conclusion, we hope that this chapter could serve as a guideline about the capabilities and potentials of the OPM technique as a synthetic method for the preparation of transition metal oxides. And that after reading it, we hope the readers can feel motivated to advance further the scope of studies involving this synthetic technique.

Acknowledgements The authors would like to thank Manhattan College for support through startup funds and Faculty Summer Grant, the Functional Materials Development Center from Federal University of Sao Carlos (UFSCar), and the Department of Chemistry from Federal Universiy of Ouro Preto (UFOP).

References

1. Camargo, E.R., Kakihana, M.: Peroxide-based route free from halides for the synthesis of lead titanate powder. Chem. Mater. **13**, 1181–1184 (2001). https://doi.org/10.1021/cm000363y
2. Wang, Y., He, Y., Lai, Q., Fan, M.: Review of the progress in preparing nano TiO2: An important environmental engineering material. J. Environ. Sci. (China) **26**, 2139–2177 (2014). https://doi.org/10.1016/j.jes.2014.09.023
3. Etacheri, V., Di Valentin, C., Schneider, J., Bahnemann, D., Pillai, S.C.: Visible-light activation of TiO_2 photocatalysts: advances in theory and experiments. J. Photochem. Photobiol. C Photochem. Rev. **25**, 1–29 (2015). https://doi.org/10.1016/j.jphotochemrev.2015.08.003
4. Zhang, J., Zhou, P., Liu, J., Yu, J.: New understanding of the difference of photocatalytic activity among anatase, rutile and brookite TiO_2. Phys. Chem. Chem. Phys. **16**, 20382–20386 (2014). https://doi.org/10.1039/c4cp02201g
5. Kumar, S.G., Devi, L.G.: Review on modified TiO_2 photocatalysis under UV/visible light: Selected results and related mechanisms on interfacial charge carrier transfer dynamics. J. Phys. Chem. A. **115**, 13211–13241 (2011). https://doi.org/10.1021/jp204364a
6. Ribeiro, C., Barrado, C.M., De Camargo, E.R., Longo, E., Leite, E.R.: Phase transformation in titania nanocrystals by the oriented attachment mechanism: the role of the pH value. Chem. A Eur. J. **15**, 2217–2222 (2009). https://doi.org/10.1002/chem.200801019

7. De Mendonça, V.R., Ribeiro, C.: Influence of TiO_2 morphological parameters in dye photodegradation: a comparative study in peroxo-based synthesis. Appl. Catal. B Environ. **105**, 298–305 (2011). https://doi.org/10.1016/j.apcatb.2011.04.018
8. Gao, Y., Masuda, Y., Peng, Z., Yonezawa, T., Koumoto, K.: Room temperature deposition of a TiO_2 thin film from aqueous peroxotitanate solution. J. Mater. Chem. **13**, 608–613 (2003). https://doi.org/10.1039/b208681f
9. Garcia, A.P., Rocha, I.C.L., Dos Santos, P.H.F., Basegio, T.M., Pereira, M.B., Clemens, F.J., Alves, A.K., Bergmann, C.P.: Low energy consumption synthesis of nanostructured TiO_2 particles by combining oxidant peroxide method and microwave-assisted hydrothermal treatment. J. Nanomater. **2016** (2016). https://doi.org/10.1155/2016/4910536.
10. de Mendonça, V.R., Lopes, O.F., Avansi, W., Arenal, R., Ribeiro, C.: Insights into formation of anatase TiO_2 nanoparticles from peroxo titanium complex degradation under microwave-assisted hydrothermal treatment. Ceram. Int. **45**, 22998–23006 (2019). https://doi.org/10.1016/j.ceramint.2019.07.345
11. Zaleska, A.: Doped-TiO_2: a review. Recent Patents Eng. **2**, 157–164 (2008). https://doi.org/10.2174/187221208786306289
12. Byrne, C., Subramanian, G., Pillai, S.C.: Recent advances in photocatalysis for environmental applications. J. Environ. Chem. Eng. **6**, 3531–3555 (2018). https://doi.org/10.1016/j.jece.2017.07.080
13. Bakar, S.A., Byzynski, G., Ribeiro, C.: Synergistic effect on the photocatalytic activity of N-doped TiO_2 nanorods synthesised by novel route with exposed (110) facet. J. Alloys Compd. **666**, 38–49 (2016). https://doi.org/10.1016/j.jallcom.2016.01.112
14. Bakar, S.A., Ribeiro, C.: Low temperature synthesis of N-doped TiO_2 with rice-like morphology through peroxo assisted hydrothermal route: Materials characterization and photocatalytic properties. Appl. Surf. Sci. **377**, 121–133 (2016). https://doi.org/10.1016/j.apsusc.2016.03.137
15. Bakar, S.A., Ribeiro, C.: Rapid and morphology controlled synthesis of anionic S-doped TiO_2 photocatalysts for the visible-light-driven photodegradation of organic pollutants. RSC Adv. **6**, 36516–36527 (2016). https://doi.org/10.1039/c6ra03819k
16. Bakar, S.A., Ribeiro, C.: A comparative run for visible-light-driven photocatalytic activity of anionic and cationic S-doped TiO_2 photocatalysts: a case study of possible sulfur doping through chemical protocol. J. Mol. Catal. A Chem. **421**, 1–15 (2016). https://doi.org/10.1016/j.molcata.2016.05.003
17. Fagan, R., McCormack, D.E., Dionysiou, D.D., Pillai, S.C.: A review of solar and visible light active TiO_2 photocatalysis for treating bacteria, cyanotoxins and contaminants of emerging concern. Mater. Sci. Semicond. Process. **42**, 2–14 (2016). https://doi.org/10.1016/j.mssp.2015.07.052
18. De Mendonça, V.R., Dalmaschio, C.J., Leite, E.R., Niederberger, M., Ribeiro, C.: Heterostructure formation from hydrothermal annealing of preformed nanocrystals. J. Mater. Chem. A. **3**, 2216–2225 (2015). https://doi.org/10.1039/c4ta05926c
19. de Mendonça, V.R., Avansi, W., Arenal, R., Ribeiro, C.: A building blocks strategy for preparing photocatalytically active anatase TiO_2/rutile SnO_2 heterostructures by hydrothermal annealing. J. Colloid Interface Sci. **505**, 454–459 (2017). https://doi.org/10.1016/j.jcis.2017.06.024
20. Avansi, W., Catto, A.C., Da Silva, L.F., Fiorido, T., Bernardini, S., Mastelaro, V.R., Aguir, K., Arenal, R.: One-dimensional V_2O_5/TiO_2 heterostructures for chemiresistive ozone sensors. ACS Appl. Nano Mater. **2**, 4756–4764 (2019). https://doi.org/10.1021/acsanm.9b00578
21. Zhu, Y., Murali, S., Cai, W., Li, X., Suk, J.W., Potts, J.R., Ruoff, R.S.: Graphene and graphene oxide: synthesis, properties, and applications. Adv. Mater. **22**, 3906–3924 (2010). https://doi.org/10.1002/adma.201001068
22. He, H., Riedl, T., Lerf, A., Klinowski, J.: Solid-state NMR studies of the structure of graphite oxide. J. Phys. Chem. **100**, 19954–19958 (1996). https://doi.org/10.1021/jp961563t
23. Stankovich, S., Piner, R.D., Chen, X., Wu, N., Nguyen, S.T., Ruoff, R.S.: Stable aqueous dispersions of graphitic nanoplatelets via the reduction of exfoliated graphite oxide in the

presence of poly(sodium 4-styrenesulfonate). J. Mater. Chem. **16**, 155–158 (2006). https://doi.org/10.1039/B512799H
24. W. Zhang, V. Carravetta, Z. Li, Y. Luo, J. Yang, Oxidation states of graphene: insights from computational spectroscopy. J. Chem. Phys. **131** (2009). https://doi.org/10.1063/1.3276339
25. Dimiev, A.M., Tour, J.M.: Mechanism of graphene oxide formation. ACS Nano **8**, 3060–3068 (2014). https://doi.org/10.1021/nn500606a
26. Mahmood, N., De Castro, I.A., Pramoda, K., Khoshmanesh, K., Bhargava, S.K., Kalantar-Zadeh, K.: Atomically thin two-dimensional metal oxide nanosheets and their heterostructures for energy storage. Energy Storage Mater. **16**, 455–480 (2019). https://doi.org/10.1016/j.ensm.2018.10.013
27. Li, Y., Wang, S., He, Y.B., Tang, L., Kaneti, Y.V., Lv, W., Lin, Z., Li, B., Yang, Q.H., Kang, F.: Li-ion and Na-ion transportation and storage properties in various sized TiO_2 spheres with hierarchical pores and high tap density. J. Mater. Chem. A. **5**, 4359–4367 (2017). https://doi.org/10.1039/c6ta08611j
28. Naguib, M., Mochalin, V.N., Barsoum, M.W., Gogotsi, Y.: 25th anniversary article: MXenes: a new family of two-dimensional materials. Adv. Mater. **26**, 992–1005 (2014). https://doi.org/10.1002/adma.201304138
29. Zhang, C.J., Pinilla, S., McEvoy, N., Cullen, C.P., Anasori, B., Long, E., Park, S.H., Seral-Ascaso, A., Shmeliov, A., Krishnan, D., Morant, C., Liu, X., Duesberg, G.S., Gogotsi, Y., Nicolosi, V.: Oxidation stability of colloidal two-dimensional titanium carbides (MXenes). Chem. Mater. **29**, 4848–4856 (2017). https://doi.org/10.1021/acs.chemmater.7b00745
30. Naguib, M., Unocic, R.R., Armstrong, B.L., Nanda, J.: Large-scale delamination of multi-layers transition metal carbides and carbonitrides "MXenes." Dalt. Trans. **44**, 9353–9358 (2015). https://doi.org/10.1039/C5DT01247C
31. Anasori, B., Xie, Y., Beidaghi, M., Lu, J., Hosler, B.C., Hultman, L., Kent, P.R.C., Gogotsi, Y., Barsoum, M.W.: Two-dimensional, ordered, double transition metals carbides (MXenes). ACS Nano **9**, 9507–9516 (2015). https://doi.org/10.1021/acsnano.5b03591
32. Wang, R., Wang, S., Zhang, Y., Jin, D., Tao, X., Zhang, L.: Graphene-coupled Ti 3 C 2 MXenes-derived TiO_2 mesostructure: promising sodium-ion capacitor anode with fast ion storage and long-term cycling. J. Mater. Chem. A. **6**, 1017–1027 (2018). https://doi.org/10.1039/c7ta09153b
33. Goh, K.H., Lim, T.T., Dong, Z.: Application of layered double hydroxides for removal of oxyanions: a review. Water Res. **42**, 1343–1368 (2008). https://doi.org/10.1016/j.watres.2007.10.043
34. Zubair, M., Daud, M., McKay, G., Shehzad, F., Al-Harthi, M.A.: Recent progress in layered double hydroxides (LDH)-containing hybrids as adsorbents for water remediation. Appl. Clay Sci. **143**, 279–292 (2017). https://doi.org/10.1016/j.clay.2017.04.002
35. He, J., Shi, H., Shu, X., Li, M.: On the nature of Ti(IV)-pillared layered metal hydroxides prepared from green, water-soluble Ti-peroxide, AIChE J. **61** (2009). https://doi.org/10.1002/aic.12029
36. Camargo, E.R., Popa, M., Frantti, J., Kakihana, M.: Wet-chemical route for the preparation of lead zirconate: an amorphous carbon- and halide-free precursor synthesized by the hydrogen peroxide based route. Chem. Mater. **13**, 3943–3948 (2001). https://doi.org/10.1021/cm010326m
37. Woodward, D.I., Knudsen, J., Reaney, I.M.: Review of crystal and domain structures in the $PbZr_xTi_{1-x}O_3$ solid solution. Phys. Rev. B Condens. Matter Mater. Phys. **72**, 1–8 (2005). https://doi.org/10.1103/PhysRevB.72.104110
38. Camargo, E.R., Frantti, J., Kakihana, M.: Low-temperature chemical synthesis of lead zirconate titanate (PZT) powders free from halides and organics. J. Mater. Chem. **11**, 1875–1879 (2001). https://doi.org/10.1039/b009096o
39. Camargo, E.R., Souza, F.L., Leite, E.R., Kakihana, M.: Structural and electrical characterization of dense lead zirconate titanate ceramics synthesized by the oxidant-peroxo wet-chemical route. J. Appl. Phys. **96**, 2169–2172 (2004). https://doi.org/10.1063/1.1765854

40. Gonçalves, M.D., Souza, F.L., Longo, E., Leite, E.R., Camargo, E.R.: Dielectric characterization of microwave sintered lead zirconate titanate ceramics. Ceram. Int. **42**, 14423–14430 (2016). https://doi.org/10.1016/j.ceramint.2016.06.035
41. Yamada, L.K.: Cerâmicas de PZT modificadas com terras raras sintetizadas usando o método dos complexos oxidantes de peróxidos (OPM) e sinterizadas por micro-ondas (2013)
42. Camargo, E.R., Kakihana, M.: Lead Hafnate ($PbHfO_3$) Perovskite powders synthesized by the oxidant peroxo method. J. Am. Ceram. Soc. **85**, 2107–2109 (2002). https://doi.org/10.1111/j.1151-2916.2002.tb00412.x
43. Camargo, E.R., Barrado, C.M., Ribeiro, C., Longo, E., Leite, E.R.: Nanosized lead lanthanum titanate (PLT) ceramic powders synthesized by the oxidant peroxo method. J. Alloys Compd. **475**, 817–821 (2009). https://doi.org/10.1016/j.jallcom.2008.08.035
44. Erdem, E., Semmelhack, H.C., Böttcher, R., Rumpf, H., Banys, J., Matthes, A., Gläsel, H.J., Hirsch, D., Hartmann, E.: Study of the tetragonal-to-cubic phase transition in $PbTiO_3$ nanopowders. J. Phys. Condens. Matter. **18**, 3861–3874 (2006). https://doi.org/10.1088/0953-8984/18/15/028
45. Pinto, A.H., Souza, F.L., Chiquito, A.J., Longo, E., Leite, E.R., Camargo, E.R.: Characterization of dense lead lanthanum titanate ceramics prepared from powders synthesized by the oxidant peroxo method. Mater. Chem. Phys. **124** (2010). https://doi.org/10.1016/j.matchemphys.2010.08.030
46. Pinto, A.H., Souza, F.L., Longo, E., Leite, E.R., Camargo, E.R.: Structural and dielectric characterization of praseodymium-modified lead titanate ceramics synthesized by the OPM route, Mater. Chem. Phys. **130** (2011). https://doi.org/10.1016/j.matchemphys.2011.06.040
47. Chen, T.Y., Chu, S.Y.: The piezoelectric and dielectric properties of Ca-additive Sm-modified $PbTiO_3$ ceramics intended for surface acoustic wave devices. J. Eur. Ceram. Soc. **23**, 2171–2176 (2003). https://doi.org/10.1016/S0955-2219(03)00017-7
48. Cho, S.B., Noh, J.S., Lencka, M.M., Riman, R.E.: Low temperature hydrothermal synthesis and formation mechanisms of lead titanate ($PbTiO_3$) particles using tetramethylammonium hydroxide: yhermodynamic modelling and experimental verification. J. Eur. Ceram. Soc. **23**, 2323–2335 (2003). https://doi.org/10.1016/S0955-2219(03)00085-2
49. Esquivel-Elizondo, J.R., Hinojosa, B.B., Nino, J.C.: Bi2Ti2O7: It is not what you have read. Chem. Mater. **23**, 4965–4974 (2011). https://doi.org/10.1021/cm202154c
50. Long, C., Du, T., Ren, W.: Significant ion conduction in Cu acceptor-substituted bismuth titanate polycrystalline ceramics. J. Mater. Sci. **55**, 5715–5729 (2020). https://doi.org/10.1007/s10853-020-04431-x
51. Nogueira, A.E., Longo, E., Leite, E.R., Camargo, E.R.: Synthesis and photocatalytic properties of bismuth titanate with different structures via oxidant peroxo method (OPM). J. Colloid Interface Sci. **415**, 89–94 (2014). https://doi.org/10.1016/j.jcis.2013.10.010
52. Zhang, Y., Chen, Z., Lu, Z.: A facile method for the preparation of colored $Bi_4Ti_3O_{12-x}$ nanosheets with enhanced visible-light photocatalytic hydrogen evolution activity, Nanomaterials **8** (2018). https://doi.org/10.3390/nano8040261
53. Chon, U., Jang, H.M., Shin, N.S., Kim, J.S., Ahn, D.C., Kim, Y.S., No, K.: Gd-substituted bismuth titanate film capacitors having ferroelectric reliability and large non-volatile charges. Phys. B Condens. Matter. **388**, 190–194 (2007). https://doi.org/10.1016/j.physb.2006.05.434
54. Megriche, A., Lebrun, L., Troccaz, M.: Materials of $Bi_4Ti_3O_{12}$ type for high temperature acoustic piezo-sensors. Sens Actuators Phys **78**, 88–91 (1999). https://doi.org/10.1016/S0924-4247(99)00223-X
55. Ranieri, M.G.A., Aguiar, E.C., Cilense, M., Simões, A.Z., Varela, J.A.: Syntheses of bismuth titanate templates obtained by the molten salt method. Ceram. Int. **39**, 7291–7296 (2013). https://doi.org/10.1016/j.ceramint.2013.02.091
56. Badge, S.K., Deshpande, A.V.: La_{3+} modified bismuth titanate (BLT) prepared by sol–gel synthesis: structural, dielectric, impedance and ferroelectric studies. Solid State Ionics **347** (2020). https://doi.org/10.1016/j.ssi.2020.115270
57. Anu, K.Y.: Optical and dielectric properties of $Bi_2Ti_2O_7/Bi_4Ti_3O_{12}$ nanocomposite, Mater. Today Proc. **28**, 153–157 (2020). https://doi.org/10.1016/j.matpr.2020.01.467

58. Sun, X., Xu, G., Bai, H., Zhao, Y., Tian, H., Wang, J., Li, X., Han, G.: Hydrothermal synthesis and formation mechanism of single-crystal Auivillius $Bi_4Ti_3O_{12}$ nanosheets with ammonium bismuth citrate ($C_6H_{10}BiNO_8$) as Bi sources. J. Cryst. Growth. **476**, 31–37 (2017). https://doi.org/10.1016/j.jcrysgro.2017.05.010
59. Nogueira, A.E., Longo, E., Leite, E.R., Camargo, E.R.: Visible-light photocatalysis with bismuth titanate ($Bi_{12}TiO_{20}$) particles synthesized by the oxidant peroxide method (OPM). Ceram. Int. **41**, 12073–12080 (2015). https://doi.org/10.1016/j.ceramint.2015.06.024
60. Wang, L., Li, H., Zhang, S., Long, Y., Li, L., Zheng, Z., Wu, S., Zhou, L., Hei, Y., Luo, L., Jiang, F.: One-step synthesis of $Bi_4Ti_3O_{12}/Bi_2O_3/Bi_{12}TiO_{20}$ spherical ternary heterojunctions with enhanced photocatalytic properties via sol-gel method. Solid State Sci. **100**, 106098 (2020). https://doi.org/10.1016/j.solidstatesciences.2019.106098
61. Hu, D., Kong, X., Mori, K., Tanaka, Y., Shinagawa, K., Feng, Q.: Ferroelectric mesocrystals of bismuth sodium titanate: formation mechanism, nanostructure, and application to piezoelectric materials. Inorg. Chem. **52**, 10542–10551 (2013). https://doi.org/10.1021/ic4015256
62. Badge, S.K., Deshpande, A.V.: Study of dielectric and ferroelectric properties of bismuth titanate (Bi_4 Ti_3O_{12}) ceramic prepared by sol-gel synthesis and solid state reaction method with varying sintering temperature. Solid State Ionics **334**, 21–28 (2019). https://doi.org/10.1016/j.ssi.2019.01.028
63. Hou, J., Qu, Y., Krsmanovic, D., Kumar, R.V.: Peroxide-based route assisted with inverse microemulsion process to well-dispersed $Bi_4Ti_3O_{12}$ nanocrystals. J. Nanoparticle Res. **12**, 1797–1805 (2010). https://doi.org/10.1007/s11051-009-9737-4
64. Nogueira, A.E., Lima, A.R.F., Longo, E., Leite, E.R. E.R. Camargo, Structure and photocatalytic properties of Nb-doped $Bi_{12}TiO_{20}$ prepared by the oxidant peroxide method (OPM), J. Nanoparticle Res. **16** (2014). https://doi.org/10.1007/s11051-014-2653-2
65. Nogueira, A.E., Lima, A.R.F., Longo, E., Leite, E.R., Camargo, E.R.: Effect of lanthanum and lead doping on the microstructure and visible light photocatalysis of bismuth titanate prepared by the oxidant peroxide method (OPM). J. Photochem. Photobiol. A Chem. **312**, 55–63 (2015). https://doi.org/10.1016/j.jphotochem.2015.07.012
66. Francatto, P., Souza Neto, F.N., Nogueira, A.E., Kubo, A.M., Ribeiro, L.S., Gonçalves, L.P., Gorup, L.F., Leite, E.R., Camargo, E.R.: Enhanced reactivity of peroxo-modified surface of titanium dioxide nanoparticles used to synthesize ultrafine bismuth titanate powders at lower temperatures, Ceram. Int. **42**,15767–15772 (2016). https://doi.org/10.1016/j.ceramint.2016.07.039
67. Nogueira, A.E., Ribeiro, L.S., Gorup, L.F, Silva, G.T.S.T., Silva, F.F.B., Ribeiro, C., Camargo, E.R.: New approach of the oxidant peroxo method (OPM) route to obtain $Ti(OH)_4$ nanoparticles with high photocatalytic activity under visible radiation, Int. J. Photoenergy. **2018** (2018). https://doi.org/10.1155/2018/6098302
68. Cardoso, F.P., Nogueira, A.E., Patrício, P.S.O., Oliveira, L.C.A.: Effect of tungsten doping on catalytic properties of niobium oxide. J. Braz. Chem. Soc. **23**, 702–709 (2012). https://doi.org/10.1590/s0103-50532012000400016
69. da Silva, G.T.S.T., Nogueira, A.E., Oliveira, J.A., Torres, J.A., Lopes, O.F., Ribeiro, C.: Acidic surface niobium pentoxide is catalytic active for CO_2 photoreduction. Appl. Catal. B Environ. **242**, 349–357 (2019). https://doi.org/10.1016/j.apcatb.2018.10.017
70. Ribeiro, L.S., Nogueira, A.E., Aquino, J.M., Camargo, E.R.: A new strategy to obtain nanoscale particles of lithium titanate ($Li_4Ti_5O_{12}$) by the oxidant peroxo method (OPM). Ceram. Int. **45**, 23917–23923 (2019). https://doi.org/10.1016/j.ceramint.2019.07.274
71. Balaya, P., Ahrens, M., Kienle, L., Maier, J., Rahmati, B., Lee, S.B., Sigle, W., Pashkin, A., Kuntscher, C., Dressel, M.: Synthesis and characterization of nanocrystalline $SrTiO_3$. J. Am. Ceram. Soc. **89**, 2804–2811 (2006). https://doi.org/10.1111/j.1551-2916.2006.01133.x
72. Wang, T.X., Chen, W.W.: Solid phase preparation of submicron-sized $SrTiO_3$ crystallites from SrO_2 nanoparticles and TiO_2 powders. Mater. Lett. **62**, 2865–2867 (2008). https://doi.org/10.1016/j.matlet.2008.01.062

73. Buscaglia, M.T., Buscaglia, V., Alessio, R.: Coating of $BaCO_3$ crystals with TiO_2: Versatile approach to the synthesis of $BaTiO_3$ tetragonal nanoparticles. Chem. Mater. **19**, 711–718 (2007). https://doi.org/10.1021/cm061823b
74. Nikolenko, N.V., Kalashnykova, A.N., Solovov, V.A., Kostyniuk, A.O., Bayahia, H., Goutenoire, F.: Peroxide-based route for the synthesis of zinc titanate powder. Arab. J. Chem. **11**, 1044–1052 (2018). https://doi.org/10.1016/j.arabjc.2016.06.018
75. Wanjun, T., Donghua, C.: Photoluminescent properties of (Ca, Zn)TiO_3:Pr, B particles synthesized by the peroxide-based route method. J. Am. Ceram. Soc. **90**, 3156–3159 (2007). https://doi.org/10.1111/j.1551-2916.2007.01929.x
76. Robert, R., Logvinovich, D., Aguirre, M.H., Ebbinghaus, S.G., Bocher, L., Tomeš, P., Weidenkaff, A.: Crystal structure, morphology and physical properties of $LaCo_{1-x}Ti_xO_{3\pm\delta}$ perovskites prepared by a citric acid assisted soft chemistry synthesis. Acta Mater. **58**, 680–691 (2010). https://doi.org/10.1016/j.actamat.2009.09.046
77. Wang, C., Wang, S., Tang, L., He, Y.B., Gan, L., Li, J., Du, H., Li, B., Lin, Z., Kang, F.: A robust strategy for crafting monodisperse $Li_4Ti_5O_{12}$ nanospheres as superior rate anode for lithium ion batteries. Nano Energy. **21**, 133–144 (2016). https://doi.org/10.1016/j.nanoen.2016.01.005
78. Yamazaki, Y., Hernandez-Sanchez, R., Haile, S.M.: High total proton conductivity in large-grained yttrium-doped barium zirconate. Chem. Mater. **21**, 2755–2762 (2009). https://doi.org/10.1021/cm900208w
79. Guo, Y., Ran, R., Shao, Z.: A novel way to improve performance of proton-conducting solid-oxide fuel cells through enhanced chemical interaction of anode components. Int. J. Hydrogen Energy. **36**, 1683–1691 (2011). https://doi.org/10.1016/j.ijhydene.2010.10.081
80. Haile, S.M.: Fuel cell materials and components. Acta Mater. **51**, 5981–6000 (2003). https://doi.org/10.1016/j.actamat.2003.08.004
81. Orera, A., Slater, P.R.: New chemical systems for solid oxide fuel cells. Chem. Mater. **22**, 675–690 (2010). https://doi.org/10.1021/cm902687z
82. Ding, H., Ge, J., Xue, X.: A ceramic-anode supported low temperature solid oxide fuel cell. Electrochem. Solid-State Lett. **15**, 15–19 (2012). https://doi.org/10.1149/2.019206esl
83. Steele, B.C.H., Heinzel, A.: Materials for fuel-cell technologies. Mater. Sustain. Energy. **414**, 224–231 (2010). https://doi.org/10.1142/9789814317665_0031
84. Iwahara, H., Asakura, Y., Katahira, K., Tanaka, M.: Prospect of hydrogen technology using proton-conducting ceramics. Solid State Ionics **168**, 299–310 (2004). https://doi.org/10.1016/j.ssi.2003.03.001
85. Xie, K., Yan, R., Xu, X., Liu, X., Meng, G.: The chemical stability and conductivity of BaCe0.9-xYxNb0.1O3-σ proton-conductive electrolyte for SOFC, Mater. Res. Bull. **44**, 1474–1480 (2009). https://doi.org/10.1016/j.materresbull.2009.02.015
86. Kreuer, K.D.: Proton-conducting oxides. Annu. Rev. Mater. Res. **33**, 333–359 (2003). https://doi.org/10.1146/annurev.matsci.33.022802.091825
87. Haile, S.M., Pintauro, P.N.: Editorial: Proton transport for fuel cells. J. Mater. Chem. **20**, 6211–6213 (2010). https://doi.org/10.1039/c0jm90035d
88. Bohn, H.G., Schober, T.: Electrical conductivity of the high-temperature proton conductor $BaZr_{0.9}$Y0.1$O_{2.95}$. J. Am. Ceram. Soc. **83**, 768–772 (2004). https://doi.org/10.1111/j.1151-2916.2000.tb01272.x
89. de Souza, E.C.C., Muccillo, R.: Properties and applications of perovskite proton conductors. Mater. Res. **13**, 385–394 (2010). https://doi.org/10.1590/S1516-14392010000300018
90. Babilo, P., Haile, S.M.: Enhanced sintering of yttrium-doped barium zirconate by addition of ZnO. J. Am. Ceram. Soc. **88**, 2362–2368 (2005). https://doi.org/10.1111/j.1551-2916.2005.00449.x
91. Kreuer, K.D., Adams, S., Münch, W., Fuchs, A., Klock, U., Maier, J.: Proton conducting alkaline earth zirconates and titanates for high drain electrochemical applications. Solid State Ionics **145**, 295–306 (2001). https://doi.org/10.1016/S0167-2738(01)00953-5
92. Yamazaki, Y., Hernandez-Sanchez, R., Haile, S.M.: Cation non-stoichiometry in yttrium-doped barium zirconate: phase behavior, microstructure, and proton conductivity. J. Mater. Chem. **20**, 8158–8166 (2010). https://doi.org/10.1039/c0jm02013c

93. Gonçalves, M.D., Muccillo, R.: Properties of yttrium-doped barium zirconate ceramics synthesized by the oxidant-peroxo method. Ceram. Int. **40**, 911–917 (2014). https://doi.org/10.1016/j.ceramint.2013.06.086
94. Gonçalves, M.D., Maram, P.S., Muccillo, R., Navrotsky, A.: Enthalpy of formation and thermodynamic insights into yttrium doped $BaZrO_3$. J. Mater. Chem. A. **2**, 17840–17847 (2014). https://doi.org/10.1039/c4ta03487b
95. Gonçalves, M.D., Maram, P.S., Navrotsky, A., Muccillo, R.: Effect of synthesis atmosphere on the proton conductivity of Y-doped barium zirconate solid electrolytes. Ceram. Int. **42**, 13689–13696 (2016). https://doi.org/10.1016/j.ceramint.2016.05.167
96. Sahraoui, D.Z., Mineva, T.: Structural properties of Y-doped BaZrO3 as a function of dopant concentration and position: a density functional study. Solid State Ionics **232**, 1–12 (2013). https://doi.org/10.1016/j.ssi.2012.11.011
97. Ishihara, T.: Development of new fast oxide ion conductor and application for intermediate temperature solid oxide fuel cells. Bull. Chem. Soc. Jpn. **79**, 1155–1166 (2006). https://doi.org/10.1246/bcsj.79.1155
98. Magrez, A., Schober, T.: Preparation, sintering, and water incorporation of proton conducting $Ba_{0.99}Zr_{0.8}Y_{0.2}O_{3-\delta}$: comparison between three different synthesis techniques. Solid State Ionics **175**, 585–588 (2004). https://doi.org/10.1016/j.ssi.2004.03.045
99. Navrotsky, A.: Progress and new directions in high temperature calorimetry. Phys. Chem. Miner. **2**, 89–104 (1977). https://doi.org/10.1007/BF00307526
100. Navrotsky, A., Simoncic, P., Yokokawa, H., Chen, W., Lee, T.: Calorimetric measurements of energetics of defect interactions in fluorite oxides. Faraday Discuss. **134**, 171–180 (2007). https://doi.org/10.1039/b604014b
101. Navrotsky, A.: Thermodynamics of solid electrolytes and related oxide ceramics based on the fluorite structure. J. Mater. Chem. **20**, 10577–10587 (2010). https://doi.org/10.1039/c0jm01521k
102. Cheng, J., Navrotsky, A.: Energetics of magnesium, strontium, and barium doped lanthanum gallate perovskites. J. Solid State Chem. **177**, 126–133 (2004). https://doi.org/10.1016/S0022-4596(03)00337-2
103. Buyukkilic, S., Shvareva, T., Navrotsky, A.: Enthalpies of formation and insights into defect association in ceria singly and doubly doped with neodymia and samaria. Solid State Ionics **227**, 17–22 (2012). https://doi.org/10.1016/j.ssi.2012.08.017
104. Aizenshtein, M., Shvareva, T.Y., Navrotsky, A.: Thermochemistry of lanthana- and yttria-doped thoria. J. Am. Ceram. Soc. **93**, 4142–4147 (2010). https://doi.org/10.1111/j.1551-2916.2010.04001.x
105. Avila-Paredes, H.J., Shvareva, T., Chen, W., Navrotsky, A., Kim, S.: A correlation between the ionic conductivities and the formation enthalpies of trivalent-doped ceria at relatively low temperatures. Phys. Chem. Chem. Phys. **11**, 8580–8585 (2009). https://doi.org/10.1039/b821982f
106. Ushakov, S.V., Navrotsky, A.: Direct measurements of water adsorption enthalpy on hafnia and zirconia. Appl. Phys. Lett. **87**, 1–3 (2005). https://doi.org/10.1063/1.2108113
107. Radha, A.V., Bomati-Miguel, O., Ushakov, S.V., Navrotsky, A., Tartaj, P.: Surface enthalpy, enthalpy of water adsorption, and phase stability in nanocrystalline monoclinic zirconia. J. Am. Ceram. Soc. **92**, 133–140 (2009). https://doi.org/10.1111/j.1551-2916.2008.02796.x
108. Gonçalves, M.D., Mielewczyk-Gryń, A., Maram, P.S., Kryścio, Ł, Gazda, M., Navrotsky, A.: Systematic water uptake energetics of yttrium-doped barium zirconate—a high resolution thermochemical study. J. Phys. Chem. C. (2020). https://doi.org/10.1021/acs.jpcc.0c01049
109. Conte, V., Bortolini, O., Carraro, M., Moro, S.: Models for the active site of vanadium-dependent haloperoxidases: insight into the solution structure of peroxo vanadium compounds. J. Inorg. Biochem. **80**, 41–49 (2000). https://doi.org/10.1016/S0162-0134(00)00038-6
110. Tan, H.L., Amal, R., Ng, Y.H.: Alternative strategies in improving the photocatalytic and photoelectrochemical activities of visible light-driven $BiVO_4$: a review. J. Mater. Chem. A. **5**, 16498–16521 (2017). https://doi.org/10.1039/c7ta04441k

111. Lopes, O.F., Carvalho, K.T.G., Macedo, G.K., De Mendonça, V.R., Avansi, W., Ribeiro, C.: Synthesis of $BiVO_4$ via oxidant peroxo-method: insights into the photocatalytic performance and degradation mechanism of pollutants. New J. Chem. **39**, 6231–6237 (2015). https://doi.org/10.1039/c5nj00984g
112. Lopes, O.F., Carvalho, K.T.G., Nogueira, A.E., Avansi, W., Ribeiro, C.: Controlled synthesis of BiVO4 photocatalysts: evidence of the role of heterojunctions in their catalytic performance driven by visible-light. Appl. Catal. B Environ. **188**, 87–97 (2016). https://doi.org/10.1016/j.apcatb.2016.01.065
113. Lopes, O.F., Carvalho, K.T.G., Avansi, W., Ribeiro, C.: Growth of BiVO4 Nanoparticles on a Bi_2O_3 surface: effect of heterojunction formation on visible irradiation-driven catalytic performance. J. Phys. Chem. C. **121**, 13747–13756 (2017). https://doi.org/10.1021/acs.jpcc.7b03340
114. Nico, C., Monteiro, T., Graça, M.P.F.: Niobium oxides and niobates physical properties: review and prospects. Prog. Mater. Sci. **80**, 1–37 (2016). https://doi.org/10.1016/j.pmatsci.2016.02.001
115. Katsumata, K.I., Cordonier, C.E.J., Shichi, T., Fujishima, A.: Photocatalytic activity of $naNbO_3$ thin films. J. Am. Chem. Soc. **131**, 3856–3857 (2009). https://doi.org/10.1021/ja900394x
116. Li, P., Ouyang, S., Xi, G., Kako, T., Ye, J.: The effects of crystal structure and electronic structure on photocatalytic H_2 evolution and CO_2 reduction over two phases of perovskite-structured $NaNbO_3$. J. Phys. Chem. C. **116**, 7621–7628 (2012). https://doi.org/10.1021/jp210106b
117. Ge, H., Hou, Y., Xia, C., Zhu, M., Wang, H., Yan, H.: Preparation and piezoelectricity of $NaNbO_3$ high-density ceramics by molten salt synthesis. J. Am. Ceram. Soc. **94**, 4329–4334 (2011). https://doi.org/10.1111/j.1551-2916.2011.04685.x
118. Dey, D., Petrykin, V., Sasaki, S., Kakihana, M.: Water soluble $Na[Nb(O_2)_3]\ 2H_2O$ as a new molecular precursor for synthesis of sodium niobate. J. Ceram. Soc. Japan **115**, 808–812 (2007). https://doi.org/10.2109/jcersj2.115.808
119. Benítez, E., Castañeda-Guzmán, R., López-Juárez, R., Depablos-Rivera, O., Gervacio-Arciniega, J.J.: Ferroelectric properties and phase transitions of high performance vertically aligned KNN nanowire-arrays grown by pulsed laser deposition. Phys. E Low-Dimensional Syst. Nanostructures **123** (2020). https://doi.org/10.1016/j.physe.2020.114143
120. Oh, Y., Noh, J., Yoo, J., Kang, J., Hwang, L., Hong, J.: Dielectric and piezoelectric properties device applications. IEEE Trans. Ultrason. Ferroelectr. Freq. Control. **58**, 1860–1866 (2011)
121. G. Stavber, B. Malič, M. Kosec, A road to environmentally friendly materials chemistry: Low-temperature synthesis of nanosized K0.5Na0.5NbO3 powders through peroxide intermediates in water. Green Chem. **13**, 1303–1310 (2011). https://doi.org/10.1039/c0gc00750a
122. Kato, H., Shimizu, K., Nakajima, K., Kobayashi, M., Kakihana, M.: Synthesis of rare earth niobate and tantalate powders via a peroxo complex route. Chem. Lett. **46**, 1515–1517 (2017). https://doi.org/10.1246/cl.170652
123. Sergienko, V.S.: Structural characteristics of peroxo complexes of group IV and V transition metals. Rev. Crystallogr. Rep. **49**, 907–929 (2004). https://doi.org/10.1134/1.1828134
124. Sergienko, V.S.: Structural chemistry of peroxo compounds of group VI transition metals: II. Peroxo complexes of molybdenum and tungsten: a review. Crystallogr. Reports. **53**, 18–46 (2008). https://doi.org/10.1134/S1063774508010045
125. Ding, Y., Wan, Y., Min, Y.L., Zhang, W., Yu, S.H.: General synthesis and phase control of metal molybdate hydrates $MMoO_4{\cdot}nH_2O$ (M = Co, Ni, Mn, n = 0, 3/4, 1) nano/microcrystals by a hydrothermal approach: magnetic, photocatalytic, and electrochemical properties. Inorg. Chem. **47**, 7813–7823 (2008). https://doi.org/10.1021/ic8007975
126. Cavalcante, L.S., Longo, V.M., Sczancoski, J.C., Almeida, M.A.P., Batista, A.A., Varela, J.A., Orlandi, M.O., Longo, E., Li, M.S.: Electronic structure, growth mechanism and photoluminescence of $CaWO_4$ crystals. CrystEngComm **14**, 853–868 (2012). https://doi.org/10.1039/c1ce05977g

127. Errandonea, D., Ruiz-Fuertes, J.: A brief review of the effects of pressure on wolframite-type oxides. Crystals **8**, 1–19 (2018). https://doi.org/10.3390/cryst8020071
128. Tablero, C.: Optical absorption and applications of the ABO4 (A = Ca, Pb and B = Mo, W) semiconductors. Chem. Phys. Lett. **635**, 190–195 (2015). https://doi.org/10.1016/j.cplett.2015.06.074
129. Dey, S., Ricciardo, R.A., Cuthbert, H.L., Woodward, P.M.: Metal-to-metal charge transfer in AWO_4 (A = Mg, Mn Co, Ni, Cu, or Zn) compounds with the wolframite structure. Inorg. Chem. **53**, 4394–4399 (2014). https://doi.org/10.1021/ic4031798
130. Ryu, J.H., Yoon, J.W., Shim, K.B.: Blue-luminescence of nanocrystalline MWO4 (M = Ca, Sr, Ba, Pb) phosphors synthesized via a citrate complex route assisted by microwave irradiation, Electrochem. Solid-State Lett. **8** (2005). https://doi.org/10.1149/1.1891626
131. Kaczmarek, A.M., Van Deun, R.: Rare earth tungstate and molybdate compounds-from 0D to 3D architectures. Chem. Soc. Rev. **42**, 8835–8848 (2013). https://doi.org/10.1039/c3cs60166h
132. Leyzerovich, N.N., Bramnik, K.G., Buhrmester, T., Ehrenberg, H., Fuess, H.: Electrochemical intercalation of lithium in ternary metal molybdates MMoO4 (M: Cu, Zn, Ni and Fe). J. Power Sources. **127**, 76–84 (2004). https://doi.org/10.1016/j.jpowsour.2003.09.010
133. Rideal VI., S.: The action of ammonia on some tungsten compounds, J. Chem. Soc. Trans. **55**, 41–45(1889). https://doi.org/10.1039/CT8895500041
134. Shariat, M.H., Setoodeh, N., Dehghan, R.A.: Optimizing conditions for hydrometallurgical production of purified molybdenum trioxide from roasted molybdenite of sarcheshmeh. Miner. Eng. **14**, 815–820 (2001). https://doi.org/10.1016/S0892-6875(99)00000-X
135. Camargo, E.R., Leite, E.R., Longo, E.: Synthesis and characterization of lead zirconate titanate powders obtained by the oxidant peroxo method. J. Alloys Compd. **469**, 523–528 (2009). https://doi.org/10.1016/j.jallcom.2008.02.008
136. Gonçalves, M.D., Camargo, E.R.: Lanthanum-doped PZT synthesized by the oxidant peroxide method and sintered by conventional and microwave routes. Ceram. Int. **43**, 3004–3009 (2017). https://doi.org/10.1016/j.ceramint.2016.11.088
137. Xie, Y., Kocaefe, D., Chen, C., Kocaefe, Y.: Review of research on template methods in preparation of nanomaterials. J. Nanomater. **2016**(2016). https://doi.org/10.1155/2016/2302595

Photoluminescence in Alkaline Earth Stannate Thin Films Grown by Physical and Chemical Methods

André Luiz Menezes de Oliveira, Mary Cristina Ferreira Alves, Jéssica Luisa Alves do Nascimento, Valérie Bouquet, Ingrid Távora Weber, Stephanie Deputier, Maximo Siu-Li, Elson Longo, Maryline Guilloux-Viry, and Ieda Maria Garcia dos Santos

Abstract Alkaline earth stannates with perovskite structure ($ASnO_3$, $A = Ca$, Sr, Ba) have been studied for a long time due to their unique structure and physicochemical properties, but few works in literature have been devoted to their thin films. The technology of thin films makes structuring materials in fettered dimensions very simple, making them useful in electronic devices. Moreover, the possibility of oriented and epitaxial growth allows a better understanding of the surface and interface properties of the films to tailor their functionalities. In this chapter, recent findings on photoluminescent properties of $ASnO_3$-type perovskites are discussed and results on polycrystalline and epitaxial thin films deposited using a physical deposition method (pulsed laser deposition, PLD) and a chemical one (chemical solution deposition, CSD) are presented. In this context, two different series were carefully investigated considering the Sr-site substitution in $SrSnO_3$ perovskite to form the $Ca_{1-x}Sr_xSnO_3$ solid solution, and the Sn-site substitution giving origin to $SrSn_{1-x}Ti_xO_3$ (x = 0, 0.25, 0.5, 0.75 and 1). The structural and microstructural characteristics of all films are

A. L. M. de Oliveira · M. C. F. Alves · I. M. G. dos Santos (✉)
Núcleo de Pesquisa e Extensão–Laboratório de Combustíveis e Materiais (NPE/LACOM), Universidade Federal da Paraíba, João Pessoa, PB, Brazil
e-mail: ieda@quimica.ufpb.br

A. L. M. de Oliveira · M. C. F. Alves · V. Bouquet · S. Deputier · M. Guilloux-Viry
ISCR UMR 6226, Univ Rennes, CNRS, 35000 Rennes, France

M. C. F. Alves · J. L. A. do Nascimento
Universidade Estadual da Paraíba, Campus I, Campina Grande, PB, Brazil

I. T. Weber
LIMA, Universidade de Brasília, Brasília, DF, Brazil

M. Siu-Li
IFSC, Universidade de São Paulo, São Carlos, SP, Brazil

E. Longo
Centro de Desenvolvimento de Materiais Funcionais—CDMF, Laboratório Interdisciplinar de Eletroquímica e Cerâmica–LIEC, Universidade Federal de São Carlos, São Carlos, SP, Brazil

F. A. La Porta and C. A. Taft (eds.), *Functional Properties of Advanced Engineering Materials and Biomolecules*, Engineering Materials,
https://doi.org/10.1007/978-3-030-62226-8_5

first presented. Then, discussion about the influence of composition, method of deposition, type of growth and short-range order/disorder related to photoluminescence properties are shown.

1 Introduction

Alkaline-earth stannates with stoichiometry $ASnO_3$ (A = Ca, Sr, Ba) are members of an important class of inorganic materials with a distinctive perovskite-type structure. They have been widely investigated owing to their unique structural characteristics, unusual dielectric, and semiconducting properties, good chemical and physical stability, besides presenting strong visible luminescence, and excellent optical properties [1–3]. Such characteristics are useful for a wide range of technological applications, such as optoelectronic devices, thermally stable capacitors, humidity and gas sensors, lithium-ion batteries, fuel cells, photocatalysts, photoelectrodes, among others [3–9]. Moreover, chemical modifications in $ASnO_3$-type materials can induce structural changes and result in the adjustment of its physical and chemical properties and to an improvement of their efficiency in such practical devices. For instance, some new optical properties of $SrSnO_3$-based material have been obtained by Sr and/or Sn substitutions by appropriate cations [3, 10–13].

In $ASnO_3$ stannates structure, Sn(IV) cations are located at the B sites coordinated to six oxygen anions, which form the BO_6 octahedra, while the alkaline-earth cations (Ca, Sr or Ba) occupy A sites and are twelve-coordinated [3, 14]. $CaSnO_3$ and $SrSnO_3$ compounds adopt a distorted *Pbnm*-type orthorhombic structure [3, 14–16], while $BaSnO_3$ has a cubic symmetry with *Pm3m* space group [14, 16]. It has been established that different magnitudes of SnO_6 octahedra tilting are closely related to the A cation size ($Ca^{2+} < Sr^{2+} < Ba^{2+}$), and are strongly associated to changes in their properties [1, 14, 17]. These stannates have previously been synthesized as bulk materials using solid-state reactions, hydrothermal synthesis or polymeric precursor method [3–5, 16–23].

In relation to stannates thin films, different studies have been reported in the literature, and some of them can be highlighted here. For instance, $SrSnO_3$ films deposited on MgO and La-YBCO substrates using radiofrequency (RF) magnetron sputtering have been used as insulating layers in the manufacturing of single flux quantum circuits (SFQ) [24, 25], whereas $CaSnO_3$:Tb-Mg films deposited on silicon and (100)$SrTiO_3$ substrates, also by rf magnetron sputtering, have been used as a phosphorescent layer in the preparation of electroluminescent devices [12]. These materials exhibited luminescent properties at room temperature after heat-treatment at high temperatures. $BaSnO_3$ films have also been deposited by sputtering on SiO_2/Si wafers, boron-glass and quartz, besides on (001)$SrTiO_3$ substrate [26, 27]. The authors have evidenced that the control of the deposition conditions is crucial for modulating the concentration of oxygen vacancies to improve the photoconductivity of the films. Pulsed laser deposition (PLD) has also been used to prepare alkaline-earth stannate thin films, especially $SrSnO_3$ and $CaSnO_3$. For instance, $SrSnO_3$

epitaxial films doped with Sb, Ta or co-doped with Sb–Nd have been grown on (100)$SrTiO_3$ substrates [25, 28–30]. Our research group has also deposited epitaxial films of $SrSnO_3$ [11, 31] and $CaSnO_3$ [10] by PLD and/or by chemical solution deposition (CSD) on various substrates at different temperatures. The preparation of thin films composed of solid solutions has not been carefully investigated in the literature.

Although $ABSnO_3$ thin films have not been studied, some alkaline-earth stannate thin films have been studied in their combined forms with $SrTiO_3$ since the discovery of the strong broadening of ferroelectric phase transitions in Ba(Ti,Sn)O_3 [32, 33]. Strontium titanate ($SrTiO_3$), which has a simple cubic perovskite symmetry with *Pm3m* space group, is an important n-type semiconductor with a bandgap of about 3.2 eV. This perovskite is considered as one of the most important multifunctional materials and it presents a diversity of useful properties, such as high dielectric constant [34–36], photoluminescence [37–41] and photocatalytic activity [42–45], and others. In this context, combining $SrSnO_3$ with $SrTiO_3$ to form $SrSn_{1-x}Ti_xO_3$ solid solution allows to tune its functionalities [11, 19, 46, 47]. As yet, few studies have been reported on thin films based on this $SrSn_{1-x}Ti_xO_3$ solid solution [11, 46]. For instance, a paper has reported the preparation of $SrSn_{1-x}Ti_xO_3$ thin films on MgO by PLD [46]. The authors have studied the optical response of the films as a function of the composition with application in solar energy devices. More recently, our research group has described the deposition of $SrSn_{1-x}Ti_xO_3$ thin films on (100)$LaAlO_3$ and R-sapphire single crystal substrates using PLD and CSD methods and the influence of Ti/Sn substitution on the structural, microstructural and photoluminescence properties has been investigated [11]. Different characteristics, including luminescent behavior, have been evidenced as a function of composition, besides the nature of growth and deposition methods. Another important study has been performed by Wang et al. [48] who investigated the ferroelectricity induced by Sn^{2+} and Sn^{4+} co-doping in $SrTiO_3$ to form $Sr_{1-x}Sn_xTi_{1-y}Sn_yO_3$ (x and y ≤ 0.10) thin films grown on (100)$SrTiO_3$ substrate using a hybrid molecular beam epitaxy (MBE) deposition method.

Among the different methods used for thin film growth, the pulsed laser deposition (PLD) and chemical solution deposition (CSD) was here chosen to be explored. The PLD is, of course, an attractive physical deposition method by which highly epitaxial thin films may be obtained (when an appropriate substrate is used) with an accurate composition and growth control. With the use of this method, it is also possible to obtain films with good adhesion on the substrates [11]. On the other hand, it is well-known that chemical solution deposition (CSD) method consists of the deposition of a solution by spin or dip coating on a substrate followed by subsequent annealing treatments of the wet film to promote the crystallization of the desired material [10, 11, 31, 49]. The CSD method is attractive, especially for economic reasons, as it is a relatively low-cost process that, in turn, allows the deposition on a large area and complex shape substrates. We can also point out that this strategy allows the preparation of films with reasonable stoichiometric control in low temperatures and does not require vacuum or atmosphere control. As such, the crystallization process based on CSD method usually is a soft process associated with a low energetic balance, in contrast with PLD which involves high energy for the evaporation step of the material as well as for the crystallization that occurs in situ at high temperature [11].

In view of the literature data cited above, it is very important to study $SrSnO_3$ perovskites in thin film form, exploring the Sr and Sn substitutions with foreign cations, as well as the influence of the substrate nature and of the method used for deposition in order to give deeper insight into the effects of the structural and morphological characteristics on their photophysical properties, including luminescence.

Particularly, photoluminescence (PL), which is one of the many forms of luminescence (light emission) is an extremely important photophysical tool used to characterize the optoelectronic properties of wide bandgap semiconductors. The principle of this technique consists of exciting electrons from the valence (VB) to the conduction band (CB) of a material using an electromagnetic light source whose energy is higher than the energy of the bandgap. As a consequence of this process, photoexcited electrons tend to relax and recombine with holes and then release energy emitting light. By analyzing the emissions, it is possible to measure the response in the desired range of the visible spectrum, accessing information about different types of defects existing in the material. For instance, applying PL associated with optical band gap characterization in thin films allow to determine film imperfections such as structural distortions in the lattice and disorder at substrate–film interface as well as to evaluate the presence of impurities and intrinsic defects in the material. All these factors induce the formation of empty energy sublevels in the bandgap and lead to different recombination mechanisms [3, 10, 11, 50–55].

In this sense, we describe in the present chapter the preparation of thin films of the $Ca_{1-x}Sr_xSnO_3$ and $SrSn_{1-x}Ti_xO_3$ series (x = 0, 0.25, 0.50, 0.75 and 1) by pulsed laser deposition (PLD) and chemical solution deposition (CSD) on (100)$SrTiO_3$ single crystal and amorphous silica substrates. The influence of parameters such as thin film composition, deposition method, and substrate nature on the photoluminescent behavior is discussed. The characterization of the films by PL response is fundamental to a full structural elucidation since it allows evaluating the presence of defects in the bandgap of the semiconductor materials, including $CaSnO_3$, $SrSnO_3$, $SrTiO_3$, and their corresponding combined forms.

2 Experimental

The thin films of the $Ca_{1-x}Sr_xSnO_3$ and $SrTi_{1-x}Sn_xO_3$ series (x = 0; 0.25; 0.50; 0.75 and 1) were prepared by chemical solution deposition (CSD) based on the polymeric precursor method and by pulsed laser deposition (PLD). Further details regarding the synthesis can be found in previous works [10, 11, 31].

2.1 Thin Films Prepared by PLD

Firstly, for the films synthesis by PLD, labmade sintered targets were prepared using solid state reaction from stoichiometric mixtures of strontium carbonate ($SrCO_3$ *Merck*), calcium carbonate ($CaCO_3$ *RP Normapur*), tin oxide (SnO_2 *Sigma Aldrich*) and titanium oxide (TiO_2 *Alfa Aesar*) powders as starting materials to obtain the desired $Ca_{1-x}Sr_xSnO_3$ and $SrTi_{1-x}Sn_xO_3$ (x = 0; 0.25; 0.50; 0.75 and 1) compositions. The mixtures of the powders were calcined in air at 1000°C for 8 h and sintered at 1100–1300 °C for 8 h. Characterizations by X-ray diffraction (XRD) and Energy dispersive spectroscopy (EDS) analyses performed on the targets confirmed that the expected perovskite single phases were obtained with the desired composition for both $Ca_{1-x}Sr_xSnO_3$ and $SrTi_{1-x}Sn_xO_3$ series.

Thin films of each composition of the solid solutions were deposited by PLD at 700 °C for 20–30 min under an oxygen pressure of 30 Pa. Deposition was performed using a KrF excimer laser (Tuilaser Excistar, $\lambda = 248$ nm, = 2 Hz, 210 mJ/pulse) with a substrate-target distance fixed at 55 mm. The laser beam has focused on the target at an incident angle of 45°, and the target was rotated in a standard vacuum chamber (background pressure of approximately 5×10^{-4} Pa). These experimental deposition conditions led to films with thicknesses varying from 200 to 300 nm.

2.2 Thin Films Prepared by CSD

For the synthesis of films by CSD, polymeric solutions were prepared according to the methodology described in our previous works [5, 10, 11, 19, 22, 31]. Briefly, the citrate solutions were prepared from calcium acetate (*Alfa Aesar* 99%), strontium nitrate (*Alfa Aesar* 99%), titanium isopropoxide (*AG hulls* 98%), and tin chloride dehydrate (*Alfa Aesar* 98%). The citric acid:cation ratio was (3:1) in mol and citric acid:ethylene glycol was (60:40) in mass. The viscosity of the resulting polymeric resins was then adjusted to a range of 25–30 cP by addition or evaporation of water and measured using a Brookfield DVII+Pro viscometer. The obtained coating solutions were deposited by spin coating at 1000 rpm during 3 s followed by rotation at 3000–4000 rpm during 10–20 s (Spin Coater model KW-4A Chemat Technology) on the substrates. Then, the wet films were subjected to two thermal treatment cycles as follows: (i) 300 °C for 4 h or 400 °C for 2 h to eliminate organic matter and (ii) 650–700 °C for 2 h for the crystallization of perovskite phase. These synthesis conditions enabled the formation of thin films with a thickness ranging from 200 to 280 nm.

2.3 Characterizations

The structural characteristics of the films were analyzed using X-ray diffraction (XRD). Standard θ-2θ scans were performed using a two-circle Bruker D8 Advance diffractometer with monochromatic Cu $K_{\alpha 1}$ radiation. For the oriented films, the ω-scans (used to evaluate the crystalline quality along the growth direction) and φ-scans (used to study the in-plane orientation) were recorded using a four-circle Bruker D8 Discover diffractometer operating with Cu $K_{<\alpha>}$ radiation. The microstructures obtained for these thin films were observed using a field emission scanning electron microscope (FE-SEM, Jeol 6301-F), which was operated at low voltage, typically 7–9 kV, to limit charge effects and to achieve high resolution without requiring surface metallization. The photoluminescence (PL) measurements were performed using a Monospec 27 monochromator (Thermal Jarrel Ash) coupled to a R446 photomultiplier (Hamamatsu). A 350.7 nm krypton ion laser (Coherent Innova 90 K) was used as the excitation source.

3 Results and Discussion

3.1 Long-Range Characterization by X-ray Diffraction

3.1.1 $Ca_{1-x}Sr_xSnO_3$ Thin Films

The θ-2θ XRD patterns of the films obtained by PLD and CSD on (100)$SrTiO_3$ substrate (STO) are shown in Figure 1. All the compositions were single-phase, and for comparison, all the peaks were indexed relative to a pseudo-cubic unit cell with $a_{pc} = 4.034$ Å for $SrSnO_3$ and $a_{pc} = 3.960$ Å for $CaSnO_3$ (Table 1).

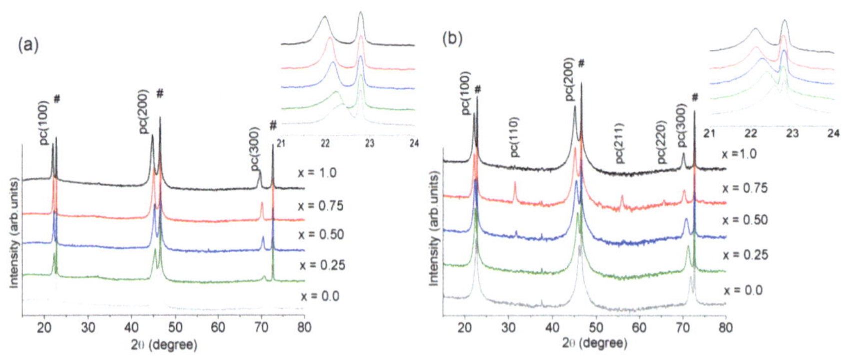

Fig. 1 θ-2θ XRD patterns of $Ca_{1-x}Sr_xSnO_3$ thin films deposited on (100)$SrTiO_3$ (STO) substrates by PLD **a** by CSD **b**. Peaks marked with # are related to the substrate

Table 1 Lattice parameters of $SrSnO_3$ and $CaSnO_3$

	Structure	Lattice parameters (Å)	Parameters in a pseudo cubic cell (Å)
$SrSnO_3$	Cubic (ICDD 22–1442)	$a = 8.068$	$a_{pc} = a/2 = 4.034$
$CaSnO_3$	Orthorhombic (ICDD 77–1797)	$a = 5.532$ $b = 5.681$ $c = 7.906$	$a_{pc} = [(a + b)\sqrt{2}/2 + c/2] / 3 = 3.960$

The films obtained using the PLD and CSD methods were highly crystalline, as indicated by the XRD patterns (Fig. 1a, b) that exhibit *h*00 preferential orientation (referring to the pseudo-cubic unit cell). The films produced using PLD presented only intense and well-defined *h00* peaks (Fig. 1a). For the films obtained using CSD, other low intensity peaks were also observed in some cases, especially for the $Ca_{0.25}Sr_{0.75}SnO_3$ film (Fig. 1b), corresponding to randomly oriented crystallites. Films grown by PLD can sometimes present higher degree of orientation compared to that of CSD films, which is strongly related to the growth mechanism of each method during the crystallization process. For the PLD, which is an *in situ* method, the crystallization occurs from the vapor phase on a heated surface and the growth takes place directly atom by atom at the substrate-film interface. For the CSD method, the deposition is performed at room temperature and the film is subsequently heated. Therefore, crystallization takes place more slowly from the randomly oriented nuclei in the film during the post-annealing treatments which can lead in some cases to a lower degree of orientation compared to PLD films (especially in the case of relatively high film—substrate mismatch) [11, 56].

As expected, a decrease of the 2θ angles of the (*h*00) reflections as a function of the Sr^{2+} content in $Ca_{1-x}Sr_xSnO_3$ was observed for films prepared using both methods, as shown by the insets of Fig. 1a (PLD) and Fig. 1b (CSD). The lattice parameters were calculated from the XRD patterns (from the (*h00*) reflections) for a better evaluation of this behavior (Fig. 2).

The monotonic increase of the a_{pc} lattice parameter when the content of Sr^{2+} increased in the solid solution could be related to the lower covalent character of Sr^{2+}, which decreases the Sr^{2+}–O^{2-} bond strength and promotes the expansion of the unit cell [23]. This behavior is in agreement with Vegard's Law [57, 58]. The films obtained using PLD had larger lattice parameters than those obtained using CSD. In both cases these values are slightly shifted from the theoretical ones obtained for the bulk materials. Difference of lattice parameters is routinely observed between films and bulk due to strain effects that can also depend on the deposition method.

The (*h00*)-oriented films obtained using PLD and CSD have a high out-of-plane and in-plane crystalline quality (Table 2).

As an example, Fig. 3 presents the ω- and φ-scans obtained for the $SrSnO_3$ and $CaSnO_3$ films synthesized using PLD.

The low values of the full-width at half-maximum (FWHM) of the ω-scans recorded for the (*h00*) reflections are representative of weak mosaicity in the films

Fig. 2 Calculated pseudo-cubic lattice parameter of $Ca_{1-x}Sr_xSnO_3$ films grown on STO substrate by PLD and CSD

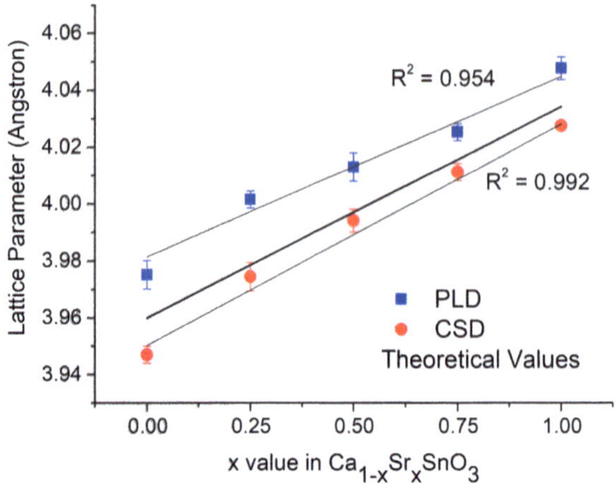

Table 2 Evaluation of the epitaxy of thin films obtained by PLD and CSD

Compositions	PLD		CSD	
	Δω-scan	Δφ-scan on (110) reflection	Δω-scan	Δφ-scan on (110) reflection
$CaSnO_3$	1.61°	3.5°*	0.90°	1.2°
$Ca_{0.75}Sr_{0.25}SnO_3$	0.34°	1.1°**	0.35°	1.4°
$Ca_{0.50}Sr_{0.50}SnO_3$	0.43°	1.3°	0.11°	1.8°
$Ca_{0.25}Sr_{0.75}SnO_3$	0.14°	1.1°	0.42°	1.2°
$SrSnO_3$	0.17°	1.4°	0.57°	1.2°

φ-scan of the samples on other plans: *pc (210) and **pc (100).

(Table 2). In the case of the $CaSnO_3$ film, the unusual form of the ω-scan pattern was related to the superposition of the contributions from the (100)$SrTiO_3$ substrate (FWHM of approximately 0.06) and of the film due to the nearest positions of the two peaks in the θ-2θ XRD patterns. Then, the FWHM value of the film was determined by deconvolution. Furthermore, the in-plane investigation revealed an epitaxial growth for all the films, as evidenced by the presence of four 90° spaced peaks in the φ-scan patterns. These results are induced by the structural similarities besides the low mismatch values between the deposited material and the substrate (3.3% and 1.4% for $SrSnO_3$ and $CaSnO_3$, respectively). Despite the lower mismatch value, the $CaSnO_3$ films exhibited relatively higher Δω values, particularly in the case of PLD compared to the ones obtained for other compositions. This behavior may be related to the covalent character of the Ca^{2+} ion, which generates a greater distortion in the lattice, as indicated in the literature [17, 23, 59, 60].

The epitaxial growth of $SrSnO_3$:Sb, $SrSnO_3$:Nd and $CaSnO_3$:Pr^{3+} films on $SrTiO_3$ substrates using PLD has also been observed by other authors [13, 28, 29, 61].

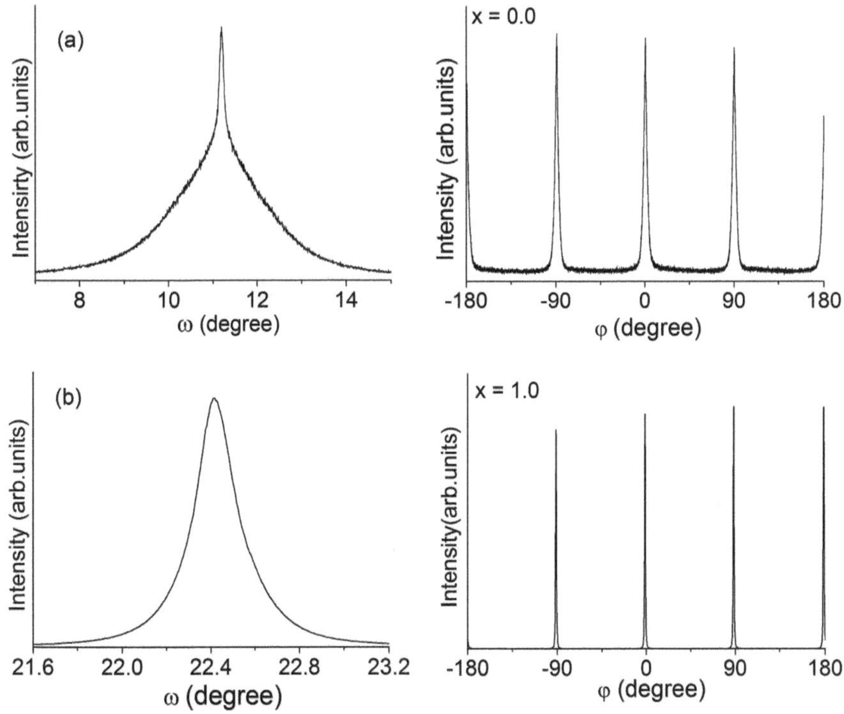

Fig. 3 In- and out-plane investigation of thin films prepared by PLD on STO: **a** ω-scan of the $CaSnO_3$ around the (100) peak and φ-scan on (210) reflection; **b** ω-scan of the $SrSnO_3$ around the (200) peak and φ-scan on (110) reflection

Although a higher crystallinity was obtained for the first two films, the epitaxial quality of the $CaSnO_3$:Pr^{3+} film was consistent with our results. The different qualities could be attributed to differences in the deposition conditions, film thicknesses and doping effects.

3.1.2 $SrSn_{1-x}Ti_xO_3$ Thin Films

To better assess the influence of the type of growth and of the cationic-site of substitution, the Ti-substituted $SrSnO_3$ films ($SrSn_{1-x}Ti_xO_3$) were grown on amorphous silica (SiO_2) substrates as described below.

Figures 4a, b show the XRD *θ-2θ* patterns of the $SrSn_{1-x}Ti_xO_3$ thin films on silica by PLD and CSD, respectively. As previously mentioned, the crystalline structures of the $SrSnO_3$ (ICDD 22–1442) is a distorted orthorhombic with *Pbnm* symmetry, whereas $SrTiO_3$ (ICDD 35–0734) presents a cubic *Pm3m* symmetry. As the other intermediary compositions in the $SrSn_{1-x}Ti_xO_3$ solid solution present no ICDD

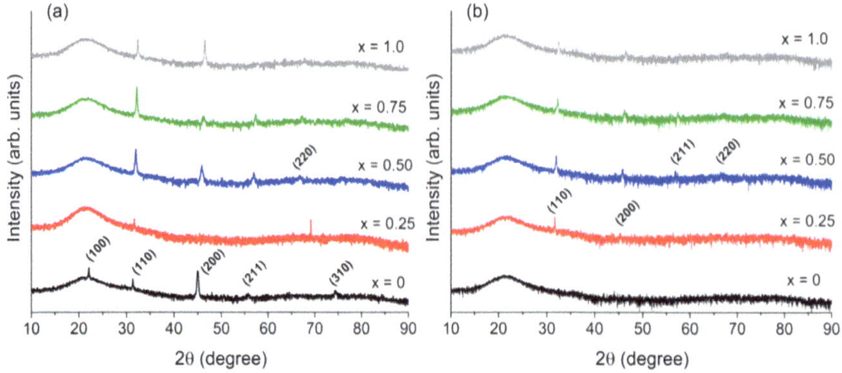

Fig. 4 θ-2θ XRD patterns of $SrSn_{1-x}Ti_xO_3$ films deposited on silica by PLD (**a**) and CSD (**b**)

indexation file, the diffraction peaks were indexed considering a pseudo-cubic unit cell.

All the films are polycrystalline and single-phase with pseudo-cubic perovskite structure. Note that the broad peak observed at around $2\theta = 20°$ is assigned to the amorphous silica substrate. As expected, a shift of the Bragg angles to higher 2θ values is observed as Ti^{4+} content increases in the solid solution, indicating that a decrease of the a_{pc} lattice parameter occurs (Fig. 5).

This decrease in the lattice parameters can be attributed to a decrease of the cation average size ($r_{Sn}{}^{4+} = 0.690$ Å; $r_{Ti}{}^{4+} = 0.605$ Å) [11, 19] which promotes a contraction of the perovskite lattice as foreseen by the Vegard Law [57, 58]. Moreover, a higher covalent character is expected for Ti^{4+}–O^{2-} bond comparing to $Sn^{4+} - O^{2-}$. This same trend was observed by Stanulis et al. [62], Lavinscky et al. [63] and by our research group [19] for the same $Sr(Sn,Ti)O_3$ solid systems in their bulk form. Liu

Fig. 5 Calculated pseudo-cubic lattice parameter of $SrSn_{1-x}Ti_xO_3$ films grown on silica by PLD and CSD

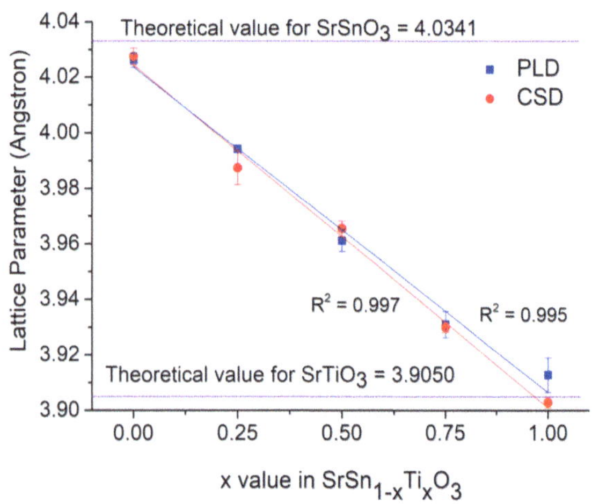

et al. [46] observed the same behavior on $SrSn_{1-x}Ti_xO_3$ thin films grown on MgO by PLD and we also observed this trend in our recent work on $SrSn_{1-x}Ti_xO_3$ films grown on (100) $LaAlO_3$ and R-sapphire single crystal substrates using both PLD and CSD methods [11].

All these results on long-range characterization by X-ray diffraction confirm that the nature of the substrate determines the type of growth: herein, we observed that the $Ca_{1-x}Sr_xSnO_3$ films when deposited on (100)$SrTiO_3$ single crystal substrates presented a highly epitaxial cube-on-cube growth, while the $SrTi_{1-x}Sn_xO_3$ thin films grown on amorphous SiO_2 (silica) substrates were polycrystalline.

3.2 *Microstructures of the Films*

3.2.1 $Ca_{1-x}Sr_xSnO_3$ Thin Films

The FE-SEM images in Fig. 6 revealed that all the $Ca_{1-x}Sr_xSnO_3$ films were homogeneous and crack-free, but the influence of deposition method and composition on the microstructure is clearly observed.

The films obtained by CSD presented smaller and smoother grains in comparison to the films deposited by PLD, regardless of the composition (Fig. 6). As previously mentioned, during CSD the nucleation of crystallites from an amorphous phase is followed by a growth process. Consequently, the crystallization process is slower and usually begins at a lower temperature than the PLD process, which leads to smaller grains. Moreover, samples deposited by CSD have generally a homogeneous surface, but porosity is also present, which could be due to the release of H2O and CO_2 during decomposition of the polymeric network.

The influence of the composition was more pronounced in PLD samples (Fig. 6). In this method, the formation of grains on the substrate surface can occur more easily at the very start of the deposition process, from which favors the growth of the grains as the deposition proceeds. For samples obtained by PLD, micrographies show small, spherical grains for $CaSnO_3$, which grow with the increase in Sr content and change to a square-like morphology for $SrSnO_3$. This behavior may be assigned to the difference in the perovskite structure, as $SrSnO_3$ exhibits less distortion between their octahedra compared to $CaSnO_3$, which facilitates the crystallization of the material [10, 23]. As a consequence, the grain growth effect is more pronounced when the Sr^{2+} content increases.

3.2.2 $SrSn_{1-x}Ti_xO_3$ Films

The FE-SEM micrographs for all the $SrSn_{1-x}Ti_xO_3$ films are presented in Fig. 7.

As stated previously, the morphology is influenced by the deposition method and the composition. In $SrSn_{1-x}Ti_xO_3$ samples the effect in morphology is less evident than in $Ca_{1-x}Sr_xSnO_3$ and relatively similar grain sizes were obtained, when the same

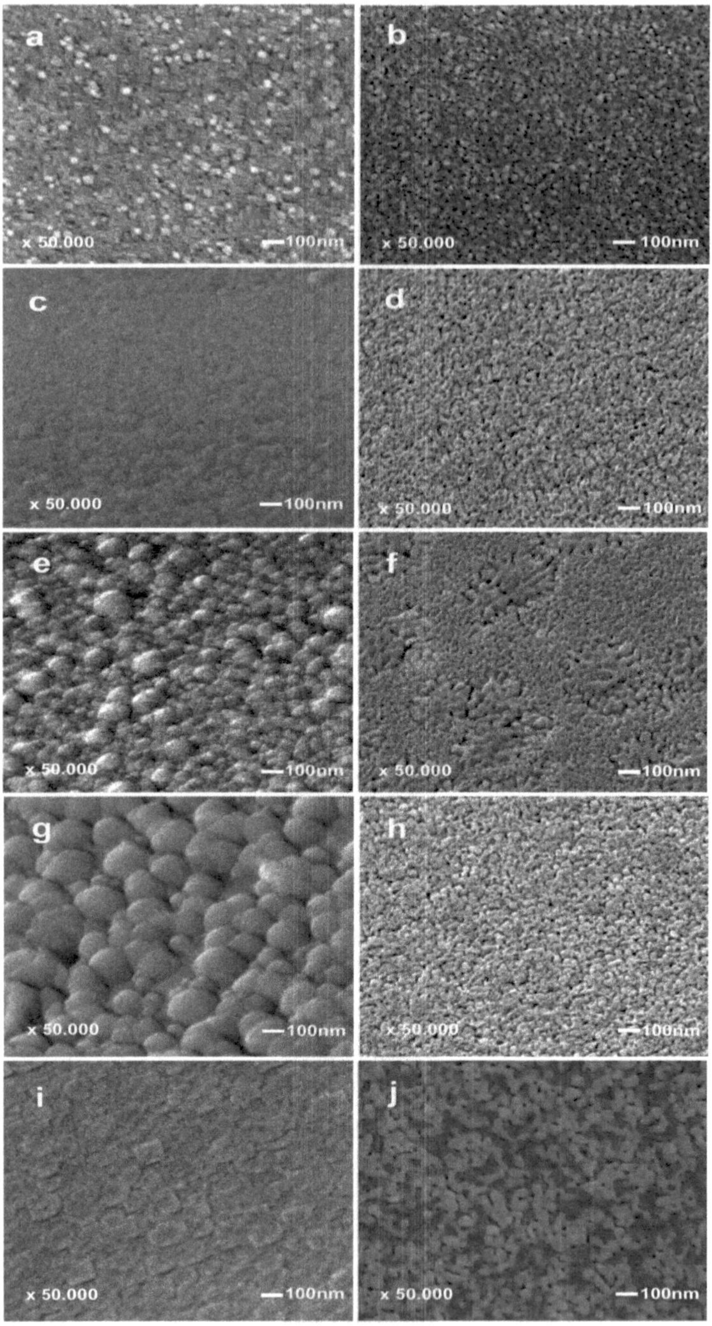

Fig. 6 FE-SEM images of the films deposited on STO prepared by PLD: **a** $CaSnO_3$; **c** $Ca_{0.75}Sr_{0.25}SnO_3$; **e** $Ca_{0.50}Sr_{0.50}SnO_3$; **g** $Ca_{0.25}Sr_{0.75}SnO_3$; **i** $SrSnO_3$; and prepared by CSD: **b** $CaSnO_3$ **d** $Ca_{0.75}Sr_{0.25}SnO_3$; **f** $Ca_{0.50}Sr_{0.50}SnO_3$; **h** $Ca_{0.25}Sr_{0.75}SnO_3$; and **j** $SrSnO_3$

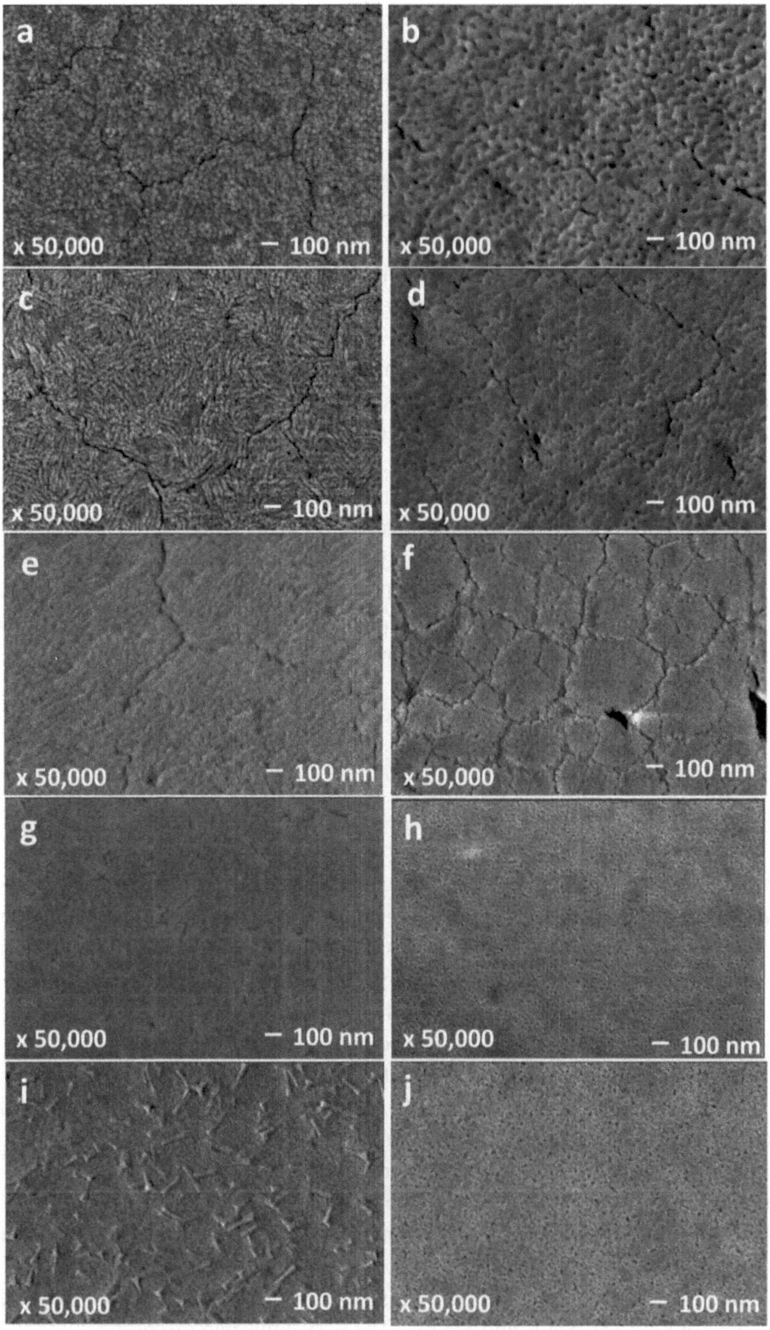

Fig. 7 FE-SEM images of the films deposited on silica prepared by PLD: **a** $SrTiO_3$; **c** $SrSn_{0.25}Ti_{0.75}O_3$; **e** $SrSn_{0.50}Ti_{0.50}O_3$; **g** $SrSn_{0.75}Ti_{0.25}O_3$; **i** $SrSnO_3$; or prepared by CSD: **b** $SrTiO_3$; **d** $SrSn_{0.25}Ti_{0.75}O_3$; **f** $SrSn_{0.50}Ti_{0.50}O_3$; **h** $SrSn_{0.75}Ti_{0.25}O_3$; **j** $SrSnO_3$

deposition method was used. This behavior is different from the results obtained by Liu et al. [46] for $SrSn_{1-x}Ti_xO_3$ thin films on MgO substrate, as well as by our research group for $SrSn_{1-x}Ti_xO_3$ films grown on $LaAlO_3$ and R-sapphire substrates [11], and may be related to the nature of the substrate and consequently to the type of growth.

As expected, films deposited by PLD were denser than those deposited by CSD, and it can be observed the presence of cracks for the Ti-richer films, mainly those synthesized by PLD. It is worth noting that cracks in thin films deposited on silica are very common. This is due to the low thermal expansion coefficient of silica substrate ($\alpha = 0.55 \times 10^{-6}$ °C^{-1} [64]) in comparison to the perovskite ones (as example, $\alpha = 7.8 \times 10^{-6}$ °C^{-1} for $SrTiO_3$ [65]). As the temperature decreases (inside the vacuum chamber after the PLD deposition, or after the crystallization heat treatment in the case of CSD method), $SrTiO_3$ tends to contract, while the dimension of silica substrate has a smaller variation. As a result, a tension in the film/substrate interface occurs and the $SrTiO_3$ film tends to crack.

As previously mentioned, the nature of the substrate determines the type of growth and consequently affects the microstructure. This fact is clearly evidenced when compared the microstructure of the epitaxial $SrSnO_3$ films obtained on (100)$SrTiO_3$ (Fig. 4i,j) and the polycrystalline ones obtained on silica (Fig. 7i,j). While a square-like morphology was observed for films deposited on (100)$SrTiO_3$ by PLD, a granular morphology with needles was obtained on silica.

3.3 Short-Range Characterization by Photoluminescence Spectroscopy

Photoluminescence may be due to intrinsic or extrinsic effects. The intrinsic photoluminescence can be divided into band-to-band luminescence and cross luminescence. Extrinsic luminescence further comprises unlocalized type such as donor–acceptor pair luminescence as well as luminescence caused by isoelectronic trap, and localized type like ionoluminescence and defect luminescence [53]. For instance, the extrinsic luminescence could be influenced by many factors, such as nature, amount, and distribution of dopants; size of the particles; synthesis method; defect concentration and degree of crystallinity [3, 11, 39, 60, 66, 67]. Specifically for films, besides crystal defects, impurities and composition, photoluminescence (PL) can also be used to evaluate electric fields and film stress [68, 69].

In this sense, PL spectroscopy was herein employed to investigate these effects on the structural order/disorder at short- and medium-range as a function of the A and B cationic substitutions in $ABSnO_3$ perovskites thin films, besides the influence of the substrate and deposition method. The PL spectra were deconvoluted into Gaussians functions for $Ca_{1-x}Sr_xSnO_3$ and $SrSn_{1-x}Ti_xO_3$ systems and discussed.

3.3.1 $Ca_{1-x}Sr_xSnO_3$ Thin Films

The PL spectra of three selected compositions in the $Ca_{1-x}Sr_xSnO_3$ series ($CaSnO_3$, $Ca_{0.50}Sr_{0.50}SnO_3$ and $SrSnO_3$) obtained using the PLD and CSD methods (Fig. 8a, b) exhibited emission bands with a profile similar to the (100)$SrTiO_3$ substrate but with a higher intensity in addition to bands at higher wavelengths for some films. A broad band emission is observed for all the materials, which characterizes a multiphonon process. This emission may be assigned to self-trapped excitons (STEs), for which relaxation occurs by several pathways and involves the participation of numerous energy sublevel states within the band gap induced by various factors as mentioned before [3, 10, 11, 37, 53, 70, 71].

The PL emission observed for the films produced using both deposition methods (Fig. 8) may be due to the epitaxial growth, which induced a short-range disorder in the films due to strains imposed by the long-range ordering. The highest short-range disorder may occur at the film/substrate interface to enable epitaxial growth. Consequently, dipoles are formed between the well-ordered SnO_6 (far from the interface) and the disordered ones (located at the interface), which leads to the photoluminescence [10, 11].

Wang and Zhang [72] reported that (0001) oriented and non-oriented ZnO films obtained using CSD presented two distinct emissions: a sharp UV emission centered at 380 nm and other visible emission characterized by a broad band centered at 560 nm. The authors attributed the UV emission to the recombination of the free excitons or near band edge emission of the wide band gap ZnO, while the broad band emission was associated to defects, especially oxygen vacancies and zinc interstitial present in the film. By calculating the ratio of these two emissions, in particular, the authors concluded that film orientation has an important effect on the increase of the broad band emission in visible range of the spectrum. Oh et al. [73] investigated the substrate dependence on structural and photoluminescence (PL) properties of $SrLaMgTaO_6$ (SLMTO) double-perovskite thin films deposited on (001)MgO and (001)$SrTiO_3$

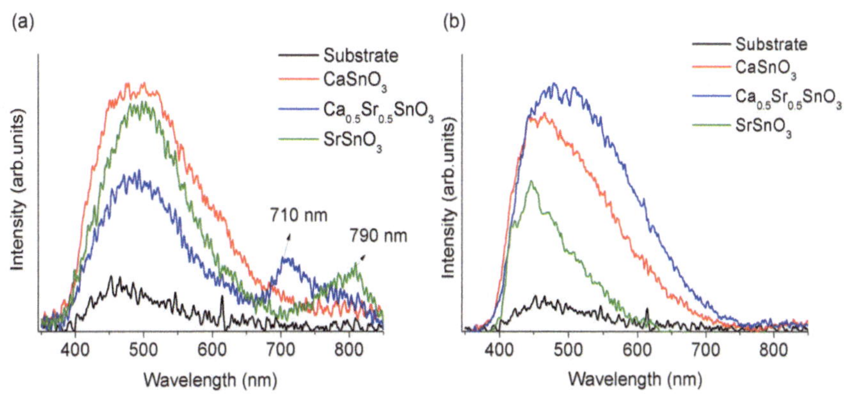

Fig. 8 PL spectra of $Ca_{1-x}Sr_xSnO_3$ films deposited on STO substrate by PLD (**a**) and CSD (**b**)

substrates by PLD. As such, the PL spectra of the polycrystalline SLMTO films on MgO were similar to those of the SLMTO bulk, which exhibited emissions in the spectral range of 400–750 nm (green-red), while the epitaxial SLMTO films on (100)$SrTiO_3$ showed emissions in the range of 475–530 nm (green). Studies based on XPS analysis revealed that Mg–O and Ta–O chemical bindings have different chemical binding states on the surfaces of the films and that the green as well as red emission bands are related to the oxygen vacancies associated to Ta–O and Mg–O environments, respectively. In a previous work from our research group concerning $SrSn_{1-x}Ti_xO_3$ films, we evidenced that the epitaxial films grown on $LaAlO_3$ single crystal substrates presented higher PL in comparison to polycrystalline films grown on R-sapphire. Hence, the short-range disorder in the films on $LaAlO_3$ lattice is responsible for the creation of energy levels within the bandgap, leading to intense PL responses occurring mostly at lower energy range.

About perovskites, Longo et al. [37, 70] observed broadband photoluminescence in disordered $SrTiO_3$ and $SrZrO_3$ materials and correlated emissions at the different regions of the visible spectrum, associated with different types of energy levels inside the band gap, due to short- and medium-range disorder in the lattice. In this sense, emissions in the violet-blue-green regions are associated with shallow defects in the band gap, whereas yellow-red emissions are due to defects inserted deeply into the band gap. Shallow defects are related to energy states close to the valence or to the conduction band originated from less distorted structures, while the deep defects are assigned to energy states close to the middle of the bandgap and are usually associated with obtained of highly distorted structures. Particularly, for perovskite films, PL intensity tends to decrease when the films attain a higher short-range order at film-substrate interface [11, 74–76]. For instance, Zhai et al. [77] observed a shift of the emission to higher energies as a higher ordering was attained for $SrTiO_3$ films on $LaAlO_3$ buffered by $SrRuO_3$ and associated the blue-light emission to intrinsic structural defects by considering that the double-oriented structure obtained in the films induced distorted clusters.

In this context, the PL curves were deconvolved to evaluate the relationship between the emission and the short-range order. As an example, Fig. 9 presents the deconvolved spectrum obtained for the $CaSnO_3$ film deposited by CSD. All the spectra exhibited four to five components centered at different ranges in the visible spectrum, which depended on the composition and on the synthesis method (Tables 3 and 4).

In relation to the synthesis method, it could be observed that the emissions are centered at lower wavelengths for the $CaSnO_3$ and $SrSnO_3$ films obtained by CSD in comparison to the films obtained by PLD. For the films synthesized by PLD, emissions in the red and infrared regions are observed, whereas the films synthesized by CSD only emit up to the orange region.

In this context, according to the reasoning above and as a consequence of the broad-band characteristic of the PL, all the emission bands observed in the present case were characterized according to the percentage of shallow defects (calculated considering the sum of the violet, blue and green emissions) and the deep defects (considered the sum of the yellow, red and infrared emissions) [37, 70]. According

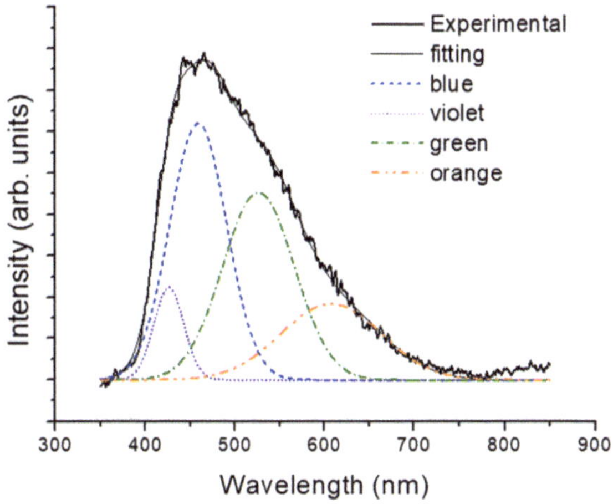

Fig. 9 Deconvolution of the PL spectrum of $CaSnO_3$ film on STO obtained by CSD

Table 3 Results of the PL deconvolution spectra of $Ca_{1-x}Sr_xSnO_3$ films grown on STO by PLD

	Shallow defects				Deep defects					
Sample	Violet		Blue-green		Yellow		Red		Infrared	
	λ_m (nm)	(%)	λ_m (nm)	(%)	λ_m (nm)	(%)	λ_m (nm)	(%)	λ_m (nm)	(%)
$CaSnO_3$	436	10	490	41	577	26	714	23	–	–
$Ca_{0.5}Sr_{0.5}SnO_3$	426	1	483	56	581	19	713	16	794	8
$SrSnO_3$	433	7	496	61	598	19	729	2	797	11

Table 4 Results of the PL deconvolution spectra of $Ca_{1-x}Sr_xSnO_3$ films grown on STO by CSD

	Shallow defects						Deep defects	
Sample	Violet		Blue		Green		y/o/o-r*	
	λ_m (nm)	(%)	λ_m (nm)	(%)	λ_m (nm)	(%)	λ_m (nm)	(%)
$CaSnO_3$	426	7	459	38	526	35	607 (o)	20
$Ca_{0.5}Sr_{0.5}SnO_3$	436	8	472	32	539	37	618 (o)	23
$SrSnO_3$	418	14	442	20	509	25	559 (y)	14
			469	27				
STO substrate	–	–	464	60	548	25	660 (o-r)	15

*y: yellow; o: Orange; o-r: Orange-red

to Figure 10, the emissions related to the shallow defects (violet, blue and green) represent 77–86% for the films obtained by CSD and 51–68 % for those obtained using PLD, which indicates a lower short-range disorder for the CSD samples.

In comparing $SrSnO_3$ and $CaSnO_3$, both methods resulted in a higher proportion of shallow defects in the $SrSnO_3$, especially for the PLD samples. This result indicates that the $SrSnO_3$ has a lower degree of short-range disorder. According to

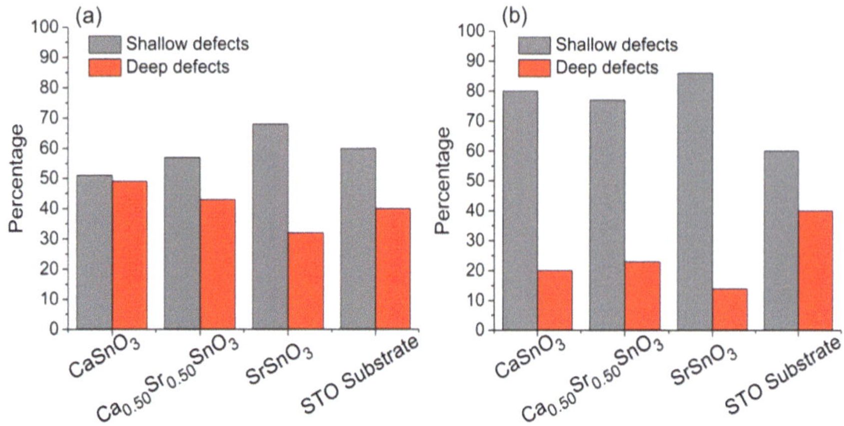

Fig. 10 Classification of the emission bands of the PL spectra into deep and shallow defects for films obtained by PLD (**a**) and CSD (**b**)

Mizoguchi et al. [78], $SrSnO_3$ and $CaSnO_3$ have distorted cubic symmetries due to the tilting among the octahedra. This tilting is different depending on the modifier cation (Sr^{2+} or Ca^{2+}), with larger tilting angles for $CaSnO_3$ than for $SrSnO_3$. Consequently, we believe that a higher short-range disorder at the film/substrate interface is required to obtain epitaxy in $CaSnO_3$. Hence, the $Ca_{0.50}Sr_{0.50}SnO_3$ film obtained by PLD confirms that a monotonic change in the short-range order occurs, with an intermediate amount of shallow defects for this composition.

For the films obtained by CSD, the evolution of the amount of shallow/deep defects with the composition was not so remarkable. Moreover, the amount of deep defects is still quite low regardless of the composition. This is in agreement with the emission in higher energy regions observed for these films, indicating that intermediate levels in the bandgap are nearer the valence band or the conduction band, being related to shallow defects. This behavior could be attributed to a less distorted structure of the perovskite, comparing the CSD with the PLD results. For CSD, the crystallization process is slower, which is characteristic of the "soft chemistry" approach with an increasing diffusion of the ions as heating occurs. This process most likely leads to less short-range disorder, for both lattice modifier cations. In addition, in contrast with PLD which promotes an *in situ* epitaxial growth, it can be assumed that CSD *ex situ* "soft chemistry" growth results in a lower short-range disorder at the film/substrate interface. Consequently, a smaller difference among the octahedra in the interface and far from the interface occurs; thus, emissions related to shallow defects (violet, blue and green emission) are more evident.

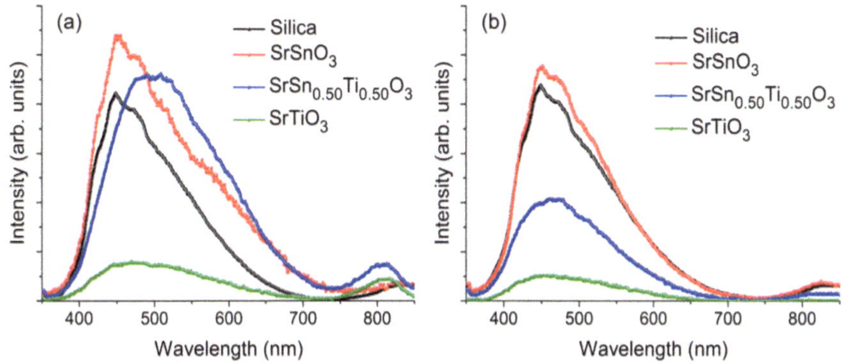

Fig. 11 PL spectra of $SrSn_{1-x}Ti_xO_3$ films deposited on silica by PLD (**a**) and CSD (**b**)

3.3.2 $SrSn_{1-x}Ti_xO_3$ Films

As in the previous study, three selected compositions ($SrSnO_3$, $SrSn_{0.50}Ti_{0.50}O_3$ and $SrTiO_3$) of the $SrSn_{1-x}Ti_xO_3$ films grown on silica substrate deposited by PLD and CSD methods were investigated by photoluminescence (PL). Their corresponding PL spectra are shown in Fig. 11a, b.

Room temperature PL was observed for $SrSn_{1-x}Ti_xO_3$ films with a broad emission band, centered at 460–520 nm (2.38–270 eV). This is a typical multiphonon processes, as previously discussed for the $Ca_{1-x}Sr_xSnO_3$ films. Multiphonon emission is, of course, characteristic of disordered materials, thus silica also displays a noticeable emission. Amorphous silica is very wide band gap ranging 7.5–9.2 eV [79]. As the energy of the emission is considerably smaller than the band gap, PL can be assigned to the presence of medium-range disorder in the material (which leads to the formation of energy levels in the band gap as stated before).

Spectra profile exhibits some changes according to the composition and to the deposition method. It is well-established that the PL efficiency at room temperature drastically decreases with the increase of structural order [37, 70, 71, 74–76]. Although all the $SrSn_{1-x}Ti_xO_3$ thin films on silica are polycrystalline, in particular, short-range disorder at the film-substrate interface can occur and it differs according to the deposition method As stated earlier, crystallization process in films deposited by PLD is more energetic as it occurs by a fast in situ crystallization at high temperature. In contrast, the crystallization in CSD occurs with lower energetic balance that facilitates the accommodation of the material on the substrate during film growth. Thus, distinct short-range disorders at the film/substrate interface are achieved for films with the same composition deposited by PLD or CSD methods. For films deposited by CSD, spectra seems to be mainly due to bare silica substrate, especially for $SrSnO_3$ (Fig. 11b), which indicates that a smaller short-range disorder is present in these films, as expected. For films deposited by PLD (Fig. 11a), the spectra profile is directly related to the composition, but all the $SrSn_{1-x}Ti_xO_3$ films have an emission above 750 nm (longer wavelengths and lower energy), which may be assigned

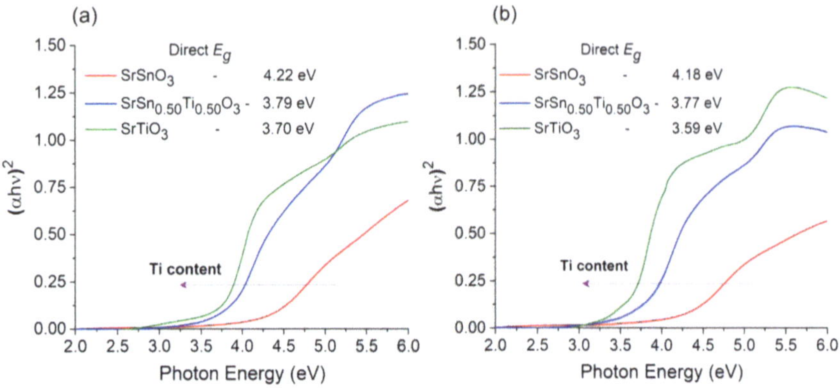

Fig. 12 UV–vis absorption spectra with the estimated direct E_g values for $SrSn_{1-x}Ti_xO_3$ thin films on silica deposited by **a** PLD and **b** CSD

to the increase of deep defects in the band gap associated to the short and medium-range structural disorder in the film-substrate interface as observed for the epitaxial $Ca_{1-x}Sr_xSnO_3$ films on (100)$SrTiO_3$ substrate deposited by the same method, which display a band at 790 nm for $SrSnO_3$.

In relation to the thin film composition, Ti clearly plays an important role on the emission profile. This behavior may in principle be related to the UV–vis spectra (Fig. 12a, b), as luminescence is closely related to the material electronic properties, especially the band gap. Similar UV–vis spectral features with a rapid increase in the absorption from 3.0 eV were observed for the films whatever the deposition method, due to the sensitivity of the method which is not able to detect interfacial disorder. Thus, the optical band gap (E_g) of the samples were also calculated from the absorbance curves according to the Tauc's method [58]. The results indicated that the E_g decreases as a function of the Ti-content, in agreement with Liu et al. [46] that observed a decrease of the band gap in $SrSn_{1-x}Ti_xO_3$ thin films grown on MgO substrates.

$SrTiO_3$ displays a low emission intensity, compared to the silica substrate (Fig. 11). This behavior may be assigned to the presence of electron traps, such as oxygen vacancies associated to the coexistence of Ti^{4+} and Ti^{3+} species in TiO_6 octahedra in $SrTiO_3$ lattice. The UV–vis spectra display an absorption tail at lower energy (<3.5 eV) for Ti-containing films. This spectral characteristic may be related to the presence of mixed Ti^{3+}/Ti^{4+} and defects such as oxygen vacancies associated to Ti^{4+} and Ti^{3+} (Ti^{4+}/Ti^{3+}–V_O), while a ligand/metal charge transfer (LMCT) is observed above 4.0 eV. The influence of Ti^{3+} on the PL spectra for Ti-containing films was also studied by other authors. For instance, Orhan et al. [80, 81] studied the dependence of structural disorder on PL of $SrTiO_3$ thin films deposited on Pt (140 nm)/Ti (10 nm)/SiO_2 (1000 nm)/Si substrates. Similarly, Mikhailik et al. [82] observed a PL quenching on Ti-doped Al_2O_3 assigned to a competition between oxygen vacancies and Ti^{4+} centers for the electron trapping. Finally, the film-substrate interface can present very particular characteristics, as observed in anatase TiO_2 thin films grown on (100)$SrTiO_3$ substrate [83]. $SrTiO_3/TiO_2$ interface presented a metallic character while the interfacial bonding presented a mixed covalent and ionic character which increased the charge confinement and then reduced PL in such films. Similar behavior was observed by Chiu et al. [84] for MoS_2, WSe_2 and MoS_2/WSe_2 heterostructured films deposited on sapphire, by Tongay et al. [85] for MoS_2/WS_2 heterostructures on silicon substrates and also by Chen et al. [86] for MoS_2/WS_2 deposited on silica. The authors demonstrated that luminescence observed in such films is due to the independent contribution of each layer. Yet, in the stacked MoS_2/WSe_2 and MoS_2/WS_2 layers negligible emission was observed, which indicated that photoexcited carriers were trapped in the material interface and quenched PL.

$SrSnO_3$ films exhibits PL profile quite similar to the bare silica substrate (Fig. 11). This behavior can be directly related to the band gap structure, as $SrSnO_3$ films have an $E_g = 4.2$ eV and a very small energy absorption up to 4.0 eV. As the energy of the laser used in the PL measurements was 3.5 eV (350.7 nm), few electrons are excited during analysis and the PL seems to be solely due to the substrate, especially for films deposited by CSD, which have a smaller short and medium-range disorder. As such, the subtraction of the spectral contribution of the substrate evidences a second emission band around 600 nm for the film obtained by PLD in comparison to the CSD one, as displayed in Fig. 13.

Hence, the $SrSn_{0.50}Ti_{0.50}O_3$ films presented an intermediary PL intensity, whatever the method. This behavior might be associated to the competitive structural ordering during crystallization process, induced by existence of both Ti^{4+} and Sn^{4+} cations sharing the octahedra sites in the perovskite lattice. For these films, a red shift is clearly observed when deposition was done by PLD (Fig. 11a).

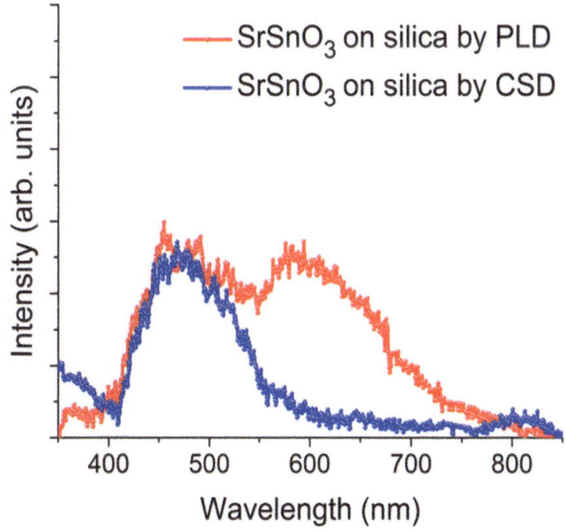

Fig. 13 Result of the spectral subtraction between the emission band of the $SrSnO_3$ and the silica substrate for the films deposited by both methods

Considering the behavior observed for $SrSnO_3$ and $SrSn_{0.50}Ti_{0.50}O_3$ films deposited on silica by PLD and taking into account that multiphonon processes during PL are generated in response to the numerous energy sublevel states created within the band gap (shallow and deep defects), the PL bands of these films were analyzed according to deconvoluted Gaussian functions as indicated by the example given for $CaSnO_3$ thin film on STO (Fig. 9). The deconvolution analysis of the PL curves shows four color components as summarized in Table 5 and the amount of shallow and deep defects was calculated according to Longo et al. [37, 70], as displayed in Fig. 14.

Comparing to bare silica, both films have more deep defects, which is related to the short- and medium-range disorder in the perovskite lattice due to the crystallization process during film formation by PLD. For $SrSn_{0.50}Ti_{0.50}O_3$ film, this is also evidenced by the red shift of the deconvoluted peaks at the violet and blue-green regions, besides the emission at 800 nm. Hence, this behavior is originated from competitive effects induced by Ti and Sn cations in the $SrSn_{0.50}Ti_{0.50}O_3$ lattice.

Table 5 Results of the deconvolution of the PL spectra of the $SrSn_{0.50}Ti_{0.50}O_3$ films grown on silica by PLD

	Shallow defects				Deep defects	
Sample	Violet		Blue-green		Yellow	
	λ_m (nm)	(%)	λ_m (nm)	(%)	λ_m (nm)	(%)
$SrSnO_3$	421	4	447/ 492	26/ 39	574	31
$SrSn_{0.50}Ti_{0.50}O_3$	432	8	472/ 521	31/ 24	580	37
Silica	429	16	459/ 511	29/ 40	585	15

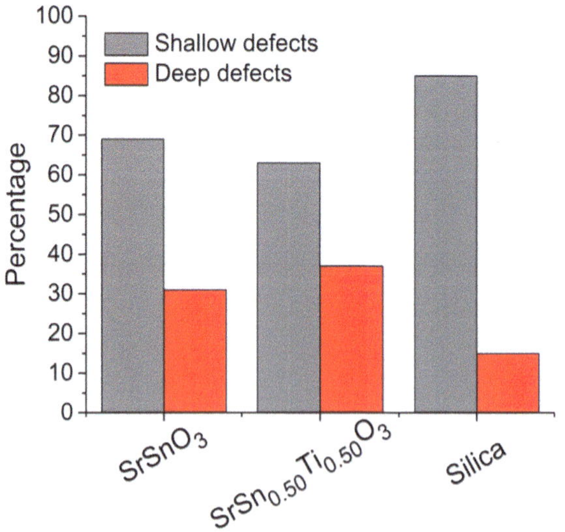

Fig. 14 Classification of the emission bands of the PL spectra into deep and shallow defects for films obtained by PLD for $SrSnO_3$ and $SrSn_{0.50}Ti_{0.50}O_3$ films

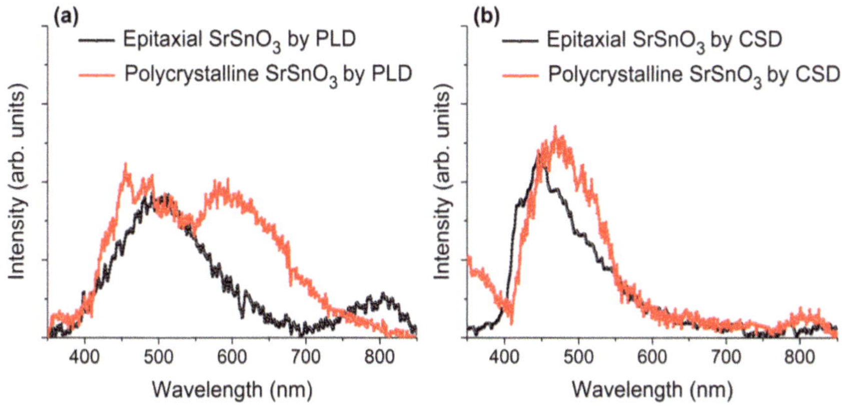

Fig. 15 PL spectra of the epitaxial $SrSnO_3$ films deposited (100)$SrTiO_3$ (STO) and polycrystalline films deposited on silica and substrates by PLD (**a**) and CSD (**b**)

Among these two films, the short-range disorder responsible for the distortion of the crystal structure of $SrSnO_3$ favors PL at lower energy range in the spectra, which are consistent with the fact of our previous reported work in which the PL intensity and the amount of shallow/deep defects were strongly associated to the type of the octahedral distortion present in the lattice.

Finally, comparison of the epitaxial and polycrystalline $SrSnO_3$ thin films deposited by both PLD and CSD was done after subtraction of the PL spectra contribution of the respective substrate (Fig. 15).

As $SrSnO_3$ has an E_g value around 4.2 eV, band-to-band emission should occur at 350 nm. So, the emission observed in the present work is related to midgap levels associated to the presence of defects in the material. In relation to films deposited by CSD, the spectral profiles are very similar with similar intensities and emission around the same region (centered between 450 and 550 nm). This behavior may in principle be assigned to the short- and mid-range disorder as consequence of the tilting among octahedra present in this type of perovskite.

In our recent study concerning $SrSnO_3$ particles, we have evidenced that multiphonon processes during PL response, which were also associated to two different type of distortions in the perovskite lattice: (i) an intraoctahedral distortion (associated with short-range disorder) that is related to the A–O and B–O interatomic distances as well as to O–Sn–O angles, and (ii) an interoctahedral distortion (associated with medium-range disorder) that is related to the cooperative tilting of the octahedral. As such, the differences in short- and medium-range disorder in $SrSnO_3$-type materials introduce intermediate more energy levels (shallow and deep defects) in the bandgap leading to different PL emissions [3].

For films deposited by PLD, emission between 450 and 550 nm is also observed, but one more emission band is present in the spectra whose position depends on the thin film growth (epitaxial or polycrystalline). As previously discussed, PL can be associated to the short- and medium-range disorder in the lattice and/or at the film/substrate interface. For the polycrystalline film deposited on silica, in particular, the second emission band is observed at 600 nm, which might be assigned to the stress induced by the difference in the thermal expansion coefficient of silica substrate and that of the perovskite one that gave rise to cracks in the microstructure of the films. For the epitaxial film deposited on STO, the emission band is observed at 800 nm and it may be assigned to the orientation process, which induces a higher disorder in the interface and leads to a higher amount of deep defects.

4 Conclusions

Herein, epitaxial $Ca_{1-x}Sr_xSnO_3$ (x = 0, 0.25, 0.50, 0.75 and 1) thin films grown on (100)$SrTiO_3$ single crystal substrates and polycrystalline $SrTi_{1-x}Sn_xO_3$ (x = 0, 0.25, 0.50, 0.75 and 1) films on amorphous silica substrates were obtained using PLD and CSD techniques. The deposition method, nature of substrate and composition played an important role on structural, microstructure and photophysical characteristics of the films. All the thin films presented room temperature photoluminescence. Broad band emissions were evidenced in different visible spectrum wavelengths according to film composition, substrate and deposition method, which indicated that short-range order/disorder can be modulated by choosing appropriate conditions for the preparation of the thin films. As such, the PL spectra were divided into color emission components that are associated with the type of defects within the band gap of the materials (shallow defects with emissions at higher energies and deep defects with emissions at lower energies. The films grown by CSD presented the smaller short-range disorder, which might be related to the nucleation and growth processes promoted by the 'soft chemistry' route. A smaller short-range disorder was attained in $SrSnO_3$ when compared to $CaSnO_3$ on STO, while films on silica, especially when deposited by CSD presented PL responses mainly induced by the amorphous substrate, which is highly disordered. In addition, the type of growth (epitaxial or polycrystalline) of $SrSnO_3$ films provoked changes in the PL responses, which were associated to the short-range disorder at the film/substrate interface. To conclude, parameters as composition, nature of substrate, deposition method have not only influence on structural properties and microstructural characteristics of the films, but also have a strong effect on observed PL behavior.

Acknowledgements The authors acknowledge the French-Brazilian doctoral college and CAPES-COFECUB (Project 644/09), INCT/CNPq/MCTI and PROINFRA/FINEP/MCTI for financial support and Francis Gouttefangeas for SEM images performed at CMEBA (SCanMat, UMS 2001 CNRS—University of Rennes 1).

References

1. Zhong, F., Zhuang, H., Gu, Q., Long, J.: Structural evolution of alkaline earth metal stannates $MSnO_3$ (M = Ca, Sr, and Ba) photocatalysts for hydrogen production. RSC Adv. 6(48), 42474–42481 (2016). https://doi.org/10.1039/C6RA05614H
2. Zhang, W.F., Tang, J., Ye, J.: Photoluminescence and photocatalytic properties of $SrSnO_3$ perovskite. Chem. Phys. Lett. **418**(1), 174–178 (2006). https://doi.org/10.1016/j.cplett.2005.10.122
3. Chantelle, L., et al.: Probing the site-selective doping in $SrSnO_3$: Eu oxides and its impact on the crystal and electronic structures using synchrotron radiation and DFT simulations. Inorg. Chem. **59**(11), 7666–7680 (2020). https://doi.org/10.1021/acs.inorgchem.0c00664
4. Tian, X., et al.: Enhanced photoluminescence and high temperature sensitivity from a novel Pr^{3+} doped $SrSnO_3/SnO_2$ composite phosphor. J. Solid State Chem. **280**, 120997 (2019). https://doi.org/10.1016/j.jssc.2019.120997
5. Teixeira, A.R.F.A., et al.: $SrSnO_3$ perovskite obtained by the modified Pechini method—insights about its photocatalytic activity. J. Photochem. Photobiol A **369**, 181–188 (2019). https://doi.org/10.1016/j.jphotochem.2018.10.028
6. Junploy, P., Phuruangrat, A., Thongtem, S., Thongtem, T.: Effect of surfactants on phase, crystal growth and photocatalysis of calcium stannate synthesized by cyclic microwave and calcination combination. Res. Chem. Intermed. **44**(10), 5981–5993 (2018). https://doi.org/10.1007/s11164-018-3470-5
7. Veerappan, G., Yoo, S, Zhang, K., Ma, M., Kang, B., Park, J.H.: High-reversible capacity of Perovskite $BaSnO_3$/rGO composite for Lithium-Ion Battery anodes. Electrochimica Acta **214**, 31–37, 2016. https://doi.org/10.1016/j.electacta.2016.07.076
8. Exner, J., Nazarenus, T., Kita, J., Moos, R.: Dense Y-doped ion conducting perovskite films of $BaZrO_3$, $BaSnO_3$, and $BaCeO_3$ for SOFC applications produced by powder aerosol deposition at room temperature. Int. J. Hydrogen Energy **45**(16), 10000–10016 (2020). https://doi.org/10.1016/j.ijhydene.2020.01.164
9. Muthukutty, B., Karthik, R., Chen, S.-M., Abinaya, M.: Designing novel perovskite-type strontium stannate ($SrSnO_3$) and its potential as an electrode material for the enhanced sensing of anti-inflammatory drug mesalamine in biological samples. New J. Chem. **43**(31), 12264–12274 (2019). doi: https://doi.org/10.1039/C9NJ02197C
10. Alves, M.C.F., et al.: Influence of the network modifier on the characteristics of $MSnO_3$ (M=Sr and Ca) thin films synthesized by chemical solution deposition. J. Solid State Chem. **199**, 34–41 (2013). https://doi.org/10.1016/j.jssc.2012.11.014
11. Menezes de Oliveira, A.L., et al.: Evolution of the structural and microstructural characteristics of $SrSn_{1-x}Ti_xO_3$ thin films under the influence of the composition, the substrate and the deposition method. Surf. Coat. Technol. **313**, 361–373 (2017). https://doi.org/10.1016/j.surfcoat.2017.01.082
12. Ueda, K., Shimizu, Y.: Fabrication of Tb–Mg codoped $CaSnO_3$ perovskite thin films and electroluminescence devices. Thin Solid Films **518**(11), 3063–3066 (2010). https://doi.org/10.1016/j.tsf.2009.09.169
13. Ueda, K., et al.: Photoluminescence from epitaxial films of perovskite-type alkaline-earth stannates. Appl. Phys. Express **1**(1), 015003 (2008). https://doi.org/10.1143/apex.1.015003
14. Prodjosantoso, A.K., Zhou, Q., Kennedy, B.J.: Synchrotron X-ray diffraction study of the $Ba_{1-x}SrSnO_3$ solid solution. J. Solid State Chem. **200**, 241–245 (2013). https://doi.org/10.1016/j.jssc.2013.01.015
15. Kennedy, B.J., Qasim, I., Knight, K.S.: Low temperature structural studies of SrSnO3. J. Phys.: Condens. Matter **27**(36), 365401 (2015). https://doi.org/10.1088/0953-8984/27/36/365401
16. Sales, H.B., et al.: $Sr_{1-x}Ba_xSnO_3$ system applied in the photocatalytic discoloration of an azo-dye. Solid State Sci. **28**, 67–73 (2014). https://doi.org/10.1016/j.solidstatesciences.2013.12.007
17. Liu, Y., et al.: Composition dependent intrinsic defect structures in $ASnO_3$ (A = Ca, Sr, Ba). J. Mater. Sci. Technol. **42**, 212–219 (2020). https://doi.org/10.1016/j.jmst.2019.10.015

18. Liu, G., et al.: Structure evolution, ferroelectric properties, and energy storage performance of $CaSnO_3$ modified $BaTiO_3$-based Pb-free ceramics. J. Alloy. Compd. **826**, 154160 (2020). https://doi.org/10.1016/j.jallcom.2020.154160
19. de Oliveira, A.L.M., et al.: Effect of the composition on the thermal behaviour of the $SrSn_{1-x}Ti_xO_3$ precursor prepared by the polymeric precursor method. J. Therm. Anal. Calorim. **114**(2), 565–572 (2013). https://doi.org/10.1007/s10973-013-3051-1
20. Sumithra, S., Jaya, N.V.: Structural, optical and magnetization studies of Fe-doped $CaSnO_3$ nanoparticles via hydrothermal route. J. Mater. Sci. Mater. Electron. **29**(5), 4048–4057 (2018). https://doi.org/10.1007/s10854-017-8348-6
21. Stanulis, A., Katelnikovas, A., Van Bael, M., Hardy, A., Kareiva, A., Justel, T.: Photoluminescence of Pr^{3+}-doped calcium and strontium stannates. J. Lumin. **172**, 323–330 (2016). https://doi.org/10.1016/j.jlumin.2015.11.021
22. Alves, M.C.F., Souza, S.C., Lima, S.J.G., Longo, E., Souza, A.G., Santos, I.M.G.: Influenceof the precursor salts in the synthesis of $CaSnO_3$ bythe polymeric precursor method. J. Therm. Anal. Calorim. **87**(3), 763–766 (2007). https://doi.org/10.1007/s10973-006-7853-2
23. Alves, M.C.F., et al.: Influence of the modifier on the short and long range disorder of stannate perovskites. J. Alloy. Compd. **476**(1), 507–512 (2009). https://doi.org/10.1016/j.jallcom.2008.09.034
24. Wakana, H., et al.: Examination of deposition conditions for $SrSnO_3$ insulating layer for single flux quantum circuits. Physica C (Amsterdam, Neth.) **426–431**, 1495–1501 (2005). https://doi.org/10.1016/j.physc.2005.01.074
25. Gao, D., Gao, X., Wu, Y., Zhang, T., Yang, J., Li, X.: Co-doped $SrSnO_3$ epitaxial thin films on MgO with tunable band gap and room-temperature ferromagnetism. Physica E **109**, 101–106 (2019). https://doi.org/10.1016/j.physe.2019.01.010
26. W. Du et al.: Bilayer $BaSnO_3$ thin film transistors on silicon substrates. J. Mater. Chem. C **8**(15), 5231–5238 2020. https://doi.org/10.1039/C9TC04815D
27. Luo, B., Hu, J.: Unraveling the oxygen effect on the properties of sputtered $BaSnO_3$ thin films. ACS Appl. Electron. Mater. **1**(1), 51–57 (2019). https://doi.org/10.1021/acsaelm.8b00007
28. Liu, Q.Z., Wang, H.F., Chen, F., Wu, W.: Single-crystalline transparent and conductive oxide films with the perovskite structure: Sb-doped $SrSnO_3$. J. Appl. Phys. **103**(9), 093709 (2008). https://doi.org/10.1063/1.2917413
29. Liu, Q., Dai, J., Zhang, X., Zhu, G., Liu, Z., Ding, G.: Perovskite-type transparent and conductive oxide films: Sb- and Nd-doped $SrSnO_3$. Thin Solid Films **519**(18), 6059–6063 (2011). https://doi.org/10.1016/j.tsf.2011.03.038
30. Liu, Q., Jin, F., Gao, G., Wang, W.: Ta doped $SrSnO_3$ epitaxial films as transparent conductive oxide. J. Alloy. Compd. **717**, 62–68 (2017). https://doi.org/10.1016/j.jallcom.2017.05.080
31. Alves, M.C.F., et al.: Synthesis of $SrSnO_3$ thin films by pulsed laser deposition: influence of substrate and deposition temperature. Thin Solid Films **519**(2), 614–618 (2010). https://doi.org/10.1016/j.tsf.2010.07.092
32. Song, S., Zhai, J., Gao, L., Yao, X.: Orientation-dependent dielectric properties of $Ba(Sn_{0.15}Ti_{0.85})O_3$ thin films prepared by sol–gel method. J. Phys. Chem. Solids **70**(8), 1213–1217 (2009). https://doi.org/10.1016/j.jpcs.2009.07.007
33. Jain, G.H., Patil, L.A., Gaikwad, V.B.: Studies on gas sensing performance of $(Ba_{0.8}Sr_{0.2})(Sn_{0.8}Ti_{0.2})O_3$ thick film resistors. Sens. Actuators B Chem. **122**(2), 605–612 (2007). https://doi.org/10.1016/j.snb.2006.07.011
34. Mochizuki, S., Fujishiro, F., Shibata, K., Ogi, A., Konya, T., Inaba, K.: Optical, electrical, and X-ray-structural studies on Verneuil-grown $SrTiO_3$ single crystal: annealing study. Physica B **401–402**, 433–436 (2007). https://doi.org/10.1016/j.physb.2007.08.205
35. Wang, Z. et al.: Effects of Sr/Ti ratio on the microstructure and energy storage properties of nonstoichiometric $SrTiO_3$ ceramics. Ceram. Int. **40**(1), 929–933 (2014). (Part A). https://doi.org/10.1016/j.ceramint.2013.06.088
36. Zhang, Z., et al.: Rod-like incipient ferroelectric $SrTiO_3$ polycrystal with crystal-axis orientation. Ceram. Int. **46**(3), 3675–3687 (2020). https://doi.org/10.1016/j.ceramint.2019.10.088

37. Longo, V.M., et al.: Structural conditions that leads to photoluminescence emission in $SrTiO_3$: an experimental and theoretical approach. J. Appl. Phys. **104**(2), 023515 (2008). https://doi.org/10.1063/1.2956741
38. Zhang, J., Hao, Y., Wang, Q., Xu, J., Guo, L., Bi, K.: Enhanced photoluminescence properties of $SrTiO_3$:Pr^{3+} nanocrystals by the TEG-sol method. APL Mater. **6**(8), 086102 (2018). https://doi.org/10.1063/1.5041821
39. da Silva, L.F. et al.: Unveiling the efficiency of microwave-assisted hydrothermal treatment for the preparation of $SrTiO_3$ mesocrystals. Phys. Chem. Chem. Phys. **21**(39), 22031–22038 (2019). https://doi.org/10.1039/C9CP02893E
40. Lim, J., Lim, H., Lee, Y.S.: Ambient dependence of visible emissions in $SrTiO_3$. Curr. Appl. Phys. **19**(11), 1177–1181 (2019). https://doi.org/10.1016/j.cap.2019.07.016
41. Trabelsi, H., et al.: Raman, EPR and ethanol sensing properties of oxygen-vacancies $SrTiO_{3-\delta}$ compounds. Appl. Surf. Sci. **426**, 386–390 (2017). https://doi.org/10.1016/j.apsusc.2017.07.128
42. Sreedhar, G., et al.: A role of lithiated sarcosine TFSI on the formation of single crystalline $SrTiO_3$ nanocubes via hydrothermal method. Mater. Lett. **133**, 127–131 (2014). https://doi.org/10.1016/j.matlet.2014.06.170
43. da Silva, L.F., et al.: An understanding of the photocatalytic properties and pollutant degradation mechanism of $SrTiO_3$ nanoparticles. Photochem. Photobiol. **92**(3), 371–378 (2016). https://doi.org/10.1111/php.12586
44. Lyu, H. et al.: An Al-doped $SrTiO_3$ photocatalyst maintaining sunlight-driven overall water splitting activity for over 1000 h of constant illumination. Chem. Sci. **10**(11), 3196–3201 (2019). https://doi.org/10.1039/C8SC05757E
45. Patial, S., Hasija, V., P. Raizada, P. Singh, A. A. P. Khan Singh, and A. M. Asiri, Tunable photocatalytic activity of $SrTiO_3$ for water splitting: Strategies and future scenario. J. Environ. Chem. Eng. **8**(3), 103791 (2020). https://doi.org/10.1016/j.jece.2020.103791
46. Liu, Q., et al.: Composition dependence of structural and optical properties in epitaxial $Sr(Sn_{1-x}Ti_x)O_3$ films. Jpn. J. Appl. Phys. **54**(3), 031101 (2015). https://doi.org/10.7567/jjap.54.031101
47. Suzuki, S., et al.: Ferroelectricity of Sn-doped SrTiO$3 perovskites with tin at both A and B sites. Physical Review B **86**(6), 060102 (2012). https://doi.org/10.1103/PhysRevB.86.060102
48. Wang, T., Pitike, K.C., Yuan, Y., Nakhmanson, S.M., Gopalan, V., Jalan, B.: Chemistry, growth kinetics, and epitaxial stabilization of Sn^{2+} in Sn-doped $SrTiO_3$ using $(CH_3)_6Sn_2$ tin precursor. APL Mater. **4**(12), 126111 (2016). https://doi.org/10.1063/1.4972995
49. Bouquet, V., et al.: Influence of two-dimensional oxide nanosheets seed layers on the growth of (100)$BiFeO_3$ thin films synthesized by chemical solution deposition. Thin Solid Films **693**, 137687 (2020). https://doi.org/10.1016/j.tsf.2019.137687
50. Yamada, Y., Kanemitsu, Y.: Band-to-band photoluminescence in $SrTiO_3$. Phys. Rev. B **82**(12), 121103 (2010). https://doi.org/10.1103/PhysRevB.82.121103
51. Kanemitsu, Y.: Multiple Exciton generation and recombination in carbon nanotubes and nanocrystals. Acc. Chem. Res. **46**(6), 1358–1366 (2013). https://doi.org/10.1021/ar300269z
52. Yuan, F., Li, S., Fan, Z., Meng, X., Fan, L., Yang, S.: Shining carbon dots: synthesis and biomedical and optoelectronic applications. Nano Today **11**(5), 565–586 (2016). https://doi.org/10.1016/j.nantod.2016.08.006
53. Li, S., Luo, J., Liu, J., Tang, J.: Self-trapped excitons in all-inorganic halide perovskites: fundamentals, status, and potential applications. J. Phys. Chem. Lett. **10**(8), 1999–2007 (2019). https://doi.org/10.1021/acs.jpclett.8b03604
54. Lakowicz, J.R.: Instrumentation for fluorescence spectroscopy. In: Lakowicz, J.R. (ed.) Principles of Fluorescence Spectroscopy, pp. 27–61. Springer, Boston, MA, (2006)
55. Mohamed, M.A., Mohd Hir, Z.A., Wan Mokthar, W.N.A., Osman, N.S. Features of metal oxide colloidal nanocrystal characterization. In: Thomas, S., Tresa Sunny, A., Velayudhan, P. (eds.) Colloidal Metal Oxide Nanoparticles, pp. 83–122. Elsevier (2020)
56. Zampieri, M., Lazaro, S.R., Paskocimas, C.A., Ferreira, A.G., Longo, E., Varela, J.A.: Structural analysis of Ti and Pb citrate using NMR and FT-Raman signals and quantum mechanics

simulations. J. Sol-Gel. Sci. Technol. **37**(1), 9–17 (2006). https://doi.org/10.1007/s10971-005-4892-2
57. Fong, C.Y., Weber, W., Phillips, J.C.: Violation of Vegard's law in covalent semiconductor alloys. Phys. Rev. B **14**(12), 5387–5391 (1976). https://doi.org/10.1103/PhysRevB.14.5387
58. Ganguly, P., Shah, N., Phadke, M., Ramaswamy, V., Mulla, I.S.: Deviation from Vegard's law: changes in the c-axis parameter in $La_{2-x}Sr_xCuO_{4-d}$ in relation to the insulator-superconductor-metal transition. Phys. Rev. B **47**(2), 991–995 (1993). https://doi.org/10.1103/PhysRevB.47.991
59. Liu, Q., Jin, F., Li, B., Geng, L.: Structure and band gap energy of $CaSnO_3$ epitaxial films on $LaAlO_3$ substrate. J. Alloy. Compd. **717**, 55–61 (2017). https://doi.org/10.1016/j.jallcom.2017.05.112
60. Song, Z., Zhao, J., Liu, Q.: Luminescent perovskites: recent advances in theory and experiments. Inorg. Chem. Front. **6**(11), 2969–3011 (2019). https://doi.org/10.1039/C9QI00777F
61. Gao, Q., et al.: Wide-range band-gap tuning and high electrical conductivity in La- and Pb-doped $SrSnO_3$ epitaxial films. ACS Appl. Mater. Interfaces. **11**(28), 25605–25612 (2019). https://doi.org/10.1021/acsami.9b07819
62. Stanulis, A., Selskis, A., Ramanauskas, R., Beganskiene, A., Kareiva, A.: Low temperature synthesis and characterization of strontium stannate–titanate ceramics. Mater. Chem. Phys. **130**(3), 1246–1250 (2011). https://doi.org/10.1016/j.matchemphys.2011.09.005
63. Lavinscky, A.B.S., Welsch, A.-M., Kennedy, B.J., Bernardi, M.I.B., Mastelaro, V.R.: Order-disorder phenomena and octahedral tilting in $SrTi_{1-X}Sn_XO_3$ perovskites—a structural and spectroscopic study. J. Solid State Chem. **269**, 521–531 (2019). https://doi.org/10.1016/j.jssc.2018.10.006
64. Chatterjee, S., Chowdhury, B.N., Das, A., Chattopadhyay, S.: Estimation of step-by-step induced stress in a sequential process integration of nano-scale SOS MOSFETs with high-kgate dielectrics. Semicond. Sci. Technol. **28**(12), 125011 (2013). https://doi.org/10.1088/0268-1242/28/12/125011
65. Taylor, T.R., et al.: Impact of thermal strain on the dielectric constant of sputtered barium strontium titanate thin films. Appl. Phys. Lett. **80**(11), 1978–1980 (2002). https://doi.org/10.1063/1.1459482
66. Ghubish, Z., Saif, M., Hafez, H., Mahmoud, H., Kamal, R., El-Kemary, M.: Novel red photoluminescence sensor based on Europium ion doped calcium hydroxy stannate $CaSn(OH)_6$:Eu^{+3} for latent fingerprint detection. J. Mol. Struct. **1207**, 127840 (2020). https://doi.org/10.1016/j.molstruc.2020.127840
67. Yang, Y., Lai, H., Xu, H., Tao, C., Yang, H.: Morphology–luminescence correlations in europium-doped ZnO nanomaterials. J. Nanopart. Res. **12**(1), 217–225 (2010). https://doi.org/10.1007/s11051-009-9598-x
68. Myhajlenko, S., et al.: Optoelectronic and microstructure attributes of epitaxial $SrTiO_3$ on Si. J. Appl. Phys. **97**(1), 014101 (2004). https://doi.org/10.1063/1.1821643
69. Amer, M.I., Moustafa, S.H., El-Hagary, M.: Enhanced band structure, optoelectronic and magnetic properties of spray pyrolysis Ni-doped SnO_2 nanostructured films. Mater. Chem. Phys. **248**, 122892 (2020). https://doi.org/10.1016/j.matchemphys.2020.122892
70. Longo, V.M., et al.: Strong violet–blue light photoluminescence emission at room temperature in $SrZrO_3$: Joint experimental and theoretical study. Acta Mater. **56**(10), 2191–2202 (2008). https://doi.org/10.1016/j.actamat.2007.12.059
71. Souza, A.E., et al.: Photoluminescence of $SrTiO_3$: influence of particle size and morphology. Cryst. Growth Des. **12**(11), 5671–5679 (2012). https://doi.org/10.1021/cg301168k
72. Wang, M., Zhang, L.: The influence of orientation on the photoluminescence behavior of ZnO thin films obtained by chemical solution deposition. Mater. Lett. **63**(2), 301–303 (2009). https://doi.org/10.1016/j.matlet.2008.10.022
73. Oh, J.H., et al.: Photoluminescence properties of $SrLaMgTaO_6$ double-perovskite thin film. J. Alloy. Compd. **755**, 67–72 (2018). https://doi.org/10.1016/j.jallcom.2018.04.210
74. Acosta-Silva, Y.J., Castañedo-Perez, R., Torres-Delgado, G., Méndez-López, A., Zelaya-Ángel, O.: Effect of annealing temperature on structural, morphological and optical properties

of CeO_2 thin films obtained from a simple precursor solution. J. Sol-Gel Sci. Technol. **82**(1), 20–27 (2017). https://doi.org/10.1007/s10971-016-4286-7
75. El Desouky, F.G., Saadeldin, M.M., Mahdy, M.A., Wahab, S.M.A.E., El Zawawi, I.K.: Impact of calcination temperature on the structure, optical and photoluminescence properties of NanocrystallineCerium oxide thin films. Mater. Sci. Semicond. Process. **111**, 104991 (2020). https://doi.org/10.1016/j.mssp.2020.104991
76. Longo, E., et al.: Density functional theory calculation of the electronic structure of $Ba_{0.5}Sr_{0.5}TiO_3$: Photoluminescent properties and structural disorder. Phys. Rev. B **69**(12), 125115 (2004). https://doi.org/10.1103/PhysRevB.69.125115
77. Zhai, Z., Xie, Q., Lin, T., Xu, J., Wu, X.: Variation in structure and photoluminescence in $SrTiO_3$ film buffered by $SrRuO_3$. J. Cryst. Growth **327**(1), 267–271 (2011). https://doi.org/10.1016/j.jcrysgro.2011.06.009
78. Mizoguchi, H., Eng, H.W., Woodward, P.M.: Probing the electronic structures of ternary perovskite and pyrochlore oxides containing Sn^{4+} or Sb^{5+}. Inorg. Chem. **43**(5), 1667–1680 (2004). https://doi.org/10.1021/ic034551c
79. Tan, G.-L., Lemon, M.F., French, R.H.: Optical properties and London dispersion forces of amorphous silica determined by vacuum ultraviolet spectroscopy and spectroscopic ellipsometry. J. Am. Ceram. Soc. **86**(11), 1885–1892 (2003). https://doi.org/10.1111/j.1151-2916.2003.tb03577.x
80. Orhan, E., et al.: Combined experimental and theoretical study to understand the photoluminescence of $Sr_{1-x}TiO_{3-x}$. J. Phys. Chem. B **108**(26), 9221–9227 (2004). https://doi.org/10.1021/jp048877y
81. Orhan, E., et al.: Origin of photoluminescence in $SrTiO_3$: a combined experimental and theoretical study. J. Solid State Chem. **177**(11), 3879–3885 (2004). https://doi.org/10.1016/j.jssc.2004.07.043
82. Mikhailik, V.B., et al.: Studies of concentration dependences in the luminescence of Ti-doped Al_2O_3. J. Appl. Phys. **109**(5), 053116 (2011). https://doi.org/10.1063/1.3552943
83. Wang, Z., Sun, R., Chen, C., Saito, M., Tsukimoto, S., Ikuhara, Y.: Structural and electronic impact of $SrTiO_3$ substrate on TiO_2 thin films. J. Mater. Sci. **47**(13), 5148–5157 (2012). https://doi.org/10.1007/s10853-012-6392-4
84. Chiu, M.-H., et al.: Determination of band alignment in the single-layer MoS_2/WSe_2 heterojunction. Nat. Commun. **6**(1), 7666 (2015). https://doi.org/10.1038/ncomms8666
85. Tongay, S., et al.: Tuning interlayer coupling in large-area heterostructures with CVD-grown MoS_2 and WS_2 monolayers. Nano Lett. **14**(6), 3185–3190 (2014). https://doi.org/10.1021/nl500515q
86. Chen, K., et al.: Electronic properties of MoS_2–WS_2 heterostructures synthesized with two-step lateral epitaxial strategy. ACS Nano **9**(10), 9868–9876 (2015). https://doi.org/10.1021/acsnano.5b03188

DFT Simulations for Heterogeneous Photocatalysis from ZnO and CuO Semiconductors

Marisa C. Oliveira, Nivaldo F. Andrade Neto, Renan Augusto Pontes Ribeiro, Mauricio R. D. Bomio, Fabiana V. Motta, and Sérgio Ricardo de Lázaro

Abstract Since the industrial revolution from century XIX, the global environment has received a charge of each more pollutant. Even with filters and catalysts to minimize some industrial residues is necessary more caution and research treat waste and other products. In such a perspective, water clean and environmental remediation are some of the most important themes for humanity. The chemical treatment for a large quantity of generated pollutant residues by industry is a great challenge. Such pollutants are in molecules or materials forms. In particular, a molecule group denominated dyes is the focus. Heterogeneous catalysis based on semiconductor oxides is a widely investigated topic as a broad and potential technology for clean water treatment. Then, we present a chapter with a comprehensive perspective of the modifications applied in CuO and ZnO to improve the efficiency in heterogeneous photocatalysis processes. The DFT approaches ally to experimental evidence showed that doping and heterojunctions are efficient tools to maximize the discovery of the advanced materials directed to water clean and environmental remediation. Mn-doped ZnO presented an exciting performance for photocatalysis on methylene blue dye from defects connected to intermediary electronic levels. Heterojunction made from CuO/ZnO is a putative candidate for photodegradation in two ways: (i) generation in situ of oxidizing molecules; and (ii) the sunlight wavelength range as an energy source. Such molecular mechanism is possible from the generation, stability, and control on the charges carriers diffusion inside semiconductor oxides. How to understand and influence the creation of the electron–hole pair is the fundamental step to establish the heterogeneous photocatalysis based on semiconductor

M. C. Oliveira · N. F. A. Neto · M. R. D. Bomio · F. V. Motta
LSQM—Laboratory of Chemical Synthesis of Materials—Department of Materials Engineering, Federal University of Rio Grande Do Norte, P.O. Box 1524, Natal, RN 59078-900, Brazil

R. A. P. Ribeiro
Department of Chemistry, State University of Minas Gerais, Av. Paraná, Divinópolis, MG 3001, 35501-170, Brazil

S. R. de Lázaro (✉)
Department of Chemistry, State University of Ponta Grossa, Av. General Carlos Cavalcanti, P.O. Box 4748, Ponta Grossa, PR 84030-900, Brazil
e-mail: srlazaro@uepg.br

F. A. La Porta and C. A. Taft (eds.), *Functional Properties of Advanced Engineering Materials and Biomolecules*, Engineering Materials,
https://doi.org/10.1007/978-3-030-62226-8_6

oxides as one of the essential applications of the advanced materials in environmental remediation.

1 Introduction

1.1 Processes Oxidative Advanced (POA)

Heterogeneous catalysis is an essential topic in chemistry. The processes connected to this area have been applied as an alternative for pollutant treatment. The advantage of the catalysis from different physical phases allows the reuse of the catalyst. Then, heterogeneous catalysis contributes genuinely to a global objective, the need to preserve the environment. Since the high growth of the organic industry for pesticides, detergents, medicines, plastic and others, the environmental remediation research should be increased on the same scale. The solution for our influence on ecosystems or environmental reached high levels. Our lifestyle and technology change nature fastly in only some years, while the natural processes delay centuries to remediation or reuse of such molecules or materials.

Among the primary pollutants studied are organic ones, because they are toxic to fauna and flora, when discarded without adequate treatments, in addition to being difficult to treat by conventional methodologies [1]. Therefore, advanced oxidative processes (POAs) have emerged as an alternative for the treatment of these contaminants. These processes aim at high efficiency, a low selectivity to the contaminant, and generate non-toxic products as products [2]. POAs can be subdivided according to the initiation step, where the main ones are Chemical (Fenton process), Photochemical (H_2O_2/UV, O_3/UV, O_3/H_2O_2/UV), photo-Fenton, Heterogeneous catalysis, Sonochemical (ultrasonic radiation (UR), O_3/UR, H_2O_2/UR, photocatalysis/UR, sono-Fenton), and Electrochemicals (anodic oxidation, electro-Fenton, photoelectron-Fenton, sonoelectrochemical, sonoeletro-Fenton) [3–5].

Another point, the POAs are based on the generation of hydroxyl radicals ($^{\bullet}OH$), which have a high oxidative capacity [6]. In addition to this high oxidative potential, hydroxyl radicals act to create chain reactions from intermediate chemical species during the process, such as oxygen (O_2), ozone (O_3), and hydrogen peroxide (H_2O_2) [7]. According to Shin et al. [8], the hydroxyl radical has a standard reduction potential (E^0) of +2.80 V, while ozone, hydrogen peroxide, and oxygen have + 2.07 V, +1.77 V, and +1.23 V, respectively. Thus, the efficiency of the increase of the generation of these $^{\bullet}OH$ radicals optimizes the oxidative process.

Heterogeneous photocatalysis from an oxide semiconductor material has been the most studied methodology today. These materials provide the generation of a high amount of $^{\bullet}OH$ radicals in the reaction environment enabling fast mineralization of organic compounds [9, 10]. The need for a semiconductor in a crystalline structure is implicit in the photocatalytic process. In fact, only an oxide semiconductor in the crystal form may absorb radiation of energy equal to or higher than the oxide semiconductor bandgap creating electron–hole pairs. The radiation absorption with the generation of the electron–hole pairs occurs only in the presence of spatially ordered atoms, as in a crystal, in which the radiation can enter in resonance. Semiconductor photocatalysis has come out as a possible technology because of its catalytic, chemical, and physical properties and the potential to utilize UV, near-UV, or visible light at ambient temperature and pressure, and in the presence of oxygen [11, 12]. From this point of view, the light should cause a decrease of the potential energy barrier relative to the dark catalytic reaction causing an acceleration in establishing the chemical equilibrium between reagent and product in the heterogeneous system [13].

2 The Molecular Mechanism for the Photogenerated Carriers

The insulating materials have the Valence Band (VB) filled with a high energy gap from the subsequent empty band. The metallic materials show a continuum of electronic states, allowing that the electrons move freely, creating the electrical conduction. Then, the semiconductors have an empty region in which there are no accessible energy levels that promote the recombination of an electron and a hole produced by photoactivation of a solid. This empty region, which extends from the top of the filled VB to the bottom of the empty Conduction Band (CB), is called the bandgap. Thus, the electronic structure of semiconductors is characterized by a filled VB and an empty CB. If the forbidden energy gap is not so high, some electrons could pass in the energetic empty band through some kinds of excitation (e.g., thermal or light), in consequence, an excited electron (e^-) moves from the VB to the CB, leaving a hole (h^+) behind [14–16]. Furthermore, the nature of the bandgap (i.e., direct or indirect) plays an essential role since it modifies the absorption coefficient of the semiconductor.

The fundamental mechanism of photocatalysis consists in the generation of the electron–hole pairs. Once this pair is separated, the redox reactions modify the molecules adsorbed on the active surface, catalyst them. The occurrence of the cycle of heterogeneous photocatalytic reactions begins from the absorption of a photon of energy ($h\nu$) by the crystalline photocatalyst. The absorbed photon energy creates stable electron–hole pairs or excitons from electronic excitations. The excited electronic state created in a single oxide semiconductor photocatalyst is fundamental to result in an efficient chemical transformation. The excitons generate oxi and redox radicals to react with other species present in the reaction environment (both organic and inorganic), reaching the formation of the products more thermodynamically stables [13, 17, 18].

The capability of an oxide semiconductor to carry out the photoinduced electron–hole transfer on surfaces is governed by the band energy positions or the oxi-redox potential for adsorbed molecules. The bottom of the CB controls the reduction potential; whereas, the top of the VB determines the oxidation potential. Then, the modulation of such values is a measure of the chemical capability associated with oxide to promote oxi- and reduction processes [18, 19]. Therefore, the light absorption processes by oxide semiconductors are essential. In a natural light absorption process, the photoinduced free excitons (the electron–hole pairs interacting through a Coulomb attraction) generate inside pure crystalline oxides [17, 20]. Extrinsic light absorption is caused by the photoexcitation from energy levels inserted into bandgap or electronic states from defects create from (i) zero-dimensional point intrinsic defects (vacancies, interstitials) and impurity atoms or ions, (ii) one-dimensional (linear defects (dislocations), and (iii) two-dimensional states (intrinsic surface states) [21, 22].

From this point of view, the photoactivity depends on the different physicochemical properties of the semiconductor, which is mainly influenced by the electron–hole recombination in a set with the generation of hydroxyl radicals ($^{\bullet}OH$). An illustration of the heterogeneous photocatalysis is illustrated in Fig. 1. The oxide semiconductor material, when receiving radiation with enough energy to excite the electron from the VB to the CB, that is, greater than the energy of the bandgap, generates electron/hole pairs (e^-/h^+). These pairs react with H_2OH_2O or O_2 generate hydroxyl radicals ($^{\bullet}OH$) and oxygen superoxide ($^{\bullet}O_2^-$), respectively. The generated radicals act to oxidate the pollutants present in the medium. The radical generation process for the species oxidation is described using the following equations [23]:

$$\text{Semiconductor} + hv \rightarrow e^-_{cb} + h^+_{vb} \tag{1}$$

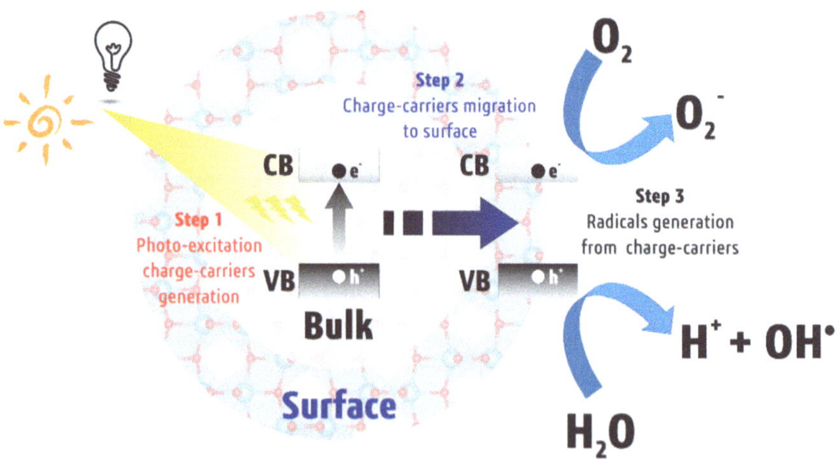

Fig. 1 Illustration of the photogeneration complex mechanism of the electron–hole pair (e^-/h^+) pairs in a hypothetical semiconductor oxide

$$O_2 + e^-_{cb} \rightarrow O_2^- \quad (2)$$

$$O_2^- + H^+ \rightarrow HO_2^{\cdot} \quad (3)$$

$$HO_2^{\cdot} + HO_2^{\cdot} \rightarrow H_2O_2 + O_2 \quad (4)$$

$$H_2O_2 + e^-_{cb} \rightarrow {}^-OH + \cdot OH \quad (5)$$

$$H_2O_2 + O_2^- \rightarrow {}^-OH + \cdot OH + O_2 \quad (6)$$

$$H_2O_{ads} + h^+_{vb} \rightarrow \cdot OH + H^+ \quad (7)$$

$$^-OH + h^+_{vb} \rightarrow \cdot OH \quad (8)$$

The initial step is the electronic excitation (e^-) from VB to CB (Fig. 1). In the VB are generated holes (h^+), which move in the material, frequently migrating on the surface, favoriting the interaction with H_2OH_2O or O_2 molecules. After this fundamental step, secondary processes generate oxidative species (as shown in Eqs. (2)–(8)), allowing reactions between them [24]. The generated oxidative species are free to migrate into the medium. The mineralization process occurs from the collision between oxidative species and organic contaminants. Then, the e-/h^+ pairs must have a low recombination rate or high stability as separated charges.

The application of the oxide semiconductor materials as a catalyst in this heterogeneous photocatalysis process because of the bandgap energy range. The modulation of this property is possible from the doping process. This technique reaches the decrease of the bandgap; however, it is essential to continue with a low recombination rate. The electronic excitation is unfeasible from the bandgap is decreased, and the recombination rate is increased [25].

A defect center creates a recombination center allowing to capture electron and hole in a recombination cycle. This recombination cycle increases the recombination rate making the photocatalytic process decrease. Then, the recombination of carriers via defects means different two-step transitions: (i) an electron is excited from VB to an empty state of the defect in the CB-band, and the electron decays from the defect state to an empty state of the VB-band; (ii) the electron is excited from the defect state of the VB-band to an empty state of the CB-band and the electron return to defect state. The excess energy of the electron is dissipated at both steps of the recombination. Therefore, the energy levels of the defect correspond to efficient recombination centers typically lie near the middle of the energy gap in oxide semiconductor materials.

3 Doped Zinc Oxide as an Alternative for Photocatalytic Properties

Among various metallic semiconductors oxide, the titanium dioxide (TiO_2) [26–29] has been the most studied compound in past decades. However, the ZnO is considered as a very good photocatalyst and an excellent supporting material for other semiconductors photocatalyst [30, 31]. Unfortunately, the ZnO has shown a low efficiency for solar-driven applications because of the high recombination between photoinduced electron–hole pairs. To address these problems, ion doping, noble metal deposition, the coupling of semiconductors, heterojunction formation, and so on approaches were adopted to enhance the activity [32–34]. For example, manuscripts reported ZnO blended with other metal oxides such as TiO_2[35], SnO_2[36], In_2O_3[37], MoS_2[38]CuO [39, 40], etc. The CuO photocatalyst is recognized as an excellent photocatalyst in hydrogen production, degradation of organic pollutants present in the wastewater, and water splitting [41–44]. Nevertheless, the photocatalytic activity of the pristine CuO was limited due to its fast electron–hole recombination, and only a meager of the photogenerated electrons were transferred to the surface of the catalyst to drive the photocatalytic activity [45]. In order to enhance the ZnO and CuO efficiencies, an effective co-catalyst between both resulted in an excellent photocatalytic activity because of the mixed oxides with n-type ZnO and p-type CuO features [46, 47].

Recently, a path to improve the adsorption capacity and photocatalytic activity is through ion doping, such as self-doping, rare-earth metal doping, and nonmetal doping [48–55], etc. The purpose of the ions doping is to introduce additional energy levels into the band structure to trap electrons or holes, creating the electron–hole pair or carriers. Thus, the carriers diffuse from bulk to surface successfully. However, different crystal surfaces have different surface energy levels for the CB and VB. Such differences in the energy levels will drive the electrons and holes to different crystal faces. Clearly, the doping can effectively modulate the electronic structures of the surfaces.

In this way, Zinc oxide (ZnO), an n-type semiconductor with wurtzite-type structure and hexagonal phase (P63mc space group), is an essential functional material widely used as sensors, photoelectrochemical cells, solar cells, photocatalysis, batteries, and mainly optical. The large bandgap energy (~3.2 eV) corresponds to the ultraviolet region with a wavelength of 387 nm [56, 57]. Then, the photocatalytic application is a consequence. The good photoactivity added a great chemically stable [58] is a way for the photocatalytic properties optimization. Usually occurs through the formation of heterojunctions or by doping, more accessible, and commonly studied, as illustrated in Fig. 2.

Paula et al. [59] studied the effects of Mn-doped ZnO and the inclusion of the metallic silver (Ag^0) using the microwave-assisted hydrothermal method. According to their results, the doping from 8% Mn increased the kinetic constancy in the order of 8x, while the ZnO/Ag^0 (30% wt.) heterostructure showed an increase close to

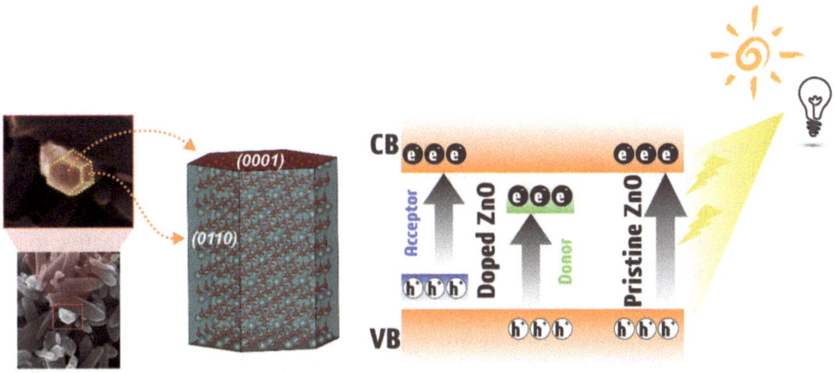

Fig. 2 In the left panel, a schematic representation of the surface models for the hexagonal shape of the ZnO nanoparticles is depicted, highlighting the chemical environment of each surface plane. In the right panel, the electronic levels for the pure and doped ZnO are compared regarding the excitation mechanism and the photogeneration of electron/hole (e^-/h^+) pairs

2×. On the other hand, the doping parallel performed between ZnO:8%Mn heterojunction with Ag^0 (ZnO:8%Mn/Ag^0 (30% wt.)) showed a fantastic increase close to 20×, indicating the presence of a synergistic effect. The creation of impediment centers to recombine the generated electron/hole pairs during the process increased the photocatalytic activity on a large scale. Andrade Neto et al. [23] investigated the application of the metallic cations, such as Fe and Pb, as an efficient way to create centers and to prevent the carriers recombination. Besides, the intermediate levels inserted in the bandgap or forbidden zone, reducing the required photon energy to excite the electron from the VB to the CB. The Fe and Pb atoms acted as electron donors creating energy levels close to the CB decreasing the gap of the forbidden zone, the intermediate levels [60].

In the previous experiments reported by Silva et al. [61], the Mn-doped ZnO material has a maximum concentration to optimize the photocatalytic activity. The ZnO:Mn2% quantity creates an excess of defects in the ZnO lattice. Then, the created recombination centers decrease photocatalytic activity. Thus, it would be expected that the recombination of the photo-generated electron/hole pairs would be difficult; however, the excess of defects motivated high recombination between the carriers. In another work, Andrade Neto et al. [62] observed similar results, where the doping of ZnO with Co and Mn at 4% decreased the photocatalytic activity drastically against the methylene blue dye. According to this study, even with doping providing an increase in superficial area form reduction of the particle size, there was a reduction in photocatalytic activity. In contrast, from the Co and Mn co-doping occurred an increase in the photocatalytic activity. This difference in photocatalytic behavior can be perceived through the more significant reduction in the energy of the prohibited band, indicating that double doping promotes the generation of intermediate levels in the forbidden band favoring the impediment of the recombination of the photogenerated pairs.

From a theoretical perspective, the understanding and controlling of photocatalytic properties depend on the charge carriers diffusion (D). The D is strictly associated with the effective mass of electrons (m_e^*) and holes (m_h^*), as well as the bandgap values associated with the localization of frontiers crystalline orbitals [63]. Indeed, D increases as the effective mass of photogenerated carriers become lighter, reflecting in enhanced photocatalytic efficiency. Furthermore, the ratio between the effective mass of charge carriers (m_h^*/m_e^*) is an alternative approach to predict the electron/hole pair stability as regards the recombination process. In particular, such an effective mass of charge carriers is extrapolated by fitting parabolic functions for the Conduction Band Minimum (CBM) and Valence Band Maximum (VBM) [63, 64].

Aiming to investigate the role of Mn-doping on the electronic structure of ZnO, Density Functional Theory (DFT) calculations at the B3LYP level of theory were carried out for three bulk models: (i) pristine ZnO; and ZnO:Mn at (ii) at 6.25 and (iii) 12.5% as illustrated in Fig. 3a–c.

For pure ZnO, the calculated band structure enables us to calculate the $m_h^* = 0.73$ and $m_e^* = 0.12$, indicating that photogenerated carriers can be separated at different surfaces. Besides, the doping with Mn promotes a decrease in the effective mass of holes (m_h^*) $= 5.08$ at 6.25% and $m_h^* = 5.69$ at 12.5% of Mn), as the effective mass of the electrons remains almost unchanged ($m_e^* = 0.1$ at 6.25 and 12.5% of Mn) [61].

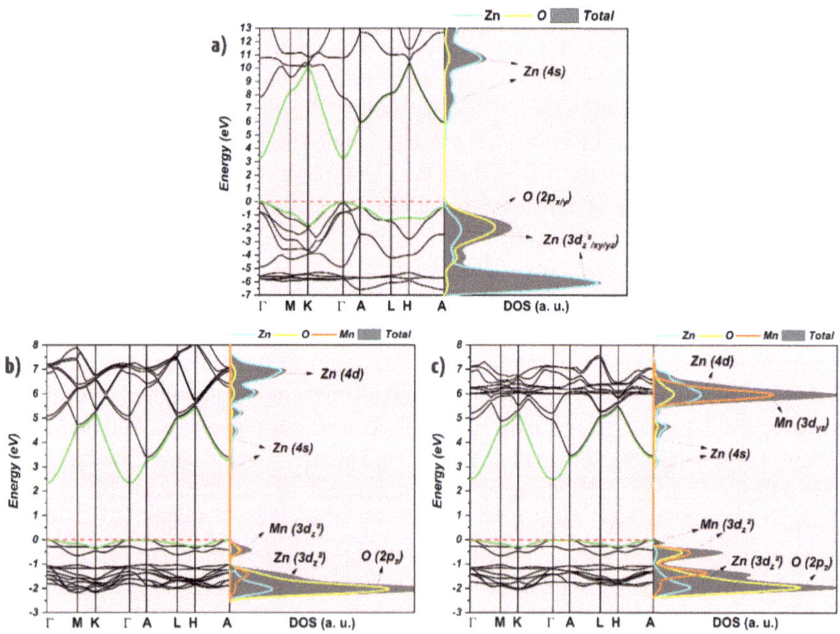

Fig. 3 Orbital-resolved Density of States and Band Structure projections for pristine **a** and Mn-doped ZnO at 6.25% **b** and 12.5% **c**, respectively

Thus, it is expected that the recombination of the photo-generated electron/hole pairs would be difficult once the m_h^*/m_e^* decrease with the amount of Mn.

However, the doping with Mn induces a significant effect in the VBM due to the localization of half-filled Mn (3d) orbitals close to the Fermi level, narrowing the bandgap energy. Moreover, the creation of intermediary energy levels in the bandgap of pure ZnO indicates an excess of defects associated with the large structural distortion resulting from doping.

Therefore, at lower Mn concentrations, the charge carriers diffusion for ZnO:Mn samples is increased, resulting in superior photocatalytic properties as compared with higher Mn concentrations, where the excess of defects contributes to enhance the recombination and reduce the photocatalytic efficiency.

4 Copper Oxide: Influence of pH on the Morphology

Many factors can influence a photocatalytic reaction. For instance, the oxygen concentration; the reaction medium conditions (pH, solvent); the catalytic surface (charge, defect, composition); the type and concentration of reactants, adsorption energy, created product, and light source (wavelength, intensity, distance). Thus, the key to the success of a catalytic process is complicated. The pH value in aqueous reacting media is an essential factor. The strong influence in the photocatalytic activity from the hydrogen activity under the protonation or deprotonation process of the substrates changes significantly the physicochemical conditions for catalysis. The key to high efficient catalytic processes is surface energy; the pH factor has a significant influence on the chemical processes connected to catalytic reactions.

In particular, the Copper oxide (CuO) is a p-type semiconductor material with small bandgap energy (close to 1.5 eV) that is found with a monoclinic crystalline structure and space group C2/c [65, 66]. It exhibits an exciting control sunlight driven photocatalytic behavior as a function of morphology. In addition to its low bandgap energy, there is high thermal and chemical stability associated with excellent electrical and optical properties [67, 68]. Among the optical properties, the CuO gains attention because it provides photogenerated charges (e^-/h^+ pairs) from the incidence of visible radiation, eliminating the application of external radiation sources, such as visible and ultraviolet LEDs [69]. On the other hand, the small barrier between the VB and CB facilitates the recombination between the photogenerated charges pairs, impairing the photocatalytic activity. Among the most used ways to improve the photocatalytic activity of the CuO, there is doping, heterostructures, or morphological control.

Andrade Neto et al. [70] showed that it is possible to control the CuO morphology modifying the pH of the chemical reaction from different surfactants synthesized by the sonochemical method. According to these results, the increase in pH (~13) is harmful to a specific morphology, tending to form only small fragments, without a defined shape. These fragments with undefined morphology have a smaller specific area than samples obtained at lower reaction pHs, indicating low surface porosity and impairing photocatalysis. On the other hand, an increase in the bandgap energy was observed in the particles fragmented, showing different exposed crystalline planes different. Each new crystalline plane has new associated surface energy changing the photocatalytic potential of the material [71]. Thus, through morphological control is possible to define the best crystalline plane to which it is exposed, optimizing the photocatalytic property [72].

Another way to control the photocatalytic properties consists of heterostructures. Such crystalline structure is the junction between two distinct solids to create an interface with molecular tension acting as a potential barrier to prevent the recombination between electrons and holes pairs. Then, the charges migration is favorable between the compounds; however, the recombination is difficult. Andrade Neto et al. [73] showed that the photocatalytic activity of the CuO/AgCl heterostructure is superior to pure CuO, especially under sunlight radiation. The optimization of the photocatalytic activity of the CuO was also observed by Oliveira et al. [74] from a CuO/ZnO heterostructure. The photocatalytic capacity increased against the methylene blue dye about 8 × higher than the pure CuO under sunlight. According to Ollis et al. [75], photocatalytic reactions can be divided into two processes: quantum confinement and photo-efficiency. Quantum confinement is connected to the absorption of photons used for photoexcitation, while the photo-efficiency deals with the observed reaction rate.

From the DFT point of view, heterostructures or coupled semiconductors show often results in photocatalytic efficiency higher than the pure or single materials. This fact is commonly attributed to heterojunctions or interfaces created between the electronic structures of the semiconductors improving the spatial charge separation. Thus, we propose a DFT-based theoretical Z-scheme to schematically represent the band edge alignment of the CuO/ZnO system, as depicted in Fig. 4.

The photocatalytic properties of the heterostructures are strictly dependent on band edge alignment associated with the crystalline frontiers orbitals of p- and n-type oxides. In the case of CuO/ZnO, the results indicate that the heterostructure exhibits a type-I junction. Indeed, the energy difference between the VBM of both oxides is 0.72 eV, while the energy difference between the CBM values was calculated as 0.77 eV, confirming the creation of the band alignment from the insertion of the intermediary energy levels.

The interpretation of the charges carriers diffusion along the p-CuO/n-ZnO heterojunction for the photocatalytic mechanism is essential. At first glance, the narrow bandgap energy of the CuO induces a sunlight-driven photoexcitation. The electrons are excited from VBM to CBM, generating an electron–hole pair. In this excited state, the electrons from p-CuO CBM are transferred to the n-ZnO CBM through the p/n junction by an electronic decay process. Now, the excited electrons on n-ZnO CBM are available to react with absorbed O_2 molecules generating hydroxyl radicals ($OH^\bullet$). This mechanism is directed by a reaction channel in which the free electrons are removed from semiconductor oxide.

On the other hand, the photogenerated holes from n-ZnO VBM are transferred along the p/n junction being captured in the p-CuO VBM, where they can react with the absorbed water (H_2O) molecules creating hydroxyl radicals ($OH^\bullet$).

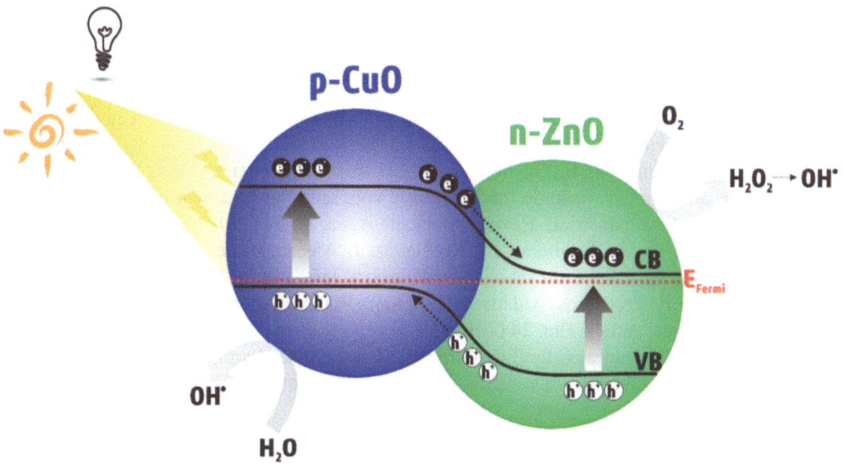

Fig. 4 Schematic band edge alignment of the p-CuO/n-ZnO heterostructure. Furthermore, the mechanism of the charge transfer under light irradiation, as well as the intermediate reactions for radical generation, are highlighted

As a consequence, the hydroxyl radicals (OH•) generated from the consumption of electrons and holes in the VBM and CBM of n- and p-type oxides, respectively, are free to react with organic molecules of dyes converting these harmful compounds to eco-friendly CO_2 and H_2O molecules.

5 Conclusion

The DFT simulations showed here to reveal one of the most promising quantum mechanical approaches to help in the knowledge of the heterogeneous catalysis based on semiconductor oxides. The molecular mechanism clarified from an in-depth theoretical analysis is a contribution to elucidate reaction of water or molecular oxygen with very singular active surfaces in semiconductor oxide. From the previous works with CuO and ZnO, more details of this quantum mechanical approach were commented. Interesting, that the high-level simulation is a pivotal point to improve the reactions to clean the water and photodegrade dyes or perhaps pesticides.

Then, the heterogeneous catalysis from doping- or heterostructure approaches are very promising to a new vision on semiconductor materials to environmental remediation. The extensive possibilities to build crystalline structures with high potential to create radicals in an aqueous medium is an essential technology to control the pollutant in the world. In this direction, simple oxides such as CuO and ZnO showed to be a suitable matrix to reach this aim. The high versatility for chemical modifications ally to new chemical synthetic routes is capable of creating crystalline morphologies more efficient for heterogeneous catalysis.

References

1. Neto, N.F.A., Dias, B.P., Tranquilin, R.L., Longo, E., Li, M., Bomio, M.R.D., et al.: Synthesis and characterization of Ag^+ and Zn^{2+} co-doped $CaWO_4$ nanoparticles by a fast and facile sonochemical method. J. Alloy. Compd. **823**, 153617 (2020)
2. Araújo, K.S.D., Antonelli, R., Gaydeczka, B., Granato, A.C., Malpass, G.R.P.: Processos oxidativos avançados: uma revisão de fundamentos e aplicações no tratamento de águas residuais urbanas e efluentes industriais. Revista Ambiente Água. **11**, 387–401 (2016)
3. Aguiar, A., Ferraz, A., Contreras, D., Rodríguez, J.: Mecanismo e aplicações da reação de fenton assistida por compostos fenólicos redutores de ferro. Quim. Nova **30**, 623–628 (2007)
4. Locatelli, M.A.F.: Processos oxidativos avançados (POA) no tratamento in sito de corpos de águas superficiais. Acesso Em. **24** (2013)
5. Santana, H.D., Bonancêa, C.E., Takashima, K.: Fotocatálise eletroquímica de atrazina sobre dióxido de titânio: efeito de diferentes parâmetros experimentais. Química Nova **26**(6), 807–11 (2003)
6. Vallejo, M., San Román, M.F., Ortiz, I., Irabien, A.: Overview of the PCDD/Fs degradation potential and formation risk in the application of advanced oxidation processes (AOPs) to wastewater treatment. Chemosphere. **118**, 44–56 (2015)
7. Pignatello, J.J., Oliveros, E., MacKay, A.: Advanced oxidation processes for organic contaminant destruction based on the Fenton reaction and related chemistry. Crit. Rev. Environ. Sci. Technol. **36**(1), 1–84 (2006)
8. Shin, S., Yoon, H., Jang, J.: Polymer-encapsulated iron oxide nanoparticles as highly efficient Fenton catalysts. Catal. Commun. **10**(2), 178–182 (2008)
9. Nascimento, G.G., Andrade Neto, N.F., Garcia, L.M.P., Li, M.S., Longo, E., Paskocimas, C.A., et al.: Effect of the Eu^{3+} (x = 0, 1, 2 and 3 mol%) doped $Zn_{2-x}TiO_4$ and $Zn_2Ti_{1-x}O_4$ obtained by complex polymerization method: photoluminescent and photocatalytic properties. J. Mater. Sci.: Mater. Electron. **30**(24), 20979–20988 (2019)
10. Santiago, A.A.G., Andrade Neto, N.F., Longo, E., Paskocimas, C.A., Motta, F.V., Bomio, M.R.D.: Fast and continuous obtaining of Eu^{3+} doped CeO_2 microspheres by ultrasonic spray pyrolysis: characterization and photocatalytic activity. J. Mater. Sci.: Mater. Electron. **30**(12), 11508–11519 (2019)
11. Khan, M.M., Adil, S.F., Al-Mayouf, A.: Metal oxides as photocatalysts. J. Saudi Chem. Soc. **19**(5), 462–464 (2015)
12. Muraro, P.C.L., Mortari, S.R., Vizzotto, B.S., Chuy, G., dos Santos, C., Brum, L.F.W., et al.: Iron oxide nanocatalyst with titanium and silver nanoparticles: synthesis, characterization and photocatalytic activity on the degradation of Rhodamine B dye. Sci. Rep. **10**(1), 3055 (2020)
13. Serpone, N., Emeline, A.V.: Semiconductor photocatalysis—past, present, and future outlook. J. Phys. Chem. Lett. **3**(5), 673–677 (2012)
14. Kumar, R.V., Coto, M.: Visible-light-active photocatalysis: nanostructured catalyst design, mechanisms, and applications (Chapter 18). In: Ghosh, S., (ed.) Visible-Light-Active Photocatalysis: Nanostructured Catalyst Design, Mechanisms, and Applications, 1st edn. Wiley (2018)
15. Boddu, S., Nishanthi, S.T., Kailasam, K.: Visible-light heterogeneous catalysts for photocatalytic CO_2 reduction (Chapter 15). In: Ghosh, S., (ed.) Visible-Light-Active Photocatalysis: Nanostructured Catalyst Design, Mechanisms, and Applications. Wiley (2018)
16. Yan, H., Wang, X., Yao, M., Yao, X.: Band structure design of semiconductors for enhanced photocatalytic activity: the case of TiO2. Prog. Nat. Sci. Mater. Int. **23**(4), 402–407 (2013)
17. Djurišić, A.B., Leung, Y.H., Ching Ng, A.M.: Strategies for improving the efficiency of semiconductor metal oxide photocatalysis. Mater. Horizons. **1**(4), 400–410 (2014)
18. Xing, M., Qiu, B., Li, X., Zhang, J.: TiO_2/graphene composites with excellent performance in photocatalysis. In: Yamashita, H., Li, H. (eds.) Nanostructured photocatalysts: advanced functional materials, pp. 23–67. Springer, Cham (2016)

19. Augugliaro, V., Palmisano, L., Soria, J.: Heterogeneous photocatalysis versus catalysis: general aspects (Chapter 1). In: Marcì, G., Palmisano, L. (eds.) Heterogeneous Photocatalysis: Relationships with Heterogeneous Catalysis and Perspectives, 1st edn. Elsevier (2019)
20. Koe, W.S., Lee, J.W., Chong, W.C., Pang, Y.L., Sim, L.C.: An overview of photocatalytic degradation: photocatalysts, mechanisms, and development of photocatalytic membrane. Environ. Sci. Pollut. Res. **27**(3), 2522–2565 (2020)
21. Zhu, S., Wang, D.: Photocatalysis: basic principles, diverse forms of implementations and emerging scientific opportunities. Adv. Energy Mater. **7**(23), 1700841 (2017)
22. Tripathi, G., Srivastava, D.K., Singh, J., Mishra, V.: Advancement and modification in photoreactor used for degradation processes (Chapter 15). In: Singh, P., Borthakur, A., Mishra, P.K., Tiwary, D. (eds.) Nano-Materials as Photocatalysts for Degradation of Environmental Pollutants, pp. 305–21. Elsevier (2020)
23. Andrade Neto, N.F., Matsui, K.N., Paskocimas, C.A., Bomio, M.R.D., Motta, F.V.: Study of the photocatalysis and increase of antimicrobial properties of $Fe^{3+}$and Pb^{2+} co-doped ZnO nanoparticles obtained by microwave-assisted hydrothermal method. Mater. Sci. Semicond. Process. **93**, 123–133 (2019)
24. Andrade Neto, N.F., Oliveira, Y.G., Nascimento, J.H.O., Carvalho, B.R., Bomio, M.R.D., Motta, F.V.: Synthesis, characterization, optical properties investigation and reusability photocatalyst capacity of AgCl-xGO composite. J. Mater. Sci. Mater. Electron. **30**(16), 15214–15223 (2019)
25. Neto, N.F.A., Nunes, T.B.O., Li, M., Longo, E., Bomio, M.R.D., Motta, F.V.: Influence of microwave-assisted hydrothermal treatment time on the crystallinity, morphology and optical properties of $ZnWO_4$ nanoparticles: photocatalytic activity. Ceram. Int. **46**(2), 1766–1774 (2020)
26. Dong, C., Xing, M., Lei, J., Zhang, J.: TiO_2/carbon composite nanomaterials for photocatalysis (Chapter 19). In: Wang, X., Anpo, M., Fu, X., (eds.) Current Developments in Photocatalysis and Photocatalytic Materials, pp. 303–21. Elsevier (2020)
27. Soltani, T., Lee, B-K.: Photocatalytic and photo-fenton catalytic degradation of organic pollutants by non-TiO_2 photocatalysts under visible light irradiation (Chapter 17). In: Wang, X., Anpo, M., Fu, X., (eds.) Current Developments in Photocatalysis and Photocatalytic Materials, pp. 267–84. Elsevier (2020)
28. Sun, P., Zhang, J., Zhang, R., Xie, Y., Liu, W., Khan, M., et al.: Controllable synthesis of TiO_2: toward an efficient photocatalyst (Chapter 4). In: Wang, X., Anpo, M., Fu, X., (eds.) Current Developments in Photocatalysis and Photocatalytic Materials, pp. 39–55. Elsevier (2020)
29. Yamazaki, S., Takaki, D., Nishiyama, N., Yamazaki, Y.: Factors affecting photocatalytic activity of TiO_2 (Chapter 3). In: Wang X, Anpo M, Fu X, (eds.) Current Developments in Photocatalysis and Photocatalytic Materials, pp. 23–38. Elsevier (2020)
30. Fujishima, A., Zhang, X., Tryk, D.A.: TiO_2 photocatalysis and related surface phenomena. Surf. Sci. Rep. **63**(12), 515–582 (2008)
31. Ishchenko, O.M., Rogé, V., Lamblin, G., Lenoble, D.: TiO_2- and ZnO-based materials for photocatalysis: material properties, device architecture and emerging concepts, semiconductor photocatalysis—materials, mechanisms and applications. In: Cao, W. (ed.) Semiconductor Photocatalysis—Materials, Mechanisms and Applications. IntechOpen (2016)
32. Gu, X., Edvinsson, T., Zhu, J.: ZnO nanomaterials: strategies for improvement of photocatalytic and photoelectrochemical activities (Chapter 15). In: Wang, X., Anpo, M., Fu, X. (eds.) Current Developments in Photocatalysis and Photocatalytic Materials, pp. 231–44. Elsevier (2020)
33. Karthikeyan, C., Arunachalam, P., Ramachandran, K., Al-Mayouf, A.M., Karuppuchamy, S.: Recent advances in semiconductor metal oxides with enhanced methods for solar photocatalytic applications. J. Alloy. Compd. **828**, 154281 (2020)
34. Bharathi, P., Harish, S., Archana, J., Navaneethan, M., Ponnusamy, S., Muthamizhchelvan, C., et al.: Enhanced charge transfer and separation of hierarchical CuO/ZnO composites: the synergistic effect of photocatalysis for the mineralization of organic pollutant in water. Appl. Surf. Sci. **484**, 884–891 (2019)

35. Sun, C., Xu, Q., Xie, Y., Ling, Y., Hou, Y.: Designed synthesis of anatase–TiO_2 (B) biphase nanowire/ZnO nanoparticle heterojunction for enhanced photocatalysis. J. Mater. Chem. A. **6**(18), 8289–8298 (2018)
36. Chen, X., Zhang, F., Wang, Q., Han, X., Li, X., Liu, J., et al.: The synthesis of ZnO/SnO_2 porous nanofibers for dye adsorption and degradation. Dalton Trans. **44**(7), 3034–3042 (2015)
37. Wang, Z., Huang, B., Dai, Y., Qin, X., Zhang, X., Wang, P., et al.: Highly photocatalytic ZnO/In_2O_3 heteronanostructures synthesized by a coprecipitation method. J. Phys. Chem. C **113**(11), 4612–4617 (2009)
38. Tian, N., Li, Z., Xu, D., Li, Y., Peng, W., Zhang, G., et al.: Utilization of MoS_2 nanosheets to enhance the photocatalytic activity of ZnO for the aerobic oxidation of benzyl halides under visible light. Ind. Eng. Chem. Res. **55**(32), 8726–8732 (2016)
39. Taufique, M.F.N., Haque, A., Karnati, P., Ghosh, K.: ZnO–CuO nanocomposites with improved photocatalytic activity for environmental and energy applications. J. Electron. Mater. **47**(11), 6731–6745 (2018)
40. Mardikar, S.P., Kulkarni, S., Adhyapak, P.V.: Sunlight driven highly efficient degradation of methylene blue by CuO-ZnO nanoflowers. J. Environ. Chem. Eng. **8**(2), 102788 (2020)
41. Yan, J-H., Yao, M-H., Zhang, L., Tang, Y-G., Yang, H-H.: Photocatalytic H_2 evolution activity of CuO/ZrO_2 composite catalyst under simulated sunlight irradiation. J. Central South Univ. Technol. **18**(1), 56–62 (2011)
42. Kum, J.M., Yoo, S.H., Ali, G., Cho, S.O.: Photocatalytic hydrogen production over CuO and TiO_2 nanoparticles mixture. Int. J. Hydrogen Energy **38**(31), 13541–13546 (2013)
43. Santos, H.L.S., Corradini, P.G., Andrade, M.A.S., Mascaro, L.H.: CuO/NiO_x thin film–based photocathodes for photoelectrochemical water splitting. J. Solid State Electrochem. (2020)
44. Raksa, P., Gardchareon, A., Chairuangsri, T., Mangkorntong, P., Mangkorntong, N., Choopun, S.: Ethanol sensing properties of CuO nanowires prepared by an oxidation reaction. Ceram. Int. **35**(2), 649–652 (2009)
45. Samad, A., Furukawa, M., Katsumata, H., Suzuki, T., Kaneco, S.: Photocatalytic oxidation and simultaneous removal of arsenite with CuO/ZnO photocatalyst. J. Photochem. Photobiol. **325**, 97–103 (2016)
46. Kandjani, A.E., Sabri, Y.M., Periasamy, S.R., Zohora, N., Amin, M.H., Nafady, A., et al.: Controlling core/shell formation of nanocubic p-Cu_2O/n-ZnO toward enhanced photocatalytic performance. Langmuir **31**(39), 10922–10930 (2015)
47. Senthil Kumar, P., Selvakumar, M., Ganesh Babu, S., Induja, S., Karuthapandian, S.: CuO/ZnO nanorods: an affordable efficient p-n heterojunction and morphology dependent photocatalytic activity against organic contaminants. J. Alloy. Compd. **701**, 562–573 (2017)
48. Liu, L., Chen, X.: Titanium dioxide nanomaterials: self-structural modifications. Chem. Rev. **114**(19), 9890–9918 (2014)
49. Cerrato, E., Gionco, C., Berruti, I., Sordello, F., Calza, P., Paganini, M.C.: Rare earth ions doped ZnO: synthesis, characterization and preliminary photoactivity assessment. J. Solid State Chem. **264**, 42–47 (2018)
50. Zong, Y., Li, Z., Wang, X., Ma, J., Men, Y.: Synthesis and high photocatalytic activity of Eu-doped ZnO nanoparticles. Ceram. Int. **40**(7, Part B), 10375–10382 (2014)
51. Alam, U., Khan, A., Ali, D., Bahnemann, D., Muneer, M.: Comparative photocatalytic activity of sol–gel derived rare earth metal (La, Nd, Sm and Dy)-doped ZnO photocatalysts for degradation of dyes. RSC Adv. **8**(31), 17582–17594 (2018)
52. Ahmad, I., Shoaib Akhtar, M., Ahmed, E., Ahmad, M., Keller, V., Qamar Khan, W., et al.: Rare earth co-doped ZnO photocatalysts: solution combustion synthesis and environmental applications. Sep. Purif. Technol. **237**, 116328 (2020)
53. Zhang, L., Zhu, X., Wang, Z., Yun, S., Guo, T., Zhang, J., et al.: Synthesis of ZnO doped high valence S element and study of photogenerated charges properties. RSC Adv. **9**(8), 4422–4427 (2019)
54. Thirukumaran, P., Atchudan, R., Parveen, A.S., Kalaiarasan, K., Lee, Y.R., Kim, S.-C.: Fabrication of ZnO nanoparticles adorned nitrogen-doped carbon balls and their application in photodegradation of organic dyes. Sci. Rep. **9**(1), 19509 (2019)

55. Marschall, R., Wang, L.: Non-metal doping of transition metal oxides for visible-light photocatalysis. Catal. Today **225**, 111–135 (2014)
56. Lynn, T.H.H., Jityen, A., Tivakornsasithorn, K., Jaisutti, R., Osotchan, T.: Nano-flower structure of indium and gallium doped zinc oxide powder. Mate Today Proc. **23**, 757–761 (2020)
57. Shelar, S.G., Mahajan, V.K., Patil, S.P., Sonawane, G.H.: Effect of doping parameters on photocatalytic degradation of methylene blue using Ag doped ZnO nanocatalyst. SN Appl. Sci. **2**(5), 820 (2020)
58. Xu, Y., Jin, J., Li, X., Han, Y., Meng, H., Wang, T., et al.: Simple synthesis of ZnO nanoflowers and its photocatalytic performances toward the photodegradation of metamitron. Mater. Res. Bull. **76**, 235–239 (2016)
59. Paula, C.H.R., Andrade Neto, N.F., Garcia, L.M.P., Nascimento, R.M., Paskocimas, C.A., Bomio, M.R.D., et al.: Increased degradation capacity of methylene blue dye using Mg-doped ZnO nanoparticles decorated by Ag0 nanoparticles. J. Electron. Mater. **48**(5), 3017–3025 (2019)
60. Green, M.A.: Intrinsic concentration, effective densities of states, and effective mass in silicon. J. Appl. Phys. **67**(6), 2944–2954 (1990)
61. Silva, J.M.P., Andrade Neto, N.F., Oliveira, M.C., Ribeiro, R.A.P., de Lazaro, S.R., Gomes, Y.F., et al.: Recent progress and approaches on the synthesis of Mn-doped zinc oxide nanoparticles: a theoretical and experimental investigation on the photocatalytic performance. New J. Chem. (2020)
62. Andrade Neto, N.F., Oliveira, Y.G., Bomio, M.R.D., Motta, F.V.: Synthesis and characterization of Co^{2+} and Mn^{2+} codoped ZnO nanoparticles obtained by the sonochemical method: photocatalytic and antimicrobial properties. J. Electron. Mater. **48**(9), 5900–5905 (2019)
63. Le Bahers, T., Rérat, M., Sautet, P.: Semiconductors used in photovoltaic and photocatalytic devices: assessing fundamental properties from DFT. J. Phys. Chem. C **118**(12), 5997–6008 (2014)
64. Ribeiro, R.A.P., Oliveira, M.C., Bomio, M.R.D., de Lazaro, S.R., Andrés, J., Longo, E.: Connecting the surface structure, morphology and photocatalytic activity of Ag_2O: an in depth and unified theoretical investigation. Appl. Surf. Sci. **509**, 145321 (2020)
65. Mnethu, O., Nkosi, S.S., Kortidis, I., Motaung, D.E., Kroon, R.E., Swart, H.C., et al.: Ultra-sensitive and selective p-xylene gas sensor at low operating temperature utilizing Zn doped CuO nanoplatelets: insignificant vestiges of oxygen vacancies. J. Colloid Interf. Sci. **576**, 364–375 (2020)
66. Andrade Neto, N.F., Oliveira, P.M., Paskocimas, C.A., Bomio, M.R.D., Motta, F.V.: Enhanced photocatalytic properties of zinc-doped CuO decorated with silver obtained by microwave-assisted hydrothermal method: statistical factorial design. J. Electron. Mater. **48**(8), 4840–4849 (2019)
67. Oosthuizen, D.N., Motaung, D.E., Swart, H.C.: In depth study on the notable room-temperature NO_2 gas sensor based on CuO nanoplatelets prepared by sonochemical method: comparison of various bases. Sens. Actuators B Chem. **266**, 761–772 (2018)
68. Siddiqui, H., Qureshi, M.S., Haque, F.Z.: Valuation of copper oxide (CuO) nanoflakes for its suitability as an absorbing material in solar cells fabrication. Optik **127**(8), 3713–3717 (2016)
69. Qin, Q., Qiao, N., Liu, Y., Wu, X.: Spongelike porous CuO as an efficient peroxymonosulfate activator for degradation of acid orange 7. Appl. Surf. Sci. **521**, 146479 (2020)
70. Andrade Neto, N.F., Oliveira, P.M., Nascimento, R.M., Paskocimas, C.A., Bomio, M.R.D., Motta, F.V.: Influence of pH on the morphology and photocatalytic activity of CuO obtained by the sonochemical method using different surfactants. Ceram. Int. **45**(1), 651–658 (2019)
71. Kusumam, T.V.A., Renuka, N.K.: Effect of crystal plane orientation in tuning the photocatalytic activity of zinc oxide particles. Mater. Today Proc. **5**(8, Part 3), 16118–16124 (2018)
72. Guo, R., Yan, A., Xu, J., Xu, B., Li, T., Liu, X., et al.: Effects of morphology on the visible-light-driven photocatalytic and bactericidal properties of $BiVO_4$/CdS heterojunctions: a discussion on photocatalysis mechanism. J. Alloy. Compd. **817**, 153246 (2020)
73. Andrade Neto, N.F., Longo, E., Matsui, K.N., Paskocimas, C.A., Bomio, M.R.D., Motta, F.V.: Photocatalytic properties under sunlight of heterostructures AgCl/CuO obtained by sonochemical method. Plasmonics **14**(1), 79–89 (2019)

74. Oliveira, M.C., Fonseca, V.S., Andrade Neto, N.F., Ribeiro, R.A.P., Longo, E., de Lazaro, S.R., et al.: Connecting theory with experiment to understand the photocatalytic activity of CuO–ZnO heterostructure. Ceram. Int. **46**(7), 9446–9454 (2020)
75. Peral, J., Ollis, D.F.: Heterogeneous photocatalytic oxidation of gas-phase organics for air purification: acetone, 1-butanol, butyraldehyde, formaldehyde, and m-xylene oxidation. J. Catal. **136**(2), 554–565 (1992)

Atomically Controlled Two-Dimensional Heterostructures: Synthesis, Characterization and Applications

Swayandipta Dey

Abstract It is well-known that information technology has been the base of our modern society. For this reason, semiconductor materials are extremely important and have a pivotal role in many technologies. Since the fantastic discovery of graphene in 2004 and the subsequent Nobel Prize for its fascinating two-dimensional (2D) properties, so far, a plethora of atomically thin 2D-layered materials and Van der Waals heterostructures has been discovered with rich material platforms; all the way from metallic (graphene, $TaSe_2$), semiconducting (WSe_2, MoS_2), superconducting ($NbSe_2$, FeSe) to topological insulators (Bi_2Se_3, Sb_2Te_3). These artificially created atomically well-controlled low dimensional electronic systems cover several exciting phenomena in the field of condensed matter physics (e.g., such as magnetism, superconductivity, topological insulation, and so on). In contrast to its 3D counterparts, these materials become strongly renormalized in the strict 2D limit, through a combination of quantum confinement and enhanced electronic interactions. As a result, these fascinating compounds exhibit enhanced quantum effects and display robust interactions with electromagnetic fields. In this chapter, several state-of-the-art fabrication methods, both top-down (such as molecular beam epitaxy, metal–organic chemical vapor deposition, mechanical exfoliation) and bottom-up (colloidal, vertical stacking, ligand displacement, etc.) will be introduced. Combining different electrical, optical and thermal measurements, the strategic characterization of complex electronic states down to atomic scales will be discussed. Finally, some of the recently studied as well as prospective applications of these constantly emerging smart materials in the field of health (bioelectronics, wearable sensors) and energy (photovoltaics, optoelectronics) industries will be further discussed.

S. Dey (✉)
Department of Physics of Complex Systems, Weizmann Institute of Science, 7610001 Rehovot, Israel
e-mail: swayandipta.dey@weizmann.ac.il

F. A. La Porta and C. A. Taft (eds.), *Functional Properties of Advanced Engineering Materials and Biomolecules*, Engineering Materials,
https://doi.org/10.1007/978-3-030-62226-8_7

1 Introduction

Recent advancements in synthesis and nanofabrication methodologies offer an exciting prospect of dimensionality control in functional materials. A revolutionary development in the field of designing new materials lies in flexible construction of two-dimensional (2D) layered materials and their heterostructures with atomic precision. Ever since the first isolation of graphene back in the year 2004, a new family of 2D-materials was born which has started growing at an accelerated rate over the last decade. As such, the graphene monolayer that was extracted successfully using a micromechanical exfoliation by Geim et al. [1, 2] triggered an enormous research interest worldwide due to its varied exotic properties, including high mechanical and tensile strength, high thermal conductivity, high charge carrier mobility along with a broadband optical absorption. With more and more researchers giving attention to developing such artificial 2D materials and their heterostructured analogs, novel layered materials are rapidly emerging with properties that are drastically different from their 3D counterparts. With relatively inexpensive production methods, these atomically thin 2D materials offer new physicochemical and tunable optoelectronic properties (effective bandgap engineering), all of which can be specifically tailored by controlling the number of layers present in these materials. With its wide variety of materials ranging from semi-metal (graphene), insulator (hexagonal boron nitride), semiconducting (WSe_2, MoS_2), superconducting ($NbSe_2$, FeSe) to topological insulators (Bi_2Se_3, Sb_2Te_3), these recently emerging plethora of 2D materials offer a tantalizing prospect for designing new device architectures ranging from integrated circuits, flexible wearable electronic devices, optoelectronics, water purification, photocatalysis to photovoltaics [1–60].

Atomically thin 2D materials are usually united by means of weak van der Waals forces (similar forces that hold layered materials together) and they potentially offer a wide combination of fabricating low dimensional heterostructured layered materials either by vertical stacking or stitching methods [3]. Vertical heterostructures based on layered 2D materials has been widely investigated by altering various parameters such as rotation (or, twisting) of different stacking layers, stacking components, layer thickness, interlayer spacing or stacking modes resulting in tunable physicochemical and optoelectronic properties. Based on the recent progress on various top-down and bottom-up strategies for nanofabrication followed by characterization techniques at hand, we will discuss the advances in the understanding of structure-property relationships for different manufactured artifical 2D heterostructures. Finally, in this chapter, we will focus on the extremely diverse and complex set of functional properties inherent of this new family of designed 2D materials [4]. Figure 1 shows the plethora of 2D materials that are available along with their electronic band structures emitting over a wide range of the electromagnetic spectrum.

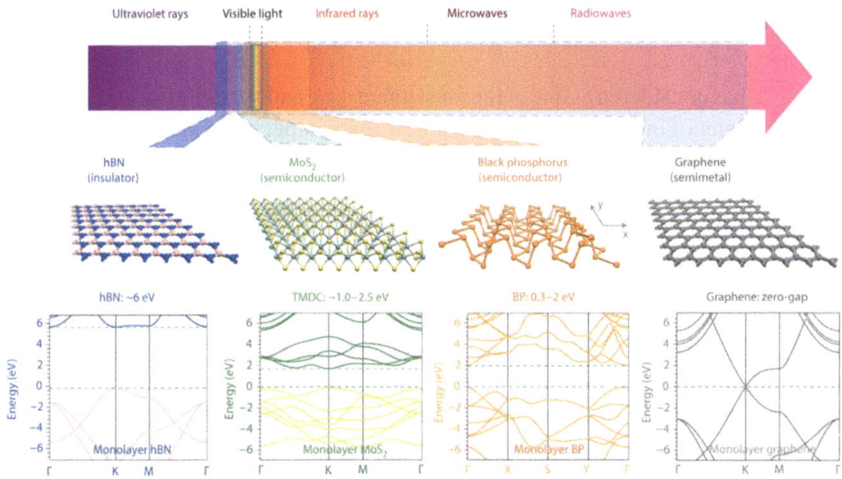

Fig. 1 Energy spectrum of various 2D materials with their atomic crystal structures (upper panel) and electronic band structures (lower panel). Reproduced with permission from reference [4]

2 Synthesis of 2D Materials

Following the first successful synthesis of layered graphene via micromechanical exfoliation on silicon substrates, the quest for fabricating low dimensional materials has accelerated ever since, thereby reducing the layered bulk solids down to few or mono-layered structures.

Vertically stacked 2D layered materials can be fabricated by mechanical stacking which provides a convenient way of preparing heterostructures by combining materials of similar morphologies. Besides, chemical vapor deposition (CVD) is another popular method of designing heterojunctions which provide a facile route of manufacturing large-area mechanical assemblies. These CVD synthesized laterally stacked heterostructures provide a relatively cleaner interface thus attributing to an enhanced performance when employed in devices.

We will first discuss the preparation methods of several pristine 2D materials. Followed by that, we then summarize existing protocols for designed van der Waals heterostructures by state of the art top-down (stepwise mechanical transfer onto a substrate, micromechanical exfoliation, chemical vapor deposition) and directed bottom-up (vapor phase deposition, reactive annealing, solution processed electrostatically self-assembly) growth approaches.

2.1 Graphene

Graphene has considered as one of the most interesting and versatile materials that has been the center of interest over the last several years, especially since the Nobel prize was awarded in 2010 to Geim and Novoselov for their "groundbreaking experiments regarding the two-dimensional material graphene" [1, 2, 5, 6]. Figure 2 shows a representation of how allotropic conversion of carbon materials can be performed. For e.g. single/multi-layered graphene to single/multi-walled carbon nanotube [7].

Whilst the atomically thin graphene as prepared by Geim et al. [1] via mechanical exfoliation displayed a very high quality layered material, this "scotch-tape" method itself has a poor output. Thus, in order to scale up its synthesis and application, it is imperative to develop an effective and inexpensive method for synthesizing large-area graphene with excellent quality and high uniformity. The quality of epitaxially fabricated graphene has been considerably improved in recent years wherein silicon carbide (SiC) has proven to be an effective substrate for growing high quality wafer-sized graphene sheets [8]. Alternatively, a large variety of materials (Cu, Ru, Pt, Co, Ir, Ag, Au, Mo, Ta, In etc.) from the periodic table has also been explored which demonstrated for the growth of graphene via a surface-reaction process or upon surface segregation of carbon due to cooling of C-metal solid solution by applying CVD method [9]. Processing parameters that determine the growth mechanism include the solubility of carbon in the metals forming the solid solution which ultimately control the quality, crystallinity, thickness and morphology of the resultant

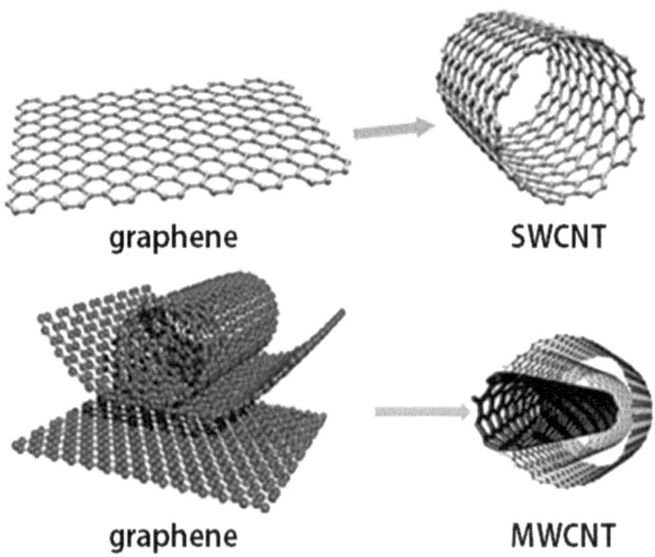

Fig. 2 Representation of atomic crystal structures of single and multi-layered layered graphene and synthetic conversion of graphene to single (SWCNT) and multi-walled carbon nanotubes (MWCNT). Reproduced with permission from reference [7]

graphene films. From the graphite oxidation, including protruding oxygen functional groups, lead to obtaining another material of great versatility for many technological applications, known as graphene oxide (GO). Yet, this functional material has interesting chemical and physical properties that in turn are usually different than those of graphene. It has been demonstrated that reducing the oxygen content through chemical, thermal, and other methods, in general, result in obtaining a form of cheaper and lower quality graphene, so-called reduced graphene oxide (abbreviated as r-GO). As such, r-GO has been mostly used for fabrication of electrodes (photoanodes) for water splitting, photocatalysis and photoelectrochemical applications [10–12].

2.2 *Hexagonal Boron Nitride (hBN)*

Boron nitride (BN) is a known synthetic material by its outstanding properties. In this regard, particularly the BN can exist in different phases: such as amorphous (a), hexagonal (h), wurtzite (w), cubic (c) and turbostratic (t). As is well-known, the hexagonal boron nitride (h-BN) is the most stable among the BN polymorphs [13–17].

As such, h-BN has been suggested as an ideal non-toxic, electrically insulating template for 2D devices due to its excellent properties. h-BN, which also is known popularly as "white graphite", has a high mechanical hardness and resilience due to its strong covalent bonding, a wide direct bandgap material (5.9–6.1 eV), a small lattice mismatch (1.7%) with graphene, high thermal conductivity with superior lubricating properties. h-BN has been employed in a multitude of device applications as a transparent membrane, encapsulation material, tunneling barrier, and dielectric substrate. As such, the atomic layers of h-BN are also considered as an ideal substrate to graphene, mainly due to their atomically smooth surface and strong ionic bonding significantly reduce the charge traps on the surface and thus improve the carrier mobility of graphene by about one order higher than other dielectric SiO_2 like substrates [33]. Similarly, the micromechanical cleavage was used to prepared atomically thin hBN from its bulk counterpart. However, it has been reported that the small size (e.g. a few microns and even less) and extremely low output of the resulting hBN flakes limit its applications. Notably, over the past few years, has been reported the synthesis of h-BN atomic layers by a CVD process. For instance, Song et al. [18] used a low-pressure assisted CVD (LPCVD) method for the large-scale growth of h-BN layers. Briefly, the authors using ammonia borane (NH_3-BH_3) as the sources of B and N and Cu as the substrate. A few months later, Shi et al. [19, 20] reported a similar approach using borazine ($B_3N_3H_6$) as the source and Ni as the substrate. Recently, the preparation of large-area uniform h-BN monolayer was usually obtained on Cu and Pt foils via a LPCVD method [20–22]. Additionally, single crystal monolayer h-BN up to several microns has also been synthesized on electropolished or pre-annealed Cu foils [33]. Jang et al. [23] introduced a high-pressure CVD growth mode and fabricated a large area thick hBN film that was used as a gate dielectric in electrical devices. Figure 3 shows the crystalline structures of various stable polymorphic forms of boron nitride (BN).

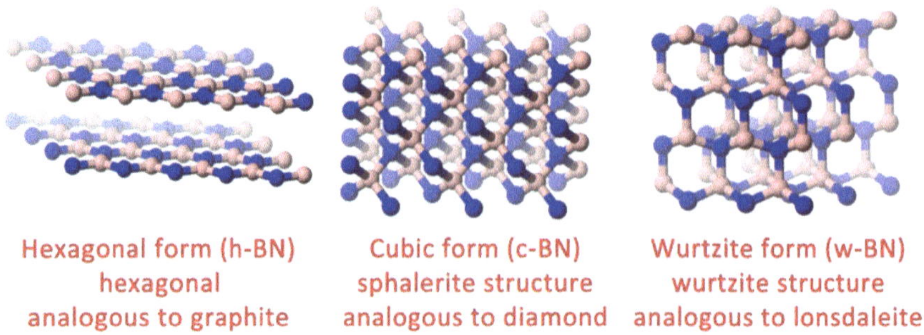

Fig. 3 Crystalline structures of various stable polymorphic forms of boron nitride (BN). The hexagonal form of h-BN is analogous to graphite, cubic form is analogous to diamond whereas the rare wurtzite form is analogous to lonsdaleite. *Source* https://www.preciseceramic.com/blog/introduction-of-pyrolytic-boron-nitride-pbn/

2.3 *Transition Metal Dichalcogenides (TMD) and MXenes*

After graphene, transition metal dichalcogenides (TMDs) are, in turn, recognized as the most studied 2D layered materials. Generally, these materials have a hexagonal or pentagonal lattice and are based on a monolayer of the interest transition metal atoms sandwiched among two layers of chalcogen atoms. Molybdenum disulfide (MoS2) is the most famous representative of this family of functional materials. As such, TMDs constitutes a diverse set of layered materials that are basically formed by more than 40 members, many of which possess a huge variety of functional properties [33]. More recently, TMDs materials and their heterostructures have attracted tremendous attention in many emerging technologies.

It should be noted that recent advances in the vapor-phase growth of TMDs can in principle be primarily divided into solid state sulfurization of the interest transition metal containing films as well as the chemical vapor growth through the reaction of sulfur and transition metal oxides [24–33]. Thus, from this perspective, MoS_2 thin films was first prepared by sulfurizing molybdenum films. In addition to silica, MoS_2 based on monolayer was also grown on nearly lattice-matching mica substrates, and the growth was proposed to follow an epitaxial growth mechanism. The preparation of wafer-size continuous MoS_2 films and large single-crystal domains is an ongoing challenge. Besides MoS_2, many other layered sulfides and selenides, such as WS_2, $MoSe_2$ and WSe_2 were also prepared by a similar vapor-phase growth method. Due to their functional properties tunable by the number of layers, in particular, TMDs materials are widely desired for potential applications in electronic devices, sensors, biomedical, water-splitting photocatalyst, and so on. Figure 4 shows the three-dimensional crystalline representation of transition metal dichalcogenides and MXenes respectively [24, 25].

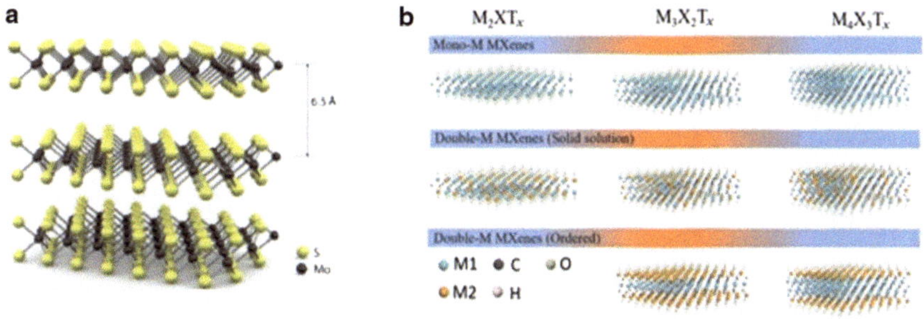

Fig. 4 **a** Three-dimensional representation of structure of transition metal dichalcogenides (MX_2). **b** Atomic crystal structures of various forms of MXenes. Reproduced with permission from reference [24, 25]

MXenes represent an emerging class of 2D materials with diverse functionalities of broad interest technological. Particularly, MXenes are based on transition metal carbides, nitrides, and carbonitrides. Notable examples include mono and double transition metal of MXenes, such as Ti_2C, Ti_4N_3, Ti_3CN, Mo_2N, Mo_2TiC_2, $Mo_2Ti_2C_3$, Cr_2TiC_2, and so on. However, in the case of MXenes with the surface terminated by Tx functional groups (e.g., O, F, OH, Cl) such as Nb_2CT_x, Ti_2CT_x, or Ti_3C_2Tx, in particular, show similar behavior of conductive clays. Also, these materials have a hydrophilic nature, mainly due to their surfaces terminate in hydroxyl or oxygen. Hence, these emerging 2D materials are promising to many technological applications, including electrochemical sensors, energy storage, chemical catalysts, field effect transistor sensors, water purification, and so on [25–28].

2.4 *Fabrication of Two-Dimensional Heterostructures*

(1) **Mechanical transfer or Manual stacking method**: Following the successful mechanical transfer of graphene onto silica substrates, all thanks to the invention of exfoliation and transfer methods of layered materials, the mechanically and chemically exfoliated 2D nanoflakes can be manually stacked forming 2D lateral heterostructures with the interlayer van der Waals force hold different layers together [29–31]. These 2D flakes can be exfoliated and transferred to different substrates either from their respective bulk forms or isolated from synthetically grown layers. The mechanical "stripping" method consists of peeling the stacked graphene sheet from bulk graphite using external forces such as normal and sheer stress. Various parameters such as stacking order, thickness of interface, stacking mode, individual thickness of layers play a critical role in modulating the electrical, conductive and optical properties of the heterostructured materials. Stepwise mechanical transfer is one of the most straightforward methods of arbitrarily stacking multiple 2D layers onto substrates via top-down manipulation of individual nanosheets one by

Fig. 5 Representation of mechanically assembled stacks of 2D crystals wherein multiple layers are held together by weak interlayer van der Waals forces. Reproduced with permission from reference [32]

one. Recent transfer methods such as carrier foil transfer and stamping are usually variations of certain basic schemes depending on if the initial layer needs to be transferred from or directly deposited on top of the substrate. In the case of carrier foil method, ideally the 2D layer is exfoliated on the carrier polymer (PMMA) layer that is deposited on the target substrate followed by an etching or peeling off the sacrificial polymer layer and subsequent thermal release of the 2D layer on the substrate. Another rather straightforward strategy is to utilize microcontact stamping from polydimethylsiloxane (PDMS)-based elastomeric stamps attached to a self-releasing layer and involving a successive dry transfer of 2D layers on a target substrate. Usually, the mechanical transfer methods lead to turbostatically disordered heterostructured stacks. Although these methods have been implemented so far for various combinations of 2D layers and nanosheets, most of the designed heterostructures in large part are restricted to two to four monolayers. Figure 5 shows the schematic representation of mechanically assembled stacks of 2D crystals in a layer by layer fashion held unitedly from the weak van der Waals forces [32].

(2) **Direct growth method**: The mechanical exfoliation method, although creates a facile route in producing high quality 2D crystals, it still remains challenging in various aspects like controlling the location, layer number or interfaces; and therefore, alternative pathways towards designing van der Waals heterostructures has been recently developed. Directed bottom-up fabrication via vapor-phase deposition methods has proven to be a versatile tool in designing such 2D heterostructures [33, 34]. CVD is one such technique that allows direct synthesis of various 2D heterojunctions with either vertically stacked or laterally stitched interfaces. Vertically stacked heterostructures with cleaner interface can be designed with CVD growth process wherein the principal growth

mechanism is via van der Waals epitaxy [35, 36]. Initial vertically stacked 2D heterostructures were based on graphene/hBN which shared similar lattice constants. Yang and co-workers [37] has implemented a plasma assisted deposition method to grow a single domain graphene on a hBN substrate. In this case, particularly the size of graphene and the preferred orientation is strongly dependent on the area of underlying h-BN. Moreover, Shi and co-workers [19] have obtained a vertically stacked MoS_2/graphene structure via thermal decomposition of inorganic Mo-based precursors on graphene surfaces. But even with higher lattice mismatch, MoS_2 has proven an efficient growth on graphene substrates since the vertical epitaxial growth can tolerate crystallite orientation during the growth process and can accommodate extra epitaxial strain. Direct synthesis of other layered materials including TMDs (MoS_2, WS_2), hBN on graphene has been fabricated using CVD wherein SiC substrate has proven to be an efficient growth platform for the underlying graphene. It has been observed that morphology of graphene plays an important role affecting the overall growth and properties of top heterostructures wherein strain, wrinkling and defects on graphene surfaces serve as the nucleation centers for the growth of other layered materials on top of it. Other methods such as thermal vaporization of powdered precursors and metal organic chemical vapor deposition (MOCVD) has been also incorporated in growing multilayered vertically stacked heterostructures such as WSe_2/MoS_2/graphene and others with varied combinations of heterojunction such as semi-metal/semiconductor (graphene/MoS_2), semi-metal/insulator (graphene/hBN) and both layer as semiconductors such as WSe_2/MoS_2 etc. Besides vertical stacking, laterally stitched heterojunctions can be also fabricated using CVD method. hBN has been grown on pre-patterned graphene surfaces by Levendorf and coworkers [38] wherein in-situ scanning tunneling microscopy (STM) has been utilized to study the lateral junction and it was observed that lattice strain can be released along the lateral direction. Early demonstration of lateral heterojunctions were based on multiple TMD materials such as MoS_2/$MoSe_2$, WS_2/WSe_2 which were fabricated by physical vapor transport with either by sequential source switching or by using mixed precursor sources [39, 40]. Figure 6 depicts the schematic representation of 2D materials using direct growth strategies like

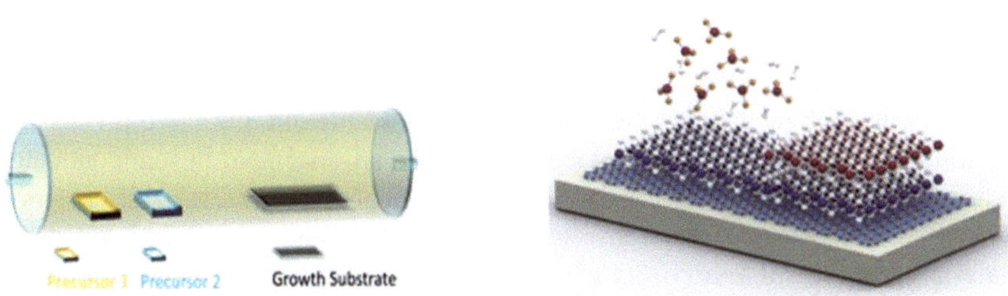

Fig. 6 Schematic representation of the synthesis of 2D layered materials using CVD and epitaxial growth on the substrate. Reproduced with permission from reference [32, 41]

CVD wherein precursors are used to grow the 2D crystals on various substrates [41].

(3) **Reactive annealing method**: van der Waals heterostructures and their superlattices can be fabricated by modulated elemental reactant method wherein elemental sources are deposited, followed by subsequent annealing, which enables growth of new binary and ternary 2D layered compounds. This design strategy, which was developed by Johnson and coworkers [42] has widely been used to obtained a series of mixed chalcogenide superlattices of the [MX2]m[MX2]n type, such as $[NbSe_2]_6[TiSe_2]_6$. As such, this technique involves controlled interfacial nucleation of specific building blocks followed by mild annealing resulting in a metastable layered heterostructures. More specifically, this is a non-epitaxial process that is usually applicable to compounds with great disproportionate chemistry and dissimilar crystal structures requiring any high growth temperatures, unlike the molecular beam epitaxy [42, 45].

(4) **Electrostatic assembly/layer-by-layer assembly method**: The bottom-up approach of self-assembly based on charged nanosheets and polyelectrolytes into artificially stacked heterostructures has widely been pursued since the early 90 s. Ionically bonded and oppositely charged materials that are amenable to the solution and colloidal processing with three main techniques that has been developed, including flocculation, electrostatic layer-by-layer (LbL) and Langmuir–Blodgett (LB) transfer to create lamellar heterostructures [43, 44]. The simplest method among them is Flocculation, which creates disordered lamellar layered heterostructures, whereas in electrostatic LbL assembly, heterostructured assemblies are created via sequential adsorption of molecular thin layered materials. Another versatile tool of fabricating molecularly thin films at air/water interface is on target substrates with precision at sub-nanoscale is via LB transfer of monolayers with amphiphilic molecules. Figure 7 depicts the schematic illustration of wet-chemistry based assembly methods for synthesis of heterostructures [45].

3 Electronic Band Structures and Electrical Properties of 2D Layered Materials

Nowadays, our societal lives have been constantly driven by interconnected electrical devices with high switching speed, low power consumption and large processing capabilities. And most of these devices rely on the advancement in semiconductor industry, which provides such capabilities in the form of transistors, variable resistors, capacitors, piezoelectric materials etc. Most of the recently developed novel 2D materials since the first discovery of graphene, the quasiparticles observed in these materials can be described as massless relativistic Dirac fermions [6]. Beside the Dirac fermions, other interesting properties including large spin–orbit coupling with gap openings at Dirac points lead to significant quantum Hall spin effects in recent

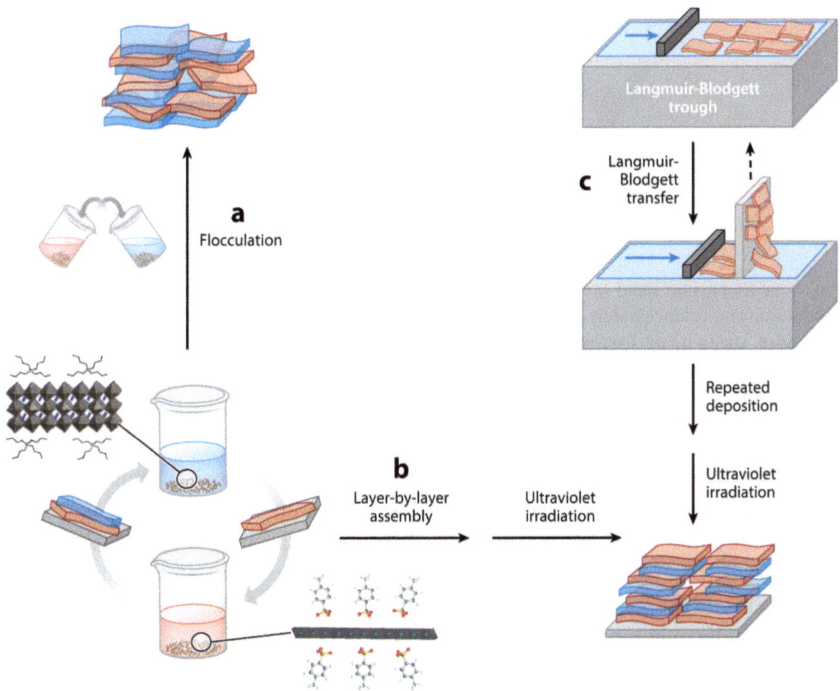

Fig. 7 Schematic illustration of three wet-chemical assembly schemes for heterostructures composed of ionic **a–c** and/or nonionic **c** nanosheet building blocks: **a** flocculation, **b** electrostatic layer-by-layer assembly, and **c** Langmuir–Blodgett transfer. Reproduced with permission from reference [45]

experiments [46, 47]. It is therefore quite obvious that the properties of 2D materials will pose remarkable changes as compared to their 3D bulk counterparts. Thus, based on their specific requirements and utilization in various innovative applications, these novel unexpected properties need to be investigated more elaborately at the nanoscale. Therefore, electrical characterization is an important tool to analyze and measure a wide variety of electrical properties in a material or device. Among them, electrical conductivity is a primordial characteristic feature which is dependent on material thickness, doping variant (and dopant concentration) and temperature. Using either direct (DC) or alternating (AC) current, the electrical properties such as electrical conductance, resistivity, Hall effect, capacitance, etc. of the materials and devices can be measured. In order to analyze electrical resistivity, few reliable methods such as two-point (on a wire-based or for uniform cross-section geometries) and four-point (also known as Hall effect) probe methods are commonly used. With the four-point probe method, it is also possible to measure other parameters such as carrier mobility, dopant concentration without actually destroying the sample. Capacitance–Voltage (C-V) characteristics is another method that provide key information of the

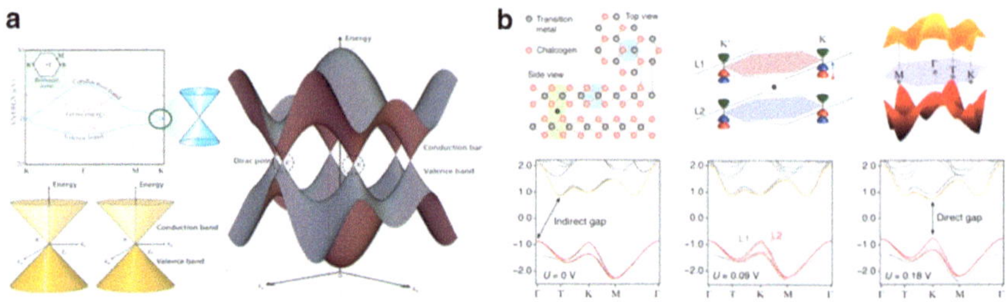

Fig. 8 **a** Electronic band structures indicating location of Dirac points with valleys present in conduction and valence bands of graphene. **b** Electronic band diagrams with theoretically calculated density of states using first principles calculation based on DFT method and type of bandgaps (direct/indirect) as observed in transition metal dichalcogenides. Reproduced with permission from reference [48, 49]

device being tested which is analyzed via several approaches such as multi-frequency capacitance, capacitance-frequency and capacitance-time.

Generally, the electrical properties of a solid has an intrinsic correlation with its electronic band structure. As such, the allowed electronic states for these structures are usually represented through valence (VB) and conduction (CB) band states. By solving the quantum mechanical equations for a periodic crystal, one can predict the electronic band structures by taking into consideration of electronic states present between the VB and CB. Typically, the energy bands of electronic materials can broadly be classified with the base on the types of wavevectors as: (1) direct bandgap wherein the electrons and holes in CB and VB both possess the same wavevector or if the momentum of the electrons and holes are the same, and (2) indirect bandgap where the energy levels of the top of VB and bottom of CB are a little misaligned having different wavevectors or dissimilar momentum values. The electrical properties can be widely tuned through proper band gap engineering thereby controlling or altering the bandgap of the material. Figure 8 shows the schematic representation of electronic band structures of graphene and transition metal dichalcogenides [48, 49].

In addition, it is well-known that the bandgap of the 2D layered materials can be also modified by strain effect. As a result, the strain on 2D structures leads to an upper tunable bandgap. An increase in electrical conductance by adding strain and bending the material with additional changes in the transport properties and bandgap has been observed in 2D like materials. This strain can in principle be converted into work function directly modifying the bandgap of the material by contributing to the energy dispersion model. As such, the electrical conductivity of 2D layered materials is directly related to the bandgap space, and in general, it can naturally occur due to a typical semiconductor behavior or being often induced by structural modifications (hence such as doping, defects, or functionalization) that change of electronic band-structure of such systems. In all cases, the smaller the bandgap value of the designed 2D structure, the greater the electrical conductivity of the material.

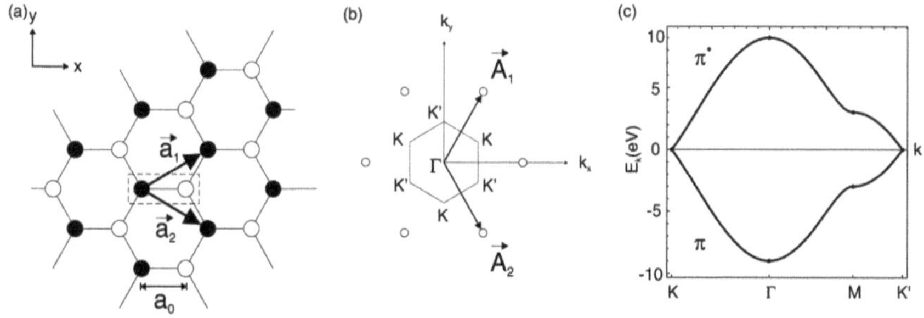

Fig. 9 **a** Representation of honeycomb lattice structure of graphene. **b** Reciprocal lattice in the first Brillouin zone with K and K′ points wherein the Dirac cones are located at the intersection of these points. **c** The graphene π-band dispersion in three-dimensional representation as calculated with the original tight-binding approach. Reproduced with permission from reference [50, 51]

Graphene, a zero-bandgap two-dimensional material consisting of single-layer C-atoms wherein the crystal structure constitutes the in-plane σ bonds between two C-atoms with sp [2] hybridization. The half-filled unhybridized π-bond along the vertical or out-plane p_z orbital axis makes the dominant contribution to electrical conductance in the transport. The two sublattice corners, $\vec{K}$ and $\vec{K'}$ exhibit different valleys in the energy contours of the first Brillioun zone condition as represented pictorially in Fig. 9 [50, 51]. By using the first order tight-binding approximation analysis, the electronic band structure of the covalent solid system can be constructed by taking into consideration of p_z orbital interaction. The energy dispersion relation can be calculated from the Bloch wave equation, for such relativistic quasiparticles, it can be simplified as:

$$E_k = \sqrt{m_0^2 + \frac{h^2k^2}{4\pi}}$$

Notably, the condition $E_k = 0$ occurs at six corners of the boundary in the first Brillioun zone. When the Fermi level E_F is located in this natural point, the π and π*-band meet each other at the $\vec{K}$ and $\vec{K'}$ points, representing a zero-gap semiconductor or a zero-overlap semimetal. Because the half-filled π-bands exhibit a linear dispersion around the natural Dirac points, the carrier density can be easily tuned by applying back gate voltage. Due to shift in the Fermi energy level in response to the back-gate voltage, V_g (V) the transport carrier is contributed by either electrons or holes depending on the sign of gate voltage (whether positive or negative) causing a change in resistivity, ρ (kΩ) with finite maximum resistance located at the natural Dirac point. For transition metal dichalcogenides (MoS_2, $MoSe_2$), the charge carrier mobility in their bulk forms show in the range of 200–500 cm^2/Vs. Recent experiments that has been carried out at room temperature observed a decrease in charge carrier mobility down to 0.1–10 cm^2/Vs due to the charge carrier traps present at the interface between the TMDC materials and the substrate. Although, WS_2 shows a much higher mobility

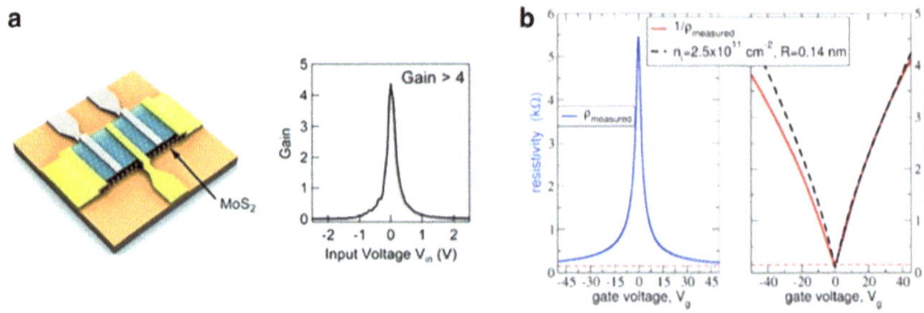

Fig. 10 **a** Schematic view of the single-layer MoS_2 based field-effect transistor (FET) device with gating response curve of the device (plotted gain vs input voltage) **b** Experimental data on graphene's conductivity (left: raw data on the measurement of resistivity of an exfoliated graphene sheet, right: fit of the conductivity curve). Reproduced with permission from reference from reference [52, 53]

due to its reduced effective mass. A recent report on WS_2 based transistor estimated its charge carrier mobility to be around 44 cm^2/Vs. It is essential to perform and measure temperature dependent electrical properties (carrier mobility) of these materials in order to understand their charge carrier conduction mechanisms. In fact, one of the most direct ways to access electrical properties is to incorporate these 2D layered materials into FET devices [52]. In this multi-terminal electronic device, the current is injected by applying the voltage between source and drain electrodes while the charge density and Fermi energy level is modulated using the gate electrode. Such electronic circuits using direct bandgap based 2D TMDCs has been demonstrated earlier by Radisavljevic et al. [53]. In 2014, Ovchinnikov et al. [54] performed temperature-dependent mobility studies of WS_2-based FET devices and the results were fitted to theoretical equations to study the conduction mechanisms. It has been observed that the current conduction mechanism in MoS_2 and WS_2 based FETs are similar. While MoS_2 is usually the n-type, p-type FET devices based on WSe_2 has been also realized with a room temperature hole mobility of 250 cm^2/Vs as well as n- type WSe_2 transistors with a room temperature mobility of 142 cm^2/Vs. Brumme et al. [55] has presented a study on field-effect doping and its influence on the electronic and transport properties of $MoSe_2$ and WSe_2 and it was observed that doping had a minor influence on the structure of the material. Figure 10 shows the schematic illustration of FET devices based on single layer MoS_2 and graphene respectively [52, 53].

4 Thermal Properties of 2D Layered Materials

With the discovery of graphene leading to an upsurge in the development of other 2D layered materials such as TMDs, MXenes, hBN and their associated heterostructures, the thermal properties of these unique artificially created materials arising from the phonon transport properties. Thus, from this perspective, a better and deeper understanding of the heat energy transfer process in such materials, in particular, has become a hot issue in chemistry and materials science research and hence can drive the emergence of new technologies in the future.

Despite its tremendous success, graphene still faces some severe problems related to the nature of their bandgap. As such, it is well-known that the current technology is based on silicon, which is a typical semiconductor material. Therefore, this can generate some incompatibilities for graphene. Alternative analogs which has been explored recently are silicene and phosphorene. Silicene, the Si counterpart of graphene, possesses a 2D structure that can smoothly solve the problems as mentioned earlier and hence has received a huge interest lately. Current research has focused on characterizing the thermal transport property of silicene structures, due to its enormous relevance in electronic applications. In comparison with the graphene, the silicene shows novel thermal transport properties. With such abnormal physical property, primarily stemming from its unique low buckling structure, silicene may enable opening up entirely new possibilities for revolutionary electronic devices and energy conversion materials. We discuss about thermal conductivity which is associated with the material's thermal properties. Graphene which has exhibited high thermal conductivity has great potential in the applications including electronic cooling and nano-/thermoelectric composites. Xie et al. [56] derived the thermal conductivity of silicone from first-principles. As such, the thermal conductivity of monolayer silicene at 300 K to be 9.4 W/mK, which is, of course, much smaller than bulk silicon. Thus, the contributions from in-plane and out-of-plane vibrations related to thermal conductivity were properly quantified in this investigation. As a result, the authors concluded that the out-of-plane vibration, in particular, contributes with the less than 10% of the thermal conductivity [56]. Hence, the difference is explained by the presence of small buckling, which breaks the reflectional symmetry of this structure. Based on this information, silicene can be considered more interesting than graphene in terms of thermoelectric application. In fact, this is due to their high electrical conductivity and low thermal conductivity of silecene compared to graphene, and can in principle be attributed to its buckled atomic structure. As yet, silicene synthesis is a considerable challenge. However, various numerical simulations suggest that the silicene's thermal conductivity range from 5 to 69 W/mK at about 300 K. In this regard, the thermal conductivity has been basically calculated by two strategies: anharmonic lattice dynamics (ALD) and molecular dynamics (MD) simulations. In particular, the results obtained by MD calculations are generally larger than those obtained by the ALD. As such, this discrepancy is probably due to the strong normal scattering near the zone center in 2D materials. Xie et al. [56] have reported the combination of first-principles-based ALD method with BTE as a strategy, which has resulted in a more accurate value to the thermal conductivity of silicene lattice. Because of the development of the calculation method, in particular, the BTE can in principle be iteratively solved. This procedure reflects the process of the redistribution of phonon modes driven by the heat flux. First principles based thermal transport modeling has also been implemented by Lindsay et al. [57] in order to analyze the lattice thermal conductivity of strained and unstrained graphene wherein the temperature and size dependence showed good agreement with the calculated parameters. Yet, in all cases, the underlying mechanisms are still not well understood. According to the iterative BET solution, the silicene's thermal conductivity for three finite sizes (i.e., 0.3, 3, and 30 μm) range from 15–30 W/mK. On the other hand, Gu et al. [58] has also calculated the silicene's thermal conductivity using the first-principles-based BTE approach. As such, the obtained results are similar in terms of magnitude for silicene's thermal conductivity. In comparison with graphene, the silicene's thermal conductivity have a logarithmic increase with respect to the sample size (mainly due to the small scattering rates of acoustic in-plane phonon modes), while that for the graphene is finite [58].

Phosphorene, the single layer counterpart of black phosphorene (BP), is another interesting 2D structure due to high carrier mobility proved by experiments and a large direct bandgap (~1.5 eV), which makes it promising for lots of nanoelectronic applications [59, 60]. There have been some experiments work on the measurement of the thermal conductivity of bulk BP and phosphorene films with different thicknesses [61, 62]. Time-resolved magneto-optical Kerr effect by Zhu et al. [58] reported results of the anisotropic thermal conductivity of BP along three primary crystalline orientations and attributed that the strong anisotropy of thermal transport in BP can be attributed to the structural-asymmetry-induced group velocity variations along different crystalline orientations, and the relaxation time variation induced by the direction of the applied temperature gradient.

The thermal conductivity of monolayer phosphorene has been theoretically reported. For instance, Liu et al. [63] studied the influence of orientation on the thermal transport properties of phosphorene sheets for a set of crystal chirality (ranging from 0.0° to 90.0°). In this study, they are using the Boltzmann transport equation (BTE) associated with the first-principles calculations. As a result, the intrinsic thermal conductivity of phosphorene sheets as a function of the crystal chirality has a sinusoidal behavior among 48.9 (0.0°) and 27.8 (90.0°) W m^{-1} K^{-1} [63]. As of yet, the exact value of the thermal conductivity of monolayer phosphorene is still unclear in the literature. However, as expected, the thermal conductivity for a monolayer of phosphorene is similar in terms of magnitude to silicene (~20 W m^{-1} K^{-1}). At the same time, this value is about two orders of magnitude lower than graphene (3000–5000 W m^{-1} K^{-1}). About 85% of the thermal conductivity is given by phonon modes in the region below the gap. Particularly, for the case of graphene, the phonon scattering processes involving odd number of FA modes that in turn are not allowed due to the symmetry-based phonon–phonon scattering selection rule. In relation to graphene, in particular, the symmetry-based phonon–phonon scattering selection rule is, of course, broken by the hinge-like structure of phosphorene, i.e., resulting in a large scattering rate of FA, among which leads to its small contribution to the thermal conductivity.

5 Optical Properties of 2D Layered Materials

In this section, a brief overview of the various optical techniques widely used in the characterization of 2D layered materials will be discussed. Depending on the frequency of excitation, it is well-known that the optical techniques can be categorized [60–84]. For example, in the terahertz (THz) frequency range, time-domain THz spectroscopy wherein the electric field (both amplitude and phase) is measured as a function of time can deduce information on intraband electronic transitions and free charge carrier mobility. By Fourier transform of the electric field in the time domain, it yields the spectrum in the frequency domain. On the other hand, Fourier-transform infra-red (FTIR) spectroscopy with having no direct access to electric field provides information on light intensity, and therefore as a result in order to obtain a complex dielectric response, Kramers–Kronig analysis is further required. FTIR spectroscopy probes the lattice vibrational modes and the low energy interband electronic transition and plasmonic resonances in 2D materials. Whereas, using Raman spectroscopy, one can probe these excitations occurring via the inelastic Raman scattering process.

One of the most exciting properties of 2D TMD materials based on the monolayer is the valley degree of freedom which arises from the non-equivalent valleys around at K and K′ points of the Brillouin zone due to breaking of inversion symmetry and spin–orbit coupling [64–66]. Due to such characteristic feature, a novel field of study such as valleytronics deals with how valley index of electrons, valley pseudospin modulates the optoelectronic behavior and could be extended in applications analogous to spintronics. Due to large momentum differences in the valleys characterized by novel optoelectronic, magnetic (valley spin) and valley electrical conductance (Hall effect) properties, these 2D monolayered materials can act as an index of carriers, e.g. in data storage and information technology-related devices.

The electronic structure of graphene was first described using a tight-binding approximation model characterized by massless Dirac fermion Hamiltonian at K and K′ points of the Brillouin zone, $\mathcal{H} = v_F \vec{\sigma}.\ \vec{k}$ where v_F is the Fermi velocity, $\vec{\sigma}$ denotes the Pauli matrices and $\vec{k}$ is the wave propagation (crystal momentum) vector. Intrinsic (undoped) graphene has its Fermi level located at the tip of the Dirac cone. The Dirac Hamiltonian leads to the distinctive half-integer quantum hall effect and the linear dispersion results in quantum hall plateaus at energies proportional to sgn (N) $\sqrt{|N|}$ where N is an integer and sgn (N) is the sign of N. Optically, the linear energy dispersion produces a constant interband contribution to the optical conductivity of $\sigma = \pi\ G_0/\ 4$ [80].

Recently discovered, the monolayer transition metal dichalcogenides MoS2, MoSe2, WS2, and WSe2 form a novel class of atomically thin materials that share many common structural, electronic, and optical properties. Similar to the bulk, the interband transitions at the K and K′ points of the Brillouin zone of monolayer TMDCs, where the valence band is split by spin–orbit coupling, give rise to the A and B peaks in the optical absorption spectra. On the other hand, the monolayers exhibit a number of important physical properties that are very distinct from the corresponding bulk materials [80]. Monolayer MoS2 is a direct-gap semiconductor, in contrast to the indirect-gap bulk MoS2. This indirect to direct bandgap transition was first observed through the shift of the indirect PL feature with layer number and a dramatic increase in the photoluminescence efficiency in the monolayer compared to the bilayer and the bulk. Heinz and co-workers [67] first observed that in atomically thin MoS2, with decreasing thickness, the indirect bandgap, among which prevails below the direct gap in the bulk material, shifts upwards in energy by more than 0.6 eV. Yet, this leads to a crossover to a direct-gap material in the limit of the single monolayer. In case of 2D MoS_2 monolayers, the exciton Bohr radius is of 9.3 Å and the binding energy is 570 meV; in general, such values are much larger than a conventional quantum well (e.g., GaAs) [68, 69]. This applies to exciton binding energies of other stable TMDs in form of monolayers in which are on the order of hundreds of meV. Due to the enhanced electron–hole interaction in 2D layered TMDs, such systems have a quasiparticle bandgap as well as the larger binding energy of exciton compared to the 3D bulk. It should note that the large binding energy for the exciton emission in such materials is specifically equal to the quasiparticle bandgap subtracted by the binding energy, and thus, it corresponds to the optical bandgap [70–72]. Figure 11 shows the exciton and bandgap dependence on optical absorption and valley dependent optical selection rules in 2D TMD materials [71, 72].

As is well-known, the MoS2 monolayer emits light strongly resulting in an increase in luminescence quantum efficiency as compared to bulk material. In the thin monolayer regime, 2D TMD materials show strong photoluminescence indicating indirect to direct bandgap transition and exhibiting quantum confinement in

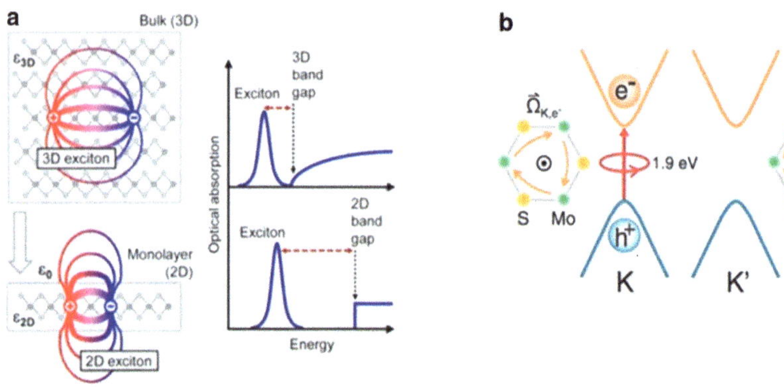

Fig. 11 **a** Real-space representation of electrons and holes bound into excitons for the three-dimensional bulk and a quasi-two-dimensional monolayer (left panel), Schematic representation on the impact of dimensionality on the electronic and excitonic properties, by optical absorption. The transition from 3D-to-2D is expected to lead to an increase of both the bandgap and the exciton binding energy (right panel) **b** Schematics of the valley-dependent optical selection rules and the electrons at the K and K′ valleys that possess opposite Berry curvatures $\vec{\Omega}$. The orange arrows represent the clockwise and counterclockwise hopping motion of K and K′ electrons. Reproduced with permission from reference [71, 72]

this d-electron system. The breaking of inversion symmetry in the case of MoS_2 monolayer gives rise to valley contrast for the Berry curvatures and orbital magnetic moment, that is, allowing the contrasted circular dichroism of K (K′) points in momentum space [73, 74]. As such, the VB splits into two bands for 2D layered TDMs, in principle, likely is due to the spin–orbit coupling. In contrast, the upper and lower bands is typical of electrons with opposite spin directions. Meanwhile, the inversion symmetry breaking and electronic spin directions (in these two bands) usually are inverted in the K and K′ valleys. Also, it should note that the VBs in the K and K′ valleys are time-reversal, from which leads to a circularly polarized dichroism [72, 75]. Thus, a new degree of freedom and the valley Hall effects of TMDs appear. Hence, in this case, the K and K′ valleys correspond to distinct electronic states with opposite momenta. The two valleys were predicted theoretically to be independently addressable by optical helicity as was then demonstrated experimentally by circular-polarization resolved photoluminescence spectroscopy [39, 76, 77]. By pumping one of the two valleys using circular polarized light, unequal transient valley populations can be created. Xiao et al. [78] showed that in monolayers of MoS2 and other group-VI dichalcogenides, interband optical transitions have frequency-dependent polarization selection rules which allow selective photoexcitation of carriers. Particularly, photoinduced spin Hall and valley Hall effects can in principle be used to produce long-lived spin and valley accumulations at boundaries on the sample, as well [78]. Second-harmonic generation (SHG) optical spectroscopy serves as a non-invasive and all-optical technique and has extensive applications in many fields of research that are ranging from detecting the structures to nonlinear optical devices. Yet, it should be noted that in case of the 2D materials (such as MoS2, WS2, and WSe2) have found an unique SHG properties, such as an odd–even layer dependent intensity, polarization effect, edge effect and also exciton resonance effect [79–81]. The strongest SHG comes from monolayer TMDs and intense SHG is observed on odd numbered layers only. The dramatic even–odd oscillation pattern of the SHG from 2D TMDs is the direct evidence for the variation of structure inversion symmetry, which is consistent with the presence (absence) of spatial inversion symmetry in even-layers (odd-layers). The third-order nonlinearities of mono- and few-layer WS2 have been explored by the two-wave mixing technique, where the optical Kerr effect plays an important role. The strong harmonic generation properties enable the nonlinear microscopic investigation of these layered materials with the significantly enhanced contrast and the reduction of the potential substrate-induced light interference and such nonlinear optical microscopic approaches can be used to probe the diverse properties associated with structural symmetries such as thicknesses, orientations, edges, stacking orders, grain boundaries, and sizes of these layered crystals and their heterostructures as represented in Fig. 12 [80, 82].

Raman and infrared (IR) spectroscopy are the major techniques to evaluate the vibrational modes. Raman spectroscopy has been established as an accurate and non-invasive technique of thickness characterization as well as a versatile method to study the phonon modes in 2D materials. As such, the Raman peak positions, intensities, and full widths at half maxima (FWHM) can directly reflect the layer number, stacking order, crystalline orientation, and applied strain of the 2D materials.

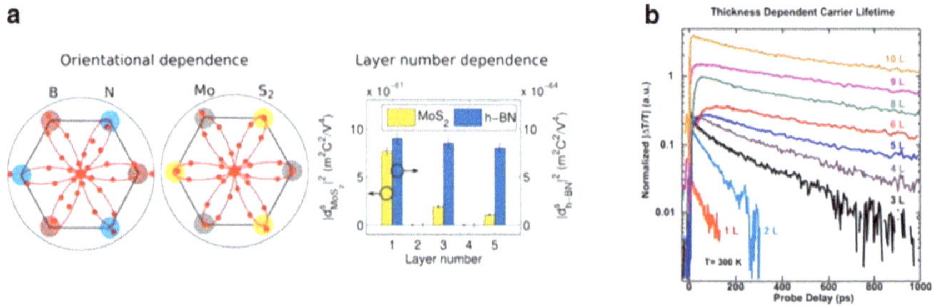

Fig. 12 **a** Schematic representation and data plots showing the dependence of optically measured SHG response on the orientation and layer numbers in atomically thin layers of hBN and MoS_2. **b** Experimentally measured layer dependent decay transient plots of photoexcited carriers in few-layers of MoS_2. Reproduced with permission from reference [80, 82]

Similarly, Raman spectroscopy has also been used to study the anisotropic in-plane structure and optical properties of 2D materials with low symmetry. Raman spectra of graphene exhibit a structure characterized by two principle bands designated as the G and 2D bands. Also, a third band, the D band may be apparent in graphene when defects within the carbon lattice are present. Usually, there are differences in the band positions and the band shapes of the G band 2D bands, as well as in the relative intensity of such bands that differ significantly. Consequently, this provides essential structural information at short-range. Note that the Raman spectra demonstrate the ability to differentiate among single-layer graphene and graphite. The huge utility of Raman spectroscopy lies in the ability to differentiate single, double, and triple-layer graphene structures. For instance, it is well-known that the Raman spectroscopy is capable of determining layer thickness for graphene of fewer than four layers. Intensive research efforts focus on four characteristic Raman active modes of TMD crystals, namely the A_{1g}, E^1_{2g}, E^2_{2g}, and B^2_{2g} modes. It is found that all the 2D materials are belonging to triclinic, monoclinic, orthorhombic, and tetragonal crystal system show polarization-dependent Raman scattering. It additionally is important to point out that the anisotropy here refers to the in-plane (xy plane) anisotropic structures of 2D materials. On the other hand, for thin samples of only a few layers, the anisotropic Raman scattering can, in principle, be widely used to identify the crystalline orientation of anisotropic 2D materials according to the angle-dependent Raman scattering efficiency. Besides, studies have suggested that periodic variation of Raman intensity of layered materials is also affected by photon-phonon interactions, and is related to the sample thickness and excitation energy [83]. Further, optical birefringence in anisotropic materials should also be considered in the polarized Raman scattering studies of anisotropic 2D materials [84]. For the polarized Raman spectroscopy of anisotropic optical crystal, the polarization states of the incident and scattering light in the Raman scattering process will both be altered due to the birefringence effect. The robust circular dichroism is the consequence of interplays among spin, valley, and layer degrees of freedoms, though the detailed mechanism is not clear yet.

6 Potential Applications of 2D Layered Materials and Their Heterostructures

Research has been fueled in the applications of 2D layered materials and their heterostructures are in general as diverse as their components. Currently, primary research directions are busy exploring (a) the miniaturization of devices based on the integration of 2D building blocks and (b) the designed of hybrid materials for a huge variety of renewable energy applications, including photocatalysis, battery materials, (super)capacitors, and thermoelectric devices. As is well-known, the functional properties of high-quality 2D material usually depend of synthetic conditions used in preparation and processesing and hence provide a superior control of interfacial electronic band, mutli-layer alignment, as well as their imperfections, which are in theory different for each desired application [45].

(a) **Optoelectronic devices**: A huge variety of 2D layered materials with tunable bandgaps (including the indirect-to-direct transition at the monolayer regime), large binding energies of the exciton, as well as, work function are widely available in the realm of nature. Also, these functional materials with versatile optical properties, it gives us the enormous possibility to design a multitude of optoelectronic devices such as photodetectors, solar cells, LEDs, single-photon sources and laser using these 2D layered materials. Based on the spatially split electrostatic gating of single TMD channel, earlier words demonstrated their optoelectronic device applications such as solar light harvesting and visible light-emission [39, 85, 86]. With the development of new fabrication methods yielding the ability to assemble 2D layers vertically, it is an important stepping stone for the realization of miniaturized, flexible/stretchable electronic and optical devices with a high density of functional components being stacked in the form of legos in both vertical and lateral dimensions. As such, the on-chip integration of various 2D materials is, of course, vital to obtaining more complex (opto)electronic circuits, opening the door to high-speed, low-power dissipation devices operating at the ultimate size scaling limit [45]. Most conventional semiconductor-based heterojunctions where the active emitting layers are either stacked or form planar arrangements have widely been used in the form of various optoelectronic devices. However, in conventional semiconducting materials, the form of heterojunction (HJ) due to limited by lattice mismatch. On the other hand, it is well-known that the layered TMD materials possess supreme advantages in altering crystalline orientations to design artificial vertically stacked heterojunctions, since such systems are formed by weak interlayer van der Waals forces and covalent bonds between the layers. In 2014, Gong et al. [87] first synthesized a TMD vertical HJ (WS_2/MoS_2). The authors reported the growth temperature of vertical WS_2–MoS_2 HJs (850 °C) is higher than that of lateral HJs (650 °C). Later, Yang et al. [88] achieved the largest lateral size (on the millimeter scale) reported in the literature for the van der Waals epitaxial growth of SnS_2/WSe_2 vertical bilayer p–n junctions with a lateral size on the millimeter scale, which represents the largest size reported.

Based on the large-scale vertical HJs, they integrated three kinds of devices on one large-scale HJ flake. As expected, the mobility and photoresponsivity performance for the HJ devices significantly improved compared to pristine WSe_2. Also, the aforementioned p-WSe_2 (p-type semiconductor)/MoS_2 (n-type semiconductor) heterojunction was the first reported light-emitting device based on heterostructured 2D layered semiconductors. Cheng et al. [89] showed that the electroluminescence (EL) occurred by hot electrons or recombination of electrons and holes injected in each layer depending on the applied voltages. Researchers at the University of Manchester [90] proposed another strategy of building 2D van der Waals heterostructures by bandgap engineering. In this study, 2D layered TMDC/h-BN/graphene heterostructures were fabricating to LEDs applications. In this device, the atomic-scale single quantum well was built by using the h-BN as a barrier and the monolayer TMDC as an emitting layer. By repeating the sandwiched h-BN/TMDC/h-BN structure several times, they managed to fabricate the MQWs or superlattice structures, which exhibited the external quantum yield of about 10%, comparable to that of the organic LEDs [90]. BP single crystals, which possess thickness independent direct bandgap with a wide range between 0.3 and 2 eV has been suggested as a potential luminescent material for nanofabrication of 2D LEDs in the mid-IR range. In a study by Xia et al. [91], wherein BP layer with a narrow bandgap was sandwiched between the electron-injecting MoS2 and hole-injecting WSe_2 layers with larger bandgap than BP, enabled both efficient injection and confinement of charge carriers. These results suggested that van der Waals 2D layered heterostructures can in principle be widely applied to build various high-efficiency LED devices.

(b) **Solar energy harvesting and conversion**: Note that the 2D layered semiconducting materials are being recently considered as potential candidates for photovoltaic devices applications due to their large surface area, free of dangling chemical bonds and efficient photoabsorbers of solar light [92–102]. TMD materials has been applied in solar cells in the form of Schottky or p–n junction that acts as an interface for separating of photogenerated charge carriers. For the Schottky junctions, the TMD semiconductors adjoins a metal or graphene electrode. In a study by Yu and co-workers [92], the MoS_2/metal (Au) based Schottky-barrier solar cell (with about 220 nm thickness for MoS_2) demonstrated a power conversion efficiency (PCE) of about 1.8%. Whereas in another study by the same group, the WS_2/graphene based photocell was able to yield a relatively higher PCE of approximately 3.3% employing a 37 nm thick WS2 nanosheet with a multilayer graphene contact [93]. In addition, TMD monolayers were predicted to absorb 5–10% of incident sunlight, which is almost one order of magnitude higher than the conventional GaAs or Si of similar thickness [41]. In fact, an independent study showed better photoabsorption and charge carrier trapping properties that could be possible in just 1 nm thick ultrathin solar cells comprising MoS_2/graphene stacked monolayers with a PCE of about 1% [94]. Particularly, the type-II electronic band alignment and van der Waals based lateral p-n heterojunctions have also been demonstrated. For a monolayer WSe_2 p–n diode, the PCE was estimated to be around 0.5%. More specifically, the solar cell based on monolayer of stacked MoS_2 and WSe2 in which exhibited PCE of 0.2% [95]. Yet, the n-type monolayer MoS2 and p-type silicon heterostructured exhibits the external quantum efficiency (EQE) exceeding 4%, [96] and a high PCE over 5% could be reached by the MoS_2/Si-based photovoltaic device [97]. As such, the application of TMDCs in the photovoltaic field is not limited to a photoactive semiconductor layer, they can also work as catalytic counter electrode materials or hole extraction buffer layers. As MoS_2 exhibits good catalytic activity for the reduction of electrolyte from triiodide(I3) to iodide(I) in a dye-sensitized solar cell (DSSC), the MoS_2 based CEs have been demonstrated in DSSCs and good photovoltaic performance was observed with PCE ranging from 5.7–7.23%. Moreover, ultrathin 2D MoS_2 nanosheets can be integrated into organic solar cells as an effective hole-extraction layer and a relatively high PCE of 8.11% can be reached [98]. Besides using as photosensitizers, various 2D layered materials has been implemented already as the electron and/or hole transport layers in solar cells. Yet, the CVD process is a bottom-up strategy compatible with the roll-to-roll method that has widely been used to produce large-area of monolayer graphene. As such, CVD-grown graphene films have widely been used as the transparent cathodes for the case of hybrid solar cells, and catalytic counter electrodes for the case of DSSCs as well as organic solar cells (OSCs). Hashim and coworkers [99] demonstrated a new structure of graphene/Si Schottky junction solar cells from introducing an interlayer of graphene oxide (GO) among graphene and Si. An optimized power conversion efficiency (PCE) of approximately 6.18% was achieved for the graphene/GO/Si solar cells. In addition, Liu et al. [100] have reported the use of monolayer graphene as the top electrode in the standard P3HT:PCBM photovoltic device. It is also feasible to incorporate layered stacked 2D materials and their heterostructures for achieving the enhanced efficiency of photovoltaic devices. Furchi et al. [95] reported a van der Waals heterojunction form the using MoS_2 and WSe_2 monolayers. Upon optical illumination, in particular, the charge transfer process occurs across the planar interface, and then the device exhibits a typical photovoltaic effect. As such, were estimated for this device a PCE of about 0.2% as well as EQE of approximately 1.5%. Further studies based on using graphene contacts for both MoS_2 and WSe_2 monolayers were done in order to improve the charge collection in the graphene-sandwiched 2D heterojunction [101]. Note that this device exhibits a high EQE value of about 2.4%. Similarly, Duan et al. [40] demonstrated a lateral WS_2/WSe_2 photodiode with EQE and internal quantum efficiencies (IQE) of about 9.9% and 43%, respectively. Phosphorene, which is the monolayer version of BP has been demonstrated to serve as the light absorber and charge transfer layer in solar cells, with enhanced light absorption and electron recombination. Deng et al. [102] reported an electrically tunable black phosphorus monolayer MoS_2 van der Waals heterojunction, which shows a maximum rectification ratio of $\sim$10, a maximum responsivity of 418 mA/W, and an EQE of 0.3% with a backgate voltage of −30 V under aser illumination (at 633 nm). Shanmugam et al. [93] demonstrated graphene/2D semiconductor Schottky barrier solar cells with different layer configurations of graphene contacts on the CVD-assembled WS_2 nanosheets. As such, the solar cell performance was significantly improved exhibiting a PCE of approximately 3.3% by incorporating multilayer graphene as the Schottky contact (Fig. 13).

(c) **Thermoelectric devices**: As is well-known, thermoelectric energy (TE) generation based on semiconductor materials has attracted considerable attention lately. As such, this physical process, in principle, is used to convert waste heat into directly usable forms of electricity. In this perspective, the efficiency of a thermoelectric material to convert heat into electrical energy is usually characterized by the dimensionless figure of merit, $ZT = \frac{\sigma\alpha 2T}{\kappa}$, where σ, α, T and κ are electrical conductivity, Seebeck coefficient, absolute temperature and thermal conductivity, respectively. Monolayer and few-layer 2D materials

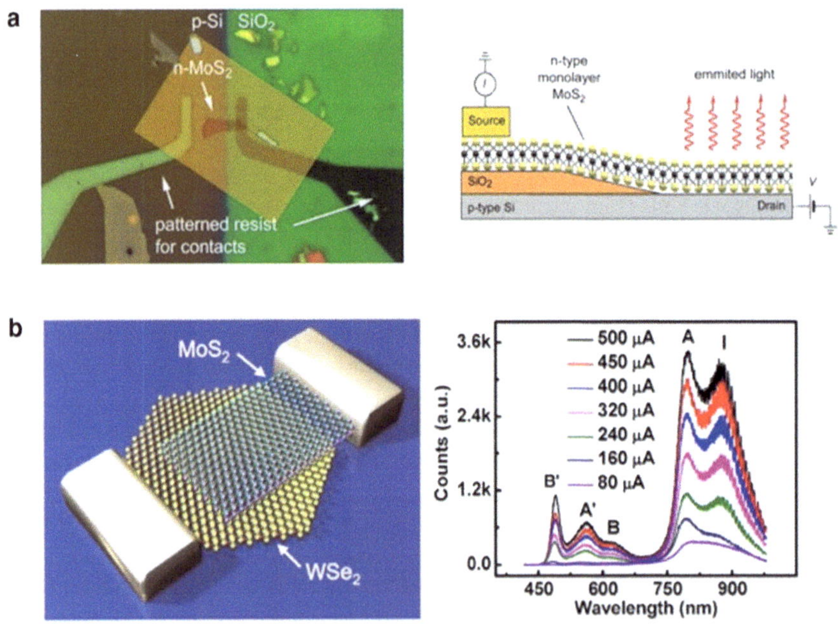

Fig. 13 **a** Geometry of the MoS_2/Si heterojunction light-emitting diode with an optical image of the device in an intermediate stage of nanofabrication. Monolayer MoS_2 is placed across the sidewall of a square window etched into a SiO_2 layer exposing the underlying p-doped silicon (left panel); Cross-sectional view of the device structure together with electrical connections used to induce light emission from the heterojunction (right panel) **b** Schematic illustration of the WSe_2/MoS_2 vertical heterojunction device shows that a transferred MoS_2 flake on synthetic WSe_2 forms a vertical heterojunction (left panel); EL spectra of the heterojunction under various injection currents (right panel). Reproduced with permission from reference [89, 96]

present nice electrical and thermal properties for such applications. It is worth mentioning that graphene isn't usually considered as an ideal material for TE devices since it has both high electric and thermal conductivities. Concerning semimetal graphene, 2D semiconductors are, in principle, expected to permit more reliable thermoelectric performance. From first-principles DFT calculation, Huang et al. [103] have investigated the thermoelectric performance of different dichalcogenide monolayers, including MoS_2, $MoSe_2$, WS_2, and WSe_2. Note that with the increase of temperature, it is well-known that the first peak values of ZT increase linearly except for the cases of monolayer n-type $WSe_2/MoSe_2$ and p-type WS_2. On the other hand, however, Naghavi et al. [104] found that monolayer Pd_2Se_3 is a promising 2D thermoelectric material with ultralow lattice thermal conductivity and high power factor. This likely is due to the anharmonicity as well as soft phonon modes that are associated with covalently bonded [Se_2] [2] dimers, resulting in an ultralow thermal conductivities for the monolayers of Pd_2Se_3 (in the range 1.5–2.9 W m^{-1} K^{-1} at 300 K). Consequently, this result is comparable to thermoelectric materials of high-performance as, for example, as bulk PbTe. Buscema et al. [105] found a large and tunable photothermoelectric effect in the monolayer MoS_2. In short, the photocurrent generation in monolayer MoS_2 usually is dominated by the photothermoelectric effect; in other words, not by the separation of photoexcited electron–hole pairs across the Schottky barriers at the MoS_2/electrode interfaces. As such, the tunable Seebeck coefficient for monolayer MoS_2 may in principle lead to new technological applications (e.g., on-chip thermoelectric devices and waste thermal energy harvesting). Later, Wu et al. [106] have reported an electrical and thermopower measurements of monolayer MoS_2 prepared by the CVD method over a wide temperature range (20–300 K) by employing microfabricated heaters thermometers. As such, they observed large values of up to ~30 mV/K at room temperature. In contrast, this result obtained is approximately two orders of magnitude larger than that in pristine graphene or one order of magnitude larger than that of bulk MoS_2. In this case, the thermopower is strongly dependent on temperature and applied gate voltage with a substantial enhancement in the vicinity of the conduction band edge. Pu et al. [107] investigated the thermoelectric properties in a FET configuration (with a channel length of ~400 μm) of monolayers of both MoS_2 and WSe_2 prepared in a large-area through the CVD method. For both monolayers, MoS_2 and WSe, using the ionic gating in the electric double-layer transistor (EDLT) configuration, has been observed a large (>200 μVK^{-1}) and power factor (>200 μWm^{-1} K^{-2}). These results show an order of magnitude enhancement in the power factor as a function of compared to their bulk counterparts. This likely is due to the quantum well-induced staircase-like density of states in 2D materials based on the TMDs. More specifically, the BP has attracted much attention for a huge variety of TE applications [108]. For instance, BP has been reported to retain a significant Seebeck coefficient (~335 μV K^{-1} at room temperature), high carrier mobility (~1000 cm^2 V^{-1} s^{-1} at room temperature), as well as a layer-dependent bandgap, which can, in principle, be tuned from 0.3 to 2 eV, i.e., with reducing thickness from bulk to monolayer [108, 109]. In this regard, Saito et al. measured the gate-tuned Seebeck coefficient at 40 nm BP [109]. As such, Seebeck value reached 510 $\mu V K^{-1}$ in the hole-depleted state, from which is much larger than the value reported for bulk phase. Thus, in this perspective, as a general trend of Seebeck value as a function of carrier concentration have found to deviate from the theoretical results. The authors concluded that the large value of Seebeck to the significant reduction of channel thickness in the depletion mode operation of the field-effect transistor. Hence, this consideration suggests that a few- or monolayer BP single-crystal might in principle be able of producing even larger thermoelectric power. Other 2D-based materials have been found to possess higher ZT than their bulky counterparts. Compared with bulk materials, the 2-D layers of SnS_2 have higher electrical conductivity and lower thermal conductivity [110]. A negative correlation has been obsrved for the enhanced electrical conductivity and reduced thermal conductivity of SnS_2. In general, this trend is directly associated with a decrease in the thickness of SnS2. Hence, this nanoscale behavior could be beneficial for applicaton as thermoelectric nanogenerators. For SnS_2 nanosheets, with a thickness of about 16 nm, the Seebeck coefficient obtained at 300 K was of 34.7 $mV K^{-1}$. The 2D layered structure materials, can therefore be cleaved down to few- or monolayer, leading to significant changes in their electrical and optical responses (such as indirect-to-direct bandgap transition). Despite the outstanding advances, a considerable amount of effort is, in turn, required in order to realize the practical applications of 2D material-based thermoelectric energy-harvesting technologies (Fig. 14) [103–110].

(d) **Capacitors**: Electrochemical capacitors are well known in the literature for next-generation energy storage devices applicaitons. This likely is due to their high specific power and long life cycle in such applications. Current efforts has driven the development of ultrathin all-solid-state hybrid supercapacitors for renewable energy storage applications. Improving specific capacitance of carbonaceous electrochemical double-layer capacitors via reversible electroadsorption of ions at the electrode/electrolyte interface, strategies including hybridization of graphene-based supercapacitors with materials based on oxides, hydroxides, and chalcogenides has been recently explored [45, 111–113]. The most important attributes to achieving high performance in hybrid supercapacitors include high intrinsic conductivity, maximum surface area with controllable pore topologies along with fast and reversible faradaic processes at the surface of redox-active nanosheets. Recently, TMDs materials have been incorporated in various energy storage devices such as (super)capacitors and batteries wherein MoS_2 with 1 T-phase is intrinsically hydrophilic and possess high electrical conductivity. Particularly, the chemically exfoliated nanosheets of MoS_2 could be efficiently intercalated with ions and achieve high capacitance [112]. As such, the high capacitance has widely been attributed to the high concentration of metallic 1 T phases generated in the process of chemical exfoliation. However, it has been reported that a nice volumetric energy

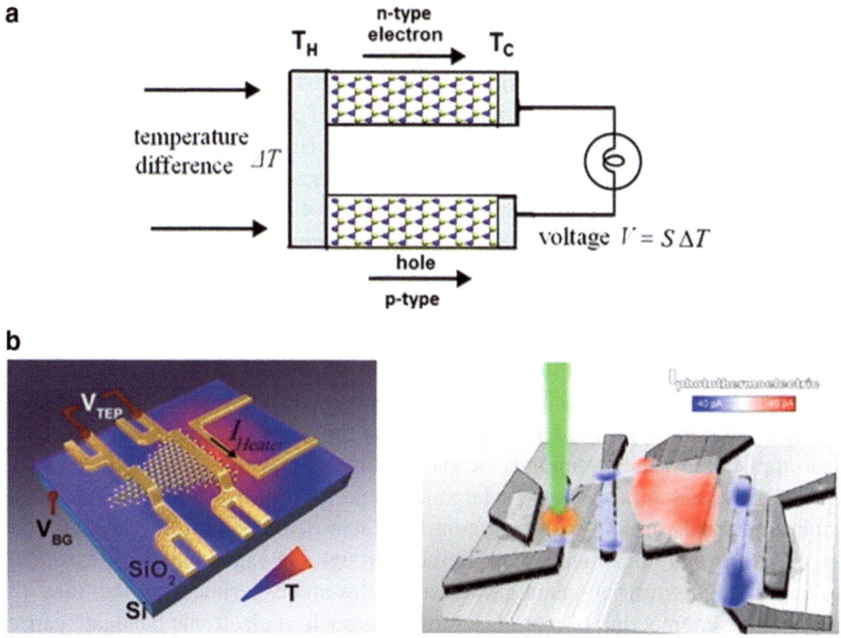

Fig. 14 **a** Schematic representation of the working principle of a thermoelectric device **b** Schematic illustration of the thermoelectric device. Two electrodes, R_{hot} and R_{cold}, contacting with the single-layer MoS_2 flake prepared by the CVD method, are used as 4-probe thermometers in order to determine the heat gradient applied by the heating current I_{heater} and also to measure the thermoelectric voltage drop. A back gate voltage V_{BG} is used to tune the carrier density of the device. (left panel); Spatial map of photocurrent response and intensity of large and tunable photothermoelectric effect observed in single layered MoS_2. Reproduced with permission from reference [105, 106, 111]

and significant power density values for the chemically exfoliated MoS_2 operating in non-aqueous organic electrolytes exhibits an excellent cycling performance with superior electrochemical properties. Also, the incorporation of carbon-based materials (such as carbon hybrid, graphene, or carbon nanotubes) into MoS2 resulting in a high energy density, good cycling stability, and high rate capability, being therefore promising to applications in high-performance lithium-ion bateries, and has been widely prepared by hydro(solvo-)thermal synthesis [19, 113, 114].

7 Conclusions and Prospective

In this chapter, we have presented a brief overview of the recent developments and research progress that has been made on the fabrication, characterization techniques, and potential applications of graphene and related 2D layered materials. Owing to the exceptional and tunable optical properties of 2D layered, combined with its simplicity of obtaining the vdW heterostructures of high quality, in general, these systems are therefore promising for applications in emerging optoelectronic technologies. Despite the current implementation of these layered materials in varied applications in the field of photovoltaics, photocatalysis, and optoelectronics, several examples have been highlighted throughout this chapter. However, additional ongoing is still required to completely study the underlying operating principles of the newly proposed devices, that is, enabling the realization of novel optoelectronic devices based on 2D layered materials with superior performance in relation to the traditional counterparts. With a wide range of 2D materials currently available for exploration, they could undoubtedly usher in several new future applications. Ever since the first discovery of graphene, substantial progress has been made over the last decade, and we anticipate further progress and significant advancement in both fundamental understanding and technological development in renewable energy harvesting devices. Furthermore, in order to move beyond the proof of concept towards practical applications, the community needs to address several critical challenges and issues: (1) scalable production of large area and reproducible strategies for high-quality materials growth with well-controlled nanoscale properties. Currently, the ongoing efforts in this direction are limited through the vague potential in scaling-up with stringent restrictions in growth substrates and experimental conditions. (2) controllable engineering of materials properties such as electronic bandgap, carrier density and strategies in improving efficiency and output performance. An extensive understanding of the mechanism of energy harvesting and charge transfer will, in turn, facilitate the optimization of device performance and output efficiency. (3) Improving stability and environmental adaptability: mainly due to the atomic thickness of the active material as well as the wide distribution of the device in the environment, thus, 2D-based material devices need to adapt to a wide variety of the environmental conditions in order to maintain stability in applications of high-performance. In this regard, the optimization of material properties, device structure, as well as the incorporation of the packaging process, in large part, would address this challenge. (4) Integrating the energy harvesters for diverse applications: based on the increasing demand for practical applications, in principle, is the main driving force for the exploration of new energy-harvesting processes. Future directions should be steered to develop devices that meet various application requirements such as flexible wearable electronics, soft bioelectronics, and sensors. Due to the flat ultrathin nature of 2D layered materials, it also allowed for integration in the vertical direction leading to a 3D architecture with novel features and boosted performance compared to planar counterparts. However, the existing issues with the electrical interconnection across layers and reliability/robustness of the manufacturing process need to be addressed. It is quite expected that with the confluence of different branches such as nanomanufacturing, data science, solid-state electronics, materials chemistry, and device physics will lead to further theoretical and experimental advancement in newer 2D materials for various renewable energy-based applications. Despite the great progress summarized in this chapter, in particular, further research are needed to clarify the self-assembly processes of 2D layered materials and their heterostructures. Undoubtedly, it is expected that such advances, in principle, will offer an outstanding level of structural-based control in order to provide the emergence of new physical phenomena and unique properties at the nanoscale. In short, the ability to design and sculpt artificial 2D materials with engineered functional properties should pave of this emerging field's progress.

References

1. Novoselov, K.S., Geim, A.K., Morozov, S.V, Jiang, D., Zhang, Y., Dubonos, S.V, Grigorieva, I.V, Firsov, A.A.: Electric field effect in atomically thin carbon films. Science (80-.). **306**(5696), 666 LP-669 (2004). https://doi.org/10.1126/science.1102896
2. Geim, A.K., Novoselov, K.S.: The rise of graphene. Nat. Mater. **6**(3), 183–191 (2007). https://doi.org/10.1038/nmat1849
3. Xu, M., Liang, T., Shi, M., Chen, H.: Graphene-like two-dimensional materials. Chem. Rev. **113**(5), 3766–3798 (2013). https://doi.org/10.1021/cr300263a
4. Xia, F., Wang, H., Xiao, D., Dubey, M., Ramasubramaniam, A.: Two-dimensional material nanophotonics. Nat. Photonics **8**(12), 899–907 (2014). https://doi.org/10.1038/nphoton.2014.271
5. Geim, A.K.: Random walk to graphene (Nobel Lecture). Angew. Chemie Int. Ed. **50**(31), 6966–6985 (2011). https://doi.org/10.1002/anie.201101174
6. Novoselov, K.S., Geim, A.K., Morozov, S.V., Jiang, D., Katsnelson, M.I., Grigorieva, I.V., Dubonos, S.V., Firsov, A.A.: Two-dimensional gas of massless dirac fermions in graphene. Nature **438**(7065), 197–200 (2005). https://doi.org/10.1038/nature04233
7. Vidu, R., Rahman, M., Mahmoudi, M., Enachescu, M., Poteca, T.D., Opris, I.: Nanostructures: a platform for brain repair and augmentation. Front. Syst. Neurosci. **8**, 91 (2014). https://doi.org/10.3389/fnsys.2014.00091
8. Mishra, N., Boeckl, J., Motta, N., Iacopi, F.: Graphene growth on silicon carbide: a review. Phys. Status Solidi **213**(9), 2277–2289 (2016). https://doi.org/10.1002/pssa.201600091
9. Bonaccorso, F., Lombardo, A., Hasan, T., Sun, Z., Colombo, L., Ferrari, A.C.: Production and processing of graphene and 2D crystals. Mater. Today **15**(12), 564–589 (2012). https://doi.org/10.1016/S1369-7021(13)70014-2
10. Li, X., Shen, R., Ma, S., Chen, X., Xie, J.: Graphene-based heterojunction photocatalysts. Appl. Surf. Sci. **430**, 53–107 (2018). https://doi.org/10.1016/j.apsusc.2017.08.194
11. Yeh, T.-F., Cihlář, J., Chang, C.-Y., Cheng, C., Teng, H.: Roles of graphene oxide in photocatalytic water splitting. Mater. Today **16**(3), 78–84 (2013). https://doi.org/10.1016/j.mattod.2013.03.006
12. Bell, N.J., Ng, Y.H., Du, A., Coster, H., Smith, S.C., Amal, R.: Understanding the enhancement in photoelectrochemical properties of photocatalytically prepared TiO2-reduced graphene oxide composite. J. Phys. Chem. C **115**(13), 6004–6009 (2011). https://doi.org/10.1021/jp1113575
13. Tay, R.Y. Synthesis of two-dimensional hexagonal boron nitride BT. In: Tay, R.Y. (ed.) Chemical Vapor Deposition Growth and Characterization of Two-Dimensional Hexagonal Boron Nitride, pp. 1–10. Springer, Singapore (2018). https://doi.org/10.1007/978-981-10-8809-4_1
14. Zedlitz, R., Heintze, M., Schubert, M.B.: Properties of amorphous boron nitride thin films. J. Non. Cryst. Solids **198–200**, 403–406 (1996). https://doi.org/10.1016/0022-3093(95)00748-2
15. Pan, Z., Sun, H., Zhang, Y., Chen, C.: Harder than diamond: superior indentation strength of Wurtzite BN and lonsdaleite. Phys. Rev. Lett. **102**(5), 55503 (2009). https://doi.org/10.1103/PhysRevLett.102.055503
16. Vel, L., Demazeau, G., Etourneau, J.: Cubic boron nitride: synthesis, physicochemical properties and applications. Mater. Sci. Eng. B **10**(2), 149–164 (1991). https://doi.org/10.1016/0921-5107(91)90121-B

17. Cazorla, C., Gould, T.: Polymorphism of bulk boron nitride. Sci. Adv. **5**(1) (2019). https://doi.org/10.1126/sciadv.aau5832
18. Song, L., Ci, L., Lu, H., Sorokin, P.B., Jin, C., Ni, J., Kvashnin, A.G., Kvashnin, D.G., Lou, J., Yakobson, B.I., et al.: Large scale growth and characterization of atomic hexagonal boron nitride layers. Nano Lett. **10**(8), 3209–3215 (2010). https://doi.org/10.1021/nl1022139
19. Shi, Y., Hamsen, C., Jia, X., Kim, K.K., Reina, A., Hofmann, M., Hsu, A.L., Zhang, K., Li, H., Juang, Z.-Y., et al.: Synthesis of few-layer hexagonal boron nitride thin film by chemical vapor deposition. Nano Lett. **10**(10), 4134–4139 (2010). https://doi.org/10.1021/nl1023707
20. Lee, K.H., Shin, H.-J., Lee, J., Lee, I., Kim, G.-H., Choi, J.-Y., Kim, S.-W.: Large-scale synthesis of high-quality hexagonal boron nitride nanosheets for large-area graphene electronics. Nano Lett. **12**(2), 714–718 (2012). https://doi.org/10.1021/nl203635v
21. Han, J., Lee, J.-Y., Kwon, H., Yeo, J.-S.: Synthesis of wafer-scale hexagonal boron nitride monolayers free of aminoborane nanoparticles by chemical vapor deposition. Nanotechnology **25**(14), 145604 (2014). https://doi.org/10.1088/0957-4484/25/14/145604
22. Kim, G., Jang, A.-R., Jeong, H.Y., Lee, Z., Kang, D.J., Shin, H.S.: Growth of high-crystalline, single-layer hexagonal boron nitride on recyclable platinum foil. Nano Lett. **13**(4), 1834–1839 (2013). https://doi.org/10.1021/nl400559s
23. Jang, S.K., Youn, J., Song, Y.J., Lee, S.: Synthesis and characterization of hexagonal boron nitride as a gate dielectric. Sci. Rep. **6**(1), 30449 (2016). https://doi.org/10.1038/srep30449
24. Radisavljevic, B., Radenovic, A., Brivio, J., Giacometti, V., Kis, A.: Single-layer MoS2 transistors. Nat. Nanotechnol. **6**(3), 147–150 (2011). https://doi.org/10.1038/nnano.2010.279
25. Liu, J., Peng, W., Li, Y., Zhang, F., Fan, X.: 2D MXene-based materials for electrocatalysis. Trans. Tianjin Univ. **26**(3), 149–171 (2020). https://doi.org/10.1007/s12209-020-00235-x
26. Gogotsi, Y., Anasori, B.: The rise of MXenes. ACS Nano **13**(8), 8491–8494 (2019). https://doi.org/10.1021/acsnano.9b06394
27. Zhang, C. (John), Ma, Y., Zhang, X., Abdolhosseinzadeh, S., Sheng, H., Lan, W., Pakdel, A., Heier, J., Nüesch, F.: Two-dimensional transition metal carbides and nitrides (MXenes): synthesis, properties, and electrochemical energy storage applications. ENERGY Environ. Mater. **3**(1), 29–55 (2020). https://doi.org/10.1002/eem2.12058
28. Jiang, X., Kuklin, A.V, Baev, A., Ge, Y., Ågren, H., Zhang, H., Prasad, P.N.: Two-dimensional MXenes: from morphological to optical, electric, and magnetic properties and applications. Phys. Rep. **848**, 1–58 (2020). https://doi.org/10.1016/j.physrep.2019.12.006
29. Bhuyan, M.S.A., Uddin, M.N., Islam, M.M., Bipasha, F.A., Hossain, S.S.: Synthesis of graphene. Int. Nano Lett. **6**(2), 65–83 (2016). https://doi.org/10.1007/s40089-015-0176-1
30. Qing, F., Zhang, Y., Niu, Y., Stehle, R., Chen, Y., Li, X.: Towards large-scale graphene transfer. Nanoscale **12**(20), 10890–10911 (2020). https://doi.org/10.1039/D0NR01198C
31. Zhou, C., Chen, S., Lou, J., Chen, Z., Xu, G., Yang, C.: Recent advances in graphene preparation methods. Mater. Sci. Forum **814**, 3–12 (2015). https://doi.org/10.4028/www.scientific.net/MSF.814.3
32. Novoselov, K.S., Mishchenko, A., Carvalho, A., Castro Neto, A.H.: 2D materials and van Der Waals heterostructures. Science (80-.). **353**(6298), aac9439 (2016). https://doi.org/10.1126/science.aac9439
33. Wang, H., Liu, F., Fu, W., Fang, Z., Zhou, W., Liu, Z.: Two-dimensional heterostructures: fabrication, characterization, and application. Nanoscale **6**(21), 12250–12272 (2014). https://doi.org/10.1039/C4NR03435J
34. Zavabeti, A., Jannat, A., Zhong, L., Haidry, A.A., Yao, Z., Ou, J.Z.: Two-dimensional materials in large-areas: synthesis, properties and applications. Nano-Micro Lett. **12**(1), 66 (2020). https://doi.org/10.1007/s40820-020-0402-x
35. Zhang, T., Fu, L.: Controllable chemical vapor deposition growth of two-dimensional heterostructures. Chem **4**(4), 671–689 (2018). https://doi.org/10.1016/j.chempr.2017.12.006
36. Kim, S., Kim, Y.C., Choi, Y.J., Woo, H.J., Song, Y.J., Kang, M.S., Lee, C., Cho, J.H.: Vertically stacked CVD-grown 2D heterostructure for wafer-scale electronics. ACS Appl. Mater. Interfaces **11**(38), 35444–35450 (2019). https://doi.org/10.1021/acsami.9b11206

37. Yang, W., Chen, G., Shi, Z., Liu, C.-C., Zhang, L., Xie, G., Cheng, M., Wang, D., Yang, R., Shi, D., et al.: Epitaxial growth of single-domain graphene on hexagonal boron nitride. Nat. Mater. **12**(9), 792–797 (2013). https://doi.org/10.1038/nmat3695
38. Levendorf, M.P., Kim, C.-J., Brown, L., Huang, P.Y., Havener, R.W., Muller, D.A., Park, J.: Graphene and boron nitride lateral heterostructures for atomically thin circuitry. Nature **488**(7413), 627–632 (2012). https://doi.org/10.1038/nature11408
39. Huang, C., Wu, S., Sanchez, A.M., Peters, J.J.P., Beanland, R., Ross, J.S., Rivera, P., Yao, W., Cobden, D.H., Xu, X.: Lateral heterojunctions within monolayer MoSe2–WSe2 semiconductors. Nat. Mater. **13**(12), 1096–1101 (2014). https://doi.org/10.1038/nmat4064
40. Duan, X., Wang, C., Shaw, J.C., Cheng, R., Chen, Y., Li, H., Wu, X., Tang, Y., Zhang, Q., Pan, A., et al.: Lateral epitaxial growth of two-dimensional layered semiconductor heterojunctions. Nat. Nanotechnol. **9**(12), 1024–1030 (2014). https://doi.org/10.1038/nnano.2014.222
41. Shi, Y., Li, H., Li, L.J.: Recent advances in controlled synthesis of two-dimensional transition metal dichalcogenides via vapour deposition techniques. Chem. Soc. Rev. **44**(9), 2744–2756 (2015). https://doi.org/10.1039/c4cs00256c
42. Alemayehu, M.B., Falmbigl, M., Ta, K., Ditto, J., Medlin, D.L., Johnson, D.C.: Designed synthesis of van Der Waals heterostructures: the power of kinetic control. Angew. Chemie Int. Ed. **54**(51), 15468–15472 (2015). https://doi.org/10.1002/anie.201506152
43. Osada, M., Sasaki, T.: Nanosheet architectonics: a hierarchically structured assembly for tailored fusion materials. Polym. J. **47**(2), 89–98 (2015). https://doi.org/10.1038/pj.2014.111
44. Xiao, F.-X., Pagliaro, M., Xu, Y.-J., Liu, B.: Layer-by-layer assembly of versatile nanoarchitectures with diverse dimensionality: a new perspective for rational construction of multilayer assemblies. Chem. Soc. Rev. **45**(11), 3088–3121 (2016). https://doi.org/10.1039/C5CS00781J
45. Lotsch, B.V.: Vertical 2D heterostructures. Annu. Rev. Mater. Res. **45**(1), 85–109 (2015). https://doi.org/10.1146/annurev-matsci-070214-020934
46. Wang, J., Deng, S., Liu, Z., Liu, Z.: The rare two-dimensional materials with Dirac cones. Natl. Sci. Rev. **2**(1), 22–39 (2015). https://doi.org/10.1093/nsr/nwu080
47. Park, M., Kim, Y., Lee, H.: Design of 2D massless dirac fermion systems and quantum spin hall insulators based on Sp–Sp2 carbon sheets. npj Comput. Mater. **4**(1), 54 (2018). https://doi.org/10.1038/s41524-018-0113-8
48. Wang, J., Ma, F., Liang, W., Sun, M.: Electrical properties and applications of graphene, hexagonal boron nitride (h-BN), and graphene/h-BN heterostructures. Mater. Today Phys. **2**, 6–34 (2017). https://doi.org/10.1016/j.mtphys.2017.07.001
49. Kang, M., Kim, B., Ryu, S.H., Jung, S.W., Kim, J., Moreschini, L., Jozwiak, C., Rotenberg, E., Bostwick, A., Kim, K.S.: Universal mechanism of band-gap engineering in transition-metal dichalcogenides. Nano Lett. **17**(3), 1610–1615 (2017). https://doi.org/10.1021/acs.nanolett.6b04775
50. Orlita, M., Potemski, M.: Dirac electronic states in graphene systems: optical spectroscopy studies. Semicond. Sci. Technol. **25**(6), 63001 (2010). https://doi.org/10.1088/0268-1242/25/6/063001
51. Janssen, T.J.B.M., Tzalenchuk, A., Lara-Avila, S., Kubatkin, S., Fal'ko, V.I.: Quantum resistance metrology using graphene. Reports Prog. Phys. **76**(10), 104501 (2013). https://doi.org/10.1088/0034-4885/76/10/104501
52. Ferreira, A., Viana-Gomes, J., Nilsson, J., Mucciolo, E.R., Peres, N.M.R., Castro Neto, A.H.: Unified description of the Dc conductivity of monolayer and bilayer graphene at finite densities based on resonant scatterers. Phys. Rev. B **83**(16), 165402 (2011). https://doi.org/10.1103/PhysRevB.83.165402
53. Radisavljevic, B., Whitwick, M.B., Kis, A.: Integrated circuits and logic operations based on single-layer MoS2. ACS Nano **5**(12), 9934–9938 (2011). https://doi.org/10.1021/nn203715c
54. Ovchinnikov, D., Allain, A., Huang, Y.-S., Dumcenco, D., Kis, A.: Electrical transport properties of single-layer WS2. ACS Nano **8**(8), 8174–8181 (2014). https://doi.org/10.1021/nn502362b
55. Brumme, T., Calandra, M., Mauri, F.: First-principles theory of field-effect doping in transition-metal dichalcogenides: structural properties, electronic structure, Hall coefficient,

and electrical conductivity. Phys. Rev. B **91**(15), 155436 (2015). https://doi.org/10.1103/PhysRevB.91.155436
56. Xie, H., Hu, M., Bao, H.: Thermal conductivity of silicene from first-principles. Appl. Phys. Lett. **104**(13), 131906 (2014). https://doi.org/10.1063/1.4870586
57. Lindsay, L., Li, W., Carrete, J., Mingo, N., Broido, D.A., Reinecke, T.L.: Phonon thermal transport in strained and unstrained graphene from first principles. Phys. Rev. B **89**(15), 155426 (2014). https://doi.org/10.1103/PhysRevB.89.155426
58. Gu, X., Yang, R.: First-principles prediction of phononic thermal conductivity of silicene: a comparison with graphene. J. Appl. Phys. **117**(2), 25102 (2015). https://doi.org/10.1063/1.4905540
59. Liu, H., Neal, A.T., Zhu, Z., Luo, Z., Xu, X., Tománek, D., Ye, P.D.: Phosphorene: an unexplored 2D semiconductor with a high hole mobility. ACS Nano **8**(4), 4033–4041 (2014). https://doi.org/10.1021/nn501226z
60. Qin, G., Hu, M.: Thermal transport properties of monolayer phosphorene: a mini-review of theoretical studies. Front. Energy **12**(1), 87–96 (2018). https://doi.org/10.1007/s11708-018-0513-y
61. Zhang, J., Liu, H.J., Cheng, L., Wei, J., Liang, J.H., Fan, D.D., Jiang, P.H., Shi, J.: Thermal conductivities of phosphorene allotropes from first-principles calculations: a comparative study. Sci. Rep. **7**(1), 4623 (2017). https://doi.org/10.1038/s41598-017-04923-y
62. Hong, Y., Zhang, J., Huang, X., Zeng, X.C.: Thermal conductivity of a two-dimensional phosphorene sheet: a comparative study with graphene. Nanoscale **7**(44), 18716–18724 (2015). https://doi.org/10.1039/C5NR03577E
63. Liu, T.-H., Chang, C.-C.: Anisotropic thermal transport in phosphorene: effects of crystal orientation. Nanoscale **7**(24), 10648–10654 (2015). https://doi.org/10.1039/C5NR01821H
64. Bragança, H., Riche, F., Qu, F., Lopez-Richard, V., Marques, G.E.: Dark-Exciton valley dynamics in transition metal dichalcogenide alloy monolayers. Sci. Rep. **9**(1), 4575 (2019). https://doi.org/10.1038/s41598-019-40932-9
65. Vitale, S.A., Nezich, D., Varghese, J.O., Kim, P., Gedik, N., Jarillo-Herrero, P., Xiao, D., Rothschild, M.: Valleytronics: opportunities, challenges, and paths forward. Small **14**(38), 1801483 (2018). https://doi.org/10.1002/smll.201801483
66. Schaibley, J.R., Yu, H., Clark, G., Rivera, P., Ross, J.S., Seyler, K.L., Yao, W., Xu, X.: Valleytronics in 2D materials. Nat. Rev. Mater. **1**(11), 16055 (2016). https://doi.org/10.1038/natrevmats.2016.55
67. Mak, K.F., Lee, C., Hone, J., Shan, J., Heinz, T.F.: Atomically thin MoS_2: a new direct-gap semiconductor. Phys. Rev. Lett. **105**(13), 136805 (2010). https://doi.org/10.1103/PhysRevLett.105.136805
68. Cheiwchanchamnangij, T., Lambrecht, W.R.L.: Quasiparticle band structure calculation of monolayer, bilayer, and bulk MoS_2. Phys. Rev. B **85**(20), 205302 (2012). https://doi.org/10.1103/PhysRevB.85.205302
69. Klots, A.R., Newaz, A.K.M., Wang, B., Prasai, D., Krzyzanowska, H., Lin, J., Caudel, D., Ghimire, N.J., Yan, J., Ivanov, B.L., et al.: Probing excitonic states in suspended two-dimensional semiconductors by photocurrent spectroscopy. Sci. Rep. **4**(1), 6608 (2014). https://doi.org/10.1038/srep06608
70. Li, Z., Xiao, Y., Gong, Y., Wang, Z., Kang, Y., Zu, S., Ajayan, P.M., Nordlander, P., Fang, Z.: Active light control of the MoS2 monolayer exciton binding energy. ACS Nano **9**(10), 10158–10164 (2015). https://doi.org/10.1021/acsnano.5b03764
71. Chernikov, A., Berkelbach, T.C., Hill, H.M., Rigosi, A., Li, Y., Aslan, O.B., Reichman, D.R., Hybertsen, M.S., Heinz, T.F.: Exciton binding energy and nonhydrogenic rydberg series in monolayer WS_2. Phys. Rev. Lett. **113**(7), 76802 (2014). https://doi.org/10.1103/PhysRevLett.113.076802
72. Mak, K.F., McGill, K.L., Park, J., McEuen, P.L.: The Valley Hall effect in MoS₂ transistors. Science (80-.) **344**(6191), 1489 LP-1492 (2014). https://doi.org/10.1126/science.1250140

73. Jiang, Z., Zhang, Y., Stormer, H.L., Kim, P.: Quantum Hall states near the charge-neutral Dirac point in graphene. Phys. Rev. Lett. **99**(10), 106802 (2007). https://doi.org/10.1103/PhysRevLett.99.106802
74. Yao, W., Xiao, D., Niu, Q.: Valley-dependent optoelectronics from inversion symmetry breaking. Phys. Rev. B **77**(23), 235406 (2008). https://doi.org/10.1103/PhysRevB.77.235406
75. Mak, K.F., He, K., Shan, J., Heinz, T.F.: Control of Valley polarization in monolayer MoS2 by optical helicity. Nat. Nanotechnol. **7**(8), 494–498 (2012). https://doi.org/10.1038/nnano.2012.96
76. Srivastava, A., Sidler, M., Allain, A.V., Lembke, D.S., Kis, A., Imamoğlu, A.: Valley Zeeman effect in elementary optical excitations of monolayer WSe2. Nat. Phys. **11**(2), 141–147 (2015). https://doi.org/10.1038/nphys3203
77. Ross, J.S., Wu, S., Yu, H., Ghimire, N.J., Jones, A.M., Aivazian, G., Yan, J., Mandrus, D.G., Xiao, D., Yao, W., et al.: Electrical control of neutral and charged excitons in a monolayer semiconductor. Nat. Commun. **4**(1), 1474 (2013). https://doi.org/10.1038/ncomms2498
78. Xiao, D., Liu, G.-B., Feng, W., Xu, X., Yao, W.: Coupled spin and valley physics in monolayers of MoS_2 and other group-VI dichalcogenides. Phys. Rev. Lett. **108**(19), 196802 (2012). https://doi.org/10.1103/PhysRevLett.108.196802
79. Zeng, H., Liu, G.-B., Dai, J., Yan, Y., Zhu, B., He, R., Xie, L., Xu, S., Chen, X., Yao, W., et al.: Optical signature of symmetry variations and spin-valley coupling in atomically thin tungsten dichalcogenides. Sci. Rep. **3**(1), 1608 (2013). https://doi.org/10.1038/srep01608
80. Li, Y., Rao, Y., Mak, K.F., You, Y., Wang, S., Dean, C.R., Heinz, T.F.: Probing symmetry properties of few-layer MoS2 and h-BN by optical second-harmonic generation. Nano Lett. **13**(7), 3329–3333 (2013). https://doi.org/10.1021/nl401561r
81. Hsu, W.-T., Zhao, Z.-A., Li, L.-J., Chen, C.-H., Chiu, M.-H., Chang, P.-S., Chou, Y.-C., Chang, W.-H.: Second harmonic generation from artificially stacked transition metal dichalcogenide twisted bilayers. ACS Nano **8**(3), 2951–2958 (2014). https://doi.org/10.1021/nn500228r
82. Wang, H., Zhang, C., Rana, F.: Surface recombination limited lifetimes of photoexcited carriers in few-layer transition metal dichalcogenide MoS2. Nano Lett. **15**(12), 8204–8210 (2015). https://doi.org/10.1021/acs.nanolett.5b03708
83. Song, Q., Wang, H., Pan, X., Xu, X., Wang, Y., Li, Y., Song, F., Wan, X., Ye, Y., Dai, L.: Anomalous in-plane anisotropic raman response of monoclinic semimetal 1 T´-MoTe2. Sci. Rep. **7**(1), 1758 (2017). https://doi.org/10.1038/s41598-017-01874-2
84. Yang, H., Jussila, H., Autere, A., Komsa, H.-P., Ye, G., Chen, X., Hasan, T., Sun, Z.: Optical waveplates based on birefringence of anisotropic two-dimensional layered materials. ACS Photonics **4**(12), 3023–3030 (2017). https://doi.org/10.1021/acsphotonics.7b00507
85. Baugher, B.W.H., Churchill, H.O.H., Yang, Y., Jarillo-Herrero, P.: Optoelectronic devices based on electrically tunable p–n diodes in a monolayer dichalcogenide. Nat. Nanotechnol. **9**(4), 262–267 (2014). https://doi.org/10.1038/nnano.2014.25
86. Pospischil, A., Furchi, M.M., Mueller, T.: Solar-energy conversion and light emission in an atomic monolayer p–n diode. Nat. Nanotechnol. **9**(4), 257–261 (2014). https://doi.org/10.1038/nnano.2014.14
87. Gong, Y., Lin, J., Wang, X., Shi, G., Lei, S., Lin, Z., Zou, X., Ye, G., Vajtai, R., Yakobson, B.I., et al.: Vertical and in-plane heterostructures from WS2/MoS2 monolayers. Nat. Mater. **13**(12), 1135–1142 (2014). https://doi.org/10.1038/nmat4091
88. Yang, T., Zheng, B., Wang, Z., Xu, T., Pan, C., Zou, J., Zhang, X., Qi, Z., Liu, H., Feng, Y., et al.: Van Der Waals epitaxial growth and optoelectronics of large-scale WSe2/SnS2 vertical bilayer p–n junctions. Nat. Commun. **8**(1), 1906 (2017). https://doi.org/10.1038/s41467-017-02093-z
89. Lopez-Sanchez, O., Alarcon Llado, E., Koman, V., Fontcuberta i Morral, A., Radenovic, A., Kis, A.: Light generation and harvesting in a van Der Waals heterostructure. ACS Nano. 8(3), 3042–3048 (2014). https://doi.org/10.1021/nn500480u
90. Withers, F., Del Pozo-Zamudio, O., Mishchenko, A., Rooney, A.P., Gholinia, A., Watanabe, K., Taniguchi, T., Haigh, S.J., Geim, A.K., Tartakovskii, A.I., et al.: Light-emitting diodes by band-structure engineering in van Der Waals heterostructures. Nat. Mater. **14**(3), 301–306 (2015). https://doi.org/10.1038/nmat4205

91. Zhu, J., Park, H., Chen, J.-Y., Gu, X., Zhang, H., Karthikeyan, S., Wendel, N., Campbell, S. A., Dawber, M., Du, X., et al.: Black phosphorus: revealing the origins of 3D anisotropic thermal conductivities of black phosphorus (Adv. Electron. Mater. 5/2016). Adv. Electron. Mater. **2**(5) (2016). https://doi.org/10.1002/aelm.201670029
92. Shanmugam, M., Durcan, C.A., Yu, B.: Layered semiconductor molybdenum disulfide nanomembrane based Schottky-Barrier solar cells. Nanoscale **4**(23), 7399–7405 (2012). https://doi.org/10.1039/C2NR32394J
93. Shanmugam, M., Jacobs-Gedrim, R., Song, E.S., Yu, B.: Two-dimensional layered semiconductor/graphene heterostructures for solar photovoltaic applications. Nanoscale **6**(21), 12682–12689 (2014). https://doi.org/10.1039/C4NR03334E
94. Bernardi, M., Palummo, M., Grossman, J.C.: Extraordinary sunlight absorption and one nanometer thick photovoltaics using two-dimensional monolayer materials. Nano Lett. **13**(8), 3664–3670 (2013). https://doi.org/10.1021/nl401544y
95. Furchi, M.M., Pospischil, A., Libisch, F., Burgdörfer, J., Mueller, T.: Photovoltaic effect in an electrically tunable van Der Waals heterojunction. Nano Lett. **14**(8), 4785–4791 (2014). https://doi.org/10.1021/nl501962c
96. Cheng, R., Li, D., Zhou, H., Wang, C., Yin, A., Jiang, S., Liu, Y., Chen, Y., Duan, X.: Electroluminescence and Photocurrent Generation from Atomically Sharp WSe2/MoS2 Heterojunction p–n Diodes. Nano. Lett. **14**(10), 5590–5597 (2014). https://doi.org/10.1021/nl502075n
97. Tsai, M.-L., Su, S.-H., Chang, J.-K., Tsai, D.-S., Chen, C.-H., Wu, C.-I., Li, L.-J., Chen, L.-J., He, J.-H.: Monolayer MoS2 heterojunction solar cells. ACS Nano **8**(8), 8317–8322 (2014). https://doi.org/10.1021/nn502776h
98. Gu, X., Cui, W., Li, H., Wu, Z., Zeng, Z., Lee, S.-T., Zhang, H., Sun, B.: A solution-processed hole extraction layer made from ultrathin MoS2 nanosheets for efficient organic solar cells. Adv. Energy Mater. **3**(10), 1262–1268 (2013). https://doi.org/10.1002/aenm201300549
99. Abdullah, M.F., Hashim, A.M.: Review and assessment of photovoltaic performance of graphene/Si heterojunction solar cells. J. Mater. Sci. **54**(2), 911–948 (2019). https://doi.org/10.1007/s10853-018-2947-3
100. Liu, Z., Li, J., Sun, Z.-H., Tai, G., Lau, S.-P., Yan, F.: The application of highly doped single-layer graphene as the top electrodes of semitransparent organic solar cells. ACS Nano **6**(1), 810–818 (2012). https://doi.org/10.1021/nn204675r
101. Lee, C.-H., Lee, G.-H., van der Zande, A.M., Chen, W., Li, Y., Han, M., Cui, X., Arefe, G., Nuckolls, C., Heinz, T.F., et al.: Atomically thin p–n junctions with van Der Waals heterointerfaces. Nat. Nanotechnol. **9**(9), 676–681 (2014). https://doi.org/10.1038/nnano.2014.150
102. Deng, Y., Luo, Z., Conrad, N.J., Liu, H., Gong, Y., Najmaei, S., Ajayan, P.M., Lou, J., Xu, X., Ye, P.D.: Black phosphorus-monolayer MoS2 van Der Waals heterojunction p–n diode. ACS Nano **8**(8), 8292–8299 (2014). https://doi.org/10.1021/nn5027388
103. Huang, W., Da, H., Liang, G.: Thermoelectric performance of MX2 (M = Mo, W; X = S, Se) monolayers. J. Appl. Phys. **113**(10), 104304 (2013). https://doi.org/10.1063/1.4794363
104. Naghavi, S.S., He, J., Xia, Y., Wolverton, C.: Pd2Se3 Monolayer: a promising two-dimensional thermoelectric material with ultralow lattice thermal conductivity and high power factor. Chem. Mater. **30**(16), 5639–5647 (2018). https://doi.org/10.1021/acs.chemmater.8b01914
105. https://thermal.ferrotec.com/technology/
106. Wu, J., Schmidt, H., Amara, K.K., Xu, X., Eda, G., Özyilmaz, B.: Large thermoelectricity via variable range hopping in chemical vapor deposition grown single-layer MoS2. Nano Lett. **14**(5), 2730–2734 (2014). https://doi.org/10.1021/nl500666m
107. Pu, J., Funahashi, K., Chen, C.-H., Li, M.-Y., Li, L.-J., Takenobu, T.: Highly flexible and high-performance complementary inverters of large-area transition metal dichalcogenide monolayers. Adv. Mater. **28**(21), 4111–4119 (2016). https://doi.org/10.1002/adma.201503872
108. Kou, L., Ma, Y., Smith, S.C., Chen, C.: Anisotropic ripple deformation in phosphorene. J. Phys. Chem. Lett. **6**(9), 1509–1513 (2015). https://doi.org/10.1021/acs.jpclett.5b00522

109. Saito, Y., Iizuka, T., Koretsune, T., Arita, R., Shimizu, S., Iwasa, Y.: Gate-tuned thermoelectric power in black phosphorus. Nano Lett. **16**(8), 4819–4824 (2016). https://doi.org/10.1021/acs.nanolett.6b00999
110. Lee, M.-J., Ahn, J.-H., Sung, J.H., Heo, H., Jeon, S.G., Lee, W., Song, J.Y., Hong, K.-H., Choi, B., Lee, S.-H., et al.: Thermoelectric materials by using two-dimensional materials with negative correlation between electrical and thermal conductivity. Nat. Commun. **7**(1), 12011 (2016). https://doi.org/10.1038/ncomms12011
111. Buscema, M., Barkelid, M., Zwiller, V., van der Zant, H., Steele, G., Castellanos-Gomez, A.: Large and tunable photothermoelectric effect in single-layer MoS2. Nano Lett. **13**(2), 358–363 (2013). https://doi.org/10.1021/nl303321g
112. Voiry, D., Salehi, M., Silva, R., Fujita, T., Chen, M., Asefa, T., Shenoy, V.B., Eda, G., Chhowalla, M.: Conducting MoS2 nanosheets as catalysts for hydrogen evolution reaction. Nano Lett. **13**(12), 6222–6227 (2013). https://doi.org/10.1021/nl403661s
113. Shi, S., Sun, Z., Hu, Y.H.: Synthesis, stabilization and applications of 2-dimensional 1T metallic MoS2. J. Mater. Chem. A **6**(47), 23932–23977 (2018). https://doi.org/10.1039/C8TA08152B
114. Zhang, G., Liu, H., Qu, J., Li, J.: Two-dimensional layered MoS2: rational design, properties and electrochemical applications. Energy Environ. Sci. **9**(4), 1190–1209 (2016). https://doi.org/10.1039/C5EE03761A

Overall Insights into Sustainable Utilization of Methane and Carbon Dioxide in Heterogeneous Catalysis

Francielle C. F. Marcos, Davi D. Petrolini, Alessandra F. Lucredio, Jose M. Assaf, and Elisabete M. Assaf

Abstract The development of society is dependent on commodities such as fuels and chemical feedstock. Most of these commodities are obtained from oil as raw material. Although the need to find a friendly solution to couple an economically viable energy model with a greener solution, it is known that technologies applying renewable sources are in an early stage of development. The conversion of methane into clean fuels or chemical feedstock with high commercial value, such as hydrogen, ethylene, or methanol is interesting from the energetic and economic point-of-view. Among the methods of methane conversion, the industrially used is the steam reforming (MSR), in which methane reacts with water to produce syngas, a mixture of CO and H_2. Nevertheless, this reaction is highly endothermic and responsible for a large volume of CO_2 emitted by the reactor burners that provide energy to the reactors. An interesting alternative process for methane conversion is the dry reforming of methane (DRM), which consists of the reaction of methane with CO_2, also yielding syngas. The advantage of this reaction is the utilization of two harmful gases to the atmosphere. The disadvantage of this reaction is due to the catalyst deactivation by carbon deposition. In heterogeneous catalysis, there is a strong relationship between catalytic performance and surface and textural properties, that are outlined by the number and distribution of available active sites. In this way, different synthetic routes may be used to design these properties and obtain products of commercial interest, such as ethylene. The commercial production of ethylene occurs by the recuperation of refinery gases, thermal cracking of light hydrocarbons, mainly ethane and propane, or a combination of both processes. An alternative process of ethylene production may be from natural gas and or biogas. This process can be performed by the syngas

F. C. F. Marcos
University of São Paulo, Escola Politécnica of University of São Paulo, Av. Prof. Luciano Gualberto, t. 3, 380, São Paulo, SP 05508-010, Brazil

D. D. Petrolini · A. F. Lucredio · E. M. Assaf (✉)
University of São Paulo, São Carlos Institute of Chemistry, Av. Trabalhador São-Carlense, 400, São Carlos, SP 13560-970, Brazil
e-mail: eassaf@iqsc.usp.br

J. M. Assaf
Federal University of São Carlos, Rod. W. Luiz, km 235, São Carlos, SP 13565-905, Brazil

F. A. La Porta and C. A. Taft (eds.), *Functional Properties of Advanced Engineering Materials and Biomolecules*, Engineering Materials,
https://doi.org/10.1007/978-3-030-62226-8_8

route or by oxidative coupling of methane (OCM). The first process is an indirect conversion of methane and involves several steps, which increases the process costs. The second one is a direct conversion, where methane is directly converted into C_2 (ethane and ethylene) hydrocarbons. Although the OCM is not yet a reaction on an industrial scale, several efforts are being made to design a catalytic system to achieve C_2 hydrocarbons yields above 30%, the minimum required. Besides that, another technology has been studied to directly produce ethylene via the oxidative coupling of methane using CO_2 as a mild oxidant (CO_2-OCM). These technological routes for the valorization of methane and CO_2 will be addressed in this review.

1 Introduction

The development of society is very dependent on commodities such as fuels and chemical feedstocks. Most of these commodities have been manufactured using petroleum as a raw material. The world population is in constant expansion, with almost 8 billion in 2020 [1]. Despite the variations in demographic density, the need for energy is common in all regions, meaning that the increase in the energy demand and consumption of fossil fuels are worldwide problems. Although the urgent need for an economically viable energy model that encompasses a greener solution, it is known that energy technologies based on renewable carbon resources are in an early stage of development. Consequently, the efficient use of the available non-renewable carbon resources, such as natural gas and biogas, arises as an eco-friendly solution to obtain energy and chemical commodities, bridging to a future 100% based on renewable energy. In 2019, the total natural gas volume worldwide was around 228 billion of m^3 [2], and strategies have been studied and developed for a conscious use of gas natural.

Most processes for energy production from carbon resources are related to catalysis, mainly heterogeneous catalysis.

Biogas is a C1 resource with a great environmental impact, produced by anaerobic fermentation of biomass, such as solid waste, wastewater, and sewage. The biogas composition depends on the feedstock and operation conditions. For instance, it should be noted that the biogas produced from sewage digesters generally comprises 55–65% CH_4, 35–45% CO_2, and 1% N_2, whereas the biogas produced from organic waste digesters in general regularly contains 60–70% CH_4, 30–40% CO_2, and 1% N_2. It implies that in both compositions also encompass traces of water vapor and other gases (e.g., H_2S). For application in C1 conversion reactions, the H_2S contaminant must be removed from the feeding biogas through an additional stage to avoid the catalyst poisoning [3–6].

Methane also is known as the principal component of natural gas. It is a harmful molecule to the atmosphere due to its contribution to the greenhouse effect. Methane is mostly consumed as a fuel, and only a small percentage is used to produce chemical feedstocks. The conversion of methane into clean fuels or value-added chemical feedstocks, such as hydrogen, methanol, or ethylene, is interesting from the energetic and economic point-of-views, considering the total volume of the global natural gas reserves and all potential of biogas production. However, most of the natural gas is flared at the extraction sites due to transportation issues [1, 7].

Currently, the production of valuable commodities, including chemicals, polymers, and fuels from methane is performed in two steps. First, methane is converted into a mixture typically of H_2 and CO, known as synthesis gas or syngas, and then, another process is usually used in order to convert syngas to more valuable products, such as ammonium, methanol, and hydrocarbons. Among the technologies to produce syngas, three processes are of industrial relevance: steam methane reforming (1), partial oxidation of methane (2), and dry reforming of methane with CO_2 (3). Basically, steam reforming of methane (SRM) and partial oxidation of methane are well-established technologies [8–12], while other reactions, namely, autothermal reforming and bi-reforming of methane, which combine an endothermic and exothermic reaction and two endothermic reactions, respectively, are still under development.

$$CH_4 + H_2O \rightleftharpoons CO + 3H_2 \quad \Delta H^\circ = 225.4\ \text{kJ/mol} \tag{1}$$

$$CH_4 + 1/2O_2 \rightarrow 2CO + 2H_2 \quad \Delta H^\circ = -22.6\ \text{kJ/mol} \tag{2}$$

$$CH_4 + CO_2 \rightleftharpoons 2CO + 2H_2 \quad \Delta H^\circ = 247\ \text{kJ/mol} \tag{3}$$

SRM is found to be the principal industrial process used for methane conversion. In the SRM, it is known that the methane reacts with water to produce syngas and CO_2. SRM is responsible for around 98% of the global H_2 supply and it is the most widespread technology used for methane conversion [13–15].

Other technologies have been studied to directly produce industrial commodities, such as ethylene, in a single step, without the use of syngas as an intermediate product. One example is shown that the oxidative coupling of methane (OCM), which involves the particular use of CO_2 as a mild oxidant (CO_2-OCM) has been recently described [13].

OCM and CO_2-OCM are eco-friendly processes that convert CH_4, O_2, and/or CO_2 into C_2 hydrocarbons (C_2H_6 and C_2H_4), thus helping reduce atmospheric emissions, that is, satisfying the industrial demand more sustainably, once CH_4 and CO_2 are greenhouses gases. Ethylene and ethane (C_2) are essential feedstocks for the manufacture of value-added chemical commodities.

2 Methane Conversion by Reforming Reactions

Low pressures favor the reforming process, but some posterior industrial synthesis gas conversion processes need a compressed synthesis gas. Thus, some processes are operated at high pressures because it is more economically viable than compressing the synthesis gas downstream. The Steam Reforming of Methane (SRM) is typically operated with H_2O/CH_4 ratio in the range of 2.5–3, to favor carbon gasification, pressure above 20 atm, and a temperature range between 500 and 900 °C [16]. These main advantages are high H_2 concentration in the product composition ($H_2/CO = 3$) and industrial maturity. The disadvantages include the energy consumption for highly endothermic reaction and a large volume of CO_2 emitted by the reactor burners producing heat by CH_4 combustion. SRM is a mature process with operating industrial plants of Topsøe Package Hydrogen Plants at Air Liquide, Belgium; plants in the USA, and Air Liquide Methane Steam Reformer in Germany [17]. Dry reforming of methane (DRM) is a process that has gained significant attention, likely because it carried out in atmospheric pressure, temperatures in the range of 650–850 °C, and $CH_4:CO_2 = 1:1$, which produces syngas with H_2/CO of 1. The main advantage is the utilization of two greenhouse gases. The disadvantages include catalysts deactivation by sintering and coke accumulation. DRM is still in development with the launch of the Linde pilot plant in Germany in 2015. On the other hand, the partial oxidation of methane (POM) is usually carried out at about 100 atm with temperature in the range of 950–1100 °C and $CH_4:O_2 = 2:1$, producing syngas with H_2/CO of 2 [17, 18]. Its main advantages are tolerance to sulfur impurities, short contact time, and more economical use of thermal energy. The disadvantages include the high pure oxygen cost, danger of explosion from $CH_4:O_2$ mixtures, hot spot formation, and operational complexity. There is a working plant Pearl GTL in Ras Laffan Industrial City, Qatar. The autothermal reforming operates with a $CH_4:H_2O:O_2$ of 1:1:0.5 and produces a variable H_2/CO based on the inlet stream. The advantages are energy conservation by the combination of endothermic and exothermic reactions, the use of compact reactors, and adjustable feed composition. The disadvantages include significant reaction limitations. There is an operating Plant in OLTIN YO'L GTL, Shurtan Gas, and Chemical Complex, Uzbekistan [17].

DRM is described to have the lowest operation cost among the reforming processes [13]. Moreover, the inauguration of a pilot plant in 2015 by the Linde Group and BASF shows the market interest in this process. Besides the advantage of using two greenhouse gases, DRM can also be used to directly convert biogas obtained from anaerobic digestion.

Atmospheric CO_2 levels has increased over the years. Before the industrial revolution, in particular, the CO_2 concentration found in the atmosphere was of 280 ppm and then, in 2017, the CO_2 levels jumped to about 405 ppm. Increases in CO_2 emission over the years have induced several environmental impacts and, therefore, the Earth's temperature has been elevated by the greenhouse effect. Worldwide efforts towards sustainable ways for reducing the atmospheric CO_2 concentration are now recognized, but the difficulty lies in reducing the CO_2 emission while meeting the energetic demand from a growing population [2]. Ways of reducing CO_2 emissions include the replacement of fossil fuels with renewable fuels, CO_2 capture and storage (CCS), and CO_2 capture and utilization (CCU). Particularly, CCU may be key for controlling the atmospheric CO_2 concentration while contributing to energy generation, and fuel and chemical production. Also, it is known that the conversion of CO_2 to fuels or chemicals is an approach that can recycle this greenhouse gas [19].

Nevertheless, the high stability of the CO_2 molecule is an issue in its conversion to fuels. CO_2 can be directly converted into methanol via H_2 hydrogenation (direct route), but the problems of this reaction lie in the low conversion rate and availability of renewable hydrogen sources. The indirect route involves the conversion of CO_2 into syngas by dry reforming, followed by Fischer–Tropsch synthesis to produce hydrocarbons. In this case, CO_2 may be captured from the industrial process or digesters and used to produce syngas by DRM [3].

Despite the environmental appeal, the impact of using atmosphere CO_2 in dry reforming of methane is low. For example, in 2013, the methanol demand was about 65 million tons/year. By assuming that all methanol derives from the CO_2 net, it will be necessary a net capture of about 89 million tons/year. This value corresponds to only 0.26% of the total CO_2 emissions in 2012, which were around 34,500 million tons. So, DRM should not be considered a solution for reducing greenhouse gases in the atmosphere, but more like a sustainable way of using anthropogenic CO_2 and biogas obtained from biomass anaerobic digestion [3, 4, 20].

Dry reforming of methane

Hence, in this case, the overall DRM process can be described by the following reactions:

Dry reforming of methane: Eq. 3

Reverse water gas shift reaction:

$$CO_2 + H_2 \rightleftharpoons CO + H_2O \quad \Delta H^\circ = 41\ kJ/mol \tag{4}$$

Methane cracking:

$$CH_4 \rightleftharpoons C + 2H_2 \quad \Delta H^\circ = 75\ kJ/mol \tag{5}$$

CO disproportionation:

$$2CO \rightleftharpoons C + CO_2 \quad \Delta H^\circ = -172\ kJ/mol \tag{6}$$

CO reduction:

$$2CO + H_2 \rightleftharpoons C + CO_2 \quad \Delta H^\circ = -131\ kJ/mol \tag{7}$$

The equilibrium conversions of CH_4 and CO_2 in DRM using a stoichiometric mixture (CH_4:CO_2 = 1:1) are presented as a function of temperature in Fig. 1a. It is possible to see that CH_4 conversion increases with temperature over all the range. This is due to the occurrence of endothermic CH_4 consuming reactions, Eqs. (3), (1) and (5). The CO_2 conversion presents a minimum at 550 °C, and then increases with temperature. At low temperatures, DRM is not favored, Eq. (3), meanwhile, the reactions leading to CO_2, water gas shift reaction (reverse of Eq. 4) and CO disproportionation (Eq. 6), are favored. Considering the component composition at equilibrium (Fig. 1b), the increase in temperature favors the production of H_2 and

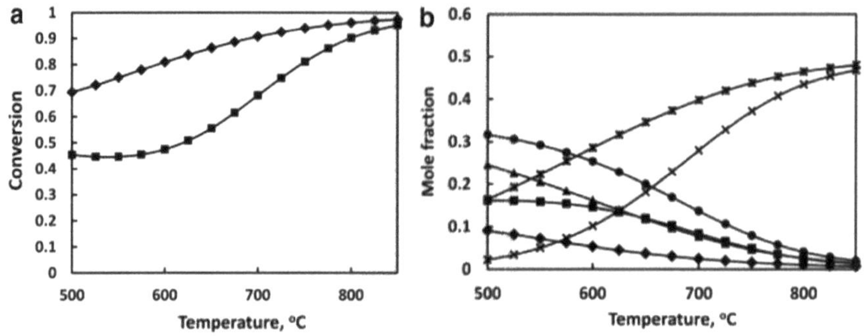

Fig. 1 (**a**) Equilibrium conversions versus temperature; (♦) methane, (■) CO_2(CH_4/CO_2 = 1, P = 1 bar); (**b**) Equilibrium mole fraction of reacting species versus temperature: (♦) CH_4, (■) CO_2, (▲) H_2O, (×) CO, (*) H_2, (●) coke (CH_4/CO_2 = 1, P = 1 bar) Copyright © 1999 Academic Press. All rights reserved (with permission) [21]

CO, and decreases the H_2/CO ratio. Yet the water gas shift reaction (reverse of Eq. 4) decreases the CO_2 conversion and H_2/CO ratio. In the range of 550–675 °C, as has widely been found the carbon deposition can occur from all the carbon producing reactions: methane cracking (Eq. 5), CO disproportionation (Eq. 6), and CO reduction (Eq. 7). This is an unfavorable temperature range for the DRM process, as a large coke deposition is expected [21]. The disadvantage of this reaction is the strong catalyst deactivation due to carbon deposition.

Catalysts containing nickel, cobalt, and noble metals have widely been tested in DRM. Among them, Ru and Rh are generally considered to be the most promising metals for SRM and DRM, followed by Ni and Ir [20, 22–25]. Yet, they are 200–2000 fold more expensive than Ni, which explains the choice of nickel for the most investigated catalytic systems [20, 22–24].

The synthesis method of supported catalysts determines the dispersion of the active phase and metal-support interactions. Catalysts with the same composition may exhibit different behaviors in the same reaction conditions depending on the preparation method. Several parameters, like the nature of the metal precursors, pH, concentration, temperature, deposition time, washing, calcination temperature, and reducing method influence on the final catalyst's properties [26]. Usually, the preparation of these catalysts is focused on the metallic phase dispersion.

The DRM mechanism is dependent on the catalyst's system and reaction conditions. The CH_4 molecule is activated on the metallic sites, such as Ni, Pt, Co, etc. The support also has a great influence on the DRM mechanism. It is known that the DRM based on inert supports (e.g., SiO_2) follows in general a monofunctional pathway with both CH_4 and CO_2 being activated on the metallic sites [19]. On acidic/basic supports, such as Al_2O_3 and MgO, DRM follows a bifunctional pathway, in which CH_4 is found to be activated on the metallic sites, while CO_2 is usually activated on the acidic/basic sites of the support, producing formate ions on the hydroxyls groups of the acidic sites and carbonate ions on the basic ones. Moreover, the reaction can also follow a bifunctional redox mechanism in supports such as CeO_2, which hence release as well as adsorb oxygen species by forming the Ce^{4+}/Ce^{3+} redox pair and creates a high concentration of oxygen vacancies that can in principle act as active sites [19]. The redox mechanism considers that CO_2 dissociates at the oxygen vacancies and the resulting active surface oxygen species react with the carbon generated from the CH_4 decomposition, thus forming CO, and regenerating the vacancy.

Lian et al. [25] investigated the catalytic properties of Ni catalysts supported on CeO_2 (Ni/CeO_2) using density functional theory (DFT) calculations. They studied the reactivity of five different sites present in the Ni/CeO_2 catalysts: metal nanoparticle (*m*), support surface (*s*), metal nanoparticle-support interface (*ms*), oxygen vacancy at the interface (*ovms*), and also oxygen vacancy at support (*ovs*). As already described, the main considered DRM mechanisms are the bifunctional and stepwise redox (bifunctional redox mechanism). Both models consider that the CH_4 activation takes place at the metal site (*m*). According to the calculations performed by the authors, the *m* and *ovms* sites present high activity in which leads to the first hydrogen abstraction from CH_4, whereas *ms* is highly active towards CH_x abstraction. Also, it can be seen that the activation of CH_4 generally occur at the support-metal interface (*ms*) and vacancies sites (*ovms*). As such, the difference among the methods relies on the sites for CO_2 activation. In the bifunctional catalysts, CO_2 activation is expected to occur at the metal-support interface (*ms*), while the redox mechanism predicts that CO_2 is activated by the interfacial oxygen vacancies (*ovms*) [25]. In addition, CO_2 may be activated on the metallic sites (*m*) and oxygen vacancies at the support (*ovs*). Based on the calculations, the authors suggested that the vacancy sites (*ov* and *ovms*) are responsible for the CO_2 activation, while the interfacial vacancy sites (*ovms*) and also metal sites (*m*) are responsible for the CH_4 activation. CO formation occurs from CO_2 dissociation and the remaining O* species fill the vacancy sites or diffuse and react with CH_x* species [25]. The interfacial site is crucial due to its balanced reactivity for the activation of CH_4 and CO_2. Moreover, the interfacial oxygen helps the active site regeneration and accelerates the elimination of coke precursors carbon species, thus contributing to the catalyst stability. This means that the metal-support interaction has great importance for example in the activity/stability of the catalyst, and approaches to increase this interface, and also to form diverse reactive sites on the catalyst surface, are crucial for developing promising catalysts for DRM [25].

Traditionally, the catalyst support materials are known to have a large specific surface area, such as alumina and silica, and in addition to favoring a high dispersion of the active metal, many of which result in increased catalytic activity significantly [26–30]. Nevertheless, most supports have small pores despite the large specific surface area, which can hinder the intra-particle diffusion of reactants and products, affecting the catalyst performance. In contrast, large pore sizes may favor intra-particle diffusion, but the low specific surface area may not favor a proper metal dispersion. Lately, the use of supports containing bimodal pore distribution (both mesopores/micropores) seemed to be promising for heterogeneous catalysis [26]. Large pores promote a fast reactant and product molecules transportation, whereas small pores yield a high specific surface area for deposition of metallic sites [27].

Shah et al. [26] studied the performance of 5 wt.% Ni catalysts supported on bimodal alumina in DRM. The bimodal porous alumina has been prepared in this study by the evaporation induced self-assembly method (EISA). Nickel was added to the support by four different methods: chemical vapors deposition (NAC), freeze-drying (NAF), wet impregnation (NAW), and urea deposition–precipitation (NAU) [26]. The authors also compared the activity and stability of the catalysts with a Ni–alumina catalyst having a unimodal porous structure. The DRM was carried out in a fixed bed down-flow using stoichiometric CH_4:CO_2 mixture (1:1) at atmospheric pressure. The catalyst stability was tested at 700 °C for 100 h at a GHSV of 10,000 mL g^{-1} h^{-1}. The particle sizes obtained by the different methods were similar, but the catalytic performances were different. None advantage of freeze-drying over the wet impregnation method was observed. NAC led to a catalyst with poor metal-support interaction as well as larger particle size after activation. The NAU-related catalyst presented the best performance, being active and stable on stream over 100 h. According to the authors, the solution pH in the urea deposition–precipitation method changed slowly, making the nickel species to precipitate progressively, which avoided the formation of large nickel ensembles. By comparing the bimodal and unimodal supports, has been shown that the unimodal catalyst with the same metal content exhibited inferior activity and stability. Hence, the authors proposed the importance of the different parameters in the following order: catalysts bimodal structure > metal-support interface nature > particle size of nickel species [26].

Reforming reactions usually require high temperatures (700 °C) to achieve an economically viable conversion yield. Moreover, high-temperature treatments usually are used to prepare catalysts by transforming the precursor in oxides and to activate the active phase. High temperatures can lead to migration and growth of crystallites and particles, reducing the final catalyst's activity. To avoid these problems, several reaction parameters can in principle be changed with the specific objective of obtaining a material with high stability at high temperatures. In this way, nanoparticles can be embedded inside controlled channels and materials with strong metal-support interactions, such as perovskite and layered double hydroxide (LDH). Recently, encapsulation techniques have attracted attention for the synthesis of catalysts. In this context, Strong Metal-Support Interaction (SMSI) and core–shell methods with different mechanisms have been used to obtain catalysts with promising results [28]. A great metal-support interaction can reduce metal mobility, avoid sintering effects, and keep the active phase highly dispersed over the reaction [19]. The core–shell methods consist of the appropriate recovery of the metal oxide with a porous layer. By varying the synthesis parameters, it is possible to obtain layers with different thickness and proper porosity for reactants and product diffusion, avoiding the sintering of the active phase at high temperatures. Catalysts with core–shell structure are an emerging class of nanomaterials that have shown promising results for DRM. Such catalysts have an active metal as the core, which is coated with a porous thermally stable shell. The shell can prevent the sintering under the high temperatures necessary for DRM [19, 28]. Core–shell structures can also avoid carbon deposition in Ni catalysts, considering that such a deposition is usually facilitated on larger nickel particles (>8 nm) [29].

Ni supported on Mg-Al_2O_3 is known as a suitable catalyst for DRM. In particular, Ni-MgAl catalysts obtained from layered double hydroxides (LDH) present a high catalytic activity and also a high resistance to carbon deposition, possibly due to the proper specific surface area, a suitable balance between basic and acid sites and homogenous dispersion of nickel species. However, aggregation of nickel species due to the reaction conditions is still a problem for long reactions. More recently, the confined active components in a mesoporous support or their desired encapsulation in supports based on the core–shell structures seems to be a good approach for minimal catalytic site segregation [30]. SiO_2 is a support with high thermal stability and large surface area, however, it presents a poor interaction with transition metals, which favors sintering processes. Du et al. [30] designed a catalyst by combining a Ni–MgO–Al_2O_3 obtained from LDH and a mesoporous SiO_2 shell (Ni–MO@SiO_2). MgO promoted a homogenous dispersion of the embedded nickels species and improved CO_2 adsorption. The recovery with a SiO_2 shell prevented the catalyst sintering. The catalyst exhibited a better dispersion of Ni species and better performance for DRM than the samples prepared via nickel impregnation on MgO–Al_2O_3@SiO_2 (IM–Ni–MO@SiO_2) and MgO–Al_2O_3 (IR–Ni–MO) supports. This difference was because the impregnation route led to a non-uniform distribution of nickel species, weaker nickel-support interaction, and blockage of support's pores. After the reaction, TG analysis indicated a weight loss of 2.5% for Ni–MO@SiO_2, 13.5% for IM–Ni–MO@SiO_2, and 43.0% for IR-Ni-MO, showing that Ni–MO@SiO_2 as catalyst is a candidate promising for enhanced DRM [30].

To avoid the sintering of Ni species, Marinho et al. [31] studied the behavior of embedded Ni nanoparticles in DRM. They prepared Ni@CeO_2 and Ni@$CeZrO_2$ catalysts by the one-step sol–gel method. These catalysts were also compared with a Ni/CeO_2 sample prepared by the support impregnation strategy. As explained above, the use of supports with high OSC (oxygen storage capacity) is interesting because they limit the metal sintering and assist in removing carbon by vacancy effects. Also, the incorporation of zirconium into the ceria structure improves the thermal stability of the support, i.e., which reduces the sintering and loss of surface area at high temperatures [31]. Also, TEM results showed that the Ni/CeO_2 catalyst prepared by the impregnation method presented larger particle size (30 nm) than the Ni@CeO_2 (13 nm) and Ni@$CeZrO_2$ (6 nm) catalysts. Raman analysis revealed typical bands of oxygen vacancies for Ni@CeO_2 and Ni@$CeZrO_2$. in-situ XRD results during the activation process with H_2 and in situ XANES experiments showed greater reduction resistance for catalysts with Ni particles embedded in the ceria matrix than the impregnated sample. These results agreed with the TEM images and in situ XANES experiments, which showed that the embedded Ni particles were less reduced, thus indicating a strong Ni-support interaction in Ni@CeO_2 and Ni@$CeZrO_2$. Considering the DRM process, all samples presented similar initial CH_4 and CO_2 conversions, with no detectable deactivation over 48 h on stream. The CO_2 conversion was higher than CH_4 conversion due to the water gas shift reaction, which led to a H_2/CO ratio of 0.8. Despite de similar conversion results, the Ni/CeO_2 catalyst presented the highest amount of deposited carbon (9.7 mgC g_{cat}^{-1} h^{-1}). Meanwhile, the catalysts containing Ni nanoparticles embedded in ceria, Ni@CeO_2, and Ni@$CeZrO_2$, were more effective in reducing the carbon deposits, presenting 1.6 and 0.0 mgC g_{cat}^{-1} h^{-1}, respectively [31].

2.1 DRM—Industrial Application

Problems related to catalyst deactivation by deposited carbon makes DRM an undeveloped industrial process. However, the Linde Group inaugurated a pilot plant for DRM in partnership with BASF and funding from the German government in 2015. This pilot plant called Linde Pilot Reformer (LPR) is located near Munich, Germany, and according to the group, after successful tests using nickel catalysts, the current focus is on the catalyst capability scale-up for commercial production [17, 32].

Besides the Linde pilot plant, two other commercial CO_2 and CH_4 conversion processes are known, namely, SPARG (Sulfur Passivated Reforming, Haldor Topsoe) and CALCOR (Caloric GMbH). In the SPARG process, part of the steam in MSR is replaced with CO_2, which results in a syngas with H_2/CO ratio of 1.8. A catalyst partially poisoned by sulfur is used to avoid carbon deposition. The final stream still has 2.7% of methane. Yet in the CALCOR process, CO is produced by CO_2 reforming. This is a multi-stage reactional process using dried CO_2, natural gas, liquid petroleum gas (LPG), and syngas to produce a highly pure CO. The exit stream contains less than 0.1% CH_4. In this process, coke deposition is prevented by packing the tube reactor with catalysts owning different shapes and activities [21, 33].

3 Oxidative Coupling of Methane

Conventional technologies adopted worldwide by oil refineries to produce ethylene are the steam and thermal cracking processes [34]. More specifically, in the case of these routes, it has been shown that the heat is necessary to favor the breaking of the C–C and C–H bonds and produce ethylene [35–37]. A drawback of this process is related to the large energy input and consequent environmental unsustainability [38]. It was established that up to three tons of CO_2, is released to the atmosphere for each ton of ethylene that is produced [39]. There are many efforts to develop a more straightforward, economical, and more sustainable alternative for ethylene production have been intensified over the last three decades [40–42].

The OCM process involves the sequential oxidation of methane to ethane and ethylene, as illustrated in Fig. 2. First, methane reacts with oxygen species at the oxide catalyst surface to produce methyl radicals ($^*CH_3$), which are further coupled in the gas phase to form ethane and water. Ethane is then dehydrogenated mainly into ethylene, in addition to existing a possibility of producing higher hydrocarbons at trace amounts [26, 27, 37]. On the other hand, any increase in oxygen concentration may lead to a significant change in the reaction, especially towards favoring the formation of CO and CO_2 [26, 27]. As such, the type of oxygen supply is another factor to be considered in these industrial processes. Performing the reaction in general with pure oxygen requires an air separation unit, i.e., making operating costs more expensive. Nonetheless, this may be mitigated by using air directly in the

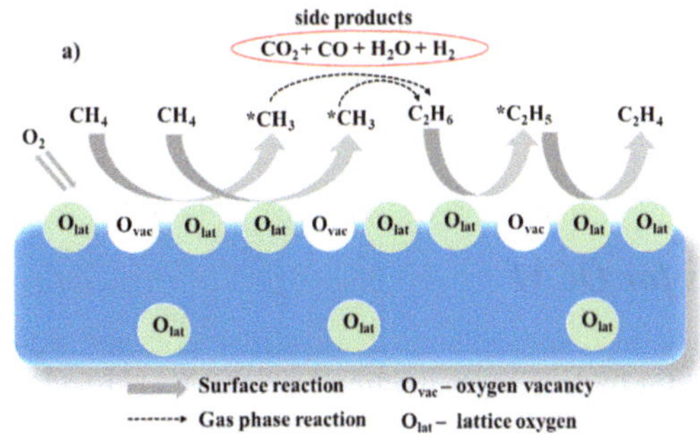

b)

Reaction steps	$\approx\Delta H^\circ_{298}$ (kJ/mol)
1) $CH_4 + 2O_2 \rightarrow CO_2 + 2H_2O$	-802
2) $2CH_4 + 0.5O_2 \rightarrow C_2H_6 + H_2O$	-177
3) $CH_4 + O_2 \rightarrow CO + H_2O + H_2$	-277
4) $CO + 0.5O_2 \rightarrow CO_2$	-283
5) $C_2H_6 + 0.5O_2 \rightarrow C_2H_4 + H_2O$	-105
6) $C_2H_4 + 2O_2 \rightarrow 2CO + 2H_2O$	-757
7) $C_2H_6 \rightarrow C_2H_4 + H_2$	+137
8) $C_2H_4 + 2H_2O \rightarrow 2CO + 4H_2$	+210
9) $CO + H_2O \rightarrow CO_2 + H_2$	-41
10) $CO_2 + H_2 \rightarrow CO + H_2O$	+41

Fig. 2 (**a**) Illustration of OCM reaction on the surface of metal oxide catalyst. (**b**) OCM reaction steps

reaction. An additional advantage of using direct air is the fact that nitrogen in the air can deal with the temperature rise due to the exothermic reactions [37].

The challenges in OCM is the development of a stable catalyst to be applied industrially and the possibility of performing this reaction at mild temperatures. The first attempts to convert methane to ethylene and other hydrocarbons are usually attributed to pioneering studies carried out by Ito and Lunsford, Hinsen and Baerns, Keller and Bhasin, throughout the 1980s [43–45]. These first studies stated that methane might be converted to ethylene and other hydrocarbons at low rates. Then, further studies focused on the understanding the effects of the structure and composition on the catalyst in addition to the reaction settings to achieve considerable conversion rates and selectivity.

Researches using numerous catalytic systems have shown that nitrous oxide offers higher performance for C_2 production (ethane and ethylene) than molecular oxygen [46], although the use of the former has the disadvantage of not being economical on an industrial scale [37]. Comparing with the single-pass system, multi-stage via methane recycling together with ethane may rise the commercial exploration of the OCM process [37]. Moreover, it is known that the configuration of the reactor in this case has a significant impact on the OCM performance [34, 35]. For this reason, the membrane reactors generally provide superior C_2 selectivity, restraining the amount of ionic oxygen species free to react with molecules of methane through selective conducting membranes [34, 47]. Though, unless operated at elevated temperatures, it is known that the use of conductive oxides in applications as membranes generally leads to less conversion of methane compared to fixed bed reactors in which operated with the use of a molecular oxygen stream. And yet, this is mainly due to the oxide thickness used, which in general is responsable to limits the oxygen ion conduction rate [48]. In this way, studies have been conducted to reveal the nature of active sites and in many other vital aspects that are directly related to the catalysis-assisted OCM. This probably includes a wide knowledge of the oxygen vacancy, their acid–base property, the oxide surface reducibility, and diverse aspects related to the oxide-support interaction. Thus, it is fundamental to discuss these improvements, and it is expected that the present literature will contribute significantly by providing a better understanding of the catalyst design for enhanced OCM applicaitons.

3.1 Potential Catalytic Sites for OCM

The potential sites in oxide surfaces for hydrogen abstraction from the C–H bond of the methane are supposed to be the reactive oxygen species available. Nonetheless, other hypotheses on the nature of these active sites have been stated, and are most likely due to presenting different oxygen species types that have widely been identified in the surface of the diverse oxide catalyst, which display very specific oxidation roles [49]. The oxygen species that have been characterized include chemisorbed oxygen (O_2^-), dissociative adsorbed oxygen (O^-), adsorbed oxygen ions (O^{2-}), and lattice oxygen (O^{2-}) [50]. Lately, the theoretical-experimental approaches are giving insights into the distinct roles of the various oxygen species in the OCM reaction [50–52].

The contributions of the lattice oxygen species on the properties of NaWMn/SiO_2 catalysts on OCM were described in a recent report [53]. The authors verified that the catalysts displayed both weak and strong bound oxygen species. They also observed that the selectivity was temperature-dependent on the reduction. Although the contribution of tightly bound oxygen species was not fully evaluated, the study disclosed that loosely bound oxygen species contributed significantly to the observed activity and stability in the OCM reaction. To clarify the nature and role of such oxygen species in OCM reaction mechanism promoted by oxide catalysts, in general, should be carried out studies with and without the presence of oxygen in the gas phase [50]. However, the study of multicomponent oxide catalysts by the systematic exchange approach may provide deep insights into the role of each oxygen species in altering the reactivity of the catalyst's active sites [53].

Furthermore, it is known that the oxygen species adsorbed from the oxygen gas phase remarkably enhanced the methane conversion for all the catalysts [50, 53]. Additional studies are still required to unravel the identity of oxygen species adsorbed in diverse oxide systems to the OCM reaction. Interestingly, the use of an infrared spectrometer as a detector is suitable for carrying out measurements with programmed temperature (e.g., TPSR, TPD, and TPO) with high peak resolution for different formed products [54]. Further studies based on this approach are great importance for an in-depth understanding of the role ofadsorbed oxygen species on the catalytic behavior of various oxides intended for OCM. Besides, it is also known that identifying the electrophilic character of the selective oxygen species in the lattice could contribute to manipulating its functional properties [37]. Several reports have disclosed the catalytic performance of oxides catalysts in OCM, as shown in Table 1.

Table 1 Catalysts, reaction conditions, conversion, and selectivity or yield data for catalysis assisted OCM reaction collected from the literature

Catalyst	Reaction condition	Conversion (%)	Selectivity (S) or Yield (Y) to C_2(%)	References
Cs/Sr/MgO	794 °C, $CH_4/O_2 = 3$	33	59 (S)	[55]
Cs/Ba/MgO	820 °C, $CH_4/O_2 = 3$	32	57 (S)	[55]
Sm_2O_3/MgO	700 °C, $CH_4/O_2 = 4$	22	52 (S)	[56]
Li/Sm_2O_3/MgO	700 °C, $CH_4/O_2 = 4$	24	64 (S)	[56]
1% Li/MgO	800 °C, $CH_4/O_2 = 2$	38	35 (S)	[57]
CaO/ZnO, Ca/Zn = 1.3	800 °C, $CH_4/O_2 = 2$	36	30 (S)	[57]
Na_2WO_4/Mn/SiO_2	850 °C, $CH_4/O_2 = 3.5$	32	57 (S)	[58]
Na_2WO_4/Mn/SiO_2 doped with 16.7 wt.% MgO	850 °C, $CH_4/O_2 = 2$	50	38 (S)	[58]
Na_2WO_4/Mn/SiO_2 doped with 40 wt.% TiO_2	850 °C, $CH_4/O_2 = 2$	39	59 (S)	[58]
10% Na_2WO_4-5% Mn/SiO_2, modified with 5% La	800 °C, $CH_4/O_2 = 32/8$,	Not reported	24 (Y)	[59]
5% Ba/La_2O_3	150 °C, in electric field, 3.0 mA, 600 V	6.3	32.3 (S)	[60]
Mn_xO_y-Na_2WO_4 supported over SBA-15	750 °C, $CH_4/O_2 = 4$	14	70 (S)	[61]
Mn_xO_y-Na_2WO_4/La_2O_3	750 °C, $CH_4/O_2 = 4$	18	34 (S)	[61]
Mn_xO_y-Na_2WO_4/CaO	750 °C, $CH_4/O_2 = 4$	18	26 (S)	[61]
Mn_xO_y-Na_2WO_4/SrO	750 °C, $CH_4/O_2 = 4$	5.4	25 (S)	[61]
Mn_xO_y-Na_2WO_4/Al_2O_3	750 °C, $CH_4/O_2 = 4$	13	25 (S)	[61]
Mn_xO_y-Na_2WO_4/ZrO_2	750 °C, $CH_4/O_2 = 4$	10	21 (S)	[61]
Mn_xO_y-Na_2WO_4/Fe_2O_3	750 °C, $CH_4/O_2 = 4$	2.3	73 (S)	[61]
Mn_xO_y-Na_2WO_4/Fe_3O_4	750 °C, $CH_4/O_2 = 4$	1.6	63 (S)	[61]

(continued)

Table 1 (continued)

Catalyst	Reaction condition	Conversion (%)	Selectivity (S) or Yield (Y) to C_2(%)	References
Mn_xO_y-Na_2WO_4/TiO_2-rutile	750 °C, $CH_4/O_2 = 4$	3.6	79 (S)	[61]
Mn_xO_y-Na_2WO_4/TiO_2-anatase	750 °C, $CH_4/O_2 = 4$	1.6	63 (S)	[61]
Mn_xO_y-Na_2WO_4/COK-12 (ordered mesoporous silica)	775 °C, $CH_4/O_2 = 4$	24	60 (S)	[62]
Mn_xO_y-Na_2WO_4/SIC (porous)	800 °C, $CH_4/O_2 = 4$	35	10 (S)	[63]
$LaInO_3$	800 °C, $CH_4/O_2 = 5$	15	54 (S)	[64]
$La_{0.9}Ba_{0.1}InO_{3-\delta}$	800 °C, $CH_4/O_2 = 5$	22	59 (S)	[64]
$La_{0.6}Ba_{0.4}InO_{3-\delta}$	800 °C, $CH_4/O_2 = 5$	21	61 (S)	[64]
$La_2Ti_2O_7$	800 °C, $CH_4/O_2 = 4$	~18	~35 (S)	[65]
$La_2Zr_2O_7$	800 °C, $CH_4/O_2 = 4$	~24	~55 (S)	[65]
$La_2Ce_2O_7$	800 °C, $CH_4/O_2 = 4$	~29	~59 (S)	[65]

3.2 *Oxygen Vacancy*

Recently, many researchers put efforts to explore the role of oxygen vacancies on the catalyst activity [66–68]. In the OCM, oxygen vacancies are generally formed in the oxide catalyst due to the release of H_2O or CO and CO_2 after CH_4 dehydrogenation [50, 69, 70]. As stated previously, the presence of oxygen species in the lattice are usually selective for the OCM reaction [50]. Also, the extent at which these oxygen species favor OCM depends on their capability of capturing hydrogen atoms from methane molecules.

Undoubtedly, the number of oxygen vacancies in oxide catalysts, particularly at high temperatures, may be converted into oxygen lattice abundance, along with the immediate availability for the reaction. Therefore, efficient catalysts for OCM must have a high amount of oxygen vacancies at the typical reaction temperatures. An iron oxide system was studied by DFT calculation as an attempt to reveal the pivotal role of the oxygen vacancies in the partial methane oxidation [69]. The study demonstrated that the oxygen vacancy amount substantially decreases the energy required for the hydrogen uptake and consequent production of methyl radicals [66]. Nonetheless, above a certain oxygen vacancy content, the methane activation barrier became

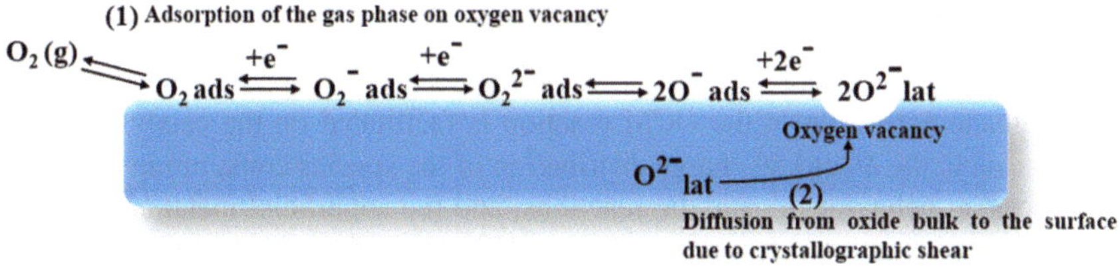

Fig. 3 Schematics of active site regeneration on the oxide surface by refilling the oxygen vacancies created during the activation of methane by OCM

independent of the vacancy distribution. This was probably due to the weak interactions between the species adsorbed on the catalyst surface and oxygen vacancies at the deep subsurface. Furthermore, the stability and location of the formed oxygen vacancies also influenced their refilling by oxygen species (from the gas phase) or via vacancy migration [70]. The refilling process results in the re-oxidation of the reduced cations and, consequently, the active sites are restored for the next catalytic cycle. Figure 3 shows the conversion of oxygen species from the gas phase into lattice oxygen species in the oxide catalyst under OCM [65, 71]. Furthermore, depending on the oxide nature, the oxygen vacancies may be restored by oxygen diffusion from the bulk lattice to the surface oxygen vacancy or by nucleation of the reduced phase with simultaneous exposure of the subjacent lattice plane [70]. However, catalysts that exhibit high activity for OCM, particularly under the oxygen-free gas phase, may have their active sites restored via diffusion of oxygen from the bulk lattice containing a crystallographic shear plane [50]. The vacancies available at the outermost oxide surface are more stable than those at the subsurface, so diffusion from the latter to the former is more energetically favorable [69, 70].

For example, Wang et al. [65] reported that the formation of active surface O_2^- species in $La_2B_2O_7$ catalysts with different structures (pyrochlore, fluorite, and perovskite) could occur by two pathways. The first is related to both $La_2Zr_2O_7$ (ordered pyrochlore cubic structure) and $La_2Ce_2O_7$ (disordered defective cubic fluorite structure), which have intrinsic oxygen vacancies 8a ($^1/_8$, $^1/_8$, $^1/_8$). The active surface species are formed by activation of oxygen species that enter the oxygen vacancies, then in the reticulated matrix, and finally, migrate to the surface. In the second pathway, active oxygen species are formed in $La_2Ti_2O_7$ (monoclinic layered perovskite structure), which does not have intrinsic oxygen vacancies, by direct transformation in the surface of O_2 molecules adsorbed. Generally, materials with many intrinsic oxygen vacancies facilitate the surface electrophilic oxygen sites. Hence, as has been shown the surface O_2^- electrophilic species determined the reactivity of the $La_2B_2O_7$ catalyst for OCM. The $La_2Ce_2O_7$ had the highest amounts of O_2^-superoxide sites, consequently having the optimal reaction performance among the three catalysts [65].

3.3 Base Properties of OCM Catalysts

The methane activation in the OCM reaction is facilitated on the catalyst surface [72]. As such, the electrical charge distribution of the species constituting the oxide influences its behavior in catalytic reactions, and the superficial metal cations and oxide anions being considered as basic sites [73]. Therefore, from this perspective, the nature and strength of the bond among cationic and anionic species determine the base property exhibited by the oxide catalyst [70]. Additionally, many oxide systems have demonstrated that their base properties are a vital factor that governs their performance in the OCM reaction [73].

The basic active surface sites with medium strength are beneficial for both CH_4 activation and C_2 selectivity, but they are inactive at low temperatures due to the adsorption of CO_2 molecules, which may be produced by side reactions. Once the reaction temperature is increased, these base sites become gradually available for methane activation due to CO_2 desorption. Though, strong base sites easily capture CO_2 to form stable carbonates, which are not active species for OCM [65, 74]. Consequently, it is coherent to believe that basic medium-strength sites (with desorption in the range 250–600 °C, according to CO_2-TPD profiles) contribute to the OCM reaction to a large extent [74].

The surface basicity is believed to be closely related to anions chemically absorbed by oxygen species, such as O_2^-, O_2^{2-}, O^- and surface defects [65, 74–76]. For example, doping the MgO lattice with Li^+ cations creates Li^+O^- electron pair defects, in which enhances the basicity of the surface and amount of active oxygen [76].

The basicity of oxide catalysts in the OCM reactions could in principle be adequately modulated via optimized doping [77]. The authors showed that strong base sites were essential for improving the OCM performance of Sr_2-doped La_2O_3 nanofibers, particularly at temperatures as high as 600 °C. It was also observed that the amount of Sr significantly increased the base site strength when compared to a pure La_2O_3. A Sr-doped La_2O_3 catalyst with 8.6 wt% of nanofiber exhibited a satisfactory OCM activity reaching a C_2 yield of about 20% at 650 °C. This catalyst presented high amounts of medium base sites, as demonstrated by CO_2-TPD experiments substantiated with XPS results.

3.4 Optimal Surface Oxide Reducibility

An additional feature that governs OCM reactions, in terms of both catalytic activity and C_2 selectivity, is the reducibility of the oxide catalyst surface at the typical reaction temperatures. Hence, the availability of selective oxygen species in the lattice relies on the extent and ease by which the oxide catalyst is reduced [70]. Using DFT calculations, Kumar et al. [78] studied several doped and non-doped oxide catalysts to describe the role of their surface reducibility in the OCM reaction [78]. It was established that the greater the oxide surface reducibility, the higher the activity for

OCM. In addition to a certain degree of reducibility, methane oxidation overloads, it is resulting in more CO and CO_2 selectivity. This suggests that the systematic adjustment of the surface reducibility, which may change the oxide activity/selectivity, is necessary for an optimal C_2 yield. Therefore, the formation of CO and CO_2 due to additional oxidation of methyl radicals ($^*CH_3$) generated during the methane activation may be limited [70, 78].

Lattice oxygen species considerably affect the reducibility of the oxide surface. At a constant temperature, more electrophilic lattice oxygen results in less surface reducibility [50, 70]. For example, lattice oxygen from $LaAlO_3$ oxide show a high electrophilic character, which results in a satisfactory C_2 selectivity. In this case, its binding energy and desorption temperature were found as being relatively high [50]. In general, the reactivity of oxide catalysts for oxidation reactions correlates strongly with the reduction temperature. Usually, oxide catalysts with low reduction temperature (high surface reducibility) are estimated to exhibit high oxidative activity, and vice versa [78]. This suggests that, for a specific oxide catalyst, there is a limit at which the reduction temperature may be adjusted downwards to obtain more C_2 selectivity in the OCM reaction. Therefore, the catalyst design for the development of efficient OCM reactions requires an in-depth understanding of the effective approaches for proper surface reducibility adjustment [70].

3.5 Prospects of Nanowire Catalysts Applied in OCM

Several innovative strategies on catalyst design and reactor configuration have widely been developed to perform OCM at mild temperatures with decent and stable C_2 yields. Up to now, the accomplished enhancement was not enough to put the OCM process in an economically viable stage (above 30% C_2 yield). The development of nanostructured catalysts with controlled and well-defined morphologies has revived the hope for enhancing the OCM yields with substantial stability [79, 80]. According to DFT calculations and experimental investigations, the C_2 yield in heterogeneous catalysts is intensely influenced by the nature of the exposed catalyst surface [81], principally for structure-sensitive OCM reactions [51]. As such, this indicates that designing new catalysts with an optimized surface composition and structure will allow OCM to be conducted mild temperatures.

The sensitivity of the $La_2O_2CO_3$ catalyst morphology at the nanoscale was studied for OCM [51]. The nanoparticle forms of the $La_2O_2CO_3$ and La_2O_3 oxide catalysts showed methane conversion lower than 2%, along with zero C_2 yields at 500 °C, demonstrating their inability for OCM at mild temperature. The variations in morphology at the nanoscale due to hydrothermal synthesis may lead to the different surface-exposed which are determinants to the activity/selectivity of the prepared oxide catalyst. Notably, the rod-shaped catalysts synthesized with similar particle sizes have displayed the highest performance (30% methane conversion and 50% C_2 selectivity at about 420 °C). It was also found that the rod-shaped catalyst had a high proportion of low-index surfaces exposed with moderate basic sites compared to the others, as evidenced by atomic density calculations of the exposed crystal planes and CO_2-TPD analysis [51].

Also, some defects (like vacancies) that display essential roles in the case of the adsorption/activation of methane and oxygen may be accessible at the exposed $La_2O_2CO_3$ surface [51]. The study further confirmed that OCM is a structure-sensitive reaction, and how this can be exploited to decrease the temperature requirement using nanowire-like catalysts. It was also reported that a rod-shaped La_2O_3 catalyst demonstrated doubled OCM performance (27.4% methane conversion and 43.4% C_2 selectivity) at temperatures lower than 500 °C than an equivalent plate-shaped catalyst.

3.6 *Synthesis of Potential OCM Catalysts*

Considering the great benefits of OCM for the direct methane conversion into platform chemicals, different methods are widely needed for designing and synthesizing new OCM catalysts, many of which could lead to an industrially feasible scale in terms of C_2 yield [82–84]. New catalysts are typically designed with basis on the kinetics of the individual OCM reaction pathways [82], as well as optimization of the catalyst properties based on fundamental theoretical/experimental studies on the structure-activity relationships [83]. Generally, the assessment of the reaction parameters are essential to the progress of kinetic models and contribute to better comprehend a diverse catalytic process [84].

Additionally, significant dispersion of the active catalyst phase on the support in the form of nanoparticles, films, and unique sites via atomic layer deposition (ALD) [85], or design of mesopores in the active phase [86] could improve the catalyst activity towards OCM. It is known that the ALD can also be applied to include promoters in the active phase, thus increasing C_2 selectivity and catalyst stability [87].

Developing new and efficient catalyst designs for supports able to establish crucial interactions with the oxide are essential to enable the use of industrial OCM process [88, 89]. The support can significantly improve the activity of the catalyst in several ways, including modification of the acid–base properties and in addition to generation of an interaction edge that could serve as a basis for leading to the formation of a new and more reactive active site. Moreover, the electrophilicity of the lattice oxygen, it might displays a fundamental role in the C_2 selectivity, could in principle be improved by the modification of the active phase electronic structure due to an appropriate support material [77, 79, 90].

Nowadays, Na_2WO_4-Mn/SiO_2 catalysts are recognized as the most promising OCM catalysts, providing high C_2 yields and stability under the OCM reaction conditions at high temperatures. However, Na_2WO_4-Mn/SiO_2 catalysts are complex systems and, therefore, the contributions of different additives, such as dopants (alkaline or alkaline earth metals), used to adjust some of their properties are difficult to be revealed. The use of mesostructured silica supports is a considerable route for achieving improved dispersion of the OCM catalyst's active phases. On the other hand, soft modeling with organic surfactants is a pronounced technique that can be used to create mesoporosity within the active phase itself, thus improving the accessibility of active sites. Furthermore, nanostructured catalysts, such as nanowires, synthesized by homogeneous precipitation with microemulsion under microwave heating can display improved activity towards OCM [70].

4 CO_2-Oxidative Coupling of Methane (CO_2-OCM)

As aforementioned, DRM and CO_2-Oxidative are two processes used to convert methane and carbon dioxide into highly valuable chemicals. Several reports have described improvements in DRM, while just a little progress on the CO_2-OCM process has been achieved. This section will discuss the knowledge update on the CH_4 and CO_2 chemistry and strategies for their conversion, mainly to ethane and ethylene, by CO_2-OCM.

Ethylene and ethane (C_2) are essential feedstocks for the manufacture of value-added chemical commodities. Currently, ethylene is produced industrially through the steam cracking of a wide range of hydrocarbon feedstocks [91, 92]. The CO_2-OCM process can be an alternative route for ethylene production using CO_2 as an oxidant, and natural gas or biogas as raw materials, both mostly composed of CH_4 and CO_2, two greenhouse gases [91–93]. In this way, the conversion of CO_2 into value-added chemicals without any separation step becomes possible. Furthermore, this process is highly advised from the environmental point of view due to the urge to suppress global warming and climate change trends.

CO_2-OCM is a quite challenging process due to the high energy required to activate both CH_4 and CO_2. Methane activation demands high temperatures due to the high stability of the C–H bonds in the gas phase (434 kJ mol^{-1}) [94, 95]. Moreover, to activate the carbon dioxide molecule, which C–O bond energy is 532 kJ mol^{-1}, it is necessary to overcome its considerable Gibbs energy of formation ($\Delta G°_{298.15\,K} = -394.4$ kJ mol^{-1}) [95, 96]. So, it is highly desirable to design effective catalytic systems for achieving optimal C_2 hydrocarbon yields.

4.1 OCM Versus CO_2-OCM

In the conventional OCM process, CH_4 and O_2 act as oxidants. Although the reaction selectivity towards ethylene and ethane may be insufficient due to the non-selective $CH_3^{\bullet}$ oxidation, especially at high oxygen concentrations, CO_2 would be an interesting oxidant to prevent the sequential reaction of C_2 products in the gas phase [95, 97–101].

Carbon dioxide has some advantages over O_2, as soft oxidant, including the formation of carbon monoxide as the only byproduct, and high selectivity towards

C_2 hydrocarbons. Moreover, CO_2 would not induce gas-phase radical reactions [95, 97–101].

The CO_2-OCM process follows two main reactions to produce ethylene and ethane, Eqs. 8 and 9, respectively [91, 93, 95, 97, 100–102]:

$$2CH_4 + 2CO_2 \rightleftharpoons C_2H_4 + 2CO + 2H_2O \quad \Delta H^\circ_{1073\,K} = 285.7\ KJ/mol \tag{8}$$

$$2CH_4 + CO_2 \rightleftharpoons C_2H_6 + CO + H_2O \quad \Delta H^\circ_{1073\,K} = 108\ kJ/mol \tag{9}$$

However, side reactions leading to CO, represented by Eqs. 3 and 10, are thermodynamically more favorable up to temperatures as high as 800 °C [93, 103]. This challenge could be overcome by the development of catalysts capable of activating both CH_4 and CO_2, as well as producing C_2 hydrocarbons [104, 105]. In this sense, some investigation efforts on the reaction conditions and catalyst composition have been carried out.

$$CH_4 + CO_2 \rightleftharpoons 2CO + 2H_2 \quad \Delta H^\circ = 247\ kJ/mol \tag{3}$$

$$CH_4 + 3CO_2 \rightleftharpoons 4CO + 2H_2O \quad \Delta H^\circ = 398\ kJ/mol \tag{10}$$

Figure 4 shows the effects of temperature and CO_2/CH_4 ratio on the equilibrium conversions of the CO_2-OCM process, as evaluated by thermodynamic calculations. It was observed that increases in both parameters favored the CH_4 conversion into C_2 hydrocarbons (C_2H_6 and C_2H_4). At 800 °C and $CO_2/CH_4 = 2$, the C_2H_6 and, C_2H_4 yields were ca. 13% and 25%, respectively. The increase in temperature to 900 °C still increased the C_2 hydrocarbon yields, with C_2H_6 higher than ca. 15% and C_2H_4 at about 40% when $CO_2/CH_4 = 2$ [104, 105].

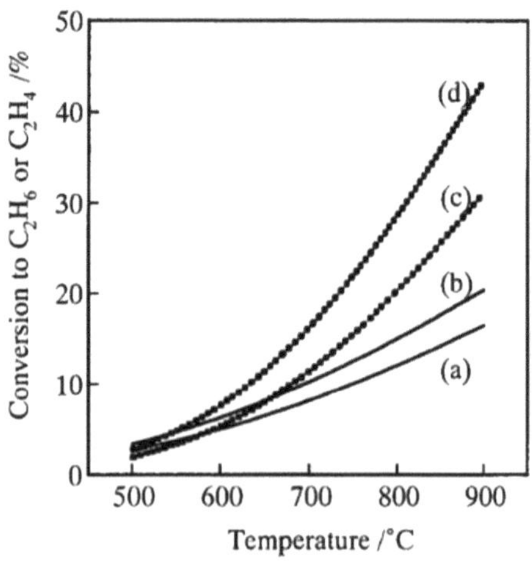

Fig. 4 Equilibrium conversions of CH_4 to C_2H_6 (solid lines) and C_2H_4 (dotted lines) using CO_2 as an oxidant. **a** and **c**: $CO_2/CH_4 = 1$. **b** and **d**: $CO_2/CH_4 = 2$ Copyright ©1999 Academic Press. All rights reserved (with permission) [104]

Wang and Ohtsuka [106] studied the effect of CO_2 on the OCM process with a binary CaO-ZnO catalyst for selective conversion of CH_4 to C_2 hydrocarbons at 875 °C. When the OCM was conducted without CO_2, the main products were H_2 and CO. At the first stage, the conversion of CH_4 was ca. 2%, and then decreased after 4 h to ca. 0.2%. However, the CH_4 conversion was higher than 5% in the presence of CO_2, with a slight change over time on stream and showing C_2 hydrocarbons as main products. The authors suggested that the oxygen species produced from CO_2 during the reaction favored the selective towards C_2.

Litawa et al. [107] evaluated the CO_2 effect on the OCM process catalyzed by CeO_2, La_2O_3/CeO_2, and CaO/CeO_2 at 800 °C. It was observed that CO_2 had a little positive influence on the CH_4 conversion and ethylene/ethane molar ratio for all the catalysts. However, the C_2 yield was enhanced in the presence of CaO/CeO_2 when compared with the process carried out without CO_2 (7% without CO_2 vs. 8.7% with CO_2).

4.2 *Catalysts Applied in CO_2-OCM*

To date, only a few papers evaluated the effect of CO_2 as a soft oxidant in the OCM process. Table 2 lists some metal oxide catalysts that have been tested for this purpose.

In 1988, Aika et al. [108] published the first report on the utilization of CO_2 in the oxidative coupling of methane using the PbO-MgO and PbO-CaO catalysts at atmospheric pressure and 750 °C. The authors observed that CO_2 raised the C_2 yield for almost all the catalysts. They also found that pure PbO and MgO did not display catalytic activity, suggesting the importance of the synergism between these oxides as active platforms for C_2 hydrocarbon production.

A wide variety of oxide catalysts have widely been investigated for CO_2-OCM, as shown in Table 2. In this case, as we can see, most of them have prepared from the impregnation method. Despite the several attempts, a CO_2-OCM catalytic system with an optimal C_2 hydrocarbon yield is still quite challenging to achieve. So far, the CaO/CeO_2 catalyst has shown the best performance, attaining C_2 yields of 9% at 800 °C [107], which still is a low yield. This means that a deep understanding of this reaction chemistry is still needed, as well as parameter optimization to design a better catalyst for attaining commercial C_2 yields in the future.

4.3 *Mechanism*

Two possible reaction mechanisms involved in the CO_2-OCM process have been proposed in the literature. The first one is related to CO_2 adsorption on the metal oxide followed by decomposition into CO and an active oxygen species (O^-), with CO being further released to the gas phase. CH_4 is oxidized by these active oxygen species to form C_2H_6 by methyl radical ($CH_3^•$) recombination. C_2H_4 can be usually produced

Table 2 Catalysts applied in the synthesis of C_2 hydrocarbons by CO_2-OCM

Catalyst	Method	T (°C)	C_2 yield (%)	References
50%PbO/MgO	Impregnation	750	4.4	[108]
15%PbO–5%Na_2O–MgO			3.9	
20% BiO–MgO			0.8	
15% K_2O–MgO			2.0	
20% Li_2O–Sm_2O_3			0.5	
Sm_2O			1.8	
20%PbO/CaO			2.6	
BaO–CaO	Impregnation	750	3.6	[109]
15%PbO/MgO	Impregnation	750	~3.6	[110]
Y, La, Sm, Ti, Zr, Hf, Nb, Cr, Mn, Fe, Co, Cu, In, Al, Si, Ge, and Bi (oxides)	–	850	–	[111]
20%La_2O_3/ZnO	Impregnation	850	2.8	[97]
La, Ce, Pr, Nd, Sm, Eu, Gd, Tb, Dy, Ho, Er, Tm, Yb, Lu, (oxides)	Commercial for all of them and two different methods for Pr: Pr(A) commercially, Pr (B) decomposition and, Pr (C) reducing it in H_2	850	1–1.5	[112]
Pr_2O_3	Thermal decomposition	600	2.3	[100]
CaO–CeO_2	Impregnation	850	6.0	[113]
Ca–Cr (1.0)	Impregnation	850	4.0	[101]
CaO/CeO_2	Impregnation	850	3.2	[104]
	Physical mixture		0.4	
CaO-ZnO	Impregnation	850	4.3	[106]
Ca–Ce	Impregnation	850	2.7	[114]
Ca–Cr			2.8	
Ca–Mn			2.7	
Ca–Zn			2.8	
Sr–Ce			1.0	
Sr–Cr			1.3	
Sr–Mn			3.3	
Sr–Zn			2.2	
Ba–Ce			0.2	
Ba–Cr			0.3	
Ba–Mn			2.6	
Ba–Zn			0.4	

(continued)

Table 2 (continued)

Catalyst	Method	T (°C)	C_2 yield (%)	References
Mn/Sr	Impregnation	875	4.5	[115]
Ca–Ce (0.5)	Impregnation	850	4.0	[116]
Ca–Cr (1)			4.0	
Ca–Mn (1)			4.1	
Sr–Ce (0.5)			1.4	
Sr–Cr (1)			1.2	
Sr–Mn (1)			4.5	
Ba–Ce (0.5)			0.4	
Ba–Cr (1)			0.4	
Ba–Mn (0.5)			3.2	
Ba–Mn (1)			2.7	
CeO_2/ZnO	Precipitation with micro-emulsion	825	4.8	[91]
12.8%CaO–6.4%MnO/CeO_2	Impregnation	854	3.9	[92]
15%CaO–5%MnO/CeO_2	Co-impregnation method	850	3.8	[93]
La_2O_3/CeO_2	Impregnation	800	5.7	[107]
CaO/CeO_2			8.7	
CeO_2/ZnO with Li, Zr, Bi, Ba, La, Co, TI and or K as promoters	Precipitation with microemulsion	700 -900	-	[117]
Ca:Na:NaCl at 950 °C	Modified polymerized complex (PC)	950	6.6 (C_{2-3})	[118]
Ca:Na:NaCl at 900 °C			2.5 (C_{2-3})	
CaO			1.6 (C_{2-3})	

by the pyrolysis and/or oxidative dehydrogenation of C_2H_6. On the other hand, the second mechanism involves the particular reaction of CH_4 with lattice oxygen of the metal oxides, that is, in order to form a $CH_3^{\bullet}$ radical, which is in general converted to C_2 hydrocarbons and then CO_2 oxidizes the partly reduced oxides [92, 95, 119].

CO_2-OCM is a complex heterogeneous (surface-catalyzed) and homogeneous (gas-phase) sequential process, in which the conversion mechanism of CH_4 and CO_2 into C_2 hydrocarbons involves the following steps [92, 95]:

I. oxygen-assisted breakage of a C–H bond in the CH_4 molecule on the solid oxide catalyst surface; or CH_4 activation by lattice oxygen in the catalyst;
II. heterogeneous decomposition of CO_2 to CO and oxygen active species on the solid oxide catalyst surface;
III. homogeneous recombination of $CH_3^{\bullet}$ radicals released from the catalyst surface;
IV. homogeneous oxidative or radical dehydrogenation of C_2H_6 to C_2H_4.

Consequently, it is essential to highlight that the CO_2-OCM process can involve two kinds of oxidant in the surface reaction: oxygen active species from CO_2 and lattice oxygen from the oxide catalyst. Some mechanisms reported in the literature from 1996 up to the present are described below.

Chen et al. [97] elucidated the CO_2-OCM mechanism on La_2O_3/ZnO, a catalyst with high selectivity towards C_2 hydrocarbons (90.6%). It was reported that the oxygen mobility and non-stoichiometry of the oxides are important to achieve the C_2 hydrocarbon selectivity. In the proposed mechanism, the surface reactive oxygen species of La_2O_3/ZnO abstract H_2 from CH_4 to produce $CH_3^{\bullet}$ radicals, which are further converted to C_2 hydrocarbons. CO_2 adsorbs on the oxygen-deficient catalyst surface (□), followed by thermal dissociation into CO, leading to surface re-oxidation, as represented in the following Eqs. (11–13).

$$\mathrm{O(surface)} + 2\mathrm{CH_4} \rightarrow 2\mathrm{CH_3^{\bullet}} + \mathrm{H_2O} + \square \tag{11}$$

$$2\,\mathrm{CH_3^{\bullet}} \rightarrow \mathrm{C_2H_6} \tag{12}$$

$$\square + \mathrm{CO_2} \rightarrow \mathrm{CO} + \mathrm{O(surface)} \tag{13}$$

Asami et al. [112] observed that the bulk phase in a Pr oxide catalyst does not correlate with the active sites to produce C_2 hydrocarbons. They also suggested that the reaction mechanism with the Pr oxide catalyst could be related to a redox reaction involving the unstable lattice oxygen. However, it was further observed that the Pr oxide has multiple oxidation states and in addition to its lattice oxygen atoms are usually labile, suggesting that the lattice oxygen instability in Pr oxides plays a pivotal role in the C_2 hydrocarbons production.

Wang et al. [104] evaluated the modification of CeO_2 with CaO, leading to solid solutions with Ca/Ce ratios of 0.1–0.5, in order to elucidate the CO_2-OCM mechanism. They observed that the CeO_2 modification achieved the highest C_2 yield of 6% at under 900 °C and $P(CO_2)$ of 70 kPa. According to the mechanism proposed by the authors, shown in Fig. 5, CO_2 molecules adsorb on the Ca^{+2} sites (step B) and the interfacial Ce^{3+} sites activate these molecules, forming CO and O^- (step C), which produces $CH_3^{\bullet}$ radicals from CH_4 (Step D). It was concluded that the formation of a solid solution (neighboring Ca^{2+} and Ce^{3+} sites), generates the active sites for adsorption and subsequent activation of CO_2, while the defects generated by the incorporation of Ca^{+2} into the CeO_2 lattice boosted the redox reactions Ce^{4+} – Ce^{3+}, thereby favoring the CH_4 conversion into C_2 hydrocarbons [104].

Wang et al. [114] also investigated the active sites and reaction mechanisms for C_2 synthesis from the CH_4 conversion using CO_2 as an oxidant over Sr–Mn and Ba–Mn binary oxide catalysts. XRD characterizations revealed peaks assigned to $SrMnO_3$ after the reaction at 850 °C, peaks typical of $SrCO_3$ + MnO and, after He treatment, peaks ascribed to $SrMnO_{2.5}$. This means that in the reaction step, the $SrMnO_3$ phase was reduced to $SrMnO_{2.5}$. It was shown by XPS that the valence state of Mn decreased from Mn^{4+} mainly to Mn^{2+}. However, after He treatment, the valence increased again

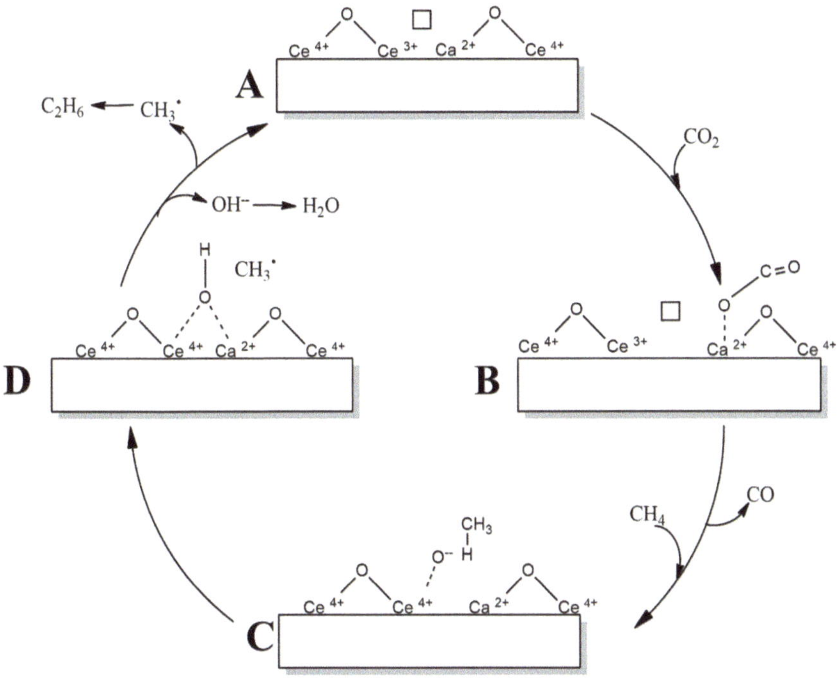

Fig. 5 Proposed mechanism for selective formation of C_2 hydrocarbons over CaO–CeO_2 catalyst Adapted from Wang et al. (with permission) (1999) [104]

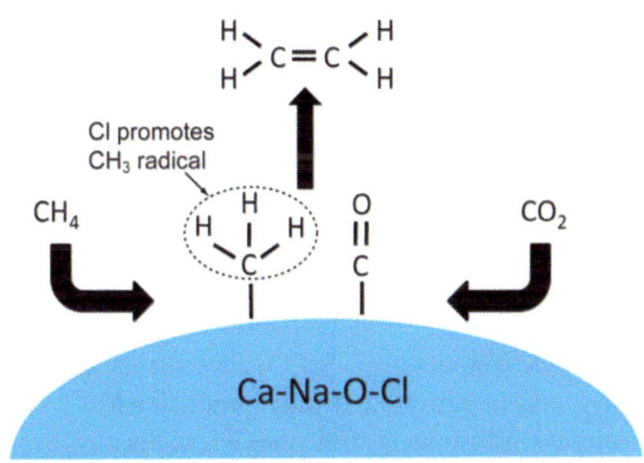

Fig. 6 Proposed mechanism for the formation of hydrocarbons over Ca–Na–O–Cl catalysts [118] (with permission) Copyright © 1999 Academic Press. All rights reserved

to Mn^{3+}, similarly to previous observations for Ba-Mn. It was considered that the decomposition of both $SrCO_3$ and $BaCO_3$ to the corresponding oxides, releasing CO_2, could take place simultaneously. Accordingly, the reaction mechanisms were elucidated, as shown in Eqs. 14–17. The Mn^{3+} ions are reduced to Mn^{2+}, and the MnO, SrO, and BaO phases are formed. In the presence of CO_2, these oxides are converted to the corresponding carbonates, which are further partly converted into $SrMnO_{2.5}$ and $BaMnO_{2.5}$ (Eqs. 14–17). The $CH_3^{\bullet}$ radicals are formed according to Eqs. 16 and 17 and they undergo coupling reactions to form C_2H_6. Though, C_2H_4 can be produced probably by the non-oxidative dehydrogenation of C_2H_6. It was concluded that the species Mn^{3+} probably catalyzed the CH_4 activation.

$$2\ SrCO_3 + 2MnO \rightarrow 2\ SrMnO_{2.5} + CO + CO_2 \quad (14)$$

$$2\ BaCO_3 + 2MnO \rightarrow 2\ BaMnO_{2.5} + CO + CO_2 \quad (15)$$

$$2\ SrMnO_{2.5} + CH_4 \rightarrow 2\ SrO + 2\ MnO + CH_3^{\bullet} + OH_{(a)} \quad (16)$$

$$2\ BaMnO_{2.5} + CH_4 \rightarrow 2\ BaO + 2\ MnO + CH_3^{\bullet} + OH_{(a)} \quad (17)$$

Cai et al. [17] reported a possible mechanism for C_2 hydrocarbon production by CO_2-OCM using Mn-$SrCO_3$ catalysts, as described by the following reactions (Eqs. 18–22) [95, 115]:

$$SrCO_3 \xrightarrow{\Delta} SrO + CO_{2ads}^{*} \quad (18)$$

$$CO_{2ads}^{*} + Mn^{2+} \rightarrow Mn^{3+} + CO + O^{-} \quad (19)$$

$$O^{-} + 2\ CH_4 + Mn^{3+} \rightarrow 2\ CH_3^{\bullet} + H_2O + Mn^{2+} \quad (20)$$

$$2\ CH_3^{\bullet} \rightarrow C_2H_6 \quad (12)$$

$$C_2H_6 \rightarrow C_2H_4 + H_2 \quad (21)$$

$$SrO + CO_{2(g)} \rightarrow SrCO_3 \quad (22)$$

At high temperatures, $SrCO_3$ dissociates into SrO and CO_{2ads}^{*}. The adsorbed CO_{2ads}^{*} may be activated by Mn^{2+} producing Mn^{+3}, CO, and O^{-}. The surface reactive species (O^{-}) abstract H_2 from the CH_4 molecules, producing methyl radicals ($CH_3^{\bullet}$), which are converted into C_2 hydrocarbons. Then, SrO may react with $CO_{2(g)}$ in the gas phase to produce $SrCO_3$ again. Based on this evidence, the satisfactory activity (C_2 yield = 5.1%, selectivity of 68.1% at 900 °C) of this catalysts could be explained by the dissociation of the $SrCO_3$ phase to form CO_{2ads}^{*}, and by the formation of the Mn^{3+}/Mn^{2+} pair that activates CO_{2ads}^{*} that finally activates CH_4.

Wang et al. [116] evaluated Ca–Ce, Ca–Cr, and Ca–Mn binary oxides catalysts in the CO_2-OCM. The proposed mechanism is presented in Eqs. 23–25. First, the chemisorption of CO_2 on basic Ca^{2+} sites takes place (Eq. 23), and this chemisorbed CO_2 ($CO_2(a)$) is activated by electron transfer of neighboring species (Ce^{3+}, Cr^{3+}, or Mn^{2+}) to form CO and O^-, possibly via CO_2^- (Eq. 24). Then, CH_4 is converted into $CH_3^{\bullet}$ by O^- species (Eq. 25), in a precursor stage for the C_2 hydrocarbons formation. The authors concluded that the formation of $Ca_xCe_{1-x}O_{2-y}$, $Ca_{0.48}Mn_{0.52}O$, and $Ca(CrO_2)_2$ involving Ca^{2+} species results in redox properties that play crucial roles in the CO_2 adsorption and activation, leading to a synergistic effect for C_2 hydrocarbons production [116].

$$CO_2 \xrightarrow{Ca^{2+}-O^{2-}} CO_2(a) \tag{23}$$

$$CO_2(a) \xrightarrow{Ce^{3+},\ Cr^{3+},\ \text{or}\ Mn^{2+}} CO_2^- \rightarrow CO + O^- \tag{24}$$

$$CH_4 + O^- \rightarrow CH_3^{\bullet} + OH^- \tag{25}$$

Zhang et al. [118] performed the CO_2-OCM using an earth-abundant CaO-based catalyst and achieved a remarkable $C_{2,3}$ yield of ~6%. They carried out an exploratory study by in situ FTIR to elucidate the hydrocarbon synthesis mechanism under the reaction conditions and evidenced the presence of carbonyl (M=C=O) and methane radical groups (–CH_3). The proposed reaction mechanism onto the Ca–Na–O–Cl catalyst is outlined in Fig. 6. First, Na_2CO_3 decomposes into $Na_2O + CO_2$ at high temperatures (above melting point). The Na_2O formed on the catalyst surface converts CH_4 into $CH_3^{\bullet}$ radicals with CaO and the assistance of Cl dopants. Then, two or more $CH_3^{\bullet}$ radicals react to produce $C_{2,3}$ hydrocarbons.

Despite the fact that some mechanisms involving binary oxides catalysts have been disclosed in the literature, the knowledge about catalytically active sites is still quite limited, and the mechanism for C_2 formation is still unclear [114]. Therefore, further investigations are needed to understand the interactions between oxides in order to develop optimized catalytic systems. This will provide insights into the CO_2-OCM activation process at the molecular level aiming at increasing the C_2 yield.

5 Concluding Remarks and Outlook

The feasibility of the C_1 process depends on the success in establishing a good conversion, maximum selectivity, and stability under the reaction conditions. Simultaneous studies on catalyst design and kinetics modeling must be carried out to obtain a suitable catalyst for the desired process.

DRM is an endothermic process that needs high temperatures to obtain good activity/stability and minimize coke deposition. Therefore, from this perspective, it is crucial to develop catalysts with high thermal stability and with a structure favorable for adsorption and desorption process of the reactants and products. The recent literature indicates that the most promising way is the design of catalysts with the active phase confined/embedded into mixed oxides, with high oxygen mobility and strong metal-support interaction. Coating the catalyst with a thermally stable shell having a controlled pore size may be a good approach to avoid metallic phase sintering and mass transport limitations.

OCM and CO_2-OCM are not conducted on an industrial scale yet. However, tremendous efforts are being put in the design of catalysts able to provide C_2 hydrocarbon yields above 30%. Despite the several attempts, a catalytic system with optimal C_2 hydrocarbon yield for the OCM and CO_2-OCM processes is still quite challenging to achieve. This means that a deep understanding of the OCM and CO_2-OCM chemistry is still needed, as well as the parameter optimization for the design of effective catalysts for reaching commercially attractive C_2 yields in the future.

Acknowledgements The authors would like to thank FAPESP (#2017/08293-8 and #2019/10980-9) for the fellowships and FAPESP Thematic Project (#2015/06246-7).

Conflict of Interest The authors declare no conflict of interest.

References

1. World population. https://www.worldometers.info/world-population. Accessed 17 Jun 2020
2. Natural gas reserves. https://www.eia.gov/international/data/world/natural-gas/dry-natural-gas-reserves. Accessed 17 Jun 2020
3. Hernández, B., Martín, M.: Optimal process operation for biogas reforming to methanol: effects of dry reforming and biogas composition. Ind. Eng. Chem. Res. **55**, 6677–6685 (2016). https://doi.org/10.1021/acs.iecr.6b01044
4. Lucrédio, A.F., Assaf, J.M., Assaf, E.M.: Reforming of a model sulfur-free biogas on Ni catalysts supported on Mg(Al)O derived from hydrotalcite precursors: effect of La and Rh addition. Biomass Bioenergy **60**, 8–17 (2014). https://doi.org/10.1016/j.biombioe.2013.11.006
5. Lucrédio, A.F., Assaf, J.M., Assaf, E.M.: Methane conversion reactions on Ni catalysts promoted with Rh: Influence of support. Appl. Catal. A Gen. **400**, 156–165 (2011). https://doi.org/10.1016/j.apcata.2011.04.035
6. Lino, A.V.P., Assaf, E.M., Assaf, J.M.: Hydrotalcites derived catalysts for syngas production from biogas reforming: effect of nickel and cerium load. Catal. Today **289**, 78–88 (2017). https://doi.org/10.1016/j.cattod.2016.08.022
7. González, A.R., Asencios, Y.J.O., Assaf, E.M., Assaf, J.M.: Dry reforming of methane on Ni–Mg–Al nano-spheroid oxide catalysts prepared by the sol-gel method from hydrotalcite-like precursors. Appl. Surf. Sci. **280**, 876–887 (2013). https://doi.org/10.1016/j.apsusc.2013.05.082
8. Usman, M., Wan Daud, W.M.A., Abbas, H.F.: Dry reforming of methane: influence of process parameters—a review. Renew. Sustain. Energy Rev. **45**, 710–744 (2015). https://doi.org/10.1016/j.rser.2015.02.026
9. Asencios, Y.J.O., Assaf, E.M.: Combination of dry reforming and partial oxidation of methane on NiO–MgO–ZrO_2catalyst: effect of nickel content. Fuel Process Technol. **106**, 247–252 (2013). https://doi.org/10.1016/j.fuproc.2012.08.004
10. Asencios, Y.J.O., Bellido, J.D.A., Assaf, E.M.: Synthesis of NiO–MgO–ZrO_2 catalysts and their performance in reforming of model biogas. Appl. Catal. A Gen. **397**, 138–144 (2011). https://doi.org/10.1016/j.apcata.2011.02.023

11. Asencios, Y.J.O., Rodella, C.B., Assaf, E.M.: Oxidative reforming of model biogas over NiO–Y_2O_3–ZrO_2 catalysts. Appl. Catal. B Environ. **132–133**, 1–12 (2013). https://doi.org/10.1016/j.apcatb.2012.10.032
12. Bezerra, D.M., Lucrédio, A.F., Valentini, A., Assaf, E.M.: Biogas reforming over Ni catalysts dispersed in different mixed oxides containing Zn^{2+}, Al^{3+} and Zr^{4+} cations. Mater. Res. Bull. **102**, 186–195 (2018). https://doi.org/10.1016/j.materresbull.2018.02.029
13. Bao, J., Yang, G., Yoneyama, Y., Tsubaki, N.: Significant advances in C1 catalysis: highly efficient catalysts and catalytic reactions. ACS Catal. **9**, 3026–3053 (2019). https://doi.org/10.1021/acscatal.8b03924
14. Lucrédio, A.F., Filho, G.T., Assaf, E.M.: Co/Mg/Al hydrotalcite-type precursor, promoted with La and Ce, studied by XPS and applied to methane steam reforming reactions. Appl. Surf. Sci. **255**, 5851–5856 (2009). https://doi.org/10.1016/j.apsusc.2009.01.020
15. De Abreu, A.J., Lucrédio, A.F., Assaf, E.M.: Ni catalyst on mixed support of CeO_2–ZrO_2 and Al_2O_3: effect of composition of CeO_2–ZrO_2 solid solution on the methane steam reforming reaction. Fuel Process Technol. **102**, 140–145 (2012). https://doi.org/10.1016/j.fuproc.2012.04.030
16. Liu, K.G.G.R.C, Song, C.P.S.U., Subramani, V.: Hydrogen and Syngas Production and Purification Technologies. Wiley, Hoboken (2010)
17. Abdulrasheed, A., Jalil, A.A., Gambo, Y., et al.: A review on catalyst development for dry reforming of methane to syngas: recent advances. Renew. Sustain. Energy Rev. **108**, 175–193 (2019). https://doi.org/10.1016/j.rser.2019.03.054
18. Asencios, Y.J.O., Marcos, F.C.F., Assaf, J.M., Assaf, E.M.: Oxidative-reforming of methane and partial oxidation of methane reactions over NiO/PrO_2/ZrO_2 catalysts: effect of nickel content. Brazilian J. Chem. Eng. **33**, 627–636 (2016). https://doi.org/10.1590/0104-6632.20160333s20150056
19. Jangam, A., Das, S., Dewangan, N., et al.: Conversion of CO_2 to C1 chemicals: catalyst design, kinetics and mechanism aspects of the reactions. Catal. Today (2019). https://doi.org/10.1016/j.cattod.2019.08.049
20. Mortensen, P.M., Dybkjær, I.: Industrial scale experience on steam reforming of CO_2-rich gas. Appl. Catal. A Gen. **495**, 141–151 (2015). https://doi.org/10.1016/j.apcata.2015.02.022
21. Jafarbegloo, M., Tarlani, A., Mesbah, A.W., Sahebdelfar, S.: Thermodynamic analysis of carbon dioxide reforming of methane and its practical relevance. Int. J. Hydrogen Energy **40**, 2445–2451 (2015). https://doi.org/10.1016/j.ijhydene.2014.12.103
22. Bellido, J.D.A., De Souza, J.E., M'Peko, J.C., Assaf, E.M.: Effect of adding CaO to ZrO_2 support on nickel catalyst activity in dry reforming of methane. Appl. Catal. A Gen. **358**, 215–223 (2009). https://doi.org/10.1016/j.apcata.2009.02.014
23. Bellido, J.D.A., Assaf, E.M.: Effect of the Y_2O_3-ZrO_2 support composition on nickel catalyst evaluated in dry reforming of methane. Appl. Catal. A Gen. **352**, 179–187 (2009). https://doi.org/10.1016/j.apcata.2008.10.002
24. Lino, A.V.P., Assaf, E.M., Assaf, J.M.: NiMgAlCe catalysts applied to reforming of a model biogas for syngas production. Catal. Lett. **148**, 979–991 (2018). https://doi.org/10.1007/s10562-018-2304-9
25. Lian, Z., Olanrele, S.O., Si, C., et al.: Critical role of interfacial sites between nickel and CeO_2 support in dry reforming of methane: revisit of reaction mechanism and origin of stability. J. Phys. Chem. C **124**, 5118–5124 (2020). https://doi.org/10.1021/acs.jpcc.9b09856
26. Shah, M., Das, S., Nayak, A.K., et al.: Smart designing of metal-support interface for imperishable dry reforming catalyst. Appl. Catal. A Gen. **556**, 137–154 (2018). https://doi.org/10.1016/j.apcata.2018.01.007
27. Bao, Z., Lu, Y., Han, J., et al.: Highly active and stable Ni-based bimodal pore catalyst for dry reforming of methane. Appl. Catal. A Gen. **491**, 116–126 (2015). https://doi.org/10.1016/j.apcata.2014.12.005
28. Otor, H.O., Steiner, J.B., García-Sancho, C., Alba-Rubio, A.C.: Encapsulation methods for control of catalyst deactivation: a review. ACS Catal. (2020). https://doi.org/10.1021/acscatal.0c01569

29. Li, Z., Li, M., Bian, Z., et al.: Design of highly stable and selective core/yolk–shell nanocatalysts—a review. Appl. Catal. B Environ. **188**, 324–341 (2016). https://doi.org/10.1016/j.apcatb.2016.01.067
30. Du, X., Zhang, D., Gao, R., et al.: Design of modular catalysts derived from NiMgAl-LDH@m-SiO_2 with dual confinement effects for dry reforming of methane. Chem. Commun. **49**, 6770–6772 (2013). https://doi.org/10.1039/C3CC42418A
31. Marinho, A.L.A., Rabelo-Neto, R.C., Epron, F., et al.: Embedded Ni nanoparticles in $CeZrO_2$ as stable catalyst for dry reforming of methane. Appl. Catal. B Environ. **268**, 118387 (2020). https://doi.org/10.1016/j.apcatb.2019.118387
32. Linde Group. https://www.linde-engineering.com/en/innovations/innovate-dry-reforming/index.html. Accessed 20 Jun 2020
33. Halmann, M.M., Steinberg, M.: Greenhouse Gas Carbon Dioxide Mitigation, 1st ed. CRC Press, Boca Raton (1998)
34. Igenegbai, V.O., Meyer, R.J., Linic, S.: In search of membrane-catalyst materials for oxidative coupling of methane: performance and phase stability studies of gadolinium-doped barium cerate and the impact of Zr doping. Appl. Catal. B Environ. **230**, 29–35 (2018). https://doi.org/10.1016/j.apcatb.2018.02.040
35. Yamaguchi, A., Jin, D., Ikeda, T., et al.: Deactivation of ZSM-5 zeolite during catalytic steam cracking of n-hexane. Fuel Process. Technol. **126**, 343–349 (2014). https://doi.org/10.1016/j.fuproc.2014.05.013
36. Wattanapaphawong, P., Reubroycharoen, P., Mimura, N., et al.: Effect of carbon number on the production of propylene and ethylene by catalytic cracking of straight-chain alkanes over phosphorus-modified ZSM-5. Fuel Process. Technol. **202**, 106367 (2020). https://doi.org/10.1016/j.fuproc.2020.106367
37. Galadima, A., Muraza, O.: Revisiting the oxidative coupling of methane to ethylene in the golden period of shale gas: a review. J. Ind. Eng. Chem. **37**, 1–13 (2016). https://doi.org/10.1016/j.jiec.2016.03.027
38. Fakhroleslam, M., Sadrameli, S.M.: Thermal/catalytic cracking of hydrocarbons for the production of olefins; a state-of-the-art review III: process modeling and simulation. Fuel **252**, 553–566 (2019). https://doi.org/10.1016/j.fuel.2019.04.127
39. Corma, A., Mengual, J., Miguel, P.J.: IM-5 zeolite for steam catalytic cracking of naphtha to produce propene and ethene. an alternative to ZSM-5 zeolite. Appl. Catal. A Gen. **460–461**, 106–115 (2013). https://doi.org/10.1016/j.apcata.2013.02.030
40. Chen, Y., Li, J., He, L.: Tradeoffs in cost competitiveness and emission reduction within micro-grid sustainable development considering price-based demand response. Sci. Total Environ. **703** (2020). https://doi.org/10.1016/j.scitotenv.2019.135545
41. Schenk, C.J., Mercier, T.J., Tennyson, M.E., et al.: Assessment of undiscovered conventional oil and gas resources of the Grand Erg/Ahnet Province, Algeria, 2018. Fact Sheet (2019). https://doi.org/10.3133/fs20193031
42. Hu, L., Pinto, D., Urakawa, A.: Active reactions and spatial gradients in oxidative coupling of methane. In: Catalysis, pp. 203–223 (2020)
43. Ito, T., Lunsford, J.H.: Synthesis of ethylene and ethane by partial oxidation of methane over lithium-doped magnesium oxide. Nature **314**, 4–5 (1985)
44. Hinsen, W., Bytyn, W., Baerns, M.: Oxidative dehydrogenation and coupling of methane. In: In: Proceedings of the 8th International Congress on Catalysis, p. 581 (1984)
45. Keller, G.E., Bhasin, M.M.: Synthesis of ethylene via oxidative coupling of methane. I. Determination of active catalysts. J. Catal. **73**, 9–19 (1982). https://doi.org/10.1016/0021-9517(82)90075-6
46. True, W.R.: Global ethylene capacity poised for major expansion. Oil Gas J. **111**, 6–90 (2013)
47. Farrell, B.L., Igenegbai, V.O., Linic, S.: A viewpoint on direct methane conversion to ethane and ethylene using oxidative coupling on solid catalysts. ACS Catal. **6**, 4340–4346 (2016). https://doi.org/10.1021/acscatal.6b01087
48. Huang, T., Guo, X., Chen, F.: Modeling transient pressure behavior of a fractured well for shale gas reservoirs based on the properties of nanopores. J. Nat. Gas Sci. Eng. **23**, 387–398 (2015). https://doi.org/10.1016/j.jngse.2015.02.020

49. Chen, S., Ma, X.: The role of oxygen species in the selective oxidation of methanol to dimethoxymethane over VOx/TS-1 catalyst. J. Ind. Eng. Chem. **45**, 296–300 (2017). https://doi.org/10.1016/j.jiec.2016.09.037
50. Kim, I., Lee, G., Bin, N.H., et al.: Selective oxygen species for the oxidative coupling of methane. Mol. Catal. **435**, 13–23 (2017). https://doi.org/10.1016/j.mcat.2017.03.012
51. Hou, Y.H., Han, W.C., Xia, W.S., Wan, H.L.: Structure sensitivity of $La_2O_2CO_3$ catalysts in the oxidative coupling of methane. ACS Catal. **5**, 1663–1674 (2015). https://doi.org/10.1021/cs501733r
52. Varghese, J.J., Trinh, Q.T., Mushrif, S.H.: Insights into the synergistic role of metal-lattice oxygen site pairs in four-centered C–H bond activation of methane: the case of CuO. Catal. Sci. Technol. **6**, 3984–3996 (2016). https://doi.org/10.1039/c5cy01784j
53. Sinev, M., Ponomareva, E., Sinev, I., et al.: Oxygen pathways in oxidative coupling of methane and related processes. case study: NaWMn/SiO_2 catalyst. Catal. Today **333**, 36–46 (2019). https://doi.org/10.1016/j.cattod.2018.06.028
54. Phung, T.K., Garbarino, G.: On the use of infrared spectrometer as detector for Temperature Programmed (TP) techniques in catalysts characterization. J. Ind. Eng. Chem. **47**, 288–296 (2017). https://doi.org/10.1016/j.jiec.2016.11.045
55. Aseem, A., Jeba, G.G., Conato, M.T., et al.: Oxidative coupling of methane over mixed metal oxide catalysts: steady state multiplicity and catalyst durability. Chem. Eng. J. **331**, 132–143 (2018). https://doi.org/10.1016/j.cej.2017.08.093
56. Elkins, T.W., Neumann, B., Bäumer, M., Hagelin-Weaver, H.E.: Effects of Li doping on MgO-supported Sm_2O_3 and TbO_x catalysts in the oxidative coupling of methane. ACS Catal. **4**, 1972–1990 (2014). https://doi.org/10.1021/cs500138j
57. Raouf, F., Taghizadeh, M., Yousefi, M.: Influence of CaO-ZnO supplementation as a secondary catalytic bed on the oxidative coupling of methane. React. Kinet. Mech. Catal. **112**, 227–240 (2014). https://doi.org/10.1007/s11144-014-0692-4
58. Lee, J.Y., Jeon, W., Choi, J.W., et al.: Scaled-up production of C_2 hydrocarbons by the oxidative coupling of methane over pelletized Na_2WO_4/Mn/SiO_2 catalysts: observing hot spots for the selective process. Fuel **106**, 851–857 (2013). https://doi.org/10.1016/j.fuel.2013.01.026
59. Ghose, R., Hwang, H.T., Varma, A.: Oxidative coupling of methane using catalysts synthesized by solution combustion method: catalyst optimization and kinetic studies. Appl. Catal. A Gen. **472**, 39–46 (2014). https://doi.org/10.1016/j.apcata.2013.12.004
60. Oshima, K., Tanaka, K., Yabe, T., et al.: Oxidative coupling of methane using carbon dioxide in an electric field over La-ZrO_2 catalyst at low external temperature. Fuel **107**, 879–881 (2013). https://doi.org/10.1016/j.fuel.2013.01.058
61. Yildiz, M., Aksu, Y., Simon, U., et al.: Enhanced catalytic performance of Mn_xO_y–Na_2WO_4/SiO_2 for the oxidative coupling of methane using an ordered mesoporous silica support. Chem. Commun. **50**, 14440–14442 (2014). https://doi.org/10.1039/c4cc06561a
62. Colmenares, M.G., Simon, U., Yildiz, M., et al.: Oxidative coupling of methane on the Na_2WO_4-Mn_xO_y catalyst: COK-12 as an inexpensive alternative to SBA-15. Catal. Commun. **85**, 75–78 (2016). https://doi.org/10.1016/j.catcom.2016.06.025
63. Wang, H., Schmack, R., Paul, B., et al.: Porous silicon carbide as a support for Mn/Na/W/SiC catalyst in the oxidative coupling of methane. Appl. Catal. A Gen. **537**, 33–39 (2017). https://doi.org/10.1016/j.apcata.2017.02.018
64. Oshima, K., Shinagawa, T., Sekine, Y.: Methane conversion assisted by plasma or electric field. J. Japan. Pet. Inst. **56**, 11–21 (2013)
65. Xu, J., Zhang, Y., Xu, X., et al.: Constructing $La_2B_2O_7$ (B = Ti, Zr, Ce) compounds with three typical crystalline phases for the oxidative coupling of methane: the effect of phase structures, superoxide anions, and alkalinity on the reactivity. ACS Catal. **7**, 4030–4045 (2019). https://doi.org/10.1021/acscatal.9b00022
66. Cheng, Z., Qin, L., Guo, M., Fan, J.A.: Methane adsorption and dissociation on iron oxide oxygen carriers: the role of oxygen vacancies. Phys. Chem. Chem. Phys. **18**, 16423–16435 (2016). https://doi.org/10.1039/C6CP01287F

67. Cheng, Z., Qin, L., Fan, J.A., Fan, L.: New insight into the development of oxygen carrier materials for chemical looping systems. Engineering **4**, 343–351 (2018). https://doi.org/10.1016/j.eng.2018.05.002
68. Cheng, Z., Baser, D.S., Nadgouda, S.G., Qin, L., Fan, J.A., Fan, L.-S.: C_2 selectivity enhancement in chemical looping oxidative coupling of methane over a Mg-Mn composite oxygen carrier by Li-Doping-induced oxygen vacancies. ACS Energy. Lett. **3**, 1730–1736 (2018). https://doi.org/10.1021/acsenergylett.8b00851
69. Cheng, Z., Qin, L., Guo, M., et al.: Oxygen vacancy promoted methane partial oxidation over iron oxide oxygen carrier in chemical looping process. Phys. Chem. Chem. Phys. **18**, 32418–32428 (2016). https://doi.org/10.1039/C6CP06264D
70. Gambo, Y., Jalil, A.A., Triwahyono, S., Abdulrasheed, A.A.: Recent advances and future prospect in catalysts for oxidative coupling of methane to ethylene: a review. J. Ind. Eng. Chem. **59**, 218–229 (2018). https://doi.org/10.1016/j.jiec.2017.10.027
71. Merino, N.A., Barbero, B.P., Eloy, P., Cadu, L.E.: $La_{1-x}Ca_xCoO_3$ perovskite-type oxides: identification of the surface oxygen species by XPS. Appl. Surf. Sci. **253**, 1489–1493 (2006). https://doi.org/10.1016/j.apsusc.2006.02.035
72. Wang, P., Zhao, G., Liu, Y., Lu, Y.: TiO_2-doped Mn_2O_3-Na_2WO_4/SiO_2 catalyst for oxidative coupling of methane: solution combustion synthesis and $MnTiO_3$-dependent LOW-temperature activity imrpovement. Appl. Catal. A Gen. **544**, 77–83 (2017). https://doi.org/10.1016/j.apcata.2017.07.012
73. Chu, C., Zhao, Y., Sun, Y.: Correlation between the acid–base properties of the La_2O_3 catalyst and its methane reactivity. Phys. Chem. Chem. Phys. **18**, 16509–16517 (2016). https://doi.org/10.1039/C6CP02459A
74. Papa, F., Luminita, P., Osiceanu, P., et al.: Acid–base properties of the active sites responsible for C^{2+} and CO_2 formation over MO-Sm_2O_3 (M=Zn, Mg, Ca and Sr) mixed oxides in OCM reaction . J. Mol. Catal. A Chem **346**, 46–54 (2011). https://doi.org/10.1016/j.molcata.2011.06.008
75. Wang, Z., Zou, G., Luo, X., et al.: Oxidative coupling of methane over $BaCl_2$–TiO_2–SnO_2 catalyst. J. Nat. Gas. Chem. **21**, 49–55 (2012). https://doi.org/10.1016/S1003-9953(11)60332-0
76. Driscoll, D.J., Mártir, W., Wang, J., Lunsford, J.H.: Formation of gas-phase methyl radicals over MgO. J. Am. Chem. Soc. **107**, 58–63 (1985). https://doi.org/10.1021/ja00287a011
77. Song, J., Sun, Y., Ba, R., Huang, S., Zhao, Y., Zhang, J., Zhu, Y.: Monodisperse Sr-La_2O_3 hybrid nanofibers for oxidative coupling of methane to synthesize C2 hydrocarbons. Nanoscale **7**, 2260–2264 (2015). https://doi.org/10.1039/b000000x
78. Kumar, G., Lap, S., Lau, J., et al.: Correlation of methane activation and oxide catalyst reducibility and its implications for oxidative coupling. ACS Catal. **6**, 1812–1821 (2016). https://doi.org/10.1021/acscatal.5b02657
79. Noon, D., Zohour, B., Bae, A., et al.: Effects of Ir-doping on the transition from oxidative CeO_2 nano fiber catalysts: spatial concentration and. RSC Adv. **7**, 26783–26789 (2017). https://doi.org/10.1039/c7ra02616a
80. Fung, V., Tao, F.F., Jiang, D.: General structure−reactivity relationship for oxygen on transition metal oxides. J. Phys. Chem. Lett. **8**, 2206–2211 (2017). https://doi.org/10.1021/acs.jpclett.7b00861
81. Liang, Z., Li, T., Kim, M., et al.: Low-temperature activation of methane on the IrO_2 (110) surface. Science (80-) **356**, 299–303 (2017)
82. Obradović, A., Thybaut, J.W., Marin, G.B.: Oxidative coupling of methane: opportunities for microkinetic model assisted process implementations. Chem. Eng. Technol. **39**, 1996–2010 (2016)
83. Li, Z.Y., Jibran, M., Sun, X., et al.: Influence of electron correlation on the electronic and magnetic structures of nitric-oxide-adsorbed manganese phthalocyanine. Chem. Phys. Lett **675**, 15–19 (2017). https://doi.org/10.1016/j.cplett.2017.02.076
84. Nabgan, W., Mat, R., Amran, T., et al.: Development of a kinetic model for hydrogen production from phenol over Ni-Co/ZrO_2 catalyst. J. Environ. Chem. Eng. **4**, 4444–4452 (2016). https://doi.org/10.1016/j.jece.2016.10.013

85. Asundi, A.S., Raiford, J.A., Bent, S.F.: Opportunities for atomic layer deposition in emerging energy technologies. ACS Energy Lett. **4**, 908–925 (2019). https://doi.org/10.1021/acsenergylett.9b00249
86. Vega, V., Gelde, L., González, A.S., et al.: Diffusive transport through surface functionalized nanoporous alumina membranes by atomic layer deposition of metal oxides. J. Ind. Eng. Chem. **52**, 66–72 (2017). https://doi.org/10.1016/j.jiec.2017.03.025
87. Neill, B.J.O., Jackson, D.H.K., Lee, J., et al.: Catalyst design with atomic layer deposition. ACS Catal. **5**, 1804–1825 (2015). https://doi.org/10.1021/cs501862h
88. Baamran, K.S., Tahir, M.: Ni-embedded TiO_2-$ZnTiO_3$ reducible perovskite composite with synergistic effect of metal/support towards enhanced H_2 production via phenol steam reforming. Energy Convers. Manage. **200**, 112064 (2019). https://doi.org/10.1016/j.enconman.2019.112064
89. Kumar, A., Singh, R., Sinha, A.S.K.: Catalyst modification strategies to enhance the catalyst activity and stability during steam reforming of acetic acid for hydrogen production. Int. J. Hydrogen. Energy **44**, 12983–13010 (2019). https://doi.org/10.1016/j.ijhydene.2019.03.136
90. Wang, P., Zhao, G., Wang, Y., Lu, Y.: $MnTiO_3$-driven low-temperature oxidative coupling of methane over TiO_2-doped Mn_2O_3-Na_2WO_4/SiO_2 catalyst. Sci. Adv. **3**, e1603180 (2017)
91. Yongjun he, Bolun Yang GC, : On the oxidative coupling of methane with carbon dioxide over CeO_2/ZnO nanocatalysts. Catal. Today **98**, 595–600 (2004)
92. Aishah, N., Amin, S.: Synergistic effect of catalyst basicity and reducibility on performance of ternary CeO_2-based catalyst for CO_2-OCM to C_2 hydrocarbons. J. Mol. Catal. A Chem. **259**, 61–66 (2006). https://doi.org/10.1016/j.molcata.2006.06.003
93. Aishah, N., Amin, S.: Optimization of process parameters and catalyst compositions in carbon dioxide oxidative coupling of methane over CaO-MnO/CeO_2 catalyst using response surface methodology. Fuel Process. Technol. **87**, 449–459 (2006). https://doi.org/10.1016/j.fuproc.2005.11.004
94. Olivos-Suarez, A.I., Sze, A., Pidko, E.A., Gascon, J.: Strategies for the direct catalytic valorization of methane using heterogeneous catalysis : challenges and opportunities. ACS Catal. **6**, 2965–2981 (2016). https://doi.org/10.1021/acscatal.6b00428
95. Cai, X., Hang, Y.: Advances in catalytic conversion of methane and carbon dioxide to highly valuable products. Energy Sci. Eng. **7**, 4–29 (2019). https://doi.org/10.1002/ese3.278
96. Homs, N., Toyir, J., Ramírez, P., Piscina, D.: Catalytic Processes for Activation of CO_2. Elsevier B.V. (2013)
97. Chen, C., Xu, Y., Li, G., Guo, X.: Oxidative coupling of methane by carbon dioxide : a highly C_2 selective La_2O_3/ZnO catalyst. Catal. Lett. **42**, 149–153 (1996)
98. Oshima, K., Tanaka, K., Yabe, T., Eiichi Kikuchi, Y.S.: Oxidative coupling of methane using carbon dioxide in an electric field over La-ZrO_2 catalyst at low external temperature. Fuel **107**, 879–881 (2013)
99. Wang, Ye., Yoshimoto Takahashi, Y.O.: Carbon dioxide-induced selective conversion of methane to C_2 hydrocarbons on CeO_2 modified with CaO. Appl. Catal. A Gen. **172**, 203–206 (1998)
100. Wang, Y., Zhuang, Q., Takahashi, Y., Ohtsuka, Y.: Remarkable enhancing effect of carbon dioxide on the conversion of methane to C_2 hydrocarbons using praseodymium oxide. Catal. Lett. **56**, 203–206 (1998)
101. Wang, Y., Yoshimoto, Y., Ohtsuka, Y.: Effective catalysts for conversion of methane to ethane and ethylene using carbon dioxide. Chem Lett **27**, 1209–1210 (1998). https://doi.org/10.1246/cl.1998.1209
102. Arora, S., Prasad, R.: An overview on dry reforming of methane: Strategies to reduce carbonaceous deactivation of catalysts. RSC Adv. **6**, 108668–108688 (2016). https://doi.org/10.1039/c6ra20450c
103. Obradovic, A., Marin, G.B.: Oxidative coupling of methane: opportunities for microkinetic model-assisted process implementations. Chem. Eng. Technol. **39**, 1996–2010 (2016). https://doi.org/10.1002/ceat.201600216

104. Wang, Y., Takahashi, Y., Ohtsuka, Y.: Carbon dioxide as oxidant for the conversion of methane to ethane and ethylene using modified CeO_2 catalysts. J. Catal. **186**, 160–168 (1999)
105. Wang, S., Zhu, Z.H.: Catalytic conversion of alkanes to olefins by carbon dioxide oxidative dehydrogenations: a review. Energy Fuels **18**, 1126–1139 (2004)
106. Ye Wang and Yasuo Ohtsuka: CaO-ZnO catalyst for selective conversion of methane to C_2 hydrocarbons using carbon dioxide as the oxidant. J. Catal. **192**, 252–255 (2000)
107. Litawa, B., Michorczyk, P., Ogonowski, J.: Influence of CO_2 on the catalytic performance of La_2O_3/CeO_2 and CaO/CeO_2 catalysts in the oxidative coupling of methane. Polish J. Chem. Technol. **15**, 22–26 (2013)
108. Aika, K., Nishiyama, T.: Utilization of CO_2 in the oxidative coupling of methane over PbO-MgO and PbO-CaO. J. Chem. Soc. Chem. Commun. pp. 70–71 (1988)
109. Ken-ichi, A., Nishiyama, T.: Utilization of CO_2 in the oxidative coupling reaction of methane over CaO based catalysts. Catal. Today **4**, 211–218 (1989)
110. Nishiyama, T., Ken-ichi, A.: Mechanism of the oxidative coupling of methane using CO_2 as an oxidant over PbO–MgO. J. Catal. **122**, 346–351 (1990)
111. Asami, K., Fujita, T., Kusakabe, K.-I., Yoshiyuki Nishiyama, Y.O.: Conversion of methane with carbon dioxide into C_2 Hydrocarbons over metal oxides. Appl. Catal. A-General **126**, 245–255 (1995)
112. Asami, K., Kusakabe, K., Ashi, N., Ohtsuka, Y.: Synthesis of ethane and ethylene from methane and carbon dioxide over praseodymium oxide catalysts. Appl. Catal. A-General **156**, 43–56 (1997)
113. Wang, Ye., Yoshimoto Takahashi, Y.O.: Carbon dioxide-induced selective conversion of methane to C_2 hydrocarbons on CeO_2 modified with CaO. App. Catal. A **172**, 203–206 (1998)
114. Wang, Y., Ohtsuka, Y.: Mn-based binary oxides as catalysts for the conversion of methane to C_2 hydrocarbons with carbon dioxide as oxidant. Appl. Catal. A Gen. **219**, 183–193 (2001)
115. Cai, Y., Chou, L., Li, S., et al.: Selective conversion of methane to C_2 hydrocarbons using carbon dioxide over Mn–$SrCO_3$ catalysts. Catal. Lett. **86**, 191–195 (2003)
116. Wang, Y., Ohtsuka, Y.: Catalysis of methane coupling with carbon dioxide over binary oxides. Top Catal. **22**, 71–75 (2003)
117. Al, M.S., Al, Z.Y., Almasry, W., et al.: Study of performance of CeO_2/ZnO nanocatalysts for the oxidative coupling of methane using carbon dioxide as oxidant. Int. J. Phys. Sci. **7**, 6270–6279 (2013). https://doi.org/10.5897/IJPS12.656
118. Zhang, Y., Cho, Y., Yamaguchi, A., et al.: CO_2 oxidative coupling of methane using an earth-abundant CaO-based catalyst. Sci. Rep. **9**, 15454 (2019). https://doi.org/10.1038/s41598-019-51817-2
119. Colmenares, J.C.: Novel trends in the utilization of CO_2 as a reagent and mild oxidant in the C–C coupling reactions. Curr. Org. Synth. **7**, 533–542 (2010)

Recent Advances in the Fabrication of $BiVO_4$ Photoanodes and $CuBi_2O_4$ Photocathodes for the Photoelectrochemical Water Splitting

Carlos Giovani Oliveira Bruziquesi, Márcio César Pereira, Tatiana Santos Andrade, and Adilson Cândido da Silva

Abstract The sun is a renewable and widely available energy source, but it needs to be converted into other types of energy to be useful. Conversion of solar energy directly to electrical energy can be done using photovoltaic cells, but solar intermittency becomes a problem in the generation of electrical energy continuously. Thus, technologies for the storage of solar energy need to be developed to make the best use of this energy source. Water splitting photoelectrochemical cells are devices that can store solar energy into chemical energy. They are formed by a photo(anode) and a photo(cathode) that can collect solar energy and carry out oxidation and reduction of water, forming O_2 in the anode and H_2 in the cathode, thus storing solar energy in the chemical bonds of the H2 molecule. In recent years, many light-collecting materials have been tested as photoanodes and photocathodes in photoelectrochemical cells. Among photoanodes, $BiVO_4$ has been widely used due to its moderate *bandgap* energy (2.4 eV), the valence band energy level suitable for water oxidation, and its theoretical solar energy conversion capacity of approximately 10%. Among photocathodes, $CuBi_2O_4$ is a promising semiconductor for use in PEC cells due to its small *bandgap* energy (1.8 V) and exceptionally large onset potential (> 1 V versus RHE) which makes it an excellent semiconductor to be coupled with wide *bandgap* photoanodes. Therefore, in this review, we examine the most current strategies used to improve the optical, electronic, and surface properties of $BiVO_4$ and $CuBi_2O_4$, including doping, heterojunction, passivation, and the use of catalysts for the oxygen and hydrogen evolution from water.

M. C. Pereira · T. S. Andrade
Institute of Science, Engineering, and Technology (ICET), Federal University of Jequitinhonha and Mucuri Valleys (UFVJM), Campus Mucuri, Teófilo Otoni, Minas Gerais 39803-371, Brazil

C. G. O. Bruziquesi
Department of Chemistry, Federal University of Minas Gerais (UFMG), Belo Horizonte, Minas Gerais 31270-901, Brazil

A. C. da Silva (✉)
Institute of Exact and Biological Sciences, Federal University of Ouro Preto (UFOP), Ouro Preto, Minas Gerais 35400-000, Brazil
e-mail: adilsonqui@ufop.edu.br

F. A. La Porta and C. A. Taft (eds.), *Functional Properties of Advanced Engineering Materials and Biomolecules*, Engineering Materials,
https://doi.org/10.1007/978-3-030-62226-8_9

Keywords $BiVO_4$ · $CuBi_2O_4$ · Photoelectrochemical cells · H_2 molecule

1 Tandem Cells

A conventional photoelectrochemical (PEC) water splitting system can store chemical energy as hydrogen bonds under light illumination. Their operation is based on the interaction of a semiconductor interface and a solution able to conduct electricity, which leads to water oxidation and hydrogen production [1–3]. Semiconductors are materials that can be photoactived if enough light energy is provided. The amount of energy needed to photoactivate the semiconductor is called *bandgap* and corresponds to the energy difference between its conduction band (CB) and its valence band (VB). In the presence of light, semiconductors generate charge carrier pairs: electrons (e^-) that goes to the CB and can act as reduction sites and holes (h^+) that stay in the VB and can act as oxidative sites [4–6].

When semiconductors are employed in PEC cells, they are called photoelectrodes. These materials can be divided into n-type and p-type semiconductors. Negative or n-type semiconductors are characterized by promoting accumulation of negative charges in the solution and accumulation of positive charges in their interface. The opposite happens with positive or p-type semiconductors in which there is an accumulation of positive charges in the solution and negative charges on its surface. In other words, for n-type semiconductor, the level of its VB is oxidative enough to promote an oxidation reaction; however, the level of its CB is not favorable for the surface flow of electrons [7–10].

During PEC operation, electrons from the photoanode (PA) are excited to the CB and moved to the cathode or photocathode (PC) where they will participate in the hydrogen-evolution reaction (HER). Meantime, holes placed in the VB of PA will take part in the water oxidation, acting in the oxygen evolution reaction (OER) [11, 12]. Equations (1) and (2) show OER and HER that takes place in the anode and cathode, respectively [13]. For PEC operation, an external bias is generally required in order to promote water splitting, which requires a difference in bias of 1.23 V plus potential losses [1]. However, a suitable pairing of two distinct light absorbers, with a photoanode being the top absorber, together an appropriate device configuration can manage the driving force to uphold water splitting reactions [14].

OER at working electrode:

$$\begin{aligned} E^0_{\text{Red}} &= 0V\ vs\ RHE \quad 2H_2O + 4h^+ \rightarrow O_2 + 4H^+ \\ E^0_{\text{Red}} &= 0V\ vs\ RHE; \quad 2H_2O + 4h^+ \rightarrow O_2 + 4H^+. \end{aligned} \tag{1}$$

HER at counter electrode:

$$\begin{aligned} E^0_{ox} &= -1.23V\ vs\ RHE \quad 4H^+ + 4e^- \rightarrow 2H_2 \\ E^0_{ox} &= -1.23V\ vs\ RHE; \end{aligned} \tag{2}$$

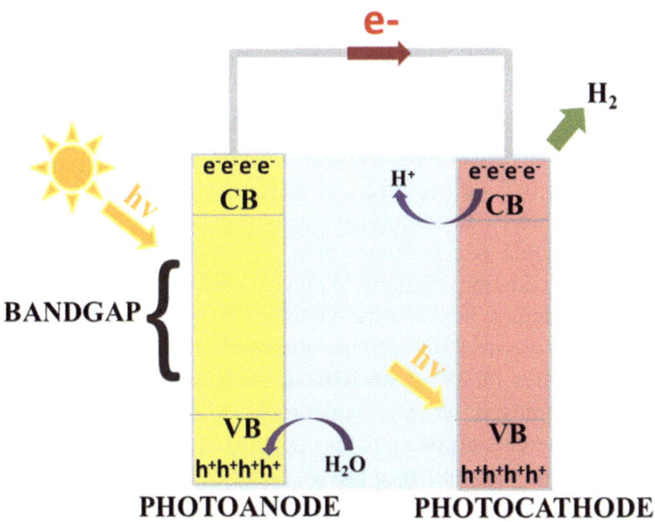

Fig. 1 Tandem cell configuration design. CB meaning conduction band, VB meaning valence band, h^+ holes and e^- electrons

Among unassisted PEC devices, tandem cells are highlighted as the most efficient configuration. In this arrangement, the photoanode is in serie with the photocathode. This way, each one of the photoelectrode creates an electron–hole pair, so photons will be absorbed to create a H2 molecule. Figure 1 shows a standard tandem cell configuration. In this approach, considering the position where the absorbers are placed, solar harvesting is optimized as the energy not absorbed by the first one is utilized by the other one. Thus, tandem cell strikes more of the spectra compared to a single absorber approach or to configurations that one absorber does not harvest the energy left from the other, such as a parallel disposition [13–16].

2 $BiVO_4$ as a Photoanode

Bismuth vanadate, $BiVO_4$, has gained increased attention along the years as a potential photoanode for solar photoelectrochemical water splitting [17–20]. Starting to become famous as a photocatalyst just in the past decade, this metal oxide has emerged as a potential alternative semiconductor mainly due to its high visible light absorption [19, 20] and favorable water splitting band structure.[18] Comparing its performance to other widespread photoanodes regarding the theoretical maximum conversion of solar energy into hydrogen (H2), $BiVO_4$ presents a conversion of 9.1% while WO_3 and TiO_2 present values of 4.8 and 2.2% respectively [1, 17, 21]. Besides that, it is environmental friendly and a low cost material based on abundant elements [1, 17–20]. In such a way, the number of publications using $BiVO_4$ for photocatalysis has increased exponentially over the years, reaching more than 800 publications up to date at Scopus database in contrast to less than 30 before 2010.

$BiVO_4$ can occur in 3 different crystalline phases: orthorrombic, tetragonal zircon and monoclinic scheelite. Monoclinic $BiVO_4$, widely abbreviated as m-$BiVO_4$, is formed in temperatures between 400 and 600 °C, and it is the most photoactive phase with *bandgap* energy of 2.4 eV [20, 22, 23]. This metal oxide presents itself as an n-type semiconductor, which means it is capable of promoting oxidation reactions and therefore acts as a photoanode. In other words, $BiVO_4$ has an adequate valence band position to promote oxidation reactions. While its conduction band is placed at 0.0 V versus RHE and therefore, not negative enough to promote reduction reactions, its valence band at 2.4 V versus RHE is greater than the bias to promote water oxidation reaction, which requires 1.23 V versus RHE. In such a way, m-$BiVO_4$ is a potential photoanode to be employed for water oxidation [1, 17–20].

Despite its high theoretical potential, the popularization of $BiVO_4$ as a photoanode spread when its limiting performance factors could be identified and overcome. Its low *bandgap*, even though it is advantageous regarding light spectrum absorption, also means that valence and conduction bands are close enough to trouble the charges to be separated and transferred to the surface, allowing their recombination. The charge recombination of the electrons and holes, both in bulk or on the surface, block them to act on the redox reactions involved and limit the efficiency of the photocatalyst [1–3, 17–20]. Therefore, studies that brought solutions to improve separation and reduce charge recombination were relevant to make $BiVO_4$ one of the most promising photoanodes. Among disseminated ways to overcome these limitations, strategies such as doping, formation of heterojunctions and deposition of oxygen catalysts are the most diffused and efficient. Besides enhancing charge transfer and separation, those approaches have also showed potential to improve optical, electronic and stability properties [24–26]. At present, many studies applying these techniques have been investigated and here we summarize and highlight some of them, and examine trends and perspectives.

2.1 Doping

Doping with different impurities, in the regular crystal of $BiVO_4$, has been proved to produce dramatic changes in charge separation. Doping provides the formation of several p-n junctions by increasing hole density and producing a more effective internal electric field. Elements such as C, Ni, F, Ce, Zn, Br, and P have been successfully employed as dopants to minimize $BiVO_4$ charge-transfer resistance [27–32]. Investigated by Jo and collaborators (2012), PO4-doped-$BiVO_4$ was first presented and reached a 30-fold enhanced photocurrent and O_2 photocatalytic rate compared to pristine $BiVO_4$ [28]. In the case of halogen dopants, although previously reported for other semiconductor [33, 34], Qin and collaborators (2020) focused for the first time to construct halogen-doped-$BiVO_4$ for photocatalytic purposes. In their work, Br dopant formed an impurity level between the valence and conduction bands that increased photo carriers' separation and boosted the photocatalytic performance of $BiVO_4$, making it a potential photoanode [32].

Even though other elements have been applied, W or Mo dopants, in amounts up to 3% molar, is, presumably, the most mature and disseminated strategy to efficiently modify $BiVO_4$ structure [35–38]. These elements are able to replace V^{5+} with M^{6+} atoms contributing to enlarge positive charge load which reduces electron–hole recombination and leads to higher $BiVO_4$ photoelectrochemical activity.[35,38] In the work of Parmar and collaborators (2012), 12 metal-ion dopants were investigated, among which just W and Mo showed improvement compared to bare $BiVO_4$ for photo water splitting. The W and Mo-doped $BiVO_4$ presented a 3–fourfold reduced charge-transfer resistance and a 5–sixfold enhanced photocurrent [35]. Similar results have been reported by many other studies [36, 37, 39, 40].

Regardless of its consolidation in the literature, this technique has still been investigated [18, 37, 41–44]. Recently, Yin and collaborators (2020) have reported a high porosity Mo-doped-$BiVO_4$ by V re-substitution using simple alkali solution dissolution. Besides an increasing in the carrier density achieved by doping, also an improving in photoelectrochemical performance has been accomplished by V enriched surface [42]. In another new report, Liu and collaborators (2019) have synthetized W-doped $BiVO_4$ single crystals. In this way, the optimization of $BiVO_4$ performance has been performed by coupling two strategies. While the formation of single-crystal optimizes the surface energy and conductivity, doping promotes greater charge separation, enhancing photoelectrochemical and photocatalytic properties of $BiVO_4$ [41]. Doped-BiVO4 is a widespread method and commonly used along other modifications, such as heterojunction formation [45] and co-catalyst deposition [46], to improve its photoactivity.

2.2 *Heterojunction Formation*

The combination of the two or more functional materials to form a new semiconductor is another strategy often explored to overwhelm the limitations of $BiVO_4$. This method, called heterojunction formation, provides materials with greater charge separation and transfer, substantially increasing the photocatalytic and photoelectrochemical efficiency [26, 47, 48]. Coupling materials with BiVO4, being n or p type, introduce different band levels in the structure, which permits better mobility and segregation of electrons and holes [26, 47, 49]. Furthermore, specific advantages can be achieved. For example, adding compounds with lower *bandgap* permits expanding in light absorption [45, 50] and materials with higher photostability can be deposited on the top to form a passivation layer [51, 52]. In this way, heterojunction formation is a potential approach to improve several $BiVO_4$ properties.

Compounds such as carbon nitrides [53–55], $g\text{-}C_3N_4$ and g-CN, and sulfides [56–58], Cu_2S, CdS, Bi_3S_3, have been successfully reported forming heterojunctions with $BiVO_4$. The different band levels between their deposited layers and the $BiVO_4$ ones has been promoted a higher separation of electrons and holes, leading to better photocatalytic activity. However, the combination of two metal oxides is the most popular sort of heterojucntion. Coupling two semiconductor metal oxides also favor the electronic transfer mechanism between the bands of the different materials promoting better separation of photocarriers. Besides that, semiconductors metal oxides have a wide distribution of band positions and low cost and especially high photoelectrochemical stability [59, 60]. In this way, several materials are listed associated with $BiVO_4$, successfully forming heterojunctions to improve $BiVO_4$ performance, such as: BiOI [61], NiO [62], Cu_2O [63], Ag_2O [64], $BiO_{0.67}F_{1.66}$ [65] and many others [66–68].

The combination of $BiVO_4$ with another widespread photoanode semiconductor such as TiO_2 and WO_3 is a popular heterojunction [52, 69–73]. At present, even studies coupling the three of them have been reported [74, 75]. These heterojunctions have been shown to be effective in combining the merits for water splitting of each of the materials. For example, TiO_2 has higher photocatalytic activity and stability, while WO_3 has higher conductivity, and both have greater electronic mobility compared to $BiVO_4$. As one of the outstanding results, reported by Pihosh and collaborators (2015), core–shell $WO_3/BiVO_4$ nanorods have achieved water splitting photocurrent of 6.7 mA at 1.23 V which corresponds to 90% of the theoretically potential for $BiVO_4$ [70].

Recently, another n-type metal oxide has been highlighted to be combined with $BiVO_4$: V_2O_5 [45, 76]. Described by Oliveira and collaborators (2018), the light spectrum absorption of $BiVO_4$ could be extended by V_2O_5, reaching a photocurrent of 7.8 mA, the greatest value reported so far for $BiVO_4$ based materials for water splitting [45]. Heterojunction formation is a widely spread strategy applied not only with doped-$BiVO_4$ but also with co-catalyst top deposition [67, 77].

2.3 *Co-Catalyst Deposition*

Top deposition of co-catalyst has been shown to be a crucial modification to provide high-performance $BiVO_4$ photoanode. The materials composing the deposited layer are both catalysts for oxygen evolution reaction as well as materials that can interact with BiVO4 surface. Co-catalysts act lowering the activation energy of OER, therefore, improving photocatalytic and photoelectrochemical activity of $BiVO_4$ [46, 78]. Moreover, they can also enhance photostability and charge separation as the case of heterojunction due the introduction of different band levels. Further, layer top co-catalyst deposition on $BiVO_4$ has produced significant improvements to enhance surface carrier transfer [1, 7, 12, 26, 78].

The problem of low charge transfer from $BiVO_4$ surface to the electrolyte has been identified by the use of solutions that act as traps to holes, such as hydrogen peroxide, methanol and sulfites. These solutions have extremely high oxidation kinetic which arrest the holes generated at the photoanode surface, therefore, the recombination of surface charges can be neglected when used [1, 46]. To overcome this deficiency of low charge transfer to the electrolyte, deposition of oxygen evolution catalysts, called co-catalysts, on the surface of $BiVO_4$ has produced expressing advances to both lower charge recombination and improve system stability [4, 46, 57, 62, 79–81]. Photodeposition of FeOOH/NiOOH is relatively simple and showed to be very efficient [82, 83]. Studies involving novel materials with top dual layer FeNiOOH on $BiVO_4$ photoanodes are still a trend topic in the literature. Recently, Yaw and collaborators (2020) employing a V_2O_5/rGO/$BiVO_4$/FeOOH/NiOOH heterojunction have achieved a 5.5 fold improved in H2 gas production rate compared to V_2O_5/$BiVO_4$ photoanodes [81].

Zhong and collaborators (2011) described a modification method that suppresses surface recombination using Co-Pi, another co-catalyst that has also been become popular. In their work, a layer of Co-Pi is deposited on the top of BiVO4, acting as a catalyst for oxygen evolution reaction and eliminating almost all the losses due to surface electron–hole recombination. The Co-Pi W-BiVO4 film yields almost double the photocurrent at 1.23 V compared to the photoanode without the surface modification [46]. This work is cited by hundreds others and along the years many studies using $BiVO_4$ modified with Co-Pi were reported [30, 39, 77, 80]. Recently, Fang and collaborators (2020) modified $BiVO_4$ with 2 co-catalysts: NiFeOOH and Co-Pi. While NiFeOOH extends the light absorption spectrum and forces charge separation, the Co-Pi layer efficiently catalyzes the oxygen evolution reaction. This strategy promotes higher active area for oxidation reactions and accelerates the surface charge transfer than when just one of the co-catalysts is applied [80]. Studies that coupled more than one strategy to improve $BiVO_4$ properties have been commonly reported and provided high-performance photoanode to be potential applied in tandem cells.

3 $CuBi_2O_4$ as Photocathode for Water Splitting

Until now, many materials have been fabricated for water reduction as photocathodes such as Cu_2O [84], $CuFeO_2$ [85], $CuGaO_2$ [86], $CuInS_2$ [87], and GaN [88] and noble metals such as Au, Pt, and Pd are currently used in the electrodes [89–91], which are expensive. The major part of these materials suffer from the drawback of poor photo charge separation, faster electron–hole recombination, photocorrosion and nowadays, novel p-type based semiconductor materials have become explored such as strategic photocathodes at photo electrochemical cells due their relative low cost of production and higher photocurrent density [92]. Thus, finding a suitable p-type photocathodic material is a task not so easy. Complementary studies are need on stable p-type materials with good photoelectrical performance.

$CuBi_2O_4$ has attracted attention recently because it is a nontoxic material, inexpensive and abundant p-type semiconductor with the potential to fill this void. Its main difference over binary copper oxides (*e.g.*, Cu_2O and CuO) second metal instead of Cu 3d, which precludes the reduction of Cu^{2+} to Cu^0 by accepting the photo generated electrons, improving photo stability [93]. This ternary metal oxide semiconductor (band gap of 1.6–1.8 eV and a photocurrent onset potential near 1 V versus RHE [10]) could be a good fit for the water splitting application, which was firstly discovered as a potential photocathode material in 2007 through a combinatorial chemistry study by the group of Sayama [94]. CBO is also studied for electrochemical and supercapacitors applications. There are several works on environmental applications, such as nanostructured composites involving $CuBi_2O_4$ composites oxides for photo degradation of dyes and oxidation of acetaldehyde to CO_2, such as other environmental employs [95–102].

Its spinel structure has a maximum predicted theoretical photocurrent of ~ 19.7 $mA.cm^{-2}$ and proper band edge position with conductive band more negative than water reduction potential, and valence band more positive than water oxidation potential, considered as a potential hydrogen producing material in PEC. Several synthesis and/or deposition strategies for CBO have been currently reported such as electrodeposition [103], hydrothermal process [104] and spray pyrolysis [105], some of them require tedious purification steps and complicated immobilization steps for electrode preparation.

The photocurrent's copper and bismuth oxide (as it is named) is still low at about 0.1 mA/cm^2 compared to that of other p-type semiconductors like Cu_2O (7.6 mA/cm^2), low charge carrier diffusion length (~10–50 nm), efficiency loss is due to the improper Ohmic contact which results in poor charge injection into the collecting electrode [106]. Then, efforts have been made to reduce its condition of poor photocurrent density including doping [107], nanostructure design, co-catalyst loading, and heterojunction construction [106, 108–114]. Other change can be the passivation layer, an important strategy for reduction in surface trap density and facilitates photogenerated semiconductor/electrolyte interface [115]. Figure 2a shows the evolution of publications related to the study of CBO, and Fig. 2b shows the different areas of study reporting the use of CBO.

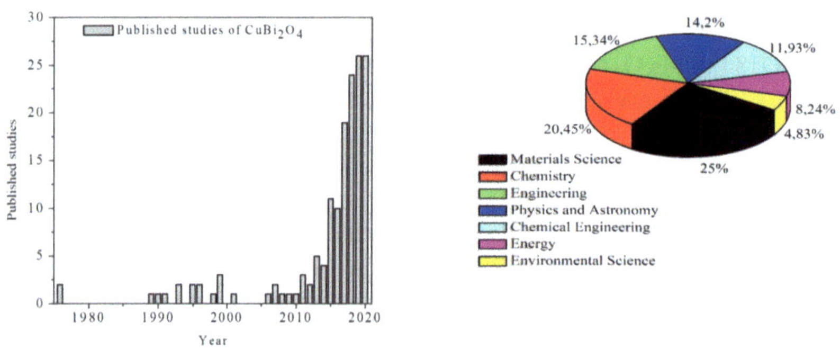

(a) (b)

Fig. 2 **a** Number of published studies on $CuBi_2O_4$ from 1976 to 2020. *Data obtained from the Scopus database;* **b** Perceptual of published studies on $CuBi_2O_4$ separated by research area. *Data obtained from the Scopus database*

3.1 Heterojunction, Passivation Layer and Co-Catalyst Deposition

Some strategies are employed to improve the charge separation of the $CuBi_2O_4$ (or any other semiconductor material). In the heterojunction, basically, an arrangement that allows direct contact between two (or more) semiconductors resulting in a new distribution of charge carriers and a band bending with the depletion region, a similar effect occurs when it puts a solid semiconductor in an aqueous solution. However, both materials must present suitably aligned energy levels and have good interfacial contact to ensure reasonable charge carriers transmission [116, 117]. Figure 3 shows a general scheme of heterojunction and the band alignments.

Some works demonstrate CBO containing cupric oxide (CuO) phase impurities can result to a phase segregation [106, 118]; however, this result cannot be a problem. A study reports the benefits of the phase segregation and its response for photo currents measurements [119]. Decreasing the Bi: Cu ratio, fabricated photocathodes by spin coating (1.97 and 1.38 with the same thickness), resulted in an increase in the photocurrent density (0.82 and 1.17 $mAcm^{-2}$ at 0.58 V_{RHE}, respectively). Once the photocathodes containing different Bi:Cu ratios but relatively the same thickness, it has been observed an increased absorption due to the presence of CuO phase (confirmed by Spectra Raman and XRD patterns). Reflectance diffuse spectra suggest a parasitic light absorption of CuO phase, where photo carriers generated in the particles do not contribute integrally to the PEC performance but present as a hole-selective contact, improving charge collection efficiency.

Combining structural information with PEC performance and band alignment analysis, the CuO particles reveal to be crucial components of the photocathode heterojunction, acting as completely selective nanoscale contacts. Thus, CuO can acts as a hole transport layer for photogenerated charge from the CBO increasing its photo current density. A photoelectrode, resulting from a CuO|CBO heterojunction, together with electrodeposited Pt, was prepared by drop casting and subsequently calcined at 500 °C. This photocathode had a photocurrent of -0.8 $mA.cm^{-2}$ (0.3 M K_2SO_4 at neutral pH). The deposition of Pt on the FTO|CuO|CBO photoelectrode

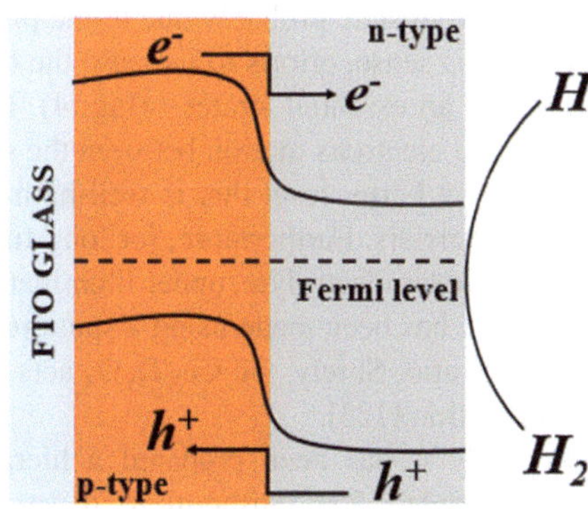

Fig. 3 General scheme of p–n junction

improves its photo instability, reducing photo degradation but it does not prevent the Cu^{2+} photo reduction.

A thin layer of Au deposited on the CBO film resulted in an improvement in the practical photocurrent density from −0.23 to −0.50 $mA.cm^{-2}$ under a potential of 0.1 V versus RHE applied [120]. In this same work, when Pt is added to the CBO and CBO/Au film, there is an increase in photocurrent density to −1.24 $mA.cm^{-2}$. There is no doping effect of Au to the structure of the CBO, and when adding a thin layer of Au, there is a uniformity of the electric field, favouring an increase in the growth of CBO crystals. In addition, there is a greater separation of the photogenerated charges and a consequent decrease in the resistance of the system, decreasing the activation energy in the electrochemical deposition [121].

Fabrication of a nanostructured n-Bi_2O_3/p-$CuBi_2O_4$/p-CuO photocathodes by the electrochemical and chemical techniques. Platelet-like BiOI nanocrystals were electrochemically deposited on FTO substrate [122]. CuI nanoparticles are deposited on the BiOI surface by successive ionic layer adsorption. Further, an oxidative heat treatment of BiOI/CuI heterostructure in air leaded to the formation of the Bi_2O_3/$CuBi_2O_4$/CuO composite[123]. In the surface, a binary oxide was formed such as result of solid-state interaction between bismuth and copper oxides. Spectral sensitization of wide-gap n-Bi_2O_3 (band gap $E_g = 2.80$ eV) with narrow-gap p-$CuBi_2O_4$ ($E_g = 1.80$ eV) and p-CuO ($E_g = 1.45$ eV) extends spectral sensitivity range up to 800 nm. The extension of the spectral range of photocurrent generation up to 800 nm is due to the spectral sensitization of wide band-gap n-Bi_2O_3 substrate (E_g = 2.80 eV) by narrow-gap p-$CuBi_2O_4$ ($E_g = 1.80$ eV) and p-CuO ($E_g = 1.45$ eV). A generous photocurrent quantum efficiency (IPCE = 70% at $\lambda = 400$ nm) occurs due a Z-scheme implementation, where the cathodic photocurrent is associated with the transition of photoelectrons from p-$CuBi_2O_4$ and p-CuO to the solution, and holes photogenerated recombine with electrons of n-Bi_2O_3 of its conduction band.

3.2 *Protection Layer*

It is recurrent for ternary oxides as $CuBi_2O_4$ some limitations related to surface electronic trap state probably due to the phase-segregation or oxygen defects [123, 124]. In this sense, efforts to increase the time-life of PEC devices, the passivation surfaces is an essential strategy (Fig. 4). In this electronic protection, it is necessary a good electrons motion between the semiconductor and the layer-passivation, also needs a Fermi level that is well-harmonized to the photocathode to decrease Schottky barriers. Furthermore, for long-time operating, the layer passivation must be stable under electrolyte, under illumination, and voltage operating conditions. A passivation has been made using a sputtered CBO/$Cu_xTi_yO_z$ photocathode varying the Cu:Ti ratio. Surely, the $Cu_xTi_yO_z$ acts as a passivation layer supressing carrier recombination [123].

Recently it has been prepared a hierarchical nanostructured FTO/CBO/PTh photoelectrode [125]. Furthermore, it was passivated with polythiophene (PTh), a

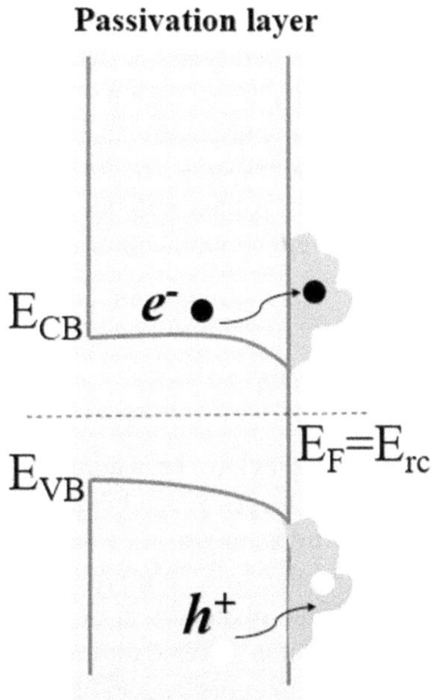

Fig. 4 Band alignments at interface of $CuBi_2O_4$ with p/n-type caused by the passivation layer

conducting polymer of thermal stability. In that electrode design, the porous CBO layers with large surface area alloys better conductivity and the same time, continuous ordered porous structure enhances light absorption. The coverage of PTh could further increase the absorption of light and furnish better stability. Pure CBO photoelectrode exhibits a photocurrent density near to 0.21 mA.cm^{-2} at 0.3 V versus RHE. In contrast, the photocurrent for the FTO/CBO/PTh reached a photocurrent of 0.41 mA cm^2 at 0.3 V (RHE), which has greatly enhanced especially under the correct voltage due to photosensitization of PTh. Moreover, the photoconversion efficiency for CBO is near to ~0.063% at 0.3 V versus RHE, and the CBO/PTh exhibits a higher photo conversion efficiency of 0.12% at 0.3 V versus RHE.

Simultaneously, an example of heterojunction and a protective layer deposition is the (CBO|ZS|P25) photocathode, both deposited by spin-coating, where CBO film modified (this, fabricated by calcination method) by the ZS and P25 semiconductors, exhibiting a photocurrent density of 0.43 mA.cm^{-2} at 0.3 V versus RHE, approximately more twice than of CBO pure [126, 127]. Due to its high electron mobility and protective layer, ZS and P25, respectively, improve the stability and hole extraction capacity of CBO. An amorphous n-TiO_2 layer by spin-coating followed of Pt (by photoelectrochemical reduction) deposited as co-catalyst altered the CBO photocathode [128]. The main reason for use of amorphous TiO_2 is its disordered structure resulting in a larger specific area and more density of active sites [128]. Due to the p-n heterojunction formed and presenting a suitable thickness of TiO_2, the layer promotes a separation of the photogenerated charges. While the $CuBi_2O_4$|Pt photocathode reaches a photocurrent density of 0.22 mA.cm^{-2}, the novel photocathode $CuBi_2O_4$|TiO_2|Pt presents a response of 0.35 mA.cm^{-2}. According to the reflectance diffuse data, the light absorption of the $CuBi_2O_4$ photocathode is not altered after modifying the TiO_2 layer, which it means that TiO_2 layer is solely used to promote the transmission of electrons generated from $CuBi_2O_4$. According to the EIS spectra, which measures the resistance between the photocathode and electrolyte interface, the resistance of the Pt|TiO_2|$CuBi_2O_4$ photocathode is smaller than that of the Pt/$CuBi_2O_4$ photocathode, allowing photogenerated electrons to transfer more efficiently upon modification with TiO_2.

3.3 Doping

Some strategies have been employed in $CuBi_2O_4$ photocathodes to reduce the photogenerated charges recombination. Ag-doped $CuBi_2O_4$ photocathodes have been prepared by adding a trace amount of Ag^+ ion. Ag^+ ions may replace Bi^{3+} ions, increasing the hole concentration in CBO photocathode. This ion exchange has improved the photocurrent density. In addition, it is largely known on the anodic photo corrosion suffered during O_2 reduction due to poor hole motion. Doping with Ag^+ seems to suppress anodic photo corrosion and, according to flat-band potentials, $CuBi_2O_4$ and Ag-doped $CuBi_2O_4$ electrodes prepared are more positive than 1.3 V versus RHE in a 0.1 M NaOH solution (pH 12.8), making these photocathodes highly promising for use in solar hydrogen production [114]. A ternary hybrid structure $CuBi_2O_4$/Au/N, Cu-C has been prepared for water splitting application [115]. N, Cu-codoped carbon layer acts a passivation layers, which decreases the surface charge recombination, and Au nanoparticles can induced between $CuBi_2O_4$ and N, Cu-codoped carbon layer improving the charge transfer. Photocurrent density is near to 0.31 mA cm^{-2} at 0.5 V (versus RHE). The N-CDs nanoparticles have been anchored onto the surfaces of $CuBi_2O_4$. A nitrogen-doped carbon dots (N-CDs) modified $CuBi_2O_4$ photocathode has been prepared using hydrothermal procedure. Results exhibited an improved hydrogen production of N-CDs/$CuBi_2O_4$ in relation to CuBi2O4 and CDs/$CuBi_2O_4$ [129]. The photocatalytic performance of N-CDs/$CuBi_2O_4$ composite can be due to the synergistic effect between N-CDs and $CuBi_2O_4$, which it increases light harvesting and electron transfer capacity.

Copper and bismuth oxide $CuBi_2O_4$ has conduction and valence bands arising from the Cu 3d and O 2p orbitals, respectively, and the most probable defects presented in $CuBi_2O_4$ are copper vacancies. These nature vacancies can be compensated by free holes, shifting the Fermi level closer to the valence band and increases the p-type character of $CuBi_2O_4$. Based on this way, gradient self-doped CBO photocathodes have been prepared by different amounts of Cu and Bi using a diffusion-assisted spray pyrolysis technique. The selfing-doping does not introduce an external dopant element, and thus the tetragonal crystal structure and carrier mobility are maintained. Decreasing the Cu/Bi ratio occurs the formation of vacancies that increase the carrier (hole) population and decrease the Fermi level. In other words, the self-doping through of decreasing of Cu/Bi ratio generates an internal electric field within $CuBi_2O_4$, facilitating the charge separation. Photocathodes containing different amounts of Cu/Bi were made by spray pyrolysis using a two-step diffusion-assisted process. The gradient self-doped CBO photocathodes produce -2.5 mA/cm^{-2} at 0.6 v versus RHE (using H2O2 as an electron scavenger at A.M 1.5) and -1.0 $mA.cm^{-2}$ at 0.0 versus RHE. These photocathodes have presented a charge separation efficiency of 34% (at 550 nm). Also for last, it is deposited a CdS/TiO_2 heterojunction layer on top of the gradient CBO photocathode along with Pt as co-catalyst.

Techniques as drop-casting and spin coating are employed in the flux-mediated one-pot solution process based on the metal–organic decomposition [130]. Precursor solutions with the various concentrations (0.5, 2, 5, and 10 mol % versus $CuBi_2O_4$) of aliovalent Li (I) flux produced were previously prepared. There is a comparison between the two deposition methods. The spin-coating method seems to produce majority secondary phase as compared to a drop-casting method. Interestingly there is a modification in the morphology for samples containing different percentages of aliovalent Li (I) varying from roughly spherical particles to a plate-like morphology probably due to a topotactic growth of the spinel crystal by the flux [130]. However, these photocathodes presented a photo current density close to -0.13 $mA.cm^{-2}$.

4 $BiVO_4$ and $CuBi_2O_4$ Cells

As $BiVO_4$ has been shown to be a potential photoanode and $CuBi_2O_4$ a promising photocathode, the combination of both photoelectrodes in a dual absorber solar tandem cell has emerged as a suitable choice. Due the highest *bandgap* of $BiVO_4$, it can act as the top absorber while $CuBi_2O_4$ can harvest the energy left. Even though this association is promising, the construction of tandem cells is highly challenging and therefore, rarely reported. Recently and first reported by Song and collaborators (2020), a tandem cell composed by W:$BiVO_4$/CoPi photoanode and a $CuBi_2O_4/CdS/TiO_2$/RuOx photocathode has been constructed and investigated regarding H2 production. The study has shown several challenges faced for the cell development such as photocorrosion and internal photo voltage, but highlight considerable factors for the design of the tandem cells composed by the 2 photoelectrodes [131].

References

1. Lianos, P.: Appl. Catal. B **210**, 235–254 (2017)
2. Jiang, C., Moniz, S.J.A., Wang, A., Zhang, T., Tang, J.: Chem. Soc. Rev. **46**, 4645–4660 (2017)
3. Abe, R.: J. Photochem. Photobiol., C **11**, 179–209 (2010)
4. Maeda, K.: J. Photochem. Photobiol., C **12**, 237–268 (2011)
5. Yerga, R.M.N., Galván, M.C.A., Del Valle, F., de la Mano, J.A.V., Fierro, J.L.G.: ChemSusChem: Chem. Sustain. Energy Mater. **2**, 471–485 (2009)

6. Li, J., Wu, N.: Catal. Sci. Technol. **5**, 1360–1384 (2015)
7. Du, H., Liu, Y.-N., Shen, C.-C., Xu, A.-W.: Chin. J. Catal. **38**, 1295–1306 (2017)
8. Bott, A.W.: Curr. Sep. **17**, 87–92 (1998)
9. Shockley, W., Queisser, H.J.: J. Appl. Phys. **32**, 510–519 (1961)
10. Mills, A., Le Hunte, S.: J. Photochem. Photobiol., A **108**, 1–35 (1997)
11. Chen, S., Takata, T., Domen, K.: Nat. Rev. Mater. **2**, 1–17 (2017)
12. Takanabe, K.: ACS Catal. **7**, 8006–8022 (2017)
13. Prévot, M.S., Sivula, K.: J. Phys. Chem. C **117**, 17879–17893 (2013)
14. Sivula, K.: Chimia **67**, 155–161 (2013)
15. Zhang, K., Ma, M., Li, P., Wang, D.H., Park, J.H.: Adv. Energy Mater. **6**, 1–16 (2016)
16. Brillet, J., Yum, J.H., Cornuz, M., Hisatomi, T., Solarska, R., Augustynski, J., Graetzel, M., Sivula, K.: Nat. Photonics **6**, 824–828 (2012)
17. Kim, J.H., Lee, J.S.: Energy Environ. Focus **3**, 339–353 (2014)
18. Kim, J.H., Lee, J.S.: Adv. Mater. **31**, 1–35 (2019)
19. Tayebi, M., Lee, B.K.: Renew. Sustain. Energy Rev. **111**, 332–343 (2019)
20. Suarez, C.M., Hernández, S., Russo, N.: Appl. Catal. A Gen. **504**, 158–170 (2015)
21. Zhang, Y., Xie, Y., Li, J., Bai, T., Wang. J.: J. Solgel. Sci. Technol. **71**, 38–42 (2014)
22. Pookmanee, P., Kojinok, S., Puntharod, R., Sangsrichan, S., Phanichphant, S.: Ferroelectrics **456**, 45–54 (2013)
23. Chen, S., Huang, D., Xu, P., Gong, X., Xue, W., Lei, L., Deng, R., Li, J., Li, Z.: ACS Catal. **10**, 1024–1059 (2019)
24. Moniz, S.J.A., Shevlin, S.A., Martin, D.J., Guo, Z.X., Tang, J.: Energy Environ. Sci. **8**, 731–759 (2015)
25. Ye, K.H., Li, H., Huang, D., Xiao, S., Qiu, W., Li, M., Hu, Y., Mai, W., Ji, H., Yang, S.: Nat. Commun. **10**, 1–9 (2019)
26. Malathi, A., Madhavan, J., Ashokkumar, M., Arunachalam, P.: Appl. Catal. A **555**, 47–74 (2018)
27. Yin, C., Zhu, S., Chen, Z., Zhang, W., Gu, J., Zhang, D.: J. Mater. Chem. A **1**, 8367–8378 (2013)
28. Jo, W.J., Jang, J.-W., Kong, K., Kang, H.J., Kim, J.Y., Jun, H., Parmar, K.P.S., Lee, J.S.: Angew. Chem. **124**, 3201–3205 (2012)
29. Regmi, C., Kshetri, Y.K., Kim, T.H., Pandey, R.P., Ray, S.K., Lee, S.W.: Appl. Surf. Sci. **413**, 253–265 (2017)
30. Pan, Q., Yang, K., Wang, G., Li, D., Sun, J., Yang, B., Zou, Z., Hu, W., Wen, K., Yang, H.: Chem. Eng. J. **372**, 399–407 (2019)
31. Abbood, H.A., Alabdie, A., Al-Hawash, A., Abbood, A.A., Huang, K.: J. Nanopart. Res. https://doi.org/10.1007/s11051-020-04792-z
32. Qin, C., Liao, H., Rao, F., Zhong, J., Li, J.: Solid State Sci. 106285 (2020)
33. Dumrongrojthanath, P., Phuruangrat, A., Thongtem, S., Thongtem, T.: Mater. Lett. **209**, 501–504 (2017)
34. Wang, Q., Zhu, S., Liang, Y., Cui, Z., Yang, X., Liang, C., Inoue, A.: Mater. Res. Bull. **86**, 248–256 (2017)
35. Parmar, K.P.S., Kang, H.J., Bist, A., Dua, P., Jang, J.S., Lee, J.S.: Chemsuschem **5**, 1926–1934 (2012)
36. Pattengale, B., Huang, J.: Phys. Chem. Chem. Phys. **18**, 32820–32825 (2016)
37 Iwase, A., Nozawa, S., Adachi, S.I., Kudo, A.: J. Photochem. Photobiol., A **353**, 284–291 (2018)
38. Huang, Z.F., Pan, L., Zou, J.J., Zhang, X., Wang, L.: Nanoscale **6**, 14044–14063 (2014)
39 Abdi, F.F., Firet, N., van de Krol, R.: ChemCatChem **5**, 490–496 (2013)
40. Shan, L., Liu, H., Wang, G.: J. Nanopart. Res. https://doi.org/10.1007/s11051-015-2996-3
41. Liu, K., Zhou, H., Wang, P., Fang, J., Li, H., Cao, M., Gao, X., Qiu, X., Liu, M.: Catal. Today **335**, 511–519 (2019)
42. Yin, X., Qiu, W., Li, W., Li, C., Wang, K., Yang, X., Du, L., Liu, Y., Li, J.: Chem. Eng. J. **389**, 124365 (2020)

43. Zhao, X., Hu, J., Yao, X., Chen, S., Chen, Z.: ACS Appl. Energy Mater. **1**, 3410–3419 (2018)
44. Zhao, L., Wei, J., Li, Y., Han, C., Pan, L., Liu, Z.: J. Mater. Sci.: Mater. Electron. **30**, 21425–21434 (2019)
45. Oliveira, A.T, Rodriguez, M., Andrade, T.S., de Souza, H.E.A., Ardisson, J.D., Oliveira, H.S., Oliveira, L.C.A, Lorençon, E., Silva, A.C., Nascimento, L.L., Patrocínio, A.O.T., Pereira, M.C.: Solar RRL. https://doi.org/10.1002/solr.201800089
46. Zhong, D.K., Choi, S., Gamelin, D.R.: J. Am. Chem. Soc. **133**, 18370–18377 (2011)
47. Tan, H.L., Amal, R., Ng, Y.H.: J. Mater. Chem. A **5**, 16498–16521 (2017)
48. Marschall, R.: Adv. Func. Mater. **24**, 2421–2440 (2014)
49. Lopes, O.F., Carvalho, K.T.G., Nogueira, A.E., Avansi, W., Ribeiro, C.: Appl. Catal. B **188**, 87–97 (2016)
50. Jiang, J., Wang, M., Li, R., Ma, L., Guo, L.: Int. J. Hydrogen Energy **38**, 13069–13076 (2013)
51. Xie, J., Guo, C., Yang, P., Wang, X., Liu, D., Li, C.M.: Nano Energy **31**, 28–36 (2017)
52. Kim, M.W., Kim, K., Ohm, T.Y., Yoon, H., Joshi, B., Samuel, E., Swihart, M.T., Choi, S.K., Park, H., Yoon, S.S.: Chem. Eng. J. **333**, 721–729 (2018)
53. Mohamed, N.A., Safaei, J., Ismail, A.F., Khalid, M.N., Jailani, M.F.A.M., Noh, M.F.M., Arzaee, N.A., Zhou, D., Sagu, J.S., Teridi, M.A.M.: Mater. Res. Bull. **125**, 110779 (2020)
54. Babu, P., Mohanty, S., Naik, B., Parida, K.: Inorg. Chem. **58**, 12480–12491 (2019)
55. Ahmed, T., Ammar, M., Saleem, A., Zhang, H., Xu, H.: RSC Adv. **10**, 3192–3202 (2020)
56. Chen, H.-Q., Lin, L.-Y., Chen, S.-L.: ACS Appl. Energy Mater. **1**, 6089–6100 (2018)
57. Bai, S., Li, Q., Han, J., Yang, X., Shu, X., Sun, J., Sun, L., Luo, R., Li, D., Chen, A.: Int. J. Hydrogen Energy **44**, 24642–24652 (2019)
58. He, B., Wang, Y., Liu, X., Li, Y., Hu, X., Huang, J., Yu, Y., Shu, Z., Li, Z., Zhao, Y.: J. Mater. Chem. A **7**, 6747–6752 (2019)
59. Yang, Y., Niu, S., Han, D., Liu, T., Wang, G., Li, Y.: Adv. Energy Mater. **7**, 1–26 (2017)
60. Gholipour, M.R., Dinh, C.T., Béland, F., Do, T.O.: Nanoscale **7**, 8187–8208 (2015)
61. Ye, K.H., Chai, Z., Gu, J., Yu, X., Zhao, C., Zhang, Y., Mai, W.: Nano Energy **18**, 222–231 (2015)
62. Zhong, M., Hisatomi, T., Kuang, Y., Zhao, J., Liu, M., Iwase, A., Jia, Q., Nishiyama, H., Minegishi, T., Nakabayashi, M., Shibata, N., Niishiro, R., Katayama, C., Shibano, H., Katayama, M., Kudo, A., Yamada, T., Domen, K.: J. Am. Chem. Soc. **137**, 5053–5060 (2015)
63. Bai, S., Liu, J., Cui, M., Luo, R., He, J., Chen, A.: Dalton Trans. **47**, 6763–6771 (2018)
64. Wu, C., Fang, Y., Tirusew, A.H., Xiang, M., Huang, Y., Chen, C.: Cuihua Xuebao/Chin. J. Catal. **38**, 192–198 (2017)
65. Feng, X., Zhao, X., Chen, L.: J. Nanopart. Res. **21**, 61 (2019)
66. Pilli, S.K., Deutsch, T.G., Furtak, T.E., Brown, L.D., Turner, J.A., Herring, A.M.: Phys. Chem. Chem. Phys. **15**, 3273–3278 (2013)
67. Chang, X., Wang, T., Zhang, P., Zhang, J., Li, A., Gong, J.: J. Am. Chem. Soc. https://doi.org/10.1021/jacs.5b04186
68. He, H., Zhou, Y., Ke, G., Zhong, X., Yang, M., Bian, L., Lv, K., Dong, F.: Electrochim. Acta **257**, 181–191 (2017)
69. Kumbhar, V.S., Lee, H., Lee, J., Lee, K.: J. Colloid Interface Sci. **557**, 478–487 (2019)
70. Pihosh, Y., Turkevych, I., Mawatari, K., Uemura, J., Kazoe, Y., Kosar, S., Makita, K., Sugaya, T., Matsui, T., Fujita, D., Tosa, M., Kondo, M., Kitamori, T.: Sci. Rep. **5**, 1–2 (2015)
71. Rao, P.M., Cai, L., Liu, C., Cho, I.S., Lee, C.H., Weisse, J.M., Yang, P., Zheng, X.: Nano Lett. **14**, 1099–1105 (2014)
72. Resasco, J., Zhang, H., Kornienko, N., Becknell, N., Lee, H., Guo, J., Briseno, A.L., Yang, P.: ACS Central Sci. **2**, 80–88 (2016)
73. Kalanoor, B.S., Seo, H., Kalanur, S.S.: Mater. Sci. Energy Technol. **1**, 49–62 (2018)
74. Yang, Y., Cheng, Y.F.: Corros. Sci. **164**, 108333 (2020)
75. Kalanur, S.S., Yoo, I.H., Park, J., Seo, H.: J. Mater. Chem. A **5**, 1455–1461 (2017)
76. Yaw, C.S., Ruan, Q., Tang, J., Soh, A.K., Chong, M.N.: Chem. Eng. J. **364**, 177–185 (2019)
77. Kim, J.H., Magesh, G., Kang, H.J., Banu, M., Kim, J.H., Lee, J., Lee, J.S.: Nano Energy **15**, 153–163 (2015)

78. Li, Y.H., Peng, C., Yang, S., Wang, H.F., Yang, H.G.: J. Catal. **330**, 120–128 (2015)
79. Arunachalam, M., Ahn, K.S., Kang, S.H.: Int. J. Hydrogen Energy **44**, 4656–4663 (2019)
80. Fang, G., Liu, Z., Han, C.: Appl. Surf. Sci. 146095 (2020)
81. Yaw, C.S., Tang, J., Soh, A.K., Chong, M.N.: Chem. Eng. J. **380**, 122501 (2020)
82. Li, L., Li, J., Bai, J., Zeng, Q., Xia, L., Zhang, Y., Chen, S., Xu, Q., Zhou, B.: Nanoscale **10**, 18378–18386 (2018)
83. Kuang, Y., Jia, Q., Nishiyama, H., Yamada, T., Kudo, A., Domen, K.: Adv. Energy Mater. **6**, 2–8 (2016)
84. Lu, X., Hart, J.N., Yao, Y., Toe, C.Y., Scott, J., Ng, Y.H.: Mater. Today Energy. https://doi.org/10.1016/j.mtener.2020.100422
85. Ferri, M, Elliott, J., Fabris, S., Piccinin, S.: Phys. Rev. B. https://doi.org/10.1103/PhysRevB.101.155201
86. Herraiz-Cardona, I., Fabregat-Santiago, F., Renaud, A., Julián-López, B., Odobel, F., Cario, L., Jobic, S., Giménez, S.: Electrochim. Acta. https://doi.org/10.1016/j.electacta.2013.09.129
87. Li, M., Chen, L., Su, Y., Yin, H., Hu, K.: J. Energy Chem. https://doi.org/10.1016/j.jechem.2020.03.070
88. Al Muyeed, S.A., Wei, X., Borovac, D., Song, R., Tansu, N., Wierer, J.J.: J. Cryst. Growth. https://doi.org/10.1016/j.jcrysgro.2020.125652
89. Babu, B., Koutavarapu, R., Shim, J., Yoo, K.: Sep. Purif. Technol. https://doi.org/10.1016/j.seppur.2020.116652
90. Li, L., Feng, K., Huang, D., Wang, K., Li, Y., Guo, Z., Ng, Y.H., Jiang, F.: Chem. Eng. J. https://doi.org/10.1016/j.cej.2020.125264
91. Shen, Z., Zhang, Y., Zhou, C., Bai, J., Chen, S., Li, J., Wang, J., Guan, X., Rahim, M., Zhou, B.: Water Res. https://doi.org/10.1016/j.watres.2019.115357
92. Thakur, A., Ghosh, D., Devi, P., Kim, K.H., Kumar, P.: Chem. Eng. J. (2020)
93. Gawande, M.B., Goswami, A., Felpin, F.X., Asefa, T., Huang, X., Silva, R., Zou, X., Zboril, R., Varma, R.S.: Chem. Rev. (2016)
94. Arai, T., Yanagida, M., Konishi, Y., Iwasaki, Y., Sugihara, H., Sayama, K.: J. Phys. Chem. C **111**, 7574–7577 (2007)
95. Klyndyuk, A.I., Petrov, G.S., Bashkirov, L.A., Akimov, A.I., Poluyan, A.F.: ChemInform **30**, (2010)
96. Ensafi, A.A., Ahmadi, N., Rezaei, B.: J. Alloy. Compd. **652**, 39–47 (2015)
97. Wang, F., Yang, H., Zhang, Y., Li, R., Wang, X.: Mater. Technol. https://doi.org/10.1080/10667857.2017.1317063
98. Deng, Y., Chen, Y., Chen, B., Ma, J.: J. Alloys Compd. **559**, 116–122 (2013)
99. Elaziouti, A., Laouedj, N., Bekka, A., Vannier, R.N.: J. King Saud Univ. Sci. **27**, 120–135 (2015)
100. Moniz, S.J.A., Shevlin, S.A., Martins, D.J., Guo, Z. Tang, J.: Energy. Environ. Sci. **8**, 731–759 (2015)
101. Chen, M., Yang, Q., Li, L., Liu, M., Xiao, P., Zhang, M.: Mater. Lett. **171**, 255–258 (2016)
102. Da Oh, W., Lua, S.-K., Dong, Z., Lim, T.T.: Nanoscale **8**, 2046–2054 (2016)
103. Zhao, L., Wang, X., Liu, Z.: Appl. Phys. A: Mater. Sci. Process. **124**, (2018)
104. Xie, Y., Zhang, Y., Yang, G., Liu, C., Wang, J.: Mater. Lett. **107**, 291–294 (2013)
105. Wang, F., Chemseddine, A., Abdi, F., van de Krol, R., Berglund, S.: J. Mater. Chem. A **5**, 12838–12847 (2017)
106. Berglund, S.P., Abdi, F.F., Bogdanoff, P., Chemseddine, A., Friedrich, D., Van De Krol, R.: Chem. Mater. **28**, 4231–4242 (2016)
107. Ma, C., Ma, D.K., Yu, W., Chen, W., Huang, S.: Appl. Surf. Sci. https://doi.org/10.1016/j.apsusc.2019.03.172
108. Patil, R., Kelkar, S., Naphade, R., Ogale, S.: J. Mater. Chem. A **2**, 3661–3668 (2014)
109. Hahn, N.T., Holmberg, V.C., Korgel, B.A., Mullins, C.B.: J. Phys. Chem. C **116**, 6459–6466 (2012)
110. Park, H.S., Lee, C.-Y., Reisner, E.: Phys. Chem. Chem. Phys. **16**, 22462–22465 (2014)

111. Sharma, G., Zhao, Z., Sarker, P., Nail, B.A., Wang, J., Huda, M., Osterloh, F.: J. Mater. Chem. A **4**, 2936–2942 (2015)
112. Cao, D., Nasori, N., Wang, Z., Mi, Y., Wen, L., Yang, Y., Qu, S., Wang, Z., Lei, Y.: J. Mater. Chem. A **4**, 8995–9001 (2016)
113. Berglund, S.P., Lee, C., Nu, P.D., Bard, J., Mullins, C.B.: 4554–4565 (2013)
114. Kang, D., Hill, J.C., Park, Y., Choi, K.S.: Chem. Mater. **28**, 4331–4340 (2016)
115. Xu, N., Li, F., Gao, L., Hu, H., Hu, Y., Long, X., Ma, J., Jin, J.: ACS Sustain. Chem. Eng. **6**, 7257–7264 (2018)
116. Hong, S.J., Lee, S., Jang, J.S., Lee, J.S.: Energy Environ. Sci. https://doi.org/10.1039/c0ee00743a
117. Saito, R., Miseki, Y., Sayama, K.: Chem. Commun. https://doi.org/10.1039/c2cc30713h
118. Wang, F., Septina, W., Chemseddine, A., Abdi, F.F., Friedrich, D., Bogdanoff, P., Van De Krol, R., Tilley, S.D., Berglund, S.P.: J. Am. Chem. Soc. https://doi.org/10.1021/jacs.7b07847
119. Zhang, Z., Lindley, S.A., Dhall, R., Bustillo, K., Han, W., Xie, E., Cooper, J.K.: ACS Appl. Energy Mater. **2**, 4111–4117 (2019)
120. Cao, D., Nasori, N., Wang, Z., Mi, Y., Wen, L., Yang, Y., Qu, S., Wang, Z., Lei, Y.: J. Mater. Chem. A **00**, 1–7 (2016)
121. Clavero, C.: Nat. Photonics. https://doi.org/10.1038/nphoton.2013.238
122. Malashchonak, M., Streltsov, E., Mazanik, A., Korolik, O., Kulak, A., Puzikova, D., Dergacheva, M., Selskis, A.: J. Solid State Electrochem. https://doi.org/10.1007/s10008-020-04494-5
123. Zhang, Z., Lindley, S.A., Guevarra, D., Kan, K., Shinde, A., Gregoire, J.M., Han, W., Xie, E., Haber, J.A., Cooper, J.K.: Adv. Funct. Mater. https://doi.org/10.1002/adfm.202000948
124. Cooper, J.K., Scott, S.B., Ling, Y., Yang, J., Hao, S., Li, Y., Toma, F.M., Stutzmann, M., Lakshmi, K.V., Sharp, I.D.: Chem. Mater. https://doi.org/10.1021/acs.chemmater.6b01994
125. Xu, N., Li, F., Gao, L., Hu, H., Hu, Y., Long, X., Ma, J., Jin, J.: Int. J. Hydrogen Energy 2064–2072 (2018)
126. Wei, S., Xu, N., Li, F., Long, X., Hu, Y., Gao, L., Wang, C., Li, S., Ma, J., Jin, J.: ChemElectroChem. https://doi.org/10.1002/celc.201900714
127. Zhu, X., Guan, Z., Wang, P., Zhang, Q., Dai, Y., Huang, B.: Cuihua Xuebao/Chin. J. Catal. **39**, 1704–1710 (2018)
128. Zywitzki, D., Jing, H., Tüysüz, H., Chan, C.K.: J. Mater. Chem. A. https://doi.org/10.1039/c7ta01614j
129. Shi, W., Guo, F., Li, M., Shi, Y., Wu, M., Tang, Y.: J. Alloy. Compd. **775**, 511–517 (2019)
130. Choi, Y.H., Yang, K.D., Kim, D.H., Nam, K.T., Hong, S.H.: Mater. Lett. **188**, 192–196 (2017)
131. Song, A., Bogdanoff, P., Esau, A., Ahmet, I.Y., Levine, I., Dittrich, T., Unold, T., Van De Krol, R., Berglund, S.P.: ACS Appl. Mater. Interfaces **12**, 13959–13970 (2020)

Photodynamic Therapy: Use of Nanocarrier Systems to Improve Its Effectiveness

Bruno Henrique Vilsinski, Renato Sonchini Gonçalves, Wilker Caetano, Paulo Ricardo de Souza, Ariel Colaco de Oliveira, Yasmin S. Gomes, Adriana P. Gerola, Alessandro Francisco Martins, Artur J. M. Valente, and Edvani Curti Muniz

Abstract Photodynamic Therapy (PDT) is a medical modality that has been applied against several types of cancer, macular degeneration, pointed condyloma, actinic keratosis as well as infections caused by fungi, viruses, and bacteria. When PDT is applied against microorganisms, the technique is called as antimicrobial photodynamic therapy (aPDT). PDT/aPDT principle involves the association of a light source (performed by a LASER, LED, and optical fiber), a non-toxic photosensitizer (PS), and molecular oxygen dissolved in the tissue of interest. The photosensitizer is excited by a light source of a specific wavelength which, in the presence of oxygen, generate high-cytotoxic reactive oxygen species (ROS) as well as superoxide anion ($O_2^{\cdot -}$), hydroxyl radical ($HO^{\cdot}$), hydrogen peroxide (H_2O_2) and singlet oxygen ($^1O_2$). These species cause damage to tumor cells and vasculatures by apoptosis, necrosis, and activating the immune responses. Among the main advantages of PDT

B. H. Vilsinski (✉) · P. R. de Souza · A. C. de Oliveira · E. C. Muniz
Grupo de Materiais Poliméricos e Compósitos (GMPC), Departamento de Química, Universidade Estadual de Maringá (UEM), Av. Colombo, 5790, Maringá, PR 87020-900, Brazil

B. H. Vilsinski · R. S. Gonçalves · W. Caetano · E. C. Muniz
Núcleo de Pesquisas em Sistemas Fotodinâmicos (NUPESF), Departamento de Química, Universidade Estadual de Maringá (UEM), Av. Colombo, 5790, Maringá, PR 87020-900, Brazil

B. H. Vilsinski · A. J. M. Valente
Department of Chemistry, University of Coimbra, CQC, 3004-535 Coimbra, Portugal

Y. S. Gomes · A. P. Gerola
INCT of Catalysis in Molecular and Nanostructured Systems, Federal University of Santa Catarina—UFSC, Florianópolis, SC 88040-900, Brazil

A. F. Martins
Laboratory of Materials, Macromolecules, and Composites (LaMMAC), Federal University of Technology (UTFPR), Apucarana, PR 86812-460, Brazil

E. C. Muniz
Universidade Tecnológica Federal do Paraná (UTFPR-LD), PPGCEM, Campus Londrina, Avenida dos Pioneiros, 3131, Londrina, PR 86036-370, Brazil

Departamento de Química, Universidade Federal do Piauí (UFPI), Campus Universitário Petrônio Portella, Bairro Ininga, Teresina 64049-550, Brazil

F. A. La Porta and C. A. Taft (eds.), *Functional Properties of Advanced Engineering Materials and Biomolecules*, Engineering Materials,
https://doi.org/10.1007/978-3-030-62226-8_10

is the specificity. This is guaranteed by the preferential accumulation of the photosensitizer in the cells of interest and the targeting of the lighting system without compromising healthy tissues. Additionally, it is a cheaper and less invasive than most known treatment, such as surgery, chemotherapy, and radiotherapy. Several classes of photosensitizers have been proposed for application in PDT treatment. It is necessary to mention phthalocyanines, porphyrins, bacteriochlorins, chlorines, chlorophyll-based compounds, phenothiazinium salts, and xanthene dyes. Generally, better PS compounds are hydrophobic once they accumulate in the interest tumors most effectively. However, the direct application of these in the body is harmful once PS can precipitate in the body, forming aggregates. Additionally, the pre-solubilization in an organic solvent before the application is not recommended once these are high toxicity in the cell. In this way, strategies have been proposed to solubilize hydrophobic PS in aqueous solutions and increase their biocompatibility. One of the most successful is the incorporation of PS in nanocarrier systems such as liposomes, copolymeric micelles, cyclodextrins, gold nanoparticles, microemulsions, self-assembled peptide-based nanomaterials, and others. Each nanocarrier has this specificity in order of its vantages and advantages. The main objective of this chapter is the description of PDT principles and understands more about formulations that have been used for PDT treatment.

Keywords Photodynamic therapy · Antimicrobial photodynamic therapy · Drug delivery systems · Photosensitizer

1 Principle

Photodynamic therapy (PDT) is a minimally invasive and emerging medical technique that has been used for the treat of a variety of diseases, such as skin diseases (condyloma acuminate, actinic keratosis, etc.), several types of cancer, atherosclerosis, macular degeneration, wound-healing, and others. Antimicrobial photodynamic therapy (aPDT) is the technique denomination when PDT is used in the inactivation of microorganisms against infections [1].

PDT principle involves the combination of a photosensitizer, light, and molecular oxygen or other substrates in the local treatment. Firstly, the PS compound accumulates in the injured tissues. The mechanism by which PS localizes selectively in the interest cells is not understood. However, it is probably due to the affinity for proliferating endothelium coupled with vascular permeability. Another possibility is the PS incorporation in the tumoral cell via receptor-mediated endocytosis of low-density proteins (LDL), primarily expressed by cancerous cells.

Afterward, PS is activated by a proper light source. The light sources used includes LEDs, LASERs, optical fibers, and near infra-red (NIR) lights. The majority of PS compounds are stimulated by a light source with wavelength comprised from 630 to 700 nm, which is the region of greater light penetration depth. The red light source, for example, penetrates in depths greater than 2 mm and is commonly used to treat thicker lesions [2]. However, PS compounds that are administered in the skin can be excited with a light source of lower wavelength and skin penetration [3]. LASERS have the advantage of allowing the medical professional to modulate the irradiance and administrate a monochromatic light. However, there are significantly more expensive than lamps and LEDs, in addition to requiring individual expertise [2]. NIR lights have the advantage of aiming to work in a deeper irradiation depth, and highly reactive [4]. Conventional lamps have been tested for PDT treatment. However, they present some disadvantages as well as low light intensity, significant thermal effect, and difficulty in controlling the light dose. However, most of these drawbacks can be solved with engineering solutions [5].

Upon activation, PS is promoted from its ground state (^{1}PS) to the first excited singlet state (^{1}PS*) (Fig. 1). Once in this state, PS can decay again to the ground state in two different forms: (I) return to its ground state by light emission or (II) by the non-radioactive process (internal conversion). Another possibility is the occurrence of an intersystem crossing process. It is characterized by the PS transition to its longer-lived triplet state (^{3}PS*). The PS in the triplet excited state can then decay to the PS_0 ground state emitting light by phosphorescence or by non-radioactive processes.

Additionally, PS molecules in the triplet excited state can transfer energy to molecular oxygen ($^3\sum_g{}^- O_2$) to form singlet oxygen ($^1\Delta_g\ ^1O_2$), the lowest excited state of oxygen [6]. The $^1O_2$ can kill tumoral cells directly or damage tumor-associated vasculature. PDT also can activate an immune response against the tumor. This

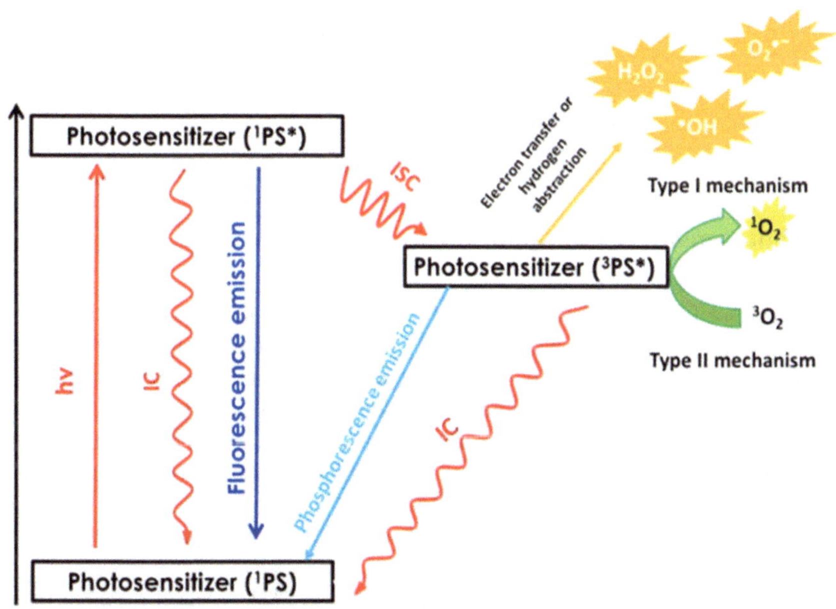

Fig. 1 Schematic illustration of the process that occurs after PS activation by a proper light source

mechanism is known as type II PDT. The oxygen singlet lifetime in the intracellular environment is 3 μs in a viable metabolically active cells. This leads to a diffusion distance of varying from 2 to 4 × 10^{-6} $cm^2.s^{-1}$ [7]. This short diffusion and lifetime of singlet oxygen make PDT a selective treatment of cancerous cells. In the type I mechanism, on the other hand, there is an electron or proton transfer from PS compound to adjacent biological substrates, forming radicals. These radicals act in photodamage of the cells or react with molecular oxygen in order to produce ROS, such as superoxide anion ($O_2^{\cdot-}$), hydroxyl radical ($OH^{\cdot}$), and hydrogen peroxide (H2O2) that causes the cellular death.

The occurrence of type I or type II mechanism depends on several factors, such as oxygen concentration, photosensitizer localization, concentration, and medium polarity [6]. For example, type I reactions are more predominant in conditions of low oxygen concentration and high PS concentration.

The advantages of photodynamic therapy include its good target selectivity, no resistance to multiple treatments, fewer harmful effects, and a cheaper technique. Additionally, PDT can be applied in regions with minimal medical instruments, is a technique increasingly used in the poorest countries around the world [8]. Moreover, aPDT is also interesting, once the increase in microbial resistance to antibiotics and other drugs is a crescent medical problem [9]. The main disadvantages include high systemic toxicity and a prolonged period of skin photosensitivity of some PS compounds. Also, it is known that pain during the irradiation is a common adverse effect during the treatment. However, these can be controlled using other optimized PS compounds or using short incubation times [2].

In addition, PS compounds can act as imaging agents in order to better tumor visualization/diagnosis. In this sense, PS can be theranostic agents against several diseases. Indocyanine green, for example, is a US-FDA approved PS that has been used in the diagnostic imaging process as a fluorescent contrast for recognization of several tumors.

2 Photosensitizer Compounds

Porphyrin and its analogs are the most useful PS classes for use in PDT. Porphyrins are characterized as an 18–22 π electron macrocycle where methine bridges linked four pyrroles ring (Fig. 2) [10]. Phthalocyanines consisted of four symmetrical macrocycles rings disposed of four isoindole units linked by nitrogen atoms (Fig. 2). These compounds have strong absorption bands around 400 nm (Soret band), and absorption bands in the region comprised from 500 to 800 nm, called Q-bands. The higher molar absorption coefficient ($\mathcal{E} \approx 5 \times 10^5\ M^{-1}\ cm^{-1}$) in the Soret band is not enough to apply PDT in deeper lesions. On the other hand, light absorption in Q bands is interesting for this purpose.

The history of photosensitizer compounds used in PDT is based on three main generations of PSs. In the first-generation of PSs, there are Photofrin® and hematoporphyrin derivatives (HPD). Photofrin® is used against early-and-late- stages of

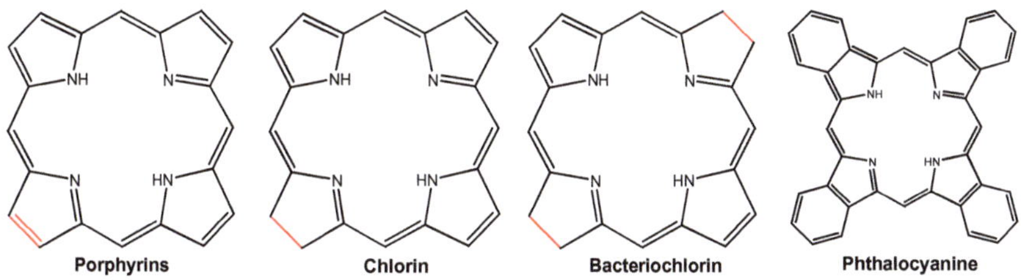

Fig. 2 Molecular structure of porphyrin and its derivatives

lung, bladder, and esophageal cancer, such as malignant, non-malignant, and early-stages of cervical cancer. It consists of a mixture of dimers and oligomers of hemato-porphyrins. In these compounds, the porphyrin unities are linked by ester, ether, and C–C bonds. The main disadvantages of this PS class are the absorption of the light source lower skin penetration, which disable the application in more deep-seated tumors [11]. Photofrin® also presents low values of singlet oxygen quantum yield (Φ_Δ). For example, the Φ_Δ is only 0.01 in the PBS medium [12]. This implicates the necessity of using a considerable amount of PS in order to obtain a phototherapeutic result.

Foscan® or Temoporfin (5,10,15,20-Tetrakis(3-hydroxyphenyl)chlorin) is a member of the second-class of PS compounds. Termoporfin presents Q_I band at 652 nm, with a high extinction coefficient and Φ_Δ in ethanol (3.0×10^4 M^{-1} cm^{-1} and 0.43, respectively). This PS has been studied against Esophageal cancer, and is this use is approved against head and neck cancer.

The PDT efficacy of Termoporfin is about 100 times greater than Photofrin and other first-generation PSs. Therefore, second-generation PS compounds present better photophysical properties than the previous generation [13].

The majority of second-generation-compounds are known for hydrophobic nature. This is a favorable characteristic since PS can permeate through the cell membrane more efficiently, further decreasing the therapeutically active concentration. However, its direct administration in aqueous medium leads to the aggregation process. The aggregation occurs because the interactions between PS molecules (by π-π stacking) are stronger than water-PS interactions. The aggregation leads to changes in the photophysical properties of PS, such as emission quenching and decreasing in the $^1O_2$ production [14]. This is the main reason by which Temoporfin is intravenously administered using a water/polyethylene glycol 400/ethanol mixture as the solvent. However, the use of an organic solvent is no desirable once it can promote several harmful effects.

Metalled phthalocyanines, chlorines, and bacteriochlorins are the new PS classes that have been synthesized for PDT treatment. The inclusion of metals as zinc (II) and aluminum (III) increases the molar absorption coefficient in the therapeutic window (650–680 nm) [15]. For comparison, the ε value for zinc phthalocyanine is 60 times larger than Photofrin® at 630 nm [16]. Additionally, these compounds present better photophysical properties than Photofrin®, as well as longer triplet state lifetime and higher singlet oxygen quantum yield [15]. However, phthalocyanines also are water-insoluble. The strategy commonly used to avoid this problem is to append biocompatible motifs at the periphery of phthalocyanines or perform the synthesis of water-soluble phthalocyanines [17]. For example, the sulphonation of phthalocyanines, chlorines, and bacteriochlorines increases the water-solubility. On the other hand, this kind of modification decreases the interaction between the PS compound and the cellular compartment, increasing the therapeutical concentration of PS [18].

Finally, hydrophilic photosensitizer classes as xanthenes and phenothiazines are known. The main advantages are that they can be easily applied intravenously and have improved tumor killing. However, they poorly accumulate in tumor cells and hardly crossing cell membranes. For example, Bengal Rose (BR) is a hydrophilic xanthene dye extensively studied for PDT purposes [19]. An alternative is to synthesize BR derivatives with a lower solubility in the aqueous medium. For example, BR ester derivatives with different alkyl chains have been synthesized. These compounds present more significant interaction with tumor cells. However, xanthene dyes have different protolytic forms. The structural pH-dependency of xanthenes is a drawback for PDT purposes, once some of these forms are no PDT active. On the other hand, phenothiazinium compounds as methylene blue (MB) can be easily irreversibly reduced to their non-active leuco-form, which is no PDT active. This is a drawback that avoids the direct MB application using PDT treatment [20].

The third-generation of PS compounds is obtained through the conjugation of PS compounds with targeting components, as an antibody directed against the tumor antigens. These PS localize and accumulate more effectively in the cellular target [21].

In addition, an attractive solution to increase the effectiveness of photodynamic therapy is the use of drug delivery systems (DDS) for hydrophobic and hydrophilic PS, as discussed in the next section.

3 Drug Delivery Systems

Drug delivery systems (DDS) are biomedical devices that make possible the solubilization of hydrophilic and hydrophobic PS compounds. These systems, besides promote the solubilization of PS compounds in the aqueous medium, also increase its bioavailability, decreasing the drug concentration necessary for the therapeutical effect. The high PS concentration in the human body leads to side-effects, in which the main is to affect the healthy cells [22]. These effects restrict the dose and duration of PDT administration, prejudicing the treatment efficacy [23]. Additionally, DDS protect the drug from degradation after systemic administration, promote the controlled release, and maintain the drug concentration in the therapeutic in the organism in a greater time. The excellent drug loading capacity of these materials is attributed to large surface areas or inner volumes into which a considerable amount of drugs can be incorporated.

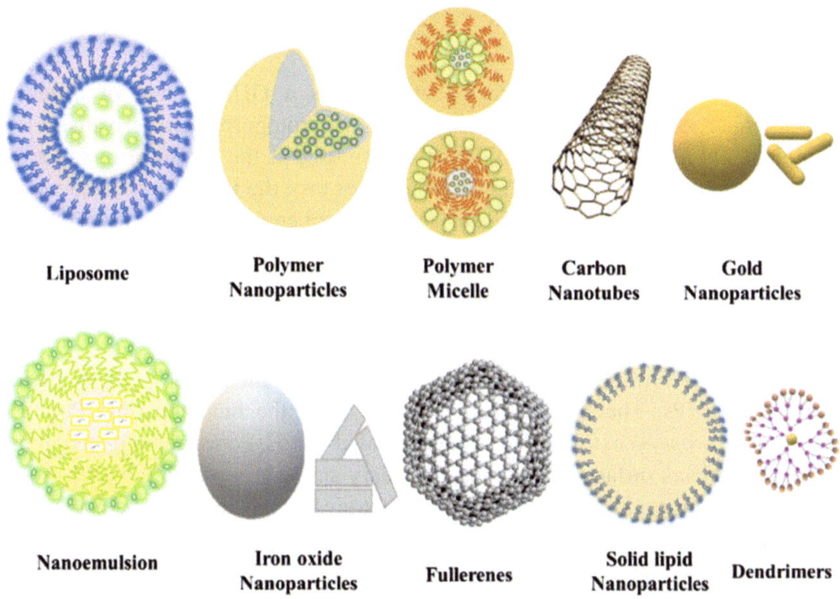

Fig. 3 Types of drug delivery systems commonly in literature. Reuse authorized [3]

DDS has been studied in a branch of science named nanomedicine, which treats the comprehension and control of organized structures with size varying from 1 to 1000 nm (Fig. 3). The nanocarriers used for DDS are classified in some subgroups, such as micelles, liposomes, dendrimers, nanogels, inorganic materials, and others. These are used in the pharmaceutical purpose in the application, diagnosis, prevention, and treatment [22].

Particularly for cancer treatment, it is desirable that the nanocarrier has size varying from 100 to 200 nm. This is due to the fact that tumor neovasculature is different from vasculature founded in healthy cells. Endothelial cells, for example, are no aligned or disordered, while basement membranes and perivascular cells are vacant or abnormal, owing to a fast angiogenesis within tumor tissue [24]. This fact results in leaky channels that aim the permeation of materials with a hydrodynamic size lower than 200 nm into the tumoral region. Additionally, the nanomaterials can accumulate into tumor cells during prolonged periods due to the fact that the tumor neovasculature possesses wide lumens and a poorly operating lymphatic drainage system. This size-motivated tumor selectivity of nanoparticles is named "*enhanced permeability and retention (EPR) effect*" [25]. Drug delivery systems widely used for PDT are discussed below.

3.1 Metalled Nanoparticles

Metalled nanoparticles have attracted attention as a DDS system of different hydrophobic PS compounds as phthalocyanines and protoporphyrin, derivatives. For example, Gold nanoparticles (AuNPs) have been used as the carrier of phthalocyanines modified with unique functional groups. For this, the PS is derivatized with a thiol moiety, attaching to the AuNPs surface. The thiol group extends a direct linkage to the AuNPs surface through self-assembly. In this research, the PS compound is not incorporated into nanocarrier but bounded to the AuNPs surface. This system aims at the solubilization of phthalocyanine in the aqueous medium and an effective PDT effect [26].

Alexeree and coworkers studied the incorporation of metal-free phthalocyanine (H_2Pc) in AuNPs. The AuNPs were reduced using the biocompatible Potatoes (*Solanum tuberosum*) extract. The results revealed that H_2Pc was conjugated to the gold nanoparticles surface through the secondary amine group of H2Pc [26]. Additionally, the conjugated Au-Pc material does not present any toxicity against buffalo epithelial cells. Thus, this work suggests the use of PS conjugated with nanoparticles, as biocompatible materials, for application in PDT.

de Freitas and coworkers combined silver nanoparticles with the PS curcumin (CUR), generating the Metal-Enhanced Singlet Oxygen effect (MEO effect) [27]. This effect is related to the distance and spectral overlapping between metallic nanoparticles and PS compounds. MEO increases some important photophysical properties, as well as the state triplet of PS compound by coupling its dipole moment with the electric field near the surface of the metallic nanoparticles. In this way, AgNPs increase the Φ_Δ of PS (from 0.01 to 0.14 in water), improving this effectiveness. In this study, the authors reveal that CUR is located on the AgNPs surface, interacting with the citrate ions. Additionally, the CUR-AgNPs conjugate leads to an increase in the fluorescence emission, which is an exciting result of applying this material as a theranostic agent in PDT studies.

Noble metal plasmonic nanoparticles, per se, also can be used as Photosensitizer compounds in PDT. For example, it is known that Gold and Silver nanoparticles present a higher extinction coefficient in comparison to organic photosensitizers [28]. They are much more stable under illumination than organic PSs. Including it is discussed that intense radiations can decompose and quickly cause the photobleaching of the organic PS compounds [29]. Finally, organic PS only can be active by UV/Vis light, while metalled nanoparticles can also absorb near-infrared light irradiation.

The human body presents a transparency window in the wavelength region from 650 to 1350 nm [28]. In metaled nanoparticles, it is possible to control the position of localized surface plasmon resonance (LSPR) depending on the shape and the size of the nanoparticles. For example, gold nanorods can shift their LSPR band position by 950 nm, changing the aspect ratio of the nanoparticles. PDT mechanism using metaled nanoparticles involves the adsorption of some O_2 molecules on the surface of the plasmonic nanoparticles. Under irradiation, the singlet oxygen is formed on the nanoparticle surface, after a plasmon mediated electron emission from the noble metal nanoparticles to surrounding media [29].

3.2 Dendrimers

Dendrimers are considered highly branched three-dimensional macromolecules. They are composed of repeating unities of polymers (monomers). These nanocarriers are synthesized by repeated chemical reactions and have a variety of peripherical functional groups. The surface of dendrimers can be functionalized by groups that have interaction with tumor cells to increase these effects.

In this carrier, the bonds emerge radially from a central core. The monomers are attached in a regular manner, layer by layer, and each layer is denominated "generation". The properties and architecture of dendrimers are related with it generation. Low-generation dendrimers are known as smaller molecules characterized by an open and asymmetric shape [30]. The higher-generation-dendrimers, on the other hand is globular and densely packed at the periphery.

PS compounds can be covalently bound to dendrimers on the surface or in the nanoparticle core [30]. The advantages of using drug-conjugated dendrimers as nanocarriers are the high drug payload, and the ability to control and modify the lipophilicity and size of the dendrimers, for optimization of cellular uptake and tissue biodistribution [31]. For example, it is reported in the literature that the pheophorbide a (Pheid a) binds covalently with the amino groups of polypropylene imine (PPI) dendrimers, forming a PS of third-generation [32].

Adittionally, PS compound can be incorporated inside dendrimer core, and interacts through non-covalent interactions. For example, non-covalent interactions between dendrimers and hydrophilic and hydrophobic PS compounds, such as xanthenes and porphyrins, have shown suitable applications in PDT. Non-covalent polyamidoamine (PAMAM)/Bengal Rose formulations (PAMAM/BR) were studied for PDT purposes. The results showed that PAMAM dendrimers presented a high drug loading of BR molecules. Additionally, the carrier systems promoted a controlled release of BR, releasing 80% of the initial PS amount during the first 72 h. Finally, PAMAM dendrimers-based nanocarriers release BR maintaining this ROS generation property upon radiation, reducing BR toxicity [33]. The last is because dendrimers can hold PS molecules in the internal cavities.

3.3 Liposomes

According to Ghosh and coworkers, liposomes are defined as synthetic or natural lipids based vesicles formed from a lipid bilayer that encloses an aqueous core and can comprise one (unilamellar) or more (multilamellar) concentric bilayers [15]. Liposomes usually constituted by lipids that have amphiphilic characters [34]. The most common liposomes synthesis is solid dispersion. Solid dispersion is considered a trivial method and consists of the solubilization of the lipids/phospholipids in an organic solvent appropriate (generally organic solvents mixture). After complete lipids/phospholipids solubilization, the organic phase is extracted by rotary evaporation. A thin film is obtained after organic solvent evaporation, with this, the thin film obtained is hydrated, also can be used buffer solutions, for obtaining liposomes.

Liposomes are formed spontaneously when certain lipids are hydrated in aqueous media. A common phospholipids used in the liposomes synthesis are: 1,2-Dipalmitoyl-*sn*-glycerol-3-phosphocholine and 1-Palmitoyl-2-linoleoylphosphatidylcholine. Both present hydrophilic and hydrophobic portions in their structural representation, as shown in Fig. 4a.

In General liposomes formation, during liposomes synthesis hydrophilic portion of the lipids interact with a hydrophilic portion of the other lipids molecule. Analogically, the hydrophobic portion of the lipids interacts with a hydrophobic portion of the other lipids molecule, as schematized in Fig. 4. Unilamellar liposomes generally have a high volume aqueous phase, which can incorporate hydrophilic photosensitizers. On the other hand, multilamellar liposomes are best for hydrophobic photosensitizer, due to the interaction with the lipidic bilayer [35]. In summary, liposomes have microenvironments with different hydrophobicity, as illustrated in Fig. 4. These

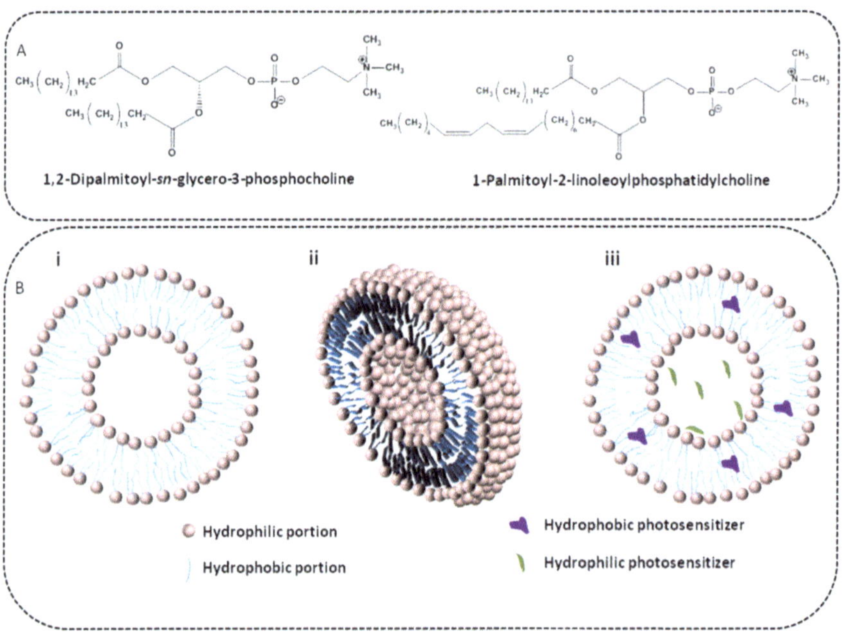

Fig. 4 **a** Structural representation of the 1,2-Dipalmitoyl-sn-glycerol-3-phosphocholine and 1-Palmitoyl-2-linoleoylphosphatidylcholine and **b** Representation schematical of the liposomes; (i) Bilayer lipid of the liposome; (ii) a cross-section of liposome; and (iii) Bilayer lipid of the liposome containing hydrophobic and hydrophilic photosensitizer

facts forward liposomes as an excellent and promissory strategy to carrier photosensitizer. Besides, this carrier is considered organic, biocompatible, and biodegradable nanosystems [36, 37]. In this way, liposomes are very useful biocompatible tools used in diverse scientific fields as pharmaceutical and biomedical applications. For example, it is extensively applied for the vehiculation and drug delivery for treatment of antimicrobial, oral cancer, breast cancer, human melanoma, among others [36, 38–40].

The main advantage of using liposomes as a drug delivery system (DDS) is that it promotes good interaction with the cell membrane and, consequently, increase the accumulation in the target site. Liposomal accumulation in the target cell or tissue stimulates the photosensitizer delivery and, consequently, it permeation for the intracellular environment. This increases therapeutic efficacy dramatically [15].

Previously reported studies had demonstrated the potential use of liposomes containing PS compounds. For example, Freitas and coworkers reported the obtention of stable liposomes incorporating erythrosine-decyl ester photosensitizer against the cancerous cell Caco-2 [41]. The low liposome stability in the aqueous medium and limitation in the sterilization procedure are considered as a drawback for liposome using as DDS system [42]. The low liposome stability has a relation with liposomes size. Liposomes higher than 100 nm presents low blood-circulation time, due to diminishing the uptake by the reticuloendothelial system, as reported in the literature [43, 44]. To minimize this drawback, researchers have used molecules, generally hydrophilic polymers, to modify the surface of the liposomes. These hydrophilic polymers are known as "imperceptible" to the endothelial reticular system [41, 45]. An example is polyethylene glycol (PEG). PEG, when linked in liposomes surface, turns liposomes imperceptive to macrophages [43].

Commercial liposomes are applied in anticancer drugs, vaccines, and ophthalmology [46]. Up to now, only ophthalmology has phototherapy applications. Visudyne® uses the lipids egg phosphatidyl glycerol (EPG) and dimyristoyl phosphatidyl choline (DMPC) (3:8 ration molar) in the liposome obtention [46]. The liposome-based on EPG/DMPC incorporates a hydrophobic photosensitizer, specifically verteporfin, for the intravenous DDS [47]. This formulation can be considered as the proof of concept of the clinical liposomes use containing photosensitizer.

Thus, liposomes can be considered as versatile platforms due to their microenvironments that stabilize hydrophobic and hydrophilic photosensitizers. Surface ligands can be modulated according to the local target, promoting interaction with cell membranes and turning invisible for some systems. Also, the interaction between liposomes and the target is the first step towards the delivery of the photosensitizer and, consequently, to the efficiency of photodynamic therapy.

3.4 Cyclodextrins

Cyclodextrins (CDs) are cyclic oligosaccharides, which have a hydrophilic exterior and a hydrophobic cavity (Fig. 5). It is obtained from bacterial digestion of cellulose

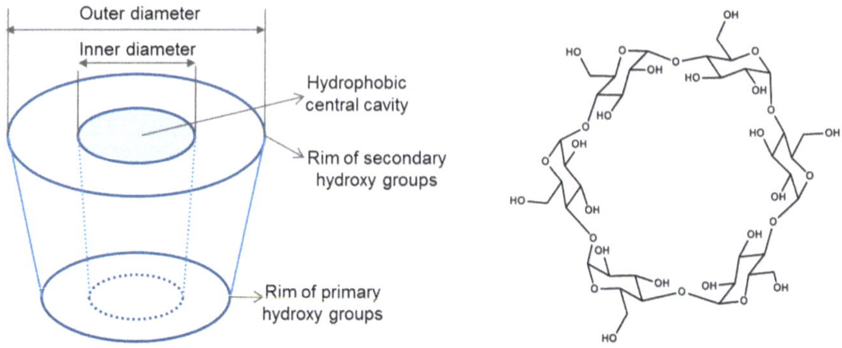

Fig. 5 Chemical structure of native α-CDs

Table 1 Main characteristics of cyclodextrins

Cyclodextrin	n	Solubility[a] (mg/mL)	MW[b] (Da)	Cavity diameter (A)	Cavity volume (A^3)
α-Cyclodextrin	6	129.5 ± 0.7	973	4.7–5.3	174
β-Cyclodextrin	7	18.4 ± 0.2	1135	6.0–6.5	262
γ-Cyclodextrin	8	249.2 ± 0.2	1297	7.5–8.3	427

Adapted from Igor Yankovsky, 2016 [51]

with units of 6 (α-CD), 7 (β-CD), and 8 (γ-CD) D-glucopyranose units linked by bonds (1–4) and each one has its characteristics, according to Table 1. This oligosaccharide allows the formation of complexes increasing the solubility and stability of hydrophobic drugs, being non-toxic, biocompatible, and used in the pharmaceutical food industry [48, 49]. Due to steric obstacles, CDs with less than six units are not obtained. On the other hand, those that have more than eight units of glucopyranose are purchased, but their applications are limited, as they are rapidly reduced to smaller products [50].

CDs interact with a large number of compounds, improving their properties. As a result, it began to be used in photodynamic therapy (PDT) as delivery systems in the mid-90 s. However, it was only in the past decade that intense research on the application of CDs in PDT began. The photosensitizers most used initially were tetrapyrroles (porphyrinoids), however, there are reports of successful use in other PSs, such as hypericin, curcumin, phthalocyanines, among others [52, 53].

With the addition of CDs to the system, some features in PSs have been improved. As an increase in water solubility, changes in photophysical characteristics, increasing the production of ROS, and improving the distribution of PSs to target cells and tissues. The literature reports three ways that are addressed to alleviate these problems. Through (i) inclusion complexes, (ii) covalent bonding (conjugated CDs-PSs) or (iii) formation of nanoassemblies between CDs and PSs (Fig. 6) [51].

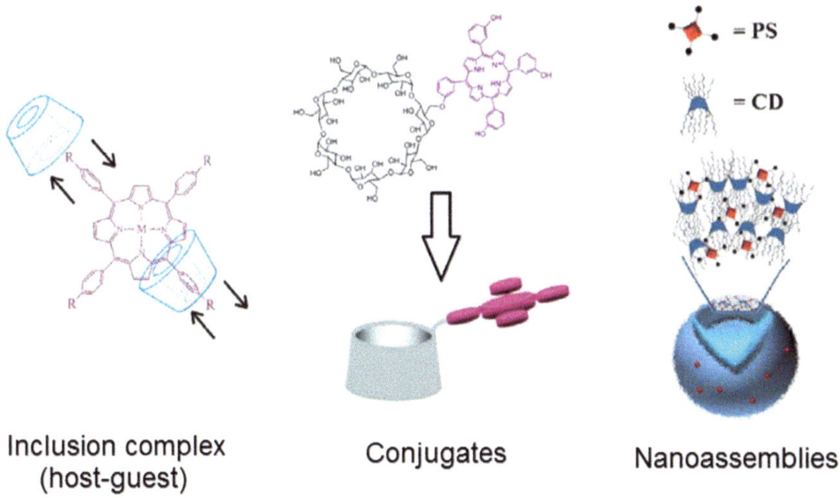

Fig. 6 Types of links between CDs with tetrapyrrole photosensitizers reported in the literature. Adapted from Igor Yankovsky (2016) [51]

The formation of inclusion complexes facilitates the solubilization of poorly soluble substances, such as PSs. In this system, the CDs act as a host and the PSs as a guest. These complexes are formed through van der Walls forces, hydrophobic interactions, and hydrogen bonds. The ability to form reversible inclusion complexes is governed by two factors: the steric effect and thermodynamic interactions. Thus, the CDs form these complexes with substances that have a size, polarity and geometric shape compatible with the size of their cavity [54]. This allows CDs and the respective molecule to interact completely, partially, or axial, depending on the host–guest stoichiometry [55]. As reported in the work of Mihoub et al. (2018), the inclusion complexes CD-porphyrinoid PSs improved the fluorescence intensity, tripled the useful life in aqueous and neutral solutions, increased the production of $^1O_2$ and the fluorescence quantum yield. Lu et al. (2014) prepared and studied a 4:1 inclusion complex of zinc phthalocyanine (ZnPc) with HP-β-CD ((HP-β-CD)4-ZnPc) to improve the PDT efficiency of ZnPc by increasing the water solubility and decreasing the aggregation in the physiological environment of ZnPc [56].

A second way used to improve the physical–chemical properties of PSs for PDT applications is through the covalent bonding between CD and PS compounds. Several studies of these phthalocyanine-CD conjugates have been carried out due to the proven effectiveness of the CD-PS set in PDT. Lourenço et al. (2014) compared the formation of conjugates between native CDs (α, β and γ) with phthalocyanine. As previously analyzed and mentioned, due to the lower solubility of β-CD, its formation of ROS, bioavailability is lower when compared to CDs α and γ. However, it is known that the presence of CD to PS improves its properties, even though it is poorly soluble like β-CD, thus improving the characteristics of phthalocyanine when applied to PDT, when compared to the drug alone [57].

The last type of connection between CDs and PSs refers to nanoassemblies. These are supramolecular colloidal systems that involve a non-specific external connection between the sites of the CDs and PSs. The changes in the edges, as in the native CDs, by several substituents, allow producing derivatives of ionic or non-ionic CDs. Through these modifications, it is possible to form supramolecular buildings that can host PSs as porphyrins. *In-vitro* applications have been reported, improving drug administration and its effectiveness in PDT [54].

3.5 Polymeric Micelles

Micelle is a designation for a self-molecular organization of amphiphilic molecules and surfactants in aqueous medium as a natural response for abrupt changes in physicochemical parameters [58]. For instance, when a molecular concentration increase in a narrow range enough to reduces the interface energy, the molecules start the formation of well-oriented colloidal aggregates (Fig. 7) [59]. Different types of block copolymers can be self-assembling as micellar systems, such as di- (AB), tri- (ABA or BAB) or tetra- (ABCA) block copolymers by end-to-end addition of a linear sequence of monomers, alternating block (ABABABAB) with two species of monomeric units distributed in alternating sequence, and random block (AAABABBAA) with monomer residues locating randomly [60–62]. On the other hand, micelles can be structured by graft copolymers, comprising a polymer chain as a backbone with two

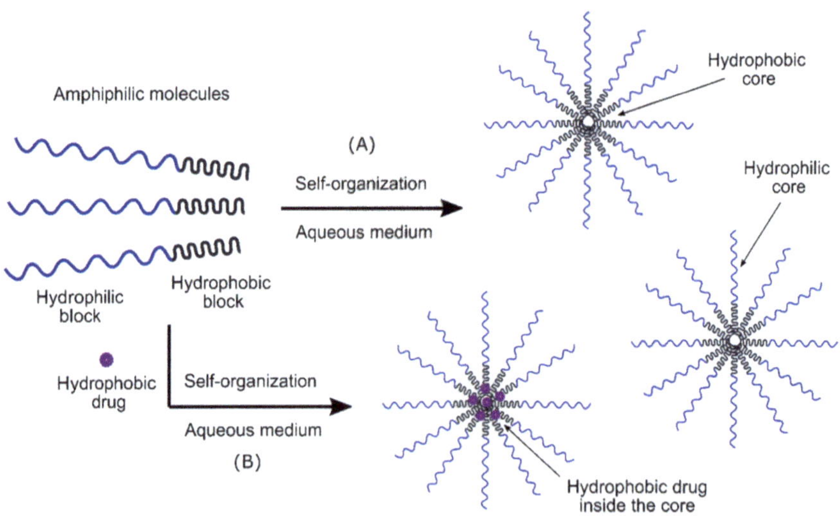

Fig. 7 The **a** Micelle formation by self-organization of amphiphilic molecules in aqueous medium; **b** Hydrophobic drugs-loaded inside the core of the micelle

monomers grafted as side parts (i.e. acrylonitrile–butadiene–styrene graft copolymer (ABS)) [63, 64].

Polimeric micelles (PM) have been extensively explored as smart and high-efficiency nanoplatforms to delivery of hydrophobic drugs in a variety of medicinal applications, especially for cancer therapy. Because the PMs show in the aqueous medium, high stability, controlled drug release, long retention time, and bioelimination, they have advantages over other micellar systems [13]. In the late 1990s, Matsumura and Maeda discovered that due to the inflamed or cancerous tissues, some macromolecular therapeutics could spontaneously accumulate via the so-called enhanced permeability and retention (EPR) effect [65]. Besides, the defects of the lymphatic system in tumors prevent these macromolecular therapeutics from being cleared from the tumor, giving to PMs a prolonged time to release their active cargo. Also, PMs have many other advantages over other nanoplatforms; these include the simple methods used in their preparation, high encapsulation efficiency and loading capacity, ideal size (20–100 nm) to in vivo applications, and controlled drug release profile.

As previously presented, most of the PSs of the first and second generation associated with the PDT for cancer therapy are hydrophobic chemical compounds, which in turn tend to aggregate in aqueous environments, losing its spectroscopic and tumor-selective properties consequently. On the other hand, PMs architected as nanoplatforms can solubilize a high concentration of PSs and protect them in the core compartment preventing the aggregation. Thus, the PMs as drug delivery systems in PDT applications, lead to an improvement of the PSs pharmacological properties, such as biodistribution, bioavailability, and pharmacokinetics [3, 66]. As a consequence, a reduced dose of the PS is required to photoinactivation of the therapeutic target, minimizing any side effects. Additionally, their small size and hydrophilic shells render PMs stealth properties facilitating their escape from the mononuclear phagocytic system, vital to increase the extension of their circulation half-life, and enhances the effects of passive targeting to make the nanoparticle more specific to a target site. Recently PMs based on ABA triblock copolymers, comprised of poly(ethylene oxide) (EO) and poly(propylene oxide) (PO) blocks), and commercially available as Pluronic®, have been employed to solubilize Hypericin (HYP), a *powerful natural PS found in plants of the genus Hypericum. In association with PDT, HYP has showed a* broad spectrum of pharmacological activities *against microorganisms and a variety of cancer cells.* Montanha et al. has reported the photodynamic potential of HYP-loaded P123 micelles to inactivation of two lineages of intestinal colon carcinoma. The authors showed that P123 copolymeric micelles not only improved the hydrosolubility of HYP but also prevented its self-aggregation, thereby keeping HYP in a monomeric stade, ensuring the photoinactivation of Caco-2 and HT29 cells [67]. In another study employing the P123/HYP system associated with PDT, Damke et al. showed that the micellar copolymer can permeate the membrane of the MCF-7 cells with the internalization of HYP in the mitochondria and endoplasmic reticulum organelles, leading a selective time- and dose-dependent phototoxic effects on cancer cells but little damage to MCF-10A cells [68].

In the last years, block copolymers linked to bioactive molecules, including chemotherapy, targeted ligands, proteins, photosensitizers, and fluorescent probes, have been engineered to produce PMs with prolonged blood-circulation times, cellular signaling effect, and targeting therapy [69]. In this approaches, the surface of PMs can be decorated with active targeting via chemical conjugation to bind specifically to target cells, resulting in the increased specificity and enhanced penetration via receptor-mediated endocytosis. The active targeting includes small molecules, proteins, peptides, antibodies, and nuclei acid-based ligands [70]. Folates, including folic acid (FA) and 5-methyltetrahydrofolate (5-MTHF), are essential cofactors for many biochemical reactions [71]. The role of this vitamin in the production of precursors for DNA synthesis and repair makes it essential for proliferating cells. Cancer cells overexpress FA receptors as folate-binding proteins and have a high affinity to uptake FA-conjugated CPMs [72]. For instance, Li et al. prepared FA-conjugated methoxypoly(ethylene glycol) (mPEG) and polylactic acid (PLA) copolymeric polymer (FA-PEG-PLA) to synthesize an active-targeting, water-soluble and pharmacomodulated hypocrellin B (HB) nanocarrier. The enhanced uptake of HB/FA-PEG-PLA micelles resulted in a more effective post-PDT killing of SKOV3 ovarian cancer cells compared to plain micelles and free drugs [73]. In the same approach, Gonçalves et al. reported a functional PMs based on Pluronic P84-linked N-(3-aminopropyl)-2-pyrrolidone (APP) for PDT-HYP treatment of melanoma. The strategy of authors has used the ability of N-alkyl-2-pyrrolidone moiety of APP as a penetration enhancer showed an improvement of the HYP potential photodamage effect in vitro models of mouse melanoma cell lines (B16-F10) [74].

3.6 Polymeric Nanoparticles

Polymeric nanoparticles have been studied for the entrapment of low soluble PSs. The nanoparticles promote the controlled PS release depending on intrinsic factors such as redox environment of cancer cells, pH gradient of tumors, or by extrinsic factors as heat, light, ultrasound, and others [62]. Between the advantages, these systems are biodegradable and biocompatible. Additionally, the controlled release of this kind nanocarrier is more controlled than nanocarriers that promotes drug delivery by a factor of hard control, such as pH, temperature, and ionic force [75].

Particularly interesting are the nanoparticles extrinsically stimulated by light incidence. These systems have vantage to circumventing the impact of microenvironments or tissue types, which restrict pathophysiology-based drug delivery. For example, the PS release by light activation can be separated into three types: chemical degradation, disruption by photothermal effect, and molecular structure change. Polycarbonate-based polymers, for example, can absolve UV light leading to a polymer degradation along the polymeric backbone [75]. After irradiation, the PS compound is controllably released from polymer, and, in a second stage, the same light is used to activate the photosensitizer compound. Finally, the PS compound promotes the photodynamic effect.

The main advantage of this kind of carrier is the possibility of the release of higher PS concentration at the tumoral site, causing fewer side-effects in other body regions. Additionally, these carriers are biocompatible biodegradable. This, allied to the small size, makes them suitable carriers for PDT purposes [76].

In addition, some nanoparticles linked to bioactive molecules for PDT applications have been investigated. In this approach, Son et al. used folate-modified poly(lactic-co-glycolic acid) nanoparticles (FA-PLGA-Pba) for the tumor-targeted delivery of Pheid a [77]. The results show a faster cellular uptake and high accumulation of the FA-PLGA-Pheid to MKN28 human gastric cancer cell line during 24 h after intravenous injection. Yoon et al. reported a tumor-targeting hyaluronic acid nanoparticles (HANPs) as the carrier of the hydrophobic photosensitizer, chlorin e6 (Ce6) for simultaneous photodynamic imaging and therapy. In this case, hyaluronic acid (HA) had shown as potential as active targeting. They found that Ce6-HANPs could be rapid uptake into tumor cells via the passive targeting mechanism and accurately enter tumor cells through the receptor-mediated endocytosis based on the interactions between HA of nanoparticles and CD44, the HA receptor on the surface of tumor cells [78].

Conjugation of NPs to iRGD peptide provides a promising strategy for tumor targeting. iRGD is a 9-amino acid cyclic peptide sequence (CRGDKGPDC) able to increase vascular and tissue permeability using a mechanism dependent on $\alpha\nu$ integrin and neuropilin-1. Wang et al. [79] reported the synthesis of iRGD-conjugated PEG-PLGA NPs for the simultaneous tumor delivery of indocyanine green (ICG) and a hypoxia-activated in both 3D tumor spheroids in vitro and orthotopic breast tumors in vivo [80].

3.7 *Peptide- and Protein-Based Nanomaterials Self-Assembled*

Recently, the use of biocompatible systems based on peptides and proteins as DDS self-assembled for PDT has gained massive attention for its bioactivity, therefore, synergic potential on treatment allied with the PS [81].

Supramolecular *self-assembly* structures build on lipids, peptides and proteins possess important function in nature, i.e., on formation of cytoskeleton and cellular membranes. Such complexes formed upon peptides and proteins have ubiquitous characteristics beyond its chemical moieties, as stability, enzymatic function and specific receptor interactions [82, 83]. The self-organization depends entirely on interactions between the amino acid residues, which can be hydrophobic, hydrophilic, aromatic, aliphatic, cationic or anionic. Thus, the structure of each amino acid enable intra and intermolecular interactions as π–π stacking, non-specific van der Waals forces, chiral dipole–dipole interactions and, most importantly, hydrogen bonding and hydrophobic interactions which maintain the system closely-knit. The many possibles interactions and any alteration of the chemical, physical and biological properties may lead to distinct materials in different sizes, as nanotubes, nanobelts, fibrils, nano-vesicles, gels and nanocages [82, 84].

Self-assembled nanomaterials can be modulated through pH, temperature, light and by the introduction of chemical and enzymatic components [85, 86]. From simple systems based on dipeptides to complex architectures which use proteins as building blocks, the infinite possibilities of these adaptable structures enable the rational design of smart drug delivery vehicles, capables of releasing the drug in specific sites, raise the PS permeation in the cellular membrane, as many other features. Still, the non-covalent bonds between amino acids and the PS may give rise to the possibility of changing and modulating its physical properties, as potencial zeta, fluorescence yield, among others.

3.7.1 Dipeptides

Dipeptides are formed by the conjugation of two amino acids through a peptidic bond. Its structure is easy to produce and, therefore, it's easy relatively simple to modulate so it can accomplish the wanted purposes. Starting with 20 amino acids, infinite possibilities are possible through its derivations, which mainly characteristic are biocompatibility, easy biodegradation and low cytotoxicity [87]. research made by Ma et al. (2016) prospected the development of a nearly 100 nm nanoparticle self-assembled from a cationic dipeptide diphenylalanine (H-Phe-Phe-NH_2.HCl, CDP), using glutaraldehyde as a covalent *crosslinker*, to deliver the PS Chlorin e6 [86]. Thereby, beyond the cationic dipeptide and the crosslinker, the nanoparticle is decorated in its surface with heparin molecules (Hep), that confer to the structure a negative potential of 25 mV, which may low its adsorption by negative charged proteins belonging to mononuclear phagocytic system (MPS) and increase its circulation time. Indeed, the results show that there are more retention in vitro when the system is used on MCF-7 tumor cells, enhancing its efficiency when comparing to the free PS. Other dipeptidic systems based on CDP and derivatives are described [88–90]. Liu et al. (2016) could observe by a batochromic shift in Ce6 spectrum that the PS actively participated in the vehicle assembly, possibly through π–π and hydrophobic interactions [91].

Abbas et al. (2017) describes the formation of nanofibers upon a diphenylalanine derivative (FF), fluorenylmethoxycarbonyl diphenylalanine (Fmoc-FF) [92]. Those nanofibers form a hydrogel upon the electrostatic interactions between the anionic dipeptide and a cation polypeptide (poly-L-*Lys*). The system was able to sustain the liberation of former encapsulated Ce6 up until 48 h, retarding tumor growth and even reducing its size when teste in mice.

3.7.2 Polypeptides

Among the use of polypeptides as major components, those with amphiphilic properties usually stand out by efficient producing nanotubes, fibers, vesicles and nanoparticles [84]. Liang et al. [93] report the development of a micellar system proper for use in PDT based on amphiphilic peptides as building blocks. These peptides are constructed through the conjugation of a hydrophobic amino acid chain and the hydrophilic amino acid *Lys,* which possess an amine moiety sensitive to pH alterations due to its pKa (VVVVVVKKGRGDS). It has been established that neoplastic tissues present different aspects in comparison with healthy tissue, as enzyme super expression, receptors, beside the acidosis ($pH < 6.0$) due to high glycolysis rate [85, 94]. Such conditions can be used by this nanocarriers to destabilize its own structure in the active site, avoiding its activation in non desired locations, hence diminishing system collateral effects. The protonation of the amine moiety promotes repulsion between polypeptidic chains and consequently extensive liberation of the PS. In this context, Chen et al. (2011) also avail the chemical properties of a polypeptide which possess 4 hydrophobic chains to enhance the stability of the system in high dissolution biological medium through increased hydrophobic interactions [95].

Both systems are capable of exploring the high malleability of those structures based in polypeptides. Still, these systems take advantage on intrinsic bioactivity of the biomolecules, since the presence of amino acid sequences also can interact with super expressed receptors presented in pathological conditions, thus raising the percentage of the PS entering the cell. Several studies use this sort of active vectorization [96–98]. A knowingly active sequence is *RGD (Arg-Gly-Asp),* added on the surface of the nanovehicles by electrostatic interactions so it can interact with $\alpha_{\upsilon}\beta_5$ e $\alpha_{\upsilon}\beta_3$ integrins, exacerbated in the membrane of certain tumors and responsible for adhesion and migration of these neoplastic cells. Xiong et al. [99] shows in its research that the functionalization of polymeric micelles with *RGD* peptide raises significantly drug permeation in melanoma cells (B16-F10) through endocytosis mechanism.

Other biomolecules super expressed in cancer cells as MMPs (metalloproteinases) also can be used as target sites for signaling sequences. Li et al. [100] proposed the utilization of an amino acid sequence *(Gly-Pro-Leu-Gly-Leu-Ala-Gly)* to transform protoporphyrin IX (PpIX) into a prodrug activated by the cleavage of the peptidic chain by MMP-2. Used as a linker, the amino acid sequence connects a cationic peptide fraction R_9 (sequence whose function is to enhance the PS cell permeation) to an anionic peptide fraction E_8. Through electrostatic interactions, the E_8 protect the R_9 fraction from interacting with receptors in the cell membrane. When the complex approaches the active site, the super expressed MMP-2 cleaves the linker and then the active portion of the drug is released.

Conveniently, peptides can also be allied to other molecules in order to optimize the self-organized complex efficiency and increase its stability. Zhu et al. [101] developed a supramolecular nanoparticulated system for use in PDT as a delivery system for hydrophobic drugs. The construction of a bimolecular complex through the alliance of pillararene derivatives and a specific designed peptidic sequence could bring thermoresponsive properties to the system, without the need for covalent linkage between the two fractions.

3.7.3 Proteins

Unlike the oligopeptides, which possess between 2 and 50 amino acids in its chain, proteins configure much more complex structure with more than 50 residues, which leads to difficult synthetic procedures. Thus, bio produced proteins as albumines, collagen and ferritins, among others, are preferably used.

Proteic structures can admit hydrophobic and hydrophilic drugs in its differents domains or empty core. Nanoparticles had been produced through the *self-assembly* of human seric albumine (HSA) by Jiang et al. [102] so it can deliver IR780, a hydrophobic PS. The cleavage of the disulfide bond in the structure exposed its hydrophobic domains, which could thereby interact with the PS. The resulting system was hydrosoluble and stable in aqueous medium.

The alliance between HSA and *crosslinkers* as glutaraldehyde can be explored so it can improve the stability of the system, as observed by Wacker et al. [103]. The use of crosslinkers is a cautious decision as it forms covalent bonding and may preclude the drug's liberation. Interestingly, the PS encapsulated in this system, 5,10,15,20-tetrakis(m-hydroxyphenyl) porphyrine (mTHPP) and 5,10,15,20-tertrakis(m-hydroxy- phenyl)chlorin (mTHPC), did not presented this drawback and the quenching of the PS when trapped inside the nanoparticles can reduce the systemic collateral effects as the PS only activates when the system is dismantled inside the tumor cell. Therefore, none the less the system could properly deliver the PS in the active site but also preserve its efficiency in producing ROS only where it's needed.

In this context, Luo et al. [104] invest in nanocages based on a hybrid protein (HPOC) developed through the joint of HSA and hemoglobin (Hb), protein responsible for carrying oxygen to the tissues.[104] Thus, the recurrent hypoxic conditions found in neoplastic tissues is outlined through oxygen supply from the system itself. The structure presented to be able to carry not only Ce6 but also doxorubicin efficiently ($93.1 \pm 3.4\%$ and $97.2 \pm 4.5\%$, respectively), enabling a multi front treatment.

Zhen et al. [105] also explored the use of nanocages based on a ferritin derivative (FRT) with its surface modified by a particular amino acid sequence, RGD4C *(Cys-Asp-CysArg-Gly-Asp-Cys-Phe-Cys)*. The complex demonstrated to be capable of encapsulating a significant percentage (up until 60 wt%) of the model PS zinc hexadecafluorophthalocyanine ($ZnF_{16}Pc$). The small size of the system (18.6 ± 2.6 nm) enables its passive targeting by EPR and active through the RGD fraction. It's possible to observe that, as many other systems presented above, this system tries to operate in different fronts in search to optimize its results.

The advantages of proteic and peptidic systems to PDT are clear, its biocompatibility and easy biodegradability, mostly of dipeptidic complexes, are notable. The countless transformation possibilities of these systems project many potentials of activation through different ambient conditions or even the modulation of the ambients itself, as O_2 supply by the own vehicle. Therefore, through basic precepts as maintenance of physical properties of the PS and low cytotoxicity in the absence of light, new *self assembly* systems can be extensively explored and available for use in PDT in a near future.

4 Conclusions and Future Prospects

Photodynamic therapy has developed as a medical modality for the treatment of different diseases caused by abnormal cell growth, with advantages ranging from greater selectivity to low cost. Although there are different PSs that are promising drugs for PDT, their intrinsic limitations can hinder the dissemination of the technique. Thus, the search for appropriate release systems for each photosensitizer is a constant challenge in different research areas. A fundamental characteristic for the development of release systems is the maintenance of the photophysical and photochemical properties of the PS, without which the effectiveness of the technique will be compromised. Therefore, hydrophobic drugs need to be monomerized in these systems during transport or after delivery to the active site. In addition, high selectivity to target cells is essential for the effectiveness of PDT and to avoid prolonged photosensitization. Among the numerous release systems investigated for PDT we highlight: liposomes, cyclodextrins, polymeric micelles, polymeric nanoparticles, peptide- and protein-based self-assembled nanomaterials, metalled nanoparticles, and dendrimers. As discussed above, these delivery systems act to increase the photodynamic effect of different drugs, thus being delivery systems suitable for PDT application. In a very close perspective, the junction of these release systems, seeking the combination of their main advantages for the assembly of mixed systems should be used. Thus, we can project the use of PSs with different targets and/or mechanisms, aiming to enhance the therapy effect.

References

1. Zhang, Y., Zhao, X., Li, Y., Wang, X., Wang, Q., Lu, H., Zhu, L.: A Fluorescent photosensitizer with far Red/near-Infrared aggregation-induced emission for imaging and photodynamic killing of bacteria. Dyes Pigm. **165**, 53–57 (2019). https://doi.org/10.1016/j.dyepig.2019.02.019
2. O'Connell, K.A., Okhovat, J.-P., Zeitouni, N.C.: Photodynamic therapy for bowen's disease (squamous cell carcinoma in situ) current review and update. Photodiagn. Photodyn. Ther. **24**, 109–114 (2018). https://doi.org/10.1016/j.pdpdt.2018.09.009
3. Shanmugapriya, K., Kang, H.W.: Engineering pharmaceutical nanocarriers for photodynamic therapy on wound healing: review. Mater. Sci. Eng., C **105**, 110110 (2019). https://doi.org/10.1016/j.msec.2019.110110
4. Yang, X., Bai, J., Qian, Y.: The investigation of unique water-soluble heptamethine cyanine dye for use as NIR photosensitizer in photodynamic therapy of cancer cells. Spectrochim.

Acta Part A Mol. Biomol. Spectrosc. **228**, 117702 (2020). https://doi.org/10.1016/j.saa.2019.117702

5. Huang, Z.: A review of progress in clinical photodynamic therapy. Technol. Cancer Res. Treat **4**(3), 283–293 (2005). https://doi.org/10.1177/153303460500400308
6. Sobolev, A.S., Jans, D.A., Rosenkranz, A.A.: Targeted intracellular delivery of photosensitizers. Prog. Biophys. Mol. Biol. **73**(1), 51–90 (2000). https://doi.org/10.1016/S0079-6107(00)00002-X
7. O'Connor, A.E., Gallagher, W.M., Byrne, A.T.: Porphyrin and nonporphyrin photosensitizers in oncology: preclinical and clinical advances in photodynamic therapy. Photochem. Photobiol. **85**(5), 1053–1074 (2009). https://doi.org/10.1111/j.1751-1097.2009.00585.x
8. Houthoofd, S., Vuylsteke, M., Mordon, S., Fourneau, I.: Photodynamic therapy for atherosclerosis. The potential of indocyanine green. Photodiagn. Photodyn. Ther. **29**, 101568 (2020). https://doi.org/10.1016/j.pdpdt.2019.10.003
9. Pertiwi, Y.D., Chikama, T., Sueoka, K., Ko, J.-A., Kiuchi, Y., Onodera, M., Sakaguchi, T.: Antimicrobial photodynamic therapy with the photosensitizer TONS504 eradicates acanthamoeba. Photodiagn. Photodyn. Ther. **28**, 166–171 (2019). https://doi.org/10.1016/j.pdpdt.2019.08.035
10. Yano, S., Hirohara, S., Obata, M., Hagiya, Y., Ogura, S., Ikeda, A., Kataoka, H., Tanaka, M., Joh, T.: Current states and future views in photodynamic therapy. J. Photochem. Photobiol., C **12**(1), 46–67 (2011). https://doi.org/10.1016/j.jphotochemrev.2011.06.001
11. Stepp, H., Kniebühler, G., Pongratz, T., Betz, C.S., Göke, B., Sroka, R., Schirra, J.: Low dose MTHPC Photodynamic Therapy for Cholangiocarcinoma. Lilge, L.D., Sroka, R. (eds.) (Munich, Germany, 2013), p 88030M. https://doi.org/10.1117/12.2032396
12. Nowak-Sliwinska, P., Karocki, A., Elas, M., Pawlak, A., Stochel, G., Urbanska, K.: Verteporfin, Photofrin II, and Merocyanine 540 as PDT photosensitizers against melanoma cells. Biochem. Biophys. Res. Commun. **349**(2), 549–555 (2006). https://doi.org/10.1016/j.bbrc.2006.08.060
13. Mesquita, M.Q., Dias, C.J., Gamelas, S., Fardilha, M., Neves, M.G.P.M.S., Faustino, M.A.F.: An insight on the role of photosensitizer nanocarriers for photodynamic therapy. An. Acad. Bras. Ciênc. **90**(1 suppl 2), 1101–1130 (2018). https://doi.org/10.1590/0001-3765201720170800
14. Li, Y., Yu, Q., Gao, P., Yang, H., Huang, T., Liu, S., Zhao, Q.: Highly efficient organic photosensitizer with aggregation-induced emission for imaging-guided photodynamic ablation of cancer cells. Tetrahedron Lett. **59**(27), 2704–2707 (2018). https://doi.org/10.1016/j.tetlet.2018.05.088
15. Ghosh, S., Carter, K.A., Lovell, J.F.: Liposomal formulations of photosensitizers. Biomaterials **218**, 119341 (2019). https://doi.org/10.1016/j.biomaterials.2019.119341
16. Ochsner, M.: Light scattering of human skin: a comparison between Zinc(II)—Phthalocyanine and Photofrin II®. J. Photochem. Photobiol., B **32**(1–2), 3–9 (1996). https://doi.org/10.1016/1011-1344(95)07209-8
17. Singh, S., Aggarwal, A., Bhupathiraju, N.V.S.D.K., Jovanovic, I.R., Landress, M., Tuz, M.P., Gao, R., Drain, C.M.: Comparing a thioglycosylated chlorin and phthalocyanine as potential theranostic agents. Bioorg. Med. Chem. **28**(3), 115259 (2020). https://doi.org/10.1016/j.bmc.2019.115259
18. Ambroz, M., Beeby, A., MacRobert, A.J., Simpson, M.S.C., Svensen, R.K., Phillips, D.: Preparative, analytical and fluorescence spectroscopic studies of sulphonated aluminium phthalocyanine photosensitizers. J. Photochem. Photobiol., B **9**(1), 87–95 (1991). https://doi.org/10.1016/1011-1344(91)80006-4
19. de Pereira, P.C.S., do Costa, P.F.A., Pellosi, D.S., Calori, I.R., Vilsinski, B.H., Estevão, B.M., Hioka, N., Caetano, W.: Photophysical properties and interaction studies of Rose Bengal derivatives with biomimetic systems based in micellar aqueous solutions. J. Mol. Liquids **230**, 674-685 (2017). https://doi.org/10.1016/j.molliq.2017.01.055
20. Tardivo, J.P., Del Giglio, A., de Oliveira, C.S., Gabrielli, D.S., Junqueira, H.C., Tada, D.B., Severino, D., de Fátima Turchiello, R., Baptista, M.S.: Methylene Blue in photodynamic

therapy: from basic mechanisms to clinical applications. Photodiagn. Photodyn. Ther. **2**(3), 175–191 (2005). https://doi.org/10.1016/S1572-1000(05)00097-9
21. Gomaa, I., Sebak, A., Afifi, N., Abdel-Kader, M.: Liposomal delivery of ferrous chlorophyllin: a novel third generation photosensitizer for in vitro pdt of melanoma. Photodiagn. Photodyn. Ther. **18**, 162–170 (2017). https://doi.org/10.1016/j.pdpdt.2017.01.186
22. van Elk, M., Murphy, B.P., Eufrásio-da-Silva, T., O'Reilly, D.P., Vermonden, T., Hennink, W.E., Duffy, G.P., Ruiz-Hernández, E.: Nanomedicines for advanced cancer treatments: transitioning towards responsive systems. Int. J. Pharm. **515**(1–2), 132–164 (2016). https://doi.org/10.1016/j.ijpharm.2016.10.013
23. Baguley, B.C.: Multiple drug resistance mechanisms in cancer. Mol Biotechnol **46**(3), 308–316 (2010). https://doi.org/10.1007/s12033-010-9321-2
24. Jeong, K., Kang, C.S., Kim, Y., Lee, Y.-D., Kwon, I.C., Kim, S.: Development of highly efficient nanocarrier-mediated delivery approaches for cancer therapy. Cancer Lett. **374**(1), 31–43 (2016). https://doi.org/10.1016/j.canlet.2016.01.050
25. Kobayashi, H., Watanabe, R., Choyke, P.L.: Improving conventional enhanced permeability and retention (EPR) effects; what is the appropriate target? Theranostics **4**(1), 81–89 (2014). https://doi.org/10.7150/thno.7193
26 Alexeree, S.M.I., Sliem, M.A., El-Balshy, R.M., Amin, R.M., Harith, M.A.: Exploiting biosynthetic gold nanoparticles for improving the aqueous solubility of metal-free phthalocyanine as biocompatible PDT agent. Mater. Sci. Eng., C **76**, 727–734 (2017). https://doi.org/10.1016/j.msec.2017.03.129
27. de Freitas, C.F., Kimura, E., Rubira, A.F., Muniz, E.C.: Curcumin and silver nanoparticles carried out from polysaccharide-based hydrogels improved the photodynamic properties of curcumin through metal-enhanced singlet oxygen effect. Mater. Sci. Eng., C **112**, 110853 (2020). https://doi.org/10.1016/j.msec.2020.110853
28. Krajczewski, J., Rucińska, K., Townley, H.E., Kudelski, A.: Role of various nanoparticles in photodynamic therapy and detection methods of singlet oxygen. Photodiagn. Photodyn. Ther. **26**, 162–178 (2019). https://doi.org/10.1016/j.pdpdt.2019.03.016
29. Lv, J., Yi, Y., Wu, G., Liu, W.: Gold nanotriangles: green synthesis and PDT & PTT effect. Mater. Lett. **187**, 148–150 (2017). https://doi.org/10.1016/j.matlet.2016.10.087
30. Klajnert, B., Rozanek, M., Bryszewska, M.: Dendrimers in photodynamic therapy. CMC **19**(29), 4903–4912 (2012). https://doi.org/10.2174/092986731209024903
31. Casas, A., Battah, S., Di Venosa, G., Dobbin, P., Rodriguez, L., Fukuda, H., Batlle, A., MacRobert, A.J.: Sustained and efficient porphyrin generation in vivo using dendrimer conjugates of 5-ALA for photodynamic therapy. J. Control. Release **135**(2), 136–143 (2009). https://doi.org/10.1016/j.jconrel.2009.01.002
32. Hackbarth, S., Ermilov, E.A., Röder, B.: Interaction of pheophorbide a molecules covalently linked to DAB dendrimers. Opt. Commun. **248**(1–3), 295–306 (2005). https://doi.org/10.1016/j.optcom.2004.11.088
33. Karthikeyan, K., Babu, A., Kim, S.-J., Murugesan, R., Jeyasubramanian, K.: Enhanced photodynamic efficacy and efficient delivery of Rose Bengal using nanostructured poly(amidoamine) dendrimers: potential application in photodynamic therapy of cancer. Cancer Nano. **2**(1–6), 95–103 (2011). https://doi.org/10.1007/s12645-011-0019-3
34. Rasti, B., Erfanian, A., Selamat, J.: Novel nanoliposomal encapsulated omega-3 fatty acids and their applications in food. Food Chem. **230**, 690–696 (2017). https://doi.org/10.1016/j.foodchem.2017.03.089
35. Bangham, A.D., Horne, R.W.: Negative staining of phospholipids and their structural modification by surface-active agents as observed in the electron microscope. J. Mol. Biol. **8**(5), 660-IN10 (1964). https://doi.org/10.1016/S0022-2836(64)80115-7
36. Eleraky, N.E., Allam, A., Hassan, S.B., Omar, M.M.: Nanomedicine fight against antibacterial resistance: an overview of the recent pharmaceutical innovations. Pharmaceutics **12**(2), 142 (2020). https://doi.org/10.3390/pharmaceutics12020142
37. Sharma, A.: Liposomes in drug delivery: progress and limitations. Int. J. Pharm. **154**(2), 123–140 (1997). https://doi.org/10.1016/S0378-5173(97)00135-X

38. Longo, J.P.F., Lozzi, S.P., Simioni, A.R., Morais, P.C., Tedesco, A.C., Azevedo, R.B.: Photodynamic therapy with aluminum-chloro-phtalocyanine induces necrosis and vascular damage in mice tongue tumors. J. Photochem. Photobiol., B **94**(2), 143–146 (2009). https://doi.org/10.1016/j.jphotobiol.2008.11.003
39. Rocha, M.S.T., Lucci, C.M., Longo, J.P.F., Galera, P.D., Simioni, A.R., Lacava, Z.G.M., Tedesco, A.C., Azevedo, R.B.: Aluminum-chloride-phthalocyanine encapsulated in liposomes: activity against naturally occurring dog breast cancer cells. J Biomed Nanotechnol **8**(2), 251–257 (2012). https://doi.org/10.1166/jbn.2012.1378
40. Barbugli, P.A., Siqueira-Moura, M.P., Espreafico, E.M., Tedesco, A.C.: In vitro phototoxicity of liposomes and nanocapsules containing chloroaluminum phthalocyanine on human melanoma cell line. J. Nanosci. Nanotech. **10**(1), 569–573 (2010). https://doi.org/10.1166/jnn.2010.1741
41. de Freitas, C.F., Montanha, M.C., Pellosi, D.S., Kimura, E., Caetano, W., Hioka, N.: Biotin-targeted mixed liposomes: a smart strategy for selective release of a photosensitizer agent in cancer cells. Mater. Sci. Eng., C **104**, 109923 (2019). https://doi.org/10.1016/j.msec.2019.109923
42. Apolinário, A.C., Pachioni-Vasconcelos, J.A., Pessoa Jr., A., Rangel Yagui, C.O.: Polimerossomos versus lipossomos: a Evolução da "Bala Mágica". Quim. Nova (2017). https://doi.org/10.21577/0100-4042.20170054
43. Bozzuto, G., Molinari, A.: Liposomes as nanomedical devices. IJN 975 (2015). https://doi.org/10.2147/IJN.S68861
44. Reshetov, V., Kachatkou, D., Shmigol, T., Zorin, V., D'Hallewin, M.-A., Guillemin, F., Bezdetnaya, L.: Redistribution of meta-Tetra(Hydroxyphenyl)Chlorin (m-THPC) from conventional and PEGylated liposomes to biological substrates. Photochem. Photobiol. Sci. **10**(6), 911 (2011). https://doi.org/10.1039/c0pp00303d
45. Immordino, M.L., Dosio, F., Cattel, L.: Stealth liposomes: review of the basic science, rationale, and clinical applications, existing and potential. Int. J. Nanomed. **1**(3), 297–315 (2006)
46. Yeh, M.-K., Chang H.-I., Cheng, M.-Y.: Clinical development of liposome based drugs: formulation, characterization, and therapeutic efficacy. IJN 49 (2011). https://doi.org/10.2147/IJN.S26766
47 Fahr, A., van Hoogevest, P., May, S., Bergstrand, N., Leigh, M.L.S.: Transfer of lipophilic drugs between liposomal membranes and biological interfaces: consequences for drug delivery. Eur. J. Pharm. Sci. **26**(3–4), 251–265 (2005). https://doi.org/10.1016/j.ejps.2005.05.012
48. Lin, C., Ng, C.: Ho. 2-Hydroxypropyl-Beta-Cyclodextrin Increases Aqueous Solubility and Photostability of All-Trans-Retinoic Acid. *J Clin Pharm Ther* **2000**, *25* (4), 265–269. https://doi.org/https://doi.org/10.1046/j.1365-2710.2000.00285.x.
49. Monteil, M., Lecouvey, M., Landy, D., Ruellan, S., Mallard, I.: Cyclodextrins: a promising drug delivery vehicle for bisphosphonate. Carbohyd. Polym. **156**, 285–293 (2017). https://doi.org/10.1016/j.carbpol.2016.09.030
50 Loftsson, T., Brewster, M.E.: Pharmaceutical applications of cyclodextrins. 1. Drug solubilization and stabilization. J. Pharm. Sci. **85**(10), 1017–1025 (1996). https://doi.org/10.1021/js950534b
51. Yankovsky, I. Evaluation of Photodynamic Activity of Chlorine-Type Photosensitizers with Bcyclodextrins Nanovectors. Human Health and Pathology. Belarussian State University, 2016. English. FfNNT : 2016LORR0171ff. Fftel-01492601v2f, Université de Lorraine, Belarussian State University, 2016.
52. Yankovsky, I., Bastien, E., Yakavets, I., Khludeyev, I., Lassalle, H.-P., Gräfe, S., Bezdetnaya, L., Zorin, V.: Inclusion complexation with β-cyclodextrin derivatives alters photodynamic activity and biodistribution of Meta-Tetra(Hydroxyphenyl)Chlorin. Eur. J. Pharm. Sci. **91**, 172–182 (2016). https://doi.org/10.1016/j.ejps.2016.06.012
53. Zhang, L., Man, S., Qiu, H., Liu, Z., Zhang, M., Ma, L., Gao, W.: Curcumin-cyclodextrin complexes enhanced the anti-cancer effects of curcumin. Environ. Toxicol. Pharmacol. **48**, 31–38 (2016). https://doi.org/10.1016/j.etap.2016.09.021

54. Ben Mihoub, A., Larue, L., Moussaron, A., Youssef, Z., Colombeau, L., Baros, F., Frochot, C., Vanderesse, R., Acherar, S.: Use of cyclodextrins in anticancer photodynamic therapy treatment. Molecules **23**(8), 1936 (2018). https://doi.org/10.3390/molecules23081936
55. Wenz, G.: Cyclodextrins as building blocks for supramolecular structures and functional units. Angew. Chem. Int. Ed. Engl. **33**(8), 803–822 (1994). https://doi.org/10.1002/anie.199408031
56. Lu, S., Ma, Y.J., Xuan, H.Y., Wang, A., Zhao, B., Li, X.D., Zhou, J.H., Lin, Y., Zhou, L., Wei, S.H.: A novel assembling complex of hydrobobic phthalocyanine-cyclodextrin: preparation, characterization, molecular modeling, and in vitro activity. RSC Adv. **4**(104), 59759–59763 (2014). https://doi.org/10.1039/C4RA12654H
57. Lourenço, L.M.O., Pereira, P.M.R., Maciel, E., Válega, M., Domingues, F.M.J., Domingues, M.R.M., Neves, M.G.P.M.S., Cavaleiro, J.A.S., Fernandes, R., Tomé, J.P.C.: Amphiphilic phthalocyanine-cyclodextrin conjugates for cancer photodynamic therapy. Chem. Commun. **50**(61), 8363–8366 (2014). https://doi.org/10.1039/C4CC02226B
58. Clint, J.H.: Micelle formation. In Surfactant Aggregation, pp 82–129. Springer Netherlands, Dordrecht (1992). https://doi.org/10.1007/978-94-011-2272-6_5
59. Lee, K.S., Lee, J.H.: Hybrid CO_2 EOR using low-salinity and smart waterflood. In: Hybrid Enhanced Oil Recovery Using Smart Waterflooding, pp 111–128. Elsevier (2019). https://doi.org/10.1016/B978-0-12-816776-2.00005-2
60. Morishima, K., Terao, K., Sato, T.: Structural analysis of hydrophobe-uptake micelle of an amphiphilic alternating copolymer in aqueous solution. Langmuir **32**(31), 7875–7881 (2016). https://doi.org/10.1021/acs.langmuir.6b01480
61. Vasey, C.E., Pearce, A.K., Sodano, F., Cavanagh, R., Abelha, T., Cuzzucoli Crucitti, V., Anane-Adjei, A.B., Ashford, M., Gellert, P., Taresco, V., Alexander, C.: Amphiphilic tri- and tetra-block co-polymers combining versatile functionality with facile assembly into cytocompatible nanoparticles. Biomater. Sci. **7**(9), 3832–3845 (2019). https://doi.org/10.1039/C9BM00667B
62. Matsen, M.W.: Block copolymers, blends, surfactants and emulsions. In: Encyclopedia of Materials: Science and Technology, pp 1–9. Elsevier (2001). https://doi.org/10.1016/B0-08-043152-6/00134-0
63. Hanafy, N., El-Kemary, M., Leporatti, S.: Micelles structure development as a strategy to improve smart cancer therapy. Cancers **10**(7), 238 (2018). https://doi.org/10.3390/cancers10070238
64. Lo, C.-L., Lin, K.-M., Huang, C.-K., Hsiue, G.-H.: Self-assembly of a micelle structure from graft and diblock copolymers: an example of overcoming the limitations of polyions in drug delivery. Adv. Funct. Mater. **16**(18), 2309–2316 (2006). https://doi.org/10.1002/adfm.200500627
65. Matsumura, Y., Maeda, H.: A new concept for macromolecular therapeutics in cancer chemotherapy: mechanism of tumoritropic accumulation of proteins and the antitumor agent smancs. Cancer Res. **46**(12 Pt 1), 6387–6392 (1986)
66. Mukherjee, B., Chakraborty, S., Mondal, L., Satapathy, B.S., Sengupta, S., Dutta, L., Choudhury, A., Mandal, D.: Multifunctional drug nanocarriers facilitate more specific entry of therapeutic payload into tumors and control multiple drug resistance in cancer. In: Nanobiomaterials in Cancer Therapy, pp 203–251. Elsevier (2016). https://doi.org/10.1016/B978-0-323-42863-7.00007-4
67. Montanha, M.C., Silva, L.L., Pangoni, F.B.B., Cesar, G.B., Gonçalves, R.S., Caetano, W., Hioka, N., Tominaga, T.T., Consolaro, M.E.L., Diniz, A., Kimura, E.: Response surface method optimization of a novel hypericin formulation in P123 micelles for colorectal cancer and antimicrobial photodynamic therapy. J. Photochem. Photobiol., B **170**, 247–255 (2017). https://doi.org/10.1016/j.jphotobiol.2017.04.008
68. Damke, G.M.Z.F., Souza, R.P., Montanha, M.C., Damke, E., Gonçalves, R.S., Cesar, G.B., Kimura, E., Caetano, W., Hioka, N., Consolaro, M.E.L.: Selective photodynamic effects on breast cancer cells provided by P123 pluronic®-based nanoparticles modulating hypericin delivery. ACAMC **18**, (2018). https://doi.org/10.2174/1871520618666181102091010
69. Torchilin, V.: Fluorescence microscopy to follow the targeting of liposomes and micelles to cells and their intracellular fate. Adv. Drug Deliv. Rev. **57**(1), 95–109 (2005). https://doi.org/10.1016/j.addr.2004.06.002

70. Yi, G., Hong, S.H., Son, J., Yoo, J., Park, C., Choi, Y., Koo, H.: Recent advances in nanoparticle carriers for photodynamic therapy. Quant. Imaging Med. Surg. **8**(4), 433–443 (2018). https://doi.org/10.21037/qims.2018.05.04
71. Doucette, M.M., Stevens, V.L.: Folate receptor function is regulated in response to different cellular growth rates in cultured mammalian cells. J. Nutr. **131**(11), 2819–2825 (2001). https://doi.org/10.1093/jn/131.11.2819
72. Henderson, G.B.: Folate-binding proteins. Annu. Rev. Nutr. **10**(1), 319–335 (1990). https://doi.org/10.1146/annurev.nu.10.070190.001535
73. Li, J., Yao, S., Wang, K., Lu, Z., Su, X., Li, L., Yuan, C., Feng, J., Yan, S., Kong, B., Song, K.: Hypocrellin b-loaded, folate-conjugated polymeric micelle for intraperitoneal targeting of ovarian cancer in vitro and in vivo. Cancer Sci. **109**(6), 1958–1969 (2018). https://doi.org/10.1111/cas.13605
74. Gonçalves, R.S., Braga, G., de Oliveira, A.C.V., César, G.B., Tominaga, T.T., Zampiere, E.H., Calori, I.R., de Morais, F.A.P., Basso, E.A., Pontes, R.M., Hioka, N., Caetano, W.: Hypericin delivery system based on p84 copolymeric micelles linked with *N* -(3-Aminopropyl)-2-Pyrrolidone for melanoma-targeted photodynamic therapy. ACS Appl. Polym. Mater. **2**(4), 1692–1701 (2020). https://doi.org/10.1021/acsapm.0c00114
75. Anderski, J., Mahlert, L., Sun, J., Birnbaum, W., Mulac, D., Schreiber, S., Herrmann, F., Kuckling, D., Langer, K.: Light-responsive nanoparticles based on new polycarbonate polymers as innovative drug delivery systems for photosensitizers in PDT. Int. J. Pharm. **557**, 182–191 (2019). https://doi.org/10.1016/j.ijpharm.2018.12.040
76. Savolainen, J., Collan, M.: How additive manufacturing technology changes business models?—Review of literature. Additive Manuf. **32**, 101070 (2020). https://doi.org/10.1016/j.addma.2020.101070
77. Son, J., Yang, S.M., Yi, G., Roh, Y.J., Park, H., Park, J.M., Choi, M.-G., Koo, H.: Folate-modified plga nanoparticles for tumor-targeted delivery of pheophorbide a in vivo. Biochem. Biophys. Res. Commun. **498**(3), 523–528 (2018). https://doi.org/10.1016/j.bbrc.2018.03.013
78. Yoon, H.Y., Koo, H., Choi, K.Y., Lee, S.J., Kim, K., Kwon, I.C., Leary, J.F., Park, K., Yuk, S.H., Park, J.H., Choi, K.: Tumor-targeting hyaluronic acid nanoparticles for photodynamic imaging and therapy. Biomaterials **33**(15), 3980–3989 (2012). https://doi.org/10.1016/j.biomaterials.2012.02.016
79. Sugahara, K.N., Teesalu, T., Karmali, P.P., Kotamraju, V.R., Agemy, L., Greenwald, D.R., Ruoslahti, E.: Coadministration of a tumor-penetrating peptide enhances the efficacy of cancer drugs. Science **328**(5981), 1031–1035 (2010). https://doi.org/10.1126/science.1183057
80. Wang, Y., Xie, Y., Li, J., Peng, Z.-H., Sheinin, Y., Zhou, J., Oupický, D.: Tumor-penetrating nanoparticles for enhanced anticancer activity of combined photodynamic and hypoxia-activated therapy. ACS Nano **11**(2), 2227–2238 (2017). https://doi.org/10.1021/acsnano.6b08731
81. Konan, Y.N., Gurny, R., Allémann, E.: State of the art in the delivery of photosensitizers for photodynamic therapy. J. Photochem. Photobiol., B **66**(2), 89–106 (2002). https://doi.org/10.1016/S1011-1344(01)00267-6
82. Abbas, M., Zou, Q., Li, S., Yan, X.: Self-assembled peptide- and protein-based nanomaterials for antitumor photodynamic and photothermal therapy. Adv. Mater. **29**(12), 1605021 (2017). https://doi.org/10.1002/adma.201605021
83. Branco, M.C., Schneider, J.P.: Self-assembling materials for therapeutic delivery. Acta Biomater. **5**(3), 817–831 (2009). https://doi.org/10.1016/j.actbio.2008.09.018
84. Zhang, S., Marini, D.M., Hwang, W., Santoso, S.: Design of Nanostructured Biological Materials through Self-Assembly of Peptides and Proteins. Curr. Opin. Chem. Biol. **6**(6), 865–871 (2002). https://doi.org/10.1016/S1367-5931(02)00391-5
85. Han, K., Ma, Z., Han, H.: Functional peptide-based nanoparticles for photodynamic therapy. J. Mater. Chem. B **6**(1), 25–38 (2018). https://doi.org/10.1039/C7TB02804K
86. Ma, K., Xing, R., Jiao, T., Shen, G., Chen, C., Li, J., Yan, X.: Injectable self-assembled dipeptide-based nanocarriers for tumor delivery and effective in vivo photodynamic therapy. ACS Appl. Mater. Interfaces **8**(45), 30759–30767 (2016). https://doi.org/10.1021/acsami.6b10754

87. van Hell, A.J., Fretz, M.M., Crommelin, D.J.A., Hennink, W.E., Mastrobattista, E.: Peptide nanocarriers for intracellular delivery of photosensitizers. J. Control. Release **141**(3), 347–353 (2010). https://doi.org/10.1016/j.jconrel.2009.09.012
88. Sun, B., Wang, L., Li, Q., He, P., Liu, H., Wang, H., Yang, Y., Li, J.: Bis(Pyrene)-doped cationic dipeptide nanoparticles for two-photon-activated photodynamic therapy. Biomacromol **18**(11), 3506–3513 (2017). https://doi.org/10.1021/acs.biomac.7b00780
89. Song, J., Xing, R., Jiao, T., Peng, Q., Yuan, C., Möhwald, H., Yan, X.: Crystalline dipeptide nanobelts based on solid-solid phase transformation self-assembly and their polarization imaging of cells. ACS Appl. Mater. Interfaces **10**(3), 2368–2376 (2018). https://doi.org/10.1021/acsami.7b17933
90. Mishra, A., Panda, J.J., Basu, A., Chauhan, V.S.: Nanovesicles based on self-assembly of conformationally constrained aromatic residue containing amphiphilic dipeptides. Langmuir **24**(9), 4571–4576 (2008). https://doi.org/10.1021/la7034533
91. Liu, K., Xing, R., Zou, Q., Ma, G., Möhwald, H., Yan, X.: Simple peptide-tuned self-assembly of photosensitizers towards anticancer photodynamic therapy. Angew. Chem. Int. Ed. **55**(9), 3036–3039 (2016). https://doi.org/10.1002/anie.201509810
92. Abbas, M., Xing, R., Zhang, N., Zou, Q., Yan, X.: Antitumor photodynamic therapy based on dipeptide fibrous hydrogels with incorporation of photosensitive drugs. ACS Biomater. Sci. Eng. **4**(6), 2046–2052 (2018). https://doi.org/10.1021/acsbiomaterials.7b00624
93. Liang, J., Wu, W.-L., Xu, X.-D., Zhuo, R.-X., Zhang, X.-Z.: PH Responsive micelle self-assembled from a new amphiphilic peptide as anti-tumor drug carrier. Colloids Surf., B **114**, 398–403 (2014). https://doi.org/10.1016/j.colsurfb.2013.10.037
94. Gao, S., Wang, G., Qin, Z., Wang, X., Zhao, G., Ma, Q., Zhu, L.: Oxygen-generating hybrid nanoparticles to enhance fluorescent/photoacoustic/ultrasound imaging guided tumor photodynamic therapy. Biomaterials **112**, 324–335 (2017). https://doi.org/10.1016/j.biomaterials.2016.10.030
95. Chen, J.-X., Wang, H.-Y., Li, C., Han, K., Zhang, X.-Z., Zhuo, R.-X.: Construction of surfactant-like tetra-tail amphiphilic peptide with rgd ligand for encapsulation of porphyrin for photodynamic therapy. Biomaterials **32**(6), 1678–1684 (2011). https://doi.org/10.1016/j.biomaterials.2010.10.047
96. Akasov, R., Zaytseva-Zotova, D., Burov, S., Leko, M., Dontenwill, M., Chiper, M., Vandamme, T., Markvicheva, E.: Formation of multicellular tumor spheroids induced by cyclic RGD-peptides and use for anticancer drug testing in vitro. Int. J. Pharm. **506**(1–2), 148–157 (2016). https://doi.org/10.1016/j.ijpharm.2016.04.005
97. Nakase, I., Akita, H., Kogure, K., Gräslund, A., Langel, Ü., Harashima, H., Futaki, S.: Efficient intracellular delivery of nucleic acid pharmaceuticals using cell-penetrating peptides. Acc. Chem. Res. **45**(7), 1132–1139 (2012). https://doi.org/10.1021/ar200256e
98. Aina, O.H., Sroka, T.C., Chen, M.-L., Lam, K.S.: Therapeutic cancer targeting peptides. Biopolymers **66**(3), 184–199 (2002). https://doi.org/10.1002/bip.10257
99. Xiong, X.-B., Mahmud, A., Uludağ, H., Lavasanifar, A.: Conjugation of Arginine-Glycine-Aspartic Acid peptides to Poly(Ethylene Oxide)-*b* -Poly(ε-Caprolactone) micelles for enhanced intracellular drug delivery to metastatic tumor cells. Biomacromol **8**(3), 874–884 (2007). https://doi.org/10.1021/bm060967g
100. Li, S.-Y., Cheng, H., Qiu, W.-X., Liu, L.-H., Chen, S., Hu, Y., Xie, B.-R., Li, B., Zhang, X.-Z.: Protease-activable cell-penetrating peptide-protoporphyrin conjugate for targeted photodynamic therapy in vivo. ACS Appl. Mater. Interfaces **7**(51), 28319–28329 (2015). https://doi.org/10.1021/acsami.5b08637
101. Zhu, H., Wang, H., Shi, B., Shangguan, L., Tong, W., Yu, G., Mao, Z., Huang, F.: Supramolecular peptide constructed by molecular lego allowing programmable self-assembly for photodynamic therapy. Nat Commun **10**(1), 2412 (2019). https://doi.org/10.1038/s41467-019-10385-9
102. Jiang, C., Cheng, H., Yuan, A., Tang, X., Wu, J., Hu, Y.: Hydrophobic IR780 encapsulated in biodegradable human serum albumin nanoparticles for photothermal and photodynamic therapy. Acta Biomater. **14**, 61–69 (2015). https://doi.org/10.1016/j.actbio.2014.11.041

103 Wacker, M., Chen, K., Preuss, A., Possemeyer, K., Roeder, B., Langer, K.: Photosensitizer loaded HSA nanoparticles I: preparation and photophysical properties. Int. J. Pharm. **393**(1–2), 254–263 (2010). https://doi.org/10.1016/j.ijpharm.2010.04.022
104. Luo, Z., Tian, H., Liu, L., Chen, Z., Liang, R., Chen, Z., Wu, Z., Ma, A., Zheng, M., Cai, L.: Tumor-targeted hybrid protein oxygen carrier to simultaneously enhance hypoxia-dampened chemotherapy and photodynamic therapy at a single dose. Theranostics **8**(13), 3584–3596 (2018). https://doi.org/10.7150/thno.25409
105. Zhen, Z., Tang, W., Guo, C., Chen, H., Lin, X., Liu, G., Fei, B., Chen, X., Xu, B., Xie, J.: Ferritin nanocages to encapsulate and deliver photosensitizers for efficient photodynamic therapy against cancer. ACS Nano **7**(8), 6988–6996 (2013). https://doi.org/10.1021/nn402199g

DFT Approaches for Smart Materials with Ferroelectric Properties

Luis Henrique da Silveira Lacerda, Renan Augusto Pontes Ribeiro, and Sérgio Ricardo de Lázaro

Abstract In the last decades, the development of new technologies is strongly related to materials development, mainly for semiconductors and smart materials. A common property usually observed in both materials is the ferroelectricity. The ferroelectric materials are largely employed on data storage and memory devices, sensors, actuators, and others. In summary, the ferroelectricity is evidenced by a high spontaneous polarization inside the crystalline structure, which can arise from crystal modifications, such as, structural distortions, chemical bond profile, or material composition. Nowadays, the more important ferroelectric materials are the $BaTiO_3$, $PbTiO_3$, $PbZr_xTi_{1-x}O_3$ (PZT), and $BiFeO_3$ materials. In addition, the search for lead-free ferroelectric materials has attracted a big interest from scientists around the world aiming at environmentally friendly alternatives. As for lead-free materials as for Pb-based ferroelectric materials, several experimental approaches are widely reported. However, the investigation of this kind of material through theoretical investigation still represents a challenge. Therefore, this work presents theoretical approaches based on the Density Functional Theory (DFT) demonstrating its potential to develop advanced materials. Thereby, the theoretical methodology has successfully employed for clarification or prediction of the ferroelectric properties raised by chemical modifications or structural deformations.

Keywords Smart materials · Semiconductors · Ferroelectricity · Structural disorder · Ferroic properties

L. H. da S. Lacerda
State University of Campinas, Campinas, São Paulo, Brazil

R. A. P. Ribeiro
State University of Minas Gerais, Divinópolis, Minas Gerais, Brazil

S. R. de Lázaro (✉)
State University of Ponta Grossa, Ponta Grossa, Paraná, Brazil
e-mail: srlazaro@uepg.br

F. A. La Porta and C. A. Taft (eds.), *Functional Properties of Advanced Engineering Materials and Biomolecules*, Engineering Materials,
https://doi.org/10.1007/978-3-030-62226-8_11

1 Towards Multifunctional Ferroelectric Materials

The ferroelectricity is a fundamental property of the solid-state materials that exhibit a spontaneous charge polarization below the Curie temperature (Tc) that can be switched by an applied electric field [1–3]. Usually, the ferroelectric materials are transition metal oxides, such as perovskite ABO_3 compounds, where the spontaneous polarization arises from ions displacement from the center-symmetrical position, which can be explained by two different factors: (i) the second-order Jahn–Teller effect (SOJT) arising from the bonding overlap between empty d shells and Oxygen (2p) orbitals; (ii) the presence of a pair cations with large stereoactivity effects. This kind of ferroelectric material is called "proper ferroelectrics" [4].

On the other hand, the polarization can be originated from a more complex lattice distortion or as a result of some electronic or magnetic ordering, originating the class of "improper ferroelectrics" [5–7]. A very important aspect of such class of ferroelectric materials is that the leading order can thus coerce the improper polarization into electronic states that would be avoided in the conventional ferroelectrics generating superior functional properties [8].

From the viewpoint of technological applications, ferroelectric materials are promising candidates in several fields. For instance, recently, the ferroelectric and piezoelectric effects were highlighted to control the optical process in emerging materials [9]. Indeed, photo-ferroelectricity defines the combination of ferroelectricity and optoelectronic properties to induce superior behavior from ferroelectric photovoltaic materials, photostriction effect, and others [9–11].

Another subject of the interest for ferroelectric materials correspond to the field of photocatalysis. In this case, the spontaneous polarization promotes naturally the separation of the photoexcited charges as well as its migration from the bulk to surface during the photocatalytic process; thus, it promotes the redox reactions. In addition, the ferroelectric materials can act as co-catalysts, inducing a large polarization electric field that accelerates the surface charge transfer of the photocatalysts [12–15].

Ferroelectric materials have been also promising candidates for the consolidation of the energy storage materials—a subject of increased attention in the past few years due to the increasing concerns regarding sustainable developments in energy topics. In this context, ferroelectric materials are receiving increased interest due to their intrinsic capability for superior energy storage density [16–18].

More recently, the interest regarding ferroelectric increased due to scientific and technological interest about multiferroic materials—a widespread class of materials that combines two or more of the primary ferroic order parameters (ferroelectricity, ferroelasticity, ferromagnetism, and others) simultaneously in the same phase. However, the main focus of recent research related to this area involves materials that combine some form of magnetic ordering (ferromagnetism, ferrimagnetism, antiferromagnetism, etc.) with the presence of the ferroelectricity, giving rise to the name of magnetoelectric materials. The interest in such materials has been strictly connected to the perspective of controlling charges through magnetic fields, as well as spins through the electric field, enabling the development of ultra-fast, dense and low energy consumption electronic devices, such as actuators, transducers, multi-state memories, FeRAMs, and others [7, 19–23].

The precise understanding of the driving force or microscopic origin of the symmetry breaking that originates the ferroelectric properties plays a fundamental role in the development of the novel high-performance candidates. In this context, the ferroelectric polarization can be visualized as a function of the orbital hybridization involved in off-centered clusters, as well as from the perspective of the exchange–correlation (XC) energy that generates the inhomogeneous electronic densities responsible for the spontaneous polarization [24]. Despite the experimental efforts devoted to explaining these effects, some answers at the atomic level remain challenging, due to the emergence of the quantum states that governs the overall ferroelectric property.

Based on previously mentioned facts, the use of computational calculations in the design of the new ferroelectric materials offers numerous complementary advantages to experimental methods, especially due to its precision. Besides, the characterization of the fundamental properties of promising materials is carried out more quickly through computer simulations than compared to experimental techniques, improving the searching for new candidates. These advantages make computer simulations an essential step in the design of new ferroelectric materials in silico before the expensive and time-consuming experimental process.

In this context, the Density Functional Theory (DFT) corresponds to the most powerful theoretical tool for studying the properties of the materials, especially the ferroelectric behavior. Indeed, the DFT can be used to calculate not only structural and dynamical properties for ferroelectrics but to investigate the polarization properties from the Berry phase method according to the modern theory of polarization [25–28]. Further, the DFT methods can also be applied for a wide range of ferroelectric materials paying special attention to their lattice dynamics, phase transitions, dielectric and piezoelectric properties, and to the study of defects and surfaces [29].

Recently, an increased number of scientists have confirmed the power of DFT calculations to understand and predict the most important properties of ferroelectric materials, contributing to the materials design from an innovative strategy that combines electronic structure calculations, symmetry analysis, and solid-state chemistry concepts [30–35].

2 DFT Approaches on Ferroelectricity

At the first moment, the theoretical approximation based on quantum mechanics plays a simple role in the materials' investigation, being applied to basic structural analysis or electronic analysis of well-known systems. Due to the strong connection between materials development and technological advances, the relevance of the theoretical methodologies increased a lot. Thereby, the theoretical methods represent important tools for helping chemists and physicists on the development of materials in a shorter time enabling the prediction of its properties and employment limitation [36–38].

Among the theoretical approaches available stand out the DFT, which has contributed largely for simulation of several materials at last thirteen five years, offering the best relationship between results precision and study time. Such methodology was firstly developed by Kohn and Hoihenberg in 1964 and posteriorly finished by Kohn and Sham in 1965. The heart of the DFT consists of the assumption that the total energy of the system is a function of the electronic density (ρ); thereby, the density functional determines exactly and completely all the properties of interest of a system in its ground state, and it is dependent only on the position (x, y, z). Another important fact about the DFT is that any tentative function for the ρ will have, therefore, energy greater or equal than the ground state for a real system [39].

The DFT formalism also includes the exchange–correlation energy, which is described according to the local (LDA, LSDA), non-local (GGA, PBE), hybrid (B3LYP, HSE06, PBE0) or meta-GGA (Massachusetts functionals series) functionals. The efficiency of each kind of exchange–correlation approximation is different and, according to extensive results reported in the literature is possible to states that the hybrid functional leads to better results when allied to localized Gaussian basis sets; whereas, the non-local approximation guarantees very reliable results using plane-wave methods [36–40]. In this chapter, all the computational simulations are based on localized Gaussian basis sets allied to hybrid functional.

Nowadays, DFT approaches are successfully employed in the study of magnetic [41–48], electronic [49–55], ferroelectric [51, 52, 56–58], optical [59, 60], morphological [61–64], and photocatalytic properties [65–67]. In particular, the ferroelectric properties investigation can be carried out by different approaches. Among this, we highlight the Berry Phase methodology [68, 69] that makes possible a direct prediction of the ferroelectric properties by Spontaneous Polarization values or piezoelectric analysis. It is noteworthy that the spontaneous polarization evaluation enables the direct comparison between theoretical and experimental results. This methodology determines the spontaneous polarization (S_P) for a crystalline material by the difference between paraelectric and ferroelectric phases and, for this reason, is known as a geometric approximation to determinate macroscopic polarization. The main idea comes from the experimental analysis since the quantity generally measured is the differential polarization, including the dielectric permittivity, Born effective charges, piezoelectricity, pyroelectricity, and ferroelectricity. Thus, the polarization observed in a unit cell is equivalent to a macroscopic current, being directly accessible to quantum calculations as a bulk property.

Furthermore, the reliability of Bery Phase results makes it the most potent approximation to ferroelectric properties prediction. However, this methodology does not evidence the anisotropic behavior of the ferroelectric properties when it is possible since it evaluates the S_P overall the material. In this context, the Perturbation Theory is very relevant predicting the dielectric constant (α) and polarizability (ε) by application of the external electric field. This kind of simulation predicts the dielectric constant by the average macroscopic electronic density as presented in Eq. 1, where ε refers to relative permissivity, E_0 refers to applied field magnitude along a specific direction of the unit cell and $\acute{E}$ is the average electric field value. In turn, the polarizability is measured as a function of the electronic density under external stimuli of an electric field.

$$\varepsilon = \frac{E_0}{E_0 + \acute{E}} \tag{1}$$

Thereby, theoretical investigations based on both methodologies have enabled a deeper understanding of the ferroelectric properties of the materials and how these properties are affected as by chemical modification as by structural factors.

3 Advanced Ferroelectric Materials

3.1 *Ferroelectricity Raised by Chemical Modification*

Lead zirconate titanate, $PbZr_xTi_{1-x}O_3$ (PZT), is a perovskite crystal well-known for its superior ferroelectric, piezoelectric, and dielectric properties, being widely used in transducers, actuators, transformers, sensors, capacitors, and other technological applications [70, 71]. The exceptional dielectric and ferroelectric properties of the PZT are deeply related to the displacement as of Ti^{4+} ion as of Zr^{4+} ion from the center-symmetrical position, which can be explained by the second-order Jahn–Teller effect (SOJT) arising from the bonding overlap between Ti/Zr (3d) and O (2p) orbitals. Moreover, the presence of large cations with lone pair $6s^2$ electrons (Pb^{2+}) in A-sites induces another SOJT effect from the stereoactivity factor of the $6s^2$ lone pair, establishing a large electric polarization on Pb–O bonds through the mixing between empty Pb(6p) and O(2p) orbitals.

The major issues involving PZT are associated with the high toxicity of the lead-based oxides, resulting in several environmental problems that contribute to increasing the scientific and technological interest in the development of new lead-free ferroelectric materials, whose ferroelectric performance is comparable to or even better than that of PZT.

Recently, several authors have dedicated experimental and theoretical efforts to elucidate the overall properties of the Sn(II)-based semiconductors owing to their potential technological importance in different fields, especially for the development of new ferroelectric devices. Indeed, in the past of a few years, $SnTiO_3$ (STO) and $SnZrO_3$ (SZO) perovskite structures were founded to have a spontaneous electric polarization larger than that of the $PbTiO_3$ material [72–76]. In addition, theoretical studies indicate that Sn-doped PZT at A-site exhibits strong ferroelectric properties and has much higher values of the piezoelectric coefficients than PZT [77].

Following this discussion, we propose the first-principle calculation of structural, electronic, and ferroelectric properties of $SnZr_{0.50}Ti_{0.50}O_3$ (SZT) material as a lead-free candidate to replace PZT. In this case, DFT/PBE0 hybrid calculations were

Table 1 Theoretical results for lattice parameters (in Å), unit cell volume (in $Å^3$), M–O bond lengths (in Å), and born effective charges (BEC in |e|) for PZT and SZT materials (A = Pb, Sn)

Models	a	c	V	M–O			BEC			
				A–O	Ti–O	Zr–O	A	Ti	Zr	O
PZT	4.007	4.239	68.062	2.511 2.872 3.513	1.758 2.020 2.244	1.958 2.056 2.517	3.534	5.561	5.637	−3.045
SZT	3.963	4.229	66.418	2.429 2.852 3.512	1.747 2.005 2.239	1.930 2.051 2.513	3.596	5.489	5.624	−3.058

carried out considering a large supercell model containing 40 atoms using CRYSTAL [78] package.

From an inspection of Table 1, it was observed that SZT has a contracted unit cell in comparison to PZT, following the smaller ionic radii for Sn atom (1.18 Å) than for Pb (1.49 Å). Regarding the atomic displacement from the central position (ideal cubic symmetry), calculated fractional coordinates indicate that both solid solutions exhibit a similar disorder along the [001] direction coinciding with the polarization axis of the tetragonal symmetry. In particular, it was noted that the unit cell contraction for SZT is directly connected with the shortening of the M–O (M = Sn, Ti, Zr) bond distances, suggesting that the isoelectronic replacement of Pb for Sn induced a local disorder on the $[AO_{12}]$ cluster, which spreads along with the crystalline structure.

The Born Effective Charge (BEC) is another fundamental property related to the dielectric and ferroelectric oxides, controlling the long-range Coulomb component of the force constants. Our calculated results show that both A- and B-site cations exhibit large dynamical charges when compared to the nominal charges in a full-ionic picture, A^{2+} and B^{4+}. These results are, in turn, a strong indicator of the existence of the covalent character for both A–O and B–O chemical bonds. Furthermore, it was observed that both A–O and B–O chemical bonds are sensitive to the isoelectronic replacement of Pb by Sn; indeed, the BEC for Sn (3.596 |e|) is higher than for Pb (3.534 |e|), indicating the enhancement of the covalent character for Sn–O, while the calculated charges for Ti/Zr at SZT are smaller than for PZT, suggesting the higher ionic component.

Similar observations can be visualized by the electron density maps along with the A–O and B–O chemical bonds, as depicted in Fig. 1. As discussed above, the electron density maps clarify the covalent component on the A–O chemical bonds, where a charge corridor was observed along the Pb–O and Sn–O bond paths. On the other hand, the plots for B–O chemical bonds indicate that the electronic density distribution exhibit a discontinuance along the [001] direction, confirming the existence of the spontaneous polarization effect for tetragonal PZT and SZT materials. In addition, the increase of the covalent character for Sn–O bonds has induced a higher charge separation along the axial plane, suggesting an enhancement of the spontaneous polarization for the lead-free SZT in comparison to PZT.

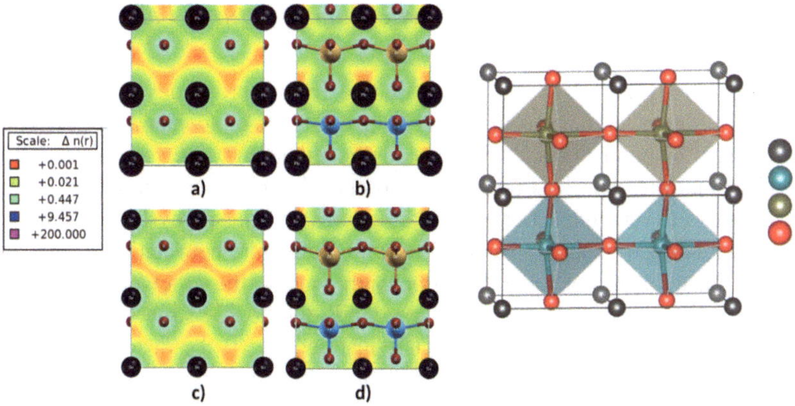

Fig. 1 Electron density maps for A–O (A = Pb, Sn) and B–O (B = Ti, Zr) chemical bonds as charge corridor or charge paths for the PZT (**a–b**) and SZT (**c–d**) materials

As a consequence of the increased spontaneous polarization, the calculated band-gap for SZT (3.03 eV) was lower than for PZT (3.79 eV), suggesting that the increased covalent character of the A–O bonds have provoked a perturbation of the frontiers crystalline orbitals, as presented in Fig. 2. Moreover, the calculated effective mass for the electrons and holes associated with the narrow band-gap suggest a potential application in photoinduced technologies, such as photovoltaic and photocatalytic devices for the ferroelectric SZT.

Therefore, combining all analysis is possible to assume that lead-free SZT is an excellent candidate for ferroelectric-based applications due to the increased spontaneous polarization associated with very high dielectric constants.

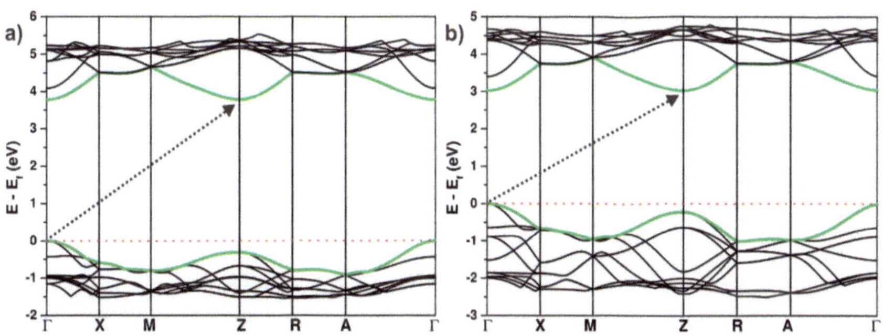

Fig. 2 Band Structure profiles indicating narrow band-gaps for **a** PZT and **b** SZT materials

3.2 Ferroelectricity Raised by Particular Structural Features

The ferroelectricity is one of the most essential ferroic properties on solid-state materials. The ferroelectricity is justified by the existence of spontaneous polarization (S_p) raised by bond character between component atoms, structural distortions, or chemical modification. It is possible to observe this phenomenon as on simple oxide structures as on complex crystalline structures, such as perovskites and other structures with ABO_3 general formula. The common feature between both kinds of materials is the existence of some distortion degree within the structures.

Among the highly ferroelectric structures stands out the R3c structures due to its intrinsic ferroelectric properties; this fact is justified by (i) the distortion degree for [AO_6] and [BO_6] clusters; (ii) the alternation between the A and B cations along the c axis; (iii) highly compacted layer composed by O atoms and (iv) the presence of intrinsic ordering vacancies [79–82]. Thus, in this section, the results for Berry methodology, Perturbation methods, and structural aspects obtained by a DFT/B3LYP calculation level applied on CRYSTAL [78] package for several R3c materials were used to determine the influence of particular structural features on ferroelectric properties. As mentioned previously, the Berry methodology considers the ferroelectric structure (R3c) regarding their respective non-polar structure (R-3c) to predict the spontaneous polarization; Fig. 3 presents both crystalline structures.

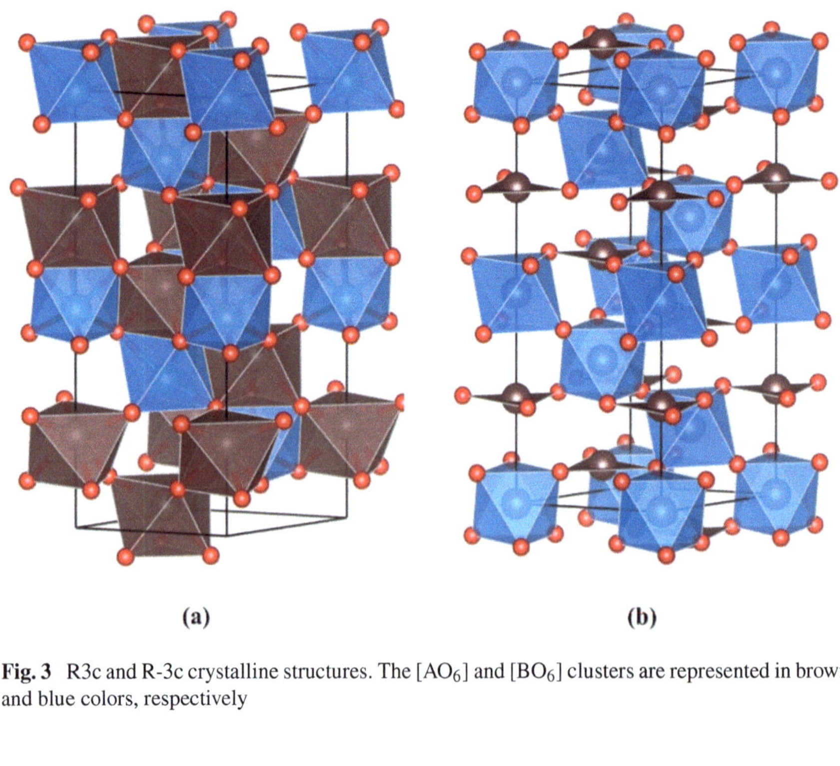

(a) (b)

Fig. 3 R3c and R-3c crystalline structures. The [AO_6] and [BO_6] clusters are represented in brown and blue colors, respectively

In summary, the ferroelectric properties of evaluated materials are related to structural aspects, such as the tilting of $[AO_6]$ and $[BO_6]$ clusters, displacement of cation located in the center of clusters and by Jahn–Teller Effects [80, 83]. Another important feature is the chemical composition of each material. To facilitate the discussion of the results, the evaluated materials were classified into four different groups according to their chemical composition. As observed in Table 2, the general evaluation of results indicates that the ferroelectric property is higher for materials containing Ti, Al, and Cu clusters, since the $TiNiO_3$, $NiTiO_3$, $FeTiO_3$, $AlFeO_3$, $FeAlO_3$, and $PbCuO_3$ materials present the highest values of the S_P in their respective groups. Furthermore, it is expected a high influence of the Pseudo Jahn–Teller (PJTE) or Jahn–Teller (JTE) Effects, structural distortion, and octahedral cluster tilting; hence, the materials present similar values. The difference in ferroelectricity for evaluated materials can be justified by the accommodation of cations in the characteristic rotated clusters of the $LiNbO_3$ structure, which are oriented between the x and y directions. This is a very particular behavior for this structure since the most common cluster type for structures is oriented along the z-direction as observed for P_4mm and Pm3m. The cation accommodation on the sites of R3c structures was assessed through the distortion degree (Δ) of the octahedra in the structure that is determined by Eq. (2), where d_i is M–O bond length and d_{ave} is the average value for bond distance in each cluster. In this context, a high disorder indicates a minor cation accommodation in the materials clusters. The obtained results indicate that the Al, Ti, Fe, and Ge clusters present the highest distortion degrees because they are observed in clusters preferentially oriented along directions different from that exhibited on the respective simple oxides and other perovskite structures [84–89]. A similar effect is found in the Pb (Ti, Zr) TiO_3 (PZT) material that presents high ferroelectric properties arising from the existence of Zr atoms in non-preferential structural sites (oriented in the z-direction) [90–93].

$$\Delta = \frac{1}{6}\sum_i \left\{ \frac{(d_i - d_{ave})^2}{d_{ave}} \right\} \tag{2}$$

Moreover, the accommodation of cations in clusters can justify the variation of ferroelectric properties for all the proposed materials in this section, since a proportional variation of ferroelectricity occurs as a function of cation comfort in the R3c clusters. The results show that, in general, the cations from the fifth period of the Periodic Table are well accommodated within the $LiNbO_3$ clusters resulting in the lowest ferroelectric properties among evaluated materials. Similarly, the disorder in the system increases for materials composed of cations from the sixth period, resulting in an increase of S_P compared to those formed by elements of the fourth period. In turn, cluster composed by second and third periods elements is responsible for the higher ferroelectricity due to their lesser adequacy in characteristic sites of R3c structures.

The ferroelectric properties of the materials were also investigated using the constant dielectric (ε) and polarizability (α) values for each material, which are

Table 2 Theoretical results for Spontaneous Polarization (S_P), dielectric constant (ε), and polarizability (α) for R3c evaluated materials

	S_P ($\mu C.cm^{-2}$)	Dielectric Constant (ε–$Bohr^3$)		Polarizability (α)		Distortion degree	
		x	z	x	z	$\Delta_{[AO]_6}$	$\Delta_{[BO]_6}$
Nickel							
$TiNiO_3$	103.63	224.03	198.27	4.67	4.25	2.14	46.91
$GeNiO_3$	88.19	149.22	146.94	3.84	3.80	1.97	17.85
$ZrNiO_3$	80.48	174.74	161.55	3.77	3.56	6.56	15.07
$SnNiO_3$	81.62	174.89	166.40	3.85	3.71	6.07	6.06
$HfNiO_3$	84.89	224.04	198.27	4.67	4.25	7.18	10.26
$PbNiO_3$	88.90	2.75×10^6	41,977.15	285.38	5.35	4.24	8.49
$NiTiO_3$	103.63	4.85	4.20	213.44	177.27	46.97	2.17
$NiGeO_3$	87.01	3.82	3.79	147.48	146.48	19.55	1.83
$NiZrO_3$	90.62	3.74	3.53	173.07	159.66	6.53	6.53
$NiSnO_3$	80.82	4.06	3.98	206.92	201.55	5.94	5.94
$NiHfO_3$	84.33	3.81	3.66	173.66	165.41	7.14	7.14
$NiPbO_3$	88.51	32,427.81	5.29	$2.12.10^6$	281.45	4.12	8.71
Iron II							
$FeTiO_3$	96.17	4.81	1239.06	241.55	78,504.35	6.23	46.50
$FeGeO_3$	66.62	4.21	3.80	175.02	152.63	8.73	13.89
$FeZrO_3$	72.62	3.92	3.50	192.72	165.01	20.57	11.79
$FeSnO_3$	64.70	4.11	3.71	199.50	174.00	18.22	4.05
$FeHfO_3$	67.48	5.18	3.66	269.52	171.58	19.24	7.47
Iron III							
$AlFeO_3$	91.29	194.16	177.59	4.59	4.28	10.50	22.62
$FeAlO_3$	91.81	184.57	161.18	4.61	4.15	22.60	10.97
$FeVO_3$	87.02	306.29	1.91×10^9	5.90	$3.05.\ 10^7$	34.68	3.40
$PbFeO_3$	76.49	2.96×10^5	425.15	4253.70	7.088	2.94	18.97
$BiFeO_3$	57.34	456.03	$1.67.\ 10^6$	7.5605	24,111.92	21.51	14.75
Lead							
$PbVO_3$	36.76	–	–	–	–	16.78	1.79
$PbCrO_3$	49.56	6.84×10^7	431.77	1.04×10^6	6.99	11.80	0.73
$PbMnO_3$	62.42	1.64×10^8	394.73	2.06×10^6	5.95	6.58	18.57
$PbFeO_3$	76.49	2.96×10^5	425.15	4253.70	7.088	2.94	18.97
$PbCoO_3$	19.73	4.93×10^9	340.94	7.21×10^7	5.98	8.24	17.57

(continued)

Table 2 (continued)

	S_P (μC.cm^{-2})	Dielectric Constant (ε–Bohr^3)		Polarizability (α)		Distortion degree	
		x	z	x	z	$\Delta_{[AO]_6}$	$\Delta_{[BO]_6}$
$PbCuO_3$	92.51	–	–	–	–	7.54	4.10

dependent on the direction. The first refers to the level of response of a material to an external electric field while the latter determines how much it is polarizable. Therefore, a material with good ferroelectric properties must present high values for S_P allied to high ε and α values in the same direction. The theoretical results obtained show that the proposed materials have high α and ε values, which are proportional to the S_P values. In addition, it is noteworthy that the systems composed of Pb are more polarizable and more susceptible to external electric fields compared to the other materials investigated. This feature is attributed to the cations size and the lesser interaction between the nucleus and the valence electrons, making the electronic cloud more polarizable. In particular, for $PbVO_3$ and $PbCuO_3$ materials, the computational simulation was not able to determine the values of ε and α due to the high covalent character for the Pb–O bonds in these materials.

It is also noteworthy that the evaluation of dielectric constant and polarizability evidences the anisotropic behavior of ferroelectric properties along the R3c unit cell. The results indicate that all investigated materials have high properties more pronounced along the x and z directions, being the S_P more pronounced along z-direction for the $FeTiO_3$, $BiFeO_3$, and $FeVO_3$ materials, which contain Fe^{3+} with a high distortion degree for a cluster within the structure. It is important to highlight the high values of S_P, α, and ε for these materials, mainly for BFO and $FeVO_3$. Whereas, the other materials possess ferroelectric properties more pronounced along the x-direction. Thus, the obtained results for evaluated materials indicate that they have high ferroelectric properties in the crystalline structure R3c, point out all materials as potential alternatives to the development of technological devices based on ferroelectricity [41–43, 49, 94].

The reliability of the theoretical methodology is evidenced by the agreement between theoretical and experimental results to S_P for $PbNiO_3$, $NiTiO_3$, and $BiFeO_3$. The experimental measurements for S_P for such materials are 100 μC.cm^{-2}, 97 μC.cm^{-2} and 60 μC.cm^{-2}, respectively [95–99]. Therefore, DFT simulations are a very useful tool for the prediction of ferroelectric properties of materials with very specific structural features.

3.3 Ferroelectricity Raised by Structural Deformations

The widespread utilization of the ferroelectric materials requires complex architectures based in nanomaterial deposited on a properly oriented substrate (thin-films), where the molecular level can be described by a combination of the surface and interface effects inducing countless physical and chemical properties [100]. In particular, in the case of ferroelectricity, the structural arrangement of the different orientations disturbs the electron density affecting the balance between Colombian (long-range) and covalent (short-range) effects associated with spontaneous polarization (S_p) [101–103].

Moreover, the film architecture originates mechanical stress in the interface region between the nanoparticle and the substrate, which is due to the lattice mismatch among the crystalline structures. In this context, strain engineering becomes a powerful and useful tool that enables the stabilization of intriguing crystalline structures for various oxides, which exhibits unusual phenomena's never observed in their unstrained states [104, 105].

Therefore, to elucidate the ferroelectric properties of strained $ATiO_3$ (A = Mn, Fe, Ni) materials, we carried out first-principles calculations based on density functional theory (DFT/PBE0) using CRYSTAL [78] package.

For this purpose, both uniaxial and biaxial strain were considered through geometry-constrained relaxation for two different models: (i) for uniaxial strain we fixed the *c* lattice parameter and relaxed the a and b lattice parameters and all atomic coordinates; (ii) meanwhile for biaxial strain we fixed the *a* and *b* lattice parameter and relaxed the c-axis and all atomic coordinates Manually changing the lattice parameter allows for the investigation of different compressive and expansive strains applied parallel to the [001] and [110] direction, as depicted in Fig. 4.

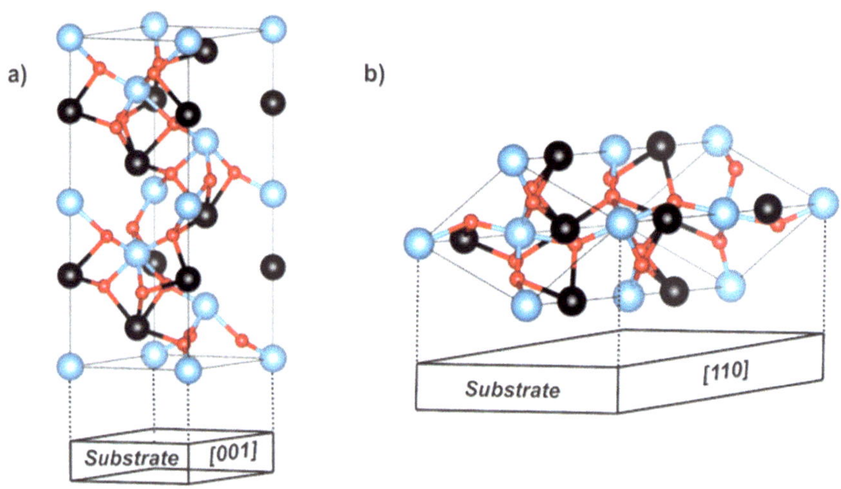

$$\%_{[001]} = \frac{(c_0 - c)}{c}\ x\ 100 \qquad \%_{[110]} = \frac{(a_0 - a)}{a}\ x\ 100$$

Fig. 4 Schematic representation of **a** [001]-oriented and **b** [110]-oriented $ATiO_3$ (A = Mn, Fe, Ni) unit cell. The black, blue, and red balls correspond to A (Mn, Fe, Ni), Ti and O atoms, respectively

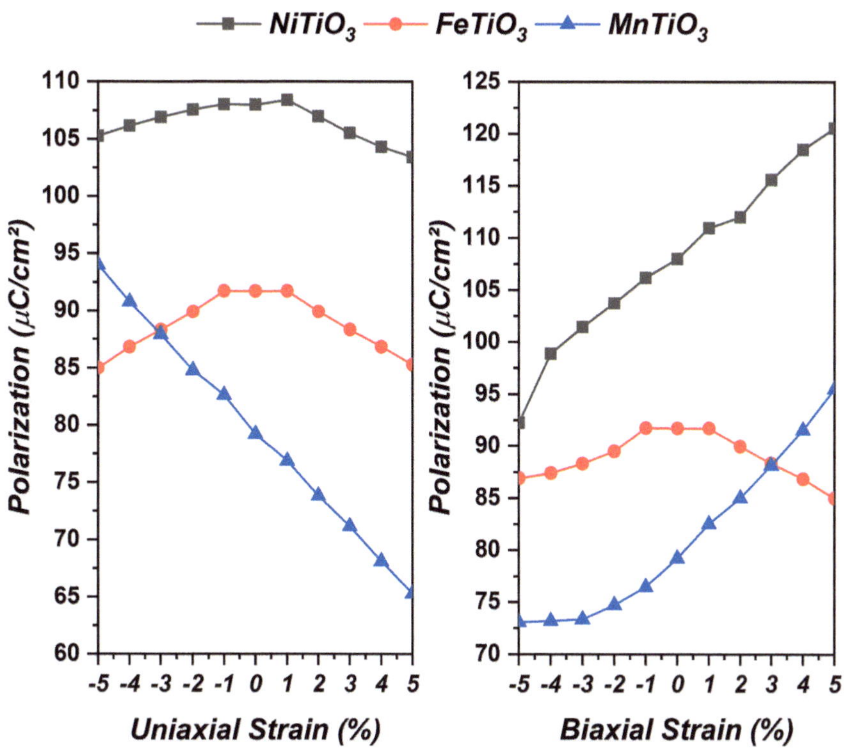

Fig. 5 Calculated spontaneous polarization for [001] uniaxial and [110] biaxial strained $ATiO_3$ (A = Mn, Fe, Ni) material

Let us now briefly discuss the especially relation among strain-induced structural disorders and the ferroelectric properties of $ATiO_3$ (A = Mn, Fe, Ni) evaluated through spontaneous polarization, as presented in Fig. 5. In this case, it was noted that [001] uniaxial strain slightly reduces the spontaneous polarization for both tensile and compressive regions in comparison to the unstrained model of $NiTiO_3$ and $FeTiO_3$. Further, it was possible to indicate that [001] uniaxial strain keeps the internal polarization because the difference between short and long A–O (A = Ni, Fe) bonds remains almost constant along the investigated region, confirming that A-site cations play a fundamental role on the description of ferroelectric properties of $ATiO_3$ materials with $LiNbO_3$-type structure.

On the other hand, the [001] uniaxial strain induces an important control of the S_p for the $MnTiO_3$, resulting in reduced/increased values for tensile/compressive regions, respectively. This fact can be associated with the singular behavior of Mn–O bonds in comparison to Fe–O and Ni–O, resulting in an intriguing ferroelectric control.

On the other hand, the biaxial [110] strain drastic change the spontaneous polarization of $MnTiO_3$ and $NiTiO_3$, where the compressive region results in a large ferroic order, whereas the tensile regime reduces the spontaneous polarization. This fact can be attributed to the increased local disorder of [(Mn/Ni)O_6] clusters that become more distorted for the compressive strain inducing superior ferroelectric properties, while the shortening of long A–O (A = Mn, Ni) bond distances makes the A center more regular resulting in reduced polarization. These results allowed us to predict that biaxial [110] strain is an interesting tool to control the ferroelectric properties of $NiTiO_3$ materials, following the experimental results reported in the literature [106–110].

Epitaxial strain can induce collective phenomena and new functionalities in complex oxide thin films. Strong coupling between strain and polar lattice modes can stabilize new ferroelectric phases from nonpolar dielectrics or enhance electric polarization as previously discussed. Furthermore, the magnetic degree of freedom can be switched or controlled by epitaxial strain being an interesting tool to develop and enhance the multiferroic properties [111–117].

4 Conclusions

In this chapter, the role of DFT calculations on ferroelectric materials was summarized. In particular, the structural, electronic, and dynamical effects associated with the raising of ferroelectricity in several oxides were discussed combining symmetry analysis, electronic structure calculations, and solid-state chemistry concepts.

As regards the origin of ferroelectric properties, main tree mechanisms were investigated: (i) in case of $PbTi_{0.5}Zr_{0.5}O_3$ and $SnTi_{0.5}Zr_{0.5}O_3$ perovskites the ferroelectric order was addressed to the role of the covalent character of A–O chemical bonds that induces an increased spontaneous polarization for SZT in comparison to PZT; (ii) moreover, for several ABO_3 perovskites with R3c crystalline structure the intrinsic polarization was related to structural aspects, such as the tilting of [AO_6] and [BO_6] clusters, off-centering cation displacement and Jahn–Teller Effects; (iii) in the third case, the strain-induced control of ferroelectric properties for $ATiO_3$ (A = Mn, Fe, Ni) materials was confirmed, suggesting that in-plane biaxial [110] strain corresponds to an interesting tool to control the ferroelectric properties.

Besides the challenge as regards the atomic-level description of the ferroelectric order, the DFT calculations have been successfully applied as a complementary tool to elucidate the origin of intrinsic spontaneous polarization, as well as can be used to predict and design innovative candidates for superior technological applications.

Acknowledgements The authors acknowledge the financial support from CAPES (PNPD Programm), CNPq, State University of Minas Gerais (UEMG), and Fundação Araucária (Financing Project 009/2017).

References

1. Cohen, R.E.: Nature **358**, 136–138 (1992)
2. Vasudevan, R.K., Balke, N., Maksymovych, P., Jesse, S., Kalinin, S.V.: Appl. Phys. Rev. **4**, 021302 (2017)
3. Li, F., Lin, D., Chen, Z., Cheng, Z., Wang, J., Li, C., Xu, Z., Huang, Q., Liao, X., Chen, L.-Q., Shrout, T.R., Zhang, S.: Nat. Mater. **17**, 349–354 (2018)
4. Seshadri, R., Hill, N.A.: Chem. Mater. **13**, 2892–2899 (2001)
5. Van Aken, B.B., Palstra, T.T.M., Filippetti, A., Spaldin, N.A.: Nat. Mater. **3**, 164–170 (2004)
6. Chen, W.-T., Wang, C.-W., Wu, H.-C., Chou, F.-C., Yang, H.-D., Simonov, A., Senn, M.S.: Phys. Rev. B **97**, 144102 (2018)
7. Cheong, S.-W., Mostovoy, M.: Nat. Mater. **6**, 13–20 (2007)
8. Nordlander, J., Campanini, M., Rossell, M.D., Erni, R., Meier, Q.N., Cano, A., Spaldin, N.A., Fiebig, M., Trassin, M.: Nat. Commun. **10**, 5591 (2019)
9. Zhang, Y., Jie, W., Chen, P., Liu, W., Hao, J.: Adv. Mater. **30**, 1707007 (2018)
10. Kreisel, J., Alexe, M., Thomas, P.A.: Nat. Mater. **11**, 260–260 (2012)
11. Yang, S.Y., Seidel, J., Byrnes, S.J., Shafer, P., Yang, C.H., Rossell, M.D., Yu, P., Chu, Y.H., Scott, J.F., Ager, J.W., Martin, L.W., Ramesh, R.: Nat. Nanotechnol. **5**, 143–147 (2010)
12. Cui, Y., Briscoe, J., Dunn, S.: Chem. Mater. **25**, 4215–4223 (2013)
13. Li, L., Salvador, P.A., Rohrer, G.S.: Nanoscale **6**, 24–42 (2014)
14. Wang, M., Wang, B., Huang, F., Lin, Z.: Angew. Chem. Int. Ed. **58**, 7526–7536 (2019)
15. Chen, F., Huang, H., Guo, L., Zhang, Y., Ma, T.: Angew. Chem. Int. Ed. **58**, 10061–10073 (2019)
16. Yao, Z., Song, Z., Hao, H., Yu, Z., Cao, M., Zhang, S., Lanagan, M.T., Liu, H.: Adv. Mater. **29**, 1601727 (2017)
17. Zhao, L., Liu, Q., Gao, J., Zhang, S., Li, J.-F.: Adv. Mater. **29**, 1701824 (2017)
18. Yang, L., Kong, X., Li, F., Hao, H., Cheng, Z., Liu, H., Li, J.-F., Zhang, S.: Prog. Mater Sci. **102**, 72–108 (2019)
19. Nat. Mater. **18**, 187–187 (2019)
20. Tokura, Y., Seki, S., Nagaosa, N.: Rep. Prog. Phys. **77**, 076501 (2014)
21. Khomskii, D.I.: J. Magn. Magn. Mater. **306**, 1–8 (2006)
22. Eerenstein, W., Mathur, N.D., Scott, J.F.: Nature **442**, 759–765 (2006)
23. Spaldin, N.A.: MRS Bull. **42**, 385–390 (2017)
24. Zhang, Y., Sun, J., Perdew, J.P., Wu, X.: Phys. Rev. B **96**, 035143 (2017)
25. Baroni, S., Giannozzi, P., Testa, A.: Phys. Rev. Lett. **58**, 1861–1864 (1987)
26. Gonze, X.: Phys. Rev. A **52**, 1096–1114 (1995)
27. King-Smith, R.D., Vanderbilt, D.: Phys. Rev. B **47**, 1651–1654 (1993)
28. Rabe, K.M., Ghosez, P.: Physics of Ferroelectrics: A Modern Perspective, pp. 117–174. Springer, Berlin Heidelberg (2007)
29. Vanderbilt, D.: Curr. Opin. Solid State Mater. Sci. **2**, 701–705 (1997)
30. Yuk, S.F., Pitike, K.C., Nakhmanson, S.M., Eisenbach, M., Li, Y.W., Cooper, V.R.: Sci. Rep. **7**, 43482 (2017)
31. Varignon, J., Bristowe, N.C., Bousquet, É., Ghosez, P.: C.R. Phys. **16**, 153–167 (2015)
32. Xu, K., Lu, X.-Z., Xiang, H.: npj Quantum Mater. **2**, 1 (2017)
33. Ye, M., Vanderbilt, D.: Phys. Rev. B **93**, 134303 (2016)
34. Varignon, J., Bristowe, N.C., Bousquet, E., Ghosez, P.: Sci. Rep. **5**, 15364 (2015)
35. Fennie, C.J.: Phys. Rev. Lett. **100**, 167203 (2008)
36. Burke, K.: J. Chem. Phys. **136**, 150901–150909 (2012)
37. Zhang, I.Y., Su, N.Q., Brémond, É.A.G., Adamo, C., Xu, X.: J. Chem. Phys. **136**, 174103–174110 (2012)
38. Hautier, G., Jain, A., Ong, S.: J. Mater. Sci. **47**, 7317–7340 (2012)
39. Lewars, E.G.: Computational Chemistry. Introduction to the Theory and Applications of Molecular and Quantum Mechanics, 2nd edn. Springer, Londres (2011)

40. Becke, A.D.: J. Chem. Phys. **140**, 18A301-18A320 (2014)
41. Lacerda, L.H.S., de Lazaro, S.R.: New J. Chem. **43**, 10610–10617 (2019)
42. Lacerda, L.H.S., de Lazaro, S.R.: J. Magn. Magn. Mater. **465**, 412–420 (2018)
43. Lacerda, L.H.S., de Lazaro, S.R.: Comput. Mater. Sci. **153**, 228–234 (2018)
44. Ribeiro, R.A.P., de Lazaro, S.R., Pianaro, S.A.: J. Magn. Magn. Mater. **391**, 166–171 (2015)
45. Ribeiro, R.A.P., Longo, E., Andrés, J., de Lazaro, S.R.: Phys. Chem. Chem. Phys. **20**, 28382–28392 (2018)
46. Rayaprol, S., Ribeiro, R.A.P., Singh, K., Reddy, V.R., Kaushik, S.D., de Lazaro, S.R.: J. Alloy. Compd. **774**, 290–298 (2019)
47. Santiago, A.A.G., Tranquilin, R.L., Oliveira, M.C., Ribeiro, R.A.P., de Lazaro, S.R., Correa, M.A., Bohn, F., Longo, E., Motta, F.V., Bomio, M.R.D.: J. Phys. Chem. C **124**, 5378–5388 (2020)
48. Manjunatha, K., Angadi, V.J., Ribeiro, R.A.P., Longo, E., Oliveira, M.C., Bomio, M.R.D., de Lázaro, S.R., Matteppanavar, S., Rayaprol, S., Babu, P.D., Pasha, M.: J. Magn. Magn. Mater. **502**, 166595 (2020)
49. Lacerda, L.H.S., Ribeiro, R.A.P., de Lazaro, S.R.: J. Magn. Magn. Mater. **480**, 199–208 (2019)
50. Lacerda, L.H.S., de Lazaro, S.R., Phys. Status Solidi B 1700119 (2017).
51. Lacerda, L.H.S., de Lazaro, S.R.: Quim. Nova **39**, 261–266 (2016)
52. Lacerda, L.H.S., de Lazaro, S.R.: Solid State Ionics **297**, 36–42 (2016)
53. Ribeiro, R.A.P., de Lázaro, S.R.: RSC Adv. **4**, 59839–59846 (2014)
54. Ribeiro, R.A.P., Camilo, A., de Lazaro, S.R.: J. Magn. Magn. Mater. **394**, 463–469 (2015)
55. Jurelo, A.R., Ribeiro, R.A.P., de Lazaro, S.R., Monteiro, J.F.H.L.: Phys. Chem. Chem. Phys. **20**, 27011–27018 (2018)
56. Ribeiro, R.A.P., de Lázaro, S.R.: Quim. Nova **37**, 1165–1170 (2014)
57. Ribeiro, R.A.P., de Lazaro, S.R.: J. Alloy. Compd. **714**, 553–559 (2017)
58. Oliveira, M.C., Ribeiro, R.A.P., Longo, E., Bomio, M.R.D., Motta, F.V., de Lazaro, S.R.: Int. J. Quantum Chem. **120**, e26054 (2020)
59. Oliveira, M.C., Ribeiro, R.A.P., Gracia, L., de Lazaro, S.R., de Assis, M., Oliva, M., Rosa, I.L.V., Gurgel, M.F.C., Longo, E., Andrés, J.: CrystEngComm **20**, 5519–5530 (2018)
60. Teixeira, M.M., de Oliveira, R.C., Oliveira, M.C., Ribeiro, R.A.P., de Lazaro, S.R., Li, M.S., Chiquito, A.J., Gracia, L., Andrés, J., Longo, E.: Inorg. Chem. **57**, 15489–15499 (2018)
61. Lacerda, L.H.S., de Lazaro, S.R.: J. Mater. Sci. **55**, 6875–6890 (2020)
62. Ribeiro, R.A.P., Lacerda, L.H.S., Longo, E., Andrés, J., de Lazaro, S.R.: J. Magn. Magn. Mater. **475**, 544–549 (2019)
63. Ribeiro, R.A.P., Oliveira, M.C., de Sousa, A.G., Bomio, M.R.D., Motta, F.V., Gracia, L., de Lazaro, S.R., Longo, E., Andrés, J.: J. Appl. Phys. **126**, 235301 (2019)
64. Ribeiro, R.A.P., de Lazaro, S.R., Gracia, L., Longo, E., Andrés, J.: J. Magn. Magn. Mater. **453**, 262–267 (2018)
65. Ribeiro, R.A.P., Oliveira, M.C., Bomio, M.R.D., de Lazaro, S.R., Andrés, J., Longo, E.: Appl. Surf. Sci. **509**, 145321 (2020)
66. Oliveira, M.C., Fonseca, V.S., Neto, N.F.A., Ribeiro, R.A.P., Longo, E., de Lazaro, S.R., Motta, F.V., Bomio, M.R.D.: Ceram. Int. **46**, 9446–9454 (2020)
67. Soares, G.B., Ribeiro, R.A.P., de Lazaro, S.R., Ribeiro, C.: RSC Adv. **6**, 89687–89698 (2016)
68. Dall'Olio, S., Dovesi, R., Resta, R.: Phys. Rev. B **56**, 10105–10114 (1997)
69. Resta, R.: Rev. Mod. Phys. **66**, 899–915 (1994)
70. Liyanage, M., Miller, R., Rajapakse, R.: Smart Mater. Struct. **28**, 034002 (2019)
71. Gopejenko, A., Piskunov, S., Zhukovskii, Y.F.: Comput. Theor. Chem. **1104**, 56–60 (2017)
72. O'Donnell, S., Chung, C.-C., Carbone, A., Broughton, R., Jones, J.L., Maggard, P.A.: Chem. Mater. (2020)s
73. Diehl, L., Bette, S., Pielnhofer, F., Betzler, S., Moudrakovski, I., Ozin, G.A., Dinnebier, R., Lotsch, B.V.: Chem. Mater. **30**, 8932–8938 (2018)
74. Fix, T., Sahonta, S.L., Garcia, V., MacManus-Driscoll, J.L., Blamire, M.G.: Cryst. Growth Des. **11**, 1422–1426 (2011)
75. Parker, W.D., Rondinelli, J.M., Nakhmanson, S.M.: Phys. Rev. B **84**, 245126 (2011)

76. Taib, M.F.M., Yaakob, M.K., Hassan, O.H., Chandra, A., Arof, A.K., Yahya, M.Z.A.: Ceram. Int. **39**, S297–S300 (2013)
77. Roy, A., Vanderbilt, D.: Phys. Rev. B **83**, 134116 (2011)
78. Dovesi, R., Erba, A., Orlando, R., Zicovich-Wilson, C.M., Civalleri, B., Maschio, L., Rérat, M., Casassa, S., Baima, J., Salustro, S., Kirtman, B.: Wiley Interdiscip. Rev.: Comput. Mol. Sci. **8**, e1360 (2018)
79. Inaguma, Y., Aimi, A., Shirako, Y., Sakurai, D., Mori, D., Kojitani, H., Akaogi, M., Nakayama, M.: J. Am. Chem. Soc. **136**, 2748–2756 (2014)
80. Benedek, N.A., Fennie, C.J.: J. Phys. Chem. C **117**, 13339–13349 (2013)
81. Silva, J., Reyes, A., Esparza, H., Camacho, H., Fuentes, L.: Integr. Ferroelectr. **126**, 47–59 (2011)
82. Roy, S., Majumder, S.B.: J. Alloy. Compd. **538**, 153–159 (2012)
83. Nassau, K., Levinstein, H.J., Loiacono, G.M.: J. Phys. Chem. Solids **27**, 983–988 (1966)
84. Wang, W., Liang, Z., Han, X., Chen, J., Xue, C., Zhao, H.: J. Alloy. Compd. **622**, 504–512 (2015)
85. Eglitis, R.I.: Appl. Surf. Sci. Part B **358**, 556–562 (2015)
86. Al-Khatatbeh, Y., Lee, K.K.M.: J. Superhard Mater. **36**, 231–245 (2014)
87. Das, S., Jayaraman, V.: Prog. Mater Sci. **66**, 112–255 (2014)
88. González, G.B., Mason, T.O., Quintana, J.P., Warschkow, O., Ellis, D.E., Hwang, J.-H., Hodges, J.P., Jorgensen, J.D.: J. Appl. Phys. **96**, 3912–3920 (2004)
89. Ramo, D.M., Gavartin, J.L., Shluger, A.L., Bersuker, G.: Phys. Rev. B **75**, 205336–205346 (2007)
90. Izyumskaya, N., Alivov, Y.I., Cho, S.J., Morkoç, H., Lee, H., Kang, Y.S.: Crit. Rev. Solid State Mater. Sci. **32**, 111–202 (2007)
91. Hoffmann, M.J., Kungl, H.: Curr. Opin. Solid State Mater. Sci. **8**, 51–57 (2004)
92. Dragan, D.: Rep. Prog. Phys. **61**, 1267–1326 (1998)
93. Setter, N., Damjanovic, D., Eng, L., Fox, G., Gevorgian, S., Hong, S., Kingon, A., Kohlstedt, H., Park, N.Y., Stephenson, G.B., Stolitchnov, I., Taganstev, A.K., Taylor, D.V., Yamada, T., Streiffer, S.: J. Appl. Phys. **100**, 051606–051651 (2006)
94. Lacerda, L.H.S., de Lazaro, S.R.: J. Magn. Magn. Mater. **500**, 166364–166374 (2020)
95. Hao, X.F., Stroppa, A., Barone, P., Filippetti, A., Franchini, C., Picozzi, S.: New J. Phys. **16**, 015030–015050 (2014)
96. Hao, X.F., Stroppa, A., Picozzi, S., Filippetti, A., Franchini, C.: Phys. Rev. B **86**, 014116–014120 (2012)
97. Xin, C., Wang, Y., Sui, Y., Wang, Y., Wang, X., Zhao, K., Liu, Z., Li, B., Liu, X.: J. Alloy. Compd. **613**, 401–406 (2014)
98. Je-Geun, P., Duc, L.M., Jaehong, J., Sanghyun, L.: J. Phys.: Condens. Matter **26**, 433202–433236 (2014)
99. Feng, H.: J. Magn. Magn. Mater. **322**, 3755–3759 (2010)
100. Wolf, S.A., Awschalom, D.D., Buhrman, R.A., Daughton, J.M., von Molnár, S., Roukes, M.L., Chtchelkanova, A.Y., Treger, D.M.: Science **294**, 1488–1495 (2001)
101. Dawber, M., Rabe, K.M., Scott, J.F.: Rev. Mod. Phys. **77**, 1083–1130 (2005)
102. Liu, S., Kim, Y., Tan, L.Z., Rappe, A.M.: Nano Lett. **16**, 1663–1668 (2016)
103. Zhang, Y., Wang, J., Sahoo, M.P.K., Shimada, T., Kitamura, T.: Phys. Chem. Chem. Phys. **19**, 26047–26055 (2017)
104. Schlom, D.G., Chen, L.-Q., Fennie, C.J., Gopalan, V., Muller, D.A., Pan, X., Ramesh, R., Uecker, R.: MRS Bull. **39**, 118–130 (2014)
105. Ederer, C., Spaldin, N.A.: Phys. Rev. Lett. **95**, 257601 (2005)
106. Varga, T., Droubay, T.C., Bowden, M.E., Colby, R.J., Manandhar, S., Shutthanandan, V., Hu, D., Kabius, B.C., Apra, E., Shelton, W.A., Chambers, S.A.: J. Vac. Sci. Technol. B **31**, 030603 (2013)
107. Varga, T., Droubay, T.C., Bowden, M.E., Kovarik, L., Hu, D., Chambers, S.A.: Adv. Condens. Matter Phys. **2015**, 9 (2015)

108. Varga, T., Droubay, T.C., Bowden, M.E., Nachimuthu, P., Shutthanandan, V., Bolin, T.B., Shelton, W.A., Chambers, S.A.: Thin Solid Films **520**, 5534–5541 (2012)
109. Varga, T., Droubay, T.C., Kovarik, L., Hu, D., Chambers, S.A.: Thin Solid Films (2018)
110. Varga, T., Droubay, T.C., Kovarik, L., Nandasiri, M.I., Shutthanandan, V., Hu, D., Kim, B., Jeon, S., Hong, S., Li, Y., Chambers, S.A.: ACS Appl. Mater. Interfaces **9**, 21879–21890 (2017)
111. Cabral, L., Aragón, F.H., Villegas-Lelovsky, L., Lima, M.P., Macedo, W.A.A., Da Silva, J.L.F.: ACS Appl. Mater. Interfaces **11**, 1529–1537 (2019)
112. Burns, S.R., Sando, D., Xu, B., Dupé, B., Russell, L., Deng, G., Clements, R., Paull, O.H.C., Seidel, J., Bellaiche, L., Valanoor, N., Ulrich, C.: npj Quantum Mater. **4**, 18 (2019)
113. Spaldin, N.A., Ramesh, R.: Nat. Mater. **18**, 203–212 (2019)
114. Agbelele, A., Sando, D., Toulouse, C., Paillard, C., Johnson, R.D., Rüffer, R., Popkov, A.F., Carrétéro, C., Rovillain, P., Le Breton, J.-M., Dkhil, B., Cazayous, M., Gallais, Y., Méasson, M.-A., Sacuto, A., Manuel, P., Zvezdin, A.K., Barthélémy, A., Juraszek, J., Bibes, M.: Adv. Mater. **29**, 1602327 (2017)
115. Wei, Y., Nukala, P., Salverda, M., Matzen, S., Zhao, H.J., Momand, J., Everhardt, A.S., Agnus, G., Blake, G.R., Lecoeur, P., Kooi, B.J., Íñiguez, J., Dkhil, B., Noheda, B.: Nat. Mater. **17**, 1095–1100 (2018)
116. Lu, W., Song, W., Yang, P., Ding, J., Chow, G.M., Chen, J.: Scientific Reports **5**, 10245 (2015)
117. Lu, X.-Z., Rondinelli, J.M.: Nat. Mater. **15**, 951 (2016)

Nanocrystalline Spinel Manganese Ferrite $MnFe_2O_4$: Synthesis, Electronic Structure, and Evaluation of Their Magnetic Hyperthermia Applications

Walmir E. Pottker, Patricia de la Presa, Mateus A. Gonçalves, Teodorico C. Ramalho, Antonio Hernando, and Felipe A. La Porta

Abstract In this paper, nanocrystalline spinel manganese ferrites $MnFe_2O_4$ (MFO) were synthesised using the co-precipitation method under different conditions and characterised by XRD, TEM, magnetic measurements and also theoretical calculations. Herein, the first-principles calculations and electron density topologic analysis at DFT level, including a comparison on the influence of relativistic effects, were carried out to obtain the geometries and the electronic parameters of the bulk and (100), (110) and (111) surfaces of the MFO structure, for the first time. The magnetisation field curves reveal a superparamagnetic behaviour for all the analysed samples. The saturation magnetisation values were determined to be 56.6, 53.3, and 54.8 emu/g at 300 K, respectively. Furthermore, the aqueous colloids were transferred to organic media to investigate the effect of particle size and aggregation degrees on the heating efficiency of the nanoparticles. Our findings demonstrate that the MFO spinel nanocrystals hold promise for innovative applications in magnetic hyperthermia.

Keywords $MnFe_2O_4$ · DFT calculations · Magnetic hyperthermia · Nanocrystals

---

W. E. Pottker · F. A. La Porta (✉)
Laboratório de Nanotecnologia E Química Computacional, Universidade Tecnológica Federal do Paraná, Londrina, PR 86036-370, Brazil
e-mail: felipelaporta@utfpr.edu.br

W. E. Pottker · P. de la Presa · A. Hernando
Instituto de Magnetismo Aplicado, UCM-ADIF-CSIC, A6 22500 Km, 28230 Las Rozas, Spain

P. de la Presa · A. Hernando
Departamento Física de Materiales UCM, Ciudad Universitaria, 28040 Madrid, Spain

M. A. Gonçalves · T. C. Ramalho
Department of Chemistry, Federal University of Lavras, Lavras, Minas Gerais, CEP 37200-000, Brazil

F. A. La Porta and C. A. Taft (eds.), *Functional Properties of Advanced Engineering Materials and Biomolecules*, Engineering Materials,
https://doi.org/10.1007/978-3-030-62226-8_12

1 Introduction

Nowadays, the heating efficiency of magnetic nanoparticles is being intensively investigated because of their potential applications in several fields such as biomedicine, catalysis, or energy harvesting [1–4]. Under certain physical conditions, the magnetic nanoparticles can convert the work of an electromagnetic field into thermal energy; therefore, they can be used as nano-heaters (e.g., for potential applications in biomedicine) [5, 6]. In this regard, it is well known that the efficiency of the energy conversion depends on intrinsic particle properties, including parameters such as coordination number, symmetry, particle size, magnetisation, magnetic anisotropy, and bond distance and disorders, as well as extrinsic properties, including parameters such as applied field, media viscosity, and aggregation degree [7–10].

Among the magnetic nanoparticles, manganese ferrite $MnFe_2O_4$ (MFO) spinel nanocrystals have been the focus of intense research in recent years, due to their low toxicity, excellent chemical stability, and small magnetic anisotropy at room temperature (RT) that enable a wide variety of biological applications, e.g. in cancer therapy [7–11]. In particular, the heating efficiency in aqueous colloids of these nanomaterials has been intensively studied [12–18]. From a structural point of view, MFO materials generally crystallise in a mixed phase between both the normal and inverse spinel structures [19, 20]. Specifically, the MFO spinel structures are conveniently described as $(Mn^{2+}_{1-\delta}Fe^{3+}_{\delta})[Mn^{2+}_{\delta}Fe^{3+}_{2-\delta}]O^{2-}_{4}$, where δ is the inversion parameter, which indicates that the divalent cations partially occupy both tetrahedral (A) and octahedral (B) sites, respectively. A broad range of synthetic strategies have been extensively employed to obtain highly crystalline MFO materials of different sizes, morphologies, and well-designed structures, utilizing several eco-friendly routes [8–11].

Overall, in this study we evaluated the crystalline structure, polydispersitivity degree, morphologies, size distributions, electronic and magnetic behaviour of MFO nanocrystals, which were synthesised using the co-precipitation method under different conditions. Also, the first-principles calculations and electron density topologic analysis at density functional theory (DFT) level with relativistic effects were carried out to obtain the geometries and the electronic parameters of the bulk and (100), (110) and (111) surfaces of the MFO structure for the first time. We also investigated the heating efficiency of these magnetic nanocrystals for hyperthermia therapeutic applications.

2 Experimental Methods

2.1 Synthesis of MFO Nanoparticles

We prepared the following MFO powders using the co-precipitation synthesis procedure: 2.5 mmol of $MnCl_2$ (Sigma Aldrich) was dissolved in 50 mL of distilled water (solution 1); 5 mmol of $FeCl_3.6H_2O$ (Sigma Aldrich) were separately dissolved in another 50 mL of distilled water (solution 2) [9, 14]. Both solutions were heated and mixed at 50 °C under vigorous magnetic stirring. During thc sequence, 100 mL of NaOH (3 M) solution at 95 °C was added dropwise. Then, the synthesis temperature was maintained at 80 °C for about 120 min under magnetic stirring (600 rpm) to obtain different particle size distributions [18]. Finally, the products obtained were magnetically separated, washed several times with water, and dried at RT. Subsequently, the obtained precipitate was dispersed in 10 mL of 0.1 M TMAOH and/or OA according to previous studies [11, 21]. Two additional samples were synthesised with a similar procedure; however, the synthesis temperature was increased up to 100 °C with different magnetic stirring times, namely 30 and 120 min. To separate the different sizes, all samples were fractioned by centrifugation [22, 23].

2.2 *Structural and Morphological Characterization*

Powder X-ray diffraction (PXRD) patterning was performed at RT on a PANalytical X'Pert Pro MPD diffractometer with Cu-Kα radiation in the 2θ range from 10 to 100 °C using a step of 0.03 min^{-1}. The PXRD patterns were also quantitatively analysed by the Rietveld method [24], using FullProf software [25]. The size and shape of the particles were observed using a JEOL JEM 2100F transmission electron microscopy (TEM) operated at 200 keV.

2.3 *Static and Dynamic Magnetic Characterization*

Magnetic characterisation of power samples was performed in a vibrating sample magnetometer (MLVSM9 MagLab 9 T, Oxford Instruments) at RT. Hysteresis loops of the samples were registered at 300 K at a maximum magnetic field of 50 kOe. The heating capacities of the MFO samples were measured in a closed circuit of water (at a constant temperature of 16 °C) using a commercial system, namely Magnetherm 1.5 (Nanotherics). The samples were characterised under a radiofrequency field with 110 kHz and 200 Oe field frequency and amplitude, respectively. More details on this methodology are described in [26].

2.4 *Computational Methods*

All of the theoretical calculations were performed using the Amsterdam Density Functional program (ADF-BAND) [27]. For all these models used in this work, were optimized at the DFT level with Perdew-Becke-Ernzerhof (PBE) functional [28] and triple-z Slater-type (TZP) basis set for all atoms, including relativistic calculations from using the zeroth-order regular approximation (ZORA) [29]. To reduce the size of the basis set, as implemented in ADF-BAND, a frozen core approximation was used for the atom cores. And so, the main characteristic of this program is that it performs "numerical" integrations for all the matrix elements.

Initially, two models were built to verify the spontaneity in forming a solid-solution with MFO phase. And so, we built the magnetite (with the space Group Fd-3 m; a = 8.3941 Å; α = 90.00; Z = 8) and MFO phases (with the space Group Fd-3 m; a = 8.4983 Å, α = 90.00; Z = 8) were evaluated based on the parameters reported in the literature [30, 31]. The theoretical results suggesting that the MFO phase has a more stable structure than pure magnetite by about 792.73 $kcal.mol^{-1}$. After the optimized bulk, the (100), (110) and (111) surfaces of MFO structure were built and re-optimized at the same theoretical level. Hence, for these surfaces, was also performed the topological analysis of the electronic density in order to better analyzed the electronic structure and chemical bonds of such structures.

3 Results and Discussion

3.1 Structure and Morphology Analysis

PXRD patterns with Rietveld analysis of the single-crystalline MFO nanoparticles are illustrated in Fig. 1a–c. The analysis results at long range suggest a cubic spinel structure belonging to the *Fd-3 m* space group (PDF # 10–0319). A good fit between

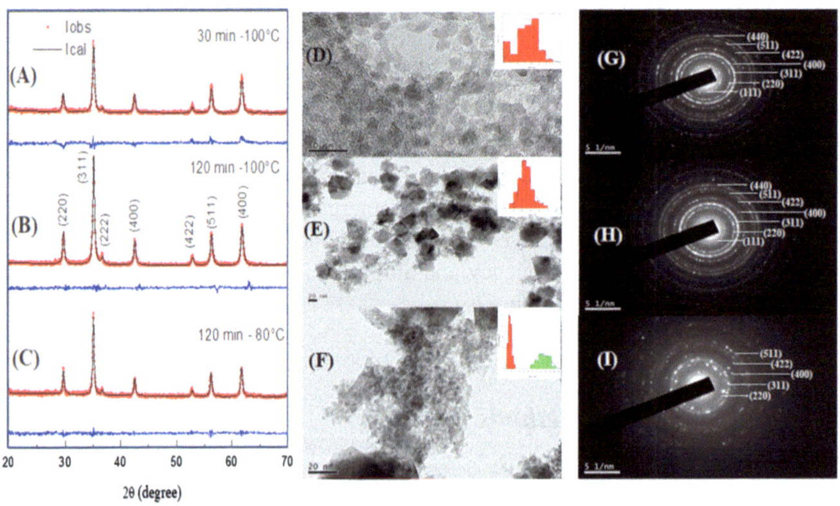

Fig. 1 **a–c** Refined PXRD pattern, **d–f** TEM images and **g–i** SAED patterns of the as-prepared MFO samples

Table 1 The refined lattice parameters (a) of the cubic phase, unit cell volume, crystalline size using Scherrer´s formula and average grain size observed by TEM for the three MFO samples

Samples	Lattice Parameters			Particle Size	
	Lattice constant a (Å)	Volume (Å^3)	χ^2 (%)	Crystallite size < D_{311} > (XRD) (nm)	Average particle size (d) (TEM) (nm)
MFO (30 min–100 °C)	8.494	612.825	1.3	19.4	17 ± 1
MFO (120 min–100 °C)	8.488	611.527	1.4	21.3	19 ± 0.3
MFO (120 min–80 °C)	8.499	613.908	1.5	25.2	9–50 $\pm$ 0.7

the observed and the calculated PXRD profiles was obtained, as shown in Table 1. A small decrease in the unit cell volume of the nanocrystalline with increasing temperature was observed, which can be related to the distribution of Mn^{2+} cations and Fe^{3+} cations in the octahedral and tetrahedral sites, respectively. In this case, the substitution of some Fe^{3+} by Mn^{2+} cations with larger ionic radius in the tetrahedral sites could be the origin of the unit-cell volume expansion for the sample prepared at 80 °C. The crystallite size is in the range of 19–25 nm; this increases with the stirring time and takes the most significant value for the lowest reaction temperature (80 °C).

The average crystallite size was estimated from an intense peak, corresponding to (311) reflection, using the Scherrer´s formula [32, 33], while the TEM images (see Fig. 1d–f) can discriminate the different particle sizes. A comparison of these results is presented in Table 1. In this case, the first one measures the coherent diffraction size, i.e., the more significant the crystallite size is, the thinner the diffraction peaks become, and the average crystallite size has different contributions from small and large particles [34]. Therefore, in case of heterogeneous distribution, the crystallite size calculated as the average of the largest and the smallest particle sizes, and thus can be larger than the smallest particle size [34, 35]. As previously reported, the particle size grows over time. Optimal stirring time for homogeneous particle size is around 30 min for synthesis at low temperature with low polydispersity degree [13]. However, the reaction at a low temperature and long stirring time produces a binomial particle size distribution, i.e., some nuclei can grow during time of synthesis.

Also, the SAED performed for the MFO samples is observed to be consistent with the phase indexed by the Rietveld method, revealing the polycrystalline nature of the as-prepared samples (see Fig. 1g–i). Results from the EDXS spectra on individual MFO samples showed that the atomic ratio of Mn:Fe for the prepared samples is 0.98:2.02 at 100 °C (30 min); 0.99:2.01 at 100 °C (120 min), and 0.86:2.09 at 80 °C (120 min). This indicates that the MFO samples have a relatively good average stoichiometry for each synthesis procedure.

3.2 Magnetic Properties

The magnetic loops of the obtained manganese ferrite are shown in Fig. 2. At 300 K, all samples exhibited small but appreciably coercivity, indicating that the particles are blocked at RT. The values of saturation magnetisation (M_S) and coercive field (H_C) of three samples are presented in Table 2.

The critical MFO size required to exhibit a superparamagnetic behaviour is around 20 nm [34]. For the three synthesis procedures, the average particle size is close to the critical size, but the largest particles in the size distribution dominate the ferromagnetic behaviour at RT. Therefore, the M_S obtained for these samples is around 55 emu/g (see Table 2). The results demonstrate an M_S value well below the expected (at 300 K) for MFO bulk (i.e., $M_s^{bulk} \approx 80$ emu/g) [33]. This difference can easily be attributed to the presence of a small particle size distribution, given that the proportion of the coupled moment carriers is relatively lower than in larger particles because of a high surface/volume ratio. It is well known that the smaller particles contribute with a lower net magnetisation than the larger ones [33]. Therefore, we can see that a variation in particle size and phase composition leads to structural heterogeneity

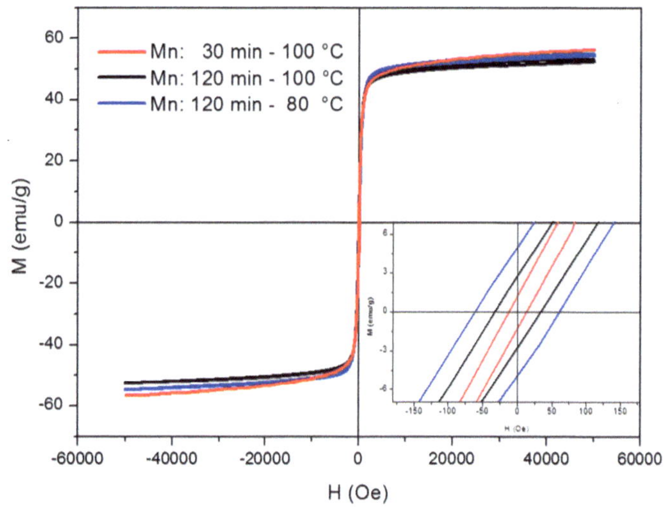

Fig. 2 Magnetization versus applied the magnetic field at 300 K with H up to 50 kOe for MFO samples

Table 2 Coercive fields (H_C) and saturation magnetization (M_S) for manganese ferrites

Samples	H_C (Oe)	M_S (emu/g)
MFO (30 min–100 °C)	13	56.6
MFO (120 min–100 °C)	32	53.3
MFO (120 min–80 °C)	62	54.8

and is responsible for modulating the movement of the magnetic moment in such particles (see Fig. 2 and Table 2).

3.3 *Heating Efficiency*

Hence, the magnetic moment of a particle suspended in a fluid can relax after magnetic field removal by two different mechanisms described by $\tau_N = \tau_0 \exp \frac{K_{eff} V_M}{k_B T}$ $\tau_B = \frac{3\eta V_h}{k_B T}$, that is, the so-called Neel and Brown relaxations, respectively [34, 35]. According to the hysteresis loops, the manganese ferrite nanoparticles are blocked at RT. Therefore, the main relaxation mechanism is due to Brownian movement [6, 26], which is governed by the hydrodynamic size. The aqueous colloids usually have large hydrodynamic sizes due to water polarity, whereas the hydrodynamic size in organic colloids is significantly reduced because the nanoparticles do not aggregate [13, 26].

It is known that, for homogeneous size distribution, coating with OA disaggregates the nanoparticles, thereby increasing the nanoparticle heating efficiency [13, 26]; however, based on these preliminary results, it is not possible to explain what would happen in the case of a heterogeneous size distribution. Therefore, the heating efficiency of the MFO sample (80 °C for 120 min) was investigated because it has a binomial particle size distribution. As can be seen from Table 3, the SAR obtained for this sample is notably small even when coated with OA. A second fractioning by centrifugation was performed on the MFO samples—denoted here as MFO (A)—coated with TMAOH and OA, respectively, as an attempt to separate once again the large particles from the small ones; this sample is called MFO (S). Figure 3 displays the temperature increased for different iron concentrations for MFO (A) and MFO (S) samples coated with OA and dispersed in hexane.

Despite the second sample fractioning, the particles coated with TMAOH do not show any improvement in the SAR. This is likely due to the fact that the centrifugation excludes large aggregates, but the small and large particles still confirm the small ones. In this case, the sizeable hydrodynamic size slows down the Brownian relaxation, thereby worsening the heating efficiency. On the other hand, the second fractioning of the particles coated with OA increases the SAR by more than 10 times. The reason for this is that, in the case of the disaggregated particles, the centrifugation allows the precipitation of the largest particles whereas the smallest ones remain in

Table 3 SAR values for MFO coated with TMOHA and OA and dispersed in water and hexane, respectively, for as-synthesized samples (A) and the second fractioning of the particles (S)

Samples	Cmg_{Fe} (mg/ml)	SAR (W/g)	
		TMAOH	OA
MFO (A)	50	7	4
MFO (S)	2	4	69

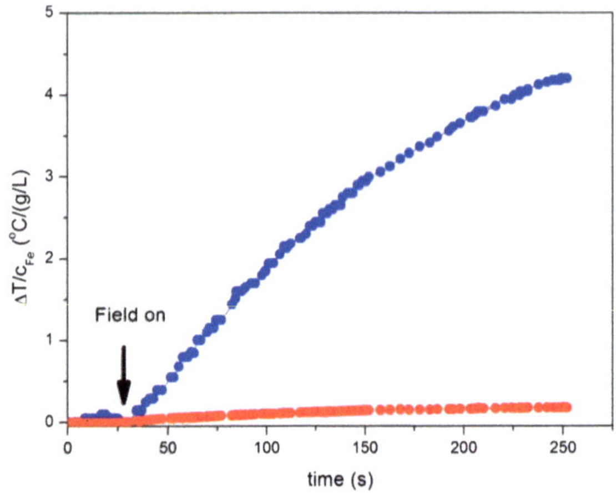

Fig. 3 Temperature increase per iron concentration as a function of time for samples MFO (A) (red curve) and MFO (S) (blue curve) with OA and dispersed in hexane at 110 kHz and 200 Oe

the colloids. Likewise, the smaller hydrodynamic size of the MFO (S) coated with OA makes the Brownian relaxation faster, thus raising the SAR. Figure 4 depicts the experimental hyperthermia curves for samples MFO (A) and MFO (S) with OA and dispersed in hexane. The best heating efficiency corresponds to as-prepared in the second fractioning of the particles, because of the MFO (A) containing the largest particles (see Table 3). However, it is well-known that the SAR values increase with the increasing concentration (TMAOH or OA) coating used in the preparation of aqueous ferrofluids [36, 37]. This is likely due that TMAOH and OA coating affects the structural and morphological features, the magnetic properties, as well as, can also lead to a high disaggregated of MFO nanoparticles, as a result, is observed an increase in SAR value [21, 36, 37].

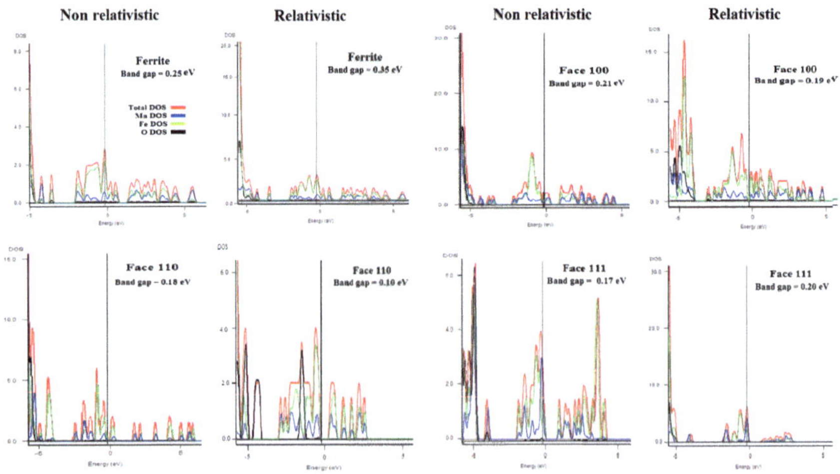

Fig. 4 Projected DOS for the bulk and (100), (110) and (111) surfaces of the MFO structure

3.4 Computational Study

Figure 4 shows the plotting the density of states (DOS) for the bulk and (100), (110) and (111) surfaces of the MFO structure, including the influence of the relativistic effects as described in the methodology. An analysis of DOS structure for both non-relativistic and relativistic calculations of the bulk MFO, in particular, reveals a typical metallic behavior. Also, these results show that main contribution of the top of the valence band (VB) is due to iron (Fe) and manganese (Mn) atoms, while that the oxygen (O) atoms has a contribution in the core level of bulk MFO structure. On the other hand, however, it is observed that Mn, Fe and O atoms present in the bulk MFO structure exhibit practically the same effect in terms of their conduction band (CB). These results are consistent with the literature [38, 39]. Our results suggest that the relativistic calculations lead to an increase in the band gap value of bulk MFO structure. Therefore, for a better description of the electronic part can be obtained from these calculations. The calculated lattice parameters for the construction of the bulk were close to the parameters found experimentally. In addition, the band gap values of the (100), (110) and (111) surfaces of MFO structure was smaller than in relation to the bulk structure. There was no significant difference in these band gap values for non-relativistic calculations. However, for the case of relativistic calculations, there is a relevant difference in the trend of surface band gap values. This tendency is inversely proportional to surface energy, which confirms that relativistic effects must be added in electronic structure calculations of MFO materials. Besides, the surface band gap values for the (110) and (111) faces were almost equal regardless of the method used. Therefore, this suggests that both surfaces likely can react similarly. These calculations also confirm that the order of stability of these surfaces is (100) > (111) > > (110). Therefore, for all subsequent theoretical analyzes, in particular, these were based on the relativistic calculations.

The electrostatic surface contours of the (100), (110) and (111) surfaces of the MFO structure in terms of Hirshfeld charges are presented in Fig. 5. In general, is well-known that the Hirshfeld charges are defined relative to the deformation density, being calculated by the difference between the relaxed and unrelaxed atomic charge

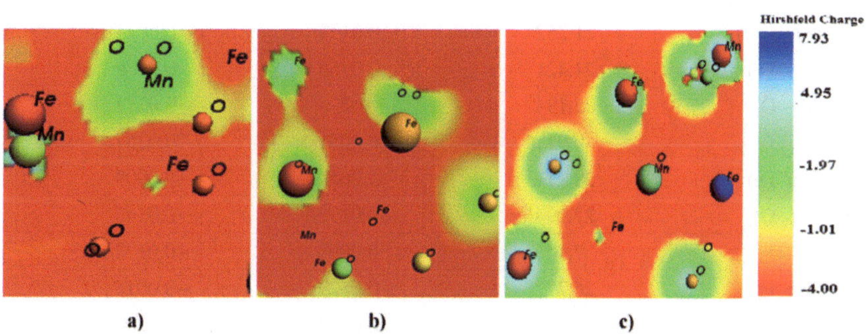

Fig. 5 Electrostatic surface contours of **a** (100), **b** (110) and **c** (111) surface of MFO models

densities [40, 41]. In Fig. 5, the region red and green indicate volumes of low electron density and high electron density, respectively. Particularly, these results reveal that the calculated charge density maps of the (110) and (111) surfaces show that there is a high charge density around the Fe and Mn atoms, but it is worth observing that there is a charge transfer from the Fe atom to the neighboring Mn atom, which contributes for the decreasing the band gap. However, for the face (100) does not evident the occurrence of this charge transfer from the Fe atom to the neighboring Mn atom (getting the atoms their respective point charges), suggesting that this effect does not change the surface band gap of the system. Thus, in general, we have that the bond Mn–O is smaller than the bond Fe–O both in the axial and equatorial position, this fact is due because the Mn has a larger atomic radius in relation to the Fe atom, consequently has a greater electrostatic attraction, decreasing the length of the bond with the O atom to which it is attached.

In order to investigate in more detail these electronic characteristics and the chemical bonding of the MFO structures, in particular, topological analysis of the electron density distribution was employed in this study. According to the quantum–mechanical concepts of QTAIM, the observable properties of a chemical system of interest are contained in its electronic density, ρ. Hence, the electronic density is in turn widely used as a quantum–mechanical observable for the execution of numerical integrations, in which the gradient vector $\nabla\rho$ is the basic condition for determining their topology [42, 43]. Thus, QTAIM calculations were performed on bulk and (100), (110) and (111) surfaces, as shown in Table 4. According to these criteria the relationship between the distance (Å) of the proton-receptors and the relocation of relationship and the electrostatic energy three considerations are important, whether $\nabla^2\rho(r) < 0$ and $H(r) < 0$ corresponds to the strong covalent bonds, whether $\nabla^2\rho(r) > 0$ and $H(r) < 0$ corresponds to partially covalent interactions and whether $\nabla^2\rho(r) > 0$

Table 4 QTAIM parameters obtained for the structures (au) for the bulk and (100), (110) and (111) surfaces of the MFO models

Structure	ρ(r)	$\nabla^2\rho(r)$	H(r)
Ferrite (Mn–Fe)	21.252	0.0625	−0.0257
Ferrite (Mn–O)	25.275	0.0592	−0.0240
Ferrite (Fe–O)	21.252	0.0179	−0.0874
100 (Mn–Fe)	24.421	0.0766	0.0179
100 (Mn–O)	25.194	0.0293	−0.0719
100 (Fe–O)	9.9111	0.0051	0.0257
110 (Mn–Fe)	29.277	0.0651	−0.0009
110 (Mn–O)	27.294	−0.0192	−0.0145
110 (Fe–O)	7.145	0.0063	−0.0172
111 (Mn–Fe)	26.277	−0.0629	−0.0018
111 (Mn–O)	23.124	0.0153	−0.0345
111 (Fe–O)	10.257	0.0073	−0.0237

and H(r) > 0 indicates a non-covalent interaction, that is the weakest type of interaction [44–46]. An analysis of Mn Fe interaction indicates that it is contributing to the stability of the system, as was observed in the electrostatic surface contours maps. Based on the Koch and Popeliers criteria [40], the interaction Mn–Fe of the face (100) represents a non-covalent interaction, while that for the face (110) and (111) corresponds to partially covalent interactions, respectively. This observed result shows that the interactions of the Mn–Fe in the (110) and (111) surfaces are more stable than face (100). Consequently, this stableness further decreases the surface band gap of such faces. Analyzing now the interaction Mn–O, for the face (100) and (111), we have $\nabla^2\rho(r) > 0$ and $H(r) < 0$ indicates partially covalent interactions, and the face (100) corresponds to the strong covalent bonds. In interaction Fe–O for the face (110) and (111) it was obtained $\nabla^2\rho(r) > 0$ and $H(r) < 0$ indicates partially covalent interactions, the face (100) it was obtained $\nabla^2\rho(r) > 0$ and $H(r) > 0$ indicates a non-covalent interaction. Thus, these analyzes show that these interactions (O–Mn–O) and (O–Fe–O) are stronger on faces (110) and (111), which collaborates with the results obtained from band gap. For the bulk (ferrite) in interaction in all analyzed (Mn–Fe, Mn–O, Fe–O) interactions we have $\nabla^2\rho(r) > 0$ and $H(r) > 0$ indicates a non-covalent interaction.

4 Conclusions

In summary, single-crystalline MFO nanoparticles with spinel structure were synthesised using the co-precipitation method under different reaction temperatures (80 and 100 °C) and stirring times (30 and 120 min). It is found that the crystallite size, which ranges from 20 to 25 nm, increases with the stirring time and takes the most significant value at the lowest synthesis temperature, i.e., 80 °C. The reaction at low temperature favours a binomial particle size distribution, centred at 9 and 50 nm. Hence, the hysteresis loops at RT show coercive fields above 15 Oe, indicating that the magnetic moments are blocked at RT as the magnetic behaviour mainly reflects the contribution of the largest particles. The saturation magnetisation is about 25% smaller than the bulk value, originated probably by the uncoupled moment carriers at the surface of the smallest particles. As the aqueous colloids are usually composed of large particle aggregates that slow down the Brownian relaxation, the hydrodynamic size is reduced by coating the magnetic nanoparticles by OA. The smaller hydrodynamic size of the MFO prepared at 80 °C for 120 min (S) coated with OA makes the Brownian relaxation faster, thus raising the SAR. These results are essential to the application of ferrite in ferrofluid technology and biomedical fields. Our theoretical findings suggest that relativist effects are most adequate for the electronic structure calculations of MFO structures. The theoretical data also unveiled fundamental aspects about the nature of chemical bonding in MFO structures, suggesting that these interactions (O–Mn–O) and (O–Fe–O) are stronger on faces (110) and (111), contributing to a greater stabilization of these faces in the crystal. Also, they have a good agreement with our experimental data. Therefore, we believe that this present study provides a deep understanding of its physical/chemical properties and may be considered as a fertile bridge to inspire future investigations on these and other complex systems.

Acknowledgements This work has been supported by the Ministerio Español de Economia y Competitividad (MINECO) projects MAT2012-37109-C02-01 and MAT2013-48009-C4-1, and Fundación Mutua Madrilena (FMM 2010). The Brazilian authors also thank to the financial support of the CNPq, CAPES, Fundação Araucária and FAPEMIG.

References

1. Hervault, A., Thanh, N.T.K.: Magnetic nanoparticle-based therapeutic agents for thermo-chemotherapy treatment of cancer. Nanoscale **6**, 11553–11573 (2014)
2. Rossi, L.M., Costa, N.J.S., Silva, F.P.: Magnetic nanomaterials in catalysis: advanced catalysts for magnetic separation and beyond. Green Chem. **16**, 2906–2933 (2014)
3. Meffre, A., Mehdaoui, B., Connord, V., Carrey, J., Fazzini, P.F., Lachaize, S., et al.: Complex nano-objects displaying both magnetic and catalytic properties: a proof of concept for magnetically induced heterogeneous catalysis. Nano Lett. **15**, 3241–3248 (2015)
4. Maier-Hauff, K., Ulrich, F., Nestler, D., Niehoff, H., Wust, P., Thiesen, B., et al.: Efficacy and safety of intratumoral thermotherapy using magnetic iron-oxide nanoparticles combined with external beam radiotherapy on patients with recurrent glioblastoma multiforme. J. Neuro-Oncol. **103**, 317–324 (2011)
5. de la Presa, P., Luengo, Y., Multigner, M., Costo, R., Morales, M.P., Rivero, G., et al.: Study of heating efficiency as a function of concentration, size, and applied field in γ-Fe_2O_3 nanoparticles. J. Phys. Chem. C. **116**, 25602–25610 (2012)
6. Cruz, M.M., Ferreira, L.P., Ramos, J., Mendo, S.G., Alves, A.F., Godinho, M., et al.: Enhanced magnetic hyperthermia of $CoFe_2O_4$ and $MnFe_2O_4$ nanoparticles. J. Alloy Compd. **703**, 370–380 (2017)
7. Salas, G., Costo, R., Morales, MDP.: In Jesus, M.d.l.F.; Grazu, V., (eds.) Frontiers of Nanoscience, vol. 4, pp. 35–79. Elsevier (2012)
8. Gonzalez-Fernandez, M.A., Torres, T.E., Andres-Verges, M., Costo, R., de la Presa, P., Serna, C.J., et al.: Magnetic Nanoparticles for Power Absorption: optimizing size, shape and magnetic properties. J. Solid State Chem. **182**, 2779–2784 (2009)
9. Sharifi, I., Shokrollahi, H., Amiri, S.: Ferrite-Based Magnetic Nanofluids Used in Hyperthermia Applications. J. Magn. Magn. Mater. **324**, 903–915 (2012)
10. Pottker, W.E., Ono, R., Cobos, M.A., Hernando, A., Araujo, J.F.D.F., Bruno, A.C.O., et al.: Influence of order-disorder effects on the magnetic and optical properties of $NiFe_2O_4$ nanoparticles. Ceram. Int. **44**, 17290–17297 (2018)
11. Mazario, E., Sánchez-Marcos, J., Menéndez, N., Cañete, M., Mayoral, A., Rivera-Fernández, S., et al.: High specific absorption rate and transverse relaxivity effects in manganese ferrite nanoparticles obtained by an electrochemical route. J. Phys. Chem. C **34**, 1196828–1196868 (2015)
12. Lemine, O.M., Omri, K., Iglesias, M., Velasco, V., Crespo, P., de la Presa, P., et al.: γ-Fe_2O_3 by sol–gel with large nanoparticles size for magnetic hyperthermia application. J Alloy Compd. **607**, 125–131 (2014)
13. Arteaga-Cardona, F., Rojas-Rojas, K., Costo, R., Mendez-Rojas, M.A., Hernando, A., de la Presa, P.: Improving the magnetic heating by disaggregating nanoparticles. J. Alloy Compd. **663**, 636–644 (2016)

14. Pereira, C., Pereira, A.M., Fernandes, C., Rocha, M., Mendes, R., Fernández-García, M.P., et al.: Superparamagnetic MFe_2O_4 (M = Fe Co, Mn) nanoparticles: tuning the particle size and magnetic properties through a novel one-step coprecipitation route. Chem. Mater. **24**, 1496–1504 (2012)
15. Laurent, S., Forge, D., Port, M., Roch, A., Robic, C., Vander Elst, L., et al.: Magnetic iron oxide nanoparticles: synthesis, stabilization, vectorization, physicochemical characterizations, and biological applications. Chem. Rev. **108**, 2064–2110 (2008)
16. Salazar, J.S., Perez, J.L., de Abril, O., Phuoc, L.T., Ihiawakrim, D., Vazquez, M., et al.: Magnetic iron oxide nanoparticles in 10–40 nm range: composition in terms of magnetite/maghemite ratio and effect on the magnetic properties. Chem. Mater. **23**, 1379–1386 (2011)
17. Chen, J.P., Sorensen, C.M., Klabunde, K.J., Hadjipanayis, G.C., Devlin, E., Kostikas, A.: Size-dependent magnetic properties of $MnFe_2O_4$ fine particles synthesized by coprecipitation. Phys. Rev. B **54**, 9288–9296 (1996)
18. Auzans, E., Zins, D., Blums, E., Massartet, R.: Synthesis and properties of Mn-Zn ferrite ferrofluids. J. Mater. Sci. **34**, 1253–1260 (1999)
19. Kim, Y.I., Kim, D., Lee, C.S.: Synthesis and characterization of $CoFe_2O_4$ magnetic nanoparticles prepared by temperature-controlled coprecipitation method. Phys. B **337**, 42–51 (2003)
20. Arelaro, A.D., Lima, E., Jr., Rossi, L.M., Kiyohara, P.K., Rechenberg, H.R.: Ion dependence of magnetic anisotropy in MFe_2O_4 (MFe, Co, Mn) nanoparticles synthesized by high-temperature reaction. J. Magn. Magn. Mater. **320**, e335–e338 (2008)
21. Massart, R.: Preparation of aqueous magnetic liquids in alkaline and acidic media. IEEE Trans. Magn. **17**, 1247–1248 (1981)
22. Tartaj, P., Morales, M.D.P., Veintemillas-Verdaguer, S.: The preparation of magnetic nanoparticles for applications in biomedicine. J. Phys. D. **36**, R182–R197 (2003)
23. Gyergyek, S., Drofenik, M., Makovec, D.: Oleic-acid-coated $CoFe_2O_4$ nanoparticles synthesized by co-precipitation and hydrothermal synthesis. Mater. Chem. Phys. **133**, 515–522 (2012)
24. Cullity, B.C., Graham, C.D.: Introduction to Magnetic Materials, 2nd ed. IEEE, Wiley (2009)
25. Natividad, E., Castro, M., Mediano, A.: Adiabatic versus non-adiabatic determination of specific absorption rate of ferrofluids. J. Magn. Magn. Mater. **32**, 1497–1500 (2009)
26. de la Presa, P., Luengo, Y., Velasco, V., Morales, M.P., Iglesias, M., Veintemillas-Verdaguer, S., et al.: Particle interactions in liquid magnetic colloids by zero field cooled measurements: effects on heating efficiency. J. Phys. Chem. C. **119**, 11022–11030 (2015)
27. Baerends EJ, Autschbach J, Bashford D, Bérces A, Bickelhaupt FM, Bo C, et al., ADF2009.01, SCM, Theoretical Chemistry,VrijeUniversiteit, Amsterdam, The Netherlands, 2009.
28. Perdew, J.P., Burke, K., Ernzerhof, M.: Generalized gradient approximation made simple. Phys. Rev. Lett. **77**, 3865–3868 (1996)
29. van Lenthe, E., Baerends, E.J., Snijders, J.G.: Relativistic regular two-component Hamiltonians. J. Chem. Phys. **99**, 4597–4610 (1993)
30. Fleet, M.E.: The structure of magnetite. Acta Crystallogr. B. **37**, 917–920 (1981)
31. Antic, B., Kremenovic, A., Nikolic, A.S., Stoiljkovic, M.: Cation distribution and size-strain microstructure analysis in ultrafine Zn−Mn ferrites obtained from acetylacetonato complexes. J. Phys. Chem. B. **108**, 12646–12651 (2004)
32. Smit, J., Wijn, H.P.J.: Eindhoven, The Netherlands: Philips Research Laboratory. Eindhoven, The Netherlands (1959)
33. Goya, G.F., Berquó, T.S., Fonseca, F.C.: Static and dynamic magnetic properties of spherical magnetite nanoparticles. J. Appl. Phys. **94**, 3520–3528 (2003)
34. Glockl, G., Hergt, R., Zeisberger, M., Dutz, S., Nagel, S., Weitschies, W.: The effect of field parameters, nanoparticle properties and immobilization on the specific heating power in magnetic particle hyperthermia. J. Phys. Condens. Matter **18**, S2935–S2949 (2006)
35. Hergt, R., Dutz, S., Muller, R., Zeisberger, M.: Magnetic particle hyperthermia: nanoparticle magnetism and materials development for cancer therapy. J. Phys. Condens. Matter **18**, S2919–S2934 (2006)

36. Raland, R.D., Saikia, D., Borgohain, C., Borah, J.P.: Heating efficiency and correlation between the structural and magnetic properties of oleic acid coated $MnFe_2O_4$ nanoparticles for magnetic hyperthermia application. J. Phys. D. Appl. Phys. 325001–325004 (2017)
37. Tourinho, F.A., Franck, R., Massart, R.: Aqueous ferrofluids based on manganese and cobalt ferrites. J. Mater. Sci. **25**, 3249–3254 (1990)
38. Ustundağ, M., Aslan, M.: Electronic and magnetic properties of Ca-doped Mn-Ferrite. Acta Phys. Pol. A. **130**, 362–364 (2016)
39. Elfalaky, A., Soliman, S.: Theoretical investigation of $MnFe_2O_4$. J Alloy Compd. **580**, 401–406 (2013)
40. Favero, P.P., de Souza-Parise, M., Fernandez, J.L.R., Miotto, R., Ferraz, A.C.: Surface properties of CdS nanoparticles. Braz. J. Phys. **36**, 1032–1034 (2006)
41. Ahuja, R., Blomqvist, A., Larsson, P., Pyykko, P., Zaleski-Ejgierd, P.: Relativity and the lead-acid battery. Phys. Rev. Lett. **106**, 018301–018305 (2011)
42. Pecul, M., Lewandiwski, J., Sadlej, J.: Benchmark calculations of the shielding constants in the water dimer. Chem. Phys. Lett. **333**, 139–145 (2011)
43. Cortesguzman, F., Bader, R.F.W.: Complementarity of QTAIM and MO theory in the study of bonding in donor-acceptor complexes. Coord. Chem. Rev. **249**, 633–662 (2005)
44. Koch, U., Popelier, P.L.: Characterization of C-H-O Hydrogen Bonds on the Basis of the Charge Density. J. Phys. Chem. A. **99**, 9747–9754 (1995)
45. Gonçalves, M.A., Peixoto, F.C., da Cunha, E.F.F., Ramalho, T.C.: Dynamics, NMR parameters and hyperfine coupling constants of the Fe_3O_4 (100)–water interface: Implications for MRI probes. Chem. Phys. Lett. **609**, 88–92 (2014)
46. La Porta, F.A., Ramos, P.H., de Resende, E.C., Guerreiro, M.C., Giacoppo, J.O.S., Ramalho, T.C., Sambrano, J.R., Andrés, J., Longo, E.: Structural, electronic and optical properties of Fe(III) complex with pyridine-2,6-dicarboxylic acid: A combined experimental and theoretical study. Inorg. Chim. Acta **416**, 200–206 (2014)

Synthesis, Properties, and Applications of Iron Oxides: Versatility and Challenges

Nathalie Danree Busti, Rodrigo Parra, and Márcio Sousa Góes

Abstract Iron (III) oxide is a compound that appears in at least four different polymorphs: α-Fe_2O_3, β-Fe_2O_3, γ-Fe_2O_3, and ε-Fe_2O_3. However, Fe^{3+} ions are also present in another form of iron oxide: Fe_3O_4, which is an iron crystal structure with both Fe^{2+} and Fe^{3+} ions. And in its turn, Fe^{2+} ions are also present in the FeO form of iron oxide. Each of these six different structures presents distinctive physical properties and, therefore, diverse applications. The different crystalline forms of iron oxide have found fertile ground in the field of nanotechnology, and therefore, became popular among researchers who have proven a wide variety of biomedicine, electronics, construction, environmental remediation, and energy harvesting applications. In this regard, the main technological challenge is related to control of its physical characteristics such as morphology, size distribution, dispersion, crystallinity, structural defects, porosity, active area, as well as impurities. All of these influence the physical and optical properties of the synthesized material and will determine its field of application. As such, the synthesized material characteristics depend on the synthesis method employed. Thereby, in this chapter, we will cover the main characteristics of iron oxides with a focus on preparation processes, physicochemical properties, and their relationship with their main applications.

1 Introduction

Iron (Fe) is, of course, the second most abundant metal on our planet, being naturally found in the oxide form. Among the different types of semiconductors that can be found, applied in the technological and scientific fields, iron oxides stand out for

N. D. Busti · M. Sousa Góes (✉)
Instituto Latino-Americano de Ciências da Vida e da Natureza (ILACVN), Universidade Federal da Integração Latino-Americana (UNILA), 85.867-900, Foz do Iguaçu, Paraná, Brazil
e-mail: marcio.goes@unila.edu.br

R. Parra
Instituto de Investigaciones en Ciencia y Tecnología de Materiales (INTEMA), CONICET - Universidad Nacional de Mar del Plata (UNMdP), B7606BWV Mar del Plata, Argentina

F. A. La Porta and C. A. Taft (eds.), *Functional Properties of Advanced Engineering Materials and Biomolecules*, Engineering Materials,
https://doi.org/10.1007/978-3-030-62226-8_13

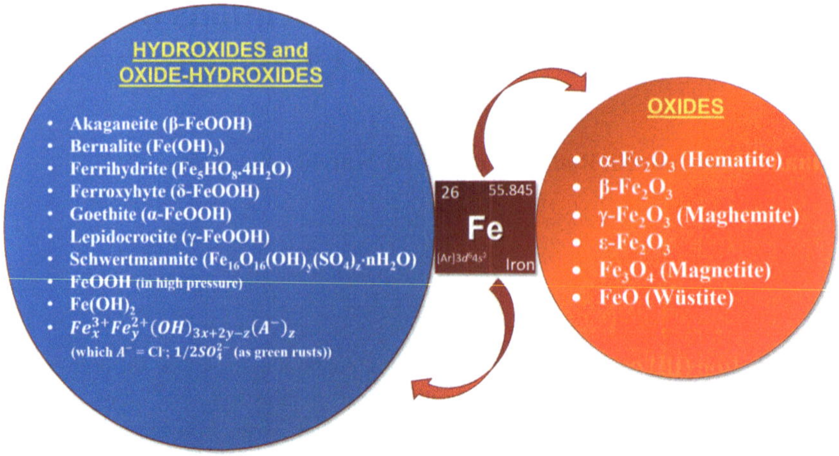

Fig. 1 Sixteen type of iron hydroxides/oxides

their diverse physical and chemical properties, in addition to their abundance. As previously reported by Cornell and Schwertmann [1], there are sixteen pure phases of iron oxides, which can be divided into the main groups described in Fig. 1.

Each of these different iron phases presents particular functional properties which are desired for a broad variety of high-performance applications. Despite the importance of all iron phases, this chapter will focus specifically on iron oxides (FeO, Fe_3O_4, α-Fe_2O_3, γ-Fe_2O_3, β-Fe_2O_3 and ε-Fe_2O_3). For further information on these and other iron compounds, we strongly recommend the books by Schwertmann and Cornell [2] and Fernández-Remolar [3].

2 Oxides Structural and Morphological Characteristics

Characterization methods such as X-ray diffraction and neutron diffraction, combined with infrared spectroscopy, electron diffraction, and high-resolution electron microscopy have widely been used by scientists to determine iron oxide structures [1, 4]. Iron oxides consist of Fe ions (range of positive oxidation states from 2^+ to 6^+) [5] and O^{2-} ions arranged in crystalline structures. As O^{2-} anions are larger than Fe cations (oxygen ionic radius is 0.14 nm while Fe^{2+} and Fe^{3+} ionic radius is 0.082 nm and 0.065 nm, respectively and with number coordination = 6) [6] the iron oxide crystal structures are governed by the anions' arrangement [1, 7]. As such, the degree at which ions are ordered and the crystal sizes vary according to the conditions in which the crystals are usually formed [8]. Table 1 shows crystal structure (made VESTA 3 software) [9] of its main oxides.

Table 1 Crystal structure of iron oxide. Brown and red spheres represent iron and oxygen, respectively

Oxide type	Crystal structure	References
α-Fe_2O_3	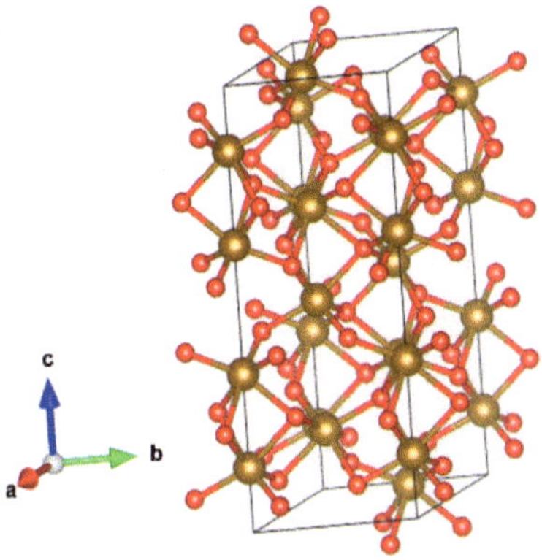	[25]
β-Fe_2O_3	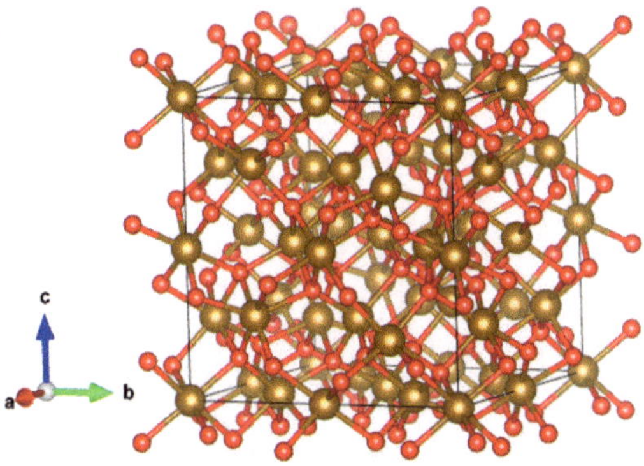	[26]
γ-Fe_2O_3	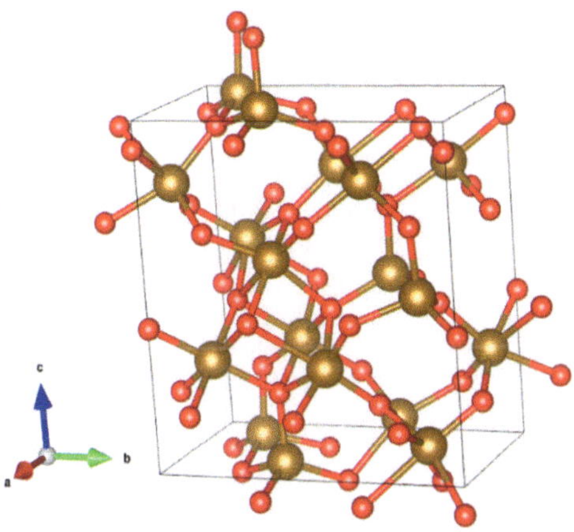	[18]

(continued)

Table 1 (continued)

Oxide type	Crystal structure	References
ε-Fe_2O_3	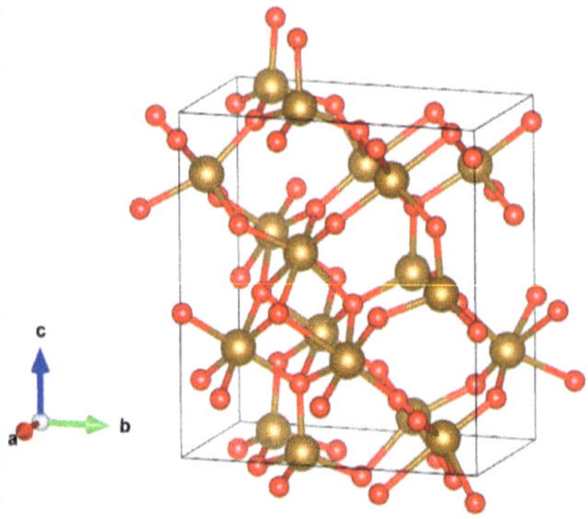	[27]
Fe_3O_4	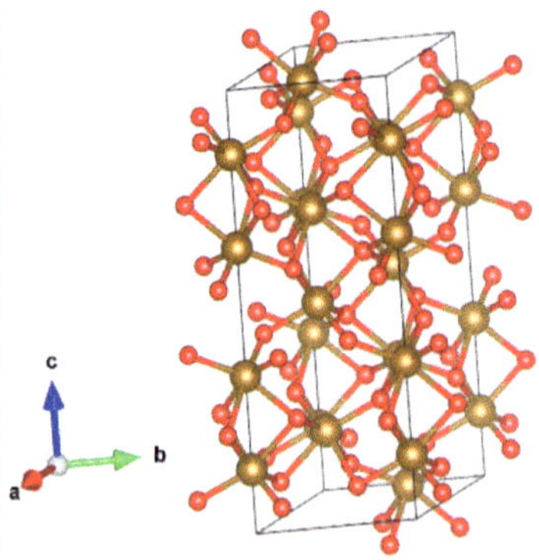	[28]
FeO	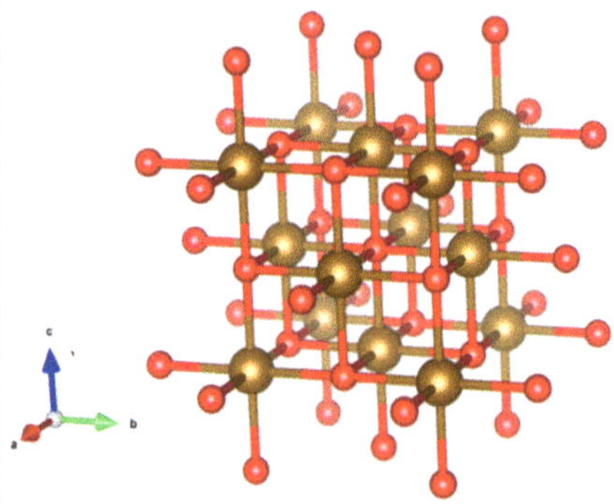	[29]

2.1 Iron (III) Oxides

Iron (III) oxide is, of course, a compound that exhibits four different crystalline polymorphs: α-Fe_2O_3, β-Fe_2O_3, γ-Fe_2O_3, and ε-Fe_2O_3. Hematite and maghemite usually appear in nature, while beta and epsilon structures are generally synthetic oxides [10, 11].

In general, the iron (III) oxides anions arrange in close-packed structures, where anions sheets (of about 0.23–0.25 nm) stack in a particular crystallographic direction, for all iron (III) oxides; cations fit into some of the interstices between the anion sheets forming different crystalline structures [1].

2.1.1 Alpha Iron Oxide

Hematite—rhombohedral corundum-type structure with iron in its 3^+ oxidation state—being the oldest known iron oxide mineral have usually found in rocks and soils [12]. Among the oxides, hematite is the more stable crystalline oxide [13]. It has a characteristic blood-red colour, and it is the most common form of iron oxide found in nature, because of its thermal and chemical stability [14]. α-Fe_2O_3 has weak ferromagnetic or antiferromagnetic behaviour at ordinary room temperature, and it is paramagnetic at temperatures above 956 K (Neel temperature) [10]. The alpha phase of iron oxide has the same trigonal structure as corundum, α-Al_2O_3, with O^{2-} anions arranged in a close-packed hexagonal crystallographic system along the [001] direction and the Fe^{3+} cations regularly occupying two-thirds of the octahedral interstices in alternate layers, forming sixfold rings [2, 10, 14]. Each of the O^{2-} anions are bonded to only two iron cations, making the structure to be electronically neutral [15].

Hematite belongs to $R\bar{3}c$ space group (rhombohedral symmetry) the unit cell is hexagonal with bulk lattice parameters a = 5.0356 Å and c = 13.7489 Å and at least six formula units per unit cell [14, 15]. Hematite may also be indexed in the rhombohedral system; in it there are two formula units per unit cell and $a_{rh} = 5.427$ Å and $\alpha = 55.3°$ [1].

2.1.2 Gamma Iron Oxide

Maghemite occurs naturally as a weathering product of magnetite or after rising the temperature from other iron oxides in addition to the presence of organic matter [1, 2]. As such, the most common procedure for the synthetic making of γ-Fe_2O_3 involves dehydrating goethite (α-FeOH) that leads to hematite (α-Fe_2O_3), which is then reducing the hematite to obtain magnetite (Fe_3O_4) and finally oxidizing to maghemite (γ-Fe_2O_3) [11]. Maghemite becomes unstable at temperatures above 775 K, transforming into hematite [10, 16].

Gamma iron oxide ($Fe_{21+x}O_{32}$, structure type) is a red-brown coloured mineral which presents a cubic spinel structure, similar to magnetite, but with cation deficient sites [1, 2, 10]. It has a cubic unit cell with lattice parameter a = 8.3474 Å, formed by 32 oxygen anions, 21 1/3 trivalent iron ions, and 2 1/3 vacancies. Its space group is $P4_332$ [1, 17, 18]. This ferromagnetic material is easily magnetized and responds easily when exposed to an external magnetic field[10].

2.1.3 Beta Iron Oxide

β-Fe_2O_3 is, of course, a rare synthetic form of the iron oxide which exhibits a body-centered cubic structure, as well. Particularly, this iron oxide has widely been obtained by the dehydroxylation process of the β-FeOOH phase under a high vacuum at about 170 °C [1, 10]. More specifically, this iron oxide form is the only that have paramagnetic properties at room temperature, being its Neel transition temperature in the range of 100 to 119 K, under which it is anti-ferromagnetic. As this is not a stable form of iron oxide, when heated, it transforms either to maghemite or hematite [10].

2.1.4 Epsilon Iron Oxide

Epsilon iron oxide phase may appear as a pure disordered form or an ordered form that might in principle be somehow associated with hematite and maghemite forms of iron oxide [1, 2]. ε-Fe_2O_3 is a not yet a fully understood iron oxide form and exhibits intermediate properties between alpha and gamma phases of iron (III) oxide [1, 10].

In 1998 Tronc et al. obtained epsilon iron oxide which presented an orthorhombic unit cell with bulk lattice parameters a = 5.095 Å, b = 8.789 Å and c = 9.437 Å, being part of $Pna2_1$ space group (orthorhombic symmetry) [19]. This iron oxide phase only exists in the form of nanoparticles and it presents a particular magnetic behaviour with different magnetic properties at different temperatures, however, its magnetic behaviour is still being discussed [10, 20, 21]. Depending on the synthesis method, it transforms into hematite between 500 and 750 °C [22].

2.2 *Iron (II)/(III) Oxide*

Magnetite (Fe_3O_4), an important iron ore, is a black ferrimagnetic material that normally appears in nature, being usually responsible for the magnetic behaviour observed in some rocks. Fe_3O_4—cubic spinel structure—contains the 3^+ and 2^+ oxidation states of iron, being the stoichiometric ratio $Fe^{2+}/Fe^{3+} = 0.5$ [1, 15]. However, it is known that the magnetite is often non-stoichiometric, resulting in a Fe^{3+} deficient layer [15].

This black iron oxide has an inverse spinel structure, with 32 O^{2-} anions arranged in a cubic close-packed along the [111] direction, forming a face-centered cubic unit cell with lattice parameter a = 8.396 Å, presenting eight formula units per unit cell. As such, it should be noted that Fe^{2+} and half of the Fe^{3+} cations occupy octahedral sites while the other half of the Fe^{3+} cations occupy tetrahedral sites [15]. Trivalent iron ions may occupy either octahedral or tetrahedral sites (see Table 1). In contrast, bivalent ions occupy only octahedral sites. This likely is because the trivalent ion has no crystal field stabilization energy (CFSE) in either coordination, while the bivalent ion presents higher CFSE for the octahedral coordination [1]. Magnetite belongs to space group $Fd\bar{3}m$ (cubic symmetry) [23].

2.3 Iron (II) Oxide

This black oxide is an essential intermediate in the reduction of iron ores. However, it is well-known that the FeO (Wüstite)—cubic rocksalt structure with iron in its 2^+ oxidation state—can only exist as a chemically stable phase at low pressure or pressures above 10 MPa [2]. Wüstite is one of the three pure iron compounds that contains Fe^{2+} ion, and it is usually oxygen deficient [1]. In the FeO iron phase, O^{2-} anions appear on a cubic centred pack structure along the [111] direction, as anions' and cations' planes appear alternately [24]. Wüstite belongs to $Fm\bar{3}m$ space group (cubic symmetry) containing four formula units per unit cell, and lattice parameter in the range of a = 4.28–4.31 Å, depending on the vacancy content, that is, created by the proportion of oxidized metal ions [24].

3 Synthesis and Properties

Over the years, diverse synthetic methods have widely been developed to obtain different forms of iron oxides. However, in the last few decades, more attention has been given to those methods that allow synthesis of materials in the nanometric scale, as researchers found an opportunity to tune material properties, including mechanical, chemical, physical, optical, electrical, and magnetic properties [10].

Nanoparticles reveal different characteristics than those presented by macroscopic scale materials, mainly related to surface effects and quantum confinement-modification in electronic structures [10]. Some metal oxide nanostructures exhibit better transport properties than those presented by their macroscopic counterparts [30]. However, the study and applications of nanomaterials strongly depend on the synthesis and preparation methods.

In this regard, diverse examples have been usually reported in the literature, such as coprecipitation, hydrothermal, solvothermal, sol–gel, spray pyrolysis, ball milling and anodization are some ways nanosized ceramic materials may be synthesized [31]. Varied mechanochemical and chemical methods have been described for the synthesis of iron oxide nanoparticle, where various iron oxides shapes have been reported after the particular using of different iron precursor salts [32, 33]. However, tailoring particle size and morphology towards a particular application still remains a challenge [15].

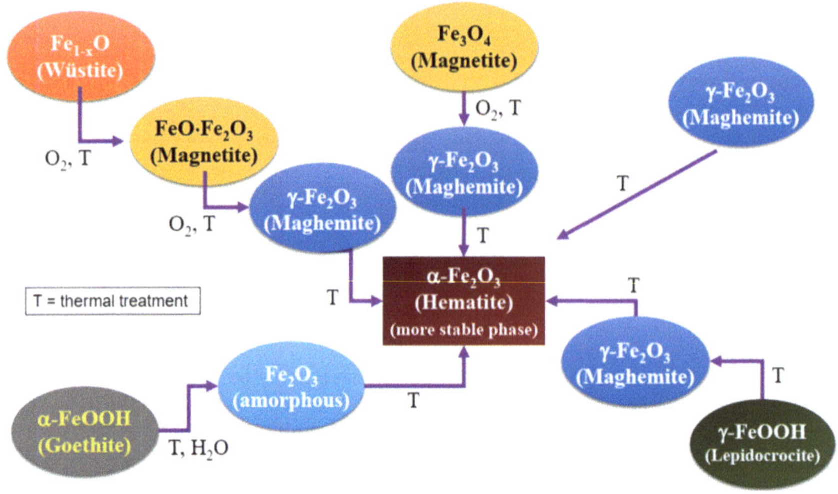

Fig. 2 Transition between iron oxide phases (Adapted from Colpas-Ruiz et al. [36])

Among all, iron(III) oxide phases easily undergo diverse phase transformations in response to heating and/or pressure treatment [34]. Figure 2 shows the main phase transitions, which depend on temperature and atmosphere (oxygen-rich atmosphere) conditions. Under standard conditions, the hematite form is the most stable of the iron oxide, and also, as it is the end-product of other iron oxides or hydroxides forms, according to Campos et al. [10] it is easier to synthesize hematite than many other forms of iron oxide (Table 2). On the other hand, the relative stability of hematite in water may be considered uncertain, since thermodynamic calculations were performed in order to evaluate its transformation into goethite [35] as [$\alpha Fe_2O_3 + H_2O \rightarrow 2\alpha FeOOH$]. However, there is a growing interest in the fields of the synthesis and applications of magnetic iron oxide (Fe_3O_4 and γ-Fe_2O_3) nanoparticles [33].

According to Ali et al. [33], for the synthesis of magnetic iron oxide nanoparticles, about 90% of the reviewed literature reported using chemical methods, while 8% used physical methods and 2% used biological ones. The variation in the synthesis

Table 2 Structure, magnetic and electronic phases of different iron oxide phases (Adapted with permission of Journal of Materials Chemistry C, Mai Hussein Hamed, David N. Muellera and Martina Müller, Copyright 2019, The Royal Society of Chemistry from [37])

Iron oxide	Iron valence	Crystal structure	Magnetic properties	Electric properties
$Fe_{1-x}O$	2^+	Rock salt	Antiferro	RT: insulator
Fe_3O_4	$2^+/3^+$	Inverse spinel	Ferri	Half metal
γ-Fe_2O_3	3^+	Defected spinel	Ferri	Insulator
α-Fe_2O_3	3^+	Corundum	Antiferro	Insulator

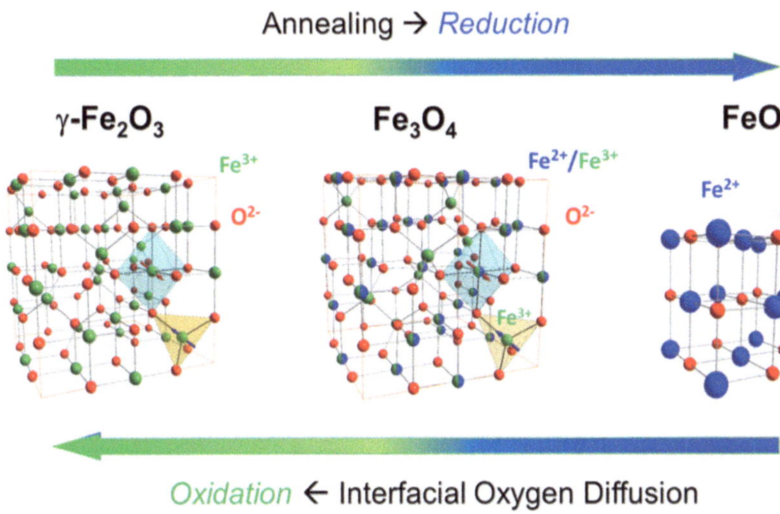

Fig. 3 Unit cells of the iron oxide phases γ-Fe_2O_3, Fe_3O_4 and FeO and the thermodynamical processes driving the oxidation–reduction reactions of the chemical phase transitions. (Reprinted with permission of Journal of Materials Chemistry C, Mai Hussein Hamed, David N. Muellera and Martina Müller, Copyright 2019, The Royal Society of Chemistry from [37])

processes promotes variation of the obtained phase (Fig. 3) and, consequently, in the final properties of the material.

Next, some of the most common synthesis methods will be described, and literature reported results concerning iron oxides would be commented and reviewed.

3.1 Spray Pyrolysis

Spray pyrolysis is a versatile processing technique that allows the preparation of metal oxide films, ceramic coatings, and powders [38]. It is a simple, low-cost method that consists of spraying a chemical precursor over a heated substrate in order to prepare films [31]. This method requires an atomizer, a precursor solution, a substrate heater and a temperature controller [38]. Three types of atomizers may be used: one is an air blast where the liquid precursor solution is exposed and carried by an air stream, the second being an ultrasonic atomizer where the liquid is atomized by small frequency ultrasonic waves, and the third type is an electrostatic atomizer, where the liquid is exposed to a high electric field [38]. According to Perednis and Gauckler [38], due to spray pyrolysis complexity, there are very few and rudimentary models that attempt to explain its mechanism, as it includes processes such as atomization, droplet transport and evaporation, spreading on the substrate, drying and precursor decomposition. Existing models divide the process into three main steps: atomization, droplet transport and precursor decomposition.

The atomization process is very dependent on the type of atomizer used and to its setting parameters, in principle, may influence the produced droplets size and size distribution directly and, consequently, on the particle sizes of the dried product [15, 38, 39]. During the transport process, the aerosol droplets, produced during atomization will be transported towards the heated substrate, and eventually solvent will evaporate. Throughout evaporation, droplet size reduces creating a concentration gradient within the droplet [38]. If droplet surface concentration exceeds the solubility limit, the precursor precipitates on the droplet surface resulting in the formation of solid microporous particles that will ultimately create dense particles or will form a porous crust increasing film roughness [31, 38].

The substrate surface temperature is critical, influencing film morphology and properties, including film roughness, porosity, cracking, and crystallinity [38]. According to its temperature, decomposition may take place either, before or after heating the substrate, influencing film adherence and solid particle formation; high substrate temperatures lead to rough and pours films, while low substrate temperatures may result in cracked films [38]. Films produced by this technique may be either dense or porous, thin or thick, and, even, single layer or multi-layered [38].

Although the experimental setting described above is common for the preparation of films and coatings, for particle production, a different experimental set may be mounted. Some of the experimental set-ups reported for the production of iron oxide in the form of nanoparticles involve the use of heated reactors, where the atomized droplets pass through different heating zones for evaporation and calcination, before being collected in a filter [40]. Other set-ups for the synthesis of iron oxide nanoparticles involve the use of a laser to heat a gaseous mixture of iron precursor materials [39].

Diverse experimental parameters may influence the quality of the synthesised material. In this regard, variables such as oxygen concentration, impurities/defects, and heating time will be key factors in the final product quality [39]. Also, the precursor solution will be an essential variable for the process, being relevant solute and solvent compositions, pH, solution concentration, as well as additives and impurities [38]. Nitrate salts and metal chlorides may be used as precursors as they present high solvability [31]. Low solubility precursors may lead to the formation of impurities [31].

The authors of this chapter have synthesized hematite thin films on conducting fluorine-doped tin oxide (FTO) glass using conventional spray pyrolysis technique. For the deposition, an ethanolic $FeCl_3{\cdot}6H_2O$ 0.2 M precursor solution was sprayed onto the substrate maintained at 450 °C. Different amounts of precursor solution were deposited, giving rise to films of varying thickness. However, all the depositions resulted in similar XRD and Raman spectra signaling the presence of the hematite phase of Fe_2O_3. The synthesized films showed enhanced photocurrent when tested under 1 sun illumination in a three-electrode configuration (using an Ag/AgCl reference electrode and a platinum counter electrode) in an electrolyte containing 1 M NaOH (pH 13) (Fig. 4).

Duret and Grätzel [41] described the deposition of 200 nm thick hematite films on conducting FTO glass, through conventional and ultrasonic spray pyrolysis. Both

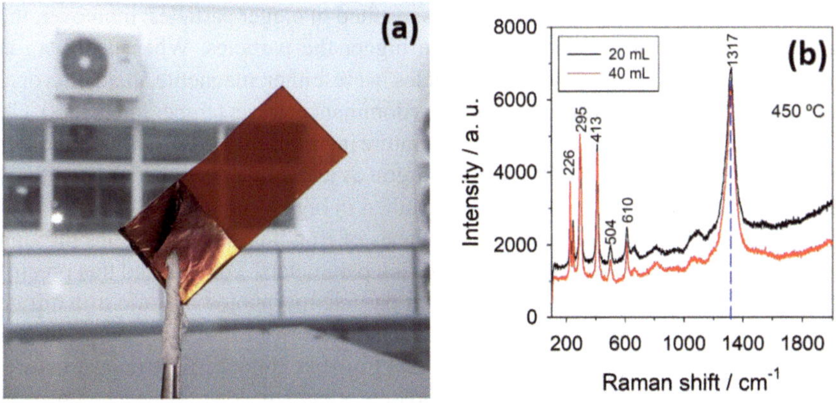

Fig. 4 **a** Hematite electrode prepared via spray-pyrolysis and **b** Raman spectra collected on films containing 20 and 40 ml of 0.2 mol·L^{-1} solution of $FeCl_3 \cdot 6H_2O$ dissolved in ethanol

depositions resulted in similar XRD and UV-visible and Raman spectra; however, they substantially differ in morphology. On the one hand, the sample deposited by conventional spray pyrolysis consisted of spherical particles with a diameter of 50–150 nm and wide size distribution. On the other hand, the sample deposited by ultrasonic spray pyrolysis consisted mainly of 100 nm platelets, 5–10 nm thick. The authors attributed this morphological difference, mainly, to the deposition rate, slower for ultrasonic spray deposition, which according to the authors possibly gave place to, at least, partial decomposition of precursor before deposition on the FTO substrate. However, it is relevant to notice that for the conventional spray pyrolysis deposition, the precursor solution consisted of a 0.05 M solution of ferric acetylacetonate in ethanol and was deposited onto a substrate maintained at 400 °C, while the precursor solution used for ultrasonic spray pyrolysis was 0.02 M ferric acetylacetonate dissolved in ethanol and the substrate temperature at steady-state conditions was of 420 °C.

Popescu et al. [42] reported the synthesis of nanostructured Fe_2O_3 films with a different number of layers using $FeCl_3$ as a precursor and nitrogen as the carrier gas. The effects of thermal treatment and the number of deposited layers was studied. The results for the absorbance of as-prepared film was investigated, and reveals a slight increase in the absorbance of visible light in their thinner films (118–212 nm thick) after thermal treatment for 1 h at 550 °C in air. In comparison, the thicker film (300 nm) presented lower light absorbance after thermal treatment. The authors concluded that thermal treatment leads to bandgap increase in thinner films, 118–212 nm thick.

Ozcelik and Ergun [43] prepared iron oxide nanoparticles from an iron nitrate ethanolic precursor solution at temperatures ranging 500–1100 °C using argon and air as carrier gases. Also, they studied the effect of collecting the synthesized particles in different parts of their experimental set-up. Their results showed that particles prepared using air as carrier gas contained pure α-Fe_2O_3 phase, and exhibit spherical uniform morphology. The average size varied according to collection location and reaction temperature; higher temperatures resulted in bigger particles; moreover, the further away from the collection site the bigger the particles. When using argon (as a carrier gas), the synthesised particles were either magnetite, hematite or a combination of both phases. Magnetite was dominant at lower temperatures (500 ºC), and hematite fraction increased with temperature increase, up to 900 ºC where another phase appeared (FeO(OH)). When using argon as a carrier gas, higher temperatures and furtherer away collection sites also resulted in bigger particles.

Kastrinaki et al. [40] reported the synthesis of iron-based nanoparticles through spray pyrolysis route under different operating parameters. They proved that precursors of different nature resulted in different particle morphology. While iron nitrate formed spherical particles, the particles synthesized from iron chloride presented irregularities. Also, their results revealed that flow rate influences particle structure, forming hollow particles at higher rates and denser ones at slower rates when solvent evaporation was slower.

3.2 Hydrothermal/Solvothermal

One of the most popular strategies to synthesize uniform crystalline size homogeneous ceramic nanoparticles is through hydrothermal reactions at relatively low temperatures [31]. This is a versatile and environmentally sound method, which does not involve organic solvents or posttreatments [39].

Particularly, the hydrothermal synthesis technique is defined as any heterogeneous reaction in the presence of aqueous solvents or mineralizers under high pressure and temperature conditions [15]. Generally, the reactions are conducted in a heated reactor or autoclave at raised pressure (up to 15 MPa) [31]. When using water as a solvent, this technique is known as hydrothermal. However, when another solvent is used, it is referred to as solvothermal [31]. This method is based on water's ability to hydrolyse and dehydrate metal salts and the low solubility of the resulting in the formation of metal oxides in the aqueous phase at the reaction conditions that supersaturate the medium [33, 39]. As such, the reaction conditions favour diffusion of reactants in water and high dehydration rates resulting in rapid nucleation and fast particle growth [39].

Morphology and particle size may be controlled by altering the synthesis conditions and optimizing the experimental parameters; this includes: adjusting solution composition and solvent or additives used, as well as reaction temperature, pressure, and time [31, 44]. High temperatures result in rapid nucleation and faster growth, leading to the formation of small-sized metastable nanoparticles [44]. Increasing reaction time and precursor concentration will increase particle size [33]. Another important variable would be the addition of a seeding agent [39]. According to Wu et al. [45] iron oxide nanoparticles obtained by the hydrothermal method are prone to present better crystallinity than those synthesised by other methods.

Ge et al. [46] synthesized highly crystalline magnetite nanoparticles by the oxidation of $FeCl_2{\cdot}4H_2O$ in basic aqueous solution under elevated temperature and pressure. By changing reaction conditions (reactants concentration or solvent composition), particle diameter was tuned (between 15 and 31 nm) and magnetic properties were modified. Depending on their size, particles presented either ferromagnetic or superparamagnetic characteristics. Smaller particles presented superparamagnetic properties, while bigger ones presented ferromagnetic behaviour.

Chen et al. [47] fabricated, through hydrothermal synthesis, magnetite dodecahedral nanostructures from ferrous chloride and ethylenediaminetetraacetic acid (EDTA), for this, the autoclave was maintained at 180 °C for 10 h. In their work, they also analyzed different reaction times and temperatures to understand the influence of these parameters in the morphology of the end product particles.

Umar et al. [48] reported the synthesis of hematite hexagonal nanoparticles from a precursor solution made from iron chloride hexahydrate and, hexamethylenetetramine through the hydrothermal synthesis in an autoclave for 3 h at 130 ºC. The hexagonal disc-shaped and highly dense nanoparticles were successfully applied as photocatalyst for the degradation of an organic dye and as mediators for the production of a highly sensitive chemical sensor for 4-nitrophenol in an aqueous medium.

3.3 Sol–Gel

The sol–gel method has typically been used to synthesize metal oxides through chemical reactions of volatile metal precursors [31]. It usually refers to the hydrolysis and condensation of metal alkoxides or alkoxide precursors, resulting in the dispersion of metal oxide particles in "sol", which is then dried or "gelled" by solvent removal or by chemical reaction [49]. In general, the solvent used is water; however, the precursors can in principle be hydrolysed by either acids or bases [49].

Notable benefits of using this strategy include the possibility of scaling up the fabrication process of creating superfine porous powders along with the possibility of controlling the particle size and product homogeneity [31]. Yet, some disadvantages of this method are well-known and include the need for product post-treatment and the possibility of contamination from reactions' by-products [49].

The sol–gel synthesis method is influenced by pH, nature, and concentration of precursor, reaction kinetics and temperature, agitation, and gel properties, affecting the size and structure of the iron oxide particles [33, 49]. Slower and more controlled hydrolysis rates tend to favour the formation of smaller particles [49].

Magnetic iron nanoparticles tend to agglomerate and form clusters due to their high surface energy (associated with their large surface-to-volume ratio) [10]. A strategy to prevent the agglomeration of magnetic iron nanoparticles is the use of coatings [10]. Different coatings have been applied including organic and inorganic materials such as polymers, biomolecules, silica, metals, etc. [45].

Chang et al. [50] reported the fabrication of Fe_3O_4 thin films via sol–gel method using iron (II) and iron (III) salts, polydentate ligands and ethanol for the preparation of the "sol". After 10 h of stirring, the films were deposited by a spinner and the xerogel films were obtained by vacuum drying at 40 °C for 1 h. The xerogel films were sintered in a nitrogen atmosphere for 3 h at 450 °C. The deposition and sintering process was repeated to obtain multilayer films. As a result, the authors obtained films with different magnetic properties, affected by grain size and film thickness, being the thinner films those which presented the best magnetic properties.

Cui et al. [51] synthesized nearly monodispersed hematite, maghemite, and magnetite nanoparticles in large quantities. According to the authors, in a single-reaction, it was possible to get about 60 g of the sample through the same sol–gel procedure and starting materials. By changing the drying conditions, different phases of iron oxide nanoparticles were obtained. Their precursor solution included iron (II) chloride tetrahydrate and propylene oxide in ethanol. After Fe_3O_4 nucleation, the nanoparticles were separated from the solution by centrifugation and dried at room temperature. The hematite nanoparticles were fabricated by drying, in the air at 150 °C, the Fe_3O_4 gel, which was previously obtained from evaporation of the Fe_3O_4 sol. Maghemite nanoparticles were fabricated by drying the sol and sintering the obtained xerogel in the air at 100 °C and 150 °C, respectively.

Akbar et al. [52] reported the sol-gel preparation of iron oxide thin films. The authors showed that varying the pH (from 1 to 13) of the sol, different phases of iron oxide are possible if they are obtained and it affects the magnetic properties of thin films.

3.4 Coprecipitation

Coprecipitation method is also one of the most frequently used methods as it presents benefits such as homogeneous mixing and good stochiometric control, which result in the formation of high purity materials [31, 33]. However, this method allows little control of particle shapes and results in wide-ranges of particle size distribution [15]. Another disadvantage of the coprecipitation method is the generation of highly basic wastewaters requiring subsequent treatment for proper disposal [45]. Phase, size, and shape of iron oxide nanoparticles synthesized by coprecipitation method depend on various parameters such as type of salt used as the iron precursor, precursors' concentration, solution's pH, ionic strength of the media, reaction temperature and stirring rate [10, 33, 45].

To reduce surface energy, caused by a large surface area to volume ratio, nanoparticles fabricated through the liquid phase coprecipitation method tend to aggregate, and anionic surfactants may be used as dispersing agents to reduce it [33, 49]. Also, for stabilization, particles' surface may be coated with proteins, starches, non-ionic detergents or polyelectrolytes [33] and, in addition, as the number of components rises, the difficulty to find a suitable precipitating agent grows [31].

Kim et al. [53] fabricated superparamagnetic magnetite nanoparticles, using the coprecipitation method, from aqueous solutions containing ferric and ferrous ions under a nitrogen atmosphere. To avoid undesired critical oxidation, the superparamagnetic magnetite nanoparticles have been coated with different biocompatible protective layers of sodium oleate, starch, and methoxy polyethylene glycol.

Magnetite and maghemite spherical nanoparticles can be fabricated using the coprecipitation method by mixing Fe^{+2} and Fe^{+3} ions in basic medium [10, 49]. Babay et al. [54] fabricated maghemite by using the coprecipitation method of ferric and ferrous ions (being the stoichiometric ratio $Fe^{+2}/Fe^{+3} = 2$) in NaOH basic solution using the pipette drop method. The authors also confirmed the transition of maghemite to the hematite phase after thermal treatment at 550 °C.

Lee et al. [55] also synthesised maghemite nanoparticles through the coprecipitation method of ferric and ferrous ions in alkali solution through the pipette drop method. However, as controlling particle size through this method is extremely difficult and depends on droplet size and dropping times they used a piezoelectric nozzle to form small size-controlled droplets and consequently control nanoparticles' size. While the size distribution of conventional pipette drop particles was 5–10 nm (pipette diameter: 2000 μm) with the use of the piezoelectric nozzle (nozzle size: 50 μm) it was possible to fabricate particles with narrow size distribution: 3–5 nm.

Farahmandjou [56] synthesized hematite nanoparticles using the coprecipitation method with iron chloride hexahydrate as a precursor and ammonia solution as the precipitator, and the precipitate was calcined at 500 ºC for 4 h.

3.5 *Anodization*

Anodization is an electrochemical synthesis method used to treat metals. In it, generally, the interesting material acts as the anode of an electrolytic cell where oxide coverings are deposited to improve surface performance. The structures and properties of the oxide layers formed are influenced by the electrolyte nature, the chemical reaction, and operational parameters [57].

Self-organized iron (III) oxide films fabricated by anodization were reported by Prakasam et al. [30], who deposited amorphous nanoporous films through potentiostat anodization of iron foil. By changing the reaction's potential, the synthesis conditions and chemical environment were modified, obtaining porous films. After annealing in a nitrogen atmosphere at 400 ºC for 30 min, the films revealed a reduction of the aspect ratio of pores and the formation of crystalline hematite, which shows an intense absorbance of light in the visible range.

Rangaraju et al. [58] also used anodization of iron foils to synthesize vertically oriented iron oxide nanotubes. In their experimental methodology, they used different reaction times (between 5 and 45 min) at a constant potential (in the range of 10–60 V) and tried different fluoride electrolyte compositions. They revealed that higher pH caused nanostructured granular oxide layers, while lower pH lead to a nanoporous oxide layer. The nanoporous structures could be modified to nanotubes by adjusting the applied potential and by using an ethylene glycol-based electrolyte with controlled water composition. After thermal treatment, the hematite samples with one and two layers of nanotubular oxide structures were tested under simulated solar illumination, at a specific potential, the two-layered structure presented 4 times higher photocurrent values than the monolayered structure.

Xie et al. [59] also obtained oriented iron oxides by anodization of iron foil at potentials ranging between 30 and 50 V. The nanotubes presented a diameter between 50 and 80 nm and thickness of 2.7–3.6 μm. Variations in the reaction conditions during the anodization process strongly influences the morphology and structure of the formed iron oxide layer. The authors revealed that the hematite nanotube shape stayed unchanged after thermal treatment at temperatures ranging between 300 and 600 °C. However, they reported that at higher temperatures, the tubular structure migrates to a porous one.

3.6 Atomic Layer Deposition

A widely used technique for material synthesis in the form of thin films is known as atomic layer deposition [60]. One of the main incentives for its mayor development has been its use for the synthesis of semiconductor materials. Yet, the main advantage related to this technique is the possibility to regulate deposition on the atomic scale [60]. Atomic layer deposition is related to chemical vapour deposition based on two reactions, the substrate is exposed to the reactants individually, and the film is generated step by step by self-limiting surface reactions [60].

Bachman et al. [61] fabricated iron oxide nanotubes by atomic layer deposition and studied the magnetic characteristics of the synthesized material. Using a porous anodic aluminum oxide membrane as substrate, they internally deposit a regular layer of Fe_2O_3 with a growth rate of 0.26 (±0.04) Å per cycle. The temperature of the deposition range was between 130 and 170 °C. After deposition, they treated the nanotubes in an atmosphere of H_2/Ar (5/95%) at 400 °C, and iron oxide was reduced to Fe_3O_4.

Rooth et al. [62] deposited iron oxide on Si(100) and anodic aluminum oxide (AAO) membrane. For this the authors used Ferrocene, $Fe(Cp)_2$, and oxygen as precursors. The authors showed effective depositions in temperature below 500 °C. Scheffe et al. [63] obtained crystalline maghemite thin films (~2.5 nm thick) on a zirconium oxide support with a high surface area. The authors use different atmospheres (mixtures of H_2, CO, and CO_2) for films reduction, showing that γ-Fe_2O_3 are reduced only to FeO. Using a similar deposition cycle, Yan et al. [64] obtained maghemite nanocomposites in a graphene oxide aluminum substrate, in a form rGO/Al@Fe_2O_3, with an average layer thickness of 3 nm, and its nanocomposite exhibited superior energy release and reduced electrostatic ignition hazard.

3.7 *Ball-Milling*

The high-energy ball milling process, although being of simple operation, has different adjustable variables, which are strongly correlated [65, 66]. During the ball milling process, the temperature can vary from ~30 to ~600 °C [67–69], which is very interesting for iron oxide synthesis.

Lemine et al. [70] analyzed the effect of high energy ball milling in obtaining the hematite phase; thus, in this study, different milling times ranging from 1 to 48 h, were used. The authors showed that the process and step affect the final characteristic (grain size, lattice parameters, etc.) of the material obtained. Already, Arbain et al. [71] showed success in getting the hematite by high energy ball milling and showed that the size of the crystal was influenced by the milling condition.

Stanhaus et al. [72] showed the influence of milling time on the phase formation, structural, and photoluminescence features of Eu-doped Fe_2O_3 to apply on energy conversion systems. The authors showed that high-energy ball milling increased structural disorder as well as decrease of crystallites sizes. Besides, for the material containing 5 wt.% of Eu_2O_3, was observed a broadband photoluminescence emission that is remarkably interesting for system energy-conversion application.

Zhao and Wen [73] prepared ε-Fe_2O_3 from by ball milling of $Fe(NO_3)_3.9H_2O$ and hydrophilic vapour phase nano-silica and post-annealing. The authors emphasize that the method is simple and suitable for large scale production of ε-Fe_2O_3 in the form of nanoparticles. ε-Fe_2O_3 nanoparticles prepared showed the magnetic hysteresis loops apparent shifts along the magnetic field axis at 300 K.

4 Applications

The usage of iron oxides by human beings goes back to prehistoric times when it was obtained from iron deposits (ochres) and used in cave and rock painting [1, 74, 75]. By around 2000 BC, already a wide range of red and brown pigments were produced by calcining raw ochres [1]. There is evidence of different civilizations, in different places and times, using iron oxide pigments either raw or calcinated, for wall murals, ceramic or pottery decoration, even as valuable commodities [1, 76]. Since the first synthetic iron was produced, around 4000 to 2000 BC, diverse applications have been developed for such materials [1]. Nowadays, iron oxides found applications in the steel and pigment industries mainly [75]. However, many other uses have appeared with the advances in nanotechnology. Nanomaterials present a wide range of possible applications, involving nanoscale optics and electronics, medicine, and, even, construction [31, 33]. More specifically, nanoscale iron oxides may serve different purposes, such as catalytic materials, wastewater treatment absorbents, pigments, flocculants, coatings, gas sensors, ion exchangers, lubrification, magnetic recording and data storage devices, toners and inks for xerography, magnetic resonance imaging, bioseparation and medicine [15].

4.1 Energy Storage

Lithium-ion batteries are very important for the development of new electrical products, portable electronic devices and electric vehicles. To improve the storage capacity of these batteries, some iron oxides with different structures and additives were synthesized and tested as anode materials in Li-ion batteries [77, 78].

Nuli et al. [79] have reported the obtained of cuboid particles and nanowires of α-Fe_2O_3 with different particle sizes by a simple hydrothermal method. As such, the anode with cuboid particles presented capacity of about 1074.5 and 609.0 mAh/g for the first and fifth cycles, respectively. However, the electrode with nanowire particles showed higher capacity: 1173.6 and 704.0 mAh/g for the first and fifth cycles, respectively. On the same year, Zeng et al. [80] used a facile solvothermal reaction and successive calcinations to synthesize porously α-Fe_2O_3 flower-like nanostructures which presented capacity of 974.43 mAh/g, 705.5 mAh/g and 548.47 mAh/g for the first, tenth a thirtieth cycle, respectively.

Pan et al. [81] synthesized α-Fe_2O_3 dendrites through the hydrothermal method and obtained an initial capacity of 1560 and 1095 mAh/g at 0.1 and 0.5 mA/cm^2 of current density, respectively, but, their cycle performance is poor.

Chen et al. [82] used a top-down method for the fabrication of α-Fe_2O_3 single-crystal nanodiscs and microparticles, the nanodisc obtained after 36 h of etching with oxalic acid presented a reversible capacity of 662 mAh/g, while the ones without oxalic acid treatment delivered a much lower capacity, 341 mAh/g, after 100 charge–discharge cycles.

Wang et al. [83] used a quasi-emulsion-templated method to synthesize α-Fe_2O_3 spheres and proved excellent cycling stability and high reversible capacity of 710 mAh/g after 100 cycles. Wang et al. [84] synthesized α-Fe_2O_3 nanotubes particles with an average size of 400–500 nm, exhibiting an opening capacity of 1200 mAh/g but only around 40% of it was maintained after 50 cycles.

Liang et al. [78] synthesized Fe_2O_3 nanocomposites prepared by oxidation of arc plasma of evaporated nano iron powders, and the nanocomposite obtained at 450 °C exhibited a capacity of about 507.6 mAh/g after 150 cycles. Xu et al. [85] studied three morphologies of α-Fe_2O_3, nanoplates, solid microspheres, and irregular nanoparticles, which presented a reversible capacity of approximately 1950 mAh/g, 790 and 680 mAh/g, respectively. Lv et al. [86] prepared composites to enhance lithium-ion batteries storage capacity by depositing γ-Fe_2O_3 nanoparticles on multi-walled carbon nanotubes by atomic layer deposition. After subsequent cycles, the synthesized material is stable, indicating no significant capacity loss and reversibility in the charge/discharge cycles. Composites with 300 and 500 cycles of atomic layer deposition presented capacities of 615 and 859.7 mAh/g after 400 charge/discharge cycles at a current density of 500 mA/g, even at a high current density of 10 A/g, the charge/discharge capacities remained 400.4 and 464.4 mAh/g, respectively [86].

4.2 Photocatalysis

While energy consumption rises, society awareness with fossil fuels environmental impacts is also increasing. Consequently, the use of renewable energy sources is becoming more and more popular every day. Solar energy is one of the most promising energy sources for environmentally friendly power generation [87, 88]. However, there are still some significant technological advances that researchers need to overcome [89].

Photosynthesis is the process by which nature collects and stores solar energy in the form of chemical bonds. The goal of many researchers in the energy field is to reproduce this process artificially [90, 91]. One desirable approach to emulate this process is by the photoelectrolysis of water using semiconductors as light absorbers and energy converters [90]. Photoelectrolysis of water, also known as solar water splitting, is a clean and sustainable form in which solar energy can be chemically stored in simple hydrogen (H_2) bonds and later consumed without the emission of pollutants either during generation or consumption processes [90, 91].

Van De Krol and Grätzel [92], mention several routes for the production of hydrogen from solar energy and water. However, several authors agree that the most popular forms are: photovoltaic cells associated with an electrochemical cell, particulate photocatalysis, and through the use of photoelectrochemical cells [93–96]. Particulate suspensions appear as a promising solution for photocatalysis. The solar production of hydrogen gas (H_2), via dissociation of water molecules using semiconductor photoelectrodes in a photoelectrochemical cell, has several advantages, among which it is highlighted that the production of hydrogen and oxygen occurs in separate electrodes at room temperature and that the separation device can be built entirely with inorganic materials at low cost: 3–5 US$/kg of hydrogen [92]. However, until now, there is limited understanding of charge and gas separation processes that limit particulate photocatalysis efficiency [97].

According to Van de Krol and Grätzel [92], the ideal material to act as a photoanode of the photoelectrochemical cell should perform several functions simultaneously: absorb light, separate and transport charges (electron–hole pairs), and on its surface, it should participate in the gas (O_2 or H_2) evolution. In addition to performing these functions at the same time, the semiconductor material for the photoelectrochemical cell should: efficiently absorb a significant portion of the solar spectrum, be abundant, low-cost and present low toxicity, however, it should also be stable when submerged for long periods in water and under concentrated lighting [88, 91].

Gibbs energy variation under standard conditions for water splitting is 237.2 kJ·mol^{-1}, which using Nernst equation ($\Delta E^{\circ} = -nF\Delta G^{\circ}$) corresponds to $\Delta E^{\circ} = 1.23$ eV [88, 92, 98]. Taking this into consideration, for solar water splitting, theoretically, the semiconductor needs to absorb photons with at least 1.23 eV [88, 92, 99]. However, in practice, this value is bigger due to the system's losses, such as overpotential for the evolution reactions or recombination of charge carriers [100, 101]. Cho et al. [93] highlight metallic oxides as the most promising materials to be considered as photoanodes in the oxidation semi-reaction of photoelectrochemical cells. One of the materials that have attracted researchers' attention is hematite due to its narrow bandgap (1.9–2.2 eV) that allows the absorption of visible sunlight (400 $\leq \lambda \leq$ 600 nm) [102].

Prakasam et al. [30] fabricated hematite nonporous films by potentiostatic anodization followed by annealing. The bang gap of the synthesized material was 2.2 eV. Under 1 sun illumination, the synthesized films exhibit photocurrent of 0.51 mA/cm^2 at 0.6 V versus Ag/AgCl in an electrolyte containing 1 M NaOH and 0.5 M H_2O_2. Townsend et al. [103] fabricated hematite nanocrystals by three different routes, including bottom-up and top-down approaches, which resulted in nanoscale crystals of various sizes that were effectively used for particulate photocatalysis in $AgNO_3$ aqueous solution. Other positive characteristics of hematite are its abundance, low toxicity, and stability, but above all, its most remarkable benefit is that it presents a theoretical solar to the hydrogen conversion efficiency of about 16.8% [104].

Hematite has been studied as a photocatalytic material since 1972 when Fujishima and Honda first used it in a photoelectrochemical cell [105]. However, the use of hematite photoanodes still faces some limitations, such as a short life of excited loads, low conductivity, and low oxygen evolution kinetics [106]. Exited charge carriers in hematite photoanodes present short diffusion lengths (2–4 nm), causing high recombination rates, which demand high overpotentials to drive the water oxidation reaction [106, 107].

Consequently, diverse efforts have consistently made in order to significantly improve the performance of this material. One of the most used techniques is the doping of hematite with metallic and non-metallic ions such as Ti, Sn, Pt, and P to improve its electrical properties [102, 104, 106, 108]. For example, Zhang et al. demonstrated that doping hematite with rare-earth ions also enhances the material's performance through photoabsorption improvement, expanding the spectrum of absorption to the infrared region [104].

4.3 *Photodegradation and Environmental Remediation*

Pollution has been a consequence of the progress and development of societies. However, to ensure appropriate living conditions for the next generations, the worldwide community has the challenge of remediating contaminated environments [109]. Environmental remediation is the reduction or removal of pollutants from water and soil for the safety of living systems [110].

Soil remediation can be accomplished by in-situ or ex-situ approaches [110], but, it is essential to identify that for remediation and pollutant removal, a combination of technologies has to be used according to specificities in environmental conditions [110].

Nanotechnology has been focusing on the remediation of poisoned soil with heavy metals and organic compounds. Heavy metals exist naturally in soil, but as that stay in it for years their toxic levels, mainly caused by mining and industrial waste disposal, make them the most challenging chemicals that need to be remediated [109]. The main technological processes for the reduction of heavy metal ions in the soil are adsorption–desorption, oxidation–reduction, and precipitation [109].

Magnetic iron oxide nanocomposites show great potential for environmental remediation applications, not only because they have large sorption capacity, but also, due to their capacity to be recycled appropriately by using a magnet, an economical and low energy consumption process [111]. Further research still is widely required in order to increase iron oxides potential, through the development of more convenient and efficient synthesis methods, coating materials, and ways to regenerate and/or reutilize the particles [111].

When considering a wastewater treatment technology, the main goal is to get the most successful treatment at the lowest rational cost [112]. Some factors to be considered include flexibility, efficiency, environmental security, and reuse capacity of the treatment's agents [112].

Some iron oxides are interesting for wastewater treatment due to their magnetic properties that are useful for the immobilisation or absorption of polluting materials; other iron oxides are attractive as photocatalysts for the fragmentation of contaminants into less toxic forms [113].

Iron oxide magnetic nanoparticles may treat great volumes of wastewater and are promising materials for water remediation by the removal of heavy metal ions such as Ag^{+}, As^{5+}, Cd^{2+}, Co^{2+}, Cr^{6+}, Cu^{2+}, Hg^{2+}, Mn^{2+}, Ni^{2+}, Pb^{2+}, Zn^{2+} cations [113–116]. Also, iron oxide magnetic particles (magnetite) may remove, by absorption, polluting organic compounds such as polycyclic aromatic hydrocarbons, a family of chemical substances that cause great damage to the environment and living species [117, 118]. However, for organic contaminant removal, the iron oxide magnetic nanoparticles need to increase their contact surface to improve their adsorption capacity [113].

As reported by Umar et al. [48] metal oxides semiconductors have a potential application in the degradation of organic dyes and as chemical sensors. Clothes and papers colouring industries use several organic compounds, which may cause serious environmental problems to human health and aquatic medium due to their toxicity and carcinogenic effect [48]. Organic dyes may be decomposed into less harmful chemicals by using semiconductor nanomaterials through photocatalysis; metal oxides are famous for this application [48]. Hematite's narrow bandgap makes it a promising photocatalysis candidate, as it might collect up to 40% of solar spectrum energy [48]. On the other hand, some of the chemicals used by these and other industries are extremely toxic and need to be detected, and their concentration needs to be quantified as it is the case of 4-nitrophenol for which hematite resulted in an effective electron mediator for the production of a highly sensitive chemical sensor in aqueous medium [48].

Laboratory research also studied catalytic properties of iron oxides nanoparticles against a variety of common environmental pollutants, including anions and radionuclides [119–121]. Magnetite nanoparticles are promising for the sorption of uranium species, even at low concentrations, from aqueous solutions with other competing anions or species [120]. Also, coated paramagnetic iron oxide nanoparticles were synthesized by Li et al. [121] aiming for stability and higher surface area, their sorption removal efficiency of uranium species was even higher than firstly reported.

4.4 Biomedical

The outstanding properties of iron oxides, in general, prompted scientists to research within the biotechnological field and, subsequently, in medicine. The magnetic properties and low toxicity have influenced the increase of recent studies in biomedical trials and cancer therapies [122].

Nanoscale iron oxides could be covered or linked to biological molecules (proteins or genes) for targeting a biological entity or tissue [123]. Magnetic iron oxide nanoparticles can be influenced by an external magnetic field, making them ideal candidates to be used as carriers for drugs or substances needed for the treatment of cancer tumours or any other localized disease [123]. Also, these particles are promising for applications such as medicine, for in vivo and in vitro studies, being useful for therapy and diagnostics. There are two types of approaches for iron oxides in medical applications: magnetic liposomes (which is a liposome with an internal ferromagnetic iron oxide shell), and "ferrofluids" (iron oxides nanoparticles covered by polymers) [123].

Iron magnetics fluids present some advantages when compared to other types of magnetic susceptible materials, e.g., such as long-term chemical stability, biocompatibility, and relatively easy surface modification and functionalization [124]. Iron oxides nanoparticles with magnetic properties are good candidates for biomedical applications as they improve the quality of magnetic resonance imaging (MRI), cell separation, hyperthermic treatment for cancer cells, targeted drug delivery, ultrasensitive bio-agent detection, gene therapy, and others [125].

4.4.1 Magnetic Resonance Imaging

MRI mechanism consists in measuring magnetization changes in hydrogen protons, present in water molecules, in a magnetic field, after a pulse of radio frequencies hit them [126]. As such, protons from different organs react differently, giving a representation of in vivo anatomical structures [127]. Magnetic resonance images can in principle be boosted by adding iron oxide nanoparticles, many of which may sharpen the contrast significantly through altering the magnetic behaviour of protons in their proximity [44, 128–130]. This contrast agents, in theory, travel by the circulatory system accessing organs, and hence growing disparity wherever they go [131].

Until now, some iron oxide nanoparticles have been produced differing in particle size, shape and type of covering material. Their size affects plasma half-life as well as biodistribution of the injected material and also their physicochemical and pharmacokinetic properties [132]. Particles greater than 50 nm, called superparamagnetic iron oxides (SPIOs), are mainly used for liver tropism. Smaller particles, called ultra-small superparamagnetic iron oxides (USPIOs) are usually used for the imaging of ganglion tropism and perfusion imaging in the brain or myocardial ischemic diseases [133–136].

USPIOs present greater blood circulation time, they provide information on cellular infiltration, related to inflammation, and also, they can be used in patients with long-lasting kidney diseases [137–139]. A more recent strategy used to increase cell uptake is the usage of anionic maghemite nanoparticles (AMNP). It has been shown that AMNPs, without any substance coating, lead to a high level of cell intake; mainly because they interact with the cell membrane due to their surface negative charges [133].

4.4.2 Hyperthermia

The eradication of cancer cells avoiding the harm of normal cells has been a desired goal to a most efficient cancer therapy. Nevertheless, the procedures used up to date, including chemotherapy, immunotherapy, radiotherapy or plus surgery, stay short in this purpose [140].

Hyperthermia is a type of cancer therapy treatment that uses temperatures ranging 40–43 °C for a small lapse of time (up to few hours) [141]. Hyperthermia as a potential strategy for cancer cells treatment first appeared after the observation that many kinds of cancer cells are widely sensitive to temperatures over 41 °C, while healthy cells resisted up to higher temperatures [142, 143]. Hyperthermia with magnetic fluids was first proposed by Gilchrist et al. [144], who presented the concept of injecting magnetic iron oxide nanoparticles, into the lymphatic system as a strategy to heat residual cancer cells from the application of an alternating magnetic field [144].

The use of magnetic fluids for hyperthermia treatment presents a great advantage due to its non-dependence of an electrical interface, which has been proved to be a severely limiting factor in radiofrequency hyperthermia [142]. Based on *in vitro* evidence that cells (ex. glioblastoma) can up-take magnetic nanoparticles more efficiently than healthy cells in certain kinds of cancers, but the mechanism is not well-known yet [145]. Improving targeting will be a step key to deliver enough quantities of magnetic fluids for selective heating of tumours [124]. Hyperthermia treatment also has the potential to be used with intermediate-risk prostate cancer patients, to control the more aggressive component of the tumour, in addition to avoid or delay the need for the use of radiotherapy or plus surgery [146].

4.4.3 Bioagent

Bacterial resistance against antibiotics is, of course, considered a primary global health concern. Iron oxides nanoparticles with surface modification or encapsulation were used as antibacterial agents efficiently in many scientific studies [147–149].

They have been proved to serve as antibacterial agents as their small size allows them to penetrate some bacterial membranes and generate reactive oxygen species, which cause severe damage to bacteria's proteins and DNA [150]. Also, there is evidence that iron oxides nanoparticles increase the production of osteoblast (bone-forming cell) without losing antibacterial action, being able to present dual function [148].

4.4.4 Drug Delivery

Nanoscale particles as an alternative strategy for delivering antibiotics have increased their use for the treatment of infectious diseases [149]. The increasing research in surface chemistry of iron oxides nanoparticles offers an opening field for the incorporation of a variety of biomolecules or antibiotics on the external surface of the nanoparticles [151]. This strategy can decrease the administration dose of antibiotics, and consequently, it can reduce resistant infections and side effects, as iron oxide magnetic nanoparticles make possible the targeting of the same group of bacteria or specifically infected tissue [152, 153]. As such, the effectiveness of magnetic drug delivery is related to some parameters, among which including field strength and gradient, as well as the volumetric and magnetic properties of iron oxide nanoparticles [44, 154].

To ensure iron oxide magnetic nanoparticles stability, in physiological conditions, a surface post adjustment is usually required. In this case, a reactive shell has frequently used for the attachment to the therapeutic compound. The covering must be carefully chosen as it must ensure stability and non-remarkable visibility of the nanoparticles in the biological media [155]. Besides, the coating can also be changed in a way that the drug release becomes stimulus-responsive of the surrounding media characteristics such as pH, temperature, or even redox states [44, 156]. The coating can be performed in two different ways: adsorption or encapsulation. In both cases, the shell will dictate the pharmacodynamic behaviour of the systems and then their efficacy as drug delivery systems [154, 155].

4.4.5 Gene Therapy

It is proposed that magnetic iron oxide nanoparticles covered with polycationic polymer polyethyleneimine may work as gene vectors and gene delivery devices in gene therapy [157]. Higher performance and activity with the presence of iron oxide magnetic nanoparticles was observed (up to 4 times) compared to traditional complexes [158]. This combination gives a novel, safe, and effective gene path as it could provide a powerful tool for treating diseases such as cancer, Alzheimer's, Huntington, and Parkinson's diseases [157, 159].

4.5 Electronics

Iron (III) oxide has been widely used as a magnetic material for magnetic audio and video recording and magnetic storage [160]. During tape recording, the iron (III) oxide is magnetized, and while playing the tape, the magnetized oxide generates an electric signal that will be the input signal for an amplifier that will reproduce sound or video [160].

Electronic and telecommunication devices have been gaining interest in the last decades as a result of human, society, and technological development [161]. The more expanded use of these devices has induced the creation of some problems of electronic pollution and electromagnetic interference [162]. Scientists have been researching new materials that may absorb and reduce electronic pollution and electromagnetic interference produced by commercial, scientific, and telecommunication devices [162, 163].

The design and use of iron oxides magnetic species as microwave absorbing materials for the protection against electromagnetic fields is a promising approach [162–166]. However, when studying and characterizing magnetic species for microwave absorption, there is still disagreement on the transmission line theory used for modelling reflection loss measurements and calculations. Liu and collaborators stated that materials scientists need a more in-depth mathematical and theoretical background for a better understanding of microwave engineering concepts [167, 168].

4.6 Electrochemical Sensing

Iron oxides nanoparticles revealed as promising functional materials in electrochemical sensing, as different techniques allow the fabrication of these materials (in various sizes and shapes) in the presence of two different oxidation states (Fe^{2+} and Fe^{3+}) [169].

New compounds created from the association of iron oxide nanoparticles with metals, organic substances, carbon nanotubes, and graphene oxide have proven to be useful resulting in better electrocatalytic efficiency for the detection of a significant number of substances, including hydroxide peroxide, hydrazine, dopamine, nitrate, nitrites, urea, hydroquinone, phenol, uric acid, glucose, ascorbic acid, L-cysteine, ciprofloxacin, salbutamol, creatinine, tetracycline, and metals [169–171]. Specifically, in the case of the carbon nanotubes and graphene, they act as nanoparticles holders, which ensure shape and size, resulting in better mechanical strength, which is reflected in efficiency gains, excellent electrical and thermal conductivities, as well as bigger surface area [169, 172].

4.7 Construction

Iron and steel industries use iron (III) oxide as one of the primary raw materials for the production of many iron and steel alloys. Iron (III) oxide reacts with aluminum to produce iron for welding processes as well as for the production of iron sculptures, weapons, and small tools [173].

Iron oxides-based nanoparticles may be used as immobilization agents of some kinds of bacterial cells, representing a promising application in bio self-healing concrete with the production of calcium carbonate in the presence of proper nutrients under specific conditions [174]. Magnetic immobilization of bacterial cells may also increase the concrete's compressive strength [175]. The strength increase is due to the properties of iron oxides nanoparticles, which act as a bind of cement hydrate and promote cement hydration [176]. Usually, bacterial cell immobilization on a solid carrier surface is accomplished by physical adsorption or covalent binding among the iron oxide nanoparticles and bacterial cell walls [175, 177].

Due to hematite nanostructures durability and capacity to absorb light in the UV region, this material is used in paints and when it is fully dispersed completely transparent paints can be obtained [15].

4.8 Other Uses

Iron (III) oxide powder, also known for its colour as rouge, is commonly used in the jewellery industry as a polishing material [1, 178]. Again, this hematite powder has been widely used as a pigment in paints and cosmetics throughout the years [14, 179]. Iron oxides potential as immobilization agent makes them ideal candidates to improve bioreactors activity, as immobilizing microbial cells can benefit their separation from the fermentation media, allowing their reuse in the fermentation process [180, 181].

5 Perspectives and Challenges

As previously discussed, iron oxide may be found in many different polymorphs, each of which presents different structural and morphological characteristics that result in distinctive properties. Iron oxides have been used by humankind for many years, firstly due to its colour, in paintings and pottery. However, nowadays, this material is being applied in many industries, mainly in the construction field. Even though iron oxides have been used for centuries, there is still plenty of research work to be done. In the last decades, nanosized iron oxides have called researchers attention as it is a low cost and environmentally friendly material which can be used in many different areas of knowledge, including electronics, agriculture, environmental remediation, energy harvesting and storage, and biomedical applications. These numerous possibilities are due to the ease of synthesis processes and the different possibilities of crystalline structures and morphologies that can be obtained. Perhaps it remains a challenge to synthesize the desired phase and morphology with the desired optical characteristics of nanosized iron oxides on a large scale. In order to expand iron oxide possibilities more research has to be done on the less stable phases, focusing either on new synthesis processes or in integrating these less stable phases with other materials, as composites, to expand their applications.

Acknowledgements The authors are gratefully acknowledges Fundação Araucária, CONICET and UNILA/Pró-Reitoria de Pesquisa e Pós-graduação (PRPPG) for financial support. Also, to B.Sc. Sandy Gonzalez Hernandez (UNILA) for his contributions and support while writing this chapter.

References

1. Cornell, R.M., Schwertmann, U.: The Iron Oxides: Structure, Properties, Reactions, Occurrences and Uses. Wiley (2003)
2. Schwertmann, U., Cornell, R.M.: Iron Oxides in the Laboratory: Preparation and Characterization, 2nd ed. Wiley-VCH (2008)
3. Fernández-Remolar, D.C.: Iron oxides, hydroxides and oxy-hydroxides BT—encyclopedia of astrobiology. In: Gargaud, M., Irvine, W.M., Amils, R., Cleaves, H.J. (Jim), Pinti, D.L., Quintanilla, J.C., Rouan, D., Spohn, T., Tirard, S., Viso, M. (eds.), pp. 1268–1270.. Springer, Berlin, Heidelberg (2015). https://doi.org/10.1007/978-3-662-44185-5_1714
4. Li, S.: Structural Design, Characterization, and Property Investigation of Iron Oxide (Nanoparticles with Visible Light Photoactivity), University of Paul Verlaine-Metz and Northeastern University (2009)
5. Atkins, P., Overton, T., Rourke, J., Weller, M., Armstrong, F.: Shriver & Atkins' Inorganic Chemistry, 5th edn. Oxford University Press, USA (2010)
6. Shannon, R.D.: Revised effective ionic radii and systematic studies of interatomic distances in halides and chalcogenides. Acta Crystallogr. Sect. A **32**(5), 751–767 (1976). https://doi.org/10.1107/S0567739476001551
7. Ahmed, M.A., Okasha, N., Nabeel, D.: The formation and characterization of nanocrystalline Mn-ferrite from magnetite. In: Modern Trends in Physics Research, pp. 104–109. WORLD SCIENTIFIC (2013). https://doi.org/10.1142/9789814504898_0015
8. Aschauer, U., Hockridge, J.G., Jones, F., Loan, M., Parkinson, G.M., Richmond, W.R.: Morphology control in the synthesis of iron oxide and oxyhydroxide nanoparticles. In: 2006 International Conference on Nanoscience and Nanotechnology, p. 1 (2006). https://doi.org/10.1109/ICONN.2006.340547
9. Momma, K., Izumi, F.: VESTA3 for three-dimensional visualization of crystal, volumetric and morphology data. J. Appl. Crystallogr. **44**(6), 1272–1276 (2011). https://doi.org/10.1107/S0021889811038970
10. Campos, E.A., Pinto, D.V.B.S., de Oliveira, J.I.S., Mattos, E.C., Dutra, R.C.L.: Synthesis, characterization and applications of iron oxide nanoparticles—a short review. J. Aerosp. Technol. Manag. **7**(3), 267–276 (2015). https://doi.org/10.5028/jatm.v7i3.471
11. Carter, C.B., Norton, M.G.: Ceramic Materials: Science and Engineering, vol. 716. Springer (2007)
12. Dissanayake, D.M.S.N., Mantilaka, M.M.M.G.P.G., Palihawadana, T.C., Chandrakumara, G.T.D., De Silva, R.T., Pitawala, H.M.T.G.A., de Silva, K.M., Amaratunga, G.A.J.: Facile and

low-cost synthesis of pure hematite (α-Fe_2O_3) nanoparticles from naturally occurring laterites and their superior adsorption capability towards acid-dyes. RSC Adv. **9**(37), 21249–21257 (2019). https://doi.org/10.1039/C9RA03756J
13. Barreca, D., Carraro, G., Gasparotto, A., Maccato, C.: 3-Metal oxide electrodes for photo-activated water splitting. In: Lin, Z., Ye, M., Wang, M.B.T.-M.P.M. Woodhead (eds.), Publishing in Materials, pp. 19–48. Woodhead Publishing (2018). https://doi.org/10.1016/B978-0-08-101977-1.00003-X
14. Sivula, K.: Nanostructured α-Fe_2O_3 photoanodes. In: van de Kro, l R., Grätzel, M. (eds.), Photoelectrochemical Hydrogen Production. Electronic Materials: Science & Technology, pp. 121–156. Springer: Boston, MA (2012). https://doi.org/10.1007/978-1-4614-1380-6_4
15. Mohapatra, M., Anand, S.: Synthesis and applications of nano-structured iron oxides/hydroxides—a review. Int. J. Eng. Sci. Technol. **2**(8), 127–146 (2010)
16. Bigham, J., Fitzpatrick, R., Schulze, D.: Iron Oxides. In: Dixon, J., Schulze, D. (eds.) Soil mineralogy with environmental applications, pp. 323–366. Soil Science Society of America, Madison (2002)
17. Van Oosterhout, G.W., Rooijmans, C.J.M.: A new superstructure in gamma-ferric oxide. Nature, 44 (1958). https://doi.org/10.1038/181044a0
18. Shmakov, A.N., Kryukova, G.N., Tsybulya, S.V., Chuvilin, A.L., Solovyeva, L.P.: Vacancy ordering in γ-Fe_2O_3: synchrotron X-ray powder diffraction and high-resolution electron microscopy studies. J. Appl. Crystallogr. **28**(2), 141–145 (1995). https://doi.org/10.1107/S0021889894010113
19. Tronc, E., Chanéac, C., Jolivet, J.P.: Structural and magnetic characterization of ε-Fe_2O_3. J. Solid State Chem. **139**(1), 93–104 (1998). https://doi.org/10.1006/jssc.1998.7817
20. García-Muñoz, J.L., Romaguera, A., Fauth, F., Nogués, J., Gich, M.: Unveiling a new high-temperature ordered magnetic phase in ϵ-Fe_2O_3. Chem. Mater. **29**(22), 9705–9713 (2017). https://doi.org/10.1021/acs.chemmater.7b03417
21. Balaev, D.A., Dubrovskiy, A.A., Yakushkin, S.S., Bukhtiyarova, G.A., Martyanov, O.N.: Temperature of the magnetic ordering of the trivalent iron oxide ε-Fe_2O_3. Phys. Solid State **61**(3), 345–349 (2019). https://doi.org/10.1134/S1063783419030053
22. Zboril, R., Mashlan, M., Petridis, D.: Iron (III) Oxides from thermal processes synthesis. Chem. Mater. **14**(III), 969–982 (2002)
23. Fleet, B.Y.M.E.: The structure of magnetite : two annealed natural magnetites, Fe_3. 00504 and. Acta Crystallogr. 40x, 1491–1493 (1984)
24. Hazen, R.M., Jeanloz, R.: Wüstite (Fe_{1-x} O): a review of its defect structure and physical properties. Rev. Geophys. **22**(1), 37–46 (1984). https://doi.org/10.1029/RG022i001p00037
25. Finger, L.W., Hazen, R.M.: Crystal structure and isothermal compression of Fe_2O_3, Cr_2O_3, and V_2O_3 to 50 Kbars. J. Appl. Phys. **51**(10), 5362–5367 (1980)
26. Danno, T., Nakatsuka, D., Kusano, Y., Asaoka, H., Nakanishi, M., Fujii, T., Ikeda, Y., Takada, J.: Crystal structure of β-Fe_2O_3 and topotactic phase transformation to α-Fe_2O_3. Cryst. Growth Des. **13**(2), 770–774 (2013)
27. Gich, M., Frontera, C., Roig, A., Taboada, E., Molins, E., Rechenberg, H.R., Ardisson, J.D., Macedo, W.A.A., Ritter, C., Hardy, V., et al.: High- and low-temperature crystal and magnetic structures of ε-Fe_2O_3 and their correlation to its magnetic properties. Chem. Mater. **18**(16), 3889–3897 (2006). https://doi.org/10.1021/cm060993l
28. Fleet, M.E.: The structure of magnetite: two annealed natural magnetites, $Fe_{3.005}O_4$ and $Fe_{2.96}Mg_{0.04}O_4$. Acta Crystallogr. Sect. C **40**(9), 1491–1493 (1984). https://doi.org/10.1107/S0108270184008489
29. Fjellvåg, H., Grønvold, F., Stølen, S., Hauback, B.: On the crystallographic and magnetic structures of nearly stoichiometric iron monoxide. J. Solid State Chem. **124**(1), 52–57 (1996)
30. Prakasam, H.E., Varghese, O.K., Paulose, M., Mor, G.K., Grimes, C.A.: Synthesis and photoelectrochemical properties of nanoporous iron (III) oxide by potentiostatic anodization. Nanotechnology **17**(17), 4285–4291 (2006). https://doi.org/10.1088/0957-4484/17/17/001

31. Ganachari, S.V., Banapurmath, N.R., Salimath, B., Yaradoddi, J.S., Shettar, A.S., Hunashyal, A.M., Venkataraman, A., Patil, P., Shoba, H., Hiremath, G.B.: Synthesis techniques for preparation of nanomaterials. Handb. Ecomater. **1**, 83–103 (2019). https://doi.org/10.1007/978-3-319-68255-6_149
32. Zia, M., Phull, A.R., Ali, J.S.: Challenges of iron oxide nanoparticles. Nanotechnol. Sci. Appl. **9**, 49–67 (2016)
33. Ali, A., Zafar, H., Zia, M., Haq, I., Phull, A., Sarfraz Ali, J., Hussain, A.: Synthesis, Characterization, Applications, and Challenges of Iron Oxide Nanoparticles. Nanotechnol. Sci. Appl. **9**, 49–67 (2016). https://doi.org/10.2147/NSA.S99986
34. Tuček, J., Machala, L., Ono, S., Namai, A., Yoshikiyo, M., Imoto, K., Tokoro, H., Ohkoshi, S., Zbořil, R.: Zeta-Fe_2O_3—a new stable polymorph in iron (III) oxide family. Sci. Rep. **5**(1), 15091 (2015). https://doi.org/10.1038/srep15091
35. Lefèvre, G., Duc, M., Fédoroff, M.: Chapter 2—accuracy in the determination of acid-base properties of metal oxides surfaces. In: Lützenkirchen, J. B. T.-I. S (eds.), Surface Complexation Modelling, vol. 11, pp. 35–66. Elsevier (2006). https://doi.org/10.1016/S1573-4285(06)80046-3
36. Colpas-Ruiz, M.A., Gnecco-Molina, C., Jiménez-Rodríguez, G.A., Pérez-Mendoza, J.A., Higuera-Cobos, Ó.F.: Synthesis of an anticorrosive pigment by thermal treatment of iron oxides from steel industry wastes. Rev. Fac. Ing. **28**(52), 44–58 (2019)
37. Hamed, M.H., Mueller, D.N., Müller, M.: Thermal phase design of ultrathin magnetic iron oxide films: from Fe_3O_4 to γ-Fe_2O_3 and FeO. J. Mater. Chem. C **8**(4), 1335–1343 (2020). https://doi.org/10.1039/C9TC05921K
38. Perednis, D., Gauckler, L.J.: Thin film deposition using spray pyrolysis. J. Electroceramics **14**(2), 103–111 (2005). https://doi.org/10.1007/s10832-005-0870-x
39. Marciello, M., Luengo, Y., Morales, M.P.: Iron oxide nanoparticles for cancer diagnosis and therapy. In: Nanoarchitectonics for Smart Delivery and Drug Targeting, pp. 667–694 (2016). https://doi.org/10.1016/B978-0-323-47347-7.00024-0
40. Kastrinaki, G., Lorentzou, S., Karagiannakis, G., Rattenbury, M., Woodhead, J., Konstandopoulos, A.G.: Parametric synthesis study of iron based nanoparticles via aerosol spray pyrolysis route. J. Aerosol Sci. **115**, 96–107 (2018). https://doi.org/10.1016/j.jaerosci.2017.10.005
41. Duret, A., Grätzel, M.: Visible light-induced water oxidation on mesoscopic α-Fe_2o_3 films made by ultrasonic spray pyrolysis. J. Phys. Chem. B **109**(36), 17184–17191 (2005). https://doi.org/10.1021/jp044127c
42. Popescu, V., Popescu, G.L., Indrea, E., Silipaş, D.T., Suciu, R.C.: Optical properties of Fe_2O_3 films obtained by spray pyrolysis. Proc. Int. Semicond. Conf. CAS **1**, 109–112 (2010). https://doi.org/10.1109/SMICND.2010.5650236
43. Ozcelik, B.K., Ergun, C.: Synthesis and characterization of iron oxide particles using spray pyrolysis technique. Ceram. Int. **41**(2), 1994–2005 (2015). https://doi.org/10.1016/j.ceramint.2014.09.103
44. Reddy, L.H., Arias, J.L., Nicolas, J., Couvreur, P.: Magnetic nanoparticles: design and characterization, toxicity and biocompatibility. Pharmaceut. Biomed. Appl. Chem. Rev. **112**(11), 5818–5878 (2012). https://doi.org/10.1021/cr300068p
45. Wu, W., He, Q., Jiang, C.: Magnetic iron oxide nanoparticles: synthesis and surface functionalization strategies. Nanoscale Res. Lett. **3**(11), 397–415 (2008). https://doi.org/10.1007/s11671-008-9174-9
46. Ge, S., Shi, X., Sun, K., Li, C., Uher, C., Baker, J.R., Banaszak Holl, M.M., Orr, B.G.: Facile hydrothermal synthesis of iron oxide nanoparticles with tunable magnetic properties. J. Phys. Chem. C **113**(31), 13593–13599 (2009). https://doi.org/10.1021/jp902953t
47. Chen, F., Gao, Q., Hong, G., Ni, J.: Synthesis and characterization of magnetite dodecahedron nanostructure by hydrothermal method. J. Magn. Magn. Mater. **320**(11), 1775–1780 (2008). https://doi.org/10.1016/j.jmmm.2008.02.117
48. Umar, A., Akhtar, M.S., Dar, G.N., Baskoutas, S.: Low-temperature synthesis of α-Fe_2O_3 hexagonal nanoparticles for environmental remediation and smart sensor applications. Talanta **116**, 1060–1066 (2013). https://doi.org/10.1016/j.talanta.2013.08.026

49. Teja, A.S., Koh, P.Y.: Synthesis, properties, and applications of magnetic iron oxide nanoparticles. Prog. Cryst. Growth Charact. Mater. **55**(1–2), 22–45 (2009). https://doi.org/10.1016/j.pcrysgrow.2008.08.003
50. Chang, H.S.W., Chiou, C.-C., Chen, Y.-W., Sheen, S.R.: Synthesis, characterization, and magnetic properties of Fe_3O_4 thin films prepared via a sol–gel method. J. Solid State Chem. **128**(1), 87–92 (1997). https://doi.org/10.1006/jssc.1996.7159
51. Cui, H., Liu, Y., Ren, W.: Structure switch between α-Fe_2O_3, γ-Fe_2O_3 and Fe_3O_4 during the large scale and low temperature sol-gel synthesis of nearly monodispersed iron oxide nanoparticles. Adv. Powder Technol. **24**(1), 93–97 (2013). https://doi.org/10.1016/j.apt.2012.03.001
52. Akbar, A., Imran, M., Riaz, S., Naseem, S.: Study of phase transition in iron oxide thin films. Mater. Today Proc. **2**(10, Part B), 5405–5409 (2015). https://doi.org/10.1016/j.matpr.2015.11.059
53. Kim, D.K., Mikhaylova, M., Zhang, Y., Muhammed, M.: Protective coating of superparamagnetic iron oxide nanoparticles. Chem. Mater. **15**(8), 1617–1627 (2003). https://doi.org/10.1021/cm021349j
54. Babay, S., Mhiri, T., Toumi, M.: Synthesis, structural and spectroscopic characterizations of maghemite γ-Fe_2O_3 prepared by one-step coprecipitation route. J. Mol. Struct. **1085**, 286–293 (2015). https://doi.org/10.1016/j.molstruc.2014.12.067
55. Lee, S.J., Jeong, J.R., Shin, S.C., Kim, J.C., Kim, J.D.: Synthesis and characterization of superparamagnetic maghemite nanoparticles prepared by coprecipitation technique. J. Magn. Magn. Mater. **282**(1–3), 147–150 (2004). https://doi.org/10.1016/j.jmmm.2004.04.035
56. Farahmandjou, M., Soflaee, F.: Synthesis and characterization of α-Fe_2O_3 nanoparticles by simple co-precipitation method. Phys. Chem. Res. **3**(3), 191–196 (2015). https://doi.org/10.22036/pcr.2015.9193
57. Ferreira, M.G.S., Zheludkevich, M.L., Tedim, J., Yasakau, K.A.: Self-healing nanocoatings for corrosion control. In: Corrosion protection and control using nanomaterials, pp. 213–263. Elsevier (2012). https://doi.org/10.1533/9780857095800.2.213
58. Rangaraju, R.R., Raja, K.S., Panday, A., Misra, M.: An Investigation on room temperature synthesis of vertically oriented arrays of iron oxide nanotubes by anodization of iron. Electrochim. Acta **55**(3), 785–793 (2010). https://doi.org/10.1016/j.electacta.2009.07.012
59. Xie, K., Guo, M., Huang, H., Liu, Y.: Fabrication of iron oxide nanotube arrays by electrochemical anodization. Corros. Sci. **88**, 66–75 (2014). https://doi.org/10.1016/j.corsci.2014.07.019
60. George, S.M.: Atomic layer deposition: an overview. Chem. Rev. **110**(1), 111–131 (2010). https://doi.org/10.1021/cr900056b
61. Bachmann, J., Jing; Knez, M., Barth, S., Shen, H., Mathur, S., Gösele, U., Nielsch, K.: Ordered iron oxide nanotube arrays of controlled geometry and tunable magnetism by atomic layer deposition. J. Am. Chem. Soc. **129**(31), 9554–9555 (2007).https://doi.org/10.1021/ja072465w
62. Rooth, M., Johansson, A., Kukli, K., Aarik, J., Boman, M., Hårsta, A.: Atomic layer deposition of iron oxide thin films and nanotubes using ferrocene and oxygen as precursors. Chem. Vap. Depos. **14**(3–4), 67–70 (2008). https://doi.org/10.1002/cvde.200706649
63. Scheffe, J.R., Allendorf, M.D., Coker, E.N., Jacobs, B.W., McDaniel, A.H., Weimer, A.W.: Hydrogen production via chemical looping redox cycles using atomic layer deposition-synthesized iron oxide and cobalt ferrites. Chem. Mater. **23**(8), 2030–2038 (2011). https://doi.org/10.1021/cm103622e
64. Yan, N., Qin, L., Hao, H., Hui, L., Zhao, F., Feng, H.: Iron oxide/aluminum/graphene energetic nanocomposites synthesized by atomic layer deposition: enhanced energy release and reduced electrostatic ignition hazard. Appl. Surf. Sci. **408**, 51–59 (2017). https://doi.org/10.1016/j.apsusc.2017.02.169
65. Rojac, T., Kosec, M., Malič, B., Holc, J.: The application of a milling map in the mechanochemical synthesis of ceramic oxides. J. Eur. Ceram. Soc. **26**(16), 3711–3716 (2006). https://doi.org/10.1016/j.jeurceramsoc.2005.11.013

66. Suryanarayana, C.: Mechanical alloying and milling. Prog. Mater. Sci. **46**(1), 1–184 (2001). https://doi.org/10.1016/S0079-6425(99)00010-9
67. Kwon, Y.-S., Gerasimov, K.B., Yoon, S.-K.: Ball temperatures during mechanical alloying in planetary mills. J. Alloys Compd. **346**(1), 276–281 (2002). https://doi.org/10.1016/S0925-8388(02)00512-1
68. Takacs, L., McHenry, J.S.: Temperature of the milling balls in shaker and planetary mills. J. Mater. Sci. **41**(16), 5246–5249 (2006). https://doi.org/10.1007/s10853-006-0312-4
69. Schmidt, R.; Scholze, H.; Stolle, A. Temperature Progression in a Mixer Ball Mill. *Int. J. Ind. Chem.* **2016**, *7*. https://doi.org/https://doi.org/10.1007/s40090-016-0078-8.
70. Lemine, O.M., Sajieddine, M., Bououdina, M., Msalam, R., Mufti, S., Alyamani, A.: Rietveld analysis and mössbauer spectroscopy studies of nanocrystalline hematite α-Fe_2O_3. J. Alloys Compd. **502**(2), 279–282 (2010). https://doi.org/10.1016/j.jallcom.2010.04.175
71. Arbain, R., Othman, M., Palaniandy, S.: Preparation of iron oxide nanoparticles by mechanical milling. Miner. Eng. **24**(1), 1–9 (2011). https://doi.org/10.1016/j.mineng.2010.08.025
72. Stanhaus, C., Alves, L.L.S., Ferrari, J.L., Padilha, J.C., Góes, M.S.; Hematite (α-Fe_2O_3) pure and doped with Eu^{3+} obtained by high-energy ball milling process. Mater. Chem. Phys. 123385 (2020). https://doi.org/10.1016/j.matchemphys.2020.123385
73. Zhao, Y., Wen, G.: Synthesis and magnetic properties of ε-Fe_2O_3 by ball milling and post annealing. J. Magn. Magn. Mater. **512**, 167039 (2020). https://doi.org/10.1016/j.jmmm.2020.167039
74. Cornell, R.M., Schwertmann, U.: The Iron Oxides (2007). https://doi.org/10.1002/9783527613229.ch01
75. Hradil, D., Grygar, T., Hradilová, J., Bezdička, P.: Clay and iron oxide pigments in the history of painting. Appl. Clay Sci. **22**(5), 223–236 (2003)
76. Schwertmann, U.: Iron Oxides (2008). https://doi.org/10.1002/9783527613229.ch01
77. Wu, H.B., Chen, J.S., Hng, H.H., Lou, X.W.: Nanostructured metal oxide-based materials as advanced anodes for lithium-ion batteries. Nanoscale **4**(8), 2526–2542 (2012). https://doi.org/10.1039/c2nr11966h
78. Liang, M., Zou, J., Zeng, X., Ding, W.: Nanostructured Fe_2O_3 based composites prepared through arc plasma method as anode materials in the lithium-ion battery. J. Nanomater. (2016).https://doi.org/10.1155/2016/1207907
79. NuLi, Y., Zhang, P., Guo, Z., Liu, H.: Shape evolution of α-Fe_2O_3 and its size-dependent electrochemical properties for lithium-ion batteries. J. Electrochem. Soc. **155**(3), A196 (2008). https://doi.org/10.1149/1.2826278
80. Zeng, S., Tang, K., Li, T., Liang, Z., Wang, D., Wang, Y., Qi, Y., Zhou, W.: Facile route for the fabrication of porous hematite nanoflowers: its synthesis, growth mechanism, application in the lithium ion battery, and magnetic and photocatalytic properties. J. Phys. Chem. C **112**(13), 4836–4843 (2008). https://doi.org/10.1021/jp0768773
81. Pan, Q., Huang, K., Ni, S., Yang, F., Lin, S., He, D.: Synthesis of α-Fe_2O_3 dendrites by a hydrothermal approach and their application in lithium-ion batteries. J. Phys. D. Appl. Phys. **42**(1), 1–6 (2009). https://doi.org/10.1088/0022-3727/42/1/015417
82. Chen, J.S., Zhu, T., Yang, X.H., Yang, H.G., Lou, X.W.: Top-down fabrication of α-Fe_2O_3 single-crystal nanodiscs and microparticles. PDF 13162–13164 (2010)
83. Wang, B., Chen, J.S., Wu, H.B., Wang, Z., Lou, X.W.: Quasiemulsion-templated formation of α-Fe_2O_3 hollow spheres with enhanced lithium storage properties. J. Am. Chem. Soc. **133**(43), 17146–17148 (2011). https://doi.org/10.1021/ja208346s
84. Wang, Z., Luan, D., Madhavi, S., Ming Li, C., Lou, X.W.: α-Fe_2O_3 nanotubes with superior lithium storage capability. Chem. Commun. **47**(28), 8061–8063 (2011). https://doi.org/10.1039/c1cc12111a
85. Xu, L., Tian, Y., Liu, T., Li, H., Qiu, J., Li, S., Li, H., Yuan, S., Zhang, S.: α-Fe_2O_3 nanoplates with superior electrochemical performance for lithium-ion batteries. Green Energy Environ. **3**(2), 156–162 (2018). https://doi.org/10.1016/j.gee.2018.01.005
86. Lv, X., Deng, J., Wang, B., Zhong, J., Sham, T.K., Sun, X., Sun, X.: γ-Fe_2O_3@CNTs anode materials for lithium ion batteries investigated by electron energy loss spectroscopy. Chem. Mater. **29**(8), 3499–3506 (2017). https://doi.org/10.1021/acs.chemmater.6b05356

87. Poizot, P., Dolhem, F., Environ, E., Poizot, P., Dolhem, F.: Clean energy new deal for a sustainable world: from non-CO_2 generating energy sources to greener electrochemical storage devices. Energy Environ. Sci. **4**(6), 2003–2019 (2011). https://doi.org/10.1039/c0ee00731e
88. Walter, M.G., Warren, E.L., McKone, J.R., Boettcher, S.W., Mi, Q., Santori, E.A., Lewis, N.S.: Solar water splitting cells. Chem. Rev. **110**(11), 6446–6473 (2010). https://doi.org/10.1021/cr1002326
89. Kannan, N., Vakeesan, D.: Solar energy for future world: a review. Renew. Sustain. Energy Rev. **62**, 1092–1105 (2016). https://doi.org/10.1016/j.rser.2016.05.022
90. Bard, A.J., Fox, M.A.: Artificial photosynthesis: solar splitting of water to hydrogen and oxygen. Acc. Chem. Res. **28**(3), 141–145 (1995). https://doi.org/10.1021/ar00051a007
91. Tachibana, Y., Vayssieres, L., Durrant, J.R.: Artificial photosynthesis for solar water-splitting. Nat. Photon. **6**(8), 511–518 (2012). https://doi.org/10.1038/nphoton.2012.175
92. Van de Krol, R., Grätzel, M.: Photoelectrochemical Hydrogen Production. In: Electronic Materials: Science & Technology, 1st ed., vol. 102 Springer US, Boston, MA (2012). https://doi.org/10.1007/978-1-4614-1380-6
93. Cho, S., Jang, J.-W.W., Lee, K.-H.H., Lee, J.S.: Research update: strategies for efficient photoelectrochemical water splitting using metal oxide photoanodes. APL Mater. **2**(1), 10703 (2014). https://doi.org/10.1063/1.4861798
94. Chang, W.J., Lee, K.-H., Ha, J.-I., Nam, K.T.: Hydrogen production via water electrolysis: the benefits of a solar cell-powered process. IEEE Electrif. Mag. **6**(1), 19–25 (2018). https://doi.org/10.1109/MELE.2017.2784632
95. Goto, Y., Hisatomi, T., Wang, Q., Higashi, T., Ishikiriyama, K., Maeda, T., Sakata, Y., Okunaka, S., Tokudome, H., Katayama, M., et al.: A Particulate photocatalyst water-splitting panel for large-scale solar hydrogen generation. Joule **2**(3), 509–520 (2018). https://doi.org/10.1016/j.joule.2017.12.009
96. Kim, J.H., Hansora, D., Sharma, P., Jang, J.-W., Lee, J.S.: Toward practical solar hydrogen production—an artificial photosynthetic leaf-to-farm challenge. Chem. Soc. Rev. **48**(7), 1908–1971 (2019). https://doi.org/10.1039/C8CS00699G
97. Pan, Z., Röhr, J.A., Ye, Z., Fishman, Z.S., Zhu, Q., Shen, X., Hu, S.: Elucidating charge separation in particulate photocatalysts using nearly intrinsic semiconductors with small asymmetric band bending. Sustain. Energy Fuels **3**(3), 850–864 (2019). https://doi.org/10.1039/C9SE00036D
98. Qi, J., Zhang, W., Cao, R.: Solar-to-hydrogen energy conversion based on water splitting. Adv. Energy Mater. **8**(5), 1701620 (2018). https://doi.org/10.1002/aenm.201701620
99. Xiao, J., Huang, H., Huang, Q., Li, X., Hou, X., Zhao, L., Ma, R., Chen, H., Li, Y.: Remarkable improvement of the turn–on characteristics of a Fe_2O_3 photoanode for photoelectrochemical water splitting with coating a fecow oxy-hydroxide gel. Appl. Catal. B Environ. **212**, 89–96 (2017). https://doi.org/10.1016/j.apcatb.2017.04.075
100. Bolton, J.R., Strickler, S.J., Connolly, J.S.: Limiting and realizable efficiencies of solar photolysis of water. Nature **316**(6028), 495–500 (1985). https://doi.org/10.1038/316495a0
101. Turner, J.A.: A realizable renewable energy future. Science **80**, *285* (5428), 687–689 (1999). https://doi.org/10.1126/science.285.5428.687
102. Maabong, K., Machatine, A.G.J., Mwankemwa, B.S., Braun, A., Bora, D.K., Toth, R., Diale, M.: Nanostructured hematite thin films for photoelectrochemical water splitting. Phys. B Condens. Matter **535**, 67–71 (2018). https://doi.org/10.1016/j.physb.2017.06.054
103. Townsend, T.K., Sabio, E.M., Browning, N.D., Osterloh, F.E.: Photocatalytic water oxidation with suspended alpha-Fe_2O_3 particles-effects of nanoscaling. Energy Environ. Sci. **4**(10), 4270 (2011). https://doi.org/10.1039/c1ee02110a
104. Zhang, M., Lin, Y., Mullen, T.J., Lin, W.-F., Sun, L.-D., Yan, C.-H., Patten, T.E., Wang, D., Liu, G.: Improving Hematite's solar water splitting efficiency by incorporating rare-earth upconversion nanomaterials. J. Phys. Chem. Lett. **3**(21), 3188–3192 (2012). https://doi.org/10.1021/jz301444a
105. Fujishima, A., Honda, K.: Electrochemical photolysis of water at a semiconductor electrode. Nature **238**(5358), 37–38 (1972). https://doi.org/10.1038/238037a0

106. Wang, J.-J.J., Hu, Y., Toth, R., Fortunato, G., Braun, A.: A facile nonpolar organic solution process of a nanostructured hematite photoanode with high efficiency and stability for water splitting. J. Mater. Chem. A **4**(8), 2821–2825 (2016). https://doi.org/10.1039/C5TA06439B
107. De Carvalho, V.A.N., Luz, R.A.D.S., Lima, B.H., Crespilho, F.N., Leite, E.R., Souza, F.L.: Highly oriented hematite nanorods arrays for photoelectrochemical water splitting. J. Power Sources **205**, 525–529 (2012). https://doi.org/10.1016/j.jpowsour.2012.01.093
108. Kim, J.Y., Magesh, G., Youn, D.H., Jang, J.-W.W., Kubota, J., Domen, K., Lee, J.S.: Single-crystalline, wormlike hematite photoanodes for efficient solar water splitting. Sci. Rep. **3**(1), 2681 (2013). https://doi.org/10.1038/srep02681
109. Kalita, E., Baruah, J.: Environmental Remediation. Elsevier Inc. (2020). https://doi.org/10.1016/b978-0-12-813357-6.00014-0
110. Yeung, A.T.: Remediation technologies for contaminated sites. Adv. Environ. Geotech. 328–369 (2010). https://doi.org/10.1007/978-3-642-04460-1_25
111. Zhu, J., Wei, S., Chen, M., Gu, H., Rapole, S.B., Pallavkar, S., Ho, T.C., Hopper, J., Guo, Z.: Magnetic nanocomposites for environmental remediation. Adv. Powder Technol. **24**(2), 459–467 (2013). https://doi.org/10.1016/j.apt.2012.10.012
112. Oller, I., Malato, S., Sánchez-Pérez, J.A.: Combination of advanced oxidation processes and biological treatments for wastewater decontamination-a review. Sci. Total Environ. **409**(20), 4141–4166 (2011). https://doi.org/10.1016/j.scitotenv.2010.08.061
113. Xu, P., Zeng, G.M., Huang, D.L., Feng, C.L., Hu, S., Zhao, M.H., Lai, C., Wei, Z., Huang, C., Xie, G.X., et al.: Use of iron oxide nanomaterials in wastewater treatment: a review. Sci. Total Environ. **424**, 1–10 (2012)
114. Hao, Y.M., Man, C., Hu, Z.B.: Effective removal of Cu (II) ions from aqueous solution by amino-functionalized magnetic nanoparticles. J. Hazard. Mater. **184**(1–3), 392–399 (2010). https://doi.org/10.1016/j.jhazmat.2010.08.048
115. Hu, H., Wang, Z., Pan, L.: Synthesis of monodisperse Fe_3O_4@silica core-shell microspheres and their application for removal of heavy metal ions from water. J. Alloys Compd. **492**(1–2), 656–661 (2010). https://doi.org/10.1016/j.jallcom.2009.11.204
116. Dave, P.N., Chopda, L.V.: Application of iron oxide nanomaterials for the removal of heavy metals. J. Nanotechnol. (2014). https://doi.org/10.1155/2014/398569
117. Zhang, S., Niu, H., Hu, Z., Cai, Y., Shi, Y.: Preparation of carbon coated Fe_3O_4 nanoparticles and their application for solid-phase extraction of polycyclic aromatic hydrocarbons from environmental water samples. J. Chromatogr. A **1217**(29), 4757–4764 (2010). https://doi.org/10.1016/j.chroma.2010.05.035
118. Abdel-Shafy, H.I., Mansour, M.S.M.: A review on polycyclic aromatic hydrocarbons: source, environmental impact, effect on human health and remediation. Egypt. J. Pet. **25**(1), 107–123 (2016). https://doi.org/10.1016/j.ejpe.2015.03.011
119. Zhang, W-X.: Nanoscale iron particles for environmental remediation: an overview. J. Nanoparticle Res. **5**(3–4), 323–332 (2003)
120. Das, D., Sureshkumar, M.K., Koley, S., Mithal, N., Pillai, C.G.S.: Sorption of uranium on magnetite nanoparticles. J. Radioanal. Nucl. Chem. **285**(3), 447–454 (2010). https://doi.org/10.1007/s10967-010-0627-0
121. Li, Y., Zhao, H., Ban, H., Yang, M.: Composites of Fe_2O_3 nanosheets with polyaniline : preparation, gas sensing properties and sensing mechanism. Sensors Actuators B. Chem. (2017). https://doi.org/10.1016/j.snb.2017.01.103
122. Dussán, K.J., Giese, E.C., Vieira, G.N.A., Lima, L.N., Silva, D.D.V.: Pharmaceutical and biomedical applications of magnetic iron-oxide nanoparticles. In: Metal Nanoparticles in Pharma, pp. 77–99. Springer (2017)
123. Nikiforov, V., Filinova, E.: Biomedical applications of magnetic nanoparticles. In: Magnetic Nanoparticles, pp. 393–455 (2009). https://doi.org/10.1002/9783527627561.ch10
124. Chang, D., Lim, M., Goos, J.A.C.M., Qiao, R., Ng, Y.Y., Mansfeld, F.M., Jackson, M., Davis, T.P., Kavallaris, M.: Biologically targeted magnetic hyperthermia: potential and limitations. Front. Pharmacol. **9**(AUG) (2018). https://doi.org/10.3389/fphar.2018.00831

125. Morber, J.R.: 1D nanowires : understanding growth and properties as steps toward biomedical and electrical application. Growth (Lakeland) (2008)
126. Berger, A.: How does it work? magnetic resonance imaging. BMJ **324**(7328), 35 (2002). https://doi.org/10.1136/bmj.324.7328.35
127. Patravale, V., Joshi, M.: Nanocarriers for imaging applications. In: Fanun, M. (ed.) Colloids in Drug Delivery, pp. 563–611. CRC Press, Boca Raton, Florida, US (2010)
128. Weissleder, R., Moore, A., Mahmood, U., Bhorade, R., Benveniste, H., Chiocca, E.A., Basilion, J.P.: In vivo magnetic resonance imaging of transgene expression. Nat. Med. **6**(3), 351–354 (2000). https://doi.org/10.1038/73219
129. Hinds, K.A., Hill, J.M., Shapiro, E.M., Laukkanen, M.O., Silva, A.C., Combs, C.A., Varney, T.R., Balaban, R.S., Koretsky, A.P., Dunbar, C.E.: Highly efficient endosomal labeling of progenitor and stem cells with large magnetic particles allows magnetic resonance imaging of single cells. Blood **102**(3), 867–872 (2003). https://doi.org/10.1182/blood-2002-12-3669
130. Lok, C.: Picture perfect. Nature **412**(6845), 372–374 (2001). https://doi.org/10.1038/35086702
131. Marghussian, V.: Nano-Glass Ceramics. Elsevier (2015). https://doi.org/10.1016/C2014-0-01294-4
132. Arami, H., Khandhar, A., Liggitt, D., Krishnan, K.M.: In vivo delivery, pharmacokinetics, biodistribution and toxicity of iron oxide nanoparticles. Chem. Soc. Rev. **44**(23), 8576–8607 (2015). https://doi.org/10.1039/C5CS00541H
133. Berry, C.C., Curtis, A.S.G.: Functionalisation of magnetic nanoparticles for applications in biomedicine. J. Phys. D. Appl. Phys. **36**(13), R198–R206 (2003). https://doi.org/10.1088/0022-3727/36/13/203
134. Pultrum, B.B., van der Jagt, E.J., van Westreenen, H.L., van Dullemen, H.M., Kappert, P., Groen, H., Sietsma, J., Oudkerk, M., Plukker, J.T.M., van Dam, G.M.: Detection of lymph node metastases with ultrasmall super paramagnetic iron oxide (USPIO)-enhanced magnetic resonance imaging in oesophageal cancer: a feasibility study. Cancer Imaging **9**(1), 19–28 (2009). https://doi.org/10.1102/1470-7330.2009.0004
135. Qin, Y., Zhuo, L., Cai, J., He, X., Liu, B., Feng, C., Zhang, L.: In vivo monitoring of magnetically labeled mesenchymal stem cells homing to rabbit hepatic VX2 tumors using magnetic resonance imaging. Mol. Med. Rep. **17**(1), 452–458 (2018). https://doi.org/10.3892/mmr.2017.7902
136. Philips, B.W.J., Stijns, R.C.H., Rietsch, S.H.G., Brunheim, S., Barentsz, J.O., Fortuin, A.S., Quick, H.H., Orzada, S., Maas, M.C., Scheenen, T.W.J.: USPIO-enhanced MRI of pelvic lymph nodes at 7-T: preliminary experience. Eur. Radiol. **29**(12), 6529–6538 (2019). https://doi.org/10.1007/s00330-019-06277-7
137. Semple, S., Alam, S.R., MacGillivray, T.J., Dweck, M.R., Shah, A.S., Richards, J., Wang, C., Lang, N., McKillop, G., Mirsadraee, S., et al.: Quantitative myocardial inflammation assessed using a novel USPIO-magnetic resonance imaging acquisition and analysis protocol. J. Cardiovasc. Magn. Reson. **15**(S1), O114 (2013). https://doi.org/10.1186/1532-429x-15-s1-o114
138. Gkagkanasiou, M., Ploussi, A., Gazouli, M., Efstathopoulos, E.P.: USPIO-enhanced MRI neuroimaging: a review. J. Neuroimag. **26**(2), 161–168 (2016). https://doi.org/10.1111/jon.12318
139. Lagan, J., Naish, J.H., Simpson, K., Zi, M., Cartwright, E.J., Foden, P., Morris, J., Clark, D., Birchall, L., Caldwell, J., et al.: Substrate for the myocardial inflammation–heart failure hypothesis identified using novel USPIO methodology. JACC Cardiovasc. Imag. 1–11 (2020). https://doi.org/10.1016/j.jcmg.2020.02.001
140. Moding, E.J., Kastan, M.B., Kirsch, D.G.: Strategies for optimizing the response of cancer and normal tissues to radiation. Nat. Rev. Drug Discov. **12**(7), 526–542 (2013). https://doi.org/10.1038/nrd4003
141. Wust, P., Hildebrandt, B., Sreenivasa, G., Rau, B., Gellermann, J., Riess, H., Felix, R., Schlag, P.: Hyperthermia in combined treatment of cancer. Lancet Oncol. **3**(8), 487–497 (2002). https://doi.org/10.1016/S1470-2045(02)00818-5

142. Jordan, A., Wust, P., Scholz, R., Tesche, B., Fähling, H., Mitrovics, T., Vogl, T., Cervós-Navarro, J., Felix, R.: Cellular uptake of magnetic fluid particles and their effects on human adenocarcinoma cells exposed to AC magnetic fields in vitro. Int. J. Hyperth. **12**(6), 705–722 (1996). https://doi.org/10.3109/02656739609027678
143. Nielsen, O.S., Horsman, M., Overgaard, J.: A future for hyperthermia in cancer treatment? Eur. J. Cancer **37**(13), 1587–1589 (2001). https://doi.org/10.1016/S0959-8049(01)00193-9
144. Gilchrist, R.K., Medal, R., Shorey, W.D., Hanselman, R.C., Parrott, J.C., Taylor, C.B.: Selective inductive heating of lymph nodes. Ann. Surg. **146**(4), 596–606 (1957). https://doi.org/10.1097/00000658-195710000-00007
145. Jordan, A., Scholz, R., Wust, P., Fähling, H., Felix, R.: Magnetic fluid hyperthermia (MFH): cancer treatment with AC magnetic field induced excitation of biocompatible superparamagnetic nanoparticles. J. Magn. Magn. Mater. **201**(1–3), 413–419 (1999). https://doi.org/10.1016/S0304-8853(99)00088-8
146. Hergt, R., Andrä, W.: Magnetic hyperthermia and thermoablation. In: Andrä, W., Nowak, H. (eds.) Magnetism in Medicine, pp. 550–570. WILEY-VCH Verlag GmbH & Co., Weinheim, Germany (2007)
147. Chudasama, B., Vala, A.K., Andhariya, N., Upadhyay, R.V., Mehta, R.V.: Enhanced antibacterial activity of bifunctional Fe_3O_4-Ag core-shell nanostructures. Nano Res. **2**(12), 955–965 (2009). https://doi.org/10.1007/s12274-009-9098-4
148. Tran, N., Mir, A., Mallik, D., Sinha, A., Nayar, S., Webster, T.J.: Bactericidal effect of iron oxide nanoparticles on staphylococcus aureus. Int. J. Nanomed. **5**(1), 277–283 (2010). https://doi.org/10.2147/ijn.s9220
149. Hajipour, M.J., Fromm, K.M., Akbar Ashkarran, A., Jimenez de Aberasturi, D., de Larramendi, I.R., Rojo, T., Serpooshan, V., Parak, W.J., Mahmoudi, M.: Antibacterial properties of nanoparticles. Trends Biotechnol. **30**(10), 499–511 (2012). https://doi.org/10.1016/j.tibtech.2012.06.004
150. Sau, T.K., Biswas, A., Ray, P.: Metal nanoparticles in nanomedicine: advantages and scope. Metal Nanopart. 121–168 (2018). https://doi.org/10.1002/9783527807093.ch6
151. Majidi, S., Zeinali Sehrig, F., Samiei, M., Milani, M., Abbasi, E., Dadashzadeh, K., Akbarzadeh, A.: Magnetic nanoparticles: applications in gene delivery and gene therapy. Artif. Cells, Nanomedicine Biotechnol. **44**(4), 1186–1193 (2016). https://doi.org/10.3109/21691401.2015.1014093
152. Dinali, R., Ebrahiminezhad, A., Manley-Harris, M., Ghasemi, Y., Berenjian, A.: Iron oxide nanoparticles in modern microbiology and biotechnology. Crit. Rev. Microbiol. **43**(4), 493–507 (2017). https://doi.org/10.1080/1040841X.2016.1267708
153. Dagan, R., Barkai, G., Givon-Lavi, N., Sharf, A.Z., Vardy, D., Cohen, T., Lipsitch, M., Greenberg, D.: Seasonality of antibiotic-resistant streptococcus pneumoniae that causes acute otitis media: a clue for an antibiotic-restriction policy? J. Infect. Dis. **197**(8), 1094–1102 (2008). https://doi.org/10.1086/528995
154. Laurent, S., Saei, A.A., Behzadi, S., Panahifar, A., Mahmoudi, M.: Superparamagnetic iron oxide nanoparticles for delivery of therapeutic agents: opportunities and challenges. Expert Opin. Drug Deliv. **11**(9), 1449–1470 (2014). https://doi.org/10.1517/17425247.2014.924501
155. Vangijzegem, T., Stanicki, D., Laurent, S.: Magnetic iron oxide nanoparticles for drug delivery: applications and characteristics. Expert Opin. Drug Deliv. **16**(1), 69–78 (2019). https://doi.org/10.1080/17425247.2019.1554647
156. Hola, K., Markova, Z., Zoppellaro, G., Tucek, J., Zboril, R.: Tailored functionalization of iron oxide nanoparticles for MRI, drug delivery, magnetic separation and immobilization of biosubstances. Biotechnol. Adv. **33**(6), 1162–1176 (2015). https://doi.org/10.1016/j.biotechadv.2015.02.003
157. Wei, W., Xu, C., Wu, H.: Magnetic iron oxide nanoparticles mediated gene therapy for breast cancer—an in vitro study. J. Huazhong Univ. Sci. Technol. **26**(6), 728–730 (2006). https://doi.org/10.1007/s11596-006-0628-y
158. Dowaidar, M., Abdelhamid, H.N., Hällbrink, M., Freimann, K., Kurrikoff, K., Zou, X., Langel, Ü.: Magnetic nanoparticle assisted self-assembly of cell penetrating peptides-oligonucleotides

complexes for gene delivery. Sci. Rep. **7**(1) (2017). https://doi.org/10.1038/s41598-017-09803-z
159. Kami, D., Takeda, S., Itakura, Y., Gojo, S., Watanabe, M., Toyoda, M.: Application of magnetic nanoparticles to gene delivery. Int. J. Mol. Sci. **12**(6), 3705–3722 (2011). https://doi.org/10.3390/ijms12063705
160. Ayers, J.W., Stephens, R.A.: Gamma Ferric Oxide for Magnetic Impulse Record Members (1962)
161. Coccia, M.: General sources of general purpose technologies in complex societies: theory of global leadership-driven innovation. Warfare Human Devel. Technol. Soc. **2015**(42), 199–226 (2014). https://doi.org/10.1016/j.techsoc.2015.05.008
162. Mederos-Henry, F., Mahin, J., Pichon, B.P., Dîrtu, M.M., Garcia, Y., Delcorte, A., Bailly, C., Huynen, I., Hermans, S.: Highly efficient wideband microwave absorbers based on zero-valent Fe@γ-Fe_2O_3 and Fe/Co/Ni carbon-protected alloy nanoparticles supported on reduced graphene oxide. Nanomaterials **9**(9) (2019). https://doi.org/10.3390/nano9091196
163. Kong, L.B., Liu, L., Yang, Z., Li, S., Zhang, T.: Magnetic nanomaterials for electromagnetic wave absorption. Magn. Nanomater. Fundam. Synth. Appl. 473–514 (2017). https://doi.org/10.1002/9783527803255.ch15
164. Ni, S., Lin, S., Pan, Q., Yang, F., Huang, K., He, D.: Hydrothermal synthesis and microwave absorption properties of Fe_3O_4 nanocrystals. J. Phys. D. Appl. Phys. **42**(5) (2009). https://doi.org/10.1088/0022-3727/42/5/055004
165. Ni, S., Sun, X., Wang, X., Zhou, G., Yang, F., Wang, J., He, D.: Low temperature synthesis of Fe_3O_4 micro-spheres and its microwave absorption properties. Mater. Chem. Phys. **124**(1), 353–358 (2010). https://doi.org/10.1016/j.matchemphys.2010.06.046
166. Yan, A., Liu, Y., Liu, Y., Li, X., Lei, Z., Liu, P.: A NaAc-assisted large-scale coprecipitation synthesis and microwave absorption efficiency of Fe_3O_4 nanowires. Mater. Lett. **68**, 402–405 (2012). https://doi.org/10.1016/j.matlet.2011.10.093
167. Liu, Y., Zhao, K., Drew, M.G.B., Liu, Y.A.: Theoretical and practical clarification on the calculation of reflection loss for microwave absorbing materials. AIP Adv. **8**(1) (2018). https://doi.org/10.1063/1.4991448
168. Liu, Y., Drew, M.G.B., Li, H., Liu, Y.: An experimental and theoretical investigation into methods concerned with "reflection loss" for microwave absorbing materials. Mater. Chem. Phys. **243** (2020). https://doi.org/10.1016/j.matchemphys.2020.122624
169. George, J.M., Antony, A., Mathew, B.: Metal oxide nanoparticles in electrochemical sensing and biosensing: a review. Microchim. Acta **185**(7), 358 (2018). https://doi.org/10.1007/s00604-018-2894-3
170. Chauhan, N., Pundir, C.S.: An amperometric biosensor based on acetylcholinesterase immobilized onto iron oxide nanoparticles/multi-walled carbon nanotubes modified gold electrode for measurement of organophosphorus insecticides. Anal. Chim. Acta **701**(1), 66–74 (2011). https://doi.org/10.1016/j.aca.2011.06.014
171. Lee, S., Oh, J., Kim, D., Piao, Y.: A sensitive electrochemical sensor using an iron oxide/graphene composite for the simultaneous detection of heavy metal ions. Talanta **160**, 528–536 (2016). https://doi.org/10.1016/j.talanta.2016.07.034
172. Zhang, Y., Cheng, Y., Zhou, Y., Li, B., Gu, W., Shi, X., Xian, Y.: Electrochemical sensor for bisphenol a based on magnetic nanoparticles decorated reduced graphene oxide. Talanta **107**, 211–218 (2013). https://doi.org/10.1016/j.talanta.2013.01.012
173. Hosford, W.F.: Iron and Steel. Cambridge University Press, Cambridge (2012). https://doi.org/10.1017/CBO9781139086233
174. Seifan, M., Samani, A.K., Berenjian, A.: Induced calcium carbonate precipitation using bacillus species. Appl. Microbiol. Biotechnol. **100**(23), 9895–9906 (2016). https://doi.org/10.1007/s00253-016-7701-7
175. Seifan, M., Sarmah, A.K., Samani, A.K., Ebrahiminezhad, A., Ghasemi, Y., Berenjian, A.: Mechanical properties of bio self-healing concrete containing immobilized bacteria with iron oxide nanoparticles. Appl. Microbiol. Biotechnol. **102**(10), 4489–4498 (2018). https://doi.org/10.1007/s00253-018-8913-9

176. Li, H., Gang Xiao, H., Ping Ou, J.: A study on mechanical and pressure-sensitive properties of cement mortar with nanophase materials. Cem. Concr. Res. **34**(3), 435–438 (2004). https://doi.org/10.1016/j.cemconres.2003.08.025
177. Kourkoutas, Y., Bekatorou, A., Banat, I.M., Marchant, R., Koutinas, A.A.: Immobilization technologies and support materials suitable in alcohol beverages production: a review. Food Microbiol. **21**(4), 377–397 (2004). https://doi.org/10.1016/j.fm.2003.10.005
178. Royal, H.: Polishing compound residues in gold jewellery alloys. Gold Bull. **22**(2), 42–47 (1989). https://doi.org/10.1007/BF03214706
179. Marcot, G.C., Cauwenberg, W.J., Lamanna, S.: A. Production of Iron Oxide Pigments. Google Patents (1951)
180. Ansari, F., Grigoriev, P., Libor, S., Tothill, I.E., Ramsden, J.J.: DBT degradation enhancement by decorating rhodococcus erythropolis IGST8 with magnetic Fe_3O_4 nanoparticles. Biotechnol. Bioeng. **102**(5), 1505–1512 (2009). https://doi.org/10.1002/bit.22161
181. Akay, G., Erhan, E., Keskinler, B.: Bioprocess intensification inflow-through monolithic microbioreactors with immobilized bacteria. Biotechnol. Bioeng. **90**(2), 180–190 (2005). https://doi.org/10.1002/bit.20376

The Importance of Factorial Design of Experiments in Functional Nanomaterials Preparation and Performance

Alexandre H. Pinto

Abstract The experimental planning and design are important parts for successful performance and result analysis in a project. Both for industry and academic settings, there is the constant need to analyze the influence of many variables in the different types of responses. Traditionally, the influence of different variables in the experimental outcome has been analyzed by changing "one factor at a time", which is usually described as univariate or OFAT approach. However, unless all variables are close to their optimum value, there is no guarantee that this approach will lead to the best optimized outcome. Additionally, the OFAT approach can lead to the implementation of an excessive number of experiments, which usually increases the expenses related to the project. Pursuing to analyze how the synergy between different variables can influence the experimental outcomes, there is the multivariate approach, where two or more variables are changed simultaneously enabling the experimentalist to analyze the beneficial or antagonistic effect of this combination of variables in the experimental outcome. Moreover, the multivariate approach may improve the chances to find the best outcome possible with the conduction of a fewer number of experiments. In this sense, this chapter introduces the concept of factorial design of experiments, a multivariate approach based on choosing two or more levels for multiple variables, calculating the effects of each variable individually and of each possible combination of variables, obtaining a model from these results, applying this model to predict untested conditions and judge the statistic significance of the model. The examples presented in the chapter will all be focused on the preparation and performance of nanomaterials. For instance, how the concentrations of different precursors can influence the particle size of colloidal silica nanoparticles. Or how different variables, such as, time, temperature and reagents concentration can influence the thickness of manganese sulfide (MnS) thin films. The chapter begins providing the definition of the basic terms underlying the factorial design, then, it presents examples from literature applying factorial designs starting with the simpler ones, such as 2^3. Then, the chapter evolves presenting optimization and

A. H. Pinto (✉)
Department of Chemistry and Biochemistry, Manhattan College, 4513 Manhattan College Parkway, Riverdale, NY 10471, USA
e-mail: alex.pinto@manhattan.edu

F. A. La Porta and C. A. Taft (eds.), *Functional Properties of Advanced Engineering Materials and Biomolecules*, Engineering Materials,
https://doi.org/10.1007/978-3-030-62226-8_14

response surface methodologies factorial designs, for instance, central composite, Box-Behken, and Doehlert designs. Finally, the chapter presents tables with references from papers published in the period from 2015 to 2020. In each one of them, the factorial design of experiments was used for the development of functional materials applied in nanoparticles preparation, drug delivery and encapsulation, wastewater remediation, and solar cells development. With this chapter, the author hopes to introduce a powerful and underexplored statistical tool to scientists, engineers, and all practitioners of nanomaterials science. Focus will be placed on how they can benefit from the concepts and examples presented, and possibly adapt them for their own projects, instead of relying on heavy mathematical notations and calculations.

1 Introduction

The development of functional materials requires accurate control of each step in their production. In this sense, to be able to test a broader range of experimental conditions by performing a lower number of experiments is desirable. So, the factorial design of experiments is a set of statistical concepts intended to optimize specific properties by taking advantage of a multivariate approach. In this approach, the interaction between different factors is interpreted holistically. The design of experiments allows the research to judge which factors are statistically significant by calculating their effects on the property to be optimized. Moreover, empirical models can be obtained, allowing the researcher to predict how that specific property will behave in untested conditions.

The ability to predict results based on an empirical model opens up an array of opportunities and saves time and resources from the researcher. So, this chapter aims to show from the scientific literature examples how the factorial design of experiments has widely been applied to the development and performance of functional materials.

This chapter is intended to fulfill the needs of readers with different knowledge levels about the design of experiments, from the beginner to the experienced reader.

That being said, readers from different familiarity levels have the opportunity to focus their attention on different parts of the chapter. For the beginner ones, it is recommended to start reading the chapter from Sect. 2. Since in this part, the essential concepts and terminology of factorial design are introduced. Then, a step by step 2^3 factorial design is performed. This 2^3 factorial design is developed very comprehensively, aiming that even a beginner reader can try to adapt the design to their own experimental situation, without using black-box programs. This part appreciates the statistical formalism whenever it is necessary. In such a way, it can sound intimidating at first sight. However, with attention and persistence, the reader can benefit from the knowledge.

Since Sect. 2 is very comprehensive. Consequently, it is also extensive. So, if a reader is already familiar with all the terminology, structure, and calculations related to factorial designs, he/she is welcome to skip this full part and start straight on Sect. 3. There, the reader will find analyses of literature examples where the factorial design of experiments and response surface methodologies were applied to the context of materials preparation and performance. This part appreciates a critical analysis of each research paper by remarking the steps followed by the authors.

Finally, in Sect. 4, readers will find references from papers published in the period from 2015 to 2020. In each one of them, the factorial design of experiments was applied to the development of four types of functional materials. They are nanoparticles preparation, drug encapsulation and delivery, wastewater remediation, and solar cell development.

2 The Fundamentals and Statistical Basis of the Factorial Design of Experiments

2.1 Factorial Design of Experiments: Initial Concepts and Terminology

The experimental planning and design are essential parts for successful performance and analysis of results in a project. Both for industry and academic settings, there is a constant need to analyze the influence of many factors in the different types of responses. Traditionally, the impact of different factors in the experimental outcome has been analyzed by changing "one factor at a time," which is usually described as a univariate or OFAT approach.

For instance, suppose that a research group is aiming to maximize the yield of a certain chemical reaction, and they know that the yield can be affected by factors such as pH, temperature, and catalyst concentration. From previous literature knowledge, the scientists figured out that this reaction has been described to occur in the following ranges for each factor, as presented in Table 1.

According to the OFAT approach, they would perform a series of experiments varying the pH and keeping temperature and catalyst concentration constant. For instance, by fixing the temperature at 85 °C, the catalyst concentration at 5×10^{-4} mol L^{-1}, and making the pH equal to 8, 10, 12, and 14 units.

Next, they would vary the temperature and keep the pH and catalyst concentration constant. For instance, by fixing pH equal to 14, catalyst concentration at 5×10^{-4} mol L^{-1}, and making the temperature equal to 25, 50, 70, and 85 °C.

Then, they would vary catalyst concentration and keep pH and temperature constant. For instance, by fixing pH equal to 14, temperature at 85 °C, and making

Table 1 Matrix of experiments showing the factors and their respective lower and upper limits

Factor	Lower limit	Upper limit
pH	8	14
Temperature (°C)	25	85
Catalyst concentration (mol L^{-1})	1×10^{-4}	5×10^{-4}

catalyst concentration equal to 1×10^{-4}, 2.5×10^{-4}, 3.5×10^{-4}, and 5×10^{-4} mol L^{-1}.

There are potential flaws associated with this OFAT approach. The first one is related to choosing the values for the factors that will remain fixed [1]. Notice that, whenever fixing the pH and catalyst concentration, the researchers always chose to fix them at the uppermost level possible. This choice could have been determined by a preconceived idea that maximizing factor values always are going to lead to a yield maximization, which is not always true.

Second, this OFAT approach could ignore potential effects related to the interaction of two or more factors being varied at the same time [1]. For instance, the yield could be maximized at pH 11, 30 °C, and catalyst concentration equal to 1.5×10^{-4} mol L^{-1}. This condition was not performed in the OFAT approach adopted, so the researchers would not be aware of the yield maximization at this set of conditions.

A third potential flaw could be related to the large number of experiments performed [1]. In general, a higher number of experiments represent higher consumption of chemicals, analysis time, production of waste, and, consequently, a higher total cost. In this sense, an approach that offers useful trends and conclusions with a lower amount of experiments is always preferred. So, in the next section, we will see how this experimental approach could be redesigned according to the factorial design of experiments. And how the factorial design could potentially decrease the chance of each of these flaws to occur.

2.2 *Planning According to the Factorial Design of Experiments*

Pursuing to analyze how the interaction between different variables can influence the experimental outcomes, there is the multivariate approach. Where two or more variables are changed simultaneously, enabling the researcher to analyze the beneficial or antagonistic effect of this interaction of variables in the experimental outcome. Moreover, the multivariate approach may improve the chances to find the best result possible with the conduction of a fewer number of experiments.

Before we start redesigning the series of experiments shown in the previous section, we should define the terms commonly related to the factorial design of experiments. The first term is **factor**. Factors, as seen in the previous section, are the variables that would be varied across the experimental design. So, according to the situation we are analyzing, three factors have been studied, they are pH, temperature, and catalyst concentration.

The second term is the **response variable**, which is the response the researchers are aiming to observe by performing the experiments. In the example presented, the yield is the response variable. It is important to say that experimental designs can have more than one response variable.

The values each factor will adopt will be called **levels**. For instance, let's suppose that the temperature will be fixed in their lowest and highest values, so, respectively, 25 and 85 °C. In this case, 25 and 85 °C are called the levels of the temperature factor. If only two levels are adopted for each factor, then the factorial design is called a **two-level factorial design**. Now, supposing that the scientist had chosen three levels for each variable, in that case, the factorial design would be called **three-level factorial design**. The levels could either be quantitative or qualitative. In case qualitative levels are chosen, it is necessary to consistently order them in such a way that a certain level is defined as the lower level, and the other one is the upper level, taking the two-level factorial design as an example.

The **effects** are how single factors or the interaction of two or more factors would influence the value obtained for the response variable. For instance, among the three factors studied, one of them, for instance, temperature, may have a higher influence on the response variable. Whereas, other factors may be concluded to be insignificant to change the response variable. The total number of possible effects will be dependent on how many factors the experimental design has.

In the studied example, as there are three factors, each factor will have its individual effect. Then, the design can have secondary effects, which will arise from the interaction of two factors at a time. So, the secondary factors would be pH-Temperature, pH-Catalyst Concentration, and Temperature-Catalyst Concentration. Finally, in models having three or more factors, tertiary effects are predicted, which are obtained by the combination of three variables at a time. In the studied example, the only tertiary effect would be pH-Temperature-Catalyst Concentration. If the design had four factors, it would have more tertiary effects and one quaternary effect. In conclusion, this three-factor, two-level experimental design has three primary effects, three secondary effects, and one tertiary effect.

A **factorial design** is the experimental design comprising all levels of each factor studied varied in a multivariate manner, in such a way that allows the interaction effects to be calculated. In general, the factorial designs are named according to how many factors they have and how many levels each factor has. A common representation for the factorial design name is $\mathbf{s^k}$, where **k** is equal to the number of factors, and **s** is how many levels each factor has. In the studied example, which has three factors, and two levels for each one of them, this can be called to be a **$\mathbf{2^3}$ factorial design**.

The s^k representation also indicates the minimum number of experiments without replicas to be performed to be considered a full s^k factorial design. To figure out the number of experiments it is necessary to solve the exponential equation shown in the s^k representation, for instance, a 2^3 factorial design would require a minimum of $2^3 = 8$ different experimental conditions to be considered complete.

Another important concept is the **coded levels**. For calculating the effects and their statistical significance, matrix calculations will be necessary. In this sense, each level, whether it is quantitative or qualitative, should be normalized according to a common scale. In general, this scale ranges from -1 to $+1$. In the case of two-level factorial design, the lower limit is normalized as -1, and thus, the upper limit

Table 2 The factors with their respective limits represented as coded (in parenthesis) and uncoded values

Factor	Lower limit	Upper limit
pH = A	8 (−1)	14 (+1)
Temperature (°C) = B	25 (1)	85 (+1)
Catalyst concentration (mol L^{-1}) = C	1×10^{-4} (−1)	5×10^{-4} (+1)

Table 3 Matrix of experiments

Exp. number	pH	Temperature	Catalyst concentration
1	−1	−1	−1
2	+1	−1	−1
3	−1	+1	−1
4	+1	+1	−1
5	−1	−1	+1
6	+1	−1	+1
7	−1	+1	+1
8	+1	+1	+1

is normalized as +1. So, Table 1 could be improved to include the coded values between parenthesis, as shown in Table 2.

Matrix of experiments is the matrix that describes each experiment to be performed, according to the combination of the coded values for each factor. For instance, for the example given, as shown in Table 3.

Translating the matrix of experiments from the coded values to the original values, one can say that the experiment number 1 is performed with pH = 8, Temperature = 25 °C, and catalyst concentration = 1×10^{-4} mol L^{-1}. Taking another example, experiment 7 is performed at pH = 8, Temperature = 85 °C, and catalyst concentration = 5×10^{-4} mol L^{-1}.

The **matrix of results** is the one that summarizes the values observed for the response variable for each one of the experiments performed, as shown in Table 4.

Table 4 Matrix of results

Exp. number	Yield (%)
1	25
2	42
3	76
4	45
5	81
6	5
7	66
8	15

After being introduced to the essential vocabulary and concepts about factorial design, we are ready to start seeing how the factorial design is structured.

2.3 *The Structure of the Factorial Design*

Factorial Design starts with its planning. In the **planning** stage, the factors and response variables to be studied should be defined. Then, within each factor is necessary to identify the upper and lower levels.

When planning a factorial design, it is essential to understand that there is no "one-size fits all" solution for all the scientific problems. However, there are some steps that the researcher can follow to better planning the design. For instance, the factors, their respective levels, and the response variables could be defined according to the need of the process that the researcher is aiming to optimize. Scientific or technical literature about the problem to be solved is a reasonable starting point. Also, the levels can be defined according to the capabilities of the pieces of equipment available for that experiment performance. The cost of using a specific piece of equipment or technique could be another decisive factor while planning the design [2].

In summary, to plan a factorial design accordingly, it is necessary: (a) to be familiar to the system to de studied, (b) to know the technical resources available, (c) to have clear goals about what is the response variable to be optimized, and (d) how the researcher is aiming to optimize this response variable, for instance, either maximizing or minimizing its value [3].

The second step in the factorial design is the **factor screening**. Due to the prior knowledge of the system, the researcher can identify variables that influence the response variable he/she is interested in optimizing. So, at this point, the researcher will perform a factor screening step. During factor screening, experiments will be performed according to some 2^k factorial design or some 2^{k-p} fractional factorial design. In either case, k is the number of factors to be studied. Then, all the experiments necessary to complete the selected factorial design will be performed. These experiments can be achieved with or without replicas, depending on the time and resources available to the researcher.

After the experiments conclusion, the researcher will analyze the data and determine which factors have significant effects and which factors do not have significant effects regarding influencing the response variable. At this point, the researcher will be able to obtain a linear polynomial equation describing how the significant factors quantitatively influence the response variable. For some situations, the model described by this equation can be enough to answer satisfactorily the question asked by the researcher. And the design can even be concluded on this point [3].

However, for other situations, the linear polynomial model obtained may not accurately describe the situation faced by the researcher. In this case, the factorial design will proceed to an **optimization** step. In this step, the insignificant factors will be disregarded, and additional experiments will be performed, considering only

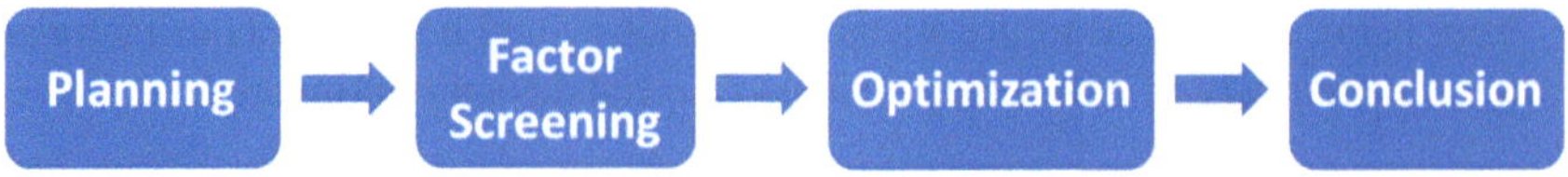

Fig. 1 The four general steps of a factorial design of experiments

the significant factors. Also, in this step, in general, more than two levels will need to be considered for every factor [3, 4].

As more than two levels will be considered for each factor, the optimization step will allow the research to obtain a model described by a polynomial equation of order higher than one, for instance, a quadratic polynomial equation. Also, as the optimization is a refinement step, some factors may prove to be insignificant under the new design conditions. Consequently, they will be excluded from the definite model equation [4].

Most of the time, after one or a few optimization steps, the researcher will have a model that describes with satisfactory precision the behavior of the system in the study. Then, the next step is the **conclusion** of the design. In this step, the researcher will interpret how the empirical model obtained translates in the context of the studied situation. Additionally, the researcher will be able to apply the model to experimental conditions initially untested and verify how the experimental result agrees with the result predicted by the model.

The successful accomplishment of the four steps shown in Fig. 1 describes what is required to complete a factorial design. In the next two sections, we will discuss how the screening and optimization steps can be performed by using different types of factorial designs and their particularities.

2.4 *Screening the Factors*

Depending on the complexity of the experimental problem to be solved, many factors can influence the response variable. In this sense, it is necessary to have a screening step, which will start with most of the factors that can hypothetically change the response variable. Then, the significance of the effects of these factors will be judged according to appropriate statistical tests, and the insignificant factors will be excluded from the optimization step. Besides deciding the significance of the effects, another goal from the screening step is to obtain a first-order polynomial equation, describing how the response variable varies according to changing the coded values of the significant factors.

The main differences between 2 and 2^{k-p} factorial designs are shown in Table 5, as adapted from Candioti et al. [5].

The fractional factorial designs 2^{k-p} are beyond the scope of this chapter. For readers interested in learning more regarding this type of design, references number

Table 5 Comparison of the features of the full factorial design (2^k) and fractional factorial design (2^{k-p})

Design	Type of factors	Type of response variable	Number of factors	Minimum number of experiments	Number of estimated effects
2^k	Quantitative or qualitative	Quantitative	Usually between 2 and 5	2^k	$2^k–1$
2^{k-p}	Quantitative or qualitative	Quantitative	Usually higher than 4	2^{k-p}	It depends on p value and how cofounded the design is

[3] and [6] are good references to learn the essential concepts of fractional factorial designs.

2.5 2^k Factorial Designs

In Sect. 2.2, we first defined what 2^k factorial designs are and how many experiments are necessary to complete the design according to the k value. A feature of a 2^k factorial design is that for each factor, only two levels are defined, a lower and an upper level. This feature holds for 2^k factorial designs used in the factors screening steps. For more comprehensive steps, like the ones used for optimization steps, more levels will be needed. In this section, we will see the particularities of a 2^k factorial design. For instance, we will discuss how to calculate the effects and the coefficients of the equation describing the design model, and how to judge the significance of these effects.

Before we start, it is important to say that to perform all the calculations and graphs presented in this section does not require any specialized software. Any software able to program matrix multiplication and graphical plotting, for instance, Microsoft Excel, is enough to perform all the calculations. However, there are software packets specialized to perform all of the design of experiments calculations, most of them are paid packages. The readers more interested to know the names and capabilities of these software packages are encouraged to read the review paper by Hibbert [7], which contains a table listing the main commercial softwares used in design of experiment calculations.

As shown in Table 6, the number of effects possible for a 2^k factorial design is equal to $2^k - 1$. In this sense, Table 6 explains how many and which are the effects, according to the k value.

Each one of the effects can be calculated according to the Eq. 1, as exemplified for Effect A [8]:

Table 6 Number and type of effects according to the number of factors (k value) in a 2^k factorial design

k value	Factors	Number of effects	Primary effects	Interaction effects
2	A, B	3	A, B	AB
3	A, B, C	7	A, B, C	AB, AC, BC, ABC
4	A, B, C, D	15	A, B, C, D	AB, AC, AD, BC, BD, CD, ABC, ABD, ADC, BDC, ABCD

$$Effect_A = \frac{\sum Y^A_{(+)}}{n/2} - \frac{\sum Y^A_{(-)}}{n/2} \tag{1}$$

where $Y^A_{(+)}$ are the values of the response variable for which A has coded value $= +1$, $Y^A_{(-)}$ are the values of the response variable for which A has coded value $= -1$. n is defined as the total number of experiments in the factorial design, for instance, n $= 8$ for a 2^3 or n $= 16$ for a 2^4. Notice that the n value does not change if replicas are performed. The only thing that changes in case replicas were performed is the fact that each $Y^A_{(+)}$ and $Y^A_{(-)}$ should be taken as the average of the replicas.

A good way to understand the Eq. 1 is to think that the effect is the difference between the average of the response variable when A has coded value $= +1$ and the average of the response variable when A has coded value $= -1$.

To minimize the chances to be mistaken during the calculation of the effects, an alternative to Eq. 1, is to use matrix multiplication for each one of the effects. In Table 3, in Sect. 2.2, we introduced the concept of the matrix of experiments. Here, we will expand that matrix to include the coded values for each one of the possible interactions.

The coded value for each interaction was obtained by simply multiplying the coded values of the factors giving rise to that respective interaction. We can convert Table 7 in a square matrix (a matrix having the same number of columns and rows) by adding a column where all entries will be +1, as the first column of the matrix.

Table 7 Matrix of experiments for the primary effects and interactions in a 2^3 factorial design

Exp. Number	A	B	C	AB	AC	BC	ABC
1	−1	−1	−1	+1	+1	+1	−1
2	+1	−1	−1	−1	−1	+1	+1
3	−1	+1	−1	−1	+1	−1	+1
4	+1	+1	−1	+1	−1	−1	−1
5	−1	−1	+1	+1	−1	−1	+1
6	+1	−1	+1	−1	+1	−1	−1
7	−1	+1	+1	−1	−1	+1	−1
8	+1	+1	+1	+1	+1	+1	+1

We can call this matrix as matrix (M1):

$$X = \begin{bmatrix} +1 & -1 & -1 & -1 & +1 & +1 & +1 & -1 \\ +1 & +1 & -1 & -1 & -1 & -1 & +1 & +1 \\ +1 & -1 & +1 & -1 & -1 & +1 & -1 & +1 \\ +1 & +1 & +1 & -1 & +1 & -1 & -1 & -1 \\ +1 & -1 & -1 & +1 & +1 & -1 & -1 & +1 \\ +1 & +1 & -1 & +1 & -1 & +1 & -1 & -1 \\ +1 & -1 & +1 & +1 & -1 & -1 & +1 & -1 \\ +1 & +1 & +1 & +1 & +1 & +1 & +1 & +1 \end{bmatrix} \quad (M1)$$

The transpose matrix (X^t) of the matrix 1 can be written as matrix (M2)

$$X^t = \begin{bmatrix} +1 & +1 & +1 & +1 & +1 & +1 & +1 & +1 \\ -1 & +1 & -1 & +1 & -1 & +1 & -1 & +1 \\ -1 & -1 & +1 & +1 & -1 & -1 & +1 & +1 \\ -1 & -1 & -1 & -1 & +1 & +1 & +1 & +1 \\ +1 & -1 & -1 & +1 & +1 & -1 & -1 & +1 \\ +1 & -1 & +1 & -1 & -1 & +1 & -1 & +1 \\ +1 & +1 & -1 & -1 & -1 & -1 & +1 & +1 \\ -1 & +1 & +1 & -1 & +1 & -1 & -1 & +1 \end{bmatrix} \quad (M2)$$

The matrix of results, shown in Table 4 can be written as matrix Y (M3):

$$Y = \begin{bmatrix} 25 \\ 42 \\ 76 \\ 45 \\ 81 \\ 5 \\ 66 \\ 15 \end{bmatrix} \quad (M3)$$

Multiplying the matrixes X^t and Y, as shown in Eq. 2, will be useful both for calculating the effects and the coefficients:

$$(X^t)Y = \begin{bmatrix} 355 \\ -141 \\ 49 \\ -21 \\ -23 \\ -113 \\ -59 \\ 73 \end{bmatrix} \tag{2}$$

The effects can be calculated by the Eq. 3:

$$Effect(i) = \frac{(X^t)Y}{n/2} \tag{3}$$

The effects calculated are shown in Fig. 2.

From Fig. 2, one can notice that except for the effect of the of the factor B, and the tertiary interaction ABC, all other effects had negative values. For the primary effects A, B, and C a negative value means that moving from the lower to the upper level of these factors, leads to a decrease in the response variable, in this case, the yield. Similarly, a positive effect means that switching from the lower to upper level leads to an increase in the response variable.

For the secondary effects, such as AB, AC, and BC, an increase in the effect means that the response variable increases when the product between the levels of the two factors switch from negative to positive. The interpretation is similar to the tertiary effect, such as ABC.

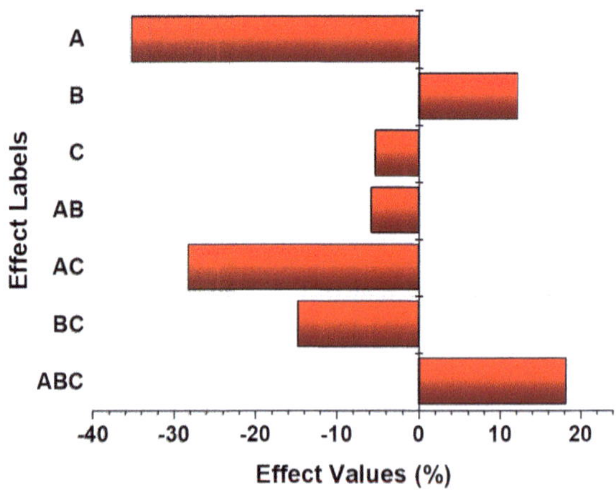

Fig. 2 Graph showing the effect value for each one of the primary effects and interactions possible for the 2^3 factorial design. The effects were calculated according to Eq. 3

2.5.1 The Empirical Model

As previously told, a 2^k factorial model can produce a polynomial equation where all the terms are linear; in general, the empirical model can be described by the Eq. 4:

$$y = b_0 + \sum_{i=1}^{k} b_i X_i + \sum_{1 \le i \le j}^{k} b_{ii} X_i X_j + \xi \tag{4}$$

where: b_0 the independent term related to the average of the response of all experiments, b_i or b_{ii} are the coefficients for each term, the X_i or X_j are the coded values of each factor, and ξ is the residual associated to the experiment.

For a factorial 2^k, the empirical model contains 2^k terms, being one term for each one of the effects plus the independent term. The residual term ξ is not explicitly written in the equation. In this sense, the empirical model equations for 2^2, 2^3, and 2^4 designs are shown, respectively, in the equations varying from 5 to 7:

$$y = b_0 + b_A X_A + b_B X_B + b_{AB} X_A X_B \tag{5}$$

$$\begin{aligned} y = b_0 &+ b_A X_A + b_B X_B + b_C X_C + b_{AB} X_A X_B \\ &+ b_{AC} X_A X_C + b_{BC} X_B X_C + b_{ABC} X_A X_B X_C \end{aligned} \tag{6}$$

$$\begin{aligned} y = b_0 &+ b_A X_A + b_B X_B + b_C X_C + b_D X_D + b_{AB} X_A X_B + b_{AC} X_A X_C \\ &+ b_{AD} X_A X_D + b_{BC} X_B X_C + b_{BD} X_B X_D + b_{CD} X_C X_D \\ &+ b_{ABC} X_A X_B X_C + b_{ABD} X_A X_B X_D + b_{ADC} X_A X_D X_C \\ &+ b_{BDC} X_B X_D X_C + b_{ABCD} X_A X_B X_C X_D \end{aligned} \tag{7}$$

Each one of the b_i or b_{ii} coefficients can be calculated by dividing the Eq. 2 by the number of experiments; in other words, the coefficients can be calculated according to Eq. 8:

$$b(i) = \frac{(X^t)Y}{n} \tag{8}$$

So, for the 2^3 factorial design, we are working on, the empirical model is shown in Eq. 9:

$$\begin{aligned} Yield(\%) = 44.4 &- 17.6 X_A + 6.1 X_B - 2.6 X_C - 2.9 X_A X_B \\ &- 14.1 X_A X_C - 7.4 X_B X_C + 9.1 X_A X_B X_C \end{aligned} \tag{9}$$

2.5.2 Judging the Significance of the Effects to Refine the Empirical Model

Equation 9 presents all the eight possible terms for the empirical model. However, not all the terms will be significant. To judge which terms will be significant we have to decide which effects are significant using statistical tests. The coefficients related to the insignificant effects will be discarded from the model.

To have a rough estimate about which effects might be significant, the effects can be squared, and then, divided by the summation of all values, and multiplying by 100%, as shown in the Fig. 3.

Figure 3 shows that effects A, AC, and ABC are the three bigger ones. The summation of these three effects accounts for about 85% of all the effects. We can affirm, almost unequivocally, that these three effects are significant. Contrastingly, we can also claim that the effect C, which is smaller than 1%, is insignificant. But how about the other four effects? Are they significant or not?

To answer this question, we need some statistical tests. One of the most popular tests is critical t-student, which is shown in Eq. 10:

$$critical - t = s_{Effect} \times t(DoF; p) \quad (10)$$

where s_{Effect} is defined as the standard deviation of the effects and $t(DoF; p)$ is the value obtained from a t-Student table with the number of degrees of freedom (DoF) from which s_{Effect} was calculated, and p is the probability. For instance, if the desired probability is 95%, the researcher should look for the t-student value with p = 0.05 in the t-Student table.

The s_{Effect} can in principle be simply calculated as the square root of the variance of the effects (${s_{Effect}}^2$), from Eq. 11 [9]:

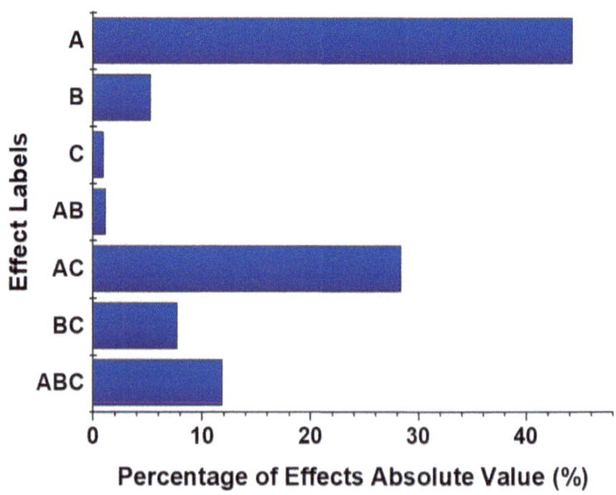

Fig. 3 Graph showing the percentage of effects absolute value for each one of the primary effects and interactions possible for the 2^3 factorial design

$$s_{Effect}{}^2 = \frac{4s_e^2}{\sum_{i=1}^{n} r_i} \tag{11}$$

where s_e^2 is an estimative of the variances, which can be calculated from Eq. 12, and r_i is the number of replicas. For instance, if each experiment was done twice, $r_i = 2$

$$s_e^2 = \frac{\sum_{i=1}^{n}\left[(r_i - 1)s^2\right]}{\sum_{i=1}^{n}[(r_i - 1)]} \tag{12}$$

In fact, in Eq. 12 the term $(r_i - 1)$ is particularly the degrees of freedom used in order to calculate the estimative of the variances, and s^2 is the variance for each experimental condition, as shown in Eq. 13:

$$s^2 = \frac{\sum_{i=1}^{r}\left(y_i - \bar{y}\right)^2}{(r_i - 1)} \tag{13}$$

where y_i is each individual result of the replica, $\bar{y}$ is the average of each individual result from the replicas, and r is the number of replicas.

After calculating critical-t using Eq. 10, the critical-t value is compared to the absolute value of each effect, and only the effects having absolute values higher than the critical-t will be considered significant, in that probability level p.

The best way to obtain all the variances necessary to calculate the critical-t is by performing all the experiments with replicas since it allows to calculate sequentially s^2, s_e^2, $s_{Effect}{}^2$, from Eqs. 11, 12, and 13, respectively.

However, if no replicas were made for any of the experiments, it is still possible to obtain the variances necessary to calculate t-value. Yet, there are distinct ways to accomplish this task, a popular and prudent one is to perform additional experimental with all the factors fixed at the central point.

For quantitative factors, the central point is defined as the average between the lower and upper levels, and it has the coded value equal to zero. Table 8 is an adaptation of Table 2, including the central point for each factor.

As such, the experiment at the central point was then performed in triplicate, and the following results were obtained for the yield, as shown in Table 9.

Table 8 The factors with their respective limits and central points represented as coded (in parenthesis) and uncoded values

Factor	Lower limit	Central point	Upper limit
pH = A	8 (−1)	11 (0)	14 (+1)
Temperature (°C) = B	25 (−1)	55 (0)	85 (+1)
Catalyst concentration (mol L^{-1}) = C	1×10^{-4} (−1)	3×10^{-4} (0)	5×10^{-4} (+1)

Table 9 Yield obtained for the experiments performed in the central point using triplicates

Factors			Yield (%)		
A	B	C	Exp 1	Exp 2	Exp 3
0	0	0	41	47	50

Table 10 Average yield, degrees of freedom, variances, standard deviation of the effect, tabled t-student value and, the critical-t value with 95% confidence obtained for the experiments performed in the central point using triplicates

Parameter	Value
Average yield	46
Degrees of freedom	2
Variance (s^2 and s_e^2)	14
S_{effect}^2	18.67
S_{effect}	4.32
t(2; 0.05)	2.92
Critical-t	12.6

The values for average, degrees of freedom, and variance (s^2 and s_e^2), S_{effect}^2, S_{effect}, t-student (2; 0.05), and critical-t are shown in Table 10.

The graph shown in Fig. 4 is called Pareto chart, it is a bar graph where in the Y-axis all the effects are categorized, and in the X-axis the absolute value of the effects are presented. Then, a vertical line is fixed on the X-axis over the critical-t value. From the Pareto chart we can see that the effects higher than the critical-t are the pH (A), the secondary interaction between pH and catalyst concentration (AC), and between temperature and catalyst concentration (BC), and the tertiary interaction among the three factors (ABC). Notice that the Pareto chart confirmed the initial guess based on Fig. 3 that the effects A, AC, and ABC are significant. Also, the

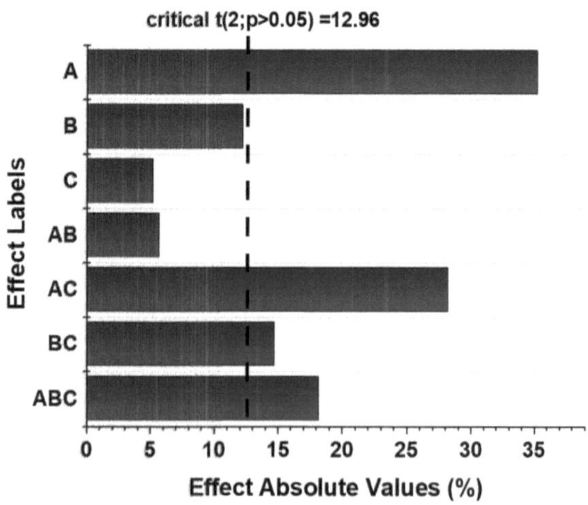

Fig. 4 Pareto chart showing which effects are higher (if so, significant) than the critical-t

Pareto chart was essential to confirm that BC is significant, whereas the effect B is insignificant, although it is very close to the critical-t value.

Now that we decided which effects are significant, based on a statistical test. We can exclude from the model (Eq. 9) the coefficients related to the insignificant values. So, the model describing the factorial design is given by Eq. 14:

$$Yield(\%) = 44.4 - 17.6X_A - 14.1X_AX_C - 7.4X_BX_C + 9.1X_AX_BX_C \quad (14)$$

2.5.3 Assessing the Quality of the Model Using Analysis of Variance (ANOVA) Table

Replacing the coded values for X_A, X_B, and X_C on Eq. 14, it was possible to calculate the yield (%) predicted by the model for each one of the experiments performed, as shown in Table 11.

A standard method to assess the quality of the model by using three parameters called the sum of squares (SS), they are: the sum of square (SS) for the mean, for the regression, and for the residuals, and they are defined according to Eqs. 15–17:

$$SS_{mean} = \sum \left(y_i - \bar{y} \right)^2 \quad (15)$$

$$SS_{regression} = \sum \left(\hat{y_i} - \bar{y} \right)^2 \quad (16)$$

$$SS_{residual} = \sum \left(y_i - \hat{y_i} \right)^2 \quad (17)$$

Table 11 Comparison between the yield (%) obtained experimentally and predicted by the model represented by Eq. 14

A	B	C	Experimental	Predicted
−1	−1	−1	25	31.4
1	−1	−1	42	42.6
-1	1	−1	76	64.4
1	1	−1	45	39.2
−1	−1	1	81	92.6
1	−1	1	5	11
−1	1	1	66	59.6
1	1	1	15	14.4

Table 12 Analysis of variance (ANOVA) Table for the 2^3 factorial design

ANOVA table			
Source of variation	Sum of square	Degrees of freedom	Mean squares
Mean	5603.88	7	800.55
Regression	5169.12	4	1292.28
Residual	421.40	3	140.47
R^2 ($SS_{regression}/SS_{mean}$)	0.9224		
$MS_{regression}/MS_{residual}$	9.20		
F-test for 4/3 degrees of freedom with 95%	5.39		

where: y_i is the value obtained experimentally, $\hat{y}_i$ being value predicted according to the model, and the $\bar{y}$ is the average of all experimental values.

Note that each sum of squares has a certain number of degrees of freedom which can be associated with them. It is known that for $SS_{residual}$ the number of degrees of freedom is found to be equal to the difference of the number of independent experiments (8, in this example) and the number of parameters contained in the equation describing the model (5, in this example), so $SS_{residual}$ has 3 degrees of freedom. For the $SS_{regression}$ the number of degrees of freedom is equal to the number of parameters (5, in this example) minus 1, so the $SS_{regression}$ has 4 degrees of freedom. Finally, the number of degrees of freedom of SS_{mean} is equal to the number of degrees of freedom for $SS_{residual}$ plus $SS_{regression}$, so 7 degrees of freedom.

Hence, the mean square (MS) for regression, residual, and mean was then calculated by dividing each sum of squares by its respective number of degrees of freedom. So, Table 12 presents the complete ANOVA table for the example analyzed.

There are two criteria used to assess the model quality through the ANOVA table. The first one is based on the coefficient of determination (R^2), which is the ratio between the $SS_{regression}$ and SS_{mean}. An R^2 as close as to 1 represents a better quality of the model. In this example, an R^2 equal to 0.9224 means that 92.24% of the variation is explained by the model, whereas only 7.76% is explained by the residuals.

The second parameter is the ratio between $MS_{regression}$ and $MS_{residual}$. This value is in general compared to the F-test value with 95% of confidence, and the number of degrees of freedom in the numerator as the same number of degrees of freedom as $MS_{regression}$. And the number of degrees of freedom in the denominator as the same number of degrees of freedom as $MS_{residual}$. As the ratio $MS_{regression}/MS_{residual}$ (9.20) is higher than the F-test with 95% confidence, and 4/3 degrees of freedom (5.39), we can conclude that the model is appropriately fitted to the data.

For most of the researchers, the model described by Eq. 14 would have enough quality, which eliminates the need to perform additional experiments for optimization to obtain a model with quadratic terms, as the ones to be discussed in Sect. 2.6.

2.6 *Optimization Steps*

Depending on the situation studied, an empirical model where all the terms in the polynomial are linear may not be satisfying. In this sense, the researcher will have to use experimental designs able to produce polynomials containing quadratic terms, such as the one generically represented by Eq. 18:

$$y = b_0 + \sum_{i=1}^{k} b_i X_i + \sum_{i=1}^{k} b_{ii} X_i{}^2 + \sum_{1 \le i \le j}^{k} b_{ii} X_i X_j + \xi \qquad (18)$$

Compare Eq. 18 with Eq. 4, and notice that there are no quadratic terms in the Eq. 4.

In this section, we will be introduced to the basics of four designs able to produce polynomial models containing quadratic terms. The four designs to be studied are 3^k Factorial Design, Central Composite Design (CCD), Box-Behnken Design, and Doehlert Design. In all these designs, all the factors are studied in more than two levels.

The group formed by these designs is called Response Surface Methodology (RSM), as the equations described by each model can be converted in a tri-dimensional graph called response surface. In none of these models there will be a detailed example, like the one we presented for 2^k factorial design. Readers interested to have a more in-depth knowledge about any one of these designs are encouraged read the references cited in this section or the ones in Sects. 3 and 4.

Also, specifically about RSM, the books: *A Comprehensive Guide to Factorial Two-level Experimentation* by Mee [6], and *Design and Analysis of Experiments* by Dean et al. [10] are good resources for readers avid for learning more about each one of the RSMs.

2.6.1 Three Level Factorial Design (3^k Design)

In the 3^k factorial design, all the factors have three levels (−1, 0, +1). Although the 3^k design implementation is not so complicated due to its similarity to 2^k designs, it requires a relatively large number of experiments. For this reason, the 3^k design is used to study a few factors, generally, two or three [11].

2.6.2 Central Composite Design (CCD)

The Central Composite Design (CCD) is made up by the combination of the following features [3, 9]:

(i) A full 2^k factorial design or a fractional factorial design 2^{k-p}.

Table 13 Matrix of experiments for a 2^2 CCD factorial design

2^2 factorial design		
X_A	X_B	2^2 factorial design
−1	−1	
+1	−1	
−1	+1	
+1	+1	
0	0	Central point experiment
−α	0	Axial point experiments
+α	0	
0	−α	
0	+α	

Table 14 Matrix of experiments for a 2^3 CCD factorial design

2^3 factorial design			
X_A	X_B	X_C	2^3 factorial design
−1	−1	−1	
+1	−1	−1	
−1	+1	−1	
+1	+1	−1	
−1	−1	+1	
+1	−1	+1	
−1	+1	+1	
+1	+1	+1	
0	0	0	Central point experiment
−α	0	0	Axial point experiments
+α	0	0	
0	−α	0	
0	+α	0	
0	0	−α	
0	0	+α	

(ii) Central point experiments, where all the factors have coded values equal to zero.

(iii) Experiments in the axial points. In these experiments, one factor will have coded value equal to $\pm\alpha$. These $\pm\alpha$ values are located in the axes of the coordinated system, and in distance $\pm\alpha$ from the origin of the system.

Tables 13 and 14 show the matrix of experiments for a CCD design having a 2^2 and 2^3 factorial as screening designs [3].

From the matrix of experiments, one can conclude that the total number of experiments for a CCD, disregarding any replicas, is equal to the Eq. 19:

Table 15 Number of experiments at axial level, axial levels, and total number of experiments for CCD factorial designs according to the number of factors

Number of factors (k)	2	3	4	5	5	6	6
p in 2^{k-p}	0	0	0	0	1	0	1
Number of Experiments in axial levels ($\pm\alpha$)	4	6	8	10	10	12	12
$+\alpha$ values	1.414	1.682	2.000	2.378	2.000	2.828	2.378
Total number of experiments	9	15	25	43	27	77	45

$$\text{NumberofExperiments} = 2^{k} + \text{CentralPoint} + 2k \tag{19}$$

where the 2^k is the number of experiments related to the 2^k factorial, the Central Point is the experiment carried out in the central point, and 2k is the number of the experiments related to the axial points $\pm\alpha$.

The coded values $\pm\alpha$ can be figured out according to Eq. 20 [9].

$$\alpha = \sqrt[4]{2^{k}} \tag{20}$$

In the CCD, any coded value can be decoded according to Eq. 21. This equation is also valid to convert the $\pm\alpha$ to their respective decoded values [9]:

$$x_i = \frac{z_i - \bar{z}}{\Delta z/2} \tag{21}$$

where x_i is the coded value for the factor i, z_i is the decoded value for the factor i, $\bar{z}$ is the decoded value for the central point for the factor i, and the Δz is the difference between the upper (+) and lower (−) levels for the factor i.

Table 15 summarizes how many experiments should be made in the axial levels $\pm\alpha$, the value of α, and the total number of experiments according to the k value in 2^{k-p}.

2.6.3 Box-Behken Designs

Similarly to the three levels factorial design the Box-Behken Desing (BBD) is a three levels (−1, 0, +1) for all factors. The number of experiments to be performed, disregarding replicas, according to the number of factors (k) is given by the Eq. 22 [12]:

$$\text{NumberofExperiments} = 2k(k-1) + \text{CentralPoint} \tag{22}$$

The conversion between the coded and uncoded values in a BBD is given by Eq. 23 [13]:

$$x_i = \left(\frac{z_i - \bar{z}}{\Delta z} \right) \quad (23)$$

where x_i is the coded value for the factor i, z_i is the decoded value for the factor i, $\bar{z}$ is the decoded value for the central point for the factor i, and the Δz is the difference between the upper (+) and lower (−) levels for the factor i.

One advantage of the BBD is that there is no experiment where the factors are taken all of them at the same time in the lower limit nor the upper limit. This fact avoids that experiments are carried out at extreme conditions and may avoid unsatisfactory or abnormal results to happen [14].

For more details about BDD matrix, readers are encouraged to read the review paper by Ferreira et al. [14].

2.6.4 Doehlert Designs

A feature differing Doehlert Designs (DD) from CCD and BBD is that DD does not have the same number of levels for all the factors. For instance, in a DD with k = 2, one factor has three levels (with coded values equal to −8.666, 0, +8.666), and the other has five levels (with coded values equal to −1, −0.5, 0, +0.5, +1). For a DD with k = 3, one factor has five levels (with coded values equal to −1, −0.5, 0, +0.5, +1), the second one has seven levels (with coded values equal to −0.866, −0.577, −0.289, 0, +0.289, +0.577, +0.866), and the third factor has 'three levels (with coded values equal to −0.866, 0, +0.866) [11, 12]. In this sense, we can conclude that DD requires the following number of experiments, disregarding replicas, according to k value as shown in Eq. 24:

$$\text{NumberofExperiments} = k^2 + \text{k} + \text{CentralPoint} \quad (24)$$

The conversion between the coded and uncoded values in a DD is given by Eq. 25 [12]:

$$x_i = \left(\frac{z_i - \bar{z}}{\Delta z} \right) \alpha \quad (25)$$

where x_i is the coded value for the factor i, z_i is the decoded value for the factor i, $\bar{z}$ is the decoded value for the central point for the factor i, the Δz is the difference between the upper (+) and lower (−) levels for the factor i, and α is the coded value limit for the factor i.

For verifying examples about DD matrix, readers are encouraged to read the review paper by Ferreira et al. [12].

2.6.5 Comparison Between the Different Response Surface Methodologies

To compare the four RSMs presented in this section, we take advantage of the efficiency parameter, which is defined by the ratio between the number of coefficients present in the model of the RSM and the number of experiments necessary to accomplish that RSM [8]. Table 16 presents the number of coefficients, the number of experiments, the efficiency, according to the number of factors for the four RSMs presented.

From Table 16, regardless of the number of factors, the DD presents a lower number of experiments and higher efficiency. Also notice that the 3^k design is not practical for $n \geq 4$ [15].

3 Factorial Design of Experiments Applied to Nanomaterials Production and Performance

Nanomaterials are recognized for presenting outstanding properties when compared with their bulk-like versions. In this sense, it is necessary to develop synthetic methods able to precisely tune properties such as nanoparticle size, surface area, bandgap, and photoluminescent emission. To be able to obtain a fine control over these properties, usually, many factors need to be considered. Most of the time, a comprehensive study would require a massive number of experiments. For making the production process more efficient, the factorial design can be used to sort the significant factors. Then, it can be used to derive models for predicting how much that response property could be in a previously untested condition. In this section, we will see five examples where the factorial design was applied to obtain as the response variables different properties of nanomaterials.

3.1 Optimization of Copper (Cu) Nanoparticle Size Using 2^2 Factorial Design

The first paper of this section deals with the synthesis of copper (Cu) nanoparticles. Granata and co-workers studied how different capping agents, such as cetyl trimethyl

Table 16 Comparison among the four RSMs presented regarding the number of experiments and efficiency

Number of factors	Number of coefficients	Number of experiments				Efficiency			
		3^k	CCD	BBD	DD	3^k	CCD	BBD	DD
2	6	9	9	N/A	7	0.67	0.67	N/A	0.86
3	10	27	15	13	13	0.37	0.67	0.77	0.77
4	15	81	25	25	21	0.19	0.60	0.60	0.71
5	21	243	43	41	31	0.09	0.49	0.51	0.68
6	28	729	77	61	43	0.04	0.36	0.46	0.65

ammonium bromide (CTAB), sodium dodecyl sulfate (SDS), and polyvinylpyrrolidone (PVP). And also, how different reducing agents, such as hydrazine and glucose, can influence the mean particle size and the stability against oxidation of the Cu nanoparticles [16].

By combining the three available capping agents with the two available reducing agents, the authors designed six different 2^2 factorial designs. The response variables analyzed were the mean and standard deviation of the particle size of the Cu nanoparticles as estimated by histograms counting 500 particles from scanning electron microscopy images. The authors were aiming to minimize the mean particle size and standard deviation since smaller Cu particles are more likely to present the surface plasmon resonance (SPR) phenomenon.

For all the designs, the temperature and initial pH were fixed at 85 °C and pH equal to 10. The levels for the surfactant and reducing agent factor for each one of the six 2^2 factorial designs are shown in Table 17 [16].

When glucose is used as the reducing agent, for design 1, only the glucose had a significant effect, and this effect was around +11.6 nm. As this effect is positive, it means that increasing glucose concentration from 0.2 to 0.5 mol L^{-1} led to an increase of 11.6 nm in the size of Cu nanocrystals. For design 2, both the capping agent and reducing agent had significant effects. The glucose had a negative effect, whereas the CTAB had a positive effect. These results mean that the Cu mean particle

Table 17 The six 2^2 factorial designs used to analyze the interactions between capping and reducing agents—data by Granata et al. [16]

Factor	Lower limit (mol L^{-1})	Upper limit (mol L^{-1})
Factorial design 1		
Capping agent	SDS 0.1	SDS 0.2 mol L^{-1}
Reducing agent	Glucose 0.2	Glucose 0.5
Factorial design 2		
Capping agent	CTAB 0.1	CTAB 0.2
Reducing agent	Glucose 0.2	Glucose 0.5
Factorial design 3		
Capping agent	PVP 0.1	PVP 0.2
Reducing agent	Glucose 0.2	Glucose 0.5
Factorial design 4		
Capping agent	SDS 0.1	SDS 0.2
Reducing agent	Hydrazine 0.2	Hydrazine 0.5
Factorial design 5		
Capping agent	CTAB 0.1	CTAB 0.2
Reducing agent	Hydrazine 0.2	Hydrazine 0.5
Factorial design 6		
Capping agent	PVP 0.1	PVP 0.2
Reducing agent	Hydrazine 0.2	Hydrazine 0.5

size increased when the glucose concentration decreased or CTAB concentration increased. Finally, for design 3, not only glucose and PVP presented significant effect, but, the interaction between these two factors was significant as well. For all cases, the effects presented were negative, which means that by increasing the level of all variables from lower to the upper limit led to a decrease in the Cu mean particle size.

When hydrazine is used as the reducing agent, for design 4, only hydrazine and SDS had significant effects, being the hydrazine positive and the SDS negative. So, the Cu particle size increased with increasing hydrazine concentration a decreasing the SDS concentration. For design 5, both hydrazine and CTAB had significant effects, both of them positive. These results mean that the Cu mean particle size increased when the hydrazine or CTAB concentrations increased. Finally, for design 6, hydrazine and PVP presented a significant effect, following the same trend observed in design 4. Interestingly, for none of the designs, the interaction between capping and reducing agents showed significant effects.

After calculating the significant effects for each factorial design, the authors did not advance on using the factorial design to estimate a model for the mean particle size or standard deviation for the nanoparticles. From effects estimation the authors concluded that the PVP was probably the best capping agent for their purpose, as it is the capping agent that, for either reducing agents, had the highest effect on changing the nanoparticle size.

To subdivide all the data in six smaller 2^2 factorial designs was a wise decision, and it simplified a lot of the data analysis. Although this simplification eliminated, to some extent, the holistic character of the factorial design. To obtain a more comprehensive model would require that each factor to have a different number of levels.

3.2 Optimization of Silica Microspheres Size Using a 2^3 Factorial Design a Linear Response Surface Model

Silica (SiO_2) microspheres are multifunctional materials, having applications in colloidal templating, photonic crystals, polymer encapsulation, to cite a few of them [17]. SiO_2 microspheres are prepared through the alkaline hydrolysis of silicon alkoxides in a water–ethanol solvent mixture. A common source of silicon is the tetraethyl orthosilicate (TEOS), whereas the reaction is usually catalyzed by water. The diameter of microspheres depends on an accurate control of the quantities of the reactants, catalysts, and solvents, for instance, TEOS, NH_3, and water. Considering that the diameter of the microspheres can determine its possible applications, to be able to control this feature systematically in a predictable way is a characteristic necessary for the synthetic method [18].

Aiming to obtain precise control over the diameter of SiO_2 microspheres, Arantes et al. used a 2^3 factorial design to study the effect of the factors number of moles

Table 18 The three factors and their levels in the 2^3 factorial designs used to control the size of SiO_2 microspheres—data by Arantes et al. [19]

Factor	Lower level	Upper level
NH_3 (mol)	3.0×10^{-2}	6.0×10^{-2}
TEOS (mol)	2.0×10^{-2}	4.0×10^{-2}
H_2O (mol)	0	1.0×10^{-1}

of ammonia, tetraethyl orthosilicate (TEOS), and water in the particle size of SiO_2 microspheres [19]. As such, the levels of each factor are shown in Table 18.

Only NH_3 and H_2O presented significant primary effects. The other significant effect was the interaction between TEOS and H_2O. After having concluded the significant effects, the authors calculated the coefficients related to each one of them and obtained a linear polynomial model to predict the SiO_2 particle size out of the initially performed experimental conditions, as shown in Eq. 26:

$$Y = 158.82 + 97.52X_1 + 47.97X_3 + 29.092X_2X_3 \tag{26}$$

where X_1 is the coded values for the number of moles of NH_3, X_2 is the number of moles of TEOS, and X_3 is the number of moles of water.

This model equation allowed the authors to plot a response surface graph by fixing NH_3 either in its lower (0.3 mol) or higher (0.6 mol) levels and varying the other two factors, as shown in Fig. 5.

From Fig. 5, we notice that both graphs presented a similar behavior. In other words, by increasing the amount of H_2O and TEOS led to a maximization of the particle size of the SiO_2 microspheres. However, when NH_3 is fixed on its upper level, the SiO_2 particle size predicted is much higher than ones predicted by fixing NH_3 on its lower level and varying H_2O and TEOS.

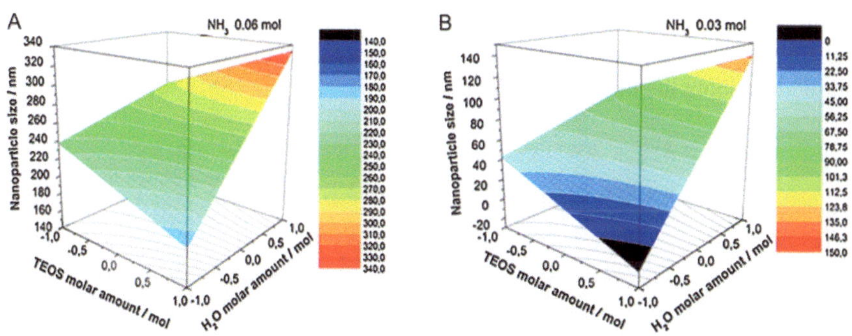

Fig. 5 Surface responses **a** by fixing the NH_3 molar amount in the level +1 (0.06 mol) and b by fixing the NH_3 molar amount in the level −1 (0.03 mol). Adapted from reference number [19], with permission from Elsevier

To confirm the quality of the model, the authors calculated the analysis of variance (ANOVA) table, and it revealed a coefficient of determination (R^2) equal to 95%, which means that 95% of the variation observed can in principle be explained by the obtained model. Finally, to experimentally confirm the applicability of the model, the authors performed an experiment where the coded values for NH_3, TEOS, and H_2O were, respectively, equal to -1, 0, and 0.5. By plugging these values into Eq. 26, it was predicted that the particle size should be around 85 nm. And in practice, the result of the experiment performed revealed that the SiO_2 particle size was very close 85 nm. This result showed a nice agreement in relation to the experimental data and the values predicted by the model.

3.3 *Optimization of Manganese Sulfide (MnS) Thin-Films Thickness Using a 2^4 Factorial Design and Dohelert Approach for Quadratic Response Surface*

To be able to control thin film thickness is an essential property of a proper deposition procedure. In this sense, Hannachi and co-workers used a 2^4 factorial to study the effects of time, temperature, initial concentrations of manganese acetate and thioacetamide on the thickness of manganese sulfide (MnS) deposited via chemical bath deposition onto glass substrates [20].

The levels were studied for each one of the factors, as shown in Table 19.

The authors found that the deposition time, temperature, and manganese acetate initial concentration had significant primary effects. All these three factors had positive effects, which means that increasing either one of them from their lower to the upper levels would lead to an increase in the film thickness. On the other hand, the thioacetamide concentration effect was not significant. The secondary interactions involving the deposition time, temperature, and manganese acetate initial concentration when taken at two by two at a time also were significant. Endly, the tertiary interaction combining the three significant factors was significant as well. After judging the significance of the effects, the authors obtained the eight terms linear-polynomial model describing the thin-film thickness:

$$Y = 483 + 204.3X_1 + 159.5X_2 + 189.4X_3 + 98.3X_1X_2 + 106.6X_1X_3 + 159.1X_2X_3 + 71.4X_1X_2X_3 \quad (27)$$

Table 19 The four factors and their levels in the 2^4 factorial designs used to control the MnS thin-film thickness—data by Hannachi et al. [20]

Factor	Lower level	Upper level
Time (h)	1	24
Temperature (°C)	Room temp	70
Mn acetate conc. (mol L^{-1})	0.5	2
Thioacetamide (mol L^{-1})	0.5	2

where X_1 is the coded values for the deposition time, X_2 is deposition temperature, and X_3 is the concentration of manganese acetate.

The conclusions obtained in the 2^4 factorial design served as the foundation to further optimize the model using a Doehlert approach, studying only the three factors that were significant in the 2^4 screening factorial design. This Doehlert design allowed the authors to obtain a quadratic model (Eq. 28), and a set of response surface graphs.

$$Y = 349.0 + 136.5X_2 + 119.2X_3 - 129.0X_1^2 - 187.0X_2^2 - 124.0X_3^2 + 98.2X_1X_2 + 148.5X_2X_3 \quad (28)$$

where X_1 is the coded values for the deposition time, X_2 is deposition temperature, and X_3 is the concentration of manganese acetate.

The first response surface graph was plot fixing the thioacetamide concentration at 0.5 mol L^{-1} and varying the deposition temperature and time. The second response surface graph was plot fixing the thioacetamide concentration at 0.5 mol L^{-1} and varying the manganese acetate initial concentration and deposition time. Finally, the third surface was plot fixing the thioacetamide concentration at 0.5 mol L^{-1} and varying the manganese acetate initial concentration and deposition temperature. The three response surfaces are shown in Fig. 6.

By analyzing the three response surface plots, it was concluded that the MnS thin-film thickness was maximized (thickness around 330–400 nm), when the deposition temperature was 60 °C, time was 20 h, manganese acetate concentration around 1.6 mol L^{-1}, and thioacetamide concentration equal to 0.5 mol L^{-1}.

To be able to indicate a narrow range of conditions that would lead to maximization of the thin-film thickness demonstrates how resourceful factorial design and response surface methodologies are. This affirmation comes from the fact that none of the conditions indicated as optimum were the same as the ones the authors initially used as levels in the 2^4 factorial.

The authors deposited a thin-film under the conditions deemed to produce the highest thickness. The structural characterization techniques indicated the MnS was

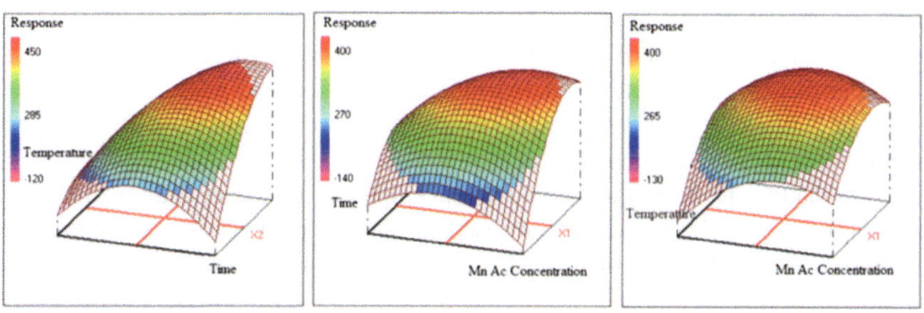

Fig. 6 Surface responses (left) varying time and temperature (middle) varying time and Mn acetate conc. (right) varying temperature and Mn acetate conc. Adapted from reference number [20], with permission from Elsevier

satisfactorily deposited onto the glass substrate. However, the authors failed to inform the thin-film thickness, which makes the readers unable to infer if the model used to produce the response surface graphs has a good agreement to the experimental data or not.

3.4 *Optimization of Cellulose and Sugarcane Bagasse Oxidation Using a 2^3 Factorial Design with Full Central Composite Design (FCCD)*

The use of functional materials derived from natural sources or obtained from wastes is a procedure aligned with the Green Chemistry Principles, as long as the obtaining of these materials does not cause any environmental harm [21]. In this sense, Martins et al. studied the oxidation of cellulose and sugarcane bagasse by a simple procedure [22]. The oxidation procedure is based on stirring either the cellulose (Cel) or the sugarcane bagasse (SB) with phosphoric acid (H_3PO_4) and sodium nitrite ($NaNO_2$) for different time intervals.

To accomplish this goal, initially, the authors designed a 2^3 factorial design, where the factors studied were the duration of the oxidation procedure (in hours), the volume of H_3PO_4 (in mL), and the mass of $NaNO_2$ (in g). The response variable was the number of carboxyl groups per gram of material, represented by the symbol n_{COOH} (in mmol/g), which can be taken as an indication of the successful oxidation of the initial material.

The Pareto plot revealed that for Cel, all the primary effects were significant, with the secondary effect between the volume of H_3PO_4 and the mass of $NaNO_2$ also being significant. For the SB, the Pareto plot revealed that only the primary effect related to the mass of $NaNO_2$ was significant. Additionally, the secondary effect between the volume of H_3PO_4 and the mass of $NaNO_2$ also was significant (Fig. 7).

Having concluded the significant factors for the oxidation of each one of the materials studied, the authors started to optimize the model by doing a full central composite design. This design was accomplished by a full central composite design (FCCD). For each one of the materials studied, this FCCD was carried out by performing the eight experiments predicted by the 2^3 design, without replicas. Plus, the experiment in the central point, in triplicates, and six more experiments. Where, each one of these six, one of the factors assumed a $+\alpha$ or $-\alpha$, whereas the two other factors were kept in the central point.

From the FCCD, the authors could obtain empirical models containing quadratic terms. These empirical models were used to plot the response surface graphs by fixing the reaction duration in 5 h, and varying the H_3PO_4 volume and $NaNO_2$ mass, as shown in Fig. 8.

The response surface graphs showed that, when the reaction duration is equal to 5 h, the n_{COOH} is maximized for H_3PO_4 volume between 11 and 19 mL, and $NaNO_2$

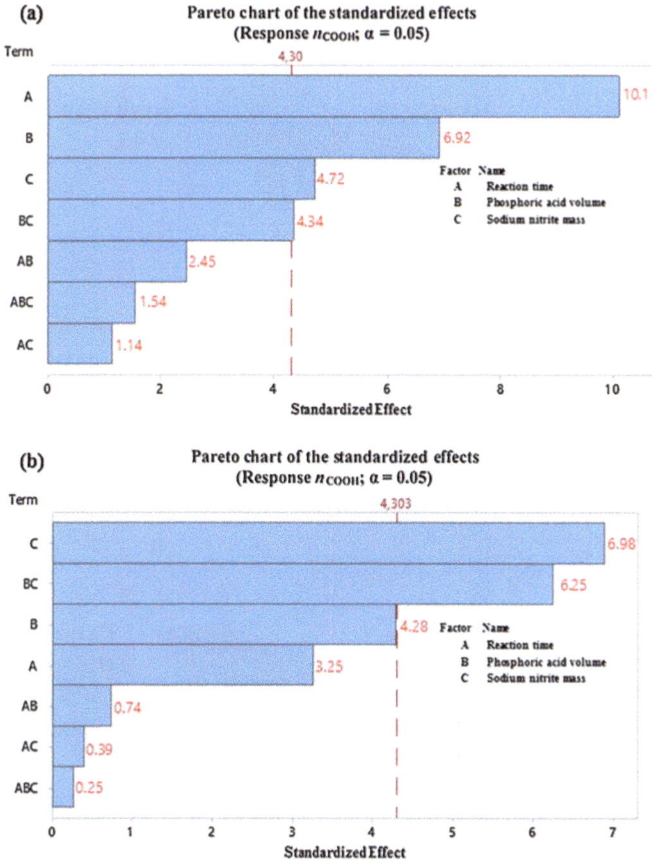

Fig. 7 Pareto charts for Cel (top) and SB (bottom). Adapted from reference number [22], with permission from Elsevier

volume between 500 and 970 mg. This conclusion agrees with the results initially obtained during the performance of the central point experiments.

After completing the full factorial design regarding the n_{COOH}, the authors used the products of Cel and SB oxidation to remove the dyes auramine O and crystal violet from simulated wastewater by adsorption. In general, the oxidated Cel presented better performance than the oxidated SB, for the removal of either of the dyes.

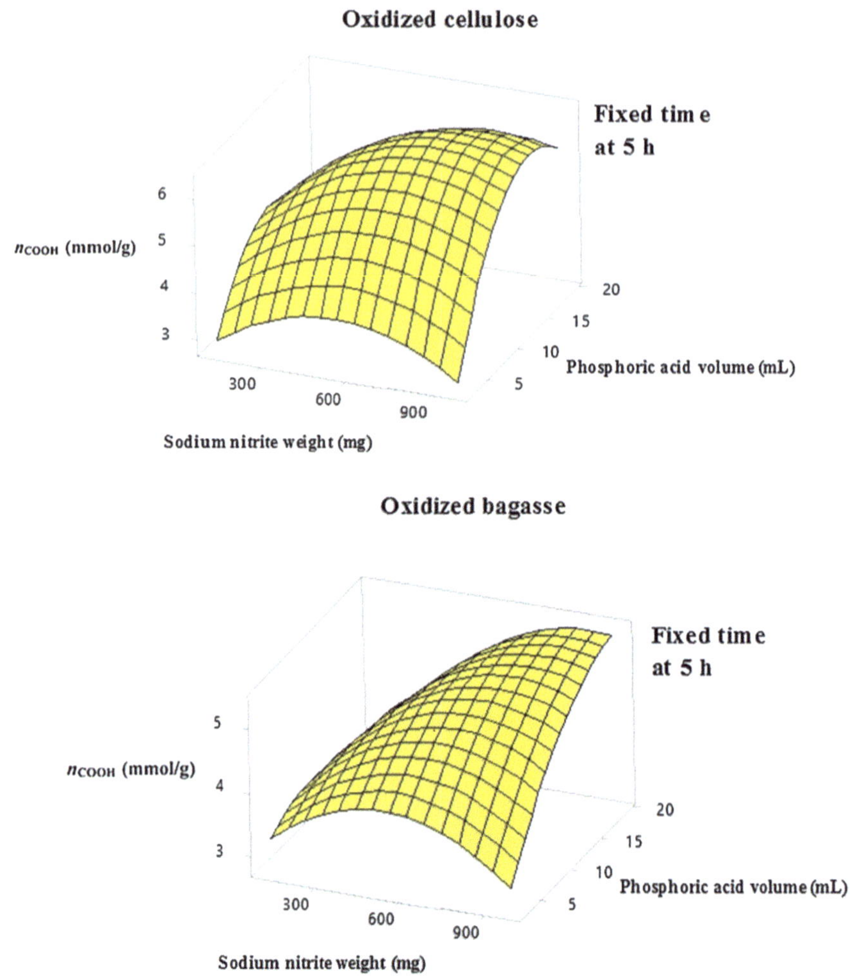

Fig. 8 Response surface graphs for Cel (top) and SB (bottom), by varying H_3PO_4 volume and $NaNO_2$ weight, the reaction duration was fixed at 5 h. Adapted from reference number [22], with permission from Elsevier

3.5 $2^{5\text{-}1}$ Fractional Factorial Design with Box-Behnken Optimization—Improving Reproducibility Between Batches of Silver Nanoparticles Using an Experimental Design Approach

Silver nanoparticles (Ag NPs) are very versatile materials since they can be applied in many different fields, such as catalysis, biological sensors, antimicrobial activities,

DNA sequencing, surface-enhanced Raman spectroscopy, and information storage [23].

Ag NPs can be produced by the reduction of Ag^+ with gallic acid (GA, $C_7H_6O_5$) in alkaline pH, by stirring the reaction mixture in the dark, at room temperature. However, this procedure lacks reproducibility since the Ag NPs tend to have different size distributions according to the reaction batch [24]. With the goal to decrease this lack of reproducibility, Núñez et al. performed a two-level factorial design to study the effects of NaOH, $AgNO_3$, and GA molar concentrations, also the effects of the reaction time and stirring speed [24]. So, the study had a total of five factors, with two-levels each. The response variable studied was a parameter called ψ, which was defined according to Eq. 29:

$$\psi = \frac{A_{max}}{\lambda_{max} \times FWHM} \tag{29}$$

In Eq. 29, A_{max} is the absorption intensity of the band corresponding to the surface plasmon resonance (SPR), λ_{max} is the wavelength of SPR band, and FWHM is the full-width at half-maximum of SPR band. The best ψ values are obtained when A_{max} is maximized, and λ_{max} and FWHM are both minimized, in other words, the ψ is better as it gets higher. And higher ψ means that Ag NPs were smaller and with a narrower size distribution [25].

As five factors were studied, a full two-level factorial design would lead to the performance of 32 experiments. To minimize the number of experiments to be done, the authors opted to do a fractional factorial design (FFD). In general, an FFD is symbolized as $2^{(k-p)}$. Where k is the number of factors, and p is the number of design generators. More specifically, in this paper, the authors carried out a 2^{5-1} factorial design. The number 5 represents the five factors ([NaOH] = A, [$AgNO_3$] = B, [GA] = C, time = D, and speed = E), and 1 represents the generator I = ABCDE.

The FFD was made up of sixteen experiments plus five experiments with all the factors in the central point. All experiments were performed with replicas. The lower and upper levels of each variable were defined, as shown in Table 20.

The Pareto chart indicated that the significant primary effects were the [GA], the reaction time, and [$AgNO_3$]. As shown in Fig. 9.

Table 20 The five factors and their levels in the 2^{5-1} factorial designs used to control the ψ in Ag NPs preparation—data by Núñez et al. [24]

Factor	Lower level	Central point	Upper level
[NaOH] (μmol L^{-1})	500	750	1000
[$AgNO_3$] (μmol L^{-1})	50	75	100
[GA] (μmol L^{-1})	5	12.5	20
Time (min)	10	30	50
Stirring speed	1	5.5	10

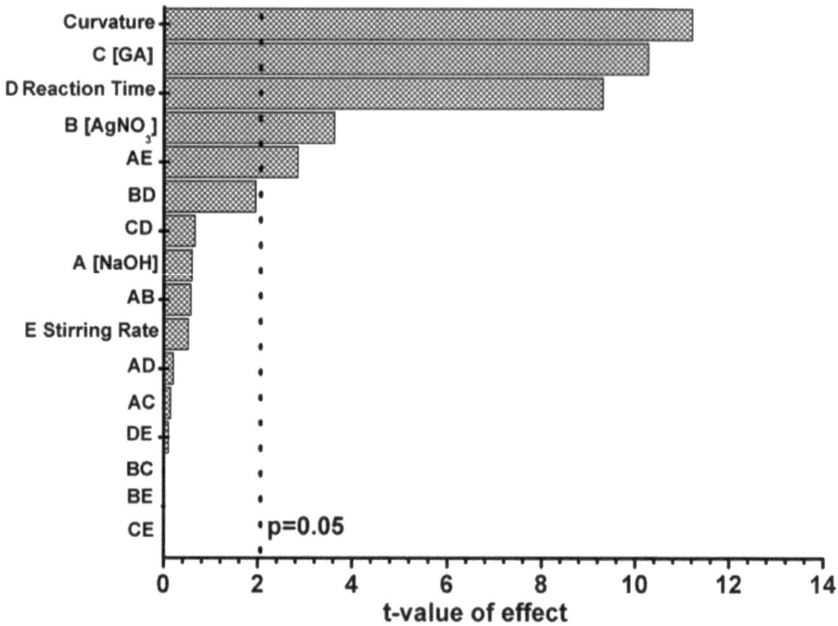

Fig. 9 Pareto chart showing the effects of ψ in Ag NPs synthesis. Adapted from reference number [24], with permission from Elsevier

After concluding the significant effects, the factorial designed continued by performing an optimization using the Box-Behnken design (BBD). The BBD is an incomplete three-level design to obtain a quadratic polynomial model. According to the number of experiments conducted is equal to $2k(k - 1) + C_0$. Where k is the number of factors, in this case, $k = 3$, and C_0 is the number of experiments in the central point, in this case, $C_0 = 5$. In summary, 17 experiments were performed on this BBD. For the BBD, instead of using ψ as the response variable, the authors used A_{max}, λ_{max}, and FWHM separately.

For each one of the response variables (A_{max}, λ_{max}, and FWHM), quadratic polynomial equations were obtained, and response surface graphs were plotted by fixing either one of the factors ([$AgNO_3$], [GA] and reaction time). For the A_{max}, it is maximized as [$AgNO_3$], [GA], and reaction time increase. The λ_{max} is minimized when [$AgNO_3$] increases, and [GA] decreases. And the FWHM became narrower when [$AgNO_3$] increased, and [GA] decreased. The response surface graphs are shown for the λ_{max} in Fig. 10.

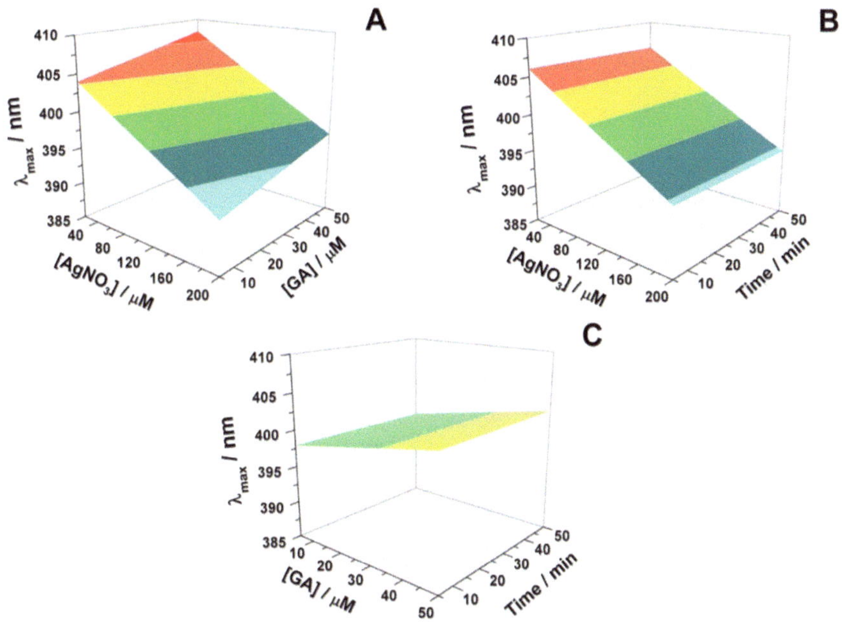

Fig. 10 Response surface graphs of λ_{max} by varying **a** [$AgNO_3$] and [GA], **b** [$AgNO_3$] and time, **c** [GA] and time. Adapted from reference number [24], with permission from Elsevier

4 Factorial Design of Experiments Applied to Nanomaterials Production and Performance

The fields of Materials Science, Materials Chemistry, and Nanotechnology are all very broad. That being said, it would be impossible to cover every single possible application in a book chapter. In this sense, this part is intended to present some references related to four areas of application in Materials Science and Nanotechnology. They are nanoparticles preparation and characterization, drug encapsulation and delivery, wastewater remediation, and solar cell development.

For each one of these areas, around ten to twenty references, are presented classified according to the Factorial Design and Response Surface Methodology used on each paper, and the goal set by the papers. Unlike Sect. 3, there will not be any further detail about any paper. Instead, the readers are encouraged to read the tables and figure out if they get interested in the statistical methods or the theme of the paper. Then, they should feel free to consult these references independently. In order to show the relevance and the insertion of the factorial design of experiments in the current scientific literature, all examples presented were published between 2015 and 2020.

4.1 Design of Experiments Applied to Materials Synthesis and Characterization

See Table 21.

4.2 Design of Experiments Applied to Drug Encapsulation and Delivery

See Table 22.

4.3 Wastewater Remediation

See Table 23.

4.4 Design of Experiments Applied to Solar Cells Design and Performance

See Table 24.

5 Concluding Remarks

This chapter presented the theoretical bases of the factorial design and applications of the design of experiments in different areas of Materials Science and Nanotechnology. The depth and extent of the technique open up the possibility for the optimization of many experimental situations.

Hopefully, the knowledge brought by this chapter can motivate the readers to see the usefulness of the factorial design of experiments and encourage them to apply this knowledge to their own experimental situations.

Table 21 List of papers related to materials synthesis and characterization

Experimental situation and year of publication	Factorial design	RSM	Goal	Refs.
Synthesis of Ag nanoparticles using plant extract, 2020	N/A	CCD for 5 factors	Maximize the intensity of SPR peak	[26]
Synthesis of $AgInS_2/ZnS$ in aqueous media using microwave heating, 2020	2^{5-1}	2^3-CCD	Photoluminescence emission peak and lifetime, quantum yield, and Ag:In:Zn ratio	[27]
Functionalization of amino-silane on superparamagnetic iron oxide nanoparticles, 2020	N/A	2^2-CCD	Atomic percentage of Fe, N, and Si	[28]
Synthesis of Au nanoparticles using *Coffea arabica* extract, 2019	2^{5-1}	3^2	Maximize yield, monodispersity, control shape and size	[29]
Synthesis of ZrO_2 nanoparticles using plant extract, 2019	2^3 with replicas in central point	N/A	Maximize the amount of tetracycline adsorbed in ZrO_2. Maximize the percentage of tetragonal ZrO_2 phase. Particle size control	[30]
Synthesis of Ni_2P in phosphonium-based ionic liquid, 2019	2^{6-3}	DD	Control particle size	[31]
Synthesis of Au nanoparticles using plant extract by sonochemical method, 2019	N/A	CCD for 4 factors, and 5 levels per factor	Control de hydrodynamical size	[32]
Synthesis of CdSe quantum dots in aqueous media, 2019	2^4 with central point	DD for 4 factors	Photoluminescence peak intensity and FWHM	[33]
Synthesis of Carbon quantum dots from chitin, chitosan, and graphite via hydrothermal carbonization, 2019	N/A	3^2	Determination of quantum dot composition and band gap	[34]

(continued)

Table 21 (continued)

Experimental situation and year of publication	Factorial design	RSM	Goal	Refs.
Synthesis of iron oxide nanoparticles from mineral coal tailings, 2019	2^2 with central point	N/A	Control particle size	[35]
Synthesis of cyano-functionalized SiO_2 nanoparticles, 2018	2^{6-2}	N/A	Minimize particle size and get narrow particle size distribution	[36]
Synthesis of 1-dimensional TiO_2 via hydrothermal method, 2018	N/A	3^3-BBD	Control the band gap energy and crystalline phases	[37]
Synthesis of multi-wall carbon nanotubes by chemical vapor deposition, 2018	2^4	N/A	Control the yield, band gap and aspect ratio of the carbon nanotubes	[38]
Growth of highly oriented Sb_2Te_3 thin-films by radio frequency sputtering, 2018	2^{5-1} with replicas in central point	BBD	Maximize the degree of thin-film preferred growth	[39]
Synthesis of $SrMoO_4$ using EDTA-Citrate complexation method, 2018	2^3 with replicas in central point	N/A	Control $SrMoO_4$ band gap	[40]
Synthesis of Au nanorods by reduction with $NaBH_4$ and ascorbic acid, 2017	2^{8-4}	CCD	Control SPR peak, particle dimensions and distribution, yield, and shape purity	[41]
Synthesis of thiol-capped CdTe quantum dots using microwave heating, 2017	N/A	2^3-CCD	Maximize photoluminescence quantum yield	[42]
Electrodeposition of Cu_2O thin-films by tuning morphology and optical properties, 2017	2^4	N/A	Deposition charge and morphology coverage degree	[43]
Synthesis of ZnO by coprecipitation in aqueous media, 2017	2^3	N/A	Control the crystallite size	[44]

(continued)

Table 21 (continued)

Experimental situation and year of publication	Factorial design	RSM	Goal	Refs.
Synthesis of curcumin-loaded albumin nanoparticles at room temperature, 2017	2^4	N/A	Control particle size, yield, and drug loading	[45]

Abbreviations: *BBD* Box-Behnken Design, *CCD* = central composite design, *DD* = Doehlert Design, *EDTA* = Ethylenediaminetetraacetic acid

Table 22 List of papers related to drug encapsulation and delivery

Experimental situation and year of publication	Factorial design	RSM	Goal	Refs.
Design of SLN as nanocarrier for the anticancer drug docetaxel, 2019	2^3 with replicas in central point	N/A	Control hydrodynamic diameter, polydispersity index, and zeta potential	[46]
Design of curcumin-loaded SLN for inhibition of post-angioplasty restenosis, 2019	2^4 with replicas in central point	N/A	Control entrapment efficiency, hydrodynamic diameter, polydispersity index, and zeta potential	[47]
Design of antiretroviral drug ritonavir-loaded SLN, 2019	N/A	3^2	Particle size minimization, entrapment efficiency and zeta potential maximization	[48]
Design of PLGA nanoparticles for co-delivery of temozolomide and O6-benzylguanine. 2019	2^{5-2} with replicas in central point	N/A	Control entrapment efficiency, hydrodynamic diameter, polydispersity index, and zeta potential	[49]
Design of furosemide loaded SLN, 2018	N/A	3^2	Entrapment and dissolution efficiencies maximization; particle size, polydispersity index, and time elapsed for 50% drug release minimization	[50]

(continued)

Table 22 (continued)

Experimental situation and year of publication	Factorial design	RSM	Goal	Refs.
Design of PLGA-chitosan nanoparticles loaded with a HIV-1 fusion inhibitor peptide, 2018	N/A	2^3 - CCD	Control entrapment efficiency, polydispersity index, and average particle size	[51]
Design of HSA nanoparticle for entrapment of anticancer agent irinotecan, 2018	N/A	3^2	Control entrapment efficiency, polydispersity index, particle size, and drug loading percentage	[52]
Design of chitosan-based nanogels for controlled release of the antibiotic gentamicin, 2018	2^{8-2} with replicas in central point	CCD	Control particle size and loading efficiency	[53]
Design of poly-caprolactone nanoparticles loaded with the antipsychotics olanzapine, 2018	N/A	3^2	Control particle size and encapsulation efficiency	[54]
Design of octanoyl chitosan nanoparticles for improved rifampicin pulmonary delivery, 2018	N/A	3^2	Particle size minimization and entrapment efficiency minimization	[55]
Design of SLN loaded with the anti-cancer drug citral, 2018	2^2 with replicas in central point	N/A	Control mean particle size, polydispersity index, and zeta potential	[56]
Design of PLGA loaded with the anti-cancer drug doxorubicin, 2017	N/A	BBD	Particle size, polydispersity index, and zeta potential minimization; entrapment efficiency maximization	[57]
Design of SLN of efavirenz for brain targeting and enhanced availability, 2017	N/A	3^2	Control mean particle size, polydispersity index, and entrapment efficiency	[58]
Design of multiple protein delivery for cardiac repair, 2016	2^{4-1}	N/A	Determine the ejection efficiency	[59]
Design of curcumin-loaded PLGA nanoparticles for colon delivery, 2016	2^3	N/A	Control particle size, zeta potential, entrapment efficiency, and in vitro release over 24 h	[60]

(continued)

Table 22 (continued)

Experimental situation and year of publication	Factorial design	RSM	Goal	Refs.
Design of levofloxacin-loaded stearic acid SLN for ocular delivery, 2016	N/A	BBD with 3 factors and 3 levels	Control particle size and entrapment efficiency	[61]
Design of PLGA/poloxamer nanoparticles loaded with zolmitriptan, 2016	2^4	N/A	Control particle size, entrapment efficiency, and cumulative drug release in 8 h	[62]
Design of carvedilol-load SLN for intranasal drug delivery, 2016	2^3	N/A	Control particle size, entrapment efficiency, and amount of drug permeated per unit area in 24 h	[63]
Design of paclitaxel-loaded folate-conjugated amphiphilic cyclodextrin nanoparticles, 2016	N/A	3^2	Control particle size, polydispersity index, entrapment efficiency, zeta potential, and paclitaxel amount per polymer	[64]
Design of methotrexate-loading in chitosan nanoparticles, 2016	N/A	3^2	Control particle size, polydispersity index, entrapment efficiency, and zeta potential	[65]

Abbreviations: *BBD* Box-Behnken Design, *CCD* central composite design, *HAS* human serum albumin, *PLGA* poly(D,L-lactic-co-glycolic acid), *SLN* solid lipid nanoparticles

Table 23 List of papers related to wastewater remediation

Experimental situation and year of publication	Factorial design	RSM	Goal	Refs.
Removal of the dye tartrazine by UV-light photocatalytic degradation using zinc oxide (ZnO) nanorods with different aspect ratios, 2020	2^4	N/A	Maximize removal efficiency and pseudo-first-order rate constant	[66]
Central composite design to mineralization of olive mill wastewater by the electro/Fe^{+2}/persulfate oxidation method, 2020	N/A	2^3-CCD	Decrease chemical oxygen demand in olive mill wastewater	[67]

(continued)

Table 23 (continued)

Experimental situation and year of publication	Factorial design	RSM	Goal	Refs.
Electrocoagulation for treating Kraft paper mill wastewater, 2020	N/A	2^2-CCD	Maximize the removal efficiency of color, chemical oxygen demand, biochemical oxygen demand, and tannin/lignin ratio. Also to maximize the biodegradability index	[68]
Microplastic pollution reduction by a carbon and nitrogen-doped TiO_2 via visible light photocatalytic degradation, 2020	2^2 with replicas in central point	N/A	Maximize high density polyethylene mass loss in microplastics	[69]
Removal of nickel and COD from wastewater from electroplating industry using Electrocoagulation and AOP, 2020	2^3	Multiple 2^2 designs until finding the optimum value for each level	Maximize the removal efficiency and COD from wastewater from electroplating industry	[70]
Pesticide Diuron removal from wastewater by photo-electrochemical oxidation with TiO_2 Nanotubes, 2020	2^3	N/A	Maximize the specific rate of removal and minimize the cell potential	[71]
Graphene oxide/TiO_2 for visible light photocatalysis to degrade eight pharmaceutical contaminants from water and wastewater, 2020	N/A	BBD with three factors	Maximize the mineralization percentage of the pharmaceutical contaminants	[72]
Solar electrocoagulation and adsorption processes with granular oil palm shell activated carbon for Pb(II) removal from aqueous solution, 2019	N/A	2^3-CCD	Maximize the removal efficiency	[73]
Performance of different coagulants in the coagulation/flocculation process of textile wastewater, 2019	N/A	3^2	Analyze the chemical oxygen demand of the coagulation/flocculation process	[74]
Electrodisinfection of real swine wastewater for water reuse, 2019	2^2	N/A	Maximize *E. coli* inactivation	[75]

(continued)

Table 23 (continued)

Experimental situation and year of publication	Factorial design	RSM	Goal	Refs.
L-Histidine (C, N) codoped-TiO_2-CdS nanocomposite for efficient visible photo-degradation of recalcitrant compounds from wastewater, 2019	N/A	CCD	Maximize methyl orange removal efficiency	[76]
Optimization of saline wastewater treatment using electrochemical oxidation process, 2019	N/A	2^4-CCD	Maximize COD and TOC removal from saline wastewaters using electrochemical oxidation process	[77]
Biological treatment of wastewater containing a mixture of polycyclic aromatic hydrocarbons using the oleaginous bacterium *Rhodococcus opacus*, 2018	2^3	N/A	Maximize polycyclic aromatic hydrocarbons removal efficiency	[78]
Degradation of paracetamol by an UV/Chlorine advanced oxidation process, 2018	N/A	2^4-CCD	Maximize paracetamol removal efficiency	[79]
Electrochemical treatment of wastewater from a bakery industry, 2018	2^3	N/A	Maximize removal of COD and minimize turbidity	[80]
Sono–photo-Fenton process for degradation of phenol derivatives in petrochemical wastewater, 2018	N/A	3^3	Maximize the removal of COD	[81]
Decolorization of Reactive Blue 235 dye by barium alginate immobilized iron nanoparticles synthesized from aluminum industry waste, 2017	N/A	BBD with 4 factors and 3 levels	Maximize the Reactive Blue 235 removal efficiency	[82]
Carbon and CNT fabricated carbon substrates for TiO_2 nanoparticles immobilization for continuous photocatalytic elimination of dye molecules, 2017	N/A	2^4-CCD	Maximize the dye and COD removal efficiency, and minimize electrical energy consumption	[83]

(continued)

Table 23 (continued)

Experimental situation and year of publication	Factorial design	RSM	Goal	Refs.
Evaluation and disposability study of actual textile wastewater treatment by electro-oxidation method using Ti/RuO_2 anode, 2017	N/A	BBD with 3 factors and 3 levels	Maximize the color and COD removal efficiency, and minimize electrical energy consumption	[84]
Treatment of real wastewater with TiO_2-films sensitized by a natural-dye obtained from *Picramnia sellowii*, 2016	2^3	N/A	Maximize the color and COD removal efficiency	[85]

Abbreviations: *AOP* advanced oxidative process, *BBD* Box-Behnken Design, *CCD* central composite design, *CNT* carbon nanotubes, *COD* chemical oxygen demand, *TOC* total organic carbon

Table 24 List of papers related to solar cells design and performance

Experimental situation and year of publication	Factorial design	RSM	Goal	Refs.
Doping of polyaniline to prepare polymer electrolytes for DSSC, 2020	N/A	3^2	Increase the efficiency of the DSSC	[86]
Improving the properties of indium tin oxide thin films by the incorporation of carbon nanotubes with solution-based techniques, 2020	2^3	N/A	Minimize the sheet resistance of the thin-film	[87]
Optimization and performance analysis of a solar concentrated photovoltaic-thermoelectric hybrid system, 2020	N/A	2^4-CCD	Increase the electrical efficiency of the solar concentrated photovoltaic-thermoelectric hybrid system	[88]
Ga-doped indium-zinc oxide films obtained by magnetron sputtering as transparent conductors for visible and solar applications, 2019	N/A	BBD	Control the chemical composition, optical, electrical, and structural properties of the GIZO film	[89]

(continued)

Table 24 (continued)

Experimental situation and year of publication	Factorial design	RSM	Goal	Refs.
Optimizing the optical properties of a dye-based luminescent down-shifter to improve the performance of organic photovoltaic devices, 2019	2^3 with replicas	N/A	Control the absorbance and photoluminescence quantum yield	[90]
Multistress testing of OPV modules for accurate predictive aging and reliability predictions, 2018	2^3	N/A	Study the degradation to 80% or 50% of the original maximum efficiency value	[91]
Improve the properties of Cu_2ZnSnS_4 films grown by spray pyrolysis, 2017	N/A	2^3-CCD	Improve the microstructural properties of Cu_2ZnSnS_4 films	[92]
Modeling of organic solar cell using response surface methodology, 2017	N/A	2^3-CCD	Improve the organic solar cell performance	[93]
Systematic optimization of boron diffusion for solar cell emitters, 2017	2^4	N/A	Control sheet resistance, lifetime, surface concentration, and junction depth of the solar cell emitter	[94]
Modeling of photogenerated charge collection of silver-based plasmonic DSSC using CCD experiments, 2017	N/A	2^5-CCD	Improve the charge collection	[95]
Systematic optimization of phosphorous diffusion for solar cell application, 2016	2^5	N/A	Control the sheet resistance, lifetime, and junction depth of solar cell p–n junction	[96]
XPS analysis of the chemical degradation of PTB7 polymers for organic photovoltaics, 2016	2^3	N/A	Control the stability of the PTB7 polymer	[97]

Abbreviations: *BBD* Box-Behnken Design, *CCD* central composite design, *DSSC* dye-sensitized solar cell, *PTB7* Poly[[4,8-bis[(2-ethylhexyl)oxy]benzo[1,2-b:4,5-b']dithiophene-2,6-diyl][3-fluoro-2-[(2-ethylhexyl)carbonyl]thieno[3,4-b]thio- phenediyl]], *OPV* organic photovoltaic

Acknowledgements The author would like to thank Manhattan College for support through startup funds and Faculty Summer Grant. The author gently dedicates this chapter to his family Olavio, Evanilde, Maria, Daniel, Caroline, Anthony, and Brianne.

References

1. Neto, B.B., Scarmino, I.S., Bruns, R.E.: Como Fazer Experimentos - Pesquisa e Desenvolvimento na Ciencia e na Industria, 4th ed., Bookman (2010)
2. Montgomery, D.C.: Experimental design for product and process design and development. J. R. Stat. Soc. Ser. D (The Stat.) **48**, 159–177 (1999). https://doi.org/10.1111/1467-9884.00179
3. Lundstedt, T., Seifert, E., Abramo, L., Thelin, B., Nyström, Å., Pettersen, J., Bergman, R.: Experimental design and optimization. Chemom. Intell. Lab. Syst. **42**, 3–40 (1998). https://doi.org/10.1016/S0169-7439(98)00065-3
4. Pereira, F., Pereira-Filho, E.: Aplicação de programa computacional livre em planejamento de experimentos: um tutorial. Quim. Nova. 1061–1071 (2018). https://doi.org/10.21577/0100-4042.20170254
5. Vera Candioti, L., De Zan, M.M., Cámara, M.S., Goicoechea, H.C.: Experimental design and multiple response optimization. Using the desirability function in analytical methods development. Talanta **124**, 123–138 (2014). https://doi.org/10.1016/j.talanta.2014.01.034
6. Mee, R.W.: A Comprehensive Guide to Factorial Two-Level Experimentation. Springer (2009)
7. Hibbert, D.B.: Experimental design in chromatography: a tutorial review. J. Chromatogr. B Anal. Technol. Biomed. Life Sci. **910**, 2–13 (2012). https://doi.org/10.1016/j.jchromb.2012.01.020
8. Ferreira, S.L.C., Lemos, V.A., de Carvalho, V.S., da Silva, E.G.P., Queiroz, A.F.S., Felix, C.S.A., da Silva, D.L.F., Dourado, G.B., Oliveira, R.V.: Multivariate optimization techniques in analytical chemistry—an overview. Microchem. J. **140**, 176–182 (2018). https://doi.org/10.1016/j.microc.2018.04.002
9. Teófilo, R.F., Ferreira, M.M.C.: Quimiometria II: planilhas eletrônicas para cálculos de planejamentos experimentais, um tutorial. Quim. Nova. **29**, 338–350 (2006). https://doi.org/10.1590/S0100-40422006000200026
10. Dean, A., Voss, D., Draguljić, D.: Design and Analysis of Experiments, 2nd ed. Springer International Publishing, Cham (2017). https://doi.org/10.1007/978-3-319-52250-0
11. Novaes, C.G., Bezerra, M.A., da Silva, E.G.P., dos Santos, A.M.P., Romão, I.L.S., Santos Neto, J.H.: A review of multivariate designs applied to the optimization of methods based on inductively coupled plasma optical emission spectrometry (ICP OES). Microchem. J. **128**, 331–346 (2016). https://doi.org/10.1016/j.microc.2016.05.015
12. Ferreira, S.L.C., Dos Santos, W.N.L., Quintella, C.M., Neto, B.B., Bosque-Sendra, J.M.: Doehlert matrix: a chemometric tool for analytical chemistry—review. Talanta **63**, 1061–1067 (2004). https://doi.org/10.1016/j.talanta.2004.01.015
13. Pasma, S.A., Daik, R., Maskat, M.Y., Hassan, O.: Application of box-behnken design in optimization of glucose production from oil palm empty fruit bunch cellulose. Int. J. Polym. Sci. (2013). https://doi.org/10.1155/2013/104502
14. Ferreira, S.L.C., Bruns, R.E., Ferreira, H.S., Matos, G.D., David, J.M., Brandão, G.C., da Silva, E.G.P., Portugal, L.A., dos Reis, P.S., Souza, A.S., dos Santos, W.N.L.: Box-Behnken design: an alternative for the optimization of analytical methods. Anal. Chim. Acta. **597**, 179–186 (2007). https://doi.org/10.1016/j.aca.2007.07.011
15. Ferreira, S.L.C., Silva Junior, M.M., Felix, C.S.A., da Silva, D.L.F., Santos, A.S., Santos Neto, J.H., de Souza, C.T., Cruz Junior, R.A., Souza, A.S.: Multivariate optimization techniques in food analysis—a review. Food Chem. **273**, 3–8 (2019). https://doi.org/10.1016/j.foodchem.2017.11.114
16. Granata, G., Yamaoka, T., Pagnanelli, F., Fuwa, A.: Study of the synthesis of copper nanoparticles: the role of capping and kinetic towards control of particle size and stability. J. Nanoparticle Res. **18**, 1–12 (2016). https://doi.org/10.1007/s11051-016-3438-6
17. Masalov, V.M., Sukhinina, N.S., Kudrenko, E.A., Emelchenko, G.A.: Mechanism of formation and nanostructure of Stöber silica particles. Nanotechnology **22** (2011). https://doi.org/10.1088/0957-4484/22/27/275718

18. Hartlen, K.D., Athanasopoulos, A.P.T., Kitaev, V.: Facile preparation of highly monodisperse small silica spheres (15 to >200 nm) suitable for colloidal templating and formation of ordered arrays. Langmuir **24**, 1714–1720 (2008). https://doi.org/10.1021/la7025285
19. Arantes, T.M., Pinto, A.H., Leite, E.R., Longo, E., Camargo, E.R.: Synthesis and optimization of colloidal silica nanoparticles and their functionalization with methacrylic acid. Colloids Surf. Physicochem. Eng. Asp. **415** (2012). https://doi.org/10.1016/j.colsurfa.2012.09.041
20. Hannachi, A., Hammami, S., Raouafi, N., Maghraoui-Meherzi, H.: Preparation of manganese sulfide (MnS) thin films by chemical bath deposition: application of the experimental design methodology. J. Alloys Compd. **663**, 507–515 (2016). https://doi.org/10.1016/j.jallcom.2015.11.058
21. Dahl, J.A., Maddux, B.L.S., Hutchison, J.E.: Toward greener nanosynthesis. Chem. Rev. **107**, 2228–2269 (2007). https://doi.org/10.1021/cr050943k
22. Martins, L.R., Rodrigues, J.A.V., Adarme, O.F.H., Melo, T.M.S., Gurgel, L.V.A., Gil, L.F.: Optimization of cellulose and sugarcane bagasse oxidation: application for adsorptive removal of crystal violet and auramine-O from aqueous solution. J. Colloid Interface Sci. **494**, 223–241 (2017). https://doi.org/10.1016/j.jcis.2017.01.085
23. Rafique, M., Sadaf, I., Rafique, M.S., Tahir, M.B.: A review on green synthesis of silver nanoparticles and their applications. Artif. Cells, Nanomed. Biotechnol. **45**, 1272–1291 (2017). https://doi.org/10.1080/21691401.2016.1241792
24. Núñez, R.N., Veglia, A.V., Pacioni, N.L.: Improving reproducibility between batches of silver nanoparticles using an experimental design approach. Microchem. J. **141**, 110–117 (2018). https://doi.org/10.1016/j.microc.2018.05.017
25. Eising, R., Signori, A.M., Fort, S., Domingos, J.B.: Development of catalytically active silver colloid nanoparticles stabilized by dextran. Langmuir **27**, 11860–11866 (2011). https://doi.org/10.1021/la2029164
26. Nikaeen, G., Yousefinejad, S., Rahmdel, S., Samari, F., Mahdavinia, S.: Central composite design for optimizing the biosynthesis of silver nanoparticles using plantago major extract and investigating antibacterial, antifungal and antioxidant activity. Sci. Rep. **10**, 1–16 (2020). https://doi.org/10.1038/s41598-020-66357-3
27. Soares, J.X., Wegner, K.D., Ribeiro, D.S.M., Melo, A., Häusler, I., Santos, J.L.M., Resch-Genger, U.: Rationally designed synthesis of bright AgInS2/ZnS quantum dots with emission control. Nano Res. **12** (2020). https://doi.org/10.1007/s12274-020-2876-8
28. Sodipo, B.K., Aziz, A.A.: Optimization of sonochemical method of functionalizing amino-silane on superparamagnetic iron oxide nanoparticles using central composite design. Ultrason. Sonochem. **64**, 104856 (2020). https://doi.org/10.1016/j.ultsonch.2019.104856
29. Keijok, W.J., Pereira, R.H.A., Alvarez, L.A.C., Prado, A.R., da Silva, A.R., Ribeiro, J., de Oliveira, J.P., Guimarães, M.C.C.: Controlled biosynthesis of gold nanoparticles with Coffea arabica using factorial design. Sci. Rep. **9**, 1–10 (2019). https://doi.org/10.1038/s41598-019-52496-9
30. da Silva, A.F.V., Fagundes, A.P., Macuvele, D.L.P., de Carvalho, E.F.U., Durazzo, M., Padoin, N., Soares, C., Riella, H.G.: Green synthesis of zirconia nanoparticles based on Euclea natalensis plant extract: optimization of reaction conditions and evaluation of adsorptive properties. Colloids Surf. Physicochem. Eng. Asp. **583**, 123915 (2019). https://doi.org/10.1016/j.colsurfa.2019.123915
31. Mora-Tamez, L., Barim, G., Downes, C., Williamson, E.M., Habas, S.E., Brutchey, R.L.: Controlled design of phase- and size-tunable monodisperse Ni 2 P nanoparticles in a phosphonium-based ionic liquid through response surface methodology. Chem. Mater. **31**, 1552–1560 (2019). https://doi.org/10.1021/acs.chemmater.8b04518
32. Usman, A.I., Aziz, A.A., Sodipo, B.K.: Application of central composite design for optimization of biosynthesized gold nanoparticles via sonochemical method. SN Appl. Sci. **1**, 1–9 (2019). https://doi.org/10.1007/s42452-019-0429-0
33. Viegas, I.M.A., Santos, B.S., Fontes, A., De Lima Pereira, G.A., Pereira, C.F.: Multivariate optimization of optical properties of CdSe quantum dots obtained by a facile one-pot aqueous synthesis. Inorg. Chem. Front. **6**, 1350–1360 (2019). https://doi.org/10.1039/c9qi00105k

34. Gomes, M.F., Gomes, Y.F., Lopes-Moriyama, A., de Barros Neto, E.L., de Souza, C.P.: Design of carbon quantum dots via hydrothermal carbonization synthesis from renewable precursors. Biomass Convers. Biorefinery **9**, 689–694 (2019). https://doi.org/10.1007/s13399-019-00387-4
35. Maass, D., Valério, A., Lourenço, L.A., de Oliveira, D., Hotza, D.: Biosynthesis of iron oxide nanoparticles from mineral coal tailings in a stirred tank reactor. Hydrometallurgy **184**, 199–205 (2019). https://doi.org/10.1016/j.hydromet.2019.01.010
36. Elazazy, M.S., Issa, A.A., Al-Mashreky, M., Al-Sulaiti, M., Al-Saad, K.: Application of fractional factorial design for green synthesis of cyano-modified silica nanoparticles: chemometrics and multifarious response optimization. Adv. Powder Technol. **29**, 1204–1215 (2018). https://doi.org/10.1016/j.apt.2018.02.012
37. Peng, T., Ray, S., Veeravalli, S.S., Lalman, J.A., Arefi-Khonsari, F.: The role of hydrothermal conditions in determining 1D TiO2 nanomaterials bandgap energies and crystal phases. Mater. Res. Bull. **105**, 104–113 (2018). https://doi.org/10.1016/j.materresbull.2018.04.021
38. Obasogie, O.E., Abdulkareem, A.S., Mohammed, I.A., Bankole, M.T., Tijani, J.O., Abubakre, O.K.: Empirical relationship between band gap and synthesis parameters of chemical vapor deposition-synthesized multiwalled carbon nanotubes. Carbon Lett. **28**, 72–80 (2018). https://doi.org/10.5714/CL.2018.28.072
39. Behera, J.K., Zhou, X., Ranjan, A., Simpson, R.E.: Sb2Te3 and its superlattices: optimization by statistical design. ACS Appl. Mater. Interf. **10**, 15040–15050 (2018). https://doi.org/10.1021/acsami.8b02100
40. Silva, M.M.S., Sena, M.S., Lopes-Moriyama, A.L., Souza, C.P., Santos, A.G.: Experimental planning of the synthesis of strontium molybdate by EDTA-citrate and its structural influence, morphology and optical bandgap. Ceram. Int. **44**, 16606–16614 (2018). https://doi.org/10.1016/j.ceramint.2018.06.087
41. Burrows, N.D., Harvey, S., Idesis, F.A., Murphy, C.J.: Understanding the seed-mediated growth of gold nanorods through a fractional factorial design of experiments. Langmuir **33**, 1891–1907 (2017). https://doi.org/10.1021/acs.langmuir.6b03606
42. Ribeiro, D.S.M., de Souza, G.C.S., Melo, A., Soares, J.X., Rodrigues, S.S.M., Araújo, A.N., Montenegro, M.C.B.S.M., Santos, J.L.M.: Synthesis of distinctly thiol-capped CdTe quantum dots under microwave heating: multivariate optimization and characterization. J. Mater. Sci. **52**, 3208–3224 (2017). https://doi.org/10.1007/s10853-016-0610-4
43. Cembrero-Coca, P., Cembrero, J., Busquets-Mataix, D., Pérez-Puig, M.A., Marí, B., Pruna, A.: Factorial electrochemical design for tailoring of morphological and optical properties of Cu2O. Mater. Sci. Technol. (United Kingdom) **33**, 2102–2109 (2017). https://doi.org/10.1080/02670836.2017.1349595
44. Machado, M.M., Savi, B.M., Perucchi, M.B., Benedetti, A., Oliveira, L.F.S., Bernardin, A.M.: Effect of temperature, precursor type and dripping time on the crystallite size of nano ZnO obtained by one-pot synthesis: 2 k full factorial design analysis. J. Nanosci. Nanotechnol. **18**, 4409–4412 (2017). https://doi.org/10.1166/jnn.2018.15043
45. Safavi, M.S., Shojaosadati, S.A., Yang, H.G., Kim, Y., Park, E.J., Lee, K.C., Na, D.H.: Reducing agent-free synthesis of curcumin-loaded albumin nanoparticles by self-assembly at room temperature. Int. J. Pharm. **529**, 303–309 (2017). https://doi.org/10.1016/j.ijpharm.2017.06.087
46. Albano, J.M.R., Ribeiro, L.N.M., Couto, V.M., Barbosa Messias, M., Rodrigues da Silva, G.H., Breitkreitz, M.C., de Paula, E., Pickholz, M.: Rational design of polymer-lipid nanoparticles for docetaxel delivery. Colloids Surf. B Biointerf. **175**, 56–64 (2019). https://doi.org/10.1016/j.colsurfb.2018.11.077
47. Akhlaghi, S., Rabbani, S., Alavi, S., Alinaghi, A., Radfar, F., Dadashzadeh, S., Haeri, A.: Green formulation of curcumin loaded lipid-based nanoparticles as a novel carrier for inhibition of post-angioplasty restenosis. Mater. Sci. Eng. C. **105**, 110037 (2019). https://doi.org/10.1016/j.msec.2019.110037
48. Pavan Rudhrabatla, V.S.A., Sudhakar, B., Suresh Reddy, K.V.N.: Ritonavir loaded surface modified stealth solid lipid nanoparticles: full factorial design and pharmacokinetic studies. Int. J. Res. Pharm. Sci. **10**, 77–89 (2019). https://doi.org/10.26452/ijrps.v10i1.1783

49. Ramalho, M.J., Loureiro, J.A., Coelho, M.A.N., Pereira, M.C.: Factorial design as a tool for the optimization of plga nanoparticles for the co-delivery of temozolomide and o6-benzylguanine. Pharmaceutics **11** (2019). https://doi.org/10.3390/pharmaceutics11080401
50. Ali, H., Singh, S.K.: Preparation and characterization of solid lipid nanoparticles of furosemide using quality by design. Part. Sci. Technol. **36**, 695–709 (2018). https://doi.org/10.1080/02726351.2017.1295293
51. Ariza-Sáenz, M., Espina, M., Calpena, A., Gómara, M.J., Pérez-Pomeda, I., Haro, I., García, M.L.: Design, characterization, and biopharmaceutical behavior of nanoparticles loaded with an HIV-1 fusion inhibitor peptide. Mol. Pharm. **15**, 5005–5018 (2018). https://doi.org/10.1021/acs.molpharmaceut.8b00609
52. Taneja, N., Singh, K.K.: Rational design of polysorbate 80 stabilized human serum albumin nanoparticles tailored for high drug loading and entrapment of irinotecan. Int. J. Pharm. **536**, 82–94 (2018). https://doi.org/10.1016/j.ijpharm.2017.11.024
53. Zabihian, A., Salouti, M., Hamidi, M.: Factorial design analysis and optimisation of chitosan-based nanogels as controlled release system for gentamicin. IET Nanobiotechnol. **12**, 12–17 (2018). https://doi.org/10.1049/iet-nbt.2017.0114
54. Joseph, E., Reddi, S., Rinwa, V., Balwani, G., Saha, R.: DoE based Olanzapine loaded poly-caprolactone nanoparticles decreases extrapyramidal effects in rodent model. Int. J. Pharm. **541**, 198–205 (2018). https://doi.org/10.1016/j.ijpharm.2018.02.010
55. Petkar, K.C., Chavhan, S., Kunda, N., Saleem, I., Somavarapu, S., Taylor, K.M.G., Sawant, K.K.: Development of novel octanoyl chitosan nanoparticles for improved rifampicin pulmonary delivery: optimization by factorial design. AAPS Pharm. Sci. Tech. **19**, 1758–1772 (2018). https://doi.org/10.1208/s12249-018-0972-9
56. Zielińska, A., Martins-Gomes, C., Ferreira, N.R., Silva, A.M., Nowak, I., Souto, E.B.: Anti-inflammatory and anti-cancer activity of citral: optimization of citral-loaded solid lipid nanoparticles (SLN) using experimental factorial design and LUMiSizer®. Int. J. Pharm. **553**, 428–440 (2018). https://doi.org/10.1016/j.ijpharm.2018.10.065
57. Shaikh, M.V., Kala, M., Nivsarkar, M.: Formulation and optimization of doxorubicin loaded polymeric nanoparticles using Box-Behnken design: ex-vivo stability and in-vitro activity. Eur. J. Pharm. Sci. **100**, 262–272 (2017). https://doi.org/10.1016/j.ejps.2017.01.026
58. Gupta, S., Kesarla, R., Chotai, N., Misra, A., Omri, A.: Systematic approach for the formulation and optimization of solid lipid nanoparticles of efavirenz by high pressure homogenization using design of experiments for brain targeting and enhanced bioavailability. Biomed Res. Int. (2017). https://doi.org/10.1155/2017/5984014
59. Awada, H.K., Johnson, L.A., Hitchens, T.K., Foley, L.M., Wang, Y.: Factorial design of experiments to optimize multiple protein delivery for cardiac repair. ACS Biomater. Sci. Eng. **2**, 879–886 (2016). https://doi.org/10.1021/acsbiomaterials.6b00146
60. Akl, M.A., Kartal-Hodzic, A., Oksanen, T., Ismael, H.R., Afouna, M.M., Yliperttula, M., Samy, A.M., Viitala, T.: Factorial design formulation optimization and in vitro characterization of curcumin-loaded PLGA nanoparticles for colon delivery. J. Drug Deliv. Sci. Technol. **32**, 10–20 (2016). https://doi.org/10.1016/j.jddst.2016.01.007
61. Baig, M.S., Ahad, A., Aslam, M., Imam, S.S., Aqil, M., Ali, A.: Application of Box-Behnken design for preparation of levofloxacin-loaded stearic acid solid lipid nanoparticles for ocular delivery: optimization, in vitro release, ocular tolerance, and antibacterial activity. Int. J. Biol. Macromol. **85**, 258–270 (2016). https://doi.org/10.1016/j.ijbiomac.2015.12.077
62. Girotra, P., Singh, S.K., Kumar, G.: Development of zolmitriptan loaded PLGA/poloxamer nanoparticles for migraine using quality by design approach. Int. J. Biol. Macromol. **85**, 92–101 (2016). https://doi.org/10.1016/j.ijbiomac.2015.12.069
63. Aboud, H.M., El komy, M.H., Ali, A.A., El Menshawe, S.F., Abd Elbary, A.: Development, optimization, and evaluation of carvedilol-loaded solid lipid nanoparticles for intranasal drug delivery. AAPS Pharm. Sci. Tech. **17**, 1353–1365 (2016). https://doi.org/10.1208/s12249-015-0440-8
64. Erdoğar, N., Esendağli, G., Nielsen, T.T., Şen, M., Öner, L., Bilensoy, E.: Design and optimization of novel paclitaxel-loaded folate-conjugated amphiphilic cyclodextrin nanoparticles. Int. J. Pharm. **509**, 375–390 (2016). https://doi.org/10.1016/j.ijpharm.2016.05.040

65. Hashad, R.A., Ishak, R.A.H., Geneidi, A.S., Mansour, S.: Methotrexate loading in chitosan nanoparticles at a novel pH: response surface modeling, optimization and characterization. Int. J. Biol. Macromol. **91**, 630–639 (2016). https://doi.org/10.1016/j.ijbiomac.2016.06.014
66. Skinner, A.W., DiBernardo, A.M., Masud, A.M., Aich, N., Pinto, A.H.: Factorial design of experiments for optimization of photocatalytic degradation of tartrazine by zinc oxide (ZnO) nanorods with different aspect ratios. J. Environ. Chem. Eng. 104235 (2020). https://doi.org/10.1016/j.jece.2020.104235
67. Görmez, F., Görmez, Ö., Yabalak, E., Gözmen, B.: Application of the central composite design to mineralization of olive mill wastewater by the electro/FeII/persulfate oxidation method. SN Appl. Sci. **2**, 1–11 (2020). https://doi.org/10.1007/s42452-020-1986-y
68. Wagle, D., Lin, C.J., Nawaz, T., Shipley, H.J.: Evaluation and optimization of electrocoagulation for treating Kraft paper mill wastewater. J. Environ. Chem. Eng. **8**, 103595 (2020). https://doi.org/10.1016/j.jece.2019.103595
69. Ariza-Tarazona, M.C., Villarreal-Chiu, J.F., Hernández-López, J.M., Rivera De la Rosa, J., Barbieri, V., Siligardi, C., Cedillo-González, E.I.: Microplastic pollution reduction by a carbon and nitrogen-doped TiO2: effect of pH and temperature in the photocatalytic degradation process. J. Hazard. Mater. **395**, 122632 (2020). https://doi.org/10.1016/j.jhazmat.2020.122632
70. Moersidik, S.S., Nugroho, R., Handayani, M., Kamilawati, Pratama, M.A.: Optimization and reaction kinetics on the removal of nickel and COD from wastewater from electroplating industry using electrocoagulation and advanced oxidation processes. Heliyon **6**, e03319 (2020). https://doi.org/10.1016/j.heliyon.2020.e03319
71. Vacca, A., Mais, L., Mascia, M., Usai, E.M., Palmas, S.: Design of experiment for the optimization of pesticide removal from wastewater by photo-electrochemical oxidation with tio2 nanotubes. Catalysts **10** (2020). https://doi.org/10.3390/catal10050512
72. Checa, M., Beltrán, F.J., Rivas, F.J., Cordero, E.: On the role of a graphene oxide/titania catalyst, visible LED and ozone in removing mixtures of pharmaceutical contaminants from water and wastewater. Environ. Sci. Water Res. Technol. (2020). https://doi.org/10.1039/d0ew00276c
73. Hussin, F., Aroua, M.K., Szlachtac, M.: Combined solar electrocoagulation and adsorption processes for Pb(II) removal from aqueous solution. Chem. Eng. Process. Process Intensif. **143**, 107619 (2019). https://doi.org/10.1016/j.cep.2019.107619
74. Dotto, J., Fagundes-Klen, M.R., Veit, M.T., Palácio, S.M., Bergamasco, R.: Performance of different coagulants in the coagulation/flocculation process of textile wastewater. J. Clean. Prod. **208**, 656–665 (2019). https://doi.org/10.1016/j.jclepro.2018.10.112
75. Simas, A., Mores, R., Steffens, J., Dallago, R.M., Kunz, A., Michelon, W., Fongaro, G., Viancelli, A.: Electrodisinfection of real swine wastewater for water reuse. Environ. Chem. Lett. **17**, 495–499 (2019). https://doi.org/10.1007/s10311-018-0782-z
76. Zangeneh, H., Zinatizadeh, A.A., Zinadini, S., Feyzi, M., Rafiee, E., Bahnemann, D.W.: A novel L-Histidine (C, N) codoped-TiO2-CdS nanocomposite for efficient visible photo-degradation of recalcitrant compounds from wastewater. J. Hazard. Mater. **369**, 384–397 (2019). https://doi.org/10.1016/j.jhazmat.2019.02.049
77. Darvishmotevalli, M., Zarei, A., Moradnia, M., Noorisepehr, M., Mohammadi, H.: Optimization of saline wastewater treatment using electrochemical oxidation process: prediction by RSM method. MethodsX. **6**, 1101–1113 (2019). https://doi.org/10.1016/j.mex.2019.03.015
78. Goswami, L., Manikandan, N.A., Dolman, B., Pakshirajan, K., Pugazhenthi, G.: Biological treatment of wastewater containing a mixture of polycyclic aromatic hydrocarbons using the oleaginous bacterium Rhodococcus opacus. J. Clean. Prod. **196**, 1282–1291 (2018). https://doi.org/10.1016/j.jclepro.2018.06.070
79. Dao, Y.H., Tran, H.N., Tran-Lam, T.T., Pham, T.Q., Le, G.T.: Degradation of paracetamol by an UV/chlorine advanced oxidation process: influencing factors, factorial design, and intermediates identification. Int. J. Environ. Res. Public Health. **15** (2018). https://doi.org/10.3390/ijerph15122637
80. de Santana, M.M., Zanoelo, E.F., Benincá, C., Freire, F.B.: Electrochemical treatment of wastewater from a bakery industry: experimental and modeling study. Process Saf. Environ. Prot. **116**, 685–692 (2018). https://doi.org/10.1016/j.psep.2018.04.001

81. Shokri, A.: Application of Sono–photo-Fenton process for degradation of phenol derivatives in petrochemical wastewater using full factorial design of experiment. Int. J. Ind. Chem. **9**, 295–303 (2018). https://doi.org/10.1007/s40090-018-0159-y
82. Natarajan, E., Ponnaiah, G.P.: Optimization of process parameters for the decolorization of reactive blue 235 dye by barium alginate immobilized iron nanoparticles synthesized from aluminum industry waste. Environ. Nanotechnol. Monit. Manag. **7**, 73–88 (2017). https://doi.org/10.1016/j.enmm.2017.01.002
83. Pajootan, E., Rahimdokht, M., Arami, M.: Carbon and CNT fabricated carbon substrates for TiO2 nanoparticles immobilization with industrial perspective of continuous photocatalytic elimination of dye molecules. J. Ind. Eng. Chem. **55**, 149–163 (2017). https://doi.org/10.1016/j.jiec.2017.06.039
84. Kaur, P., Kushwaha, J.P., Sangal, V.K.: Evaluation and disposability study of actual textile wastewater treatment by electro-oxidation method using Ti/RuO2 anode. Process Saf. Environ. Prot. **111**, 13–22 (2017). https://doi.org/10.1016/j.psep.2017.06.004
85. Fernández, J.A., Suan, A., Ramírez, J.C., Robles, J., Salcedo, J.C., Pedroza, A.M., Daza, C.E.: Treatment of real wastewater with TiO2-films sensitized by a natural-dye obtained from Picramnia sellowii. J. Environ. Chem. Eng. **4**, 2848–2856 (2016). https://doi.org/10.1016/j.jece.2016.05.037
86. Patni, N., Pillai, S.G.: Optimization of the doping of polyaniline via response surface method to prepare polymer electrolytes for dye sensitized solar cells. J. Environ. Chem. Eng. **8**, 103709 (2020). https://doi.org/10.1016/j.jece.2020.103709
87. Gökçeli, G., Karatepe, N.: Improving the properties of indium tin oxide thin films by the incorporation of carbon nanotubes with solution-based techniques. Thin Solid Films **697**, 137844 (2020). https://doi.org/10.1016/j.tsf.2020.137844
88. Rejeb, O., Shittu, S., Ghenai, C., Li, G., Zhao, X., Bettayeb, M.: Optimization and performance analysis of a solar concentrated photovoltaic-thermoelectric (CPV-TE) hybrid system. Renew. Energy. **152**, 1342–1353 (2020). https://doi.org/10.1016/j.renene.2020.02.007
89. Solís-Cortés, D., Navarrete-Astorga, E., Costa-Krämer, J.L., Salguero-Fernandez, J., Schrebler, R., Leinen, D., Dalchiele, E.A., Ramos-Barrado, J.R., Martín, F.: Ga-doped IZO films obtained by magnetron sputtering as transparent conductors for visible and solar applications. Ceram. Int. **45**, 5577–5587 (2019). https://doi.org/10.1016/j.ceramint.2018.12.017
90. Fernandes, R.V., Cordeiro Junior, N.A., Pardo Perdomo, A.D., Duarte, J.L., Laureto, E.: Optimizing the optical properties of a dye-based luminescent down-shifter to improve the performance of organic photovoltaic devices. Dye. Pigment. **169**, 1–6 (2019). https://doi.org/10.1016/j.dyepig.2019.04.074
91. Stoichkov, V., Kumar, D., Tyagi, P., Kettle, J.: Multistress testing of OPV modules for accurate predictive aging and reliability predictions. IEEE J. Photovolt. **8**, 1058–1065 (2018). https://doi.org/10.1109/JPHOTOV.2018.2838438
92. Ramírez, E.A., Ramírez, A., Gordillo, G.: Cu2ZnSnS4 films grown in one-step process by spray pyrolysis with improved properties. Mater. Sci. Semicond. Process. **67**, 110–117 (2017). https://doi.org/10.1016/j.mssp.2017.05.024
93. Suliman, R., Mitul, A.F., Mohammad, L., Djira, G., Pan, Y., Qiao, Q.: Modeling of organic solar cell using response surface methodology. Results Phys. **7**, 2232–2241 (2017). https://doi.org/10.1016/j.rinp.2017.04.037
94. Ebrahimi, P., Kolahdouz, M., Iraj, M., Ganjian, M., Aghababa, H., Asl-Soleimani, E., Radamson, H.H.: Systematic optimization of boron diffusion for solar cell emitters. J. Electron. Mater. **46**, 4236–4241 (2017). https://doi.org/10.1007/s11664-017-5364-5
95. Buda, S., Shafie, S., Rashid, S.A., Jaafar, H., Khalifa, A.: Response surface modeling of photogenerated charge collection of silver-based plasmonic dye-sensitized solar cell using central composite design experiments. Results Phys. **7**, 493–497 (2017). https://doi.org/10.1016/j.rinp.2017.01.011

96. Kermaniha, M., Kolahdouz, M., Manavizadeh, N., Aghababa, H., Elahi, M., Iraj, M., Asl-Soleimani, E., Radamson, H.H.: Systematic optimization of phosphorous diffusion for solar cell application. J. Mater. Sci. Mater. Electron. **27**, 13086–13092 (2016). https://doi.org/10.1007/s10854-016-5452-y
97. Kettle, J., Ding, Z., Horie, M., Smith, G.C.: XPS analysis of the chemical degradation of PTB7 polymers for organic photovoltaics. Org. Electron. **39**, 222–228 (2016). https://doi.org/10.1016/j.orgel.2016.10.016

Preparation and Implementation of a High Throughput Virtual Screening Protocol on a Shared Memory GPU Supercomputer

Hortensia Gomes Leal, Michelle A. Arya, Richard Anderson, and Kimberly Stieglitz

Abstract A novel computational protocol for drug discovery is presented. Pre-screening binding sites of selected targets with high accuracy and precision using different molecular docking programs is combined with a virtual drug discovery platform (VDDP) to screen a large number of small molecules. Several relevant proteins were identified as targets for this protocol. The protein coordinates were downloaded for docking studies from the RCSB PDB for each target. FBPase was used as a case study for the complete implementation of the VDDP. Compounds with high affinity predicted scores were identified and molecular dynamics studies were performed on the 10 highest scoring protein/small molecule complexes focusing on the allosteric binding sites. Studies were carried out using NAMD. Dissociation constants (Ki's) for the protein–ligand complex were calculated throughout the NAMD runs and respective conformational changes in the FBPase binding pocket were observed. The inhibitor binding pocket was evaluated based on pocket shape and volume. Predicted interfacial structural changes produced in response to small molecule binding at two allosteric sites was generated in NAMD. Novel hydrogen bonding and hydrophobic interaction networks of residues in the 3D structures were identified to connect the active sites to the allosteric binding sites using an NAMD protocol.

H. G. Leal
Departamento de Genética, Ecologia e Evolução, Universidade Federal de Minas Gerais, 6627 Presidente Antônio Carlos Ave, Belo Horizonte, MG 31270-901, Brazil

M. A. Arya
Independent Researcher, 104 Rock Glen Road, Medford, MA 02155, USA

R. Anderson · K. Stieglitz (✉)
Roxbury Community College STEM Biotechnology Division, 1234 Columbus Ave, Roxbury, MA 02120, USA
e-mail: kastieglitz@rcc.mass.edu

R. Anderson
Symmetric Computing, Venture Development Center, University of Massachusetts Boston, 100 Morrissey Blvd, Boston, MA 02125, USA

F. A. La Porta and C. A. Taft (eds.), *Functional Properties of Advanced Engineering Materials and Biomolecules*, Engineering Materials,
https://doi.org/10.1007/978-3-030-62226-8_15

1 Prescreening the Binding Sites of Selected Targets and Precision with Different Molecular Docking Programs

Docking programs can be distinguished by their different ligand placement approaches, which directly interfere with the scores results [1]. Additionally, it has been shown that the docking scores can vary according to the target and the ligand molecule [2, 3]. Thus, the choice of the docking software program is an important factor for the automated docking protocol. Keyes et al. [4] have demonstrated that the geometric shape of docking sites from various protein structures can be used as a protocol to choose the best performing docking program.

2 Monoamine Oxidases in Cancer and Parkinson's Disease

Monoamine Oxidase (MAO) enzymes are expressed in two isoforms MAO-A and MAO-B and their amino acid sequences share 70% identity. MAO enzymes are present at the outer membrane of the mitochondria and catalyze the oxidative deamination of amines neurotransmitters in the brain and peripheral tissues [5]. MAO enzymes have dopamine, noradrenaline, adrenaline, tryptamine and tyramine as common substrates. MAO-A also acts on serotonin, while MAO-B has phenylethylamine and benzylamine as substrates, presenting an important role in degrading neurotransmitters. Both enzymes have been suggested as mediator molecules in tumorigenesis, psychiatric and neurodegenerative disorders [6–8].

MAO-A has been shown to be highly expressed in prostate cancer, glioma, lung cancer, and classical Hodgkin's Lymphoma, while its down-regulation has been identified to play an important role in multiple cancer types, including breast cancer, hepatocellular cancer and nasopharyngeal carcinoma [9–15]. MAO-A is inhibited by clorgyline and phenelzine and shown suppressive effects in the growth of prostate carcinoma and restore sensitivity in anti-androgen enzalutamide resistant prostate cancer [16–18]. The use of clorgyline has also been investigated for brain tumor and Hodgkin's Lymphoma, showing in vitro and in vivo reduction of tumor cell proliferation, migration, and angiogenesis [12, 13].

MAO-B is predominant in the human brain and elevated levels are associated with neurodegenerative disorders, such as Alzeimer's and Parkinson's disease [6]. MAO-B converts dopamine to hydrogen peroxide, which is essential for the pathomechanism processes of oxidative stress and oxidative damage that occur in Parkinson's disease [19]. Parkinson's disease therapy mainly includes the used of dopaminergic regulators (i.e. Levodopa) and selective MAO B inhibitors, such as selegiline (L-deprenyl). These may lead to an improvement of the motor symptoms, protect against neuronal damage, potentialize dopaminergic effects, and are shown to reduce oxidative stress and glutamate-induced toxicity in in vitro studies. However MAO-B inhibitors also stimulate cardiovascular side effects and action of the neurotoxic metabolites might compromise its neuroprotective effect at high doses they interfere with MAO-A activity [19–21].

3 Beta Secretase 1 and Alzheimer's Disease

β-site APP-cleaving enzyme (BACE-) 1 and 2 are membrane-bound aspartic proteases expressed in peripheral tissues of the brain, sharing 59% of their amino acid sequences, BACE-1, being the most abundant isoform [22, 23]. BACE-1 has been reported as an important enzyme for the control of nerve myelination by cleaving neuregulin 1 (NRG1), and in synaptic function by using Seizure protein 6 (Sez6) and Seizure 6-like protein (Sez6L) as substrates [24–26]. BACE-1 also cleaves the transmembrane amyloid precursor protein (APP) and, together with y-secretase, generates amyloid β-peptide (Aβ) that aggregates and deposits in the brain in Alzheimer's disease (AD) [23, 27].

BACE-1 concentration is increased in the neocortex of Alzheimers' disease patients [28, 29]. In this view, BACE-1 has been posited as a potential target to reduce Aβ levels in AD patients. However, all clinical trials with BACE-1 inhibitors were discontinued due to infectivity or safety reasons, such as liver toxicity and cognitive worsening in the active treatment group [30].

4 Chloride Protein Channel and Cystic Fibrosis

The cystic fibrosis transmembrane conductance regulator (CFTR) is an ATP-binding cassette (ABC) family transporter and its structure includes two transmembrane domains, two nucleotide-binding domains and a regulatory domain with multiple phosphorylation sites [31]. ATP hydrolysis, cyclic AMP-dependent protein kinase A (PKA) and protein kinase C (PKC) are used to control its activity [32].

CFTR plays a fundamental role mediating the transport of chloride and other anions across the membrane of many epithelial organs including the lungs, pancreas, and gastro-intestinal tract. Mutations in the gene encoding CFTR causesCystic Fibrosis (CF) [33]. The major impact of CFTR dysfunction in CF is on the respiratory system. Dehydration and obstruction of airway surfaces by mucus accumulation, acidification of surface fluids, frequent opportunistic pathogenic colonization of the airways that trigger intense inflammatory response, and irreversible progressive destruction of the epithelial lung cells lead to declining lung functions [34–37].

Over 2,000 CFTR variants) identified in CF patients. CFTR gene mutations have particularly been classified into six classes and the mutations can compromise protein expression, processing, channel function and stability [38, 39]. The F508del is the most prevalent, accounting for 60% of all CF mutations [40]. Due to it's main role in CF pathophysiology CFTR is an ideal target for small molecules that can improve and rescue the mutant protein's function. Modulators that directly target CFTR mutants aiming to restore CFTR function have shown promise for a subgroup of patients with CF [41]. Drugs that act as correctors and potentiators, such as Lumacaftor (VX-809) and Ivacaftor (VX-770) have been used to restore the CFTR channel gating activity in patients bearing class II F508del mutation and reduce chloride levels, improving forced expiratory volume to one second in patients that presents at least one allele of class III mutations [42–44].

5 FBPase as a Case Study: A Target in Drug Discovery for Lowering Blood Glucose Levels in Type II Diabetes

To facilitate understanding of the process of using the VDDP, the remainder of this chapter will focus on one target as a case study: FBPas. eFructose 1,6 Bisphosphatase (FBPase) is an allosteric enzyme found in two isotypic forms in mammals, sharing 77% sequence identity [45]. FBP1 is mainly expressed in the liver and kidneys and has a regulatory role in gluconeogenesis, while FBP2 is primarily found in muscle, and is involved in the glycolytic and gluconeogenesis pathways and is indicated as a thermogenic regulatory enzyme in mice [46–49].

FBPase is a catalyst in the rate controlling step of this pathway. FBPase FBPase cleaves the terminal P-O phosphate bond of fructose 1,6 bisphosphate. (F1, 6) to fructose 6-phosphate and inorganic phosphate [50]. The physiological form of this enzyme is a homotetramer composed of two dimers which changes conformational state to either the active "R" state which is stabilized by FBPase bound with fructose 1,6-bisphosphate at the active site and a metal cation (Mg^{2+}, Mn^{2+}, or Zn^{2+}), or the inactive "T" state where the enzyme is in complex with adenosine monophosphate (AMP). Each monomer of FBPase has one fructose-1,6-bisphosphate binding site located at the active site of the enzyme and an AMP binding domain at the allosteric regulatory site[51–53].

FBP1 is synergistically inhibited by AMP and fructose 2,6-bisphosphate [54]. Binding of AMP at the allosteric site promotes a conformational change in the tetrameric structure, stabilizing the inactive state. [52]. Binding of Fructose 2,6-bisphosphate (Fru 2,6-P2) at the active site inhibits enzyme activity further [54]. A schematic picture of the FBPase biological unit is shown in Fig. 1 showing the AMP allosteric binding site (1), active site (2), and binding site at the tetramer interface (3) [55].

Inhibitors of the active site would need to target the phosphatase activity of FBPase. Phosphatase activity is common to many other enzymes and phosphate binding sites are highly conserved. Hence, inhibitors of the active site may not be specific enough and may inhibit more than one enzyme in vivo, causing unwanted off-target effects. Additionally, FBPase competitive inhibitors are often highly polarized molecules and interact with residues in the active site primarily through H-bonding networks and polar interactions since the active site is highly hydrophilic.

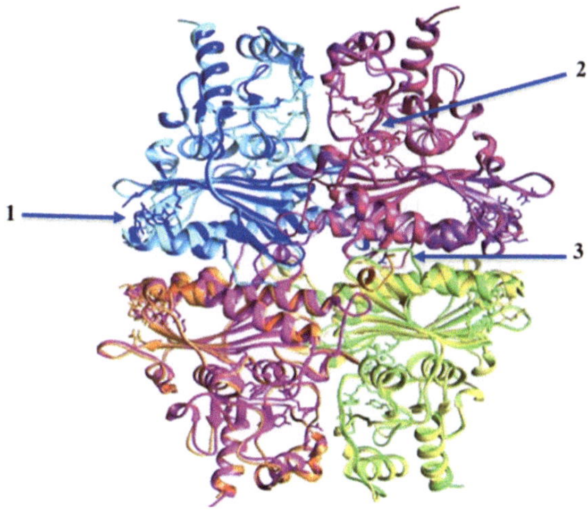

Fig. 1 Overlay of human and pig kidney FBPase (PDB codes 1FTA and 1KZ8). Identification of the three possible targets for drug discovery: The AMP allosteric binding site (1); the active site (2); and the tetramer interfacial PFE binding site (3) (PDB 1FTA [56], 1KZ8 [55])

A highly charged potential drug may have trouble passing through the lipophilic cell membrane.

In contrast, the AMP binding site contains ~ 65% hydrophobic residues and hydrophobic interactions dominate for AMP and other small molecules that may bind the site with high affinity. As the AMP binding site plays an important role in the negative regulation of FBPase, researchers have been working on FBPase as a drug target with a focus on the AMP binding site [57]. Uuncompetitive and noncompetitive FBPase inhibitors that interact with the AMP allosteric site or act as AMP mimics, such as Anilinoquinazoline, ZMP (AICAr 5-Aminoimidazole-4-carboxamide-1-β-D-ribofuranosyl 5′-monophosphate), MB05032 are of interest. [55, 58, 59].

In this perspective FBPase is an attractive target for inhibitory drugs since it would directly control glycemic levels while avoiding side effects such as hypoglycemia which is associated with conventional Type II Diabetes drug therapies [60]. However, challenges of toxicity and/or bioavailability have prevented a definitive promising lead compound until this time.

In our approach, FBPase small-molecule inhibition is achieved by a two pronged approach where large numbers of small molecules are screened with in silico docking and molecular dynamics on a high speed computer using a Virtual Drug Discovery Platform (VDDP). Once docking and molecular dynamics protocols have produced predicted lead compounds or “hits” they are then validated with assays in the lab. An overview of this process is shown in Fig. 2.

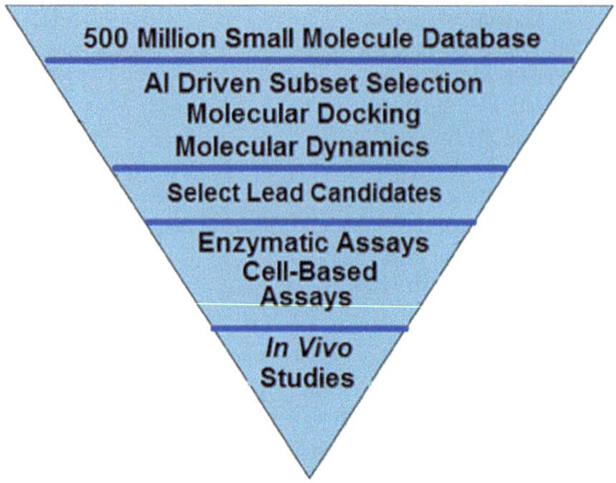

Fig. 2 Schematic flowchart incorporating the Virtual Drug Discovery Platform (VDDP) in the drug discovery process. Starting with ~ 500 million molecules a subset of molecules is selected followed by molecular docking and molecular dynamics. Analysis of scores leads to candidate selection and validation (enzymatic/cell-based assays) and finally in vivo studies

6 The Virtual Drug Discovery Platform (VDDP): Design and Implementation

Selection of Small Molecules and Ideal Docking Program:

The molecules or "hits" for FBPase were selected from the original ZINC database [61] and downloaded onto a departmental supercomputer "Ada" for these studies. A description of Ada's architecture is available at the company website: (https://www.symmetriccomputing.com/products.html). A subset of natural products purchasable from IBScreen Bioactives selected from the original ZINC cited above (~85,000 molecules) were identified based on similar LogP, pH and molecular weight as AMP (AI driven subset selection from Fig. 2). Autodock Vina [62], Autodock 4 [63] and Dock 6.9 [64], were compared to identify the best docking program whose docking scores best matched literature binding constants before large scale searches were performed (Table 1) [4]. In addition, the position of the docked ligands in the crystal structure was also examined in a graphics program to check the orientation of the docked ligands was correct like the crystallographic ligand. Table 1 shows that for the targets discussed in this chapter, the best predicted binding scores are achieved with Autodock Vina. This is consistent with a previously described method of correlating the shape of the binding pocket to a solvent channel. The ratio of solvent accessible volume to total molecular surface volume is high for binding pockets on these targets and they tend to be enclosed on ~ 3 sides with an "open mouth" to a solvent channel [4] which makes them amenable to AutoDock program predictions.

Kd: Estimated dissociation constant. Ki: Inhibition constant. Km: Michaelis constant. IC_{50}: inhibitory concentration 50%.

Molecular docking of FBPase was performed by first targeting the AMP binding site using Autodock Vina [62] with rigid protein geometric docking of small molecules (flexible torsions) and then implementing key flexible residues (Lys112, Tyr113, Arg140, Thr31 and Val17). As shown below in Fig. 3 with PDB 1FTA AMP

Table 1 Comparison of binding scores among docking programs

			Binding energy (kcal/moL)			Kd (estimated)(M)			In Vitro		
PDB Code	Target	Ligand	Autodock Vina	Autodock 4	Dock 6.9	Autodock Vina	Autodock 4	Dock 6.9	K_d/K_i	IC_{50}	K_m
2BXR [65]	MAO_A	Clorgyline	−13.4	−11.2	−9.2	2.77×10^{-9}	7.81×10^{-7}	3.04×10^{-6}		2.99 nM	
2BYB [65]	MAO_B	Deprenyl	−12.1	−10.5	−10.0	3.36×10^{-8}	2.51×10^{-7}	1.11×10^{-7}		7.04 nM	
602P [66]	CFTR	Ivacaftor	−12.8	−13.8	−8.5	1.05×10^{-9}	5.30×10^{-9}	9.77×10^{-5}	6.6 nM		
601 V [66]	CFTR	GLPG1837	−11.9	−12.8	−7.9	2.43×10^{-8}	1.05×10^{-9}	3.69×10^{-5}		~ 0.12 μM	
1FTA [56]	FBP1	AMP	−9.2	−8.5	−6.2	3.04×10^{-6}	9.77×10^{-5}	2.34×10^{-4}	~ 9 μM		~ 1.6 μM
IKZ8 [55]	FBP1	Anilinoquinazoline inhibitors	−13.2	−10.2	−8.8	2.00×10^{-9}	1.54×10^{-7}	1.59×10^{-6}		0.64 μM	
3VF3 [67]	BACE1	NVP-BQQ711 (60p)	−13.9	−10.7	−12.8	6.24×10^{-9}	3.47×10^{-7}	1.05×10^{-9}		250 nM	

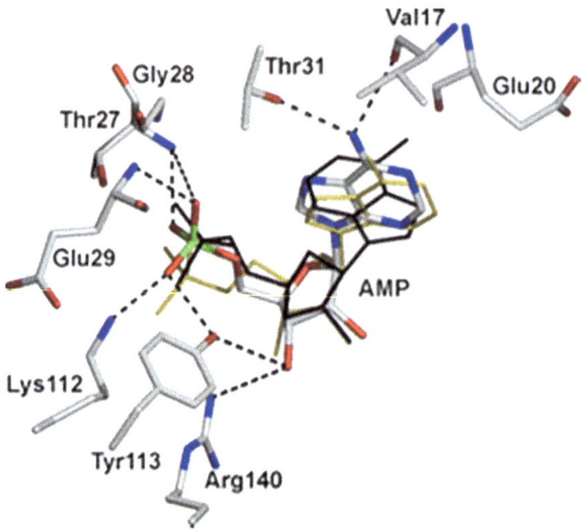

Fig. 3 The crystallographic AMP binding site of human FBPase compared between Autodock Vina and Autodock 4.2.6, respectively. The best performing programs that agreed with actual (experimental) Ki values for AMP as black [68] and Autodock Vina as gold [62] (AMP docked in black) compared to actual crystallographic AMP co-crystallized with the protein in the 1FTA structure (gray) [4, 56]. The RMSD's for the docked AMPs were < 1.5 angstroms squared compared to the crystallographic AMP in 1FTA

binding site with AMP inhibitor, AutoDock 4 and Autodock Vina docked AMP were in close proximity to the actual AMP co-crystallized in the x-ray structure:

Once the most accurate predictive docking program was identified, based on orientation and Ki, the large-scale docking was executed within the VDDP. Immediately after docking the data was harvested from the database and small molecules with affinity scores that produced calculated Ki's exhibiting ~ tenfold or tighter binding (lower) values than natural or known inhibitors were selected to continue on to molecular dynamics protocols. A protein–ligand complex was prepared for Nanoscale Molecular Dynamics (NAMD) [69, 70] by generating the protein–ligand complex protein structure file (psf) using standard topology files for the protein and a Visual Molecular Dynamics (VMD) plugin called Molefacture which allows building parameter and topology files for small molecules in VMD [71, 72]. A water box was generated using molecular dimensions and periodic boundary conditions (PBC) defined as previously described [73, 74]. The protein–ligand complex was ionized to physiological ionic strength (0.150 M) and files then generated for input to NAMD. Molecular dynamics scripts were tested for ΔG calculations utilizing two different approaches to calculate the overall ΔG of the protein first alone then the protein–ligand complex. The ΔG of binding was the difference between these two values and calculated as previously described [75]. Once each ΔG for the undocked positionally randomized potential inhibitor was calculated and deemed insignificant (< 1% of the ΔG of the system), the overall ΔG of the protein alone in the water box was subtracted from the protein/ligand complex ΔG which estimates the ΔG of binding for the protein and the ligand.

Briefly, the two molecular dynamics methods compared the overall ΔG of the system with or without small molecules using thermodynamic integration or free energy perturbation (FEP) [75]. For each molecule both methods were run and a % difference of < 10–15% between the two resultant estimated ΔG's of binding for each molecule was deemed adequate to proceed to laboratory experimental validation (as long as calculated Ki's for the small molecule potential inhibitors were tenfold lower than predicted Ki for AMP). Calculated ΔG of binding values were used in the $\Delta G = -RT\ln K$ equation to calculate the Ki. The calculated Ki for docked AMP was compared with the literature for experimental Ki values to assure the computational methods were sound and a % error of 5% was deemed acceptable. See Table 2 for top 3 ZINC Molecules Predicted Ki's from AutoDock Vina and NAMD. As the data indicates, these molecules are predicted to bind to the AMP binding site with ~10 and 100 fold higher affinity than AMP, the natural allosteric inhibitor of FBPase.

In Fig. 4 the top three scoring molecules from the search of the natural products collection from the VDDP binding to the AMP binding site of human FBPase [56] are shown. Two are piperazine derivatives that bound with ~10 and ~100-fold higher predicted affinity over natural inhibitor AMP and a carboxamide carboline derivative

Table 2 Comparison of predicted binding affinities (kJ/mol) and Ki's (nm) of best ZINC molecules docked to human FBPase with autoDock vina

Ligand	Autodock predicted binding constant (kCal/mol)	Predicted Ki (nM)	NAMD Potential Energy of binding difference (kCal/mol)	NAMD Predicted Ki (nM)
AMP	−7.7	3.33×10^{-6}[a]	−10.7	2.67×10^{-6}
ZINC_1	−10.4	4.69×10^{-8}	−13.2	1.28×10^{-8}
ZINC_2	−9.2	3.29×10^{-7}	−8.80	6.29×10^{-7}
ZINC_3	−9.2	3.29×10^{-7}	−10.2	2.5×10^{-7}

[a]Actual AMP binding constant (Ki) is 1.60×10^{-6}

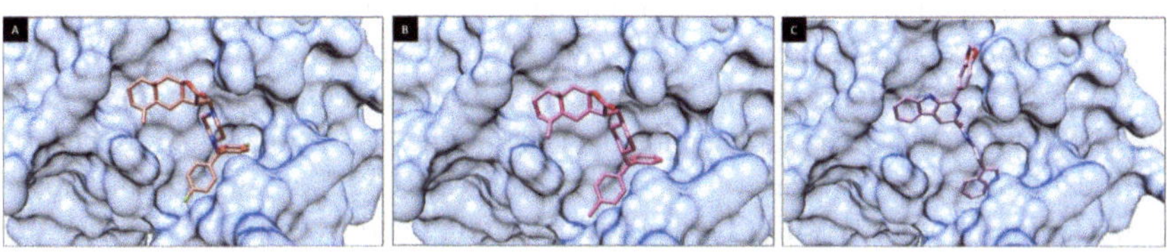

Fig. 4 Surface maps of the AMP binding site of human FBPase PDB code 1FTA [56] VDDP Autodock Vina Docked (a) ZINC_1 (**ID# 70687957**) [3S,3aR,4aS,8aR,9aR)-3-[[4-[bis(4-fluorophenyl)methyl]piperazin-1-yl]methyl]-8a-methyl-5-methylene-3], (b) ZINC_2 (**ID# 20412773**) [3S,3aR,8aR,9aR)-3-[[4-[(R)-(4-chlorophenyl)-phenyl-methyl]piperazin-1-yl]methyl]-5,8a-dimethyl-3,3a, and (c) ZINC_3 (**ZINC02161329**) 1-(1,3-benzodioxol-5-yl)-N-[2-(1H-indol-3-yl)ethyl]-9H-$b-carboline-3-carboxamide molecules

with ~ tenfold lower predicted Ki thanAMP. Notably the fluorophenyl piperazine derivative fluorine atom is predicted to be stabilized by a hydrogen bonding network at the mouth of the AMP binding site (Fig. 4a). It is also striking that all 3 inhibitors completely fill the space of the AMP binding site cavity. Within the natural products catalogue of the VDDP, the next highest predicted affinity inhibitor was only ~2–threefold better than AMP.

The AMP binding site is ~20 angstroms away from the active site. Through hydrogen bonding and hydrophobic interaction networks the AMP binding site connects to the active site. After docking the ZINC molecules with AutoDock Vina in the VDDP, top scoring ZINC/FBPase complexes were then run through NAMD to confirm the predicted Ki's by performing molecular dynamics on the protein/ligand complex. The results for the estimated Ki's calculated from NAMD runs using the overall potential energy as an estimate of the difference of binding energy of protein/ligand complex (subtracting overall potential energy of protein alone) as described above as shown in Table 2.

The active site architecture before and after the binding of the ZINC_1 molecule is significantly different (Fig. 5).

More specifically, in Fig. 5c, the AMP binding site is 460 and 203 Å^3 after ZINC_1 binding indicating the binding site changes shape when in complex with the ZINC molecule. In contrast, the active site is in an open solvent accessible confirmation conformation before ZINC binding at 323 Å^3, but only 224 Å^3 after the ZINC molecule has bound to the allosteric binding site. In the absence of a crystal structure of the FBPase/ZINC complex, the conformations generated from NAMD are compelling and to predict that the ZINC molecules may indeed act as allosteric regulators of the activity of the FBPase, by changing the active site to a closed conformation. In addition, the predicted closed active site conformation upon ZINC_1 binding at the allosteric site, has been shown to prevent successful docking of the substrate fructose 1,6-bis-phosphate into the active site cavity. The predicted change in the active site architecture after ZINC molecule binding to the AMP binding strongly indicative of allosteric inhibition.

Many FBPase/ligand structures have been solved and the data deposited in the Rutgers Consortium for Structural Biology (RCSB) including all three sites on the protein indicated in Fig. 1 (PDBs 1FTA [56], 1KZ8 [55] and 1CNQ [76]), the global conformational changes of the native protein tetramer from the T or inactive to the R or active conformational state have been studied and characterized in detail. [77, 78].

As a result of studying the crystal structures of the FBPase/ligand complexes in the PDB and the many site directed mutagenesis studies of the pig kidney and nearly identical human liver FBPase in the active site, AMP binding site and interfacial residues linking the two sites, the hydrogen bonding and hydrophobic interactions networks have been identified from the AMP binding site to the active site and the tetramer binding site the active site [55, 79–81].

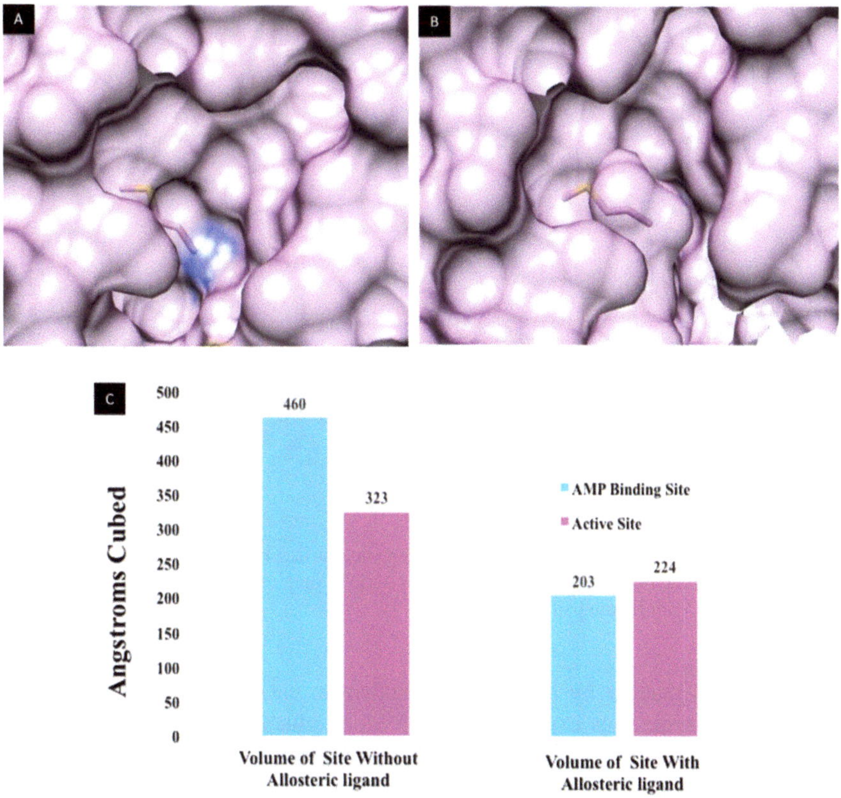

Fig. 5 Comparison of surface maps of human FBPase active site before (A) and after (B) NAMD with ZINC_1 docked in AMP binding site and (C) graph of volume of active site and AMP binding site before and after NAMD

7 Future Studies and New Directions

Now that the molecular underpinnings of specific amino acid interactions are known that lead to allosteric inhibition of FBPase, the identification and development of potent allosteric inhibitors of FBPase continues to be a viable research program for many investigators to pursue. Issues of toxicity and bioavailability currently need to be worked out. In addition, mutations in the FBPase gene of patients with hypoglycemia have been discovered in the clinical area. Two critical residues Met177 and Tyr164 are mutated to alanine in some patients with hypoglycemia. As shown in Fig. 6, Met177 is adjacent to Asp178 and probably stabilizes the conformation of this residue at the interface of the AMP binding site and the active site.

Tyr164 and Asp178 are critical for hydrogen bonding networks connecting the allosteric binding site to the active site. These residues may display a pivotal role in allosteric regulation of FBPase. In addition, it may be that some patients with elevated blood glucose levels (impaired glucose tolerance) or type II diabetes also

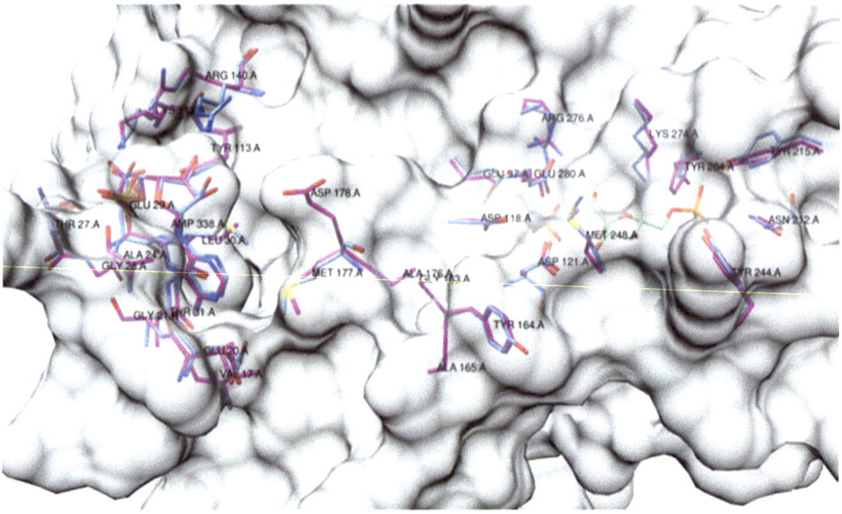

Fig. 6 Connecting the active site to the AMP binding site. Surface map of PDB codes 1FTA/1KZ8 [55, 56] overlay showing crucial conserved residues Met177 (adjacent to Asp178) and Tyr164 that link active site to AMP binding site

have mutations that act by a unique mechanism to uncouple AMP regulation of the FBPase enzyme thus exasperating high blood glucose levels in patients with metabolic disorders [81]. Once mutations in the FBPase gene have been identified in patients with blood glucose disorders, specific drugs that bind tightly to the mutant FBPase present in these patients may be developed to help alleviate their maladies.

References

1. Pagadala, N.S., Syed, K., Tuszynski, J.: Software for molecular docking: a review. Biophys. Rev. **9**(2), 91–102 (2017). https://doi.org/10.1007/s12551-016-0247-1
2. Vieira, T.F., Sousa, S.F.: Comparing autodock and vina in ligand/decoy discrimination for virtual screening. Appl. Sci. **9**(21), 4538 (2019). https://doi.org/10.3390/app9214538
3. Wang, D., Ding, X., Cui, C., et al.: Improving the virtual screening ability of target-specific scoring functions using deep learning methods. Front Pharmacol. **10**, 924 (2019). https://doi.org/10.3389/fphar.2019.00924
4. Keyes, R.M., Pejo, E., Katagiri, K., et al.: Shape Matters: Improving Docking Results by Prior Analysis of Geometric Attributes of Binding Sites. JSM Chem **4**(1), 1020 (2016)
5. Grimsby, J., Chen, K., Wang, L.J., et al.: Human monoamine oxidase A and B genes exhibit identical exon-intron organization. PNAS **88**(9), 3637–3641 (1991). https://doi.org/10.1073/pnas.88.9.3637
6. Youdim, M.B., Bakhle, Y.S.: Monoamine oxidase: isoforms and inhibitors in Parkinson's disease and depressive illness. Br. J. Pharmacol. **147**(1), 287–296 (2006). https://doi.org/10.1038/sj.bjp.0706464

7. Shih, J.C.: Monoamine oxidase isoenzymes: genes, functions and targets for behavior and cancer therapy. J. Neural. Transm. Suppl. **125**(11), 1553–1566 (2018). https://doi.org/10.1007/s00702-018-1927-8
8. Ziegler, C., Wolf, C., Schiele, M.A. et al.: Monoamine oxidase a gene methylation and its role in posttraumatic stress disorder: first evidence from the South Eastern Europe (SEE)-PTSD Study. Int. J. Neuropsychopharmacol. **21**(5), 423–432 (2018). https://doi.org/10.1093/ijnp/pyx111
9. True, L., Coleman, I., Hawley, S., et al.: A molecular correlate to the Gleason grading system for prostate adenocarcinoma. PNAS **103**(29), 10991–10996 (2006). https://doi.org/10.1073/pnas.0603678103
10. Li, J., Yang, X.M., Wang, Y.H., et al.: Monoamine oxidase A suppresses hepatocellular carcinoma metastasis by inhibiting the adrenergic system and its transactivation of EGFR signaling. J. Hepatol. **60**(6), 1225–1234 (2014). https://doi.org/10.1016/j.jhep.2014.02.025
11. Wu, J.B., Shao, C., Li, X., et al.: Monoamine oxidase A mediates prostate tumorigenesis and cancer metastasis. J. Clin. Invest. **124**(7), 2891–2908 (2014). https://doi.org/10.1172/JCI70982
12. Kushal, S., Wang, W., Vaikari, V.P. et al.: Monoamine oxidase A (MAO A) inhibitors decrease glioma progression. Oncotarget **7**(12), 13842 (2016). https://doi.org/10.18632/oncotarget.7283
13. Li, P.C., Siddiqi, I.N., Mottok, A., et al.: Monoamine oxidase A is highly expressed in classical Hodgkin lymphoma. J. Pathol. **243**(2), 220–229 (2017). https://doi.org/10.1002/path.4944
14. Liu, F., Hu, L., Ma, Y., et al.: Increased expression of monoamine oxidase A is associated with epithelial to mesenchymal transition and clinicopathological features in non-small cell lung cancer. Oncol Lett. **15**(3), 3245–3251 (2018). https://doi.org/10.3892/ol.2017.7683
15. Lee, H.M., Sia, A.P., Li, E., et al.: Monoamine oxidase A is down-regulated in EBV-associated nasopharyngeal carcinoma. Sci. Rep. **10**(1), 1–10 (2020). https://doi.org/10.1038/s41598-020-63150-0
16. Gordon, R.R., Wu, M., Huang, C.Y., et al.: Chemotherapy-induced monoamine oxidase expression in prostate carcinoma functions as a cytoprotective resistance enzyme and associates with clinical outcomes. PLoS ONE **9**(9), e104271 (2014). https://doi.org/10.1371/journal.pone.0104271
17. Xu, S., Adisetiyo, H., Tamura, S., et al.: Dual inhibition of survivin and MAOA synergistically impairs growth of PTEN-negative prostate cancer. Br. J. Cancer **113**(2), 242–251 (2015). https://doi.org/10.1038/bjc.2015.228
18. Wang, K., Luo, J., Yeh, S., et al.: The MAO inhibitors phenelzine and clorgyline revert enzalutamide resistance in castration resistant prostate cancer. Nat. Commun. **11**(1), 1–14 (2020). https://doi.org/10.1038/s41467-020-15396-5
19. Dezsi, L., Vecsei, L.: Monoamine oxidase B inhibitors in Parkinson's disease. CNS Neurol. Disord. Drug. Targets **16**(4), 425–439 (2017). https://doi.org/10.2174/1871527316666170124165222
20. Finberg, J.P., Gillman, K.: Selective inhibitors of monoamine oxidase type B and the "cheese effect". In: International Review of Neurobiology. Vol. 100, pp. 169-190. Academic press, (2011)
21. Fowler, J.S., Logan, J., Volkow, N.D., et al.: Evidence that formulations of the selective MAO-B inhibitor, selegiline, which bypass first-pass metabolism, also inhibit MAO-A in the human brain. Neuropsychopharmacology **40**(3), 650–657 (2015). https://doi.org/10.1038/npp.2014.214
22. Sinha, S., Anderson, J.P., Barbour, R., et al.: Purification and cloning of amyloid precursor protein β-secretase from human brain. Nature **402**(6761), 537–540 (1999). https://doi.org/10.1038/990114
23. Stockley, J.H., O'neill C, : The proteins BACE1 and BACE2 and β-secretase activity in normal and Alzheimer's disease brain. Biochem. Soc. Trans. **35**(3), 574–576 (2007). https://doi.org/10.1042/BST0350574
24. Willem, M., Garratt, A.N., Novak, B., et al.: Control of peripheral nerve myelination by the ß-secretase BACE1. Science **314**(5799), 664–666 (2006). https://doi.org/10.1126/science.1132341

25. Osaki, G., Mitsui, S., Yuri, K.: The distribution of the seizure-related gene 6 (Sez-6) protein during postnatal development of the mouse forebrain suggests multiple functions for this protein: an analysis using a new antibody. Brain Res. **1386**, 58–69 (2011). https://doi.org/10.1016/j.brainres.2011.02.025
26. Pigoni, M., Gunnersen, J.M., Lichtenthaler, S.F.: Seizure-6 proteins highlight BACE1 functions in neurobiology. Oncotarget **8**(5), 7214 (2017). https://doi.org/10.18632/oncotarget.13801
27. Zhao, J., Fu, Y., Yasvoina, M., et al.: β-Site amyloid precursor protein cleaving enzyme 1 levels become elevated in neurons around amyloid plaques: implications for Alzheimer's disease pathogenesis. J. Neurosci. **27**(14), 3639–3649 (2007). https://doi.org/10.1523/JNEUROSCI.4396-06.2007
28. Fukumoto, H., Cheung, B.S., Hyman, B.T. et al.: β-Secretase protein and activity are increased in the neocortex in Alzheimer disease. Arch. Neurol. **59**(9), 1381–1389 (2002). https://doi.org/10.1001/archneur.59.9.1381
29. Yang, L.B., Lindholm, K., Yan, R., et al.: Elevated β-secretase expression and enzymatic activity detected in sporadic Alzheimer disease. Nat Med **9**(1), 3–4 (2003). https://doi.org/10.1038/nm0103-3
30. Hampel, H., Vassar, R., De Strooper, B., et al.: The β-secretase BACE1 in Alzheimer's disease. Biol. Psychiatry (2020). https://doi.org/10.1016/j.biopsych.2020.02.001
31. Gadsby, D.C., Vergani, P., Csanády, L.: The ABC protein turned chloride channel whose failure causes cystic fibrosis. Nature **440**(7083), 477–483 (2006). https://doi.org/10.1038/nature04712
32. Chappe, V., Irvine, T., Liao, J., et al.: Phosphorylation of CFTR by PKA promotes binding of the regulatory domain. EMBOJ **24**(15), 2730–2740 (2005). https://doi.org/10.1038/sj.emboj.7600747
33. Mall, M.A., Hartl, D.: CFTR: cystic fibrosis and beyond. Eur. Respir. J. **44**(4), 1042–1054 (2014). https://doi.org/10.1183/09031936.00228013
34. Bhatt, J.M.: Treatment of pulmonary exacerbations in cystic fibrosis. Eur Respir Rev **22**(129), 205–216 (2013). https://doi.org/10.1183/09059180.00006512
35. Cowley, E.S., Kopf, S.H., LaRiviere, A., et al.: Pediatric cystic fibrosis sputum can be chemically dynamic, anoxic, and extremely reduced due to hydrogen sulfide formation. MBio **6**(4), e00767-e815 (2015). https://doi.org/10.1128/mBio.00767-15
36. Shah, V.S., Meyerholz, D.K., Tang, X.X., et al.: Airway acidification initiates host defense abnormalities in cystic fibrosis mice. Science **351**(6272), 503–507 (2016). https://doi.org/10.1038/s41598-019-42751-4
37. Tang, X.X., Ostedgaard, L.S., Hoegger, M.J., et al.: Acidic pH increases airway surface liquid viscosity in cystic fibrosis. J. Clin. Invest. **126**(3), 879–891 (2016). https://doi.org/10.1172/JCI83922
38. Quon, B.S., Rowe, S.M.: New and emerging targeted therapies for cystic fibrosis. BMJ **352**, i859 (2016). https://doi.org/10.1136/bmj.i859
39. Veit, G., Avramesc, R.G., Chiang, A.N., et al.: From CFTR biology toward combinatorial pharmacotherapy: expanded classification of cystic fibrosis mutations. Mol. Biol. Cell **27**(3), 424–433 (2016). https://doi.org/10.1091/mbc.E14-04-0935
40. Sosnay, P.R., Siklosi, K.R., Van Goor, F., et al.: Defining the disease liability of variants in the cystic fibrosis transmembrane conductance regulator gene. Nat Gene **45**(10), 1160–1167 (2013). https://doi.org/10.1038/ng.2745
41. Gentzsch, M., Mall, M.A.: Ion channel modulators in cystic fibrosis. Chest **154**(2), 383–393 (2018). https://doi.org/10.1016/j.chest.2018.04.036
42. De Boeck, K., Munck, A., Walker, S., et al.: Efficacy and safety of ivacaftor in patients with cystic fibrosis and a non-G551D gating mutation. J. Cyst. Fibros **13**(6), 674–680 (2014). https://doi.org/10.1016/j.jcf.2014.09.005
43. McKone, E.F., Borowitz, D., Drevinek, P., et al.: Long-term safety and efficacy of ivacaftor in patients with cystic fibrosis who have the Gly551Asp-CFTR mutation: a phase 3, open-label extension study (PERSIST). Lancet Respir. Med. **2**(11), 902–910 (2014). https://doi.org/10.1016/S2213-2600(14)70218-8

44. Ratjen, F., Hug, C., Marigowda, G., et al.: Efficacy and safety of lumacaftor and ivacaftor in patients aged 6–11 years with cystic fibrosis homozygous for F508del-CFTR: a randomised, placebo-controlled phase 3 trial. Lancet Respir. Med. **5**(7), 557–567 (2017). https://doi.org/10.1016/S2213-2600(17)30215-1
45. Tillmann, H., Bernhard, D., Eschrich, K.: Fructose-1, 6-bisphosphatase genes in animals. Gene **291**(1–2), 57–66 (2002). https://doi.org/10.1016/s0378-1119(02)00627-3
46. Tillmann, H., Eschrich, K.: Isolation and characterization of an allelic cDNA for human muscle fructose-1, 6-bisphosphatase. Gene **212**(2), 295–304 (1998). https://doi.org/10.1016/s0378-1119(98)00181-4
47. Herzog, B., Walttner-Law, M., Scott, D.K. et al.: Characterization of the human liver fructose-1, 6-bisphosphatase gene promoter. Biochem. J. **351**(2), 385-392 (2000)https://doi.org/10.1042/bj3510385
48. van Poelje, P.D., Potter, S.C., Erion, M.D.: Fructose-1, 6-bisphosphatase inhibitors for reducing excessive endogenous glucose production in type 2 diabetes. In: Diabetes-Perspectives in Drug Therapy, pp. 279–301. Springer, Berlin, Heidelberg (2011)
49. Park, H.J., Jang, H.R., Park, S.Y., et al.: The essential role of fructose-1, 6-bisphosphatase 2 enzyme in thermal homeostasis upon cold stress. Exp Mol Med **52**(3), 485–496 (2020). https://doi.org/10.1038/s12276-020-0402-4
50. Benkovic, S.J., de Maine, M.M.: Mechanism of action of fructose 1, 6-bisphosphatase. Adv. Enzymol. Relat. Areas Mol. Biol. **53**, 45–82 (1982). https://doi.org/10.1042/BSR20190124
51. Pilkis, S.J., El-Maghrabi, M.R., Claus, T.H.: Hormonal regulation of hepatic gluconeogenesis and glycolysis. Annu. Rev. Biochem. **57**(1), 755–783 (1988).https://doi.org/10.1146/annurev.bi.57.070188.003543
52. Ke, H.M., Zhang, Y.P., Lipscomb, W.N.: Crystal structure of fructose-1, 6-bisphosphatase complexed with fructose 6-phosphate AMP and magnesium. PNAS **87**(14), 5243–5247 (1990). https://doi.org/10.1073/pnas.87.14.5243
53. Barciszewski, J., Wisniewski, J., Kolodziejczyk, R., et al.: T-to-R switch of muscle fructose-1, 6-bisphosphatase involves fundamental changes of secondary and quaternary structure. Acta. Cryst. D **72**(4), 536–550 (2016). https://doi.org/10.1107/S2059798316001765
54. Pilkis, S.J., Granner, D.K.: Molecular physiology of the regulation of hepatic gluconeogenesis and glycolysis. Annu. Rev. Physiol. **54**(1), 885–909 (1992). https://doi.org/10.1146/annurev.ph.54.030192.004321
55. Wright, S.W., Carlo, A.A., Carty, M.D., et al.: Anilinoquinazoline inhibitors of fructose 1, 6-bisphosphatase bind at a novel allosteric site: synthesis, in vitro characterization, and X-ray crystallography. J. Med. Chem. **45**(18), 3865–3877 (2002). https://doi.org/10.1021/jm010496a
56. Gidh-Jain, M., Zhang, Y., Van Poelje, P.D. et al.: The allosteric site of human liver fructose-1, 6-bisphosphatase. analysis of six AMP site mutants based on the crystal structure. J. Biol. Chem. **269**(44), 27732–27738 (1994)
57. Erion, M.D., Dang, Q., Reddy, M.R., et al.: Structure-guided design of AMP mimics that inhibit fructose-1, 6-bisphosphatase with high affinity and specificity. J. Am. Chem. Soc. **129**(50), 15480–15490 (2007). https://doi.org/10.1021/ja074869u
58. Song, X.M., Fiedler, M., Galuska, D., et al.: 5-Aminoimidazole-4-carboxamide ribonucleoside treatment improves glucose homeostasis in insulin-resistant diabetic (ob/ob) mice. Diabetologia **45**(1), 56–65 (2002). https://doi.org/10.1007/s125-002-8245-8
59. Erion, M.D., van Poelje, P.D., Dang, Q., et al.: MB06322 (CS-917): A potent and selective inhibitor of fructose 1, 6-bisphosphatase for controlling gluconeogenesis in type 2 diabetes. PNAS **102**(22), 7970–7975 (2005). https://doi.org/10.1073/pnas.0502983102
60. Kaur, R., Dahiya, L., Kumar, M.: Fructose-1, 6-bisphosphatase inhibitors: a new valid approach for management of type 2 diabetes mellitus. Eur. J. Med. Chem. **141**, 473–505 (2017). https://doi.org/10.1016/j.ejmech.2017.09.029
61. Irwin, J.J., Sterling, T., Mysinger, M.M., et al.: ZINC: a free tool to discover chemistry for biology. J. Chem. Inf. Model **52**(7), 1757–1768 (2012). https://doi.org/10.1021/ci3001277
62. Morris, G.M., Huey, R., Lindstrom, W., et al.: Autodock4 and AutoDockTools4: automated docking with selective receptor flexiblity. J. Comput. Chem. **16**, 2785–2791 (2009). https://doi.org/10.1002/jcc.21256

63. Trott, O., Olson, A.J.: AutoDock Vina: improving the speed and accuracy of docking with a new scoring function, efficient optimization and multithreading. J. Comput. Chem. **31**(2), 455–461 (2010). https://doi.org/10.1002/jcc.21334
64. Allen WJ, Balius T, Brozell S et al (2018) DOCK 6.9 Users Manual. University of California, California, 27 November 2018
65. Guang, H.M., Du, G.H.: High-throughput screening for monoamine oxidase-A and monoamine oxidase-B inhibitors using one-step fluorescence assay. Acta. Pharmacol. Sin. **27**(6), 760–766 (2006). https://doi.org/10.1111/j.1745-7254.2006.00336.x
66. Liu, F., Zhang, Z., Levit, A. et al.: Structural identification of a hotspot on CFTR for potentiation. Science **364**, 1184–1188 (2019). https://doi.org/10.1126/science.aaw7611
67. Rueeger, H., Lueoend, R., Rogel, O., et al.: Discovery of cyclic sulfone hydroxyethylamines as potent and selective β-site APP—cleaving enzyme 1 (BACE1) inhibitors: structure-based design and in vivo reduction of amyloid β peptides. J. Med. Chem. **55**, 3364–3386 (2012). https://doi.org/10.1021/jm300069y
68. Huey, R., Morris, G.M., Olson, A.J., et al.: A semiempirical free energy force field with charge-based desolvation. J. Comput. Chem. **28**, 1145–1152 (2007). https://doi.org/10.1002/jcc.20634
69. Kalé, L., Skeel, R., Bhandarkar, M., et al.: NAMD2: Greater scalability for parallel molecular dynamics. J. Comp. Phys. **151**, 283–312 (1999). https://doi.org/10.1006/jcph.1999.6201
70. Phillips, J.C., Braun, R., Wang, W., et al.: Scalable molecular dynamics with NAMD. J. Comput. Chem. **26**(16), 1781–1802 (2005). https://doi.org/10.1002/jcc.20289
71. Humphrey, W., Dalke, A., Schulten, K.: VMD—visual molecular dynamics. J. Molec. Graph. **14**(1), 33–38 (1996). https://doi.org/10.1016/0263-7855(96)00018-5
72. Strumpfer J, Freddolino P, Chipot C (2012) Molefacture: A tutorial to build and edit molecules. Centre National de la Recherche Scientifique, University of Illinois, Urbana–Champaign, 19 March 2012
73. Helms, V., Wade, R.C.: Thermodynamics of water mediating protein-ligand interactions in cytochrome P450cam: a molecular dynamics study. Biophys. J. **69**(3), 810–824 (1995). https://doi.org/10.1016/S0006-3495(95)79955-6
74. Bernardi, M., Bhandarkar, A., Bhatele, E. et al.: Boundary Conditions. In: NAMD User's Guide, Version 2.13. Theoretical and Computational Biophysics Group, Beckman Institute, University of Illinois, 9 November (2018)
75. Gumbart, J., Benoît, R., Chipot, R.C.: Protein: ligand standard binding free energies: A tutorial for alchemical and geometrical transformations. Centre National de la Recherche Scientifique, University of Illinois, Urbana, Champaign 5 November (2018)
76. Choe, J.Y., Poland, B.W., Fromm, H.J., et al.: Role of a dynamic loop in cation activation and allosteric regulation of recombinant porcine fructose-1, 6-bisphosphatase. Biochemistry **37**(33), 11441–11450 (1998). https://doi.org/10.1021/bi981112u
77. McIninch, J.K., Kantrowitz, E.R.: Use of silicate sol-gels to trap the R and T quaternary conformational states of pig kidney fructose-1, 6-bisphosphatase. BBA Protein Struct. Mol. Enzymol. **1547**(2), 320–328 (2001). https://doi.org/10.1016/S0167-4838(01)00203-5
78. Kelley-Loughnane, N., Kantrowitz, E.R.: Binding of AMP to two of the four subunits of pig kidney fructose-1, 6-bisphosphatase induces the allosteric transition. Proteins **44**(3), 255–261 (2001). https://doi.org/10.1002/prot.1090
79. Lu, G., Stec, B., Giroux, E.L., et al.: Evidence for an active T-state pig kidney fructose 1, 6-bisphosphatase: Interface residue Lys-42 is important for allosteric inhibition and AMP cooperativity. Protein Sci. **5**(11), 2333–2342 (1996). https://doi.org/10.1002/pro.5560051120
80. Stec, B., Abraham, R., Giroux, E., et al.: Crystal structures of the active site mutant (Arg-243→ Ala) in the T and R allosteric states of pig kidney fructose-1, 6-bisphosphatase expressed in Escherichia coli. Protein Sci. **5**(8), 1541–1553 (1996). https://doi.org/10.1002/pro.5560050810
81. Topaz, G., Epiter-Smith, V., Robalo, C. et al.: Characterization of recombinant fructose-1, 6-bisphosphatase gene mutations: evidence of inhibition/activation of FBPase protein by gene mutation. Biosci. Rep. **39**(2), BSR20180960 (2019). https://doi.org/10.1042/BSR20180960

Key Aspects for Achieving Hits by Virtual Screening Studies

Leonardo Bruno Federico, Mariana Pegrucci Barcelos, Gulherme Martins Silva, Isaque Antonio Galindo Francischini, Carlton A. Taft, and Carlos Henrique Tomich de Paula da Silva

Abstract Virtual screening studies consists of applying successive filters to large groups of molecules, called virtual libraries, in order to obtain a small number of hits. These hits, after going through activity-proving studies, generally in vitro enzymatic activity assays, can be followed by optimization studies aiming to increase activity and generating lead compounds. Computational approaches used in virtual screening studies are based on previously reported information about ligand or macromolecular receptor structures, and thus are called *Ligand or Structure-Based Drug Design,* respectively. The starting point for a virtual screening study is to obtain and prepare the virtual library to be used. The number of compounds present in virtual libraries, which can reach hundreds of thousands, have different characteristics, and depend on the study to be performed. In order to have a good efficiency of a virtual screening study, the compounds of a library must follow the criteria of representativeness and diversity, since these studies aim not only to obtain compounds with biological activity, but also to broaden the knowledge of different chemical classes of compounds able of interacting with a given target. In this chapter, we will address the fundaments of virtual screening studies, as well as the emergence and how these studies are currently conducted. From the initial choice of strategies that can be adopted, followed by the choice and preparation of databases, techniques that can be adopted, and ending with studies of hits optimization.

L. B. Federico · M. P. Barcelos · I. A. G. Francischini · C. H. T. de P. da Silva (✉)
Faculdade de Ciências Farmacêuticas de Ribeirão Preto, Universidade de São Paulo, Av. do Café, s/n, Ribeirão Preto 14040-903, Brazil
e-mail: tomich@fcfrp.usp.br

G. M. Silva · C. H. T. de P. da Silva
Departamento de Química, Faculdade de Filosofia, Ciências e Letras de Ribeirão Preto, Universidade de São Paulo, Av. Bandeirantes 3900, Ribeirão Preto 14090-901, Brazil

C. A. Taft
Centro Brasileiro de Pesquisas Físicas, Rio de Janeiro, Brasil

F. A. La Porta and C. A. Taft (eds.), *Functional Properties of Advanced Engineering Materials and Biomolecules*, Engineering Materials,
https://doi.org/10.1007/978-3-030-62226-8_16

1 Introduction

The process of developing new drugs is a lengthy and costly activity. For this reason, there is a high added value to the product, which accounted for ca. of 17.5% in estimated profit in the year 2017 in the USA [1]. The cost data for the development of a new drug is not accurate, but the sales from 2000 to 2018 generated an accumulated revenue of USD 11.5 trillion for the 35 major pharmaceutical companies and a gross profit of USD 8.6 trillions [2]. Kiu Tay-Teo and collaborators studied the income from sales of cancer drugs, comparing research and development costs, and concluded that there is an average of accumulated sales revenue of USD 14.50 per dollar invested in R&D, which is obtained in an average time return of 5 years [3]. Experts estimate that for a new drug to be launched on the market there is financing in the range of billions of dollars, about USD 2.56 billion [4, 5]. These expenses account for failures and advertising, invested over 10 to 15 years, [6, 7] and there are still those who say that this process requires an even longer period ranging from 12 to 24 years [8].

Unlike costs, the trajectory for the development of new pharmacological entities is well defined. As such, this starts with a medical need and moves through the pre-clinical studies phase, determined by the discovery of new chemical entities, which are led to the development phase, comprised of clinical studies until the drug is approved. More precisely, the FDA segments the drug development process in 5 steps: (1) Discovery and Development, (2) Preclinical Research, (3) Clinical Research, (4) FDA Review and (5) FDA Post-Market Safety Monitoring [9].

The first years of research by new pharmaceutical entities (Discovery and Development) are aimed at the discovery phase of bioactive molecules, chemical substances called hits, which are the starting point for the development of new substances with potential biological activities. This process usually starts with new ideas that aim to interrupt or reverse the effects of a disease. As such, it is necessary to have prior knowledge of the biochemistry of the disease and to determine a target to be studied, which is validated to combat the pathology, called biomacromolecules that are usually proteins, enzymes, ion channels or DNA / RNA. After choosing the target, one may seek to identify small molecules capable of modulating their action, which ultimately will have this capacity evaluated by in vitro studies and later, depending on the inhibition rate, by in vivo studies. If the inhibition rate reached is not satisfactory, but the molecule under study is promising, additional studies on drug optimization can be developed, aiming on increasing the activity of these molecules [8, 10–12] Molecular optimization studies can still be designed at later stages, whereby improvements in the pharmacokinetic characteristics of the compounds may be necessary.

The identification of hits usually occurs through real or virtual screenings. Real screening, commonly called high-throughput screening (HTS), was introduced by the pharmaceutical industry in the 1990s. This drug development technique is a scientific experimentation methodology that involves a wide variety of chemical libraries in addition to testing biological techniques, process engineering platforms, and computer procedures to track and analyze the results in order to eliminate the intrinsic noises to the method and elucidate complex biological processes. Thus, large libraries of chemical compounds can be analyzed against specific biological targets through the use of automation, miniaturized assays, and large-scale data analysis. [13, 14].

On the other hand, virtual or in silico methods that mean "computer-aided", are methods that have been widely, and continue to be, driven by the advancement of computational technology. These drug development methods are able to reduce the time and funding spent on traditional screenings, in addition to using a greater diversity and quantity of compounds, present in libraries, which can currently reach up to hundreds of millions. Finally, these selected compounds can be purchased commercially and tested in vitro to confirm the required activity.

The use of computational chemistry in drug development is already a well established process present in the largest pharmaceutical industries around the world [15]. As an example of the importance of using computational techniques in the development of drugs, in 2015, five out of twenty drugs that were undergoing pre-clinical trials at the pharmaceutical giant Bayer, were only possible to be discovered through the use of computational chemistry. We can observe the growing consensus within large companies of the importance given to the process of development and improvement of new drugs through computational chemistry and that this activity is essential in the current scenario of drug development [16].

The growing importance of using computational methods in the process of drug development and optimization is due to the ability of these methods to allow researchers to obtain as much information as possible about a biological target, such as conformational, topographic data and interaction sites, among others. In addition to information about ligands, such as the locations and modes of interactions with a macromolecular target under study and pharmacokinetic and toxicological predictions [17].

The development and planning of drugs through computational chemistry, commonly referred to by the term Computer-Aided Drug Desing (CADD) has achieved its expansion and popularization thanks to advances in computer science. The increase in the processing and storage capacity of current computers was essential for the development and execution of the new software used for this activity. In addition to these computational advances, the increase in chemical and experimental data, such as the number of databases containing chemical compounds with biological activity and protein structures resolved by crystallography studies, are essential for the quality of the information obtained [18, 19].

Currently, CADD has been used for the development of compounds that have potential biological activity (hits compounds) and in the optimization of these compounds (leading compounds), which aim to improve the pharmacokinetic and pharmacodynamic properties (ADMET) [20]. However, there are limiting factors in the use of CADD. The process and methodology adopted will depend on the amount of information and data on a specific pathology to be studied, the work dynamics of each researcher, the power of the computational apparatus of the research center, and the time available.

Currently, CADD studies are divided into two approaches that can be adopted individually or together called structure or ligand-based drug design (SBDD and

LBDD, respectively). SBDD studies may be carried out in view of the existence of data regarding the therapeutic target under investigation, which can be elucidated by means of X-ray crystallography, nuclear magnetic resonance imaging [21], and the Protein Data Bank (PDB) is deposited in the protein structure repository [22] or even by studies of structural protein homology [23]. The use of crystallographic structures is the most common choice. From it, it is possible to identify the target's interactions with the ligand, through the drug-protein complex, allowing this information to be used in virtual structure-based screening (SBVS), in addition to docking simulation studies [24]. In conjunction with SBDD, techniques can be applied to better elucidate the behavior of the drug-receptor complex, such as molecular dynamics, which allows assessing the conformational changes that will be caused by the interaction among the target protein and the drug under study, thus being possible optimize such interactions [25].

If the structural data are not available, it is possible to use techniques that are based on previous knowledge of ligands with activity already reported. These techniques are known by the term development of drugs based on ligand (of (ligand based drug design; LBDD). These techniques mainly address studies of pharmacophoric patterns (pharmacophore), studies of similarity (shape and electrostatics) and studies related to the quantitative structure–activity relationship (QSAR) [26, 27]. In addition to being used in the early stages of drug development, LBDD techniques are also commonly used in the optimization process of hits in leading compounds or in the optimization of existing leaders, for which energy optimization studies, bioisosteric modifications are common and calculation of electronic, geometric and physicochemical properties [28].

The use of CADD techniques can be approached from three different perspectives: (1) Perform the identification of hits with potential biological activity in extensive databases of chemical compounds using LBDD or SBDD techniques, called virtual screens (VS) that when based in ligands or the structure of the receptor they are named LBVS and SBVS, respectively; (2) Optimization of leading compounds [29]; (3) New development, approach based on the construction of new molecules from functional groups found in key positions within the active site of a specific target, which may be fragmented or not [30].

Therefore, given the versatility of CADD techniques, we intend in this chapter to highlight the main virtual screening techniques given the importance of these studies in the current scenario. We aim to clarify the concepts adopted by the main existing VS techniques, in order to assist the reader since the initial choice of strategies that can be adopted in a VS study, as well as to make this study more efficient with the preparation of databases and validations.

1.1 Virtual Screening Studies

Several researchers, when describing virtual screening studies, in order to exemplify and facilitate the understanding of this complex area, make an analogy with the expression “look for a needle in a haystack”. According to them, the haystack would be the representation of vast chemical space, and the needle would be the desired chemical structure, which would become a medicine in the future. However, even for a didactic representation, this analogy oversimplifies virtual screening tasks and corresponding difficulty in finding an active molecule.

In order to better illustrate the difficulty and conduct of VS studies, we can make another analogy. Imagine a seamstress in her craft and she needs a specific button, which can be found in a drawer that contains tens of millions or even hundreds of millions of buttons. What is the most efficient way for this seamstress to find such a button? In fact, does she really need a button exactly as she imagines or does a button with similar characteristics already satisfy her need? Therefore, an efficient way for her to find such a button would be to use filters in her button drawer, initially removing the very large and the tiny ones, later with different shapes or colors to the intended one, and so on.

Now, imagine that the button box is a database of chemical compounds that can contain hundreds of millions of compounds, and that in this database there is not only one molecule of interest, that is, that for a given study there may be some molecules that can inhibit a pharmacological target. What is the best approach to find these compounds? Applying successive drug discovery techniques. So, didactically, we can refer to VS studies as successive filters applied to large groups of molecules (drug library) in order to obtain, at the end, a small amount of hits that can later be optimized and become leading compounds [31].

The filters in question are LBDD and SBDD techniques and each filter adopted must contain a specific methodology in order to minimize the number of compounds present, seeking the selection of the most viable molecules [32]. Additionally, worth mentioning that there is no rule to be followed for the application of filters, and only one technique, or more than one, can be applied depending on the previous data, the computational capacity of the study center, the time required, or even the experience of the researcher.

In VS studies, it is most convenient to use methods with less computational detachment in the first stages of screening, such as similarity studies, for example, and the more robust methodologies that require more processing capacity are for the final stages of the study, in which they will already have a reduced number of compounds. At the end of the screening stages, after selecting the hits, they can have their activities confirmed by enzyme inhibition assays and, if necessary, the selected hits can be subjected to optimization studies foreseeing the increase in power and efficiency. Additional studies of molecular dynamics and QSAR studies (depending on the compounds tested) are great suggestions for understanding the mode of action and physicochemical characteristics important for the activity of the compounds. A suggestion of virtual screening studies and optimization studies can be seen in Fig. 1.

Filters adopted in VS studies can be applied in different ways, being called sequential VS, parallel VS and integrated VS. In sequential methods, the different approaches are adopted, as so referenced, in sequential forms, where the output of one filter becomes the input of the next, so it is possible to use methodologies with low computational detachment at the beginning, and end with more efficient robust

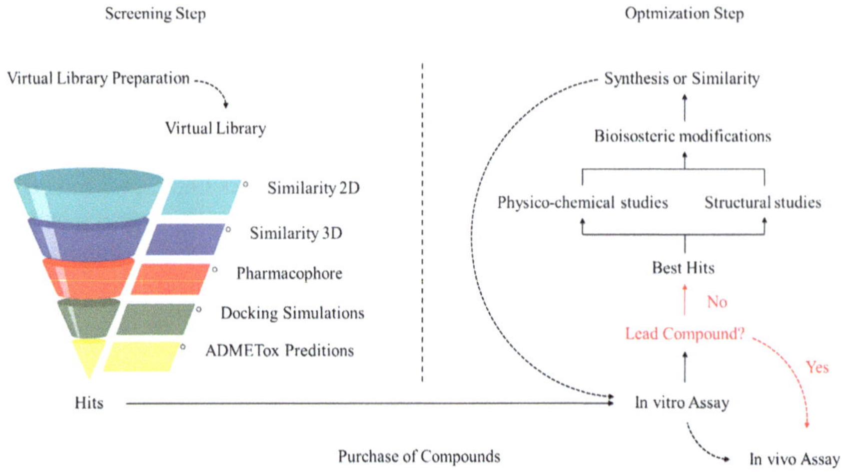

Fig. 1 Schematic representation of a virtual screening study, in which examples of screening can be observed, as well as examples of hit optimization studies aiming to obtain lead compounds

methods. In the parallel VS methodology, several methods are executed independently and the final results are analyzed. Finally, integrated VS methods are methods developed in conjunction with experimental HTS techniques, for example [33–35].

In addition to identifying new inhibitors, VS studies have the property of investigating and expanding the chemical diversity of known ligands. These studies have become increasingly attractive in recent decades due to the ability to investigate billions of compounds contained in a single database, such as the database Drug Like of the company ZINC15 [36, 37].

Currently, the need for supercomputers, clusters or a computer lab with a large processing capacity, desirable for the use of extensive databases or long simulations, is a panorama that has been changing. Today, some researchers in the field of drug development can opt for computer services available in high performance systems (HPC) in the cloud, offered at low costs, which greatly reduce the costs of infrastructure and maintenance of these centers processing [38–40]. In addition to this option, which will be discussed in more detail in the final sub item of this chapter, there are still computer programs available on online servers, as many examples mentioned in the Click2Drug (www.click2drug.org) and VLS3D (www.vls3d.com) repositories [41, 42].

In the following sections of this chapter, we will cover some of the main and most used techniques in virtual screening studies, as well as some processes that are extremely important to the development of an efficient study in the area.

2 Choice and Preparation of Databases

After determining the strategies that will be adopted in the virtual screening process, the common starting point for any study consists of choosing and preparing the databases. This process is extremely important, since the final results will depend on the appropriate choice of the compounds under study and they will directly influence the final result of the same [43]. The size of the library to be used will depend on the strategy and the computational apparatus available, reaching hundreds of millions of compounds, as in the current compilation of bases made by ZINC15 [37], a relevant one used in VS studies, in addition to others compiled bases made available by PubChem [44], NCI [44], eMolecules (https://www.emolecules.com/), ChemSpider (https://www.chemspider.com/), ChEMBL (https://www.ebi.ac.uk/chembl/), or bases made available by the Enamine suppliers themselves (https://enamine.net/), and MolPort (https://www.molport.com/), for example. It is worth mentioning that these mentioned bases have, in their entirety, compounds that are commercially available, which, after acquisition, can now proceed directly to in vitro studies.

The number of compounds contained in each of the currently available bases is not relevant at this time, as these values are updated periodically. It is noteworthy that there are various bases available, Gopal Pawar and his collaborators identified in a study 80 databases of chemical structures available for drug-design [45], some bookstores being able to reach hundreds of thousands, or even hundreds of millions, of compounds.

There are studies that relate the influence of the size of the database to its effectiveness, so that the success rate is proportional to the chemical space covered [46] and suggest that applicability, efficiency and diversity are dependent on the extent of such bases [45]. The total chemical space, which covers all possible small organic molecules (about 500 Daltons) is estimated at 10^{60}, [47] which makes it impossible, with current computational resources, to apply any methodology capable of exploring this number of compounds.

On the other hand, considering that the total number of small molecules present in our organism, called "biologically relevant chemical space", it presents only a few thousand compounds and mediates all reactions that occur in our organism [48], in such a manner that we notice that approaching as close as possible to 10^{60} it is not a task of extreme necessity. Thus, the vast chemical space can be restricted to libraries of specific and accessible compounds [49], determined by studies of refinements in existing libraries [50], or previous studies for the development of bases that can direct them to a specific target, as in the case of study of combinatorial analyzes for the development of libraries for specific targets or studies [51]. Thus, the library will not be as extensive, but in all cases the base under study must ensure that the structures meet criteria of diversity and representativeness [52], that is, the bases need a wide variety of compounds that must not only diverge in small portions of the structure as radical changes, these must also have a wide variation in their scaffolds [53].

According to the study in progress, bases can be previously prepared and there are computer programs to meet this demand. Many bases due to the size can be made available only in 2D format, or in extensions that are devoid of formal loads or, depending on the sorting to be performed, it may be necessary to have a database that presents a library of conformers, as they can be used primarily by programs that do not perform the step of generating conformers. It may still be necessary to correct the size of the bonds and the dihedral angles, check the presence of tautomers and protonation states; additionally, programs that standardize the database can be used, eliminating molecules outside drug-like or lead-like profiles, as per example excessive number of rotatable bonds, heavy atoms or even carbon atoms [54, 55].

In addition to correction methods, another preparation method for the bases can be applied with the intention of previously reducing the compounds present with undesirable characteristics. Some physicochemical properties may be irrelevant to the study in question or premature withdrawal of compounds with toxic, reactive and undesirable groups, or even pan-assay interference compounds (PAINS) [56].

The use of specific bases may also be desired or indicated. There are bases previously filtered for studies of Fragment Based Drug Design (FBDD), bases with drug-like or lead-likes compounds, bases pre-filtered by the rule of 5 of Lipinsky [57, 58] or rule of 3 [59], for screening compounds used as pesticides, [60] specific bases such as those of natural products or metabolites, drugs approved by the FDA or other global regulatory agencies, used in studies of drug repositioning, compounds in the testing phase and still specific to certain pathologies or targets.

As stated earlier, success in virtual screening studies does not depend directly on the use of supercomputers, robust methods or the use of bases with hundreds of millions of compounds. These factors can, and do, add significance to the study, however, the correct choice of the applicable methods and their parameterization, in addition to the database to be used, as well as the validation of the methods before starting the screening studies itself, are fundamental to the success rate. Therefore, even before we approach the main VS methodologies, we will describe some of the main validation methods in VS.

3 Validation Methods Applied in VS

Even before starting the virtual screening studies, it is advisable to validate the methods adopted in the study of interest. These validation studies are necessary to assess the ability of these methods to distinguish active from inactive ligands, and in the case of SBDD whether the method is capable of reproducing the crystallographic pose of a ligand in a complex. These previous studies generate reliability to the work to be developed and guarantee that they are able to obtain adequate performances, that is, that the adopted parameters are adjusted to the study and that they are efficient in the identification / recovery of the active ligands against a diverse base and whether the pose generated in an SBDD study is reliable.

Recovery of active ligands

Bearing in mind that virtual screening studies aim to identify hits in the middle of a large set of different molecules, it is advisable to check if the methodology to be used for in a given study is able to distinguish ligands with and without activity. One way to carry out this verification, and evaluate the performance of the method to be used, is to develop a previous study of recovery of active ligands. This validation study is nothing more than a simple screening using a database with information on reported biological activity, composed of active and inactive ligands, thus, it is possible to analyze whether the adopted method is able to recover the ligands with activity, that is, if the ranking generated by the program is able to score active ligands better than inactive ones.

In order to develop a more rigorous evaluation of the performance of a given method, the base used may contain, in addition to the active compounds, ligands with great similarity to these, called decoys. Decoys are molecules compatible in terms of physicochemical properties, such as molecular weight, logP, number of rotating bonds and hydrogen bond acceptor donors, in relation to the active ligands. This base that will be used to validate a VS study can be developed by the online server Directory of Useful Decoys–Enhanced (DUD-E), which is capable of generating 50 decoys for each added asset [61], or DecoyFinder [62]. These sets of decoys are commonly used to validate docking simulations, LBVS methodologies, or to compare / evaluate functions with different scores (in the same program, or not).

After the generation of a subset of molecules, this may be used in the screening study that one wants to evaluate, either by methods based on ligands or on structure. As such, it is possible to assess the recovery capacity of these ligands by analyzing the scores projected on enrichment factor (EFs), receiver operating characteristic (ROC), the area under the curve (AUC) [63], among other forecast models and metrics [64].

One of the most used metrics, the ROC curve, is applied in several and different areas of knowledge, in which it is intended to evaluate the rate of true unanimous and false positives in an event of binary occurrence, that is, if the expected positive event was really positive and not false positive, and if the expected negative event was really negative and not false negative. For this, the expected events are represented by a graph correlating true positive fractions, called the sensitivity axis (y-axis), with false positive fractions, called the specificity axis (x-axis); where a curve is expected (optimal event) passing through the upper left corner of the graph, where the true positive fraction is 1 (perfect sensitivity) and the false negative fraction is 0 (perfect specificity).

Still in relation to the obtained ROC graph, the area under the curve (AUC) that determines the accuracy of the method can be calculated, by the probability that a randomly chosen active compound has a higher score than a randomly chosen inactive compound, or by the factor enrichment (EF) fraction of active compounds obtained when a certain percentage of inactive compounds is found [63, 64].

The development of a ROC, AUC and EF curve study can be used for any virtual screening event, including similarity, pharmacophore, docking or QSAR, in which it is intended to assess the system's ability to recover active ligands. The process starts by creating a simple table where the expected and observed results are placed, that is, if the compound that is known to be active was well ranked (true positive) and if the compound that is known to be inactive was badly ranked (true negative). Still, there will be compounds that are known to be active and have been poorly rated (false negatives) and compounds that are known to be negative and have been well ranked (false positives). A threshold will define what will be called "good" or "bad" ranked, and from this, the false positive fractions (False positive = False Positive / True Negative) x-axis and the positive fractions (True) Positive = True Positive / False Negative) y-axis. Changing the threshold will generate several points and thus determine the curve.

Redocking and Cross-Docking

Despite the increasing use of molecular docking in the discovery of new drugs, some problems to determine the success of such involved algorithms still occur [61, 65]. Some reasons to these failures can be mentioned, such as the reconstruction of the known position of a ligand [66, 67], the accuracy of the binding affinity capable of differentiating the ligands from the decoys [61, 66, 67] or even the identification of the receptor when ligands are coupled to different conformations of that same protein [68].

Several studies have shown that one of the most influential factors in the success of molecular docking is the choice of the ideal receptor/protein structure to be used as a reference [69–71]. To solve the flaws presented in the success of an algorithm, two techniques can be used: redocking and cross-docking.

Redocking consists of the coupling of a ligand extracted from a resolved 3D structure and embedded in that same structure. Cross-docking, on the other hand, consists of extracting a ligand from a 3D structure and fitting it into another shape of the same receptor [68]. In both cases, success is relative to the mean square root deviation (RMSD) between the embedded pose and the respective complex formed [72]. Therefore, cases in which redocking is used are easier to be solved since the structure of the receptor is in its ideal shape. The cases involving cross-docking are more challenging and simulate the type of link predictions to which the algorithms are intended [68].

Fischer randomization for pharmacophore validation

Fischer's randomization is a validation methodology often used in studies of the development of pharmacophoric patterns. This methodology aims to verify the existence of a strong correlation between the chemical structure of the selected molecule and its biological activity, and is developed based on the random reassignment of the activity data of a set of compounds with known biological activity. Several hypotheses of pharmacophores are generated randomly using the same parameters used to develop the original hypothesis and this is confirmed apt if the correlation and cost coefficients are higher in it, than the pharmacophoric models generated randomly [73–76].

4 Approaches Used in Virtual Screening

4.1 Similarity Studies

Molecular similarity: 2D structure

A widely used approach in VS studies, not only, but mainly when there are no resolved receptor structures, are the ligand-based methodologies, LBVS. These methodologies are linked to previously reported activity data and are developed from molecularly similar aspects between the ligands under study and ligands with known activity [53, 77]. Therefore, being based on the structure of the ligand, it allows the possibility of establishing a relationship between structural similarity and its biological activity, as already proposed by Koshland in 1958 [78].

LBVS studies can be developed from structural similarity data, and thus carry out studies of proposals and analyzes of pharmacophores, or of quantitative structure–activity relationships (QSAR), which allow the possibility of analyzing the molecular determinants for activity of compounds, among them, the specific arrangement of functional groups or even physicochemical properties. Molecular similarity analyzes derive from the concept / term ‘molecular similarity property’ (SPP), which employs a general molecular view of the structure of compounds, being directly related to the concept of structure–activity relationship (SAR), and ligand-based VS (LBVS) approaches [79].

The SPP concept is based on the theory called “key-lock” proposed by Fischer, [80] in 1894. Fischer was the first researcher to relate the complementarity between the ligand and its receptor, which was later, even better, exemplified by Koshland in 1958 [78] suggesting the “induced fit” theory. In fact, such theories help us until current days, since they allow to propose the correspondence between similarity and biological activity between molecules. Thus, a simple LBVS strategy can be applied based only on the calculation of similarity between a reference structure, with known activity, and other molecules present in a database, resulting in the method known as VS based on ‘2D fingerprint’. This method uses only the “fingerprint” that each molecule has in its structure, and thus, it makes it a simple, fast technique, with low computational detachment and widely used in the initial stages of the VS process [81, 82].

There are three important components related to similarity measures that need to be considered in an LBVS study based on structural similarity: the structural representation of molecules, the weighting scheme that is used to determine values of relative importance of structural representation, and the similarity coefficient [82].

In order to define structural representations in 2D, the concept of “fingerprints” was developed, which makes it possible the computational analysis of the complex and subjective comparison by similarity. Fingerprints are representations of the molecule in the form of bits, which in turn can be analyzed objectively by computational methods. There are several methods for generating 2D fingerprints, among them, the two most used are the one based on key substructures in the molecule, and the topological or based on the molecular path.

The substructure-based method uses the 10-bit format. This method assigns 1 bit for parts of molecule that contains specifics functional groups, and 0 bits for parts of the molecule that do not contain functional groups. Thus, a fingerprint in 10 bits is obtained for each molecule, based on the structure of the compound under analysis. The topological method also makes use of 10 bits, but it consists of performing the analysis of all fragments of a molecule. This method starts from a specific point, usually a functional group, and follows a linear path in its structure, forming fragments according to the number of pre-established connections, in this way, fragments are generated according to the analyzed path and the related bits are analyzed [83].

After the structural representation and weighting steps, to perform the structural similarity analysis, it is necessary to choose a mathematical method that makes these results obtained into quantitative results. There are several coefficients that can be used for this analysis, the most consensual being the Tanimoto coefficient which is defined by the following equation:

$$\mathrm{TC} = \frac{\mathrm{NAB}}{\mathrm{NA} + \mathrm{NB} - \mathrm{NAB}}$$

Therefore, NA corresponds to the number of characteristics present in molecule A, (bits = 1), NB corresponds to the number of characteristics present in molecule B (bits = 1), and NAB being the number of characteristics shared by the two molecules under analysis (bits = 1). Thus, the Tanimoto coefficient has its value calculated between 0 and 1, with 0 corresponding to a structure that has no similarity with the reference molecule, and 1 corresponds to a greater similarity between the analyzed molecule and the reference molecule [81, 82, 84–86].

Therefore, in spite of a simpler methodology, 2D fingerprint studies have some peculiarities, for which this method still remains in use even after the emergence of more robust methods. Among them, we can highlight the low need for computational processing to perform the screenings and their rapid development, in addition to an easy execution and the need for a minimum amount of structural information, which makes this method still in use in studies concerning planning and development of drugs [81, 83, 87].

3D Similarity Studies–Shape

Methodologies that address the structural similarity between ligands with reported activity and different structures present in databases are frequently used in LBVS studies. These studies stand out for presenting a high efficiency and effectiveness, mainly when using data of two and three dimensions (2D and 3D) [88]. Processing capacity, memory and time are factors that must be taken into account when choosing the method used, as the amount of information used in a given study is directly proportional to their consumption. Thus, simpler methodologies like those based on 2D, omit information such as the spatial conformation of the compounds under study and require low computational processing and storage capacity. 3D studies, when using this information, require greater computational performance [89].

The increase in amount of information present in similarity studies has occurred due to 3D similarity methods. From structural and mechanistic information of molecules, absent in two-dimensional methods, the diversification of similarity techniques in LBVS studies improved by incorporating pharmacophore modeling, 3D fingerprint, shape similarity and electrostatic similarity, among others [90]. These studies take into account the conformation and volume of molecules, and are based on the assumption that molecules that have similar shapes to an active molecule will also have a great propensity to bind in a similar way in the same receptor [78].

Comparison of structural similarity between molecules can be performed by methods that use 3D information, and such methods employ approaches that may, or may not, use structural alignment or overlap between the molecules. Both methods allow the comparison between molecular surface properties, such as hydrophobicity and polarity, to be determined satisfactorily. However, methods that do not require overlap are, to a large extent, faster methods and they are generally used in VS for large libraries of compounds. On the other hand, those that perform alignment or overlap are highly effective methods for determining similarity, but they are computationally more expensive methods, as they require more computational processing and storage [90].

Approaches that do not use structural overlap to perform similarity calculation between molecules, use methods that are based on atomic distance. These methods are often used in similarity searches based on shape, due to their ability to perform scans at high speed, which allows the use of this methodology in extensive databases, in addition to their use as pre-screening [91, 92].

The methodologies based on atomic distance make use of geometric descriptors and, thus, describe the shape of a molecule through the relative position of the corresponding atoms. Therefore, similarity is calculated from the comparison of the distribution corresponding to atomic distances. For performing a similarity calculation based on atomic distance, each molecule is considered to be a collection of 3 triatomic sub molecules, and each molecule is capable of generating between 300 and 500 atomic triplets. The atomic triplet triangles the distance perimeter with the other two sub molecules, and in this way, comparison histograms are generated, which are used in the comparison of similarity [90, 93].

Another approach, frequently used in LBVS studies for calculating shape similarity without using structural overlap, is the USR methodology (Ultra Speed Recognition) [94–96]. USR is based on the method of descriptors based on atomic distance and is capable of calculating the molecular form from the atomic distance distribution in four reference positions: the molecular centroid being the atom closest to the centroid, the atom furthest from the centroid and the most distant atom to the furthest atom from the centroid. These positions represent the center of the molecule, and its extremes, and thus, these measures allow the shape of the molecule to be calculated, due to the understanding that the number of atomic distances is proportional to the number of atoms in the molecule [90–92, 97].

A method often used in shape similarity studies that uses structural alignment or overlap to calculate similarity is the atom-based Gaussian form method. This method is based on the molecular volume of the structure to perform the calculation of shape similarity, because it is based on the assumption that two molecules will have similar shapes if they have similar volumes. Gaussian method is used regularly in VS methodologies in large databases even though it is not as fast as the methods based on atomic distance. The wide use of this method in LBVS studies is, due to the fact, that it allows the graphic visualization of the similarity between the molecules under study. This functionality makes it extremely useful for understanding SAR and optimizing structures [90].

The two most adopted models for representing the volumes used in the methods that use the atom-based Gaussian form are those of Gaussian sphere and hard sphere. The hard sphere model represents the molecule as a set of spheres, each corresponding to an atom of the molecule, and for this purpose, use the size relative to the Van de Walls radius. In this way, the volume of the molecule can be obtained from calculations that correlate the number of sets of spheres that make up the structure and its interactions. Gaussian model of spherical representation, on the other hand, presents the molecule using a set of overlapping Gaussian spheres, and the calculation of the volume of the structure is carried out from the integral volume of all overlapping spheres [90, 98].

Similarity methods, described above, are often used in LBVS approaches. Similarities in shape are procedures usually performed in the first stages of screening, but the literature also reports their use in single screenings, with performance compared to SBVS studies [99, 100]. In fact, these methods have their own characteristics, which allow them to be widely used for scans in extensive databases, such as the fact that they do not need a large amount of computational processing, in addition to not requiring a large storage capacity. Thus, similarity methods become increasingly versatile, which depending on the study, may have their results used later in other approaches to VS, such as electrostatic similarity or pharmacophore, for example.

3D Similarity Studies - Electrostatic potential

As mentioned in the previous items of this chapter, the formation of the drug-receptor complex is a dynamic process, whereby the structure of the ligand and the macromolecular target interact in order to form a stable and low-energy complex. For the formation of the complex, the ligand and the residues of the receptor interaction site adopt conformations that aim to lower the energy of this system, thus assuming conformations that can efficiently direct the functional groups present in their structures. Therefore, not only the shape of the ligand is important for interaction with the receptor, but also, how its functional groups are arranged, that is, how the electrostatic potential is distributed over their surface.

In this way, similarly that the form and shape of ligands are explored in virtual screening studies, the possibility of identifying potential maps quickly and effectively makes this measure an essential tool for the development of new drugs in virtual screening studies. This study, called electrostatic similarity, is developed from obtaining the value of molecular electrostatic potential (MEP), a value that makes it possible to compare two different molecules, based on their electronic properties [101, 102].

An efficient possibility of exploring similarity in LBVS studies is the association of similarities by shape and by electrostatic potential as proposed by the company OpenEye, through the programs ROCS and EON [103–108]. ROCS software is used initially to perform a search for similarity in 3D shape and it is based on the shape-based overlay method. This program performs the alignment of the structures by maximizing the volume of overlap between a reference structure used as query and all structures/conformers contained in the analyzed database. The structural similarity of shape between two molecules is calculated by Tanimoto indices [104–106].

The molecules obtained from the ROCS that present greater shape similarity to the reference molecule are used as input in the EON program, which is used to calculate the electrostatic similarity of these molecules, using the electrostatic coefficient of Tanimoto (ET) defined by the equation:

$$\mathrm{ET} = \frac{A * A}{A * A + B * B - A * B}$$

where A is the electrostatic potential around the reference molecule and B is the electrostatic potential around the molecule under analysis.

The electrostatic potential of the molecules is calculated internally in the software, through the Zap program, which solves the Poisson-Boltzmann equation, calculating new partial charges using the MMFF94 force field in addition to having the ability to establish a neutral pH for the molecules in analyze. Finally, EON classifies the molecules by means of the ET value, which varies between -0.3 and 1, being -0.3 those molecules that have similar potential but opposite, and at 1, those that present same electrostatic potential [103–106].

4.2 *Pharmacophore Studies*

The first concept of pharmacophore formulated in the literature was suggested in 1909 by the German bacteriologist Paul Ehrlrich. Ehrlrich stated that pharmacophore is a molecular structure, composed of certain structural patterns that are fundamental to the biological activity in a given molecule. In short, the etymology of the word pharmacophore derives from the terms *phoros*, related to molecular structure, and *pharmacon*, related to the essential molecular patterns for biological activity [109]. A century after the first elucidation of a pharmacophore model, we can see that the current concept has not undergone many changes but there has been an expansion as to its meaning and its applications. Currently, according to IUPAC (International Union of Pure and Applied Chemistry), a pharmacophoric model is “a set of steric and electronic characteristics, which are necessary to guarantee the ideal molecular interaction with a specific biological target, to be able to activate and/or block the biological response of that particular target" [109, 110].

In other words, a pharmacophoric model is a hypothesis obtained through a group of molecules, which share common aspects and interact with the same target. Such characteristics may be based on atoms present in the structure of molecules, on their topology, or even on the chemical functions present in their structure [111, 112]. Therefore, a pharmacophoric model can be obtained both by investigating points of interaction and by investigating structural properties between the target and a group of overlapping molecules [109]. It is noteworthy that pharmacophore is a model and not a representation of a real molecule or an association of functional groups, but rather, a model that describes a set of points that favor the ligand-receptor interaction, a "common denominator" between a group of molecules [113].

Currently, programs aimed at the study and generation of a pharmacophore model face two difficulties. The first is due to the generation of conformers from the group of compounds used to generate a pharmacophore in a three-dimensional field that involves flexible structures and to find a correlation between these structures [109, 111, 114]. If conformational data like crystallized complex data exist, this step can be neglected and it is called generation directly [112, 114, 115]. The second difficulty present in the study of generating a pharmacophoric model is the efficiency of the algorithm in aligning unusual structures [109, 111, 114].

Based on the informations related to presence or absence of the resolved structure under study, there are two possibilities for working with the derivation of pharmacophoric patterns. A pharmacophoric model can be established based only on structural data of ligands, whereby a set of molecules with known biological activity on a given target are superimposed allowing the identification of chemical features that are essential for its activity, called LBP (ligand based pharmacophore). If the structure of the molecular target is available in repositories of crystallographic structures, the study can be improved using the SBP (structure based pharmacophore) approach, in which points of essential interactions between target-ligand can be analyzed, defining a pattern of chemical interactions [116].

SBP approach can be used in two manners: if the structure of the protein-inhibitor complex is present, or if only the protein structure is present. These approaches use an interaction site analysis of the biological target and from this, it is possible to derive chemical patterns complementary to the cavity. With this information, a pharmacophoric model is then generated, which can be used to investigate the complementarity of the evaluated cavity [109, 117].

For the development of a pharmacophore pattern using LBP it is necessary to select a set of active molecules presenting important chemical patterns for the interaction with the biological target. These molecules are aligned, in their bioactive conformations, when they exist, or when nonexistent, the program tries to strategically position the functional groups present in the ligands in such a way to represent their interactions with the organic functions present in the receptor. In this way, it can be deduced how such functions are correlated with the formation of a drug-receptor complex and, if possible, carry out the identification of common pharmacophoric patterns between them, when they are overlapping [117]. The next step is to refine these patterns, obtaining in the end a consensus pattern that suits most of the molecules, for which, virtual screening steps will be performed in databases [109].

As an example for generations of pharmacophore models the program *Catalyst* uses two algorithms for LBP studies: *HipHop* [114] and *HypoGen* [118] algorithm. In the *HipHop* algorithm, molecular patterns are identified through an exhaustive search, which begins with the identification of common features between molecules and spatially locates them. After identifying the common features among the molecules that make up the training set, a simplistic pharmacophore model is proposed. This model is prepared for each molecule and all possible combinations of two common aspects between them are aligned. Subsequently, the research continues with all possible combinations of three aspects and so on, until it is no longer possible to generate further combinations [111–114, 119].

Another possible way to generate pharmacophoric patterns is through the inclusion of biological activity data (IC_{50}). *Hypogen* algorithm uses this information to distinguish pharmacophoric patterns belonging to active compounds from those belonging to inactive compounds. The generation of the pharmacophore model using this algorithm is divided into three phases, being the first one intended for construction, i.e. associated with formation of the pharmacophore. At this phase, the active molecules are identified and all the pharmacophoric patterns of the two most active molecules are stored. This is followed by the subtraction phase that is intended to refine the selected pharmacophore, in which pharmacophores that fit at least half of the inactive molecules are removed and then finalized by the optimization phase. The quality of the pharmacophore hypothesis generated by *Hypogen* can be described using statistical parameters like fixed cost, null cost and total cost. As such, the best model selected is usually the one with the highest correlation coefficient (R [2]), low total cost, high cost, largest cost difference, and low RMSD (*root mean square deviation)* values [118, 120].

Result generated by studies involving pharmacophoric patterns, are represented by pharmacophoric models constituted of features. For instance, the BIOVIA Discovery Studio program generates features that are represented either by a sphere or by two spheres and an arrow. Features composed of two spheres represent the region of the ligand that is coupled by the arrow, which represents the direction of the charge donation, followed by the second sphere that is called the sphere of exclusion, representing regions of steric hindrance. In order to distinguish different features the spheres have different colors, as shown in Fig. 2 that exemplifies all possible features generated by the program [121].

In addition to methods named as substructure pattern-based, widely used and made available by programs PHASE [122], LigandScout [123] and Catalyst, there are still other methodologies used to derive a pharmacophore pattern, such as feature-based and molecular field-based [117]. We emphasize that generation of pharmacophore models is very useful because they may serve as a basis for the search for ligands in 3D databases (virtual screening), as well as they can also serve as an extremely helpful tool for synthetic chemicals, in an attempt to synthesize a new inhibitor through visual inspection of the generated model [115].

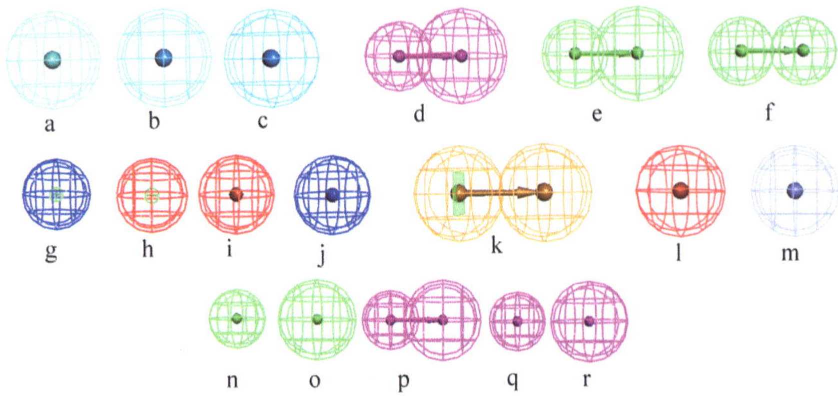

Fig. 2 Examples of features generated by the BIOVIA Discovery Studio program. **a** Hydrophobic, **b** Hydrophobic, aromatic, **c** Hydrophobic, aliphatic, **d** Hydrogen bond donor, **e** Hydrogen bond acceptor, **f** Hydrogen bond acceptor lipidic, **g** Negative charge, **h** Positive charge, **i** Positive ionizable, **j** Negative ionizable, **k** Ring aromatic, **l** Positive ionizable 1, **m** Zn binder, **n** Hydrogen bond acceptor heavy, **o** Hydrogen bond acceptor projection, **p** Hydrogen bond donor 1, **q** Hydrogen bond donor heavy and **r** Hydrogen bond donor projection

4.3 Molecular Docking

One of the major goals during the drug discovery process is to obtain drugs that perform interactions with the chosen target and one of the most used strategies is the experimental method High-throughput screening [124]. However, as mentioned in introduction, it is an expensive and time consuming technique [125]. Thus, it becomes necessary to use techniques that has lower cost and does not consume a large amount of time [126]. Obtaining resolved crystallographic structures from therapeutic targets associated with the development of robust and sophisticated computational techniques allowed the development of techniques such as strucuture based drug design (SBDD) [127]. The fact of dealing with the disease at a molecular level associated with a 3D structure of the corresponding protein target, made SBDD become a more specific, efficient and fast process in the discovery and optimization of leads [125].

Among the most common methods used in SBDD, one can mention molecular docking simulations applied to structure-based virtual screening (SBVS). It is additionally possible to evaluate the interactions that occurs in the formation of the ligand-receptor complex, analysis of the binding energy and evaluation of the conformational changes that occur during the coupling process through this technique [128]. The great advantage when using SBDD techniques in VS studies is the possibility of identification and direct evaluation of the probability of molecules, present in databases, to bind to the therapeutic target in the study [129].

The structural description between the ligand-receptor complex is useful for the discovery of key intermolecular interactions, investigation of binding conformations, study mechanisms, discovery of new binding sites and of conformational changes induced by ligands [130].

One main aspect in SBVS studies is the knowledge of the three-dimensional structure of the biological target. This data is extremely important due that it allows the study of the binding site, which in turn makes it possible to identify the optimization of the binding between the ligand and the receptor [131]. Crystallographic structures of macromolecules are obtained by methods such as X-ray crystallography or NMR spectroscopy [132], but sometimes that is not possible. If the experimental structure is not available, it is possible to create a model by homology based on experimentally resolved structures of similar proteins [133].

A SBVS study begins with the preparation of the receptor and the library of compounds to be used in the docking stage. Molecular docking is intended to predict the complex formed between the molecules present in the library and the protein target [134] with certain precision [135]. The simplest way to describe these studies is by means of combining two components: a conformation algorithm and a scoring function [136]. Thus, the fit between the ligand and the active site will be scored, and this value indicates the degree of this interaction. Then, the compounds with the best score will proceed to new filters or even an experimental evaluation [136].

The conformation algorithm consists of exploring the conformational space of the receptor and/or the ligand and obtaining the pose (conformation and orientation) of the ligand [137], while the scoring function evaluates each docking mode [138]. Despite all the technology used, the true potential of SBVS is questioned due to the great challenge that no single tool is able to consistently predict the correct bioactive conformation of all compounds from a chemical library [139].

Molecular docking can be performed in 3 different ways depending on the flexibility of the ligand and receptor. In rigid docking, both the receptor and the ligand are considered rigid. This type of docking works with small molecules that have great complementarity with the receptor's site and the simplest algorithms such as DOCK are used [140]. This algorithm works with geometric correspondence to overlap the ligand within the connection pocket and although it reduces the computational load, it is not the best methodology for fitting solutions [141].

Flexible docking can be performed with the flexible ligand and the rigid receptor, or vice versa, or with both flexible. Generally, in the case of the receptor, only the residues that make up the binding site are used in flexible form. In cases where only the ligand is flexible, its pose generation can be determined using conformational search algorithms. These algorithms must be able to explore the different energy scenarios within a reasonable period of time, regardless of the method used, which can be systematic, stochastic or simulations [142].

Systematic search method consists of analyzing the degrees of freedom of the ligand, often using incremental construction based on fragments for this task [143]. The ligand grows in the active site and is then fragmented into a rigid core and flexible side chains. The core is positioned and side chains are included incrementally, considering their degrees of freedom [143] and thus avoiding the combinatorial explosion [144]. One of the disadvantages presented by this method is that, although it is effective in exploring conformational space, results can converge to a local and not a global minimum. However, this can be solved by conducting simultaneous searches from different conformations [145].

The stochastic method performs conformational search by randomly modifying the ligand parameters [146], generating a set of molecular conformations and filling a wide range of energy scenarios. These random modifications are accepted or rejected depending on the probability functions, which are based on Monte Carlo and genetic algorithms [142] which apply concepts from the Theory of Evolution and Natural Selection. These algorithms increase the probability of finding the global minimum, however the computational cost is a key limitation factor [145, 146].

Finally, the simulation algorithm is integrated with energy minimization in order to reach local minimums. This type of algorithm together with Monte Carlo is widely used in flexible protein docking [147, 148]. The algorithms available in such a method are Newton–Raphson, least squares and conjugated gradient [142].

After the step of generating conformers, the second problem to be solved by docking studies is the classification of these by means of a score, which is determined by means of a scoring function. The scoring function is intended to help the molecular docking program used in the study to calculate the binding affinity between the ligand molecule and the receptor [149], evaluating the molecular interactions, desolvation and entropic effects involved in ligand-receptor binding [150].

The larger the number of physicochemical parameters evaluated, the greater will be the accuracy of a scoring function. It must offer a balance between precision and speed, a critical aspect when working with a large library of compounds. Next, we describe the four classes of scoring functions: empirical, force field, knowledge based and machine learning [137, 138, 151].

Empirical scoring functions analyze the affinity of the ligand and receptor based on their number of atoms, [152] and each function term describes an event involved in the formation of the complex, such as hydrogen bonds, ionic and non-polar interactions, solvation and entropic effects [153]. A disadvantage of this scoring function is the dependence on the accuracy of data used [153]. However, the energy terms used are simple, making the empirical scoring function faster than the force field based scoring [153].

The calculation of the force field score is based on the energy of the complex plus intermolecular interactions, such as electrostatic and Van der Waals forces between the objects of study [154]. A limitation of this function is the precision in the estimation of entropic contributions, in addition to the solvent not being considered, which makes complicate to estimate desolvation energies [155].

A knowledge-based scoring function depends on the statistical potential of intermolecular interactions. This method is based purely on the assumption that frequent functional groups or a certain type of atom are energetically favorable and contribute to the binding affinity [156]. This type of scoring function offers an appropriate balance between accuracy and speed, as it does not depend on the reproduction of binding affinities or ab initio calculations [157], as also present in the empirical and force field methods respectively.

The last method of scoring functions is machine learning, which does not restrict analysis to a single predefined functional form [158]. This method takes into account interactions among the ligand and the receptor, ignoring error-prone interactions and working with nonlinear dependence between binding interactions [149]. Machine learning method is becoming more and more popular [159] and has presented a better performance than other methods in calculating binding energy [160].

Hence, docking programs are able to successfully predict the formation of the ligand at the receptor's binding site, however, most do not reproduce the complex's absolute interaction energy satisfactorily. Some issues like entropic effects and desolvation represent significant challenges that must be overcome by available docking algorithms [161, 162]. Scoring functions are often compared in several studies [153] and present different rates of accuracy depending on the conditions, showing that one cannot replace the other. Thus, the use of different scores applied simultaneously has been increasing, in order to obtain a consensus score [163].

Consensus score has been used in order to combine the advantages presented by each different score aiming to minimize their errors [164]. It additionally is based on the fact that the true value is closest to the average value of replicated experiments [165]. Therefore, consensus scoring function can perform better than a single scoring function since it presents results with a greater probability of true hits and a false positive rate, besides being widely used in several studies and applications.

Finally, SBVS studies have proven to be the most robust method for drug discovery. Molecular docking programs successfully predict binding modes between ligands and receptors, proving their ability to track compounds in large databases and making it a low-cost alternative when compared to experimental techniques such as HTS. Current search algorithms are efficient, but the scoring functions do not estimate with satisfaction the absolute energy of intermolecular interactions since they ignore issues such as solvent effects and entropy calculations. Therefore, successful SBVS studies require knowledge of the fundamentals applied since its comprehension is essential in producing significant results.

5 Pharmacokinetic and Toxicological Predictions

In silico studies to estimate pharmacokinetic and toxicological properties (Absorption, Distribution, Metabolism, Excretion and Toxicity–ADMET or ADMETox) aim to predict the in vivo behavior of drug candidate molecules, bringing together the kinetic processes in an inclusive model. These studies are extremely important in the development of new pharmacological entities, as the early discovery of this information can reduce failures in clinical trials, which are part of the costliest steps in the development of drugs and are, generally, result of inappropriate ADMETox properties. Thus, it becomes important to elucidate, as soon as possible, adequate pharmacokinetic and toxicological indices, as this should result in a saving factor regarding time and money during the development of new bioactive substances [166–169].

In silico methods for the prediction of ADMETox properties at initial stages of researching new pharmacological entities have assured satisfactory results, and have become increasingly acceptable methods. Besides being relatively inexpensive and quick to execute, which allows their use in initial stages of virtual screening campaigns when the number of compounds is high [170], these methods are also commonly used in initial stages of preparing databases, in final selection of hits or even in the drug optimization process, as they are able to highlight physicochemical aspects that can provide important information when optimizing lead compounds [167, 169, 171]. Therefore, one could state that these methods go along since the discovery phase until clinical trial phase [172].

However, a negative point of such methods is the low accuracy, which can, at times, overestimate some values [167]. Nevertheless, it must be taken into account that depending on the ADMETox property that is intended to be predicted, and also on the computational approach used for this, low predictive power or, sometimes, false negative results can be verified. Thus, it can be seen that these approaches are extremely useful and important for predicting /estimating ADMETox properties, especially in the planning steps mentioned above, but they are limited and dependent on experimental studies.

Currently, there are several methodologies and computer programs designed for ADMETox calculations, for instance there are several methodologies and programs available at the vls3d.com repository (www.vls3d.com), which contains approximately 200 programs or free online servers with the theme "ADMET and physchem predictions". An interesting point to highlight are the physicochemical prediction models and the molecular descriptors, which are models of rapid execution and the properties analyzed have significant effect on pharmacokinetic behavior. From the various examples, we can highlight the calculation of aqueous solubility and dissolution rate, which are important factors for the rate of oral bioavailability, the pKa calculation to determinations of solubility, lipophilicity, permeability and absorption, and also, calculations of descriptors that predict the permeability being oral, gastro intestinal or blood–brain barrier [168]. There are several ways to calculate each one of these descriptors and physicochemical properties and such models are generally based on statistical models that relate structural characteristics of the compounds, measured in a given test to their biological properties, whereby each developer holds their equations, parameters, validations and distinct methodologies [171], which can lead to different results [173].

A major milestone for the use of molecular parameters relating them to oral bioavailability (good water solubility and intestinal permeability) was the widely cited work by Lipinski et al., which is referred to as the Lipinski's rule of 5. By correlating more than 2000 drugs with their physicochemical characteristics, Lipinski and collaborators identified that for good absorption or permeability, they should have no more than 5 hydrogen bond donors (HBD), more than 10 hydrogen bond acceptors (HBA), molecular weight greater than 500 daltons and calculated logP (ClogP) greater than 5 [57, 58]. Although there are exceptions to the rule of five, this is still commonly used to assess druglikeness and is widely used in drug development studies.

After the rule of 5, other rules emerged in order to assist researchers in the drug development process, by correlating physicochemical and structural data with oral bioavailability. Oprea and collaborators studies differentiated characteristics of drugs series with other compounds that gave rise to the same lead compounds, concluding that there is an increase in the lipophilicity and complexity of a lead compounds, originating the Rule of 4 that is used for lead development.

According to the authors of Rule 4, these rules are guidelines for lead development, structures of less structural complexity with affinities for the structural target and subject to optimization, which must have a molecular weight $\leq$ 450, ClogP between -3.5 and 4.5, no more than 10 terminal methylene groups, HBD $\leq$ 5 and HBA $\leq$ 8 [174]. A further study on this theme conducted by Yvonne C. Martin attributed the dependence on bioavailability to the predominant charge in the physiological pH because neutral species, negatively or positively charged, differ in the behavior in the pH range between 6 and 7. Hence, it was concluded that the polar surface area is a property that should be considered for anions, while for cations and neutral compounds the rule of 5 has good predictive capacity, thus creating a bioavailability score [175].

Veber and colleagues identified a positive influence of increasing molecular rigidity in association with the count of rotatable bonds, i.e. $<$ 10, as well as a negative impact of increasing polar surface area, i.e. $\leq$ 140. They alert for the rigidity of molecules being attributed only to the low number of rotatable bonds, and emphasize that rings are not rigid systems, as addressed by them [176]. Lu and colleagues, when reproducing Veber's studies using two different methodologies, found a significant difference in results for calculations of polar surface area and rotatable bonds, which made them state to be aware of using generalized rules depending on the classes of drugs under study [173].

Depending on the study, there are interesting rules to be used beforehand, ensuring greater effectiveness, as in the case of the Rule of 3 used in studies Fragment-based lead Discovery, which is characterized by structures with molecular mass $\leq$ 300; ClogP $\leq$ 3, HBD $\leq$ 3 and HBA $\leq$ 6 or the Tice Rules for herbicides 150 $\geq$ MM $\leq$ 500; alogP $\leq$ 5; HBD $\leq$ 3; 2 $\geq$ HBA $\leq$ 12 and RTB $\leq$ 12 and insecticides 150 $\geq$ MM $\leq$ 500; 0 $\geq$ alogP $\leq$ 6.5; HBD $\leq$ 2; 1 $\geq$ HBA $\leq$ 8 and RTB $\leq$ 12 [60, 177, 178].

The use of a specific druglikeness rule in VS studies is an important factor in increasing the probability that a selected hit will have biological activity with favorable pharmacokinetic properties [179]. Methods for determining these properties are simple and are used more often in the initial stages of screening, in which it is intended to eliminate a large number of compounds that are not interesting for the study of interest, such as molecules with high polar surface area, excessive number of rotatable bonds, hydrogen bond donors or acceptors. However, to ensure that the study generated a selection of relevant compounds they can be evaluated by such rules in final stages of obtaining the hits.

More robust and specific studies are generally used in advanced stages of screening, or even optimization stages. In addition to calculation of descriptors and filters for estimating ADMETox properties, different studies may also be applied, such as machine learning, Artificial Neural Networks Ensemble, and also use of statistical models, often QSAR, QSAR-3D, similarity, and structure-based methods.

In this way, considering the above exposed concerning the use of in silico methods to predict ADME/Tox properties, it is evident that such stage/filter plays a crucial role in the development of robust virtual screening studies. These, together with a rational selection of most appropriate guideline/criteria to select compounds with a greater tendency to become hit, lead or drug candidate, in the context of interest, should be considered as essential to achieve such goal.

6 Use of Cloud Platforms in VS: Future Perspectives

Although we have portrayed a coherent and effective way of working with VS, which is possible to be developed independently of large installations with high computational processing capacity, there is a great interest on the part of medicinal chemists to identify and cover the effective rate for development of drugs present in the vast chemical space composed of small molecules. In fact, there is still a great deal of difficulty in working with ultra-large libraries that can easily contain up to hundreds of millions of compounds [50], however, this difficulty has been overcome by following a strong worldwide trend, driven by the development and use of programs, aimed at the development of drugs using high efficiency computing (HPC) redirected to computational clouds.

Some virtual screening services are already offered on cloud computing platforms, with the proposal to cover billions of molecules in virtual screening studies. The pioneer of Big Data analytics was Google with MapReduce designed to analyze 20 pentabytes of data per day, [180] through the open source implementation with Apache Hadoop [181] and later with Apache Spark [182]. The implementation and success of these systems for drug development studies has been demonstrated by the development of several studies using robust LBDD and SBDD techniques, as a supervised machine learning technique Support Vector Machines (SVM) [183] and even docking simulations with the popular AutoDock4 docking program (AutoDock-Cloud) [184–186] also reported being carried out in extensive databases with millions of molecules [187].

In addition to the platforms offered by Google, new proposals for implementing programs aimed at drug development are also being developed in other HPC cloud infrastructures such as Amazon Web Servives (AWS). Kainrad and collaborators, implemented the LigandScout [123] intended for the generation of conformers and LBVS based on pharmacophore models directly into the Graphical User Interface (GUI) applications, which they named LigandScout Remote. The tests developed by the authors estimate the approximate time of 211 min for the screening of a base containing 1 million compounds [188].

Another cloud plataform available to virtual screening study is Orion, OpenEye's cloud platform [189] also hosted on Amazon Web Services (AWS). This server houses several programs from the company OpenEye and these softwares are mostly offered as toolkits versions, which are simplified versions of corresponding original softwares, with equivalent efficacy. Among toolkits and softwares included in Orion, we can cite FastROCS, OMEGA, EON, BROOD, OEDocking, and others.

For instance, FastROCS is able to develop 3D similarity studies by shape overlaying and it has been applied in virtual screening studies with large databases of ca. 10^{10} conformers, showing a great performance, being able to run that amount of compounds in minutes [40].

In fact, hosting servers on highly efficient cloud computing platforms will change the landscape of drug development in a few years. The costs for the maintenance and use of an HPC center are extremely high and these factors make it impossible for many researchers to carry out large-scale screening or to develop simulations over long periods of time. There are still few services offered but their low cost will enable, in addition to greater efficiency in research in the field of drug development, a greater number of researchers using equipment and programs with high efficiency. It is possible that the drug development scenario will change in a short period of time, as it is possible to observe the growing interest of large companies that develop pharmaceutical programs and industries that are increasingly closer to these services.

7 Conclusion

The stages of proposals and development of hits and leads compounds are critical steps for the success of the discovery of a new drug. In view of the high financial expense and time spent in the process of developing a new drug, the selection of inappropriate molecules in initial stages of this process can carry ineffective molecules to future stages, causing enormous financial losses. Thus, aiming at a better use of the initial stages of drug development, CADD studies become increasingly essential stages for the pharmaceutical industries.

CADD techniques stand out from the others for being comparatively cheaper, in addition to presenting efficient results for the initial stages of drug development. These studies being directly allied with computational progress evolve together with it, both in terms of software and hardware, and demonstrate themselves as an efficient, fast, cheap and versatile technique for the initial stages of discovering hits and leads compounds. Regardless of the computational need required in CADD studies, some previous steps can guarantee their quality, especially if they are intended for VS studies such as the choice and preparation of databases and the most appropriate method for the study. The methods can vary between LBVS, SBVS, or combined, and must be in accordance with the intended study, as well as validated, so that they can guarantee the reliability of the study.

Finally, considering that VS studies evolve every day and become more and more efficient in the development and search for new bioactive molecules, we have pushed our efforts to show some of the techniques used in these studies, as well as the processes necessary for a good development of the same.

Bibliography

1. Office, U. S. G. A. Drug Industry: Profits, Research and Development Spending, and Merger and Acquisition Deals. (2017)
2. Ledley, F.D., McCoy, S.S., Vaughan, G., Cleary, E.G.: Profitability of large pharmaceutical companies compared with other large public companies. JAMA-J. Am. Med. Assoc. **323**, 834–843 (2020)
3. Tay-Teo, K., Ilbawi, A., Hill, S.R.: Comparison of sales income and research and development costs for FDA-approved cancer drugs sold by originator drug companies. JAMA Netw. Open **2**, e186875 (2019)
4. DiMasi, J.A., Grabowski, H.G., Hansen, R.W.: Innovation in the pharmaceutical industry: new estimates of R&D costs. J. Health Econ. **47**, 20–33 (2016)
5. Costantini, S., Walensky, R.P.: The costs of drugs in infectious diseases: branded, generics, and why we should care. J. Infect. Dis. **221**, (2019)
6. Tomar, V., Mazumder, M., Chandra, R., Yang, J., Sakharkar, M.K.: Small molecule drug design. Encycl. Bioinforma. Comput. Biol. 741–760 (2019). https://doi.org/10.1016/b978-0-12-809633-8.20157-x
7. Issa, N.T., Badiavas, E.V., Schürer, S.: Research techniques made simple: molecular docking in dermatology—a foray into in silico drug discovery. J. Invest. Dermatol. **139**, 2400-2408.e1 (2019)
8. Lombardino, J.G., Lowe, J.A.: The role of the medicinal chemist in drug discovery--then and now. Nat. Rev. Drug Discov. **3**, 853–62 (2004)
9. U. S. Food & Drug Administration. The Drug Development Process (2018). https://www.fda.gov/patients/learn-about-drug-and-device-approvals/drug-development-process
10. Andricopulo, A., Salum, L., Abraham, D.: Structure-based drug design strategies in medicinal chemistry. Curr. Top. Med. Chem. **9**, 771–790 (2009)
11. Guido, R.V.C., Andricopulo, A.D., Oliva, G.: Biotecnologia e química medicinal: aplicações em doenças infecciosas. **24**, 81–98 (2010)
12. Khurshid Ahmad, M.H.: Drug discovery and in silico techniques: a mini-review. Enzym. Eng. **04**, 1–3 (2014)
13. Cronk, D.: High-throughput screening. Drug Discovery and Development: Technology in Transition. Elsevier Ltd, (2013). https://doi.org/10.1016/B978-0-7020-4299-7.00008-1
14. Begley, T.P., Inglese, J., Auld, D.S.: High throughput screening (HTS) techniques: applications in chemical biology. Wiley Encycl. Chem. Biol. (2008). https://doi.org/10.1002/9780470048672.wecb223
15. Cobb, K.: Dock this : insilico drug design feeds drug development. Biomed. Comput. Rev. 20–30 (2007)
16. Hillisch, A., Heinrich, N., Wild, H.: Computational chemistry in the pharmaceutical industry: from childhood to adolescence. ChemMedChem **10**, 1958–1962 (2015)
17. Jorgensen, W.L.: The many roles of computation in drug discovery. Science (80-.). **303**, 1813–1818 (2004)
18. Song, C.M., Lim, S.J., Tong, J.C.: Recent advances in computer-aided drug design. Brief. Bioinform. **10**, 579–591 (2009)
19. Baig, M.H., Ahmad, K., Rabbani, G., Danishuddin, M., Choi, I.: Computer aided drug design and its application to the development of potential drugs for neurodegenerative disorders. Curr. Neuropharmacol. **16**, 740–748 (2017)
20. Xiang, M., Cao, Y., Fan, W., Chen, L., Mo, Y.: Computer-aided drug design: lead discovery and optimization. Comb. Chem. High Throughput Screen. **15**, 328–337 (2012)
21. Chaudhary, K.K., Mishra, N.: A review on molecular docking : novel tool for drug discovery. JSM Chem. **4**, 1–4 (2016)
22. Berman, H.M., et al.: The protein data bank. Nucleic. Acids Res. **28**, 235–242 (2000)
23. M.P., B., L.B., F., C.A., T., C.H.T, de P. da S.: Prediction of the three-dimensional structure of phosphate-6-mannose PMI present in the cell membrane of xanthomonas citri subsp. citri

of interest for the citrus canker control. In: F., L. P., C., T. (eds.) Emerging Research in Science and Engineering Based on Advanced Experimental and Computational Strategies. Engineering Materials. pp. 259–276. Springer, Cham, (2020)
24. Anderson, A.C.: The process of structure-based drug design. Chem. Biol. **10**, 787–797 (2003)
25. Kalyaanamoorthy, S., Chen, Y.P.P.: Modelling and enhanced molecular dynamics to steer structure-based drug discovery. Prog. Biophys. Mol. Biol. **114**, 123–136 (2014)
26. Bacilieri, M., Moro, S.: Ligand-based drug design methodologies in drug discovery process: an overview. Curr. Drug Discov. Technol. **3**, 155–165 (2007)
27. Geppert, H., Vogt, M., Bajorath, J.: Current trends in ligand-based virtual screening: molecular representations, data mining methods new application areas, and performance evaluation. J. Chem. Inf. Model. **50**, 205–216 (2010)
28. Ou-Yang, S.-S., et al.: Computational drug discovery. Acta Pharmacol. Sin. **33**, 1131–1140 (2012)
29. Gimeno, A. et al.: The light and dark sides of virtual screening: what is there to know? Int. J. Mol. Sci. **20**, (2019)
30. Mortenson, P.N., Erlanson, D.A., De Esch, I.J.P., Jahnke, W., Johnson, C.N.: Fragment-to-lead medicinal chemistry publications in 2017. J. Med. Chem. **62**, 3857–3872 (2018)
31. Gimeno, A., Tom, S.: In: The Light and Dark Sides of Virtual Screening : What Is There to Know ? (2019). https://doi.org/10.3390/ijms20061375
32. Reddy, A.S., Pati, S.P., Kumar, P.P., Pradeep, H.N., Sastry, G.N.: Virtual screening in drug discovery—a computational perspective. Curr. Protein Pept. Sci. **8**, 329–351 (2007)
33. Kumar, A., Zhang, K.Y.J.: Hierarchical virtual screening approaches in small molecule drug discovery. Methods **71**, 26–37 (2015)
34. Ferreira, R.S., Oliva, G., Andricopulo, A.D.: Integrating virtual and high-throughput screening: Opportunities and challenges in drug research and development. Quim. Nova **34**, 1770–1778 (2011)
35. Tanrikulu, Y., Krüger, B., Proschak, E.: The holistic integration of virtual screening in drug discovery. Drug Discov. Today **18**, 358–364 (2013)
36. Irwin, J.J., Shoichet, B.K.: ZINC—a free database of commercially available compounds for virtual screening ZINC—a free database of commercially available compounds for virtual screening. J. Chem. Inf. Model. **45**, 177–182 (2005)
37. Sterling, T., Irwin, J.J.: ZINC 15-ligand discovery for everyone. J. Chem. Inf. Model. **55**, 2324–2337 (2015)
38. Ol\ugaç, A., Türe, A., Ol\ugaç, S., Möller, S.: Cloud-based high throughput virtual screening in novel drug discovery. In: Kołodziej, J., González-Vélez, H. (eds.) High-Performance Modelling and Simulation for Big Data Applications: Selected Results of the COST Action IC1406 cHiPSet, pp. 250–278. Springer International Publishing, (2019). https://doi.org/10.1007/978-3-030-16272-6_9
39. McCarthy, A.: Drug discovery in the clouds. Chem. Biol. **19**, 1–2 (2012)
40. Grebner, C. et al.: Virtual screening in the cloud: how big is big enough? J. Chem. Inf. Model (2019).https://doi.org/10.1021/acs.jcim.9b00779
41. Singh, N., Chaput, L., Villoutreix, B.O.: Virtual screening web servers: designing chemical probes and drug candidates in the cyberspace. Brief. Bioinform. **00**, 1–29 (2020)
42. Artimo, P., et al.: ExPASy: SIB bioinformatics resource portal. Nucleic Acids Res. **40**, 597–603 (2012)
43. Shi, Y., von Itzstein, M.: How size matters: diversity for fragment library design. Molecules **24**, 1–10 (2019)
44. Kim, S.: Getting the most out of pubchem for virtual screening. Expert Opin Drug Discov **11**, 843–855 (2016)
45. Pawar, G., Madden, J.C., Ebbrell, D., Firman, J.W., Cronin, M.T.D.: In silico toxicology data resources to support read-across and (Q)SAR. Front. Pharmacol. **10**, 1–26 (2019)
46. Lavecchia, A., Giovanni, C.D.: Virtual screening strategies in drug discovery: a critical review. Curr. Med. Chem. **20**, 2839–2860 (2013)

47. Bohacek, R.S., McMartin, C., Guida, W.C.: The art and practice of structure-based drug design: a molecular modeling perspective. Med. Res. Rev. **16**, 3–50 (1996)
48. Odling-smee, L., Dobson, C.M.: Insight Chemical space and biology. **432**, 824–828 (2004)
49. Shoichet, B.K.: Virtual screening of chemical libraries problems with virtual screening. Nature **432**, 862–865 (2006)
50. Lyu, J., et al.: Ultra-large library docking for discovering new chemotypes. Nature **566**, 224–229 (2019)
51. Wu, C., et al.: Analysis of therapeutic targets for SARS-CoV-2 and discovery of potential drugs by computational methods. Acta Pharm. Sin. B (2020). https://doi.org/10.1016/j.apsb.2020.02.008
52. Sheisi, F.L., Rocha, S., Olanda, C.G., Fokoue, H.H., Anna, C.M.R.S.: Virtual screening techniques in drug discovery: review and recent applications. 1751–1767 (2019). https://doi.org/10.2174/1568026619666190816101948
53. Johnson, M.A., Maggiora, G.M.: Concepts and applications of molecular similarity. Wiley (1990). https://doi.org/10.1002/jcc.540130415
54. Ropp, P.J., et al.: Gypsum-DL: an open-source program for preparing small-molecule libraries for structure-based virtual screening. J. Cheminform. **11**, 1–13 (2019)
55. Hawkins, P.C.D., Nicholls, A.: Conformer generation with OMEGA: Learning from the data set and the analysis of failures. J. Chem. Inf. Model. **52**, 2919–2936 (2012)
56. Lagarde, N., et al.: A free web-based protocol to assist structure-based virtual screening experiments. Int. J. Mol. Sci. **20**, 1–15 (2019)
57. Lipinski, C.A., Lombardo, F., Dominy, B.W., Feeney, P.J.: Experimental and computational approaches to estimate solubility and permeability in drug discovery and development settings. Adv. Drug Deliv. Rev. **23**, 3–25 (1997)
58. Lipinski, C.A., Lombardo, F., Dominy, B.W., Feeney, P.J.: Experimental and computational approaches to estimate solubility and permeability in drug discovery and development settings1. Adv. Drug Deliv. Rev. **46**, 3–26 (2001)
59. M, C., R, C., C, M., H, J. A 'Rule of Three' for fragment-based lead discovery? Drug Discov. Today **8**, 876–7 (2003)
60. Tice, C.M.: Selecting the right compounds for screening: Does Lipinski's rule of 5 for pharmaceuticals apply to agrochemicals? Pest Manag. Sci. **57**, 3–16 (2001)
61. Mysinger, M.M., Carchia, M., Irwin, J.J., Shoichet, B.K.: Directory of useful decoys enhanced (DUD-E): Better ligands and decoys for better benchmarking. J. Med. Chem. **55**, 6582–6594 (2012)
62. Cereto-Massagué, A., et al.: DecoyFinder: An easy-to-use python GUI application for building target-specific decoy sets. Bioinformatics **28**, 1661–1662 (2012)
63. Bertrand, H., Triballeau, N.: Virtual screening workflow development guided by the ROC curve approach . application to high-throughput docking on metabotropic glutamate receptor subtype 4 accepting the uncertainty ? Receptor 2534–2547 (2005)
64. Empereur-Mot, C., et al.: Predictiveness curves in virtual screening. J. Cheminform. **7**, 1–17 (2015)
65. Jain, A.N., Nicholls, A.: Recommendations for evaluation of computational methods. J. Comput. Aided. Mol. Des. **22**, 133–139 (2008)
66. Kim, R., Skolnick, J.: Assessment of programs for ligand binding affinity prediction. J. Comput. Chem. **29**, 1316–1331 (2008)
67. Koes, D.R., Baumgartner, M.P., Camacho, C.J.: Lessons learned in empirical scoring with smina from the CSAR 2011 benchmarking exercise. J. Chem. Inf. Model. **53**, 1893–1904 (2013)
68. Wierbowski, S.D., Camacho, C.J., Wingert, B.M., Zheng, J.: Cross-docking benchmark for automated pose and ranking prediction of ligand binding. 298–305 (2020). https://doi.org/10.1002/pro.3784
69. Baumgartner, M.P., Camacho, C.J.: Choosing the optimal rigid receptor for docking and scoring in the CSAR 2013/2014 experiment. J. Chem. Inf. Model. **56**, 1004–1012 (2016)

70. Ye, Z., Baumgartner, M.P., Wingert, B.M., Camacho, C.J.: Optimal strategies for virtual screening of induced-fit and flexible target in the 2015 D3R grand challenge. J. Comput. Aided. Mol. Des. **30**, 695–706 (2016)
71. Wingert, B.M., Oerlemans, R., Camacho, C.J.: Optimal affinity ranking for automated virtual screening validated in prospective D3R grand challenges. J. Comput. Aided. Mol. Des. **32**, 287–297 (2018)
72. Shamsara, J.: CrossDocker : a tool for performing cross-docking using Autodock Vina. Springerplus 0–4 (2016). https://doi.org/10.1186/s40064-016-1972-4
73. Kavitha, R., Karunagaran, S., Subhash, S., Woo, K., Meganathan, C.: BioSystems Pharmacophore modeling, virtual screening, molecular docking studies and density functional theory approaches to identify novel ketohexokinase (KHK) inhibitors. BioSystems **138**, 39–52 (2015)
74. Ece, A., Sevin, F.: Chemistry The discovery of potential cyclin A / CDK2 inhibitors : a combination of 3D QSAR pharmacophore modeling virtual screening and molecular docking studies. (2013). https://doi.org/10.1007/s00044-013-0571-y
75. Tai, W., Lu, T., Yuan, H.: Pharmacophore modeling and virtual screening studies to identify new c-Met inhibitors. 3087–3100 (2012). https://doi.org/10.1007/s00894-011-1328-5
76. John, S., Thangapandian, S., Arooj, M., Hong, J.C., Kim, K.D.: Development evaluation and application of 3D QSAR Pharmacophore model in the discovery of potential human renin inhibitors. **12**, (2011)
77. Screening, L.V., Himmat, M., Salim, N., Al-dabbagh, M.M., Saeed, F.: Adapting document similarity measures for ligand-based virtual screening. 1–13 (2016). https://doi.org/10.3390/molecules21040476
78. Koshland, D.E.: Application of a theory of enzyme specificity to protein synthesis. Proc. Natl. Acad. Sci. **44**, 98–104 (1958)
79. Eckert, H., Bajorath, J.: Molecular similarity analysis in virtual screening: foundations, limitations and novel approaches. Drug Discov. Today **12**, 225–233 (2007)
80. Fischer, E.: Einfiuss der Configuration aUf die Wirkung der Enzyme. Ber Dtsch. Chem Gesellschaft 3479–3483 (1898)
81. Willett, P.: Similarity-based virtual screening using 2D fingerprints. Drug Discov. Today **11**, 1046–1053 (2006)
82. Bero, S.A., Muda, A.K., Choo, Y.H., Muda, N.A., Pratama, S.F.: Similarity measure for molecular structure: a brief review. J. Phys. Conf. Ser. **892**, (2017)
83. Cereto-Massagué, A., et al.: Molecular fingerprint similarity search in virtual screening. Methods **71**, 58–63 (2015)
84. Hu, G., et al.: Performance evaluation of 2D fingerprint and 3D shape similarity methods in virtual screening. J. Chem. Inf. Model. **52**, 1103–1113 (2012)
85. Sastry, M., Lowrie, J.F., Dixon, S.L., Sherman, W.: Large-scale systematic analysis of 2D fingerprint methods and parameters to improve virtual screening enrichments. J. Chem. Inf. Model. **50**, 771–784 (2010)
86. Almeida, D. et al.: Artigo Estratégias de Triagem Virtual no Planejamento de Fármacos Estratégias de Triagem Virtual no Planejamento de Fármacos. **4**, 739–776 (2012)
87. Muegge, I., Mukherjee, P.: An overview of molecular fingerprint similarity search in virtual screening. Expert Opin. Drug Discov. **11**, 137–148 (2016)
88. Miyao, T., Jasial, S., Bajorath, J., Funatsu, K.: Evaluation of different virtual screening strategies on the basis of compound sets with characteristic core distributions and dissimilarity relationships. J. Comput. Aided. Mol. Des. (2019). https://doi.org/10.1007/s10822-019-00218-8
89. Performance, H.: A review of ligand-based virtual screening web tools and screening algorithms in large molecular databases in the age of big data. **10**, 2641–2658 (2018)
90. Kumar, A., Zhang, K.Y.J.: Advances in the development of shape similarity methods and their application in drug discovery. **6**, 1–21 (2018)
91. Ballester, P.J., Finn, P.W., Richards, W.G.: Ultrafast shape recognition : Evaluating a new ligand-based virtual screening technology. J. Molecular Graph. Modell. **27**, 836–845 (2009)

92. Ballester, P.J. et al.: Ultrafast shape recognition in molecular databases. **463**, 1307–1321 (2014)
93. Kumari, V., Tripathi, R., Patel, S., Raj, U., Varadwaj, P.K.: DUSR (Distributed Ultrafast Shape Recognition): a Hadoop based tool to identify similar shaped ligand molecules. **51**, 6–9 (2017)
94. Kerezsi, C.R.: Prospective virtual screening for novel p53–MDM2 inhibitors using ultrafast shape recognition. pp. 89–97 (2014). https://doi.org/10.1007/s10822-014-9732-4
95. Bommu, U.D., Konidala, K.K., Pabbaraju, N.: Ligand-based virtual screening , molecular docking , QSAR and pharmacophore analysis of quercetin- associated potential novel analogs against epidermal growth factor receptor. **9893**, (2017)
96. Friedrich, L., Byrne, R., Treder, A., Singh, I., Bauer, C.: Shape similarity by fractal dimensionality : an application in the de novo design of (À) -englerin A mimetics. 1–6 (2020). https://doi.org/10.1002/cmdc.202000017
97. Li, H., Leung, K., Wong, M., Ballester, P.J.: USR-VS : a web server for large-scale prospective virtual screening using ultrafast shape recognition techniques. **2**, 1–6 (2016)
98. Finn, P.W., Morris, G.M.: Shape-based similarity searching in chemical databases. **00**, 1–16 (2012)
99. Venhorst, J., Núñez, S., Terpstra, J.W., Kruse, C.G.: Assessment of scaffold hopping efficiency by use of molecular interaction fingerprints. J. Med. Chem. **51**, 3222–3229 (2008)
100. Hawkins, P.C.D., Skillman, A.G., Nicholls, A.: Comparison of shape-matching and docking as virtual screening tools. J. Med. Chem. **50**, 74–82 (2007)
101. Marín, R.M., Aguirre, N.F., Daza, E.E.: Graph theoretical similarity approach to compare molecular electrostatic potentials. J. Chem. Inf. Model. **48**, 109–118 (2008)
102. Puertas-Martín, S., Redondo, J.L., Pérez-Sánchez, H,. Ortigosa, P.M.: Multi-objective evolutionary algorithm for evaluation of shape and electrostatic similarity. In: AIP Conference Proceedings **2070**, (2019)
103. Borges, N.M., et al.: Similarity search combined with docking and molecular dynamics for novel hAChE inhibitor scaffolds. J. Mol. Model. **24**, 41 (2018)
104. Wiggers, H.J., Rocha, J.R., Cheleski, J., Montanari, C.A.: Integration of ligand- and target-based virtual screening for the discovery of cruzain inhibitors. 565–578 (2011). https://doi.org/10.1002/minf.201000146
105. Simo, C.J.V., Mukherjee, T., Brito, R.M.M., Jackson, R.M.: Toward the discovery of functional transthyretin amyloid inhibitors : application of virtual screening methods. 1806–1820 (2010)
106. Tresadern, G., Bemporad, D., Howe, T.: A comparison of ligand based virtual screening methods and application to corticotropin releasing factor 1 receptor. J. Mol. Graph. Model. **27**, 860–870 (2009)
107. Omprakash, T.K., Selvan, A.T., Hameed, A.S., Geetha, S.P.: Identification of novel human immunodeficiency virus-1 integrase inhibitors by shape-based virtual screening. **1**, 151–156 (2010)
108. Silva, R.A.Í.C., Poiani, J.G.C., Ramos, R.S., Costa, J.S.: Ligand- and structure-based virtual screening of 16- ((diiso- a compound with potential anti-prostate cancer activity. **83**, 2018–2019 (2019)
109. Yang, S.-Y.: Pharmacophore modeling and applications in drug discovery: challenges and recent advances. Drug Discov. Today **15**, 444–450 (2010)
110. Wermuth, C.G., Ganellin, C.R., Lindberg, P., Mitscher, L.A.: Glossary of terms used in medicinal chemistry (IUPAC Recommendations 1998). Pure Appl. Chem. (1998).https://doi.org/10.1351/pac199870051129
111. Leach, A.R., Gillet, V.J., Lewis, R.A., Taylor, R.: Three-dimensional pharmacophore methods in drug discovery. J. Med. Chem. **53**, 539–58 (2010)
112. Dror, O., Shulman-Peleg, A., Nussinov, R., Wolfson, H.J:. Predicting molecular interactions in silico: I. a guide to pharmacophore identification and its applications to drug design. Curr. Med. Chem. **11**, 71–90 (2004)
113. Langer, T., Hoffmann, R.D.: In: Pharmacophores and Pharmacophore Searches (2006)

114. Drive, H.R.: Automated chemical hypothesis generation and database searching with Catalyst ®. **3**, (1995)
115. Tintori, C., Corradi, V., Magnani, M., Manetti, F., Botta, M.: Targets looking for drugs: a multistep computational protocol for the development of structure-based pharmacophores and their applications for hit discovery. J. Chem. Inf. Model. **48**, 2166–2179 (2008)
116. Langer, T., Wolber, G.: Pharmacophore definition and 3D searches. Drug Discov. Today Technol. **1**, 203–207 (2004)
117. Schaller, D. et al.: Next generation 3D pharmacophore modeling. Wiley Interdiscip. Rev. Comput. Mol. Sci. 1–20 (2020). https://doi.org/10.1002/wcms.1468.
118. Ma, Y., et al.: 3D QSAR Pharmacophore based virtual screening for identification of potential inhibitors for CDC25B. Comput. Biol. Chem. **73**, 1–12 (2018)
119. Güner, O., Clement, O., Kurogi, Y.: Pharmacophore modeling and three dimensional database searching for drug design using catalyst: recent advances. Curr. Med. Chem. **11**, 2991–3005 (2004)
120. Sutter, J., et al.: New features that improve the pharmacophore tools from accelrys. Curr. Comput. Aided-Drug Des. **7**, 173–180 (2011)
121. Dassault Systèmes BIOVIA. Discovery Studio Modeling Environment (2016)
122. Dixon, S.L. et al. PHASE: a new engine for pharmacophore perception, 3D QSAR model development, and 3D database screening: 1. Methodology and preliminary results. J. Comput. Aided. Mol. Des. **20**, 647–671 (2006)
123. Wolber, G., Langer, T.: LigandScout: 3-D pharmacophores derived from protein-bound ligands and their use as virtual screening filters. J. Chem. Inf. Model. **45**, 160–169 (2005)
124. Searls, D.B.: Data integration: challenges for drug discovery. Nat. Rev. Drug Discov. **4**, 45–58 (2005)
125. Lionta, E., Spyrou, G., Vassilatis, D. K., Cournia, Z.: Structure-based virtual screening for drug discovery : principles. Appl. Recent Adv. 1923–1938 (2014)
126. Swinney, D.C., Anthony, J.: How were new medicines discovered? Nat. Rev. Drug Discov. **10**, 507–519 (2011)
127. Anderson, A.C.: The process of structure-based drug design. **10**, 787–797 (2003)
128. Dixon, S.J., Stockwell, B.R.: Identifying druggable disease-modifying gene products. Curr. Opin. Chem. Biol. **13**, 549–555 (2009)
129. Rollinger, B.J.M., Stuppner, H., Hofbauergasse, C.M.: Virtual screening for the discovery of bioactive natural products. **65**, (2008)
130. Kahsai, A.W., et al.: Multiple ligand-specific conformations of the β2-adrenergic receptor. Nat. Chem. Biol. **7**, 692–700 (2011)
131. Waring, M.J., et al.: An analysis of the attrition of drug candidates from four major pharmaceutical companies. Nat. Rev. Drug Discov. **14**, 475–486 (2015)
132. Kontoyianni, M.: Docking and virtual screening in drug discovery. **1647**, 255–266 (2017)
133. Barcelos, M.P., Federico, L.B., Taft, C.A., de Paula da Silva, C.H.T.: Prediction of the three-dimensional structure of Phosphate-6-mannose PMI present in the cell membrane of Xanthomonas citri subsp. Citri of Interest for the Citrus Canker Control. in 259–276 (2020). https://doi.org/10.1007/978-3-030-31403-3_10
134. Stahura, F., Bajorath, J.: New methodologies for ligand-based virtual screening. Curr. Pharm. Des. **11**, 1189–1202 (2005)
135. Meng, X.-Y., Zhang, H.-X., Mezei, M., Cui, M.: Molecular docking: a powerful approach for structure-based drug discovery. Curr. Comput. Aided-Drug Des. **7**, 146–157 (2011)
136. Li, Q., Shah, S.: Structure-based virtual screening. In: Wu, C.H., Arighi, C.N., Ross, K.E. (eds.) Protein Bioinformatics: From Protein Modifications and Networks to Proteomics, vol. 1558, pp. 111–124. Springer, New York, (2017)
137. Moitessier, N., Englebienne, P., Lee, D., Lawandi, J., Corbeil, C.R.: Towards the development of universal, fast and highly accurate docking/scoring methods: a long way to go. Br. J. Pharmacol. **153**, S7–S26 (2009)
138. Huang, S.-Y., Grinter, S.Z., Zou, X.: Scoring functions and their evaluation methods for protein–ligand docking: recent advances and future directions. Phys. Chem. Chem. Phys. **12**, 12899 (2010)

139. Schneider, G.: Virtual screening: an endless staircase? Nat. Rev. Drug Discov. **9**, 273–276 (2010)
140. Kuntz, I.D., Blaney, J.M., Oatley, S.J., Langridge, R., Ferrin, T.E.: A Geometric approach to macromolecule-ligand interactions. J. Mol. Biol. **161**, 269–288 (1982)
141. Siavoush, D., Maryam Hamzeh-Mivehroud, B.S.: In: Applied Case Studies and Solutions in Molecular Docking-based Drug Design. IGI Global, (2016)
142. Sousa, S.F., Fernandes, P.A., Ramos, M.J.: Protein-ligand docking: current status and future challenges. Proteins Struct. Funct. Bioinforma. **65**, 15–26 (2006)
143. Rarey, M., Kramer, B., Lengauer, T., Klebe, G.: A fast flexible docking method using an incremental construction algorithm. J. Mol. Biol. **261**, 470–489 (1996)
144. Dias, R., de Azevedo Jr., W.: Molecular docking algorithms. Curr. Drug Targets **9**, 1040–1047 (2008)
145. Zsoldos, Z., Reid, D., Simon, A., Sadjad, S.B., Johnson, A.P.: eHiTS: A new fast, exhaustive flexible ligand docking system. J. Mol. Graph. Model. **26**, 198–212 (2007)
146. Gorelik, B., Goldblum, A.: High quality binding modes in docking ligands to proteins. Proteins Struct. Funct. Bioinforma. **71**, 1373–1386 (2007)
147. Oshiro, C.M., Kuntz, I.D., Dixon, J.S.: Flexible ligand docking using a genetic algorithm. J. Comput. Aided. Mol. Des. **9**, 113–130 (1995)
148. Hart, T.N., Read, R.J.: A multiple-start Monte Carlo docking method. Proteins Struct. Funct. Genet. **13**, 206–222 (1992)
149. Batool, M., Ahmad, B., Choi, S.A.: In: Structure-Based Drug Discovery Paradigm (2019)
150. Jain, A.N.: In: Scoring Functions for Protein-Ligand Docking. pp. 407–420. (2006)
151. Ain, Q.U., Aleksandrova, A., Roessler, F.D., Ballester, P.J.: Machine-learning scoring functions to improve structure-based binding affinity prediction and virtual screening. Wiley Interdiscip. Rev. Comput. Mol. Sci. **5**, 405–424 (2015)
152. Guedes, I.A., Pereira, F.S.S., Dardenne, L.E.: Empirical scoring functions for structure-based virtual screening: applications, critical aspects, and challenges. Front. Pharmacol. **9**, (2018)
153. Ferreira, L.G., Santos, R.N., Oliva, G., Andricopulo, A.D.: Molecular docking and structure-based drug design strategies. (2015)https://doi.org/10.3390/molecules200713384
154. Ponder, J.W., Case, D.A.: In: Force Fields for Protein Simulations. pp. 27–85. (2003). https://doi.org/10.1016/S0065-3233(03)66002-X
155. Kitchen, D.B., Decornez, H., Furr, J.R., Bajorath, J.: Docking and scoring in virtual screening for drug discovery: methods and applications. Nat. Rev. Drug Discov. **3**, 935–949 (2004)
156. Muegge, I.: PMF Scoring revisited. J. Med. Chem. **49**, 5895–5902 (2006)
157. Gohlke, H., Hendlich, M., Klebe, G.: Knowledge-based scoring function to predict protein-ligand interactions. J. Mol. Biol. **295**, 337–356 (2000)
158. Li, H., et al.: The impact of protein structure and sequence similarity on the accuracy of machine-learning scoring functions for binding affinity prediction. Biomolecules **8**, 12 (2018)
159. Hecht, D., Fogel, G.: Computational intelligence methods for docking scores. Curr. Comput. Aided-Drug Des. **5**, 56–68 (2009)
160. Cheng, T., Li, Q., Zhou, Z., Wang, Y., Bryant, S.H.: Structure-based virtual screening for drug discovery : a problem-centric review. **14**, (2012)
161. Mysinger, M.M., Shoichet, B.K.: Rapid context-dependent ligand desolvation in molecular docking. J. Chem. Inf. Model. **50**, 1561–1573 (2010)
162. Ruvinsky, A.M.: Role of binding entropy in the refinement of protein-ligand docking predictions: analysis based on the use of 11 scoring functions. J. Comput. Chem. **28**, 1364–1372 (2007)
163. Charifson, P.S., Corkery, J.J., Murcko, M.A., Walters, W.P.: Consensus scoring: a method for obtaining improved hit rates from docking databases of three-dimensional structures into proteins. J. Med. Chem. **42**, 5100–5109 (1999)
164. Feher, M.: Consensus scoring for protein-ligand interactions. Drug Discov. Today **11**, 421–428 (2006)
165. Wang, R., Wang, S.: How does consensus scoring work for virtual library screening? an idealized computer experiment. J. Chem. Inf. Comput. Sci. **41**, 1422–1426 (2001)

166. Lagorce, D. et al.: In Silico ADME/Tox Predictions. ADMET Med. Chem. A Pract. Guid. 29–124 (2010). https://doi.org/10.1002/9780470915110.ch2
167. Ren, H.C., Sai, Y., Chen, T.: Evaluation of generic methods to predict human pharmacokinetics using physiologically based pharmacokinetic model for early drug discovery of tyrosine kinase inhibitors. Eur. J. Drug Metab. Pharmacokinet. **44**, 121–132 (2019)
168. Alqahtani, S.: In silico ADME-Tox modeling: progress and prospects. Expert Opin. Drug Metab. Toxicol. (2017)
169. Amashita, F.Y., Ashida, M.H.: Review in silico approaches for predicting ADME properties of drugs. **19**, 327–338 (2004)
170. Daina, A., Michielin, O., Zoete, V.: SwissADME: A free web tool to evaluate pharmacokinetics, drug-likeness and medicinal chemistry friendliness of small molecules. Sci. Rep. **7**, 1–13 (2017)
171. Paul Gleeson, M., Hersey, A., Hannongbua, S.: In-Silico ADME Models: a general assessment of their utility in drug discovery applications. Curr. Top. Med. Chem. **11**, 358–381 (2011)
172. Hosea, N.A., Jones, H.M.: Predicting pharmacokinetic profiles using in silico derived parameters. Mol. Pharm. **10**, 1207–1215 (2013)
173. Lu, J.J., et al.: Influence of molecular flexibility and polar surface area metrics on oral bioavailability in the rat. J. Med. Chem. **47**, 6104–6107 (2004)
174. Teague, S.J., Davis, A.M., Leeson, P.D., Oprea, T.: The design of leadlike combinatorial libraries. Angew. Chemie-Int. Ed. **38**, 3743–3748 (1999)
175. Martin, Y.C.: A bioavailability score. J. Med. Chem. **48**, 3164–3170 (2005)
176. Veber, D.F., et al.: Molecular properties that influence the oral bioavailability of drug candidates. J. Med. Chem. **45**, 2615–2623 (2002)
177. Makara, G.M.: On sampling of fragment space. J. Med. Chem. **50**, 3214–3221 (2007)
178. Piccirillo, E., Do Amaral, A.T.: Virtual screening of bioactive compounds: concepts and aplications. Quim. Nova**41**, 662–677 (2018)
179. Sotriffer, C., Mannhold, R., Kubinyi, H., Folkers, G.: Virtual Screening: Principles, Challenges, and Practical Guidelines. Wiley, 2011
180. Dean, J., Ghemawat, S.: MapReduce: Simplified data processing on large clusters. In: OSDI 2004-6th Symposium Operations Systems Design Implement **51**, 137–149 (2004)
181. Apache Software Foundation. Hadoop.
182. Zaharia, M., et al.: Apache spark: a unified engine for big data processing. Commun. ACM **59**, 56–65 (2016)
183. Ahmed, L., Edlund, A., Laure, E., Spjuth, O.: Using iterative MapReduce for parallel virtual screening. In: Proceedings International Conference Cloud Computing Technology Science CloudCom vol. 2, pp. 27–32. (2013)
184. Zhao, J., Zhang, R., Zhao, Z., Chen, D., Hou, L.: Hadoop mapreduce framework to implement molecular docking of large-scale virtual screening. In: Proceedings-2012 IEEE Asia-Pacific Services Computing Conference APSCC 2012, pp. 350–353 (2012). https://doi.org/10.1109/APSCC.2012.67
185. Sid, K., Batouche, M.C.: Big data analytics techniques in virtual screening for drug discovery. In: ACM Interantional Conference Proceeding Series Part F1294 (2017)
186. Ellingson, S.R., Baudry, J.: High-throughput virtual molecular docking: hadoop implementation of AutoDock4 on a private cloud. In: ECMLS'11-Proceedings 2nd International Working Emerging Computing Methods Life Science pp. 33–38 (2011). https://doi.org/10.1145/1996023.1996028.
187. Capuccini, M., Ahmed, L., Schaal, W., Laure, E., Spjuth, O.: Large-scale virtual screening on public cloud resources with Apache spark. J. Cheminform. **9**, 4–9 (2017)
188. Kainrad, T., Hunold, S., Seidel, T., Langer, T.: LigandScout remote: a new user-friendly interface for HPC and cloud resources. J. Chem. Inf. Model. **59**, 31–37 (2019)
189. Orion. https://www.eyesopen.com/orion (2020)

High-Throughput-Based Virtual Screening via Molecular Docking for Oxidative Stress Mediated by ROS Enzyme

Williams J. C. Macêdo, Rai C. Silva, Carlton A. Taft, Carlos H. T. P. Silva, Anselmo F. R. Rodriguez, Joaquín M. Campos, and Cleydson Breno Rodrigues dos Santos

Abstract The regulation of redox homeostasis and the reduction of oxidative stress is one of several strategies used in the development of anticancer drugs. Understanding from in silico studies, how a particular molecule binds to the receptor, allows the selection of promising compounds that may be used in antineoplastic pharmacotherapy. Protein–ligand coupling can use the study and relationship between the protein–ligand complex or that is the origin of the ligand-target interaction. The docking algorithms used present a high complexity; however, currently, the systems

W. J. C. Macêdo (✉) · C. B. R. dos Santos
Posgraduate Program of Pharmaceutical Innovation, Federal University of Amapá, Macapá, AP 68902-280, Brazil
e-mail: williams.macedo@ufra.edu.br

Laboratory of Molecular Modeling and Simulation System, Federal Rural University of Amazônia, Rua João Pessoa, 121 Capanema, Pará 68700-030, Brazil

R. C. Silva · C. H. T. P. Silva
Departamento de Química, Faculdade de Filosofia, Ciências e Letras de Ribeirão Preto, University of Sao Paulo, Av. do Café s/n, Monte Alegre, Ribeirão Preto, São Paulo 14040-903, Brazil

Computational Laboratory of Pharmaceutical Chemistry, University of Sao Paulo, Av. do Café s/n, Monte Alegre, Ribeirão Preto, São Paulo 14040-903, Brazil

C. A. Taft
Brazilian Center for Physics Research, R. Dr. Xavier Sigaud, 150-Urca, Rio de Janeiro 22290-180, Brazil

C. H. T. P. Silva
Faculty of Pharmaceutical Sciences of Ribeirão Preto, University of Sao Paulo, Av do Cafe, s/n Ribeirão Preto, São Paulo 14040–903, Brazil

A. F. R. Rodriguez
Biodiversity and Biotechnology Network of the Legal Amazon, Biotechnology Department, Federal University of Acre, Road BR 364, Km 04, Rio Branco CEP 69915–900, Brazil
e-mail: ruiz@ufac.br

J. M. Campos
Department of Pharmaceutical and Organic Chemistry, University of Granada, Campus of Cartuja, Granada 18071, Spain
e-mail: jmcampos@ugr.es

F. A. La Porta and C. A. Taft (eds.), *Functional Properties of Advanced Engineering Materials and Biomolecules*, Engineering Materials,
https://doi.org/10.1007/978-3-030-62226-8_17

used to perform such studies present a friendly interface. A comparative study of various coupling algorithms can provide us with useful information to select the appropriate algorithm for drug research, design, and selection using new computational techniques. Hence, from this perspective, the purpose of this chapter is to provide new information about how it is possible to study via docking molecular Reactive Oxygen Species (ROS) enzymes against antineoplastic agents and to associate them with antitumor pharmacotherapy. The performed molecular docking results will be shown both the lower binding affinity (ΔG) values for the receptor-ligand, as well as interactions in the two enzymes, obtained after validation of the molecular docking protocols for the receptors: cytochrome P450 (CYP450) and NADPH oxidase (NOX).

Keywords Binding site · Free radicals · Anticancer drugs · Inhibition constant

1 Introduction

The system of redox reactions at the cellular level helps in the regulation and maintenance of redox homeostasis in living organisms. The existence of a group of enzymes that act in conjunction with the mechanisms of aerobic respiration is known to act on reactive oxygen species (ROS). In the living organism, causing oxidative damage and leading to a condition known as cellular oxidative stress [1, 2].

Several studies indicate that chronic illnesses, such as diabetes, neurodegenerative and cardiovascular diseases, and cancer itself may be associated with an oxidative stress condition [3]. This condition is due to the activation of transcription factors, which are capable of expressing a great variability of different genes, among them growth factors and cytokines linked to inflammation, which in turn activate inflammatory pathways modifying a normal cell for the state of a cancer cell [4]. It must be considered that the maintenance of homeostasis of the redox system and the reduction of cellular oxidative stress are dependent on the antioxidant efficiency present in the cell [5]. Enzymes such as cytochrome P450 (CP450), lipoxygenase (LOX), myeloperoxidase (MPO), NADPH oxidase (NOX) and xanthine oxidase (XOX), originate ROS during arachidonic acid metabolism and which, when inhibited, intervene in the ROS production cycle leading to the reduction of oxidative stress and promoting the maintenance of cellular redox homeostasis [6, 7].

C. B. R. dos Santos
Laboratory of Modeling and Computational Chemistry, Department of Biological Sciences, Federal University of Amapá, Rod. Juscelino Kubitschek, Km 02, s/n, Macapá, AP 68902-280, Brazil

1.1 Virtual Screening Versus and High-Throughput-Based Virtual Screening HTBVS

The initial search for compounds that have a promising biological activity on a given target carried out from an in silico study is known as virtual screening VS. The identification of hits that can be studied as drug candidates from a library of compounds from the perspective of a theoretical model will support a more advanced study based on an experimental approach at the molecular biochemical level and its substrate enzyme reactivity, and how this set of larger molecules selected in public and commercial databases can be studied applied to a specific target is a technique called High-Throughput-based Virtual Screening HTBVS. Therefore, analyzing these enzymes in the light of the drug-receptor coupling study, using molecular docking techniques, is relevant in the observation and evaluation of the receptor-ligand interaction, as well as in the inhibition of enzymes related to antioxidant activity. The use of these techniques can assist in the selection of molecules as well as point out binding sites and which amino acid residues are involved in the interactions, the type of bond, and atoms that form the complex [7]. The exploration of an extensive online platform in the search for compounds based on ligands, with access to databases commercial and public is available on the Web, using a Bayesian learning method, being able to create virtual screening models due to their noise tolerance, speed, and efficiency in extracting knowledge from a database which is widely used in the world of virtual research [8].

In this study, we will aim to observe the following aspects: (1) visualizing the structural model of the five ROS-generating proteins (CYP450C9 and NOX), (2) investigating a structural analysis of the proteins using control molecules in the studies to confirm the target site (3) perform sorting and virtual anchoring to identify inhibitory molecules against the appropriate target.

2 Materials and Methods

2.1 Molecules selection

Based on a large number of methods exist to molecular structures map, the choice of a group of initial molecules for study depends on several factors, such as the definition of the molecular target under research and the types of filters that will be used in the virtual search, which can be from 2D similarity, pharmacophoric or based on the format and anchoring. In general, the main steps employed in preparing the database for study involve the use of filters the "garbage filters"; it is the removal of compounds with unwanted functional groups, removal of compounds with unwanted physical–chemical or structural properties, correction of structures, and obtaining different conformations. Allied to this, the Lipinsk rule of five RO5 is used when looking for promising compounds that present possible oral solubility [18] and the

Tanimoto index [18, 19], which is an appropriate method in a few specific scenarios for fingerprint-based similarity calculations, as shown in Table 1.

2.2 Public and Commercial Database Collections

Currently, the world of research with the aid of computers has a great variability of virtual databases of compounds, public or commercial, the data banks, to be used in HTBVS [9–14]. Table 1 A large number of molecules are available on the websites in different formats according to each database, among which they may present similarities regarding the availability of the other formats, using international codifications such as: International Chemical Identifier InChI, which is a textual identifier for chemical substances, created by IUPAC [14], whose purpose is to facilitate the use of molecular databases, the Smile Simplified Molecular Input Line Entry Specification [15], a representation of chemical structures encoded using ASCII characters. The Chemical table file (CT File) is a chemical file formats that describe molecules, lists each atom in a molecule the x–y-z coordinates atomics, and the bonds among the atoms. This format was created by MDL Information Systems (MDL), which is a file format widely used for holding information, e.g., regarding the atoms, bonds, connectivity, and coordinates of a molecule [16]. However, such information are currently acquired by Symyx Technologies, that now so-called BIOVIA (a subsidiary of Dassault Systemes of Dassault Group) [16]. The SDF is one of a family of chemical-data file formats developed by MDL. Another important format file used in drug design is The Protein Data Bank (pdb) file format PDB [17]. Note that this text file format describing the three-dimensional structures of molecules held on Protein Data Bank. The pdb format provides description and annotation and structures, including atomic coordinates, secondary structure assignments of protein, atomic connectivity. Also, various experimental metadata are stored.

Table 1 List of available databases compounds for HTBVS

Database	Website	Availability since	Current size in 2020
PubChem	https://pubchemdocs.ncbi.nlm.nih.gov/	Public 2004	102,404,298
ChEMBL	https://www.ebi.ac.uk/chembl/	Public 2009	1,897,206
BindingDB	https://www.bindingdb.org/bind/index.jsp	Public 2000	804,949
ZINC	https://zinc.docking.org/	Public	over 230 million
DrugBank	https://www.drugbank.ca/	Public	13,491
ChemBridge	https://www.chembridge.com/index.php	Commercial 1995	Over 1.3 million
Maybridge	https://www.maybridge.com	Commercial	53,000

2.2.1 PubChem

PubChem is a famous free chemistry database, which is maintained by the National Institutes of Health (NIH) https://pubchem.ncbi.nlm.nih.gov. Launch in 2004, PubChem is the greatest chemical information resource for scientists, students, and the general public [20–23]. PubChem contains a vast colections of data with about of 102,404,298 compounds with unique chemical structures deposited in this database. The PubChem Substance records, are approximately 251,658,675 substances. As such, information about chemical entities provided by PubChem contributors, 1,067,849 biological experiments also are provided. About 268,275,849 data related to biological activity are made available in this database. Other relevant information in this database are: genes targes tested (~58,029); protein targets (~17,847); taxonomy of organisms (~3746); scientific publications with links in PubChem (~30,474,654); and so on. Additionally, this database provides all records for a specific molecule into an aggregate record, thus making searching more efficient. The Bioassay database provides descriptions of biological experiments on the tested compounds, particularly from HTS. It should be pointed out that PubChem represents the largest base of molecular structures, including data of spectral information available of 13C NMR, 1H NMR, 2D NMR, ATRIR, FT-IR, MS, GC–MS, Raman, UV–Vis, vapor-phase IR [24–26].

2.2.2 ChEMBL

The ChEMBL25 is the current realese, [26–29] [3, 8–10] Provided under a Creative Commons Attribution-ShareAlike 3.0 Unported license. This database holds to date about 12.482 targets, 1.897.206 distints compunds, 15.404.603 activities, 72.221 publications band 54 deposites datasets., including information on assays, binding constants, pharmacology, absorption, distribution, metabolism, excretion, and toxicity ADMET data. All data are derived from the literature extensive from multiple screening resources, PubChem bioassays, GSK (Glaxo SmithKline) deposited data, and the BindingDB database [30].

A widely used public database, is BindingDB web interface for query, download and virtual compound screening, which is based on providing online binding data, the existing data sets are 2291 crystalline structures of protein ligands with BindingDB affinity measurements and 5816 mostly crystalline structures with 100% of its structure elucidated. This database is completely accessible via the web for measurement connections, focusing mainly on the interactions of chemical substances and drug targets with small molecule compounds similar to drugs. The BindingDB contains 1,819,720 binding data, for 7,470 protein targets and 804,949 small molecules in September 2017 [31–34].

2.2.3 ZINC

ZINC is a famous free database of commercially-available compounds to perform virtual screening, for use in ligand Discovery and phamacophore screening. This data base provides access to other resources. Applications in this web is find Endogenous human metabolites and epigenetic targets, currently ZINC contains over 230 million purchasable compounds in ready-to-dock, 3D formats. ZINC also contains over 750 million purchasable compounds you can search for analogs in under a minute. INC is provided by the Irwin and Shoichet Laboratories in the Department of Pharmaceutical Chemistry at the University of California, San Francisco (UCSF). We thank NIGMS for financial support (GM71896) [35].

2.2.4 DrugBank

DrugBank The latest release oin january 2020, is a data bank wath contains 13,491 drug entries including 2641 approved small molecule drugs, 1364 approved biologics, informations of proteins, peptides, vaccines, and some allergenics, 130 nutraceuticals and over 6,347 experimental (discovery-phase) drugs. Additionally, 5183 non-redundant protein (i.e. drug target/enzyme/transporter/carrier). Note that each entry usually holds more than 200 data fields, as well various sequences are linked to these drug entries; in this case, half of the information has been devoted to drug/chemical data, while that the other half are in turn related to drug target or protein [36, 37].

2.2.5 ChemBridge and MayBridge

ChemBridge https://www.chembridge.com is a comercial data base what has been a provider of high quality screening compounds and libraries since 1995, in last years continues to produce novel druglike and leadlike screening compounds. Over 1,3 milion of compounds indata base canbe used in requirements of drug discovery and chemical biology researchers in industry and academia in all world. The EXPRESS-Pick Collection and CORE Library,in respectively Collection stock compounds, are selected using novelty, diversity, druglike properties, offering diverse classes of compounds with analogs to support initial SAR work. This data base in CORE Library of small molecule screening compounds offer more than 735,000. MayBridge avaliable in https://www.maybridge.com as ChemBrige is commercially available screening Collections libraries complementary, with highly diverse set of over 53,000 hit-like and lead-like molecules widely acknowledged as a critical tool in screening campaigns. The core of the Maybridge building block collection, these pharmacophorically rich intermediates are specifically designed for medicinal chemistry, allowing logical SAR development and Hit-to-Lead optimization. 30,000 small compounds (mw <300) taken from the complete Maybridge collection for use in fragment screening to further accelerate the identification of lead compounds.

3 Result

3.1 Selected Molecules

Four compounds were used in redocking on receptors cited according to Table 2. In silico simulations molecules with potential epitelial anticancer activity via inhibition ROS have been studied from the zinc database by HTBVS virtual screening, ZINC000000004840, ZINC000003872277, ZINC000029562604, ZINC3984017 and molecules controls [38] 5-Fluorouracil (FLU), Dextromethorphan (DEX).

Table 2 Ligands ZINC data base promissor ROS enzyme inhibitor and molecules control used

Molecules	Assignment
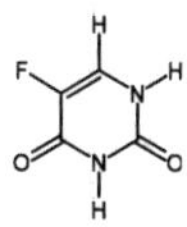 5-Fluorouracil (FLU)*	Available on inhibitory activity at the CP450 receptor [39, 40]
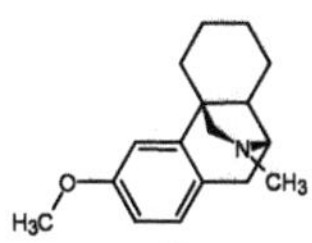 Dextromethorphan (DEX)*	Available on inhibitory activity at the NO receptor [40–42]
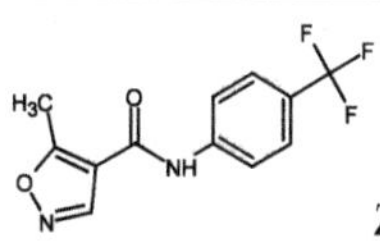 Z40	ZINC000000004840**
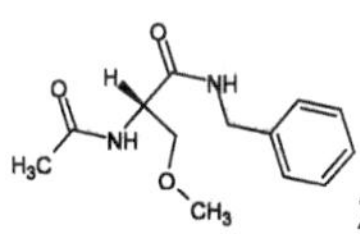 Z04	ZINC000029562604**
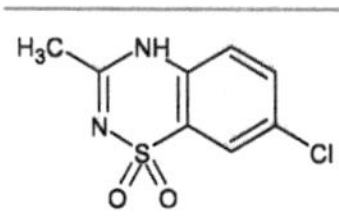 Z77	ZINC000003872277**
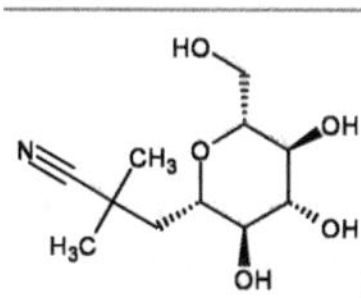 Z17	ZINC3984017**

*Molecule control commercially available, **Molecule test commercially available

3.2 Protein Preparation

To explore a protein complex at the level of molecular structural biochemistry and observe drug-receptor interactions using molecular docking, initially the crystallized structure of the appropriate protein containing the complex for studies is selected. Proceed with the removal of the crystallization water molecules and some ionic artifacts that are not capable of influencing the interaction of the complex and its ligands. In some cases, it becomes necessary to optimize the protein and to reconstitute some lost loop which is observed with the sequence of the incomplete or interrupted chain. This can be done with Chimera or Pymol easily. In this paper, the Discovery Studio Visualizer 2019 software was used, using the receptor-Ligand Interactions tool, for more accurate visualization of the residues that form the pocket binding site of all proteins.

3.2.1 Cytochrome P450

A superfamily with a very large group is cytochrome P450 (CYP450s) proteins. The haem proteins promote the biotransformation of many physiologically important compounds, they are present not only in mammals in the isoforms CYP1A2, CYP2C9, CYP2C19, CYP2D6 and CYP3A4, they contribute to the oxidative metabolism of over 90% of drugs in current clinical use but are also found in certain species of microorganisms, plants and animals [43, 44] Fig. 1.

After properly choosing the CYP2C9 protein with a resolution of 2.55 Å, which catalyzes the 6-and 7-hydroxy hydroxylation of the active enantiomer of warfarin, S-warfarin [45, 46], so that it is currently considered that a good resolution should have up to 3.0 Å. For the proposed study, it must contain a complexed molecule on the active site. The cytochrome P450 of ID 1OG5 obtained from the link https://www.rcsb.org/structure/1OG5, this protein has a record in the literature of the following residues that make up the active site: a bag of hydrophobic nature coated with residues: Arg 97, Gly 98, Ile 99, Phe 100, Leu 102, Ala 103, Val 113, Phe 114, Asn 217, Thr 364, Ser 365, Leu 366, Pro 367 e Phe 476 [47] Fig. 2.

3.2.2 NADPH Oxidase NOX

NADPH oxidase is an enzyme complex of oxyrredutase systems expressed in mammalian cells. Its isoforms are called NOX1, NOX2, NOX3, NOX4, NOX5, DOUX1 and DOUX2. The inflammatory cell NADPH oxidase enzyme consists of a number of protein subunits including the catalytic subunit Nox2 [48]. They are capable of generating in a oxidative stress via ROS as superoxide. Found in plasmatic membranes as well as in phagosomes and are used by neutrophils. Yet, it is well-known that the NADPH oxidase is the primary source of superoxide production by inflammatory cells [49, 50]. The ID 2CDU Crystal Structure of Nad(P)H Oxidase

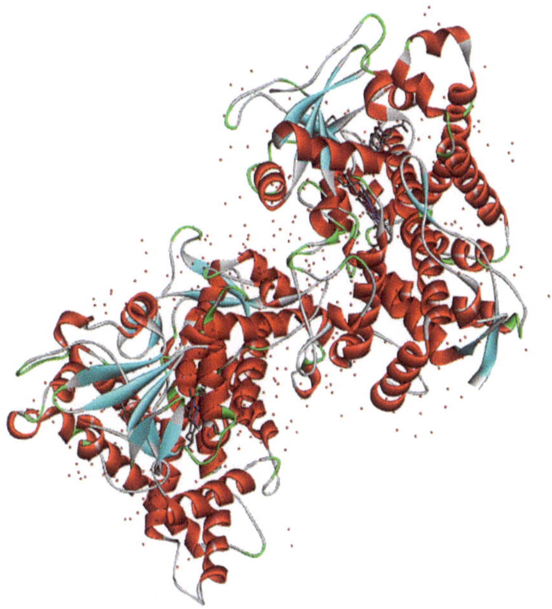

Fig. 1 Structure of human cytochrome P450 CYP2C9 ID 1OG5 chains A and B, HEME C complex and *S*-Warfarin substrates

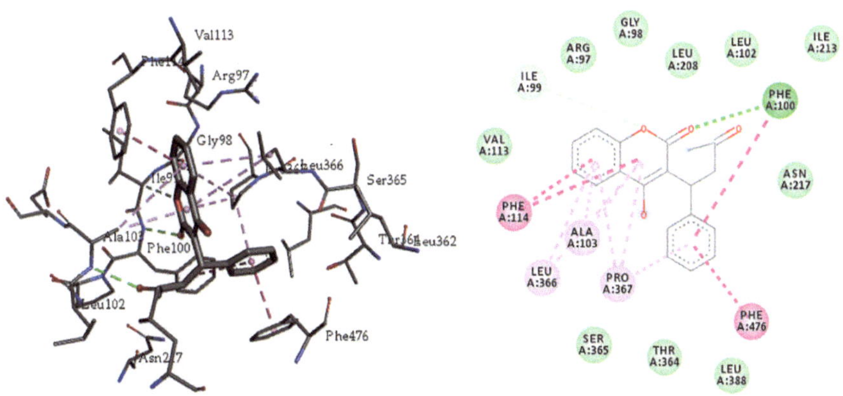

Fig. 2 Interactions S-Warfarin bindin site pocket CYP450: □ van der Walls; ■ Conventional H-Bond; □ Carbon H-Bond; ■ Pi-Pi Stacked and Pi-Alkyl □

from Lactobacillus Sanfranciscensis Insights Into the Conversion of O(2) into two Water Molecules by the Flavoenzyme was deposited in the data bank https://www.rcsb.org/structure/2CDU with ligand adenosine-5′-diphosphate ADP [51] Fig. 3.

Phosphate groups of the ADP are H-bonded with the Ile160N, Tyr159N and Tyr188OH. Additional hydrogen bonds are suggested between the ADP ribose

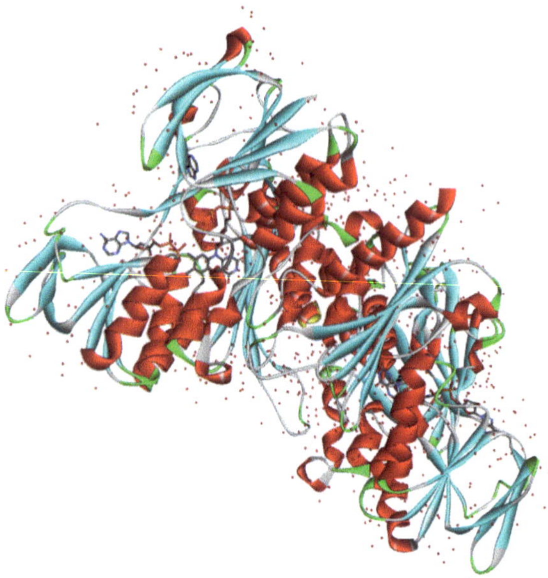

Fig. 3 Crystal Structure of Nad(P)H Oxidase from *Lactobacillus Sanfranciscensis* ID 2CDU chains A and B with adenosine-5′-diphosphate and flavin-adenine dinucleotide complex

hydroxyl groups and His181N.The active site pocket is surrounded by Gly 156, Tyr 159, Ile160, Asp 179, His 181, Tyr 188, Lys 213, Val 214,, Ile 243, Gly 244, Phe 245 residues, which interact with the ligand adenosine-5′-diphosphate Fig. 4.

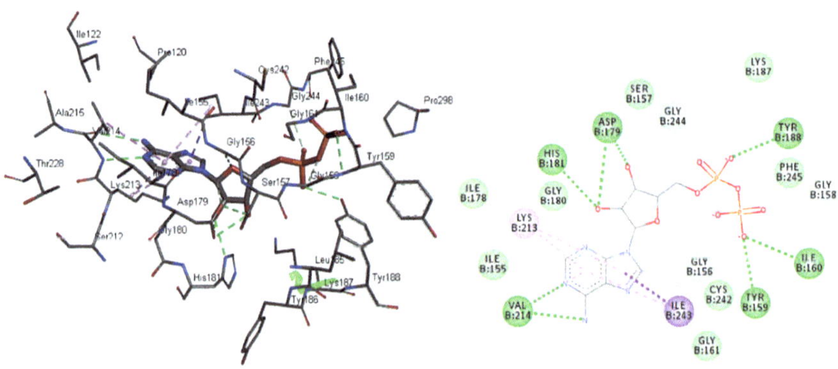

Fig. 4 Interations adenosine-5′-diphosphate bindin site pocket NADPH oxidase: van der Walls; Conventional H-Bond; Carbon H-Bond; Pi-Sigma and Pi-Alkyl

3.3 *Workflow*

3.3.1 Hardware and Software

Computer: Linux, Macintosh or Windows PC, Internet access, pyrxvirtual, Autodock, Autodock Vina, OpenBabel and Python 2.5, Discovery Studio and Text editor.

The typical inputs of HTBVS workflow are selection the target protein, either experimentally solved or computationally modeled and a compound library of small molecules available via purchase in data base or synthesis.

PyRx 0.8 [52, 53] downloaded in https://pyrx.sourceforge.net/downloads it is a free powerfull GUI with AutoDock Vina [54], AutoDock 4.2 [55], Mayavi, Open Babel and Python to docking molecular and virtual screening studies. Two fitting methods can be performed using PyRx to meet different needs. Start with Pyrx to run the AutoDock 4.2, continue sequentially observing the results compared to the complete study. This initial procedure is called validation, in this point will be observed docking for the estimation of ligand-receptor affinity with complex. Then, perform the molecular coupling of the entire test set with autodock vina to VS using the parameters obtained by grid validation.

3.3.2 Protocol

Prerequisites: PyRx 0.8, Autodock Vina 1.1.1, Autodock 4.2, Python 2.7 and Discovery Studio Visualizer 2019 [56].

Files Supplied: Targets-(CYP450 ID 1OG5, MP ID 1DNU, LOX ID NQ8, XOR ID 3NRZ and NOX ID 2CDU; ligands- ZINC000029562604(Z04), ZINC000000004840 (Z40),ZINC000003872277(Z77), ZINC3984017(17) and molecules controls 5-Fluorouracil (FLU), Dextromethorphan (DEX).

This protocol establish how to performe a virtual screening for potential inhibitors of ROS enzyme, using docking applications PyRx. The procedure will be to demonstrate how to run the docking to generate a protein receptor grid, set ligands to dock in target grid and analyze results.

The PyRx procedure is summarized in 10 steps:

1. Double-click in PyRx Icon: 

   Initially the structures com be imported from PDB directly, or from directory designed workspace added in Browse of prefernces.
2. Go to **Edit > Preferences** (Find the dirtectory and choose Set workspace Browse) Fig. 5.
3. File > Load Molecule
   This botton load your **protein** and **ligand (s)** into PyRx workspace (Fig. 6).
4. Right Button Click > Autodock > Make Macromolecula (Fig. 7)

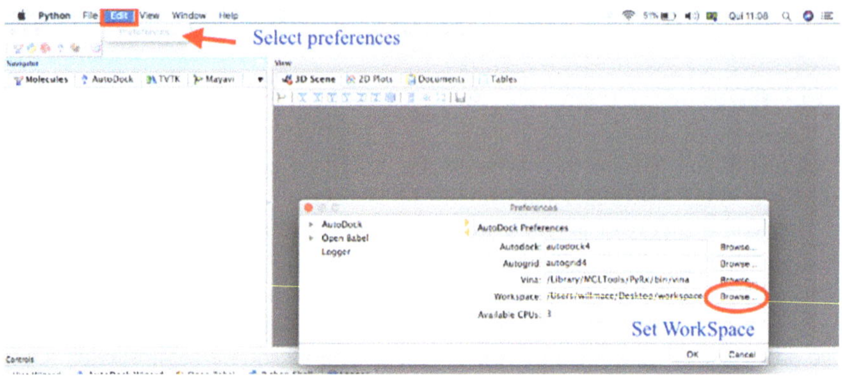

Fig. 5 Select a preferences directory

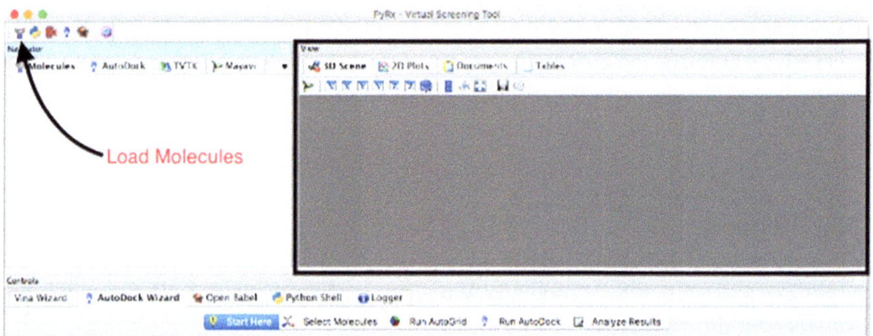

Fig. 6 Load molecules

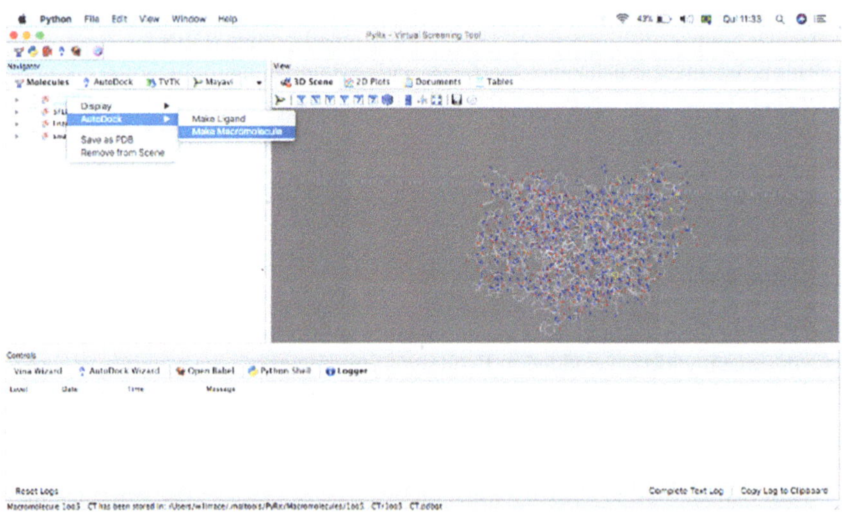

Fig. 7 Convert macromolecule to PDBQT format

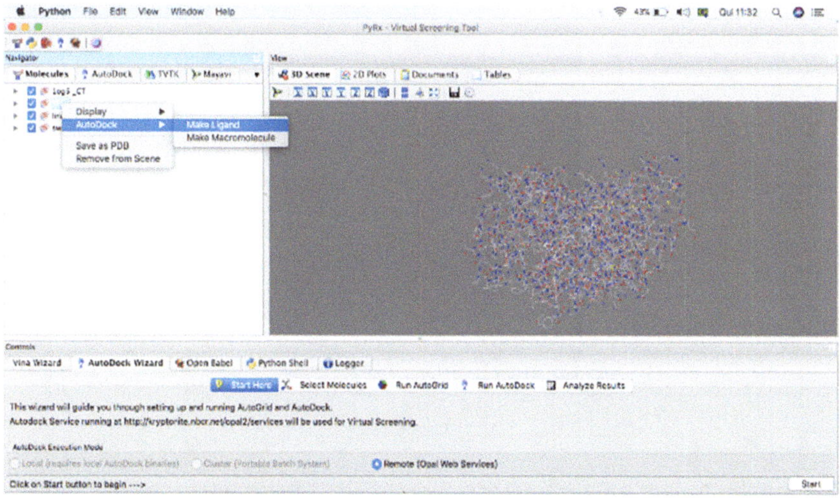

Fig. 8 Convert ligand to PDBQT format

Select protein and convert to PBBQT autodock input format. In Make Macromolecule will be add polar hydrogens atmom, Atom types, charges in macromolecule and parameters necessary to proced docking.

5. Right Button Click > Autodock > Make Ligand
   Now select ligand and convert to PQDQT format. This conversion is realized for Open Babel directly included in Pyrx software (Fig. 8).
6. Select the Ligand and Macromolecule for view
   In this window you can visualize all loading molecules (Fig. 9).
7. In Vina Wizard select start here and leave the option checked: Local (using /Library/MLGTools/PyRx/bin/vina) and click start.
   In this step, you can choose to perform docking calculations on your own PC or perform on a server externally. this protocol demonstrates how to perform in local mode. To perform on an external server, one must consult the correct address which may change according to the updates (Fig. 10).
8. Then This Screen Will Appear Noting that the Grid Box Must Be Involve All the Residues of the Active Site in 3D Scene (Fig. 11).
9. The bottom Toggle Selection Spheres.

The sphere selection button will help to make it possible to dimension the grid box with a minimum size which must contain the binding pocket. Select the residues Arg 97, Gly 98, Ile 99, Phe 100, Leu 102, Ala 103, Val 113, Phe 114, Asn 217, Thr 364, Ser 365, Leu 366, Pro 367 and Phe 476 (Fig. 12).

10. Click **forward** button to start Vina calculations (Fig. 13).
    In this step autodock vina is running all simullations to set compunds ou just one compond. When the calculation is then finished, in particular, the obtained results will be available in the workbook selected in the preferences.

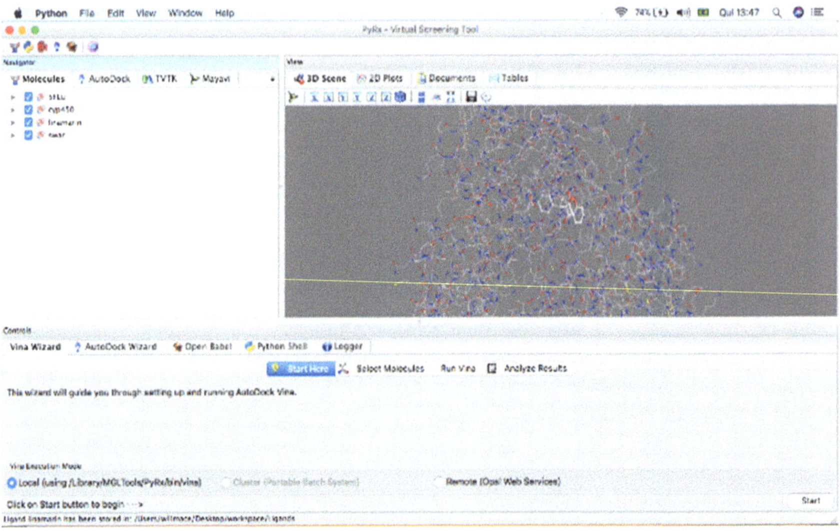

Fig. 9 Visualization loading molecules

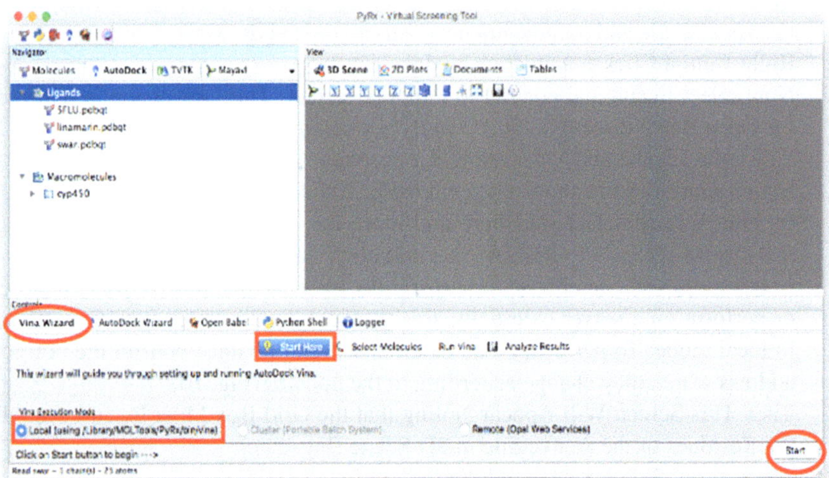

Fig. 10 Configuation wizard to local calculations

11. Finished calculations

When the calculations are completed, the results will be filled in as shown below in the table with the Binding affinity values BA (kcal / mol). The Most significant parameter is the most negative value regarding the BA, which must be associated with better orientation binder at the pocket (Fig. 14).

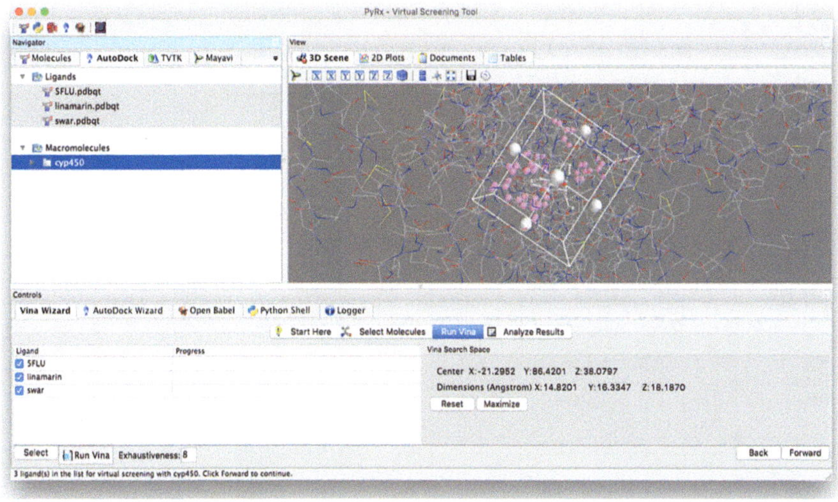

Fig. 11 Preview grid box extension

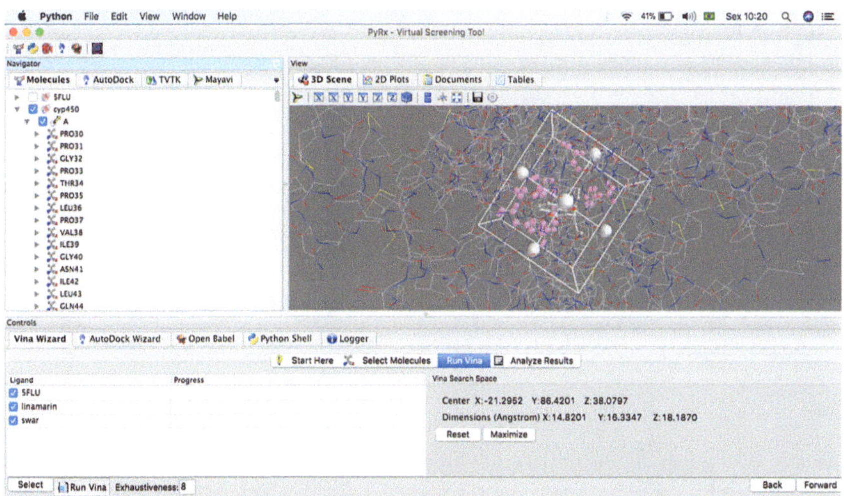

Fig. 12 Visualize the grid box without outlines residues

4 Evaluation of Molecular Docking

The molecular docking calculations take into account the contributions of intra and intermolecular forces (TROTT; OLSON, 2010) [57], and require a prior specification of the search spaces and the elaboration of the grid box containing the specifications

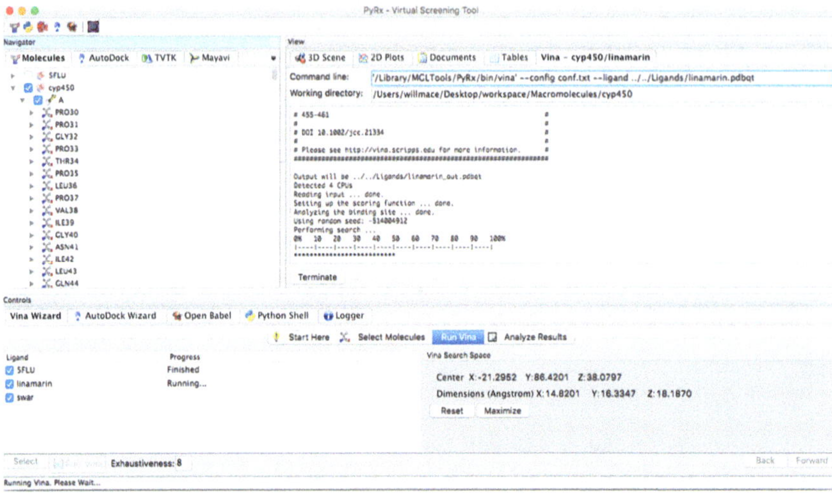

Fig. 13 Run autodock vina

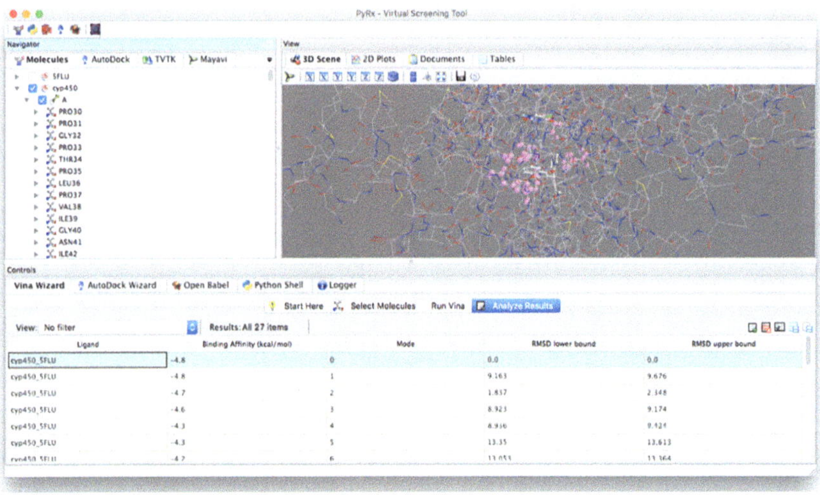

Fig. 14 Results obtained to analyses in Discovery Studio or another molecule visualization software

regarding the size of the binding site that involves the enzyme site of interest is assigned to information from the X, Y, Z coordinates of the search point.

At this point there is a translation problem to solve, the translation in the pocket and its relationship of effective interaction with the binder that must have the conditions in terms of a rotational solution for their rotational degrees of freedom according to statistical thermodynamics. Then when these conditions are satisfied, we will typically have the optimization of the space of translation and rotation of the studied system, associated with the accuracy of connection distances and angles present in the receptor-ligand relationship (Wang et al. [58]).

The use of the Lamarckian algorithm allows quite an assessment in that task. The 10 best results obtained are ordered based on the lowest energy values and Root Mean Square deviation RMSD, such results that should be inspected visually with the overlapping of the 3D ligands in the target site containing the ligand. The ideal is to obtain a good crystallographic structure containing the co-crystallized ligand (Aggarwal et al. [59]). The RMSD value essential help in analyzing the difference between coordinates when theoretical versus experimental observed, as the difference is essentially the distance here between those coordinates which are calculated as an average quadratic function of the square root on these who pondered the distances between observations and calculated between theoretical and experimental values. Usually, the observed or incorrect initial limit is used as compounds that attribute their analytical quality to values less than <0.5 angstroms. In that case, it is considered really good. however, treating the result as <1.5 angstroms is ideal for the accuracy of the study and with values around <20 being acceptable. However, there are some problems with the RMSD that should be noted. The dimensions of the ligand, in this case, it will be possible for the study to have a higher RMS which may be larger can be tolerated as opposed to a small fragment molecule and sometimes there may even be symmetry artifacts, depending on the programs that you use and your implementations.

The validation protocol for molecular coupling was performed with Pyrx using Autodock wizard (Autodock 4.2), where the results of the compound overlap at the binding site. Literatures express the RMSD values corresponding to the relationship between the x, y and z coordinates of crystallographic data calculated from the complexed ligand with acceptable resolution up to 2.0 Å [14–16]. The visualization of the superposition of crystallographic poses obtained shows the similarities regarding orientation, calculated conformation, obtained by means of molecular coupling which exhibits a low RMSD value, characterizing good results (see Fig. 15). These results suggest that the protocols used in molecular coupling analyzes are very useful, to assess the relationship between the selected receptors and the ligands.

CP450 complexed with S-warfarin shows the active site of an interaction is constituted by ARG97, GLY98, ILE99, PHE100, LEU102, ALA103, VAL113, PHE114, ASN217. Waste THR364, SER365, LEU366, PRO367 and PHE476. Specific interactions as described in PDB 1og5. Pi–pi interaction), PHE100 and ALA103, hydrogen bonding are present [17], showing similarity of interactions with the CP450 receptor between the tested and control molecules Table 3.

The values of BA (Z04 = -6.7 kcal mol^{-1}, Z40 = -7.4 kcal mol^{-1} and Z77 = -6.6 kcal mol^{-1}, suggests antioxidant capacity for these molecules.

Figure 15 shows the 2D interactions of the tested molecules linamarin (Z17), Z40, Z04, Z77) and control (FLU) with the CyP450 receiver. The amino acid residues of the binding site (PRO367, ALA 103, LEU366 and PHE 114) are present in the connections of Z04, Z40, Z77 and FLU. suggesting that these molecules may have antioxidant potential. The molecules respectively) described BA (-6.7, -7.4 and $-$ 6.6 kcal mol^{-1}, respectively) greater than control 1 (FLU) with BA = -4.7 kcal.mol^{-1}.

Complex	Grid data File
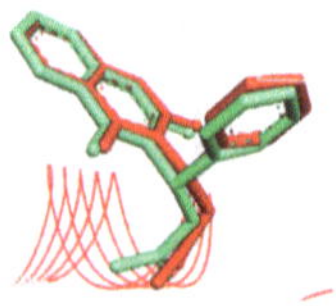 *S*-warfarin in Complex CYP450 ID 1OG5 Resolution 2.55 Å ΔG= -9.04 kcal mol^{-1} RMSD 0.47 Å	Grid Point Spacing =0.375 Angstroms Coordinates of Central Grid Point of Maps (-20.025, 87.282, 38.599) Even Number of User-specified Grid Points 48 x-points 50 y-points 50 z-points
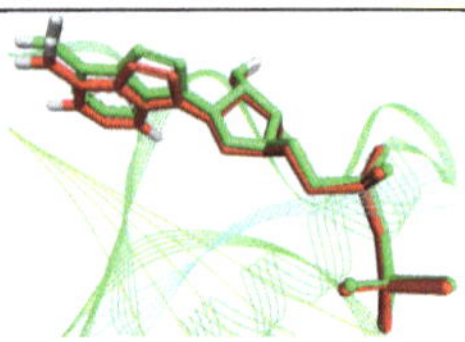 adenosine-5′-diphosphate in Complex NOX ID 2CDU Resolution 1.80 Å ΔG=-4.49 kcal mol^{-1} RMSD 0.48 Å	Grid Point Spacing =0.375 Angstroms Coordinates of Central Grid Point of Maps (0.380, 8.526, 51.245) Even Number of User-specified Grid Points 48 x-points 40 y-points 64 z-points

Fig. 15 Results of the validation protocol: in green experimental pose and redocking in red representation calculated for CYP450 and NDPH oxidase NOX

The antioxidant capacity is evidenced by the similarity of interactions and site assets [18]. The local asset is surrounded by residues ILE160, ILE243, ASP179, LYS213, VAL214 and TYR188, which interact with the adenosine-5′-diphosphate ligand. (ILE243, ASP179, LYS213 and VAL214) (see Fig. 16).

Using the same analytical methodology to study the NOX enzyme against ligands, it appears that the molecules do not bind in a set of at least three residues to compose the catalytic triad. In this case, even the studied compounds had a satisfactory BA Z04, Z40, Z77 and Z17 energy when compared to the control energy, it clearly demonstrates that the NOX enzyme is not suitable to satisfy a minimal condition to classify such compounds as promising. In cases like this, one should look for an appropriate test set or analyze the tale of compounds in another binding site with another control and thoroughly study the possibilities of the enzyme in question Table 4.

In this way, the understanding of the aspects related to the activation and inhibition of ROS enzymes can provide the means for the development of the molecular design of possible modifications in the ligands, which result in a better thermodynamic and conformational system more favorable to docking with the active enzyme–substrate site, which may serve as a starting point for the rational proposition of promising new drugs that can treat diverse metabolic disorders Fig. 17.

Table 3 Binding energies of ligands components in blind docking against Cyp 450

Ligand	Free binding Energy in Kcal/mol*	Hydrogen bonding interacting residues		π-π Staked/ π-Alkil interactions	van der Waals interactions
		Conventional H-Bond	Carbon H-Bond		
	CYP450 ID 1OG5				
S-warfarin	10.1	–	Phe 100	Phe 114, Phe 476 / Ala 103, Leu 366, Pro 367	Arg 97, Val 113, Leu 102, Leu 208, Ile 213, Asn 217, Ser 365, Thr 364, Leu 388
5-fluorouracil	4.7	Gly 98, Phe100	–	Phe114/Leu 366, Pro 367, Ala 103	Arg 97, Val 113, Ile 99
ZINC000029562604(Z04)	6.7	Asn 217	–	Phe114/Leu 366, Pro 367, Ala 103	–
ZINC000000004840 (Z40)	7.4	Phe 476	Pro 367, Arg 97	Ala 477*/ Ala 103, Leu 366, Phe 114	–
ZINC000003872277(Z77)	6.6	–	–	Arg 97*, Val 113*, Leu 366*, Phe 114, Ala 103, Leu 366. Pro 367	–
ZINC3984017(17)	6.3	Arg 97, Cys 436, His 368	–	–	–

*Alkyl, **π-donor H, ***π-Cation

5 Final Considerations

HTBVS using molecular docking on crystallographic data from biological receptors with activities available in the Protein Database, allows to evaluate the possible protocols for coupling the studied molecules (Z04, Z40, Z77, Z17, FLU and DEX).

In this protocol the two receivers CYP450 and NO,have all been validated by with RMSD values (experimental x theoretical) less than 2.0 Å. A binding affinity (BA) was assessed for values with low ΔG values. A good one correspondence of negative BA values for CP450 and NO) associated with interactions of the Z04, Z40 and Z77, molecules in relation to the control molecules based on the Show good antioxidant capacity for the tested molecules.

Overall, the results obtained revealed that Z04, Z40 and Z77 have at least antioxidant potential at least three receivers (CP450 and NO) via ROS generation. Thus,

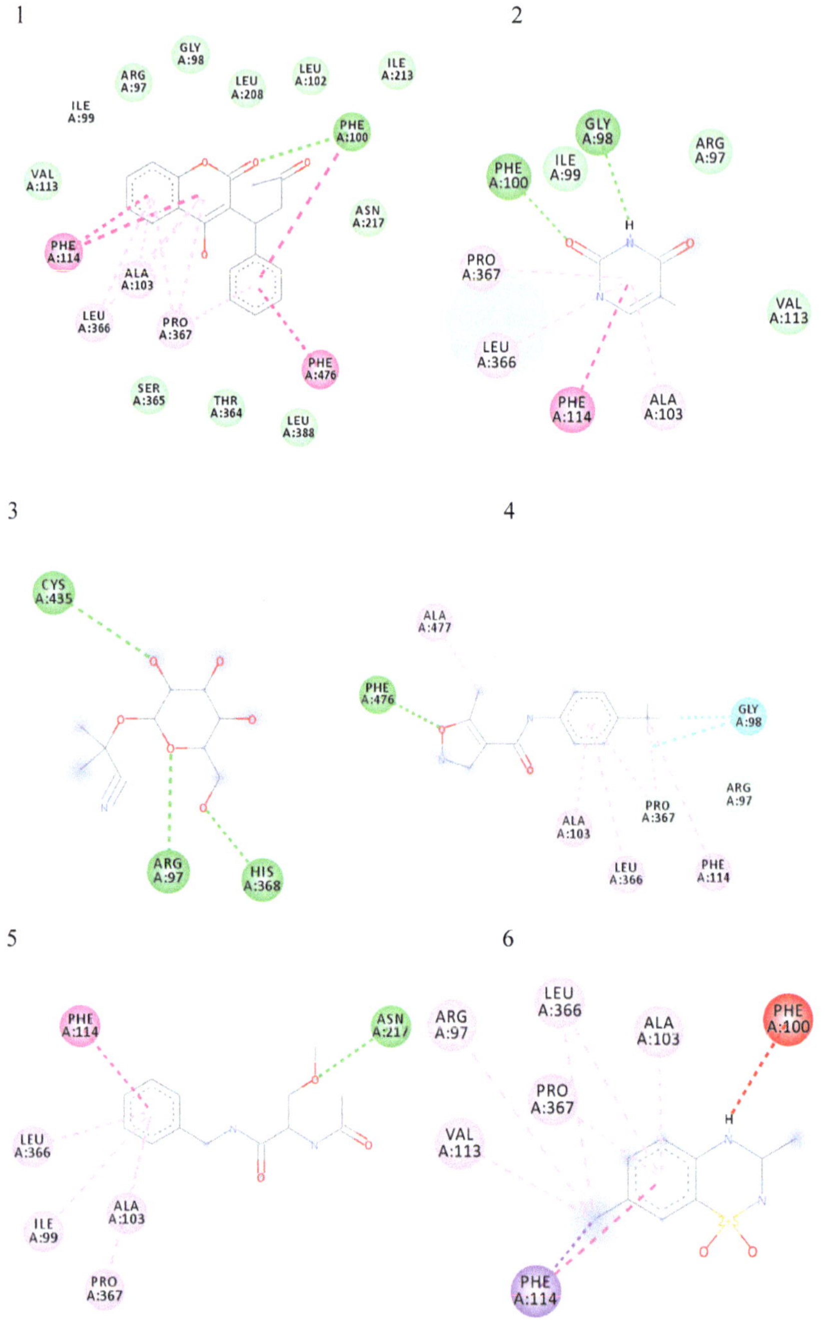

Fig. 16 Two-dimensional binding representation of Cyp P450 with different ligands: (1) S-warfarin, (2) fluoracil (control), (3) linamarin (Z17), (4) Z40, (5) Z04, (6) Z77. □ van der Walls; □ Conventional H-Bond; □ Carbon H-Bond; □ Pi-Pi Stacked and □ Pi-Alkyl

Table 4 Binding energies of ligands components in blind docking against NOX

Ligand	Free binding Energy in Kcal/mol*	Hydrogen bonding interacting residues		π-π Staked/ π-Alkil interactions	van der Waals interactions
		Conventional H-Bond	Carbon H-Bond		
	CYP450 ID 1OG5				
	NOX ID 2CDU				–
Aadenosine-5′-diphosphate	10.8	Tyr 159, Asp 179, His 181, Val 214, Tyr 188	Gly 156, Gly 144	Lys 213, Ile 243	–
Dextromethorphan	6.7	–	–	Phe 429/Leu 424, His 396	–
ZINC000029562604(Z04)	7.7	Thr 9	Thr 113	Lys 134***	–
ZINC000000004840 (Z40)	8.4	Asn 34, ser 41, Csx 42	–	Ala 303*	–
ZINC000003872277(Z77)	7.1				–
ZINC3984017(17)	6.9	Asp 397, Ser 401	Tyr 435	Met 423*, Leu 424*/Thr9**	–

*Alkyl, **π-donor H, ***π-Cation

results can be used to more analysis to better assess its efficiency in reducing oxidative stress and to be used as possible antioxidant to be used in the production or replacement of drugs within the pharmaceutical industry.

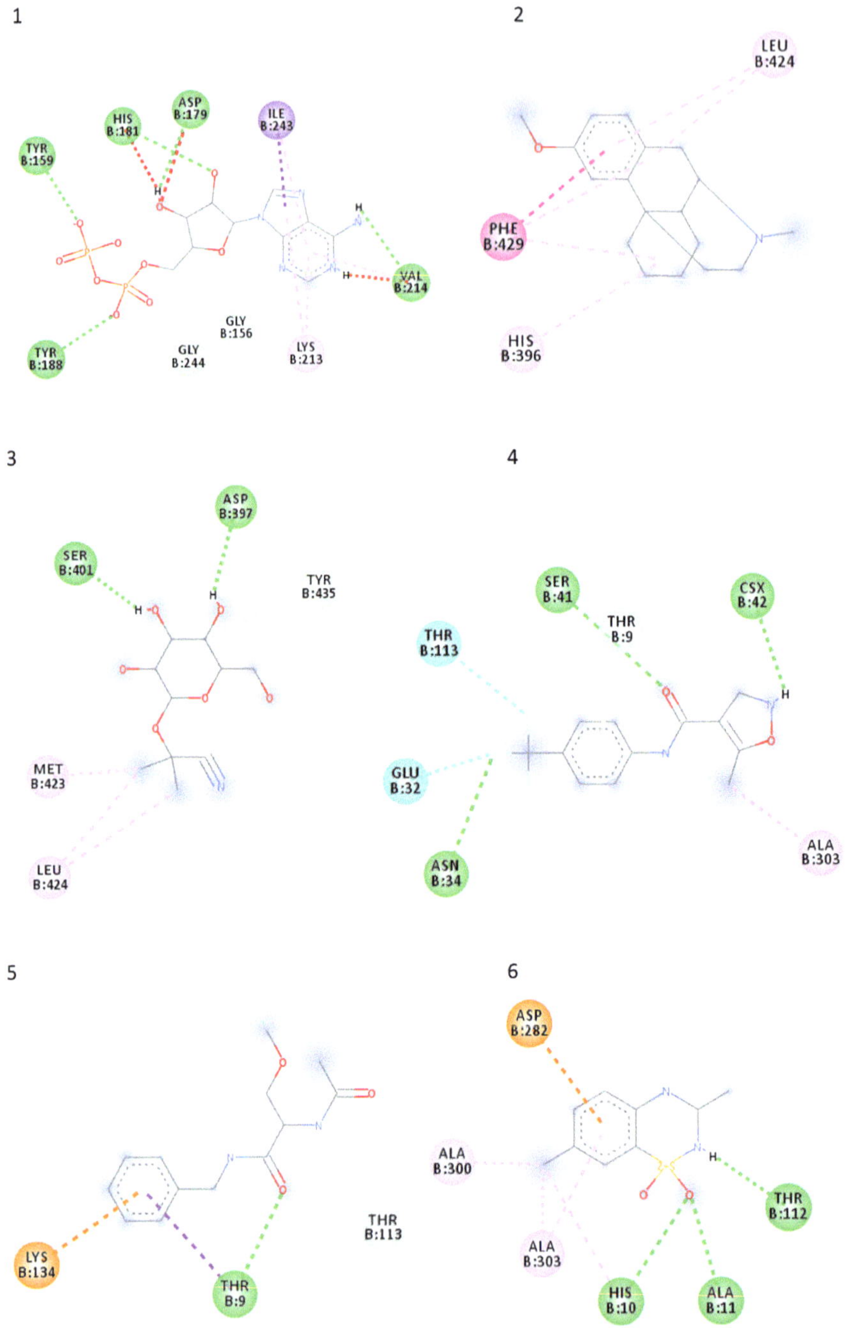

Fig. 17 Two-dimensional binding representation of NOX with different ligands: (1) adenosine-5′ -diphophate, (2) DEX (control), (3) Linamarin (Z17), (4) Z40, (5) Z04, (6) Z77. ☐ van der Walls; ☐ Conventional H-Bond; ☐ Carbon H-Bond; ☐ Pi-Sigma, ☐ Pi- Cation and ☐ Pi-Alkyl

References

1. Silva, A.A., Gonçalves, R.C.: Espécies reativas de oxigênio e as doenças do trato respiratório de grandes animais. Ci ê ncia Rural **40**, 994–1002 (2010). [CrossRef]
2. Yun, J., Mullarky, E., Lu, C., Bosch, K.N., Kavalier, A., Rivera, K., Roper, J., Chio, I.I., Giannopoulou, E.G., Rago, C. et al.: Vitamina C mata seletivamente kras e braf mutante colorretal células cancerosas, visandoGAPdh. Ciência **350**, 1391–1396 (2015). [CrossRef] [PubMed]
3. Branco, P.A., Oliveira, R.C., Oliveira, A.P., Serafini, M.R., Araújo, A.A., Gelain, D.P., Moreira, J.C., Almeida, Jr. R., Quintans, J.S., Quintans-Junior, L.J., et al.: Atividade antioxidante e mecanismos de ação de compostos naturais isolados de líquens: Uma revisão sistemática. Moléculas **19**, 14496–14527 (2014). [CrossRef][PubMed]
4. Reuter, S., Gupta, S.C., Chaturvedi, M.M., Aggarwal, B.B.: Estresse oxidativo, inflamação e câncer: Como eles estão ligados? Radic livre. Biol. Med. **49**,1603–1616 (2010). [CrossRef] [PubMed]
5. Cadenas, E.: Mecanismos básicos de atividade antioxidante. BioFatores **6**, 391–397 (1997). [CrossRef] [PubMed]
6. Dharmaraja, A.T.: Papel das espécies reativas de oxigênio (ROS) na terapêutica e resistência a medicamentos em câncer e bactérias. Química. **60**, 3221–3240 (2017). [CrossRef] [PubMed]
7. Gupta, M., Sharma, R., Kumar, A.: Técnicas de acoplamento em farmacologia: Quanto promissor? Comp. Biol. Chem. **76**, 210–217 (2018). [CrossRef] [PubMed]
8. Zheng, M., Liu, Z., Yan, X., Ding, Q., Gu, Q., Xu, J.: LBVS: an online platform for ligand-based virtual screening using publicly accessible databases. Mol. Diversity **18**(4), 829–840 (2014). https://doi.org/10.1007/s11030-014-9545-3
9. Bento, A.P., Gaulton, A., Hersey, A., Bellis, L.J., Chambers, J., Davies, M., Krüger, F.A., Light, Y., Mak, L., McGlinchey, S., Nowotka, M., Papadatos, G., Santos, R., Overington, J.P.: Nucleic Acids Res. **42**, D1083 (2014)
10. Sterling, T., Irwin, J. J., J. Chem. Inf. Model. **55**, 2324 (2015)
11. Kim, S., Thiessen, P.A., Bolton, E.E., Chen, J., Fu, G., Gindulyte, A., Han, L., He, J., He, S., Shoemaker, B.A., Wang, J., Yu, B., Zhang, J., Bryant, S.H.: Nucleic Acids Res. **44**, D1202 (2016)
12. Wang, R., Fang, X., Lu, Y., Wang, S., J. Med. Chem. **47**, 2977 (2004)
13. Olah, M., Mracec, M., Ostopovici, L., Rad, R., Bora, A., Hadaruga, N., Olah, I., Banda, M., Simon, Z., Mracec, M., Oprea, T.I.: In: Oprea, T. I. (ed.) Em Chemoinformatics in Drug Discovery, Wiley-VCH Verlag GmbH & Co. KGaA: Weinheim, cap. 9 (2005)
14. Wishart, D.S.: Nucleic Acids Res. **34**, D668 (2006)
15. Weininger, D.J.: SMILES, a chemical language and information system. J. Chem. Inf. Comput. Sci. **28**, 31–36 (1988)
16. Dalby, A., Nourse, J.G., Hounshell, W.D., Gushurst, A.K.I., Grier, D.L., Leland, B.A., Laufer, J.: Description of several chemical structure file formats used by computer programs developed at molecular design limited. J. Chem. Inf. Model. **32**(3), 244 (1992). https://doi.org/10.1021/ci00007a012
17. Berman, H.M.: The protein data bank: a historical perspective. Acta Crystallogr. A **64**(1), 88–95 (2007)
18. Bajusz, D., Rácz, A., Héberger, K.: Why is Tanimoto index an appropriate choice for fingerprint-based similarity calculations? J. Cheminf. **7**(1), 20 (2015)
19. Lipinski, C.A.: Drug-like properties andthe causes of poor solubility and poor permeability. J. Pharmacol Toxicol Methods **44**(1), 235–249 (2000)
20. Kim, S., Thiessen, P.A., Bolton, E.E., Chen, J., Fu, G., Gindulyte, A., Han, L., He, J., He, S., Shoemaker, B.A., et al.: PubChem substance and compound databases. Nucleic Acids Res. **44**, D1202–D1213 (2016)
21. Wang, Y., Bryant, S.H., Cheng, T., Wang, J., Gindulyte, A., Shoemaker, B.A., Thiessen, P.A., He, S., Zhang, J.: PubChem bioAssay: 2017 update. Nucleic Acids Res. **45**, D955–D963 (2017)

22. Kim, S.: Getting the most out of PubChem for virtual screening. Expert Opin. Drug Discov. **11**, 843–855 (2016)
23. Hahnke, V.D., Kim, S., Bolton, E.E.: PubChem chemical ¨structure standardization. J. Cheminform. **10**, 36 (2018)
24. Kim, S., Thiessen, P.A., Cheng, T., Yu, B., Shoemaker, B.A., Wang, J.Y., Bolton, E.E., Wang, Y.L., Bryant, S.H.: Literature information in PubChem: associations between PubChem records and scientific articles. J. Cheminform. **8**, 32 (2016)
25. Wang, Y., Xiao, J., Suzek, T.O., Zhang, J., Wang, J., Zhou, Z., Han, L., Karapetyan, K., Dracheva, S., Shoemaker, B.A., Bolton, E., Gindulyte, A., Bryant, S.H.: PubChem's bioassay database. Nucleic Acids Res. **40**(Database issue): D400–D412 (2012). https://doi.org/10.1093/nar/gkr1132
26. Kim, S., Chen, J., Cheng, T., Gindulyte, A., He, J., He, S., Li, Q., Shoemaker, B.A., Thiessen, P.A., Yu, B., Zaslavsky, L., Zhang, J., Bolton, E.E.: PubChem 2019 update: improved access to chemical data. Nucleic Acids Res. **47**(D1), D1102-1109 (2019). https://doi.org/10.1093/nar/gky1033. [PubMed PMID: 30371825]
27. Gilson, M.K., Liu, T., Baitaluk, M., Nicola, G., Hwang, L., Chong, J.: BindingDB in 2015: a public database for medicinal chemistry, computational chemistry and systems pharmacology. Nucleic Acids Res. **44**(D1), D1045–D1053 (2016). https://doi.org/10.1093/nar/gkv1072
28. Davies, M., Nowotka, M., Papadatos, G., Dedman, N., Gaulton, A., Atkinson, F., Bellis, L., Overington, J.P.: ChEMBL web services: streamlining access to drug discovery data and utilities. Nucleic Acids Res. **43**(W1), W612–W620 (2015). https://doi.org/10.1093/nar/gkv352
29. Liu, T., Lin, Y., Wen, X., Jorissen, R.N., Gilson, M.K.: BindingDB: a web-accessible database of experimentally determined protein-ligand binding affinities. Nucleic. Acids Res. **35**(Database), D198–D201 (2007). https://doi.org/10.1093/nar/gkl999
30. Gaulton, A., Hersey, A., Nowotka, M., Bento, A.P., Chambers, J., Mendez, D., Mutowo, P., Atkinson, F., Bellis, L.J., Cibrián-Uhalte, E., Davies, M., Dedman, N., Karlsson, A., Magariños, M.P., Overington, J.P., Papadatos, G., Smit, I., Leach, A.R.: The ChEMBL database in 2017. Nucleic Acids Res. **45**(D1), D945–D954 (2016). https://doi.org/10.1093/nar/gkw1074
31. Gilson, M.K., Liu, T., Baitaluk, M., Nicola, G., Hwang, L., Chong, J.: BindingDB in 2015: A public database for medicinal chemistry, computational chemistry and systems pharmacology. Nucleic Acids Res. **44**, D1045–D1063 (2016)
32. Liu, T., Lin, Y., Wen, X., Jorrisen, R.N., Gilson, M.K.: BindingDB: a web-accessible database of experimentally determined protein-ligand binding affinities. Nucleic Acids Res. **35**, *D198-D201* (2007)
33. Chen,X., Lin,Y., Gilson, M.K.: The binding database: overview and user's guide. Biopolym. Nucleic Acid Sci. **61**,127-141 (2002)
34. Chen, X., Lin, Y., Liu, M., Gilson, M.K.: The binding database: data management and interface design. Bioinformatics **18**, 130–139 (2002)
35. Chen, X., Liu, M., Gilson, M.K.: Binding DB: a web-accessible molecular recognition database. J. Combi. Chem. High-Throughput Screen **4**, 719–725 (2001)
36. Sterling and Irwin, J. Chem. Inf. Model, (2015). http://pubs.acs.org/doi/abs/https://doi.org/10.1021/acs.jcim.5b00559
37. . Wishart, D.S., Feunang, Y.D., Guo, A.C., Lo, E.J., Marcu, A., Grant, J.R., Sajed, T., Johnson, D., Li, C., Sayeeda, Z., Assempour, N., Iynkkaran, I., Liu, Y., Maciejewski, A., Gale, N., Wilson, A., Chin, L., Cummings, R., Le, D., Pon, A., Knox, C., Wilson, M.: DrugBank 5.0: a major update to the DrugBank database for 2018. Nucleic Acids Res. 2017 Nov 8 (2017). https://doi.org/10.1093/nar/gkx1037
38. Costa, J.S., Ramos, R.S., Costa, K.S.L., Brasil, D.S.B., Silva, C.H.T.P., Ferreira, E.F.B., Borges, R.S., Campos, J.M., Macêdo, W.J.C., Santos, C.B.R.: An In silico study of the antioxidant ability for two caffeine analogs using molecular docking and quantum chemical methods. Molecules **23**(11), 2801 (2018)
39. Gunes, A., Coskun, U., Boruban, C., Gunel, N., Babaoglu, M.O., Sencan, O., Bozkurt, A., Rane, A., Hassan, M., Zengil, H., et al.: Inhibitory effect of 5-fluorouracil on cytochrome P450 2C9 activity in cancer patients. Basic Clin. Pharmacol. Toxicol. **98**, 197–200 (2006). [CrossRef] [PubMed]

40. Drugbank (2020). Available online: https://www.drugbank.ca/unearth/(accessed on 3 mach 2020)
41. Liu, S.L., Li, Y.H., Shi, G.Y., Tang, S.H., Jiang, S.J., Huang, C.W., Liu, P.Y., Hong, J.S., Wu, H.L.: Dextromethorphan reduces oxidative stress and inhibits atherosclerosis and neointima formation in mice.Cardiovasc. Res. **82**, 161–169 (2009). [CrossRef] [PubMed]
42. Wu, T.C., Chao, C.Y., Lin, S.J., Chen, J.W.: Low-dose dextromethorphan, a NADPH oxidase inhibitor, reduces blood pressure and enhances vascular protection in experimental hypertension. PLoS ONE 9, 1–12 (2012). [CrossRef] [PubMed]
43. Anzenbacher, P., Anzenbacherova, E.: Cytochromes P450 and metabolism of xenobiotics. Cell. Mol. Life Sci. **58**, 737–747 (2001)
44. Hodgson, J.: ADMET–turning chemicals into drugs. Nature Biotechnol. **19**, 722–726 (2001)
45. Takahashi, H., Echizen, H.: Pharmacogenetics of warfarin elimination and its clinical implications. Clin. Pharmacokinet. **40**, 587–603 (2001)
46. Kaminsky, L.S., Zhang, Z.Y.: Human P450 metabolism of warfarin. Pharmacol. Ther. **73**, 67–74 (1997)
47. Williams, P.A., Cosme, J., Ward, A., Angove, H.C., Matak Vinković, D., Jhoti, H.: Crystal structure of human cytochrome P450 2C9 with bound warfarin. Nature **424**(6947), 464–468 (2003). https://doi.org/10.1038/nature01862
48. Vlahos, R., Stambas, J., Bozinovski, S., Broughton, B.R.S., Drummond, G.R., Selemidis, S.: Inhibition of Nox2 oxidase activity ameliorates influenza A Virus-Induced Lung Inflammation. PLoS Pathog. **7**(2), e1001271 (2011). https://doi.org/10.1371/journal.ppat.1001271
49. Selemidis, S., Sobey, C.G., Wingler, K., Schmidt, H.H., Drummond, G.R.: NADPH oxidases in the vasculature: molecular features, roles in disease and pharmacological inhibition. Pharmacol Ther **120**, 254–291 (2008)
50. Bedard, K., Krause, K.H.: The NOX family of ROS-generating NADPH oxidases: physiology and pathophysiology. Physiol Rev **87**, 245–313 (2007)
51. Lountos, G.T., Jiang, R., Wellborn, W.B., Thaler, T.L., Bommarius, A.S., Orville, A.M.: The crystal structure of NAD(P)H oxidase fromLactobacillus sanfranciscensis: insights into the conversion of O2into two water molecules by the flavoenzyme†,‡. Biochemistry **45**(32), 9648–9659 (2006). https://doi.org/10.1021/bi060692p
52. Dallakyan, S., Olson, A.J.: Small-molecule library screening by docking with PyRx. Methods Mol. Bio. **1263**, 243–250 (2015)
53. Morris, G.M., Goodsell, D.S., Halliday, R.S., Huey, R., Hart, W.E., Belew, R.K., Olson, A.J.: Automated docking using a Lamarckian genetic algorithm and an empirical binding free energy function. J. Comp. Chem. **19**, 1639–1662 (1998). [CrossRef]
54. Trott, O., Olson, A.J.: AutoDock Vina: improving the speed and accuracy of docking with a new scoring function efficient optimization and multithreading. J. Comput. Chem. **31**, 455–461 (2010)
55. Morris, G.M., Huey, R., Lindstrom, W., Sanner, M.F., Belew, R.K., Goodsell, D.S., Olson, A.J. (2009)
56. Dassault Systèmes BIOVIA, Discovery Studio Modeling Environment. Release 2017. Dassault Systèmes: San Diego, CA, USA, (2019)
57. Trott, O., Olson, A.J.: AutoDock Vina: improving the speed and accuracy of docking with a new scoring function efficient optimization and multithreading. National Institute of Health **31**(2), 455–461 (2010)
58. Wang, R., Lai, L., Wang, S.: Further development and validation of empirival scoring functions for structure-based binding affinity prediction. J. Comput. Aided Mol. Des. **16**(1), 11–26 (2002)
59. Aggarwal, R., Koes, D.R.: Learning RMSD to improve protein-ligand scoring and pose selection. ChemRxiv (2020). Preprint. https://doi.org/10.26434/chemrxiv.11910870.v2

Syntheses and Biological Applications of Fluorescent Probes

Suzane Quintana Gomes, Ismael Raitz, Mariana Pegrucci Barcelos, Carlton Anthony Taft, and Carlos Henrique Tomich de Paula da Silva

Abstract Fluorescent probes are powerful tools with vast potential for application in chemical biology. The specific characteristics of the main group of fluorophores coupled with the development of new techniques, have boosted their investigation in various research areas. For instance, the necessity of fluorescent tags applicable in different studies of subcellular localization and mechanisms of action of bioactive compounds has increased the development of fluorophores and new synthetic protocols toward the application in medicinal chemistry. This chapter discuss the first syntheses as well as modern synthetic methods, and some biological applications of the main fluorescent probes for drug discovery.

1 Introduction

The study of the molecular structure and its interactions encouraged the development of methods to identify and detect biomolecules such as enzymes, proteins, antibodies or amino acids [1]. One of the techniques developed was the fluorescent probe, a molecule that contains a fluorescent functional group, which changes its fluorescence emission when binding to a specific region of the target biomolecule, in response to a chemical reaction or alteration in their environment [2].

S. Q. Gomes · C. H. T. de Paula da Silva (✉)
Departamento de Química, Faculdade de Filosofia, Ciências e Letras de Ribeirão Preto (FFCLRP), Universidade de São Paulo, Av. Bandeirantes, 3900, Ribeirão Preto 14090-901, Brazil
e-mail: tomich@fcfrp.usp.br

I. Raitz · M. P. Barcelos · C. H. T. de Paula da Silva
Faculdade de Ciências Farmacêuticas de Ribeirão Preto (FCFRP), Universidade de São Paulo, Av. do Café, s/n, Ribeirão Preto 14040-903, Brazil

C. A. Taft
Centro Brasileiro de Pesquisas Fisicas, Rua Dr. Xavier Sigaud, 150-Urca, Rio de Janeiro 22290-180, Brazil

F. A. La Porta and C. A. Taft (eds.), *Functional Properties of Advanced Engineering Materials and Biomolecules*, Engineering Materials,
https://doi.org/10.1007/978-3-030-62226-8_18

These molecules are being widely used in drug discovery, cell imaging, environmental analysis, and various medical applications [3]. One of the main reasons for its use at scale is that the fluorescence emission can be measured with great sensitivity, versatility, and quantitative capacity. In addition, the composition of the probe can be varied in order to control the excitation and emission of wavelengths, binding affinity to the molecular target, chemical reactivity, among others [4]. Although fluorescent probes have only been used recently, the discovery of fluorescence has been going on for a long time.

1.1 The Discovery of Fluorescence and Its Mechanism

The first method used to detect and identify molecular compounds was the use of radioisotopes, however, over time, safer methods have emerged, based on fluorescence [4].

The first reports on fluorescence appeared in 1560 and 1565 by Bernadino de Sahagún and Nicolás Monardes, respectively. Both described an infusion of water from a small Mexican tree, used to treat kidney and urinary problems, which had a bluish opalescence [5].

Although several other scientists have described fluorescence over the years, it was not until 1852 that the term was coined [6, 7]. George Gabriel Stokes was responsible for coining the term in his article "On the Change of Refrangibility of Light" in which he describes the ability of fluorite and uranium glass to transform invisible light, in addition to the violet end of the visible spectrum, into light blue [8].

Fluorescence is a form of luminescence in which light is emitted by a substance that has absorbed electromagnetic radiation. In this case, the light emission occurs from an excited singlet state, in which the excited electron does not change the spin orientation, remaining unpaired [9]. During the short excitation period, part of the energy is dissipated by molecular collisions or transferred to a nearby molecule, while the remaining energy is emitted, in the form of a photon, returning the electron to its fundamental state [10].

The entire cycle of excitation and emission described above is presented in the Jablonski Diagram, designed by the Polish physicist Aleksander Jablonski and published in Nature in 1933, in the article entitled "Efficiency of anti-Stokes fluorescence in dyes". The diagram created by him is simple and effective, being divided into three energy levels and highlights the behavior of the functional group in the process of excitation of the electron from the fundamental to the excited state and its return with the emission of a photon with the length of the longer wave [11].

The chemical group associated with fluorescence is called fluorophore and was named by Richard Meyes in 1897 [12]. Fluorophores are divided into two major classes: intrinsic and extrinsic. The former corresponds to the functional groups that emit light naturally while the latter refers to those added to the sample to perform the probe function [9].

Each fluorophore has an excitation and emission wavelength, which is the fundamental aspect in the detection of fluorescence in biological applications. As such, the first fluorophores used in research emitted only in the visible range of the electromagnetic spectrum, mainly due to limitations in the detection of the photon [13]. Also, the entire fluorescence process is cyclical and it can be repeatedly performed until it is irreversibly damaged, allowing these molecules to be used in a wide range of research applications [14].

For the use of fluorophores in bioimaging, properties such as maximum absorption (λ_{max}) and maximum emission (λ_{em}), fluorescence quantum yield (Φ) and extinction coefficient (ε) must be considered, as well as solubility, the tendency to aggregate and photobleaching [15]. Fluorophores absorbs and emits in a range of wavelength, with a λ_{max} and λ_{em} values. Usually, the λ_{em} has a longer wavelength then the λ_{max} and the difference between them is called the Stokes shift. Quantum yield (Φ) is related to the brightness of emission and can be defined with the base in the ratio among the number of emitted photons by the number of absorbed photons. Extinction coefficient (ε) is also related to fluorophore brightness emission and measures the absorbed light by the dye. So, the highest extinction coefficient the greater is the amount of light will be absorbed by the dye. Photobleaching is the permanently loss of fluorescence upon continuous light excitation, caused by irreversible structural modifications in the fluorophore [15].

1.2 Types of Fluorophores

Several types of fluorophores have been developed in recent years due to technical advances and developments in fluorescence chemistry, although this property has been explored in biological research in the last 100 years [16].

Since each fluorophore has different characteristics and there is a vast amount on the market, these compounds have greater flexibility, performance, and variation for applications in the research area. Below, two general groups of these molecules may be explored: organic dyes and biological fluorophores [13].

Organic dyes

As is well-known, the first fluorescent compounds used in biological research were the synthetic organic dyes. Their small size allow them to participate in bioconjugation strategies with biomolecules without interfering with their biological functions. There is a wide variety of compounds comercially available, and derivatives of the original compounds were created to improve their photostability and solubility [17].

Biological fluorophores

The first biological fluorophore used in the research was the green fluorescent protein (GFP). This protein was cloned from the living water *Aequorea victoria* and used as a probe for gene expression [18]. After this use, several GFP derivatives were created and other proteins were designed, making the use of biological fluorophores common in research [19]. The advantage of this type of fluorophore over others is that the expression plasmids can be induced in other organisms and allow the expression of that fluorophore alone or fused to a protein of interest. On the other hand, its disadvantage is due to its size, the alteration of the biological function of the protein of interest is possible in addition to providing a sensitivity and photostability equal to synthetic organic dyes [17].

2 Synthesis of Fluorophores and Biological Applications

2.1 2,1,3-Benzoxadiazole

The history of the small-molecule probe 2,1,3-benzoxadiazole (**2**, benzofurazan) started in the late 1960s. The reduction reaction of benzoxadiazole 1-oxide (**1**, benzofuraxan) using hydroxylamine was the first synthesis to obtain benzofurazan (**2**) many decades before, in 1899 (Scheme 1) [20].

Fluorogenic property was revealed after efforts to synthesized 4,7-substituted 2,1,3-benzoxadiazoles (Scheme 2). 1,3-Dichloro-2-nitrosobenzene (**3**)—previously prepared by oxidation of the respective aniline—reacted with sodium azide to produce spontaneous 4-chlorobenzofurazan (**5**) via attack from the nitrogen of the

Scheme 1 First synthesis of 2,1,3-benzoxadiazole (**2**)

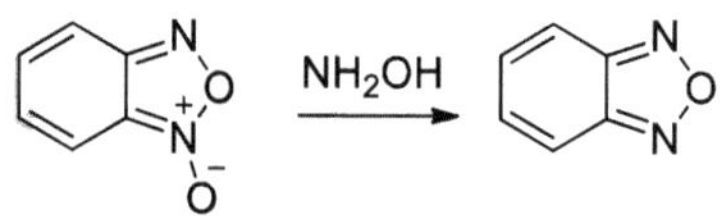

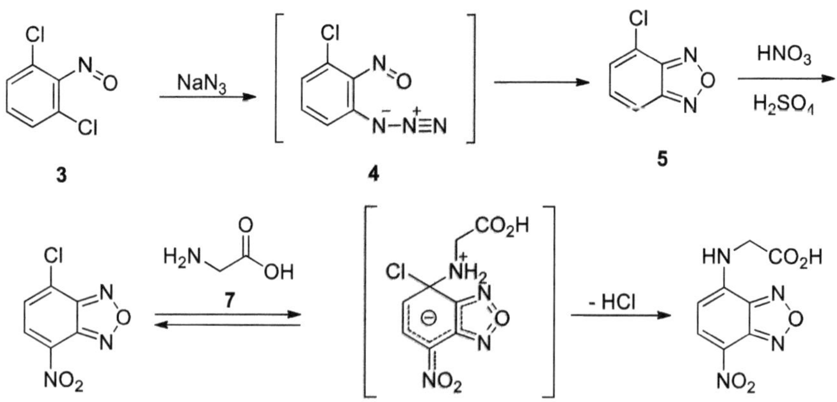

Scheme 2 Synthesis of 4-chloro-7-nitro-2,1,3-benzoxadiazole (**6**) and further amination

azide moiety to the oxygen on nitroso group with the loss of nitrogen gas [21]. Nitration reaction was performed to obtain the 4-chloro-7-nitro-2,1,3-benzoxadiazole (**6**) [22]. The high fluorescence was serendipity discovered after nucleophilic aromatic substitution of the chlorine atom of **6** to amino groups (Scheme 2) [23].

Benzofurazan ring was also prepared using *N*-(O-nitrophenyl)carbamates in high temperatures [24] and the propose mechanism is described in Scheme 3 [25].

Oxygen from the nitro group of the compound **10** attacks the electron-deficient carbonyl group from carbamate to remove the alkoxide ion which reacts with the proton on the nitrogen in an acid–base reaction toward carbon dioxide elimination. After that, cyclization of the 2,1,3- benzoxadiazole (**2**) is conclude when nitroso substituent attacks the nitrogen atom on *ortho* position [25]. Benzofurazan derivatives was also prepared in mild conditions reacting 4-substituted 2-nitroaniline with potassium hydroxide and sodium hypochlorite solution [26]. These methods make feasible different groups on benzene ring including those to produce fluorescent compounds.

Fluorescence observed in 4,7-substituted benzofurazans derivatives occurs due to system named "push–pull" wherein one substituent is electron withdrawal group (e.g. NO_2) and another one is an electron donor group (e.g. amino-alkyl) [27]. The substituent effects at the 4- and 7-positions were investigated evaluating seventy compounds which was classified into two groups according to difference of maximum excitation and emission wavelengths: excitation related to intramolecular charge transfer transitions and $\pi \rightarrow \pi^*$ transitions [28]. A new highly fluorescent compound precursor was synthesized based on this study considering Hammett substituent constant as parameter, the 4-phenylaminosulfonyl-7-fluoro-2,1,3-benzoxadiazole (**15**), which could be react with amines to produce the very intense chromophore **16** (Scheme 4), a useful tool to investigate amino acids [28]. Furthermore, molecular

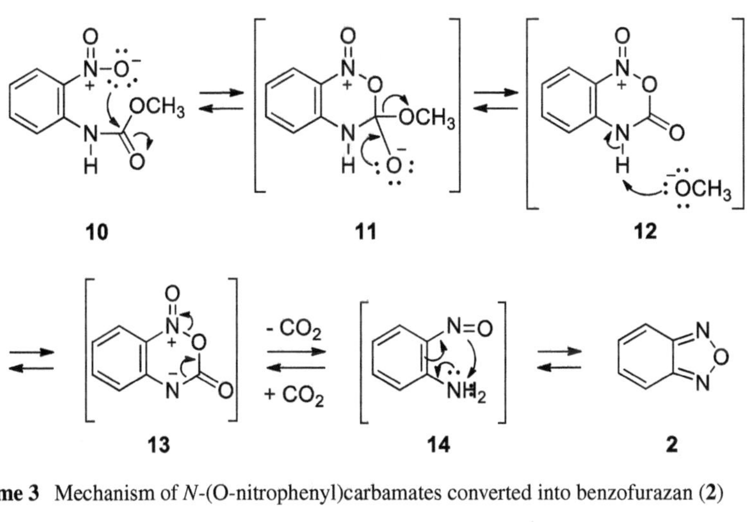

Scheme 3 Mechanism of *N*-(O-nitrophenyl)carbamates converted into benzofurazan (**2**)

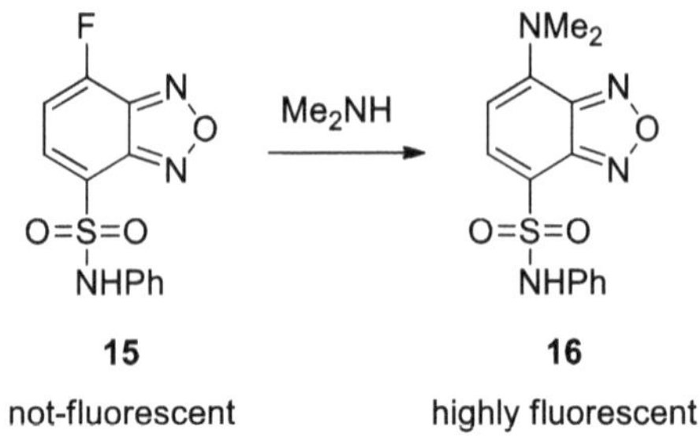

Scheme 4 Amination of 4-phenylaminosulfonyl-7-fluoro-2,1,3-benzoxadiazole (**15**)

orbitals study [29] and elucidation of fluorescence on–off switching mechanism—associated to fast intersystem crossing or fast photoreaction—of 4,7-substituted benzofurazans [30] was able to sum at the previous findings.

A small collection of 5-substituted benzofurazans had a similar evaluation and its results indicated the photoreaction is responsible for the relaxation process [26]. The quantum yield of these compounds was investigated according to excited states and ground states and that information is a start to develop new fluorogenic 2,1,3-benzoxadiazoles [26].

4-Chloro-7-nitro-2,1,3-benzoxadiazole (**6**) reacts with nucleophiles, however the 7-sulfonium salt **17** is water soluble and specific for thiols in protic solvents (Scheme 5) [31].

Compound **17** was synthesized via sulfonation reaction of the 4-chloro-2,1,3-benzoxadiazole (**5**) and neutralised with ammonium hydroxide. Next, in a competition reaction using product **17** and similar amine and thiol compounds, **17** was specific for thiol in aqueous medium [31]. The investigation continued and it proved in a similar fashion the same behaviour using different amino acids and peptides: only those bearing a free thiol groups had reaction with the 4-chloro-7-sulfobenzofurazan **17**. All products are highly fluorescents [31]. More fluorogenic sulfo-2,1,3-benzoxadiazoles **19** and **20** were specific for thiols—bearing fluorine rather than chlorine atom—and recommended as a tool for bioimaging (Fig. 1) [32].

The synthetic chemistry involved in nitrobenzoxadiazoles (NBD) were explored in bioimaging applications. Modern eletrophilic NBD derivatives could react with cysteine, homocysteine, and reduced glutathione (GSH) in different manners. Ether-NBD **21** was cleaved by thiol group and changed the fluorescence spectrum

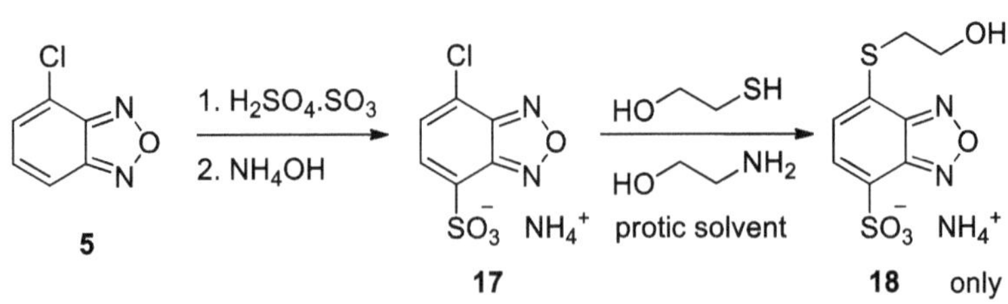

Scheme 5 Synthesis of 4-Chloro-7-sulfobenzofurazan ammonium salt **17** and its preferential thiolation

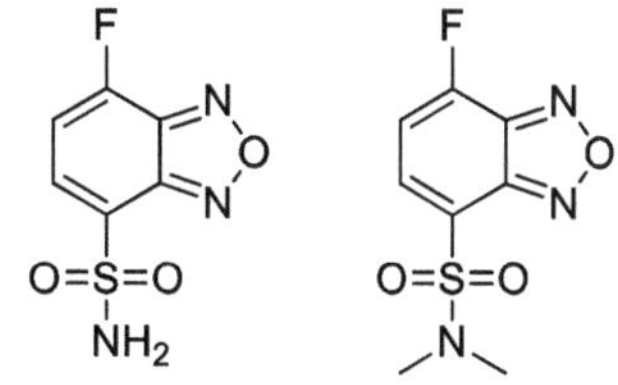

19 20

Fig. 1 Fluorogenic sulfo-2,1,3-benzoxadiazoles specific for thiols

(Scheme 6) [33]. Compound **24** was specific for cysteine and homocysteine because the thiol group replaced the thioether-pyrimidine and further it was converted to a higher fluorescent amino compound via intramolecular nucleophilic aromatic substitution reaction—Smiles rearrangement—due to proximity of amino residue which is not possible for glutathione (Scheme 7) [34].

Reactivity of benzofurazan derivatives was utilized for different species rather than amines or thiols. Compound **27** is an example of fluorescence probe for reactive oxygen species (Scheme 8) [35] and the structure **31** could identify formaldehyde (**32**) in living cells via aza-Cope rearrangement (Scheme 9) [36].

Fluorescent probes based on 2,1,3-benzoxadiazol (**2**) are an important tool to label biologically active substances. NBD derivatives **36** and **37** are tagged on ligands for mitochondria membrane receptor [37, 38], **38** on the Tigeclyne [39]—a newest

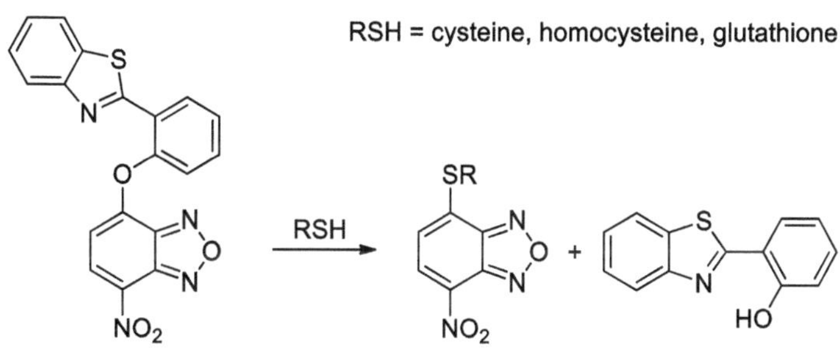

21 22 23

Scheme 6 Cleavage of NBD **21** by biological molecules

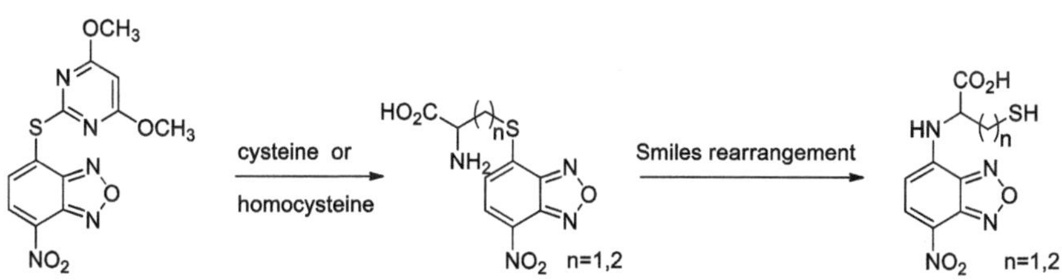

24 25 26

Scheme 7 Selective reaction of NBD **24** for cysteine and homocysteine

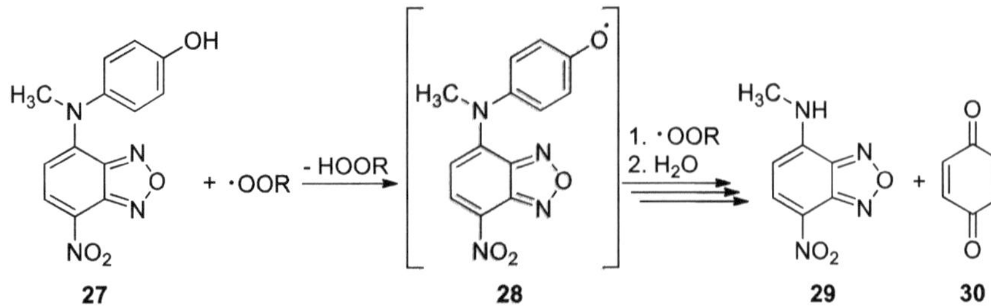

Scheme 8 Reaction of NBD **27** with reactive oxygen species

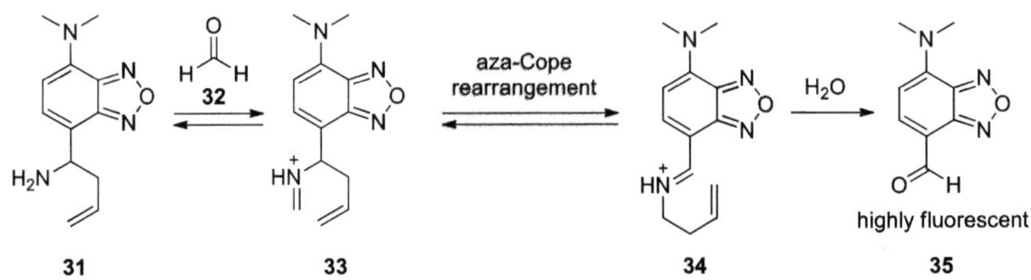

Scheme 9 2,1,3-Benzoxadiazole specific for formaldehyde

tetracycline derivative antibiotic with low toxicity—and **39** on the *D*-gluocose for the anti-diabetic investigation in zebrafish [40] (Fig. 2).

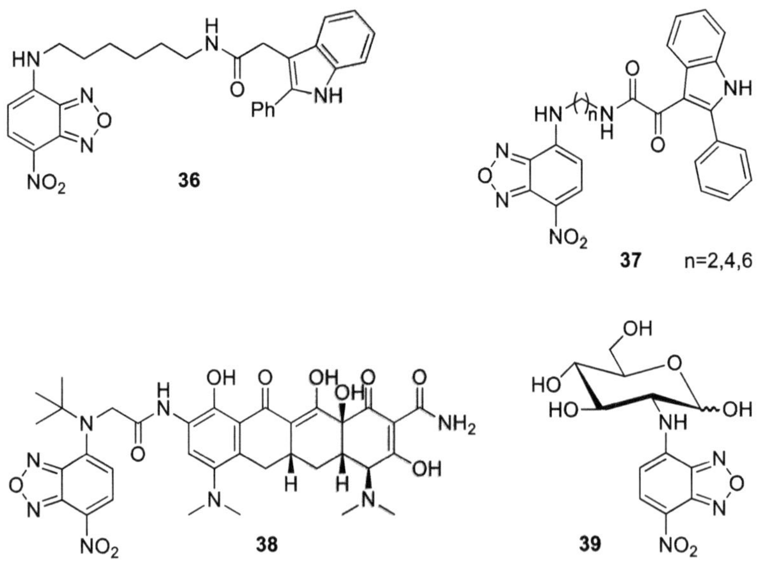

Fig. 2 NBDs structures tagged on biologically active molecules

2.2 2,1,3-Benzothiadiazole

Heterocycle 2,1,3-benzothiadiazole (**41**, BTD; piazthiole is the old name) is isostere of 2,1,3-benzoxadiazole (**2**). The basic synthetic pass to prepare it uses *ortho*-phenylenediamine (**40**) and a sulphur source like sulfur dioxide (SO_2), *N*-sulfinylaniline (PhNSO), bis(benzenesulfonyl)sulfodiimide [$(PhSO_2N)_2S$] or thionyl chloride ($SOCl_2$) [41]. The latter reactant is more common (Scheme 10) [42].

Each nitrogen atom of **40** attacks the sulphur atom in the thionyl chloride and, after aromatization process, BTD (**41**) is formed eliminating triethylammonium chloride and water.

An alternative method to synthesize BTD (**41**) uses 1-piperidinosulfenyl chloride (**42**) whose proposed mechanism is described in Scheme 11 what is like the thionyl chloride. However, dihydro-2,1,3-benzothiadiazole **44** is also isolated after purification [42].

The interest in the fluorescence of BTD (**41**) provided a detailed report about its spectral properties in 1975 [43]. Substituents on the benzene ring of the BTD (**41**) causes different brightness—as well as it works for benzofurazan (**2**). Recently, the synthesis of 4,7-diaryl-2,1,3-benzothiadiazoles **48** (Scheme 12) displayed new compounds with high fluorescent quantum yields justified by the π-extended conjugated structures [42].

Bromination reaction of BTD (**41**) to obtain 4,7-dibromo-2,1,3-benzothiadiazole (**47**) was not so feasible. BTD (**41**) and bromine mixture produced 4,5,6,7-tetrabromo-4,5,6,7-tetrahydro-2,1,3-benzothiadiazole (**46**) in an electrophilic addition reaction—and probably with the quinoid system intermediate **45** (Scheme 12) [44]. Product **47** was isolated only after it had reacted with potassium hydroxide [44]. However, a solution of BTD (**41**) and 47% hydrobromic acid, and droplet addition of bromine allowed prepare compound **47** in almost quantitative yield [44]. Dibromo-BTD derivative **47** is the ideal starting material to produce π-conjugated

Scheme 10 Synthesis of 2,1,3-benzothiadiazole (**41**, BTD)

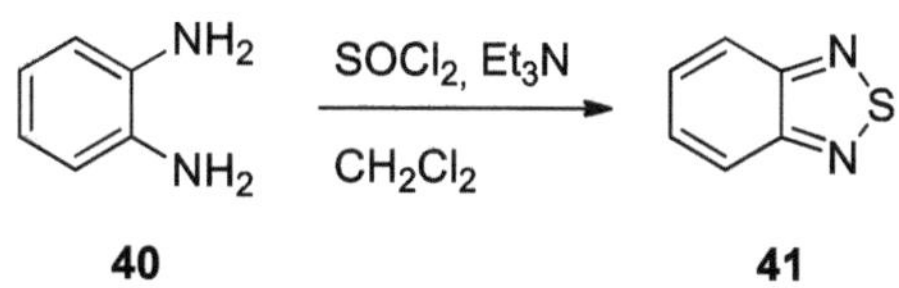

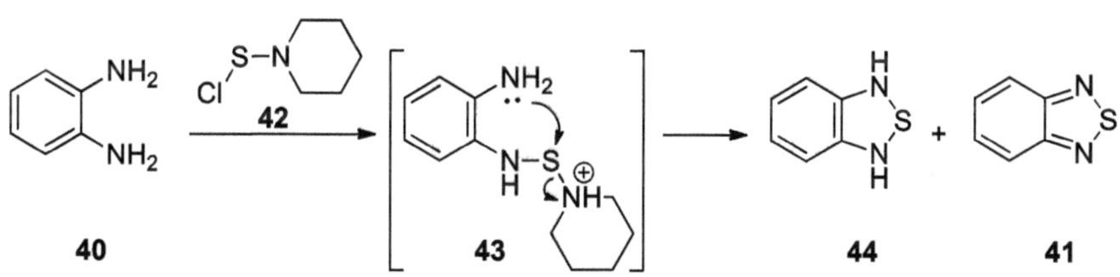

Scheme 11 Alternative method to synthesize BTD (**41**)

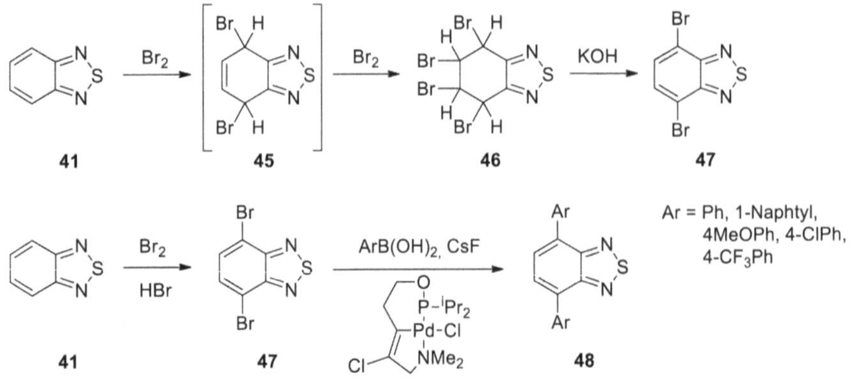

Scheme 12 Bromination of BTD (**41**) further Suzuki coupling

fluorescent BTDs by means the synthetic protocols of carbon–carbon couplings like Suzuki (Scheme 12) and Sonogashira reactions (Scheme 13) [42].

Fluorescence properties of 4,7-disubstituted-2,1,3-benzothiadiazole—aryl **48** and ethynyl **51**—were applied in biological assays in the last years. Compound **52** (Fig. 3) were a sensitive probe for DNA [45, 46]. Fluorescence of **52** occurs due to charge-transfer process from electron-donor 4-methoxyphenyl group to electron-acceptor BTD core [45]. Alkynyl moiety had the linker function—and π-extension conjugation to improve the brightness—for the DNA binding site [45]. BTD derivatives **53** and **54** were incorporated into DNA oligonucleotides showing a fluorescence dependence of the neighbourhood. Such structures were considered building blocks for optical DNA-materials owing π-π stacking interactions [47].

Other views to the BTD applications are a non-fluorescent cytotoxic nitro-derivative **55** converted to light-up amino-derivative **56** by nitroreductase in cancer cells (Scheme 14) [48], a simple push–pull fluorophore BTD **57**—4-(*N,N*-dimetilamino)phenyl group as electron donor—for lipid droplets in living and fixed cells (Fig. 4) [49], and a BTD derivative **58** specific to plasma membrane—

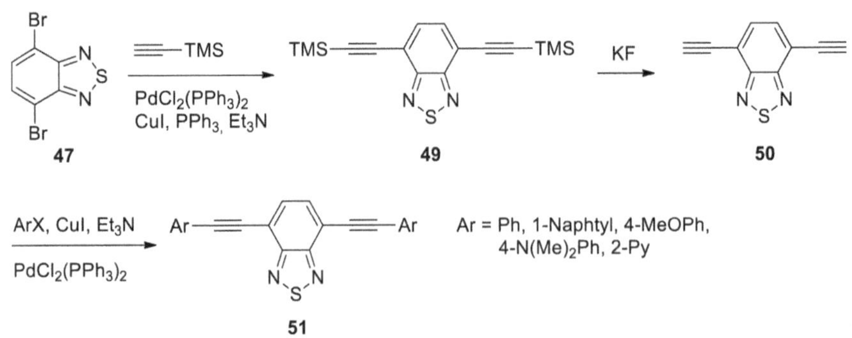

Scheme 13 Sonogashira coupling of dibromo-BTD (**47**)

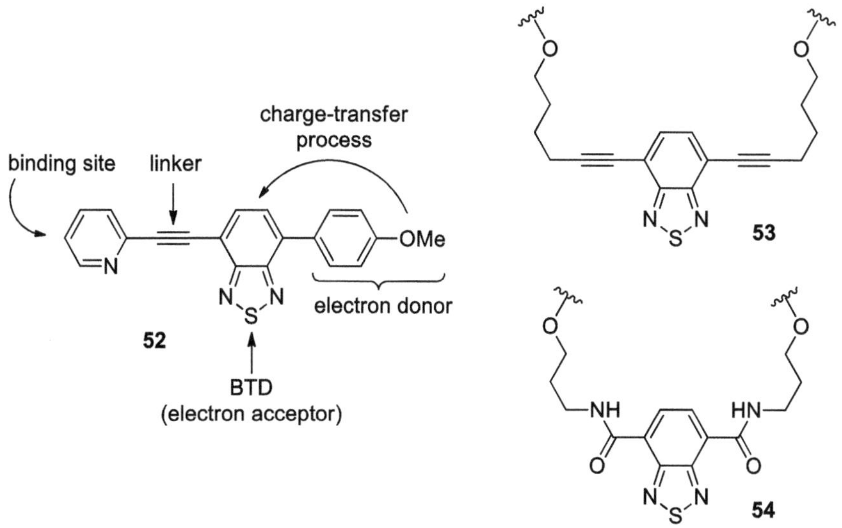

Fig. 3 Fluorescent BTD derivatives for DNA

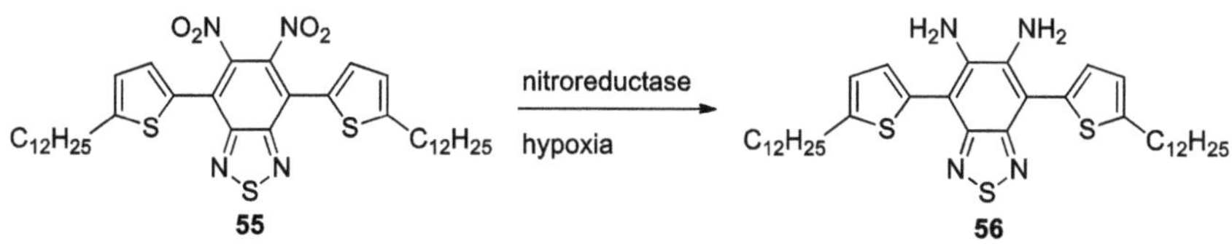

Scheme 14 Light-up process by action of nitroreductase enzyme

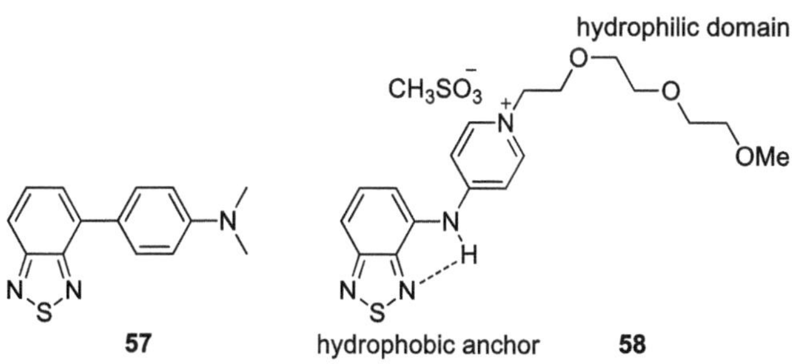

Fig. 4 BTD derivatives for lipids and plasma membrane

water soluble compound with a hydrophilic domain and a lipophilic anchor due to hydrophobic preferences of BTD core (Fig. 4) [49, 50].

Mitochondria is an important organelle to study because their disfunctions are responsible for many diseases [51]. Then, selective fluorescent probes for mitochondria—like some BTD derivatives—help to investigate cellular mechanisms of action

[52]. Previous presented molecule **58** has not aryl or ethynyl group directly attached to the heterocycle. Besides that, it has only one substituent. Structures like **58** were synthesized via direct Buchwald–Hartwig amination, a palladium-catalysed coupling reaction of amines with aryl halides in mild conditions and ample scope (Scheme 15) [51, 52].

2,1,3-Benzothiadiazole **60** displayed an excited state intramolecular proton transfer (ESIPT) effect that stabilize the molecule during the energy emission in addition to intramolecular charge transfer (ICT) [51]. Compound **60** stained only mitochondria in different tumor cell lines [51]. Backing to aryl- and ethynyl-BTD, substance **52** was also specific for mitochondria in bioimaging experiments in living cells [53].

A positional isomer of **60** had an additional study extending the molecule with alkyl groups on nitrogen atom of the pyridine moiety providing singly and doubly charged structures (Fig. 5) [52]. In biological assays, structures like **61** was selective for mitochondria and **62** was an unexpected fluorescent dye for plasma membrane [52].

Biologically active molecules may be a target to tag BTD probes. (*R*)-Goniothalamin (**63**) is a natural product has presented antitumor activity [54], and its structure allowed synthesize the racemic hybrid compound **64** with a fluorescent BTD indicating a preferential mitochondrial localization in cancer cell lines to this styryl lactone (Fig. 6) [55] Lapachone (**65**) is also a natural product with medicinal interest, and a fluorescent BTD core was tagged to the nor-β-lapachone by Click reaction (compound **66**) [56]—a 1,3-dipolar cycloaddition in mild conditions between azides and terminal alkynes with large application as synthetic tool for bioconjugation, including *in situ* reactions (Fig. 6) [57]. Quinone moiety was the potential

Scheme 15 Buchwald–Hartwig amination of 4-bromo-BTD

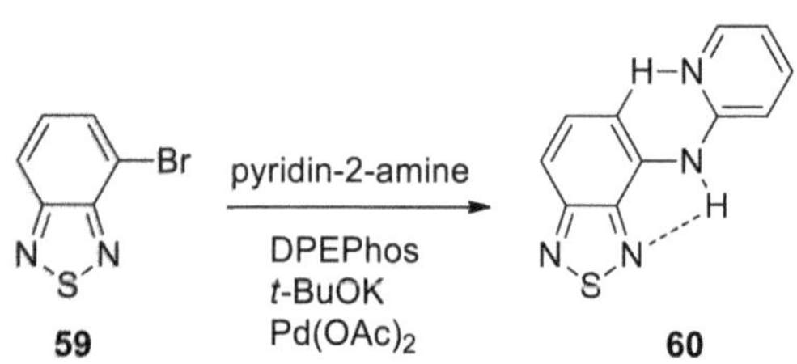

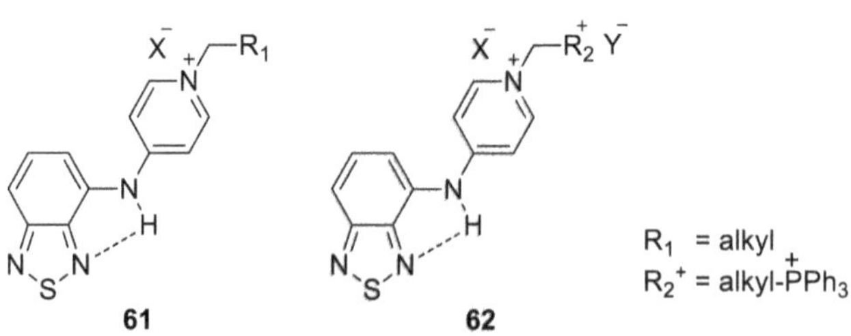

Fig. 5 Singly and doubly charged BTDs

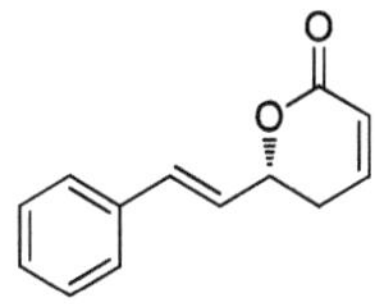

63

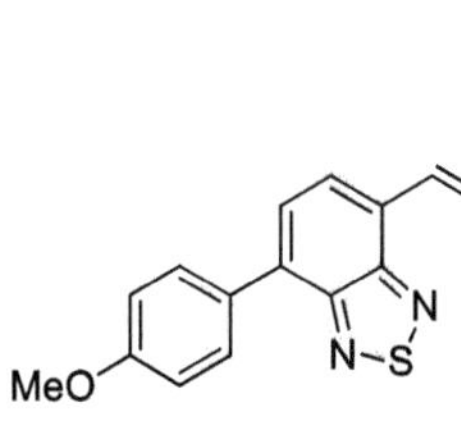

64

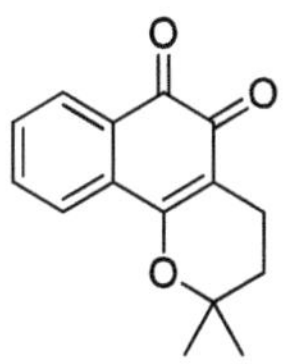

65

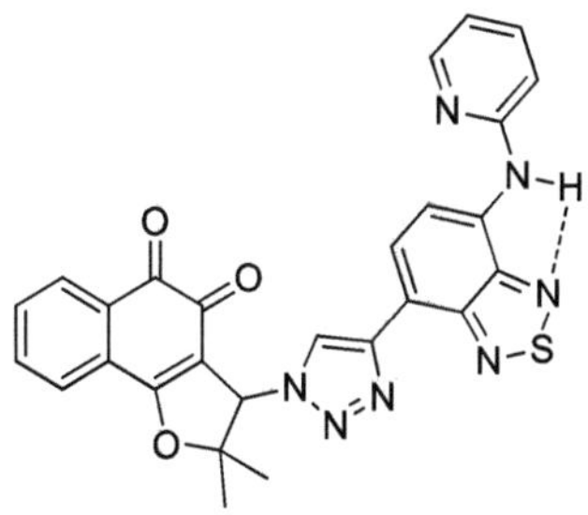

66

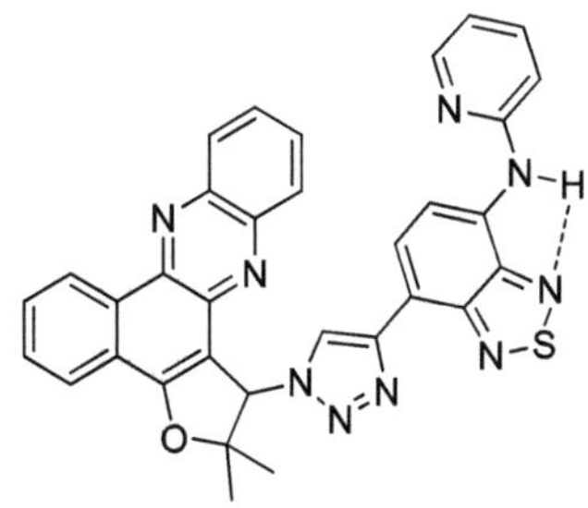

67

Fig. 6 BTD tagged into natural products derivatives

reason for the quenching in **66** which was solved synthesizing the compound **67**—by reacting **66** with *o*-phenylenediamine (**40**, Fig. 6) [56]. Even though nor-β-lapachone derivative **66** had low fluorescence intensity, it was active against tumor cell lines, what was not observed for **67** [56].

2.3 *Cyanine*

The first report of cyanine dye occurred in 1856 by Williams who synthesized a blue cyanine using quinoline, amyl iodide and ammonia as chemical reagents [58]. The general structure of cyanine dye (**68**) consists of two heterocyclic nitrogen rings, one of which is a positively charged, conjugated through an odd number of carbon atoms (Fig. 7). Cyanine dyes can be classified according to the number of carbons chain as monomethine (Cy1), trimethine (Cy3), pentamethine (Cy5), and heptamethine (Cy7) [59].

Fig. 7 General structure of cyanine dye

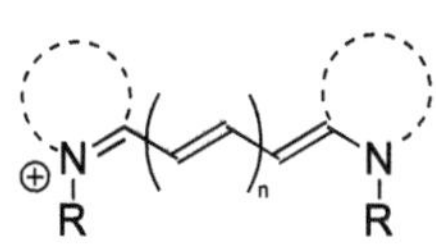

68

Cy3 n=3
Cy5 n=5

General synthesis of Cy3 (**71**), Cy5 (**73**) and other cyanine dyes involves the condensation of aromatic quaternary ammonium salts and condensing agents [59]. For the formation of trimethine (**71**), it is necessary the presence of methyl group that can be activated by base, resulting in the formation of a carbanion. It is also necessary the presence of a condensing agent as trimethyl othoester [59]. In this case, the formation of the condensing agent occurs in the reaction medium, with the protonation of trimethyl orthoformate (**70**), resulting in an aldehyde intermediate. This intermediary can undergo a condensation reaction. The cycle is repeated, resulting in the formation of trimethine (**71**, Scheme 16). The traditional synthesis of pentamethine (**73**) is similar to the synthesis of Cy3 (**71**). For Cy5 (**73**) synthesis, it is necessary the use of an unsaturated bis-aldehyde or equivalent compound, usually as Schiff base in the presence of the sodium acetate or other catalyst, for condensation reaction (Scheme 16) [59].

In 1995, Narayanan and Patonay reported a new non-catalyzed synthesis of heptamethine cyanine dyes. The protocol involved a quaternary salt of nitrogen heterocyclic rings and 2-chloro-1-formyl-3-(hydroxymethylene)-1-cyclohexene **75** in 1-butanol and benzene as solvent, resulting in a chloro dye intermediate **76**. The intermediate **76** can be reacted with other nucleophiles from phenols, in presence of sodium hydride, to form substitution products in high yields (Scheme 17). Additionally, it is essential to notice this protocol is not limited to synthesize symmetrical cyanine dyes due its slower rate reaction, but it is also not limited to the formation of pure nonsymmetric dye alone [60]. This protocol avoided some disadvantages of the traditional synthesis of heptamethine cyanine dyes, such as difficulty in purifying the product, the need to use a catalyst, problems during the synthesis of non-symmetrical dyes and low yield during reaction scale up [61].

Initially, the new cyanine dyes synthesized for use in biological systems exhibited disadvantages in photochemical and photophysical properties which precluded their application. However, the possibilities for change of heteroaromatic moiety and insertion of new groups in both heterocycles and nitrogen boosted the discovery of new cyanine derivatives with wide application in bioimaging studies [62].

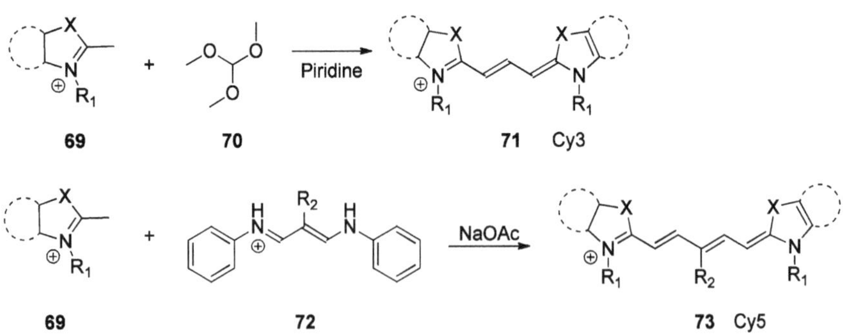

Scheme 16 General synthesis of Cy3 (**71**) and Cy5 (**73**)

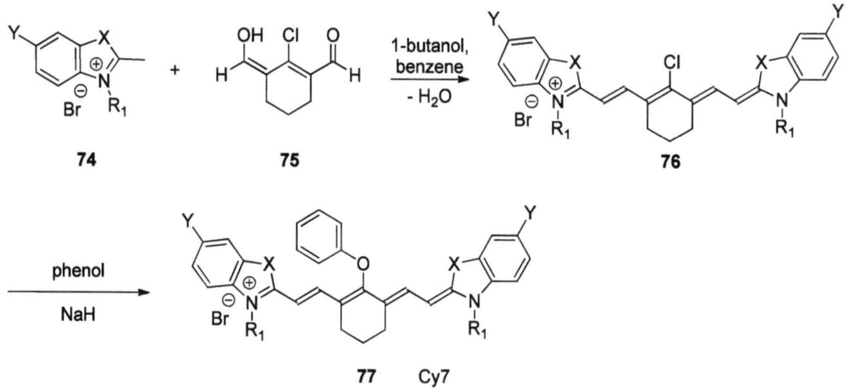

Scheme 17 General synthesis heptamethine cyanine dyes (**77**)

Fig. 8 General structure of the sulfoindocyanine dyes **78**

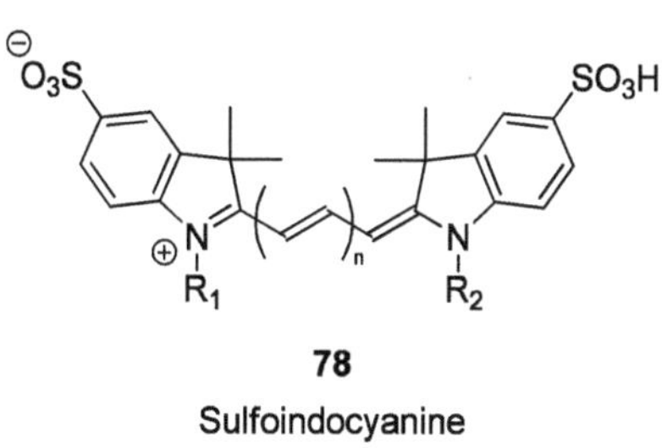

In 1993, Waggoner et al. synthesized a series of symmetrical and unsymmetrical sulfoindocyanine dyes **78**. The presence of negatively charged sulfonate group on the aromatic nucleus increased the water soluble, being used to prepare brightly fluorescent materials varying the visible to the near-infrared region (Fig. 8) [62]. Nowadays, sulfo-cyanine dyes with modifications in the structure (heteroaromatic moiety and *N*, *N'*-substituents) are still widely used in fluorescence applications.

In 2000, Lilley et al. could define the effective position of Cy3 (**71**) in the DNA. It stills commonly used in applications such as nucleic acid labeling, which attaches to 5′ end of a DNA duplex being stacked onto the end of the helix [63].

In general, Cy5 (**73**) and Cy7 dyes are more favorable for developing chemosensors. The Cy5 (**73**) and Cy7 fluoresces in greenish yellow and red region, respectively. Both have less photodamaging, low background interference, and less light scattering [64]. Almost all recent cyanine chemosensors developed for small molecule in biological image are based on heptamethine cyanine (Cy7) dyes. Modifications on the nitrogen of 3*H*-indolenine and substituents in the central chlorine atom on the cyclohexene ring improve the photostability, fluorescence and solubility of these dyes [65]. For instance, a highly selective and sensitive cyanine derivative **79** was developed for detection of arylamine *N*-acetyltransferase (NAT2). In this case, NAT2 transfers an acetyl group to the aromatic amine of the **79** causing an effective PET quenching effect in **80** (Scheme 18). The probe could monitor the NAT2 activity in

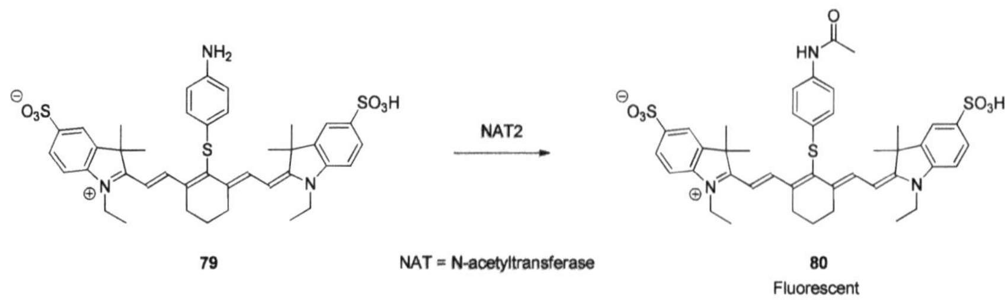

Scheme 18 Structure of derivative cyanine dye **79** used for detection of NAT2

enzymatic system, living mice and tissue homogenate samples [66].

A different application of heptamethine was the development of two cyanine-based fluorescent probes able to detect thiols such as glutathione, cysteine, and homocysteine in living cells. The new structures contain 2,4-dinitrobenzenesulfonamide group in **81** and 5-(dimethylamino)naphthalenesulfonamide group in **82** as substituents in the central chlorine atom on the cyclohexene ring (Fig. 9). The probe **81** exhibited a reversible response to thiols when treated with a thiol-blocking reagent *N*-methylmaleimide, while probe **82** showed high selectivity for glutathione over cysteine and homocysteine (Hcy), with reversible response when used *N*-methylmaleimide. Probe **82** proved to be able to monitor glutathione in a mouse model, with strong fluorescence emitted from various tissues, and, monitoring the depletion of glutathione in mouse tissue cells promoted by excessive administration of acetaminophen [67].

Other uses of cyanine dyes derivatives can be found in literature. Hemicyanine **83** and squarines **84** and **85** are members of the cyanine family, but biological applications still have a lack (Scheme 19) [64]. For instance, relatively short excitation and emission wavelengths of hemicyanines **83** and generation of aggregates in aqueous media of squaraines **84** and **85** are not desirable properties in biological research

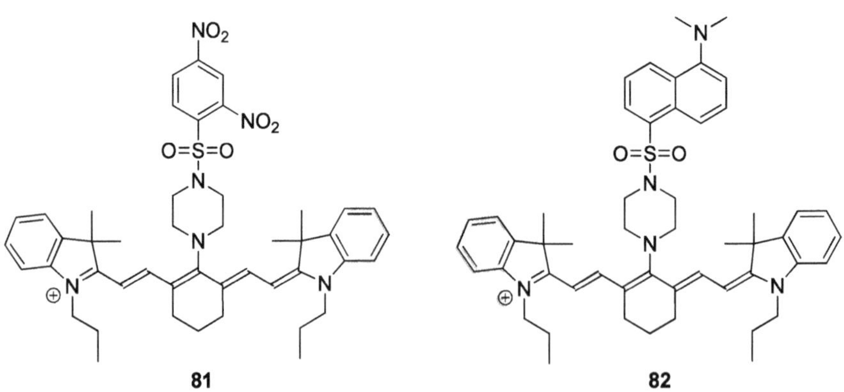

Fig. 9 Structure cyanine-based fluorescent probes **81** and **82**

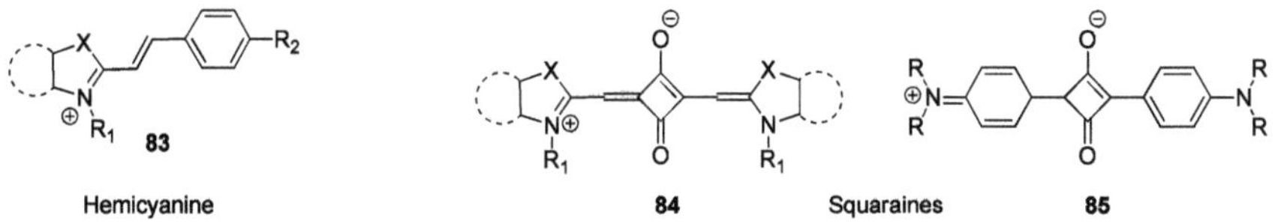

Scheme 19 Hemicyanines and squaraines cyanine family members

[64]. As mentioned, photochemical and photophysical properties can be improved by changing the heteroaromatic moiety and insertion of new groups in the structure of cyanine. Maybe, these strategies can also be applied to hemicyanine **83** and squaraines **84** and **85** for expand their use in biological systems.

2.4 Dansyl

In 1970, Mendel published a new protocol to synthesized dansyl chloride (**87**) [68]. Mendel complained that the commercially available dansyl chloride (**87**) had a variable purity and the current protocol was tediously because the need to melt the reagents, resulting in 25 to 40% yield. So, he described a new synthesis using phosphorus oxychloride in the presence of 5-dimethylamino-1-naphthalenesulfonic acid (**86**) and phosphorus pentachloride in a room temperature solution, in 85% yield (Scheme 20) [68]. Dansyl chloride or 1-dimethylaminonaphthalene-5-sulfonyl chloride (**87**) is a typical aromatic sulfonyl chloride, which reacts with a wide variety of bases, and, nowadays, is commercially available with 98% purity.

Hartley and Massey's proposed the use of dansyl structure in studies of the α-amino and other reactive groups of proteins and peptides, but was Gray who described the formation and stability of dansyl amino acids [69]. When dansyl chloride (**87**) reacted with primary and secondary amino groups, it formed a fluorescent sulfonamide adduct, while tertiary nitrogen bases catalyzed the reaction of dansyl chloride (**87**) with other bases [70]. In 1975, the amino acid sequence of natural enkephalin, a natural ligand for opiate receptors, was determined using dansyl-Edman procedure [71].

Jiang et al. reported a new dansyl based probe **88** for detection of cysteine *in vivo* based on a d-PeT switching mechanism (Fig. 10). The strategy was based on electron

Scheme 20 Dansyl chloride (**87**) synthesis

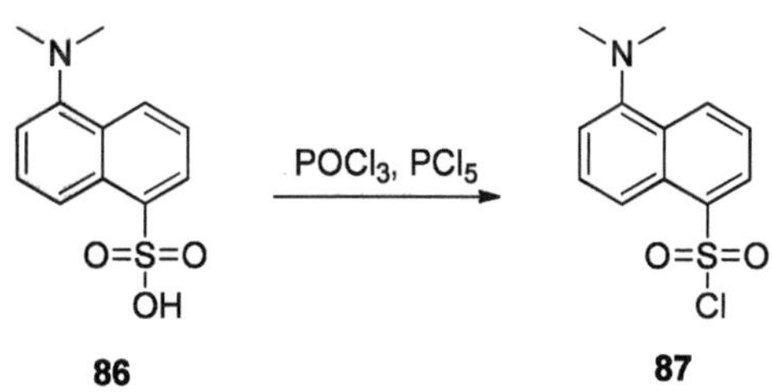

Fig. 10 Dansyl-based fluorescent probe **88** for detection of cysteine

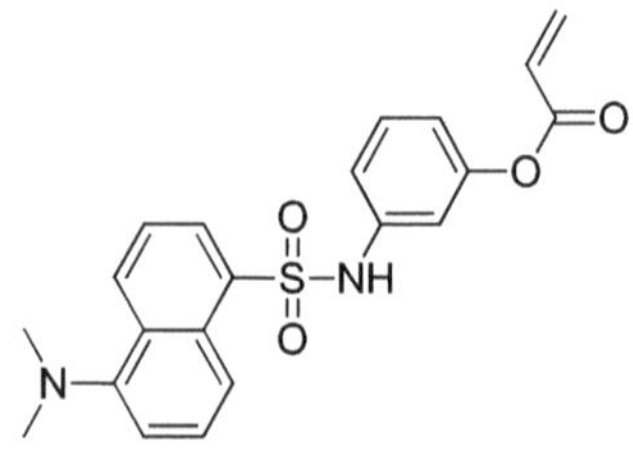

88

transfer from the dansyl moiety to the LUMO of acryloyl group, resulting in the lack of fluorescence. When acrylate is cleaved by the cysteine, the electron transfer and the fluorescence is restored with bright green fluorescence. The probe showed selectivity and sensitivity for cysteine *in vitro* over other various biologically relevant species, and could detect Cys in living cells with potential use in real-time cysteine monitoring [72].

Despite dansyl fluorophore is widely used in amino acids and protein, a few dansyl based probes for the detection of metal ion have been reported. For instance, a sulfonamide-based probe containing a dansyl fluorophore was developed to detection of copper in living cell. Ion Cu^{2+} can strongly quench the fluorescence of fluorophores by electron or energy transfer. Then, when the dansyl based probe **89** interacts with the metal ion and the complex **90** is formed, the fluorescence is extinguished (Scheme 21). The probe **89** exhibited low detect limit for Cu^{2+} (2.98×10^{-8} M), good water-solubility, good selectivity, high sensitivity, good cell permeability, with potential to be employed as biomarker to monitor the level of Cu^{2+} in living cells [73].

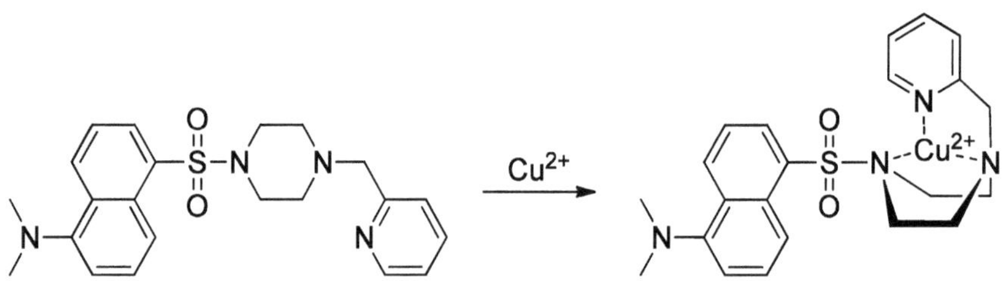

89 **90**

Scheme 21 Dansyl based probe **89** selective for copper(II) ion

2.5 *Fluo-4*

Fluo-4 (**97**) is a fluorescent dye used to quantifying cellular Ca^{2+} in living cells. Fluo-4 (**97**) is similar in structure and spectral properties of Fluo-3. The structural modification consisted in the substitution of two fluorines for two chlorine in the fluorophore which conferred certain advantages over Fluo-3. So, Fluo-4 (**97**) absorbs more efficiently than Fluo-3 in 488nm, generating more intense fluorescence when used with argon-ion laser sources or in microscopes equipped with standard fluorescein filter sets [74].

The synthesis of Fluo-4 (**97**) is similar to that used for Fluo-3, reacting **91** with *tert*-butyl bromoacetate which protects the amino groups for further treatment with bromine and *tert*-butyllithium. The organolithium intermediate **94** reacts with **95** to synthesize the intermediate **96** which is treated with BF_3 etherate in acetic acid to remove the *t*-butyl groups, resulting in the Fluo-4 (**97**, Scheme 22) [75].

Fluo-4 (**97**) can be used in the non-fluorescent acetoxymethyl ester form, Fluo-4 AM (**98**), which usually is cleaved inside the cell in order to furnish the free fluorescent Fluo-4 (**97**) (Fig. 11) [74]. Fluo-4 AM (**98**) is commercially available for use in fluorescence and confocal microscopy, flow cytometry, and microplate screening applications.

The use of Fluo-4 (**97**) in calcium image neuronal network was successfully applied in mouse brain tissue. The method provided a simple and robust approach for imaging Ca^{2+} in cortical networks with single-cell resolution *in vivo* [76]. Fluo-4 (**97**) also showed applicability in the erythrocytes measurements of intracellular calcium. Compared to popular calcium indicators such as Fura-2 and Indo-1, Fluo-4 (**97**) exhibited visible light excitation more suitable for calcium imaging in

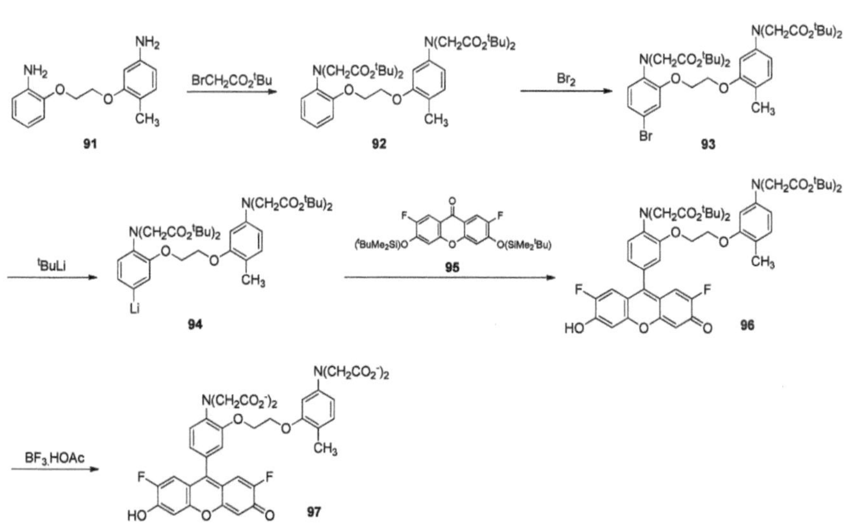

Scheme 22 Synthesis of Fluo-4 (**97**)

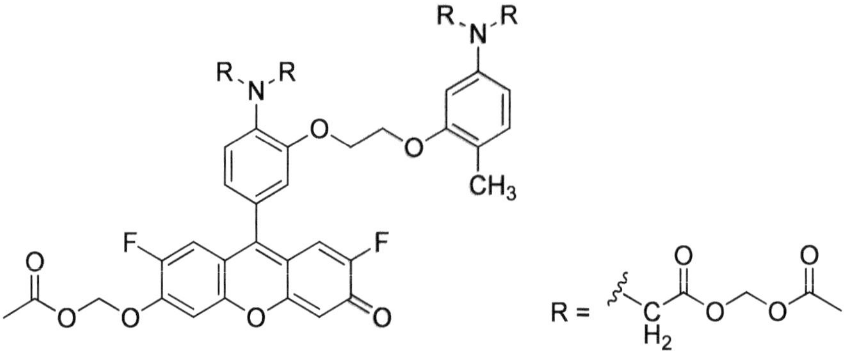

98

Fig. 11 Structure of Fluo-4 AM (**98**)

individual erythrocytes, with minimal influence by hemoglobin and no significant auto-fluorescence of the erythrocytes [77].

2.6 Fluorescein

Fluorescein (**101**) was first synthesized by Adolf von Baeyer in 1871 from phthalic anhydride (**99**) and resorcinol (**100**) in the presence of zinc chloride (Scheme 23)

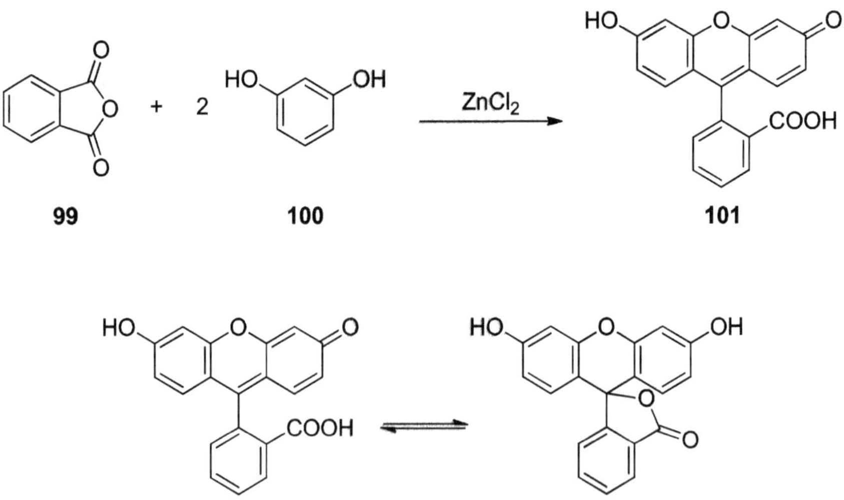

99 100 101

"open form" 101 "lactone form"

Scheme 23 First synthesis of fluorescein (**101**) and its open-closed equilibrium

[78]. Fluorescein (**101**) can exist in two major different forms: ring opened fluorescent form or lactone non-fluorescent form (Scheme 23) [79]. Interesting features of fluorescein (**101**) as good water solubility, absorption maximum at 494 nm and emission maximum of 521 nm (in water) and high quantum yield can be explored in different biological and biochemical applications [79].

Fluorescein isothiocyanate (**104**, FITC), a derivative of fluorescein (**101**), was first synthesized in 1958. Fluorescein isothiocyanate (**104**) was developed to substitute the fluorescein isocyanate used to labeled antibodies due to disadvantages in the synthesis of fluorescein isocyanate and the low stability of product. The substitution of isocyanate for isothiocyanate resulted in a mild synthesis by the reaction of aminofluorescein diacetate **102** and thiophosgene (**103**) in acetone, resulting in stable solid with satisfactory fluorescence when labeling antibodies (Scheme 24). The use of FITC (**104**) showed a great advanced compared with previous fluorescein isocyanate, which demanded use of highly toxic phosgene gas, besides the instability of the isocyanates [80].

In 1990, Confalone found that succinyl fluorescein dye could be synthesized in mild conditions using phthalic anhydride (**99**) and resorcinol (**100**) with methanesulfonic acid as both solvent and a Lewis acid catalyst [81]. In 1997, Sun et al. prepared a series of novel fluorinated fluorescein based on Confalone protocol, resulting in Oregon Green carboxylic acid **107** and **108**, which could be excited with the 488 nm spectral line of the argon-ion laser (Scheme 25) [82]. The higher photo-

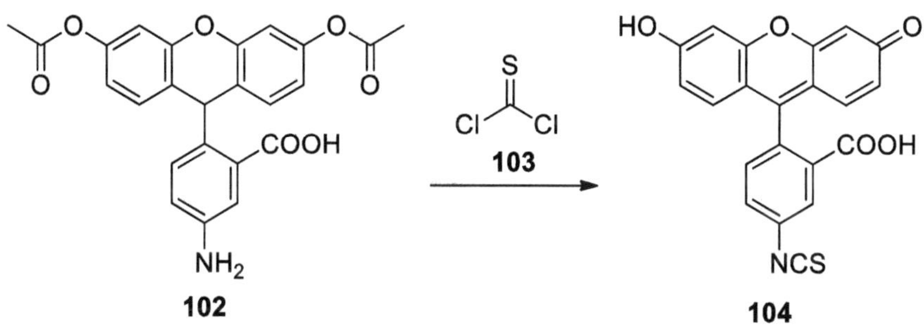

Scheme 24 Synthesis of fluorescein isothiocyanate (**104**, FITC)

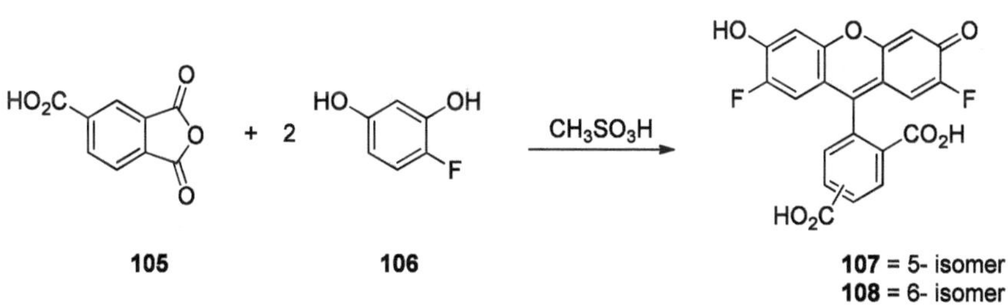

Scheme 25 Synthesis of the isomers of the Oregon Green® carboxylic acid **107** and **108**

stability, ionization at a lower pH (insensitive in the pH range), and a high quantum yield make these fluorinated fluoresceins very popular. Nowadays, Oregon Green® carboxylic acid, 5-isomer (**107**) and Oregon Green® carboxylic acid, 6-isomer (**108**) are commercially available.

Ma et al. reported a highly selective and sensitive spiro fluorescein hydrazide probe **109** for Cu^{2+}. Experiments showed that spiro fluorescein hydrazide **109** has nearly no absorption in the visible region. According to the authors, hydrazide recognizes and binds to Cu^{2+}, and the complexation of Cu^{2+} results in the hydrolytic cleavage of the amide bond, causing the release of fluorophore **112**, and recovering the yellowish green fluorescence characteristic of fluorescein (Scheme 26). In 0.01 M Tris–HCl buffer (pH 7.2), the probe exhibited a highly selective fluorescence response only to Cu^{2+} allowing the probe to be used for the direct detection of Cu^{2+} in some biological systems, as demonstrate in preliminary applications in fluids such as human serum and cerebrospinal fluid [83].

Yoon and coworkers have developed a fluorescein-based fluorescent probe **113** for the detection of thiol-containing molecules with high selectivity and sensitivity. They observed changes in the fluorescence spectra of probe in the absence or presence of glutathione (0–100 mM) in HEPES buffer. The probe **113** is essentially non-fluorescent in the absence of glutathione, however, the addition of glutathione causes probe fluorescence enhancements (**115**) with a strong new emission peak at 520 nm.

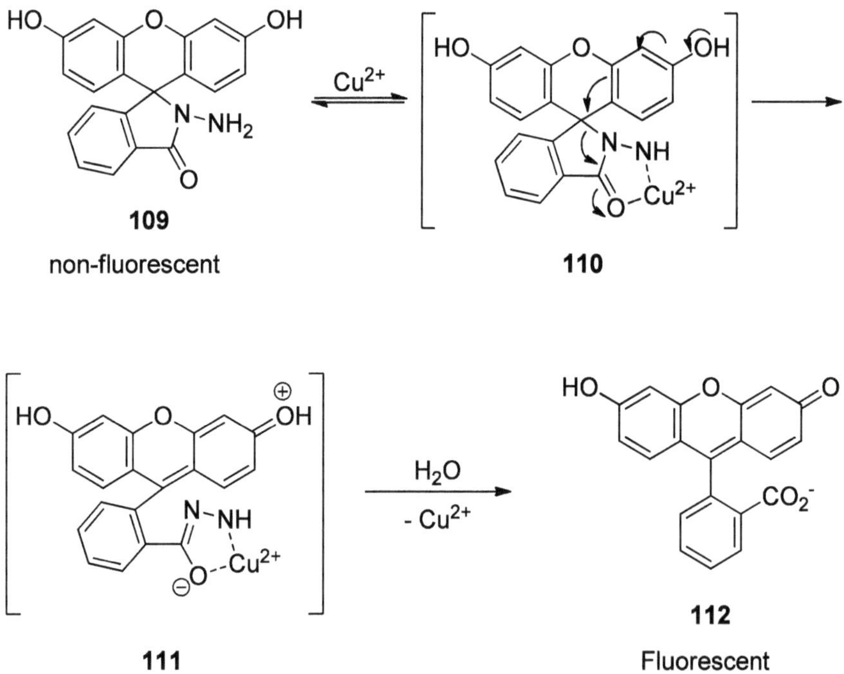

Scheme 26 Spiro fluorescein hydrazide probe **109**, and hydrolytic cleavage of the amide bond by Cu^{2+}

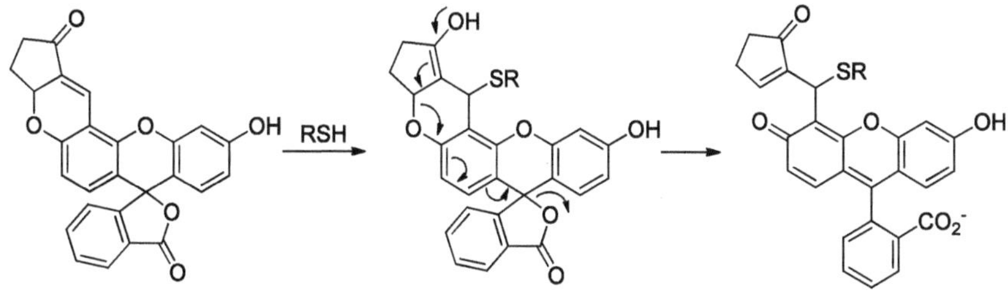

Scheme 27 The opening of the fluorescein **113** by 1,4-addition of thiol specie

Murine P19 embryonic carcinoma cells and zebrafish exhibited high fluorescence when treated with **69**. However, when they were pretreated with *N*-methylmaleimide (NMM) and, then, incubated with probe **113**, they observed a decrease in fluorescence intensity, confirming the presence of thiol species. Experimental evidence suggests that the 1,4-addition of thiol to the unsaturated ketone results in the opening of the fluorescein spiro ring, increasing the probe's fluorescence in the aqueous medium (Scheme 27) [84].

2.7 *Phenoxazine*

Phenoxazine (**118**) is an oxazine fused to two benzene rings. The first synthesis of phenoxazine was reported in 1887 by Bernthsen by pyrolytic condensation of *o*-aminophenol (**116**) with cathecol (**117**, Scheme 28) [85] Almost 30 years later, Kehrmann and Neil synthesized phenoxazine (**118**) heating an equimolar mixture of *o*-aminophenol (**116**) and *o*-aminophenol hydrochloride (**119**, Scheme 28). Later, it was demonstrated that the catechol (**117**) acts only as proton donor, and the evidence was obtained by the isolation of the intermediate 2,2′-dihydroxydiphenylamine [86, 87]. A different procedure for the synthesis of unsubstituted phenoxazine (**118**) involves the auto condensation of *o*-aminophenol (**116**) in the presence of iodine, with elimination of ammonia and water (Scheme 28) [88]. However, for substituted phenoxazines, the Turpin's reaction has been widely used. This reaction is based on the condensation of *o*-aminophenol (**116**) with picryl chloride (**120**), resulting in an intramolecular nucleophilic displacement of the 2-nitro group by the phenoxide group in alkaline medium to obtain **122** (Scheme 28) [89].

Phenoxazine derivatives most used in cell biology are the dyes Nile Blue (**125**) and Nile Red (**126**). The first reported synthesis of Nile Blue dates to 1896 by Möhlau and Uhlmann by condensing 1-naphthylamine (**124**) with 4-nitroso-*N*,*N*-diethyl-3-aminophenol (**123**) to obtain a blue dye **125** (Scheme 29) [90]. An alternative synthesis of Nile Blue (**125**) involves the use of perchloric acid instead of acetic acid for condensation (Scheme 29) [91].

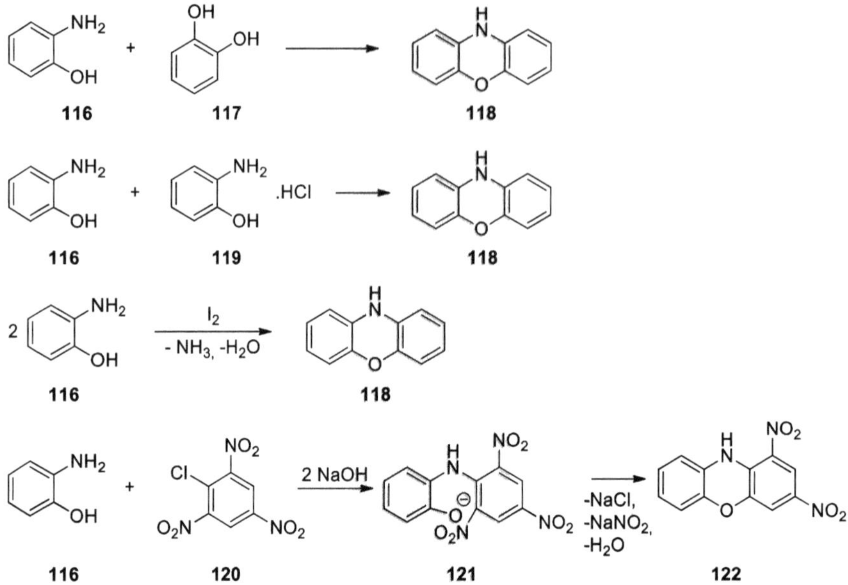

Scheme 28 Syntheses of phenoxazine phenoxazines **118** and **122**

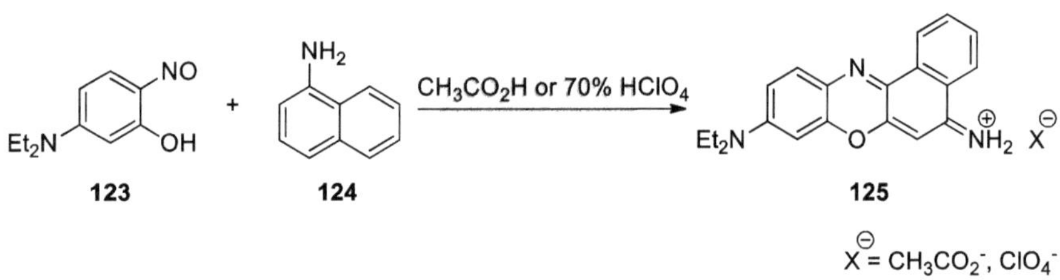

Scheme 29 Original and modified synthesis of Nile Blue (**125**)

Nile Red (**126**) can be synthesized via hydrolysis of Nile Blue (**125**) with sulfuric acid (Scheme 30) [90]. The most common method reported for preparing Nile Red dyes is the use of 5-(dialkylamino)-2-nitrosophenol like **123** with 2-naphthol (**128**) through an acid-catalyzed condensation reaction (Scheme 30) [92]. Nile Red (**126**) has a neutral oxidized phenoxazine structure and exhibits highly fluorescent properties. It is poor water soluble, but in other solvents its fluorescence varies, consequently, Nile Red (**126**) is particularly useful for studying lipids and events that involve impregnation of the dye in apolar medium [93]. For instance, the first report of Nile Red (**126**) for the detection of intracellular lipid droplets by fluorescence microscopy and flow cytofluorometry was in 1985. It was observed that the cytoplasmic lipid droplets could be performed with yellow-gold fluorescence [94].

Chang and coworkers reported a phenoxazine **129** and fluorescein **130** derivatives as a peroxysensor (Fig. 12). The new probes are ROS selective, membrane

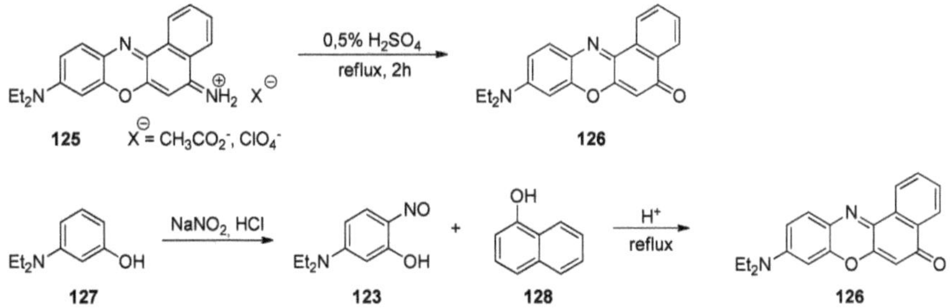

Scheme 30 Syntheses of Nile Red (**126**)

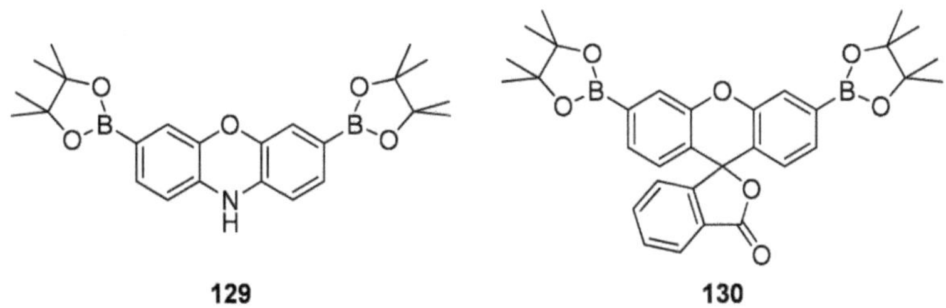

Fig. 12 Boronate based fluorescent probes

permeable and respond to H_2O_2 by an increase in red, green, and blue fluorescence. The hydrolytic deprotection of the boronates with H_2O_2 recover the fluorescence of phenoxazine and fluorescein derivatives. These boronate dyes are all membrane-permeable and can be used to measure micromolar changes in intracellular H_2O_2 concentrations in living cells using confocal microscopy and two-photon fluorescence microscopy as well as nanomolar concentrations of H_2O_2 *in vitro* [95].

Wu et al. reported a study with phenoxazine derivative fluorescent probe **131** for detection of the biomarker of melanoma and tyrosinase activity. The probe structure consists of (4-hydroxyphenyl) urea group as a substrate for the enzyme, and a phenoxazine moiety as chromophore. Tyrosinase can recognize the hydroxyphenylurea and reacts with the urea linkage by oxidation followed by hydrolysis, activating the fluorescence (Scheme 31). The probe **131** was able to detect sensitively and selectively the endogenous tyrosinase in live cells and in zebrafish [96].

2.8 *Rhodamine B*

Classical rhodamines are prepared from 3-dialkylaminophenols condensed with phthalic anhydrides under harsh conditions in different protocols described in the literature [75, 80, 97]. A generic synthesis of Rhodamine B (**135**) is shown in the Scheme 32.

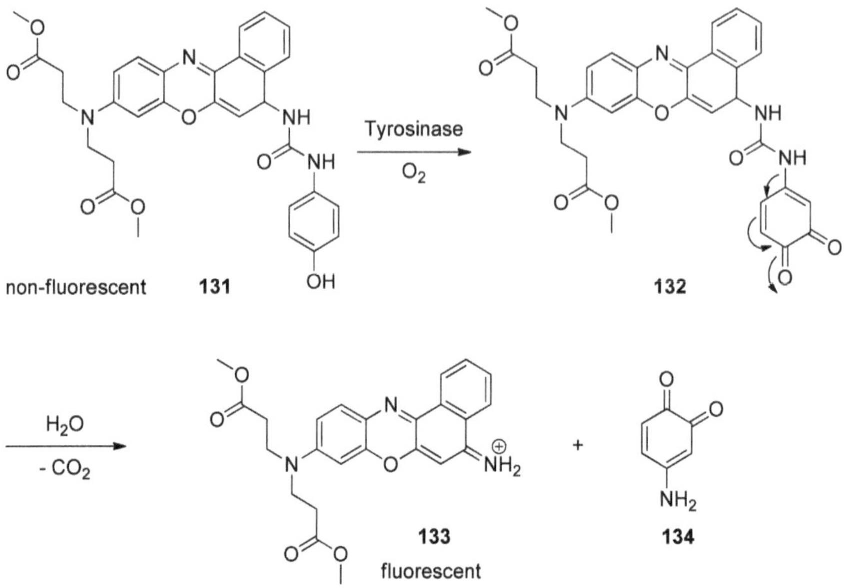

Scheme 31 Phenoxazine derivative fluorescent probe **131** for detecting tyrosinase

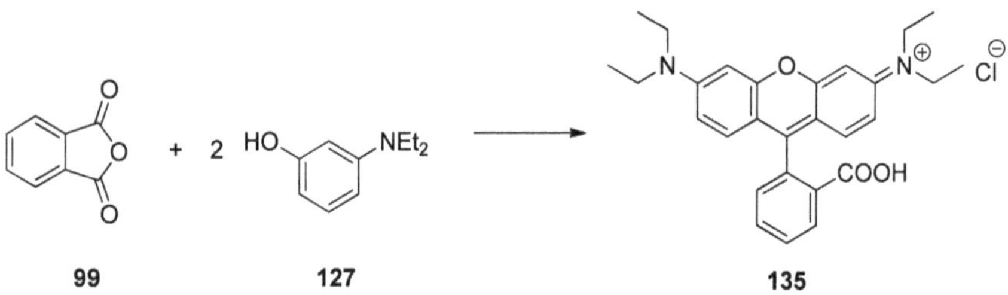

Scheme 32 Synthesis of Rhodamine B (**135**)

Rhodamine B (**135**) was first reported by Noelting and Dziewonsky in 1905 [97], however, it was only in 1997 that the Rhodamine B derivative and its ring-opening reaction received more attention [98]. Czarnik and coworkers designed the Rhodamine B hydrazide probe **136** selective for Cu^{2+} in water. The reaction with Cu^{2+} in water resulted in redox hydrolysis, restoring the fluorescence of rhodamine (Scheme 33) [98].

Ma et al. reported a *N*-benzoyl rhodamine B–hydrazide probe **140** designed for hypochlorite anion (OCl^-) detection. The non-fluorescent probe **140** interacts with hypochlorite anion selectively, resulting in the oxidation of dibenzoylhydrazine, which promotes the opening of the closed spirolactam ring (Scheme 34). They observed that the pink color of Rhodamine B (**135**) is largely restored due to the

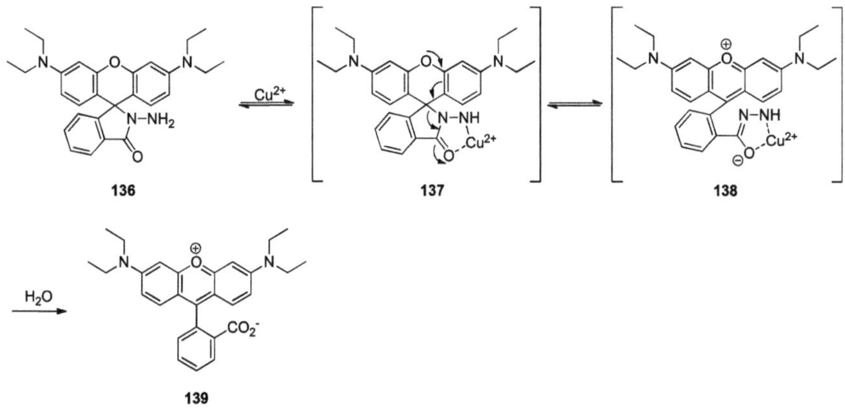

Scheme 33 Mechanism of interaction of Rhodamine B hydrazide **136** with Cu^{2+}

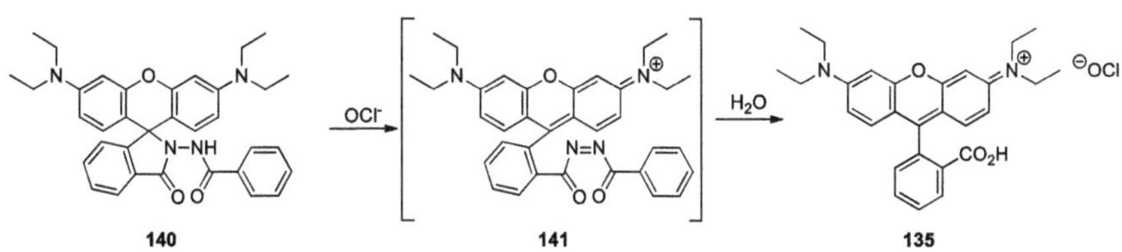

Scheme 34 Mechanism of interaction of *N*-benzoyl rhodamine B–hydrazide **140** with hypochlorite ion

increase of the OCl^- concentration. The fluorescence-on reaction has shown high sensitivity and high selectivity for OCl^- over other common ions and oxidants [99].

2.9 *Alexa Fluor®*

Alexa Fluor® is the basic name of a series of commercially available fluorescent dyes widely used in many laboratories around the world. Before that, the research group managed by Richard P. Haugland and his wife Rosaria P. Haugland synthetized various other fluorescent probes for decades until the breakthrough which it was patented and produced by his Company—Molecular Probes, nowadays part of Thermo Fischer Scientific.

The first Alexa Fluor® compounds reported were Alexa 350 (**142**), Alexa 430 (**143**), Alexa 488 (**144**), Alexa 532 (**145**), Alexa 546 (**146**), Alexa 568 (**147**), and Alexa 594 (**148**, Fig. 13)—each number is related to excitation wavelength maximum in nanometres—with significantly quantum yield and photostability in bioconjugate forms [100].

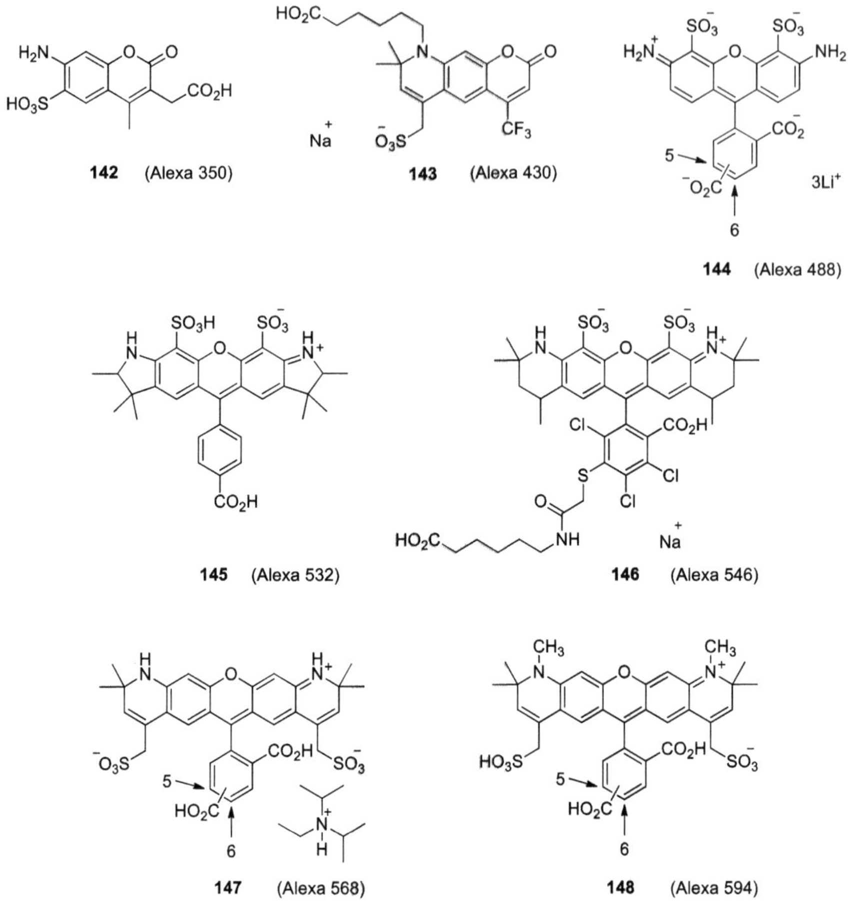

Fig. 13 Structures of Alexa 350 (**142**), Alexa 430 (**143**), Alexa 488 (**144**), Alexa 532 (**145**), Alexa 546 (**146**), Alexa 568 (**147**) and Alexa 594 (**148**)

Those structures of Alexa dyes are the sulfonated version of aminocoumarins and rhodamines derivatives. Sulfonation were largely tested in many hydrophobic compounds to increase the solubility [100]. The sulfonyl group had improved the hydrophilicity and furthermore the brightness of the fluorescence for the Alexa dyes. Synthesis of Alexa 488 (**144**) is described in Scheme 35 [101].

5(6)-Carboxyrhodamine 110 was prepared like presented before—for rhodamines—and the classic sulfonation was performed using fuming sulfuric acid in an ice bath. The product was obtained as a triethylammonium salt after neutralization and purification. This protocol is carried out to prepare other Alexa dyes, including the starting materials (e.g. 7-hydroxy-1,2,2,4-tetramethyl-1,2-dihydroquinoline and a phthalic anhydride (**99**)). Carboxylic acid substituent may

Scheme 35 Synthesis of Alexa 488 (**144**)

be converted to succinimidyl ester, esters containing alkyl chains and amides for future reactions [100].

Other Alexa Fluor® compounds were reported and compared to cyanines when conjugated to antibodies and specific proteins [102]. The conclusion was Alexa 555, Alexa 633, Alexa 647, Alexa 660, Alexa 680, Alexa 700 and Alexa 750—some of them are sulfonated cyanines and other structures has not been disclosed by the suppliers—were more fluorescent when observed properties like resistance to photobleaching and labelling [102].

Applications of Alexa dyes are numerous. Alexa 647 was tagged to an effector caspase recognition sequence and a quencher compound what it was possible to observe a quenching in baseline conditions and the imaging of apoptosis after cleavage of fluorescent probe [103].

Alexa 488 (**144**) was fluorescent probe for bioorthogonal labelling by copper-free click chemistry using difluorinated cyclooctyne promoting the imaging in mouse [104] and zebrafish [105]. Bicyclo[6.1.0]nonyne derivative was functionalised with Alexa 555 for the three-dimensional visualization of living melanoma cells demonstrating a novel alkyne derivative for biorthogonal chemistry [106]. Likewise, tetrazines carrying Alexa 750 or Alexa 568 (**147**) were efficient for biorthogonal conjugation via Diels-Alder cycloaddition reaction in cancer cells [107].

A major application of Alexa 488 (**144**) is as donor fluorophore in Förster (fluorescence) resonance energy transfer (FRET) which was useful in the study of linker length and rigidity for the determination of distance of nucleic acids [108]. RNA molecules also were labelled with Alexa 488 (**144**) and Alexa 594 (**148**) in specific sites in a long chain [109]. Synthetic carbohydrates could be monitored in a single-molecule level when tagged with Alexa 488 (**144**), an example of the study of glycosaminoglycan fragments [110].

In the G protein-coupled receptors (GPCRs) field, Alexa 647 was tagged in a human adenosine A_{2A} receptor (hA2AAR) antagonist for pharmacological studies of this drug target [111]. Cell surfaces could be investigated via semisynthetic fluorescent sensor proteins by FRET using the pair Alexa 488 (**144**)/Alexa 594 (**148**) [112].

2.10 Coumarin

Coumarin (**152**)—further its derivatives—is a versatile natural product first isolated in 1820 from tonka beans, widely diffused in plants and applicable in biology, medicine, materials and common products like cosmetics and beverages [113, 114]. Perkin was responsible for first synthesis of coumarin (**152**) [115]. Essentially, it was first reported the use of the sodium phenolate salt as starting material aiming the acetylation of the oxygen, but additional investigations by Perkin—because he had unexpected obtained coumarin (**152**)—determinate salicylaldehyde (**150**), acetic anhydride (**151**) and sodium acetate as reactants (Scheme 36) [116, 117]. Studies involving the synthesis of coumarin (**152**) lead Perkin to produce cinnamic acids from aromatic aldehydes and acid anhydrides what is known as Perkin Reaction [117].

Pechmann reaction is a classic method to synthesize coumarin derivatives like umbelliferone (**155**, 7-hydroxycoumarin), described in Scheme 37 [118].

Initially malic acid (**153**) is decarboxylated and dehydrated by sulfuric acid medium to form formylacetic acid (**154**) which reacts with resorcinol (**100**) to produce umbelliferone (**155**) after three steps (attack on the carbonyl group, electrophilic aromatic substitution followed by dehydration for aromatization) [118]. β-Ketoester could be used in this protocol [118]. Knoevenagel condensation is a well know reaction and it is also relevant for synthesis of coumarins whose mechanism resemble Perkin reaction [113, 119].

Coumarins were first identified after isolation of the product through the odour and its fluorescence, what was the case of umbelliferone (**155**) [118]. 7-Substituted coumarins with electron donor groups are fluorescent [120]. This luminescence is increased with an electron acceptor substituent at 3-position of the heterocycle (Fig. 14) [121].

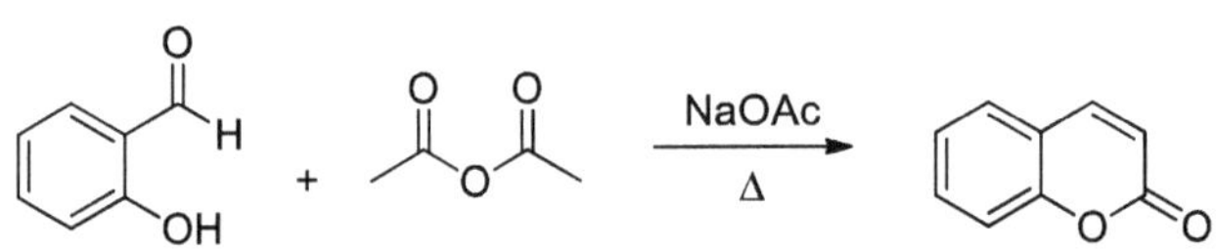

150 **151** **152**

Scheme 36 Synthesis of coumarin (**152**) by Perkin

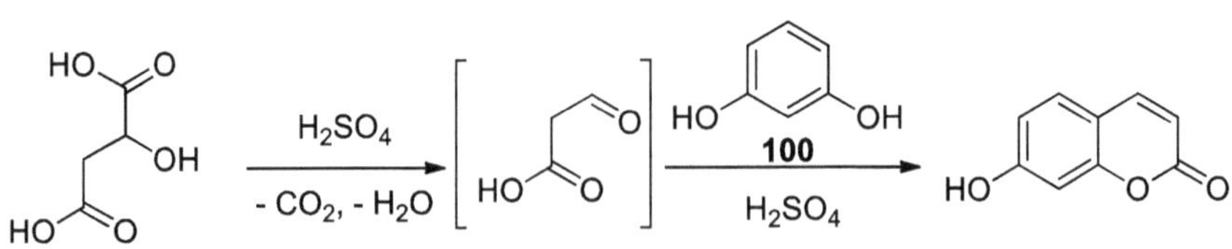

153 **154** **155**

Scheme 37 Synthesis of umbelliferone (**155**) by Pechmann reaction

Fig. 14 Design for fluorescent coumarins

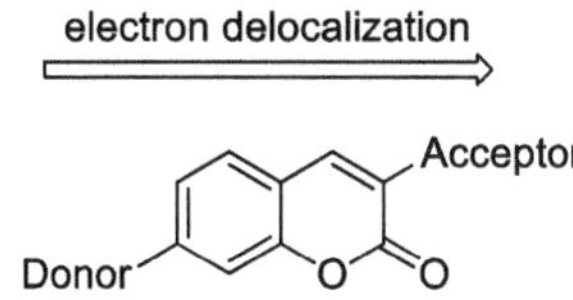

156

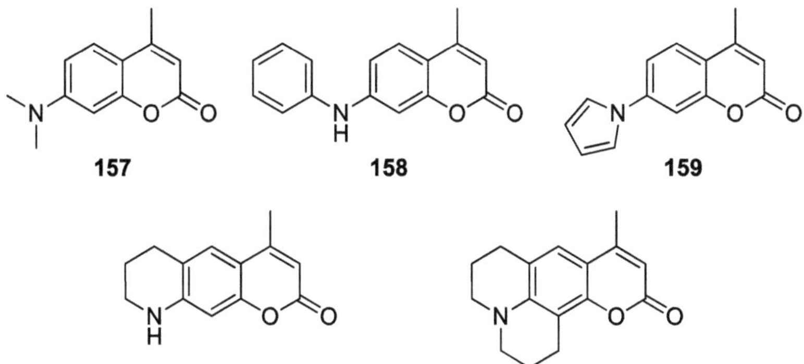

157

158

159

160

161

Fig. 15 7-Aminocoumarins; fused carbazole ring had highest quantum yield

In this context, 7-aminocoumarins are a significant class among coumarins in fluorescent field. A series of this amino compounds were evaluated, and their results indicated fused carbazole ring had highest quantum yield (Fig. 15) [120].

The advantageous electron acceptors in 3-position of coumarin derivatives are cyano, benzothiazoles and benzimidazoles groups. Then, coumarin analogue **166** was synthesized (Scheme 38), and it presented red shift in absorption and emission assigned to rigidity of the nitrogen at 7-position what enhanced the electron donation efficiency [121].

Fluorescent 3,7-substituted coumarins could be prepared through one-pot reaction using basic reactants to form coumarin ring as presented before plus *ortho*-substituted aniline to construct heterocycle at 3-position (Scheme 39) [122].

Post-functionalization of coumarin derivative **170** was realized via palladium-catalysed carbon–carbon cross-coupling reaction when two equivalents of 3-bromocoumarin (**170**) were converted into 3-(benzofuran-2-yl)-coumarin (**172**, Scheme 40) [123].

Plausible mechanism to produce **172** contains a halo-reduction of one molecule of 3-bromocoumarin (**170**), and a decarboxylation ring-contraction to form benzofuran ring [123]. There are other examples of post-functionalization of coumarin derivatives aiming different purposes [124].

Coumarins have a countless biological application as such as active molecule for diseases and fluorescent probe in bioassays [123]. Chromones **174** and **176** were designed as fluorescent false neurotransmitters (FFNs) based on serotonin—an

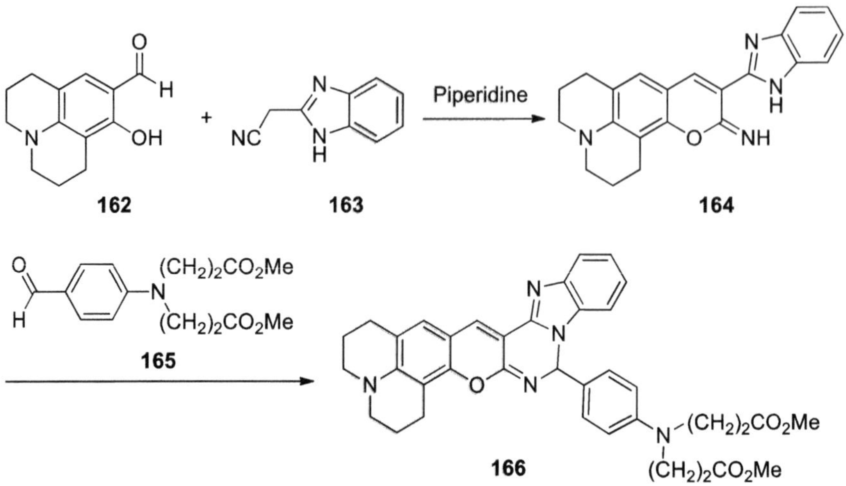

Scheme 38 Synthesis of coumarin analogue **166**

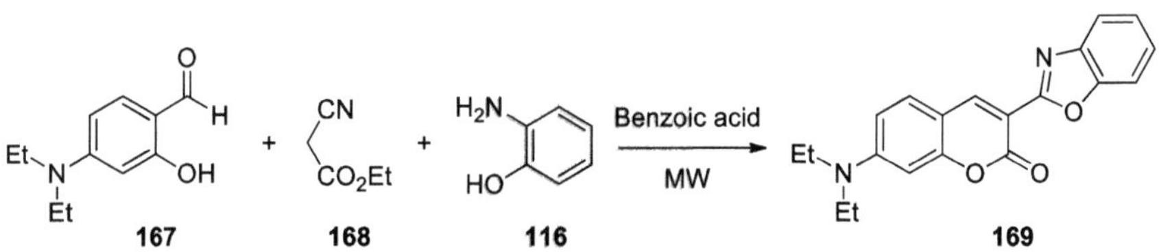

Scheme 39 Synthesis of the 3,7-substituted coumarin **169**

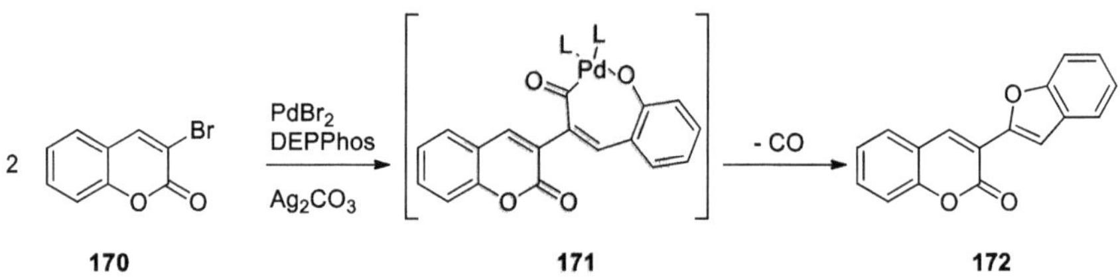

Scheme 40 Synthesis of 3-(benzofuran-2-yl)-coumarin (**172**)

endogenous neurotransmitter requiring a fluorescent core—and a previous reported fluorescent coumarin derivative—demanding a recognition element for synaptic activity [125]. These molecules could be prepared by same method using either chloro(triphenylphosphine)gold(I) or platinum(IV) tetrachloride with latter being less sensitive (Scheme 41) [125].

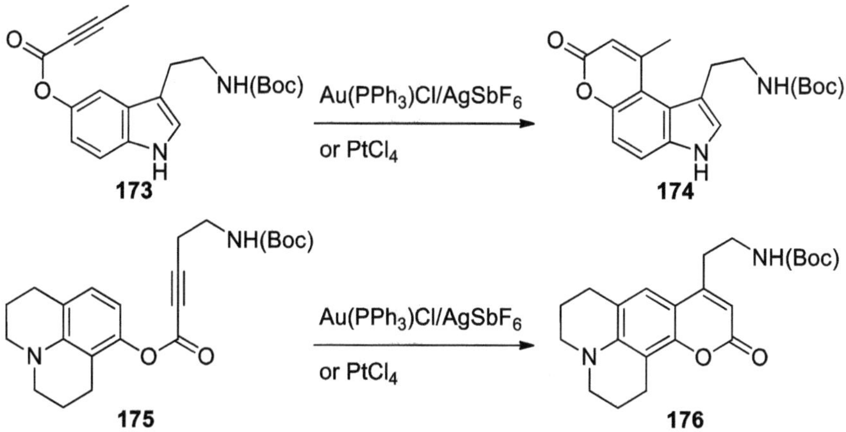

Scheme 41 Synthesis of potential fluorescent false neurotransmitters

One of the main biological tools for tag fluorescent probes is the copper-catalyzed azide-alkyne cycloaddition (CuAAC; Click) reaction. Azido-coumarin **177** is a non-fluorescent compound though triazole moiety bounded in proteins after Click reaction has revealed the fluorogenic substance **179** (Scheme 42) [126].

Bioorthogonal chemistry has applied coumarin dyes in many important studies. Scheme 43 describes Staudinger reaction between the phosphorous-derivative of coumarin **180** and enzyme bearing azido-group to form the highly fluorescent

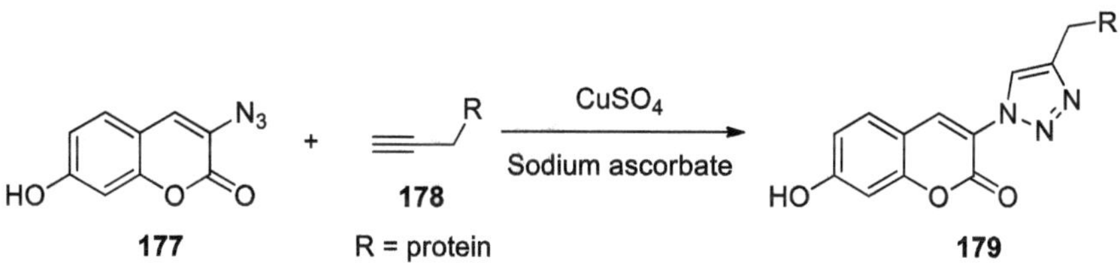

Scheme 42 Fluorescent coumarin after Click reaction

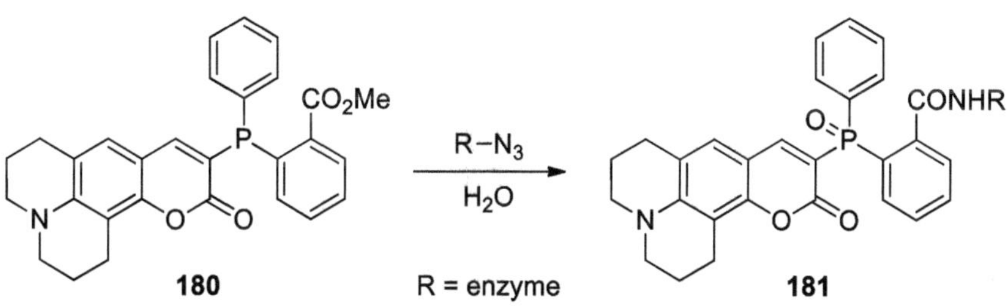

Scheme 43 Fluorescent coumarin **181** after biorthogonal Staudinger reaction

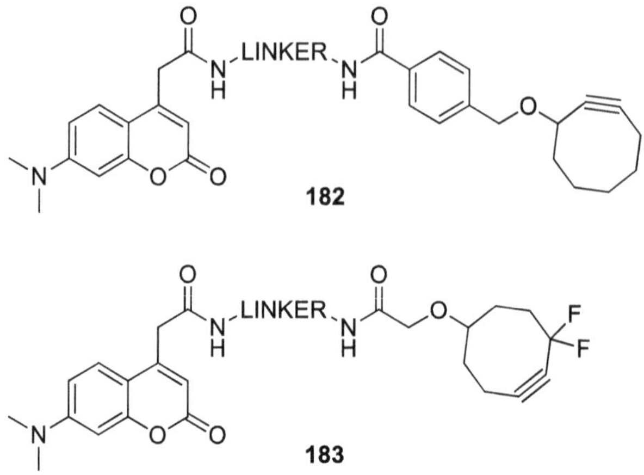

Fig. 16 Coumarins bearing cyclooctyne group for biorthogonal chemistry

compound **181**—because lone pair of electrons on phosphorous atom trigger the quenching [127].

Likewise, coumarins were tagged into cyclooctynes, the major class of compounds to bioassays (Fig. 16) in living cells in biorthogonal protocol [128, 129].

Coumarin derivatives were utilized as probes for biological metal detections like magnesium(II) (**184**) [130], zinc(II) (**185**) [131] and copper(II) (**186**), latter being selectively detectable in LLC-MK2 cultured cells (Fig. 17) [132].

Nucleic acids could be monitored by coumarin derivatives [133]. Chromone **187** remained fluorescent in DNA after incorporation by oligonucleotide synthesis methods, and it was possible observe dynamics in the interior of DNA (Fig. 18) [134].

Determination of thiol groups in biomolecules inside the cells was achieved by coumarins, in general not fluorescent before reaction with cysteine or reduced glutathione (Scheme 44) [135–137] These structures were designed to be either cleaved by thiols or suffer nucleophilic attack or electrophilic addition [136, 138].

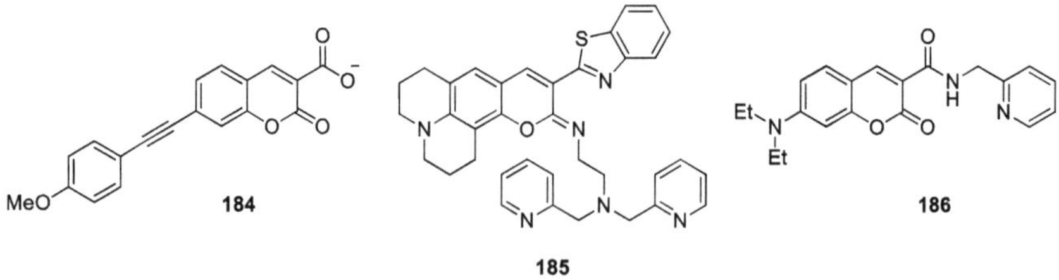

Fig. 17 Coumarins for metal detection

Fig. 18 Coumarin derivative **187** to monitor DNA

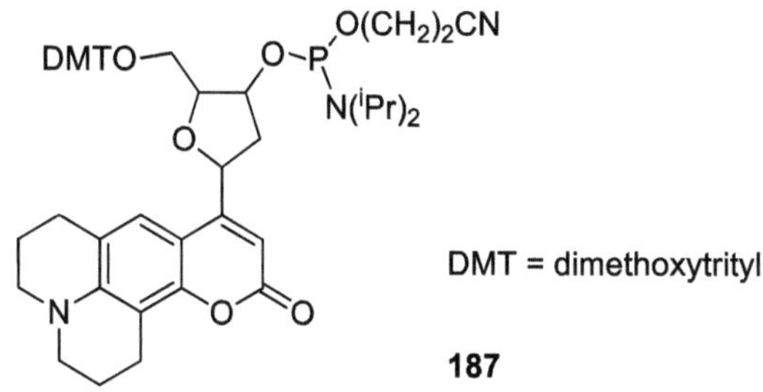

187

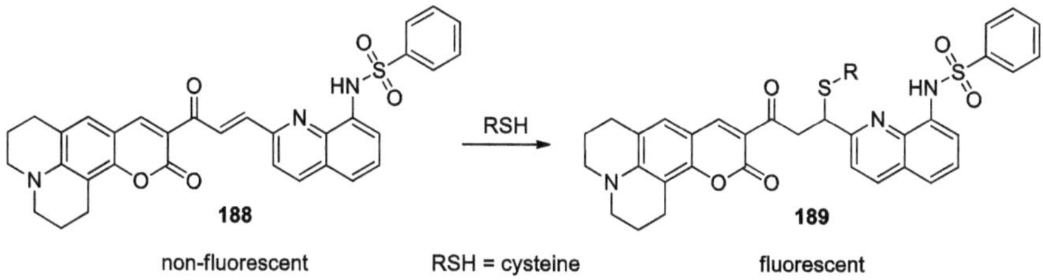

Scheme 44 Coumarin derivative **188** selective for cysteine

Lastly, some commercially available coumarins acted as fluorescent reporter in drug release investigations for visualization of cellular uptake of the nanoparticles containing the active ingredient [139, 140].

2.11 BODIPY

BODIPY is an acronym name for boron dipyrromethene (or boron dipyrrin), a usual denomination for the fluorescent 4-bora-3a,4a-diaza-*s*-indacene derivatives whose synthesis of the basic structure **192** is described in Scheme 45 [141].

BODIPY **192** was prepared reacting pyrrole (**190**) and its 2-aldehyde analogue **191** with boron trifluoride diethyl etherate in a one-pot two-steps reaction in basic medium [141]. This method was reported more than forty years after the synthesis of the first BODIPY [142] (Scheme 46) probably because that simple structure has

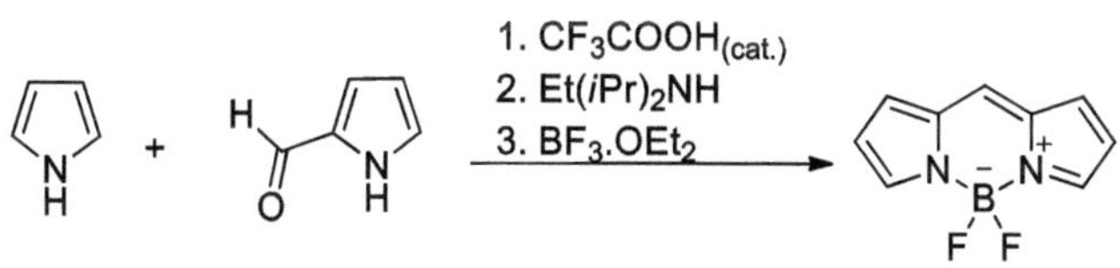

190 **191** **192**

Scheme 45 Synthesis of 4,4-difluoro-4-bora-(3a,4a)-diaza-*s*-indacene (**192**)

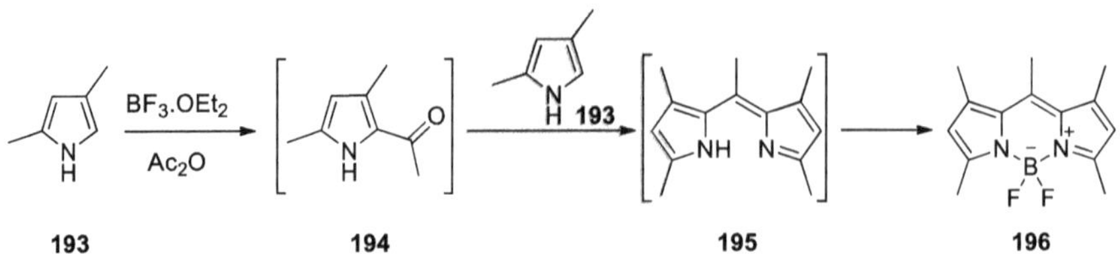

Scheme 46 Serendipity formation of BODIPY core

low stability in post-functionalization protocols [141].

BODIPY's breakthrough was an unexpected reaction aiming the acetylation of 2,4-dimethylpyrrole (**193**) with boron trifluoride diethyl etherate and acetic anhydride (**151**) under reflux (Scheme 46). Further investigations—derivatives and functionalization—were made in the same work on account of new molecule was highly fluorescent [142].

Preparation of symmetric BODIPY **201** has the following general reaction using two equivalents of pyrrole **197** (Scheme 47) [143].

Initially two molecules of pyrrole **197** react with the aldehyde **198** via acid catalysis to produce **199** eliminating water. This compound is oxidised to dipyrrin **200** usually with 2,3-dichloro-5,6-dicyanobenzoquinone (DDQ) or *p*-chloranil as oxidant [143]. Compound **200** could be prepared directly—no oxidation step—using acyl chloride rather than aldehyde [144]. Boron complex **201** is formed as presented before [143]. It was just recently reported results about the mechanism of complexation of boron trifluoride by dipyrrin which they indicated a first attack of pyrrolenine nitrogen on the boron atom further high interaction of hydrogen from pyrrole with the fluorine atom to produce hydrofluoric acid [145]. The basic protocol to synthesize asymmetric BODIPY **203** is described in Scheme 48 [143].

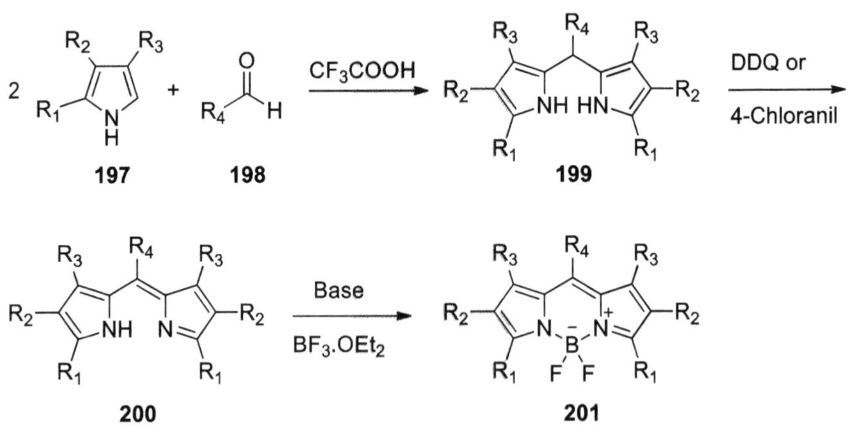

Scheme 47 General reaction for symmetric BODIPYs

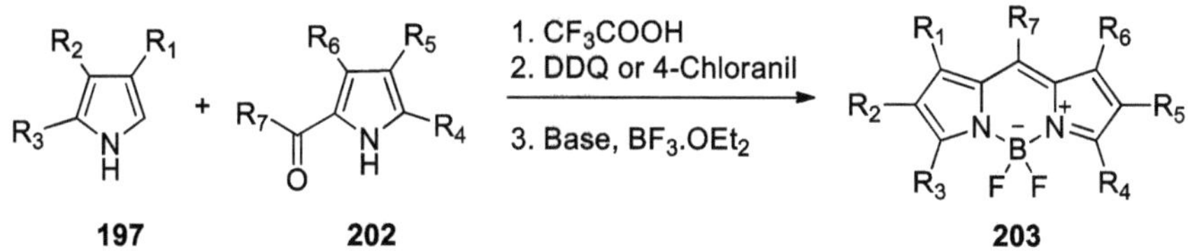

Scheme 48 General reaction for asymmetric BODIPYs

Mechanism of reaction to prepare asymmetric BODIPY **203** is like that to symmetric ones. It only difference is one pyrrole ring—aldehyde or ketone **202**—has different substituents from the other heterocycle and may be prepared in a previous step using formylation—like Vilsmeier-Haack reaction—or acetylation methods [143].

As noted, functionalised BODIPYs may be prepared with substituted pyrroles. Another option is the post-functionalisation of BODIPY core which it is summarized in Fig. 19 [146].

There is a myriad of methods of post-functionalisation of BODIPYs. Basically, reactions on the carbon atom are related to reactivity of pyrrole rings like nucleophilic and electrophilic aromatic substitution, modern carbon–carbon cross couplings—using halogens—and C–H activation, and radical C–H arylation with aryldiazonium salts. Knoevenagel condensation occurs on methyl substituent. Fluorine atom can be suffering nucleophilic substitutions by carbon and oxygen nucleophiles [146].

Last strategies to post-functionalisation of BODIPY core can be reviewed as follow (Fig. 20). Arylation on the methyl group at the *meso* position was achieved through palladium-catalysed direct $C_{(sp3)}$-H arylation (compound **204**) [147]. A styryl group at the *meso* position obtained by Suzuki cross-coupling reaction (compound **205**) could suffer cycloaddition reaction with trienamine—via amino-catalysis—wherein BODIPY moiety act as electron withdrawing group (compound **206**) [148].

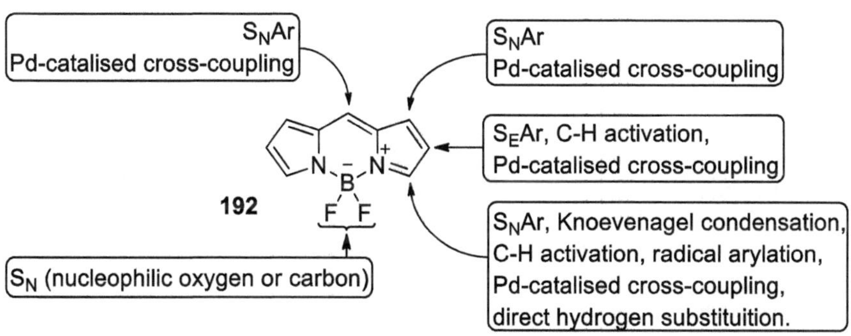

Fig. 19 Resume of post-functionalization reactions of BODIPY core

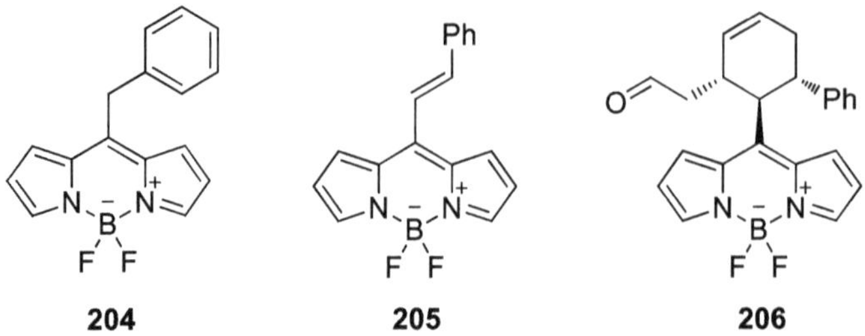

Fig. 20 Examples of post-functionalisation of BODIPY at the *meso* position

All these protocols are important to improve the fluorescence properties of BODIPY (**192**). In general, BODIPYs have emission wavelength up to 600 nm. Beyond that, BODIPY is poor water soluble. Solutions for these drawbacks could be rationalized in different research and it was compiled in recent reviews [149, 150]. In the main term, aryl and styryl groups at 2,6- or 3,5-positions like in **210**, and fused aromatic rings like in **208** help to obtain a near infrared spectrum (Fig. 21) [150]. Hydrophilicity might be reached by ethylene glycol chains as observed in **207**, carboxylates, sulfonates (e.g. **209**), and phosphonates (Fig. 21) [150].

Since there are numerous BODIPY derivatives—in addition to many synthetic possibilities—a major application of this kind of fluorescent compounds is bioassays because their structures have high fluorescence quantum yields, photostability and they are biologically active [141, 151]. In order to use BODIPY as fluorescent label for receptor ligands, an investigation about linkers has shown different alkyl groups at *meso* position of **211**–**214** were stable in distinct conditions without quenching (Fig. 22) [152] Biorthogonality was aimed synthesizing the BODIPY attached to bicyclo[6.1.0]nonyne **215** for bioimaging of (azido-tagged) organelles in living cells [153], or it was prepared with tetrazine moiety to obtain **216** which made it highly fluorescent after reaction with *trans*-cyclooctene [154], and the resembling structure **217** was applied for labelling of DNAs (Fig. 22) [155].

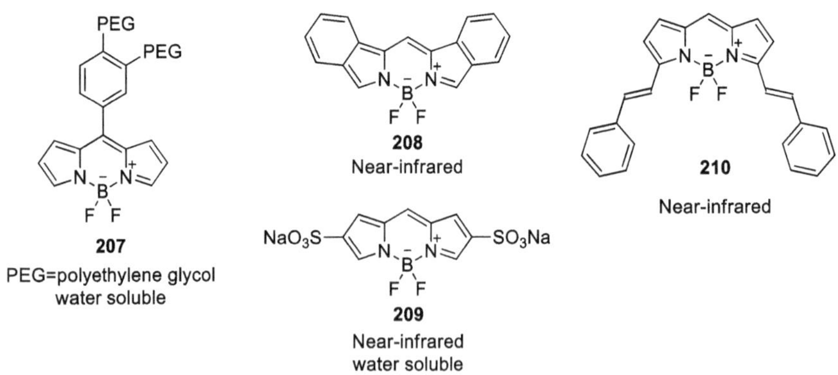

Fig. 21 Hydrophilic and/or with near-infrared shift BODIPY derivatives

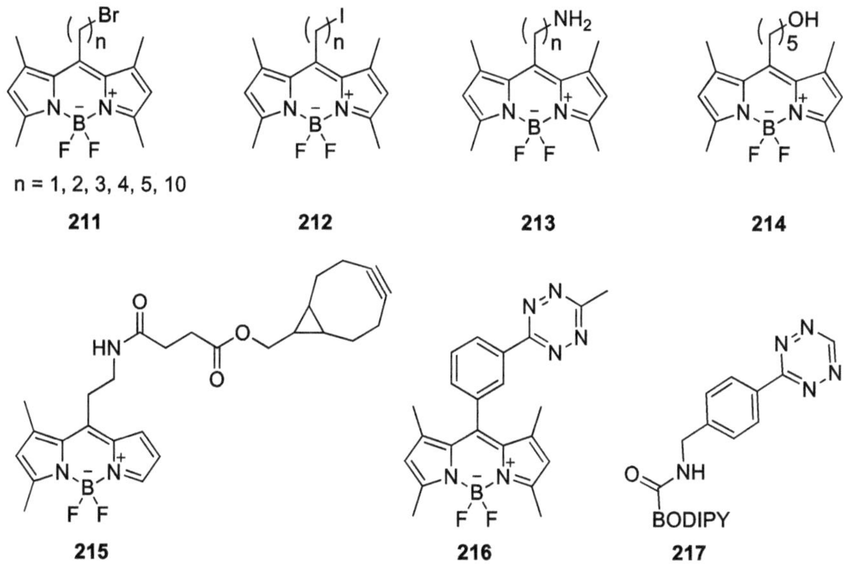

Fig. 22 BODIPY designed for receptor ligands (**211–214**) and for biorthogonal assays (**215–217**)

BODIPYs bearing chelating agents were useful tools for monitor metals like zinc(II) using **218**, [156] copper(II) and mercury(II) testing **219** in biological experiments (Fig. 23) [157]. Indeed, coordination complexes with biological interest were label with fluorescent BODIPY as observed in **220** (Fig. 23) [158].

Amino acids and proteins also are targets for BODIPY labelling mainly reacting with biothiols [159, 160]. Fluorescent probe **221** (Fig. 24) was selective for cysteine

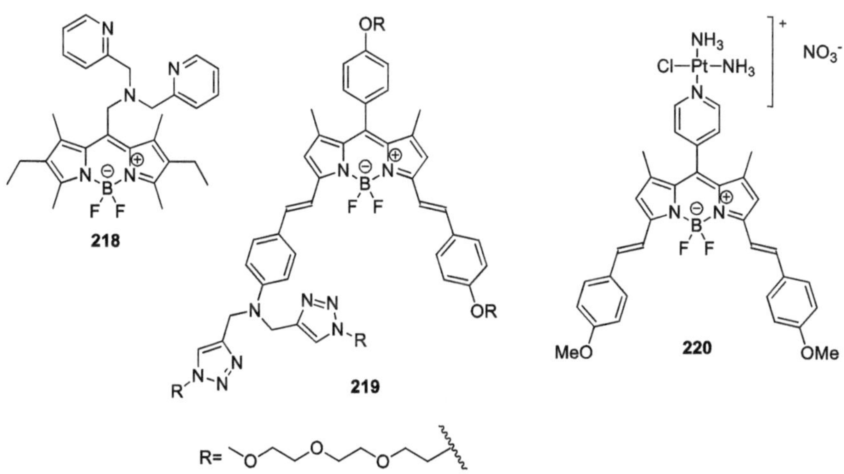

Fig. 23 BODIPYs for metal investigations

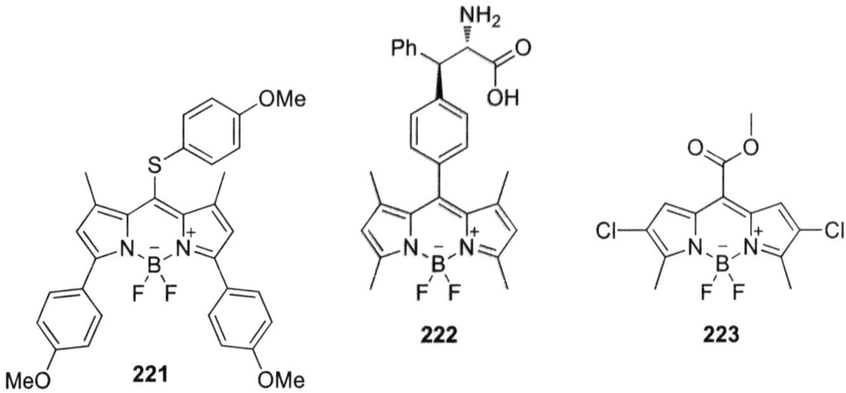

221 222 223

Fig. 24 BODIPYs for amino acids, peptides, or proteins

detection in lysosome—leaving behind homocysteine and reduced glutathione—through amino substitution of the thioether moiety [161]. Structural modification via palladium-catalysed $C_{(sp3)}$-H activation of BODIPY probe **222** (Fig. 24) was efficient for experiments in peptides [162]. Carboxylesterase 1 reacted with not-fluorescent *meso*-ester BODIPY **223** resulting in a very brightness carboxylate anion compound (Fig. 24) [163].

BODIPY **224** and **225** were examples of selective probe for endoplasmic reticulum—in low and no cytotoxic concentration [164]—and for membrane potential imaging—with increase water solubility [165]—, respectively (Fig. 25). Fluorescent dyes were designed for *in vivo* evaluations using BODIPY [166] like the compound **226** whose "light up" response was observed in zebrafish after sulfonate cleavage by cysteine (Fig. 25) [167].

Lastly, a simple mention for application of fluorescent BODIPY core for detection of reactive oxygen/nitrogen species, and gaseous molecules in living cells [168].

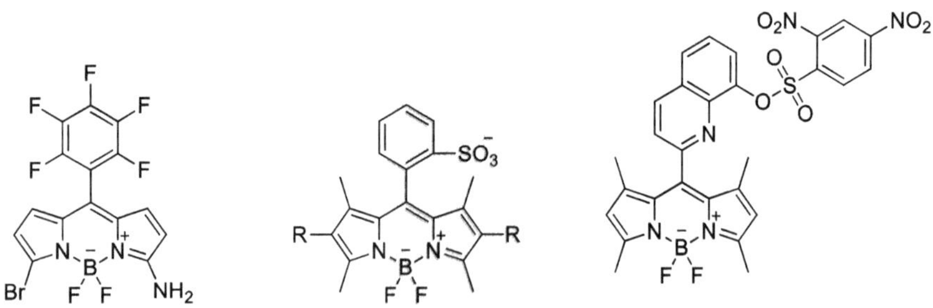

224 225 226

R= -H, -Et, $-CO_2^-$, -CONHR, -CN.

Fig. 25 BODIPYs for biological assays

There are many other utilizations for BODIPY derivates, but they did not fit in this document.

2.12 *Carbazole*

Carbazole (**230**) is a substance mainly found in coal tar whose first known synthesis date from 1872 when aniline passing through glowing tube had resulted in a mixture of compounds where the required heterocycle was isolated after purification [169, 170]. In the same opportunity, it was confirmed the structure of carbazole (**230**), a tricyclic aromatic ring [170].

Shortly thereafter, other (nowadays acknowledged) protocols to prepare carbazoles were reported like Graebe-Ullmann synthesis (Scheme 49) [171].

Graebe-Ullmann synthesis of carbazole (**230**) involves the intermediate of diazotization **228**, and one step to formation of benzotriazole ring **229**. Product **230** is formed after elimination of nitrogen gas [171].

The Borsche–Drechsel cyclization produce tetrahydrocarbazoles using phenylhidrazine (**231**) and cyclohexanone (**232**) via acid catalysis with removal of ammonia and, in a second procedure, an oxidation reaction with lead oxide results in carbazoles (Scheme 50) [172].

There are also modern methods to synthesize carbazoles. This heterocycle was prepared in a one-pot palladium-catalysed reaction based on Suzuki–Miyaura carbon–carbon coupling and on intramolecular amination (Scheme 51) [173].

3-Substituted carbazoles were synthesized in three steps: first, Ullmann cross-coupling between enone **239** and bromo-aryl **240** followed by reduction—to isolate tetrahydrocarbazoles—and lastly an oxidation with palladium on carbon (Scheme 52) [174].

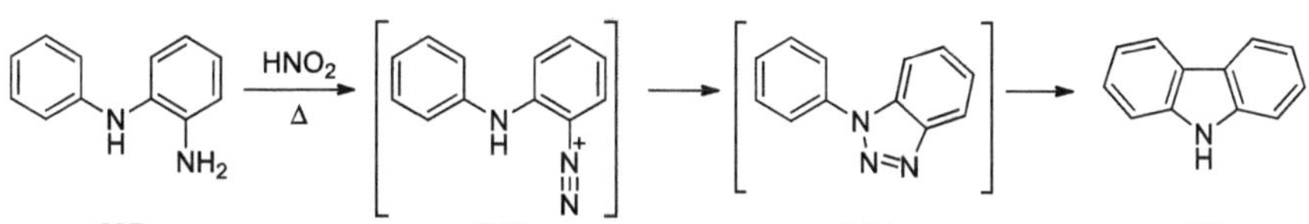

227 **228** **229** **230**

Scheme 49 Graebe-Ullmann synthesis to produce carbazole (**230**)

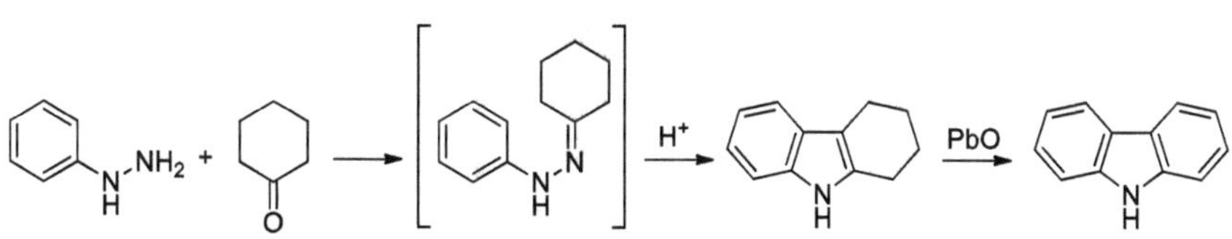

231 **232** **233** **234** **230**

Scheme 50 Borsche–Drechsel cyclization to produce carbazole (**230**)

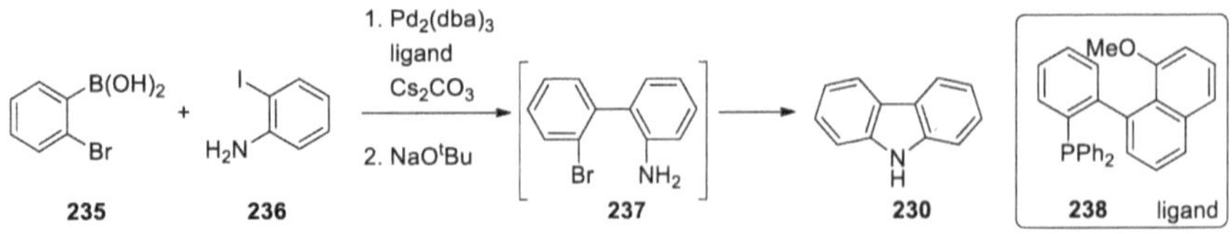

Scheme 51 Tandem Suzuki–Miyaura coupling and intramolecular amination

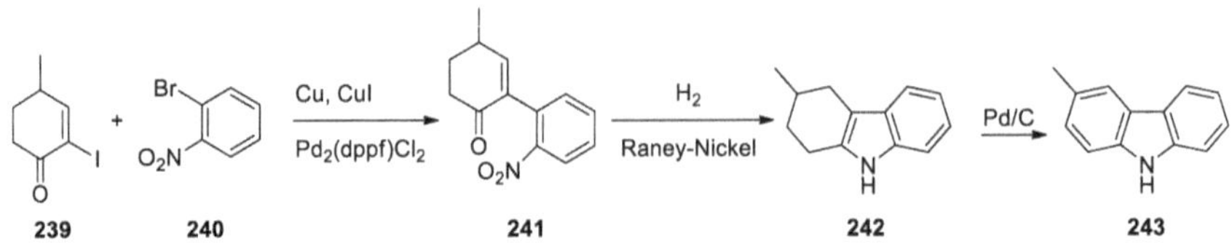

Scheme 52 Synthesis of carbazole **243** via Ullmann cross-coupling

Backing to free-metal protocols, carbazoles could be prepared via (*meta*-) C-H activation using hypervalent iodine (Scheme 53) [175].

Post-functionalization of carbazole ring is summarized in Fig. 26 [169]. There are three major positions able to add new substituents: 3-,6- and 9-position. Hydrogen on nitrogen (9-position) is weakly acidic which it may be removed by strong bases to

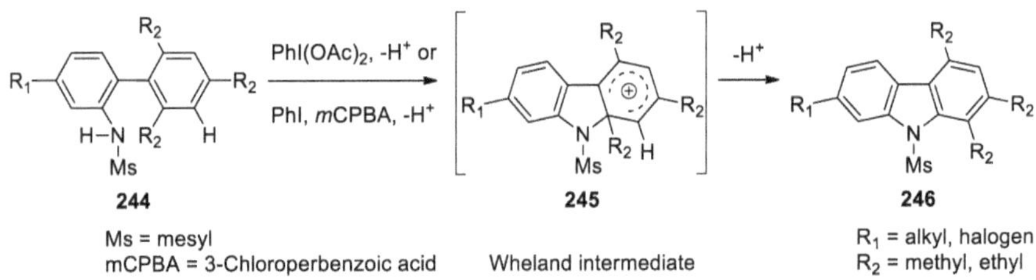

Scheme 53 Synthesis of carbazole **246** using hypervalent iodine

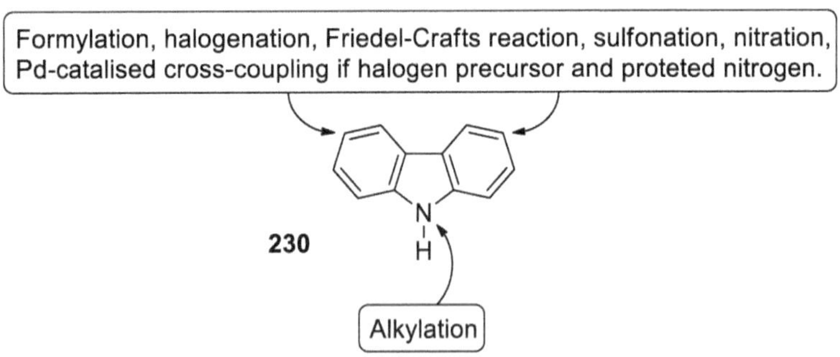

Fig. 26 Main post-functionalization reactions of carbazole ring

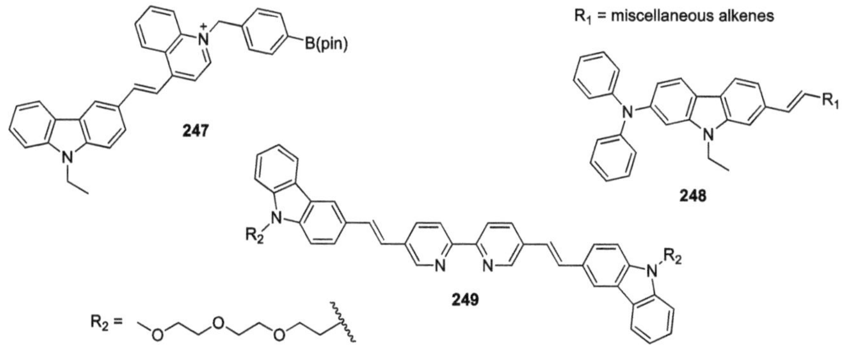

Fig. 27 Examples of carbazoles as fluorescent probes for biological application

realize nucleophilic attacks. Several characteristic reactions are possible at electron-rich 3- and 6-positions. Halogens in those positions work as precursor for palladium cross-coupling reactions since nitrogen had been protected [169].

Carbazoles had its electronic and fluorescent properties investigated since the late 1950s [176, 177]. The main carbazole dyes bear electron donating substituents in 9-position and electron withdrawing groups at 3- and 6-positions [169], however this is not a verdict [178].

Some examples of biological application of carbazoles as fluorescent probes are resumed as follows (Fig. 27): carbazole **247** targeted mitochondria to react with hydrogen peroxide [179]; derivatives of compound **248** are selective for mitochondria, lipid droplets or tissues depending on substituent [180]; copper(II) and zinc(II) could be detected including *in vivo* analysis (compound **249** for zinc) [181, 182]; lastly carbazoles stained peptides [183] and nucleic acids [184].

References

1. Gonçalves, M.S.T.: Fluorescent labeling of biomolecules with organic probes. Chem. Rev. **109**, 190–212 (2009)
2. Sahoo, H.: Fluorescent labeling techniques in biomolecules: a flashback. RSC Adv. **2**, 7017–7029 (2012)
3. Terai, T., Nagano, T.: Small-molecule fluorophores and fluorescent probes for bioimaging. Pflügers Arch. Eur. J. Physiol. **465**, 347–359 (2013)
4. Valeur, B., Berberan-santos, N.: A brief history of fluorescence and phosphorescence before the emergence of quantum theory. J. Chem. Educ. **88**, 731–738 (2011)
5. Muyskens, M.: Phantastic fluorescence. J. Chem. Educ. **83** 768A (2006)
6. Brewster, D., Lond, K.H., Edin, V.: On the colours of natural bodies. Earth Environ. Sci. Trans. R. Soc. Edinburgh **12**, 538–545 (1834)
7. Herschel, J.F.W., Bart, K.H.: On a case of superficial colour presented by a homogeneous liquid internally colourless. Philosofical Trans. R. Soc. London pp. 147–153 (1845)
8. Stokes, G.G.: On the change of refrangibility of light. Philos. Trans. R. Soc. London (1852)
9. Lakowics, J.R.: Principles of Fluorescence Spectroscopy. Kluwer Academic/Plenum Publishers (1999)

10. Fu, Y., Finney, N.S.: Small-molecule fluorescent probes and their design. RSC Adv. **8**, 29051–29061 (2018)
11. Frackowiak, D.: The Jablonski diagram. J. Photochem. Photobiol. **2**, 399–401 (1988)
12. Kricka, L.J., Fortina, P.: Analytical ancestry : "Firsts" in fluorescent labeling of nucleosides, nucleotides, and nuclei acid. Clin. Chem. **55**, 670–683 (2009)
13. Martynov, V.I., Pakhomov, A.A., Popova, N.V, Deyev, I.E., Petrenko, A.G.: Synthetic fluorophores for visualizing biomolecules in living systems. Acta Naturae **8**, 33–46 (2016)
14. Ghisaidoobe, A.B.T., Chung, S.J.: Intrinsic tryptophan fluorescence in the detection and analysis of proteins: a focus on Förster resonance energy transfer techniques. Int. J. Mol. Sci. **15**, 22518–22538 (2014)
15. Morris, M.: Fluorescence-Based biosensors. Academic Press Inc. (2013)
16. Kobayashi, H., Ogawa, M., Alford, R., Choyke, P.L., Urano, Y.: New strategies for fluorescent probe design in medical diagnostic imaging. Chem. Rev. **110**, 2620–2640 (2010)
17. Demchenko, A.P.: Introduction to Fluorescence Sensing. Springer Science & Business Media (2008)
18. Shimoura, O., Johnson, F.H., Saiga, Y.: Extraction, purification and properties of aequorin, a bioluminescent protein from the luminous Hydrimedusan, Aequorea. J. Cell. Comp. Physiol. pp. 223–239 (1962)
19. Chalfie, M., Tu, Y., Euskirchen, G., Ward, W.W., Prasherf, D.C.: Green fluorescent protein as a marker for gene expression. Science **263**, 802–805 (1994)
20. Uchiyama, S., Santa, T., Okiyama, N., Fukushima, T., Imai, K.: Fluorogenic and fluorescent labeling reagents with a benzofurazan skeleton. Biomed Chromatogr. **15**, 295–318 (2001)
21. Boulton, A.J., Ghosh, P.B., Katritzky, A.R.: A new benzofurazan synthesis. Tetrahedron Lett. **25**, 2887–2888 (1966)
22. Boulton, A.J., Ghosh, P.B., Katritzky, A.R.: Rearrangement of 4-Arylazo- and 4-Nitroso-benzofuroxans: new syntheses of the benzotriazole and benzofurazan ring systems. J. Chem. Soc. B. pp. 1004–1011 (1966)
23. Ghosh, P.B., Whitehouse, M.W.: 7-Chloro-4-nitrobenzo-2-oxa-1,3-diazole: a new fluorigenic reagent for amino acids and other amines. Biochem. J. **108**, 155–156 (1968)
24. Prokipcak, J.M., Forte, P.A., Lenno, D.D.: Thermal decomposition of alkyl *N*-(*o*-nitrophenyl)carbamates: a novel synthesis of benzofurazan. Can. J. Chem. **47**, 2482–2484 (1969)
25. Prokipcak, J.M., Forte, P.A.: New preparation of benzo-2,1,3-oxadiazoles (benzofuazans). Can. J. Chem. **48**, 3059–3063 (1970)
26. Uchiyama, S., Takehira, K., Kohtani, S., Santa, T., Nakagaki, R., Tobitaband, S., Imaia, K.: Photophysical study of 5-substituted benzofurazan compounds as fluorogenic probes. Phys. Chem. Chem. Phys. **4**, 4514–4522 (2002)
27. Norris, S.R., Warner, C.C., Lampkin, B.J., Bouc, P., VanVeller, B.: Synthesis and spectral properties of push-pull dyes based on isobenzofuran scaffolds. Org. Lett. **21**, 3817–3821 (2019)
28. Uchiyama, S., Santa, T., Fukushima, T., Homma, H., Imai, K.: Effects of the substituent groups at the 4- and 7-positions on the fluorescence characteristics of benzofurazan compounds. J. Chem. Soc. Perkin Trans. **2**, 2165–2173 (1998)
29. Uchiyama, S., Santa, T., Imai, K.: Fluorescence characteristics of six 4,7-disubstituted benzofurazan compounds: an experimental and semi-empirical MO study. J. Chem. Soc. Perkin Trans. **2**, 2525–2532 (1999)
30. Uchiyama, S., Takehira, K., Kohtani, S., Imai, K., Nakagaki, R., Tobitab, S., Santa, T.: Fluorescence on–off switching mechanism of benzofurazans. Org. Biomol. Chem. **1**, 1067–1072 (2003)
31. Andrews, J.L., Ghosh, P., Ternai, B., Whitehouse, M.W.: Ammonium 4-chloro-7-sulfobenzofurazan: A new fluorigenic thiol-specific reagent. Arch. Biochem. Biophys. **214**, 386–396 (1982)
32. Toyooka, T., Imai, K.: New fluorogenic reagent having halogenobenzofurazan structure for thiols: 4-(aminosulfonyl)-7-fluoro-2,1,3-benzoxadiazole. Anal. Chem. **56**, 2461–2464

(1984); Toyooka, T., Suzuki, T., Saito, Y., Uzu, S., Imai, K.: Fluorogenic reagent for thiols: 4-(*N*,*N*-dimethylaminosulphonyl)-7-fluoro-2,1,3-benzoxadiazole. Analyst **114**, 413–419 (1989)
33. Zhu, Z., Liu, W., Cheng, L., Li, Z., Xi, Z., Yi, L.: New NBD-based fluorescent probes for biological thiols. Tetrahedron Lett. **56**, 3909–3912 (2015)
34. Wang, J., Niu, L., Huang, J., Yan, Z. Wang, J.: A novel NBD-based fluorescent turn-on probe for the detection of cysteine and homocysteine in living cells. Spectrochim. Acta Part A **192**, 52–58 (2018)
35. Heyne, B., Beddie, C., Scaiano, J.C.: Synthesis and characterization of a new fluorescent probe for reactive oxygen species. Org. Biomol. Chem. **5**, 1454–1458 (2007)
36. Yang, M., Fan, J., Du, J., Long, S., Wang, J., Peng, X.: Imaging of formaldehyde in live cells and *daphnia magna* via Aza-cope reaction utilizing fluorescence probe with large stokes shifts. Front. Chem. **6**, 488 (2018)
37. Kozikowski, A.P., Kotoula, M., Ma, D., Boujrad, N., Tückmantel, W., Papadopoulos, V.: Synthesis and biology of a 7-nitro-2,1,3-benzoxadiazol-4-yl derivative of 2-phenylindole-3-acetamide: a fluorescent probe for the peripheral-type benzodiazepine receptor. J. Med. Chem. **40**, 2435–2439 (1997)
38. Taliani, S., Simorini, F., Sergianni, V., La Motta, C., Da Settimo, F., Cosimelli, B., Abignente, E., Greco, G., Novellino, E., Rossi, L., Gremigni, V., Spinetti, F., Chelli, B., Martini, C.: New fluorescent 2-phenylindolglyoxylamide derivatives as probes targeting the peripheral-type benzodiazepine receptor: design, synthesis, and biological evaluation. J. Med. Chem. **50**, 404–407 (2007)
39. Salman, B.I., Ali, M.F.B., Marzouq, M.A., Hussein, S.A.: Utility of the fluorogenic characters of benzofurazan for analysis of tigecycline using spectrometric technique; application to pharmacokinetic study, urine and pharmaceutical formulations. Luminescence **34**, 175–182 (2019)
40. Wang, L., Kong, H., Jin, M., Li, X., Stoika, R., Lin, H., Liu, K.: Synthesis of disaccharide modified berberine derivatives and their anti-diabetic investigation in zebrafish using a fluorescence-based technology. Org. Biomol. Chem. **18**, 3563–3574 (2020)
41. Kresze, G., Wucherpfennig, W.: New methods of preparative organic chemistry V. Organic syntheses with imides of sulfur dioxide. Angew. Chem. Int. Ed. **6**, 149–167 (1967)
42. Neto, B.A.S., Lopes, A.S., Ebeling, G., Gonçalves, R.S., Costa, V.E.U., Quina, F.H., Dupont, J.: Photophysical and electrochemical properties of π-extended molecular 2,1,3-benzothiadiazoles. Tetrahedron **61**, 10975–10982 (2005)
43. Henry, B.R., Morrison, J.D.: Studies of the electronic absorption and emission spectra of 2,1,3-benzothiadiazole. J. Mol. Spectrosc. **55**, 311–318 (1975)
44. Kurt Pilgram, K., Zupan, M., Skiles, R.: Bromination of 2,1,3-benzothiadiazoles. J. Heterocycl. Chem. **7**, 629–633 (1970)
45. Neto, B.A., Lapis, A.A., Mancilha, F.S., Vasconcelos, I.B., Thum, C., Basso, L.A., Santos, D.S., Dupont, J.: New sensitive fluorophores for selective DNA detection. Org. Lett. **9**, 4001–4004 (2007)
46. Neto, B.A., Lapis, A.A., Mancilha, F.S., Batista, E.L., Jr., Netz, P.A., Rominger, F., Basso, L.A., Santos, D.S., Dupont, J.: On the selective detection of duplex deoxyribonucleic acids by 2,1,3-benzothiadiazole fluorophores. Mol. Biosyst. **6**, 967–975 (2010)
47. Garo, F., Häner, R.: 2,1,3-Benzothiadiazole-modified DNA. Eur. J. Org. Chem. pp. 2801–2808 (2012)
48. Jiang, Q., Zhang, Z., Lu, J., Huang, Y., Lu, Z., Tan, Y., Jiang, Q.: A novel nitro-substituted benzothiadiazole as fluorescent probe for tumor cells under hypoxic condition. Bioorg. Med. Chem. **21**, 7735–7741 (2013)
49. Appelqvist, H., Stranius, K., Börjesson, K., Nilsson, K.P.R., Dyrager, C.: Specific imaging of intracellular lipid droplets using a benzothiadiazole derivative with solvatochromic properties. Bioconjug. Chem. **28**, 1363–1370 (2017)
50. Carvalho, P.H.P.R., Correa, J.R., Paiva, K.L.R., Machado, D.F.S., Scholten, J.D., Neto, B.A.D.: Plasma membrane imaging with a fluorescent benzothiadiazole derivative. Beilstein J. Org. Chem. **15**, 2644–2654 (2019)

51. Neto, B.A.D., Carvalho, P.H.P.R., Santos, D.C.B.D., Gatto, C.C., Ramos, L.M., de Vasconcelos, N.M., Corrêa, J.R., Costa, M.B., de Oliveira, H.C.B., Silva, R.G.: Synthesis, properties and highly selective mitochondria staining with novel, stable and superior benzothiadiazole fluorescent probes. RSC Adv. **2**, 1524–1532 (2012)
52. Carvalho, P.H.P.R., Corrêa, J.R., Paiva, K.L.R., Baril, M., Machado, D.F.S., Scholten, J.D., de Souza, P.E.N., Veiga-Souza, F.H., Spencer, J., Neto, B.A.D.: When the strategies for cellular selectivity fail. Challenges and surprises in the design and application of fluorescent benzothiadiazole derivatives for mitochondrial staining. Org. Chem. Front. **6**, 2371–2384 (2019)
53. Carvalho, P.H.P.R., Correa, J.R., Guido, B.C., Gatto, C.C., de Oliveira, H.C.B., Soares, T.A., Neto, B.A.D.: Designed benzothiadiazole fluorophores for selective mitochondrial imaging and dynamics. Chem. Eur. J. **20**, 15360–15374 (2014)
54. Barcelos, R.C., Pastre, J.C., Caixeta, V., Vendramini-Costa, D.B., de Carvalho, J.E., Pilli, R.A.: Synthesis of methoxylated goniothalamin, aza-goniothalamin and γ-pyrones and their in vitro evaluation against human cancer cells. Bioorg. Med. Chem. **20**, 3635–3651 (2012)
55. Raitz, I., de Souza Filho, R.Y., de Andrade, L.P., Correa, J.R., Neto, B.A.D., Pilli, R.A.: Preferential mitochondrial localization of a goniothalamin fluorescent derivative. ACS Omega. **2**, 3774–3784 (2017)
56. Cruz, E.H.G., Carvalho, P.H.P.R., Correa, J.R., Silva, D.A.C., Diogo, E.B.T., Filho, J.D.D.S., Cavalcanti, B.C., Pessoa, C., de Oliveira, H.C.B., Guido, B.C., Filho, D.A.D.S., Neto, B.A.D., Junior, E.N.D.S.: Design, synthesis and application of fluorescent 2,1,3-benzothiadiazole-triazole-linked biologically active lapachone derivatives. New J. Chem. **38**, 2569–2580 (2014)
57. Dheer, D., Singh, V., Shankar, R.: Medicinal attributes of 1,2,3-triazoles: current developments. Bioorg. Chem. **71**, 30–54 (2017)
58. Williams, C.H.G.: Researches on Chinoline and its Homologues. Trans. R. Soc. Edinburg **21**, 377–401 (1857)
59. Mishra, A., Behera, R.K., Behera, P.K., Mishra, B.K., Behera, G.B.: Cyanines during the 1990s: a review. Chem. Rev. **100**, 1973–2011 (2000); Henary, M., Paranjpe, S., Owens, E.A.: Syazole containing cyanine dyes Heterocyclic Communications, 19, 1–11 (2013)
60. Narayanan, N., Patonay, G.: A new method for the synthesis of heptamethine cyanine dyes: synthesis of new near-infrared fluorescent labels. J. Org. Chem. **60**, 2391–2395 (1995)
61. Reynolds, G.A., Drexhage, K.H.: Stable heptamethine pyrylium dyes that absorb in the infrared. J. Org. Chem. **42**, 885–888 (1977)
62. Mujumdar, R.B., Ernst, L.A., Mujumdar, S.R., Lewis, C.J., Waggoner, A.S.: Cyanine dye labeling reagents: sulfoindocyanine succinimidyl esters. Bioconjug. Chem. **4**(2), 105–111 (1993)
63. Norman, D.G., Grainger, R.J., Uhrín, D., Lilley, D.M.: Location of cyanine-3 on double-stranded DNA: importance for fluorescence resonance energy transfer studies. Biochemistry **39**, 6317–6324 (2000)
64. Sun, W., Guo, S., Hu, C., Fan, J., Peng, X.: Recent development of chemosensors based on cyanine platforms. Chem. Rev. **116**, 7768–7817 (2016)
65. Mojzych, M., Henary, M.: Synthesis of cyanine dyes. In: Strekowski, L. (ed.) Heterocyclic Polymethine Dyes, pp. 1–9. Springer, Berlin Heidelberg (2008)
66. Wang, X., Cui, L., Zhou, N., Zhu, W., Wang, R., Qian, X., Xu, Y.: A highly selective and sensitive near-infrared fluorescence probe for arylamine N-acetyltransferase 2 in vitro and in vivo. Chem. Sci. **4**, 2936–2940 (2013)
67. Yin, J., Kwon, Y., Kim, D., Lee, D., Kim, G., Hu, Y., Ryu, J.-H., Yoon, J.: Cyanine-based fluorescent probe for highly selective detection of glutathione in cell cultures and live mouse tissues. J. Am. Chem. Soc. **136**, 5351–5358 (2014)
68. Mendel, A.: Improved preparation of 5 dimethylamino-1-naphthalenesulfonyl chloride. J. Chem. Eng. Data **15**(2), 340–341 (1970)
69. Hartley, B.S., Massey, V.: The active centre of chymotrypsin: I. Labelling with a fluorescent dye Biochim. Biophys. Acta **21**, 58–70 (1956); Gray, W.R., Hartley, B.S.: A fluorescent end group reagent for peptides and proteins. Biochem. J. **89**, 59p (1963)

70. Gray, W.R.: Dansyl chloride procedure. Methods Enzymol. **11**, 139–151 (1967)
71. Hughes, J., Smith, T., Kosterlitz, H., Fothergill, L.A., Morgan, B.A., Morris, H.R.: Identification of two related pentapeptides from the brain with potent opiate agonist activity. Nature **258**, 577–579 (1975)
72. Xiao, Y., Guo, Y., Dang, R., Yan, X., Xu, P., Jiang, P.: A dansyl-based fluorescent probe for the highly selective detection of cysteine based on a d-PeT switching mechanism. RSC Adv. **7**, 21050–21053 (2017)
73. Cao, M., Jiang, L., Hu, F., Zhang, Y., Yang, W.C., Liu, S.H., Yin, J.: A dansyl-based fluorescent probe for selectively detecting Cu^{2+} and imaging in living cells. RSC Adv. **5**, 23666–23670 (2015)
74. Gee, K.R., Brown, K.A., Chen, W.-N.U., Bishop-Stewart, J., Gray, D., Johnson, I.: Chemical and physiological characterization of Fluo-4 Ca^{2+}-indicator dyes. Cell Calcium **27**, 97–106 (2000)
75. Mintaz, A., Kao, J.P.Y., Tsieng, R.Y.: Fluorescent indicators for cytosolic calcium based on rhodamine and fluorescein chromophores. J. Biol. Chem. **264**, 8171–8178 (1989)
76. Stosiek, C., Garaschuk, O., Holthoff, K., Konnerth, A.: In vivo two-photon calcium imaging of neuronal networks. PNAS **100**, 7319–7324 (2003)
77. Kaestner, L., Tabellion, W., Weiss, W., Bernhardt, I., Lippa, P.: Calcium imaging of individual erythrocytes: problems and approaches. Cell Calcium **39**, 13–19 (2006)
78. von Baeyer, A.: Uber ein neue Klasse von Farbstoffen. Chem. Ber. **4**, 555–558 (1871)
79. Klonis, N., Sawyer, W.H.: Spectral properties of the prototropic forms of fluorescein in aqueous solution. J. Fluoresc. **6**, 147–157 (1996)
80. Riggs, J.L., Seiwald, R.J., Burckhalter, J.H.: Isothiocyanate compounds as fluorescent labeling agents for immune serum. Am. J. Pathol. **34**, 1081–1097 (1958)
81. Confalone, P.N.: The use of heterocyclic chemistry in the synthesis of natural and unnatural products. J. Heterocycl. Chem. **27**, 31–46 (1990)
82. Sun, W.C., Gee, K.R., Klaubert, D.H., Haugland, R.P.: Synthesis of fluorinated fluoresceins. J. Org. Chem. **62**, 6469–6475 (1997)
83. Chen, X., Ma, H.: A selective fluorescence-on reaction of spiro form fluorescein hydrazide with Cu(II). Anal. Chim. Acta **575**, 217–222 (2006)
84. Chen, X., Ko, S.-K., Kim, M.J., Shin, I., Yoon, J.: A thiol-specific fluorescent probe and its application for bioimaging. Chem. Commun. **46**, 2751–2753 (2010)
85. Berthsen, A.: Ueber ein neues Chromogen, das Phenazoxin. Chem. Ber. **20**, 942–944 (1887)
86. Kehrmaiin, F., Neil, A.A.: Synthese in der Gruppe der Azoxine. Ber. Dtsch. Chem. Ges. **47**, 3102–3109 (1914)
87. Katritzky, A.R., Boulton, A.J.: Advances in Heterocyclic Chemistry. Academic Press (1967
88. Müller, P., Buu-Hoï, N.P., Rips, R.: Preparation and some reactions of phenoxazine and phenoselenazine. J. Org. Chem. **24**, 37–39 (1959)
89. Turpin, G.S.: The action of picric chloride on amines in presence of alkali. J. Chem. Soc. **59**, 714–725 (1891)
90. Möhlau, R., Uhlmann, K.: Zur Kenntniss der Chinazin- und Oxazinfarbstoffe. Justus Liebigs Ann. Chem. **289**, 90–130 (1896)
91. Kanitz, A., Hartmann, H.: Preparation and characterization of bridged naphthoxazinium salts. Eur. J. Org. Chem. **4**, 923–930 (1999)
92. Martinez, V., Henary, M.: Nile red and nile blue: applications and syntheses of structural analogues. Chem. Eur. J. **22**, 13764–13782 (2016)
93. Jose, J., Burgess, K.: Benzophenoxazine-based fluorescent dyes for labeling biomolecules. Tetrahedron **62**, 11021–11037 (2006)
94. Greenspan, P., Mayer, E.P., Fowler, S.D.: Nile red: a selective fluorescent stain for intracellular lipid droplets. J. Cell Biol. **100**, 965–973 (1985)
95. Miller, E.W., Albers, A.E., Pralle, A., Isacoff, E.Y., Chang, C.J.: Boronate-based fluorescent probes for imaging cellular hydrogen peroxide. J. Am. Chem. Soc. **127**, 16652–16659 (2005)
96. Zhan, C., Cheng, J., Li, B., Huang, S., Zeng, F., Wu, S.: A fluorescent probe for early detection of melanoma and its metastasis by specifically imaging tyrosinase activity in a mouse model. Anal. Chem. **90**, 8807–8815 (2018)

97. Noelting, E., Dziewonski, K.: Zur Kenntniss der Rhodamine. Ber. Dtsch. Chem. Ges. **38**, 3516–3527 (1905); Ehrlich, P., Benda, L.: Über die Einwirkung von Cyankalium auf Pyronin- und Acridinium-Farbstoffe. Ber. Dtsch. Chem. Ges. **46**, 1931–1951 (1913)
98. Dujols, V., Ford, F., Czarnik, A.W.: A long-wavelength fluorescent chemodosimeter selective for Cu(II) ion in water. J. Am. Chem. Soc. **119**, 7386–7387 (1997)
99. Chen, X., Wang, X., Wang, S., Shi, W., Wang, K., Ma, H.: A highly selective and sensitive fluorescence probe for the hypochlorite anion. Chem. Eur. J. **14**, 4719–4724 (2008)
100. Panchuk-Voloshina, N., Haugland, R.P., Bishop-Stewart, J., Bhalgat, M.K., Millard, P.J., Mao, F., Leung, W.Y., Haugland, R.P.: Alexa dyes, a series of new fluorescent dyes that yield exceptionally bright, photostable conjugates. J. Histochem. Cytochem. **47**, 1179–1188 (1999)
101. Mao, F., Leung, W-L., Haugland, R.P.: Sulfonated Xanthene Derivatives. U. S. Patent 6,130,101, Oct. 10, 2000
102. Berlier, J.E., Rothe, A., Buller, G., Bradford, J., Gray, D.R., Filanoski, B.J., Telford, W.G., Yue, S., Liu, J., Cheung, C.-Y., Chang, W., Hirsch, J.D., Beechem, J.M., Haugland, R.P., Haugland, R.P.: Quantitative comparison of long-wavelength Alexa Fluor dyes to Cy dyes: fluorescence of the dyes and their bioconjugates. J. Histochem. Cytochem. **51**, 1699–1712 (2003)
103. Bullok, K., Piwnica-Worms, D.: Synthesis and characterization of a small, membrane-permeant, caspase-activatable far-red fluorescent peptide for imaging apoptosis. J. Med. Chem. **48**, 5404–5407 (2005)
104. Baskin, J.M., Prescher, J.A., Laughlin, S.T., Agard, N.J., Chang, P.V., Miller, I.A., Lo, A., Codelli, J.A., Bertozzi, C.R.: Copper-free click chemistry for dynamic in vivo imaging. Proc. Natl. Acad. Sci. U. S. A. **104**, 16793–16797 (2007)
105. Laughlin, S.T., Baskin, J.M., Amacher, S.L., Bertozzi, C.R.: *In vivo* imaging of membrane-associated glycans in developing zebrafish. Science **320**, 664–667 (2008)
106. Dommerholt, J., Schmidt, S., Temming, R., Hendriks, L.J.A., Rutjes, F.P.J.T., van Hest, J.C.M., Lefeber, D.J., Friedl, P., van Delft, F.L.: Readily accessible bicyclononynes for bioorthogonal labeling and three-dimensional imaging of living cells. Angew. Chem. Int. Ed. **49**, 9422–9425 (2010)
107. Karver, M.R., Weissleder, R., Hilderbrand, S.A.: Synthesis and evaluation of a series of 1,2,4,5-tetrazines for bioorthogonal conjugation. Bioconjug. Chem. **22**, 2263–2270 (2011)
108. Sindbert, S., Kalinin, S., Nguyen, H., Kienzler, A., Clima, L., Bannwarth, W., Appel, B., Müller, S., Seidel, C.A.M.: Accurate distance determination of nucleic acids via Förster resonance energy transfer: implications of dye linker length and rigidity. J. Am. Chem. Soc. **133**, 2463–2480 (2011)
109. Someya, T., Ando, A., Kimoto, M., Hirao, I.: Site-specific labeling of RNA by combining genetic alphabet expansion transcription and copper-free click chemistry. Nucleic Acids Res. **43**, 6665–6676 (2015)
110. Dalton, C.E., Quinn, S.D., Rafferty, A., Morten, M.J., Gardiner, J.M., Magennis, S.W.: Single-molecule fluorescence detection of a synthetic heparan sulfate disaccharide. ChemPhysChem **17**, 3442–3446 (2016)
111. Comeo, E., Kindon, N.D., Soave, M., Stoddart, L.A., Kilpatrick, L.E., Scammells, P.J., Hill, S.J., Kellam, B.: Subtype-selective fluorescent ligands as pharmacological research tools for the human adenosine A_{2A} receptor. J. Med. Chem. **63**, 2656–2672 (2020)
112. Brun, M.A., Griss, R., Reymond, L., Tan, K.-T., Piguet, J., Peters, R.J.R.W., Vogel, H., Johnsson, K.: Semisynthesis of fluorescent metabolite sensors on cell surfaces. J. Am. Chem. Soc. **133**, 16235–16242 (2011)
113. Sethna, S.M., Shah, N.M.: The chemistry of coumarins. Chem. Rev. **36**, 1–62 (1945)
114. Trenor, S.R., Shultz, A.R., Love, B.J., Long, T.E.: Coumarins in polymers: from light harvesting to photo-cross-linkable tissue scaffolds. Chem. Rev. **104**, 3059–3077 (2004)
115. Perkin, W.H.: On the artificial production of coumarin and formation of its homologues. J. Chem. Soc. **21**, 53–63 (1868)
116. Perkin, W.H.: On the hydride of aceto-salicyl. J. Chem. Soc. **21**, 181–186 (1868)

117. Perkin, W.H.: On the formation of coumarin and of cinnamic and of other analogous acids from the aromatic aldehydes. J. Chem. Soc. **31**, 388–427 (1877)
118. von Pechmann, H., Welsh, W.: Ueber einige neue Cumarine. Ber. Dtsch. Chem. Ges. **17**, 1646–1652 (1884)
119. Knoevenagel, E.: Condensationen zwischen Malonester und Aldehyden unter dem Einfluss von Ammoniak und organischen Aminen. Ber. Dtsch. Chem. Ges. **31**, 2585–2595 (1898)
120. Kitamura, N., Fukagawa, T., Kohtani, S., Kitoh, S., Kunimoto, K.-K., Nakagaki, R.: Synthesis, absorption, and fluorescence properties and crystal structures of 7-aminocoumarin derivatives. J. Photochem. Photobiol. A Chem. **188**, 378–386 (2007)
121. Chemate, S.B., Sekar, N.: Novel iminocoumarin derivatives: synthesis, spectroscopic and computational studies. J. Fluoresc. **25**, 1615–1628 (2015)
122. Nourmohammadian, F., Gholami, M.D.: Microwave-promoted one-pot syntheses of coumarin dyes. Synth. Commun. **40**, 901–909 (2010)
123. Galvani, G., Reddy, K.H.V., Beauvineau, C., Ghermani, N., Mahuteau-Betzer, F., Alami, M., Messaoudi, S.: Conversion of 3-bromo-*2H*-coumarins to 3-(Benzofuran-2-yl)-*2H*-coumarins under palladium catalysis: synthesis and photophysical properties study. Org Lett. **19**, 910–913 (2017)
124. Zou, Q., Fang, Y., Zhao, Y., Zhao, H., Wang, Y., Gu, Y., Wu, F.: Synthesis and in vitro photocytotoxicity of coumarin derivatives for one- and two-photon excited photodynamic therapy. J. Med. Chem. **56**, 5288–5294 (2013)
125. Vadola, P.A., Sames, D.: Catalytic coupling of arene C-H bonds and alkynes for the synthesis of coumarins: substrate scope and application to the development of neuroimaging agents. J. Org. Chem. **77**, 7804–7814 (2012)
126. Hong, V., Presolski, S.I., Ma, C., Finn, M.G.: Analysis and optimization of copper-catalyzed azide-alkyne cycloaddition for bioconjugation. Angew. Chem. Int. Ed. **48**, 9879–9883 (2009)
127. Lemieux, G.A., De Graffenried, C.L., Bertozzi, C.R.: A fluorogenic dye activated by the Staudinger ligation. J. Am. Chem. Soc. **125**, 4708–4709 (2003)
128. Beatty, K.E., Fisk, J.D., Smart, B.P., Lu, Y.Y., Szychowski, J., Hangauer, M.J., Baskin, J.M., Bertozzi, C.R., Tirrell, D.A.: Live-cell imaging of cellular proteins by a strain-promoted azide-alkyne cycloaddition. ChemBioChem **11**, 2092–2095 (2010)
129. Yao, J.Z., Uttamapinant, C., Poloukhtine, A., Baskin, J.M., Codelli, J.A., Sletten, E.M., Bertozzi, C.R., Popik, V.V., Ting, A.Y.: Fluorophore targeting to cellular proteins via enzyme-mediated azide ligation and strain-promoted cycloaddition. J. Am. Chem. Soc. **134**, 3720–3728 (2012)
130. Yin, H., Zhang, B., Yu, H., Zhu, L., Feng, Y., Zhu, M., Guo, Q., Meng, X.: Two-photon fluorescent probes for biological $Mg^{(2+)}$ detection based on 7-substituted coumarin. J. Org. Chem. **80**, 4306–4312 (2015)
131. Komatsu, K., Urano, Y., Kojima, H., Nagano, T.: Development of an iminocoumarin-based zinc sensor suitable for ratiometric fluorescence imaging of neuronal zinc. J. Am. Chem. Soc. **129**, 13447–13454 (2007)
132. Jung, H.S., Kwon, P.S., Lee, J.W., Kim, J.I., Hong, C.S., Kim, J.W., Yan, S., Lee, J.Y., Lee, J.H., Joo, T., Kim, J.S.: Coumarin-derived $Cu^{(2+)}$-selective fluorescence sensor: synthesis, mechanisms, and applications in living cells. J. Am. Chem. Soc. **131**, 2008–2012 (2009)
133. Zhou, B., Liu, W., Zhang, H., Wu, J., Liu, S., Xu, H., Wang, P.: Imaging of nucleolar RNA in living cells using a highly photostable deep-red fluorescent probe. Biosens. Bioelectron. **68**, 189–196 (2015)
134. Brauns, E.B., Madaras, M.L., Coleman, R.S., Murphy, C.J., Berg, M.A.: Measurement of local DNA reorganization on the picosecond and nanosecond time scales. J. Am. Chem. Soc. **121**, 11644–11649 (1999)
135. Jung, H.S., Han, J.H., Pradhan, T., Kim, S., Lee, S.W., Sessler, J.L., Kim, T.W., Kang, C., Kim, J.S.: A cysteine-selective fluorescent probe for the cellular detection of cysteine. Biomaterials **33**, 945–953 (2012)
136. He, L., Xu, Q., Liu, Y., Wei, H., Tang, Y., Lin, W.: Coumarin-based turn-on fluorescence probe for specific detection of glutathione over cysteine and homocysteine. ACS Appl. Mater. Interfaces **7**, 12809–12813 (2015)

137. Yue, Y., Huo, F., Ning, P., Zhang, Y., Chao, J., Meng, X., Yin, C.: Dual-site fluorescent probe for visualizing the metabolism of Cys in living cells. J. Am. Chem. Soc. **139**, 3181–3185 (2017)
138. Wu, L., Tian, X., Groleau, R.R., Wang, J., Han, H.-H., Reeksting, S.B., Sedgwick, A.C., He, X.-P., Bull, S.D., James, T.D.: Coumarin-based fluorescent probe for the rapid detection of peroxynitrite 'AND' biological thiols. RSC Adv. **10**, 13496–13499 (2020)
139. Win, K.Y., Feng, S.S.: Effects of particle size and surface coating on cellular uptake of polymeric nanoparticles for oral delivery of anticancer drugs. Biomaterials **26**, 2713–2722 (2005)
140. Lin, Q., Huang, Q., Li, C., Bao, C., Liu, Z., Li, F., Zhu, L.: Anticancer drug release from a mesoporous silica based nanophotocage regulated by either a one- or two-photon process. J. Am. Chem. Soc. **132**, 10645–10647 (2010)
141. Schmitt, A., Hinkeldey, B., Wild, M., Jung, G.: Synthesis of the Core Compound of the BODIPY Dye Class: 4,4′-Difluoro-4-bora-(3a,4a)-diaza-s-indacene. J. Fluoresc. **19**, 755–758 (2009)
142. Treibs, A., Kreuzer, F.: Difluorboryl-Komplexe von Di- und Tripyrrylmethenen. Justus Liebigs Ann. Chem. **718**, 208–223 (1968)
143. Squeo, B.M., Gregoriou, V.G., Avgeropoulos, A., Bayse, S., Allard, S., Scherf, U., Chochos, C.L.: BODIPY-based polymeric dyes as emerging horizon materials for biological sensing and organic electronic applications. Prog. Polym. Sci. **71**, 26–52 (2017)
144. Zhang, M., Hao, E., Xu, Y., Zhang, S., Zhu, H., Wang, Q., Yua, C., Jiao, L.: One-pot efficient synthesis of pyrrolylBODIPY dyes from pyrrole and acyl chloride. RSC Adv. **2**, 11215–11218 (2012)
145. Rumyantsev, E.V., Marfin, Y.S.: Boron Dipirrins: mechanism of formation, spectral and photophysical properties, and directions of functionalization. Russ. J. Gen. Chem. **89**, 2682–2699 (2019)
146. Boens, N., Verbelen, B., Dehaen, W.: Postfunctionalization of the BODIPY core: synthesis and spectroscopy. Eur. J. Org Chem. pp. 6577–6595 (2015)
147. Chong, H., Fron, E., Liu, Z., Boodts, S., Thomas, J., Harvey, J.N., Hofkens, J.H., Dehaen, W.V., der Auweraer, M., Smet, M.: Acid-sensitive BODIPY dyes: synthesis through Pd-catalyzed direct C(sp3)-H arylation and photophysics. Chem. Eur. J. **23**, 4687–4699 (2017)
148. Guerrero-Corella, A., Asenjo-Pascual, J., Pawar, T.J., Díaz-Tendero, S., Martín-Sómer, A., Gómez, C.V., Belmonte-Vázquez, J.L., Ramírez-Ornelas, D.E., Peña-Cabrera, E., Fraile, A., Cruz, D.C., Alemán, J.: BODIPY as electron withdrawing group for the activation of double bonds in asymmetric cycloaddition reactions. Chem. Sci. **10**, 4346–4351 (2019)
149. Lu, H., Mack, J., Yang, Y., Shen, Z.: Structural modification strategies for the rational design of red/NIR region BODIPYs. Chem. Soc. Rev. **43**, 4778–4823 (2014)
150. Liu, M., Ma, S., She, M., Chen, J., Wang, Z., Liu, P., Zhang, S., Li, J.: Structural modification of BODIPY: improve its applicability. Chin. Chem. Lett. **30**, 1815–1824 (2019)
151. Bodio, E., Goze, C.: Investigation of B-F substitution on BODIPY and aza-BODIPY dyes: development of B-O and B-C BODIPYs. Dyes Pigm. **160**, 700–710 (2019)
152. Heisig, F., Gollos, S., Freudenthal, S.J., El-Tayeb, A., Iqbal, J., Müller, C.E.: Synthesis of BODIPY derivatives substituted with various bioconjugatable linker groups: a construction kit for fluorescent labeling of receptor ligands. J. Fluoresc. **24**, 213–230 (2014)
153. Alamudi, S.H., Satapathy, R., Kim, J., Su, D., Ren, H., Das, R., Hu, L., Alvarado-Martínez, E., Lee, J.Y., Hoppmann, C., Peña-Cabrera, E., Ha, H.-H., Park, H.-S., Wang, L., Chang, Y.-T.: Development of background-free tame fluorescent probes for intracellular live cell imaging. Nat. Commun. **7**, 11964 (2016)
154. Carlson, J.C., Meimetis, L.G., Hilderbrand, S.A., Weissleder, R.: BODIPY-tetrazine derivatives as superbright bioorthogonal turn-on probes. Angew. Chem. Int. Ed. **52**, 6917–6920 (2013)
155. Reisacher, U., Ploschik, D., Rönicke, F., Cserép, G.B., Kele, P., Wagenknecht, H.-A.: Copper-free dual labeling of DNA by triazines and cyclopropenes as minimal orthogonal and bioorthogonal functions. Chem. Sci.**10**, 4032–4037 (2019)

156. Gorbatov, S.A., Kozlov, M.A., Zlobin, I.E., Kartashov, A.E., Zavarzin, I.V., Volkova, Y.A.: Highly selective BODIPY-based fluorescent probe for Zn^{2+} imaging in plant roots. Mendeleev Commun. **28**, 615–617 (2018)
157. Huang, Y., Li, C.-F., Shi, W.-J., Tan, H.-Y., He, Z.-Z., Zheng, L., Liu, F., Yan, J.-W.: A near-infrared BODIPY-based fluorescent probe for ratiometric and discriminative detection of Hg^{2+} and Cu^{2+} ions in living cells. Talanta. **198**, 390–397 (2019).
158. Bertrand, B., Passador, K., Goze, C., Denat, F., Bodio, E., Salmain, M.: Metal-based BODIPY derivatives as multimodal tools for life sciences. Coord. Chem. Rev. **358**, 108–124 (2018)
159. Marfin, Y.S., Solomonov, A.V., Timin, A.S., Rumyantsev, E.V.: Recent advances of individual BODIPY and BODIPY-based functional materials in medical diagnostics and treatment. Curr. Med. Chem. **24**, 2745–2772 (2017)
160. Zhang, J., Wang, N., Ji, X., Tao, Y., Wang, J., Zhao, W.: BODIPY-based fluorescent probes for biothiols. Chem. Eur. J. **26**, 4172–4192 (2020)
161. Gao, J., Tao, Y., Zhang, J., Wang, N., Ji, X., He, J., Si, Y., Zhao, W.: Development of lysosome-targeted fluorescent probes for Cys by regulating the boron-dipyrromethene (BODIPY) molecular structure. Chem. Eur. J. **25**, 11246–11256 (2019)
162. Wang, W., Lorion, M.M., Martinazzoli, O., Ackermann, L.: BODIPY peptide labeling by late-stage C(sp3)-H activation. Angew. Chem. Int. Ed. **57**, 10554–10558 (2018)
163. Tian, Z., Ding, L., Li, K., Song, Y., Dou, T., Hou, J., Tian, X., Feng, L., Ge, G., Cui, J.: Rational design of a long-wavelength fluorescent probe for highly selective sensing of carboxylesterase 1 in living systems. Anal. Chem. **91**, 5638–5645 (2019)
164. Sekhar, A.R., Mallik, B., Kumar, V., Sankar, J.: A cell-permeant small molecule for the super-resolution imaging of the endoplasmic reticulum in live cells. Org. Biomol. Chem. **17**, 3732–3736 (2019)
165. Franke, J.M., Raliski, B.K., Boggess, S.C., Natesan, D.V., Koretsky, E.T., Zhang, P., Kulkarni, R.U., Deal, P.E., Milleret, E.W.: BODIPY fluorophores for membrane potential imaging. J. Am. Chem. Soc. **141**, 12824–12831 (2019)
166. Bodio, E., Denat, F., Goze, C.: BODIPYS and aza-BODIPY derivatives as promising fluorophores for in vivo molecular imaging and theranostic applications. J. Porphyrins. Phthalocyanines **23**, 1159–1183 (2019)
167. Wang, L., Qian, Y.: A novel quinoline-BODIPY fluorescent probe for fast sensing biothiols via hydrogen bonds assisted-deprotonation mechanism and its application in cells and zebrafish imaging. J. Photochem. Photobiol. A. Chem. **372**, 122–130 (2019)
168. Kolemen, S., Akkaya, E.U.: Reaction-based BODIPY probes for selective bio-imaging. Coord. Chem. Rev. **354**, 121–134 (2018)
169. Yin, J., Ma, Y., Li, G., Peng, M., Lin, W.: A versatile small-molecule fluorescence scaffold: carbazole derivatives for bioimaging. Coord. Chem. Rev. **412**, 213257 (2020)
170. Graebe, C.: Ueber Synthesen des Carbazols. Ber. Dtsch. Chem. Ges. **5**, 376–378 (1872)
171. Graebe, C., Ullmann, F.: Ueber eine neue Carbazolsynthese. Justus Liebigs Ann. Chem. **291**, 16–17 (1896)
172. Borsche, W.: Ueber Tetra- und Hexahydrocarbazolverbindungen und eine neue Carbazolsynthese. Justus Liebigs Ann. Chem. **359**, 49–80 (1908)
173. Kitamura, Y., Yoshikawa, S., Furuta, T., Kan, T.: One-pot synthesis of carbazole via tandem suzuki–miyaura and amination reaction. Synlett **3**, 377–380 (2008)
174. Yan, Q., Gin, E., Wasinska-Kalwa, M., Banwell, M.G., Carr, P.D.: A palladium-catalyzed ullmann cross-coupling/reductive cyclization route to the carbazole natural products 3-methyl-*9H*-carbazole, glycoborine, glycozoline, clauszoline K, mukonine, and karapinchamine A. J. Org. Chem. **82**, 4148–4159 (2017)
175. Bal, A., Maiti, S., Mal, P.: Iodine(III)-enabled distal C-H functionalization of biarylsulfonanilides. J. Org. Chem. **83**, 11278–11287 (2018)
176. Weber, G., Teale, F.W.J.: Fluorescence excitation spectrum of organic compounds in solution. Part 1. - Systems with quantum yield independent of the exciting wavelength. Trans. Faraday Soc. **54**, 640–648 (1958)

177. van Duuren, B.L.: The fluorescence spectra of aromatic hydrocarbons and heterocyclic aromatic compounds. Anal. Chem. **32**, 1436–1442 (1960)
178. Adhikari, R.M., Neckers, D.C., Shah, B.K.: Photophysical study of blue, green, and orange-red light-emitting carbazoles. J. Org. Chem. **74**, 3341–3349 (2009)
179. Xu, J., Zhang, Y., Yu, H., Gao, X., Shao, S.: Mitochondria-targeted fluorescent probe for imaging hydrogen peroxide in living cells. Anal. Chem. **88**, 1455–1461 (2016)
180. Zheng, Z., Zhang, T., Liu, H., Chen, Y., Kwok, R.T.K., Ma, C., Zhang, P., Sung, H.H.Y., Williams, I.D., Lam, J.W.Y., Wong, K.S., Tang, B.Z.: Bright near-infrared aggregation-induced emission luminogens with strong two-photon absorption, excellent organelle specificity, and efficient photodynamic therapy potential. ACS Nano **12**, 8145–8159 (2018)
181. Li, D., Sun, X., Huang, J., Wang, Q., Feng, Y., Chen, M., Meng, X., Zhu, M., Wang, X.: A carbazole-based "turn-on" two-photon fluorescent probe for biological Cu^{2+} detection vis Cu^{2+}-promoted hydrolysis. Dyes Pigm. **125**, 185–191 (2016)
182. Divya, K. P., Sreejith, S., Ashokkumar, P., Yuzhan, K., Peng, Q., Maji, S.K., Tong, Y., Yu, H., Zhao, Y., Ramamurthy, P., Ajayaghosh, A.: A ratiometric fluorescent molecular probe with enhanced two-photon response upon Zn^{2+} binding for *in vitro* and *in vivo* bioimaging. Chem. Sci. **5**, 3469–3474 (2014)
183. Haohan Song, H., Zhang, J., Wang, X., Zhou, Y., Xu, C., Pang, X., Peng, X.: A novel "turn-on" fluorescent probe with a large stokes shift for homocysteine and cysteine: performance in living cells and zebrafish. Sens. Actuators, B. **259**, 233–240 (2018)
184. Liu, X., Sun, Y., Zhang, Y., Miao, F., Wang, G., Zhao, H., Yu, X., Liu, H., Wong, W.-Y.: A 2,7-carbazole-based dicationic salt for fluorescence detection of nucleic acids and two-photon fluorescence imaging of RNA in nucleoli and cytoplasm. Org. Biomol. Chem. **9**, 3615–3618 (2011)

An Overview of New Strategies Based on Functional Nanoscale Materials to the Treatment of Tuberculosis

Letícia C. Assis, Alexandre A. de Castro, Teodorico C. Ramalho, Carlton A. Taft, and Felipe A. La Porta

Abstract Tuberculosis (TB) is a serious infectious disease of chronic evolution caused by Mycobacterium tuberculosis (MTB). In general, TB control depends on many factors, among which a fast and accurate diagnosis is essential, which in turn makes it possible to carry out a complete treatment involving the proper administration of medications by patients with active disease, that is, to prevent the transmission and evolution of this disease. Despite efforts to TB control, only in 2018, about 1.5 million people died for causes attributed to TB. This likely is due to the appearance of multidrug-resistant strains to known drugs, as well as individuals with HIV/AIDS who are more susceptible to TB. Thus, in this perspective, it is fundamentally important to develop new therapeutic options for the treatment of TB. Hence, this chapter aims to identify new therapeutic alternatives available in the scientific literature based on the use of functional nanoscale materials as strategies to control the increase in bacterial resistance to drugs commonly used in the treatment of TB.

L. C. Assis · A. A. de Castro · T. C. Ramalho
Department of Chemistry, Federal University of Lavras, CEP 37200-000 Lavras, Minas Gerais, Brazil

T. C. Ramalho
Department of Chemistry, Faculty of Science, University of Hradec Kralove, Rokitanskeho 62, 500 03 Hradec Kralove, Czech Republic

C. A. Taft
Centro Brasileiro de Pesquisas Físicas, Rua Dr. Xavier Sigaud, 150, Urca, 22290-180 Rio de Janeiro, Brazil

F. A. La Porta (✉)
Laboratório de Nanotecnologia e Química Computacional, Universidade Tecnológica Federal do Paraná, 86036-370 Londrina, PR, Brazil
e-mail: felipelaporta@utfpr.edu.br

F. A. La Porta and C. A. Taft (eds.), *Functional Properties of Advanced Engineering Materials and Biomolecules*, Engineering Materials,
https://doi.org/10.1007/978-3-030-62226-8_19

1 Introduction to Tuberculosis

Tuberculosis (TB) has plagued humankind for millennia and to remains considered one of the most lethal infectious diseases in the modern world [1–3]. Along this line, there are reports of its occurrence, e.g., in Greece and the Rome empire. However, it is believed that the disease has existed since the time of ancient Egypt since researchers have found characteristic TB lesions in mummified bodies [4]. Despite this, the etiologic agent responsible for this disease was isolated only in 1882, by the scientist Robert Koch, known as Mycobacterium tuberculosis (MTB) or Koch's bacillus (BK) [4–6].

As such, the MTB bacterium belongs to the Mycobacterium genus of the Mycobacteriaceae family, which has great clinical relevance [7]. This species of saprophytic bacteria is a mandatory, or opportunistic, parasite that generally does not develop in the living organism. On the other hand, they feed mainly on the waste products present in these organisms. Also, they may have free life in soil/water, but it is widely known that most of these microorganisms prefer to parasitize the tissues of living hosts [8].

The MTB species has curved or straight bacilli, not sporulated and immobile, in general, its chemical structure is similar, varying only in terms of its thickness and length, which is, measuring about 0.2–0.6 μm in thickness, with approximately 1–10 μm in length [9]. In addition, it is well-known that MTB has a high resistance to acid-alcohol due to a cell wall absorbed by high lipid content, such as microorganisms, which causes a hydrophobic barrier it difficult for acid-acid discoloration and, consequently contributing to its high resistance to antibiotics [9, 10].

The first antibiotics started to appear around 1940, when there was a hope that this severe disease could eventually be eradicated. Unfortunately, this optimism was premature, as TB is currently the bacterial infection that kills most people worldwide [8, 11, 12]. In 2018 alone, this disease was responsible for about 1.5 million deaths, with approximately 10 million new cases reported worldwide, according to the World Health Organization (WHO) [2, 13]. The frightening spread of TB is mainly associated with society's living conditions. Like most infectious diseases, TB also proliferates in areas of high human concentration, especially with poor infrastructure services (such as sanitation and housing), where, for example, hunger and misery exist. For obvious reasons, particularly its incidence is more significant in the peripheries of large urban centers, and can also infect anyone in rural areas [14].

Data released in 2018 (Fig. 1) by WHO shows that the majority of incidence of TB cases occurred mainly in Africa (24% of the total), Southeast Asia region (44%), and Western Pacific (18%), encompassing countries like Bangladesh, China, India, Indonesia, Nigeria, Pakistan, Philippines, and South Africa [13]. In this scenario, Brazil occupies the 20th position in the classification of the incidence of this disease. The latest Ministry of Health report, released in 2018, reported records of about 75,717 new tuberculosis cases and 4614 deaths attributed to the TB disease, with higher mortality rates in the states of Amazonas and Rio de Janeiro [15].

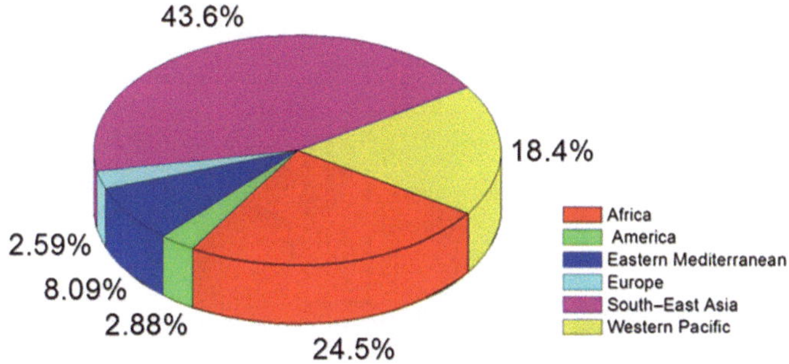

Fig. 1 Incidence of tuberculosis cases in the world in 2018. Reproduced with permission from reference [13]

Usually, transmission occurs when an individual infected with BK eliminates it to the external environment, by sneezing or coughing, contaminating a healthy person [15, 16]. As such, the most important factors to prevent the MTB transmission are to avoid agglomerations in closed spaces, cover your mouth with your arm or handkerchief when coughing and keep the environment well ventilated [4, 9]. Another prominent factor is the delay in visiting a doctor and to diagnosing of this disease, which can trigger the intensification of the spread of the disease, since, in each positive case, several people may also be infected before it has been diagnosed in locations with a high prevalence of the disease [9].

Yet, a peculiar fact related to TB transmission is that not all individuals infected with MTB progress to the active form of the disease. As such, there are latent and active TB infections. In particular, latent TB infection usually is when the patient has no symptoms of the disease, as the body is able to fight the infection, it is assumed that more than 2.0 billion people have latent TB [9, 17, 18]. These patients are at risk of some point reactivating the disease, because factors such as aging, poor diet, stress, and hormonal disorders and people infected with HIV/AIDS can be important factors in decreasing the effectiveness of the immune system, leading to the appearance of the active TB [17, 19]. Other routes of transmission of the TB bacillus, e.g., such as the skin or the placenta, are rare and hence have no considerable epidemiological importance [4].

As for TB infection, it usually starts when the bacilli reach the alveoli through the individual's airways, as illustrated in Fig. 2. In particular, BKs are ingested by alveolar macrophages that most often can kill the bacteria, that is, likely due to the host's immune response. If BKs manage to survive through this first line of defense of the immune system, in particular, the bacteria begin to multiply within macrophages and spread to neighboring cells, such as endothelial and epithelial cells. In addition, through the lymphatic and circulatory systems, they can also migrate to other organs, contaminating them and therefore reaching a high bacterial load in a few weeks [12, 20].

Thus, after this initial inflammatory reaction of the innate immune system, in particular, the adaptive immune system leads to the dispersion of neutrophils,

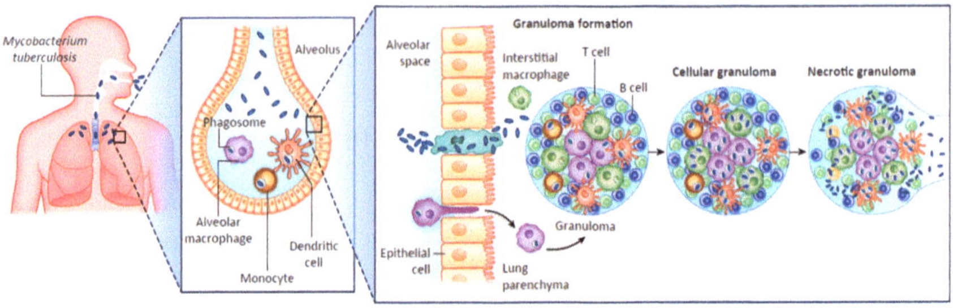

Fig. 2 Process of infection by *M. tuberculosis*. Reproduced with permission from reference [21]

lymphocytes, as well as other immune cells to the first infection site (pulmonary), naturally originating a cellular infiltrate that, at some point, exhibits the structure of a granuloma, with fibrotic components, surrounding it and making it calcified [12]. Then, the bacilli are protected and asleep within this granuloma. For reasons still not well-defined, the immune system fails and the bacilli uncontrollably start to reproduce, making the TB disease active, with manifestations resulting from symptoms [9, 12].

As previously mentioned, TB can affect any organ of the infected individual. Yet, most infections are restricted to the lungs. Outside the lungs, the places where TB is located, in order of frequency, are ganglia, pleura, urinary system, bones and joints, meninges, and peritoneum, but virtually all organs and systems can be affected by TB [4].

The detection of TB is an essential step towards its control, since, one of the leading causes of high TB load in low-income countries are the cases of TB undetected. Thus, the early detection of TB implies the reduced transmission of the disease significantly [22]. Generally, the set of symptoms, which involve a chronic cough, fever, night sweat, chest pain, anorexia and adynamia (muscle weakness) can raise the suspicion of TB [5, 23]. However, confirmation only occurs through the definitive diagnosis that is made by identifying the bacilli in a biological sample of the patient, which can be bronchial sputum, bronchoalveolar lavage, as well as other samples related to the respiratory tract, through bacilloscopic, culture or molecular methods. Other tests such as blood count, biochemical and radiological tests assist in the diagnosis, leading to solicitation of more specific tests [9]. Hence, the main advantages and disadvantages of some TB detection tests are summarized in Table 1.

These tests shown in Table 1 are widely used to verify the immune response developed in the host against MTB, since the latent infection generates a late hypersensitivity reaction to the type, as opposed to various components of the bacterium [30, 31]. After diagnosis, the current treatment for controlling the spread of TB usually consists of the use of antibiotics such as isoniazid, rifampicin, ethambutol and pyrazinamide (Fig. 3), which involves ingesting these antibiotics in its combination for the first 2 (two) months of treatment [6, 32, 33]. In view that this initial phase consists of the rapid elimination of the bacilli. While the second phase consists of preventing relapses and is called the continuation phase, where treatment with isoniazid and rifampicin is carried out for 4 (four) months [22]. The chemical structure of these drugs used to treat TB are illustrated in the Fig. 3.

In addition, among these measures available for the prevention of the disease are the intervention to assess latent infection of contacts and BCG vaccination (Bacillus

Table 1 TB detection tests and their advantages and disadvantages

TB detection test	Advantages	Disadvantages	Explanation	References
Chest X-ray	Simple procedure, identify atypical changes in the lungs	Small peripheral nodules may not be seen and risk of cancer due to excessive radiation exposure	Excessive exposure to X-rays affects the location of exposure	[24]
TB skin test/Mendel Mantoux test	Highly requested exam to detect TB and replace multiple puncture tests	Test that indicates only infection and is not sufficient for the diagnosis of the disease, in addition to generating false-positive or false-negative results	The result depends on knowledge about the antigen used, immunology of the reaction to the antigen and administration and reading techniques, as well as on the result of epidemiological and clinical experiences	[25]
TB culture Test	It is a detection method of high specificity and sensitivity for pulmonary and extrapulmonary TB In addition to indicating cases of suspected bacterial drug resistance	Expensive procedure that requires infrastructure and levels of biosafety	Due to the slow growth of mycobacteria, this procedure takes time, taking a long time to obtain the results	[26]
TB interferon gamma release assays (IGRAs)	The result of this test is obtained in 24 h, it is a blood test and vaccination does not give false results	Results do not show selectivity between latent tuberculosis and active tuberculosis	In this test, the white blood cells of the patient who has already had contact with the TB mycobacterium, when mixed with the Mtb antigen, release a large amount of IFN-g (interferon-gamma)	[9]

(continued)

Table 1 (continued)

TB detection test	Advantages	Disadvantages	Explanation	References
Bacilloscopy	Most widely used identification method in the world, because it is simple, fast and cheap	In the case of positive results, thousands of bacilli are needed per ml of sputum, in addition, it has low sensitivity and specificity	As the sample is sputum, it cannot detect extra-pulmonary TB	[26]
Serological tests	Method considered simple, economical and minimally invasive	The antibody response to M. tuberculosis in chronic infection is highly complex and variable, not precise and quite difficult	Detection of antibodies to TB in a blood sample	[27]
GeneXpert test	The automated GeneXpert method identifies the complex, also checks resistance to rifampicin	Constant power and air conditioning is required	Detection method based on the identification of specific sequences related to antibiotic resistance	[28]
TrueNat TB	The method is partially automated and only 0.5 ml of sputum sample is needed	Rapidly detects Tb resistance to rifampicin	Polymerase chain reaction-based testing identifies Mtb infection and antibiotic resistance	[29]

of Calmette-Guérin) [34]. In particular, the BCG vaccine was developed from a strain of Mycobacterium bovis by researchers Léon Charles Albert Calmette and Jean-Marie CamilleGuérin, in Paris, between the years 1906 and 1919 [34–36]. As such, the BCG vaccine is became widely used successfully in humans and prevents the development of lethal disease in children and young people; however, its effectiveness in protecting against lung disease in adults is still questionable [4, 34]. Therefore, there are huge efforts to the development of new and more efficient vaccines, being that some of them are underway in the phase of clinical trials [22].

On the other hand, however, patients' lack of adherence to the available therapeutic regimens, the subsequent increase in multidrug-resistant strains of TB (with high tolerance to some drugs) and also, co-infection with the HIV/AIDS virus, decrease the effectiveness of treatment [4–22]. Particularly, we had reached an alarming point where some bacteria exhibit high resistance to the use of the main antibiotics. This likely is due to the fact that, in many cases, these drugs are used in large excess (often wrongly and without any medical consent). This could lead to a global health crisis, which could lead us to the time when antibiotics did not exist [37–41]. Thus,

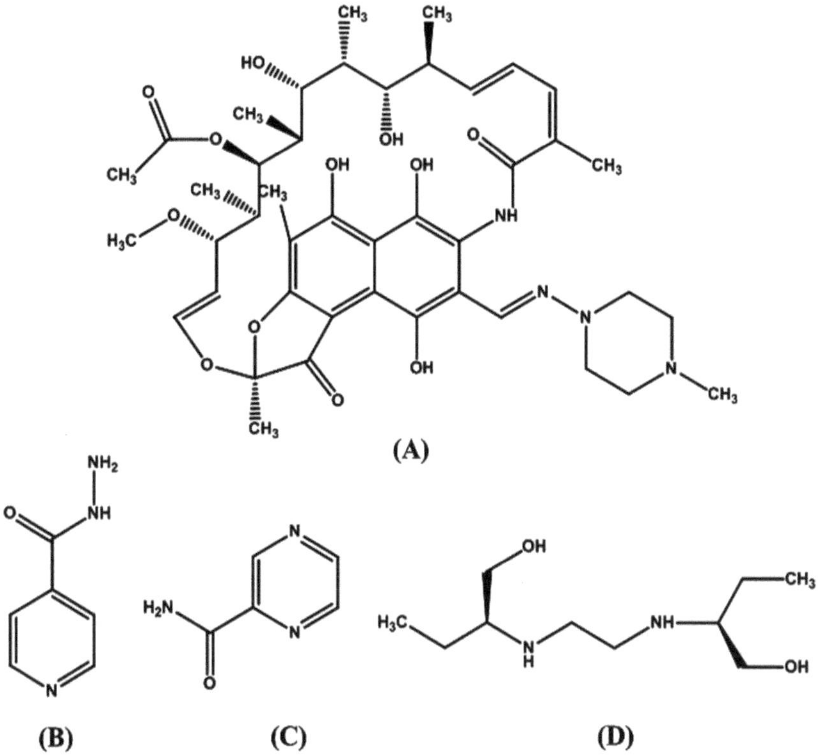

Fig. 3 Cocktail of drugs used to treat TB: Rifampicin (**a**), isoniazid (**b**), pyrazinamide (**c**) and Itambutol (**d**)

in this perspective, the functional nanoscale materials emerging as a new and very promising strategy for overcoming bacterial resistance [4, 9].

There are many recent studies emphasizing the use of nanotechnology as a therapeutic strategy [42–44]. Significant advances have been made in the design of novel functional nanoscale materials, which can be obtained in a controlled manner by different preparation methods. Hence, the major goal of this chapter is to introduce the potential benefits related to the potential use of functional nanoscale materials to the treatment of multidrug-resistant TB strains. In addition, we believed that this therapeutic approach based on nanomaterials could, in principle, contribute to reducing the overall incidence of death and new TB cases in the future.

2 Therapeutic Strategies Based on Functional Nanoscale Materials

The employment of nanomaterials in medical applications, such as for the TB treatment, has increasingly grown in the last decade [45]. It is important to highlight that the conventional antibacterial drugs present some limitations, such as short-term efficacy, a poor release profile along with cumulative toxicity, and even the development of resistance. Hence, the fundamental concepts of nanotechnology may aid in solving these drug-related problems [46]. Also, these functional nanoscale materials can protect antibiotics from deactivating enzymes, increasing the therapeutic efficacy of the medication and are easy to handle, portable, efficient and inexpensive [22]. Particularly the use of nanomaterials for drug delivery purposes aims at maximizing the dose into specific target organs, thus resulting in the reduction of systemic side effects [47–49]. Thus, it is important a careful assessment of the in vivo biodistribution, which for nano-sized materials, due to their behavior is deeply dependent on the size, charge, stability, in addition to other biophysical features [50]. In this context, the formation of the biomolecular corona is known to determine the successful application of nanoparticles in vivo (Fig. 4). As such, the adsorption process of biomacromolecules (like proteins), in principle, is an important and necessary step that occurs immediately by biding with the surface of nanomaterials when circulating in physiological fluids (such as the blood). More specifically, these strategies are needed in order to improve the functional properties of the nanoparticles significantly and hence allow for targeted stimuli-responsive drug delivery [51]. In view that the biomolecular corona has a crucial role in determining immune system clearance and targeting when nanomaterials are injected through the intravenous route [52], particularly for anti-TB treatment since macrophages stand for the leading target site [53, 54].

Some methods for the preparation of polymer-related nanoparticles can be observed in Fig. 5. Nano-delivery-based systems developed for both purposes of targeting and treating TB, without the use of conventional drugs, have been designed [46]. These include species like liposomes, which boost the intracellular killing of MTB [55], as well as the employment of liposomal formulations containing

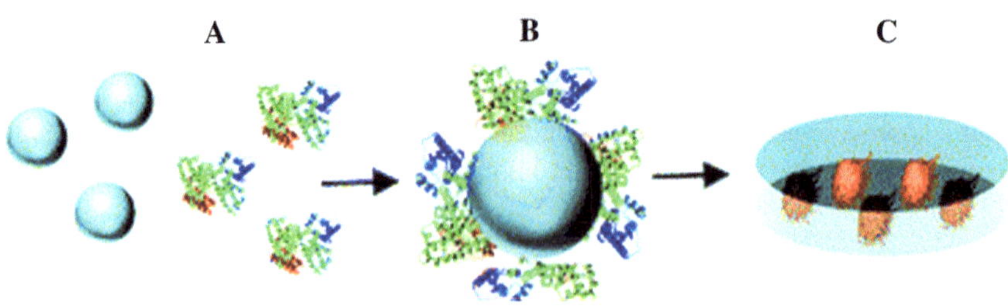

Fig. 4 Representation of the corona formation process. A = pre-coating, B = corona formation and C = cell uptake. Reproduced with permission from reference [51]

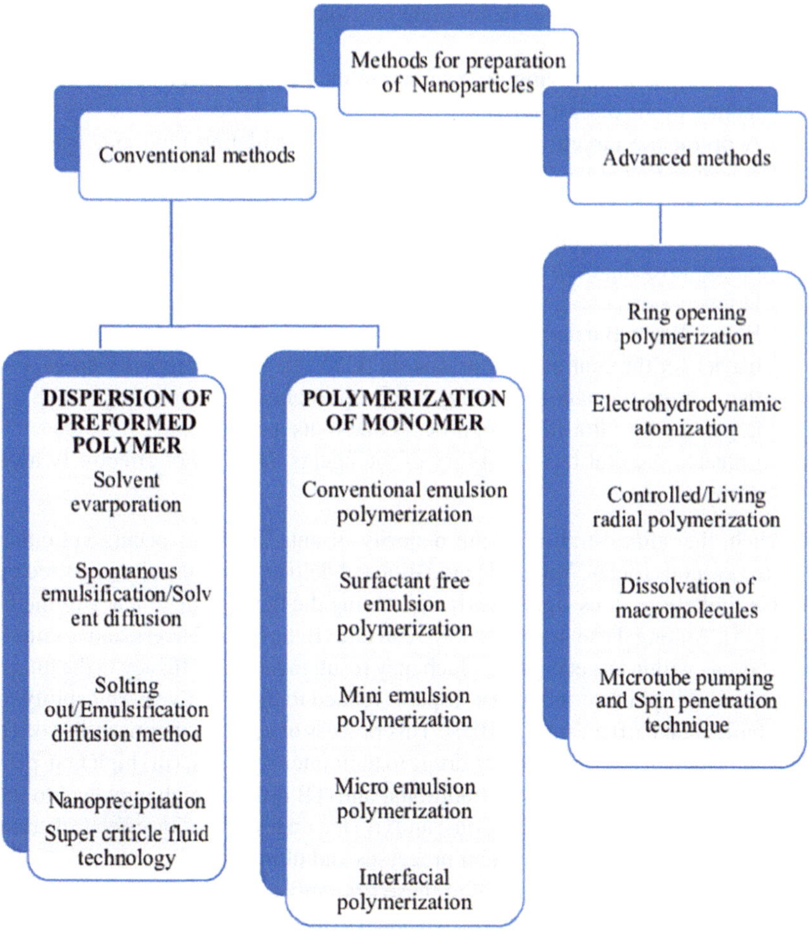

Fig. 5 Representation of conventional methods for the preparation of nanoparticles. Reproduced with permission from reference [57]

antisense oligonucleotides [56]. Faced with the exposed, some advantages of the nanoformulation are as follows, according to Patel et al. [57]:

(i) The use of nanomaterials-based delivery systems can potentially protect drugs from degradation;
(ii) Tunable physicochemical properties due to the alteration in composition, particle size, and morphology;
(iii) Reduction in dose frequency preventing problems related to misuse and in excess;
(iv) Economic and patient compliance;
(v) The delivery of insoluble drugs can also take place by using nanomaterials-based delivery systems;

(vi) These systems can also incorporate previously rejected drugs or drugs with administration issues;
(vii) Refining and improving drug targeting due to specific pathophysiological feature of the diseased tissues;
(viii) A promising targeting system have longer circulating time and optimum concentration at target sites;
(ix) Its pharmacological activity is not affected by longer circulating time;
(x) Enhanced permeability and retention effect are characteristic aspects of the tumor, favoring drug delivery;
(xi) Drugs passively target macrophages (liver and spleen);
(xii) Blood–Brain Barrier (BBB) is the most efficient and challenging natural barrier for the central nervous system (CNS) targeting lipophilic drugs. For these drugs, the nanotechnology is a simple solution as such drugs reach the target site via ultra-filtration process due to its nano-size;
(xiii) Enhance the oral bioavailability of the agents that are not effectively used orally.

As such, the administration of the majority of anti-TB drugs occurs via either oral or intravenous routes. Yet, these modalities of delivery have usually impaired by extensive limitations, thus significantly decreasing the therapeutic efficacy of medications [45]. Among these limitations, we can cite (i) the poor delivery and accumulation of drugs within target organs, which may result in reduced efficacy in the lungs, urinary tract and spinal vertebrae; (ii) aspects related to the selective permeability of diverse biological barriers such as BBB. This process of selective permeability often inhibits the efficient delivery of these drugs to their intended sites; (iii) high first-pass metabolism and short half-lives of traditional anti-TB drugs, which may lead to the need of a prolonged treatment durations; and (iv) the undesirable side effects of some medicines result in reduced treatment processes and ultimately, drug resistance and disease recurrence [45, 58, 59]. Due to these limitations mentioned above, particularly the nanocarriers have widely been designed to manage these deficiencies, that is, thus improving TB drug delivery. Some strategies that have demonstrated enhancing in vivo drug performance consist in the employment of nanoparticulate-based drug delivery systems, with improved plasma membrane permeability, tissue selectivity, and even controlled drug release [45, 58, 59]. As such, the use of nanotechnology in medicine is undoubtedly revolutionizing the current forms of treatment, allowing for and making the treatment process more efficient and customized for a range of diseases. Figure 6 shows the representation of some functional nanoscale materials can be widely used in diverse medical applications.

It is additionally biodegradable and biocompatible polymeric nanoparticles have been used since long in the cancer treatment as drug vehicles for enhancing anticancer drug delivery so that it would be possible to maximize in vivo drug efficacy [61]. Guided by the prosperous results demonstrated with anticancer drugs, these polymeric nanoparticles have currently been used to improve anti-TB drug delivery [62, 63]. Some of the approaches, which have been adopted, as well as notable achievements, are highlighted in Tables 2 and 3.

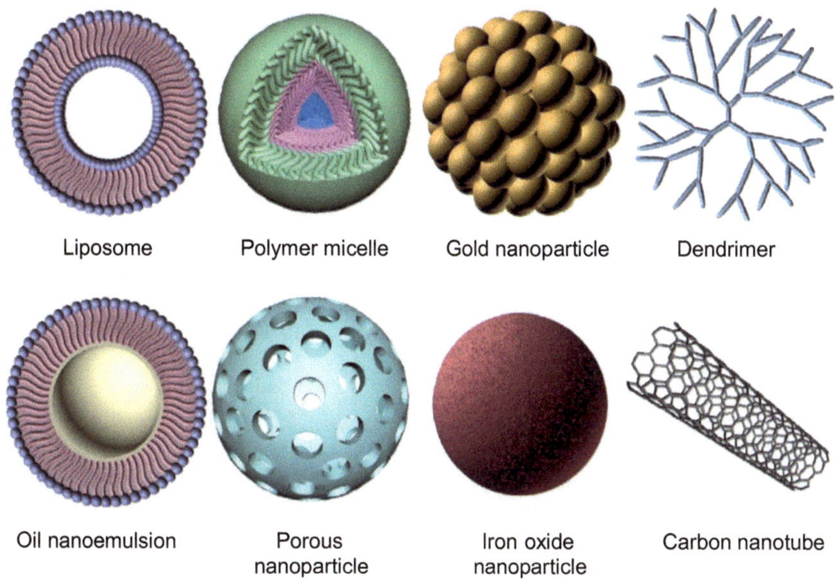

Fig. 6 Schematic representation of some nanoparticulate materials. Reproduced with permission from reference [60]

A range of polymeric nanoparticles have been studied for the purpose of enhancing drug delivery, and among these species, the poly(lactide-coglycolide) (PLGA) in form of nanoparticles is one of the most deeply investigated. In view of their biocompatibility and good safety profiles, the PLGA nanoparticles have particularly been approved by the US Food and Drug Administration (FDA), for application in a variety of situations. Efforts have been made in order to design PLGA nanoparticles for the delivery of anti-TB drugs. These strategies are, therefore, essential to address legitimate clinical needs [45].

In work from Pandey et al. [64], were developed PLGA nanoparticles (diameter 186–290 nm), aiming at improving anti-TB drug delivery. In addition, these nanoparticles have been applied to encapsulate first-line anti-TB drugs, such as rifampicin (RMP), isoniazid (INH), and pyrazinamide (PZA) (Table 2). Interestingly, the encapsulation efficiencies (EE) reached 56%, 66%, and 68%, respectively. On the other hand, these PLGA-based nanoparticles also demonstrated good potential in prolonging the half-life of anti-TB drugs, as well as reducing the dosing frequency of these medications. Pharmacokinetic investigations in mice models were capable of detecting the presence of RMP in plasma, and surprisingly, even up to 6 days after oral administration. In the case of INH and PZA, up to 9 days. These outcomes are totally in contrast with the shorter half-lives of the unadulterated drugs, being 2 h for RMP and 2–3 h for INH [64]. Furthermore, sustained drug levels, i.e., above minimum inhibitory concentrations, were shown in target organs, such as lungs, for a period of time of up to 9–11 days [45].

Table 2 Nanoparticle formulations employed in enhancing in vivo TB-drug delivery

Formulation	Drug	Drug EE [%]		Animal/dose/routes
PLGA nanoparticles (186–290 nm)	RMP INH PZA	56 66 68	a	Mice/5 doses (RMP, INH and PZA, 12, 10, and 25 mg kg^{-1}; once every 10 days)/oral
			b	Mice/single dose (5 mg of drug loaded PLGA nanoparticles)/subcutaneous
			c	Guinea pigs/5 doses (once every 10 days administered via nebulization)/inhaled
PLGA nanoparticles (316 nm)	RMP		d	Zebrafish/intravenous
Lectin functionalized PLGA nanoparticles (350–400 nm)	RMP INH PZA	54 64 66	e	Guinea pigs/3 doses (RMP, INH, and PZA, 12, 10, and 25 mg kg^{-1}; once every 15 days)/oral
			f	Guinea pigs/3 doses (RMP, INH, and PZA, 12, 10, and 25 mg kg^{-1}; once every 15 days)/inhaled
Alginate nanoparticles (235 nm)	RMP INH PZA	80–90 70–90 70–90	g	Mice/Guinea pigs/3 doses (RMP, INH, and PZA, 12, 10, and 25 mg kg^{-1}; once every 15 days)/oral
			h	Guinea pigs/3 doses (RMP, INH, and PZA, 12, 10, and 25 mg kg^{-1}; once every 15 days)/inhaled
Liposomes	RMP INH	44–49 8–10	i	Mice/Guinea pigs/12 doses (free drug RMP, INH, 12 and 10 mg kg^{-1}; RMP and INH loaded liposomes, 3 and 4 mg kg^{-1}; twice every week)/intravenous
Liposomes	RMP INH	40–45 8–12	j	Guinea pigs/1 dose (RMP, INH, 46.5 and 23.25 mg kg^{-1}) /inhaled
Asymmetric apoptotic body like liposomes (APL)			k	Mice/intranasal
Solid lipid nanoparticles (SLN)	RMP INH PZA	51 45 41	l	Mice/5 doses (5 mg of SLN containing 0.24 mg of RMP, 0.2 mg of INH and 0.5 mg of PZA), once every 10 days/oral

(continued)

Table 2 (continued)

Formulation	Drug	Drug EE [%]		Animal/dose/routes
			m	Guinea pigs/7 doses (free drug RMP, INH, 12 and 10 mg kg − 1; RMP and INH loaded liposome, 3 and 4 mg kg − 1; once every week)/inhaled

Reproduced with permission from reference [45]

In the process of free drug administrations, it is verified an initial burst release within hours, subsequently being quickly metabolized and excreted. This situation explains, e.g., the need for high dosing frequencies for TB treatment. According to a study with M. tuberculosis-infected mice, 5 daily doses of drug-loaded nanoparticles, which were administered orally, were enough to totally treat the infection. In contrast, conventional free drugs in clinical use provide the same result with 46 daily [64].

The delivery process via inhalation of these nanoparticles also provided similarly remarkable outcomes [63]. For example, in a study with M. tuberculosis-infected guinea pigs, showed that drug-loaded nanoparticles led to complete clearance of infection over only 4–5 treatment sessions, as well as, resulting in a significantly improve the drug bioavailability concerning their orally administered equivalents (i.e., 12.7, 32.8, and 14.7 folds for RMP, INH, and PZA, respectively) [63].

Note that the results obtained indicate potential implications in order to minimize dosing frequencies, and equally important, they can reduce total drug concentrations. These aspects are important to make it easier for the treatment administration, reduce side effects, and circumvent multidrug resistance [45]. As said previously, nanoparticles have demonstrated to be effective in boosting the therapeutic efficacy of conventional anti-TB drugs. This trend is widely mediated through their endocytosis by phagocytic cells, such as macrophages [65]. This effect from phagocytic cells is notably most pronounced for nanoparticles within a specific size range of 50–200 nm [66–69]. Due to the trend of preferentially targeting macrophages, drug-loaded nanocarrier systems have shown efficacy in reducing bacterial loads at the target organ sites, explaining the superior performance of these systems. In a work from Griffith et al. [70], in which BCG-infected mice models were used, they demonstrated the differential localization of the BCG and PLGA nanoparticles within mouse primary macrophages. When infecting macrophages, BCG predominantly resided within phagosomes. On the contrary, when drug-loaded PLGA nanoparticles were phagocytosed by macrophages, they instead localized to the phagolysosome [70].

With noninvasive imaging techniques, it was possible to visualize nanoparticles being taken up by both infected and noninfected macrophages. One important point to highlight is the fact that noninfected macrophages are actively recruited to the infection site, where they phagocytose their infected counterparts. The investigators observed that drug-loaded nanoparticles within the phagolysosomes degraded

Table 3 Achievements of nanoparticle formulations employed in enhancing in vivo TB-drug delivery

Formulation		Achievements
PLGA nanoparticles (186–290 nm)	a	TB bacilli was completely cleared from the lungs and spleens of mice that received 5 doses of oral treatment
	b	A single subcutaneous dose led to sustained drug plasma levels in the lungs (32 days) and spleen (36 days); M. tuberculosis-infected mice receiving single dose of treatment showed undetectable bacteria counts in the lungs and spleen
	c	A single nebulization dose led to sustained drug plasma levels for 6–8 days; prolonged half-life and enhanced bioavailability of drugs; M. tuberculosis infected Guinea pigs that received five doses of treatment showed undetectable bacteria counts in the lungs
PLGA nanoparticles (316 nm)	d	RMP-loaded NPs significantly increased embryo survival and decreased bacterial load
Lectin functionalized PLGA nanoparticles (350–400 nm)	e	Extended plasma life of drugs up to 7–13 days; M. tuberculosis infected Guinea pigs that received 3 doses of oral treatment showed undetectable bacteria counts in various organs (lungs, liver, and spleen)
	f	Extended plasma life of drugs up to 6–14 days; sustained therapeutic drug levels in tissues for up to 15 days; administration of nebulized formulations to M. tuberculosis infected guinea pigs led to undetectable CFUs in various organs (lungs, liver, and spleen)
Alginate nanoparticles (235 nm)	g	Prolonged plasma life of drugs for up to 8–11 days; sustained drug levels up to 15 days in organs; oral administration of drug-loaded NPs to infected guinea pigs resulted in undetectable CFUs in various organs (lungs, liver, and spleen)
	h	Prolonged plasma life of drugs for up to 10–14 days; sustained drug levels for up to 15 days in organs; administration of nebulized formulations to infected Guinea pigs led to undetectable CFU in various organs (lungs, liver, and spleen)
Liposomes	i	Extended plasma life of drugs for up to 5–7 days; drugs-loaded liposome demonstrated higher efficacy than free drugs against tuberculosis

(continued)

Table 3 (continued)

Formulation		Achievements
Liposomes	j	Extended plasma life of drugs for up to 2 days; sustained drug levels for up to 5 days in alveolar macrophages
Asymmetric apoptotic body like liposomes (APL)	k	APL enhanced intracellular killing of M. tuberculosis in bronchoalveolar lavage cells obtained from patients with active pulmonary tuberculosis; Infected mice receiving ABL treatment with or without INH dramatically reduced lung, liver and spleen mycobacterial burden
Solid lipid nanoparticles (SLN)	l	Extended plasma life of drugs for up to 8 days; sustained drug levels for up to 10 days in the lungs, liver, and spleen; oral administration of drug loaded SLN (5 doses) to infected mice led to undetectable CFUs in various organs (lungs and spleen)
	m	Prolonged plasma life of drugs for up to 5 days; sustained drug levels for up to 7 days in the lungs, liver and spleen; nebulization of drug-loaded SLN (7 doses) to infected guinea pigs led to undetectable CFUs in various organs (lungs and spleen)

Reproduced with permission from reference [45]

through hydrolysis. This trend led the antibiotics to be first released into the cytoplasm (with the intracellular killing of pathogens) and then into the systemic blood circulation. The RMP-loaded PLGA nanoparticles once more showed their utility for treating TB, significantly lowering bacterial loads in target organs [45].

Along with PLGA nanoparticles, alginate nanoparticles have also been loaded with anti-TB drugs and delivered to their target organs, also providing good results [59]. In the study with TB-infected guinea pigs with inhaled INH, PZA, and RMP-loaded alginate nanoparticles, it was possible to achieve full clearance of infection, being observed in the lungs and spleen after 45 days of treatment. Interestingly, the plasma drug concentration levels were comparable to those observed on day 1 post-drug administration, even after 14 days of treatment [61]. This outcome reinforces the fact that dosing frequencies can significantly be reduced with the employment of drug-loaded nanoparticles.

On the other hand, it is well-known that functional nanoscale materials as a promising platform for biosensors applications. For instance, Khoder et al. [42], who in 2020 used polypyrrole nanostructure to develop a biosensor to detect the genomic DNA of the MTB and the mutant that is resistant to RMP. According to the experiment, through electrochemical polymerization, polypyrrole nanowires were formed, then they were chemically modified by electrochemical oxidation of ethylenediamine or PAMAM dendrimers to obtain an aminated surface [42]. Sharma et al. [71] studied the nano-carbons as bio-sensing tools to detect variations in the concentrations of bacteria. In this work, the authors pointed out the limitations and argued strategies to improve bacterial sensors based on nano-carbon, especially for monitoring infections, such as TB [71].

Carbon-based nanomaterials (denoted here as CNMs) stand for a novel class of functional materials, which have been broadly explored in many biomedical applications (e.g., such as drug delivery, tissue engineering, diagnosis, cancer therapy, among others) [72]. These functional materials have shown to present a broad-spectrum antibacterial efficacy [73]. In this line, they have recently so-called huge attention for the development of more promising therapeutic strategies against drug-resistant species. As such, CNMs are species that differ from shape, size and therefore exhibit specific interactions with eukaryotic and prokaryotic cells. Moreover, the carbon skeleton of CNMs is usually functionalized with groups that lead to peruse better solubility, beyond defining interactions with large macromolecules. Note that the bactericidal activity of CNMs involves a combination of physical and chemical mechanisms, as well. These nanomaterials designed can induce critical damages to the cell wall and membrane of the target microorganisms. Interestingly in other cases, the physical action of CNMs is related to the isolation of bacteria from the external environment [74], affecting bacterial nutrient uptake, reducing the number of colonized and infected cells as well as improving host immune response and drug activity against the pathogen. Chemical interactions between CNMs and the microorganism surface may lead to the generation of oxidative stress by electron transfer phenomena, resulting in the production of known reactive oxygen species [73]. Hence, these effects are deeply dependent on the particle size of CNM and can be used to eliminate antibiotic-resistant microorganisms [46]. Some types of CNM are represented in Fig. 7.

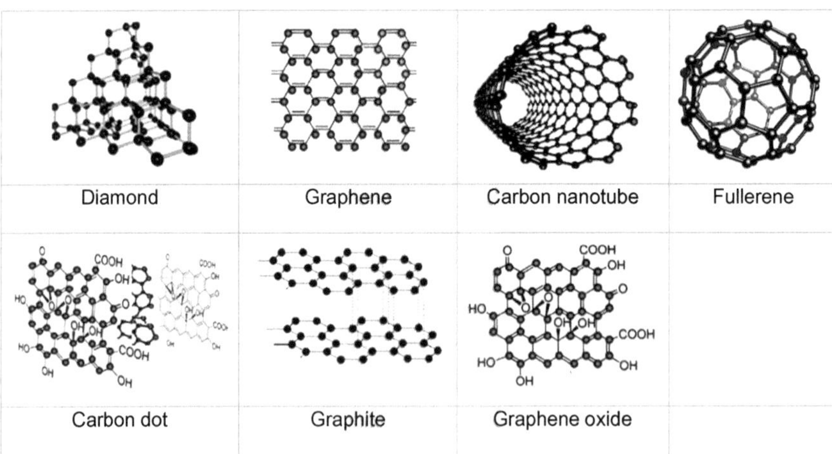

Fig. 7 Representation of some carbon-based nanomaterials. Reproduced with permission from reference [75]

As such, CNMs display an intrinsic antibacterial action, which can strikingly enhance anti-TB drugs effects in order to overcome microbial drug resistance. Notably, the synergistic effects of the combination of graphene oxide (GO) with anti-TB drugs, in principle, can reduce drug resistance and allow for a constant drug release. In this context, De Maio et al. [46] showed that administration of a low dose of GO could, in principle, be enough to induce the trapping of extracellular mycobacteria in the lungs, blocking the infection of forthcoming cells and increasing the anti-TB efficacy. It is worth mentioning that the trapping effect may be an essential ally to confine mycobacteria, thus avoiding their spread in other sites.

Functional nanoscale materials, in the current scenario, have opened novel approaches in medicine, diagnosis, and advanced therapeutics. For the case of multi- and extensively drug-resistant TB is a global threat to human health. Thus, we mention here the importance of phytogenic metal nanoparticles to inhibit mycobacteria. As is well-known, the silver nanoparticles (AgNPs) have a high antimicrobial activity with a broad spectral action [76]. Thus, the scientists extended the investigation to screen phytogenic AgNPs, gold (AuNPs), and gold–silver bimetallic (Au-AgNPs) nanoparticles for application as antitubercular agents (antimycobacterial activity) [77]. For instance, Singh et al. [77] studied the biogenic-synthesized from three medicinal plants (Barleria prionitis, Plumbago zeylanica, and Syzygium cumini) AgNPs, AuNPs, and Au-AgNPs nanoparticles and their promising application against MTB and M. bovis BCG. The investigators studied macrophage infection models in order to determine the minimum inhibitory concentration (MIC) and half-maximal inhibitory concentration of nanoparticles. Techniques such as microscopic analyses were performed to show intracellular uptake of nanoparticles in macrophages. Other parameters were also established in relation to human cell lines, such as biocompatibility, specificity, and selectivity of nanoparticles [77]. The bimetallic Au–AgNPs displayed excellent efficiency to inhibit mycobacteria. Although RMP presents very high selectivity, the relevance of this study lies in the fact that microorganisms are quickly obtaining resistance against conventional TB drugs, and researches on alternative strategies are of great importance for efficient treatments. In this report, Au–AgNPs from S. cumini showed deep efficiency, specificity, and selectivity to kill mycobacteria, suggesting a good potential as antitubercular agent [77]. Au–AgNPs demonstrated the highest antitubercular activity, with MIC of 2.56 μg/mL, followed by AgNPs. In addition, AuNPs did not show such activity at concentrations of up to 100 μg/mL. The bimetallic Au–AgNPs were able to enter macrophage cells and show up to 45% cytotoxicity at 30 μg/mL after 48 h. Furthermore, these bimetallic nanomaterials were found to be more specific toward mycobacteria, with their selectivity index in the range of 94–108 [77]. In this scenario, a specific drug should be able to eliminate the target microorganism at concentrations significantly below its cytotoxic level [78]. Currently, there are reports about bimetallic nanoparticles comprising of gold and silver, which present higher antimicrobial activity than monometallic ones [79, 80]. On the other hand, the biogenic synthesis of different metallic nanoparticles has called the attention of the scientific community due to the clean, nontoxic, and eco-friendly procedures involved [81]. Despite higher selectivity index, it is known that conventional TB drugs are losing their functional value. This fact comes from the development of resistance among mycobacteria, and in this case, it makes necessary high doses for treatment, which can lead to the emergence of side effects in TB patients [82]. Differently from antibiotics, nanoparticles do not present a single specific way to act on microorganisms. Diverse mechanisms have been proposed to unveil the killing action of nanoparticles. In this context, there are the morphology disruption, DNA cleavage, inhibition of DNA replication, as well as enzyme inactivation [83]. These mentioned mechanisms become the survival of microorganisms difficult in the presence of nanoparticles.

Despite the nanomedicine advances in diagnostic and therapeutic systems, the potential effect on human health due to prolonged exposure to these species has not yet been established. The limited amount of information about the toxic effects of these nanomaterials causes the nanomedicine to confront several regulatory issues in a complex and costly process, but necessary, independent on the nanomaterial or application. One of the main concerns is how nanoparticles can penetrate cell membranes more easily than several other molecules or drugs [84, 85]. However, could this greater ease be an indication of toxic effect? These questions are the fuel of nanotoxicology studies, that is, it is of interest to answer the most considerable number of questions about the interaction between nanomaterials and biological systems, such as the limitations and mechanisms associated to short, medium and long term exposures.

Classic analytical and biological methodologies are being adapted for toxicological investigations [86–88]. Among them are flow cytometry, cell proliferation, DNA fragmentation, cell cycle, histology, bioavailability and biodispersion. In vitro and in vivo analyzes are the most used, however the information available today is still contradictory and inaccurate. To assist the search for results in this area, other tools have been proposed as models, computer simulations or instrumentations at the atomic and/or molecular level [89].

There are, however, new problems in comparing in vitro and in vitro tests, for instance, the differences in concentrations and doses administered for assays [90]. Very high doses in in vitro assays do not match the doses administered in in vivo trials, not counting the exposure time or use of culture media in in vitro tests, making nanoparticles with physical–chemical properties different from those administered in in vivo tests, due to the adsorption of some components of the medium on the surface of nanoparticle [91]. Parameters such as in vitro—in vivo correlation become of fundamental importance.

3 Conclusion

In this chapter, we summarize the fundamental background on the application of novel functional nanoscale materials to the treatment of multidrug-resistant TB strains. In short, for applications with high sensitivity and selectivity in diagnostic and therapeutic systems, nanomaterials should be combined with even more specific biomolecules, for example, antigen/antibody, enzyme/substrate, receptors or membrane proteins, DNA sequences, RNA, cofactors, that is, a variety of biomolecules [92, 93]. In addition to these factors, an important parameter that must be considered in the case of the development of a diagnostic and/or therapeutic system is the bioavailability in the biological environment.

Due to their small size, functional nanoscale materials can easily enter the individual's body, for example, when used as a skin cream, orally, or also by inhalation [94]. However, this entry into the biological system depends on factors, such as the chemical composition of the surface, physical–chemical properties, size and shape, capacity for aggregation or agglomeration, or even functionalization with specific biomolecules [95]. Another important factor is directly related to the doses. In this regard, some questions should be considered, for example, what is the dosage limit that can be administered without demonstrating toxic effects, or how long this nanomaterial remains available inside the organism until it is completely eliminated. As the nanomaterials may have different sizes and structural functionalization, each one may have a distinct cell entry mechanism. A study performed with polystyrene-based nanosphere (dimensions of about 20 nm) suggests that they enter cells directly through the membrane without the need for specific transport mechanisms, followed by distribution throughout the cytoplasm and binding to several cytoskeleton biostructures [96]. Understanding the cell entry and exit mechanisms, as well as ensuring the stability of functional nanoscale materials, especially those surface-related properties, will be critical to ensure the safety as well as better use of their biological functions.

References

1. McHenry, M.L., Williams, S.M., Stein, C.M.: Genetics and evolution of tuberculosis pathogenesis: new perspectives and approaches. Infect. Genet. Evol. **81**, 104204 (2020)
2. Li, J., Zhao, A., Tang, J., Wang, G., Shi, Y., Zhan, L., Qin, C.: Tuberculosis vaccine development: from classic to clinical candidates. Eur. J. Clin. Microbiol. Infect. Dis. **(2020)**
3. Santos, N.C. de S., Scodro, R.B. de L., Sampiron, E.G., Ieque, A.L., Carvalho, H.C. de, Santos, T. da S., Ghiraldi Lopes, L.D., Campanerut-Sá, P.A.Z., Siqueira, V.L.D., Caleffi-Ferracioli, K.R. et al.: Minimum Bactericidal Concentration Techniques in Mycobacterium tuberculosis : a systematic review. Microb. Drug Resist. (2020)
4. Nogueira, A., Facchinetti, V., Souza, M., Vasconcelos, T.: Universidade Federal Fluminense. Faculdade de Farmácia. Rio de Janeiro, R.; Fundação Oswaldo Cruz. Instituto de Tecnologia em Fármacos. Rio de Janeiro, R.; Fundação Oswaldo Cruz. Instituto de Tecnologia em Fármacos. Rio de Janeiro, R.; Universidade Federal Fluminense. Instituto de Química. Rio de Janeiro, R. Tuberculose: uma abordagem geral dos principais aspectos. **2012**, 93, 3–9
5. De Souza, M.V.N., Vasconcelos, T.R.A.: Fármacos no combate à tuberculose: Passado, presente e futuro. Quim. Nova **28**, 678–682 (2005)
6. Silva, D.R., Mello, F.C. de Q., Kritski, A., Dalcolmo, M., Zumla, A., Migliori, G.B.: Tuberculosis series. J. Bras. Pneumol. **44**, 71–72 (2018)
7. Wang, Y.N., Chi, C.Q., Cai, M., Lou, Z.Y., Tang, Y.Q., Zhi, X.Y., Li, W.J., Wu, X.L., Du, X.: Amycolicicoccus subflavus gen. nov., sp. nov., an actinomycete isolated from a saline soil contaminated by crude oil. Int. J. Syst. Evol. Microbiol. **60**, 638–643 (2010)

8. Casal, M., Gutierrez, J., Vaquero, M.: Comparative evaluation of the mycobacteria growth indicator tube with the BACTEC 460 TB system and Lowenstein-Jensen medium for isolation of mycobacteria from clinical specimens. Int. J. Tuberc. Lung Dis. **1**, 81–84 (1997)
9. Osório Ferri, A., Aguiar, B., Mörschbächer Wilhelm, C., Schmidt, D., Fussieger, F., Ulrich Picoli, S.: Diagnóstico da tuberculose: uma revisão. Rev. Lib. **15**, 145–154 (2014)
10. Sekaggya-Wiltshire, C., Lamorde, M., Kiragga, A.N., Dooley, K.E., Kamya, M.R., Kambugu, A., Fehr, J., Manabe, Y.C., Castelnuovo, B.: The utility of pharmacokinetic studies for the evaluation of exposure-response relationships for standard dose anti-tuberculosis drugs. Tuberculosis **108**, 77–82 (2018)
11. Berthel, S.J., Cooper, C.B., Fotouhi, N.: Tuberculosis. Annu. Rep. Med. Chem. **52**, 1–25 (2019)
12. Griffiths, G., Nyström, B., Sable, S.B., Khuller, G.K.: Nanobead-based interventions for the treatment and prevention of tuberculosis. Nat. Rev. Microbiol. **8**, 827–834 (2010)
13. ORGANIZATION, W.H. Global tuberculosis report 2019
14. Ministério da Saúde-BRASIL Secretaria de Vigilância em Saúde - Departamento de Vigilância, Prevenção e Controle das Infecções Sexualmente Transmissíveis, do HIV/Aids e das Hepatites Virais
15. Ministério da Saúde; Panorama da Tuberculose no Brasil: indicadores epidemiologicos e operacionais 92 (2014)
16. Silva, D.B. da, Costa, G.S. da, Rosa, L.F.B., Guilherme, M. dos S., Oliveira, S.A. de, Cavalcanti, R.L. de S.: Assistência Farmacêutica a Pacientes Com Tuberculose Pulmonar: Uma Revisão Integrativa. Rev. Presença **2**, 83–106 (2017)
17. Cadena, A.M., Flynn, J.L., Fortune, S.M.: The importance of first impressions: early events in *Mycobacterium tuberculosis* infection influence outcome. MBio **7**, e00342-e416 (2016)
18. World Health Organization (WHO) Latent tuberculosis infection: updated and consolidated guidelines for programmatic management. Geneva
19. Dodd, C.E., Schlesinger, L.S.: New concepts in understanding latent tuberculosis. Curr. Opin. Infect. Dis. **30**, 316–321 (2017)
20. Murray, P.R., Rosenthal, K.S., Pfaller, M.A.: Microbiologia Médica. Elsevier, São Paulo, pp. 1038–1056 (2009)
21. Koch, A., Mizrahi, V.: Mycobacterium tuberculosis. Trends Microbiol. **26**, 555–556 (2018)
22. Yadav, J., Verma, S., Chaudhary, D., Jaiwal, P.K., Jaiwal, R.: Tuberculosis: current status, diagnosis, treatment and development of novel vaccines. Curr. Pharm. Biotechnol. **20**, 446–458 (2019)
23. Ministério da Saúde-BRASIL Tuberculose: o que é, causas, sintomas, tratamento, diagnóstico e prevenção (2020)
24. Sehgal, V.N., Srivastava, G.: Fixed drug eruption (FDE): changing scenario of incriminating drugs. Int. J. Dermatol. **45**, 897–908 (2006)
25. Cailleaux-Cezar, M.: Diagnóstico e Tratamento da Tuberculose Latente Diagnosis and Treatment of Latent Tuberculosis. Pulmão RJ **21**, 41–45 (2012)
26. Costa, R.R. da, Silva, M.R., Gonçalves, I.C.: Diagnóstico laboratorial da tuberculose : Revisão de literatura. Rev Med Minas Gerais **28**, 197–206 (**2018**)
27. Steingart, K.R., Flores, L.L., Dendukuri, N., Schiller, I., Laal, S., Ramsay, A., Hopewell, P.C., Pai, M.: Commercial serological tests for the diagnosis of active pulmonary and extrapulmonary tuberculosis: an updated systematic review and meta-analysis. PLOS Med. **8**, e1001062 (2011)
28. Lima, T.M. de, Belotti, N.C.U., Nardi, S.M.T., Pedro, H. da S.P.: Teste rápido molecular GeneXpert MTB/RIF para diagnóstico da tuberculose. Rev. Pan-Amazônica Saúde **8**, 65–76 (2017)
29. Lee, D.J., Kumarasamy, N., Resch, S.C., Sivaramakrishnan, G.N., Mayer, K.H., Tripathy, S., Paltiel, A.D., Freedberg, K.A., Reddy, K.P.: Rapid, point-of-care diagnosis of tuberculosis with novel Truenat assay: cost-effectiveness analysis for India's public sector. PLoS ONE **14**, e0218890 (2019)
30. Ferraz, J.C., Melo, F.B.S., Albuquerque, M.F.P.M., Montenegro, S.M.L., Abath, F.G.C.: Immune factors and immunoregulation in tuberculosis. Brazilian J. Med. Biol. Res. **39**, 1387–1397 (2006)

31. Lu, L.L., Smith, M.T., Yu, K.K.Q., Luedemann, C., Suscovich, T.J., Grace, P.S., Cain, A., Yu, W.H., McKitrick, T.R., Lauffenburger, D., et al.: IFN-γ-independent immune markers of Mycobacterium tuberculosis exposure. Nat. Med. **25**, 977–987 (2019)
32. Iseman, M.D.: Tuberculosis therapy: past, present and future. Eur. Respir. J. **20**, 87S LP-94s (2002)
33. Controlled trial of 2, 4, and 6 months of pyrazinamide in 6-month, three-times-weekly regimens for smear-positive pulmonary tuberculosis, including an assessment of a combined preparation of isoniazid, rifampin, and pyrazinamide: results at 30 months. Am. Rev. Respir. Dis. **143**, 700–706 (1991)
34. Pereira, S.M., Dantas, O.M.S., Ximenes, R., Barreto, M.L.: BCG vaccine against tuberculosis: its protective effect and vaccination policies. Rev. Saude Publica **41**, 1–7 (2007)
35. Stefanova, T.: Quality control and safety assessment of BCG vaccines in the post-genomic era. Biotechnol. Biotechnol. Equip. **28**, 387–391 (2014)
36. Daniel, T.M.: Leon Charles Albert Calmette and BCG vaccine. Int. J. Tuberc. Lung Dis. **9**, 944–945 (2005)
37. Rios, A.C., Moutinho, C.G., Pinto, F.C., Del Fiol, F.S., Jozala, A., Chaud, M.V., Vila, M.M.D.C., Teixeira, J.A., Balcão, V.M.: Alternatives to overcoming bacterial resistances: State-of-the-art. Microbiol. Res. **191**, 51–80 (2016)
38. Rather, I.A., Kim, B.-C., Bajpai, V.K., Park, Y.-H.: Self-medication and antibiotic resistance: crisis, current challenges, and prevention. Saudi J. Biol. Sci. **24**, 808–812 (2017)
39. Stewart, P.S., William Costerton, J.: Antibiotic resistance of bacteria in biofilms. Lancet **358**, 135–138 (2001)
40. Gagneux, S., Long, C.D., Small, P.M., Van, T., Schoolnik, G.K., Bohannan, B.J.M.: The competitive cost of antibiotic resistance in *Mycobacterium tuberculosis Science (80-.)* **312**, 1944 LP–1946 (2006)
41. Sandgren, A., Strong, M., Muthukrishnan, P., Weiner, B.K., Church, G.M., Murray, M.B.: Tuberculosis drug resistance mutation database. PLOS Med. **6**, e1000002 (2009)
42. Khoder, R., Korri-Youssoufi, H.: E-DNA biosensors of M. tuberculosis based on nanostructured polypyrrole. Mater. Sci. Eng. C **108**, 110371 (2020)
43. Sharma, R., Gupta, B., Singh, N., Acharya, J.R., Musilek, K., Kuca, K., Ghosh, K.: Development and structural modifications of cholinesterase reactivators against chemical warfare agents in last decade: a review. Mini-Reviews Med. Chem. **15**, 58–72 (2014)
44. Van Giau, V., An, S.S.A., Hulme, J.: Recent advances in the treatment of pathogenic infections using antibiotics and nano-drug delivery vehicles. Drug Des. Devel. Ther. **13**, 327–343 (2019)
45. Xu, K., Liang, Z.C., Ding, X., Hu, H., Liu, S., Nurmik, M., Bi, S., Hu, F., Ji, Z., Ren, J. et al.: Nanomaterials in the prevention, diagnosis, and treatment of mycobacterium tuberculosis infections. Adv. Healthc. Mater. **7** (2018)
46. De Maio, F., Palmieri, V., De Spirito, M., Delogu, G., Papi, M.: Carbon nanomaterials: a new way against tuberculosis. Expert Rev. Med. Devices **16**, 863–875 (2019)
47. Volpedo, G., Costa, L., Ryan, N., Halsey, G., Satoskar, A., Oghumu, S.: Nanoparticulate drug delivery systems for the treatment of neglected tropical protozoan diseases. J. Venom. Anim. Toxins Incl. Trop. Dis. **25** (2019)
48. Yang, Y., Wang, S., Wang, Y., Wang, X., Wang, Q., Chen, M.: Advances in self-assembled chitosan nanomaterials for drug delivery. Biotechnol. Adv. **32**, 1301–1316 (2014)
49. Gelperina, S., Kisich, K., Iseman, M.D., Heifets, L.: The potential advantages of nanoparticle drug delivery systems in chemotherapy of tuberculosis. Am. J. Respir. Crit. Care Med. **172**, 1487–1490 (2005)
50. Palmieri, V., Perini, G., De Spirito, M., Papi, M.: Graphene oxide touches blood: in vivo interactions of bio-coronated 2D materials. Nanoscale Horiz. **4**, 273–290 (2019)
51. Simon, J., Müller, L.K., Kokkinopoulou, M., Lieberwirth, I., Morsbach, S., Landfester, K., Mailänder, V.: Exploiting the biomolecular corona: pre-coating of nanoparticles enables controlled cellular interactions. Nanoscale **10**, 10731–10739 (2018)
52. Papi, M., Caracciolo, G.: Principal component analysis of personalized biomolecular corona data for early disease detection. Nano Today **21**, 14–17 (2018)

53. Weiss, G., Schaible, U.E.: Macrophage defense mechanisms against intracellular bacteria. Immunol. Rev. **264**, 182–203 (2015)
54. Russell, D.G., Cardona, P.-J., Kim, M.-J., Allain, S., Altare, F.: Foamy macrophages and the progression of the human tuberculosis granuloma. Nat. Immunol. **10**, 943–948 (2009)
55. Greco, E., Quintiliani, G., Santucci, M.B., Serafino, A., Ciccaglione, A.R., Marcantonio, C., Papi, M., Maulucci, G., Delogu, G., Martino, A., et al.: Janus-faced liposomes enhance antimicrobial innate immune response in *Mycobacterium tuberculosis* infection. Proc. Natl. Acad. Sci. U. S. A. **109**, E1360–E1368 (2012)
56. Dasari Shareena, T.P., McShan, D., Dasmahapatra, A.K., Tchounwou, P.B.: A review on graphene-based nanomaterials in biomedical applications and risks in environment and health. Nano-Micro Lett. **10**, 53 (2018)
57. Khanna, K., Khanna, S., Goel, B., Patel, A., Xavier, G.: Polymeric nano-particles for tumor targeting—a review. Int. J. Drug Dev. Res. (2017)
58. Sosnik, A., Carcaboso, A.M., Glisoni, R.J., Moretton, M.A., Chiappetta, D.A.: New old challenges in tuberculosis: potentially effective nanotechnologies in drug delivery. Adv. Drug Deliv. Rev. **62**, 547–559 (2010)
59. Andrade, F., Rafael, D., Videira, M., Ferreira, D., Sosnik, A., Sarmento, B.: Nanotechnology and pulmonary delivery to overcome resistance in infectious diseases. Adv. Drug Deliv. Rev. **65**, 1816–1827 (2013)
60. Heo, D.N., Min, K.H., Choi, G.H., Kwon, I.K., Park, K., Lee, S.C.: Scale-up production of theranostic nanoparticles (Chap.24). In: Chen, X., Wong, S.B.T.-C.T. (eds.) pp. 457–470. Academic Press, Oxford (2014) ISBN 978-0-12-407722-5
61. Ahmad, Z., Sharma, S., Khuller, G.K.: Inhalable alginate nanoparticles as antitubercular drug carriers against experimental tuberculosis. Int. J. Antimicrob. Agents **26**, 298–303 (2005)
62. Pandey, R., Khuller, G.K.: Solid lipid particle-based inhalable sustained drug delivery system against experimental tuberculosis. Tuberculosis (Edinb). **85**, 227–234 (2005)
63. Pandey, R., Sharma, A., Zahoor, A., Sharma, S., Khuller, G.K., Prasad, B.: Poly (DL-lactide-co-glycolide) nanoparticle-based inhalable sustained drug delivery system for experimental tuberculosis. J. Antimicrob. Chemother. **52**, 981–986 (2003)
64. Pandey, R., Zahoor, A., Sharma, S., Khuller, G.K.: Nanoparticle encapsulated antitubercular drugs as a potential oral drug delivery system against murine tuberculosis. Tuberculosis (Edinb). **83**, 373–378 (2003)
65. Andrade, F., Videira, M., Ferreira, D., Sarmento, B.: Nanocarriers for pulmonary administration of peptides and therapeutic proteins. Nanomedicine (Lond). **6**, 123–141 (2011)
66. Patton, J.S., Byron, P.R.: Inhaling medicines: delivering drugs to the body through the lungs. Nat. Rev. Drug Discov. **6**, 67–74 (2007)
67. Garcia-Contreras, L., Fiegel, J., Telko, M.J., Elbert, K., Hawi, A., Thomas, M., VerBerkmoes, J., Germishuizen, W.A., Fourie, P.B., Hickey, A.J. et al.: Inhaled large porous particles of capreomycin for treatment of tuberculosis in a guinea pig model. Antimicrob. Agents Chemother. **51**, 2830 LP–2836 (2007)
68. Hwang, S.M., Kim, D.D., Chung, S.J., Shim, C.K.: Delivery of ofloxacin to the lung and alveolar macrophages via hyaluronan microspheres for the treatment of tuberculosis. J. Control. Release **129**, 100–106 (2008)
69. Moretton, M.A., Chiappetta, D.A., Andrade, F., das Neves, J., Ferreira, D., Sarmento, B., Sosnik, A.: Hydrolyzed galactomannan-modified nanoparticles and flower-like polymeric micelles for the active targeting of rifampicin to macrophages. J. Biomed. Nanotechnol. **9**, 1076–1087 (2013)
70. Kalluru, R., Fenaroli, F., Westmoreland, D., Ulanova, L., Maleki, A., Roos, N., Paulsen Madsen, M., Koster, G., Egge-Jacobsen, W., Wilson, S. et al.: Poly(lactide-co-glycolide)-rifampicin nanoparticles efficiently clear *Mycobacterium bovis* BCG infection in macrophages and remain membrane-bound in phago-lysosomes. J. Cell Sci. **126**, 3043 LP–3054 (2013)
71. Sharma, A., Sharma, N., Kumari, A., Lee, H.J., Kim, T.Y., Tripathi, K.M.: Nano-carbon based sensors for bacterial detection and discrimination in clinical diagnosis: A junction between material science and biology. Appl. Mater. Today **18**, 100467 (2020)

72. Maiti, D., Tong, X., Mou, X., Yang, K.: Carbon-based nanomaterials for biomedical applications: a recent study. Front. Pharmacol. **9**, 1401 (2019)
73. Al-Jumaili, A., Alancherry, S., Bazaka, K., Jacob, M.V.: Review on the antimicrobial properties of carbon nanostructures. Mater. (Basel, Switzerland) **10** (2017)
74. Palmieri, V., Bugli, F., Lauriola, M.C., Cacaci, M., Torelli, R., Ciasca, G., Conti, C., Sanguinetti, M., Papi, M., De Spirito, M.: Bacteria meet graphene: modulation of graphene oxide nanosheet interaction with human pathogens for effective antimicrobial therapy. ACS Biomater. Sci. Eng. **3**, 619–627 (2017)
75. Sudha, P.N., Sangeetha, K., Vijayalakshmi, K., Barhoum, A.: Nanomaterials history, classification, unique properties, production and market (Chap. 12). In: Barhoum, A., Makhlouf, A.S.H.B.T.-E.A. of N. A.N. (eds.) Micro and Nano Technologies, pp. 341–384. Elsevier (2018) ISBN 978-0-323-51254-1
76. Singh, R., Nawale, L.U., Arkile, M., Shedbalkar, U.U., Wadhwani, S.A., Sarkar, D., Chopade, B.A.: Chemical and biological metal nanoparticles as antimycobacterial agents: a comparative study. Int. J. Antimicrob. Agents **46**, 183–188 (2015)
77. Singh, R., Nawale, L., Arkile, M., Wadhwani, S., Shedbalkar, U., Chopade, S., Sarkar, D., Chopade, B.A.: Phytogenic silver, gold, and bimetallic nanoparticles as novel antitubercular agents. Int. J. Nanomedicine **11**, 1889–1897 (2016)
78. Pritchett, J.C., Naesens, L., Montoya, J.: Treating HHV-6 infections: the laboratory efficacy and clinical use of anti-HHV-6 agents. In: Flamand, L., Lautenschlager, I., Krueger, G.R.F. (eds.) Ablashi HHV-6B & HHV-7, 3rd ed., D.V.B.T.-H.H.H.-6A, pp. 311–331. Elsevier, Boston (2014) ISBN 978-0-444-62703-2
79. Salunke, G.R., Ghosh, S., Santosh Kumar, R.J., Khade, S., Vashisth, P., Kale, T., Chopade, S., Pruthi, V., Kundu, G., Bellare, J.R., et al.: Rapid efficient synthesis and characterization of silver, gold, and bimetallic nanoparticles from the medicinal plant *Plumbago zeylanica* and their application in biofilm control. Int. J. Nanomed. **9**, 2635–2653 (2014)
80. Ghosh, S., Jagtap, S., More, P., Shete, U.J., Maheshwari, N.O., Rao, S.J., Kitture, R., Kale, S., Bellare, J., Patil, S., et al.: *Dioscorea bulbifera* mediated synthesis of novel $Au_{core}Ag_{shell}$ nanoparticles with potent antibiofilm and antileishmanial activity. J. Nanomater. **2015**, 562938 (2015)
81. Iravani, S.: Green synthesis of metal nanoparticles using plants. Green Chem. **13**, 2638–2650 (2011)
82. Gandhi, N.R., Nunn, P., Dheda, K., Schaaf, H.S., Zignol, M., van Soolingen, D., Jensen, P., Bayona, J.: Multidrug-resistant and extensively drug-resistant tuberculosis: a threat to global control of tuberculosis. Lancet **375**, 1830–1843 (2010)
83. Singh, R., Shedbalkar, U.U., Wadhwani, S.A., Chopade, B.A.: Bacteriagenic silver nanoparticles: synthesis, mechanism, and applications. Appl. Microbiol. Biotechnol. **99**, 4579–4593 (2015)
84. Warheit, D.B.: How meaningful are the results of nanotoxicity studies in the absence of adequate material characterization? Toxicol. Sci. **101**, 183–185 (2008)
85. Dukhin, A.S., Ulberg, Z.R., Karamushka, V.I., Gruzina, T.G.: Peculiarities of live cells' interaction with micro- and nanoparticles. Adv. Colloid Interface Sci. **159**, 60–71 (2010)
86. Oberdörster, G.: Nanotoxicology: in vitro-in vivo dosimetry. Environ. Health Perspect. **120**, A13; author reply A13 (2012)
87. Oberdörster, G.: Safety assessment for nanotechnology and nanomedicine: concepts of nanotoxicology. J. Intern. Med. **267**, 89–105 (2010)
88. Stone, V., Johnston, H., Schins, R.P.F.: Development of in vitro systems for nanotoxicology: methodological considerations. Crit. Rev. Toxicol. **39**, 613–626 (2009)
89. Åberg, C., Kim, J.A., Salvati, A., Dawson, K.A.: Theoretical framework for nanoparticle uptake and accumulation kinetics in dividing cell populations. EPL **101**, 38007 (2013)
90. Maurer-Jones, M.A., Haynes, C.L.: Toward correlation in in vivo and in vitro nanotoxicology studies. J. Am. Soc. Law, Med. Ethics **40**, 795–801 (2012)
91. Han, X., Corson, N., Wade-Mercer, P., Gelein, R., Jiang, J., Sahu, M., Biswas, P., Finkelstein, J.N., Elder, A., Oberdörster, G.: Assessing the relevance of in vitro studies in nanotoxicology by examining correlations between in vitro and in vivo data. Toxicology **297**, 1–9 (2012)

92. Chen, X., Gambhir, S.S., Cheon, J.: Theranostic nanomedicine. Acc. Chem. Res. **44**, 841 (2011)
93. Zhang, L., Gu, F.X., Chan, J.M., Wang, A.Z., Langer, R.S., Farokhzad, O.C.: Nanoparticles in medicine: therapeutic applications and developments. Clin. Pharmacol. Ther. **83**, 761–769 (2008)
94. Couvreur, P.: Nanoparticles in drug delivery: past, present and future. Adv. Drug Deliv. Rev. **65**, 21–23 (2013)
95. Rivera Gil, P., Oberdörster, G., Elder, A., Puntes, V., Parak, W.J.: Correlating physico-chemical with toxicological properties of nanoparticles: the present and the future. ACS Nano **4**, 5527–5531 (2010)
96. Edetsberger, M., Gaubitzer, E., Valic, E., Waigmann, E., Köhler, G.: Detection of nanometer-sized particles in living cells using modern fluorescence fluctuation methods. Biochem. Biophys. Res. Commun. **332**, 109–116 (2005)

An in Silico Study of Natural Compounds as Potential MAO-B Inhibitors for the Treatment of Parkinson's Disease

Bianca Liffey Brito Marino, Kessia Priscila Araújo Sousa, Cleydson Breno Rodrigues dos Santos, Carlton Anthony Taft, Carlos Henrique Tomich de Paula da Silva, and Lorane Izabel da Silva Hage-Melim

Abstract Parkinson's Disease (PD) is a neurodegenerative disease that causes damage to the cognitive and motor system due to the death of dopaminergic neurons, which are responsible for the synthesis of the neurotransmitter dopamine. The study aimed to compare the monoamine oxidase B (MAO-B) inhibitory activity of natural molecules described in the literature with Selegiline, as potential drugs for the treatment of PD through molecular modeling, molecular docking and prediction of ADME/Tox properties. Thus, it was found the structure of the four natural molecules, Amburoside A, Harman, Harmaline and Harmalol, showed antiparkinsonian biological activity. Maps Electrostatic Potential showed similar regions between the molecules, except for Amburoside A, and Harmaline had a greater similarity in the positive potential with Selegiline. Molecular docking demonstrated that the studied molecules interact with 4–6 amino acids from the active site of the MAO-B enzyme, indicating that it has an inhibitory action on the enzyme, through hydrogen bonding and hydrophobic interactions. For ADME property predictions, most of the molecules showed good human oral absorption, all showed average permeability in Caco-2 cells, most showed average permeability in MDCK cells, showed low binding to plasma proteins, and for permeability in the blood-brain barrier, they were between good and medium. Overall, Harmaline has more properties similar to Selegiline. For toxicological properties, all molecules including Selegiline showed a positive result for the possibility of mutagenicity, whereas for the parameter of carcinogenicity in rats only the molecules Harmaline and Harmalol were positive, but no molecule was

B. L. B. Marino · K. P. A. Sousa · L. I. da S. Hage-Melim (✉)
Laboratory of Pharmaceutical and Medicinal Chemistry (PharMedChem), Federal University of Amapá, Macapá, Amapá, Brazil

C. B. R. dos Santos
Laboratory of Modeling and Computational Chemistry, Federal University of Amapá, Macapá, Amapá, Brazil

C. A. Taft
Brazilian Center for Physics Research, Rio de Janeiro, Brazil

C. H. T. de P. da Silva
Computational Laboratory of Pharmaceutical Chemistry, School of Pharmaceutical Sciences of Ribeirão Preto, University of São Paulo, Ribeirão Preto, São Paulo, Brazil

F. A. La Porta and C. A. Taft (eds.), *Functional Properties of Advanced Engineering Materials and Biomolecules*, Engineering Materials,
https://doi.org/10.1007/978-3-030-62226-8_20

positive for carcinogenicity in mice. Therefore, the molecule that presented the best results was Harmaline, opening perspectives for the execution of in vitro studies.

Keywords Parkinson · Monoamine oxidase B · Natural compounds · Molecular modeling

1 Introduction

Parkinson's disease (abbreviated as PD) is a progressive neurodegenerative disease caused by necrosis of dopaminergic neurons located in substantia nigra, the brain region responsible for the synthesis of the neurotransmitter dopamine (DA) [4-(2-aminoethyl)benzene-1,2-diol]. This degradation leads to a decrease in DA in the synaptic cleft, consequently, a movement control disorder occurs, since this neurotransmitter is related to motor and cognitive functions. Nowadays, the possible causes of this neurodegeneration are studied and generated the following hypotheses: One related to brain damage due to excitotoxicity, where there is an accumulation of glutamate in the region where the dopaminergic neurons are, and another related to oxidative stress caused by monoamine oxidase B (MAO-B) when it degrades dopamine, releasing free radicals [1, 2]. Alterations in α-synuclein levels create filamentous clusters that are added to the cells of the central nervous system, causing a decrease in the substantia nigra [3]; with the genetic mapping of the PD, it can be noticed that the alterations, mainly, in the SNCA, GBA, LRRK2, PARK2, UCHL1, PINK1, ATP13A2, HTRA2 genes [4, 5]; another hypothesis is the change that occurs in the proteasome-ubiquitin system, in which Cul-1 protein makes use of ubiquitin ligase and a desubiquitinator protein associated with the proteasome Pad-1, lead to the reactivation of the cell cycle and apoptosis in subsets of postmitotic neurons [6].

Studies in European populations indicate that PD affects 1.8% of individuals over 65 years and can reach 2.6% of individuals over 85 years old. Additionally, some authors have affirmed that there is no difference in prevalence between the sexes, but studies in American populations indicate a higher prevalence in white males [7]. This increase in PD is closely associated with an increase in the longevity of the world population and, as a result, the appearance of neurodegenerative diseases, but with changes in the exposure of risk factors, possible multidisciplinary care, early diagnosis and improvement in the registration of the disease, it is possible to improve the quality of life of these patients, resulting in longer life after the diagnosis of the disease [8]. The diagnosis of PD is clinical and is mainly done through the identification of four cardinal signs: bradykinesia, rest tremor, postural instability and rigidity. Among PD patients without dementia, approximately 25–30% have mild cognitive impairment, diagnosed in 10–20% of patients; PD can also be confused with Lewy bodies dementia (LBD), and for the definitive diagnosis, the 1-year strategy is used, which, if more evident appearances occur after one year, will be diagnosed with PD [9].

Pharmacological treatment of PD consists of symptomatic control. The drugs are classified as dopaminergic agonists whose example is L-DOPA [(S) -2-amino-3-(3,4-dihydroxyphenyl) propanoic acid]; monoamine oxidase B (MAO-B) inhibitors, selegiline and rasagiline; catechol-o-methyltransferase (COMT) inhibitors such as entacapone and tolcapone, and anticholinergic drugs such as triexiphenidyl, which are described in Table 1 [10]. Many MAO-B and COMT inhibitors used in the clinic are irreversible, by covalently binding to the enzyme and inactivating it, which results in important side effects. Besides, irreversible MAO-B inhibitors (MAOI) may exhibit several other undesirable side effects in the central nervous system (insomnia, irritability, agitation, hypomania, REM sleep suppression), as well as cardiovascular dysfunction (orthostatic hypotension) or sexual disorders. Thus, the search for natural molecules is a promising proposal for the development of drugs with fewer adverse events [2, 11, 12].

And one of the methodologies used in the development of new drugs is computational chemistry, with the characteristics of high affinity and specificity for the target site and with greater intrinsic activity, increasing the effectiveness of the treatment of this pathology [13]. In association with computational chemistry, molecular modeling helps to understand the most attractive targets in the genesis representativeness or progression of the pathological process. Through virtual screening techniques, data management and mapping of the 3D structure of molecular targets and ligands can be done with less time and costs than conventional techniques for the identification and optimization of new drug candidates [14].

This study approaches natural compounds isolated from plants in the North and Northeast of Brazil, *Amburana cearensis* and *Passiflora incarnata*, used as an alternative treatment for PD, according to the authors, but its mechanism of action has not been still fully clarified [15, 16]. The objective of the study was to compare the natural molecules found in the literature with Selegiline, which is a drug currently used in the treatment of PD, through calculations of descriptors, molecular docking and prediction of pharmacokinetic and toxicological properties, to determine natural compounds with inhibitory activity of Monoamine Oxidase B (MAO-B) as drug candidates for PD.

2 Materials and Methods

2.1 Studied Compounds

For this study, a literature search was made through Scielo and PubMed databases, in which it can be found scientific articles about studies of natural compounds from plants of North and Northeast regions of Brazil used for the treatment of PD, preferably MAO-B inhibitors. Three criteria for choosing the molecules were used: to have more than one study on the same molecule; to have at least one experimental study in animals and to be a plant from Brazil, to improve the studies of plants of

Table 1 Classes of drugs used to treat Parkinson's disease [17–19]

Class	Action	Drugs	Contraindications	Adverse effects
Doparminergic agonists	Increased synthesis, release and inhibition of dopamine reuptake in striatum in parkinsonian patients	Amantadine	Epilepsy, heart failure, mental confusion	Nausea Dizziness Insomnia
		Bromocriptine	Pregnancy, lactating, hypersensitivity to other alkaloids, psychic disorders	
		Cabergoline	History of cardiac changes	Depression Irritability Hypotension Fibrosis Liver failure Psychiatric disorders Sweating Infections Changes in the urinary tract Xerostomia Tachycardia Benign neoplasms Anorexia
		Apomorphine		
Central anticholinergics	It blocks cholinergic transmission and decreases the level of cerebral acetylcholine, restoring balance with dopamine	Biperiden	Glaucoma and changes in the digestive system	
		Triexiphenidyl	Glaucoma and pregnancy	
MAO-B inhibitors	Selective and irreversible inhibition of monoamine oxidase B enzyme	Selegiline	Involuntary and abnormal movements, severe psychosis, liver ulcer, extremities, pregnancy and lactation	
		Rasagiline	Use with other MAOIs, cyclobenzoprim, *Hypericum perforatum*, narcotic analgesics, liver failure and general anesthesia	
COMT inhibitors	Selective catechol-O-methyl transferase enzyme inhibitors	Tolcapone	Liver disorders, neuroleptic syndrome, hyperemia, non-selective MAOIs	
		Entacapone	Liver disorders, cardiac disorders, neuroleptic syndrome, use with entacapone and non-selective MAOIs	

(continued)

Table 1 (continued)

Class	Action	Drugs	Contraindications	Adverse effects
Dopamine precursor	Unlike DA, levodopa crosses the blood–brain barrier, then is converted by the enzyme aromatic L-amino acid decarboxylase	Levodopa	Non-selective MAOIs, endocrine problems, glaucoma, severe psychiatric illness, pregnancy and lactation	

the national flora. The keywords used were "antiparkinsonian natural molecules". Subsequently, a search was done looking for the most used drug and, then, it was used as a standard molecule for physico-chemical, pharmacokinetic, toxicological and experimental properties on docking simulation, through Molecular Modeling.

Selegiline (L-Deprenyl®; Eldery®) [(R)-N-methyl-N-(1-phenylpropan-2-yl)prop-1-yn-3-amine] is a non-competitive and irreversible MAO-B inhibitor, which was initially developed as an antidepressant and soon became a success in the treatment of PD. In PD, Selegiline irreversibly inhibits MAO-B through a reaction between the flavin unit of the enzyme and the acetylenic unit of the drug [2]. *Amburana cearensis* A.C. Smith (sin. *Torresea cearenses* Fr. All) belongs to Leguminoseae Papilionoideae (Fabaceae) family. Popularly, it is known by various names such as "imburana-de-cheiro" and "cumaru" [20]. A compound of *Amburana cearense* presented neuroprotective activity by the antioxidant action, being able to be a medicine in the future, is the Ambruroside A [21, 22]. *Passiflora incarnata* L., popularly known as "maracujá", belongs to the genus *Passiflora*, which some native species are from Brazil, and distributed by Asia, Australia and North America. The main constituents of *P. incarnata* leaves are flavonoids (0.25%), such as vitexin, isovitexin, orientin, isoorientin, apigenin and kaempferol. Alkaloids based on viz system ß-carboline ring, Harman, Harmine, Harmaline, Harmalol. Harmine and Harmaline alkaloids are said to be effective anti-Parkinson compounds [15, 16].

2.2 *Molecular Modeling of Studied Compounds and Descriptor Calculations*

The compounds studied were constructed as follows: initially, the structure of Selegiline, Amburoside A, Harman, Harmaline and Harmalol which were drawn in ChemSketch 12.0 program [23], and was established with Gauss View 5.0 program [24] and optimized using Density Functional Theory (DFT) method and B3LYP/6-31G** basis sets implemented with Gaussian 03 program [25]. After the structures were determined in three dimensions and various descriptors for each molecule were calculated. They represent different sources of chemical information (features) regarding the molecules and include geometric, electronic, quantum-chemical, physical-chemical and topological descriptors, among others [26]. Note that the computation of the descriptors has been performed employing the following softwares: Gaussian 03 program [25], Molekel [26] and HyperChem 6.02 [28].

2.2.1 Quantum Chemical Descriptors

In this study 11 reactivity descriptors were calculated, with the aid of the GaussView 5.0 [29], using a base set B3LYP/6-31G** implemented with Gaussian 03 [24] and the DFT method: total energy (TE), Mulliken electronegativity (χ), molecular hardness (η), molecular softness ($1/\eta$), chemical potential (μ), global electrophilicity index (ω), highest occupied molecular orbital (HOMO), one level below of highest occupied molecular orbital (HOMO-1), lowest unoccupied molecular orbital (LUMO), a level above of lowest unoccupied molecular orbital (LUMO +1), and the difference of energy between HOMO and LUMO (GAP = HOMO–LUMO). These descriptors are indicators of stability and chemical reactivity [28, 29]. The construction of the frontier orbitals (HOMO and LUMO) was visualized with the aid of the Molekel software [27].

2.2.2 Structure Activity Relationship Descriptors

The following SAR descriptors were calculated for the natural molecules reported in the literature: Total Surface Area (TSA), Molecular Volume (MV), Molar Refractivity (MR), Molecular Polarizability (MP), Lipophilicity coefficient (logP), Molar Mass (MM) and Hydration Energy (HE) [30] using the HyperChem 6.02 software [27]. HE descriptor was used to calculate the Pearson correlation matrix as indicator variable.

2.3 Statistical Analysis

With the determination of all the molecular descriptors, it was possible to construct the data matrix. For the analysis was necessary to standardize the data matrix X = (n, m) containing the 5 molecules studied and the descriptors calculated in columns, where n is the number of molecules and m is the number of variables. The objective of standardization is to give each variable an equal weight in mathematical terms, each variable was centered in the mean and scaled to the unit variance. The selection of the variables was started by analyzing the Pearson correlation matrix [31] with two tables, one relating the variables to the total energy (TE) and another relating to the hydration energy (HE). The Person's correlation analysis was performed with the Pirouette 3.01 [32] and Statistica 6.1 softwares [33]. The final objective of the analysis was to evaluate the descriptors that best correlate with the inhibitor Selegiline and the natural molecules.

2.4 Molecular Electrostatic Potential Maps (MEP)

An important concept explored in this study was the correlation of structure-activity of the inhibitors studied through the characteristics of the electrostatic potential map. In particular, the MEP allows the use of qualitative analysis to locate reactive sites in a molecule. As such, MEP results determine the roles played by both electronic and steric effects (size/shape) on its potency. It is of fundamental importance that the visualization of MEP provides qualitative information about the molecules, such as the behavior of the interaction between a ligand and the receptor. The MEP at a given point (x, y, z) in the neighborhood of a molecule is defined in terms of the interaction energy between the electric charge generated from electrons of the molecules, nuclei and a positive test charge (a proton) located in r. The MEPs were generated from the atomic charge at the DFT B3LYP/6-31G** [34] calculated in the Gaussian 03 [25] software, and the results will be visualized with the help of the Molekel software [27].

2.5 Molecular Docking Study Using GOLD 4.1 Software

Molecular docking is an intensive and prominent computational method in the drug discovery process. The benefit of docking is to identify the ligand binding mode at the enzyme or receptor binding site through specific key interactions and to predict the binding affinity between the protein-ligand complexes [35].

The GOLD software (GeneticOptimisation for LigandDocking) uses the genetic algorithm for purposes of flexible docking experiments of ligands within protein binding sites. A population of potential solutions (possible anchorage orientations of ligands) is randomly configured [36]. In this regard, each member of the population is, of course, encoded as a chromosome that contains information about the hydrogen bonds mapping of ligand atoms to (complementary) hydrogen bonds of protein atoms, hydrophobic points mapping on the ligand to hydrophobic points on the protein and conformation around the ligand flexible bonds and hydroxyl groups of the protein. It is difficult to predict how a small molecule will bind to a protein and no software can guarantee success. The next step to be undertaken is to measure as accurately as possible the reliability of the software, that is, the chance that it will make a good prediction in a given instance. For this reason, GOLD has been tested with a large number of complexes extracted from PDB. The overall conclusion of these tests was that the solutions best classified by GOLD were correct in 70–80% of cases [36].

Given the crystallographic structure of the target, the molecular docking automatically shows samples of the ligands conformations with a specific region of receptor binding. It has been used successfully to identify active substances by filtering out those that do not fit into the active site. The binding site of the studied human enzymes was defined with all residues within a sphere of 10 Å radius located at the center of the active site. For the initial stage of docking experiments, some parameters of the genetic algorithm (GA) were applied to search for binding modes of the selected natural products keeping them rigid, since there was a need to preserve the similar pose (conformation + bioactive orientation) of the inhibitor originally complexed using as the reference molecule. Ten docking solutions were calculated for each selected natural product and GoldScore was analyzed. In addition, the energy score function was used to evaluate the free binding energy of the interactions between the receptor and ligand. The visual analysis of poses was also take into account the choice of the best free energy (ΔG) [35, 36].

A maximum efficiency search was performed (100% efficiency self-configuration). The best natural products selected in this step were classified according to the PLPFitness scoring function which assessed the binding modes of these substances to the binding site. A visual inspection was performed to select the best poses of selected natural products that revealed interactions with the amino acid residues of the active site of the enzyme studied.

2.6 *Pharmacokinetic and Toxicological Properties*

The human organism consists of components such as transporters, channels, receptors and enzymes in which are related to absorption, distribution, metabolism, excretion and toxicity (ADME/Tox) of a molecule to be administered in an individual. Observing and understanding the interactions that occur between molecules and their macromolecular targets should enhance the ability to predict the toxicological consequences that are responsible for the removal of many marketed drugs and failures in the final development of the drug. By using the PreADMET webserver, (https://preadmet.bmdrc.org/) it was possible to determine the pharmacokinetic properties of the structures, where is predicted human intestinal absorption (HIA), cellular permeability in Caco-2 (pCaco-2) cells in vitro, cell permeability in Maden Darby Canine Kidney cells (pMDCK), plasma protein binding (PPB) and penetration of the blood-brain barrier (BBB), and toxicological properties as mutagenicity and carcinogenicity. Pharmacokinetic and toxicological properties were calculated for the inhibitor selegiline and natural compounds [37–39].

3 Results and Discussions

3.1 Studied Compounds and Molecular Modeling

In Fig. 1, the molecules described in the literature and used in this study are observed: Selegiline (1), Amburoside A (2), Harman (3), Harmaline (4), Harmalol (5), which were drawn in ChemSketch 12.0 program [23]. Selegiline is a drug used in PD therapy and has antidepressant activity in high dose; it does not cause hypertension; neuroprotection against MPTP (1-methyl-4-phenyl-1,2,3,6-tetrahydropyridine); sympathomimetic effects related to methamphetamine metabolites [2]. The phenolic glycosides found in *A. cearensis*, amburoside A and B, had their neuroprotective properties evaluated in cultures of cells exposed to neurotoxin 6-hydroxydopamine (6-OHDA). 6-OHDA is a hydroxylated derivative of dopamine that is possibly endogenously formed in patients with PD and is related to various signs of the disease [16, 20]. For the studies of *P. incarnata*, the aqueous extract with dry leaves was used and later the effects of the extract were evaluated in animals that were induced experimentally by neuroleptics to develop the catalepsy of PD, as this is an animal model used for

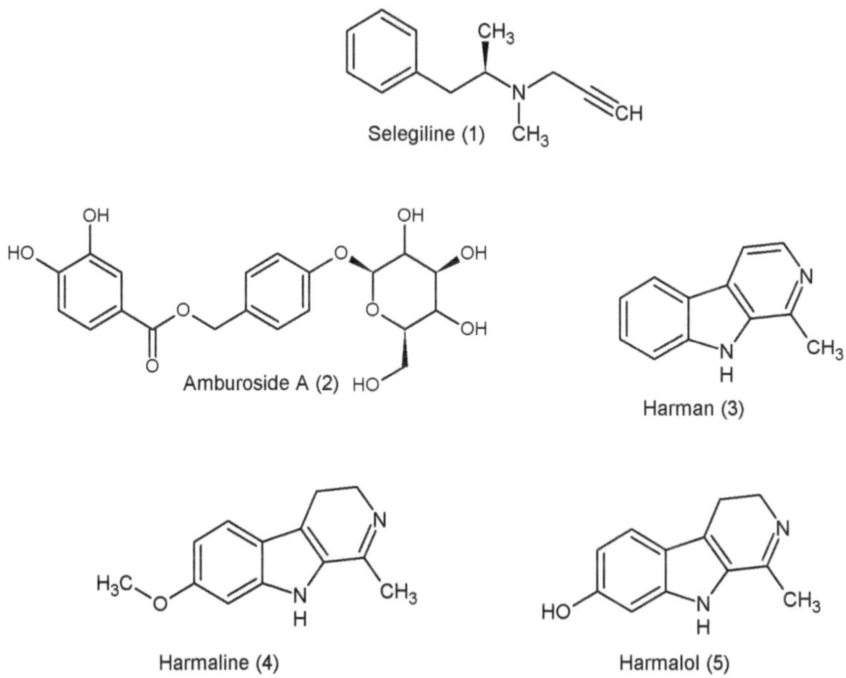

Fig. 1 Structure of Selegiline (1) and the natural compounds Amburoside A (2), Harman (3), Harmaline (4), Harmalol (5)

drug screening [15, 16]. Studies of natural molecules indicate that their antiparkinsonian activity is due to antioxidant action, but the mechanisms of action of the molecules have not been proven, since, to reduce the oxidative stress, the molecules must have the capacity to bind to reactive oxygen species or inhibit MAO-B that degrades substances that generate oxidative stress. With this, it is believed that the natural molecules can be used as MAO-B inhibitors.

3.2 Chemical Reactivity and Statistical Analysis of Studied Compounds

With the computational advance, softwares that identify chemical reactions are being incremented in research for discoveries of new drugs, because it is possible to obtain information on the probable regions of interaction between molecules, more deeply about which atoms these interactions may occur, facilitating the decision regarding the molecular modifications to improve the activity of the new drug candidate.

Table 2 shows the results for Selegiline (1), Amburoside A (2), Harman (3), Harmaline (4), Harmalol (5) compounds of E_{HOMO}, E_{HOMO-1}, E_{LUMO}, E_{LUMO+1}, gap energy in B3LYP/6-31G** [20] level and Pearson correlation matrix in function of Total Energy in which 1 represents a Positive Absolut Correlation, 0 to no Correlation, and −1 to Negative Total Correlation [31].

In Fig. 2, the images of HOMO orbital energy are shown. That is the property that informs the values related to the electrons affinity of the molecules studied, characterized by the ability to perform nucleophilic attacks [40], as shown in Table

Table 2 Calculated E_{HOMO}, E_{HOMO-1}, E_{LUMO}, E_{LUMO+1} and gap energy of Selegiline (1), Amburoside A (2), Harman (3), Harmaline (4) and Harmalol (5) compounds in B3LYP/6-31G** level and Pearson correlation matrix

Compounds	E_{HOMO} (eV)	E_{HOMO-1} (eV)	E_{LUMO} (eV)	E_{LUMO+1} (eV)	*Gap* (eV)	Total Energy (a.u)
1	−8.55	−8.80	4.12	4.18	−12.67	−556.55
2	−5.74	−6.41	−0.67	−0.59	−0.36	−152.75
3	−7.72	−8.23	2.63	4.02	−10.35	−569.15
4	−7.20	−8.28	2.76	4.56	−9.96	−645.15
5	−7.74	−8.71	3.95	4.68	−11.70	−646.31
E_{HOMO}	1.00	0.94	−0.94	−0.85	0.95	0.79
E_{HOMO-1}		1.00	−0.99	−0.96	0.99	0.93
E_{LUMO}			1.00	0.94	−0.98	−0.91
E_{LUMO+1}				1.00	−0.96	−0.99
Gap					1.00	0.93
TE						1.00

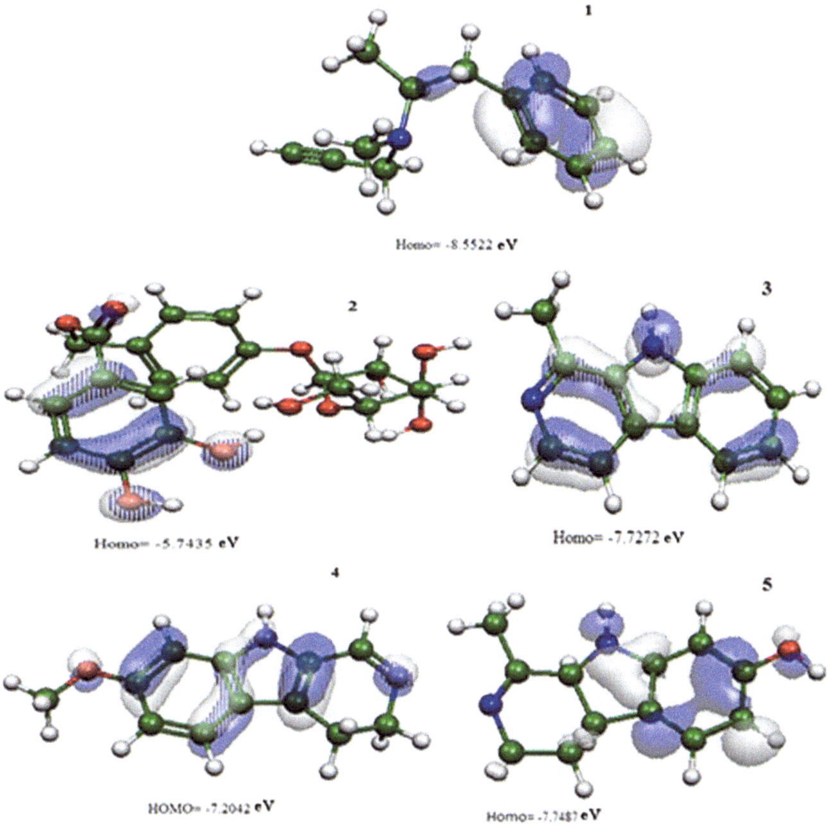

Fig. 2 Representation of the HOMO orbitals in the Selegiline (1), Amburoside A (2), Harman (3), Harmaline (4), Harmalol(5) natural compounds

2, HOMO values ranged from −8.55 eV to −5.74 eV, lower and higher for the molecules Selegiline and Amburoside A, respectively. The Harmalol molecule has −7.75 eV, being the closest value to the standard molecule Selegiline. For HOMO-1, The values ranged from −8.8086 eV and −6.42 eV, lower and higher for the Selegiline and Amburoside A molecules, respectively, with the closest value to the standard molecule, Harmalol with −8.71 eV. The Pearson correlation value for HOMO and HOMO-1 with total energy was 0.80 and 0.94, respectively, demonstrating that there is a positive correlation.

In Fig. 3, presents images of LUMO energy which is the property that informs the values related to electrons affinity of the molecules studied, characterized by the susceptibility of the compound relative to nucleophilic attack, same concept is for other orbital LUMO, for example, LUMO + 1 [40], as shown in Table 2, LUMO values ranged from −0.67 eV and

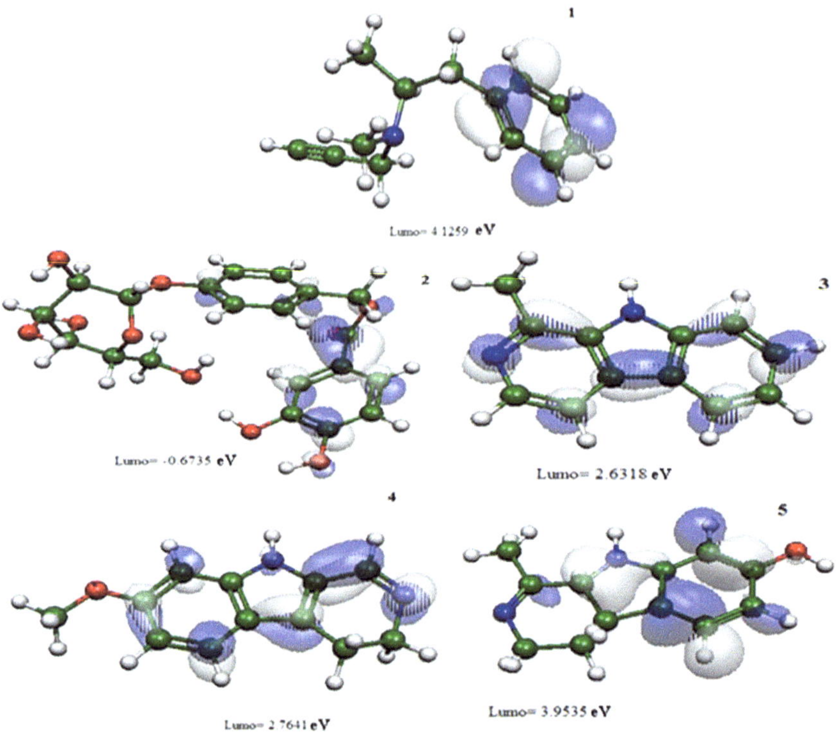

Fig. 3 Representation of the LUMO orbitals in the Selegiline (1) and natural compounds Amburoside A (2), Harman (3), Harmaline (4), Harmalol (5)

4.12 eV, lower and higher for the molecules Amburoside A and Selegiline, respectively, being Harmalol molecule with 3.95 eV the closest value to the standard molecule Selegiline. For LUMO+1, the values varied from −0.59 eV and 4.69 eV, lower and higher for the molecules Amburoside A and Harmalol, respectively, with the value closest to the standard molecule Harmane with a value of 4.03 eV. Table 2 shows the Pearson correlation value for LUMO and LUMO +1 with the total energy, being −0.99 and −0.99, respectively, showing a negative correlation.

In another study, Elshaflu and collaborators [41] performed in silico calculations of HOMO and LUMO for synthesized MAO inhibitors, to identify the substituent that had the highest degree of energy. The results of the energies of the HOMO orbitals varied from −5.51 to −5.20 eV, being possible to observe a similar result with Amburoside A. While for LUMO the results varied from −2.79 and −1.53 eV for Harman and Harmaline, respectively.

The difference between HOMO–LUMO energies is identified as GAP and it is an important indicator of molecular stability and chemical reactivity. The lower the GAP values are, the more reactive will be the molecules, while higher values indicate high

Table 3 Molecular descriptors of chemical reactivity and molecular stability of Selegiline (1), Amburoside A (2), Harman (3), Harmaline (4) and Harmalol (5) compounds and Pearson correlation matrix

Compounds	X (eV)	μ (eV)	η (eV)	1/η (eV)	HE (Kcal/mol)
1	0.08	−0.08	0.23	4.27	−0.25
2	0.00	−0.00	0.00	147.60	−33.35
3	0.09	−0.09	0.19	5.25	−4.91
4	0.08	−0.08	0.18	5.45	−6.27
5	0.06	−0.06	0.21	4.65	−8.73
χ	1.00	0.99	−0.92	−0.97	0.96
μ		1.00	−0.92	−0.985	0.96
η			1.00	0.913	−0.97
1/η				1.00	−0.97
HE					1.00

molecular stability [29]. As shown in Table 2, GAP values ranged from −12.6781 eV and −0.3687 eV, lower and higher, for the molecules Selegiline and Amburoside A, respectively, being Harmalol molecule with −11.7022 eV as the closest value to the standard molecule Selegiline. The Pearson correlation value for GAP with the total energy was 0.9386, demonstrating that there is a positive correlation and being the best correlation of the matrix.

The total energy (TE) is the property that correlates the structure with the chemical reactivity, corresponding to the sum of the energy of nuclear repulsion and the electronic energy, in which the latter determines the fixed position of nuclei of the molecule to estimate its stability [42]. As shown in Table 2, TE values varied from -646.3129 a.u and −152.7570 a.u, lower and higher for the molecules Harmalol and Amburoside A, respectively, where Harman molecule was -569.1508 a.u, being the closest value to the standard Selegiline molecule.

In Table 3 are shown the molecular properties of the Selegiline (1), Amburoside A (2), Harman (3), Harmaline (4), Harmalol (5) compounds in B3LYP/6-31G** level and Pearson correlation matrix. The correlation between pairs of molecular properties is less than 0.99408, while the correlation between the molecular properties and hydration energy (HE) value is less than −0.97380.

The electronegativity (χ) is the property of the atom that measures its force of attracting electrons for itself, in which it makes a connection with another atom. This property can be used to estimate the capacity of one molecule to attract electrons to another when there is an interaction between these two molecules [43]. As shown in Table 3, χ values ranged from 0.0041 eV and 0.0936 eV, lower and higher, for the molecules Amburoside A and Harman, respectively, having the Harmaline molecule with 0.0816 eV, being the closest value to the standard molecule Selegiline. The Pearson correlation value for electronegativity with HE was 0.9941 demonstrating that there is a positive correlation.

The chemical potential (μ) is the property related to the increase of Gibbs free energy that came from the addition of an infinitesimal number of moles of one component of the added substance mixture per mole, under conditions of T, P, number of moles of other constant components. Note that this property can in principle be expressed as a function of other thermodynamic functions (E, H, A, S), since the parameters are changed [43] as shown in Table 3, values ranged from −0.0936 eV and −0.0041 eV lower and higher for the Harman and Amburoside A molecules, respectively, having the Harmaline molecule with −0.0815 eV being the closest value to the standard molecule Selegiline. The Pearson correlation value for chemical potential with HE was 0.9885 demonstrating that there is a positive correlation.

The concept of molecular hardness (η) and softness ($1 / \eta$) was formulated according to the Lewis concept of acids and bases. The molecules that have a high ionization potential and high electronegativity have high hardness. Therefore, hardness can be said to represent the resistance of a molecule to deformation, and softness represents the ease with which a molecule is deformed [42]. As shown in Table 3, hardness values varied from 0.0067 eV and 0.2337 eV, lower and higher value for the molecules Amburoside A and Selegiline, respectively, having the Harmalol molecule with 0.2150 eV being the closest value to the standard molecule Selegiline. For softness, the values varied from 4.2772 eV and 147.6014 eV, lower and higher for the molecules Selegiline and Amburoside A, respectively, having the Harmalol molecule with 4.6506 eV being the closest value to the standard molecule Selegiline. The Pearson correlation values with HE for hardness were −0.9738 and for softness −0.9599 demonstrating that both have a negative relation.

Hydration Energy (HE) is related to the stability of different molecular conformations in aqueous solution, providing information regarding the type of interaction that can occur between the compounds and their respective biological receptor and also indicates the separation between active and inactive compounds [43], as shown in Table 3, the values varied from –33.35 (Kcal/mol) and −0.25 (Kcal/mol), lower and higher for the Amburoside A and Selegiline molecules, respectively, having the Harman molecule with 4.91(Kcal/mol) being the closest value to the standard molecule Selegiline.

3.3 Molecular Electrostatic Potential Maps

The MEP view is shown in Fig. 4, for the compounds investigated. The regions highlighted in red represent negative regions that the molecule is more likely to perform a nucleophilic attack, and blue regions are positive, evidence of regions where the compound is most likely to undergo nucleophilic attack [34, 45]. To identify the main characteristics related to the electron donor and acceptor regions of Selegiline (1), Amburoside A (2), Harman (3), Harmaline (4) and Harmalol (5), qualitative evaluations were performed resulting in amine group region and ester highlighted by red in the MEP as an electron donor, while the amide and hydroxyl region is highlighted by blue in the MEP as an electron acceptor.

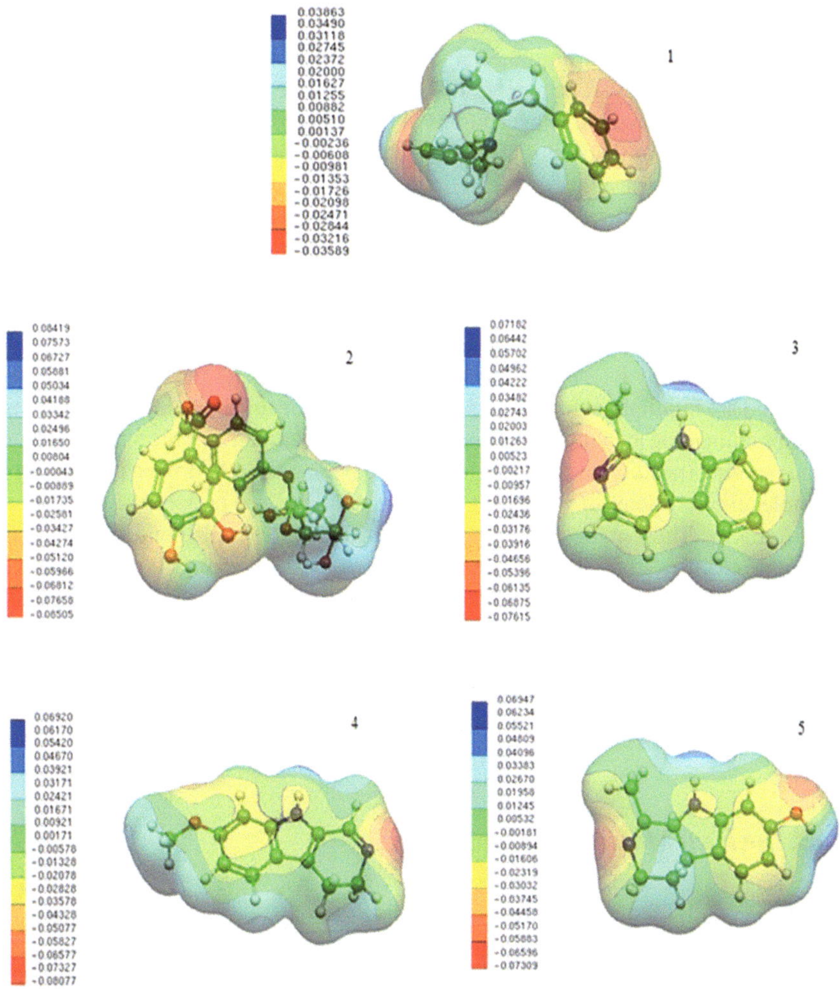

Fig. 4 Molecular electrostatic potential maps of natural compounds Selegiline (1), Amburoside A (2), Harman (3), Harmaline (4) and Harmalol (5)

It was verified that the negative molecular electrostatic potential is distributed in the nitrogen atoms present in the tertiary amine group (–C=N–C) and carbonyl (C=O). The positive electrostatic potential in the investigated compounds was evenly distributed along the aromatic rings, being more evident in the secondary amine group (–C–NH–C). The lowest and highest values of the positive electrostatic potential were 0.036 and 0.084 ua, for the molecules Selegiline and Amburoside A, respectively. While the molecule with the closest value to the standard Selegiline molecule was the Harmaline with a value of 0.069 ua. The near and far values corresponding to the negative electrostatic potential were −0.08505 ua and −0.035 ua, for the Amburoside and Selegiline molecules, respectively. While the Harmalol molecule with a value of −0.076 ua has a similarity to the standard Selegiline molecule.

The region of electrostatic potential is negative due to the nitrogen atoms present in the amine groups and the oxygen atoms in the carbonyl, which is the most notable feature of MEP. The distribution of the electron density around these functional groups may be responsible for the activity of the compounds. The presence of a negative surface near the nitrogen atoms present in the amine groups and the oxygen atoms in the carbonyl suggests that these compounds have a reactive site for electrophilic attack. Thus, in the case of an electrophilic attack against an electronegative zone, there is a preference to happen in these regions [43, 44].

Through MEP analysis (Fig. 4), the selection of inactive molecules can be avoided. For the Selegiline molecule, regions in the aromatic ring and in the carbon of the triple bond are highlighted in red (negative molecular electrostatic potential), having 2 regions, indicating that these regions are propitious for a nucleophilic attack, and the regions in blue (positive molecular electrostatic potential) in hydrogens throughout the molecule, in a total of 8 regions, indicate that they are protected from nucleophilic attacks. Thus, in the 4 natural molecules, there is no triple bond and negative molecular electrostatic potential distributed in the aromatic ring, but the positive molecular electrostatic potential is found in the hydrogens bound to atoms that have strong bonds with other atoms, being able to release this hydrogen for a possible interaction with another molecule. Nevertheless, it could be said that the MEP of the natural molecules has similarities with the MEP of the standard molecule Selegiline [43, 44].

3.4 *Docking Molecular Simulation*

For this study, the docking simulation between the ligands and the active site of the enzyme was performed with the Selegiline and natural molecules Amburoside A (1), Harman (2), Harmaline (3), Harmalol (4) and MAO-B respectively. For this purpose, the selection was first made from the Protein Data Bank (PDB) of the MAO-B crystallographic structure deposited in the complex form under code 4A79 at 1.89 Å resolution. The GOLD software was used, which uses the genetic algorithm for purposes of flexible docking experiments of ligands within protein binding sites [36]. And this was used to identify the binding site of MAO-B to simulate the interaction between the ligands and the enzyme.

The validation of the program for docking simulation consists in the replication of the experimental results of the positions of the ligand atoms about the active site of the enzyme, acquired from the crystallographic complex, and through the GOLD 4.1, several coupling tests were carried out in an attempt to obtain the same positions of the atoms of the experimental result. The most used method to analyze if the program succeeded in repeating the experimental result is the calculation of Mean Square Deviation (RMSD) in which previous studies showed results smaller than 2 Å have a high success rate for a replication of the experimental result [45]. Through

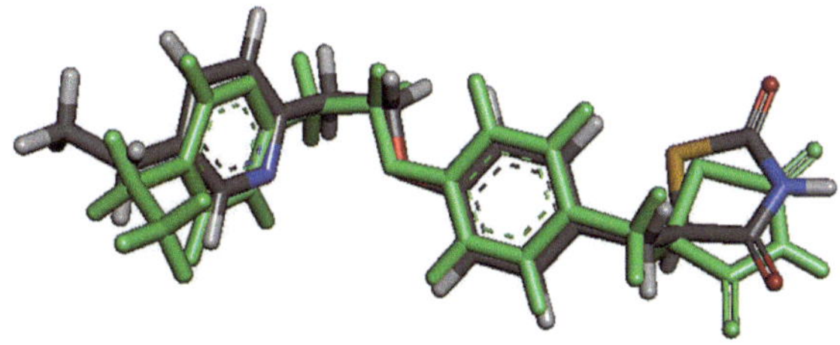

Fig. 5 Result of docking simulation with the lowest RMSD for the validation of the complex 4A79

the Discovery Studio Visualizer software [46] it was possible to visualize the test with the lowest RMSD of 1.3814 Å compared to the position of the ligand from the experimental result of the 4A79 complex, shown in Fig. 5.

In Fig. 6 shows the interaction of the amino acids active sites that were selected as described in the literature: TYR398, TYR435, PHE343, LEU171, TYR326, ILE316, PRO104, PRO104, ILE199, LEU171, CYS172, TRP119 [47]. The analyzes demonstrated that the bonds were mostly hydrophobic and hydrogen bond interactions, the 5 molecules had only one common amino acid, CYS172, but all molecules had interaction with more than 3 amino acids among the 12 active sites.

Hydrophobic interactions occur between apolar regions such as aromatic rings and methyl rings. Conventional hydrogen bonds are characterized when a proton acceptor molecule interacts with a proton donor molecule because of the difference in electronegativity. This type of coupling occurs between electronegative atoms such as fluorine, oxygen and chlorine [48]. Hydrogen bonds with a distance less than or equal to 3 Å are relevant to assert that there is interaction, whereas the hydrophobic bonds have a distance less than or equal to 5 Å so that an interaction between the amino acids and the binding molecule occur [49].

In Table 4 shows the bonds between the amino acids and the studied molecules, with their respective types of interactions. The docking simulation between molecule 1 and MAO-B amino acids showed 6 bonds in 5 different amino acids of the active site. Half of the bonds are hydrogen bonds between the nitrogen atom (N) and oxygen (O) with hydrogen, and the other half consists of hydrophobic interactions between aromatic rings and hydrophobic regions. The simulation between molecule 2 and the MAO-B amino acids showed 17 bonds in 11 different amino acids. Most of the bonds are hydrogen bonds between hydrogen (H) and oxygen (O) atoms, and hydrophobic interactions between aromatic rings also occurred. In the simulation between molecule 3 and the amino acids of MAO-B with 5 bonds in 4 different amino acids, most of the bonds are hydrogen bonds between hydrogen (H) and oxygen (N) atoms, and hydrophobic interactions between aromatic rings. In the simulation between molecule 4 and MAO-B amino acids with 9 bonds in 7 different amino acids, most of the bonds are hydrogen bonds between hydrogen (H) and nitrogen (N) atoms, and hydrophobic interactions have also occurred between aromatic rings. In the simulation between molecule 5 and the MAO-B amino acids with 7 bonds in 5 different amino acids, most of the bonds are hydrogen bonds between hydrogen (H) and oxygen (N) atoms, and hydrophobic interactions between aromatic rings.

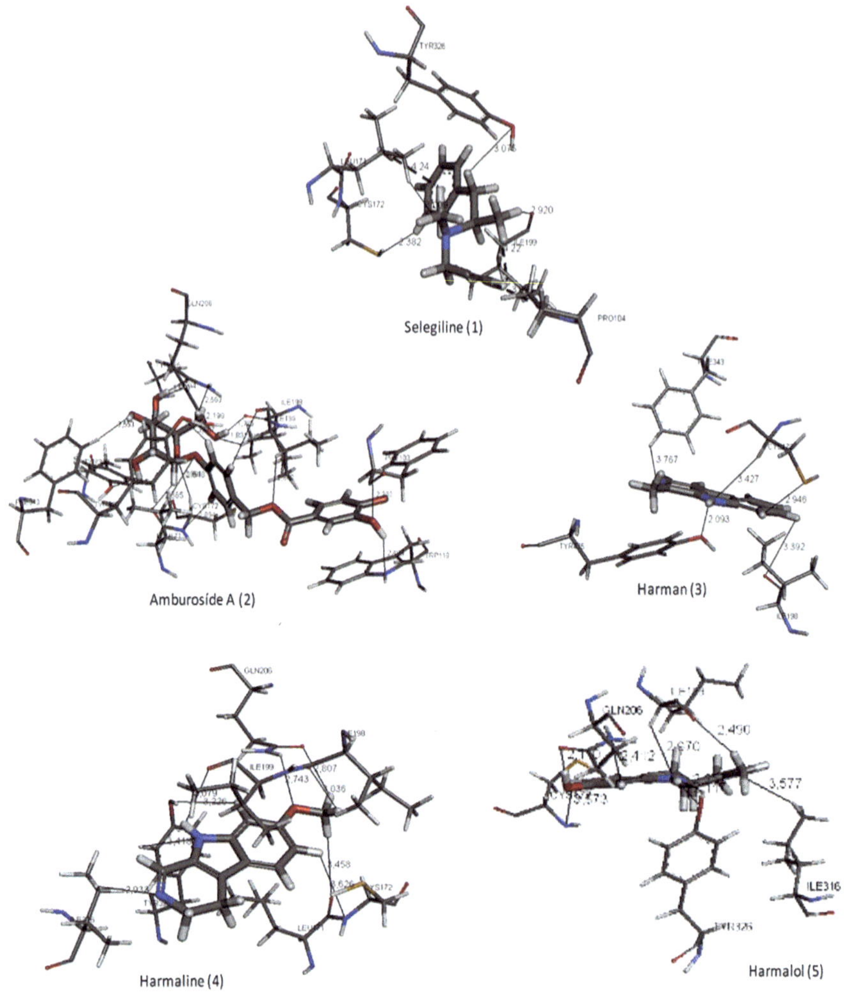

Fig. 6 Docking simulation between the active site of amino acids and natural molecules

The docking simulation between Selegiline and MAO-B showed 8 bonds in 6 different amino acids from the active site: CYS172, TYR326, PRO104, ILE316, ILE199, LEU171. All interactions are hydrophobic, with the amino acid CYS172 a van der Waals interaction was obtained between the sulfur atom of the amino acid and the aromatic ring of the molecule with a distance of 4.73 Å; with the amino acid TYR326, two hydrophobic interactions were observed, one between the aromatic ring of the amino acid and the carbon of the methyl radical of the molecule, and the other between the aromatic rings of both the amino acid and the molecule, with distances of 5.28 Å and 5.46 Å, respectively. With the amino acid PRO104, a hydrophobic interaction was obtained between an alkyl group of the amino acid and the hydrogen

Table 4 Result of the docking simulation between the studied molecules and the amino acids of the MAO-B enzyme active site

Molecules	Amino acids active site	Atoms	Interaction types	Distance (Å)
Selegiline	PRO104	Alkyl–H	Hydrophobic	3.99
	LEU171	1343 C–Ph	Hydrophobic	4.24
	CYS172	S–Ph	Hydrophobic	4.73
	ILE199	Alkyl–C	Hydrophobic	4.22
	ILE316	Alkyl–C	Hydrophobic	4.64
		Alkyl–C	Hydrophobic	4.84
	TYR326	Ph–C	Hydrophobic	5.28
		Ph–Ph	Hydrophobic	5.46
Amburoside A	LEU171	1343 C–Ph	Hydrophobic	3.97
	CYS172	H–O	Hydrogen bond	2.65
		Alkyl–Ph	Hydrophobic	5.06
	ILE199	Alkyl–Ph	Hydrophobic	3.83
		Alkyl–Ph	Hydrophobic	4.89
	TYR398	O–H	Hydrogen bond	2.11
	TYR435	OH–H	Hydrogen bond	2.51
Harman	LEU171	C1343–Ph	Hydrophobic	4.52
	CYS172	Alkyl–Ph	Hydrophobic	4.83
	TYR398	Ph–Ph	Hydrophobic	4.51
		Ph–Pirrol	Hydrophobic	5.36
		Ph–C14	Hydrophobic	4.06
	TYR435	Ph–C14	Hydrophobic	3.84
		O–H	Hydrogen bond	2.09
Harmaline	LEU171	C1343–Ph	Hydrophobic	4.21
	CYS172	S–Ph	Hydrophobic	4.76
	ILE199	O–H	Hydrogen bond	3.08
		Alkyl–C11	Hydrophobic	4.79
		Alkyl–Ph	Hydrophobic	4.86
		Alkyl–Ph	Hydrophobic	4.22
	ILE316	Alkyl–C11	Hydrophobic	4.87
	TYR326	Ph–Ph	Hydrophobic	3.36
Harmalol	LEU171	C1343–Ph	Hydrophobic	4.19
	CYS172	S–Ph	Hydrophobic	4.88
	ILE199	Alkyl–C11	Hydrophobic	4.20
	ILE316	Alkyl–C11	Hydrophobic	5.28
		Alkyl–C15	Hydrophobic	4.72
	TYR326	Ph–Ph	Hydrophobic	5.34

(continued)

Table 4 (continued)

Molecules	Amino acids active site	Atoms	Interaction types	Distance (Å)
		Ph–C15	Hydrophobic	4.92

atom of the molecule with a distance of 3.99 Å. With the amino acid ILE316, two hydrophobic interactions were observed, between an alkyl group of the amino acid and the carbon of the methyl radical, and another interaction between the alkyl group and the tertiary carbon, with distances of 4.84 Å and 4.64 Å, respectively. With the amino acid ILE199, a hydrophobic interaction was obtained between an alkyl group of the amino acid and the carbon of the methyl radical of the molecule with a distance of 4.22 Å. With the amino acid LEU171, a hydrophobic bond was obtained between an alkyl group of the amino acid and the aromatic ring of the molecule with a distance of 4.24 Å.

The simulation between Amburoside A and MAO-B showed 7 bonds in 5 different amino acids: LEU171, CYS172, TYR398, TYR435 and ILE199. Most of the bonds are of the hydrophobic type with four interactions, and there have also been three hydrogen interactions between the hydrogen (H) and oxygen (O) atoms. With the amino acid LEU171, a hydrophobic interaction was obtained between the alkyl group of the amino acid and the aromatic ring of the molecule with a distance of 3.97 Å. CYS172 presented two interactions, one of the hydrogen bond type between the H of the amino acid and O of the molecule with a distance of 2.65 Å, and the other of the hydrophobic type between an alkyl group of the amino acid and the aromatic ring of the molecule, with a distance of 5.06 Å. With the amino acid residue TYR398, a hydrogen bond interaction was obtained between the O of the amino acid and the H of the molecule with a distance of 2.11 Å. The amino acid residue TYR435 showed a hydrogen bond interaction between the hydroxyl O of the amino acid and the H of the molecule, with a distance of 2.51 Å. The amino acid residue ILE199 showed two hydrophobic interactions, between alkyl groups of the amino acid and the aromatic ring of the molecule, with distances of 3.83 Å and 4.89 Å.

In the simulation between Harman and MAO-B, it was possible to observe 7 bonds in 4 different amino acids: TYR435, TYR 398, LEU171 and CYS172. Most bonds are hydrophobic and only one of the hydrogen bond type. With the amino acid residue TYR435, two interactions were observed, one of the hydrogen bond type between the O of the amino acid and the H of the molecule, and another of the hydrophobic type between the aromatic ring of the amino acid and the alkyl group of the molecule, with distances of 2.09 Å and 3.84 Å, respectively. With the amino acid residue TYR398, three hydrophobic interactions were observed, between the aromatic ring of the amino acid and the aromatic ring, pyrrolidine ring and the alkyl group of the molecule, with distances of 4.51 Å, 5.36 Å, 4.06 Å, respectively. The amino acid residue LEU171 showed a hydrophobic interaction between the alkyl group of the amino acid and the aromatic ring of the molecule with a distance of 4.52 Å. The amino acid residue CYS172 showed a hydrophobic interaction between the alkyl group of the amino acid and the aromatic ring with a distance of 4.83 Å.

In the simulation between Harmaline and MAO-B, 8 interactions were observed with 5 different amino acid residues: ILE199, CYS172, TYR326, ILE316 and LEU171. Most bonds are hydrophobic, with only one hydrogen bond type. With the amino acid residue ILE199, a hydrogen bond interaction was obtained between the O atom of the amino acid and the H atom of the molecule, and three hydrophobic interactions, one between the alkyl groups of the amino acid and the molecule, and two between the alkyl group of the amino acid and the aromatic ring of the molecule, with distances of 3.08 Å, 4.79 Å, 4.86 Å, 4.22 Å, respectively. With the amino acid residue CYS172, a hydrophobic interaction was obtained between the sulfur atom (S) of the amino acid and the aromatic ring of the molecule with a distance of 4.76 Å. With the amino acid residue TYR326, a hydrophobic interaction was obtained between the aromatic rings of the amino acid and the molecule with a distance of 3.36 Å. The amino acid residue ILE316 showed a hydrophobic interaction between alkyl groups of the amino acid and the molecule with a distance of 4.87 Å. The amino acid residue LEU171 showed two hydrophobic interactions, one between the alkyl group of the amino acid and the aromatic ring of the molecule with a distance of 4.21 Å.

In the simulation between Harmalol and MAO-B, it was possible to observe 7 bonds with 5 different amino acids. All interactions are hydrophobic. With CYS172, a hydrophobic interaction was obtained between the S atom of the amino acid and the aromatic ring of the molecule with a distance of 4.88 Å. With the amino acid residue TYR326, two interactions were observed, one between the aromatic ring of the amino acid and the aromatic ring and the alkyl group of the molecule, with distances of 5.34 Å and 4.92 Å, respectively. With the amino acid residue LEU171, a hydrophobic interaction was obtained between the alkyl group of the amino acid and the aromatic ring of the molecule with a distance of 4.19 Å. The amino acid residue ILE199 showed an interaction between the alkyl groups of the amino acid and the molecule with a distance of 4.20 Å. The amino acid residue ILE316 showed two interactions between the alkyl groups of the amino acid and the molecule, with distances of 5.28 Å and 4.72 Å.

However, the standard molecule Selegiline showed 8 interactions with six different amino acid residues being CYS172, TYR326, PRO104, ILE316, ILE199, LEU171, and the closest result was from Harmaline for presenting 8 interactions with the same amino acids which Selegiline also interacted with, except PRO104. Therefore, it can be deduced that Harmaline has a higher probability among the four natural products to present inhibitory activity of the MAO-B enzyme.

The regions where the interactions took place were confirmed with the MEP since all molecules interacted in the positive (blue) and negative (red) regions shown on the map. Selegiline interacted in two (aromatic ring and hydrogen of the triple bond) of the four regions. Amburoside A interacted in four (hydroxyls and ester groups) of the six regions. Harman, Harmaline and Harmalol interacted in 3 regions (ketone, the nitrogen of the pyrrolidine ring and aromatic ring) of the four presented in MEP.

In the study by Farahnaz and Jahan [50], in silico evaluations of chalcones synthesized as possible MAO-B inhibitors were performed. One of the tests performed was molecular docking, in which the four synthesized proposals interacted with the MAO-B binding site through the amino acids Tyr326, Tyr435 and Tyr398 [50]. These results were similar to this study, in which the natural molecules Amburoside A, Harman, Harmaline and Harmalol also interacted.

Hagenow and collaborators [51] docked two compounds for MAO-A and MAO-B. With that, it was observed that the compounds had interactions with the amino acid residues LEU164, PHE168, PRO102, GLN206 and CYS172, the latter also interacting with the molecules of this study.

3.5 Pharmacokinetic and Toxicological Properties

Using PreADMET, it was possible to determine the pharmacokinetic properties of the compounds Selegiline (1), Amburoside A (2), Harman (3), Harmaline (4) and Harmalol (5) through the ADME prediction, where it predicts human intestinal absorption (HIA), Cell permeability in Caco-2 (pCaco-2) and Madin-Darby Canine Kidney (pMDCK) cells, plasma protein binding (PPB) and blood-brain barrier (BBB). According to Yamashita [23, 24], these cells have been recommended as a reliable in vitro model for the prediction of oral absorption of the drug candidate. In addition, it can predict the toxicological properties that the natural compounds can cause, such as mutagenicity and carcinogenicity [24, 25].

ADME in silico models, compared to traditional experimental assays, have greater applicability to meet the huge demand generated in the large-scale screening of new molecules. Also, in vitro and in vivo assays have drawbacks that limit their use on a large scale: they are complex and costly in terms of materials, infrastructure and skilled personnel [21, 22]. For these reasons, there is great industry interest in the generation of ADME in silico models that can quickly aid in the selection of promising molecules and guide the elimination of compounds with an inappropriate pharmacokinetic profile. On the other hand, integration of ADME (in silico, in vitro and in vivo) models seems to be an essential pathway to be followed at all stages of the drug discovery process. The results of the ADME prediction are presented in Table 5.

Human intestinal absorption data are found from the sum of the bioavailability and the assessed absorption of the proportion of cumulative excretion or excretion in

Table 5 Pharmacokinetics prediction of Selegiline (1), Amburoside A (2), Harman (3), Harmaline (4) and Harmalol (5) compounds

Compounds	Absorption			Distribution	
	HIA (%)	P_{Caco2} (nm/s)	P_{MDCK} (nm/s)	PPB (%)	BBB (C_{Brain}/C_{Blood})
1	100.00	31.94	164.08	53.20	4.43
2	38.14	14.72	0.92	67.97	0.03
3	94.10	23.99	371.07	100.00	5.59
4	91.53	36.08	345.82	52.36	4.59
5	90.95	20.80	82.50	58.37	1.11

urine, bile and feces. The patterns are classified as: bad absorbed between 0 and 20%, moderately absorbed 20–70%, well absorbed 70–100% [21, 22]. Table 5 shows the results for HIA, the molecules were in the range of well-absorbed molecules, except for the molecule Amburoside A that was in the margin of moderately absorbed molecules. The values of the range of minimum and maximum were 38.14% and 100% for the molecules Amburoside A and Selegiline respectively, indicating that molecule 2 has a higher percentage of absorption. The Harman molecule with 94.10% has the closest value to the standard molecule Selegiline.

Caco-2 cells are derived from human colon adenocarcinoma and have several routes of drug transport through the intestinal epithelium. Although there are some differences in the in vivo experimental values of compounds or their metabolisms, they can be classified as: low permeability if less than 4 nm/s, medium permeability between 4 and 70 nm/s, high permeability if greater than 70 nm/s [21, 22]. For Caco-2 cells, the molecules were in the medium permeability range, with the range of minimum and maximum variation of 14.72 nm/s and 36.08 nm/s for Amburoside A and Harmaline molecules, respectively, indicating that the molecule Harmaline has a higher Permeability in the body by Caco-2 cells (Table 5). The Harmaline molecule has the closest value to the standard molecule Selegiline.

The advantage of MDCK cells is that their growth period is shorter than that of Caco-2 cells. The correlation of MDCK and Caco-2 cells for high permeability is R2 = 0.79. These results prove that the MDCK cell system can be used as a good tool for the rapid screening of permeability. It can be classified as: low permeability if less than 25 (nm/s), medium permeability between 25 and 500 (nm/s), high permeability if greater than 500 (nm/s) [21, 22]. For MDCK cells, the molecules were in the medium permeability range, except molecule Amburoside A that is in the low permeability range. The range of variation for the minimum and maximum range of 0.92 (nm/s) and 371.07 (nm/s) for the molecules Amburoside A and Harman, respectively, indicates that the Harman molecule has better permeability in the body by MDCK cells (Table 5). The Harmalol molecule with 82.50 (nm/sec) has the closest value to the standard Selegiline molecule.

As is well-known, the degree of plasma protein binding of a molecule influences its bioavailability and effectiveness, where molecules considered to be strongly bound to proteins have more than 90% probability and weakly bound to plasma proteins have less than 90% probability [21, 22]. According to the PPB prediction, the molecules showed medium permeability, except for Harman who had high permeability. The range of variation for minimum and maximum was 52.36% and 100% for the Harmaline and Harman molecules, respectively, indicating that the Harman molecule has low bioavailability in the human body, with the Harmaline molecule with 52.36% being the highest value close to the standard Selegiline molecule.

Blood-Brain Barrier Penetration (BBB) is represented as BB = [Brain]/[Blood], where [Brain] and [Blood] are the steady-state concentration of radiolabeled compounds in the brain and peripheral blood, respectively. Penetration at BBB means predicting whether the compounds will pass through the blood-brain barrier. This is crucial in the pharmaceutical field because compounds acting on the CNS must pass through it and inactive compounds in the CNS must not pass in order

Table 6 Mutagenicity (Ames test) and carcinogenicity properties (rats and mouse) of Selegiline (1), Amburoside A (2), Harman (3), Harmaline (4) and Harmalol (5) compounds

Compounds	Ames test	Carcinogenicity	
	Mutagenicity	Rats	Mouse
1	Mutagen	Negative	Negative
2	Mutagen	Negative	Negative
3	Mutagen	Negative	Negative
4	Mutagen	Negative	Positive
5	Mutagen	Negative	Positive

to avoid CNS side effects. It can be put in general categories, as follows, although there are some differences in the in vivo experimental values of compounds or their metabolites: high absorption if greater than 2.0 (CBrain/CBlood), medium absorption between 2.0 and 0.1 (CBrain/CBlood), low absorption less than 0.1 (CBrain/CBlood) [21, 22]. According to the BBB prediction, the molecules were in the high permeability range, except Amburoside A and Harmalol which had low permeability. The range of variation for minimum and maximum was 0.037 (CBrain/CBlood) and 5.59 (CBrain/CBlood) for the molecules Amburoside A and Harman respectively, indicating that the Harman molecule is easier to cross the blood-brain barrier (Table 5). The Harmaline molecule with 4.59 (CBrain/CBlood) has the closest value to the standard molecule Selegiline.

Dhiman et al. [52] performed the ADMET prediction of natural products derived from piperine that have potential inhibitory activity for MAO. In this study, natural products showed between medium and good results for the analyzed parameters, and there was no violation of Lipinski's Ro5. Jiang et al. [53] performed a pharmacokinetic study of Harma and Harmaline in vitro and in vivo using the rat animal model, with intravenous administration, natural products were detected in the striatum 300 min after administration, indicated that they have good blood-brain barrier permeability, confirming the result in silico.

Table 6 shows the results for mutagenicity and carcinogenicity. Mutagenicity is predicted by the Ames test, which is a simple method for testing mutagenicity of a compound, as suggested by Ames, where several strains of the bacteria *Salmonella typhimurium* are used with mutations in the genes involved in the histidine synthesis being fundamentally important for histidine growth. Carcinogenicity is the toxicity that causes cancer in the body and generally the test requires a long time (on average two years), the test uses mouse or rats exposing them to a compound [22, 23]. PreADMET predicts the outcome of this model, constructed from data from the National Toxicology Program (NTP) and FDA in the United States. The 5 molecules showed mutagenicity. All the compounds showed clear evidence of carcinogenic activity for rats (negative prediction), whereas in mouse only Harmaline and Harmalol were non-carcinogenic (positive prediction).

Is and collaborators [54] presented a study with MAO inhibitors in which the toxicological properties were evaluated in silico, the two molecules were subjected to 26 different QSAR models of toxicity, such as mutagenicity, carcinogenicity, cytotoxicity, anemia, genotoxicity, hepatotoxicity, hepatic necrosis and neurotoxicity, in which they presented low and negative results for the QSAR models, indicating that they may not be toxic.

4 Conclusion

The Harmaline molecule presented significant results, and of the four natural products studied, this structure obtained more satisfactory results regarding the descriptors of global reactivity and pharmacokinetic and toxicological predictions compared to the results of Selegiline. The natural compound studied showed good intestinal absorption and medium plasma proteins binding. The values of the frontier orbitals (HOMO and LUMO), GAP and MEP were significant to evaluate the regions of probable sites of biological activity, as well as the regions of electrophilic and nucleophilic attacks of the compounds studied. Molecular docking indicated that the interaction between the amino acids of the active sites and the natural molecules studied were mostly of the hydrophobic and hydrogen bonding type. The Ames test showed a non-carcinogenic prediction in mouse, a factor that contributed to the satisfactory nature of the study in taking the research forward, aiming to improve the findings, so that, in the future, natural products aid in the treatment of PD. In order to better confirm the inhibitory activity of natural *n* molecules, additional studies should be carried out to evaluate the conformational characteristics of the enzyme to better identify the interaction regions of the natural molecules and the target molecule, using other methods of study of computational chemistry.

Acknowledgments We thank the Institute of Exact and Natural Sciences of the Federal University of Pará for the use of the GaussView and Gaussian softwares and support from Conselho Nacional de Desenvolvimento Cientifico e Tecnologico (CNPq).

Conflict of Interest The authors confirm that there is no conflict of interest.

References

1. Almeida, R.D.: Universidade Federal da Bahia, Faculdade de Medicina da Bahia, Salvador. Thesis (2013)
2. Follmer, C., Netto, H.J.C.B.: Quim. Nova **36**(2), 306–313 (2013)
3. Emmanouilidou, E., Papagiannakis, N., Kouloulia, S., Galaziou, A., Antonellou, R., Papadimitriou, D., Athanasiadou, A., Bozi, M., Koros, C., Maniati, M., Vekrellis, K., Ioannou, P.C., Stefanis, L.: Parkinsonism & Disorders **73**, 35–40 (2020)
4. Hirotaka, I., Cornelis, B., Hampton, L.L., Ganqiang, L., Maple-Grødem, J., Corvol, J.-C., Lasse, P., Nimwegen, M.V., Hutten, S.J., Nguyen, K.-D.H., Rick, J., Eberly, S., Faghri, F., Auinger, P., Scott, K.M., Wijeyekoon, M.R., Van Deerlin, V.M., Hernandez, D.G., Day-Williams, A.G., Brice, A., Alves, G., Noyce, A.J., Tysnes, O.-B., Evans, J.R., Breen, D.P., Estrada, K., Wegel, C.E., Danjou, F., Simon, D.K., Ravina, B., Toft, M., Heutink, P., Bloem, B.R., Weintraub, D., Barker, R.A., Williams-Gray, C.H., Warrenburg, B.P.V., Van Hilten, J.J., Scherzer, C.R., Singleton, A.B., Nalls, M.A.: Neurol.: Genetics, V.5, N.4 (2019)

5. Lautier, C., Goldwurm, S., Durr, A., Giovannone, B., Tsiaras, W.G., Pezzoli, G., Brice, A., Smith, R.J.: Am. J. Hum. Genet. **82**, 822–833 (2008)
6. Fishman-Jacob, T., Reznichenko, L., Youdim, M.B.H., Mandel, S.A.: J. Biol. Chem. **20**, v.284, n. 47 (2009)
7. Savica, R., Grossardt, B.R., Rocca, W.A., Bower, J.H.: Mov Disord. **33**(4), 537–543 (2018)
8. Darweesh, S.K.L., Raphael, K.G., Brundin, P., Matthews, H., Wyse, R.K., Chen, H., Bloem, B.R.: J. Parkinsons Dis. **8**(4), 495–498 (2018)
9. Aarsland, D., Creese, B., Politis, M., Chaudhuri, K.R., Ffytche, D.H., Weintraub, D., Ballard, C.: Nat. Rev. Neurol. **13**(4), 217–231 (2017)
10. Maiti, P., Manna, J., Dunbar, G.L.: Transl. Neurodegenerat. **6**, 28 (2017)
11. Csoti, I., Jost, W.H., Reichmann, H.: J. Neural Transm. **123**, 3–17 (2016)
12. Marino, B.L.B., Souza, L.R., Sousa, K.P.A., Ferreira, J.V., Padilha, E.C., Silva, C.H.T.P., Taft, C.A., Hage-Melim, L.I.S.: Mini-Rev. Med. Chem. **20** (2020)
13. Andrade, C.H., TrossinI, G.H.G., Ferreira, E.I.: Revista Eletrônica De Farmácia **7**, 1–23 (2010)
14. Schlessinger, A., Welch, M.A., Vlijmen, H.V., Korzekwa, K., Swaan, P.W., Matsson, P.: Clin. Pharmacol. Ther. **104**(5), 818–835 (2018)
15. Campos, H.C., da Rocha, M.D., Viegas, F.P., Nicastro, P.C., Fossaluzza, P.C., Fraga, C.A., Barreiro, E.J., Viegas Junior, C.: CNS & Neurol. Disord. Drug Targets **10**(2), 239–250 (2011)
16. Almeida, J.R.G.S., Guimarães, A.G., Siqueira, S.S., Santos, M.R.V., Lima, T.J., Nunes, X.P., Quintans-Junior, L.J.: Scientia Plena **6**, 11 (2010)
17. Ferraz. Rev. Neurociências. n.7, v.1, 06–12 (1999)
18. Bravo, P.A.F., Nassif, M.C.: Infarma **18**(9/10) (2006)
19. Saito, T.C.: Curso de Especialização em Saúde Coletiva e Saúde da Família do Centro Universitário Filadélfia – UniFil (2011)
20. Leal, L.K., Nobre Júnior, H.V., Cunha, G.M., Morae, S.M.O., Pessoa, C., Oliveira, R.A., Silveira, E.R., Canuto, K.M., Viana, G.: S. Neurosci. Lett **388**(2), 86–90 (2005)
21. Canuto, K.M., Silveira, E.R.: Quim. Nova **29**(6), 1241–1243 (2006)
22. Ingale, S.P., Sanjay, B.K.: Orient. Pharm. Exp. Med. (2014)
23. ACD/Structure Elucidator, version 12.01, Advanced Chemistry Development, Inc., Toronto, ON, Canada. www.acdlabs.com (2014)
24. Dennington, R., Keith, T., Millam, J.: GaussView, Version 5, Semichem Inc., Shawnee Mission KS (2009)
25. Frisch, M.J.G., Trucks, W., Schlegel, H.B., Scuseria, G.E., Robb, M.A., Cheeseman, J.R., Montgomery, J.A., Vreven, Jr. T., Kudin, K.N., Burant, J.C., Millam, J.M., Iyengar, S.S., TomasI, J., Barone, V., Mennucci, B., Cossi, M., Scalmani, G., Rega, N., Petersson, G.A., Nakatsuji, H., Hada, M., Ehara, M., Toyota, K., Fukuda, R., Hasegawa, J., Ishida, M., Nakajima, T., Honda, Y., Kitao, O., Nakai, H., Klene, M., Li, X., Knox, J.E., Hratchian, H.P., Cross, J.B., Adamo, C., Jaramillo, J., Gomperts, R., Stratmann, R.E., Yazyev, O., Austin, A.J., Cammi, R., Pomelli, C., Ochterski, J.W., Ayala, P.Y., Morokuma, K., Voth, G.A., Salvador, P., Dannenberg, J.J., Zakrzewski, V.G., Dapprich, S., Daniels, A.D., Strain, M.C., Farkas, O., Malick, D.K., Rabuck, A.D., Raghavachari, K., Foresman, J.B., Ortiz, J.V., Cui, Q., Baboul, A.G., Clifford, S., Cioslowski, J., Stefanov, B.B., Liu, G., Liashenko, A., Piskorz, P., KomaromI, I., Martin, R.L., Fox, D.J., Keith, T., AL-Laham, M.A., Peng, C.Y., Nanayakkara, A., Challacombe, M., Gill, P.M.W., Johnson, B., Chen, W., Wong, W., Gonzalez, C., Pople, J.A.: Gaussian 03, Revision A.1; Gaussian, Inc., Pittsburgh, PA (2003)
26. Chemplus: Modular Extensions for HyperChem Release 6.02, Molecular Modeling for Windows, HyperClub, Inc., Gainesville (2000)
27. Plukiger, P., Lutthi, H.P., Portmann, S., Webber, J.: MOLEKEL 4.1, Swiiss Center for Scientific Computing, Switzerland (2001)
28. Chattaraj, P.K., GirI, S.: J. Phys. Chem **111**(43), 11116–11121 (2007)
29. Padmanabhan, J., Parthasarathi, R., Subramanian, V., Chattaraj, P.K.: J. Phys. Chem. **111**(7), 1358–1361 (2007)
30. Silva, B.F.A., Arroio, A., Honório, M.: K. Quim. Nova **33**(3), 694–699 (2010)

31. Santos, C.B.R., Vieira, J.B., Lobato, C.C., Hage-Melim, L.I.S., Souto, R.N.P., Lima, C.S., Costa, E.V.M., Brasil, D.S.B., Macêdo, W.J.C., Carvalho, J.C.T.: Molecules **19**, 367–399 (2014)
32. Pirouette 3.01, Infometrix Inc., Seattle, WA (2001)
33. Statistica (Data Analysis Software System); Version 6.1, Statsoft Inc. (2004)
34. Politzer, P., Murray, J.S.: Theor. Chem. Acc. **108**, 134–142 (2002)
35. Gupta, S., Mohan, C.G.: Biomed. Res. Int. **21** (2014)
36. Verdonk, M.L., Cole, J.C., Hartshom, M.J., Murray, C.W., Taylor, R.D.: Proteins **52**(4), 609–623 (2003)
37. Yamashita, S., Furubayashi, T., Kataoka, M., Sakane, T., Sezaki, H.: Eur. J. Pharm. **10**, 195–204 (2000)
38. Silva, N.S.R., Santos, C.F., Gonçalves, L.K.S., Braga, F.S., Almeida, J.R., Lima, C.S., Brasil, D.S., Silva, C.H.T.P., Hage-Melim, L.I.S., Santos, C.B.R.: British J. Pharm. Res. **7**(4), 247–263 (2015)
39. Ames, R.E., Marwell, G.: Am. J. Sociol. **84**(6), 1335–1360 (1979)
40. Hage-Melim, L.I., Santos, C.B.R., Poiani, J.G., Vaidergom, M.M., Manzolliand, E.S.: Paula Da Silva. C. H. T. Cur. Bioac. Comp. **10**, 147–157 (2014)
41. Elshaflu, H., Todorović, T.R., Nikolić, M., Lolić, A., Višnjeva, A., Hagenow, S., Padrón, J.M., García-Sosa, A.T., Djordjević, I.S., Grubišić, S., Stark, H., Filipović, N.R.: Front Chem. **6**, 247 (2018)
42. Soldi, C.: Universidade Federal de Santa Catarina, Programa de Pós-graduação em Química.Thesis (2011)
43. Cunha, E.L., Santos, C.F., Braga, F.S., Costa, J.S., Silva, R.C., Favacho, H.A.S., Hage-Melim, L.I.S., Carvalho, J.C.T., da Silva, C.H.T.P., Santos, C.B.R.: J. Comput. Theor. Nanosci. **12**(10), 3682–4369 (2015)
44. Solomons, G.: Química Orgânica, 10ª Ed. LTC (2012)
45. Cole, J.C., Murray, C.W., Nissink, J.W.M., Taylor, R.D., Taylor, R.: Proteins: Struc. Func. Bioinf. **60**, 325–332 (2005)
46. Accelrys Discovery Studio. v. 4.0, T.C., San Diego, CA, USA (2007)
47. Binda, C., Aldeco, M., Geldenhuys, W.T., Tortorici, M., Mattevi, A., Edmondson, D.E.: Acs. Med. Chem. Lett **3**, 39–42 (2012)
48. Picanço, L.C.S., Casstro, L.L., Pinheiro, A.A., Silva, K.R., Souza, L.R., Braga, F.S., Silva, C.H.T.P., Santos, C.B.R., Hage-Melim, L.I.S.: British J. Pharm. Res. **7**(3), 152–175 (2015)
49. Rodrigues, J.A.R.: Quim. Nova **23**, 6 (2000)
50. Farahnaz, R.M., Jahan, B.G.: Curr. Neuropharmacol. **16**, 664–725 (2018)
51. Hagenow, J., Hagenow, S., Grau, K., Khanfar, M., Proschak, H., Stark, H.: Drug Design, Development and Therapy, vol. 14, pp. 371–393 (2020)
52. Dhiman, P., Malik, N., Khatkar, A.: BMC Chem. **14**(12) (2020)
53. Jiang, B., Meng, L., Zou, N., Wang, H., Li, S., Cheng, X., Wang, Z., Chen, W., Wang, C.: Phytomedicine **62**, 152–967 (2019)
54. Is, Y.S., Durdagi, S., Aksoydan, B., Yurtsever, M.: Pharmacokinetic and toxicity properties. ACS Chem. Neurosci. **9**, 1768–1782 (2018)

Advances Toward the Development of New Therapeutic Strategies Targeting Acetylcholinesterase and Its Remediation Processes

Alexandre A. de Castro, Letícia C. Assis, Giovanna C. Gajo, Teodorico C. Ramalho, and Felipe A. La Porta

Abstract Acetylcholinesterase (AChE) is a serine protease, responsible for finalising the transmission of nerve impulses at cholinergic synapses by hydrolysis of the acetylcholine (ACh) neurotransmitter. Therefore, from this perspective, it is well-known that the irreversible or prolonged inhibition of AChE elevates the synaptic Ach level, resulting in severe central and peripheral adverse effects that fall under the cholinergic syndrome spectra. Certain AChE inhibitors (AChEI) with reactivator effects stand out more specifically to combat the possible toxic effects, such as denominated oximes that are widely designed substances. Current investigations focus on searching for new and more effective broad-spectrum reactivators of the inhibited AChE (same in its aged form) against diverse organophosphorus agents. Thus, the objective of this chapter is to present a more complete understanding of new therapeutic strategies targeting AChE and its remediation processes. Through the bioremediation techniques employing degrading enzymes also show advances as a promising approach of degrading toxic organophosphorus compounds, that is, preventing the individuals from undergoing the toxic effects of the AChE inhibition. It is also important to mention that AChE is a significant therapeutic target for the treatment of certain disorders, particularly neurodegenerative diseases, such as Alzheimer's disease. The employment of nanotechnology and biosensors represents a promising alternative, with the potential to boost the forms of treatment and diagnosis.

A. A. de Castro · L. C. Assis · G. C. Gajo · T. C. Ramalho
Department of Chemistry, Federal University of Lavras, Lavras, Minas Gerais, CEP 37200-000, Brazil

T. C. Ramalho
Department of Chemistry, Faculty of Science, University of Hradec Kralove, Rokitanskeho 62, 500 03 Hradec Kralove, Czech Republic

F. A. La Porta (✉)
Laboratório de Nanotecnologia e Química Computacional, Universidade Tecnológica Federal do Paraná, Londrina, PR 86036-370, Brazil
e-mail: felipelaporta@utfpr.edu.br

F. A. La Porta and C. A. Taft (eds.), *Functional Properties of Advanced Engineering Materials and Biomolecules*, Engineering Materials,
https://doi.org/10.1007/978-3-030-62226-8_21

1 Introduction to Acetylcholinesterase

Anatomically, the cholinergic system corresponds to an automatic portion of the peripheral nervous system (PNS). It additionally is characterised by the presence of synapses, containing acetylcholine (ACh) as a neurotransmitter, which is responsible for the transmission of nerve stimuli from a neurone to another [1]. Cholinergic synapses could also be found in the central nervous system (CNS) and the ganglionic junctions of the adrenergic system, as well [2].

Cholinesterase synapses are present in cholinesterase, are part of a class of enzymes that catalyse ACh hydrolysis into acetate and choline in the synaptic cleft (Fig. 1), allowing the active cholinergic neurone to return into its resting state. There are usually two types of cholinesterases: butyrylcholinesterase (denoted as BuChE) and acetylcholinesterase (denoted as AChE) [1].

These enzymes differ from each other in tissue distribution, specificity towards their substrates, as well as in their kinetic properties, although they are evolutionarily similar, presenting amino acids with a homology of approximately 50%. For this reason, other amino acids that exhibit 50% heterogeneity show differences in the selectivity towards both the substrates and inhibitors of these enzymes [3]. Particularly, AChE is mostly present in the CNS, skeletal muscles, and erythrocyte membrane. It primarily catalyses ACh. BuChE is abundant in blood plasma and is less selective during catalysis, hydrolysing both ACh and butyrylcholine (BuCh) in comparable amounts [1].

In the synaptic cleft, e.g., AChE is a regulatory enzyme responsible for terminating the transmission of nerve impulses through the ACh hydrolysis, a simple molecule with an ester group and a quaternary amine [1, 4, 5]. In the presynaptic neurones, ACh is synthesised from choline and acetyl-coenzyme A (acetyl-CoA) through choline acetyltransferase catalysis. After its formation, it remains stored in vesicles and it is released into the synaptic cleft only when a neuronal stimulus occurs [5].

Upon its release, ACh binds to the postsynaptic receptor, transmitting the information. After propagating the message, the ACh molecule disconnects from the postsynaptic receptor. It then returns to the synaptic cleft, where it undergoes hydrolysis catalysed by AChE, giving rise to acetate and choline [1]. Yet, during the hydrolysis process, a tetrahedral intermediate of a choline ester carbonyl group is formed, allowing the development of transition analogue bio-esters (Fig. 2) [6].

AChE has three branches joined by disulphide bonds and is fixed to the cell membrane by collagen. Each branch is an enzymatic unit composed of four protein subunits, each with an active site. As such, this site consists of four domains. The first

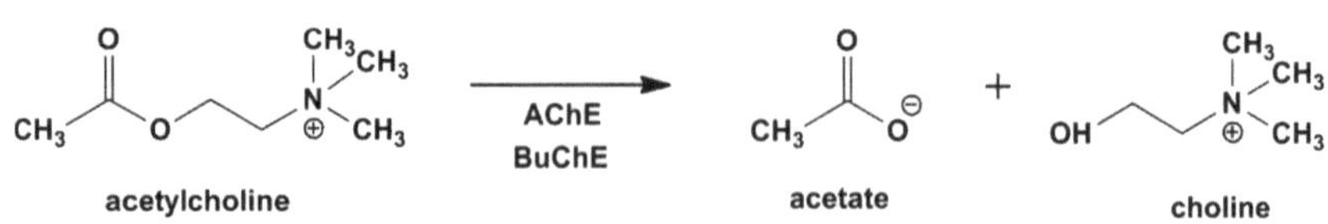

Fig. 1 ACh hydrolysis using cholinesterase, AChE and BuChE enzymes

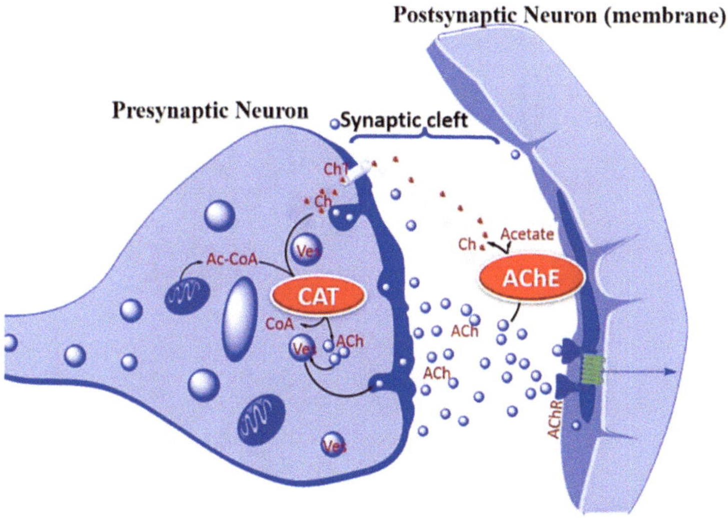

Fig. 2 Scheme of the transmission process and nervous control at synapses. Reproduced with permission from reference [7]

is formed by serine and histidine residues, parts of the catalytic triad (Ser203, Glu334, and His447), as shown in Fig. 3 [6, 8]. The second domain consists of an anionic bag containing the ACh quaternary ammonium group, which, in turn, interacts electrostatically with Glu334 [2, 9]. The third domain consists of a hydrophobic region that is extremely important for bonding with cyclic substrates. The fourth domain is responsible for the interaction with cationic and some other neutral ligands. Yet, this domain is distant from the active site and is thus called a peripheral anionic site, consisting of Asp74 and Trp286 residues as a common nucleus (Fig. 3) [10].

It is noteworthy that the occupation of ligands at the peripheral site often influences the active site conformation. The interactions of ligands with these residues may be the key to the AChE catalytic activity allosteric modulation [8, 12, 13]. Based on these emplacements, AChE is an important molecular target for a variety of bioactive molecules as it would be highly relevant to evaluate the potential of these molecules in inhibiting or reactivating AChE for the development of neurotoxic agents and pesticides, as well as for the research of new drugs against yet-uncurable diseases, such as Parkinson's disease (PD) [14] and Alzheimer's disease (AD) [2, 8].

2 AChE Inhibition Processes

Several drugs target the AChE enzyme, either through inhibition or reactivation. Most of these drugs are, in turn, directed to cholinergic synapses. A drug that inhibits AChE is generally referred to as anticholinesterase (AChEI) or an indirect cholinergic [15].

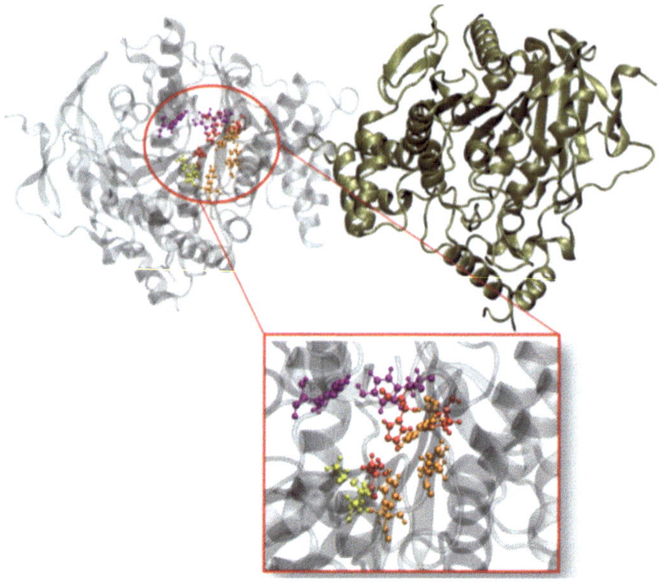

Fig. 3 Three-dimensional structure of human AChE (PDB code = 5HF9). The catalytic triad is highlighted in red (active site), the oxyanion hole in green and peripheral anionic site (PAS) in yellow. Reproduced with permission from reference [11]

AChEI prolongs the ACh duration and intensity in the synaptic cleft. Generally, the action mechanism of such inhibitors involves the competitive blocking of the AChE enzyme, which could in principle be reversible, irreversible, and also pseudo-irreversible [5, 8].

Based on this information, the irreversible AChEI agents are usually compounds that have a central pentavalent phosphorus atom in their structure, directly linked to fluorine groups or an organic group [16], as shown in Fig. 4. Spontaneously, these agents phosphorylate AChE, making poisoning by these agents quite dangerous. These AChEI pesticides display high solubility and hence are well absorbed by all respiratory, digestive, and dermal routes. After absorption, they are quickly distributed to various tissues and organs, targeting mainly the liver and the kidneys. In addition, many organophosphorus-based compounds (OPs) are highly lipophilic.

Fig. 4 General structure of organophosphorus compounds (OP)

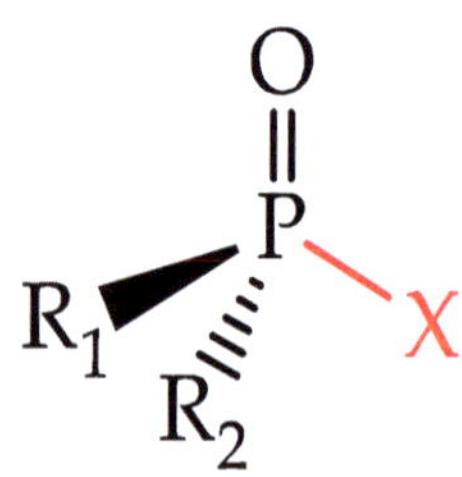

Consequently, they are deposited in the adipose tissue and after several days of exposure are gradually released. Certain studies focused on the enzymatic biodegradation of these compounds by using different degrading enzymes [17–23].

Other well-known AChE inhibitors are carbamates (Fig. 5) that, along with OP, represent a significant class of insecticides. These compounds inhibit the activity of the AChE and BuChE enzymes, resulting in the ACh accumulation and the uncontrolled activation of cholinergic synapses [15, 24]. As such, the influence of these manifestations is directly involved in the administration of the dose and exposure to these compounds. Also, unlike OPs that readily cross the blood–brain barrier (BBB), carbamates do not effectively penetrate the CNS. Under these circumstances, the use of carbamates results in less neurological toxicity and, in addition, these substances do not accumulate in the body [24].

Nervous or neurotoxic agents are the most lethal group among the OPs [25]. The most prominent examples are presented in Fig. 6. They have a chiral phosphorus

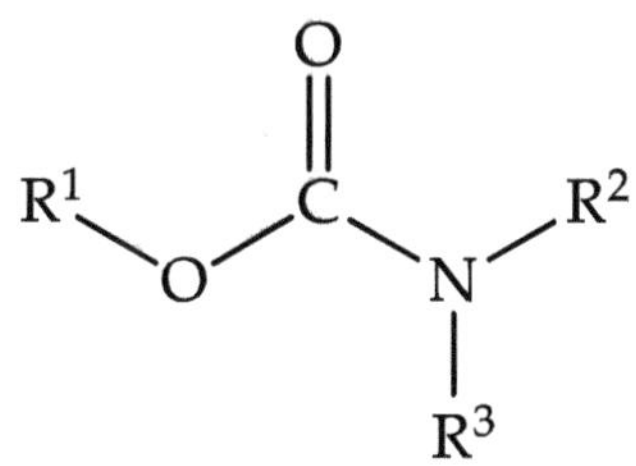

Fig. 5 General structure of carbamates

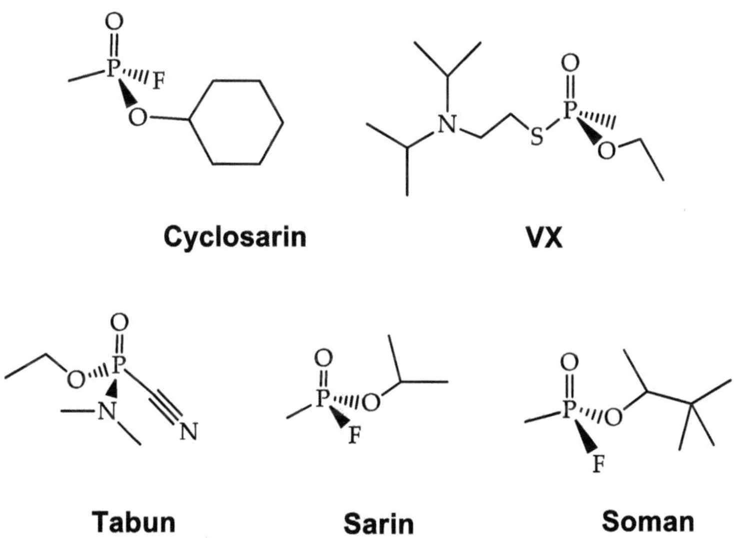

Fig. 6 Chemical structures of the main warfare nerve agents

atom that results in a pair of optical isomers in equal proportions. Among the neurotoxic drugs in Fig. 6, soman, with an optically active carbon in the pinacolyl group, generates another pair of enantiomers [26]. As much as this agent has two stereocenters [27], it is known that the dominant chiral centre is formed by the phosphorus atom since it determines the huge potential toxicity of this compound. Thus, from this perspective, the soman has been extensively studied in many theoretical and experimental investigations [28].

In addition to insecticides, chemical weapons represent an even more toxic OP class [8, 17–23]. These compounds were applied for the first time in World War II and later on a variety of compounds were developed that present high toxicity and risk [24]. Due to these properties, many chemical weapons have moved from a strictly military interest to that of terrorist organisations and represent a significant threat to national security [8, 24]. Currently, numerous synthetic pathways for various nerve agents have been reported, and huge stocks are still available. However, these chemical warfare agents have particularly been banned by the International Chemical Weapons Convention (CWC) [24]. These OP agents could potentially cause toxic damage to the nervous system, leading to the emergence of serious, life-threatening diseases due to their long-term chemical reactivity of human exposure [29, 30]. As such, the mechanisms of inhibition and ageing are shown in Fig. 7.

Figure 7 shows that the reaction is guided by the phosphorylation of the hydroxyl functional group [32, 33]. Where in the AChE active site, inhibition occurs through a biomolecular nucleophilic substitution reaction (SN2), resulting in the formation of a covalent bond between the serine oxygen atom (Ser203) and the central phosphorus atom of the OP [34]. AChE could be inhibited in an irreversible process known as ageing (Fig. 7), after a period that varies according to the neurotoxic agent, characterised by a dealkylation process [31]. Spontaneous enzyme reactivation could also occur, but at an almost negligible rate [35].

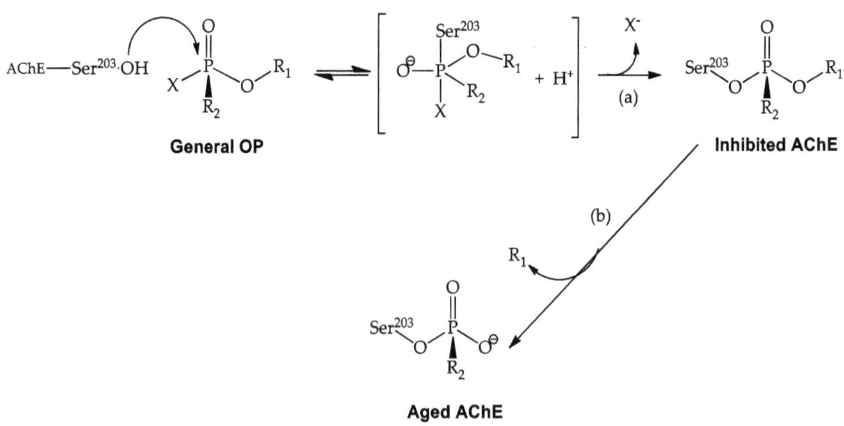

Fig. 7 General representation of the **a** inhibition and **b** ageing mechanisms. Reproduced with permission from reference [31]

OP-derived neurotoxic agents are compounds well-known for their high toxicity, lipophilicity, and volatility. Due to these properties, neurotoxic agents are rapidly and widely absorbed through the air and topical routes and could penetrate the BBB. Besides, the low cost and ease of manufacture favour the use of certain compounds of this class as agrochemicals [36]. Among the commercially available agrochemicals (Fig. 8), OP agents were the mostly preferred due to their broad-spectrum bioactivity and easy availability [24].

Therefore, in this perspective, the treatment adopted against neurotoxic poisoning is generally based on atropine, diazepam (Fig. 9), and oxime [37, 38]. This process will be discussed in the next topic. In addition to these AChE inhibitors, the use of bioactive compounds with therapeutic potential, such as AD, stands out in treating

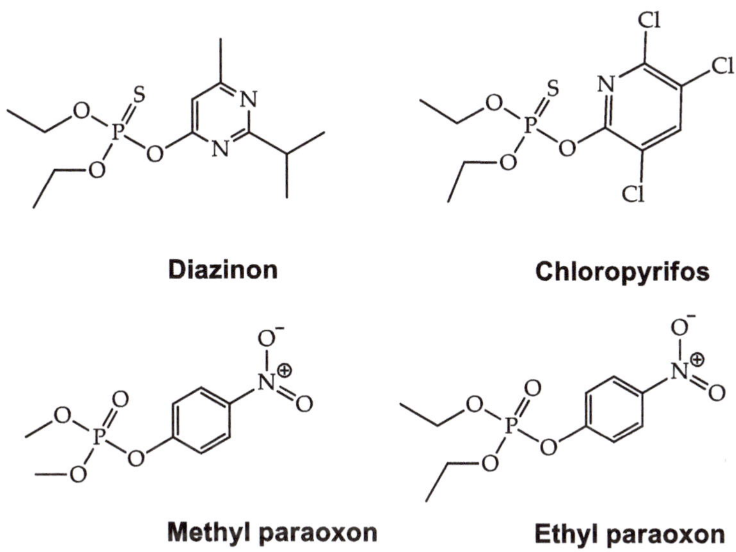

Fig. 8 Chemical structures of some important pesticides

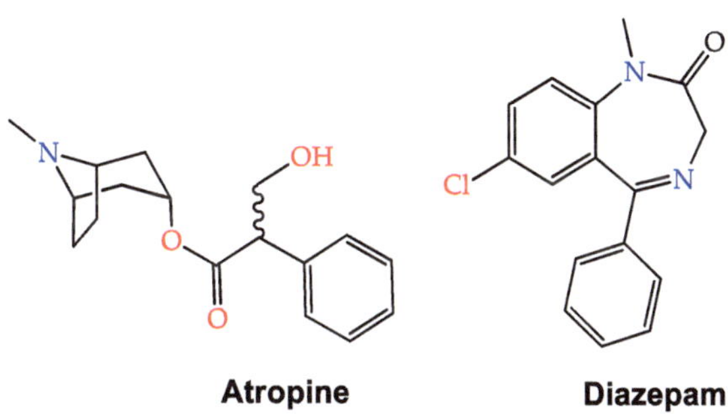

Fig. 9 Representation of the chemical structures of atropine and diazepam

certain diseases. This is likely due to the ability of this class of AChEI drugs to enhance cholinergic function, inducing improvement in the cognitive profile [1].

3 AChE Reactivation Processes

Concerning the treatment processes currently available for the remediation of the OP-induced intoxication, the well-known reactivators are the most important species. Reactivators are nucleophilic substances capable of breaking the newly formed covalent bonds between OP and AChE in the inhibition process [15]. Yet, due to the action of these reactivators, the catalytic activity of the enzyme could be restored. An antidote named pralidoxime, discovered in 1955, is known as the first reactivator to exhibit efficacy in the treatment of OP intoxication. Note that so far this drug is one of the most broadly applied reactivators [39–41]. Thus, the oxime-based compounds present characteristic structures, with one or two functional groups –C = N–OH at one end [39, 42], as shown in Fig. 10. Such compounds, e.g., are usually found in the form of a conjugate base (i.e., oximate ion) in biological environments [7].

In structural terms, the enzyme active site reportedly undergoes specific modifications depending on the type of the bound OP. This fact leads to the need for developing reactivators with the potential to bind to the modified enzymatic cavity [43, 44]. Regarding the reactivation mechanism, the two stages involved are as follows:

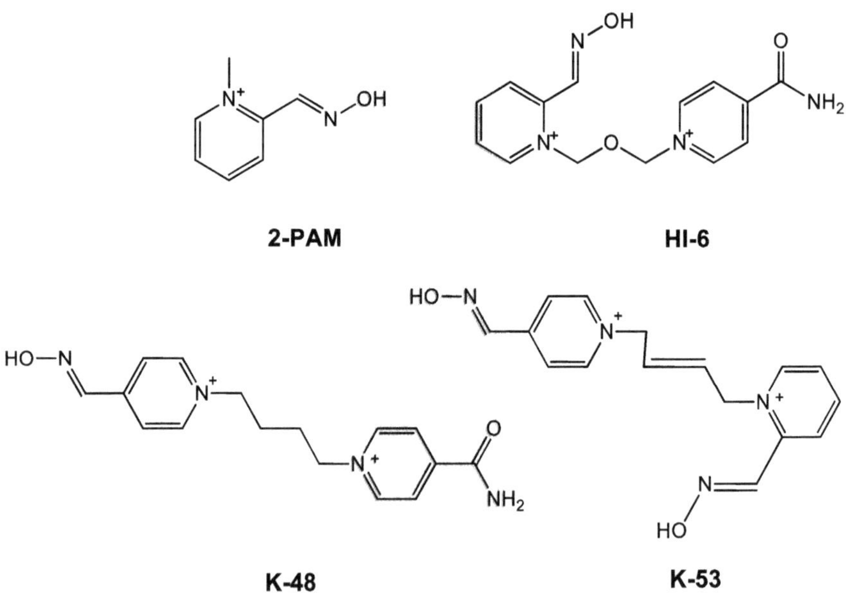

Fig. 10 Representation of the chemical structures of the oximes 2-PAM, HI-6, K-48 and K-53. Reproduced with permission from reference [26]

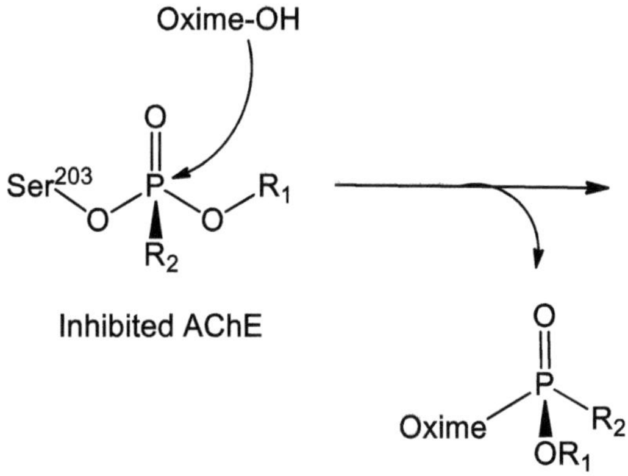

Fig. 11 General representation of the reactivation process of the inhibited AChE. Reproduced with permission from reference [26]

the approximation of the reactivator to the phosphoric centre of the AChE-OP adduct, giving rise to a pentacoordinate transition state, then the release of the OP-reactivator conjugate, leading to the restoration of the enzymatic function (Fig. 11) [45, 46].

The reactivators commonly have up to three or no pyridine rings. Generally, quaternary pyridine rings interact with the AChE peripheral anionic site (PAS). These interactions stabilise the reactivator in the cavity, thus giving rise to conformations and orientations that favour the nucleophilic attack. The reactivator efficiency could be influenced by factors such as the presence of quaternary nitrogen in the reactivator structure, the presence of the oxime functional group, the position of the oxime group on the pyridinium ring, as well as the number of nucleophilic groups in the reactivator structure. For bis-pyridinium oximes, the length and rigidity of the connection chain between them also influence the reactivator efficiency [47]. Some of these aspects were further investigated and more information could be found in Tables 1, 2 and 3.

As shown in Table 1, the length of the connecting chain between both pyridinium rings (bisquaternary pyridinium reactivators) is directly related to the antidote potency in reactivating nerve agent-inhibited AChE [47, 48]. For the *n*-methylene connecting chain, Table 1 shows that there is a clear dependence between the length of the connecting chain of the reactivator and the type of nerve agent present in the inhibitor-AChE adduct. According to these results, the ideal length of the connecting chain concerning the reactivation of Tabun or VX-inhibited AChE is 3 or 4 methylene groups (Table 1). However, this rule is satisfied only for *n*-methylene connecting chains. These trends change in the case of oxygen, Sulphur, or other structural fragments incorporated into the connecting chain [46, 47].

In the presence of electronegative atoms in the connecting chain, this might lead to different interaction profiles due to the presence of free electrons, allowing further interactions between this part of the reactivator and the enzyme active site. Another crucial structural factor influencing the reactivation process (potency) is the "rigidity" of the connecting chain. The as-prepared compounds showed a trend with a certain

Table 1 Reactivation potency of diverse bisquaternary pyridinium AChE reactivators (concentration of oxime 0.001 M). Reproduced with permission from reference [47]

		Reactivation potency [%]			
AChE Reactivator	Connecting Chain	Sarin	Cyclosarin	VX	Tabun
Trimedoxime	$CH_2CH_2CH_2$	7	0	85	41
K074	$CH_2CH_2CH_2CH_2$	9	0	72	46
TO 057	$CH_2CH_2SO_2CH_2CH_2$	8	8	26	8
TO 058	$CH_2CH_2S+(CH_3)CH_2CH_2$	5	0	46	9
Obidoxime	CH_2OCH_2	4	0	79	0
TO 052	CH_2COCH_2	2	3	71	8
K075	$CH_2CH = CHCH_2$	13	0	87	34

Table 2 Reactivation potency of several bisquaternary pyridinium AChE reactivators (concentration of oxime 0.001 M). Reproduced with permission from reference [47]

	Reactivation potency [%]		
Connecting Chain	**2#**	**3#**	4#
$CH_2CH_2CH_2$	19	16	79
$CH_2CH_2CH_2CH_2$	24	6	50
$CH_2CH{=}CHCH_2$	4	13	25
$CH_2C{\equiv}CCH_2$	0	0	0
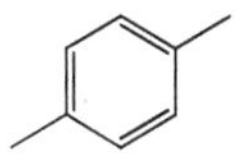	41	8	34
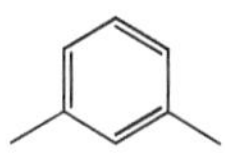	26	5	37
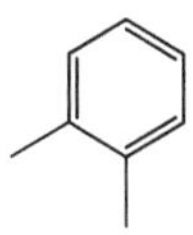	19	0	20

#Position of the oxime group at the pyridinium rings

level of rigidity in the linking chain. This aspect is essential as the rigidity of this chain might influence the spatial orientation of the pyridinium rings in the enzyme cavity, making it more limited. In this investigation, alkene, alkyne and xylene moieties were inserted into the connecting chain [47]. Table 2 shows that the incorporation of rigidity in employing but-2-ene or but-2-ine quickly reduces the reactivation potency of the AChE reactivators. As mentioned previously, another essential structural factor is related to the position of oxime group in the quaternary pyridinium ring. As shown in Table 2, relevant differences in the reactivation potency of the oximes regarding the oxime group position. In this case, the oxime group in the *para*-position showed the best reactivation results, based on biological tests [47].

However, on reactivators, the number of oxime groups in the reactive molecule is another key structural factor discussed in this chapter. The AChE reactivator might contain one to four oxime groups in its chemical structure at one, two, or all pyridinium rings. According to Table 3, note that the number of oxime groups does not enhance the reactivation potency. This trend might also be related to the increasing size of the reactivator.

The class of oximes has been intensely investigated over the past 50 years. It additionally is known that certain molecules are efficient against some specific OP and there are thus no broad-spectrum antidotes available in the market so far. Concerning structural terms, it is necessary to mention that quaternary oximes commonly possess

Table 3 Reactivation potency of different reactivators with different number of oxime groups at a different number of pyridinium rings (concentration of oxime 0.001 M)

Structure of AChE reactivator	Description	Reactivation potency (%)
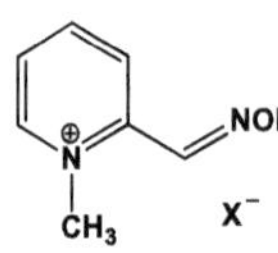 Pralidoxime	• One oxime group at one pyridinium ring • One pyridinium ring	4
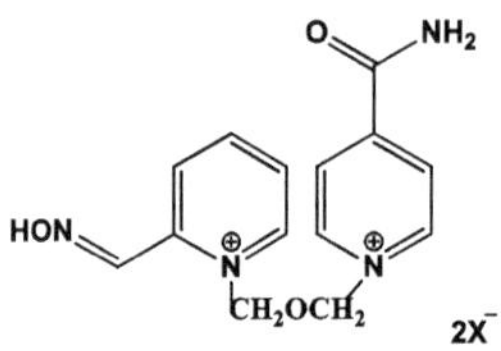 HI-6	• One oxime group at one pyridinium ring • Two pyridinium rings	6
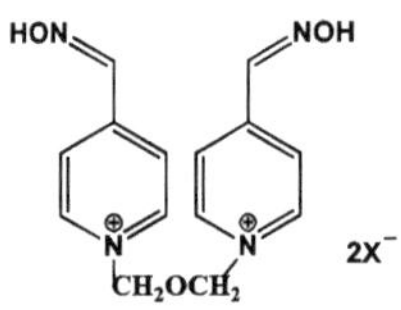 Obidoxime	• Two oxime groups at two pyridinium rings • Two pyridinium rings	37
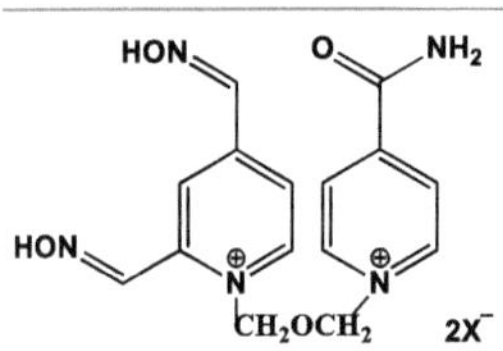 HLö-7	• Two oxime groups at one pyridinium ring • Two pyridinium rings	3
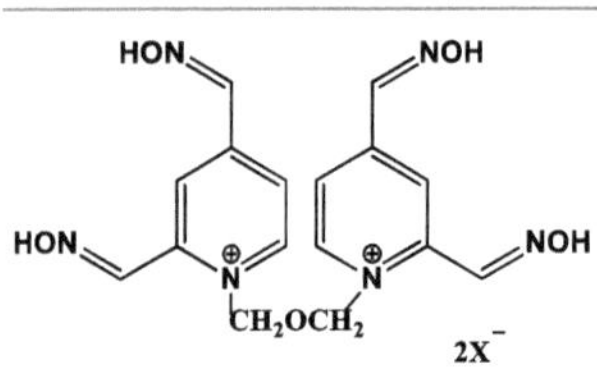 Tetroxime	• Four oxime groups at two pyridinium rings • Two pyridinium rings	1
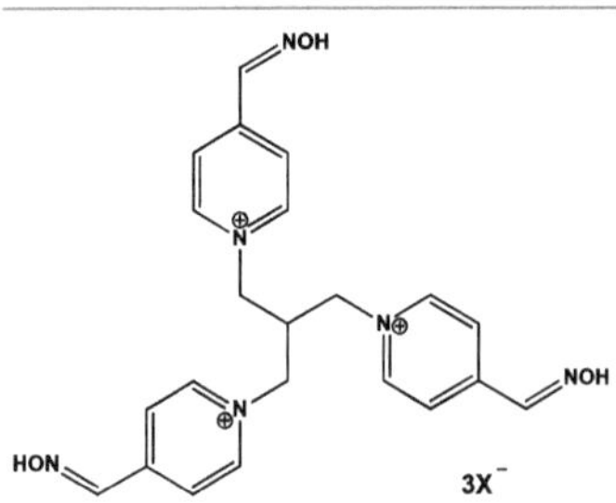 TO055	• Three oxime groups at three pyridinium rings • Three pyridinium rings	2

Reproduced with permission from reference [47]

a permanent positive charge, which interacts with PAS, thus enhancing the reactivation potency. However, this charge makes it difficult for the molecule to pass through the BBB. This trend results in a divergent remark between promising in vitro test results and unsatisfactory in vivo test results [26, 44].

Currently, diverse studies are ongoing, improving the efficiency of antidotes, and decreasing related side effects. For instance, Kuca et al. [49] have demonstrated satisfactory results in in vitro tests with aryl oximes and some neutral derivatives, by analysing the effect of substituents capable of improving the therapeutic performance of existing antidotes. In the same context, it was shown in an investigation with bisoxime (a three positive charge oxime) that even with its hydrophilicity, this reactivator was able to stabilise the sarin and VX-inhibited AChE cavity. Note that this happened through interactions with some amino acid residues, such as Tyr124 and Tyr337 [48]. However, as the three positive charges, this reactivator would not penetrate the BBB. In this context, this reactivator could be a good option for employment in BChE reactivation. BChE is an important biomacromolecule that could function as a bioscavenger for the detoxification of free nerve agents in the body [50].

Further studies showed the influence of the non-nucleophilic end position on isomeric oximes, revealing a significant difference in the inhibited AChE reactivation rate through each isomer [51]. More precisely, the change in position of the carbamoyl group led to differences in the reactivation rate for different enzyme-OP adducts. Theoretical studies have suggested that the structural changes between the isomeric oximes result in different interactions with the cavity residues [51, 52]. Another aspect approached by the same authors is the employment of oximes as pre-treatment prior to acute paraoxon intoxication. The study revealed that the investigated oximes, K-48 and K-53 (Fig. 10), can decrease the risk of death of individuals if applied preventively before pesticide intoxication. Therefore, this is a viable alternative as pretreatment to complement the available treatment protocol [53].

According to a range of studies in this area regarding the benefit of oximes for the treatment of OP poisoning, there is a concern regarding the balance between applied concentrations and efficiency, with the absence of tissue damage to the treated organisms. In this regard, Kuca et al. [48] investigated the effects of applying different concentrations of reactivators on certain living tissues. As found in the literature, the reactivators investigated at high dosages led to damage to both visceral tissues, and generally, to the individual's health. These studies were relevant to show the importance of adjusting the doses for different reactivator structures for newly developed drugs [54, 55]. The reactivation process through a non-oxime reactivator is shown in Fig. 12.

In the past decade, some patents were produced aiming at applications in the reactivation processes of the aged-AChE complex. The patent from Quinn and Topczewski [31] brings about molecules and methods that help combat the poisoning caused by toxic OP agents. In the case of oximes, if administered immediately after exposure, the reactivators could displace the bound OP. Thus, the serine residue could be released. On the other hand, however, if the administration is delayed, a process called ageing takes place through a solvolytic loss of an alkyl group from the AChE-OP adduct. The newly formed aged adduct is stabilised by various interactions with

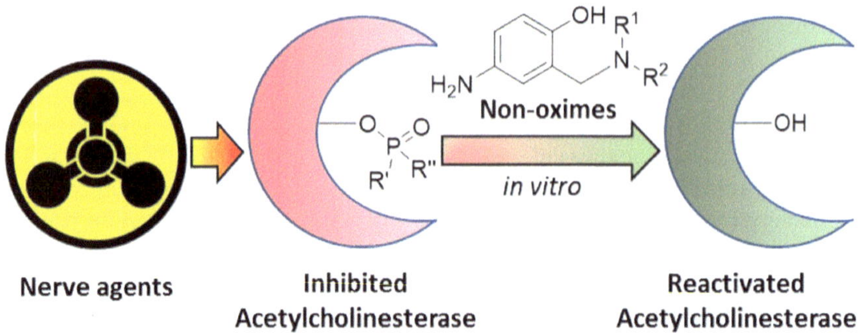

Fig. 12 Representation of the inhibition and reactivation processes of the AChE enzyme. Reproduced with permission from reference [56]

the active site, turning out to be incapable of undergoing oxime reactivation [57]. Unfortunately, there is a lack of antidotes with potential action against aged AChE-OP adducts, and scientists around the world are investing efforts and time to develop new potent antidotes [58].

The ageing process was considered irreversible, but scientists have tried to convert the phosphorylated oxyanion aged form to a re-alkylated form of AChE. According to Franjesevic et al. [44], the reactivation or re-alkylation processes of aged AChE are challenging. For this process to take place, a re-alkylating agent should bind selectively and efficiently into the AChE active site, thus allowing for reactivation of the re-alkylated serine [7, 43]. Thus, the aged state of AChE is a thermodynamically stable form of the enzyme. Therefore, the resurrection of the enzyme might presents some difficulties in surpassing the conformational changes as well as hydrogen-bonding networks formed within the active site of the aged adduct [44]. Consequently, this could reduce the strength of the intermolecular interactions, leading to the reactivity of the phosphorylated oxyanion. This makes the desired transition state for electrophilic re-alkylation. Figure 13 shows the reaction scheme of the resurrection process of the aged AChE, proposed by Quinn et al. [59].

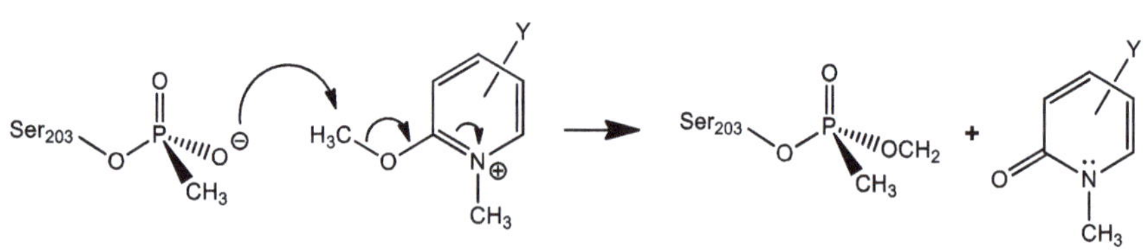

Fig. 13 Proposed resurrection process for aged AChE. Reproduced with permission from reference [59]

4 Enzymatic Biodegradation

Chemical weapons are a major concern around the world as they are relatively cheap, with the possibility of production on a large scale. These substances are also difficult to detect and control. Among the chemical warfare agents, OP constitutes an important class of molecules. As explained previously, these toxic agents affect the CNS and could lead to death. A range of works invests time to design novel efficient antidotes to treat the intoxication caused by them. The standard treatment makes use of an anticholinergic, which is usually administered in combination with a CNS depressor and an oxime. As observed in the previous sections, the oximes are important compounds whose function consists in the reactivation of the inhibited AChE enzyme, a regulatory enzyme responsible for the transmission of nerve impulses. Note that AChE is one of the molecular targets most vulnerable to neurotoxic agents [20]. Currently, enzymatic treatment is an increasingly promising alternative as a remediation method, called bioremediation. A variety of enzymes have been widely investigated for applications in OP degradation, such as phosphotriesterase (PTE) from bacterias and diisopropyl fluorophosphatase (DFPase) [20]. Hence, biomacromolecules have demonstrated significant performance in OP detoxification. Understanding the mechanisms by which these enzymes act is essential for the development of antidotes against warfare nerve agents and pesticides. The proposed mechanisms for PTE and DFPase during the hydrolysis process are shown in Fig. 14. Computational chemistry is important, so it is possible to investigate biological systems at the molecular level, obtaining diverse chemical parameters with reduced time and costs for the process. Molecular docking, molecular dynamics and quantum-mechanics/molecular-mechanics (QM/MM) are widely applied to study molecular interactions between ligands and proteins, as well as reaction mechanisms [20].

The enzymatic treatment has become a promising alternative as conventional methods of OP detoxification are expensive and require severe conditions. Enzymes are a class of biomolecules known for their diversity and ability to perform chemical reactions at high-speed rates [23, 60]. Research areas usually consist of enzyme catalysis, regarding protein stability, stereoselectivity and substrate preference [61].

5 AChE as a Therapeutic Target

Neurodegenerative diseases lead to dementia or locomotion problems, reaching the point of total disability [63]. These neurological disorders result from progressive degeneration and death of nerve cells, affecting the central and peripheral nervous system [64]. The genetic, environmental, and endogenous factors are the main processes that cause neurodegeneration and are amplified with increasing age. Thus, life expectancy increase means that more people could be affected [65].

These diseases share pathophysiological patterns, such as oxidative stress, neuroinflammation, toxic proteins, mitochondrial dysfunction, and neuronic death

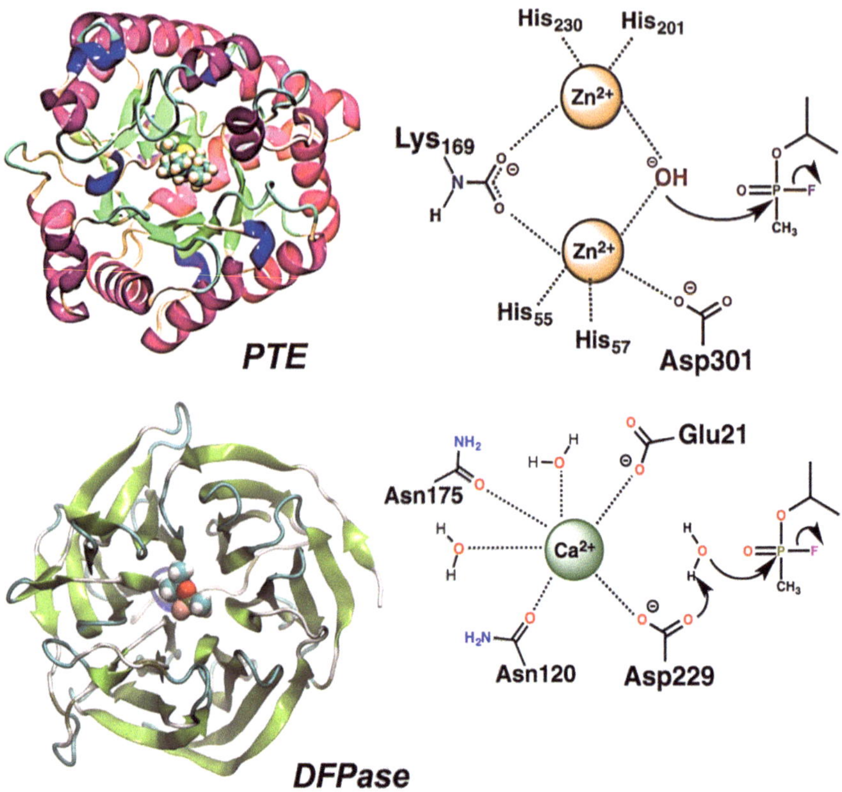

Fig. 14 Crystallographic structure of some degrading enzymes. Reproduced with permission from reference [62]

[66]. Because of this, they are considered to be multifaceted diseases with the involvement of multiple factors combined [67]. In each neurodegenerative disease, there is an abnormal replication followed by aggregation with the underlying protein. As such, this gradual accumulation of these aggregates in different regions of the brain explains the onset and characteristic symptoms of each disease [64].

The most common neurodegenerative diseases are Huntington's disease (HD), PD and AD [63]. Genetic factors are related to specific mutations or genetic polymorphisms such as the mutation in the amyloid precursor protein (APP) leading to AD, A53T mutation in α-synuclein leading to DP and polyQ expansions in the huntingtin protein leading to HD [64, 67]

HD causes loss of motor ability a cognitive, is hereditary, induced by the trinucleotide repeat expansion of the CAG codon in exon-1, the DNA segment, in the Huntingtin (HTT) gene on chromosome 4, which encodes the huntingtin protein (Htt), which plays an important role in neogenesis, transcriptional regulation, and synaptic connectivity [68]. Moreover, the deficiency of cholinergic pathways in the brain could lead to the clinical condition of cognitive decline and behavioural changes in patients with neurodegenerative disorders [69]. As stated earlier, cholinergic synapses are responsible for several human physiological functions. The enzyme AChE hydrolyses ACh. When AChE is inhibited, cholinergic transmission increases due to the presence of ACh in the synaptic cleft [70].

PD causes degenerative changes in several regions of the nervous system, such as the brainstem nuclei, olfactory bulb, hippocampus amygdala, and neocortex, in addition to triggering the progressive loss of dopaminergic neurones in the substantia nigra pars compacta (SNc) [71]. PD affects an individual's motor control, progressively eliminating it. Furthermore, it causes characteristic symptoms, such as motor slowness (bradykinesia); stiffness between the wrist, elbow, shoulder, thigh and ankle joints, rest tremors and imbalance; as well as decreased smell; or intestinal and sleep changes [72]. PD patients with dementia have a significant cholinergic loss with deficient cholinergic neurone terminals affecting various brain regions [67].

Despite extensive research, there is still no definitive treatment for PD. Most used drugs act in a dopaminomimetic manner, as precursors of dopamine (such as levodopa). The chemical structures of the drugs that inhibit the enzyme AChE, such as Biperiden, Methixene, and Triexiphenidyl are shown in Fig. 15.

According to Fig. 15, these central anticholinergic compounds exert their effects by blocking the central cholinergic transmission of ACh. As a result, the level of

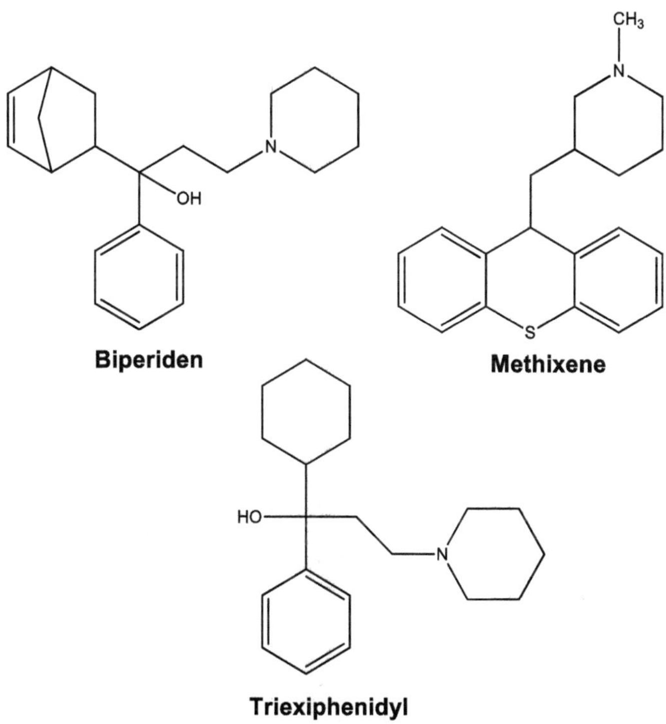

Fig. 15 Anticholinesterase compounds against Parkinson's disease

cerebral ACh decreases, restoring balance with dopamine. Table 4 lists the main anti-parkinsonian drugs [73].

Unfortunately, treatment with these drugs is palliative, more research would thus be needed to identify new therapies delaying or, ideally, stopping dopaminergic neurone degeneration. Another degenerative disease with AChE therapeutic target is AD, one of the most significant known dementia-causing neurodegenerative disease [73].

AD is characterised as a process of progressive and irreversible decline in cognitive functions, AD is the most common form of dementia and extends to disorganised behaviours and psychotic symptoms [74]. The clinical condition begins with the loss of recent memory, evolves damage in the nerve cells of the CNS, causing language

Table 4 Antiparkinsonian drugs

First line	
Compounds	Action
Levodopa	Converted to dopamine by the enzyme aromatic L-amino acid decarboxylase, being an effective drug in PD treatment, since it crosses the blood–brain barrier
Bromocriptine, Lissuride, Pergolide and Paramipexol	They are dopamine (D-2) receptor agonist agents. They can be used alone or in combination with levodopa
Cardidopa	Administered in association with levodopa, which blocks the systemic aromatic L-amino acid decarboxylase, allowing a higher concentration of levodopa to reach the brain using a much lower dose than administered alone
Second line	
Biperiden, Metixene and Triexifenidil	Blocks AChE
Third line	
Benserazide, Entacapona and Tolcapona	It inhibits the enzyme catechol-o-methyl transferase (COMT), responsible for the o-methylation of levodopa. When combined with L-DOPA they can increase the effect of each dose, resulting in a reduction in the administration of cardidopa/levodopa
Selegiline	It slows down the dopamine degradation by irreversible inhibition of the MAO type B enzyme, in addition, it inhibits the reuptake of dopamine from the synaptic space
Amantadina	It is an antiviral agent that is prescribed against PD. It increases the release of dopamine from the nerve endings and inhibits dopamine reuptake

Reproduced with permission from reference [73]

disorders, confusion, delusions, and in a critical phase, causing changes in muscle tone leading to a total loss of movement [64].

Disease progression is associated with a cholinergic synaptic reduction in the cortex and hippocampus, since ACh is a neurotransmitter associated with learning, memory, and cognition [75]. Cognitive deficits and dementia are caused by low ACh levels and extracellular amyloid-beta peptide deposits, and TAU protein hyperphosphorylation that accumulates in the cytoplasm, associated with the loss of synapses and neurones [76]. Patients with AD exhibit severe ACh defects at the cortical level, an early diagnosis could thus help minimising the devastating effects of such diseases.

The diagnosis of AD is made primarily through pre-established clinical criteria along with the exclusion of other possible causes of dementia [77, 78]. The exclusion of other causes was done through a set of clinical tests, laboratory tests and cerebral neuroimaging. In the clinical examination, the patient's previous history must be addressed, such as pre-existing diseases, trauma, surgery, alcohol use, medication use, among other factors that could cause cognitive impairment and even the dementia syndrome itself [79].

The physical examination aims to identify focal neurological deficits, such as pairs and paresthesias, signs of hydrocephaly, such as changes in gait and urinary incontinence, changes in motor skills, slowing and tremors, and signs of hypothyroidism, among other changes consistent with the most common differential diagnoses with AD dementia [79]. Among the differential diagnoses, severe depression found in up to 12% of patients with dementia stands out [80], vitamin B12 deficiency, common among the elderly, and hypothyroidism [81]. In addition, the disease is so costly that the condition of dementia has a variable evolution, moving to a vegetative state within 10 to 15 years from the onset of symptoms [82].

The current treatment for AD is only palliative, aiming at improving cognition, delaying the evolution of the disease and mitigating symptoms and behavioural changes [6]. There are two drug categories used for treating patients with dementia: AChEI drugs, targeting Ach replacement, as memory mediator; and anti-glutamatergic drug, targerting calcium overload reduction. Both directly interfere with learning and memory. Moreover, they might show good results when used in combination, or even separated [83].

Currently, only four drugs are approved by the US Food and Drug Administration (FDA) and commercially available for the treatment of AD: tacrine (Cognex®, 1), galantamine (Reminyl®, 2), rivastigmine (Exelon®, 3), memantine (Namenda®, 4) and donepezil (Aricept®, 5) (Fig. 16). These drugs aim at delaying or ameliorating the cholinergic deficit by partially inhibiting AChE activity [6, 8].

Tacrine, an aminoacridine, is a non-classical cholinesterase inhibitor that binds to AChE or BuChE. It was the first drug approved by the FDA for palliative AD treatment, exhibiting an improvement in the cognitive performance [5, 84, 85], but due to its hepatotoxicity, its use has significantly decreased since the recent development of safer AChEIs [86]. This compound has a quinolinic ring that performs π-stacking interactions with the Trp84 residue of the enzyme and a protonated nitrogen atom

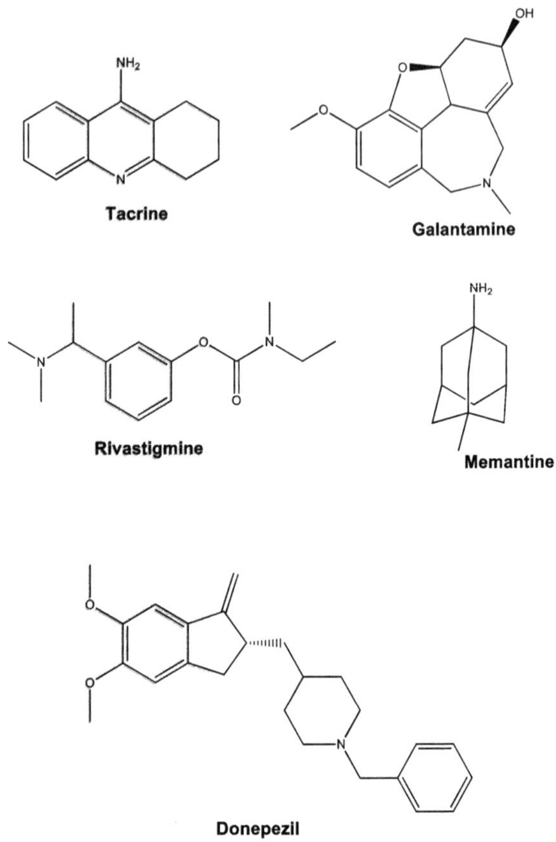

Fig. 16 Chemical structure of drugs approved for the treatment of Alzheimer's: tacrine, galantamine, rivastigmine), memantine and donepezil

that forms hydrogen bonds with His440 of the Catalytic Triad. In addition, the quinolinic ring is disposed in parallel and contact with the group phenyl residue Phe330, establishing strong interactions with the AChE active site.

Galantamine is an alkaloid that acts as a nicotinic receptor agonist and has been widely studied as a therapeutic target for the development of new candidates for AChEI drugs. This natural compound is a reversible AChE inhibitor [6]. As it is a tertiary amine and could cross the blood–brain barrier, it is indicated for the treatment of mild to moderate AD and dementia. Galantamine differs from other cholinesterase inhibitors as it binds allosterically to nicotinic receptors, resulting in a double-cholinergic action [87, 88]. As for its behaviour at the AChE active site, galantamine through its hydroxyl group, performs hydrogen bonding with Glu199 and a second interaction is possible between the hydrogen of Ser200 and the oxygen-methoxy group of Galantamine. In addition, a bifurcated link could also be formed with His440 of the catalytic triad [89, 90]. Rivastigmine is a pseudo-irreversible carbamate-type compound that inhibits AChE, contributing to an Ach increase. As such, the carbamyl and NAP groups of the rivastigmine interact with the amino acids Trp84 and Phe330 [90, 91].

Memantine is an N-methyl-D-aspartate (NMDA) type glutamate receptor antagonist. This FDA-approved inhibits the excessive influx of calcium (Ca^{2+}), reported in the case of patients in moderate and advanced stages of AD. Memantine administration improves communication skills, dressing and bathing [92]. Its adverse effects are mainly related to confusion, nephrotoxicity, constipation, migraines, and other effects such as high blood pressure, drowsiness and visual hallucinations [93]. Despite decreasing the cytotoxic activity of excessive glutamate release at neuronic endings, memantine has no effective action in paralysing or curing AD [92].

The donepezil compound is a reversible, non-competitive AChE inhibitor, less toxic than tacrine, more selective for AChE than for BuChE as it interacts with the Trp279 and Phe330 residues, absent amino acids in BuChE. This drug increases the availability of intra-synaptic ACh, with few side effects. Donepezil shows good blood–brain barrier transposition, and due to its pharmacokinetic profile and low toxicity, it has been used as a drug of the first choice in the treatment of patients with AD [6, 94]. Donepezil interacts with the active site of the enzyme from the benzyl, piperidine nitrogen, and methoxyindane groups. Performing π-stacking interactions with aromatic residues, as well as hydrophobic interactions with the amino acids Trp84, Phe330, and Trp279. This drug does not interact directly with the catalytic triad (Ser200, His440, and Glu327) [89, 90].

According to the complex and multifactorial nature of AD, where a set of biochemical events and several receptors are involved, it is necessary to search for new strategies allowing a higher planning and prospection efficiency of drug candidates [95–97]. In this context, nanomaterial-based technologies and biosensors are promising strategies for enhancing detection, drug delivery and release at the target sites.

6 Nanotechnology Applications in Remediation Processes

As such, the BBB is an efficient against exogenous substances that maintain the chemical composition of the neuronic "milieu" for the appropriate functioning of neuronic circuits and synaptic communication. This barrier is the most important interface between the blood and the brain. The most significant aspect limiting the design of new drugs for acting on the CNS is the BBB permeability. In general, various medications do not cross the BBB [98]. Over the past decade, researchers have tried to deal with this fundamental problem by developing different strategies that could facilitate drug passage through the BBB. In this context, we could cite nanotechnology-based therapeutic strategies, which have gained huge importance as they could overcome the limitations inherent to BBB passage. Among these, diverse kinds of lipidic, polymeric, inorganic, and other types of nanoparticles (NPs) for regulated drug delivery and release have been developed [99–102]. Some of these strategies are shown in Fig. 17.

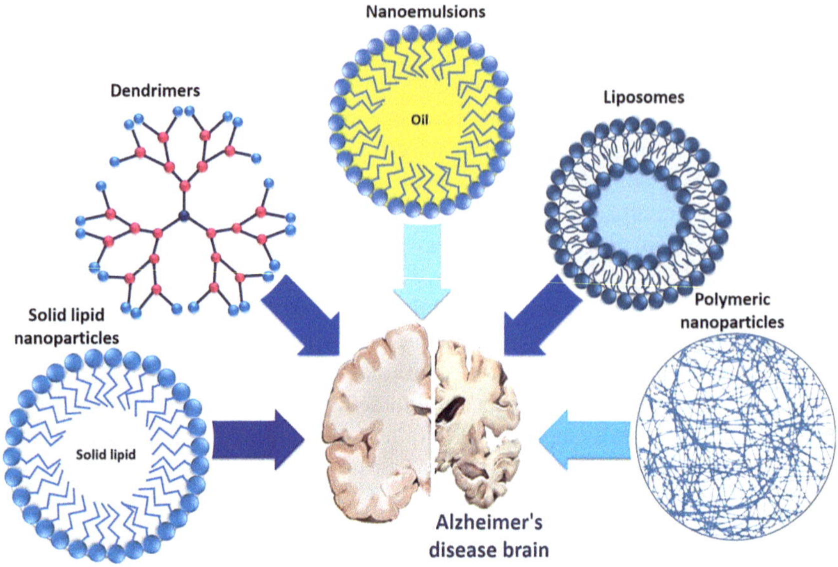

Fig. 17 Strategies of nanotechnology application in AD pathology. Reproduced with permission from reference [103]

It is noteworthy that the rise of nanotechnology could potentially provide a solution to overcome the challenges and difficulties encountered in AD diagnosis and neurotherapeutic stages. Nanotechnology uses engineered materials or tools, which have a functional organisation on a surprising nanometre scale (1–100 nm) and could interact with biological systems at the molecular level. Nanoparticles could penetrate the BBB in case of both in vitro and in vivo models [104–107]. Therefore, nanotechnology could be applied to design diagnostic devices and nano-enabled delivery systems that might penetrate the BBB, thus making it easier to use traditional and novel neurotherapeutic interventions, such as drug therapy, gene therapy, and tissue regeneration [108]. Today, nanotechnology is being applied in various aspects, such as the detection of biomarkers, molecular diagnostics, drug discovery, or drug delivery, and is clearly of importance in AD therapy [102, 109].

The majority of nanotechnology-based drug delivery systems for the treatment of AD and other neurodegenerative diseases, such as PD, are in the form of polymeric nanoparticles. Polymeric nanoparticles are important for the remediation of AD and PD due to their ability to cross the BBB [109]. There are some interesting features of these nanosystems in targeted CNS drug delivery. Thus, it is possible to cite their chemical properties that could be potentially altered to reach organ-, tissue-, or cell-specific and selective drug delivery. This is possible by disguising the physicochemical features and thus enhancing therapeutic agent delivery across the BBB. Finally, there is the protection of incorporated drugs against enzymatic degradation, and equally important, with fewer side effects [101, 109]. Nanotechnology could also be the basis of novel techniques that aim to detect AD and assist in the treatment stage [102, 110].

7 Biosensors in the AChE Detection

Currently, biosensors are being widely used in several areas, in the food and beverage sector, in energy matrices, as well as in the development of pesticides and chemicals, among others [111, 112]. Particularly, the biosensors could provide quantitative or semi-quantitative analytical information. For this, they using a biological recognition element (for example, enzymes, controls, proteins, tissues, microorganisms) in contact with medical or chemical transducers, which can be electrochemical, optical, thermal, and others. In this perspective, enzymes are the most used biological substances in the development of several biosensors, and could usually be found in their purified form or present in targeting microorganisms or in parts of the animal and plant tissue [108–111], as shown in Fig. 18.

Khaled Elmory et al. [110] developed a simple potentiometric biosensor to detect drugs for the treatment of AD, using donepezil and galantamine. As a result, the proposed procedures were successfully applied in the determination of drugs in pharmaceutical formulations and biological samples with sensitivity and precision comparable to the official method. The edge shown can be suggested to test the toxicity of the pharmaceutical preparation against cholinesterase enzymes in vitro [113]. Liu et al. [114] proposed a biosensor highly sensitive to the determination of pesticides based on tin disulfide (SnS_2) and chitosan (CHIT) decorated with a fly exclusive AChE British household. As such, the biosensor had an ultra-high sensitivity and a wide linear detection range from 0.02 nM to 20,000 nM with a detection limit of about 0.02 nM for the detection of chlorpyrifos as a pesticide model.

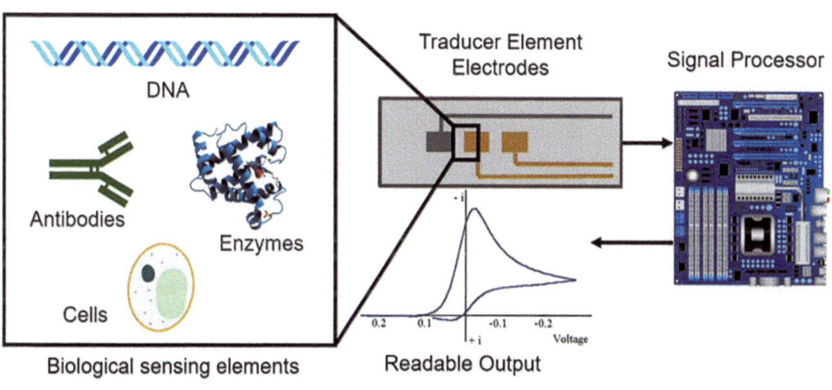

Fig. 18 Scheme of the main components of a sensor. Reproduced with permission from reference [115]

In addition, AChE decorated CHIT/SnS_2 exhibited acceptable storage stability, good reproducibility, and selectivity [114].

References

1. Araújo, C.R.M., Santos, V.L.A., Gonsalves, A.A.: Acetilcolinesterase—AChE: Uma Enzima de Interesse Farmacológico. Rev. Virtual Quim. **8**, 1818–1834 (2016)
2. Quinn, D.M.: Acetylcholinesterase: enzyme structure, reaction dynamics, and virtual transition states. Chem. Rev. **87**, 955–979 (1987)
3. Pezzementi, L., Nachon, F., Chatonnet, A.: Evolution of acetylcholinesterase and butyrylcholinesterase in the vertebrates: an atypical butyrylcholinesterase from the medaka oryzias latipes. PLoS One **6** (2011)
4. Doytchinova, I., Atanasova, M., Valkova, I., Stavrakov, G., Philipova, I., Zhivkova, Z., Zheleva-Dimitrova, D., Konstantinov, S., Dimitrov, I.: Novel hits for acetylcholinesterase inhibition derived by docking-based screening on ZINC database. J. Enzyme Inhib. Med. Chem. **33**, 768–776 (2018)
5. Petronilho. C.E., P.C.A.V.F.D.J.: Acetilcolinesterase: Alzheimer e guerra química. Rev. Mil. Ciência e Tecnol. **28**, 3–14 (2011)
6. Dias, K.S.T., De Paula, C.T., Riquiel, M.M., Do Lago, S.T.L., Costa, K.C.M., Vaz, S.M., Machado, R.P., De Lima, L.M.S., Viegas, C.: Aplicações recentes da abordagem de fármacos multialvo para o tratamento da doença de Alzheimer. Rev. Virtual Quim. **7**, 609–648 (2015)
7. Habtemariam, S.: Natural products in Alzheimer's disease therapy: would old therapeutic approaches fix the broken promise of modern medicines? Molecules **24**, 1519 (2019)
8. De Giacoppo, J.O.S., De Lima, W.E.A., Kuca, K., Da Cunha, E.F.F., França, T.C.C., De Ramalho, T.C.: Guerra química: Perspectivas no estudo de reativadores da enzima acetilcolinesterase inibida por organofosforados. Rev. Virtual Quim. **6**, 653–670 (2014)
9. Giacoppo, J.O.S., Mancini, D.T., Guimarães, A.P., Gonçalves, A.S., Da Cunha, E.F.F., França, T.C.C., Ramalho, T.C.: Molecular modeling toward selective inhibitors of dihydrofolate reductase from the biological warfare agent Bacillus anthracis. Eur. J. Med. Chem. **91**, 63–71 (2015)
10. Dvir, H., Silman, I., Harel, M., Rosenberry, T.L., Sussman, J.L.: Acetylcholinesterase: from 3D structure to function. Chem. Biol. Interact. **187**, 10–22 (2010)
11. Franklin, M.C., Rudolph, M.J., Ginter, C., Cassidy, M.S., Cheung, J.: Structures of paraoxon-inhibited human acetylcholinesterase reveal perturbations of the acyl loop and the dimer interface. Proteins **84**, 1246–1256 (2016)
12. Patočka, J., Cabal, J., Kuča, K., Jun, D.: Oxime reactivation of acetylcholinesterase inhibited by toxic phosphorus esters: in vitro kinetics and thermodynamics. J. Appl. Biomed. **3**, 91–99 (2005)
13. Darvesh, S.: Butyrylcholinesterase as a Diagnostic and Therapeutic Target for Alzheimers Disease. Curr. Alzheimer Res. **13** (2016)
14. de Azevedo, L.L., Cardoso, F.: Ação da levodopa e sua influência na voz e na fala de indivíduos com doença de Parkinson. Rev. da Soc. Bras. Fonoaudiol. **14**, 136–141 (2009)
15. Colovic, M.B., Krstic, D.Z., Lazarevic-Pasti, T.D., Bondzic, A.M., Vasic, V.M.: Acetylcholinesterase Inhibitors: pharmacology and toxicology. Curr. Neuropharmacol. **11**, 315–335 (2013)
16. Castro, A.A. de, Prandi, I.G., Kuca, K., Ramalho, T.C.: Enzimas degradantes de organofosforados: Base molecular e perspectivas para biorremediaÃ§Ã\poundso enzimÃ¡tica de agroquÃ\-micos. CiÃ\textordfemeninencia e Agrotecnologia **41**, 471–482 (2017)
17. Pereira, A.F., de Castro, A.A., Soares, F.V., Soares Leal, D.H., da Cunha, E.F.F., Mancini, D.T., Ramalho, T.C.: Development of technologies applied to the biodegradation of warfare nerve

agents: Theoretical evidence for asymmetric homogeneous catalysis. Chem. Biol. Interact. **308**, 323–331 (2019)
18. de Castro, A.A., Soares, F.V., Pereira, A.F., Silva, T.C., Silva, D.R., Mancini, D.T., Caetano, M.S., da Cunha, E.F.F., Ramalho, T.C.: Asymmetric biodegradation of the nerve agents Sarin and VX by human dUTPase: chemometrics, molecular docking and hybrid QM/MM calculations. J. Biomol. Struct. Dyn. **37**, 2154–2164 (2019)
19. Soares, F.V., de Castro, A.A., Pereira, A.F., Leal, D.H.S., Mancini, D.T., Krejcar, O., Ramalho, T.C., da Cunha, E.F.F., Kuca, K.: Theoretical studies applied to the evaluation of the DFPase bioremediation potential against chemical warfare agents intoxication. Int. J. Mol. Sci. **19** (2018)
20. de Castro, A.A., Assis, L.C., Silva, D.R., Corrêa, S., Assis, T.M., Gajo, G.C., Soares, F.V., Ramalho, T.C.: Computational enzymology for degradation of chemical warfare agents: promising technologies for remediation processes. AIMS Microbiol. **3**, 108–135 (2017)
21. Sartorelli, J., de Castro, A.A., Ramalho, T.C., Giacoppo, J.O.S., Mancini, D.T., Caetano, M.S., da Cunha, E.F.F.: Asymmetric biocatalysis of the nerve agent VX by human serum paraoxonase 1: molecular docking and reaction mechanism calculations. Med. Chem. Res. **25**, 2521–2533 (2016)
22. de Castro, A.A., Caetano, M.S., Silva, T.C., Mancini, D.T., Rocha, E.P., da Cunha, E.F.F., Ramalho, T.C.: Molecular docking, metal substitution and hydrolysis reaction of chiral substrates of phosphotriesterase. Comb. Chem. High Throughput Screen. **19**, 334–344 (2016)
23. Ramalho, T.C., de Castro, A.A., Silva, D.R., Silva, M.C., Franca, T.C.C., Bennion, B.J., Kuca, K.: Computational enzymology and organophosphorus degrading enzymes: promising approaches toward remediation technologies of warfare agents and pesticides. Curr. Med. Chem. **23**, 1041–1061 (2016)
24. Sharma, R., Gupta, B., Singh, N., Acharya, J.R., Musilek, K., Kuca, K., Ghosh, K.: Development and structural modifications of cholinesterase reactivators against chemical warfare agents in last decade: a review. Mini-Reviews Med. Chem. **15**, 58–72 (2014)
25. De Giacoppo, J.O.S., De Lima, W.E. a; Kuca, K., Da Cunha, E.F.F., França, T.C.C., De Ramalho, T.C.: Chemical warfare: perspectives on reactivating the enzyme acetylcholinesterase inhibited by organophosphates. Rev. Virtual Quim. **6**, 653–670 (2014)
26. de Castro, A.A., Assis, L.C., Soares, F. V; Kuca, K., Polisel, D.A., da Cunha, E.F.F., Ramalho, T.C.: Trends in the recent patent literature on cholinesterase reactivators (2016–2019). Biomolecules **10** (2020)
27. Benschop, H.P., De Jong, L.P.A.: Nerve agent stereoisomers: analysis, isolation, and toxicology. Acc. Chem. Res. **21**, 368–374 (1988)
28. Melzer, M., Chen, J.C.H., Heidenreich, A., Gäb, J., Koller, M., Kehe, K., Blum, M.M.: Reversed enantioselectivity of diisopropyl fluorophosphatase against organophosphorus nerve agents by rational design. J. Am. Chem. Soc. **131**, 17226–17232 (2009)
29. Marimuthu, P., Lee, Y.-J., Kim, B., Seo, S.S.: In silico approaches to evaluate the molecular properties of organophosphate compounds to inhibit acetylcholinesterase activity in housefly. J. Biomol. Struct. Dyn. **37**, 307–320 (2019)
30. Wong, P.T., Bhattacharjee, S., Cannon, J., Tang, S., Yang, K., Bowden, S., Varnau, V., O'Konek, J.J., Choi, S.K.: Reactivity and mechanism of α-nucleophile scaffolds as catalytic organophosphate scavengers. Org. Biomol. Chem. **17**, 3951–3963 (2019)
31. Quinn, D.M., Topczewski, J.J.: Compounds and methods to treat organophosphorus poisoning, 1–10 (2016)
32. Da Petronilho, E.C., Figueroa-Villar, J.D.: Agents for defense against chemical warfare: reactivators of the inhibited acetylcholinesterase with organophosphorus neurotoxic compounds. Rev. Virtual Quim. **6**, 671–686 (2014)
33. Kim, K., Tsay, O.G., Atwood, D.A., Churchill, D.G.: Destruction and detection of chemical warfare agents. Chem. Rev. **111**, 5345–5403 (2011)
34. Ordentlich, A., Barak, D., Sod-Moriah, G., Kaplan, D., Mizrahi, D., Segall, Y., Kronman, C., Karton, Y., Lazar, A., Marcus, D., et al.: Stereoselectivity toward VX is determined by interactions with residues of the acyl pocket as well as of the peripheral anionic site of AChE. Biochemistry **43**, 11255–11265 (2004)

35. Alvim, R.S., Vaiss, V.S., Leitão, A.A., Borges, I.: Theoretical chemistry at the service of the chemical defense: degradation of nerve agents in magnesium oxide and hydroxide surface. Rev. Virtual Quim. **6**, 687–723 (2014)
36. Cavalcanti, L.P.A.N., De Aguiar, A.P., Lima, J.A., Lima, A.L.S.: Organophosphorous poisoning: treatment and analytical methodologies applied in evaluation of reactivation and inhibition of acetylcholinesterase. Rev. Virtual Quim. **8**, 739–766 (2016)
37. Zilker, T.: Medical management of incidents with chemical warfare agents. Toxicology **214**, 221–231 (2005)
38. Dichtwald, S., Weinbroum, A.A.: Bioterrorism and the anaesthesiologist's perspective. Best Pract. Res. Clin. Anaesthesiol. **22**, 477–502 (2008)
39. WILSON, I.B. Acetylcholinesterase. XI. Reversibility of tetraethyl pyrophosphate. *J. Biol. Chem.* **1951**, *190*, 111–7.
40. Wilson, I.B., Ginsburg, S.: A powerful reactivator of alkylphosphate-inhibited acetylcholinesterase. Biochim. Biophys. Acta **18**, 168–170 (1955)
41. Petroianu, G.A.: The history of pyridinium oximes as nerve gas antidotes: the British contribution. Pharmazie **68**, 916–918 (2013)
42. Lundy, P.M., Raveh, L., Amitai, G.: Development of the bisquaternary oxime HI-6 toward clinical use in the treatment of organophosphate nerve agent poisoning. Toxicol. Rev. **25**, 231–243 (2006)
43. Worek, F., Thiermann, H.: The value of novel oximes for treatment of poisoning by organophosphorus compounds. Pharmacol. Ther. **139**, 249–259 (2013)
44. Franjesevic, A.J., Sillart, S.B., Beck, J.M., Vyas, S., Callam, C.S., Hadad, C.M.: Resurrection and reactivation of acetylcholinesterase and butyrylcholinesterase. Chemistry **25**, 5337–5371 (2019)
45. Wang, J., Gu, J., Leszczynski, J., Feliks, M., Sokalski, W.A.: Oxime-induced reactivation of sarin-inhibited AChE: a theoretical mechanisms study. J. Phys. Chem. B **111**, 2404–2408 (2007)
46. Artursson, E., Akfur, C., Hörnberg, A., Worek, F., Ekström, F.: Reactivation of tabun-hAChE investigated by structurally analogous oximes and mutagenesis. Toxicology **265**, 108–114 (2009)
47. Kuca, K., Jun, D., Musilek, K.: Structural requirements of acetylcholinesterase reactivators. Mini-Reviews Med. Chem. **6**, 269–277 (2006)
48. Pang, Y.-P., Kollmeyer, T.M., Hong, F., Lee, J.-C., Hammond, P.I., Haugabouk, S.P., Brimijoin, S.: Rational design of alkylene-linked bis-pyridiniumaldoximes as improved acetylcholinesterase reactivators. Chem. Biol. **10**, 491–502 (2003)
49. Kitagawa, D., Cavalcante, S., de Paula, R., Rodrigues, R., Bernardo, L., da Silva, M., da Silva, T., dos Santos, W., Granjeiro, J., de Almeida, J., et al.: In vitro evaluation of neutral aryloximes as reactivators for electrophorus eel acetylcholinesterase inhibited by paraoxon. Biomolecules **9**, 583 (2019)
50. Kuca, K., Nepovimova, E., Wu, Q., de Souza, F.R., Ramalho, T. de C., Franca, T.C.C., Musilek, K.: Experimental hydrophilic reactivator: bisoxime with three positive charges. Chem. Pap. **73**, 777–782 (2019)
51. Polisel, D.A., de Castro, A.A., Mancini, D.T., da Cunha, E.F.F., França, T.C.C., Ramalho, T.C., Kuca, K.: Slight difference in the isomeric oximes K206 and K203 makes huge difference for the reactivation of organophosphorus-inhibited AChE: theoretical and experimental aspects. Chem. Biol. Interact. **309**, 108671 (2019)
52. Musilek, K., Holas, O., Kuca, K., Jun, D., Dohnal, V., Opletalova, V., Dolezal, M.: Synthesis of monooxime-monocarbamoyl bispyridinium compounds bearing (E)-but-2-ene linker and evaluation of their reactivation activity against tabun- and paraoxon-inhibited acetylcholinesterase. J. Enzyme Inhib. Med. Chem. **23**, 70–76 (2008)
53. Lorke, D.E., Nurulain, S.M., Hasan, M.Y., Kuča, K., Petroianu, G.A.: Oximes as pretreatment before acute exposure to paraoxon. J. Appl. Toxicol. jat.3835 (2019)
54. Jaćević, V., Nepovimova, E., Kuča, K.: Toxic injury to muscle tissue of rats following acute oximes exposure. Sci. Rep. **9**, 1–13 (2019)

55. Jaćević, V., Nepovimova, E., Kuča, K.: Acute toxic injuries of rat's visceral tissues induced by different oximes. Sci. Rep. **9**, 1–13 (2019)
56. de Koning, M.C., Horn, G., Worek, F., van Grol, M.: Discovery of a potent non-oxime reactivator of nerve agent inhibited human acetylcholinesterase. Eur. J. Med. Chem. **157**, 151–160 (2018)
57. Carletti, E., Colletier, J.-P., Dupeux, F., Trovaslet, M., Masson, P., Nachon, F.: Structural evidence that human acetylcholinesterase inhibited by tabun ages through O-dealkylation. J. Med. Chem. **53**, 4002–4008 (2010)
58. Kalisiak, J., Ralph, E.C., Zhang, J., Cashman, J.R.: Amidine−oximes: reactivators for organophosphate exposure. J. Med. Chem. **54**, 3319–3330 (2011)
59. Quinn, M.D., Topczewski, J., Yasapala, N., Lodge, A.: Why is Aged Acetylcholinesterase So Difficult to Reactivate? Mol **22** (2017)
60. Iyer, R., Iken, B., Damania, A.: A comparison of organophosphate degradation genes and bioremediation applications. Environ. Microbiol. Rep. **5**, 787–798 (2013)
61. Iyer, R., Iken, B.: Protein engineering of representative hydrolytic enzymes for remediation of organophosphates. Biochem. Eng. J. **94**, 134–144 (2015)
62. Field, M.J., Wymore, T.W.: Multiscale modeling of nerve agent hydrolysis mechanisms: a tale of two Nobel Prizes. Phys. Scr. **89**, 108004 (2014)
63. Wood, L.B., Winslow, A.R., Strasser, S.D.: Systems biology of neurodegenerative diseases. Integr. Biol. (United Kingdom) **7**, 758–775 (2015)
64. Gontijo, V.S., Viegas, F.P.D., Ortiz, C.J.C., de Freitas Silva, M., Damasio, C.M., Rosa, M.C., Campos, T.G., Couto, D.S., Tranches Dias, K.S., Viegas, C.: Molecular hybridization as a tool in the design of multi-target directed drug candidates for neurodegenerative diseases. Curr. Neuropharmacol. **18**, 348–407 (2020)
65. Ortiz, C.J.C.: Síntese e Avaliação Farmacológica de Novos Ligantes Multialvo Visando ao Tratamento de Doenças Neurodegenerativas, Universidade Federal de Alfenas (2020)
66. Mathew, B., Parambi, D.G.T., Mathew, G.E., Uddin, M.S., Inasu, S.T., Kim, H., Marathakam, A., Unnikrishnan, M.K., Carradori, S.: Emerging therapeutic potentials of dual-acting MAO and AChE inhibitors in Alzheimer's and Parkinson's diseases. Arch. Pharm. (Weinheim). **352**, 1–13 (2019)
67. Farah, R., Haraty, H., Salame, Z., Fares, Y., Ojcius, D.M., Said Sadier, N.: Salivary biomarkers for the diagnosis and monitoring of neurological diseases. Biomed. J. **41**, 63–87 (2018)
68. McColgan, P., Tabrizi, S.J.: Huntington's disease: a clinical review. Eur. J. Neurol. **25**, 24–34 (2018)
69. Ceschi, M., Pilotti, R., Lopes, J., Dapont, H., da Rocha, J., Afolabi, B., Guedes, I., Dardenne, L.: An expedient synthesis of tacrine-squaric hybrids as potent, selective and dual-binding cholinesterase inhibitors. J. Braz. Chem. Soc. **31**, 857–866 (2020)
70. Araújo, C.R.M., Santos, V.L.A., Gonsalves, A.A.: Acetylcholinesterase—AChE: a pharmacological interesting enzyme. Rev. Virtual Quim. **8**, 1818–1834 (2016)
71. Toker, L., Tran, G.T., Sundaresan, J., Tysnes, O.: Genome-wide dysregulation of histone acetylation in the Parkinson ' s disease brain 1–47 (2020)
72. Klein, C., Westenberger, A.: Genetics of Parkinson's disease. Cold Spring Harb. Perspect. Med. **2**, a008888 (2012)
73. BRAVO, Patrício Andres Fuentes; NASSIF, M.C.: Doença de Parkinson: terapêutica atual e avançada. Pharm. Bras. **55**, 25–29 (2006)
74. Chierrito, T.P.C., Mantoani, S.P., Roca, C., Requena, C., Sebastian-Perez, V., Castillo, W.O., Moreira, N.C.S., Pérez, C., Sakamoto-Hojo, E.T., Takahashi, C.S., et al.: From dual binding site acetylcholinesterase inhibitors to allosteric modulators: a new avenue for disease-modifying drugs in Alzheimer's disease. Eur. J. Med. Chem. **139**, 773–791 (2017)
75. Silva, T., Reis, J., Teixeira, J., Borges, F.: Alzheimer's disease, enzyme targets and drug discovery struggles: from natural products to drug prototypes. Ageing Res. Rev. **15**, 116–145 (2014)
76. Grøntvedt, G.R., Schröder, T.N., Sando, S.B., White, L., Bråthen, G., Doeller, C.F.: Alzheimer's disease. Curr. Biol. **28**, R645–R649 (2018)

77. Gabriel, U., Cunha, D.V., Marino, C.G., Balabram, K., Marquete, C.R.: Uso de inibidores da colinesterase em idosos com comorbidades clínicas. 162–166
78. Nitrini, R., Caramelli, P., Bottino, C.M.D., Damasceno, B.P., Brucki, S.M.D., Anghinah, R.: Diagnosis of Alzheimer's disease in Brazil: cognitive and functional evaluation. Recommendations of the scientifc department of cognitive neurology and aging of the Brazilian academy of neurology. Arq. Neuropsiquiatr. **63**, 720–727 (2005)
79. Aprahamian, I., Martinelli, J.E., Yassuda, M.S.: Doença de Alzheimer : revisão da epidemiologia e diagnóstico * Alzheimer ' s disease : an epidemiology and diagnosis review. Rev Bras Clin Med. **7**, 27–35 (2009)
80. Forsell, Y., Winblad, B.: Major depression in a population of demented and nondemented older people: prevalence and correlates. J. Am. Geriatr. Soc. **46**, 27–30 (1998)
81. Pennypacker, L.C., Allen, R.H., Kelly, J.P., Matthews, L.M., Grigsby, J., Kaye, K., Lindenbaum, J., Stabler, S.P.: High prevalence of cobalamin deficiency in elderly outpatients. J. Am. Geriatr. Soc. **40**, 1197–1204 (1992)
82. Gilmore, J.H., Castillo, M., Rojas, M.: Early onset schizophrenia in a patient with premature birth, germinal matrix hemorrhage and periventricular leukomalacia. Schizophr. Res. **44**, 158–160 (2000)
83. Canineu, P.R., Canineu, R.F.B., Canineu, P.R.B., Silva, M.C.: Terapia Multidisciplinar: uma proposta de tratamento global do idoso. Mundo Saúde. **29**, 662–665 (2005)
84. Giacobini, E.: Invited review cholinesterase inhibitors for Alzheimer's disease therapy: from tacrine to future applications. Neurochem. Int. **32**, 413–419 (1998)
85. Chand, K., Alsoghier, H.M., Chaves, S., Santos, M.A.: Tacrine-(hydroxybenzoyl-pyridone) hybrids as potential multifunctional anti-Alzheimer's agents: AChE inhibition, antioxidant activity and metal chelating capacity. J. Inorg. Biochem. **163**, 266–277 (2016)
86. Foye, W.O.: Foye's Principles of Medicinal Chemistry. In: Wilkins, L.W. (ed.) (2008). ISBN 9780781768795.
87. Dias Viegas, F.P., de Freitas Silva, M., Divino da Rocha, M., Castelli, M.R., Riquiel, M.M., Machado, R.P., Vaz, S.M., Simoes de Lima, L.M., Mancini, K.C., Marques de Oliveira, P.C., et al.: Design, synthesis and pharmacological evaluation of N-benzyl-piperidinyl-aryl-acylhydrazone derivatives as donepezil hybrids: Discovery of novel multi-target anti-alzheimer prototype drug candidates. Eur. J. Med. Chem. **147**, 48–65 (2018)
88. Tiseo, P.J., Rogers, S.L., Friedhoff, L.T.: Pharmacokinetic and pharmacodynamic profile of donepezil HCl following evening administration. Br. J. Clin. Pharmacol. **46**(Suppl 1), 13–18 (1998)
89. Cheung, J., Rudolph, M.J., Burshteyn, F., Cassidy, M.S., Gary, E.N., Love, J., Franklin, M.C., Height, J.J.: Structures of human acetylcholinesterase in complex with pharmacologically important ligands. J. Med. Chem. **55**, 10282–10286 (2012)
90. Almeida, J.R.: De Estudos de modelagem molecular e relação estrutura-atividade da acetilcolinesterase e inibidores em Mal de Alzheimer Estudos de modelagem molecular e relação estrutura-atividade da acetilcolinesterase e inibidores em Mal de Alzheimer 126 (2011)
91. Bar-On, P., Millard, C.B., Harel, M., Dvir, H., Enz, A., Sussman, J.L., Silman, I.: Kinetic and structural studies on the interaction of cholinesterases with the anti-Alzheimer drug rivastigmine. Biochemistry **41**, 3555–3564 (2002)
92. Geldenhuys, W.J., Van der Schyf, C.J.: Rationally designed multi-targeted agents against neurodegenerative diseases. Curr. Med. Chem. **20**, 1662–1672 (2013)
93. Chaves, M.B., Ferreira, A.A.F.: Terapia medicamentosa da doença de Alzheimer. Rev. Eletrônica Farmácia **1**, 1–7 (2008)
94. Viegas, C., Bolzani, V.S., Pimentel, L.S.B., Castro, N.G., Cabral, R.F., Costa, R.S., Floyd, C., Rocha, M.S., Young, M.C.M., Barreiro, E.J., et al.: New selective acetylcholinesterase inhibitors designed from natural piperidine alkaloids. Bioorg. Med. Chem. **13**, 4184–4190 (2005)
95. Martínez, A., Zahran, M., Gomez, M., Cooper, C., Guevara, J., Ekengard, E., Nordlander, E., Alcendor, R., Hambleton, S.: Novel multi-target compounds in the quest for new chemotherapies against Alzheimer's disease: an experimental and theoretical study. Bioorganic Med. Chem. **26**, 4823–4840 (2018)

96. Jalili-Baleh, L., Babaei, E., Abdpour, S., Nasir Abbas Bukhari, S., Foroumadi, A., Ramazani, A., Sharifzadeh, M., Abdollahi, M., Khoobi, M.: A review on flavonoid-based scaffolds as multi-target-directed ligands (MTDLs) for Alzheimer's disease. Eur. J. Med. Chem. **152**, 570–589 (2018)
97. de Freitas Silva, M., Dias, K.S.T., Gontijo, V.S., Ortiz, C.J.C., Viega, C.: Multi-target directed drugs as a modern approach for drug design towards alzheimer's disease: an update. Curr. Med. Chem. **25**, 3491–3525 (2018)
98. Potschka, H.: Targeting the brain--surmounting or bypassing the blood-brain barrier. Handb. Exp. Pharmacol. 411–431 (2010)
99. Silva, G.A.: Nanotechnology applications and approaches for neuroregeneration and drug delivery to the central nervous system. Ann. N. Y. Acad. Sci. **1199**, 221–230 (2010)
100. Kreuter, J.: Nanoparticulate systems for brain delivery of drugs. Adv. Drug Deliv. Rev. **47**, 65–81 (2001)
101. Brambilla, D., Le Droumaguet, B., Nicolas, J., Hashemi, S.H., Wu, L.-P., Moghimi, S.M., Couvreur, P., Andrieux, K.: Nanotechnologies for Alzheimer's disease: diagnosis, therapy, and safety issues. Nanomedicine **7**, 521–540 (2011)
102. de Castro, A.A., Soares, F.V., Pereira, A.F., Polisel, D.A., Caetano, M.S., Leal, D.H.S., da Cunha, E.F.F., Nepovimova, E., Kuca, K., Ramalho, T.C.: Non-conventional compounds with potential therapeutic effects against Alzheimer's disease. Expert Rev. Neurother. **19**, 375–395 (2019)
103. Wen, M.M., El-Salamouni, N.S., El-Refaie, W.M., Hazzah, H.A., Ali, M.M., Tosi, G., Farid, R.M., Blanco-Prieto, M.J., Billa, N., Hanafy, A.S.: Nanotechnology-based drug delivery systems for Alzheimer's disease management: technical, industrial, and clinical challenges. J. Control. Release **245**, 95–107 (2017)
104. Kostarelos, K., Miller, A.D.: Synthetic, self-assembly ABCD nanoparticles; a structural paradigm for viable synthetic non-viral vectors. Chem. Soc. Rev. **34**, 970–994 (2005)
105. Borm, P.J.A., Muller-Schulte, D.: Nanoparticles in drug delivery and environmental exposure: same size, same risks? Nanomedicine (Lond). **1**, 235–249 (2006)
106. Xu, Z.P., Zeng, Q.H., Lu, G.Q., Yu, A.B.: Inorganic nanoparticles as carriers for efficient cellular delivery. Chem. Eng. Sci. **61**, 1027–1040 (2006)
107. Foged, C., Nielsen, H.M.: Cell-penetrating peptides for drug delivery across membrane barriers. Expert Opin. Drug Deliv. **5**, 105–117 (2008)
108. Leary, S.P., Liu, C.Y., Apuzzo, M.L.J.: Toward the emergence of nanoneurosurgery: part III–nanomedicine: targeted nanotherapy, nanosurgery, and progress toward the realization of nanoneurosurgery. Neurosurgery **58**, 1009–1026 (2006)
109. Modi, G., Pillay, V., Choonara, Y.E.: Advances in the treatment of neurodegenerative disorders employing nanotechnology. Ann. N. Y. Acad. Sci. **1184**, 154–172 (2010)
110. Nazem, A., Mansoori, G.A.: Nanotechnology solutions for Alzheimer's disease: advances in research tools, diagnostic methods and therapeutic agents. J. Alzheimers. Dis. **13**, 199–223 (2008)
111. Pohanka, M., Jun, D., Kalasz, H., Kuca, K.: Cholinesterase biosensor construction—a review. Protein Pept. Lett. **15**, 795–798 (2008)
112. Porfírio, F.F.O., Giarola, J.D.F., Pereira, A.C.: Biosensor and beverages—review. Rev. Virtual Quim. **8**, 1366–1391 (2016)
113. Khaled, E., Hassan, H.N.A., Ahmed, M.A., El-Attar, R.O.: Highly sensitive method for analysis anticholinesterase drugs. Egypt. J. Chem. **62**, 441–449 (2019)
114. Liu, X., Sakthivel, R., Liu, W.C., Huang, C.W., Li, J., Xu, C., Wu, Y., Song, L., He, W., Chung, R.J.: Ultra-highly sensitive organophosphorus biosensor based on chitosan/tin disulfide and British housefly acetylcholinesterase. Food Chem. **324**, 126889 (2020)
115. Hernandez-Vargas, G., Sosa-Hernández, J.E., Saldarriaga-Hernandez, S., Villalba-Rodríguez, A.M., Parra-Saldivar, R., Iqbal, H.M.N.: Electrochemical biosensors: a solution to pollution detection with reference to environmental contaminants. Biosensors **8**, 29 (2018)

In Silico Methods to Predict Relevant Toxicological Endpoints of Bioactive Substances

Guilherme Martins Silva, Leonardo Bruno Federico, Vinicius Medeiros Alves, and Carlos Henrique Tomich de Paula da Silva

Abstract Toxicity assessment is an essential step in the development of drugs, agrochemicals and other bioactive substances. In silico toxicity prediction, or computational toxicology, has been growing fast during the last years mainly due to advantageous cost and labor reduction, and avoidance of experiments using animals. Considering the vast number of software available for toxicity prediction, one should be aware to pick up the right software for the right task. Herein, we aim to describe the main software as well as corresponding methodologies behind them, that are able to provide reliable and accurate results for diverse toxicological endpoints. We give special emphasis to software based on quantitative structure–activity relationships, expert systems and machine learning methodologies. With this in mind, we hope that this chapter content may serve as a valuable guide to select toxicity prediction software, and also to reinforce their usefulness towards researches concerning the development of bioactive substances.

1 Introduction

Evaluation of the toxicity potential of chemical substances is an essential step associated with the most diverse branches of activities performed in modern and industrialized times [1–3]. Such evaluation can be performed by in vivo or in vitro

G. M. Silva · C. H. T. de Paula da Silva (✉)
Departamento de Química, Faculdade de Filosofia, Ciências E Letras de Ribeirão Preto (FFCLRP), Universidade de São Paulo (USP), Av. Bandeirantes, Ribeirão Preto 3900, 14090-901, Brazil
e-mail: tomich@fcfrp.usp.br

L. B. Federico · C. H. T. de Paula da Silva
Faculdade de Ciências Farmacêuticas de Ribeirão Preto (FCFRP), Universidade de São Paulo (USP), Av. do Café, s/n, Ribeirão Preto 14040-903, Brazil

V. M. Alves
Laboratory for Molecular Modeling, UNC Eshelman School of Pharmacy, University of North Carolina At Chapel Hill, Chapel Hill, NC 27599, USA

F. A. La Porta and C. A. Taft (eds.), *Functional Properties of Advanced Engineering Materials and Biomolecules*, Engineering Materials,
https://doi.org/10.1007/978-3-030-62226-8_22

experimental procedures, but computational (in silico) methods can provide accurate predictions that guide researchers to make better decisions [4, 5]. In this introduction, we are going to briefly run through general aspects of toxicity assessment, and then emphasize how in silico predictions turned out to be such an important tool in his context.

Understanding how, and to which extent, the contact (ingestion, handling, development, or production) with chemicals may entail toxicity damages and produce harm is critical in regulatory risk assessment of chemicals [6, 7]. Approved drugs and agrochemicals need to provide limited/acceptable adverse outcomes at the therapeutical biological responses [8].

During the process of drug development—i.e., since its initial discovery as a bioactive substance until it reaches the market—there are several causes that may lead to its failure, among which the presence of dose-limiting toxicity stands out [1, 9]. For this reason, the study and design of a novel drug candidate must consider predicting its toxicity potential [10, 11]. In the last few decades, regulatory agencies, such as the US FDA (Food and Drug Administration), EMA (European Medicines Agency), and Brazilian ANVISA *(Agência Nacional de Vigilância Sanitária)* have become more strict on approval of chemicals in order to guarantee their safety profile [5].

Agrochemicals development also requires the support of studies that ensure its safety considering the conditions of exposed dose, that is, to have minimal risks of causing damage to the environment. In general, agrochemicals should be assessed in relation to their environmental fate, to the potential harms to humans who will handle them, as well as to final consumers of vegetables and grains that will be treated with such chemicals [7, 12]. To this end, regulatory agencies control the use of agrochemicals, according to their hazard potentials, such as the US EPA (United States Environmental Protection Agency) and ECHA (European Chemicals Agency). Albeit, here, we accentuate that we will be partially biased to deal with the context of drug toxicity.

In order to perform toxicity prediction of these bioactive compounds, computational methodologies have been exhaustively employed, either by pharmaceutical companies or by academic groups, with the main goals to save time and financial expenditures and also to avoid animal-based tests [1, 11, 13, 14]. Worth mentioning that such methodologies are nowadays widely accepted and considered for evaluating the initial toxicological profile of a substance under the process of approval. In fact, many supporting agencies, programs and set of laws (e.g., Registration, Evaluation, Authorization, and Restriction of Chemicals (REACH), Toxic Substances Control Act (TSCA), US National Toxicology Program (US NTP), Office of Pollution Prevention and Toxics (OPPT), European Center for the Validation of Alternative Methods, etc.) strongly encourage the use of them [3, 5, 15]—or actually require them to be included on studies -, in such a way that these agencies indeed provide updated-guidelines to support the use of computational methodologies to predict the toxicity profile of chemicals (e.g., US EPA—(Quantitative) Structure Activity Relationship [(Q)SAR] Guidance Document, 2012 [16]).

The use of computational methodologies/strategies to predict toxicity—also commonly found being mentioned by the terms computational toxicology or in silico methods to predict toxicity or predictive toxicology—is a must, when it comes to studies concerning toxicity potential of compounds [1, 7, 11]. It allows the prediction of specific toxicological endpoints with good reliability by well-established and validated methods [8, 15]. Moreover, it provides useful information to elucidate the mechanisms through which a compound provoke harm [17].

Computational toxicology is a multidisciplinary field that put together researchers with various backgrounds, such as chemistry, biology, biochemistry, toxicology, pharmacotoxicology, computer science, statistics, also proteomics, genomics, metabolomics, among others [1]. The importance of computational toxicology and its applications is presented and discussed in this chapter by highlighting strategies to develop toxicity models that have been developed over the years by many groups and many toxicity endpoints, such as mutagenicity [18–23], acute oral and lethal toxicity [24–31], carcinogenicity [32–38], and cardiotoxicity [39–43].

As can be seen, among principal endpoints that are considered within toxicity prediction studies, one can highlight mutagenicity, hepatotoxicity, carcinogenicity, inhalation and oral toxicity, skin sensitization, and developmental toxicity potential. Such endpoints are fundamental to clarify if a compound may be safe and move forward in the drug and agrochemical development pipeline [6, 7]. Computational models are built using experimental data from in vitro or in vivo assays [3, 44]. For instance, mutagenicity can be evaluated by the in vitro Ames test, which consists of using strains of histidine dependent Salmonella, with different mutations in histidine operon, and then growing them in agar plates containing a minimal amount of histidine to check which one can form colonies [45–47].

Normally, in order to provide a reasonable amount of data to allow prediction by in silico model, there must be enough tested compounds originated from in vitro/in vivo toxicological assays and the data must have good quality. Therefore, one can notice that there is this straight and synergistic relation in-between in silico and the experiment, that is, the experiment furnishes data to train, generate models and validate in silico methodologies, whereas in silico software predict future experimental endpoints for compounds with unknown toxicity [15, 44]. The assurance of good data quality can be performed during the process of data curation, which is a mandatory step to develop predictive models [48, 49].

There are different types of methodologies implemented on the main software used in computational toxicology, usually coming from studies related to molecular modeling and cheminformatics [8]. In this chapter, we aim to describe the most popular software to predict the toxicity potential of bioactive compounds, according to their corresponding methodologies. We split and opt to describe these software in most relevant methodologies, namely QSAR/QSTR (Quantitative Structure–Activity Relationship / Quantitative Structure–Toxicity Relationship), expert knowledge-based, and machine learning. Moreover, we mention a few studies concerning different methodologies such as 3D-QSAR, docking, pharmacophore and molecular dynamics. We believe this shall be much useful for better understanding the software's operation and, thus, shall facilitate the pickup of most adequate software to the desired end, and with respect to toxicity predictions.

Since the historical trajectory and most of the software that are used in computational toxicology to predict toxicity endpoints had their origin by means of the approach of QSAR, we will dedicate a special topic in this context to some theoretical aspects of such relevant methodology.

2 Overview of Aspects Concerning QSAR Theory

The effect that a chemical substance exerts over a biological system can be elucidated by analysis of their in-between interactions, as well as by inspection of their relationships depicted by corresponding structural and biological properties. When establishing such relationships in a quantitative manner, one may denote this as Quantitative Structure–Activity Relationship (QSAR) [50–53].

QSAR studies are extended to different fields of application including the design/discovery/optimization of drugs, the prediction of toxicity potential for compounds with biological activity, and also to the evaluation of mechanisms of action of drugs and agrochemicals [54–56].

The pioneer publication of QSAR was in 1964 by Hansch and Fujita [57], whose proposition became known as the extra-thermodynamic approach, which consists of a mathematical model relating the biological activity of a set of molecules with their physicochemical and structural properties expressed by parameters. This classical extra-thermodynamic approach was based on the extension of organic chemistry general concepts and the principles of LFER (Linear Free Energy Related), which permits analysis of possible relations between kinetic properties (expressed by biological parameters, e.g., IC_{50}) and experimental/computed related thermodynamic parameters attributed to group substituents or fragments of compounds [53, 58].

The so-called Hansch & Fujita QSAR approach was limited to small sets of analogs. Also, to the use of a few measured/calculated physicochemical parameters (such as *logP*, σ, π, and others) that could be possibly related with the biological activity. Hence, evaluation of the relative contributions of each one of them to the activity should only be established by simple and linear mathematical models [59–61]. But soon emerged the proposition of more complex relationships through nonlinear models, such as the parabolic [57] and the bilinear [53]. We take this context to mention some literature with historical information and evolutionary timeline of QSAR [62–65].

In fact, in this brief section we do not intend to provide an extensive description of QSAR history and best practices procedures, but rather provide an overview of QSAR concepts, with additional suggestions of some relevant papers in literature. Additional information concerning these topics can be found in two seminal reviews [62, 66].

With regard to physicochemical and structural parameters, these can be used as descriptors, i.e. a quantitative description of molecular structure, in respective QSAR models [67]. Nowadays, there is a great variety of descriptors due to increasing interest in identifying additional ways of describing molecules, as well as to the huge expansion of data generated in the last decades [68, 69].

As mentioned, there are currently numerous molecular descriptors such as topological, steric, electrostatic indices, and also indices that encode 3D geometric information about the molecule. These descriptors can be easily and quickly calculated [70–73], therefore being suitable for QSAR studies and similarity/diversity analyzes of extensive databases. It is noteworthy that one should prefer to use molecular descriptors that are meaningful and interpretable, along with discerning selection methods, in order to significantly represent the information related to structural/physicochemical properties of the dataset of compounds under study [55, 74, 75].

Considering the great amount of data available to manage to build QSAR models, sophisticated statistical and chemometrics tools have been employed to analyze such data and also to establish strong relationships and significant correlations—in corroboration with the topic of the principle of unambiguous algorithm, as denoted by [75]. For instance, MLR (Multiple Linear Regression) [76, 77], OLS (Ordinary Least Squares), PCA (Principal Component Analysis), PCR (Principal Component Regression), DA (Discriminant Analysis), RF (Random Forest) [78], PLS (Partial Least-Square) [79], k-NN (*k* nearest neighbor), Genetic Algorithms (GA), Neural Networks (ANN) [80], and so on [67, 81].

In fact, this seems to be a topic of interest of researchers with a chemical background, who tend to focus their studies on molecular descriptors, statistical methods for the selection of compounds/descriptors, and model validation. On the other hand, there is this topic where researchers with biological background tend to form a group more interested in studying modes of action, and biological endpoints/mechanisms. The former is known as "predictive QSAR" and the latter as "descriptive QSAR", and this division of QSAR modeling has been commonly mentioned in the literature [82].

When it comes to validation methods applied to QSAR models, there are many guidelines and criteria available on the literature [62, 75, 81, 83–85]. According to Gramatica [65], a robust and reliable QSAR model, besides being useful, should be the right one for the purposed aim.

First, and ideally, compounds from the whole dataset should be originated from the same source and have their biological activities measured by the same assays and experimental protocols, under the same conditions. So there should be a concern with the quality and the well definition of the endpoint of interest. This is part of the dataset curation, which is critical to assure data quality and guarantee the predictivity of the models [49, 86, 87].

The validation of QSAR models composes another critical step, which should proceed by performing a correct split of a dataset into training and test sets; and internal and external validations. Putting in other words, splitting of the whole dataset into two different sets, i.e., training and test sets, should be executed in order to permit both internal and external validation, respectively. The training set consists of ~ 80% of molecules from the dataset and must be considered in model development (fitting and internal validation), and the test set consists of molecules that originally came from the same whole dataset, but corresponds to ~ 20% of the total number of compounds and will be used to after model development to evaluate its external predictivity. There are some works reporting on different strategies to split training/test sets [65, 88].

After checking the goodness-of-fit of the model (i.e. verifying the R^2), internal validation of a model's performance, also represented by its robustness, may be done by evaluation of the statistical parameters Leave-One-Out (LOO) cross-validation, Q^2_{LOO}, and by the overall error of the model, Root Mean Square Error (RMSE). The predictivity, that is the ability of the model to predict outside the training set, is performed by using the test set for external validation and analysis of the parameter Q^2_{EXT}, that can be obtained by different manners and have different meanings. This validation overview, here described, is general and there's much more to be aware of, so that we cite the following papers—please also consider those cited three paragraphs above [60, 83, 89–91].

Over time, QSAR studies have shown that, in addition to specific interactions between the ligand and the biological target, there should be an influence of field effects by non-covalent intermolecular interactions on the spatial vicinity of the ligands [62]. This led to the deepening of studies related to three-dimensional molecular descriptors and, thus, to the emergence of 3D-QSAR [92], which incorporates three-dimensional information on ligands, but requires their alignment and conformational choice [93].

Traditionally, 3D-QSAR has been subdivided into the CoMFA (Comparative Molecular Field Analysis) and CoMSIA (Molecular Similarity Indexes in a Comparative-Analysis) approaches [94]. The CoMFA method positions the binding molecules with a steric (shape or volume) or electrostatic test group, within a grid of equally spaced points (the grid). In CoMSIA, both electrostatic and steric characteristics are considered, in addition to the hydrogen bridge donor, hydrogen bridge acceptor, and hydrophobic fields [95].

Worth to mention that other developments in the modeling field of QSAR have led to the disclosure of different approaches, such as HQSAR [96], 4D-QSAR [97], 5D-QSAR [98], multi-target and multi-objective QSAR [99, 100], aiQSAR [52], among others.

3 QSAR Software

Theoretical concepts behind QSAR, including the fact that it is based on the principles of LFER, allows one to deduce that it may be applied not only to biological parameters (such as IC_{50}, EC_{50}, etc.) but also to toxicological endpoints/parameters (LD_{50}, LOAEL, carcinogenicity, etc.) [101]. These studies are commonly named as QSTR, i.e. Quantitative Structure–Toxicity Relationships, and represents the fundamental basis for almost every QSAR software that is commonly employed in computational toxicology [102, 103]. We highlight that here, in this chapter, we are employing the terms QSAR or QSTR as similar terms, since we are dealing with the context of toxicity prediction.

In literature, after the initial development of QSAR studies, many studies have been published with different kinds of datasets, and respective different kinds of descriptors, aiming to correlate them with different kind of toxicological endpoints [15, 56, 101]. Therefore, since then, one can find several QSAR models, which, in turn, allowed the improvement and development of different QSAR software that are currently available to evaluate the toxicity of chemical substances [3].

Worth to note that many software are defined as using QSAR methodologies, once they are statistics based and/or employ regression models. However, since that current recommendation from official agencies dealing with computational toxicology evaluations, suggests that not only one methodology should be applied for a chemical in order to predict its toxicological endpoints, many software are equipped with a mix of methodologies. Anyhow, here we will mention those software that are most recognized for employing the QSAR approach.

3.1 TopKat

One of the first software ever launched as a toxicity predictor based on the approach of QSTR was TopKat (Toxicity Prediction by Komputer Assisted Technology). Since its initial development, this software has been under the domain of different companies and, nowadays, its commercial version can be used as an implemented tool in ADME/Tox applications package by BIOVIA/Dassault Systemes [4].

Toxicity prediction by this software utilizes cross-validated QSTR models. In order to make the use of its implemented models more versatile, TopKat allows checking model applicability and features contributions, as well as employs the patented Optimal Predictive Space (OPS) validation method to assist in interpreting the results, putting in other words, reports similar compounds to assist in results interpretation. If a prediction generated for an input structure stays outside of the OPS this will be considered unreliable [104]. Furthermore, it is possible to extend models and improve coverage of chemistry space by adding new training data.

According to information displayed on owner's webpage https://www.3dsbiovia.com/products/collaborative-science/biovia-discovery-studio/qsar-admet-and-predictive-toxicology.html, all extensible TopKat models have been published as Toxicity Models QSAR Model Report Format (QMRF) reports on the European Commission Joint Research Center (JRC) [105].

QSTR models implemented in TopKat are built upon data provided by in vitro experimental assays and animal models, being, therefore, suitable to predict the following toxicological endpoints: Ames mutagenicity, rodent carcinogenicity (NTP and FDA data), the weight of evidence carcinogenicity, carcinogenic potency TD_{50}, developmental toxicity potential, rat oral LD_{50}, rat maximum tolerated dose, rat inhalation toxicity LC_{50}, rat chronic LOAEL, skin irritancy and sensitization, eye irritancy, aerobic biodegradability, fathead minnow LC_{50}, *Daphnia magna* EC_{50} and *logP*.

3.2 ADMET Predictor

ADMET Predictor is a commercial software provided by SimuationsPlus that consists of several built-in modules such as Physicochemical & Biopharmaceutical, Metabolism, ADMET Modeler, HTPK Simulation, MedChem Studio, and Toxicity (https://www.simulations-plus.com/software/admetpredictor/toxicity/). Besides predicting properties, as well as quantitative and qualitative values and endpoints of interest, it works as a QSAR model-building application, since it is based on literature databases and robust QSAR/QSTR models, but not only. Just to mention, two important works that developed relevant QSAR studies are employed/cited by ADMET Predictor, see references [106, 107].

In order to provide predictive cardiac toxicity (binding affinity towards hERG-encoded potassium channels), this software employs classification and regression neural network models to assess the chances of a compound to block hERG-encoded K^+ channels, which are responsible for the normal repolarization of the cardiac action potential.

ADMET Predictor uses the information of a subset of 490 molecules from the Spontaneous Reporting System (SRS) database, concerning human liver adverse effects of many popular drugs, to model hepatotoxicity. Moreover, it provides the following individual models corresponding to individual enzymes used in hepatotoxicity diagnostics: alkaline phosphatase increase, SGOT increase, SGPT increase, LDH increase and GGT increase.

It also allows to predict reproductive (developmental) toxicity; chronic carcinogenicity and mutagenicity endpoints, since it contains models built using data from the Carcinogenic Potency Database (CPDB); phospholipidosis; endocrine toxicity; skin and respiratory sensitization; acute rat toxicity (by oral administration); chromosomal aberrations; maximum recommended therapeutic dose; and environmental toxicity (in terms of bioconcentration factor).

3.3 VEGA

VEGA is a computational software to predict toxicity potential that consists of a platform including several validated QSAR models as well as an independent tool that helps the user in the evaluation of results through an Applicability Domain Index (https://www.vegahub.eu/portfolio-item/vega-qsar/).

This software is freely available and allows the use of several implemented QSAR models, simply by providing the structure of the chemical compound (SMILES or sdf formats) as an input. It is an independent JAVA application and does not require any network communication—chemicals are processed on the local machine and information is not shared over the Internet.

Its methodology is a completely independent algorithm used to assess the reliability of the model's prediction through the "Applicability Domain Index". This algorithm works on all QSAR models, separately. Among the models incorporated to predict specific toxicity endpoints we highlight the CAESAR, which used to be a separate platform and has migrated to VEGA, as well as IRFMN, ISS, ISSCAN-CGX, SarPy, and others.

At a glance, this algorithm shows similar and structurally related compounds (three as default), evaluates QSAR results for similar compounds, and analyzes some relevant chemical characteristics between the input compound and its associated compounds. For the utilization of VEGA software, the following endpoints can be predicted: carcinogenicity (also through oral or inhalation exposures), developmental toxicity, reproductive toxicity, hepatotoxicity, mutagenicity (Ames test) and skin sensitization, among others.

3.4 *OECD QSAR Toolbox*

The software OECD QSAR Toolbox performs toxicity and hazard assessment of chemical substances, it is a freely available program, and it is emphasized that, by using such software, one can avoid making use of animal testing, without reducing the safety of humans and environment. Using this software allows the possibility to search among all available data either for endpoints of interest or for compounds with specific experimental results, through more than 50 databases containing more than 80,000 chemicals and with almost 2.5 million data measured and implemented (https://qsartoolbox.org/about/).

Moreover, the performance functionality of generated alerts informs the user about the predictive power of specific alerts or structural features in profilers, thus enabling its applicability to create categories of analogs, based on consistent experimental results. In addition, there are 902 QSAR models to predict different properties that are available in QSAR Toolbox 4.3.1 (latest release), of which 145 QSAR models are available with information pertinent to human health risks.

In addition, toxicological data reported in the literature for the analyzed compounds can be retrieved or found using the following 5 predefined databases implemented in the software: Database Affiliation, Inventory Affiliation, OECD HPV Chemical Categories, Substance Type and US EPA New Chemical Categories. Some of the endpoints that may be estimated include: Carcinogenicity, Developmental Toxicity/Teratogenicity, Genetic Toxicity, DNA binding (by OASIS and OECD), in vitro mutagenicity (Ames test) alerts by ISS, Respiratory Sensitization, Immunotoxicity, and Neurotoxicity.

3.5 QSARINS

QSARINS, also known as QSAR-INSubria, is a software provided by the Italian research group from the University of Insubria. It allows the development and validation of multiple linear regression models by Ordinary Least Squares (OLS) and Genetic Algorithm (GA) for variable selection, following the best practices to perform statistical and chemometric approaches, and thus enhancing the robustness of QSAR models generation [83, 89, 91]. Worth to mention that it includes an add-on to calculate PaDEL descriptors (Yap 2011) for an input set of compounds, and its handling is very intuitive and simple.

Academic licenses can be acquired under registration and request through their official website https://www.qsar.it/ and, also, as it is stated on their website, the aim of the developers with such a helpful tool "is to propose the use of validated QSAR predictions for the screening and prioritization of hazardous compounds and the 'benign by design' approach of Green Chemistry".

Besides being a very useful software to perform QSAR, or QSTR, studies it comes along with a dataset containing more than 3,000 3D chemical structures (and experimental data) as well as 45 ready-to-use QSAR models, in a module named QSARINS-Chem. These models were developed to estimate environmental endpoints for organic pollutants, such as: benzo-triazole *D. magna* toxicity [108, 109]; endocrine disruptor chemicals estrogen receptor binding [109, 110]; global half-life index [109, 111]; human biotransformation *logHLT_h* [112]; personal care products aquatic toxicity index-ATI [113]; *logKoc* of pesticides [109, 114]; nitrated polycyclic aromatic hydrocarbon mutagenicity [109, 115]; persistence bioaccumulation toxicity index [116]; perfluorinated chemicals mouse inhalation tox [117]; pharmaceuticals *D.magna* acute tox [118]; and others (see reference [109]).

3.6 TEST

The Toxicity Estimation Software Tool (TEST) is freely available on US EPA website https://www.epa.gov/chemical-research/toxicity-estimation-software-tool-test and it was developed to provide an estimation of compound's toxicity by means of QSAR methodologies. It does not require any use of external programs and the user must input a molecule of interest to predict its toxicological profile. It also allows the prediction of physicochemical properties such as boiling and melting point, density and water solubility [119].

In order to perform QSAR analyzes it employs one (or more) of the following methodologies: hierarchical, FDA, single-model, group contribution, nearest neighbor, consensus and mode of action [86, 120, 121]. We emphasize the consensus method, that takes into account applying each of the above QSAR methodologies and uses an average of their predicted toxicities to estimate the toxicity of a compound.

According to the software user's guide from TEST [119], it includes models to predict the endpoints: 96-h fathead minnow 50 percent lethal concentration (LC_{50}), 48-h *Daphnia magna* 50 percent lethal concentration (LC_{50}), *Tetrahymena pyriformis* 50 percent growth inhibition concentration (IGC_{50}), Oral rat 50 percent lethal dose (LD_{50}), Bioconcentration Factor (BCF), Developmental Toxicity (DevTox), Ames Mutagenicity (Mutagenicity).

3.7 EPI Suite

This set of models and tools known as EPI (Estimation Programs Interface) Suite predicts physical/chemical property and environmental fate, which are predominantly built by means of QSAR [122].

United States Environmental Protection Agency (US EPA) together with Syracuse Research Corp. (SRC) have developed such predictive models and tools, along with assessment methods and databases (over 40 thousand of compounds), that has been widely used by governmental and industry organizations to support the evaluation of new chemical substances, especially to screen those under the Toxic Substances Control Act (TSCA). The idea is to provide predictive knowledge upon what happens, in terms of hazard effects, to humans and to the environment when using and being exposed to these substances. Models can be freely accessed or downloaded through the official website: https://www.epa.gov/tsca-screening-tools/epi-suitetm-estimation-program-interface

Among the models and tools available within EPI Suite, one can mention: Ecological Structure–Activity Relationships Program (ECOSAR), and WSKOWWIN. The first one mentioned, as an e.g., estimates toxicity potential for a compound based on short and long-term exposure for fish, invertebrates, and aquatic plants along with a limited number of methods for terrestrial and marine organisms [123]. WSKOWWIN is useful to predict 1-octanol/water partition coefficient *(logP)* and also water solubility.

3.8 Lazar

Lazar is the nomenclature for Lazy Structure–Activity Relationships, and it was primarily developed around 2008 [124, 125]. It used to work based on the use of two main methods, which are classification (*k*-nearest neighbors and kernel models) and regression (multiple linear regression and kernel models), and treating chemical similarities as toxicity dependent values, so that it detains most relevant fragments for the toxicity endpoint under analysis.

Its current version, an open-source software from the small company in silico *toxicology* from the University of Freiburg, can be easily run on the website https://lazar.in-silico.ch/predict. The practical version of Lazar generates local QSAR models for each compound submitted to toxicity prediction, in an equivalent manner to read across procedure in toxicological risk assessment, allowing the user to appoint between a large variety of algorithms for descriptor calculation and selection, chemical similarity indices, and model building [126].

It is possible to apply several models that are available to predict toxicity endpoints by regression or classification methods, in the software's read-across approach. The following endpoints can be predicted: acute toxicity (*Daphnia magna* / *Fathead minnow*), blood–brain barrier penetration in humans, carcinogenicity (rodents, rat, mouse—TD_{50}), lowest observed adverse effect level in rat, maximum recommended daily dose in human, and mutagenicity in *Salmonella typhimurium*.

3.9 MultiCASE

CASE Ultra is the name of the QSAR software afforded commercially by MultiCASE that aims to model and predict toxicity for chemical compounds, by automatically deriving alerts from data applying statistical methods or use alerts obtained applying expert knowledge. Models generated by CASE Ultra consists of a set of key alerts, which present a statistical relationship with the endpoint of interest, once it extracts alerts based on fragments or chemical structures [127].

It makes the use of physicochemical such as *logP*, solubility, E-states, surface descriptors, and others, to generate local QSAR models for alerts. Furthermore, one should emphasize that predictions are done, considering alerts found in the compound submitted to test, as well as that this software performs domain applicability evaluation. It also allows the user to add in-house chemical structure–activity data to an existing model, so that model's capabilities can be improved (https://multicase.com/case-ultra).

The software CASE ultra includes a high-quality collection of in silico toxicity models, covering the toxicological endpoints of genotoxicity, renal (kidney and urinary tract) toxicity, mammalian reproductive toxicity, adverse effects, carcinogenicity, cardiotoxicity, acute toxicity, endocrine receptor models, skin and eye toxicity, hepatotoxicity, ecotoxicity and environmental hazard, and bacterial mutagenicity. For the latter, it provides a pair of built-in QSAR methodologies, which are complementary (i.e. statistics and expert rule-based,) to estimate bacterial mutagenicity of chemical substances in agreement with ICH M7 compliance. Also, worth mentioning that many of the above-mentioned endpoints are predicted by using models based on FDA data [128, 129].

4 Expert Software

An expert software or system may be defined as a system able to emulate a human expert in making predictions, starting from the point of using generalized knowledge to achieve such goal, and hence it is commonly also described by the term knowledge-based systems [130]. In fact, one can describe its operation by a knowledge base that contains subject matter as a set of rules originated from experimental data, and an inference engine that makes logical deductions by assembling those rules [4, 46].

When it comes to software developed to toxicity prediction, in general, these set of rules are usually made of structure–toxicity relationships collected by toxicologists and, in this way, the expert system infers the compound's structure making a comparative analysis with those knowledge-based rules. Finally, the inference engine assesses the toxicity potential for such compound qualitatively in order to report its conclusions [131].

There are some outcomes of using expert software mainly regarding the quality and variation of toxicophores included in knowledge bases, which depending on the endpoint of interest there are many variables to describe mechanisms and experiments. Nevertheless, when comparing to traditional statistical models used as in the case of QSAR software, expert software can justify their estimations with considerations to the mechanisms of action provided by the knowledge database provided [132, 133]. These assays have received prestige due to the easy of interpretation [134]. However, they are often associated with poor accuracy, since many structural alerts, i.e., fragments associated with the mechanism of toxicity are also found in non-toxic compounds [135].

Our intention here is not to focus on the comparison between one methodology or another but to present a brief description of them, and their corresponding software, as well as their options of toxicological endpoints to be predicted. So, next, we will introduce some of the most valuable expert software that have been used by the computational toxicology scientific community in late years.

4.1 DEREK

DEREK is one of the pioneering software when it comes to expert knowledge-based methodologies to be used with the intention of predicting toxicity endpoints. It is a commercial product offered by the company Lhasa Limited, www.lhasa-limited.org, and it is an expert knowledge-based software that provides predictions for various toxicological parameters [136]. The term DEREK is actually an abbreviation for Deductive Estimation of Risk from Existing Knowledge.

What is attractive in such software, it is the fast, accurate and simple usage, where the user must submit a given structure as input to DEREK, which will be subsequently standardized and then analyzed in comparison to the certified Lhasa knowledge base implemented in the software, followed by a toxicity prediction as 'alerts'. Prediction using DEREK includes a general conclusion about the probability of toxicity for a structure and detailed information justifying such probability.

Standardization of structures is made by a series of transformation rules, including, but not limited to: aromaticity perception, transformation of pentavalent nitrogen and removal of specific stereochemistry. The goal of standardization is to interpret structures more accurately to optimize predictions. Systems that are rule-based present “if–then-else” rules that combine toxicological knowledge, expert judgment and fuzzy logic [137].

The DEREK database contains data from published sources and data donated by partner organizations. These data are collected and verified by scientists, members of the Lhasa company, before they are used to establish new rules and relationships. The latter two, in turn, are based on empirical observations that are supported by understanding the toxicity mechanism [133]. For the use of the DEREK software, the following endpoints can be predicted: mutagenicity, genotoxicity, neurotoxicity, developmental toxicity, carcinogenicity, HERG channel inhibition in vitro, teratogenicity, skin sensitization and irritation, hepatotoxicity in mammal, and others [138].

In general, the alert that is given by DEREK can be translated into a toxicophore, i.e. a substructure that is known or thought to be the main reason for its toxicity. Worth mentioning that many of the alerts refers to data for different kind of species, such as humans, other mammals and bacteria. DEREK uses a glossary, which contains a hierarchical order (from more certain to more doubtful) of alerts that will be fired, or not, as results for predictions. These are: certain, probable, plausible, equivocal, doubted, improbable, impossible, open and contradicted.

4.2 *HazardExpert*

HazardExpert was one of the first ever released expert software, in the 80 s, and today consists of a module of the Pallas System suite integrated within CompuDrug package of products (https://www.compudrug.com/hazardexpertpro) [139]. This software predicts a range of toxicity endpoints including carcinogenicity, irritation, teratogenicity, neurotoxicity, mutagenicity and immunotoxicity.

It allows the toxicity prediction of compounds and its symptoms in animals and humans, based on toxic fragments through a rule based system and open knowledge base, containing fragments derived from structure–toxicity relationships available on the literature and from US EPA data. Moreover, the user can change, expand or optimize data fragments where the most probable toxic estimation stands for. Further interesting resources included in HazardExpert is the ability to attenuate predictions by factors such as dose level, duration of exposure and route of administration; and also calculates bioavailability based on logP and pKa and bioaccumulation.

The software HazardExpert is commercially available to Windows and Linux platforms and its latest release include an upgrade with improved prediction of cytotoxicity, in which an artificial neural network based approach using atomic fragmental descriptors has been developed and implemented to categorize molecules according to their in vitro human cytotoxicity.

Worth mentioning that within Pallas System suite, offered by CompuDrug, there is the software MetabolExpert, which can work as an extension to initial toxicological predictions by parental compound by HazardExpert, allowing the estimation of the structural formula of its metabolites, which might be produced by such parental compound in humans, animals and plants.

Furthermore, another integrated tool in the same suite is ToxAlert, which function in a similar fashion as HazardExpert, indicating compounds with possible hazards for specific pharmacophores and, in addition, outputs probability percentages for endpoints under investigation [104].

4.3 *Toxtree*

Toxtree is a freely available platform that can be ran online or in its standalone version for PC, full-featured and flexible user-friendly open source application, which is able to estimate toxic hazard by applying a decision tree approach (https://toxtree.sourceforge.net/). It is cited in the official website from the European Union https://ec.europa.eu/jrc/en/scientific-tool/toxtree-tool as scientific computer tool able to estimate assess chemical toxicity, since it was developed by the JRC in collaboration with various consultants [137].

This software includes classification schemes for mutagenicity and carcinogenicity by the Benigni-Bossa rulebase, protein and DNA binding alerts, SMARTCyp (cytochrome P450 mediated drug metabolism and metabolites prediction), skin and eye irritation, as well as Cramer decision tree for the allocation of substances of known structure into class I, II or III based on their potential to present oral toxicity. Further details on such schemes are to be found on its main reference publication by Patlewicz and colaborators [140].

4.4 *Oncologic*

OncoLogic™ is a predictive computer system to evaluate the carcinogenic potential of chemical substances [141], by means of mimicking the judgment of human experts and following sets of knowledge rules based on studies of how such substances cause cancer in animals and humans (https://www.epa.gov/tsca-screening-tools/oncologictm-computer-system-evaluate-carcinogenic-potential-chemicals).

This software is freely available and allows the estimative of carcinogenicity (as its only predictable endpoint) in potential by compounds, by means of applying structure–activity relationships and considering their corresponding mechanisms of action, as well as supply a database with toxicological data to assess carcinogenicity endpoint [142].

It is an expert software developed by US EPA Office of Pollution Prevention and Toxics in collaboration with LogiChem, inc. One of its most interesting features is that, besides more than 50 classes of organic compounds, it is able to predict carcinogenicity for polymers, metals and fibers.

4.5 ToxPHACTS

ToxPHACTS is a software developed by the *Department für Pharmazeutische Chemie, Universität Wien*, supported by INiTS (Innovation into Business), the high-tech business incubator of the Business Agency of Vienna and Vienna's universities, within the AplusB Scale-up program, and commercially offered by the company Phenaris GmbH (https://www.phenaris.com/products/toxphacts/). It is also freely available as webserver for non-commercial use, hosted by University of Vienna, and in the format of ToxPHACTS light, that contains its basic features (https://toxphacts.univie.ac.at/).

The software ToxPHACTS works as an expert system for toxicological read across that can be split in three main tasks. At first, it performs a similarity search after the user inputs its query molecular structure, by automated read across utilizing five similarity search algorithms and consensus scoring, including the chemical space of ChEMBL. Then, it performs a data extraction by simultaneously retrieving several data sources with toxicity endpoints and including in-house data bases, subsequently leading to data analysis, which display results in standardized and aggregated form to guide the user. In addition, results may be exported as heat-map (CSV file) and it furnishes detailed knowledge based target-toxicity associations [143].

5 Machine Learning Models

Machine learning is a subfield of artificial intelligence in which computer algorithms learn and adapt to data. The artificial intelligence refers to a broader concept, in which computers can make better decisions. As in the previous topics, a QSAR model does not necessarily need the use of computers to be developed [144, 145]. Yet, with the increase in the amount of available data, such models have become impractical to be developed manually, with simple regression equations. Currently, QSAR/QSPR models are generated using thousands of compounds and molecular descriptors, applying the most variable algorithms machine learning [62]. For this reason, machine learning methods have become essential in cheminformatics. Machine learning methods can be roughly classified as unsupervised and supervised.

In unsupervised learning, predictions are made to characterize the data. In this case, only independent variables, i.e., those inherent to the data, are used in the development of the model. In cheminformatics, unsupervised methods are widely used to determine common or different structural characteristics between molecules, which can be used to design new bioactive molecules [146, 147]. Algorithms such as principal component analysis (PCA), hierarchical clustering analysis (HCA) and self-organizing maps (SOM) are commonly used to perform these tasks [148].

Supervised methods are those in which an input variable (dependent variable) is known and is intended to be predicted. In the case of QSAR, the biological activity or toxicity of a molecule is the dependent variable (represented on the **Y**-axis) and the molecular descriptors make up the independent variables (represented on the **X**-axis) [84]. The QSAR model is composed of a mathematical equation that establishes a relationship between molecular descriptors (chemical structure) and biological activity/property [149]. This equation is not always simple and observable. Modern machine learning algorithms such as *random forest* [150], *support vector machine* [151], *neural networks* [152], and *deep learning* [153] generate complex equations and, very often, difficult to interpret. For this reason, machine learning methods of learning are often described as "black boxes". One of the ways to interpret QSAR models is through the analysis of the molecular descriptors identified by the equation as the most important [135, 154]. In this process, it is possible to determine which physical–chemical, electrostatic characteristics or which molecular fragments are most important for the modeled activity. More recently, deep learning algorithms have become a major approach to develop QSAR models [155]. Neural networks are machine learning methods that were inspired by the connections of biological neural networks. The architecture of this algorithm has "neurons" or interconnected nodes that compute input data and release transformed output data. Deep learning, or deep neural networks, consisting of neural networks with more than one layer [153].

Computational tools to predict toxicity endpoints have become widely spread. In recent years, the National Institute for Environmental Health Sciences has promoted several global collaborations for the development of consensus QSAR models for the prediction of major toxicity endpoints, such as estrogen [156] and androgen [157] receptor activity. These models have been implemented into the OPERA software [158]. The UL REACHAcross™ a fee-based commercial application that implement machine learning models through a new approach named read-across structure–activity relationships (RASAR) [159]. Unfortunately, the authors did not provide information regarding data curation and many aspects of proper model validation. Many of these deficiencies have emphasized that the reported high accuracy of the models could be the consequence of inadequate data curation leaving many duplicate compounds in the modeling datasets [160].

The Pred-Skin is a freely-available web application that implements multiple QSAR models developed using machine learning for the evaluation of skin sensitization based on multiple assays, including data on humans, the mice, and in vitro assays [161]. Another tool developed by the same group, Pred-hERG [39], was generated by implementing models developed with random forest on cardiac toxicity data collected from ChEMBL [162]. The admetSAR is another free tool that implement machine learning models for many endpoints related to the absorption, distribution, metabolism, excretion, and toxicity (ADMET) of drugs [163]. The Pro-Tox II [164] is a web server that implements multiple molecular modeling approaches, such as similarity search, pharmacophores, and machine learning for prediction of toxicity endpoints, such as Ames mutagenicity, hepatotoxicity, and carcinogenicity. The Vienna LiverTox Workspace has a series of machine learning models for many liver transporters and hepatotoxicity-related endpoints, such as hyperbilirubinemia, cholestasis, and drug-induced liver injury [165]. The CarcinoPred-EL implements models based on ensemble support vector machine, random forest, and XGBoost algorithms to predict carcinogenicity of chemicals [34].

6 Use of Other Molecular Modeling Methodologies to Assess Toxicity

The development and improvement of methodologies associated with molecular modeling unveil more and more software and algorithms with useful applications in different contexts, including computational toxicology [8]. Here, we will briefly mention a few case studies involving toxicity prediction, in which it was used software employing some of the methodologies we consider more relevant, that is 3D-QSAR, docking, pharmacophore and/or molecular dynamics.

As we mentioned in the subtopic of this chapter regarding QSAR theory, the evolution of QSAR studies naturally achieved the use three-dimensional and conformational data in order to establish relationships, giving birth to 3D-QSAR [92]. This approach has also shown to be very useful in studies concerning toxicity prediction, e.g. [166] developed CoMFA and CoMSIA predictive models to show that the toxicity of substituted benzenes is related with electrostatic and H-bonding interactions, besides hydrophobic effects, and they could also get a picture of the respective toxicity mechanism involved, by analyzing corresponding contour maps.

Another work [167], proposed to design polychlorinated naphtalenes derivatives with less risk to the environment and human health, by building CoMSIA, 3D-QSAR, predictive models, and showing that the biodegradability mechanism of such compounds are related with electrostatic property. Moreover, they could show by docking simulations, that their proposed derivatives bind more efficiently with oxidase (degrading enzyme used as target), thus showing an increased degradability of such compounds. Finally, they performed molecular dynamics simulation to observe that during binding of derivatives with oxidase, the change rate of the electrostatic energy was highest, indicating that the electrostatic energy plays a key role on the change of binding energy and, thus, the electrostatic interaction is indeed the main factor affecting biodegradation of polychlorinated naphtalenes.

In the study of [168], they performed docking simulations using the FRED [169] software to evaluate interactions between potential molecular toxicants (environmentally relevant chemicals) and a library of macromolecular targets of toxicity (including rat estrogen receptors), to evaluate estrogenicity potential. Moreover, they noticed that using a pharmacophore filter, previously to the docking, is advantageous and improve the ability to identify potential chemicals active towards receptors of toxicity.

As one can see, docking studies may help to elucidate binding and intermolecular interactions between potential toxic compounds and amino acid residues from the binding site of targets—mostly enzymes related to toxicity mechanisms. More details concerning several available crystal structures of enzymes (e.g. CYP2C8, CYP2D6, P450, etc.) related to toxic mechanisms can be found in revision of Blaney and colaborators [170], as well as further research articles regarding docking studies [171, 172]. We also cite further literature regarding studies using pharmacophore [173, 174] and molecular dynamics [175, 176].

7 Concluding Remarks

Prediction of toxicity potential for bioactive substances is an important step associated with most of research projects involving drugs and agrochemicals development. To this point, one can note that most of the used computational toxicology software, employ the QSAR/QSTR approach, which is still the predominant methodology in computational toxicology. This traditional approach is reliable and easy to interpret. But that does not mean that is the preferable one, since we observe a great increase on popularity and reliability of predictions by expert software, and also by those employing machine learning methodologies. Another thing to note is that the tendency of using toxicity prediction software is related to their license availability, i.e. if they are either freely available or if they are only obtained on commercial versions.

Regardless, software here described provide a wide range of options, both in terms of methodologies and in terms of expanding possibilities to predict different toxicological endpoints, that spans from dermal sensitization to carcinogenicity, among others. The software described in this chapter are among the most robust options currently available, and guarantees a safe assessment of toxicological profile in studies that aim harnessing the potential of chemicals of interest.

In our opinion, we emphasize that the use of toxicological prediction has been gaining ground in the scientific community, and consequently in the world, considering that one of the greatest advantages is the fact that they avoid and reduce experimental procedures with laboratory animals, as well as reduce environmental impacts. Furthermore, the use of predictive toxicology software is very important in computer-aided drug design (in pharma industry or academia) because it allows to filter out toxic compounds during virtual screening campaigns, rationalize the synthesis and optimization of drug candidates, diminishing significantly the cost and time spent on drug discovery/development process. We also highlight that the use of other molecular modeling methodologies, such as docking and molecular dynamics, greatly helps in the mechanistic elucidation of the toxic action of compounds.

This chapter has aimed to describe the main software and methodologies that have been used to provide accurate and reliable toxicity prediction to a variety of toxicological endpoints. This overview of computational toxicology software should be useful to help on choosing the most adequate software for those who intend to improve their toxicity predictions on their research projects. To conclude, we hope that our efforts to this end, facilitate the work of gathering information on this subject to researchers who act on drugs and agrochemicals development and seek useful predictive toxicology tools. It is worth remembering that, in order to obtain more detailed information on how certain software applies a certain model, or on their database, or even on which species of animals does the experimental data come from, all references have been respectively assigned.

References

1. Reisfeld, B., Mayeno, A.N.: What is computational toxicology? in 3–7 (2012). https://doi.org/10.1007/978-1-62703-050-2_1
2. Helma, C., King, R.D., Kramer, S., Srinivasan, A.: The predictive toxicology challenge 2000–2001. Bioinformatics **17**, 107–108 (2001)
3. Alberga, D., Trisciuzzi, D., Mansouri, K., Mangiatordi, G.F., Nicolotti, O.: Prediction of acute oral systemic toxicity using a multifingerprint similarity approach. Toxicol. Sci. **167**, 484–495 (2019)
4. Gombar, V.K., Mattioni, B.E., Zwickl, C., Deahl, J.T.: Computational approaches for assessment of toxicity: a historical perspective and current status. In: Computational Toxicology, pp. 183–215. Wiley (2006). https://doi.org/10.1002/9780470145890.ch7
5. Steger-Hartmann, T., Pognan, F.: Improving the safety assessment of chemicals and drug candidates by the integration of bioinformatics and chemoinformatics data. Basic Clin. Pharmacol. Toxicol. **123**, 29–36 (2018)
6. Combs, A. B., Acosta, D. An Introduction to Toxicology and Its Methodologies. In: Computational Toxicology, pp. 1–20. Wiley (2006). https://doi.org/10.1002/9780470145890.ch1
7. Greene, N., Pennie, W.: Computational toxicology, friend or foe? Toxicol. Res. (2015). https://doi.org/10.1039/c5tx00055f
8. Goldsmith, M.R. et al.: Informing mechanistic toxicology with computational molecular models. in 139–165 (2012). https://doi.org/10.1007/978-1-62703-050-2_7
9. Schuster, D., Laggner, C., Langer, T.: Why drugs fail—a study on side effects in new chemical entities. Curr. Pharm. Des. **11**, 3545–3559 (2005)
10. Pollard, C.E., et al.: An introduction to QT interval prolongation and non-clinical approaches to assessing and reducing risk. Br. J. Pharmacol. (2010). https://doi.org/10.1111/j.1476-5381.2009.00207.x
11. Kar, S., Sanderson, H., Roy, K., Benfenati, E., Leszczynski, J.: Ecotoxicological assessment of pharmaceuticals and personal care products using predictive toxicology approaches. Green Chem. (2020). https://doi.org/10.1039/c9gc03265g
12. Sánchez-Bayo, F., Tennekes, H.A.: Environmental risk assessment of agrochemicals—a critical appraisal of current approaches. In: Toxicity and Hazard of Agrochemicals (InTech, 2015). https://doi.org/10.5772/60739
13. Insel, P.A., Amara, S.G., Blaschke, T.F., Meyer, U.A.: Introduction to the theme new methods and novel therapeutic approaches in pharmacology and toxicology. Annu. Rev. Pharmacol. Toxicol. **57**, 13–17 (2017)
14. Balls, M. et al.: Development and validation of non-animal tests and testing strategies: The identification of a coordinated response to the challenge and the opportunity presented by the sixth amendment to the cosmetics directive (76/768/EEC). The Report and Recommendatio. In: ATLA Alternatives to Laboratory Animals (1995)
15. Mangiatordi, G.F., et al.: Mind the gap! a journey towards computational toxicology. Mol. Inf. (2016). https://doi.org/10.1002/minf.201501017
16. Manibusan, M., Paterson, J., Kent, R., Chen, J.: (Quantitative) structure activity relationship [(Q)SAR] guidance document. US EPA Archive Document—NAFTA—TWG 186 (2012). Available at: https://www.epa.gov/sites/production/files/2016-01/documents/qsar-guidance.pdf. (Accessed: 30th May 2020)
17. US EPA.: Computational Toxicology Communities of Practice. US EPA (2020). Available at: https://www.epa.gov/chemical-research/computational-toxicology-communities-practice. Accessed 30 May 2020
18. Vian, M., Raitano, G., Roncaglioni, A., Benfenati, E.: In silico model for mutagenicity (Ames test), taking into account metabolism. Mutagenesis **34**, 41–48 (2019)
19. Kovacic, P., Somanathan, R.: Nitroaromatic compounds: Environmental toxicity, carcinogenicity, mutagenicity, therapy and mechanism. J. Appl. Toxicol. (2014). https://doi.org/10.1002/jat.2980

20. Kazius, J., McGuire, R., Bursi, R.: Derivation and validation of toxicophores for mutagenicity prediction. J. Med. Chem. **48**, 312–320 (2005)
21. Debnath, A.: The importance of the hydrophobic interaction in the mutagenicity of organic compounds. Mutat. Res. Mol. Mech. Mutagen. **305**, 63–72 (1994)
22. Valencia, A., Prous, J., Mora, O., Sadrieh, N., Valerio, L.G.: A novel QSAR model of Salmonella mutagenicity and its application in the safety assessment of drug impurities. Toxicol. Appl. Pharmacol. (2013). https://doi.org/10.1016/j.taap.2013.09.015
23. Villemin, D., Cherqaoui, D., Cense, J.: Neural networks studies: quantitative structure-activity relationship of mutagenic aromatic nitro compounds. J. Chim. Phys. **90**, 1505–1519 (1993)
24. do Amaral, A. et al.: Physicochemical parameters involved in the lethal toxicity of N, N--4-substituted benzoate hydrochlorides: a QSAR study. Eur. J. **32**, 433–443 (1997)
25. Fan, T., Sun, G., Zhao, L., Cui, X., Zhong, R.: QSAR and Classification study on prediction of acute oral toxicity of N-nitroso compounds. Int. J. Mol. Sci. **19**, 3015 (2018)
26. Gissi, A.: Alternative acute oral toxicity assessment under REACH based on sub-acute toxicity values. ALTEX 353–361 (2017). https://doi.org/10.14573/altex.1609121
27. Russo, D.P., et al.: Nonanimal models for acute toxicity evaluations: Applying data-driven profiling and read-across. Environ. Health Perspect. (2019). https://doi.org/10.1289/EHP3614
28. Ballabio, D., Grisoni, F., Consonni, V., Todeschini, R.: Integrated QSAR models to predict acute oral systemic toxicity. Mol. Inform. **38**, 1800124 (2019)
29. Roy, P.P., Banjare, P., Verma, S., Singh, J.: Acute rat and mouse oral toxicity determination of anticholinesterase inhibitor carbamate pesticides: a QSTR approach. Mol. Inform. **38**, 1800151 (2019)
30. Hoffmann, S., et al.: Acute oral toxicity: Variability, reliability, relevance and interspecies comparison of rodent LD50 data from literature surveyed for the ACuteTox project. Regul. Toxicol. Pharmacol. **58**, 395–407 (2010)
31. Strigari, L., et al.: A new model for predicting acute mucosal toxicity in head-and-neck cancer patients undergoing radiotherapy with altered schedules. Int. J. Radiat. Oncol. Biol. Phys. (2012). https://doi.org/10.1016/j.ijrobp.2012.02.004
32. Vainio, H., Coleman, M., Wilbourn, J.: Carcinogenicity evaluations and ongoing studies: the IARC databases. Environ. Health Perspect. **96**, 5–9 (1991)
33. Petrescu, A.-M., Ilia, G.: Potential carcinogenicity predicted by computational toxicity evaluation of thiophosphate pesticides using QSTR/QSCarciAR model. Drug Chem. Toxicol. **40**, 263–272 (2017)
34. Zhang, L., et al.: CarcinoPred-EL: novel models for predicting the carcinogenicity of chemicals using molecular fingerprints and ensemble learning methods. Sci. Rep. **7**, 2118 (2017)
35. Niculescu-duvâz, I., Craescu, T., Tugulea, M., Croisy, A., Jacquignon, P.C.: A quantitative structure-activity analysis of the mutagenic and carcinogenic action of 43 structurally related heterocyclic compounds. Carcinogenesis (1981). https://doi.org/10.1093/carcin/2.4.269
36. Richard, A.M., Benigni, R.: AI and SAR approaches for predicting chemical carcinogenicity: survey and status report. SAR QSAR Environ. Res. **13**, 1–19 (2002)
37. Kar, S., Roy, K.: Development and validation of a robust QSAR model for prediction of carcinogenicity of drugs. Indian J. Biochem, Biophys (2011)
38. Li, Z.: Risk-based principles and incompleteness theorems for linear dose-response extrapolation for carcinogenic chemicals. Chemosphere **247**, 125934 (2020)
39. Braga, R.C., et al.: Pred-hERG: a novel web-accessible computational tool for predicting cardiac toxicity. Mol. Inform. **34**, 698–701 (2015)
40. Magdy, T., Burmeister, B.T., Burridge, P.W.: Validating the pharmacogenomics of chemotherapy-induced cardiotoxicity: what is missing? Pharmacol. Ther. (2016). https://doi.org/10.1016/j.pharmthera.2016.09.009
41. Cavalli, A., Poluzzi, E., De Ponti, F., Recanatini, M.: Toward a pharmacophore for drugs inducing the long QT syndrome: insights from a CoMFA study of HERG K+ channel blockers. J. Med. Chem. (2002). https://doi.org/10.1021/jm0208875

42. Keserü, G.M.: Prediction of hERG potassium channel affinity by traditional and hologram qSAR methods. Bioorg. Med. Chem. Lett. **13**, 2773–2775 (2003)
43. Ekins, S., Balakin, K.V., Savchuk, N., Ivanenkov, Y.: Insights for human ether-a-Go-Go-related gene potassium channel inhibition using recursive partitioning and kohonen and sammon mapping techniques. J. Med. Chem. **49**, 5059–5071 (2006)
44. Xu, J.J.: In vitro Toxicology: bringing the In silico and in vivo worlds closer. In: Computational Toxicology, pp. 21–32. Wiley (2006). https://doi.org/10.1002/9780470145890.ch2
45. Mortelmans, K., & Zeiger, E. The Ames Salmonella/microsome mutagenicity assay. Mutat. Res. Fundam. Mol. Mech. Mutagen. (2000). https://doi.org/10.1016/S0027-5107(00)00064-6
46. Gadaleta, D., Manganelli, S., Manganaro, A., Porta, N., Benfenati, E.: A knowledge-based expert rule system for predicting mutagenicity (Ames test) of aromatic amines and azo compounds. Toxicology (2016). https://doi.org/10.1016/j.tox.2016.09.008
47. Ames, B.N., McCann, J., Yamasaki, E.: Methods for detecting carcinogens and mutagens with the salmonella/mammalian-microsome mutagenicity test. Mutat. Res. Mutagen. Relat. Subj. **31**, 347–363 (1975)
48. Fourches, D., Muratov, E., Tropsha, A.: Curation of chemogenomics data. Nat. Chem. Biol. **11**, 535–535 (2015)
49. Fourches, D., Muratov, E., Tropsha, A.: Trust, but verify: on the importance of chemical structure curation in cheminformatics and QSAR modeling research. J. Chem. Inf. Model. **50**, 1189–1204 (2010)
50. Worth, A., Fuart-Gatnik, M., Lapenna, S., Serafimova, R.: SCIENTIFIC REPORT submitted to EFSA applicability of QSAR analysis in the evaluation of developmental and neurotoxicity effects for the assessment of the toxicological relevance of metabolites and degradates of pesticide active substances for dietary risk 174 (2011)
51. Andrews, P.R., Craik, D.J., Martin, J.L.: Functional group contributions to drug-receptor interactions. J. Med. Chem. **27**, 1648–1657 (1984)
52. Vukovic, K., Gadaleta, D., Benfenati, E.: Methodology of AiQSAr: a group-specific approach to QSAR modelling. J. Cheminform. (2019). https://doi.org/10.1186/s13321-019-0350-y
53. Kubinyi, H. QSAR: Hansch Analysis and Related Approaches. Wiley (1993)
54. Rusyn, I., Sedykh, A., Low, Y., Guyton, K.Z., Tropsha, A.: Predictive modeling of chemical hazard by integrating numerical descriptors of chemical structures and short-term toxicity assay data. Toxicol. Sci. **127**, 1–9 (2012)
55. Baumann, D., Baumann, K.: Reliable estimation of prediction errors for QSAR models under model uncertainty using double cross-validation. J. Cheminform. **6**, 1–19 (2014)
56. Hansch, C., Leo, A.: Exploring QSAR: Fundamentals and Applications in Chemistry and Biology. American Chemical Society—ACS (1995)
57. C. Hansch, T.F.: p-sigma-pi analysis. A Method for the Correlation of Biological Activity and Chemical Structure. Correl. Biol. Act. Chem. Struct. **86**, 1616–1626 (1964)
58. Gertzen, C.G.W., Gohlke, H.: From Hansch-Fujita analysis to AFMoC: a road to structure-based QSAR. Mol. Inform. **31**, 698–704 (2012)
59. Hansch, C., Maloney, P., Fujita, T., Muir, R.: Correlation of biological activity of phenoxy-acetic acids with Hammett substituent constants and partition coefficients. Nature **194**, 178–180 (1962)
60. Alexander, D.L.J., Tropsha, A., Winkler, D.A.: Beware of R^2: simple, unambiguous assessment of the prediction accuracy of QSAR and QSPR models. J. Chem. Inf. Model. **55**, 1316–1322 (2015)
61. Willett, P.: Chemoinformatics: a history. Wiley Interdiscip. Rev. Comput. Mol. Sci. **1**, 46–56 (2011)
62. Cherkasov, A., et al.: Perspective QSAR modeling: where have you been ? Where are you going to? QSAR modeling: where have you been? Where are you going to? J. Med. Chem. **57**, 4977–5010 (2014)
63. Kubinyi, H.: The changing landscape in drug discovery. In: Discovery, pp. 24–46 (2007). https://doi.org/10.1039/9781847557964-00024

64. Hansch, C.: The advent and evolution of QSAR at Pomona College. J. Comput. Aided. Mol. Des. **25**, 495–507 (2011)
65. Gramatica, P.: Principles of QSAR modeling. Int. J. Quant. Struct. Relationships **5**, 1–37 (2020)
66. Muratov, E.N., et al.: QSAR without borders. Chem. Soc. Rev. (2020). https://doi.org/10.1039/D0CS00098A
67. Sliwoski, G., Mendenhall, J., Meiler, J.: Autocorrelation descriptor improvements for QSAR: 2DA-Sign and 3DA-Sign. J. Comput. Aided. Mol. Des. **30**, 209–217 (2016)
68. Tropsha, A., Golbraikh, A.: Predictive QSAR modeling workflow, model applicability domains, and virtual screening. Curr. Pharm. Des. **13**, 3494–3504 (2007)
69. Kausar, S., Falcao, A.O.: An automated framework for QSAR model building. J. Cheminform. **10**, 1–23 (2018)
70. Tetko, I.V., Tanchuk, V.Y.: Application of associative neural networks for prediction of lipophilicity in ALOGPS 2.1 program. J. Chem. Inf. Comput. Sci. **42**, 1136–1145 (2002)
71. Schwab, C.H.: Conformations and 3D pharmacophore searching. Drug Discov. Today Technol. **7**, e245–e253 (2010)
72. Yap, C.W.: PaDEL-descriptor: an open source software to calculate molecular descriptors and fingerprints. J. Comput. Chem. (2011). https://doi.org/10.1002/jcc.21707
73. Todeschini, R., Consonni, V.: Handbook of Molecular Descriptors. Wiley (2000). https://doi.org/10.1002/9783527613106
74. Scotti, M.T., et al.: Variable-selection approaches to generate QSAR models for a set of antichagasic semicarbazones and analogues. Chemom. Intell. Lab. Syst. **154**, 137–149 (2016)
75. OECD.: Guidance Document on the Validation of (Quantitative) Structure-Activity Relationship [(Q)SAR] Models. **69**, OECD, (2007)
76. Martin, T.M., Lilavois, C.R., Barron, M.G.: Prediction of pesticide acute toxicity using two-dimensional chemical descriptors and target species classification. SAR QSAR Environ. Res. (2017). https://doi.org/10.1080/1062936X.2017.1343204
77. Enslein, K., Craig, P.N.: A toxicity estimation model. J. Environ. Pathol. Toxicol. (1978)
78. Lei, T. et al.: ADMET evaluation in drug discovery: 15. Accurate prediction of rat oral acute toxicity using relevance vector machine and consensus modeling. J. Cheminform. (2016). https://doi.org/10.1186/s13321-016-0117-7
79. Sazonovas, A., Japertas, P., Didziapetris, R.: Estimation of reliability of predictions and model applicability domain evaluation in the analysis of acute toxicity (LD50). SAR QSAR Environ. Res. (2010). https://doi.org/10.1080/10629360903568671
80. Hamadache, M., et al.: A quantitative structure activity relationship for acute oral toxicity of pesticides on rats: validation, domain of application and prediction. J. Hazard. Mater. (2016). https://doi.org/10.1016/j.jhazmat.2015.09.021
81. Roy, K., Kar, S., Das, R.N.: Statistical methods in QSAR/QSPR. In: A Primer on QSAR/QSPR Modeling, pp. 37–59. Springer (2015). https://doi.org/10.1007/978-3-319-17281-1_2
82. Fujita, T., Winkler, D.A.: Understanding the roles of the 'two QSARs.' J. Chem. Inf. Model. (2016). https://doi.org/10.1021/acs.jcim.5b00229
83. Gramatica, P.: Principles of QSAR models validation: internal and external. QSAR Comb. Sci. **26**, 694–701 (2007)
84. Tropsha, A.: Best practices for QSAR model development, validation, and exploitation. Mol. Inform. **29**, 476–488 (2010)
85. Dearden, J.C., Cronin, M.T.D., Kaiser, K.L.E.: How not to develop a quantitative structure–activity or structure–property relationship (QSAR/QSPR). SAR QSAR Environ. Res. **20**, 241–266 (2009)
86. Young, D., Martin, T., Venkatapathy, R., Harten, P.: Are the chemical structures in your QSAR correct? QSAR Comb. Sci. **27**, 1337–1345 (2008)
87. Castillo-González, D., et al.: Harmonization of QSAR best practices and molecular docking provides an efficient virtual screening tool for discovering new G-quadruplex ligands. J. Chem. Inf. Model. **55**, 2094–2110 (2015)

88. Golbraikh, A., Shen, M., Xiao, Z., Xiao, Y., Lee, K.: Rational selection of training and test sets for the development of validated QSAR models. J. Comput. Aided. Mol. Des. **17**, 241–253 (2003)
89. Tropsha, A., Gramatica, P., Gombar, V.K.: The importance of being earnest: validation is the absolute essential for successful application and interpretation of QSPR models. QSAR Comb. Sci. **22**, 69–77 (2003)
90. Kubinyi, H.: From narcosis to hyperspace: the history of QSAR. Quant. Struct. Act. Relat. (2002). https://doi.org/10.1002/1521-3838(200210)21:4%3c348::AID-QSAR348%3e3.0.CO;2-D
91. Gramatica, P., Sangion, A.: A historical excursus on the statistical validation parameters for QSAR models: a clarification concerning metrics and terminology. J. Chem. Inf. Model. **56**, 1127–1131 (2016)
92. Cramer, R.D., Patterson, D.E., Bunce, J.D.: Comparative molecular field analysis (CoMFA). 1. Effect of shape on binding of steroids to carrier proteins. J. Am. Chem. Soc. (1988). https://doi.org/10.1021/ja00226a005
93. Doweyko, A.M.: 3D-QSAR illusions. J. Comput. Aided. Mol. Des. **18**, 587–596 (2004)
94. Klebe, G., Abraham, U., Mietzner, T.: Molecular similarity indices in a comparative analysis (CoMSIA) of drug molecules to correlate and predict their biological activity. J. Med. Chem. **37**, 4130–4146 (1994)
95. Verma, J., Khedkar, V., Coutinho, E.: 3D-QSAR in drug design—a review. Curr. Top. Med. Chem. **10**, 95–115 (2010)
96. Lowis, D.R.: HQSAR: a new, highly predictive QSAR technique. Tripos Tech. Notes **1**, 1–17 (1997)
97. Vedani, A., Briem, H., Dobler, M., Dollinger, H., McMasters, D.R.: Multiple-conformation and protonation-state representation in 4D-QSAR: the neurokinin-1 receptor system. J. Med. Chem. (2000). https://doi.org/10.1021/jm000986n
98. Vedani, A., Dobler, M.: Multidimensional QSAR: moving from three- to five-dimensional concepts. Quant. Struct. Relationships **21**, 382–390 (2002)
99. Prado-Prado, F.J. et al.: Unified QSAR approach to antimicrobials. 4. Multi-target QSAR modeling and comparative multi-distance study of the giant components of antiviral drug–drug complex networks. Bioorg. Med. Chem. **17**, 569–575 (2009)
100. Lambrinidis, G., Tsantili-Kakoulidou, A.: Challenges with multi-objective QSAR in drug discovery. Expert Opin. Drug Discov. **13**, 851–859 (2018)
101. Schultz, T.W., Cronin, M.T.D., Walker, J.D., Aptula, A.O.: Quantitative structure–activity relationships (QSARs) in toxicology: a historical perspective. J. Mol. Struct. THEOCHEM **622**, 1–22 (2003)
102. Roy, K., Ghosh, G.: QSTR with extended topochemical atom indices. 2. Fish toxicity of substituted benzenes. J. Chem. Inf. Comput. Sci. (2004). https://doi.org/10.1021/ci0342066
103. He, L., et al.: Insights into pesticide toxicity against aquatic organism: QSTR models on Daphnia Magna. Ecotoxicol. Environ. Saf. (2019). https://doi.org/10.1016/j.ecoenv.2019.02.014
104. Fuart Gatnik, M., Worth, A.P.: Review of software tools for toxicity prediction. JRC Sci. Tech. Rep. (2010). https://doi.org/10.2788/60101
105. Sild, S., Piir, G., Neagu, D., Maran, U.: Storing and using qualitative and quantitative structure–activity relationships in the era of toxicological and chemical data expansion. In: Issues in Toxicology, pp. 185–213 (2019). https://doi.org/10.1039/9781782623656-00185
106. Zhu, H., et al.: Combinatorial QSAR Modeling of chemical toxicants tested against tetrahymena pyriformis. J. Chem. Inf. Model. **48**, 766–784 (2008)
107. Matthews, E., Kruhlak, N., Benz, R., Contrera, J.: Assessment of the health effects of chemicals in humans: I. QSAR estimation of the maximum recommended therapeutic dose (MRTD) and no effect level (NOEL) of organic chemicals based on clinical trial data1. Curr. Drug Discov. Technol. **1**, 61–76 (2004)
108. Cassani, S., et al.: Daphnia and fish toxicity of (benzo)triazoles: validated QSAR models, and interspecies quantitative activity-activity modelling. J. Hazard. Mater. (2013). https://doi.org/10.1016/j.jhazmat.2013.04.025

109. Gramatica, P., Cassani, S., Chirico, N.: QSARINS-chem: insubria datasets and new QSAR/QSPR models for environmental pollutants in QSARINS. J. Comput. Chem. (2014). https://doi.org/10.1002/jcc.23576
110. Li, J., Gramatica, P.: The importance of molecular structures, endpoints' values, and predictivity parameters in QSAR research: QSAR analysis of a series of estrogen receptor binders. Mol. Divers. (2010). https://doi.org/10.1007/s11030-009-9212-2
111. Gramatica, P., Papa, E.: Screening and ranking of POPs for global half-life: QSAR approaches for prioritization based on molecular structure. Environ. Sci. Technol. (2007). https://doi.org/10.1021/es061773b
112. Papa, E., Sangion, A., Arnot, J.A., Gramatica, P.: Development of human biotransformation QSARs and application for PBT assessment refinement. Food Chem. Toxicol. **112**, 535–543 (2018)
113. Gramatica, P., Cassani, S., Sangion, A.: Aquatic ecotoxicity of personal care products: QSAR models and ranking for prioritization and safer alternatives' design. Green Chem. (2016). https://doi.org/10.1039/c5gc02818c
114. Gramatica, P., Giani, E., Papa, E.: Statistical external validation and consensus modeling: a QSPR case study for Koc prediction. J. Mol. Graph. Model. (2007). https://doi.org/10.1016/j.jmgm.2006.06.005
115. Gramatica, P., Pilutti, P., Papa, E.: Approaches for externally validated QSAR modelling of nitrated polycyclic aromatic hydrocarbon mutagenicity. SAR QSAR Environ. Res. (2007). https://doi.org/10.1080/10629360601054388
116. Papa, E., Gramatica, P.: QSPR as a support for the EU REACH regulation and rational design of environmentally safer chemicals: PBT identification from molecular structure. Green Chem. (2010). https://doi.org/10.1039/b923843c
117. Bhhatarai, B., Gramatica, P.: Per-and polyfluoro toxicity (LC50 inhalation) study in rat and mouse using QSAR modeling. Chem. Res. Toxicol. (2010). https://doi.org/10.1021/tx900252h
118. Sangion, A., Gramatica, P.: Ecotoxicity interspecies QAAR models from Daphnia toxicity of pharmaceuticals and personal care products $. SAR QSAR Environ. Res. **27**, 781–798 (2016)
119. US EPA.: User's Guide for T.E.S.T. (version 4.2) (Toxicity Estimation Software Tool): A Program to Estimate Toxicity from Molecular Structure. US EPA (2016). Available at: https://www.epa.gov/chemical-research/users-guide-test-version-42-toxicity-estimation-software-tool-program-estimate. Accessed 30 May 2020
120. Zhu, H., et al.: Quantitative structure-activity relationship modeling of rat acute toxicity by oral exposure. Chem. Res. Toxicol. (2009). https://doi.org/10.1021/tx900189p
121. Martin, T.M., Harten, P., Venkatapathy, R., Das, S., Young, D.M.: A hierarchical clustering methodology for the estimation of toxicity. Toxicol. Mech. Methods (2008). https://doi.org/10.1080/15376510701857353
122. US EPA. EPI Suite^{TM}-Estimation Program Interface. US EPA (2020). Available at: https://www.epa.gov/tsca-screening-tools/epi-suitetm-estimation-program-interface. Accessed 30 May 2020
123. US EPA.: OPERATION MANUAL for the ECOlogical structure-activity relationship model (ECOSAR) class program. US EPA (2020). Available at: https://www.epa.gov/sites/production/files/2017-11/documents/operationmanual_v2.0.pdf. Accessed 30 May 2020
124. Helma, C.: Lazy structure-activity relationships (lazar) for the prediction of rodent carcinogenicity and Salmonella mutagenicity. Mol. Divers. (2006). https://doi.org/10.1007/s11030-005-9001-5
125. Maunz, A., Helma, C.: Prediction of chemical toxicity with local support vector regression and activity-specific kernels. SAR QSAR Environ. Res. **19**, 413–431 (2008)
126. Maunz, A. *et al.* lazar: a modular predictive toxicology framework. Front. Pharmacol. **4** (2013)
127. Chakravarti, S.K., Saiakhov, R.D., Klopman, G.: Optimizing predictive performance of CASE ultra expert system models using the applicability domains of individual toxicity alerts. J. Chem. Inf. Model. **52**, 2609–2618 (2012)

128. Saiakhov, R., Chakravarti, S., Klopman, G.: Effectiveness of CASE ultra expert system in evaluating adverse effects of drugs. Mol. Inform. (2013). https://doi.org/10.1002/minf.201200081
129. Chakravarti, S.K., Alla, S.R.M.: Descriptor free QSAR modeling using deep learning with long short-term memory neural networks. Front. Artif. Intell. (2019). https://doi.org/10.3389/frai.2019.00017
130. Dearden, J.C. et al.: The Development and Validation of Expert Systems for Predicting Toxicity. In: The report and recommendations of an ECVAM/ECB workshop (ECVAM workshop 24). ATLA (1997)
131. Judson, P.: Knowledge-Based Expert Systems in Chemistry. The Royal Society of Chemistry. Royal Society of Chemistry (2009). https://doi.org/10.1039/9781847559807
132. Judson, P.N.: Expert systems. In: Computational Toxicology, pp. 521–543. Wiley, (2006). https://doi.org/10.1002/9780470145890.ch18
133. Van Bossuyt, M., et al.: (Q)SAR tools for priority setting: a case study with printed paper and board food contact material substances. Food Chem. Toxicol. (2017). https://doi.org/10.1016/j.fct.2017.02.002
134. Cronin, M.T.D.: Evaluation of categories and read-across for toxicity prediction allowing for regulatory acceptance. Issues Toxicol. (2013)
135. Alves, V.M., et al.: Alarms about structural alerts. Green Chem. **18**, 4348–4360 (2016)
136. Sanderson, D.M., Earnshaw, C.G.: Computer prediction of possible toxic action from chemical structure; the DEREK system. Hum. Exp. Toxicol. (1991). https://doi.org/10.1177/096032719101000405
137. Worth, A. et al.: The Use of Computational Methods in the Toxicological Assessment of Chemicals in Food: Current Status and Future Prospects (2011). https://doi.org/10.2788/6234
138. Barratt, M.D.: Prediction of toxicity from chemical structure. Cell Biol. Toxicol. (2000). https://doi.org/10.1023/A:1007676602908
139. Smithing, M.P.: Hazard expert: an expert system for predicting chemical toxicity. In: Food Safety Assessment (1992). https://doi.org/10.1021/bk-1992-0484.ch019
140. Patlewicz, G., Jeliazkova, N., Safford, R.J., Worth, A.P., Aleksiev, B.: An evaluation of the implementation of the Cramer classification scheme in the Toxtree software. SAR QSAR Environ. Res. **19**, 495–524 (2008)
141. Woo, Y.T., Lai, D.Y.: Oncologic: a mechanism-based expert system for predicting the carcinogenic potential of chemicals. In: Predictive Toxicology (2005)
142. US EPA. OncoLogic^{TM}—A Computer System to Evaluate the Carcinogenic Potential of Chemicals. US EPA (2005). Available at: https://www.epa.gov/tsca-screening-tools/oncologictm-computer-system-evaluate-carcinogenic-potential-chemicals. Accessed: 30 May 2020
143. Vienna BioCenter. News—ToxPHACTS.: (2020). Available at: https://www.viennabiocenter.org/about/news/toxphacts-available-for-purchase/. Accessed 30 May 2020
144. Tetko, I. V, Engkvist, O., Chen, H.: Does 'Big Data' exist in medicinal chemistry, and if so, how can it be harnessed? Future Med. Chem. **8**, 1801–1806 (2016)
145. Fourches, D.: Cheminformatics: at the crossroad of eras. In: Gorb, L., Kuz'min, V., Muratov, E. (eds.) Application of Computational Techniques in Pharmacy and Medicine, pp. 539–546. Springer, Netherlands, (2014). doi:https://doi.org/10.1007/978-94-017-9257-8
146. Downs, G.M., Barnard, J.M.: Clustering methods and their uses in computational chemistry. In: Lipkowitz, K.B., Boyd, D.B. (eds.) Reviews in Computational Chemistry, vol. **18**, pp. 1–40. Wiley (2003)
147. Mercier, D.: Clustering large datasets. Electronic review—Linacre College (2003)
148. Miller, J.N., Miller, J.C.: Statistics and Chemometrics for Analytical Chemistry. Ashford Colour Press Ltd. (2010)
149. Welling, M.: A First Encounter with Machine Learning. University of California (2011)
150. Breiman, L.E.O.: Random forests. Mach. Learn. **45**, 5–32 (2001)
151. Cortes, C., Vapnik, V.: Support-vector networks. Mach. Learn. **20**, 273–297 (1995)
152. Dreiseitl, S., Ohno-Machado, L.: Logistic regression and artificial neural network classification models: a methodology review. J. Biomed. Inform. **35**, 352–359 (2002)

153. LeCun, Y., Bengio, Y., Hinton, G.: Deep learning. Nature **521**, 436–444 (2015)
154. Kuz'min, V.E., Polishchuk, P.G., Artemenko, A.G., Andronati, S.A.: Interpretation of QSAR models based on random forest methods. Mol. Inform.**30**, 593–603 (2011)
155. Smith, J.S., Roitberg, A.E., Isayev, O.: Transforming computational drug discovery with machine learning and AI. ACS Med. Chem. Lett. **9**, 1065–1069 (2018)
156. Mansouri, K., et al.: CERAPP: collaborative estrogen receptor activity prediction project. Environ. Health Perspect. **124**, 1023–1033 (2016)
157. Mansouri, K., et al.: CoMPARA: collaborative modeling project for androgen receptor activity. Environ. Health Perspect. **128**, 027002 (2020)
158. Mansouri, K., Grulke, C.M., Judson, R.S., Williams, A.J.: OPERA models for predicting physicochemical properties and environmental fate endpoints. J. Cheminform. **10**, 10 (2018)
159. Luechtefeld, T., Marsh, D., Rowlands, C., Hartung, T.: Machine learning of toxicological big data enables read-across structure activity relationships (RASAR) outperforming animal test reproducibility. Toxicol. Sci. **165**, 198–212 (2018)
160. Alves, V.M., et al.: Oy Vey! a comment on machine learning of toxicological big data enables read-across structure activity relationships outperforming animal test reproducibility. Toxicol. Sci. **167**, 3–4 (2019)
161. Braga, R.C., et al.: Pred-Skin: a fast and reliable web application to assess skin sensitization effect of chemicals. J. Chem. Inf. Model. **57**, 1013–1017 (2017)
162. Gaulton, A., et al.: The ChEMBL database in 2017. Nucleic Acids Res. **45**, D945–D954 (2017)
163. Cheng, F., et al.: admetSAR: a comprehensive source and free tool for assessment of chemical ADMET properties. J. Chem. Inf. Model. **52**, 3099–3105 (2012)
164. Banerjee, P., Eckert, A.O., Schrey, A.K., Preissner, R.: ProTox-II: a webserver for the prediction of toxicity of chemicals. Nucleic Acids Res. **46**, W257–W263 (2018)
165. Montanari, F., et al.: Vienna livertox workspace—a set of machine learning models for prediction of interactions profiles of small molecules with transporters relevant for regulatory agencies. Front. Chem. **7**, 899 (2020)
166. Salahinejad, M., Ghasemi, J.B.: 3D-QSAR studies on the toxicity of substituted benzenes to tetrahymena pyriformis: CoMFA, CoMSIA and VolSurf approaches. Ecotoxicol. Environ. Saf. **105**, 128–134 (2014)
167. Gu, W., Li, Q., Li, Y.: Law and mechanism analysis of biodegradability of polychlorinated naphthalenes based on principal component analysis, QSAR models, molecular docking and molecular dynamics simulation. Chemosphere **243**, 125427 (2020)
168. Rabinowitz, J.R., Little, S.B., Laws, S.C., Goldsmith, M.R.: Molecular modeling for screening environmental chemicals for estrogenicity: use of the toxicant-target approach. Chem. Res. Toxicol. (2009). https://doi.org/10.1021/tx900135x
169. OpenEye. FRED (2009)
170. Blaney, F.E., Tehan, B.G.: Crystal structures of toxicology targets. In: Computational Toxicology, pp. 469–519. Wiley (2006). https://doi.org/10.1002/9780470145890.ch17
171. Hirst, W.D., et al.: Differences in the central nervous system distribution and pharmacology of the mouse 5-hydroxytryptamine-6 receptor compared with rat and human receptors investigated by radioligand binding, site-directed mutagenesis, and molecular modeling. Mol. Pharmacol. (2003). https://doi.org/10.1124/mol.64.6.1295
172. Aureli, L., et al.: Predicting human serum albumin affinity of interleukin-8 (CXCL8) inhibitors by 3D-QSPR approach. J. Med. Chem. (2005). https://doi.org/10.1021/jm049227l
173. Ekins, S., et al.: Three-dimensional quantitative structure-activity relationships of inhibitors of P-glycoprotein. Mol. Pharmacol. (2002). https://doi.org/10.1124/mol.61.5.964
174. Lewis, D.F.V.: Molecular modeling of human cytochrome P450–substrate interactions. Drug Metab. Rev. **34**, 55–67 (2002)
175. Thorsteinson, N., et al.: In silico identification of anthropogenic chemicals as ligands of zebrafish sex hormone binding globulin. Toxicol. Appl. Pharmacol. (2009). https://doi.org/10.1016/j.taap.2008.07.014

176. Cheng, X., Ivanov, I.: Molecular dynamics. In: Reisfeld, B., Mayeno, A.N. (eds.) Computational Toxicology: Volume I, Methods in Molecular Biology, vol. 929, pp. 243–285. Springer, LLC (2012)

Anticancer Lead Compounds that Prevent DNA Binding to hnRNP K

Vinicius Barreto da Silva, Flavia Amoroso Matos e Silva, Cristiana Bernadelli Garcia, Andreia Machado Leopoldino, Carlos Henrique Tomich de Paula da Silva, and Carlton Anthony Taft

Abstract hnRNP K is an important constitutive protein in which is found in the nucleus, cytoplasm, and mitochondria of cells. As such, this protein interacts in turn with various molecules, which are directly involved in gene expression as well as signal transduction. However, it is well-known that its aberrant expression is related to the development of most commonly diagnosed cancers, including prostate, lung, breast, and colorectal. Hence, the binding to nucleotides is the main molecular event responsible for triggers the biological activity of hnRNP K and in which is mediated by its K homology (KH) domains. Using the structure of KH3 domain, virtual screening simulations were then performed by docking using GOLD software to select small molecules that could compete with nucleotides by the binding site of the domain, intending to block the protein activity and discover new lead compounds against cancer. In vitro assays revealed the discovery of a benzimidazole and a phenylbenzamide derivative able to prevent DNA binding to hnRNP K. The molecular interaction fields computed for hydrophobic and polar interactions for KH3 structure and molecular dynamics simulations with docked compounds revealed energetically viable binding modes for these derivatives, where arginine protein residues should play a central role in molecular recognition. The design of benzimidazole and phenylbenzamide derivatives enriches the knowledge of lead compounds in the search for a novel class of anticancer drugs able to down-regulate hnRNP K.

Keywords Heterogeneous ribonucleoprotein K · KH domain · Antineoplastic · Virtual screening · Molecular modeling

---

V. B. da Silva (✉)
School of Medical, Pharmaceutical and Biomedical Sciences, Pontifícia Universidade Católica de Goiás, Goiânia, Brazil
e-mail: vbarreto@pucgoias.edu.br

F. A. M. e Silva · C. B. Garcia · A. M. Leopoldino · C. H. T. de P. da Silva
Faculdade de Ciências Farmacêuticas de Ribeirão Preto, Universidade de São Paulo, Ribeirão Preto, Brazil

C. A. Taft
Centro Brasileiro de Pesquisas Físicas, Rio de Janeiro, Brazil

F. A. La Porta and C. A. Taft (eds.), *Functional Properties of Advanced Engineering Materials and Biomolecules*, Engineering Materials,
https://doi.org/10.1007/978-3-030-62226-8_23

1 Introducution

Currently, cancer has been a leading cause of death worldwide, and the number of cases is also expected to overgrow rapidly as populations grow, age, and adopt lifestyle behaviors that in general increase cancer risk significantly [1]. Thus, from this perspective, it is well-known that the cancer is particularly caused by diverse deregulated cell signaling cascades, leading more than 200 neoplastic diseases. As such, the disease develops from the result of the dysregulation of as high as approximately 500 different genes, among which may leads to over a very long duration of time untill the symptoms become apparent [2, 3]. From a histogenetic viewpoint, cancers have been classified into carcinomas and sarcomas. Carcinomas are originated from epithelial tissues and sarcomas from connective tissues. Around 90% of clinical tumors are carcinomas and in case of lung cancer have widely been reported to be the most commonly diagnosed followed by female breast cancer, prostate cancer and also colorectal cancers [2, 4–6].

Mechloretamine was the first anticancer drug introduced in market in 1949, when the knowledge of cancer biology was limited. Mechloretamine has the ability to bind to purine bases of DNA and generate cytotoxic effects against tumor cells. The lack of selectivity is the major limitation and restricts the wide usage because of serious drug adverse events. Currently, cancer biology is better understood because of the scientific advances made in the past decades, especially in molecular biology, biochemistry, immunology and medicinal chemistry fields that notably changed cancer treatment paradigms since mechloretamine. Current anticancer drug design focuses on more selective molecules that are safer and better tolerated than first drugs. Such molecules are designed to inhibit cancer-related proteins have displayed a pivotal role in a variety of tumors [7–10].

Molecular modeling provides numerous tools that can be applied in the drug design process with several reported successful cases [11–16]. Conceptually, since coordinates of 3D structures of attractive molecular targets for pharmacological intervention are available it is possible to apply structure-based drug design techniques, where interactions of known ligands can be modeled in order to derive structure–activity relationships [17–21]. On the other hand, when a database of ligands for a known target is available one can apply ligand-based approaches, where the concept of similarity guide the design of novel candidates, specially in the absence of 3D structures of drug targets [22–24]. Usually, a protocol combining strategies from both approaches is the best choice.

Interest in developing drug-likes compounds from computational methods and techniques is a growing trend, mainly due to the wide variety of success cases, which saves equally the cost and time [25]. The major contribution of the computational chemistry to cancer drug design is the development of drug candidates for new anticancer pathways discovered recently. Actual computational approaches that facilitate anticancer drug discovery include designing drug-like libraries, multi-targeted drug discovery, ligand–protein inverse docking, virtual screening of drug-like molecules and multi-target QSAR [25].

Significant progress has been made in hit discovery, lead optimization and drug candidates designing in anticancer research. Computational methods have been applied to speed up the discovery of molecules that interfere with promising drug targets linked with cancer initiation, progression and metastasis, including kinases, transcription factors and receptors of growth factors [26–29].

Lastly, it is known that cancer researches mainly arise from the drug design, leading to the development of new therapeutic options for a broad variety of illnesses, as has widely been shown in the literature. Despite major advances, the current approaches in general target single cancer pathways with single molecules that will not likely lead to cancer cure. Drugs combinations against several cancer hallmarks in a way similar to that done with AIDS seems to be a promising therapeutic strategy. Hence, in this context, it is well-known that the design of lead compounds for novel molecular targets continues to be important in order to enrich the variety of therapeutic options contributing to refreshed the expectations for development of improved treatments [7, 30].

New attractive molecular targets with known 3D structures must be studied by computational methods in order to stimulate the design of new anticancer drug candidates. A promising drug target against cancer is hnRNP K (heterogeneous ribonucleoprotein K), an interesting protein involved in the genesis of most prevalent cancers, including lung, breast, and colorectal [31–33]. Considering that coordinates of protein domain related to DNA interactions are available at Protein Data Bank, the molecular basis for DNA recognition can be used to drive the design of small molecules capable of blocking its action in cancer.

2 hnRNP K Protein

hnRNP K is, of course, an incredibly multifunctional protein that usually found in the nucleus, cytoplasm, and mitochondria of cells that influences multiple steps that compose gene expression, including chromatin remodeling, transcriptional, splicing, and mRNA stability through its capacity to bind RNA and single-stranded (ss) DNA via its KH domains. Aberrant hnRNP K expression has already been related to the development of melanoma, breast, prostate, colorectal, lung, hepatocellular, and esophageal cancers [31–39].

Specific interactions had already been illustrated for case of hnRNP K with both RNA, single-stranded (*ss*), and DNA, double-stranded (*ds*) [40–42], as for example with CT element. Additionally, it is known that the hnRNP K is able to bind to homopurine (GGGG) and homopyrimidine (CCCC) sequences present at the CT element, which is located within the *c-myc* promoter P1 [43]. As is well-known, the inadequate activation of *c-myc* gene can in principle triggers the cancer [44]. Hence, from this perspective, the pivotal role of hnRNP K marker related to the manifestations of diverse cancer, in general, it can then classify as an attractive molecular target for potential pharmacological intervention.

Hence, the interactions of hnRNP K with nucleotides are usually mediated by its K homology (KH) domains. It must be noted that the KH3 domain also shows the ability to bind to nucleic acids as an isolated domain, nevertheless with reduced affinity, i.e., when compared with hnRNP K [45–47]. However, in some cases the structural complexes between KH3 and specific oligonucleotide sequences [40–48], are available in the Protein Data Bank (PDB), helping to better understand the binding of nucleotides to KH domains.

It is possible to observe that the *ss*DNA oligonucleotides are usually situated in the binding cleft of the domain, which is predominantly hydrophobic, and base contacts are made predominantly by the tetrad TCCC. As such, the binding site of KH3 is located at the surface of the domain and because of being narrow it favors the accommodation of pyrimidine bases, which are composed by only one central ring, especially cytosine-rich sequences. The central region of the binding site is composed by hydrophobic amino acids (I29, I36, I47, I49) while it is surrounded by polar residues (K31, K37, R40, S46, K48, R59) [Fig. 1]. R40 e R59 are usually considered the key polar residues involved in the recognition process of nucleotide bases [40, 46, 48].

The identification of the structural properties involved in molecular recognition of nucleotides by KH3 domain of hnRNP K compose substantial information in order to help the design and search for small organic molecules able to prevent DNA binding, thus guiding the discovery of novel anticancer lead compounds. Based on the knowledge of KH3-DNA complex, it was designed new anticancer candidates targeting KH domains of hnRNP K.

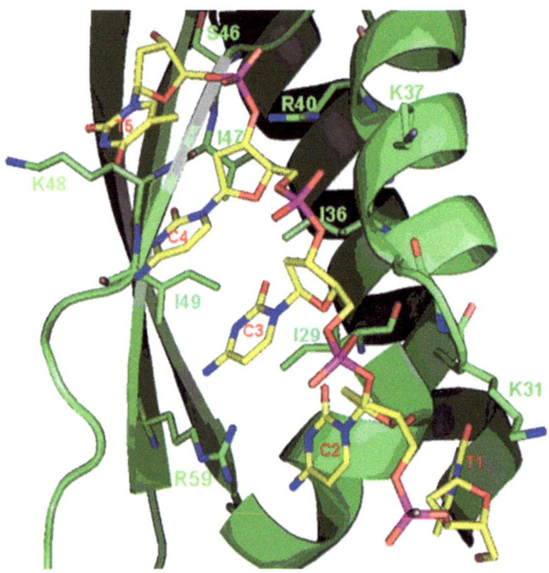

Fig. 1 Binding mode of T1-C2-C3-C4-T5 *ss*DNA in the binding site of KH3 domain (PDB code: 1J5K), highlighting the main protein residues that perform interactions and recognize the oligonucleotide

3 Design of Anticancer Lead Compounds Targeting hnRNP K

The design of novel anticancer lead compounds targeting KH domains of hnRNP K was made after the use of a structure-based virtual screening protocol with docking simulations that filtered small molecules ligands validated subsequently by in vitro assays. As such, the binding modes suggested by docking were evaluated with the application of molecular interactions fields e molecular dynamics simulations.

4 Methods

Virtual screening simulations were then performed using GOLD 3.1.1 [49] software based on docking simulations targeting tridimensional structure of KH3 domain of hnRNP K available in PDB (PDB code: 1J5K). Simulations were carried out with CNSset, DIVERSet and MWset virtual compound collections from EXPRESS-Pick database, inside a sphere of 15 Å radius centered at delta 1 carbon atom of I49 residue side chain. In the case of the KH3 domain present in the 1J5K complex was then elucidated by NMR [40] and their hydrogen orientations were considered for simulations. In this case, the genetic algorithm parameters of GOLD software were set up in 10,000 operations, 100 mutations and 100 crossovers and a population size of 100. The top-ranked orientation for each compound was generated and the 50 top-ranked ones for each database collection were filtered. The rescore docking procedure was applied to the 150 compounds filtered by virtual screening in order to select the most promising molecules able to interact with the KH3 domain binding site. Rescoring was performed for each compound inside the same sphere defined for virtual screening simulations. The 15 top-ranked orientations for each compound were obtained. Also, the genetic algorithm parameters for rescoring were set up in 100,000 operations, 95 mutations and 95 crossovers and a population size of 100.

Molecular interaction fields were computed with GRID22b [50] software for KH3 structure in order to map the interaction capabilities of the domain. The virtual interaction sites obtained for each molecular probe (DRY or hydrophobic, aromatic carbon, amide nitrogen and carbonyl oxygen) were superimposed with KH3 domain structure and the rescored docked compounds. The ten top-ranked compounds by docking that presented orientations with functional groups within favorable virtual interaction sites in 3D space were selected for in vitro experiments with hnRNP K.

Selected screened compounds from EXPRESS-Pick database were commercially acquired and solubilized in DMSO 99.9%. The ability of compounds in preventing DNA binding to hnRNP K protein was observed by EMSA (Electrophoretic Mobility Shift Assay) using 7% polyacrylamide gel [42]. The human hnRNP K cDNA (transcript 2) was amplified and cloned into PCR®2.1-TOPO® (Invitrogen) cloning vector. After DNA sequencing, in particular, the selected clone was digested using *NheI* and *XhoI* restriction enzymes and thus, subcloned into pET-28a expression vector (Novagen). Based on the plasmid was transformed into *E. coli* Rosetta (DE3) and the recombinant hnRNP K protein was then purified by *affinity chromatography* using a *Ni* Sepharose™ *High Performance resin* (GE HealthCare). As such, the proteins bound to the column *were eluted* step-wise *using* 25 mM phosphate buffer at a pH 7.4, in which containing 60–500 mM*imidazole and 500 mM NaCl. One aliquot of each fraction were* analyzed using SDS-PAGE and Coomassie brilliant blue staining. This pooled fractions containing purified hnRNP K protein were dialyzed against 100 mM Tris buffer at a pH 7.4 containing about 1 mM of EDTA (ethylenediamine tetraacetic acid). Immunoblotting was performed using antibody anti-hnRNP K and the immunoreactivity of the human recombinant hnRNP K protein was confirmed. The protein quantification was determined by Bradford method (BioRad). Reactions were performed incubating together 3 μL of water, 2 μL of EMSA binding buffer, 1 μL (630 ng/12.25 μM) of human purified hnRNP K and 2 μL of each compound in two different concentrations (16.78 mM and 50 μM for compound 294890-39-4; 12.88 mM and 50 μM for compound 693832-60-9) and 2 μL (at ca. 60 ng) of CT2-30 oligonucleotide (AGCTAACCCTCCCCAGATCAAGCCTGC-GATGATTTATACTCACAGCCCTCCCCATAG) [42] following homogenization. Oligonucleotide control was set up with 6 μL of water, 2 μL of DMSO 99.9% and 2 μL (at ca. 60 ng) of CT2-30, following homogenization. hnRNP K-oligonucleotide binding control was prepared with 5 μL of water, 2 μL of EMSA binding buffer, 1 μL of hnRNP K (at ca. 12.25 μM) e 2 μL of CT2-30 (at ca. 60 ng). The protein control was prepared with 7 μL of water, 2 μL of EMSA binding buffer and 1 μL (at ca. 12.25 μM) of hnRNP K. After preparation of each reaction and respective control samples all of them were homogenized and stored at 0 °C for a period of about 20 min. Next, reactions and controls were transferred to 7% polyacrylamide gel and submitted to electrophoresis for approximately 40 min. The oligonucleotides were stained with SYBR® Gold (Invitrogen) and thus, revealed through ultraviolet light.

Molecular dynamics simulations were performed with compound orientations identified among the 10 top-ranked ones during the rescore docking procedure and the KH3 structure. Simulations were set up with Discover_3 module from Insight II package [51]. First, the complexes were solvated with 8 water layers and then submitted to 1500 steps energy minimization procedure combining steepest descent/conjugated gradient algorithms using the CVFF force field. For each complex it was generated a 2 ns trajectory at 298 K temperature. The system coordinates were saved every 0.001 ns, generating 2000 snapshots for each trajectory.

5 Search and Validation of KH3 Domain Ligands

Virtual screening simulations were performed for KH3 domain (available at PDB, code: 1J5K) using small molecule virtual collections from EXPRESS-Pick database from ChemBridge Corporation, including CNSset, DIVERSet and MWset. Thus, the virtual simulations with KH3 domain covered about 150,000 compounds.

Firstly, using the Goldscore function of the GOLD software [49], the 50 top-ranked molecules for each compound collection were selected by docking simulations, filtering a total of 150 compound structures. These 150 structures were submitted to a rescore docking procedure where only the 15 top-ranked orientations for each compound were retained. These 15 orientations were superimposed and the ones that presented RMSD values lower than 2 Å each were considered equivalent. All different orientations obtained for each compound in the binding site of KH3 were analyzed inside the receptor site together with the virtual binding sites computed for KH3 domain structure.

For virtual binding sites generation, the most relevant molecular probes able to generate typical protein–ligand interactions were used to identify selected regions inside the receptor site space that could favor the binding of a specific chemical group of a small ligand molecule. Thus, amide nitrogen, representing hydrogen bond donor and acceptor groups, carbonyl oxygen, aromatic carbon, representing aromatic rings, and the DRY probe, representing hydrophobic groups, were chosen in order to identify where these chemical features could be accommodated within the KH3 binding site. In this way, only compounds that presented docked orientations positioning at least one chemical group in favored position were selected for further analysis. The ten top-ranked compounds indicated by the Goldscore function that fitted virtual interaction sites in the space within KH3 domain were selected for purchase (Fig. 2).

Experimental in vitro assays were performed with purified hnRNP K and then 10 compounds have been selected through virtual screening (Fig. 2) in order to evaluate their ability to prevent protein-DNA binding. In vitro assays revealed that only compounds 294890-39-4 and 693832-60-9 were able to prevent DNA binding to hnRNP K. This effect was observed during the EMSA (Electrophoretic Mobility Shift Assay) procedure (Fig. 3), where the DNA staining shows the position of the DNA oligonucleotide in each sample. In this experimental analysis, the objective is to determine the amount of *ss*DNA that is able to move down or stay retained at the top of the gel. Considering that hnRNP K protein has a high molecular weight, once present in any sample it should be retained by the gel pores. When in complex with the protein, the DNA should be retained as well, and when not in complex, the DNA should move at the bottom of the gel due to the absence of protein or even the presence of a molecule able to prevent DNA-protein binding in a sample.

Samples 2 and 3 (Fig. 3) were set as control samples of *ss*DNA oligonucleotide without and with hnRNP K protein, respectively. For sample 2, only DNA is present and it moves at the bottom of the gel because it is not possible to form any complex with the protein, which is absent in this sample. However, in sample 3, the DNA stain remains at the top of the gel, because hnRNP K was present, indicating that DNA is bounded to hnRNP K, which stays retained by the gel pores.

For samples 4–7, where compounds 294890-39-4 and 693832-60-9 were added in two concentrations each (16.78 mM and 50 μM for compound 294890-39-4; 12.88 mM and 50 μM for compound 693832-60-9), with DNA and hnRNP K, the oligonucleotide was present at the bottom of the gel. This means that in the presence of compounds 294890-39-4 and 693832-60-9 the expected binding of DNA to hnRNP K is prevented. In sample 8, only the protein was added, and no stain was observed,

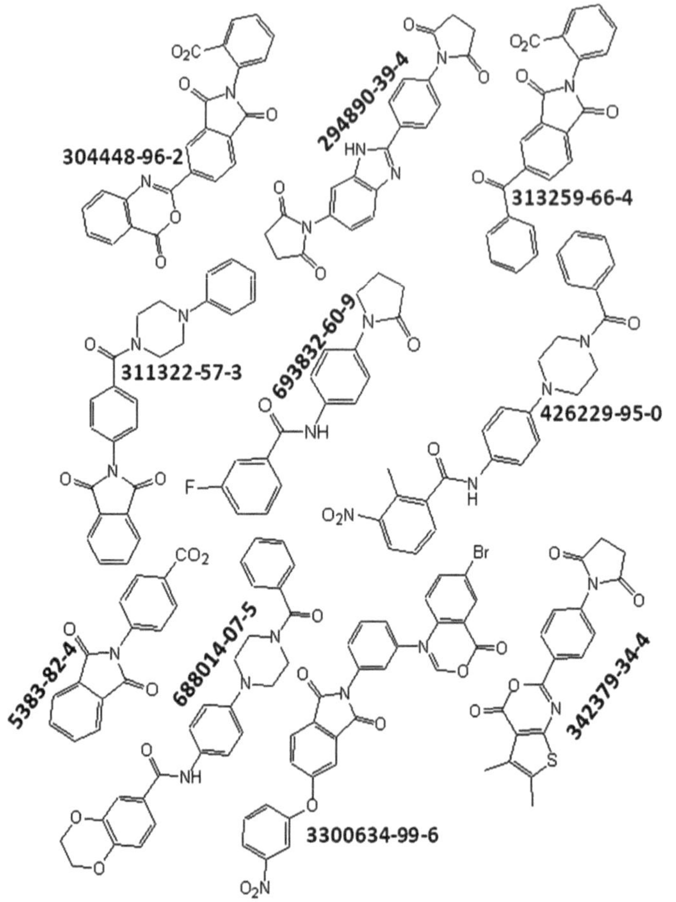

Fig. 2 Structure of the all compounds selected by virtual screening simulations with their respective CAS numbers

as expected. When the other eight compounds (Fig. 2) were tested, no effect in DNA mobility was observed, maintaining the same pattern of DNA-hnRNP K binding control (Fig. 3).

The rescored docked models were used to propose the binding modes for bioactive compounds, 294890-39-4 and 693832-60-9, in the binding site of the KH3 domain. All the orientations obtained for each compound were superimposed with the virtual interaction sites computed for KH3 domain structure. Three docked orientations that presented fit with virtual interaction sites were obtained (Fig. 4) for compound 294890-39-4. Regarding compound 693832-60-9, only one orientation fitted the virtual interaction sites (Fig. 5). It is expected that one of these theoretical orientations represent the real binding mode of each compound. The main polar interactions proposed for both compounds are hydrogen bonds performed with R40 and R59

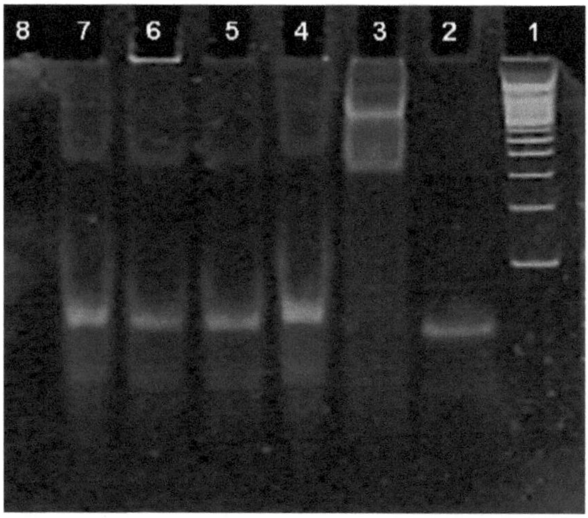

Fig. 3 Eletroforetic mobility shift assay with human hnRNP K, *ss*DNA oligonucleotide and small organic molecules 294890-39-4 and 693832-60-9. Samples: 1- molecular weight marker; 2- *ss*DNA CT2-30 (2 μl/60 ng); 3- hnRNP K (1 μl/12.25 μM) + CT2-30 (2 μl/60 ng); 4- hnRNPK (1 μl/12.25 μM) + CT2-30 (2 μl/60 ng) + compound 294890-39-4 (2 μl/16.78 mM); 5- hnRNPK (1 μl/12.25 μM) + CT2-30 (2 μl/60 ng) + compound 294890-39-4 (2 μl/50 μM); 6- hnRNPK (1 μl/12.25 μM) + CT2-30 (2 μl/60 ng) + compound 693832-60-9 (2 μl/12.88 mM); 7- hnRNPK (1 μl/12.25 μM) + CT2-30 (2 μl/60 ng) + compound 693832-60-9 (2 μl/50 μM); 8- hnRNP K (1 μl/12.25 μM)

protein residues. For compound 294890-39-4 (Fig. 4), only orientation 3 was able to interact simultaneously with R40 and R59. For compound 693832-60-9, the proposed orientation performed an additional hydrogen bond with K22 residue (Fig. 5).

These proposed orientations for compounds 294890-39-4 and 693832-60-9 show a good fit with virtual interaction sites computed for aromatic carbon and carbonyl oxygen (Figs. 6 and 7), confirming that these orientations are energetically favored in the binding site molecular environment and could represent the binding modes of such compounds.

Attempting to evaluate the theoretical binding modes for bioactive compounds, molecular dynamics simulations were performed in order to define the most probable binding mode for compound 294890-39-4 and determine whether the unique orientation proposed by compound 693832-60-9 inside the binding site of KH3 is viable. Considering the three orientations proposed for compound 294890-39-4, orientation 1 seems to be unstable (Fig. 8). During the trajectory, achieving lower energy levels (Fig. 8A), it is possible to observe a higher degree of conformational variability inside the KH3 binding site (Fig. 8C) when compared with orientations 2 and 3 (Figs. 9 and 10, respectively). Furthermore, the polar interaction proposed with R40 by docking was broken just at the beginning of the trajectory, observed by interatomic distances of about 4.72 Å (Fig. 8B).

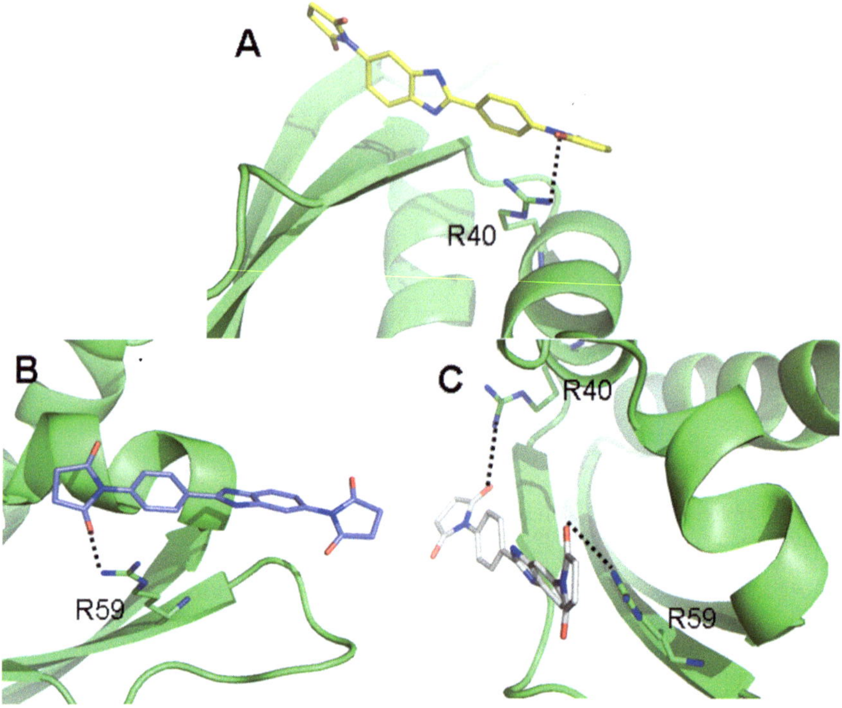

Fig. 4 Top-ranked orientations obtained by docking simulations for compound 294890-39-4 within the binding site of hnRNP K KH3 domain, highlighting the suggested polar interactions (dashed lines). **a** Orientation 1; **b** orientation 2; **c** orientation 3

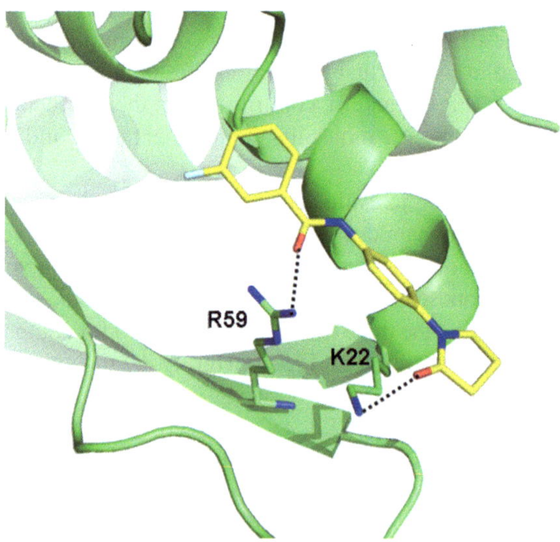

Fig. 5 Orientation obtained by docking simulations for compound 693832-60-9 within the binding site of the KH3 domain, highlighting the proposed polar interactions (dashed lines) of the complex

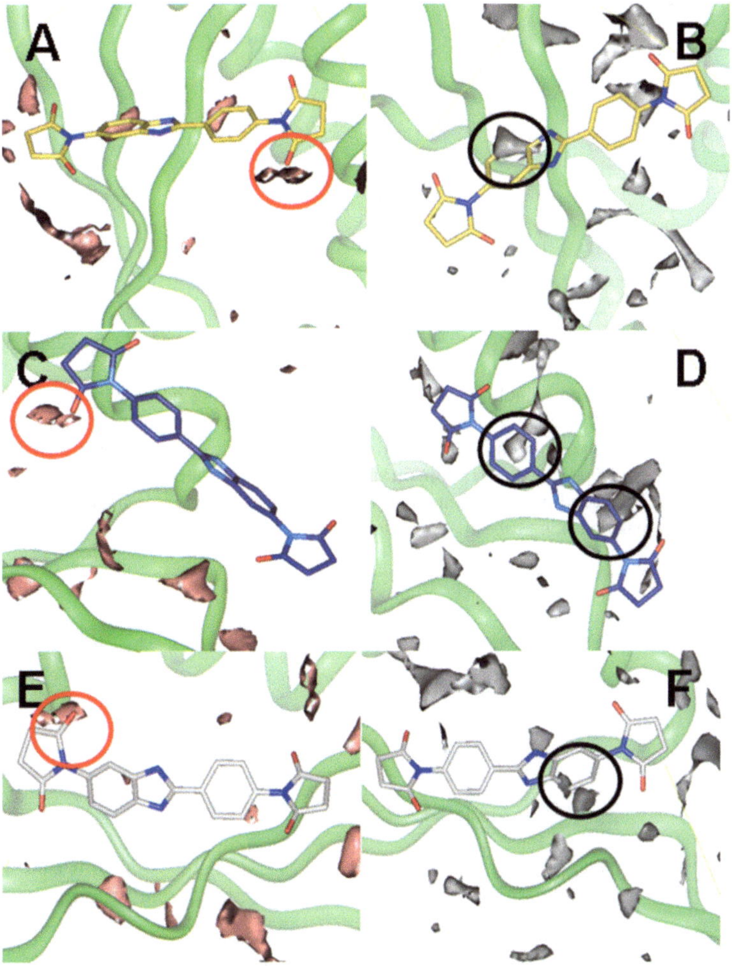

Fig. 6 Orientations of compound 294890-39-4 obtained by docking posed with the isoenergetic contours of the virtual interaction sites computed inside KH3 domain structure, highlighting the adequate posing of such compound fragments (red and black circles). **a** Orientation 1 inside KH3 domain and isoenergetic contours for carbonyl oxygen probe at −5 kcal/mol. **b** Orientation 1 inside KH3 domain and isoenergetic contours for aromatic carbon probe at − 2.5 kcal/mol. **c** Orientation 2 inside KH3 domain and isoenergetic contours for carbonyl oxygen probe at −5 kcal/mol. **d** Orientation 2 inside KH3 domain and isoenergetic contours for aromatic carbon probe at −2.5 kcal/mol. **e** Orientation 3 inside KH3 domain and isoenergetic contours for carbonyl oxygen probe at − 5 kcal/mol. **f** Orientation 3 inside KH3 domain and isoenergetic contours for aromatic carbon probe at −2.5 kcal/mol

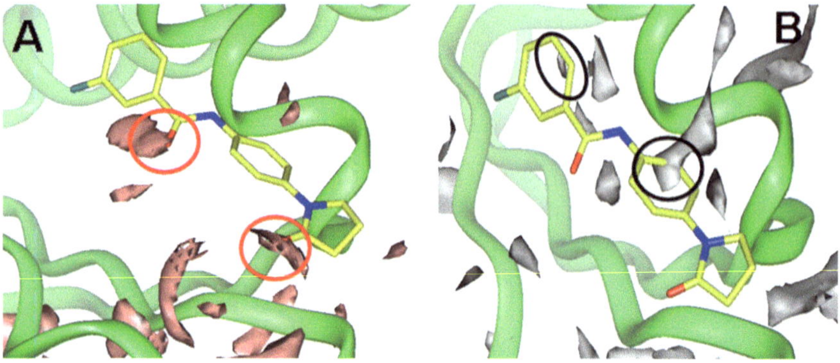

Fig. 7 Orientation of compound 693832-60-9 obtained by docking posed with the isoenergetic contours of the virtual interaction sites computed inside KH3 domain structure, highlighting the adequate posing of such compound fragments (red and black circles, respectively). **a** Docked pose of compound 693832-60-9 inside KH3 domain and isoenergetic contours for carbonyl oxygen probe at −5 kcal/mol. **b** Docked pose of compound 693832-60-9 inside KH3 domain and isoenergetic contours for aromatic carbon probe at −2.5 kcal/mol

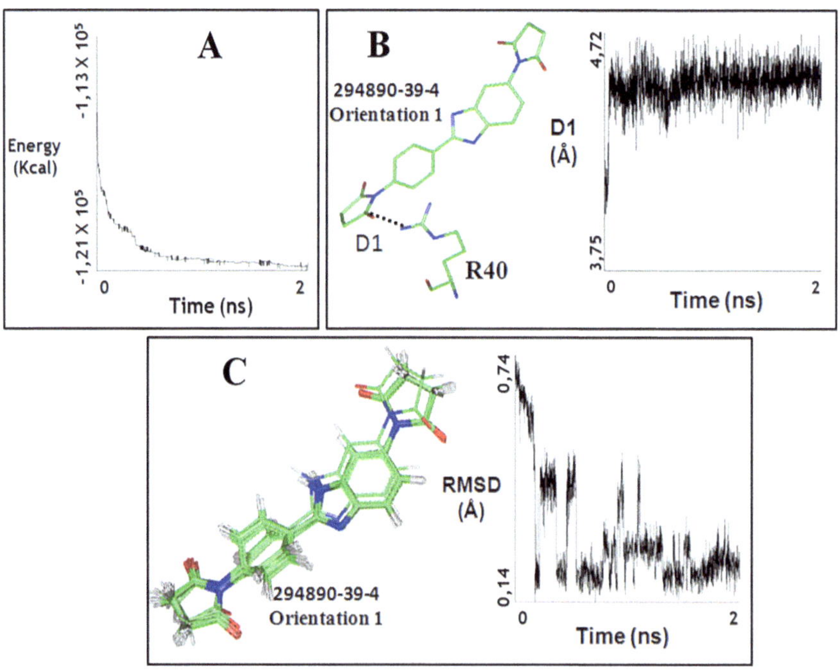

Fig. 8 Molecular dynamics trajectory. **a** Total energy of the complex KH3-orientation 1 (compound 294890-39-4). **b** Interatomic distance variation (D1) between the carbonyl oxygen of compound 294890-39-4 (orientation 1) and nitrogen atom of R40. **c** Snapshot samples and RMSD (Å) of the conformations obtained for orientation 1 (compound 294890-39-4)

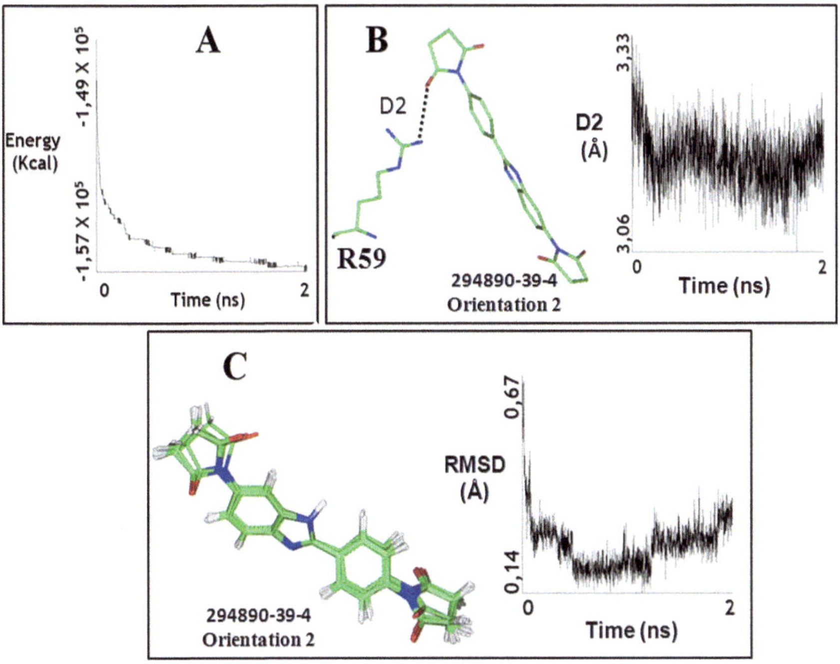

Fig. 9 Molecular dynamics trajectory. **a** Total energy of the complex KH3-orientation 2 (compound 294890-39-4). **b** Interatomic distance variation (D2) between the carbonyl oxygen of compound 294890-39-4 (orientation 2) and nitrogen atom of R59. **c** Snapshot samples and RMSD of the conformations obtained for orientation 2 (compound 294890-39-4)

For compound 693832-60-9, its unique evaluated orientation indicates stability inside the KH3 domain (Fig. 11), resulting in a viable binding mode for such hnRNP K ligand. The polar interaction proposed with R59 was maintained during all the simulated trajectory (Fig. 11B) and the conformation proposed did not suffer substantial variation inside the KH3 domain (Fig. 11C).

Compounds 294890-39-4 and 693832-60-9 were searched during virtual screening simulations as possible KH3 domain binders. Since the three KH domains of hnRNP K share sequential and structural homology and all bind to C-rich nucleotide sequences, it's reasonable to consider that a molecule able to interact with the KH3 domain could also interact with KH1 and KH2. Nevertheless, if a compound only binds to KH3 it is expected that, in biological systems, the hnRNP K biological activity might suffer interference because of the simultaneous binding of the three KH domains of hnRNP K to *ss*DNA is a cooperative process [45]. Therefore, the design of small molecule compounds as selective ligands of KH3 domain may also be relevant to reduce the aberrant function of hnRNP K in tumorigenesis.

Considering that the biological activities of hnRNP K associated with cancer are related with transcription and translation processes and involve the binding with DNA, the ability of compounds 294890-39-4 and 693832-60-9 in preventing DNA

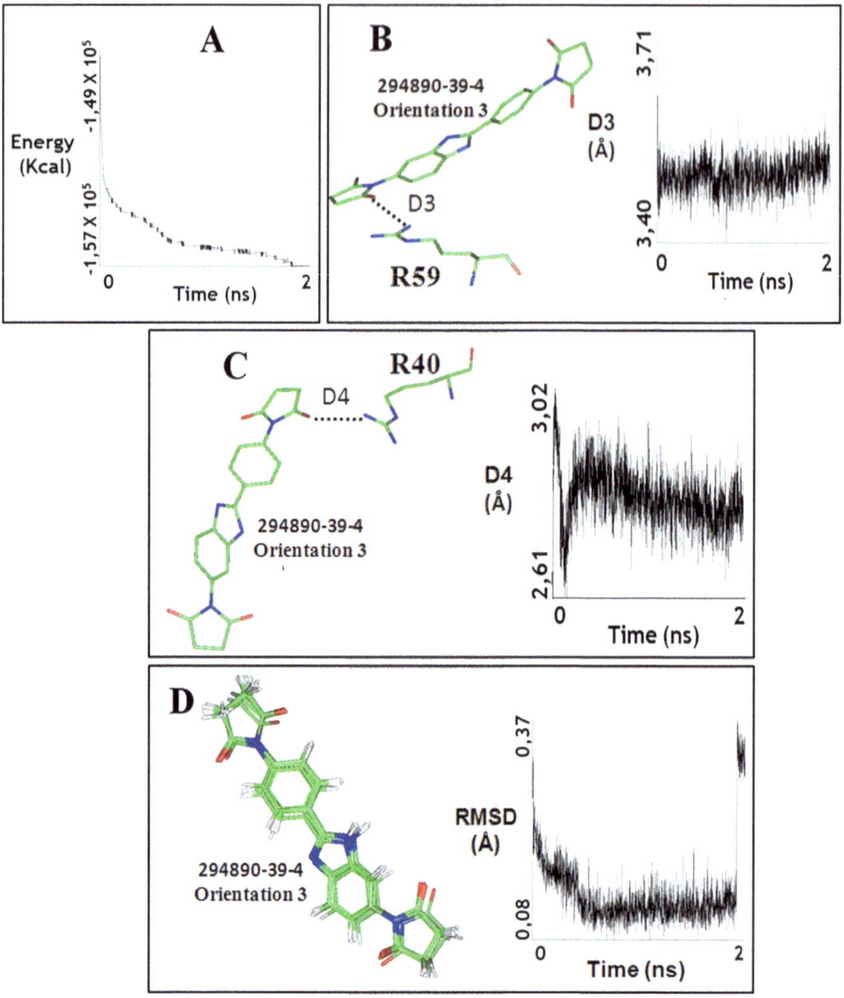

Fig. 10 Molecular dynamics simulations. **a** Total energy of the complex KH3-orientation 3 (compound 294890-39-4). **b** Interatomic distance variation (D3) between the carbonyl oxygen of compound 294890-39-4 (orientation 3) and nitrogen atom of R59. **c** Interatomic distance variation (D4) between the carbonyl oxygen of compound 294890-39-4 (orientation 3) and nitrogen atom of R40. **d** Snapshot samples and RMSD (Å) of the conformations obtained for orientation 3 (compound 294890-39-4)

binding to hnRNP K could be useful to design drug candidates able to reduce the protein biological effects related to cancer, becoming valuable lead compounds in order to investigate the treatment types of cancer associated with over-expression of hnRNP K. Huth et al [52] have already reported the discovery of benzoylanthranilic acid inhibitors of FBP (Far Upstream Element Binding Protein). FBP also binds

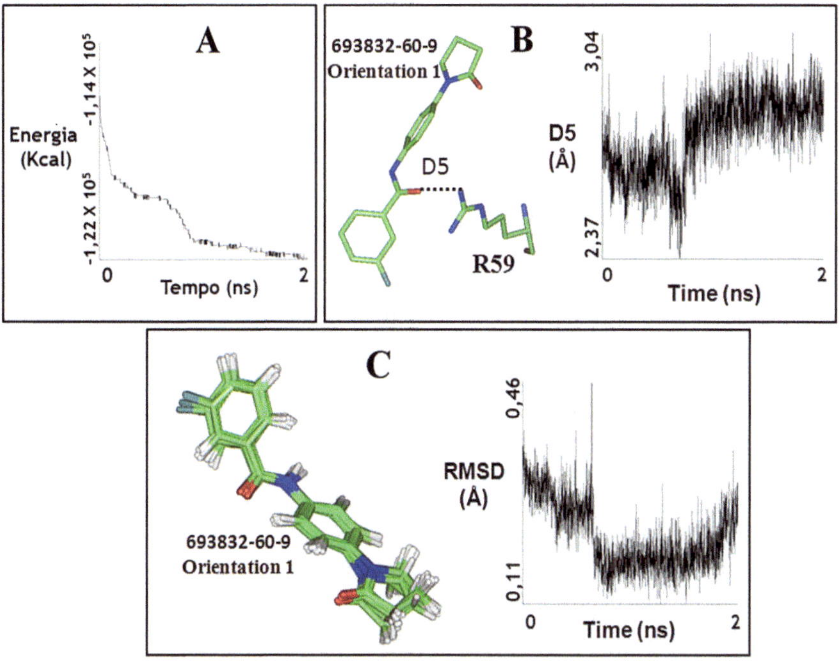

Fig. 11 Molecular dynamics trajectory. **a** The total energy of the complex KH3-compound 693832-60-9. **b** Interatomic distance variation (D5) between the carbonyl oxygen of compound 693832-60-9 and the nitrogen atom of R59. **c** Snapshot samples and RMSD of the conformations obtained for compound 693832-60-9

DNA molecules through KH domains and is involved in cancer genesis through *c-myc* pathway, as hnRNP K. In this way, the design of ligands targeting KH domains of different molecular markers could become a new trend in cancer drug discovery and development by down-regulating the *c-myc* pathway. The prevention of hnRNP K-nucleotide binding may contribute to inhibit cellular events that trigger cancer development mediated by hnRNP K activity.

Compounds 294890-39-4 and 693832-60-9 correspond to benzimidazole and phenylbenzamide derivatives, respectively. Such compounds have molecular structures with carbonyl groups, which are able to interact with R40 and R59 residues. Also, it was concluded that these two arginines are in general considered key residues to the recognition process of nucleotides by KH3 domain of hnRNP K [40, 48]. Since hydrogen bonds or ion–dipole interactions can be established between carbonyl oxygen atoms of active compounds and the guanidinium group of R40 and R59, it is expected that in the biological environment the natural recognition of nucleotides by hnRNP K could be prevented, explaining the activity of compounds 294890-39-4 and 693832-60-9 in vitro.

6 Conclusions

The discovery of hnRNP K ligands that prevent DNA binding to KH domains is the first step to design novel anticancer agents targeting this attractive molecular target, which is already associated with the development of most prevalent cancers worldwide, including breast, prostate and colorectal. The main activity of hnRNP K in the cellular processes is mediated by its interaction with DNA through KH domains. After small ligand molecules were discovered, like reported here for compounds 294890-39-4 and 693832-60-9, their interactions with the KH3 domain were proposed through computational methods, including docking, molecular interaction fields, and molecular dynamics. The 3D structure of KH domains is composed of a hydrophobic binding cleft surrounded by polar residues, especially R40 and R59 that are important to recognize approaching DNA molecules. Interesting ligands should be composed of heterocyclic compounds, able to be accommodated within the KH hydrophobic binding cleft, also presenting polar substituents able to interact with both R40 and R59 by means of hydrogen bonding or ionic interactions. Thus, the benzimidazole and phenylbenzamide derivatives proposed here share such chemical characteristics and are now classified as novel anticancer lead compounds available for guiding the design of novel analogues that can be further optimized. The discovery of such compounds contributes to enrich the knowledge of small molecules as lead compounds for designing a novel class of anticancer agents that could down-regulate hnRNP K.

References

1. Torre, L.A., Siegel, R.L., Ward, E.M., Jemal, A.: Global cancer incidence and mortality rates and trends—an update. Cancer Epidemiol Biomarkers Prev. **25**(1), 16–27 (2016)
2. Kunnumakkara, A.B., Bordoloi, D., Sailo, B.L., Roy, N.K., Thakur, K.K., Banik, K., Shakibaei, M., Gupta, S.C., Aggarwal, B.B.: Cancer drug development: the missing links. Exp. Biol. Med. **244**(8), 663–689 (2019)
3. Hasima, N., Aggarwal, B.B.: Cancer-linked targets modulated by curcumin. Int. J. Biochem. Mol. Biol. **3**(4), 328–351 (2012)
4. Sonnenschein, C., Soto, A.: Carcinogenesis explained within the contexto of a theory of organisms. Prog. Biophys. Mol. Biol. **122**(1), 70–76 (2016)
5. Sonnenschein, C., Davis, B., Soto, A.M.: A novel pathogenic classification of cancers. Cancer Cell. Int. **14**(1), 113–117 (2014)
6. Bray, F., Ferlay, J., Soerjomataram, I., Siegel, R.L., Torre, L.A., Jemal, A.: Global Cancer statistics 2018: GLOBOCAN estimates of incidence and mortality worldwide for 36 cancers in 185 countries. CA Cancer J. Clin. **68**, 394–424 (2018)
7. Magalhães, L.G., Ferreira, L.L.G., Andricopulo, A.D.: Recent advances and perspectives in cancer drug design. Acad. Bras. Cienc. **90**(1 Suppl 2), 1233–1250 (2018)
8. Polavarapu, A., Stillabower, J.A., Stubblefield, S.G.W., Taylor, W.M., Baik, M.: The mechanism of guanine alkylation by nitrogen mustards: a computational study. J. Org. Chem. **77**(14), 5914–5921 (2012)
9. Chen, Y., Jia, Y., Song, W., Zhang, L.: Therapeutic potential of nitrogen mustard based hybrid molecules. Front. Pharmacol. **9**, 1453–1464 (2018)

10. Geromichalos, G.D.: Importance of molecular computer modeling in anticâncer drug development. J. BUON **12**(Suppl 1), S101–S118 (2007)
11. Taft, C.A., Silva, V.B., Silva, C.H.T.P.: Current topics in computer-aided drug design. J. Pharm. Sci. **97**, 1089–1098 (2008)
12. Tramontano, A.: The role of molecular modeling in biomedical research. FEBS Lett. **580**, 2928–2934 (2006)
13. Stahl, M., Guba, W., Kansy, M.: Integrating molecular design resources within modern drug discovery research: the Roche experience. Drug. Discov. Today **11**, 326–333 (2006)
14. Macalino, S.J.Y., Gosu, V., Hong, S., Choi, S.: Role of computer-aided drug design in modern drug discovery. Arch. Pharm. Res. **38**(9), 1686–1701 (2015)
15. Macalino, S.J.Y., Billones, J.B., Organo, V.G., Carillo, M.C.O.: In silico strategies in tuberculosis drug discovery. Molecules **25**(3), 665 (2020)
16. Loganathan, L., Muthusamy, K.: Current scenario in structure and ligand-based drug design on anti-colon câncer drugs. Curr. Pharm. Des. **24**(32), 3829–3841 (2018)
17. Kapetanovic, I.M.: Computer-aided drug discovery and development (CADDD): In silico-chemico-biological approach. Chem-Biol. Interact. **171**, 165–176 (2008)
18. Kitchen, D.B., Decornez, H., Furr, J.R., Bajorath, J.: Docking and scoring in virtual screening for drug discovery: methods and applications. Nat. Rev. Drug Discov. **03**, 935–949 (2004)
19. Forli, S., Huey, R., Pique, M.E., Sanner, M., Goodsell, D.S., Olson, A.J.: Computational protein-ligand docking and virtual drug screening with the autodock suíte. Nat. Protoc. **11**(5), 905–919 (2016)
20. Li, Q., Shah, S.: Structure-based virtual screening. Methods Mol. Biol. **1558**, 111–124 (2017)
21. dos Santos, R.N., Ferreira, L.G., Andricopulo, A.D.: Practices in molecular docking and structure-based virtual screening. Methods Mol. Biol. **1762**, 31–50 (2018)
22. Hawkins, P.C.D., Stahl, G.: Ligand-based methods in GPCR computer-aided drug design. Methods Mol. Biol. **1705**, 365–374 (2018)
23. Seidel, T., Schuetz, D.A., Garon, A., Langer, T.: The pharmacophore concept and its applications in computer-aided drug design. Prog. Chem. Org. Nat. Prod. **110**, 99–141 (2019)
24. Acharya, C., Coop, A., Polli, J.E., Mackerell, A.D., Jr.: Recent advances in ligand-based drug design: relevance and utility of the conformationally sampled pharmacophore approach. Curr. Comput. Aided. Drug. Des. **7**(1), 10–22 (2011)
25. Hameed, R., Khan, A., Khan, S., Perveen, S.: Computational approches towards kinases as attractive targets for anticancer drug Discovery and development. Anticancer Agents Med. Chem. **19**(5), 692–598 (2019)
26. Luo, Y., Wang, L.: Discovery and development of ATP-competitive mTOR inhitors using computational approaches. Curr. Pharm. Des. **23**(29), 4321–4331 (2017)
27. Lambert, M., Jambon, S., Depauw, S., David-Cordonnier, M.: Targeting transcription factors for cancer treatment. Molecules **23**(6), 1479 (2018)
28. Gutierrez, A.N., McDonald, P.H.: GPCRs: emerging anticancer drug targets. Cell Signal **41**, 65–74 (2018)
29. Moody, T.W., Ramos-Alvarez, I., Jensen, R.T.: Neuropeptide G protein-coupled receptors as oncotargets. Front. Endocrinol. **9**, 345 (2018)
30. Zugazagoitia, J., Guedes, C., Ponce, S., Ferrer, I., Molina-Pinelo, S., Paz-Ares, L.: Current challenges in câncer treatment. Clin. Therap. **38**, 1551–1566 (2016)
31. Ostrowski, J., Bomsztyk, K.: Nuclear shift of hnRNP K protein in neoplasms and other states of enhanced cell proliferation. Br. J. Cancer **89**, 1493–1501 (2003)
32. Barboro, P., Repaci, E., Rubagotti, A., Salvi, S., Boccardo, S., Spina, B., Truini, M., Introini, C., Puppo, P., Ferrari, N., Carmignani, G., Boccardo, F., Balbi, C.: Heterogeneous nuclear ribonucleoprotein K: altered pattern of expression associated with diagnosis and prognosis of prostate cancer. Br. J. Cancer **100**, 1608–1616 (2009)
33. Carpenter, B., Mckay, M., Dundas, S.R., Lawrie, L.C., Telfer, C., Murray, G.I.: Heterogeneous nuclear ribonucleoprotein K is over expressed, aberrantly localised and is associated with poor prognosis in colorectal cancer. Br. J. Cancer **95**, 921–927 (2006)

34. Bomsztyk, K., Denisenko, O., Ostrowski, J.: hnRNP K: one protein multiple processes. BioEssays **26**, 629–638 (2004)
35. Chen, L.C., Chung, I.C., Hsueh, C., Tsang, N.M., Chi, L.M., Liang, Y., Chen, C.C., Wang, L.J., Chang, Y.S.: The antiapoptotic protein, FLIP, is regulated by heterogeneous nuclear ribonucleoprotein K and correlates with poor overall survival of nasopharyngeal carcinoma patients. Cell Death Differ **17**, 1463–1473 (2010)
36. Ciarlo, M., Benelli, R., Barbieri, O., Minghelli, S., Barboro, P., Balbi, C., Ferrari, N.: Regulation of neuroendocrine differentiation by AKT/hnRNP K/AR/β-catenin signaling in prostate câncer cells. Int. J. Cancer **131**, 582–590 (2012)
37. Wen, F., Shen, A., Shanas, R., Bhattacharyya, A., Lian, F., Hostetter, G., Shi, J.: Higher expression of heterogeneous ribonucleoprotein K in melanoma. Annals Surg. Oncol. **17**, 2619–2617 (2010)
38. Gallardo M, Hornbaker MJ, Zhang X, Hu P, Bueso-Ramos C, Post SM. Aberrant hnRNP K expression.: all roads lead to câncer. Cell Cycle **15**(12), 1552–1557 (2016)
39. Kedzierska, H., Piekielko-Witkowska, A.: Splicing factors of SR and hnRNP families as regulators of apoptosis in câncer. Cancer Lett. **396**, 53–65 (2017)
40. Braddock, D.T., Baber, J.L., Levens, D., Clore, G.M.: Molecular basis of sequence-specific single-stranded DNA recognition by KH domains: solution structure of a complex between hnRNP K KH3 and single-stranded DNA. EMBO J. **21**, 3476–3485 (2002)
41. Bomsztyk, K., Seuningen, I.V., Suzuki, H., Denisenko, O., Ostrowski, J.: Diverse molecular interactions of the hnRNP K protein. FEBS Lett. **403**, 113–115 (1997)
42. Tomonaga, T., Levens, D.: Heterogeneous nuclear ribonucleoprotein is a DNA-binding transactivator. J. Biol. Chem. **270**, 4875–4881 (1995)
43. Takimoto, M., Tomonaga, T., Matunis, M., Avigan, M., Krutzsch, H., Dreyfuss, G., Levens, D.: Specific binding of heterogeneous ribonucleoprotein particle protein K to the human c-myc promoter, in vitro. J. Biol. Chem. **268**, 18249–18258 (1993)
44. Prochownik, E.V.: c-Myc as a therapeutic target in cancer. Expert Rev. Anticanc. **4**, 289–302 (2004)
45. Paziewska, A., Wyrwics, L.S., Bujnicki, J.M., Bomsztyk, K., Ostrowski, J.: Cooperative binding of the hnRNP K three KH domains to mRNA targets. FEBS Lett. **577**, 134–140 (2004)
46. Baber, J.L., Libutti, D., Levens, D., Tjandra, N.: High precision solution structure of the C-terminal KH domain of heterogeneous nuclear ribonucleoprotein K, a c-*myc* transcription factor. J. Mol. Biol. **289**, 949–962 (1999)
47. Nicastro, G., Taylor, I.A., Ramos, A.: KH-RNA interactions: back in the groove. Curr. Opin Struct. Biol. **30**, 63–70 (2015)
48. Backe, P.H., Messias, A.C., Ravelli, R.B.G., Sattler, M., Cusack, S.: X-ray crystallographic and NMR studies of the third KH domain of hnRNP K in complex with single-stranded nucleic acids. Structure **13**, 1055–1067 (2005)
49. Verdonk, M.L., Cole, J.C., Hartshorn, M.J., Mulrray, C.W., TayloR, R.D.: Improved protein-ligand docking using GOLD. Proteins **52**, 609–623 (2003)
50. Goodford, P.J.: A Computational procedure for determining energetically favorable binding sites on biologically important macromolecules. J. Med. Chem. **28**, 849–857 (1985)
51. Insight, I.I., software, : Version 2005. CA, USA, Accelrys (2005)
52. Huth, J.R., Yu, L., Collins, I., Mack, J., Mendoza, R., Isaac, B., Braddock, D.T., Muchmore, S.W., Comess, K.M., Fesik, S.W., Clore, G.M., Levens, D., Hajduk, P.J.: NMR-driven discovery of benzoylanthranilic acid inhibitors of far upstream element binding protein binding to the human oncogene c-myc promoter. J. Med. Chem. **47**, 4851–4857 (2004)

Surfactant/Alkali Stress Effect in Exopolysaccharide Production by *Xanthomonas* and *Enterobacter* Strains

Igor Carvalho Fontes Sampaio, Pedro Jorge louro Crugeira, Joalene de Azevedo Santos Ferreira, Jamile de Almeida Santos, Jacson Nunes dos Santos, Elias Ramos-de-Souza, Antônio Luiz Barbosa Pinheiro, Josilene Borges Torres Lima Matos, Fabio Alexandre Chinalia, and Paulo Fernando de Almeida

Abstract New innovative technologies are required today for lowering the production costs of highly demanded microbial exopolysaccharides (EPS). The physiological "stress" caused by the presence of alkaline and non-cationic surfactants (Triton X-100) in the media (alone or in combination) may significantly increase *Xanthomonas campestris* and *Enterobacte*r sp. Production of EPS. A mineral media (MSM) was supplemented with sucrose and crude glycerin (GB) and prepared with produced water either diluted (PW) or dialyzed (DPW) for removing mineral salts. Under conditions of alkaline stress (pH 9.5), the microbial population of *X. campestris* showed an increase in EPS production of 16.7% ($p < 0.01$). The addition of Triton X-100, however, increased the production and also the viscosity of the *X. campestris* EPS. On the other hand, this former substance was toxic to *Enterobacter* sp. in the lowest concentration tested (0.1%). The association of alkaline stress with Triton X-100 increased the production and quality of the EPS produced by *X. campestris*, with maximum values of 88.72% of production and 190.35% of viscosity in the medium prepared with DPW ($P < 0.0001$). The technology developed "surfactant/alkali stress" is an innovative way of reducing production costs and increasing the quality of EPS xanthan gum.

Keywords Alkalinization · Detergent · Biopolymer · By-products · Recycling

---

I. C. F. Sampaio · P. J. Crugeira · J. de Azevedo Santos Ferreira · J. de Almeida Santos · J. N. dos Santos · J. B. T. L. Matos · F. A. Chinalia · P. F. de Almeida (✉)
Laboratory of Biotechnology and Ecology of Micro-organisms, Institute of Health Sciences, Federal University of Bahia, Reitor Miguel Calmon Ave, S/N, Salvador, BA 40110-100, Brazil
e-mail: pfatk@yahoo.com.br

P. J. Crugeira · A. L. B. Pinheiro
Center of Biophotonics, Federal University of Bahia, 62, Araujo Pinho Ave, Canela, Salvador, BA 40110-150, Brazil

E. Ramos-de-Souza
Federal Institute of Education, Science and Technology of Bahia—Campus Salvador, Rua Emídio Santos, S/N Barbalho, Salvador, BA 40300-010, Brazil

F. A. La Porta and C. A. Taft (eds.), *Functional Properties of Advanced Engineering Materials and Biomolecules*, Engineering Materials,
https://doi.org/10.1007/978-3-030-62226-8_24

1 Introduction

Exopolysaccharides (EPS) are a class of biopolymers produced by bacteria through the use of simple or complex substrates and which are located outside the cell [1]. They are composed of repeated units of monosaccharides, and may or may not be linked to glycoproteins, glycolipids, acids or even extracellular DNA [2] and other constituents of the exopolymeric matrix, such as inorganic ions [3]. EPS that are initially produced during the log phase of the microbial growth are associated with the membrane capsule, and those are produced in the stationary phase when compared to the ones that accumulate in the culture medium [4]. The former is an example of the most industrially produced such as xanthan gum [5]. The excretion of biopolymers by bacteria is related to protection from adverse conditions in the environment such as toxic compounds, desiccation, stresses (including nutritional imbalance), antibiotics, metals and alcohol [6], providing an increase in their ability to compete and survive, as well as adhesion mechanisms, biofilm formation and/or symbiotic relationships [7].

A physiological stress during culturing is the exposure of microorganisms to alkaline environments, which can cause damage to the outer membrane, ribosomes, proteins and DNA [8]. Additionally, the dissociation of bases, such as NaOH, in sodium and hydroxyl ions can disturb the energy metabolism of bacteria. An excess of sodium could modify the gradient maintained through a Na^+/H^+ antiport, and with an increase in extracellular pH, antiport activity would also increase [9]. At a high pH, antiports can saturate with excess Na^+ entering the cell, therefore, the ion gradient may not be sustained and the membrane proteins involved in energy synthesis are then inhibited. It is believed that in high pH environments (e.g. pH 12) there may be saponification of membrane lipids and destabilization of such proteins, which is directly depend on the integrity of the lipid double layer [10, 11]. With the reduction of energy production, stress can inhibit growth [12] if the microorganism does not produce an emergency biological response to the stimulus experienced [13, 14].

Studies report a higher production of exopolysaccharides in the presence of high pH [6, 15, 16]. Although the mechanisms of how this production is induced have not yet been fully elucidated, but alkaline stress is used to optimize the production of EPS. Detergents in isolation has also been tested as a means of increasing the production of EPS [17–19]. They can increase oxygen transfer and improve the production of xanthan gum and its rheological characteristics, such as its viscosity [17, 20]. Studies indicate that surfactants can infer the catabolic process by increasing the production of hydrolytic enzymes, but the metabolic processes involved have not yet been clarified [21]. It is known that there is an increase in the permeability of the cell membrane, altering the lipid metabolism that assist in releasing enzymes. These can positively interfere in the secretion of exopolysaccharides. This metabolic imbalance may develop a higher production of EPS [22].

This work aims to evaluate the production of EPS (*Xanthomonas campestris* and *Enterobacter sp.*) in produced water from the oil industry and crude glycerin under the stressing presence of surfactant and alkaline substance.

2 Material and Methods

2.1 Produced Water and Crude Glycerin

The produced water has collected from a carbonate oil field in primary recovery stage in Brazil, naturally originating from the reservoir, with no application of injection fluids. Crude glycerin (GB) was obtained from a Biodiesel production plant at SENAI/CIMATEC (Salvador, BA, Brazil), from the transesterification of soy oil and animal fat (8: 2) using sodium methoxide (0, 6%) and methanol (17% in excess).

2.2 Produced Water Dyalysis

As a pretreatment, the dialysis of the produced water was carried out using artificial collagenic tubular membranes (DEVRO, Miami, USA) of 45 mm of thickness and pores of 10 nm in diameter was carried out against distilled water. Water was changed every 8 h until obtaining a conductivity value of about 300 μS/cm (FiveEasy Benchtop FP30, conductivity Meter, Powai Mumbai, India).

2.3 Microrganisms and Inoclum Preparation

The strain *Xanthomonas campestris* IBSBF 2103 was kindly supplied by the Faculty of Pharmacy, Federal University of Bahia, Salvador, Brazil. Thus, the preparation of the microbial inoculum consisted of inoculating a cryotube with the cryopreserved strain in 20 mL of YM medium [23] contained in a 125 mL Erlenmeyer flask. It was incubated at 28 °C and 150 rpm (New Brunswick Scientific, Edison, New Jersey, US) and after 32 h, a 10% (v/v) raise was performed and after 16 h of incubation the culture was ready for production tests.

The *Enterobacter* sp. was obtained from the Culture Collection of the Laboratory of Biotechnology and Ecology of Microorganisms—LABEM—of the Institute of Health Sciences of the Federal University of Bahia, Salvador, Brazil. The preparation of the microbial inoculum consisted of inoculating a cryoprobe of the cryopreserved strain in 20 mL of Trypticase Soy Broth—TSB medium (Becton Dickinson GmbH, Heidelberg, Heidelberg, Germany) in a 125 mL Erlenmeyer flask and were incubated for a period of 24 h at 30 °C and 150 rpm orbital agitation (New Brunswick Scientific, Edison, New Jersey, US). After 24 h, a 10% (v/v) raise was performed and after 16 h of incubation at 30 °C and 150 rpm the culture was ready for production tests.

2.4 Production Protocol

X. campestris and *Enterobacter* sp. EPS productions were carried out in modified MSM [24] medium, containing (w/v) 0.277% Na_2HPO_4, 0.1% KH_2PO_4, 0.00713% Ca (NO_3) $2.4H_2O$, 0, 1% (NH_4) $2SO_4$, 0.02% $MgSO_4.7H_2O$, 0.05% yeast extract, 2.5% Sucrose and 2.0% crude glycerin. The media were prepared with 3 different solute base: distilled water (dW); water produced from the oil industry diluted in concentrations of 15 and 25% for *X. campestris* and *Enterobacter* sp., respectively, and dialyzed produced water (DPW). The culture media were pH adjusted to 7.0 with the addition of 5 M NaOH solution and then were pasteurized at about 65 °C/30 min.

Culturing was carried out in 250-mL Erlenmeyer flasks containing 45 mL of production medium and 5 mL of inoculum, 250 rpm orbital shaking and temperature of 28 ± 0.25 °C for *X. campestris* and 30 ± 0.25 °C for *Enterobacter* sp. in an I26 Incubator Shaker Series rotational incubator (New Brunswick Scientific, Edison, New Jersey, US). The incubation period was of 120 h for *X. campestris* and 44 h for *Enterobacter* sp. After production, centrifugation was then performed at 10,000 × g for 30 min at a constant temperature of 4 °C (Eppendorf 5804 R, Westbury-NY, USA). Subsequently, 150 mL of 99.5% ethyl alcohol at 4 °C was added to the supernatant and stored at 4 °C for 16 h to insolubilize the biopolymers. The insolubilized material was then recovered by centrifugation at 10,000 × g for 10 min at 4 °C. Hence, the biopolymers were then arranged for drying in an oven at 30 °C until they reached constant weight, measured with an ATY224 analytical balance (Shimadzu, Kyoto, Japan).

2.5 Alkaline and Surfactant Physiological Stress

The alkaline stress was introduced after the start of EPS production, that is at 96 and 20 h of incubation for the *X. campestris* and *Enterobacter* sp., respectively. The pH was adjusted to 9.5 using sterilized 5 M NaOH solution. A pH meter (430 pH meter, Corning, New York, USA) was used, disinfected with 2.0% sodium hypochlorite and later with 70% alcohol during the procedure. Firstly, the effect of this treatment on the microbial concentration of the cultures was verified in MSM medium prepared with dW before the application of the alkaline stress, 1 h after the application and at the end of the production process. Subsequently, it was also tested the effect of alkaline stress on the production of EPS in media prepared with dW, PW and DPW.

Microbial quantification for carried out using the most probable number technique—MPN [25]. Briefly, triplicates of 100 μL aliquots were inoculated into deep-well plates, with 96 autoclavable wells and with a capacity of 1.6 mL (Axygen, Inc.), containing 900 μL of sterile YM or TSB medium in each well for serial dilution up to 10–12. Iodonitrote chloride (Sigma-Aldrich, San Luis, US) was used as an indicator of microbial growth, added after sterilization of the culture media to obtain a concentration of 0.01% [26]. The dilutions that showed growth were detected by

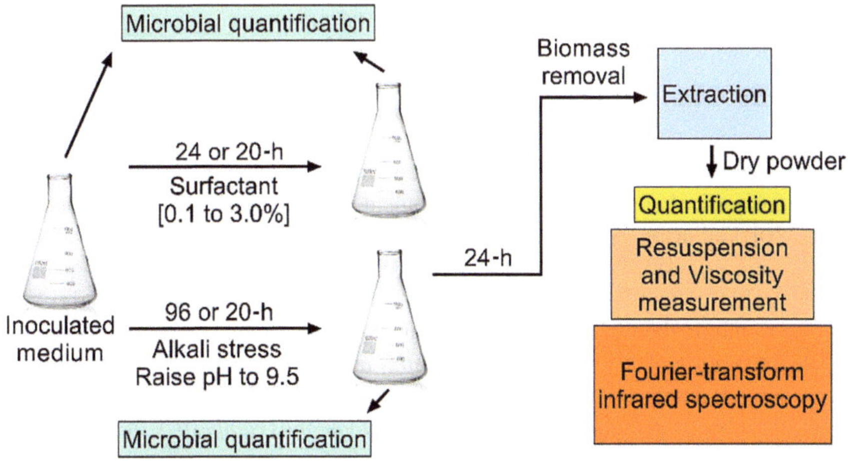

Fig. 1 Surfactant and alkali stress application scheme in inoculated production medium

changing the color from yellow to pink or purple were counted as positive and used to estimate the microbial concentration [27].

A screening test with different concentrations was carried out with Triton X-100 before determining the working solution. The evaluation was carried out in MSM medium prepared with dW and with the final concentrations (%) of Triton X-100 of 0.1, 0.5, 1.0, 2.0, 3.0, 4.0. The surfactant was introduced at 24 and 20 h of culturing. After determining the best concentration, by the production and viscosity parameters, PW and DPW tests were carried out, as well as the simultaneous association with alkaline stress (Fig. 1).

2.6 EPS Characterization (FT-IR)

An infrared spectrometer with Fourier transform model VERTEX 70v (Bruker Corporation, Billerica, Massachusetts) was used to obtain the characterization of the EPS produced [28, 29]. The dry samples were placed on the ATR diamond (model A225/Q Platinum ATR unit, Bruker Corporation, Billerica, Massachusetts) and the spectra obtained in a spectral range from 400 to 4000 cm^{-1} with acquisition of spectra through 32 scans with a resolution of 4 cm^{-1}, at a temperature of 22 ± 2 °C and controlled humidity.

3 Results and Discussion

3.1 The Alkaline Stress

Bacteria can produce EPS under various conditions. However, the EPS structure, arrangements and concentrations are dependent on nutritional and culturing conditions [30]. Culturing conditions can affect production in two different ways: it can impose stress causing alteration of cell numbers or it can regulate distinct metabolic pathways [31]. The study presented is based on the second principle mentioned above, aiming to mimic the tools of bacteriological adaptation to industrial production.

Quantification of *Xanthomonas campestris* and *Enterobacter* sp. under alkaline stress (Fig. 2) demonstrated that after alkalinization (pH 9.5) there was no reduction in the microbial population. In the culture of *Xanthomonas campestris*, the population

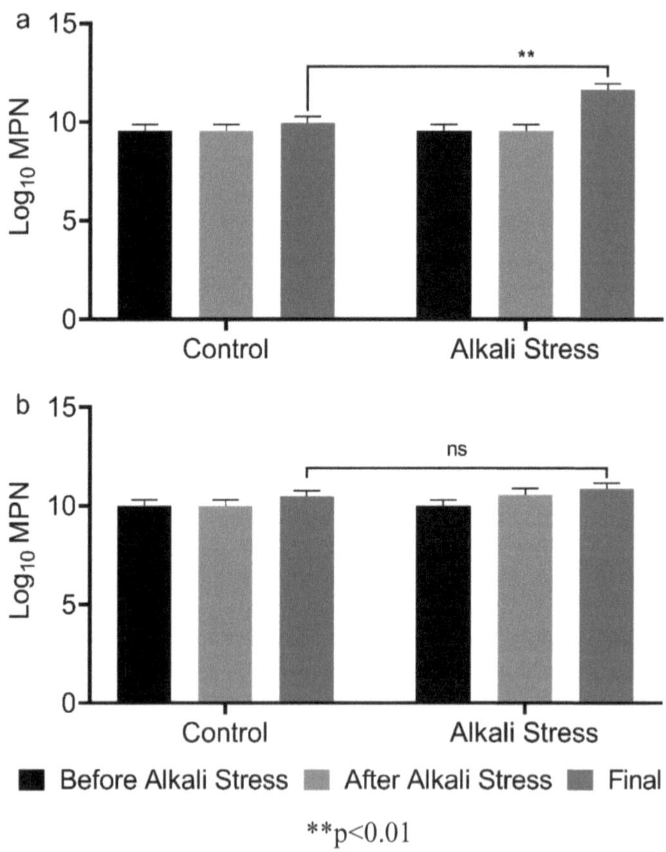

Fig. 2 Microbial quantification of before, after the alkali stress treatment and in the end of the batch. **a** *Xanthomonas campestris*; **b** *Enterobacter* sp

of viable cells remained the same after 1 h of alkaline stress and was 16.7% higher ($P < 0.01$) at the end of production (Fig. 2a). For *Enterobacter* sp. there was an increase in the microbial population after 1 h of stress, but without statistical significance (Fig. 2b).

The growth of a population that is already in a stationary phase suggests a significant change in their physiological state. Bacteria can increase the production of fatty acids as an attempt to respond to alkaline stress [14, 32–43]. It is not yet clear whether this process has an effect on reducing the cytoplasmic pH, external or both. Overexpression of Na^+ and H^+ antiports allows for a series of physiological roles, including the export of undesirable Na^+ concentrations, pH homeostasis, osmoregulation and signaling [44]. Molecular chaperones expressed between 15 and 30 min after exposure to alkalis are believed to interact with other proteins as a way of protecting against denaturation and to maintain the high level of intracellular organization, as well as essential chaperones for damaged protein refolding processes [45, 46], proteins involved in DNA metabolism and repair, cell wall catabolism, peptidoglycan and lipoteichoic acid biosynthesis and membrane proteins, stress genes and genes involved in detoxification and adaptation [13].

Publications report the existence of different genetic profiles analyzed by microarray according to different periods of alkaline stress, suggesting the existence of a regulatory mechanism as a response to alkalization [13, 14]. The maximum number of genes that have been altered in a strain of *Listeria monocytogenes* occurred in the first 15 min, indicating that the reorganization of gene expression patterns may also occurs quickly after the application of stress conditions [14]. After 30–60 min, the results showed that the cells begin to recover from their initial stress, shifting from the emergency response to a more sustainable long-term pattern to deal with the alkali challenge. It is not clear how the adaptation to the alkaline stress occurs (pH 9.0) [14].

The most significant changes in *Listeria monocytogenes* occurs in the regulation of genes related to carbohydrate metabolism during alkaline stress. In the short term (15–30 min) the metabolic pathways are found that lead to the rapid production of acids may be a compensatory mechanism to respond to the effect of high pH [13, 14]. In addition, it is known that the induction of genes like maltodextrin transporters implies that the uptake, as well as initial processing of certain sugars, is favored under alkaline stress, as they are easily and quickly fermented to produce acid products [33]. Many enzymes induced by alkaline stress are associated with the metabolism of pyruvate, alanine dehydrogenase, lactate dehydrogenase and pyruvate lyase format, enzymes related to biosynthesis and degradation of valine, leucine and isoleucine, glycerolipid metabolism, as well as the citric acid cycle, proteins transport and linkage, including Na^+/H^+ antiports and ATP-binding cassette-type carriers [14].

In the adaptation phase, most genes related to carbohydrate metabolism are at normal levels of expression [13], however, the phosphotransferase-dependent phosphotransferases show high levels of expression (in some cases $30\times$). These genes are involved with the uptake of sugars and transporters that import H^+ during oxidative respiration. These are highly regulated under alkaline stress [14]. Such changes

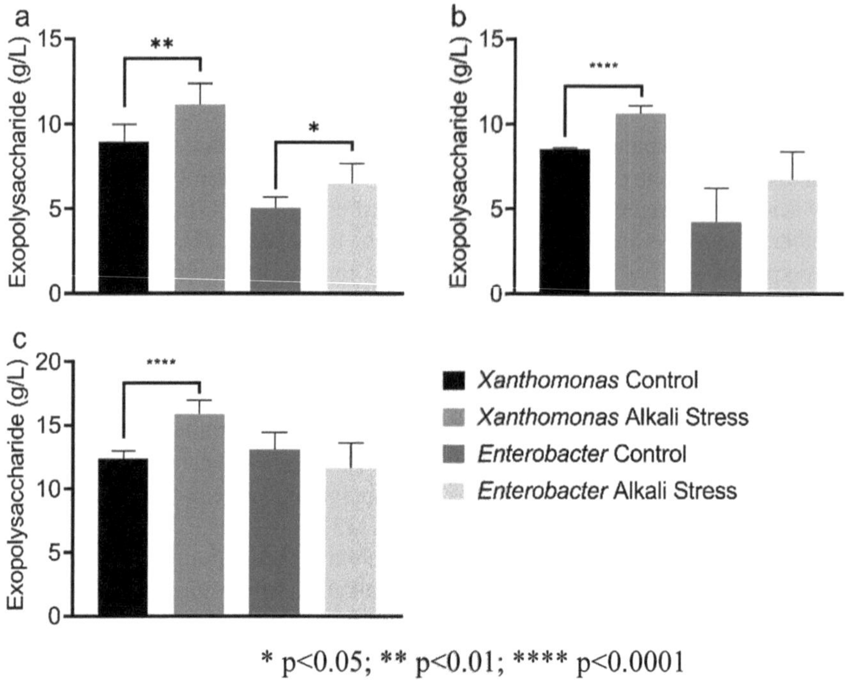

Fig. 3 Exopolysaccharide production of *Xanthomonas campestris* and *Enterobacter* sp. with alkali-stress. **a** distilled water; **b** produced water; **c** dialyzed produced water

in genetic expression may explain the higher production of exopolysaccharides by *Xanthomonas campestris* and *Enterobacter* sp.

It was observed that in a medium prepared with distilled water, there was obtained a higher recovery of exopolysaccharides by both strains of *X. campestris* (increase from 8.97 to 11.15 g/L) and *Enterobacter* sp. (from 5.0 to 6.5 g/L), representing an increase of 24.37 and 30%, respectively (Fig. 3a). In the study groups with PW or DPW, there was an increase only in the production of Xanthan gum, with no significant difference between the production with the *Enterobacter* sp. (Fig. 3b, c).

The analysis of viscosity of EPS produced with alkaline stress was lower (Fig. 4a, c) for most tests, except in tests that used produced water with *X. campestris*. In this case there was an increase in viscosity (Fig. 4b). The production of *X. campestris* in DPW showed better values of viscosity (from 390.9 to 1030.7 mPa.s) when this microbe was subjected to alkaline stress. It resulted in an increase of 163% (Fig. 4b).

Significant production of xanthan was achieved using produced water and alkaline stress, 15.9 g/L (Fig. 3c). Some studies report production of xanthan gum under alkaline conditions, with changes in the quantity and quality of the exopolysaccharide [7, 15, 16, 47–49]. Using 5% sucrose as a C source, production of 16.5 g/L has been reported, representing an increase of 74.8% in xanthan gum after 24 h in a bioreactor and with alkalinization to pH 12.0 [15], another This study used

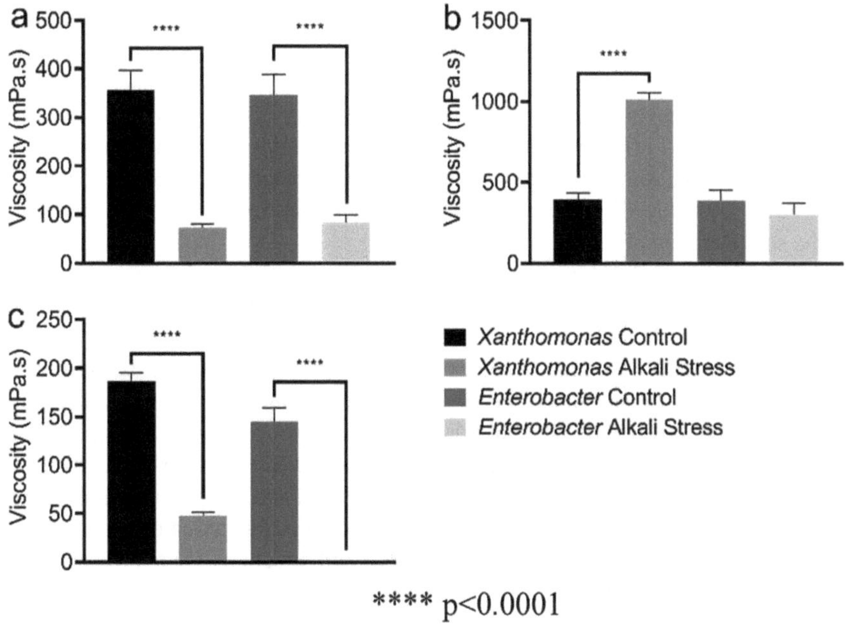

Fig. 4 Viscosity of 1.0% exopolysaccharide solutions of *Xanthomonas campestris* and *Enterobacter* sp. with alkali-stress. **a** distilled water; **b** dyalised produced water; **c** produced water

2.5% sucrose and 2.5% crude glycerin as a source of C and alkalinization of pH 11.0 resulted in the production of 10.78 g/L in Erlenmeyer flasks, representing an increase of 164.86% [16]. Although there was observed high production, their viscosity has reduced when compared to the control. Low viscosity may be due to the produced xanthan low molecular weight at high pH [50]. This may be caused by electrostatic repulsion or positively charged ions [15, 51] that caused the hydrolysis of the polymer. The authors suggest that alkaline stress can induce the cell to produce a large amount of exopolysaccharides as a protective mechanism against adverse conditions, however, the presence of vacuolated cytoplasm and discontinuity of membranes after alkalization processes, indicated lysis and microbial death [15] in such a high pH.

Our results differed from the literature when obtained xanthan gum with higher viscosity using DPW at alkaline stress—pH 9.5. Constituents present in the produced water can act as enzymatic cofactors (metals) or stress agents (hydrocarbons), which is why productions using DPW have higher yields than those with dW. Our results suggest that the association of double stress is causing the increase in production and in the viscosity of the xanthan gum. It is believed that the increase in viscosity with increasing pH [52] is due to the high level of ionization of the EPS carboxylic groups. On the other hand, the reduction in viscosity is possibly related to the accumulation of alkali in the negatively charged polysaccharide. An interesting aspect is that EPS solutions that exhibit more viscosity, show considerable tolerance to high temperatures, increase in pH, and a wide range of salinity [53]. Chemical characterization of the microbial community present in a hyperalkaline environment [54], showed that its composition is based on an exopolymeric substance containing different types of carbohydrates (represented mainly by mannose), salts, lipids, proteins and eDNA, creating a microenvironment (pH 10.4) with chemistry different from the external environment (pH 11.0). This determines the success of this community in that environment [54–56]. Another study suggested [57] that biopolymers rich in mannose, such as those produced by *Xanthomonas* sp. and *Enterobacer* sp. are directly involved in the sequestration of calcium.

In addition, the eDNA (external DNA), present in an exopolymeric substance, also plays a role in calcium sequestration due to its thermodynamically favorable interaction with calcium ions [54, 58]. Calcium is believed to be a bioflocculation-provoking agent [59], forming floccules up to 240 μm in size and their sequestration by the exopolymeric substance serves as a structural support to the flocculant and as a pH buffer [54]. Acid phospholipids are also associated with alkaline bacterial membranes and can act as a buffer at external pH [60]. Proteins can cover the carbohydrate fractions of the exopolymeric substance and the production of extracellular proteins is related to flocculation [61] and buffering action due to its acidic nature [62]. A summary scheme of the approach and direction of exopolysaccharide production is shown in Fig. 5.

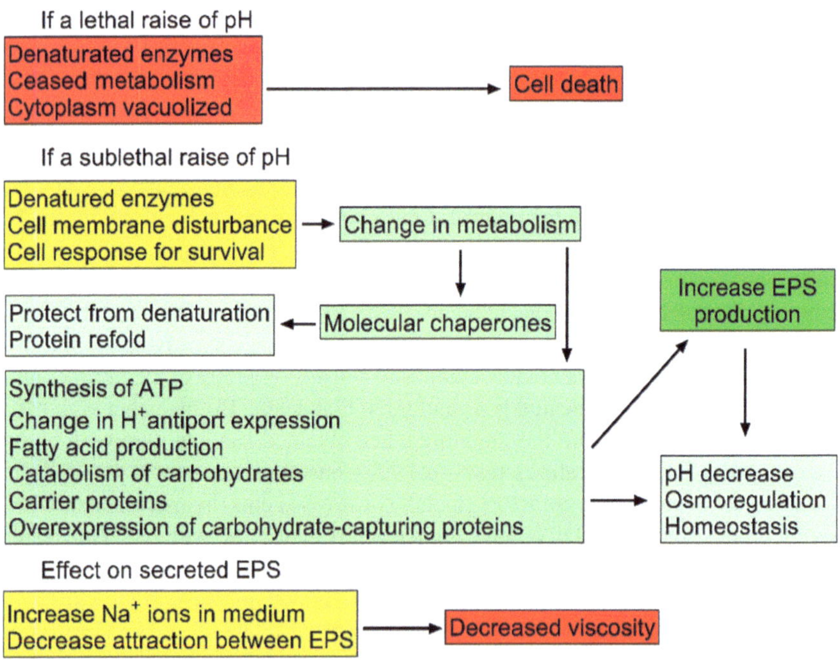

Fig. 5 Graphical abstract of the mechanisms of alkali-stress in bacteria and exopolysaccharide production

3.2 Impact of Surfactant Supplementation in EPS Production

It is believed that surfactants may assist with the increase of exopolysaccharide production [17–19]. Detergents can increase oxygen transfer and therefore improve the production of exopolysaccharides and their rheological characteristics, such as viscosity [18, 20] and consequently increase the production of hydrolyzing enzymes [63, 64]. Increased membrane permeability, modification of lipid metabolism and elevated enzyme production are among the factors that contribute to greater xanthan gum secretion [63, 64]. Surfactants affect bacterial membranes causing an increase in xanthan gum release [17, 20, 65]. It can cause an expansion of membrane pores [66], as well as moistening the cell envelope and, thus, it improves the absorption of nutrients [67, 68]. It is believed that this mechanism may be better applicable to fungi and other organisms containing thick and condensed membranes.

In Gram negative bacteria, such as *Xanthomonas* sp. and *Enterobacter* sp., the interaction with surfactants and cell barriers (inner and outer membranes) results in cell damage and increased level of leakage, as in the case of the release of reduced ATP and sugars into the culture medium [66]. It is speculated that after membrane damage and the leakage of low molecular weight compounds (purines, pyrimidines, pentoses and inorganic phosphates) are followed by the leakage of larger molecules, such as ATP [19]. Changes in the morphology from bacillus to coccobacillus [18] and reduction in size [18, 66] have also been reported, as well as expansion of the membrane pores and formation of micelles [19, 69], showing cell leakage. It is believed that this reduction in size and change in morphology may result in a higher rate of oxygen absorption, given the greater surface area of the cells [18]. In addition, it is hypothesized that detergents may interact with the cell membrane and interfere with the process of polymerization or excretion of the xanthan molecule [70].

Some studies report that Tritons surfactants are toxic to bacterial growth [19, 66, 71], although they increase membrane permeability [65], while other studies show that this type of surfactant is beneficial to the production of bioactives [18, 20]. It was shown that the application of surfactants in the idiophase resulted in a positive effect, both in concentration—with increases of about 10% [66], 36% [20] 50% [18] and viscosity—10% [66], 200% [18]. However, a considerable decrease in microbial biomass—20% [66], 45% [18, 20].

Xanthomonas campestris strain can tolerate significant concentrations of the Triton X-100 surfactant (Fig. 6a), with no statistical difference when compared to the control. The EPS viscosity was higher in the 0.5% Triton X-100 surfactant treatment increasing from 357.1 to 906 mPa.s (Fig. 6b).

On the other hand, the *Enterobacter* sp. showed sensitivity to the surfactant, with a reduction in EPS production at the initial concentration of 0.1% and absence of microbial growth in other concentrations (Fig. 7a) as well as a reduction in viscosity (Fig. 7b).

When *Xanthomonas campestris* productions were tested with 0.5% Triton X-100 in produced water (Fig. 8), it was observed a reduction in production with 15% PW and an increase with DPW (Fig. 8a).

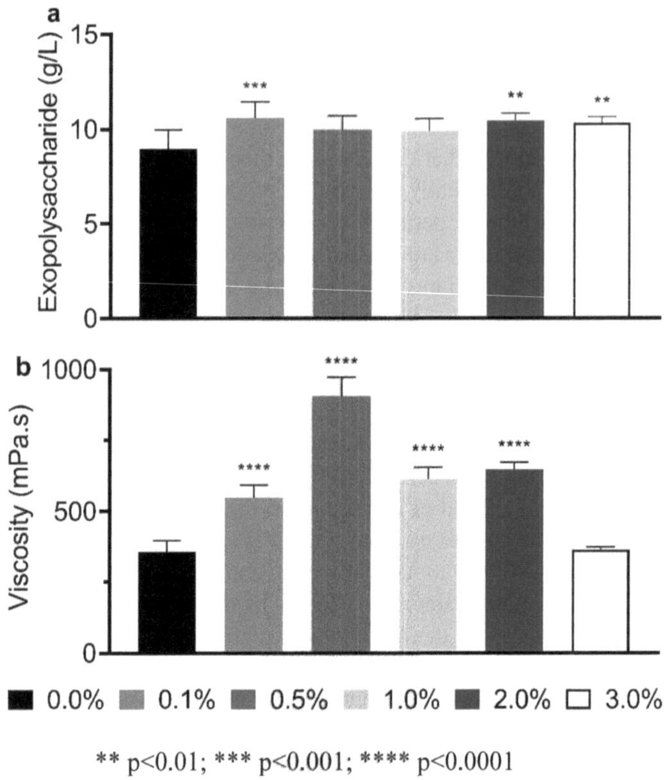

Fig. 6 Surfactant effect in *Xanthomonas campestris* in **a** exopolysaccharide production and **b** viscosity in distilled water

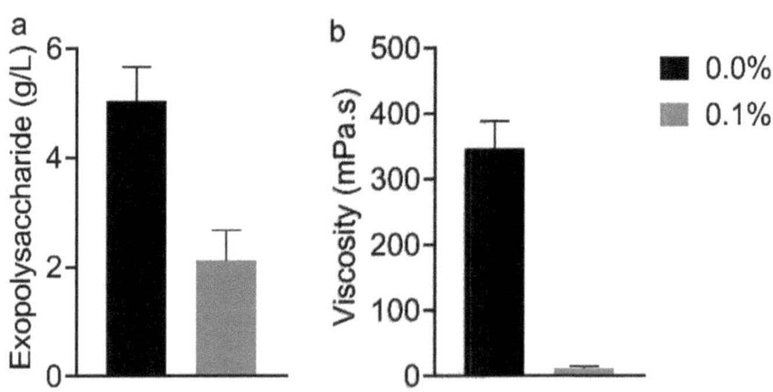

Fig. 7 Surfactant effect in *Enterobacter* sp. exopolysaccharide production (**a**) and viscosity (**b**) in distilled water

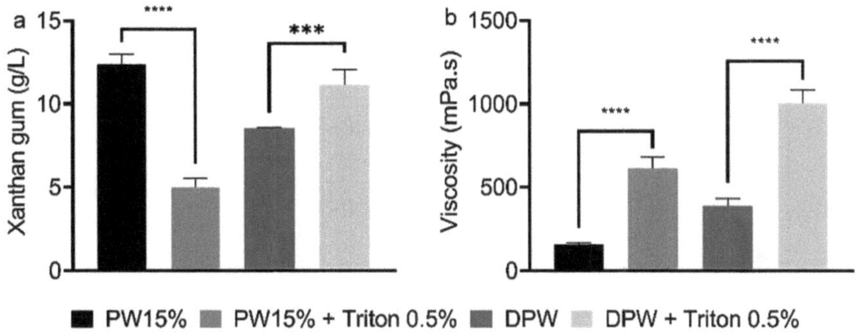

Fig. 8 Surfactant/alkali-stress effect in **a** *Xanthomonas campestris* exopolysaccharide production and **b** viscosity in produced water and dialyzed produced water

In both cases, the product generated had a higher viscosity (Fig. 8b). It is hypothesized that the disturbance caused by Triton X-100 in the *Xanthomonas* membrane caused the bacteria to lose homeostasis in the presence of the salts of PW15%, thus reducing their processes of energy acquisition and production of xanthan gum, since many of the enzymes responsible for polymerization are located on the cell membrane.

3.3 The Physiological Stress Caused by Alkaline and Surfactant Supplementation

Enterobacter sp. production of EPS under alkaline stress increased. The EPS production also increased with Triton X-100 and 5 M NaOH in dW from 8.96 to 13.53 g/L (Fig. 9a). Viscosity increased from 357.1 to 597.6 mPa.s (Fig. 9b). Thus, the association of alkaline stress and Triton X-100 resulted in significant approach for increasing EPS production and its viscosity.

Surfactant supplementation generated a small amount of foam during the preparation of solutions for analyzing the viscosity of xanthan gum. It was necessary to infer and discuss the possible effects of this detergent on the interactions in the xanthan molecule. The interactions between polymers and surfactants are mainly electrostatic and hydrophobic [72–75]. The interactions between a polyelectrolyte with a surfactant of the same charge, for example, can happen if the polyelectrolyte contains hydrophobic groups, and thus, considering that the attractive forces between them overcome their mutual repulsion [76–78]. Xanthan gum can acts as an electrolyte and its viscosity is relatively highly dependent on attractive/repulsive interactions [79]. However, it has been shown that the addition of the Triton X-100 (a surfactant), in xanthan solutions, resulting in a significant alteration of the ionic strength. In

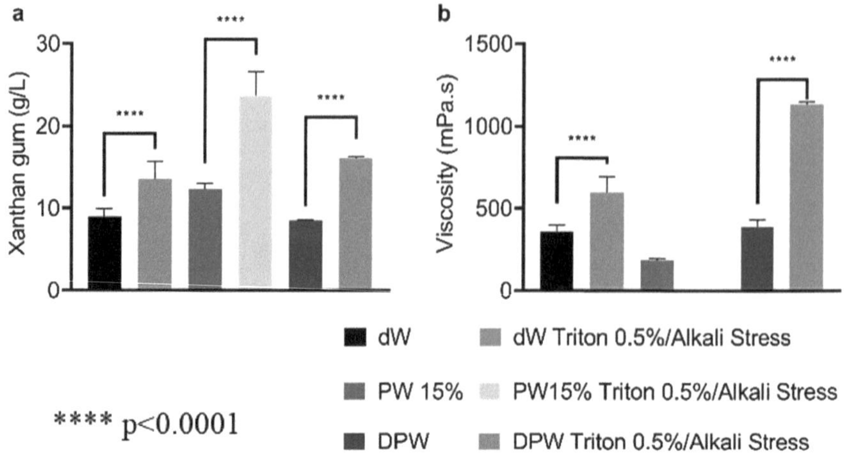

Fig. 9 Effect of simultaneous surfactant/alkali-stress in *Xanthomonas* **a** exopolysaccharide production and **b** viscosity in distilled water, produced water and dialyzed produced water

this study, we propose the existence of possible interactions: (I) the nonpolar region of the surfactant interacts with the nonpolar region of xanthan gum, hydrophilic tracking of the free surfactant to capture Na^+ ions and thereby reduce their action on xanthan OH^- groups, restoring the attraction between the polymer molecules; (II) interaction of the nonpolar segment of Triton X-100 with the nonpolar region of a xanthan molecule, and the hydrophilic group interacting with OH- radicals of another xanthan molecule, acting as an electrostatic "glue" and thus increasing the interactions between the xanthan molecules.

Surfactants such as SDS and Tween 80 interact with polysaccharide changing the viscosity. Viscosity increases when increases the attraction between molecules of the EPS [79, 80]. Given the low viscosity of xanthan gum produced only with the use of alkaline stress, we propose that Triton X-100, in addition to acting directly on improving the production of exopolysaccharide, also has a beneficial effect in restoring the attractive electrostatic forces normally present in xanthan gum solutions, but reduced when using alkalinization, given the high concentration of Na^+ ions.

A summary scheme of the effect of surfactants on bacteria and their implications for the production of exopolysaccharides are shown in Fig. 10.

3.4 *EPS Chemical Characterization (FT-IR)*

The EPS FT-IR characterization was performed with the results of the alkaline stress trials due to the sensitivity of *Enterobacter* sp. for the surfactant (Fig. 11). It was observed the final spectra common to most polysaccharides, as is the case in the 3301 cm^{-1} region; which corresponds to the presence of hydroxyl groups, an

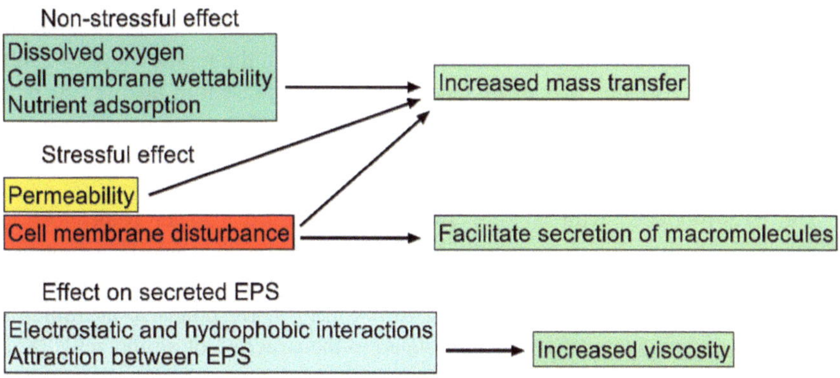

Fig. 10 Graphical abstract of the effect of surfactants in bacteria and exopolysaccharide production

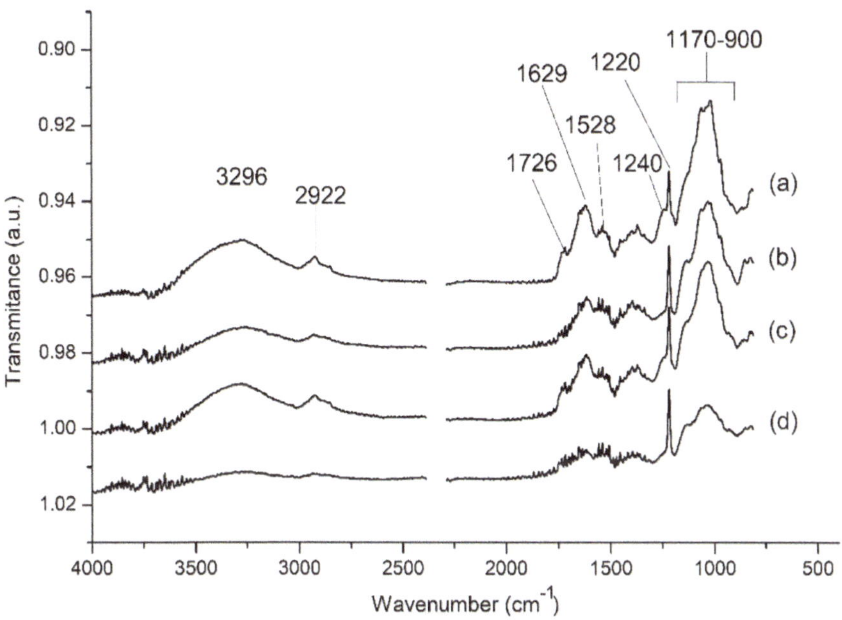

Fig. 11 EPS FTIR spectra of *Enterobacter* sp. obtained from productions in different conditions. **a** distilled water; **b** dialyzed produced water; **c** distilled water with alkaline stress; **d** produced water dialyzed with alkaline stress

absorbance band characteristic of the carbohydrate ring, responsible for conferring water solubility to EPS [81]. As well as the 2929 cm^{-1}, which correspond to CH peak of methyl and methylene groups, present in hexoses such as glucose, galactose or deoxyhexoses, such as rhamnose and fucose [82] (reference to the total sugar content).

Comparing distilled water yields (Fig. 11a, c), similarity is observed in the band close to 3200 cm^{-1} for the total carbohydrate content; in the region close to 2900, representative for methyl and methylene—CH groups, present in hexoses such as glucose; the 1720 band corresponding to the C = The elongation of EPS acetate groups and the band close to 1620, indicative of the monosaccharide skeleton -OH and/or carboxylate ions [82] and carboxyl groups [4].

The EPS produced with alkaline stress it is observed a lower peak close to 1520 cm^{-1}, which is due to protein vibration, associated with amide II, acting as a support for polypeptide structures [83, 84]. Likewise, the absorbance region around 1401 cm^{-1}, observed in EPS containing fucose [29], corresponds to asymmetric stretches of carboxylates—COO being related to the structure of amino acids [85]. The strong peak at 1220 cm^{-1} is due to the presence of asymmetric stretches of the CO ether group, in addition to alcohols, phenols, lactone and aromatic or cyclic ketones [86], being less intense in the sample produced in water produced with and without alkaline stress (Fig. 11c). In addition, the next absorbance band for 1200 to 800 cm^{-1} represents the fingerprint region, which is crucial for the identification of different polysaccharides [86], showing greater similarity in the production with distilled water, distilled water and alkaline stress and dialyzed produced water (Figs. 11a, c, b). Within this region, the range 1000–1200 cm^{-1} is due to elongations of COC, CO groups as well as the presence of carbohydrates [87] and also, the peaks between 1000–1125 cm^{-1}, more specifically, suggest the presence of uronic acid, groups o -acetyl and ester bonds [83]; glycosidic bonds produce an absorption peak of around 1020 cm^{-1} as a result of the aforementioned C–O–C vibrations. Although there are noises in the curves of some spectra, mainly with dialyzed produced water (Fig. 11b, c), it is possible to observe similarity in the conformation of the curves and in the absorbance bands analyzed, showing that they are exopolysaccharides. The sample produced in distilled water (Fig. 11c) and with alkaline stress spectral was more similar to the control production in distilled water (Fig. 11a).

FT-IR spectra of the samples have been analyzed in this study to identify the functional groups of the produced biopolymers (Fig. 12). The spectrum of xanthan obtained with DPW using Triton X-100 and alkaline stress includes the spectral bands characteristic of the polymer (Fig. 12a).

Bands referring to the sugar content are present in the four samples, such as the 3300 cm^{-1} region corresponding to OH- groups of the Triton X-100 and the carbohydrate ring, present in most polysaccharides and responsible for providing solubility to them [81].

The result observed in the 2928 cm^{-1} region is indicative of CH of methyl groups present in hexoses such as glucose and mannose [82], just as the 1723 cm^{-1} region represents the carboxyl stretch or stretching of C = O of acetate groups of xanthan gum, as well as lipids. The region close to 1601 cm^{-1} may be related to the presence of the functional group COO^- [88, 89] or to the elongation of C = O of Triton X-100 and in the pyruvate group of xanthan gum [90–92]. The 1406 and 1379 cm^{-1} regions indicate asymmetric elongation of xanthan carboxymethylates. According to data consulted in the literature [93], there is no evidence that the modification and definition of the bands are due to the spectral interference of the surfactant.

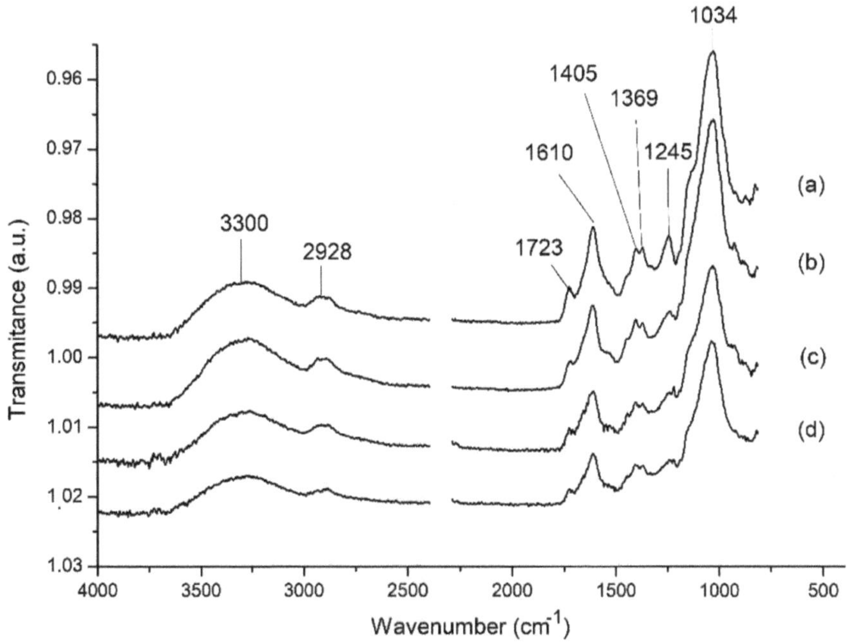

Fig. 12 FT-IR spectra obtained with the EPS samples produced with *Xanthomonas campestris* with dialyzed produced water. **a** the product obtained with alkaline stress and the addition of surfactant; **b** without alkaline stress and without Triton X; **c** surfactant supplementation and **d** alkaline stress

A significant similarity is observed in the bands profile obtained with the commercial products of xanthan gum 99% and the biopolymers produced with either alkaline stress and Triton in dialyzed produced water. It can be inferred from FT-IR data that the alkaline stress associated with Triton X-100 is beneficial and results in an extract with a better defined spectral composition than the other tests.

In order to determine whether the interaction of the surfactant and exopolysaccharide alone would explain the increase viscosity, tests EPS mixed with surfactant were carried as suggested previously [18]. For this reason, it is proposed that the exopolysaccharide interaction starts at the polymerization stage and is reduced after the polymer is already assembled. The association of alkaline stress and Triton X-100 resulted in the development of a technology to increase EPS production and viscosity.

4 Conclusions

It was observed an increase in the production of exopolysaccharides under conditions of alkaline stress of both strains under study. We have observed that the addition of about 0.5% Triton X-100 also increased the viscosity of the xanthan gum produced with produced water and crude glycerin. On the other hand, this compound was toxic to *Enterobacter* sp.

The association of alkaline stress with Triton X-100 increased the production and quality of EPS produced by *X. campestris*. It was observed an increase of 88% in production and 190% in viscosity using the medium prepared with DPW ($p < 0.0001$). This report shows an innovative approach to increase microbial EPS production by means of supplemented the media with a "surfactant/alkali" substance that causes physiological stress during growth.

Funding This work was financed by Petrogal Brasil, SA, through the project "Sustainable biotechnological alternatives to increase the oil recovery factor of carbonate reservoirs" under the management of ISPG—Instituto de Petróleo e Gás.

Compliance with Ethical Standards All authors were associated with a grant from the project "Sustainable biotechnological alternatives to increase the oil recovery factor of carbonate reservoirs" under the management of ISPG—Instituto de Petróleo e Gás and National Agency of Petroleum, Natural Gas and Biofuels—ANP.

References

1. Nwodo, U.U., Green, E., Okoh, A.I.: Bacterial exopolysaccharides: functionality and prospects. Int. J. Mol. Sci. **13**, 14002–14015 (2012). https://doi.org/10.3390/ijms131114002
2. Mishra, A., Jha, B.: Microbial Exopolysaccharides. In: The Prokaryotes, pp. 179–192. Springer Berlin Heidelberg, Berlin, Heidelberg (2013)
3. Meisen, S., Wingender, J., Telgheder, U.: Analysis of microbial extracellular polysaccharides in biofilms by HPLC. Part I: development of the analytical method using two complementary stationary phases. Anal. Bioanal. Chem. **391**, 993–1002 (2008). https://doi.org/10.1007/s00216-008-2068-y
4. Kumar, M.A., Anandapandian, K.T.K., Parthiban, K.: Production and characterization of exopolysaccharides (EPS) from biofilm forming marine bacterium. Braz. Arch. Biol. Technol. **54**, 259–265 (2011). https://doi.org/10.1590/S1516-89132011000200006
5. Lopez, B.D.M., Lessa, V.L., Lacerda, L.G.: Xanthan gum: properties, production conditions, quality and economic perspective. J. Food Nutr. Res. **54**, 185–194 (2015)
6. Almansoory, A.F., Al-Baldawi, I.A., Hazaimeh, M.: Optimization of the EPS production of a bacterial floc consortium using different parameters. Biocatal. Agric. Biotechnol. **23**, 101466 (2020). https://doi.org/10.1016/j.bcab.2019.101466
7. Nichols, C.A.M., Guezennec, J., Bowman, J.P.: Bacterial exopolysaccharides from extreme marine environments with special consideration of the Southern Ocean, Sea Ice, and Deep-Sea hydrothermal vents: a review. Mar. Biotechnol. **7**, 253–271 (2005). https://doi.org/10.1007/s10126-004-5118-2
8. Rowbury, R.J., Lazim, Z., Goodson, M.: Regulatory aspects of alkali tolerance induction in Escherichia coli. Lett. Appl. Microbiol. **22**, 429–432 (1996). https://doi.org/10.1111/j.1472-765X.1996.tb01196.x
9. Karpel, R., Alon, T., Glaser, G., et al.: Expression of a sodium proton antiporter (NhaA) in escherichia coli is induced by Na^+ and Li^+ Ions. J. Biol. Chem. **266**, 21753–21759 (1991)
10. Mendonca, A.F., Amoroso, T.L., Knabel, S.J.: Destruction of gram-negative food-borne pathogens by high pH involves disruption of the cytoplasmic membrane. Appl Environ Microbiol **60**, 4009–4014 (1994)

11. Giotis, E.S., Blair, I.S., McDowell, D.A.: Effects of short-term alkaline adaptation on surface properties of Listeria monocytogenes 10403S. Open Food Sci. J. **3**, 62–65 (2014). https://doi.org/10.2174/1874256400903010062
12. Vasseur, C., Baverel, L., Hébraud, M., Labadie, J.: Effect of osmotic, alkaline, acid or thermal stresses on the growth and inhibition of Listeria monocytogenes. J. Appl. Microbiol. **86**, 469–476 (1999). https://doi.org/10.1046/j.1365-2672.1999.00686.x
13. Giotis, E.S., Muthaiyan, A., Natesan, S., et al.: Transcriptome analysis of alkali shock and alkali adaptation in Listeria monocytogenes 10403S. Foodborne Pathog. Dis. **7**, 1147–1157 (2010). https://doi.org/10.1089/fpd.2009.0501
14. Giotis, E.S., Muthaiyan, A., Blair, I.S., et al.: Genomic and proteomic analysis of the alkali-tolerance response (AlTR) in Listeria monocytogenes 10403S. BMC Microbiol. **8**, 1–11 (2008). https://doi.org/10.1186/1471-2180-8-102
15. De, M.M., Dellinghausen, C., De, O.D., et al.: Structure of xanthan gum and cell ultrastructure at different times of alkali stress. Brazilian J Microbiol **47**, 102–109 (2015). https://doi.org/10.1016/j.bjm.2015.11.006
16. Trindade, R.A., Munhoz, A.P., Burkert, C.A.V.: Impact of a carbon source and stress conditions on some properties of xanthan gum produced by Xanthomonas campestris pv. mangiferaeindicae. Biocatal. Agric. Biotechnol. **15**, 167–172 (2018). https://doi.org/10.1016/j.bcab.2018.06.003
17. Ghashghaei, T., Soudi, M.R., Hoseinkhani, S.: Optimization of xanthan gum production from grape juice concentrate using Plackett-Burman design and response surface methodology. Appl. Food Biotechnol. 3, 15–23 (2016). https://doi.org/10.22037/afb.v3i1.9984
18. Galindo, E., Salcedo, G.: Detergents improve xanthan yield and polymer quality in cultures of Xanthomonas campestris. Enzym. Microb. Technol. **19**, 145–149 (1996)
19. Johnston, M.D., Hanlon, G.W., Denyer, S.P., Lambert, R.J.W.: Membrane damage to bacteria caused by single and combined biocides. J. Appl. Microbiol. **94**, 1015–1023 (2003). https://doi.org/10.1046/j.1365-2672.2003.01923.x
20. Janas, P., Gustaw, W., Mleko, S., Pielecki, J.: Effect of detergents on xanthan production during batch and continuous cultivation of Xanthomonas campestris NRRL B-1459. Acta Sci. Pol. Technol. Aliment **02**, 125–133 (2003)
21. Stutzenberger, F.J.: Component-specific stimulation of cellulase secretion in Thermomonospora curvata by the surfactant Tween 80. J. Appl. Bacteriol. **63**, 239–244 (1987). https://doi.org/10.1111/j.1365-2672.1987.tb04942.x
22. Alakomi, H.-L., Skyttä, E., Saarela, M., et al.: Lactic acid permeabilizes gram-negative bacteria by disrupting the outer membrane. Appl. Environ. Microbiol. **66**, 2001–2005 (2000). https://doi.org/10.1128/AEM.66.5.2001-2005.2000
23. Jeanes, A., Rogovin, P., Cadmus, M.C., et al.: Polysaccharide (Xanthan) of Xanthomonas campestris NRRL B-1459: Procedures for Culture Maintenance and Polysaccharide Production, Purification, and Analysis. US Department of Agriculture, Washigton, DC (1976)
24. Dorn, E., Hellwig, M., Reineke, W., Knackmuss, H.J.: Isolation and characterization of a 3-chlorobenzoate degrading pseudomonad. Arch. Microbiol. **99**, 61–70 (1974). https://doi.org/10.1007/BF00696222
25. Johnsen, A.R.: Introduction to microplate MPN-enumeration of hydrocarbon degraders. In: Timmis, K.N. (ed.) Handbook of Hydrocarbon and Lipid Microbiology, pp. 4159–4172. Springer Berlin Heidelberg, Berlin, Heidelberg (2010)
26. Chinalia, F.A., Paton, G.I., Killham, K.S.: Physiological and toxicological characterization of an engineered whole-cell biosensor. Bioresour. Technol. **99**, 714–721 (2008). https://doi.org/10.1016/j.biortech.2007.01.041
27. Jarvis, B., Wilrich, C., Wilrich, P.T.: Reconsideration of the derivation of most probable numbers, their standard deviations, confidence bounds and rarity values. J. Appl. Microbiol. **109**, 1660–1667 (2010). https://doi.org/10.1111/j.1365-2672.2010.04792.x
28. Nejadmansouri, M., Shad, E., Razmjooei, M., et al.: Production of xanthan gum using immobilized Xanthomonas campestris cells: Effects of support type. Biochem. Eng. J. **157**, 107554 (2020). https://doi.org/10.1016/j.bej.2020.107554

29. Freitas, F., Alves, V.D., Torres, C.A.V., et al.: Fucose-containing exopolysaccharide produced by the newly isolated Enterobacter strain A47 DSM 23139. Carbohydr. Polym. **83**, 159–165 (2011). https://doi.org/10.1016/j.carbpol.2010.07.034
30. Xu, R., Ma, S., Wang, Y., et al.: Screening, identification and statistic optimization of a novel exopolysaccharide producing Lactobacillus paracasei HCT. Afr. J. Microbiol. Res. **4**, 783–795 (2010)
31. Liu, Y., Zhu, J., Feng, M.G., et al.: Six-helix bundle assembly and analysis of the central core of mumps virus fusion protein. Arch. Biochem. Biophys. **421**, 143–148 (2004). https://doi.org/10.1016/j.abb.2003.09.037
32. Kaneda, T.: Iso- and anteiso-fatty acids in bacteria: biosynthesis, function, and taxonomic significance. Microbiol. Rev. **55**, 288–302 (1991)
33. Maurer, L.M., Yohannes, E., Bondurant, S.S., et al.: pH regulates genes for flagellar motility, catabolism, and oxidative stress in Escherichia coli K-12. J. Bacteriol. **187**, 304–319 (2005). https://doi.org/10.1128/JB.187.1.304-319.2005
34. Giotis, E.S., McDowell, D.A., Blair, I.S., Wilkinson, B.J.: Role of branched-chain fatty acids in pH stress tolerance in Listeria monocytogenes. Appl. Environ. Microbiol. **73**, 997–1001 (2007). https://doi.org/10.1128/AEM.00865-06
35. Giotis, E.S., Blair, I.S., McDowell, D.A.: Morphological changes in Listeria monocytogenes subjected to sublethal alkaline stress. Int. J. Food Microbiol. **120**, 250–258 (2007). https://doi.org/10.1016/j.ijfoodmicro.2007.08.036
36. Neumann, G., Kabelitz, N., Heipieper, H.J.: The regulation of the cis-trans isomerase of unsaturated fatty acids in Pseudomonas putida: correlation between cti activity and K+-uptake systems. Eur. J. Lipid Sci. Technol. **105**, 585–589 (2003). https://doi.org/10.1002/ejlt.200300803
37. Kim, B.H., Kim, S., Kim, H.G., et al.: The formation of cyclopropane fatty acids in Salmonella enterica serovar Typhimurium. Microbiology **151**, 209–218 (2005). https://doi.org/10.1099/mic.0.27265-0
38. Brown, J.L., Ross, T., McMeekin, T.A., Nichols, P.D.: Acid habituation of Escherichia coli and the potential role of cyclopropane fatty acids in low pH tolerance. Int. J. Food Microbiol. **37**, 163–173 (1997). https://doi.org/10.1016/S0168-1605(97)00068-8
39. Yuk, H.-G., Marshall, D.L.: Adaptation of Escherichia coli O157:H7 to pH alters membrane lipid composition, verotoxin secretion, and resistance to simulated gastric fluid acid. Appl. Environ. Microbiol. **70**, 3500–3505 (2004). https://doi.org/10.1128/AEM.70.6.3500-3505.2004
40. Fozo, E.M., Quivey, R.G.: Shifts in the membrane fatty acid profile of Streptococcus mutans enhance survival in acidic environments. Appl. Environ. Microbiol. **70**, 929–936 (2004). https://doi.org/10.1128/AEM.70.2.929-936.2004
41. Fozo, E., Kajasz, J., Quiveyjr, R.: Low pH-induced membrane fatty acid alterations in oral bacteria. FEMS Microbiol. Lett. **238**, 291–295 (2004). https://doi.org/10.1016/j.femsle.2004.07.047
42. Fozo, E.M., Quivey, R.G.: The fabM Gene product of Streptococcus mutans is responsible for the synthesis of monounsaturated fatty acids and is necessary for survival at low pH. J Bacteriol **186**, 4152–4158 (2004). https://doi.org/10.1128/JB.186.13.4152-4158.2004
43. Padan, E., Bibi, E., Ito, M., Krulwich, T.A.: Alkaline pH homeostasis in bacteria: new insights. Biochim. Biophys. Acta – Biomembr. **1717**, 67–88 (2005). https://doi.org/10.1016/j.bbamem.2005.09.010
44. Krulwich, T.A.: Alkaliphiles: 'basic' molecular problems of pH tolerance and bioenergetics. Mol. Microbiol. **15**, 403–410 (1995). https://doi.org/10.1111/j.1365-2958.1995.tb02253.x
45. Liu, S., Graham, J.E., Bigelow, L., et al.: Identification of Listeria monocytogenes genes expressed in response to growth at low temperature. Appl Environ Microbiol **68**, 1697–1705 (2002). https://doi.org/10.1128/AEM.68.4.1697-1705.2002
46. Folio, P., Chavant, P., Chafsey, I., et al.: Two-dimensional electrophoresis database of Listeria monocytogenes EGDe proteome and proteomic analysis of mid-log and stationary growth phase cells. Proteomics **4**, 3187–3201 (2004). https://doi.org/10.1002/pmic.200300841

47. Esgalhado, M.E., Roseiro, J.C., Collaço, M.T.A.: Interactive effects of pH and temperature on cell growth and polymer production by Xanthomonas campestris. Process. Biochem. **30**, 667–671 (1995). https://doi.org/10.1016/0032-9592(94)00044-1
48. Rye, A.J., Drozd, J.W., Jones, C.W., Linton, J.D.: Growth efficiency of Xanthomonas campestris in continuous culture. Microbiology **134**, 1055–1061 (1988). https://doi.org/10.1099/00221287-134-4-1055
49. Lee, Y., Seo, H., Yeom, J., Park, W.: Molecular characterization of the extracellular matrix in a Pseudomonas putida dsbA mutant: Implications for acidic stress defense and plant growth promotion. Res. Microbiol. **162**, 302–310 (2011). https://doi.org/10.1016/j.resmic.2010.11.002
50. Borges, C.D., de Paula, R.C.M., Feitosa, J.P.A., Vendruscolo, C.T.: The influence of thermal treatment and operational conditions on xanthan produced by X. arboricola pv pruni strain 106. Carbohydr. Polym. **75**, 262–268 (2009). https://doi.org/10.1016/j.carbpol.2008.07.013
51. Carrington, S., Odell, J., Fisher, L., et al.: Polyelectrolyte behaviour of dilute xanthan solutions: Salt effects on extensional rheology. Polymer (Guildf) **37**, 2871–2875 (1996). https://doi.org/10.1016/0032-3861(96)87653-1
52. Liu, S.-B., Qiao, L.-P., He, H.-L., et al.: Optimization of fermentation conditions and rheological properties of Exopolysaccharide produced by deep-sea bacterium Zunongwangia profunda SM-A87. PLoS ONE **6**, e26825 (2011). https://doi.org/10.1371/journal.pone.0026825
53. Patel, A., Zhang, H.J., Ke, M., et al.: Lubricants and drag reducers for oilfield applications—chemistry, performance, and environmental impact. In: SPE International Symposium on Oilfield Chemistry, pp. 1–12. Society of Petroleum Engineers, The Woodlands, Texas, USA (2013)
54. Charles, C.J., Rout, S.P., Patel, K.A., et al.: Floc formation reduces the pH stress experienced by microorganisms living in Alkaline environments. Appl. Environ. Microbiol. **83**, 1–12 (2017). https://doi.org/10.1128/AEM.02985-16
55. Dang, H., Lovell, C.R.: Microbial surface colonization and biofilm development in marine environments. Microbiol. Mol. Biol. Rev. **80**, 91–138 (2016). https://doi.org/10.1128/MMBR.00037-15
56. Katharios-Lanwermeyer, S., Xi, C., Jakubovics, N.S., Rickard, A.H.: Mini-review: microbial coaggregation: ubiquity and implications for biofilm development. Biofouling **30**, 1235–1251 (2014). https://doi.org/10.1080/08927014.2014.976206
57. Perry, T.D., IV., Klepac-Cera-J, V., Zhang, X.V., et al.: Binding of harvested bacterial exopolymers to the surface of calcite. Environ. Sci. Technol. **39**, 8770–8775 (2005). https://doi.org/10.1021/es0508368
58. Das, T., Sehar, S., Koop, L., et al.: Influence of calcium in extracellular DNA mediated bacterial aggregation and biofilm formation. PLoS ONE **9**, 1–11 (2014). https://doi.org/10.1371/journal.pone.0091935
59. Endo, T., Nakamura, K., Takahashi, H.: Pronase-susceptible floc forming bacteria: relationship between flocculation and calcium ion. Agric. Biol. Chem. **40**, 2289–2295 (1976). https://doi.org/10.1080/00021369.1976.10862394
60. Enomoto, K., Koyama, N.: Effect of growth pH on the phospholipid contents of the membranes from alkaliphilic bacteria. Curr. Microbiol. **39**, 270–273 (1999). https://doi.org/10.1007/s002849900458
61. Higgins, M.J., Novak, J.T.: Characterization of exocellular protein and its role in bioflocculation. J. Environ. Eng. **123**, 479–485 (1997). https://doi.org/10.1061/(ASCE)0733-9372(1997)123:5(479)
62. Tsujii, K.: Donnan equilibria in microbial cell walls: A pH-homeostatic mechanism in alkaliphiles. Colloids Surf. B: Biointerfaces **24**, 247–251 (2002). https://doi.org/10.1016/S0927-7765(01)00244-2
63. Fischer, B.E., Kleber, H.-P.: Isolation and characterization of the extracellular lipase of Acinetobacter calcoaceticus 69 V. J. Basic Microbiol. **27**, 427–432 (1987). https://doi.org/10.1002/jobm.3620270807
64. Asther, M., Corrieu, G., Drapron, R., Odier, E.: Effect of tween 80 and oleic acid on ligninase production by Phanerochaete chrysosporium INA-12. Enzym. Microb. Technol. **9**, 245–249 (1987). https://doi.org/10.1016/0141-0229(87)90024-X

65. Alakomi, H.-L., Skytta, E., Saarela, M., et al.: Lactic acid Permeabilizes gram-negative bacteria by disrupting the outer membrane. Appl. Environ. Microbiol. **66**, 2001–2005 (2000). https://doi.org/10.1128/AEM.66.5.2001-2005.2000
66. Ghashghaei, T., Soudi, M.R., Hoseinkhani, S., Shiri, M.: Effects of nonionic surfactants on xanthan gum production: a survey on cellular interactions. Iran J. Biotechnol. **16**, 60–66 (2018). https://doi.org/10.21859/ijb.1483
67. Desai, J.D., Patel, H.C., Desai, A.J.: Effect of tween series surfactants on alkaloid production by submerged cultures of Claviceps species. J. Ferment. Technol. **64**, 499–501 (1986). https://doi.org/10.1016/0385-6380(86)90073-7
68. Mizrahi, A., Miller, G.: Role of glycols and tweens in the production of ergot alkaloids by Claviceps paspali. J. Bacteriol. **97**, 1155–1159 (1969)
69. García-Ochoa, F., Santos, V., Casas, J., Gómez, E.: Xanthan gum: production, recovery, and properties. Biotechnol. Adv. **18**, 549–579 (2000). https://doi.org/10.1016/S0734-9750(00)00050-1
70. Ielpi, L., Couso, R.O., Dankert, M.A.: Sequential assembly and polymerization of the polyprenol-linked pentasaccharide repeating unit of the xanthan polysaccharide in Xanthomonas campestris. J. Bacteriol. **175**, 2490–2500 (1993). https://doi.org/10.1128/jb.175.9.2490-2500.1993
71. Makhdoumi-Kakhki, A.: Suitable nonionic surfactants for the erythromycin production by Saccharopolyspora erythraea. J. Sci. (University Tehran) **32**, 41–46 (2006)
72. El Aferni, A., Guettari, M., Tajouri, T.: Effect of polymer conformation on polymer-surfactant interaction in salt-free water. Colloid Polym Sci **294**, 1097–1106 (2016). https://doi.org/10.1007/s00396-016-3869-8
73. Deo, P., Deo, N., Somasundaran, P.: Complexation of hydrophobically modified polyelectrolytes with surfactants: anionic poly(maleic acid/octyl vinyl ether)/anionic sodium dodecyl sulfate. Langmuir **21**, 9998–10003 (2005). https://doi.org/10.1021/la050539g
74. Deo, P., Jockusch, S., Ottaviani, M.F., et al.: Interactions of hydrophobically modified polyelectrolytes with surfactants of the same charge. Langmuir **19**, 10747–10752 (2003). https://doi.org/10.1021/la035038e
75. Rub, M.A., Asiri, A.M., Khan, J.M., et al.: Interaction of gelatin with promethazine hydrochloride: conductimetry, tensiometry and circular dichroism studies. J Mol Struct **1050**, 35–42 (2013). https://doi.org/10.1016/j.molstruc.2013.07.010
76. Anthony, O., Zana, R.: Interactions between water-soluble polymers and surfactants: effect of the polymer hydrophobicity. 1. Hydrophilic polyelectrolytes. Langmuir **12**, 1967–1975 (1996). https://doi.org/10.1021/la950817j
77. Miyazawa, K., Winnik, F.M.: Isothermal titration calorimetry and fluorescence spectroscopy studies of the interactions between surfactants and a phosphorylcholine-based polybetaine. In: Aqueous Polymer—Cosolute Systems, pp. 149–156. Springer Berlin Heidelberg, Berlin, Heidelberg (2003)
78. Winnik, F.M., Regismond, S.T.A., Goddard, E.D.: Interactions of an anionic surfactant with a fluorescent-dye-labeled hydrophobically-modified cationic cellulose ether. Langmuir **13**, 111–114 (1997). https://doi.org/10.1021/la9607159
79. Krstonošić, V., Milanović, M., Dokić, L.: Application of different techniques in the determination of xanthan gum-SDS and xanthan gum-tween 80 interaction. Food Hydrocoll. **87**, 108–118 (2019). https://doi.org/10.1016/j.foodhyd.2018.07.040
80. Zhang, R., Zhang, Z., Zhang, H., et al.: Influence of emulsifier type on gastrointestinal fate of oil-in-water emulsions containing anionic dietary fiber (pectin). Food Hydrocoll. **45**, 175–185 (2015). https://doi.org/10.1016/j.foodhyd.2014.11.020
81. Karbowiak, T., Ferret, E., Debeaufort, F., et al.: Investigation of water transfer across thin layer biopolymer films by infrared spectroscopy. J. Memb. Sci. (2011). https://doi.org/10.1016/j.memsci.2010.12.037
82. Castellane, T.C.L., Campanharo, J.C., Colnago, L.A., et al.: Characterization of new exopolysaccharide production by Rhizobium tropici during growth on hydrocarbon substrate. Int. J. Biol. Macromol. **96**, 361–369 (2017). https://doi.org/10.1016/j.ijbiomac.2016.11.123

83. Bramhachari, P.V., Dubey, S.K.: Isolation and characterization of exopolysaccharide produced by Vibrio harveyi strain VB23. Lett. Appl. Microbiol. **43**, 571–577 (2006). https://doi.org/10.1111/j.1472-765X.2006.01967.x
84. Helm, D.: Identification of some bacterial cell components by FT-IR spectroscopy. FEMS Microbiol. Lett. **126**, 75–79 (1995). https://doi.org/10.1016/0378-1097(94)00529-Z
85. Maddela, N.R., Zhou, Z., Yu, Z., et al.: Functional determinants of extracellular polymeric substances in membrane biofouling: experimental evidence from pure-cultured sludge bacteria. Appl. Environ. Microbiol. **84**, e00756-e818 (2018). https://doi.org/10.1128/AEM.00756-18
86. Vélez, P.R.C. Nanoparticulas de prata na presença de ácido húmico em meio aquoso: caracterização físico-química e avaliação toxicológica em modelo zebrafish (Danio rerio). Universidade de Brasília (2017)
87. Copikova, J., Barros, A., Smidova, I., et al.: Influence of hydration of food additive polysaccharides on FT-IR spectra distinction. Carbohydr. Polym. **63**, 355–359 (2006). https://doi.org/10.1016/j.carbpol.2005.08.049
88. Faria, S., De Oliveira Petkowicz, C.L., De Morais, S.A.L., et al.: Characterization of xanthan gum produced from sugar cane broth. Carbohydr. Polym. **86**, 469–476 (2011). https://doi.org/10.1016/j.carbpol.2011.04.063
89. Anjum, F., Bukhari, S.A., Siddique, M., et al.: Microwave irradiated copolymerization of xanthan gum with acrylamide for colonic drug delivery. BioResources **10**, 1434–1451 (2015). https://doi.org/10.15376/biores.10.1.1434-1451
90. Wang, Z., Wu, J., Zhu, L., Zhan, X.: Characterization of xanthan gum produced from glycerol by a mutant strain Xanthomonas campestris CCTCC M2015714. Carbohydr. Polym. **157**, 521–526 (2017). https://doi.org/10.1016/j.carbpol.2016.10.033
91. Kang, Y., Li, P., Zeng, X., et al.: Biosynthesis, structure and antioxidant activities of xanthan gum from Xanthomonas campestris with additional furfural. Carbohydr. Polym. **216**, 369–375 (2019). https://doi.org/10.1016/j.carbpol.2019.04.018
92. Wang, Z., Chen, P., Tao, N., et al.: Anticancer activity of polysaccharides produced from glycerol and crude glycerol by an endophytic fungus Chaetomium globosum CGMCC 6882 on human lung cancer A549 cells. Biomolecules **8**, 171 (2018). https://doi.org/10.3390/biom8040171
93. Banu, V.R.N., Babu, V.R., Rajendran, S.: Investigating the corrosion inhibition efficiency of surgical carbon steel instruments used in medical field. Int. Res. J. Pharm. **8**, 79–90 (2018). https://doi.org/10.7897/2230-8407.0812254

ADME/Tox Study and Molecular Dynamics Simulations Applied in the Design of New Potential GABA-AT Inhibitors

Rai C. Silva, Irlon M. Ferreira, Leonardo Bruno Federico, Lorane Izabel da Silva Hage-Melim, Williams J. C. Macêdo, André L. M. Porto, Carlton A. Taft, Carlos H. T. P. Silva, and Cleydson Breno Rodrigues dos Santos

Abstract Chalcones derivatives compounds possess several biological activities. 08 molecules were selected from the literature. PASS Server was used to predict biological activities of chalcones (**1–4**) and dihydrochalcones (**5–8**). Pharmacokinetic and toxicological properties (ADME/Tox) of chalcones derivatives were calculated via preADMET Server. We investigated the interaction mode of selected compounds against GABA aminotransferase (GABA-AT) -PDB code 4Y0I- target indicated by PASS, using GOLD program in our molecular docking study. Our analysis was based

R. C. Silva · C. H. T. P. Silva (✉)
Departamento de Química, Faculdade de Filosofia, Ciências e Letras de Ribeirão Preto, Universidade de São Paulo, Av. Prof. Do Café, s/n Monte Alegre, Ribeirão Preto, São Paulo 14040-903, Brazil
e-mail: tomich@fcfrp.usp.br

I. M. Ferreira · C. B. R. dos Santos
Laboratory of Modeling and Computational Chemistry, Department of Biological and Health Sciences, Federal University of Amapá, Macapá, AP 68902-280, Brazil

L. B. Federico · L. I. da S. Hage-Melim · C. H. T. P. Silva
Computational Laboratory of Pharmaceutical Chemistry, University of Sao Paulo, Av. Prof. Do Café, s/n Monte Alegre, Ribeirão Preto, São Paulo 14040-903, Brazil

A. L. M. Porto
Laboratório de Química Orgânica E Biocatálise, Institute of Chemistry of São Carlos, Universidade de São Paulo, Av. João Dagnone, 1100 Santa Angelina, São Carlos, São Paulo 13563-120, Brazil

W. J. C. Macêdo
Laboratory of Molecular Modeling and Simulation System, Federal Rural University of Amazônia, Rua João Pessoa, 121, Capanema, PA 68700-030, Brazil

C. A. Taft
Brazilian Center for Physical Research, R. Dr. Xavier Sigaud, 150 Urca, Rio de Janeiro 22290-180, Brazil

L. I. da S. Hage-Melim
Laboratory of Pharmaceutical and Medicinal Chemistry (PharMedChem), Federal University of Amapá, Macapá, Amapá, Brazil

F. A. La Porta and C. A. Taft (eds.), *Functional Properties of Advanced Engineering Materials and Biomolecules*, Engineering Materials,
https://doi.org/10.1007/978-3-030-62226-8_25

on the interactions with Glu270, Arg445, Lys329, and Arg192 present in the active site, and indicated compounds **2**, **5,** and **6** as the most promising inhibitors. Molecular dynamics (MD) simulations were carried out via Biovia Discovery Studio software, based on the CHARMm force-field engine, using as input top-ranked docking poses obtained for the selected **2**, **5,** and **6**, regarding the same crystallographic GABA-AT enzyme structure. Our results showed these compounds could bind strongly to plasma proteins, with a variation in the range of 90.21% to 100%. Compound **5** exhibited $C_{Brain}/C_{Blood} = 1.07505$ considered active in the CNS, while compounds **2** and **6** showed low values 0.34815 and 0.28086, respectively, therefore regarded as inactive. Analyses of RMSDs obtained for such three chalcones reveal a higher stabilization for the complex GABA-AT enzyme-chalcone **6**, since a lower fluctuation of RMSD values is for it observed (from 4.0 to 7.0 Å), in comparison with the values obtained for GABA-AT enzyme-chalcone **2** (RMSD from 2.5 to 8.5 Å, the less stable) and GABA-AT enzyme-chalcone **5** (RMSD from 2.5 to 7.0).

Keywords Chalcones · Dihydrochalcones · Molecular docking · Bioactivity predictions · Central nervous system

1 Introduction

Numerous types of chalcones and dihydrochalcones have attracted attention because of reports that they possess various biological activities, i.e., cytotoxic [1, 2], antileishmanial [3], antibacterial [4], antitumor [5], anti-Trypanosoma cruzi [6] and antioxidant [7]. Indicating a variety of pharmacological activities, chalcone compounds, of medicinal interest, have been developed.

The production of dihydrochalcones based on the use of microorganisms is, of course, more effective than the extraction of plants or the traditional organic synthesis [8, 9]. Particularly, the development of the modified microbial strains has allowed the use of a relatively low-cost cultivation medium for the biosynthesis of a variety of dihydrochalcones [10]. Consequently, this leads to a significant reduction in processing costs, making this methodology attractive for the production of a large scale of dihydrochalcones-related compounds [8–10].

The disorderly increase in neuronal excitation can cause seizures, this happens when the cerebral concentration of γ-aminobutyric acid (GABA), an inhibitory neurotransmitter, decreases below a critical level. This neurotransmission disorder can be resolved by inhibiting the activity of γ-aminobutyric acid aminotransferase (GABA-AT), an enzyme that favors the conversion of GABA to l-glutamic acid, an arousal neurotransmitter. Another finding is that the elevation of GABA levels can be antagonistic in the processes of rapid elevation and release of dopamine in the nucleus accumbens, responsible for the satisfaction response in addiction [11].

In this paper, we selected 8 compounds (chalcones 1–4 and dihydrochalcones 5–8), from the previous study performed by Ferreira et al. [12]. These compounds had pharmacokinetic, toxicological and biological activity predictions, followed by docking and molecular dynamics simulations.

2 A Theoretical and Computational Approach to Predictions of Biological Target, Pharmacokinetics, and Toxicity from Chalcone Derivatives

2.1 Selection of the Compounds of Interest for Preliminary Evaluations

We have selected compounds **1–8**, from the literature for this study [12]. In general, the chemical structures of all compounds is shown in Fig. 1. As such, Ferreira et al. have synthesized chalcone compounds by aldol condensation and dihydrochalcones derivatives via biotransformation from fungus Penicillium citrinum CBMAI 1186 resulting in satisfactory yields [12].

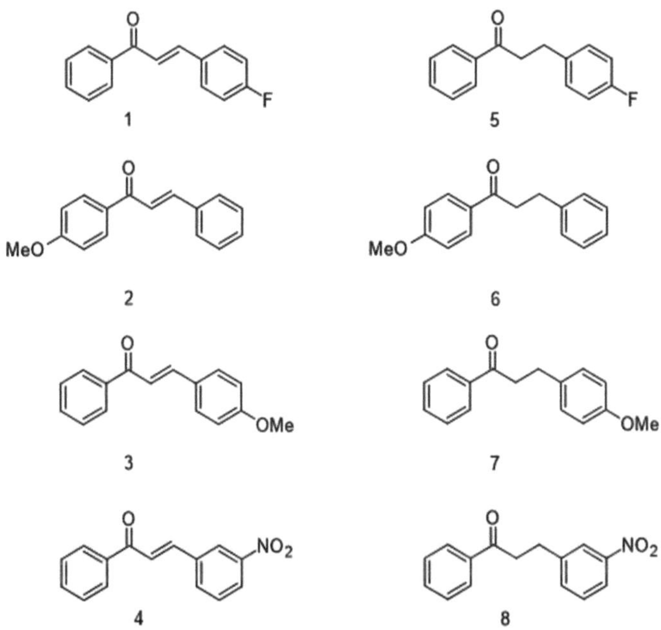

Fig. 1 Chalcones (1–4) and dihydrochalcones (5–8) compounds selected for study

2.2 *Biological Target Identification, Absorption, and Toxicity Assessment, and Molecular Dynamics Simulations*

PASS (Prediction of Activity Spectra of Substances) server [13] was then used to predict biological activities of eight compounds, chalcones (1–4) and dihydrochalcones (5–8). The server indicated high efficiency, which predicts with high accuracy (70–80%) up to 2000 chemical compounds for biological activities. In addition, in this study, the PASS served to select biological targets for performing docking and molecular dynamics simulations.

Pharmacokinetic and toxicological (ADME/Tox) properties of the chalcones derived compounds have been calculated through the preADMET [14– 16].

2.3 *Molecular Docking Studies and Molecular Dynamics Procedures*

We investigated the interaction mode of chalcones and dihydrochalcones against the target indicated by the PASS server using GOLD program [17] in our molecular docking study. The crystallographic GABA-AT enzyme structure obtained from the complex with ligand4-[({3-hydroxy-2-methyl-5-[(phosphonooxy)methyl]pyridin-4-yl}methyl)amino]thiophene-2-carboxylic acid (PDB **code 4Y0I**) with 1.66 Â resolution [18] was then selected in this study.

Note that we used protocol with CHEMPLP (Piecewise Linear Potential) scoring function, population size = 100, selection-pressure = 1.1, number of operations = 10.000, islands = 1, niche = 2, weights operator to migrate = 0, number of mutations = 100, crossover = 100 and 10 docking runs, a sphere of size 10 Å with center coordinates x = 33.11, y = 0.06 and z = 54.64. We used Maestro (*version 11*) program [19] to analyze results and visualize interactions in active site.

Molecular Dynamics (MD) is an essential computational tool that has widely been used in Medicinal Chemistry for drug design. More specifically, the classic dynamic behavior of a system is simulated as an extension of methods based on molecular mechanics from the numerical integration of the displacement equations. Thus, the MD simulations can widely be used to help in designing new drug design as well as an understanding of the mechanism for a therapeutic target.

In this case, MD simulations were carried out using Biovia Discovery Studio software 2017 [20], based on the CHARMm force-field engine. As such, in this study, we have used the standard dynamics cascade as a strategy. In particular, no restraints were used in the MD calculations. The input for the MD simulations in GABA-AT active site are based on the top-ranked docking poses for the selected chalcones **2**, **5** and **6**.

In order to energy minimization, it was carried out using maximum steps of 100 and an RMS gradient of 1.0, respectively. Next, the MD calculations were setting (using a conjugated gradient as an algorithm) using a maximum steps of 500 and an RMS gradient of about 0.1. In this case, we have used a time of 50 ps (with each step of 2 ps) for the heating among 50 and 300 K, respectively. According to these simulations, the equilibrium phase at 300 K was obtained with a time of 100 ps. On the production step, a time of about 1,000,000 ps (or 1000 ns) was considered in MD simulation, and these results were saved considered intervals of 1,000 ps (snapshots). Finally, we have used an implicit solvent dielectric constant of 80 (simulation in water), with a non-bond higher cutoff distance of 12.0 and a non-bond lower cutoff distance of 10.0 leapfrog Verlet dynamics integrator.

3 Bioactivity Predictions and Intermolecular Aspects Between Compounds Derived from Chalcones and Gaba Aminotransferase Enzyme

PASS online predicts with high accuracy (>80%) up to 4000 kinds of biological activities, including pharmacological effects, mechanisms of action, toxic and adverse effects, interaction with metabolic enzymes and transporters, influence on gene expression, and so on [21, 22]. Hence, for a start is necessary, only the 2D structure in mol format as input. It additionally can easily be used for a huge number of activities in different projects. The great success of this program is related to its ability to provide preliminary results of great utility. As such, this strategy is extremely interesting for in silico screening [21, 22].

Biological activity prediction spectra were performed for those chemical structures previously selected (chemical structures **1–8**). The prediction search was focused on GABA-AT inhibitors. The parameters related to the activity of GABA-AT inhibitor with Pa > Pi were selected, see Table 1.

Note that the interpretation of these results is usually realized in terms of the Pa values. Hence, a high biological activity is expected for cases in which Pa assumes values greater than 0.7. Yet, the probability of finding biological activity during an experiment decreases significantly in the range of $0.5 < Pa < 0.7$ [23]. Thus, compounds **1**, **2** and **3** are the most promising GABA-AT inhibitors. There

Table 1 Biological activity prediction of the compounds via *PASS* server

Compounds	Biological activity	Pa	Pi
1	GABA-AT inhibitor	0.796	0.003
2	GABA-AT inhibitor	0.734	0.004
3	GABA-AT inhibitor	0.734	0.004
4	GABA-AT inhibitor	0.533	0.016
5	GABA-AT inhibitor	0.653	0.006
6	GABA-AT inhibitor	0.581	0.011
7	GABA-AT inhibitor	0.581	0.011
8	GABA-AT inhibitor	0.604	0.009

is strong evidence that GABA display a fundamental role in epilepsy treatment, being hence the main inhibitory neurotransmitter in the cerebral cortex, in which counterbalances neuronal excitation [23].

Bark extract Ficus religiosa L. (*Moraceae*) showed in practice an interesting anti-convulsant mechanism [24]. In order to information on the biological spectrum of all the reported bioactive metabolites found in the bark extract has been performed an investigation on PASS software. These findings suggest a high activity score for the GABA-AT inhibitory effect of the identified bioactive metabolites in the bark extract [24], in good agreement with this study.

At the molecular level, it is known that a conveyor system is, in general, coordinates via channels, receptors, and also by enzymes. Additionally is this complex system could particularly affects the pharmacokinetics properties (such as absorption, distribution, metabolism, excretion, and toxicity) of a drug-like molecule in humans. Hence, from this perspective, it is evident that an in-depth understanding of the interactions between small molecules and their molecular targets, in principle, have a pivotal role predicting toxic effects that are in general responsible for the removal of several marketed drugs and late-stage failures of drug development [25, 26].

ADME prediction results are shown in Tables 2 and 3. Also, it is known that the absorption processes are mainly due to the permeation of compounds into biological membranes under the influence of the physicochemical features [27, 28].

Analyzing ADME results of compounds shown in Table 2, it was observed that compounds presented HIA% values $\geq$ 98.21, whereas compounds **1–3** and **5–7** showed absorption equal to 100%. However, it is known that orally administered drugs do not exhibit total absorption.

Cell permeability *in vitro* Caco-2 is, of course, an essential test to assess the intestinal absorption of drugs. Caco-2 cells are, therefore, derived from the human colon adenocarcinoma, owning various transport properties through the intestinal

Table 2 Absorption properties for chalcones (**1–4**) and dihydrochalcones (**5–8**) compounds

Compounds	Absorption			
	HIA(%)[a]	P_{Caco-2} (nms^{-1})[b]	P_{MDCK} (nms^{-1})[c]	P_{skin}[d]
1	100.0000	55.4962	36.8439	−2.0081
2	100.0000	55.4311	52.6245	−1.9653
3	100.0000	55.1765	52.6243	−1.9675
4	98.4270	21.2869	8.9107	−2.2539
5	100.0000	55.5253	36.7073	−1.8322
6	100.0000	55.4487	52.3492	−1.4846
7	100.0000	55.2021	52.3495	−1.9191
8	98.2065	21.2986	8.7008	−2.3264

[a]percentage of human intestinal absorption; [b]cell permeability (Caco-2 in nms^{-1}); [c]cell permeability Maden Darby Canine Kidney in $nm.s^{-1}$; [d]skin permeability

Table 3 Distribution in percentages of PPB and penetration of (C_{Brain}/C_{Blood}) for chalcone derived compounds (**1**–**8**)

Compounds	Distribution	
	PPB(%)[a]	C_{Brain}/C_{Blood}[b]
1	96.64789	1.42624
2	91.40194	0.34815
3	91.74429	0.34966
4	93.49657	0.01787
5	100.00000	1.07505
6	92.04427	0.28086
7	91.10597	0.28841
8	90.20775	0.01891

[a]percentage of plasma protein binding; [b]penetration of the blood brain barrier

epithelium [16, 29]. Table 2 showed an average permeability of 46.85818. For the compounds calculated, it was found that PCaco-2 (nms^{-1}) was moderate, ranging from 21.2869 to 55.5253 nms^{-1}.

Cell permeability *in vitro* in the MDCK system in general utilizes canine kidney cells and has a relatively shorter growth than Caco-2 cells. As such, this system has been used explicitly as a tool for rapid prediction of the permeability [16, 30]. Analyzing data in the MDCK system for inhibitors shown in Table 2 has been verified that permeability can be classified as low (25 nms^{-1}) and medium (25–500 nms^{-1}), being low in the compounds **4** and **8** with values of 8.70085 nms^{-1} to 8.91071 nms^{-1}, respectively, and medium in the other compounds varying between 36.7073 and 52.6245 nms^{-1}.

On the other hand, knowing the permeability of the skin is an extremely important parameter for transdermal drug delivery. Remarkably, in the pharmaceutical industry, this parameter is widely used to assess the risk potential of accidental contact involved new chemical products with skin [31]. All compounds studied showed negative permeability values varying between −2.3265 to −1.4846 not indicating the importance of the drug by this type of administration (see Table 2).

Note that the drug has two forms in the blood, such as plasma protein bond form and non-bond form. It is known that bond directly depends on their affinity for bond portion and free portion. Hence, the binding to these proteins can in principle alter the half-life of the drug in the body of the individual [16, 32, 33].

The distribution properties of PPB and penetration of (C_{Brain}/C_{Blood}) for all compounds studied, as shown in Table 3, suggests a bind strongly to plasma proteins, with a variation in the range of 90.20775 to 100%.

Compounds **1** and **5** have strength in the bond with these proteins due to high values. As such, this is directly related to the small amount of free drug that interacts with its receptors. Consequently, it ends up triggering the formation of this pharmacological response [10]. As is well-known, the PPB also modifies renal excretion, mainly due to unbound drug, i.e., increasing in large part its excretion as well as leading a complete decreasing of the half-life [16, 34].

Blood–brain barrier C_{Brain}/C_{Blood} is, of course, an essential component of a communication network that connects the central nervous system and peripheral tissues [16, 35]. Consequently, this fact can in principle lead to a limiting/regulation the exchange of substances that occurs among the central nervous system and blood [16, 35].

It is known that compounds in which penetration value ($C_{Brain}/C_{Blood} > 1$) are usually considered active in CNS may in theory cause collateral effects [36]. As shown in Table 3, the compounds analyzed showed that all have fewer penetration values than 1 (hence are inactive in the CNS), except for the compounds **1** and **5** that exhibited values of about 1.42624 and 1.07505, respectively, whereas the compound **4** have a low value of $C_{Brain}/C_{Blood} = 0.017875$.

Particularly, the Ames test is a famous biological assay based on the Salmonella typhimurium strains' use to assesses the mutagenicity of such compounds. In this regard, the Salmonella typhimurium strains have modifications identified in their genes, which are responsible by the histidine synthesis [16]. As such, it is well-known that histidine is important for growth. Therefore, the test assessment allowas evaluating the mutagenic agent responsible for the growth inhibition in a medium without histidine [37].

Based on this test, the compounds (**1–8**) have a positive predictionin sense of being mutagen, see Table 4. As such, the carcinogenicity test has the objective of identifying the tumorigenic potential in animals and risk assessment in humans [16, 38]. In general, these tests require considerable study time (>2 years), whereas PreADMET online server may predict results from a set data of National Toxicology Program (NTP) and Food and Drug Administration (FDA), both of USA, which coordinates toxicology research and *in vivo* testing in rats and mice [38].

Analyzing carcinogenicity in mice models, for the **3**, **7** and **8** compounds were predicted as negative, and in case of **1**, **2**, **4**, **5** and **6** compounds have a positive prediction, i.e., shows non-carcinogenicity for mice. By analyzing rat carcinogenicity, **2**, **3**, **4** and **8** compounds were predicted positive, that is, showing non-carcinogenic, and

Table 4 Mutagenicity (Ames Test) and carcinogenicity (mouse and rat) for chalcones derived compounds (**1–8**)

Compounds	Ames test	Carcinogenicity	
	Mutagenicity	Mice	Rat
1	Mutagen	Positive	Negative
2	Mutagen	Positive	Positive
3	Mutagen	Negative	Positive
4	Mutagen	Positive	Positive
5	Mutagen	Positive	Negative
6	Mutagen	Positive	Negative
7	Mutagen	Negative	Negative
8	Mutagen	Negative	Positive

1, **5**, **6** and **7**, presented negative predictions, suggesting evidence of carcinogenic, according to Table 4.

Finally, we have also extended our studies with molecular docking to evaluate the binding mode with the 3D structure of the GABA-AT receptor deposited in PDB **code 4Y0I** [18]. Initially, we did the redocking study and the result obtained was an RMSD value of 0.8, which ensured us the quality of the model, demonstrating that parameters used for redocking are optimal. Validation was accepted by presenting a small deviation of RMSD, because both are possible crystallographic poses and pose found not invalidate the other. According to Cole et al. (2005), for binding mode predictions using "docking", RMSD values smaller than 2 Å with crystallographic poses of ligand, is considered successful [39].

The results of the molecular docking study showed that it is favorable for our compounds to interact with residues of the active site of GABA-AT in the same way as some compounds reported that interacted in the same region, presented biological activity [18, 23, 24]. Our analysis was based on the interactions with Glu270, Arg445, Lys329 and Arg192 present in the active site. As such, these residues are, in turn, essential for the interaction of the ligands with GABA-AT [40].

Figures 3 and 4 show the results of the docking study in 3D and 2D dimensions, respectively. Thus, we can observe the possibilities of chalcone **1–4** interacting with GABA-AT enzyme. In these figures, it is possible to correlate the results with the frontier orbitals highest occupied molecular orbital (HOMO) and Lowest Unoccupied Molecular Orbital (LUMO).

Through visual inspection, we observe that for HOMO and LUMO orbitals of chalcones **1–4** are in their aromatic ring X and HOMO regions that allow the transfer of electrons, see Fig. 2. Docking studies exploit this property, reveal that the aromatic ring X of compounds **1**, **2** and **4**, perform pi-stacking interaction with the ring present in amino acid residue Phe19, depending on the positioning of benzene ring X. There is also the possibility of interacting with Arg445 through a Cation-Pi interaction, observed in compounds **1**, **3**, and **4**. This fact can be elucidated by the HOMO electron donation to Phe189, since this residue can present positive inductive effects on the aromatic ring.

When analyzing the hypothesis of binding mode according to the LUMO orbital in interaction with Phe189, we highlight that the inductive effect can, in principle, be understood it due to the presence of the carbonyl group in the chemical structure, as well HOMO orbital will promote inductive effect on the molecule. Another important region of our compounds is the region having a carbonyl. Note that this is enables of making a hydrogen bond with Lys329. Finally, NO_2 group present in compound **4** is

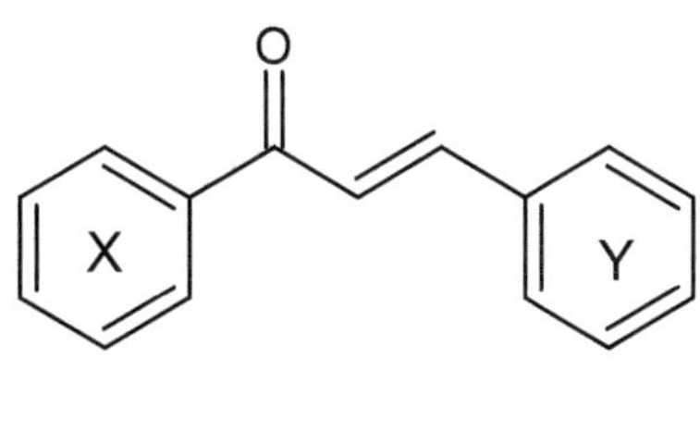

Fig. 2 Generic molecular structure representation of derived compound chalcones

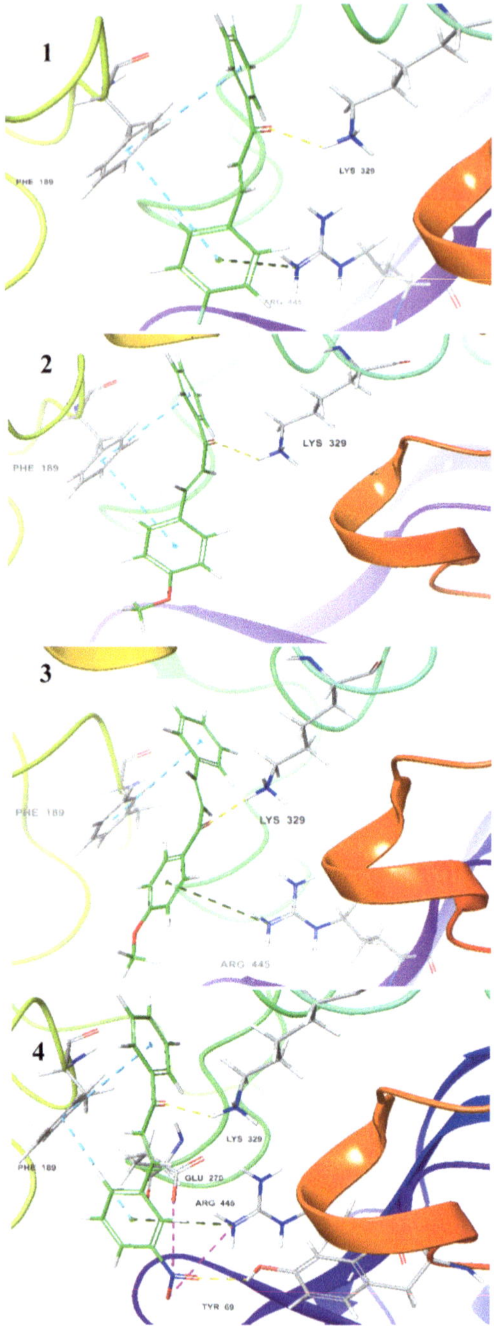

Fig. 3 3D representation of interactions between chalcone compounds (1–4) and GABA-AT enzyme resulting from the results of docking simulations

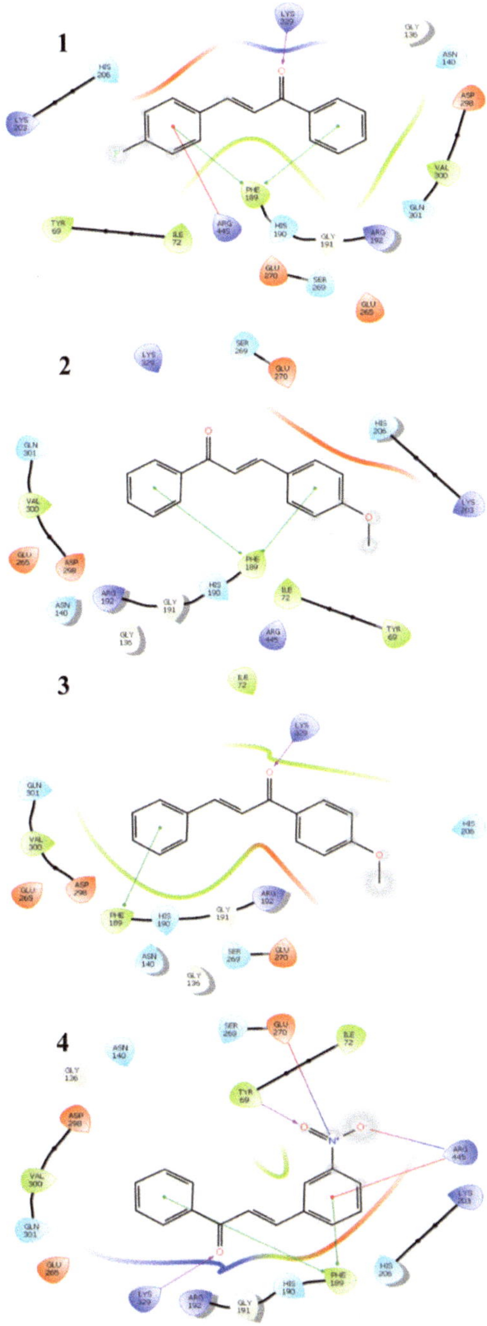

Fig. 4 2D representation of interactions between chalcone compounds (1–4) and GABA-AT enzyme resulting from the results of docking simulations

an essential group for these molecules, since it is capable of performing 3 significant interactions with GABA-AT.

We observe the possibility of the occurrence of two salt bridges between the N and O- of the compound with Glu270 and Arg445, respectively. Thus, the same grouping can still interact with Tyr69 through an H-bond accept with its carbonyl.

The greater spacing between the rings present in dihydrochalcone structures generated similar results with those observed by the chalcones, as can be observed in Figs. 5 and 6.

Pi–Pi interaction between the ring of the Phe189 and at least one of the dihydrochalcone rings is common to all structures under study. We further observe compounds **6**, **7**, and **8** making a Cation-Pi interaction with Arg445, and also, the compound **5** and **8** by making a hydrogen bond with Lys329 and further compound **6** by making a hydrogen bond with Gln301 and compound **7** with Arg192.

In the comparison between the ligand of the PDB **code 4Y0I** crystallographic complex and compounds studied, we note that the conservation of the direct interaction with Phe189 and Arg445 in regions of the large electron cloud, as in aromatic rings. Lys329 interacts directly with carbonyl and indirectly with electrons located in alpha and beta carbon unsaturation and may influence the electrophilic and hydrophobic character of the region.

When we compare the types of interactions with the electrostatic potential, we observed that the negative potentials confirm the hydrogen bonding interactions "positive potential", for example, the interaction of the carbonyl with the residue Lys329 and Gln301, as well as, in the Compound 4 that presented interaction in the -NO_2 region with Tyr69 and compound **6** that showed interaction with the residue Gln301.

This type of analysis is a proposal that provides additional information on features of compounds to get a more in-depth comprehension on the chemical behavior for new drugs. For each very long MD trajectory here obtained, the root mean square deviations (RMSDs) of the enzyme-inhibitor conformations were analyzed along the respective trajectories (1000 conformations/snapshots were produced) and regarding the initial pose of the respective complex enzyme-inhibitor (Figs. 7, 8 and 9).

Compounds **2**, **5** and **6** were chosen for MD based on the best results of PPB, C_{Brain}/C_{Blood}, pharmacokinetics and docking analysis. The total energies of the three systems were also calculated along the trajectory (1,000,000 ps—Figs. 8a-c), as well as the target temperature was maintained constant at 300 K (Figs. 7a-c) in all the simulations (at the production step).

Results obtained with molecular dynamics procedures indicate that all the three chalcones here investigated remain relatively stable inside the docked GABA-AT enzyme active site after 1000 ns of simulation time, using the top-ranked GOLD solutions as input, since energy stabilization is guaranteed for all the chalcones after the earlier stages of the respective MD production steps, at a constant temperature, theoretically corroborating and enabling the chalcone poses suggested by docking procedures. However, analyses of the RMSDs obtained for such three chalcones reveal a higher stabilization for the complex GABA-AT enzyme-chalcone **6**, since a lower fluctuation of the RMSD values is for it observed (from 4.0 to 7.0 Å), in

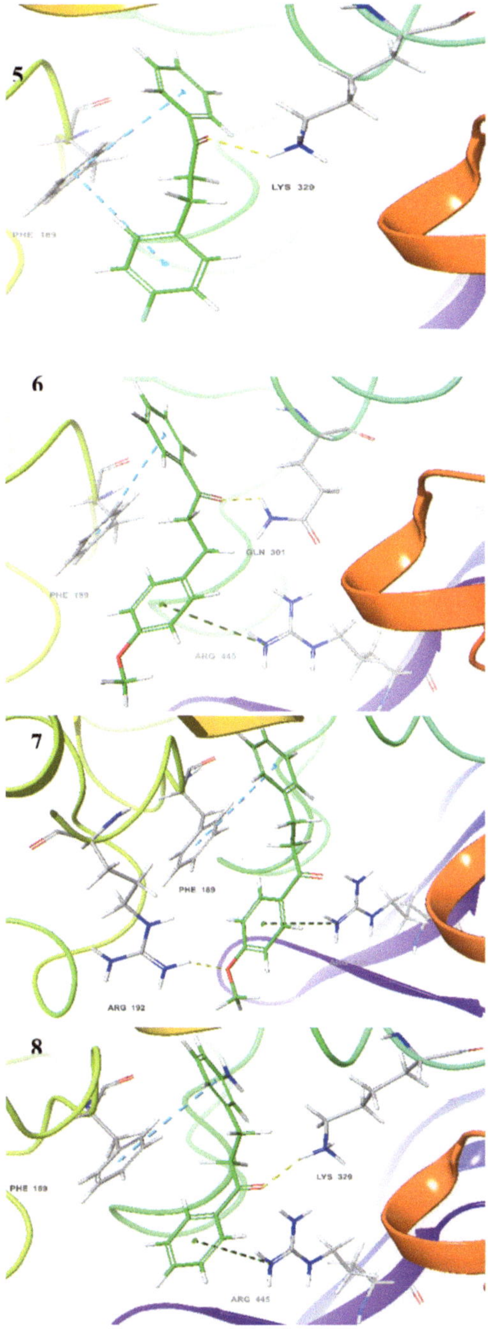

Fig. 5 3D representation of interactions between dihydrochalcone compounds (**5**–**8**) and GABA-AT enzyme resulting from the results of docking simulations

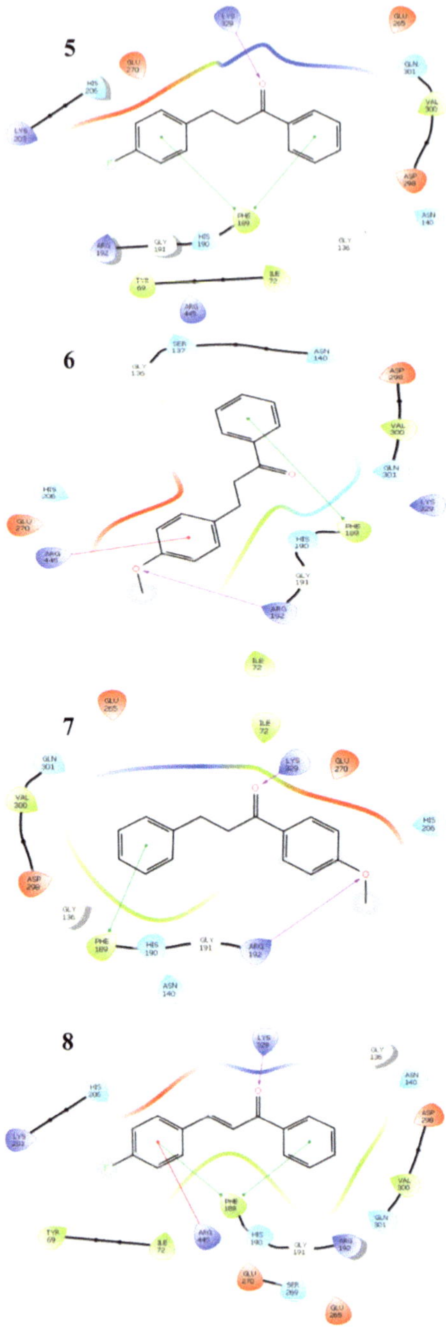

Fig. 6 2D representation of interactions between dihydrochalcone compounds (**5**–**8**) and GABA-AT enzyme from docking simulations

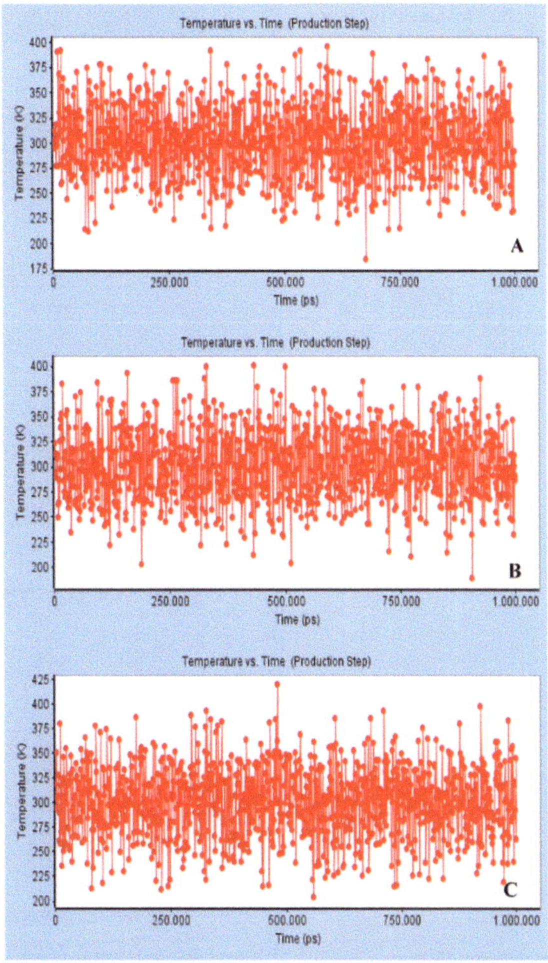

Fig. 7 Molecular dynamics results (Temperature versus Time see text for discussion) obtained for the following inhibitors in complex with GABA-AT enzyme: in **a**, chalcone **2**; in **b**, chalcone **5**; in **c**, chalcone **6**

comparison with the values obtained for GABA-AT enzyme-chalcone **2** (RMSD from 2.5 to 8.5 Å, the less stable) and GABA-AT enzyme-chalcone **5** (RMSD from 2.5 to 7.0).

4 Conclusions

The investigated compounds showed a nice values of HIA > 98% for chalcones and dihydrochalcones, being that compounds **1**, **2**, **3**, **5**, **6** and **7** showed absorption to 100%. However, compound **4** showed a low value of $C_{Brain}/C_{Blood} = 0.017875$. A detailed analysis of distribution properties for these compounds reveals that all have

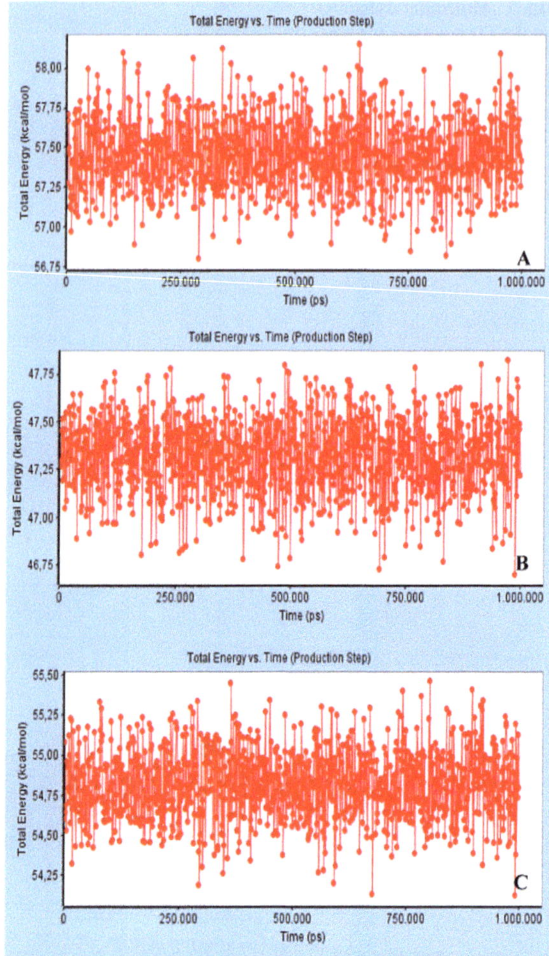

Fig. 8 Molecular dynamics results (Total energy versus Time) obtained for the following inhibitors in complex with GABA-AT enzyme: in **a**, chalcone **2**; in **b**, chalcone **5**; in **c**, chalcone **6**

fewer penetration values than 1, with the exception of compounds **1** and **5** have a little amount of free drugs in order to interact with their respective receptors. As such, these results indicating the possibility of favoring the formation of this pharmacological response.

Docking studies using Gold program indicate interaction of our compounds with essential residues of the active site of GABA-AT. Hence, MD procedures indicated that the chalcones investigated remain stable inside the docked GABA-AT enzyme active site.

Three chalcone compounds showed satisfactory results for promising GABA aminotransferase inhibitors and good results for the pharmacokinetic and toxicological properties in silico. These compounds are suggested as promising candidates

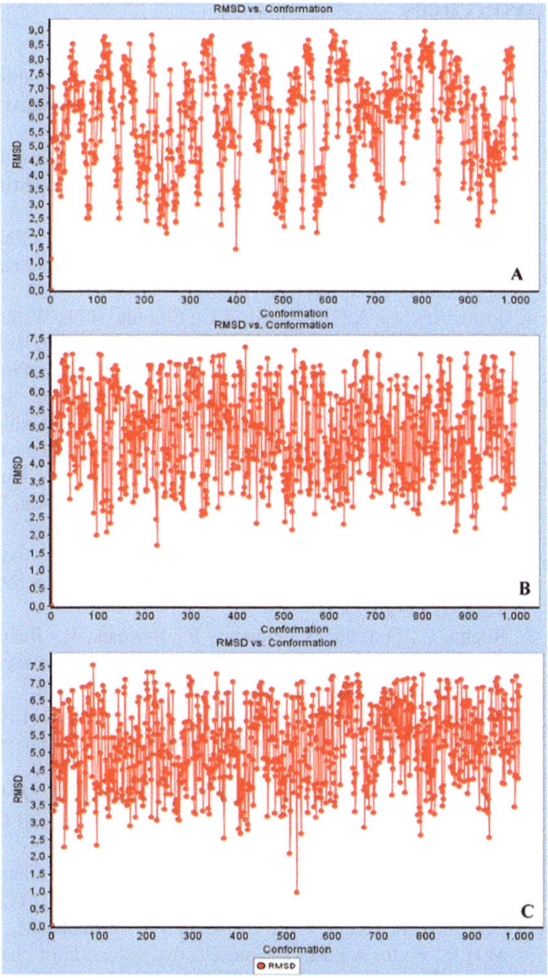

Fig. 9 Molecular dynamics results (RMSD versus Conformation) obtained for the following inhibitors in complex with GABA-AT enzyme: in **a**, chalcone **2**; in **b**, chalcone **5**; in **c**, chalcone **6**

for GABA-AT inhibitors. Therefore, in this perspective, we believed that a biological evaluation (*in vitro* and *in vivo*) of these compounds might help to validate these findings related to designing new ligands to be used as GABA aminotransferase inhibitors agents. Nevertheless, further and still more accurate studies need to be performed in order to elucidate their mechanisms actions, toxicity as well as use potential for the treat of diverse pathologies related to the cerebral cortex.

References

1. Anto, R.J., Sukumaran, K., Kuttan, G., Rao, M.N.A., Subbubbaraju, V., Kuttan, R.: Anticancer and antioxidant activity of synthetic chalcones and related compounds. Cancer Lett. **97**, 33–37 (1995)
2. Ducki, S., Forrest, R., Hadfield, J.A., Kendall, A., Lawrence, N.J., McGown, A.T., Rennison, D.: Potent antimitotic and cell growth inhibitory properties of substituted chalcones. Bioorg. Med. Chem. Lett. **8**, 1051–1056 (1998)
3. Boeck, P., Bandeira, F.C.A., Leal, P.C., Yunes, R.A., Filho, V.C., Torres-Santos, E.C., Rossi-Bergmann, B.: Synthesis of chalcone analogues with increased antileishmanial activity. Bioorg. Med. Chem. **14**, 1538–1545 (2006)
4. Joshi, A.S., Li, X.C., Nimrod, A.C., ElSohly, H.N., Walker, L.A., Clark, A.M.: Dihydrochalcones from Piper longicaudatum. Planta Med. **67**, 186–188 (2001)
5. Cabrera, M., Simoens, M., Falchi, G., Lavaggi, M.L., Piro, O.E., Castellano, E.E., Vidal, A., Azqueta, A., Monge, A., de Ceráin, A.L., Sagrera, G., Seoane, G., Cerecetto, H., González, M.: Synthetic chalcones, flavanones, and flavones as antitumoral agents: biological evaluation and structure-activity relationships. Bioorg. Med. Chem. **15**, 3356–3367 (2007)
6. Aponte, J.C., Verastegui, M., Malaga, E., Zimic, M., Quiliano, M., Vaisberg, A.J., Gilman, R.H., Hammond, G.B.: Synthesis, Cytotoxicity, and Anti-Trypanosoma cruzi Activity of New Chalcones. J. Med. Chem. **51**, 6230–6234 (2008)
7. Dong, J., Zhang, M., Lu, L., Sun, L., Xu, M.: Nitric oxide fumigation stimulates flavonoid and phenolic accumulation and enhances antioxidant activity of mushroom. Food. Chem. **135**, 1220–1225 (2012)
8. Rocha, L., Ferreira, H., Pimenta, E., Berlinck, R., Rezende, M., Landgraf, M., Seleghim, M., Sette, L., Porto, A.: Biotransformation of α-Bromoacetophenones by the Marine Fungus Aspergillus sydowii. Mar. Biotec. **12**, 552–557 (2010)
9. Ricci, L.C., Comasseto, J.V., Andrade, L.H., Capelari, M., Cass, Q.B., Porto, A.L.M.: Biotransformations of aryl alkyl sulfides by whole cells of white-rot Basidiomycetes. Enz. Micro Tec. **36**, 937–946 (2005)
10. Stompor, M., Potaniec, B., Szumny, A., Zielinski, P., Zolnierczyk, A., Aniol, M.: Microbial Synthesis of dihydrochalcones using Rhodococcus and Gordonia species. J. Mol. Catal. B. Enz. **97**, 283–288 (2013)
11. Silverman, R.B.: Design and mechanism of GABA aminotransferase inactivators. Treatments for epilepsies and addictions. Chem. Rev. **7**, 4037–4070 (2018)
12. Ferreira, I.M., Rocha, L.C., Yoshioka, S.A., Nitschke, M., Jellera, A.H., Pizzuti, L., Seleghim, M.H.R., Porto, A.L.M.: Chemoselective reduction of chalcones by whole hyphae of marine fungus Penicillium citrinum CBMAI 1186, free and immobilized on biopolymers. Biocatal. Agric. Biotec. **3**, 358–364 (2014)
13. Filimonov, D.A., Poroikov, V.V.: PASS: computerized prediction of biological activity spectra for chemical substances. In: Bioactive Compound Design: Possibilities for Industrial Use, pp. 47–56. BIOS Scientific Publishers (1996)
14. Yamashita, S., Furubayashi, T., Kataoka, M., Sakane, T., Sezaki, H., Tokuda, H.: Optimized conditions for prediction of intestinal drug permeability using Caco-2 cells. Eur. J. Pharm. Sci. **10**, 195–204 (2000)
15. Hou, T., Wang, J.: Structure-ADME relationship: still a long way to go? Expert Opin. Drug Met. Tox. **4**, 759–770 (2008)
16. Silva, N.S.R., Gonçalves, D.J.L., Silva, J.S., Santos, C.F., Braga, F.S., Silva, R.C., Costa, J.S., Hage-Melim, L.I.S., Santos, C.B.R.: Computational analysis of physicochemical, pharmacokinetic and toxicological properties of deoxyhypusine synthase inhibitors with antimalarial activity. Comp. Mol. Biosc. **4**, 47–57 (2014)
17. Jones, G., Willett, P., Glen, R.C., Leach, A.R., Taylor, R.: Development and validation of a genetic algorithm for flexible docking. J. Mol. Biol. **267**, 727–748 (1997)

18. Lee, H.V., Dustin, D.H., Wu, R., Emma, D., Sanishvili, R., Liu, D., Neil, K., Silverman, R.B.: GABA-aminotransferase inactivated by conformationally-restricted inactivator. (In Press, 2016). https://doi.org/10.2210/pdb4Y0I/pdb
19. Schrödinger Release 2017–2: Maestro, Schrödinger, LLC, New York, NY, (2017)
20. Dassault Systèmes BIOVIA: BIOVIA discovery studio 2017 R2: A comprehensive predictive science application for the life sciences. San Diego, CA, USA (2017). https://accelrys.com/products/collaborative-science/biovia-discovery-studio/.
21. Filimonov, D.A., Poroĭkov, V.V., Karaicheva, E.I., Kazarian, R.K., Budunova, A.P., Mikhaĭlovskiĭ, E.M., Rudnitskikh, A.V., Goncharenko, L.V., Burov, I.U.V.: Computer-aided prediction of biological activity spectra of chemical substances on the basis of their structural formulae: computerized system PASS. Exp. Clin. Pharm. (Rus) **8**, 56–62 (1995)
22. Stepanchikova, A.V., Lagunin, A.A., Filimonov, D.A., Poroikov, V.V.: Prediction of biological activity spectra for substances: evaluation on the diverse sets of drug-like structures. Curr. Med. Chem. **10**, 225–233 (2003)
23. Treiman, D.M.: GABAergic mechanisms in epilepsy. Epilepsia **42**, 8–12 (2001)
24. Singh, D., Gawande, D.Y., Singh, T., Poroikov, V.V., Goel, R.K.: Revealing pharmacodynamics of medicinal plants using in silico approach: a case study with wet lab validation. Comput. Biol. Med. **47**, 1–6 (2014)
25. Bittencourt, J.A.H.M., Oliveira, N.K.S., Cabral, M.S., Ribeiro, J.R., Henriques, S.V.C., Picanço, L.C.S., dos Santos, C.B., Stien, D., Carvalho, J.C.T., Silva, J.O.: Antiophidian activity of Brosimum guianense (AUBL) HUBER. Am. J. Pharm. Tox. **9**, 148–156 (2014)
26. Santos, C.B.R., Lobato, C.C., Braga, F.S., Costa, J.S., Favacho, H.A.S., Carvalho, J.C.T., Macedo, W.J.C., Brasil, D.S.B., Silva, C.H.T.P., Hage-Melim, L.I.S.: Rational design of antimalarial drugs using molecular modeling and statistical analysis. Curr. Pharm. Des. **21**, 4112–4127 (2015)
27. Balimane, P.V., Chong, S., Morrison, R.A.: Current methodologies used for evaluation of intestinal permeability and absorption. J. Pharm. Tox. Met. **44**, 301–312 (2000)
28. Zhao, Y.H., Le, J., Abraham, M.H., Hersey, H., Eddershaw, P.J., Luscombe, C.N., Butina, D., Beck, G., Sherborne, B., Cooper, I., Platts, I.J.A.: Evaluation of human intestinal absorption data and subsequent derivation of a quantitative structure-activity relationship (QSAR) with the Abraham descriptors. J. Pharm. Sci. **90**, 749–784 (2001)
29. Yazdanian, M., Glynn, S.L., Wright, J.L., Hawi, A.: Correlating partitioning and caco-2 cell permeability of structurally diverse small molecular weight. Pharm. Res. **15**, 1490–1494 (1998)
30. Irvine, J.D., Takahashi, L., Lockhart, K., Cheong, J., Tolan, J.W., Selick, H.E., Grove, J.R.: MDCK (Madin–Darby canine kidney) cells: A tool for membrane permeability screening. J. Pharm. Sci. **88**, 28–33 (1999)
31. Singh, S., Singh, J.: Transdermal drug delivery by passive diffusion and iontophoresis: A review. Med. Res. Rev. **13**, 569–621 (1993)
32. Godin, D.V.: Pharmacokinetics: disposition and metabolism of drugs. In: Munson, P.L., Mueller, R.A., Breese, G.R., (eds.) Principles of Pharmacology: Basic Concepts and Clinical Applications, pp. 39–84, 2nd edn. Chapman & Hall, New York (1995)
33. Pratt, W.B., Taylor, P.: Principles of Drug Action: The Basis of Pharmacology, pp. 209–276, 3rd edn. Churchill Livingstone, New York (1990)
34. Brunton, L.L., Goodman, G.: As Bases Farmacológicas da Terapêutica, pp. 636–637, 12th edn. McGraw-Hill, Rio de Janeiro (2012)
35. Banks, W.A.: Blood-brain barrier as a regulatory interface. In: Langhans, W., Geary, N. (eds.) Frontiers in Eating and Weight Regulation, vol. 63, pp. 102–110. Forum Nutr. Basel, Karger (2010)
36. Ma, X., Chen, C., Yang, J.: Predictive model of blood-brain barrier penetration of organic compounds. Acta. Pharm. Sinic. **26**, 500–512 (2005)
37. Ames, B.N., Gurney, E.G., Miller, J.A., Bartsch, H.: Carcinogens as frameshift mutagens: metabolites and derivatives of 2-acetylaminofluorene and other aromatic amine carcinogens. Proc. Natl. Acad. Sci. U.S.A. **69**, 3128–3132 (1972)

38. Woo, Y., Lai, D.Y.: Mechanisms of action of chemical carcinogens and their role in structure-activity relationships (SAR) analysis and risk assessment. In: Benigni, R. (ed.) Quantitative Structure-Activity Relationship (QSAR) Models of Mutagens and Carcinogens, pp. 41–80. CRC, Boca Raton (2003)
39. Cole, J.C., Murray, C.W., Nissink, J.W.M., Taylor, R.D., Taylor, R.: Comparing protein-ligand docking programs is difficult. Proteins Struc. Funct. Bioinf. **60**, 325–332 (2005)
40. Lee, H., Doud, E.H., Wu, R., Sanishvili, R., Juncosa, J.I., Liu, D., Kellehe, N.L., Silverman, R.B.: Mechanism of inactivation of γ-aminobutyric acid aminotransferase by (1S,3S)-3-Amino-4-difluoromethylene-1-cyclopentanoic acid (CPP-115). J. Am. Chem. Soc. **137**, 2628–2640 (2015)

Phytotechnological and Pharmaceutical Potential of *Myrciaria cauliflora* Fruits

Leonardo Luiz Borges, Clayson Moura Gomes, Vinicius Barreto da Silva, Edemilson Cardoso da Conceição, Damaris Silveira, and Carlton Anthony Taft

Abstract *Myrciaria cauliflora* (jabuticaba) is a Brazilian native species of the Atlantic forest region that produces fruits enriched with outstanding antioxidant content, such as anthocyanins, polyphenols, tannins and flavonoids. It additionally is known that jabuticaba fruits are attractive sources of bioactive molecules that drive phytotechnological research worldwide. Several works with jabuticaba were developed in last years covering mainly the agricultural, pharmaceutical and food science fields. However, there is still a gap regarding technological development. Here, we make an approach over the technological products obtained form jabuticaba fruits, their biological potential, the drug-like properties of chemical markers and the main techniques that would be employed in future development of other products with commercial value.

1 Introduction

The use of medicinal plants derivatives in general to the treatment of diseases is a practice that has accompanied humanity for millennia in various cultures around the

L. L. Borges · C. M. Gomes · V. B. da Silva (✉)
School of Medical, Pharmaceutical and Biomedical Sciences, Pontifícia Universidade Católica de Goiás, Goiânia, Brazil
e-mail: vbarreto@pucgoias.edu.br

L. L. Borges
Câmpus Anápolis de Ciências Exatas e Tecnológicas – Henrique Santillo, Universidade Estadual de Goiás, Anápolis, Brazil

E. C. da Conceição
Faculdade de Farmácia, Universidade Federal de Goiás, Goiânia, Brazil

D. Silveira
Faculdade de Ciências da Saúde, Universidade de Brasília, Brasília, Brazil

C. A. Taft
Centro Brasileiro de Pesquisas Físicas, Rio de Janeiro, Brazil

F. A. La Porta and C. A. Taft (eds.), *Functional Properties of Advanced Engineering Materials and Biomolecules*, Engineering Materials,
https://doi.org/10.1007/978-3-030-62226-8_26

world and is therefore related to primary health care today in several countries [1–6]. Several active substances have widely been obtained from plants, i.e., reinforcing the importance of natural products as sources that are still little explored in terms of chemical characterization [7].

In recent decades, the World Health Organization (WHO) has encouraged research related to the use of natural products, mainly of plant origin, considering the use of medicinal plants and their derivatives as part of Traditional and Complementary Medicine continues to play a role of extreme relevance in public health in several countries [8]. For example, in compliance with WHO strategies, Brazil has published policies for the insertion of complementary practices in the Unified Health System (SUS), including Phytotherapy [9]. However, some challenges related to the effectiveness, safety, and quality of plant derivatives persist.

Obtaining products from natural sources, which have a constant quality and yield, is subject to several variables that constitute difficulties to be overcome. In natural products obtained from plants, the following can be mentioned: cultivation conditions, collection time, extractive methods used, drying, and storage conditions [10]. Due to these various factors, it is necessary to use chemical markers to monitor the quality and standardize them [11]. A chemical marker consists of the substance or set of substances present in a plant species and that are preferably responsible for the biological activity (such as flavonoids, alkaloids, fatty acids, etc.). Chemical markers can also be used to monitor the plant's derivatives quality and must be capable of being analyzed by analytical techniques, such as spectrophotometry and chromatography, gas chromatography (GC), as well as high-performance liquid chromatography (HPLC) [12].

Brazilian Cerrado is, in turn, classified as one of the biomes with the highest extent of biodiversity in the world, having a high number of species that only occur in that location. It is located mainly in the central-west region of Brazil. Through its physical, hydrographic, and rich biodiversity characteristics, Cerrado provides an excellent scenario for developing agricultural practices [13]. Studies demonstrate a varied number of plants from Cerrado or adapted that can be valorized and exploited commercially, one of them is jabuticaba (*Myrciaria cauliflora*). Thus, the fruit jabuticaba (*Myrciaria cauliflora*) belonging to the Myrtaceae, originally from the south-central region of Brazil. Its main characteristics are globose berry up to 3 cm in diameter, reddish skin almost black, whitish mucilaginous pulp with bittersweet flavor. It presents from one to four seeds [14]. The species *M. cauliflora* is popularly known as "jabuticabeira", "jabuticaba", "jabuticaba paulista", "jabuticaba-açu", "jabuticaba-do-mato" and "jabuticaba-sabará" [15]. It is a fruit species that arouses high interest among rural producers in several regions in Brazil, as it presents high productivity, rusticity, and fruit utilization in several different ways [16].

M. cauliflora fruits are still collected in an extractive manner and freely traded, always with the tremendous widespread acceptance of the population. The commercialization period is relatively short, and studies have indicated that, in just two days after harvest, there is a rapid change in appearance and flavor due to the intense loss of water, which results in pulp deterioration and fermentation [16] (Fig. 1).

Fig. 1 The general aspects of aerial parts and fruits of *Myrciaria cauliflora*

2 Pharmaceutical Potential of *Myrciaria cauliflora* Species

Several biological activities are reported for *M. cauliflora* fruits, such as antioxidant activity, which can be attributed mainly to anthocyanins present in the epicarp [17, 18], hypoglycemic and hypolipidemic effects [19] and anti-inflammatory activity [20]. As such, its leaves and bark of *M. cauliflora* are well known as astringent and the decoction is widely used in the treatment of diarrhea and skin irritations [21]. Macêdo-Costa and coworkers [22] investigated the antimicrobial activity of *M. cauliflora* leaf extract against bacteria that form a dental biofilm (*Streptococcus mitis*, *S. mutans*, *S. sanguinis*, *S. oralis*, *S. salivarius* and *Lactobacillus casei*). Finally, these results found showed the effectiveness of the extract when compared to antimicrobial chlorhexidine.

The biological properties found for *M. cauliflora* can be attributed mainly to the vast amount of polyphenolic compounds present in this species. Among these compounds, in addition to anthocyanins (penidinic glycosides) and aglycones (cyanidins), tannins, isoquercitrin, myricetin and ellagic acid stand out [23, 24].

Ellagic acid and ellagitanins can be found in several other edible fruits [25, 26] and are considered promising constituents of the diet to prevent of various degenerative diseases. These compounds have several biological activities such as: antioxidant, also acting in the prevention of lipid peroxidation [27, 28], vascular protection [29], anti-inflammatory activity [30], anticancer [31], anti-atherosclerotic [32, 33], anti-hepatotoxic [34] and antibacterial [35]. Ellagic acid is a substance originating from the hydrolysis (acidic or basic) of elagitanins, promoting the formation of the intermediate hexahydroxydiphenic acid, which undergoes a spontaneous lactonization reaction, resulting in ellagic acid[25]. Also, elagitanins can be metabolized by the microbiota of the human gastrointestinal tract and other animals resulting in the release of ellagic acid, as shown [36]. *Myrciaria cauliflora* also contains other antioxidant substances, in addition to polyphenolic compounds, such as ascorbic acid [24, 37].

Wu and colleagues [38] showed that *Myrciaria cauliflora* extract stabilized the plasma glucose and indirectly improved diabetic mice's insulin sensitivity. Besides, diabetes-caused glomerular atrophy, as well as accumulation of saccharide, in addition to the formation of collagen IV has been significantly reduced from the use of Myrciaria cauliflora extract rich in anthocyanins and polyphenols has potential benefits to the treatment of diabetic mice, mainly due to inhibiting Ras/PI3K/Akt and kidney fibrosis-related proteins [38].

Fidelis and colleagues [39] evaluated the solvent effects on the extraction of desired bioactive compounds from seeds of jabuticaba. In this case, they found that a mixture among water and propanone (in range of 52:48 v/v) leads to a significantly increased, therefore, extraction yield of compounds biologically actives. Also, the optimized extract shows a relevant antiproliferative and cytotoxic effects against both A549 and HCT8 cell lines, inhibited α-amylase/α-glucosidase activities, antimicrobial and antihemolytic effects, as well antihypertensive effect. The main phenolic constituents determined in this study were vescalagin, castalagin and ellagic acid. Additionally, no cytotoxicity in a human cell model (IMR90) has been found for the optimized extract. *Myrciaria cauliflora* extract activity is also observed as a microbicidal activity in bacteria and fungi of dental importance [22, 40], and antimicrobial activity of jabuticaba bark extract in Gram-negative bacteria and Gram-positive [41]. Hence, the bioactive compounds that act as bacterial inhibition are phenolic compounds, mainly cyanidin chloride, catechin and epicatechin, which have inhibitory activity mainly in *S. aureus* [42].

3 Technological Potential of *Myrciaria cauliflora*

Several authors have reported the industrial importance of jabuticaba fruits. It is possible to find them benefited from juices, ice creams, fermented, distillates, jellies, jams, liquor and vinegar. They can also be used in the manufacture of fermented and distilled [37].

In industrialization, the peels and seeds are neglected, representing approximately a 50% loss of fruit. This waste is usually used as animal feed, but most are discarded or used in composting. Considering the variety of useful substances with the high added value present in this residue, better use of these by-products would add commercial value and avoid environmental problems generated by its disposal in the soil [43]. The peels rich in natural dyes could be used as raw material by the food, pharmaceutical and cosmetic industries [44], as a raw material for the isolation of phytopharmaceuticals, or in the form of standardized extracts, from just as it is possible to find on the market several suppliers of pomegranate extracts standardized in ellagic acid, for example. Other technological approaches have been investigated, such as the application of chitosan- coating jabuticaba peel extract (85% ethanol acidified with about 15% of a stock solution of 1.5 mol L^{-1} hydrochloric acid), more specifically, with the purpose of avoiding microorganism growth, that is, for enhanced the product's shelf life.

Some innovative ways of the use of *M. cauliflora* extracts have been developed, such as the work conducted by Santos and colleagues [45], which employed the liquid extract as an antimicrobial photosensitizer against *Staphylococcus aureus*. In this work, were used residue of the processed fruit bark for the obtaing of the aqueous extract in ratio of 1:3 (residue:water). Briefly, the pigment extraction was carried out under constant mechanical agitation (at 6500 rpm) for a period of six hours in the dark. The obtained fluid was then filtered and, next, was then concentrated to 1/3 of its initial volume using a rotary evaporator at 60 °C coupled to a vacuum pump operating at about 600 mm Hg, generating the concentrated extract.

Batista and colleagues [46] fed rats based on a high-fat diet supplemented with lyophilized jabuticaba peel (FJP). Thus, the lyophilization method consists of other technological approaches to obtain standardized products from jabuticaba fruits. The authors of the study found that all diets with different concentrations (1, 2 and 4%) of jabuticaba extract contribute to prevent the occurrence of lipid peroxidation in the liver, as well as, minimized the oxidative stress in which likely can be associated as a consequence of its high polyphenols content (such as anthocyanins, ellagic acid, gallic acid, in addition to quercetin derivatives).

Microencapsulation of liquid extract from jabuticaba peels using spray drying technique have also showed a high anthocyanins content (cyanidin-3-glucoside) [47]. These products were applied in mortadella to improve their stored characteristics. Also, this strategy in general provide a reduced of redness, yellowness, pH value, aroma and also color acceptance significantly. Therefore, the study authors have also reported an improvement in sensory analysis in terms of texture and flavor. And yet, no significant changes in the chemical composition, lightness, lipid oxidation, or overall sensory quality was observed [47].

Asquieri and coworkers [48] developed a method of reuse of bark and jabuticaba sludge that underwent a fermentation process for the production of brandy. Silva and coworkers [49] carried out a study with jabuticaba residue for the production of pigments with functional properties using the spray-drying technique, using maltodextrin as a drying aid, and the resulting product showed antioxidant, antimicrobial, and also arginase inhibitory activity, an enzyme produced by the protozoan *Leishmania amazonensis.*

Morales and colleagues [50] showed that the by-products of *M. cauliflora* fruits are of large importance to obtaining the valuable bioactive compounds. As such, the effect of the fermented versus non-fermented jabuticaba pomaces was studied in detail. The authors investigated their hydrophilic/lipophilic compounds, as well as their oustanding antioxidant properties. Concerning chemical composition, the study shows the sugars, organic acids and tocopherols, phenolics and anthocyanins, and fatty acids. Also, non-fermented and fermented jabuticaba pomaces represent a rich source of compounds biollogically actives (tocopherols, polyunsaturated fatty acid, and phenolic compounds) and have a high and well-known potential antioxidant. Therefore, from this perspective, it should be noted that the use of jabuticaba pomaces from the industries could lead to novel products with added value, e.g., in the human and animal feed manufacturing process [50].

The residues of jabuticaba fruits also present technological potential, as showed by Borges and colleagues [51], in which the waste powder fruits were employed to obtain a liquid hydroethanolic extract (55%) and the spray drying process was optimized using response surface methodology to obtain a dried intermediate product with increase levels of ellagic acid and other phenolic compounds.

Through the characterization and standardization of the dry extract, the potential of the residues generated by the fermentation of jabuticaba can be used more rationally, thus valuing all the links of the productive chain. In this way, it is necessary and immediate to valorize the potentialities and the visibility of the possibilities of using and economically exploiting agroindustrial residues, which indicate a wide spectrum of use and which are not yet being commercially exploited rationally. In this context, it is important to develop and optimize extraction and drying processes to obtain standardized dry extracts with a wide spectrum of applications in the pharmaceutical, food and cosmetic areas. Thus, jabuticaba fruits represent a raw material with a high potential to obtain standardized products from different techniques.

4 Drug-Like Properties of *Myrciaria* Molecules

Borges and coworkers [24] compiled the active compounds already detected for *Myrciaria* genus species. They describe 53 molecular structures with vast diversity and biological importance, showing that *Myrciaria* is an interesting source for collecting useful molecules for drug design projects. To filter the orally active drug-like compounds from *Myrciaria* genus the physicochemical parameters defined in the rule-of-five (RO5) [52] were calculated for each molecule. According to this RO5, the most drug-like compounds have lipophilicity (Log P) ≤ 5, molecular weight ≤ 500, number of hydrogen bond donors ≤ 5 and number of hydrogen bond acceptors ≤ 10. Most orally active molecules (or in phase II clinical trials) in general no violate more than one of these parameters [52, 53].

The physicochemical parameters defined by RO5 are, of course, associated with its solubility in aqueous medium and the intestinal permeability, as well as, comprise the first steps related to their oral bioavailability. If a molecule violates more than one parameter and fails the RO5, it will encounter certain oral activity limitations. On the other hand, passing the RO5 is no guarantee that a compound is drug-like but indicates a high probability for oral bioavailability [52, 53].

Around 64% of molecules described by Borges and coworkers [24] fit the RO5 criteria, revealing that *Myrciaria* genus provides a useful list of chemical entities for drug design interest with a high probability to be orally active (Table 1). The rule of five parameters was calculated with the Calculation of Molecular Properties and Bioactivity Score tool from Molinspiration Cheminformatics Software (https://www.molinspiration.com/cgi-bin/properties).

Table 1 RO5 parameters calculated for molecules encountered in *Myrciaria* genus species that are approved by Lipinski's rule

Myrciaria Molecules	Molecular weight	miLogP	H-bond donors	H-bond acceptors
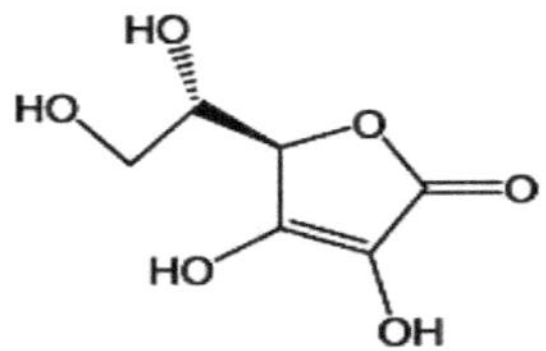 Ascorbic acid	178.14	– 2.99	4	6
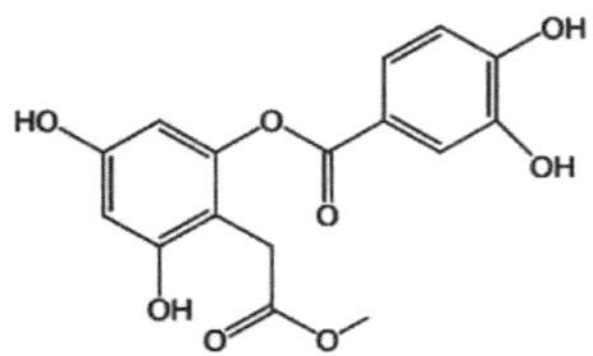 Jaboticabin	334.28	2.24	4	8
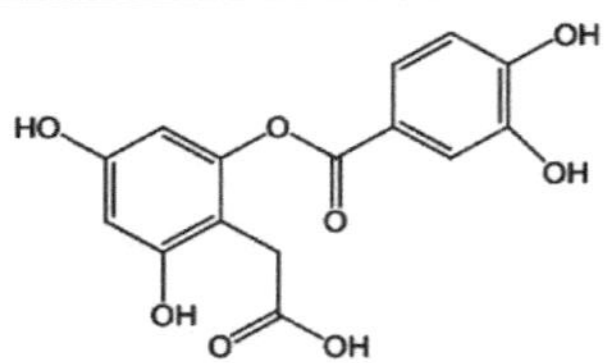 2-O-(3,4-dihydroxybenzoyl)-2,4,6-trihydroxyphenylacetic acid	320.25	1.63	5	8

(continued)

Table 1 (continued)

Myrciaria Molecules	Molecular weight	miLogP	H-bond donors	H-bond acceptors
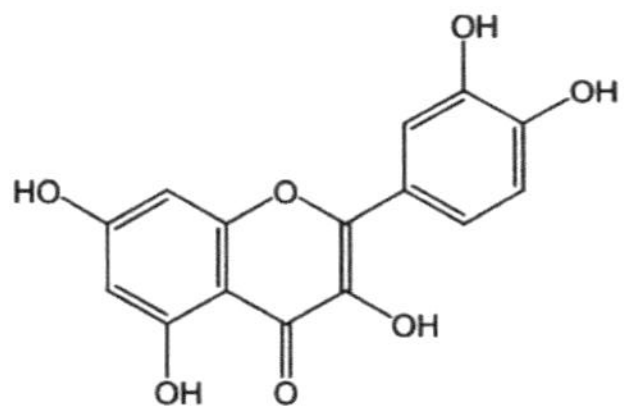 Quercetin	302.24	1.68	5	7
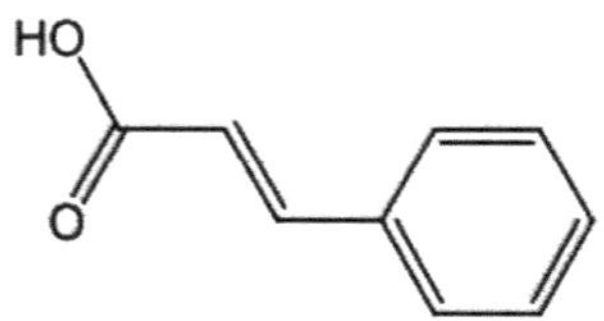 Cinnamic acid	148.16	1.91	1	2
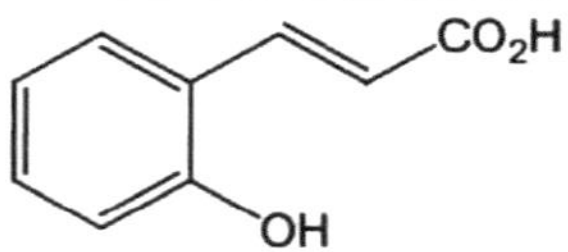 *O*-cumaric acid	164.16	1.67	2	3

(continued)

Table 1 (continued)

Myrciaria Molecules	Molecular weight	miLogP	H-bond donors	H-bond acceptors
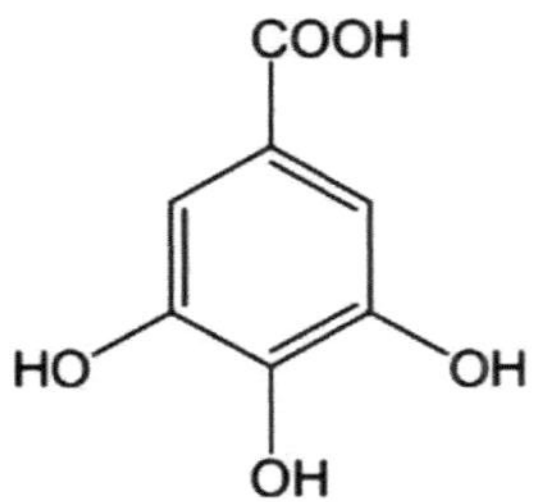 Gallic acid	170.12	0.59	4	5
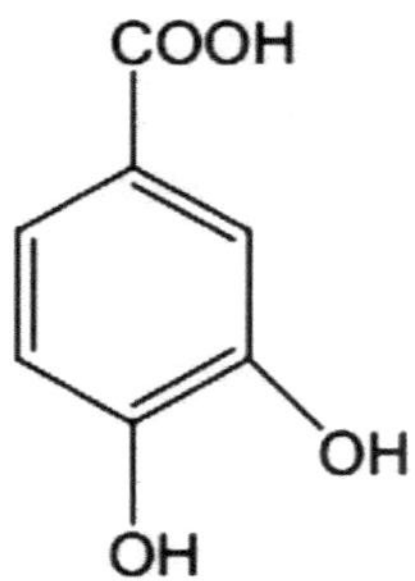 Protocatechuic acid	154.12	0.88	3	4

(continued)

Table 1 (continued)

Myrciaria Molecules	Molecular weight	miLogP	H-bond donors	H-bond acceptors
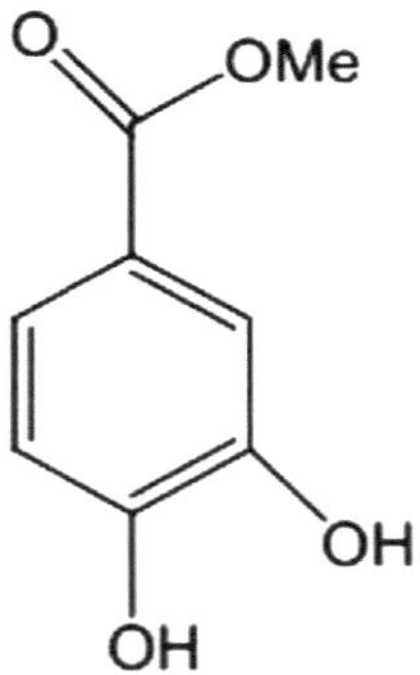 Methylprotocatechuate	168.15	1.14	2	4
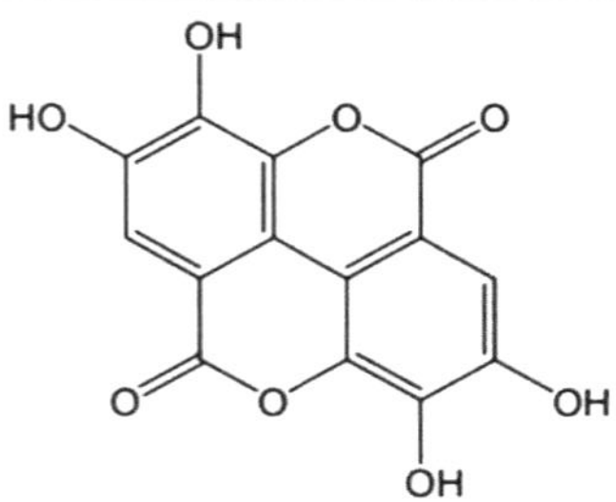 Ellagic acid	302.19	0.94	4	8

(continued)

Table 1 (continued)

Myrciaria Molecules	Molecular weight	miLogP	H-bond donors	H-bond acceptors
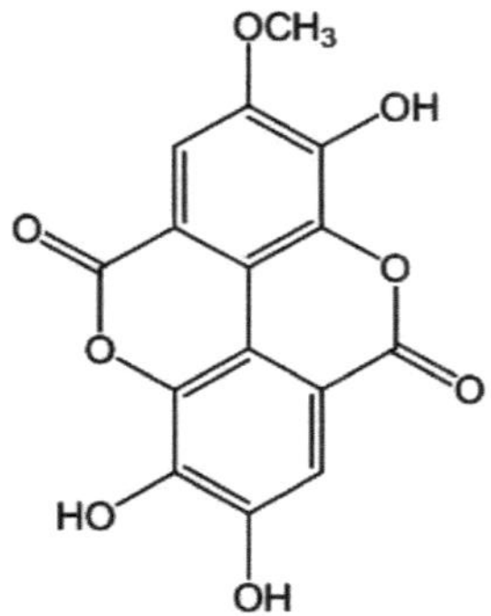 *O*-methylellagic acid	316.22	1.25	3	8
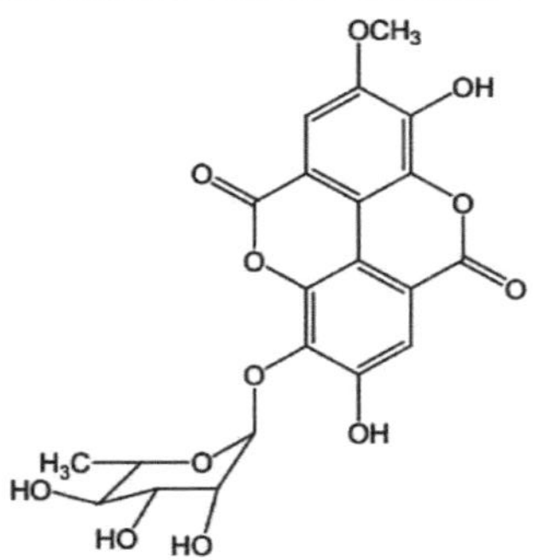 4-alpha-rhamnopyranosyl ellagic acid	462.36	0.21	5	12

(continued)

Table 1 (continued)

Myrciaria Molecules	Molecular weight	miLogP	H-bond donors	H-bond acceptors
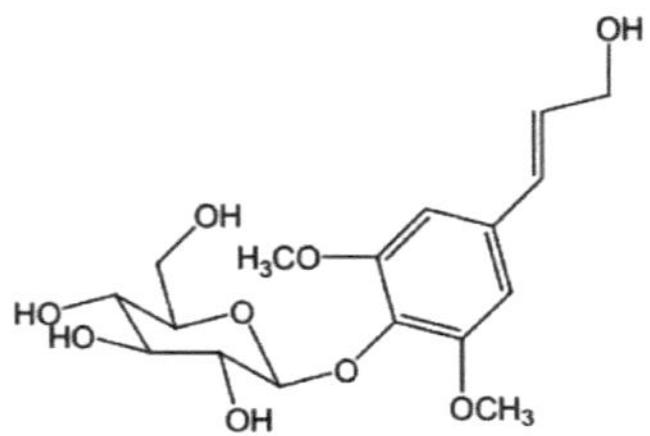 Syringin	372.37	– 0.66	5	9
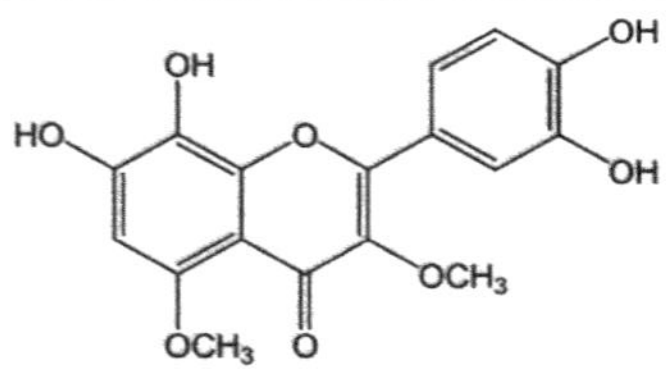 2-(3,4-dihydroxyphenyl)-7,8-dihydroxy-3,5-dimethoxychromen-4-one	346.29	1.98	4	8

(continued)

Table 1 (continued)

Myrciaria Molecules	Molecular weight	miLogP	H-bond donors	H-bond acceptors
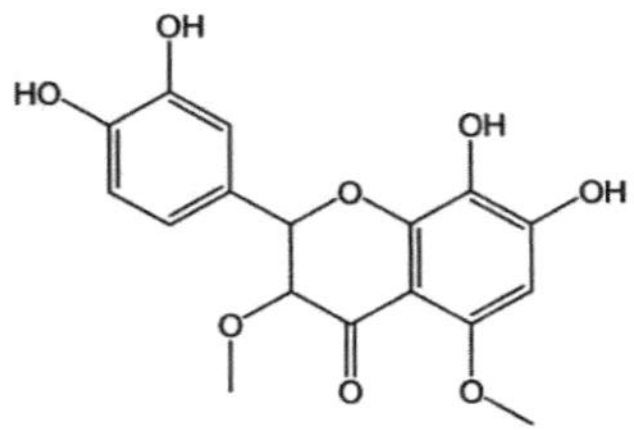 Grossypetin-3,5-dimethyl ether	348.31	1.14	4	8
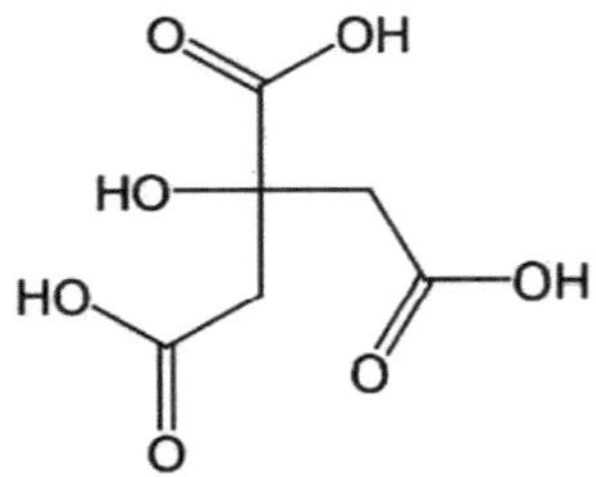 Citric acid	192.12	– 1.98	4	7

(continued)

Table 1 (continued)

Myrciaria Molecules	Molecular weight	miLogP	H-bond donors	H-bond acceptors
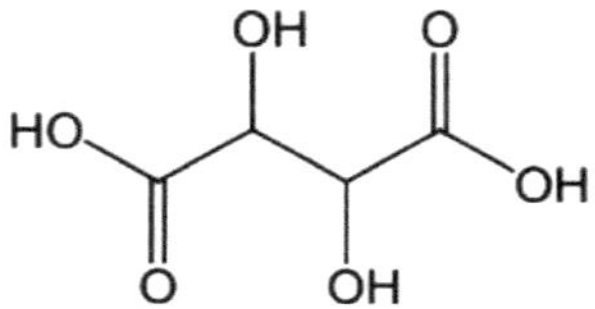 Tartaric acid	150.09	– 2.49	4	6
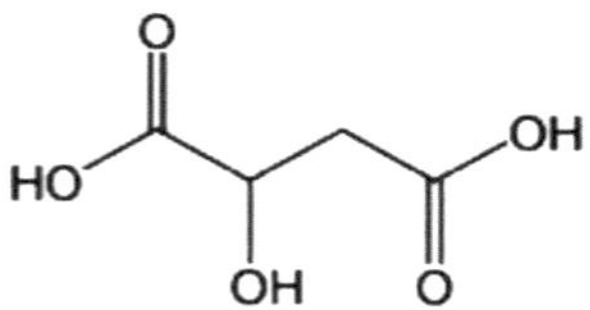 Malic acid	134.09	– 1.57	3	5
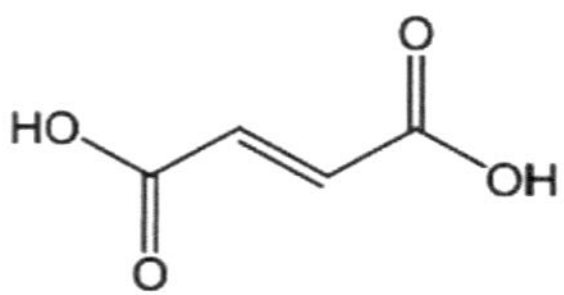 Fumaric acid	116.07	– 0.68	2	4

(continued)

Table 1 (continued)

Myrciaria Molecules	Molecular weight	miLogP	H-bond donors	H-bond acceptors
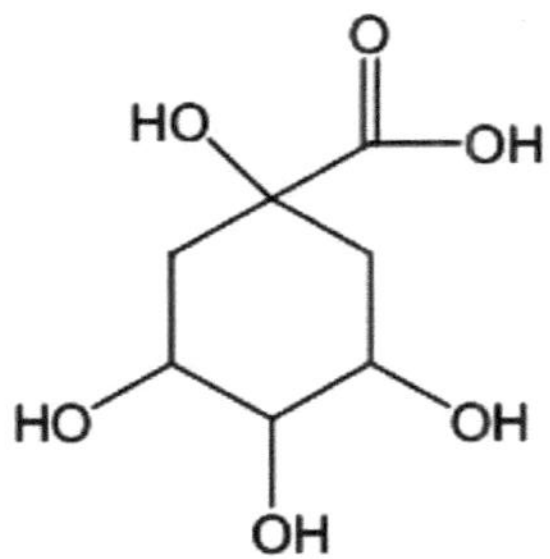 Quinic acid	192.17	– 2.33	5	4
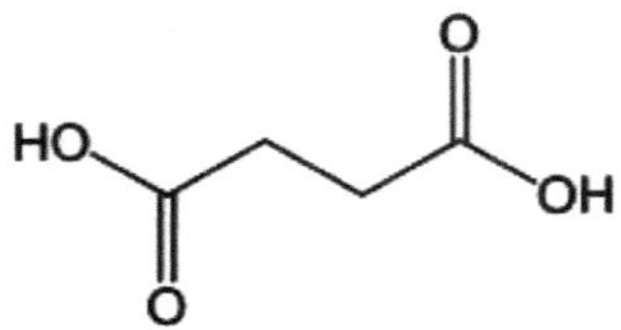 Succinic acid	118.09	– 0.66	2	4

(continued)

Table 1 (continued)

Myrciaria Molecules	Molecular weight	miLogP	H-bond donors	H-bond acceptors
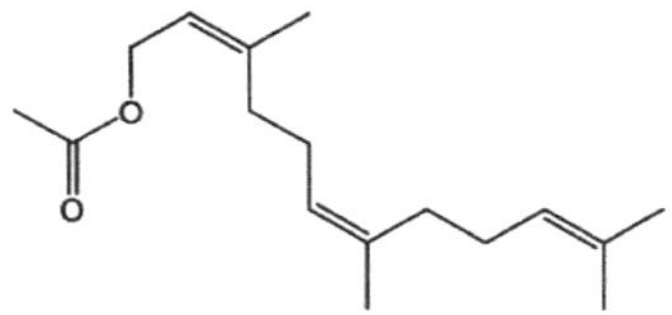 (2E,6E)-farnesyl acetate	264.41	5.75	0	2
 1,4-cineole	154.25	2.72	0	1
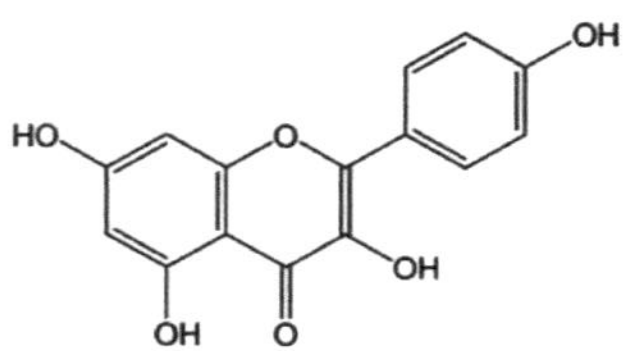 Kaempferol	286.24	2.17	4	6

(continued)

Table 1 (continued)

Myrciaria Molecules	Molecular weight	miLogP	H-bond donors	H-bond acceptors
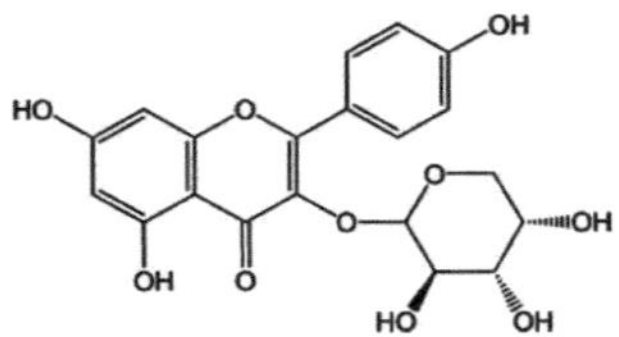 Kaempferol-3-O-alpha-arabinofuranoside	412.44	4.51	3	7
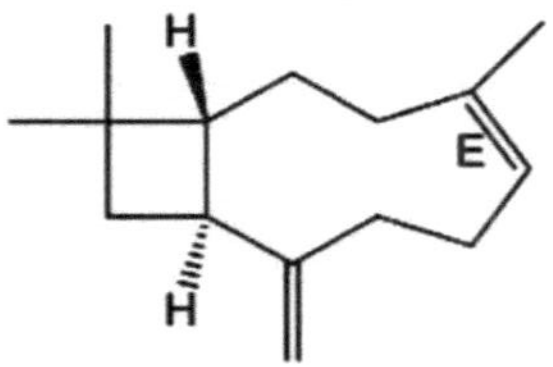 β-caryophyllene	204.36	5.17	0	0

(continued)

Table 1 (continued)

Myrciaria Molecules	Molecular weight	miLogP	H-bond donors	H-bond acceptors
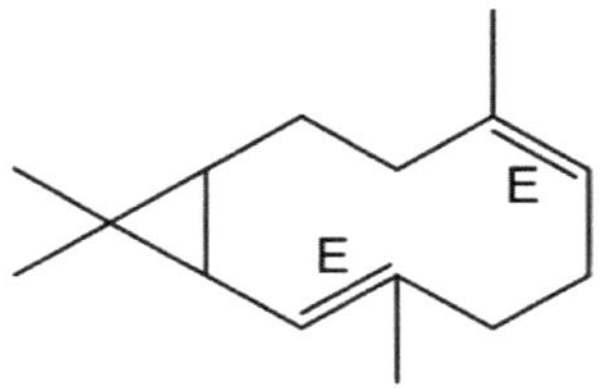 Bicyclogermacrene	204.36	5.29	0	0
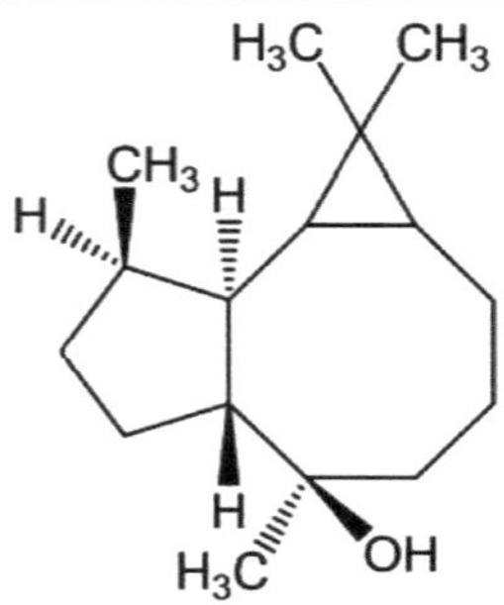 Globulol	236.40	4.58	1	1

(continued)

Table 1 (continued)

Myrciaria Molecules	Molecular weight	miLogP	H-bond donors	H-bond acceptors
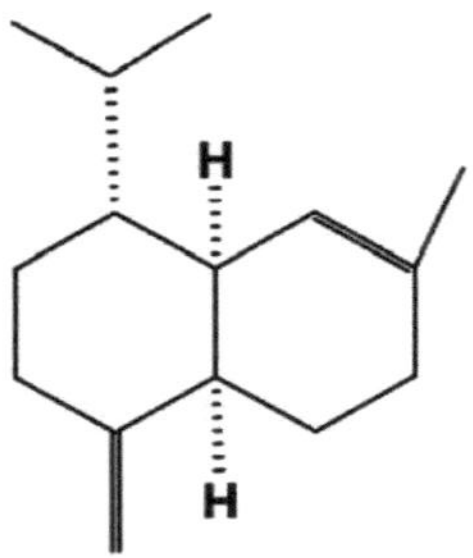 Gamma-muurolene	204.36	5.75	0	0
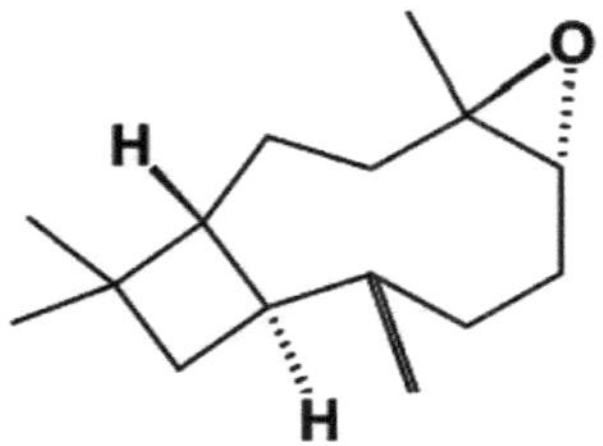 β-caryophyllene oxide	220.36	4.14	0	1

(continued)

Table 1 (continued)

Myrciaria Molecules	Molecular weight	miLogP	H-bond donors	H-bond acceptors
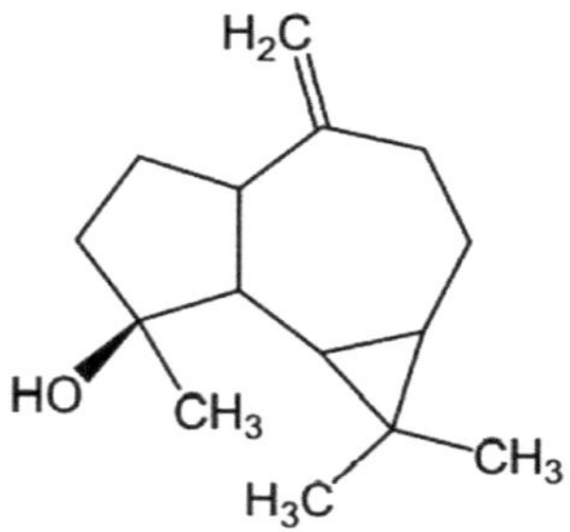 Spathulenol	220.36	3.91	1	1
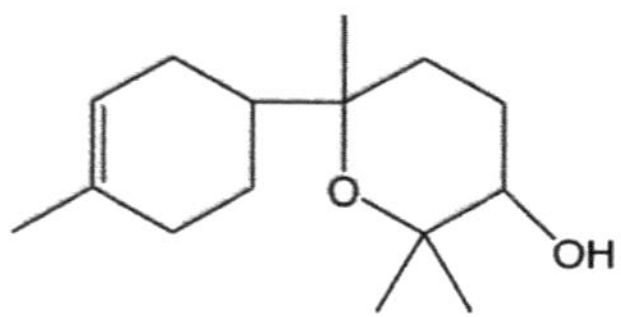 α-bisabolol oxide A	238.37	3.41	1	2

(continued)

Table 1 (continued)

Myrciaria Molecules	Molecular weight	miLogP	H-bond donors	H-bond acceptors
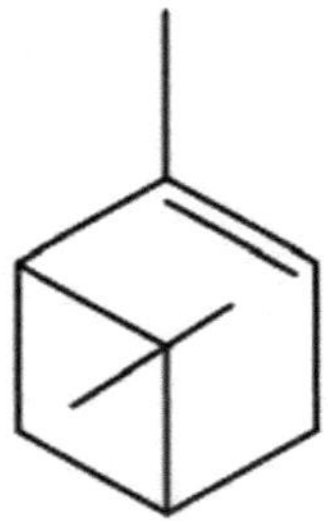 α-pinene	136.24	3.54	0	0
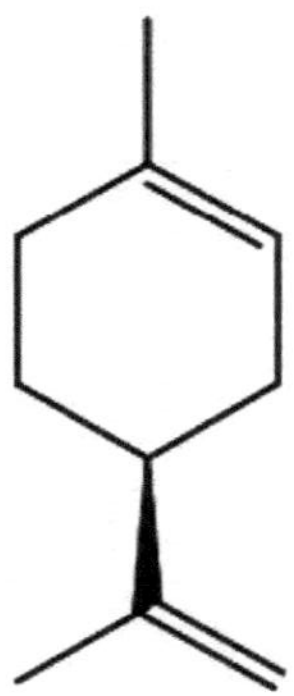 d-limonene	136.24	3.62	0	0

5 Conclusion

Myrciaria cauliflora fruits are an attractive source of molecules with biological activities that should be used as starting points in drug design projects. The plant extracts are rich in polyphenols and anthocyanins that reveal antioxidant, hypoglycemic, hypolipidemic, antimicrobial and anti-inflammatory properties. During industrialization, some parts of the fruits are neglected as waste, including peels and seeds. Such material is rich in valuable compounds for developing pharmaceutical products and is an interesting source for isolating biologically optimized lead compounds for drug design. Such an argument is reinforced by the fact that many bioactive molecules described in the literature for *Myrciaria* genus, around 60%, pass through Lipinskis's rule-of-five. Thus, molecules from *Myrciaria cauliflora* have a high probability of being orally active, providing a valuable list of bioavailable bioactive molecules.

References

1. Lehmann, H.: Medicinal plants in France, between pharmacy and herb trade: historical and legislative aspects. Ann. Pharm. Fr. **73**(5), 391–398 (2015)
2. Cavero, R.T., Calvo, M.I.: Medicinal plants used for ophthalmological problems in Navarra (Spain). J. Ethnopharmacol. **190**, 212–218 (2016)
3. Marmitt, D.J., Bitencourt, S., Silva, A.C., Rempel, C., Goettert, M.I.: The healing properties of medicinal plants used in the Brazilian public health system: a systematic review. J. Wound Care **27**(Sup6), S4–S13 (2018)
4. Jaradat, N.A., Shawahna, R.: Ethnopharmacological survey of medicinal plants used by patients with psoriasis in the west bank of Palestine. BMC Complement. Altern. Med. **17**(1), 4 (2017)
5. Nega, S.S., Bekele, H.M., Meles, G.G., Nordeng, H.: Medicinal plants and concomitant use with pharmaceutical drugs among pregnant women. J. Altern. Complement. Med. **25**(4), 427–434 (2019)
6. Calixto, J.B.: Efficacy, safety, quality control, marketing and regulatory guidelines for herbal medicines (phytotherapeutic agents). Braz. J. Med. Biol. Res. **33**, 179–189 (2000)
7. Siani, A.C.: Desenvolvimento tecnológico de fitoterápicos: plataforma metodológica, 1st edn. Scriptorio, Rio de Janeiro (2003)
8. WHO: Traditional Medicine Strategy: 2014–2023. World Health Organization, Geneve (2013)
9. BRASIL: Política Nacional de Plantas Medicinais e Fitoterápicos, p. 189. Ministério da Saúde, Brasília (2016)
10. Castro, H.G., Ferreira, F.A., Silva, D.J.H., Mosquim, P.R.: Contribuição ao estudo de plantas medicinais: metabólitos secundários. 2nd edn. (2004)
11. Chaves, J.S.: Obtenção e caracterização de um fitoterápio a partir de *Tanacetum parthenium* Shultz-Bip (Asteraceae): caracterização química e física do pó das partes aéreas de *T. parthenium* e obtenção e caracterização química, física e físico-química dos extratos hidroalcoólico e seco e de comprimidos de dissolução entérica. p. - Faculdade de Ciências Farmacêuticas de Ribeirão Preto, Universidade de São Paulo, Ribeirão Preto, 2005.
12. BRASIL. Resolução RE nº 27, de 27 de maio de 2012. Guia para validação de métodos analíticos e bioanalíticos. Agência Nacional de Vigilância Sanitária, Brasilia (2012)
13 De Brito, B.N., Da Silva, E.B.: Análise multitemporal de uso e cobertura da terra na Reserva da Biosfera do Cerrado. Ateliê Geográfico **13**(2), 73–91 (2019)
14. Gomes, R.P.: Fruticultura Brasileira, 13th edn. Nobel, São Paulo (2007)
15 Citadin, I., Danner, M.A., Sasso, S.A.Z.: Jabuticabeiras. Rev. Bras. Fruticultura **32**, 1 (2010)

16. Brunini, M.A., Oliveira, A.L., Salandini, C.A.R., Bazzo, F.R.: Influência de embalagens e temperatura no armazenamento de jabuticabas (*Myrciaria jabuticaba* (Vell) Berg) cv 'sabará' . Ciênc. Tecnol. Aliment. **24**(3), 378–383 (2004)
17. Reynertson, K.A., Basile, M.J., Kennelly, E.J.: Antioxidant potential of seven myrtaceous fruits. Ethnobotany Res. Appl. **3**, 25–35 (2005)
18. Leite, A.V., Malta, L.G., Riccio, M.F., Eberlin, M.N., Pastore, G.M., Marostica, M.R.: Antioxidant potential of rat plasma by administration of freeze-dried jaboticaba peel (*Myrciaria jaboticaba* Vell Berg). J. Agric. Food Chem. **59**(6), 2277–2283 (2011)
19. Lenquiste, S.A., Batista, A.G., Marineli, R.D.S., Dragano, N.R.V., Maróstica, M.R.: Freeze-dried jaboticaba peel added to high-fat diet increases HDL-cholesterol and improves insulin resistance in obese rats. Food Res. Int. **49**(1), 153–160 (2012)
20. Reynertson, K.A., Wallace, A.M., Adachi, S., Gil, R.R., Yang, H., Basile, M.J., D'Armiento, J., Weinstein, I.B., Kennelly, E.J.: Bioactive depsides and anthocyanins from jaboticaba (*Myrciaria cauliflora*). J. Nat. Prod. **69**(8), 1228–1230 (2006)
21. Souza, T. M. Estudo farmacognóstico e avaliação da atividade antimicrobiana e citotóxica de preparações cosméticas contendo o extrato de folhas de *Myrciaria cauliflora* O. Berg. (Myrtaceae) e de casca de *Stryphnodendron adstringens* (Mart.) Coville (Leguminosae-Mimosoidae). 171 p. - Faculdade de Ciências Farmacêuticas, Universidade Estadual Paulista, Araraquara (2007)
22 Macedo-Costa, M.R., Diniz, D.N., Carvalho, C.N., Pereira, M.S.V., Pereira, J.V., Higino, J.S.: Eficácia do extrato de Myrciaria cauliflora (Mart.) O. Berg. (jabuticabeira) sobre bactérias orais. Braz. J. Pharm. **19**(2B), 565–571 (2009)
23. Reynertson, K.A., Yang, H., Jiang, B., Basile, M.J., Kennelly, E.J.: Quantitative analysis of antiradical phenolic constituents from fourteen edible Myrtaceae fruits. Food Chem. **109**(4), 883–890 (2008)
24. Borges, L.L., Conceição, E.C., Silveira, D.: Active compounds and medicinal properties of *Myrciariagenus*. Food Chem. **153**, 224–233 (2014)
25. Landete, J.M.: Ellagitannins, ellagic acid and their derived metabolites: a review about source, metabolism, functions and health. Food Res. Int. **44**(5), 1150–1160 (2011)
26. Berni, A., Grossi, M.R., Pepe, G., Filippi, S., Muthukumar, S., Papeschi, C., Natarajan, A.T., Palitti, F.: Protective effect of ellagic acid (EA) on micronucleus formation induced by N-methyl-N′-nitro-N-nitrosoguanidine (MNNG) in mammalian cells, in in vitro assays and in vivo. Mutat. Res. **746**, 60–65 (2012)
27. Priyadarsini, K.I., Khopde, S.M., Kumar, S.S., Mohan, H.: Free radical studies of ellagic acid, a natural phenolic antioxidant. J. Agric. Food Chem. **50**, 2200–2206 (2002)
28. Borges, L.L., Lucio, T.C., Gil, E.S., Barbosa, E.F.: Uma abordagem sobre métodos analíticos para determinação da atividade antioxidante em produtos naturais. Enciclopédia Biosfera **7**, 01–20 (2011)
29. Larrosa, M., García-Conesa, M.T., Espín, J.C., Tomás-Barberán, F.A.: Ellagitannins, ellagic acid and vascular health. Mol. Aspects Med. **31**(6), 513–539 (2010)
30. Marín, M., Giner, R.M., Ríos, J., Recio, M.C.: Intestinal anti-inflammatory activity of ellagic acid in the acute and chronic dextrane sulfate sodium models of mice colitis. J. Ethnopharmacol. **150**(3), 925–934 (2013)
31. Larrosa, M., Tomás-Barberán, F.A., Espín, J.C.: The dietary hydrolysable tannin punicalagin releases ellagic acid that induces apoptosis in human colon adenocarcinoma Caco-2 cells by using the mitochondrial pathway. J. Nutr. Biochem. **17**, 611–625 (2006)
32. Yu, Y.M., Chang, W.C., Wu, C.H., Chiang, S.Y.: Reduction of oxidative stress and apoptosis in hyperlipidemic rabbits by ellagic acid. J. Nutr. Biochem. **16**, 675–681 (2005)
33. Aviram, M., Volkova, N., Coleman, R., Dreher, M., Reddy, M.K., Ferreira, D., Rosenblat, M.: Pomegranate phenolics from the peels, arils, and flowers are antiatherogenic: studies in vivo in atherosclerotic apolipoprotein e-deficient (E0) mice and in vitro in cultured macrophages and lipoproteins. J. Agric. Food Chem. **56**, 1148–1157 (2008)
34. Lin, C.-C., Hsu, Y.-F., Lin, T.-C., Hsu, H.-Y.: Antioxidant and hepatoprotective effects of punicalagin and punicalin on acetaminophen-induced liver damage in rats. Phytother. Res. **15**(3), 206–212 (2001)

35. Akiyama, H., Fujii, K., Yamasaki, O., Oono, T., Iwatsuki, K.: Antibacterial action of several tannins against *Staphylococcus aureus*. J. Antimicrob. Chemother. **48**, 487–491 (2001)
36. Cerdá, B., Llorach, R., Cerón, J.J., Espín, J.C., Tomás-Barberán, F.A.: Evaluation of the bioavailability and metabolism in the rat of punicalagin and antioxidant polyphenol from pomegranate juice. Eur. J. Nutr. **42**, 18–28 (2003)
37. Vieites, R.L., Daiuto, É.R., Moraes, M.R., Neves, L.C., Carvalho, L.R.: Caracterização físico-química, bioquímica e funcional da jabuticaba armazenada sob diferentes temperaturas. Rev. Bras. Fruticultura **33**(2), 362–375 (2011)
38. Wu, C.-C., et al.: *Myrciaria cauliflora* extracts attenuate diabetic nephropathy involving the Ras signaling pathway in streptozotocin/nicotinamide mice on a high fat diet. J. Food Drug Anal. **24**(1), 136–146 (2016)
39. Fidelis, M., do Carmo, M.A.V., Azevedo, L., Cruz, T.M., Marques, M.B., Myoda, T., Sant'ana, A.S., Furtado, M.M., Wen, M., Zhang, L., Rosso, N.D., Genovese, M.I., Oh, W.Y., Shahidi, F., Pap, N., Granato, D.: Response surface optimization of phenolic compounds from jabuticaba (Myrciaria cauliflora [Mart.] O.Berg) seeds: antioxidant, antimicrobial, antihyperglycemic, antihypertensive and cytotoxic assessments. Food Chem. Toxicol. **142**, (2020)
40. Diniz, D.N., Macêdo-Costa, M.R., Pereira, M.S.V., Pereira, J.V., Higino, J.S.: Efeito antifúngico in vitro do extrato da folha e do caule de *Myrciaria cauliflora* berg. sobre microorganismos orais. Rev. Odontologia UNESP **39**(3), 151–156 (2010)
41. Albuquerque, B.R., et al.: Jabuticaba residues (*Myrciaria jaboticaba* (Vell.) Berg) are rich sources of valuable compounds with bioactive properties. Food Chem. **309**(2019), 125735 (2020)
42. Oliveira, F.C., Marques, T.R., Machado, G.H.A., Carvalho, T.C.L., Caetano, A.A., Batista, L.R., Corrêa, A.D.: Jabuticaba skin extracts : phenolic compounds and antibacterial activity. Braz. J. Food Technol. **21**, 1–11 (2018)
43. Soong, Y.Y., Barlow, P.J.: Antioxidant activity and phenolic content of selected fruit seeds. Food Chem. **88**(3), 411–417 (2004)
44. Oliveira, A.L., Brunini, M.A., Salandini, C.A.R., Bazzo, F.R.: Caracterização tecnológica de jabuticabas Sabará provenientes de diferentes regiões de cultivo. Rev. Bras. Fruticultura **25**(3), 397–400 (2003)
45 Santos, D.P., Lopes, D.P.S., De Melo Calado, S.P., et al.: Efficacy of photoactivated *Myrciaria cauliflora* extract against *Staphylococcus aureus* infection—A pilot study. J. Photochem. Photobiol. B. **191**, 107–115 (2019)
46 Batista, Â.G., et al.: Intake of jaboticaba peel attenuates oxidative stress in tissues and reduces circulating saturated lipids of rats with high-fat diet-induced obesity. J. Funct. Foods **6**, 450–461 (2014)
47. Baldin, J.C., Munekata, P.E.S., Michelin, E.C., Polizer, Y.J., Silva, P.M., Canan, T.M., Pires, M.A., Godoy, S.H.S., Fávaro-Trindade, C.S., Lima, C.G., Fernandes, A.M., Trindade, M.A.: Effect of microencapsulated jabuticaba (Myrciaria cauliflora) extract on quality and storage stability of mortadella sausage. Food Res. Int. **108**, 551–557 (2018)
48. Asquieri, E.R., Silva, A.G.M., Candido, M.A.: Aguardente de jabuticaba obtida da casca e borra da fabricação de fermentado de jabuticaba. Ciênc. Tecnol. Aliment. **29**(4), 896–904 (2009)
49. Silva, M.C., Souza, V.B.D., Thomazini, M., da Silva, E.R., Smaniotto, T., Carvalho, R.A.D., Genovese, M.I., Favaro-Trindade, C.S.: Use of the jabuticaba (*Myrciaria cauliflora*) depulping residue to produce a natural pigment powder with functional properties. LWT Food Sci. Technol. **55**(1), 203–209 (2014)
50 Morales, P., Barros, L., Dias, M.I., Santos-Buelga, C., Ferreira, I.C.F.R., Asquieri, E.R., Berrios, J.J.: Non-fermented and fermented jabuticaba (Myrciaria cauliflora Mart.) pomaces as valuable sources of functional ingredients. Food Chem. **208**, 220–227 (2016)
51. Borges, L.L., Martins, F.S., Conceição, E.C., Silveira, D.: Optimization of the spray-drying process for developing Jabuticaba waste powder employing response surface methodology. J. Food Process Eng. **40**(1), (2017). Article: e12276

52. Lipinski, C.A., Lombardo, F., Dominy, B.W., Feeney, P.J.: Experimental and computational approaches to estimate solubility and permeability in drug discovery and development settings. Adv. Drug Deliv. Revisions **23**, 4–25 (1997)
53. Lipinski, C.A.: Lead- and drug-like compounds: the rule-of-five revolution. Drug Discov. Today: Technol. **1**(4), (2004)

Natural Product Bioactive as Antiviral Agents Against Zika Virus

Nerilson M. Lima, Teresinha de Jesus A. S. Andrade, and Marcone A. L. de Oliveira

Abstract Zika virus (ZIKV) is a Flavivirus belonging to the Flaviviridae family transmitted by an infected *Aedes aegypti* mosquito's bite. Although it is not a highly lethal infection; however, it is associated with several morbidities such as damage to the testicles, eye damage, Guillain–Barre syndrome, fetal microcephaly, and other severe neurological complications. Due to the complications associated with the ZIKV infection and because it is also transmitted through sexual, blood, and perinatal routes, this disease becomes a serious public health problem. Since there are still no effective drugs and a licensed vaccine to control and combat the virus, many therapeutic resources from biodiversity have stood out searching for anti-ZIKV agents. In this context, plants are important sources of raw materials and prototypes for the synthesis of new drugs with therapeutic properties. Thus, this study brings a full review from classes of natural products such as phenolic compounds (flavonoids, phenolic acids, tannins, and quinones), alkaloids and terpenes (including triterpenes, saponins, and steroids), and plant species and their bioproducts as essential oils (mono, sesquiterpenes, and phenylpropanoids), with antiviral potential against ZIKV strains evidenced by in vitro and in *vivo* tests.

1 Introduction

The Zika virus (ZIKV) belongs to the family of Flaviviruses and it was discovered around 1947 in the Zika forest in Uganda. Other family members are known to cause infections like dengue, yellow fever, and West Nile fever. In Brazil, the viral infection caused by ZIKV was reported in 2015 [1–5]. The effects of virulence, rapid spread and its serious neurological complications in general include Guillain-Barré syndrome and microcephaly that occurs especially in fetuses/newborns caused

N. M. Lima (✉) · M. A. L. de Oliveira
Analytical Chemistry Department, Federal University of Juiz de Fora, Juiz de Fora, MG, Brazil

T. de Jesus A. S. Andrade
Nucleus of Applied Research To Sciences, Federal Institute of Maranhão, Maranhão, Brazil

F. A. La Porta and C. A. Taft (eds.), *Functional Properties of Advanced Engineering Materials and Biomolecules*, Engineering Materials,
https://doi.org/10.1007/978-3-030-62226-8_27

alarm and spurred research in the discovery of new antiviral agents [6, 7]. Usually, most people infected with ZIKV do not become sick, but can transmit the disease through sexual intercourse and vertical transmission that can result in serious birth defects, including serious cases of microcephaly [1–7]. We emphasizing that there is no therapeutic agent approved for the treatment of ZIKV infection in humans, only patient support medications, and the anti-ZIKV drug candidates have been tested only in animals [8–11]. As such, the incipient research on herbal medicines to treat viruses and viral infections is still quite limited. Yet, in recent years, there is a large number of molecules with therapeutic applications derived from natural products, mainly from plant species and their associated fungi [12, 13]. For instance, many secondary metabolites, like polyphenols, possess outstanding antiviral activity against different type of viruses (such as influenza, DENV, coronavirus, HIV-1 and hepatitis B viruses) and hence they have tested against the NS2B-NS3 protease, and as a result, has been shown to inhibit the activity of ZIKV protease, as well [14]. Several polyphenols belonging to the flavonoid class, e.g., such as myricetin, quercetin, luteolin, isorhamnetin, apigenin, and the natural phenolic compounds curcumin showed inhibition ZIKV protease [14–17]. Based on this information, the natural products isolated from seaweed have demonstrated a great antiviral potential against flavivirus as those that cause the dengue. Other bioassays such as the drug addiction time trial have shown that isoquercitrin acts on the entry of ZIKV, preventing mainly the internalization of viral particles in the host cell [14]. In field of the natural product research, all of the efforts are directed in the search of new organic molecules with antivirals or virucide properties to treat of infection caused by the ZIKV, since there are no vaccines or medication available has effectively treated recent cases of infection. Therefore, given the urgent need to develop new therapeutic agents with antiviral action against ZIKV, this research brings a survey of the latest studies published in the scientific literature on discovering of new prototypes of drugs from natural products against ZIKV.

2 Phytochemical Bioactive as Novel Antiviral Agents Against ZIKV

In the years 2015–2016, Brazil began to face a major epidemiological challenge of viral infection caused by ZIKV, which occurs due to the bite from the mosquito vector *Aedes Aegypti*, and can also be transmitted through sexual intercourse [1, 2].

As such, the epidemic caused by the ZIKV infection had been reported in more than 60 countries worldwide [3]. Despite having a low lethality rate, this epidemic has raised many concerns because of the morbidity associated with viral infection and its action triggered on the human biological system, such as damage to the testicles [4, 5], eye damage [6], Guillain–Barre syndrome, fetal microcephaly [7], and other severe neurological complications.

Given this scenario, many scientific studies aiming the discovery of new molecular targets in the therapy of the viral infection caused by ZIKV and in the control of the mosquito vector have been developed. However, no drugs selected in pharmacological screenings from the drug repositioning strategy have shown sufficient efficacy to treat the ZIKV infection. In this way, many natural products derived from plants have been tested to combat the virus, control the *Aedes Aegypti* vector and identify the mode of action of these products, whether through repellency or biocidal action [1–17].

Although some platforms committed to the development of ZIKV vaccines and encouraging results have been obtained, the protection through this immunity has only been tested in rats and monkeys [8–10]. Thus, in this perspective, while there are no safe and licensed vaccines to prevent and combat the ZIKV, the infection treatment usually has been carried out using antivirals only to relieve symptoms, mitigate morbidities, as well as, interrupt the chain of transmission [11]. In this regard, many natural compounds of non-plant origin, such as the fungal lipopeptide named cavinafungin, have potent activity and are selectively active against ZIKV as well as all four dengue virus, named DEN-1, DEN-2, DEN-3, and DEN-4, serotypes acting against the signal peptidase (enzyme found in eukaryotic cells) and interfere in the flavivirus replicative cycle [12], but this book chapter will focus on small molecules from plant origin.

Therefore, the investigation and discovery of new phytochemicals with anti-Zika activity has become effective and safe therapeutic tools, since a large part of the medicinal plants species used in the in vitro and in vivo assays are used in popular phytotherapy to treat several other diseases, in addition to providing prototypes for studies of molecular variation in order to enhance their effectiveness and reduce their toxicity to the host tissues. Thus, this book chapter provides general information about plant species and their phytochemicals with antiviral action against ZIKV. Vegetable macromolecules with antiviral activity against ZIKV were divided into the three main groups of secondary metabolites: **phenolic compounds, alkaloids** and **terpenes** (Fig. 1).

2.1 Phenolic Compounds

Phenolic compounds are well-known by their antioxidant potential that has hydroxyl groups attached to an aromatic ring, widely distributed in the vegetable kingdom. They are usually present in fruits and vegetables, and have several properties in the plant and animal organisms, as protection against ultraviolet radiation and aggression of pathogens, including antiviral [13].

Flavonoids are one of the largest classes of phenolic compounds, and have a chemical diversity of structures and marked chemical-pharmacological activities [13–17]. This secondary metabolite class has an inexhaustible reservoir of molecules with significant antiviral properties, including arbovirus inhibitory activity such as

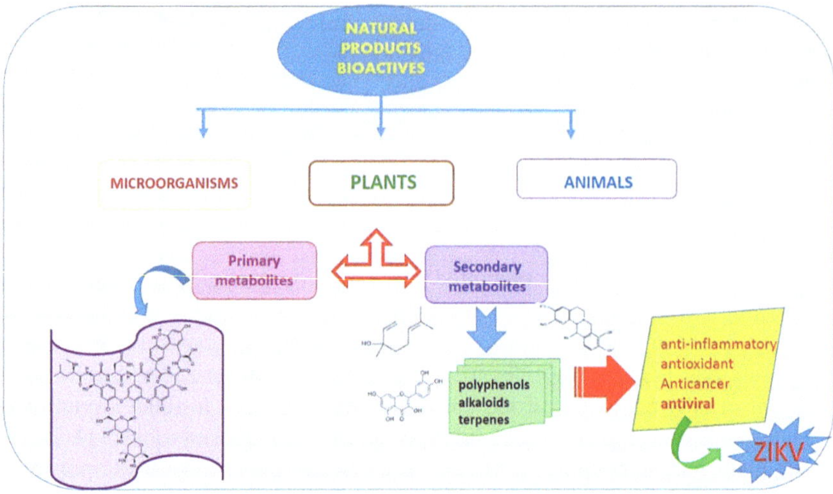

Fig. 1 Overview from plant natural product research to obtain antiviral agents against ZIKV

the Zika virus, which can play an important role in preventing and combating of ZIKV infection [14–17].

In studies of qualitative and quantitative evaluation of flavonoids with anti-ZIKV activity, many polyphenols such as flavonol isoquercitrin (quercetin-3-*O*-glucoside or Q3G) and its glycosylated fraction had been shown to be effective inhibitors of African and Asian epidemic ZIKV viral strains in human hepatoma, epithelial and neuroblastoma cell lines. Drug addiction time trials concluded that isoquercitrin acts on the entry of the ZIKV, preventing the internalization of viral particles in the host cell [14]. Other recent studies highlight the efficacy of isoquercitrin flavonol and Epigallocatechin Gallate polyphenols and curcumin against ZIKV strains [15–17].

Epigallocatechin Gallate is formed by an ester of epigallocatechin and gallic acid, has a high antioxidant potential. It is found in green tea, which can inhibit the entry of the virus into host cells in Vero E6 cell assay [16]. To elucidate their mechanism of action of this compound, were performed molecular modeling calculations to reveal the interactions of the active ingredient with the viral envelope protein. These results showed that the compound interacts with various residues on the receptor, in addition to maintained stability the conformations of the protein up to 50 ns of simulations [16].

A phenolic complex named Silymarin, which consists mainly of seven flavonolignans and one flavonoid, demonstrated significant antiviral activity for arboviruses and acts to combat and treat many other pathologies. This complex, which is found mainly in seeds of the species *Silybum marianum* L. Gaertn, achieves anti-ZIKV activity by reducing the virus titer, viral RNA and plaque-forming units. Also, it can inhibit the adsorption and penetration of ZIKV-like particles [18].

Flavonoids found in citrus fruits (such as flavanone naringenin), in particular, demonstrating an alternative promising to combating viral ZIKA infections. Pharmacological studies have shown that this compound prevents ZIKV infection in human A549 cells and other cell lines (such as primary dendritic cells derived from human monocytes infected by the virus). For this reason, flavanone naringenin acts in the replication of the virus or the assembly of viral particles. Also, molecular coupling studies indicate that there may be a potential interaction between the compound and the ZIKV protease, whose mechanism would explain the antiviral action of the target compound [19].

In addition to fruits and vegetables, it is well-known that flavonoids are widely found in other commercial natural products (e.g., honey, tea, and red wine). As such, this flavonoid compounds exhibited a significant antiviral effect against ZIKV in JEG-3 cells from human placenta ($IC_{50} = 17.4\ \mu M$), whose anti-ZIKV effect occurs through the post-entry process of the virus replication cycle that inhibition of RNA production and in the synthesis of viral envelope proteins [20].

Despite the numerous experimental studies in vitro and in vivo involving anti-ZIKV activity of flavonoids, the mechanisms by which these compounds act in inhibiting the infection are still not completely elucidated. However, it is believed that it can happen within interactions between phenolic rings of the molecule with viral proteins and RNA or the ability to interfere in the defense of host cells, that is, more specifically in regulating MAP kinase signaling [13–21].

Based on this information, studies of the structure activity relationship (SAR) have been used successfully and have helped in the discovery of new bioactive and innovative molecules. For example, Bhargava et al. [21] using simulation-based QSAR method, it was applied in order to know the molecular factors of the chemical structure of the compound for viral inhibitory activity, whose studies carried out with the flavonoids amentoflavone, fisetin, isorhamnetin and theaflavin-3-gallate demonstrated greater activity predicted inhibitory and more validated by the fitting analysis.

Chemical structure-biological activity relationship studies aiming to screen for new viral inhibitors for ZIKV infection were performed with 18 flavonoids (6 flavonol, 8 flavanol, 2 flavone, and 2 flavanone) and in addition to four other compounds phenolic (such as pyrogallol, pyrocatechol, caffeic acid, and gallic acid) [22]. The results of this research highlight that the compounds myricetin ($IC_{50} = 22 \pm 0.2\ \mu M$), astragalin ($IC_{50} = 112 \pm 5.5\ \mu M$), rutin ($IC_{50} = 105 \pm 2.9\ \mu M$), epigallocatechin gallate ($IC_{50} = 87 \pm 1.2\ \mu M$), epicatechin gallate ($IC_{50} = 89 \pm 1.6\ \mu M$), gallocatechin gallate ($IC_{50} = 99 \pm 1.8\ \mu M$) and luteolin ($IC_{50} = 53 \pm 1.3\ \mu M$) inhibited more than 40% of viral activity [22]. As such, the antiviral activity evaluation was based on the molecular structure of the compounds, and showed that myricetin showed greater potential; luteolin showed 2.8 times greater potential than chrysin, likely due to the presence of two OH groups in C3′ and C5′ in its B ring. The methoxy group on carbon C–4′ in ring B and the prenyl group at position C8 in ring A from the icaritin may have negatively affected to its lower antiviral potential ZIKV strain. Thus, this study concluded that the OH group's presence in C3′, C4′

and C5′ in ring B might be an important requirement for the inhibitory activity of ZIKV [22].

The flavonoid Sophoraflavenone G is a highly oxygenated flavanone with C-prenyl groups, found in Chinese medicine plants such as the *Sophora Flavecens* species, known for its antiviral property. This flavanone demonstrated significant antiviral activity against Dengue and ZIKV flaviviruses in pre- and post-infection treatments, the mechanism of which occurs through the inhibition of the virus RNA polymerase enzyme [23].

Other flavonoid candidates for prototype drugs against flavivirus are the flavones baicalein and baicalin, which showed the ability to negatively regulate the virus's replication up to 10 h after ZIKV infection and pronounced prophylactic action in pretreated cells [24]. Baicalein, whose structure has three hydroxyls in ring A, showed greater antiviral effect during intracellular replication of ZIKV, while the glycosylated flavone baicalin showed greater efficiency in preventing the entry of the virus. However, theoretical studies investigating the mechanism of action of both compounds indicate that they have a higher affinity of binding to ZIKV NS5 than to the viral envelope glycoprotein [24].

Highly hydroxylated molecules such as anthocyanin delfinidine and the tannin epigallocatechin have also been shown to be effective flavivirus inhibitors such as those of ZIKV and dengue [25], as well as plant species with a high anthocyanin content have also shown anti-ZIKV activity [26].

Although flavonoids, phenolic acids, and tannins are the main polyphenols reported in the literature with antiviral activity against ZIKV. Other phenolic compounds stand out in vitro anti-arbovirus activity, such as gossypol, curcumin, digitonin and conessin pigments. Gao et al. [27] showed that gossypol exhibited strong inhibitory activity against almost all ten ZIKV strains assessed, which can neutralize viral infection by reaching the ZIKV envelope III (EDIII) domain. Hence, other constituents evaluated inhibited the infection by inhibiting the virus's entry into the cell or by inhibiting viral replication. There was also a synergistic effect of gossypol with any of these metabolites and with the anti-ZIKV bortezomib [27].

Plant species used in folk medicine such as *Phyllanthus phillyreifolius*, have been shown to be an important source of the polyphenol geraniin which has antiviral activity against flavivirus strains including ZIKV [28]. Another active metabolite is the lignan (Dibenzylbutanes) named nordihydroguaiaretic acid found in the leaves of *Larrea tridentata* (Zygophyllaceae) with potent antiviral effect in vitro against the ZIKA virus [29] (Fig. 2).

2.2 Alkaloids

Alkaloids are secondary metabolites formed by heterocyclic rings containing the nitrogen heteroatom, with a broad spectrum of pharmacological activities and invaluable therapeutic potential due to their antimicrobial property and their ability to bind to genetic material [30]. Thus, many studies have been directed to the isolation and

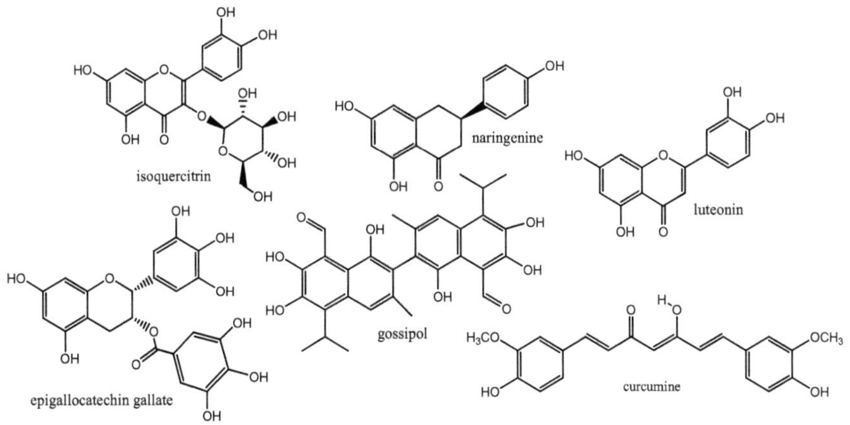

Fig. 2 Phenolic compounds (isoquercitrin, naringenin, luteolin, epigallocatechin gallate, gossypol and curcumin) with anti-ZIKV activity

pharmacological screening of alkaloids with antiviral potential for flavivirus such as ZIKV.

Considering that they are compounds of restricted distribution to some plant families, the search for new alkaloids with therapeutic potential for viral infections has been directed to these taxa, such as Amaryllidaceae family, which produces the alkaloid R430 that has an inhibitory action on the viruses of RNA and DNA, including ZIKV [31].

The alkaloids of this family are divided into subclasses, in the narciclasin subclass are antiviral alkaloids trans-dihydronarciclasine, trans-dihydrolycoricidine, narciclasine, 7-deoxy analog (lycoricidine), pancratistatin and 7-deoxy. Of these, the alkaloid trans-dihydronarciclasine has an inhibitory effect on some types of arbovirus, including ZIKV ($IC_{50} = 0.1\ \mu M$ and Therapeutic index = 7), therefore, a good candidate for therapeutic agent against ZIKV infection [32].

Another striking example of an alkaloid with anti-ZIKV activity in vitro and in vivo is the benzylphenethylamine alkaloid named licorine, whose antiviral action occurs by inhibiting RNA replication and protein synthesis. Licorine inhibited ZIKV RNA-dependent RNA polymerase activity, increased the survival of AG6 mice in ZIKV-induced in vivo assays by decreasing viral load in the blood and reduced viremia. In addition, this alkaloid also inhibited plaque formation partially at a concentration of about 0.5 μM and completely at 5.0 μM [33].

Medicinal plants from traditional Chinese medicine are an important source of the alkaloid with antiviral properties of the isoquinoline class named palmatine, which may in principle be directly isolated from species such as *Fibraurea recisa, Phellodendron Amurense, Enantia Chlorantha, Corydalis yanhusuo,* and *Coptis Chinensis.* This substance can inhibit ZIKV infection and resist the cytopathic effects induced by this virus, whose antiviral action mechanism occurs by interrupting arbovirus binding, entry and stability [34].

Another antiviral phytochemical from similar class is the alkaloid berbamine, a bisbenzylisoquinoline alkaloid found in species used in traditional Chinese medicine. This alkaloid was able to inhibit viral infection by ZIKV at an IC_{50} of approximately 2.17 mM in virus titration assays and also inhibited significantly the production of viral particles from the infectious progeny of the virus [35].

Other alkaloids of plants from traditional Chinese medicine have shown to be powerful antiviral agents. A notable example of this is the alkaloid cephalotaxin isolated from the species *Cephalotaxus drupacea*, which proved to be a potent inhibitor of ZIKV activity in cell lines Vero and A549. Yet, the mechanism of action of this molecule involves the decrease in viral replication by reducing the expression of virus RNA and viral protein, interrupting the virus's life cycle by preventing its replication and reaching virucidal activity [36].

In the screening of biologically active alkaloids against ZIKV infection, the isoquinolinic alkaloid emetine stands out, which is widely known and used in the treatment of several pathologies. This compound is already approved by the FDA for amebiasis treatment and has antiviral activity against other viruses, including ZIKV. The effectiveness of emetin in inhibiting ZIKV infection has been demonstrated at extremely low inhibitory concentrations, of the order of nanomolar in in vitro and in vivo assays [37].

The mechanism of inhibition of the alkaloid emetine as well as its bioactive analogue cefhaelina were determined. This study concluded that emetin acts against ZIKV by inhibiting the activity of the polymerase enzyme ZIKV NS5 and disrupting lysosomal function. However, its desmethyl analog, cefhaelina, has also been tested against the ZIKV. As result, both compounds have shown a very similar efficacy against infections caused by ZIKV and hence can contribute to a better tolerated in patients, being an excellent therapeutic alternative for infection. This research also showed that the NS1 protein level decreased in an emetine-dependent manner in HEK293 cell lines infected with African ZIKV (MR766) ($IC_{50} = 52.9$ nM) [37].

The alkaloid conessine, classified as a steroidal alkaloid, found in plant species from Apocynaceae family, showed important anti-ZIKV activity with IC_{50} 9.75 μM. The results showed that this alkaloid inhibited ZIKV infection by 80–96% using a method of co-incubation or post-incubation of cells with the compound and the virus. However, in the method of pretreating cells with before virus incubation, conessine blocked only 38% of viral infection [27]. Another steroidal alkaloid that has anti-ZIKV activity is tomatidine, whose compound is an aglycone metabolite of tomatin, which can be extracted from the skin and leaves of tomatoes [38]. In addition to alkaloids with molecular structures of alkaloids linked to steroids, there are those with structures linked to terpenoids, such as indole monoterpenoid alkaloids, for example, compounds of the Vallesiachotaman class, which have a wide range of bioactivities, including antiviral. In this class are found the alkaloids (−)-angustoline and (+)-angustoline found in several species of the genera *Strychnos*, *Mitragyna, Nauclea, Uncaria* and *Vinca*, which exhibit weak inhibitory activity against ZIKV, but which can be useful in models molecular studies for structural modification studies aiming to increase its therapeutic potency [39].

Concerning the alkaloids with the pyrrole group that have anti-ZIKV activity, the anisomycin alkali stands out, an important inhibitor of the replication of Asian and African ZIKV strains in Vero cells. This compound acts by preventing the virus from multiplying in the host cell as well as by inhibiting viral protein expression. In addition, the results from in vivo assays demonstrated that anisomycin significantly reduced viremia levels and the animals (mice with ZIKV morbidity and mortality) treated with the compound died much later than the control group. Thus, the studies concluded that the alkaloid anisomycin is a potent and selective inhibitor in vitro that impairs a post-entry stage of viral replication and exerted a protective effect in animals infected with ZIKV [40].

For being metabolites of restricted distribution, the alkaloids can be used in chemosystematic studies. They are compounds known since antiquity in ancient species such as *Atropa belladonna* or Belladonna (Solanaceae), which is an important source of the widely known bioactive alkaloids, scopolamine and hyoscyamine. Thus, a research using Belladonna has been promising in reducing infection by the Japanese encephalitis virus, whose treatment can be tried in the case of ZIKV, since both belong to the same family. In homeopathy, e.g., Belladonna alkaloids are usually used in the treatment of infectious diseases, just as the homeopathic pharmaceutical product *Eupatorium perfoliatum* can be used as a prophylactic as part of the ZIKV, since this medicine is the one that comes to the closest match symptoms exhibited by the disease caused by ZIKV [41] (Fig. 3).

2.3 *Terpenes*

Terpenes or terpenoids are a very diverse class of secondary metabolites and they are known for pleasant aromas extracted from plant essences, which have been used for thousands of years for their medicinal effect and raw material for the production of cosmetics. These metabolites can be found in various plant organs such as seeds, flowers, leaves, roots and wood from higher plants as well as in fungi and animals, which perform various biological activities in the animal organism, including antiviral [13].

Several terpenes have recently been the target of pharmacological studies aiming to discover new antiviral agents against arborivores. In investigations of new biologically active molecules against ZIKV infection, the terpenes bisabolol and andrographolide have been suggested as potential antiviral ZIKV therapy [42]. In these studies, the plant species that showed the highest therapeutic index were *Momordica charantia, Psidium guajava, Vitex negundo* and *Blumea balsamifera,* whose metabolites in the aqueous extracts were saponin, terpenes and terpenoids and anthocyanin [26].

Terpenes can be found linked to sugar molecules, such as the sesquiterpene glycoside named plumieride found in species of the genus *Himatanthus* that was active against ZIKV lineage (IC_{50} 14.49 ± 0.72) and showed an SI increase to 15.97 compared to the crude ethanolic extract of the species from which it was isolated [43].

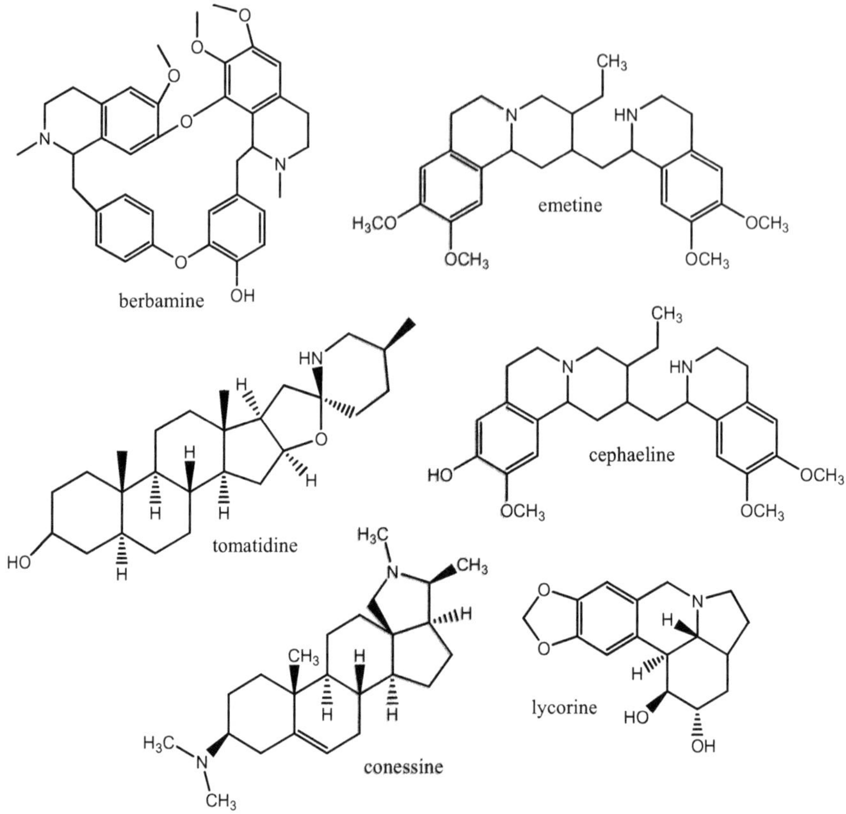

Fig. 3 Alkaloids (berbamine, emetine, cephaeline, tomatidine, lycorine, conessine), with antiviral activity against ZIKV

Other potent viral inhibitors against ZIKV from terpene class are the diterpenes from the medicinal plant *Stillingia loranthacea*: one loranthones A, four tigliane diterpenes and the flexibilene diterpene named tonantzitlolones C. The most active compounds were the loranthones B terpenes and the new tigliane diterpene, which reduced 1.7 and 1.8 log10 TCID50/mL in viral replication, respectively [44].

Pentacyclic triterpenes identified from leaf and branch extracts of *Tontelea micrantha* (friedelin, 3β-friedelinol, 3,4-*seco*-friedelanoic acid, 28-hydroxyfriedelan-3-one, friedelan-3,21-dione, 21β-hydroxyfriedelan-3-one, 21α-hydroxyfriedelan-3-one, 30-hydroxyfriedelan-3-one, 3-oxo-friedelan-28-al, friedel-1-en-3,21-dione, friedelan-1,3,21-trione, 3-oxo-friedelan-28-oic acid, 28,29-dihydroxyfriedelan-3-one, 21α,30-dihydroxyfriedelan-3-one, 28,30-dihydroxyfriedelan-3-one, tingenin B, netzahualcoyone, netzahualcoyondiol, salaquinone A and the orthosphenic acid) may be responsible for the potent anti-ZIKV effect reported in the species, whose leaf extracts showed inhibition of viral penetration at concentrations lower than 15.6 μg/mL [45].

Most assays to assess the anti-ZIKV activity of terpenes are performed with essential oils or tests of larvicidal effect against larvae of the vector mosquito, *Aedes aegypti* L. (Diptera: Culicidae) [46, 47]. In general, the essential oil from a plant species can contain 20–80 compounds [46]. Studies of antiviral activity against ZIKV using essential oils and its major constituent characterization have been widely carried out in this research field. In this context, the analysis of the essential oil of the medicinal herbs *Ayapana triplinervis*, which have as principal component the dimethylhydroquinone-dimethyl ether (abbreviated as THQ), was carried out, prospecting compounds with potential for prevent ZIKV infection in human cells. This research was carried out by means of virological assays using human A549 epithelial cells infected with the GFP reporter ZIKV or the epidemic viral strain, whose results showed that the essential oil ($IC_{50} = 38$ μg/mL) and its major component (THQ) ($IC_{50} = 45$ μg/mL) inhibit ZIKV infection in human cells. In addition, both reduced the production of 3-log virus progenies at non-cytotoxic concentrations and showed that THQ could inhibit virus entry into the cell [47].

A widely known species and already used in herbal preparations is the medicinal herb *Lippia alba*. This plant's essential oil have showed significant antiviral activity to concentrations in the range of 8.02 μg/mL to 20.88 μg/mL and, therefore, has been evaluated as a antiviral against ZIKV [48].

One of ZIKV's control and combat strategies is to eliminate the vector responsible for transmitting of the virus. Thus, much research involving suppressing the life cycle of the *A. aegypti* mosquito has been developed [49]. Natural products with significant larvicidal activity are essential oils, a mixture of mono, sesquiterpenes and also phenylpropanoids. In this context, substances present in these products such as monoterpenes geraniol and citronellol and the phenylpropanoids are explored in biological studies. Among these compounds, we can mention eugenol, which has an insecticidal effect that can be useful in the control and spread of the mosquito. In addition, essential oils can be manipulated using different concentrations as well as combinations of growth regulators and elicitors [49].

Concerning the inhibition of the vector mosquito, the terpenes δ-Cadinene, Calarene and δ-4-Carene isolated from *Kadsura heteroclita* essential oil were effective against *A. aegypti* ($CL_{50} = 9.03$, 13.33 and 17.91 μg/mL) [50].

In this regard, the compound *p*-Cymene ($LC_{50} = 51$ ppm) also showed potent toxicity against larvae of the mosquito *A. aegypti*, exceedingly greater than the monoterpene 1, 8-cineole ($LC_{50} = 1419$ ppm). The study concluded that the presence of lipophilic groups resulted in greater larvicidal potential, while hydroxyl groups resulted in less toxic compounds. As such, the lipophilicity of the substituents appears to play a crucial role in the larvicidal activity, since more potent compounds were obtained by more lipophilic molecules, whose methylation of hydroxyls led to a general increase in potency [51].

The evaluation of the larvicidal effect of the citrus terpenes R- and S-limonene showed excellent larvicidal activity ($LC_{50} = 27$ and 30 ppm, respectively), as well as the γ-terpinene monoterpenes ($LC_{50} = 56$ ppm) and R, S-carvone ($LC_{50} = 118$ ppm) [52]. The lower larvicidal potency of the evaluated compounds may be related to the presence of heteroatoms while conjugated and exo double bonds seem to increase the larvicidal activity. However, it was found that the substitution of double bonds by more reactive epoxies reduces this effect as well as the presence of hydroxyls in the cyclic structure. Thus, the lipophilicity appears to display a pivotal role in increasing larvicidal activity [52].

Major compounds (99.05%) of the essential oil of *Syzygium aromaticum* may be usually related to their larvicidal potential ($LC_{50} = 62.3$ ppm), whose main component, the eugenol terpene, also showed a potent larvicidal effect against vector mosquito larvae *A. aegypti* ($LC_{50} = 71.9$ ppm) [53].

The larvicidal activity of essential oils can be achieved in different ways and act on specific biochemical routes and various action sites on the mosquito, including changes that lead insects to intoxication, whether lethal or sub-lethal. The most common action sites involve the nervous system, specifically at the synapses and in the chlorine-GABA channels where the larvicidal compounds can act by blocking the nervous signal between neurons. Some constituents present in essential oils such as terpinen-4-ol, 1, 8-cineole, S-carvone, linalool, fenchone, γ-terpinene and geraniol act on the enzyme acetylcholinesterase [54].

According to Pavela [55] the lethal concentration to kill 50% of the population (LC_{50}) of culicids $\leq$ 100 ppm come from the main botanical families Lamiaceae, Cupressaceae, Rutaceae, Apiaceae and Myrtaceae, being their most common constituents, the 1,8-cineol, thymol, ρ-cymene, α- and β-pinene, δ-3-carene, sabinene, limonene and trans-anethole.

Figure 4 shows the chemical structure of some terpenes present in essential oils with larvicidal activity against the ZIKV vector-transmitting mosquito, since

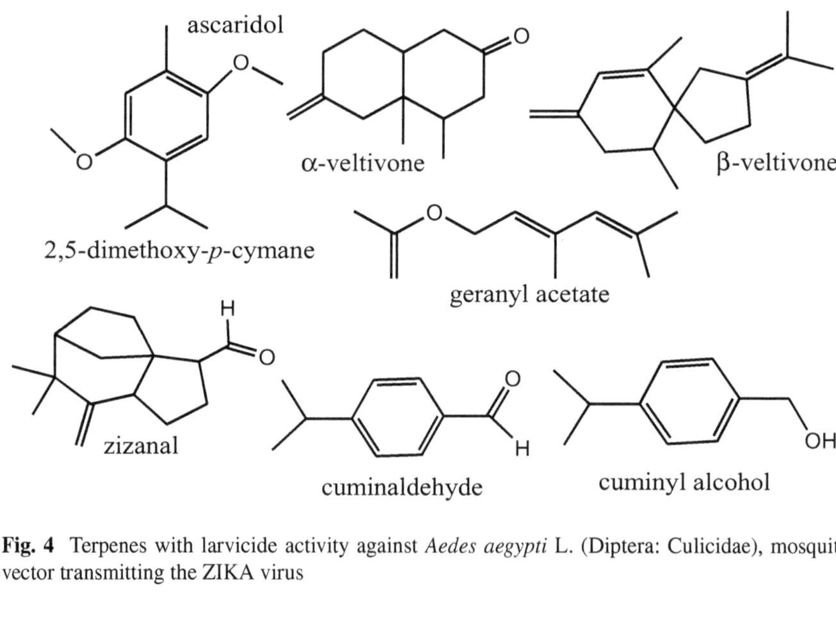

Fig. 4 Terpenes with larvicide activity against *Aedes aegypti* L. (Diptera: Culicidae), mosquito vector transmitting the ZIKA virus

the chemical composition of these natural products is formed mainly by hydrocarbons such as monoterpenes, accompanied by sesquiterpenes, phenylpropanoids, oxygenated and aromatic compounds [49, 56].

3 Conclusion

The prospection of new biologically active molecules from natural sources such as plants and microorganisms is a powerful and promising tool for developing new drugs for the treatment of ZIKV infectious. Hence, the screening of these biocompounds from plants secondary metabolism was carried out through in vitro and in vivo tests in order to prove their effectiveness in several pathologies such as infection caused by Flavivirus, which represents a severe public health problem due to its mortality and the morbidity associated with the disease. Thus, this study showed several phytochemicals and their pharmacological action mechanism with antiviral effect against strains of ZIKV, whose metabolites classified into phenolic compounds, alkaloids and terpenes provide molecular models for the technological development of new antiviral drugs. Many chemical constituents such as flavonoids (isoquercitrin, naringenin, pinocembrin, myricetin, and luteolin), alkaloids (trans-dihydro narciclasine, licorice, palmatine, and emetine), terpenes (bisabolol, andrographolide) and phytochemicals as essential oils have shown significant antiviral potential against ZIKV infection. Overall, we believed that such compounds due to their therapeutic potential against this virus, in principle, could be widely used in prophylactic treatment, as well as to the control of the mosquito A. aegypti.

References

1. Musso, D., Roche, C., Robin, E., Nhan, T., Teissier, A., CaoLormeau, V.M.: Potential sexual transmission of Zika virus. Emerg. Infect. Dis. **21**(2), 359–361 (2015)
2. D'Ortenzio, E., Matheron, S., Yazdanpanah, Y., de Lamballerie, X., Hubert, B., Piorkowski, G., Maquart, M., Descamps, D., Damond, F., Leparc-Goffart, I.: Evidence of sexual transmission of Zika virus. N. Engl. J. Med. **374**, 2195–2198 (2016)
3. WHO (World Health Organization): Cancer. Fact sheet N° 297, Update February 2015. Disponível em: Acesso em: 24 May 2016
4. Govero, J., Esakky, P., Scheaffer, S., et al.: Zika vírus infection damages the tests in mice. Nature **540**, 438–442 (2016)
5. Ma, W., Li, S., Ma, S., et al.: Zika virus causes testis damage and leads to male infertility in mice. Cell **167**, 1511-1524.e15 (2016)
6. Miner JJ, Sene A, Richner JM, et al.: A infecção do vírus Zika em camundongos provoca panuveíte com derramamento de vírus em lágrimas. **16**, 3208-3218 (2016)
7. Calvet, G., Aguiar, R.S., Melo, A.S., Sampaio, S.A., De Filippis, I., Fabri, A., Tschoeke, D.A.: Detection and sequencing of Zika virus from amniotic fluid of fetuses with microcephaly in Brazil: a case study. Lancet Infect. Dis. **16**(6), 653–660 (2016)
8. Abbink, P., Larocca, R.A., De La Barrera, R.A., Bricault, C.A., Moseley, E.T., Boyd, M., Kirilova, M., Li, Z., Ng'ang'a, D., Nanayakkara, O., Nityanandam, R., Mercado, N.B.,

Borducchi, E.N., Agarwal, A., Brinkman, A.L., Cabral, C., Chandrashekar, A., Giglio, P.B., Jetton, D., Jimenez, J., Lee, B.C., Mojta, S., Molloy, K., Shetty, M., Neubauer, G.H., Stephenson, K.E., Peron, J.P.S., Zanotto, P.M.D.A., Misamore, J., Finneyfrock, B., Lewis, M.G., Alter, G., Modjarrad, K., Jarman, R.G., Eckels, K.H., Michael, N.L., Thomas, S.J., Barouch, D.H.: Protective efficacy of multiple vaccine platforms against Zika virus challenge in rhesus monkeys. Science 353, 1129, (2016)
9. Larocca, R.A., Abbink, P., Peron, J.P.S., De A Zanotto, P.M., Iampietro, M.J., Badamchi-Zadeh, A., Boyd, M., Ng'ang'a, D., Kirilova, M., Nityanandam, R., Mercado, N.B., Li, Z., Moseley, E.T., Bricault, C.A., Borducchi, E.N., Giglio, P.B., Jetton, D., Neubauer, G., Nkolola, J.P., Maxfield, L.F., De La Barrera, R.A., Jarman, R.G., Eckels, K.H., Michael, N.L., Thomas, S.J., Barouch, D.H.: Vaccine protection against Zika virus from Brazil. Nature 536, 474–478, (2016)
10. Pardi, N., Weissman, D.: Nucleoside modified mRNA vaccines for infectious diseases. In: Kramps, T., Elbers, K. (eds.), RNA Vaccines: Methods and Protocols, pp. 109–121. Springer New York, New York, NY, (2017)
11. Chutaputti, A.: Adverse effects and other safety aspects of the hepatitis C antivirals. J. Gastroenterol. Hepatol. **15**, E156-163 (2000)
12. Estoppey, D., Lee, C.M., Janoschke, M., Lee, B.H., Wan, K.F., Dong, H., Mathys, P., Filipuzzi, I., Schuhmann, T., Riedl, R.: The natural product cavinafungin selectively interferes with Zika and Dengue virus replication by inhibition of the host signal peptidase. Cell Rep. **19**, 451–460 (2017)
13. Simões, C.M.O., Schenkel, E.P., Gosman, G., Palazzo de Melo, J., Mentz, L.E.P.: PR (org.) Farmacognosia: da Planta ao Medicamento. Editora da UFRGS/Editora da UFSC. Porto Alegre/Florianópolis (1999)
14. Gaudry, A., Bos, S., Viranaicken, W., Roche, M., Krejbich-Trotot, P., Gadea, G., El-Kalamouni, C.: The flavonoid isoquercitrin precludes initiation of Zika virus infection in human cells. Int. J. Mol. Sci. **19**(4), 1093 (2018)
15. Mounce, B.C., Cesaro, T., Carrau, L., Vallet, T., Vignuzzi, M.: Curcumin inhibits Zika and chikungunya virus infection by inhibiting cell binding. Antivir. Res. **142**, 148–157 (2017)
16. Sharma, N., Murali, A., Singh, S.K., Giri, R.: Epigallocatechin gallate, an active green tea compound inhibits the Zika virus entry into host cells via binding the envelope protein. Int. J. Biol. Macromol. **104**, 1046–1054 (2017)
17. Wong, G., He, S., Siragam, V., Bi, Y., Mbikay, M., Chretien, M., Qiu, X.: Antiviral activity of quercetin-3-beta-O-d-glucoside against Zika virus infection. Virol. Sin. **32**, 545–547 (2017)
18. da Silva, T.F., Ferraz, A.C., Almeida, L.T., da Silva Caetano, C.C., Camini, F.C., Lima, R.L.S., de Magalhães, J.C.: Antiviral effect of silymarin against Zika virus in vitro. Acta Tropica, 105613, (2020)
19. Cataneo, A.H.D., Kuczera, D., Koishi, A.C., Zanluca, C., et al.: The citrus flavonoid naringenin impairs the in vitro infection of human cells by Zika virus. Sci. Rep. **9**(1), 1–15 (2019)
20. Lee, J.L., Loe, M.W.C., Lee, R.C.H., Chu, J.J.H.: Antiviral activity of pinocembrin against Zika virus replication. Antiviral Res. **167**, 13–24 (2019)
21. Bhargava, S., Patel, T., Gaikwad, R., Patil, U.K., Gayen, S.: Identification of structural requirements and prediction of inhibitory activity of natural flavonoids against Zika virus through molecular docking and Monte Carlo based QSAR simulation. Nat. Prod. Res. **33**(6), 851–857 (2019)
22. Lim, H.J., Nguyen, T.T.H., Kim, N.M., Park, J.S., Jang, T.S., Kim, D.: Inhibitory effect of flavonoids against NS2B-NS3 protease of ZIKA virus and their structure activity relationship. Biotechnol. Lett. **39**, 415–421 (2017)
23. Sze, A., Olagnier, D., Hadj, S.B., Han, X., Tian, X.H., et al.: Sophoraflavenone G restricts Dengue and Zika virus infection via RNA polymerase interference. Viruses **9**(10), E287 (2017)
24. Oo, A., Teoh, B.T., Sam, S.S., Bakar, S.A., Zandi, K.: Baicalein and baicalin as Zika virus inhibitors. Arch. Virol. **164**(2), 585–593 (2019)
25. Vázquez-Calvo, Á., Jiménez de Oya, N., Martín-Acebes, M.A., Garcia-Moruno, E., Saiz, J.C.: Antiviral properties of the natural polyphenols delphinidin and epigallocatechin gallate against the flaviviruses West Nile virus, Zika virus, and dengue virus. Front. Microbiol. **8**, 1314 (2017)

26. Vista, F.E.S., Dalmacio, L.M.M., Corales, L.G.M., Salem, G.M., Galula, J.U., Chao, D.Y.: Antiviral effect of crude aqueous extracts from ten Philippine medicinal plants against Zika virus. Acta Med. Philippina, **54**(2), (2020)
27. Gao, Y., Tai, W., Wang, N., Li, X., Jiang, S., et al.: Identification of novel natural products as effective and broad-spectrum anti-Zika virus inhibitors. Viruses **11**(11), 1019 (2019)
28. Haddad, J.G., Grauzdytė, D., Koishi, A.C., Viranaicken, W., Venskutonis, P.R., et al.: The Geraniin-rich extract from reunion island endemic medicinal plant Phyllanthus phillyreifolius inhibits Zika and Dengue virus infection at non-toxic effect doses in Zebrafish. Molecules **25**(10), 2316 (2020)
29. Merino-Ramos, T.; de Oya, N.J.; Saiz, J.-C.; Martín-Acebes, M.A.: Antiviral activity of nordihydroguaiaretic acid and its derivative tetra-o-methyl nordihydroguaiaretic acid against West Nile virus and Zika virus. Antimicrob. Agents Chemother. **61**, (2017)
30. Jayasurriya, H., Koonchanok, M.K., Geahlen, R.L., Mclanghlin, J.L., Chang, C.J.: Emodine, a protein kinase inhibitor from polygonum Cuspidatum. J. Nat. Pro. **55**, 696–698 (1991)
31. D'Aiuto, L., McNulty, J., Hartline, C., Demers, M., Kalkeri, R., Wood, J., et al.: R430: a potent inhibitor of DNA and RNA viruses. Sci. Rep. **8**, 16662 (2018)
32. Revu, O., Zepeda-Velázquez, C., Nielsen, A.J., McNulty, J., Yolken, R.H., Jones-Brando, L.: Total synthesis of the natural product (+)-trans-Dihydronarciclasine via an asymmetric organocatalytic [3+3]-Cycloaddition and discovery of its potent anti-Zika virus (ZIKV) Activity. ChemistrySelect **1**(18), 5895–5899 (2016)
33. Chen, H., Lao, Z., Xu, J., Li, Z., Long, H., Li, D., Li, G.: Antiviral activity of lycorine against Zika virus in vivo and in vitro. Virology, (2020)
34. Ho, Y.J., Lu, J.W., Huang, Y.L., Lai, Z.Z.: Palmatine inhibits Zika virus infection by disrupting virus binding, entry, and stability. Biochem. Biophys. Res. Commun. **518**(4), 732–738 (2019)
35. Huang, L., Li, H., Yuen, T.T.T., Ye, Z., et al.: Berbamine inhibits the infection of SARS-CoV-2 and flaviviruses by compromising TPRMLs-mediated endolysosomal trafficking of viral receptors, (2020)
36. Lai, Z.Z., Ho, Y.J., Lu, J.W.: Cephalotaxine inhibits Zika infection by impeding viral replication and stability. Biochem. Biophys. Res. Commun. **522**(4), 1052–1058 (2020)
37. Yang, S., Xu, M., Lee, E.M., Gorshkov, K., Shiryaev, S.A., He, S., Lu, B.: Emetine inhibits Zika and Ebola virus infections through two molecular mechanisms: inhibiting viral replication and decreasing viral entry. Cell Discov. **4**(1), 1–14 (2018)
38. Diosa-Toro, M., Troost, B., Van De Pol, D., Heberle, A.M., Urcuqui-Inchima, S., Thedieck, K., Smit, J.M.: Tomatidine, a novel antiviral compound towards dengue virus. Antiviral Res. **161**, 90–99 (2019)
39. Peng, X., Fu, M., Ou, J., Cao, R., Song, H., Liu, X.Y., Qin, Y.: Total synthesis of angustine and angustoline. Tetrahedron Lett. 151757, (2020)
40. Quintana, V.M., Selisko, B., Brunetti, J.E., Eydoux, C., et al.: Antiviral activity of the natural alkaloid anisomycin against dengue and Zika viruses. Antiviral Res. **176**, 104749 (2020)
41. Saxena, S.K., Elahi, A., Gadugu, S., Prasad, A.K.: Zika virus outbreak: an overview of the experimental therapeutics and treatment. VirusDisease **27**(2), 111–115 (2016)
42. Feranchuk, S., Potapova, U., Belikov, S.: Virtual screening of inhibitors for the Zika virus proteins. bioRxiv, 060798, (2016). https://doi.org/10.1101/060798
43. da Silva, M.L., Stehmann, J.R., Serafim, M.S.M., Vale, V.V., Gontijo, D.C., Brandão, G.C., de Oliveira, A.B.: Himatanthus bracteatus stem extracts present anti-flavivirus activity while an isolated sesquiterpene glucoside present only anti-Zika virus activity in vitro. Nat. Prod. Res. 1–5, (2019)
44 Abreu, L.S., do Nascimento, Y.M., Costa, R.D.S., Guedes, M.L.S., et al.: Tri-and diterpenoids from Stillingia loranthacea as inhibitors of Zika virus replication. J. Nat. Prod. **82**(10), 2721–2730 (2019)
45. Ferreira, F.L., Hauck, M.S., Duarte, L.P., de Magalhães, J.C., da Silva, L.S.M., et al.: Zika virus activity of the leaf and branch extracts of Tontelea micrantha and its hexane extracts phytochemical study. J. Braz. Chem. Soc. **30**(4), 1–11 (2019)

46. Regnault-Roger, C., Vincent, C., Arnason, J.T.: Essential oils in insect control: low-risk products in a high-stakes world. Rev. Bras. Entomol. **57**, 405–424 (2012)
47. Haddad, J.G., Picard, M., Bénard, S., Desvignes, C., Desprès, P., Diotel, N., El Kalamouni, C.: Ayapana triplinervis essential oil and its main component thymohydroquinone dimethyl ether inhibit Zika virus at doses devoid of toxicity in Zebrafish. Molecules **24**(19), 3447 (2019)
48. Quispe-Bravo, B.E., Drozdek, L.A.S., Jara, J.H., Díaz, I.E.C., et al.: In vitro activity evaluation of Lippia alba essential oil against Zika virus. bioRxiv, 2020. doi: https://doi.org/10.1101/2020.06.25.170720
49. Dias, C.N., de Mesquita, L.S.S., Coutinho, D.F., Malik, S.: Plant derived essential oils against Aedes aegypti L. and their biotechnological production. In: Medicinal Plants and Environmental Challenges, pp. 345–357. (Springer, Cham, 2017)
50. Govindarajan, M., Rajeswary, M., Benelli, G.: δ-Cadinene, calarene and δ-4-carene from Kadsura heteroclita essential oil as novel larvicides against malaria, dengue and filariasis mosquitoes. Comb. Chem. High Throughput Screening **19**(7), 565–571 (2016)
51. Santos, S.R., Silva, V.B., Melo, M.A., Barbosa, J.D., Santos, R.L., de Sousa, D.P., Cavalcanti, S.C.: Toxic effects on and structure-toxicity relationships of phenylpropanoids, terpenes, and related compounds in Aedes aegypti larvae. Vector-Borne Zoonotic Dis. **10**(10), 1049–1054 (2010)
52. Santos, S.R., Melo, M.A., Cardoso, A.V., Santos, R.L., de Sousa, D.P., Cavalcanti, S.C.: Structure–activity relationships of larvicidal monoterpenes and derivatives against Aedes aegypti Linn. Chemosphere **84**(1), 150–153 (2011)
53. Barbosa, J.D., Silva, V.B., Alves, P.B., Gumina, G., Santos, R.L., Sousa, D.P., Cavalcanti, S.C.: Structure–activity relationships of eugenol derivatives against Aedes aegypti (Diptera: Culicidae) larvae. Pest Manag. Sci. **68**(11), 1478–1483 (2012)
54. Priestley, C.M., Williamson, E.M., Wafford, K.A., Satelle, D.B.: Thymol, a constituent of thyme essential oils, is a positive modulator of human GABA and a homooligosteric GABA receptor from Drosophila melanogaster. Br. J. Pharmacol. **140**, 72–1363 (2003)
55. Pavela, R.: Essential oils for the development of eco-friendly mosquito larvicides: A review. Ind. Crops Prod. **76**, 174–187 (2015)
56. Nerio, L.S., Verbel, J.O., Stashenko, E.: Repellent activity of essential oils: a review. Biores. Technol. **101**, 372–378 (2010)

MIX
Papier aus verantwortungsvollen Quellen
Paper from responsible sources
FSC® C105338

If you have any concerns about our products,
you can contact us on
ProductSafety@springernature.com

In case Publisher is established outside the EU,
the EU authorized representative is:
Springer Nature Customer Service Center GmbH
Europaplatz 3, 69115 Heidelberg, Germany

Printed by Libri Plureos GmbH
in Hamburg, Germany